for my teacher who left before his time, Richard O. Stone

to my parents Martin C. Sara, and Nancy Sara,

and to Martin James, Nicole, Amy, Amanda and most of all Terrie

PREFACE

This book is the result of many years of trying to evaluate the physical reality of ground-water flow in geologic materials. This text concentrates on practical aspects of hydrogeology and was designed as a working tool used to evaluate the heterogenities of subsurface units. The key to understanding the conceptual basis of a particular geologic environment is building the geologic models from known regional systems and adding site facility knowledge to the regional picture. I have tried to build a conceptual basis for understanding geologic and hydrogeologic systems using traditional methods, some dating back over fifty years. These traditional evaluation methods provide a systematic basis for sorting out some very difficult field problems. Rather than taking a total cook book approach I have also tried to keep professional insights as an integral part of the site assessment process.

One should not lose sight that excessive drilling to evaluate a particular property should not be substituted for a knowledgeable, targeted cost effective evaluation. Experience in conducting site investigations in a specific geologic environment must always be an important consideration. Since assessment methods may be dramatically different, I have tried to provide illustrative guidance on the procedural aspects of site investigations for both primary and secondary porosity systems. Low hydraulic conductivity geologic environments provide special challenges to the investigative hydrogeologist. No hydrogeologic system provides so many problems (and have been so thoughly miss-evaluated) as these low hydraulic conductivity geologic environments. This book provides a number of evaluation methods for these geologic materials often representing important confining units for waste disposal facilities.

The case examples used in the text are presented for illustrative purposes and the actual locations are generally irrevelant. As such I have removed locational references, as much as, possible and wish to only focus on the technical aspects of the presentations. These examples are presented as evidence that one can use these illustrations for working out the conceptual understanding of geologic environments.

I have used, whenever possible, a form of structured graphics that I refer to as pictographics. These technical 'organization charts' provide a systematic view of a particular group of evaluation methods with inter-method relationships and references preserved within the pictographic.

ACKNOWLEDGMENTS

The generation of a text such as this requires the efforts of many individuals that have provide both illustrative and textural materials for which I am indebted. Versions of the original Site Assessment Manual (SAM) that served a the root document for this text were reviewed by three of the most capable professionals I have ever had the pleasure to work with: Dr. Lee C. Atkinson, Professor Allen W. Hatheway and Dr. Travis H. Hughes. Their continued support and comment were critical to the concepts and guidance presented herein.

The concepts presented in Chapter 11 relative to statistical evaluation of ground-water data is the product of the work of Professor Robert Gibbons, his contributions to the science of statistical evaluation of ground-water parameters have made detection monitoring possible for solid and hazardous waste disposal facilities. Special recognition is made to John Baker who provided many insights to the detection and assessment monitoring methods presented in this text. Special recognition must also be given to the following individuals: Richard C. Benson for the concepts associated with geophysics and scale relationships of site investigations; Michael R. Noel for the conceptual process as applied to numerous geologic environments; Harry Morris for the three dimensional conceptual process as applied to detection monitoring; James F. Quinlan for Karst hydrogeology and monitoring; Frank Jarke for text describing concepts of MCL's in Chapter 11; Richard C. F. King for several pump testing discussions and especially for fractured rock illustrative figures; John F. Clerici for development of procedures for testing of confining units; Professor F. D. I. (Frank) Hodgson for conceptual development of ground-water flow in fractured rock; David D. Slaine for the many photographs in Chapter 3 that illustrate geophysical field applications; John E. Scaife for illustrative presentations in Chapter 3 on GPR and reflection seismic techniques; Dave Burt for the flow diagrams on Subtitle D, and Jack Dowden for fast track Superfund investigative processes.

Many consulting companies provided both their expertise and professional support in the example materials presented within the various sections. These contributions are referenced below the illustration or in the text describing the figure. Special recognition is given for the following organizations: Canonie Environmental, Dames & Moore, Donohue (now RUST Environment & Infrastructure), Eckenfelder Inc., EMCON Associates, Gartner Lee, Golder Associates Inc., GZA GeoEnviromental, Kerfoot and Associates, Meredith/Boli & Associates, Inc., multiVIEW Geoservices Inc., P. E LaMoreaux & Associates, Patrick Engineering Inc., Simon Hydro-Search, Inc., Solinst, Technos, Warzyn, Wehran EnviroTech.

The value of the text has been greatly increased by the contributions of the following individuals: Florin Gheorghiu, Jeff Shanks, Lawrence E. Annen, David Nielsen, Les G. McMillion, Sam Brown, Phil Wagner, Louis Lindsay, Henry B. Kerfoot, Paul Sanborn, A. S. (Tony) Burgess, Donald J. Miller, Richard S. Williams, Dennis G. Fenn, Lorne G. Everett, Charles O Riggs, Robert V. Colangelo, Douglas R. Fraser, Dirk Kassenaar, John Luttinger, Pedro Fierro, Dennis Goldman, John V. A. Sharp, Mike D. Shotton, Russell H. Plumb, Jr., Joe D'Lugosz, Michael J. Mann, David Nielsen, David B. Kaminski, Bashir A. Memon, J. W. LaMoreaux, Lori C. Huntoon, Timothy D. Lynch, Ronald Schalla, and George Gillespie. This work represents a particular perspective of site investigations and the individuals above are exempt from any responsibility for whatever omissions or errors that may be present within this document. Source information credit has been provided, whenever a source was known. If omissions are present, they are unintentional and will be corrected in later printings. I would like readers to bring to my attention any such credits, where considered significant.

Special thanks are extended to Gary A. Williams, Leonard J. Butler, Patrick R. Spooner, Bob Barber, Ron DeBattista, and Vito Galante for their constant striving for technical excellence.

Finally this book could not have been completed without the high technical goals set by Donald Wallgren, Peter Vardy, Phil Rooney and Dean Buntrock as part of the Waste Management Inc. (WMX) extensive Environmental Monitoring Program.

Errata
Standard Handbook for Solid and Hazardous Waste Facilty Assessments
Martin N. Sara
ISBN: 0-87371-318-4
Lewis Publishers

Page 10-83

Equation 10-22 should read:

$$CDI = \frac{Cw \cdot IR \cdot EF \cdot ED}{BW \cdot AT \cdot Days}$$

Equation 10-23 should read:

$$IEC = Ca \cdot \frac{ET}{24} \cdot \frac{EF}{365} \cdot \frac{ED}{AT}$$

The cover art was produced by Ramesh Venkatakrishnan.

For more information, contact CRC Press, Inc., 2000 Corporate Blvd. NW, Boca Raton, FL 33431.

TABLE OF CONTENTS

APPENDICES

LIST OF FIGURES

LIST OF FIGURES (continued)

LIST OF FIGURES (continued)

LIST OF FIGURES (continued)

LIST OF FIGURES (continued)

LIST OF FIGURES (continued)

LIST OF FIGURES (continued)

LIST OF FIGURES (continued)

LIST OF FIGURES (continued)

LIST OF FIGURES (continued)

LIST OF FIGURES (continued)

LIST OF FIGURES (continued)

LIST OF FIGURES (continued)

LIST OF FIGURES (continued)

LIST OF FIGURES (continued)

LIST OF FIGURES (continued)

LIST OF FIGURES (continued)

LIST OF FIGURES (continued)

LIST OF FIGURES (continued)

LIST OF FIGURES (continued)

LIST OF FIGURES (continued)

LIST OF TABLES

LIST OF TABLES (Continued)

LIST OF TABLES (Continued)

LIST OF TABLES (Continued)

STANDARD HANDBOOK
for
SOLID and HAZARDOUS WASTE
FACILITY ASSESSMENTS

CHAPTER 1

INTRODUCTION

1.1 PURPOSE AND SCOPE

This manual is intended to assist those individuals planning, performing, and interpreting investigations for the facility suitability for disposal of solid or hazardous waste and the selection of appropriate locations for ground water monitoring the effectiveness of the landfill's leachate collection and containment systems. This document also address technical issues of remedial assessments for land disposal facilities.

Many factors influence the applicability of a site for use as a landfill: rainfall, runoff, air quality, and public concerns. Two of the most critical are geology and hydrology. This manual concentrates on these two factors. However, each factor must be carefully examined to the extent necessary to assess the overall suitability of a site. We must be fully aware of the factors that make a site acceptable for land disposal and of the mitigating measures available to minimize environmental impacts. This manual proposes a holistic program as illustrated on Figure 1-1, to examine the physical, chemical, and environmental factors that effect the suitability of a site for waste disposal. This program is extended to also address technical aspects of site assessments that deal with evaluation and remediation of contaminated ground water.

There is always a need for subsurface information and geotechnical data during the planning and development stages of land disposal facilities. A detailed understanding of the site geology is necessary for any project that has major components supported on or in the earth and underlying rock. The geologic and hydrogeologic features that will affect design and construction of the waste disposal facility must be investigated and evaluated to define the ultimate potential of the site for land disposal activities. This text stresses the acquisition of both geologic and geotechnical information that allow the prediction of ground water flow adjacent to a proposed, active, or closed land disposal facility.

A major part of the prediction of ground-water flow

adjacent to land disposal of solid or hazardous waste is the development of an understanding of the heterogeneous nature of the subsurface. Geologists have to use the method of multiple working hypotheses because they work with incomplete data and because several different processes may have contributed to the final site situation. The emphasis of this book is on the variability of conceptual models and on the complexity of subsurface environments. It is only by concentrating on differences

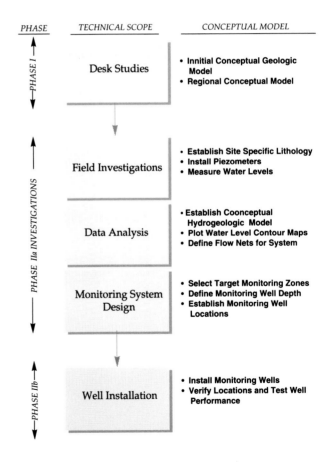

Figure 1-1 Summary Ground-Gas Monitoring Procedures

rather than on similarities that the importance of various processes which combine to make a particular geologic and hydrogeologic environment.

This text in many ways represents a philosophy of geologic understanding, perhaps more than a manual on techniques to develop that understanding. This book is all about sorting out geologic heterogeneity of specific sites (see Figure 1-2 for an example of glacial material with heterogeneity). These sites may represent facilities proposed for solid or hazardous waste land disposal or those areas that have received waste disposal and require some form of remediation or assessment. The technical literature of the late 1980s described the difficulty of predicting movement of contaminants in ground water due to heterogeneities in the subsurface (see Kazman 1989, Anderson, 1989). These problems with the inability to assess movement of contaminants due to small scale heterogeneity has been cited (OTA 1989) as one of the major hindrance in aquifer remediation under the Superfund program.

This text presents an organized but flexible approach based on traditional site investigation technology that can define the important criteria of ground water flow in the subsurface environment. These traditional site investigation methods have been developed over the last eighty years. Unfortunately, traditional data collection and field analysis methods are commonly ignored in extensive site assessment Superfund projects due to drive to collect samples for wide-spectrum organic parameter testing at

analytical laboratories. One can observe assessment projects under Comprehensive Environmental Response, Compensation, and Liability Act of 1980 (CERCLA) Superfund programs that expend 50% of project costs to obtain laboratory analytical results that have 95% of the wide spectrum chemical parameters lists as non-detects. Of the remaining 50% of project costs, 40% is spent on the field (drilling and sampling activities) collection of the samples for the analytical tests. The remaining 10% of project cost is typically used to develop the understanding of the geologic environment and the heterogeneities that can greatly complicate the understanding of how the analytical results relate to the site as a whole.

This shotgun approach is a tremendous waste of time, money and effort that could be used to develop a focused site investigation that would point directly to the optimum remediation that would be most effective in reducing risk to human health and the environment. This manual look at assessment programs that would comply with both Resource, Conservation & Recovery Act of 1976 and CERCLA remediations.

The 1980s saw the increased awareness of protection of ground water resources. Federal regulations associated with CERCLA, RCRA, sole source aquifers and many additional state and federal regulations directly address protection of the ground water resource.

This manual supports the protection of the environment through assessment of the performance of land disposal facilities. The assessment of the true risk to the

Source: A. Hathaway

Figure 1-2 Variations in Hydraulic Conductivity of Seven Orders of Magnitude Can Occur Within a Few Feet in Glacial Units

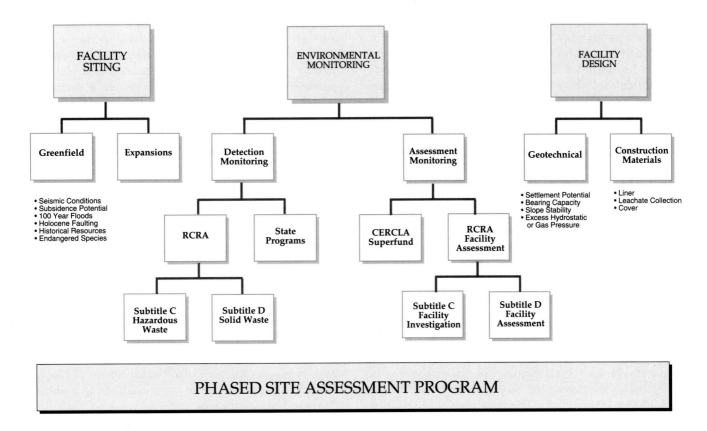

Figure 1-3 Categories of Site Investigations for Waste Disposal

environment of these facilities has been traditionally based on chemical indicator parameter monitoring. The performance of ground water monitoring wells in defining facility performance are highly sensitive to the specific location and depth of the well screens. These screens must be placed in the direct hydraulic flow paths of the facility to be monitored.

Monitoring wells can be compared to real estate with the analogy, "There are three important criteria for an effective monitoring well system; location, location, location." Federal regulations such as RCRA and CERCLA are directly keyed into water quality monitoring results; either by statistical trigger tests or by maximum contaminant limits (MCLs) as codified by water quality standards. These statistical and MCL triggers are directly dependent on ground-water monitoring. If the monitoring wells are located outside of the hydraulic flow path from the facility to be monitored, the true risk of that facility to the environment will never be evaluated properly.

An important purpose of this manual is to describe the various procedures for subsurface investigations applicable to the understanding of the near surface flow regime that

serves as the pathway for potential or actual movement of leachate from land disposal of solid or hazardous waste. An outline of a sequence of operations for conducting an investigation is presented throughout this manual for a wide spectrum of geologic environments. Data obtained at each operational step should be interpreted and the findings applied to optimize each successive work step. These geologic and geotechnical data should be considered as influential or even critical in all planning, design and construction stages of the land disposal project.

Site investigations can be divided into three main categories; facility siting, facility design, and environmental monitoring. These categories are illustrated on Figure 1-3. By far the largest segment and cost associated with site investigations is environmental monitoring. Federal and state regulatory programs are particularly directed toward facility performance, detection monitoring (RCRA) or assessment monitoring of rate and extent of contamination as codified by the CERCLA.

The primary purpose of this manual is to describe the technical components necessary to develop monitoring and assessment procedures for each of the three types of programs shown on Figure 1-3.

The manual addresses the increasing demand for detailed hydrogeologic and geotechnical information which has initiated extensive and costly surface and subsurface explorations for siting, design and monitoring of land disposal facilities. The level of investigation appropriate to a particular project must be given careful consideration during planning efforts. Though the additional information gathered in these waste disposal investigations will generally decrease possible unknowns and environmental risks, a balance must be maintained between the costs of the investigation program and the level of produced information.

Widely diverse geologic environments, local equipment, personal preferences, experience, time and budget constraints can all contribute to the development of investigative approaches. Subsurface exploration procedures cannot be reduced to a few guidelines that fit all conditions. The effects of specific geologic and hydrogeologic conditions on the type of proposed waste disposal facility or assessment project must be evaluated for each project. Therefore, it is imperative that one understands the critical elements of site assessment in selecting and developing landfill sites as well as addressing remedial investigations. This manual is intended as a guide to characterizing site hydrogeology and developing an effective ground water monitoring for both detection and assessment programs.

The conceptual hydrogeological model is the primary mechanism for accomplishing this task. Development of this model is an iterative process of increasing detail and is, therefore, developed in phases. This manual defines:

- the developmental phases of the conceptual model;
- the key elements of each phase and their purpose; and
- the methodology for defining the key elements.

1.1.1 Scope

This manual provides guidance on site characterization procedures basic to all site investigations. However, this manual does not detail the additional elements required for procedural planning efforts associated with Superfund activities and similar State level enforcement actions. This text does, however, address technical components of assessment monitoring programs that define both rate and extent of contamination and provide design criteria for ultimate aquifer remediation. The various technical tools of site assessments are discussed in detail to provide sufficient background techniques such as flow net constructions, cross section instructions, and documentation standards.

This manual is provided as guidance to technical staff with prior geotechnical and hydrogeologic training and background. Care has been taken throughout this manual to describe the recommended procedures as fully as possible within the physical constraints of the document. Where necessary, additional source references are cited that develop the concepts to the point where a full site investigation can be implemented from initial planning to final report. The assessment methods presented require technical evaluation to be most effective. Hence, those intrusted with such investigations should have a strong founding in geology, hydrogeology, geotechnical and soils engineering.

1.1.2 Historic Background to Site Assessments

Geology has long been recognized as a key factor in the successful construction and performance of engineered works. Although "designed" land disposal facilities have developed only over the last twenty years, by the early 1900s, "earthwork engineers," the forerunner of geotechnical engineers, saw that a better understanding of geology was needed, particularly in the context of geology as related to the solution of many engineering problems that existed during the time.

At the beginning of the twentieth century, the technology of the earthwork engineer consisted primarily of empirical practice (guidance developed through past successes and failures in construction) supported by early theories in soil mechanics. The earthwork engineer relied upon geologic observations that were made without the benefit of today's investigative methods (Terzaghi, 1960). Geologic investigations were commonly made prior to construction of the project, but not in a cognizant scientific manner, and not always with satisfactory results. The typically academic description of geologic conditions were considered not relatable to the solution of engineering problems of the day. Terzaghi first recognized this in 1911. By 1918, Terzaghi determined that the successful integration of geology into earthwork engineering required the quantification of geologic data. At this point, Terzaghi began to develop the investigative methods, which includes the systematic evaluation and measurement of the physical properties of earth materials, that are the basis of today's site investigation (Kent, 1981).

Site characterizations (Figure 1-4) have gradually evolved over the last forty years from soil boring programs for roads, dams, and buildings (which actually carried a relatively high risk factor in human lives) in the 1950s to literally hundreds of borings and monitoring wells for hazardous and solid waste investigations in the late 1980s. This trend of more data and lower real risk will probably continue through the 1990's with the passage of additional components and technical guidance under RCRA.

The development of systematic evaluation and measurement of geologic materials required that predictions made through surface geologic investigations must be checked by actual penetration of the subsurface to obtain records of the subsurface materials and to collect samples for later laboratory testing. No matter how sophisticated the evaluation, the field data must be gathered through drilling, sample collection and logging of the observations using methods and procedures that have changed little in the last 50 years. The basic technology of field collection of subsurface samples for site investigations was described by Hvorslev in his masterly post-World War II report (1949). This report was reprinted in 1968 and is still regarded as probably the best guide on the subject (Legget & Hatheway 1988). Figure 1-5, taken from Hvorslev (1949), still represents the basis for many site investigations performed today. Although this site investigation methodology was described by Hvorslev over fifty years ago, the basic phased approach was correctly structured for subsurface explorations and stratigraphical surveys. Hvorslev directed his discussion toward foundation studies, however, the technique described would be fully applicable for assessment studies for landfill design, ground-water detection monitoring well locating and the majority of rate and extent assessment monitoring investigations. The application and extent of use of these techniques represents the current challenge for those intrusted with site assessment responsibilities.

Geotechnical practice increased dramatically with the advent of the major construction projects from the 1940s to the present. From disasters such as the 1928 St. Francis Dam to more recent fault investigations associated with Preliminary Safety Analysis Reports (PSARs) for nuclear power station siting, regulatory driven site investigations accelerated the development of geotechnical practice. The initiation of Superfund in 1980 directly affected the practice of site investigations by the requirement of rigid planning documents for all phases of remedial investigations (RIs). These heavily planned projects unfortunately have relegated collection of geotechnical and hydrogeologic field data secondary to sample collection for chemical analytical programs. The percentage proportion of cost devoted to a wide variety of site investigations is given in Table 1-1. Typical site investigations rarely exceeded 2% of construction costs even for geotechnically sensitive tunnel projects, until the advent of the Superfund RI/FS investigation. The average of over 120 RI/FS projects represented 8% of the estimated remedial project costs.

THE EVOLUTION IN SITE CHARACTERIZATIONS: 1960-90

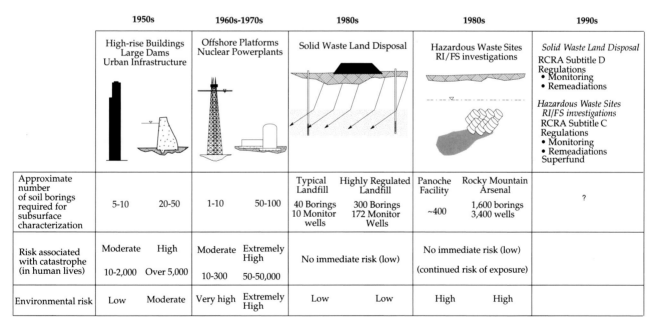

SOURCE: MODIFIED FROM DUPLANCIC & BUCKLE (1989)

Figure 1-4 Evolution of Site Characterizations

TABLE 1-1 Proportional Cost of Geotechnical Site Investigations

TYPE OF PROJECT	RANGE OF COSTS (% OF TOTAL COSTS)
Tunnels	0.3-2.0
Dams	0.3-1.6
Bridges	0.3-1.8
Roads	0.2-1.5
Buildings	0.2-0.5
Solid waste landfill	1.2-2.4
Superfund RI/FS	8.0*

*U.S. EPA Figures (1988)

Source: Legget & Karrow (1983)

The obvious jump in the percentage of investigation costs associated with the Superfund project is indicative of the drive to reduce perceived environmental risk associated with these sites. Actual risks of infra-structure facilities carries much higher real risk to human life than the majority of Superfund sites. However, perceived environmental risk is very difficult to fully assess in an investigation that primarily is directed toward sample collection and chemical analysis. For this reason, the RI/FS investigation becomes very expensive when compared to more traditional site investigations for structures. A case will be made in Chapter 10 for redirecting the Superfund RI back toward the more traditional site investigation procedures developed over the last eighty years. These methodologies include reliance on defining ground-water flow conditions before installation of monitoring wells. These procedures also include use of

GENERAL SITE ASSESSMENT CLASSIFICATIONS

Stability of subsurface materials Deformation and consolidation Pressure on supporting structures	Foundation Problems	FOUNDATION INVESTIGATIONS	Stratigraphic Survey	Exploration Sampling Identification	Stratigraphical Profiles Rock-line, rough, or detailed soil profiles, limited physical profiles
Excavation of subsurface material Use of excavated material	Construction Problems		Physical Survey	Laboratory Tests Major Field Tests	Physical Profiles Rough or detailed data on variation of physical properties with depth
Flow and action of ground water Quantity and use of ground water	Ground water Problems		Evaluation of Data	Empirical Rules Soil Mechanics	Final Design Based on stratigraphical and physical profiles, latter often estimated

Soil from various strata mixed or some soil constituents removed or foreign materials added to sample	Non-Representative Sampling	SAMPLES AND SAMPLERS	Exploration Samplers	Augers, bailers , sandpumps; slit and cup samplers or sampling tubes with a longitudial or circumferential slit. Non-representative and representative samples
Soil structure disturbed or change in water content or void ratio but no change in the soil constituents	Representative Samples		Drive Samplers	Sampling tubes driven without rotation or chopping; displaced soil pushed aside. Open drive samplers and piston samplers. Represent. and undisturbed samples
No disturbance of soil structure No change in water content, void ratio, or chemical composition	Undisturbed Samples		Core Boring Samplers	Rotation or chopping action of sampler; displaced material ground up and removed by circulating water or drilling fluid. Representative of undisturbed samples

Geophysical methods, soil sounding, yielding limited physical profiles, Borings without taking representative samples of principal strata	Indirect Methods	STRATIGRAPHICAL SURVEY OR SUBSURFACE EXPLORATIONS	Fact Finding and Geological Survey	Gathering of data on foundation experence and results of previous explorations at site and vicinity. In case of large projects, site survey by an engineer-geologist
Borings- Changes in material detected by action of tools or non-representative samples. Identification by representative samples	Semi-Direct Methods		Reconnaissance Explorations	Semi-direct methods with or without indirect methods. Rough ground water observations in permeable strata. Required for all projects. Sufficient for many standard structures, small loads, simple soil conditions
			Detailed Explorations	Continuous undist. samples of small diameter or large repres. samples of pertinent strata. Ground water observations in fairly pervious strata. Sufficient for common foundation problems, all construction problems
Test pits, trenches, drifts, other accessible explorations or borings yielding fairly continuous, representative or undisturbed samples	Direct Methods		Special Explorations Undisturbed Sampling	Large undisturbed samples in boreholes or in accessible explorations. Ground water levels and pressures in impermeable strata. Required for structures unusual size and character and for difficult foundation conditions

M.J. HVORSLEV (1949)

Figure 1-5 Hvorslev's Site Assessment Classifications

indicator parameters before wide spectrum analytical testing of the affected environment. The purpose of this manual is to provide a technically correct methodology for assessing true risk of land disposal to human health and the environment.

The philosophy of the classic site investigation was described by Ralph B. Peck in 1969 as the observational method. The complete application of the observational method embodies six general steps.

- Conduct an investigation of sufficient scope to establish the general characteristics of a site.

- Assess the most probable site conditions and the deviations from these probable conditions.

- Develop a design based on the most probable site conditions.

- Determine what courses of action should be taken if the conditions deviate from predictions.

- Measure and evaluate actual conditions during construction.

- Modify the design as needed to suit actual conditions.

Although the goals of the investigation may be to define the proper location for a detection monitoring system or to establish rate and extent as part of a remedial investigation, the observational method is presented in various forms throughout this manual. Since this approach represents a flexible methodology, accommodations must be made within the rigid planning associated with Superfund-type investigations to allow for reasonable adjustments to the program based on observed conditions. The adjustments to the flow of these assessment projects are addressed in Chapter 10 of this document.

1.1.3 Objectives

The objectives of this manual's approach is defining an effective and efficient environmental monitoring system for land disposal facilities that includes assessments of:

- ground water;
- potential subsurface leachate migration;
- potential gas migration;
- climatology/meteorology; and
- surface water.

A second important objective is to define the site characteristics that effect landfill design and facility remediation. The third major technical objective of facility site assessments is to quantify and qualify landfill construction materials.

This manual's technical approach to site assessments consists of first establishing regional data and gradually identifying and refining the understanding of the site characteristics as more information is gathered through a structured approach. This process is illustrated in Figure 1-6 where we begin with a regional perspective and home-in on local and specific borehole data assessment to develop a full picture of the site conditions. This process is used in selecting locations for boreholes through regional background literature review (Regional) supported by site geologic and geophysical mapping (local setting) to select the target drilling locations. This process of regional conceptualizations from background literature review to develop site assessment procedures and specific site conceptualization is a reoccuring thiem in this manual.

This structured approach usually consists of the following:

- A thorough review of existing literature and technical information.

- Preliminary site aerial and ground reconnaissance.

- Development of initial regional and site specific geologic conceptual models.

- Design of the field investigation is then based on the initial conceptual models.

This investigation must be sufficient to establish target monitoring zones and identify limiting design criteria and/or quantify availability of construction materials. This process of conceptualization, design, and well placement by a phased investigation is illustrated in Figure 1-1. This phased approach is consistent with both detection and assessment monitoring programs such as codified by CERCLA (Superfund) and many state remedial clean-up programs.

Conceptual Site Understanding

The conceptual reconstruction of geologic environments requires the following:

(1) A thorough field description of the unconsolidated and rocks with additional laboratory data obtained from samples collected to answer specific questions. Since time is always limited, rock description is inevitably selective, emphasizing

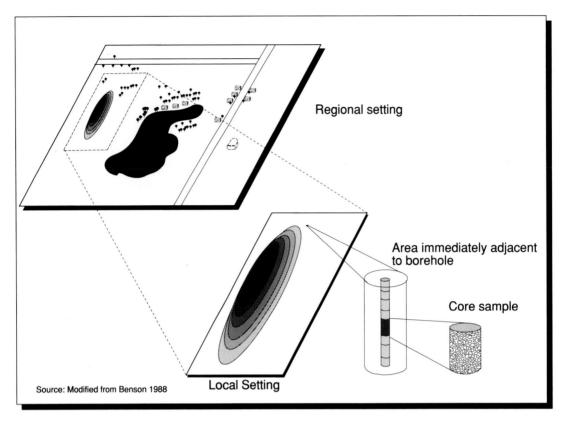

Source: Modified from Benson 1988

Figure 1-6 Regional to Site Specific Data

some features, underplaying others and rejecting yet others as quite unimportant. The selection depends on the judgement, experience and purpose of the investigator.

(2) An awareness of processes so that, simultaneously with rock description, the investigator develops conceptually a working basis of the depositional environment. We also have to consider the later alteration, diagenetic and fracturing processes which may have changed not only the porosity of the rocks but also their cementing composition.

(3) An understanding of the relationship of rocks to their vertical and lateral neighbours, shape of permeable bodies and their contacts with neighboring units. These relationships are used as constraints to eliminate certain environments and thereby reduce the number of options.

(4) A knowledge of present-day environments and the processes which operate within them. Structural features and permeable units observed in recent sediments have direct analogs to bedrock heterogeneities.

Thus the emphasis in this book will be on: (1) conceptual *environments through field evaluation techniques*, (2) *processes,* concentrating on those that cause ground water to move along specific pathways in the environment and showing how they relate to the resultant ground-water monitoring issues. They will not be discussed on their own as there are already several good textbooks on processes and the genesis of sedimentary rocks and structures, *(3) Field classification of unconsolidated deposites and bedrock,* stressing field data, facies relationships, sequences and associations, (4) *geological applications,* illustrating how sedimentary rocks are related to their geological background and how the recognition of sedimentary processes and environments illuminates our understanding of ground-water flow, the chemistry of ground water, and the environmental processes that control containment of land disposed waste.

1.2 ANALYSIS OF EXISTING INFORMATION

Utilizing the criteria and definitions contained in the Site Assessment Manual, site investigation always begins with an assessment of regional geology with existing available information to design a site investigation plan. Site specific information (as described in Section 2.0) may be available from published sources or previous site studies. In some cases, existing information may be adequate to sufficiently characterize the site for environmental monitoring purposes. Additional field investigation for detection monitoring programs should only be undertaken when existing information must be supplemented in order to obtain adequate assurance that aquifers or subsurface migration pathways are sufficiently defined and monitored. Assessment monitoring programs should also first fully incorporate available geologic and hydrogeologic information before proceeding with additional field drilling programs. Existing geologic and hydrogeologic data may be available to provide the following information:

- Characterization of subsurface geology and geomorphology.

- A regional location map and a specific map of the facility showing site topography and locations of all borings, piezometers, and monitoring wells.

- Location of the uppermost aquifer and aquitard(s).

- Hydrogeologic characterizations of each water-bearing formation down to the uppermost aquifer.

- Hydrogeologic characterization of the area.

- Water quality characteristics of the water-bearing formations.

- Formulation of preliminary conceptual model of site geology.

This preliminary phase can be expanded to account for planning associated with remedial site assessments.

EXAMPLE OF GENERATIONS OF ENVIRONMENTAL DATA

FIRST GENERATION DATA   SECOND GENERATION DATA   THIRD GENERATION DATA   FOURTH GENERATION DATA

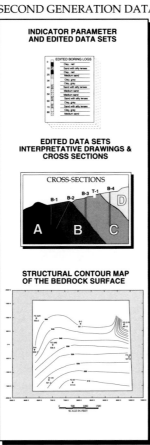

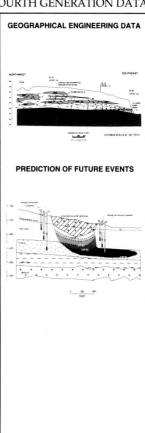

Figure 1-7 Levels of Environmental Data

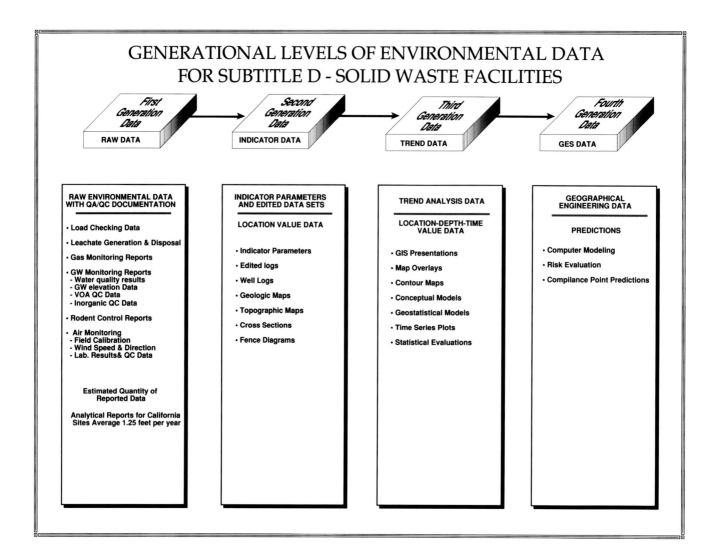

Figure 1-8 Subtitle D Levels of Environmental Data

However, the planning documents should not be massive, highly structured documentation efforts, and should only point toward the goals of acquiring sufficient understanding of the chemical, geologic and hydrogeologic conditions to evaluate the sites' environmental conditions for design of the site remediations.

Site assessment projects generate a lot of data. The collection evaluation and documentation of environmental data can quickly become a technical hurdle that many project do not successfully negotiate. Technical data can become even more intractable when collected from disconnected projects by different staff members working under variable (or non-existent) documentation standards. Environmental data from detection and assessment monitoring analytical programs generate some of the most intensive and difficult to interpret data sets. Many

operational problems with simple data handling and validation the massive volumes of information can be expected before this water quality data can be used within a monitoring program. Environmental monitoring data is often considered as program specific, (i.e., Superfund, State permit requirement, RCRA), however, these data have many similarities between both state and federal regulatory programs. These data can be categorized into "generation" of data. Figure 1-7 illustrates the various generations beginning with the first generation "RAW" data with QA/QC documentation. This first generation also includes planning documentation raw field data, unedited logs, and tabular analytical data. First generational data tends to be very specific to the program. For example, Superfund projects require very specific format standards for planning, execution and reporting of

" SIX LAWS OF ENVIRONMENTAL DATA"

1) THE MOST IMPORTANT DATA IS THAT WHICH IS USED IN MAKING DECISIONS.

   ∴ ONLY COLLECT DATA THAT IS PART OF THE DECISION-MAKING PROCESS.

2) THE COST OF COLLECTING, INTERPRETING, AND REPORTING IS ABOUT EQUAL TO THE COST OF ANALYTICAL SERVICES.

3) ABOUT 90% OF DATA DOES NOT PASS FROM THE FIRST GENERATION LEVEL TO THE NEXT.

4) THERE IS SIGNIFICANT OPERATIONAL AND INTERPRETIVE SKILL REQUIRED IN MOVING UP THE DATA GENERATION LADDER.

5) DATA INTERPRETATION IS NO BETTER THAN THE QUALITY CONTROL USED TO GENERATE THE ORIGINAL DATA.

6) SIGNIFICANT ENVIRONMENTAL DATA SHOULD BE APPARENT.

Figure 1-9 Laws of Environmental Data

the project results to define the optimum remedial methods for site clean-ups. Samples must be collected according to federal guidance and analyzed according to well defined quality standards. RCRA within the Subtitle C regulations also have many "generation one" raw data than are required by the specific permit. Subtitle D within the solid waste regulations of RCRA. As with the other Federal (and many state regulatory programs) the first generation data is specific to the individual regulatory standard. Higher generation data sets are illustrated in Figure 1-8 that move from indicator parameters and edited data set (2nd generation) to third generation data that is based on point in space (x-y-z) value at a specific time. Second generation data includes interpretative drawings, cross sections and maps obtained form the raw data. Third generational data includes statistical data evaluations and geographical information systems that overlies data sets to aid in the interpretation. Fourth generation data permits the prediction of future events by using the lower order

generational data to model the coming environmental conditions. Inspection of the two figures will illustrate the similarity of required data once moving past the first generation. Hence, the presentations and interpretations tend to be very similar for the higher order data. This makes the execution of assessment monitoring and ultimate aquifer remediation very similar for any of the federal or state remediation programs once moving past the first generation raw data gathering. There have also been many similar observations to environmental data gathered to meet regulatory requirements These observations have been formulated into a series of environmental "laws"as shown in Figure 1-9.

These "laws" provide operation guidance in generating and handling the various generations of environmental data. These "laws" are used throughout this manual to direct and target important data for environmental monitoring systems.

SITE COMPLEXITY INDICATORS
FOR SELECTION OF ASSESSMENT TECHNIQUES

CONCEPTUAL HYDROGEOLOGY   EXAMPLE ASSESSMENT TECHNIQUES

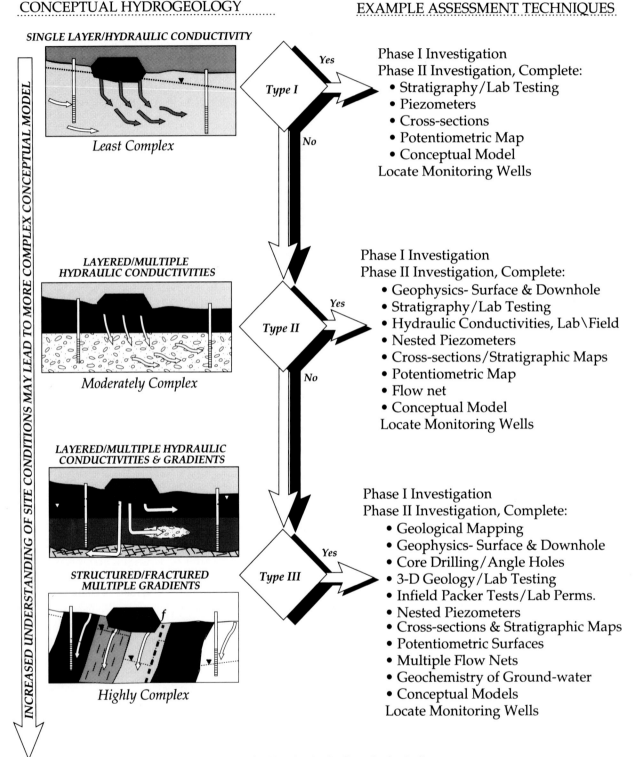

SINGLE LAYER/HYDRAULIC CONDUCTIVITY

Least Complex

Type I — *Yes*

Phase I Investigation
Phase II Investigation, Complete:
• Stratigraphy/Lab Testing
• Piezometers
• Cross-sections
• Potentiometric Map
• Conceptual Model
Locate Monitoring Wells

No

*LAYERED/MULTIPLE
HYDRAULIC CONDUCTIVITIES*

Moderately Complex

Type II — *Yes*

Phase I Investigation
Phase II Investigation, Complete:
• Geophysics- Surface & Downhole
• Stratigraphy/Lab Testing
• Hydraulic Conductivities, Lab\Field
• Nested Piezometers
• Cross-sections/Stratigraphic Maps
• Potentiometric Map
• Flow net
• Conceptual Model
Locate Monitoring Wells

No

*LAYERED/MULTIPLE HYDRAULIC
CONDUCTIVITIES & GRADIENTS*

*STRUCTURED/FRACTURED
MULTIPLE GRADIENTS*

Highly Complex

Type III — *Yes*

Phase I Investigation
Phase II Investigation, Complete:
• Geological Mapping
• Geophysics- Surface & Downhole
• Core Drilling/Angle Holes
• 3-D Geology/Lab Testing
• Infield Packer Tests/Lab Perms.
• Nested Piezometers
• Cross-sections & Stratigraphic Maps
• Potentiometric Surfaces
• Multiple Flow Nets
• Geochemistry of Ground-water
• Conceptual Models
Locate Monitoring Wells

INCREASED UNDERSTANDING OF SITE CONDITIONS MAY LEAD TO MORE COMPLEX CONCEPTUAL MODEL

Figure 1-10   Site Geologic Complexity Indicators

1.3 SITE INVESTIGATIONS (PHASE II)

When more information is necessary to establish an adequate monitoring well network or to evaluate the facility's risk to the environment, a field study designed to produce supplemental data should be conducted. In advance of any field work, the staff involved in the project need to have a clear understanding of the necessity for the work, and the objective of each task or activity. Phase II field work may include the following tasks:

- Remote imagery and aerial photography
- Geologic mapping
- Geophysics (surface and downhole)
- Soil borings, rock coring, trenching, and test site installation
- In-field hydraulic conductivity tests
- Water-level measurements
- Sample collection for assessment purposes

The actual techniques used in the assessment depends on how complex is the geologic environment under investigation. A conceptual basis for this selection of field assessment techniques is given in Figure 1-10. As more data is gathered on a particular site, the conceptual understanding may change sufficiently to cause an investigation to add additional field and office assessment techniques. Only after a full understanding of the site is known should monitoring wells as shown on Figure 1-11 be installed at a facility for ground-water monitoring activities.

A full discussion of the field collection of data for "typical" site assessments is presented in Section 3.0 to section 5.0 of this manual with fractured rock covered in Chapter 6.0. Interpretation and analysis guidelines for this information is given in Section 7.0. Section 8.0 describes the process of developing conceptual models and flow nets. Section 9.0 defines the use of the collected data to make decisions on design of both detection and assessment monitoring systems. Section 10.0 looks at assessment programs that comply with technical aspects of both RCRA and CERCLA assessments. Section 11.0 reviews the presentation of water quality data for both detection and assessment monitoring programs. Each of the presentation techniques covered in these chapters provide a system approach toward the overall understanding of the final results of the performance of monitoring and assessment programs.

A number of sections contained in this manual provide more detail than would be traditionally included in a guidance manual. This is a result of local needs for detailed information on specific issues such as holocene faulting assessments.

1.4 CLIENT / OWNER RELATIONSHIPS

The relationships between owners or clients and a consultant can have a significant effect on the performance of a project. In the "good old days" a client would often pose a problem to a consultant for scoping and ultimate solution. These days are now past in the majority of cases. Projects are often scoped by the client and competitively bid. The following chapters are presented in the hope that the guidance provided can reduce technical differences in scoping site investigations. Most consultant firms enjoy a 60 to 80 percent repeat client base; its erosion through poor client relationships can have a devastating effect on both morale and marketing costs. This text provides technical answers to many environmental questions associated with waste disposal. Following the guidance provided can aid in the evaluation of facilities for the general benefit of society and specifically to the benefit of the facility owner. Nevertheless even with a correct scope of work for a project, relationships do go sour, and more that likely it's generally due to people-problems.

Common causes of "people-problems":

- Changing personnel without the client's prior knowledge or approval. "Changing" includes switching horses between proposal or interview and starting the work. We care about who's doing our work, and consultant selection is heavily influenced by personal credentials and chemistry.

- NOT changing personnel in the course of a job when it's obvious there is a personality conflict with the client.

Figure 1-11 Monitoring Wells for Waste Disposal Require Accurate Placement in a Three-Dimensional Context

- Not hearing problems or ignoring what you hear. Not listening is a major client complaint. "We get very frustrated when we ask our consultant to explore one concept, and the person comes back with something completely different."

- Not following client procedures and specifications. Large commercial clients must use systematic methodology for their field activities and designs to provide continuity between various sites and regions. Many clients also require consultants to follow specifications, manuals and procedures while working at their facilities. The consultant must train his field staff properly in the required techniques before coming on site to do work.

- The marketer or project data-collector isn't the doer. When the person collecting project background information, for example, is not involved in the execution of the work, a lot of listening can get mistranslated. This can also happen when the principal is the sole contact with the client and depends on the back room to produce the work.

- Silence when things aren't going well. When there's a problem, tell the client as soon as possible. Do not delay in the hope that it will go away or be solved without the client knowing about it. If there are budget or schedule difficulties, say so. Most clients can handle bad news, but they almost always hate surprises.

- Leaving your client to struggle alone when there's a tough decision to be made. It may be the client's decision, and you may have provided all the necessary technical material; now offer some personal support.

- Poor communications skills. Too many professionals can't articulate their concepts and plans; they stumble through public agency approvals; they flop at community hearings; they can't be convincing at presentations to the client's client (prospective site owners, boards of directors, etc.). Hone your public speaking skills.

- Making promises you can't keep -- the quickest way to lose a client's trust. When you agree to do something, not only perform -- also give the client progress reports and advise when it's done.

- Poor writing skills. Often the only contact regulatory agencies have with client sites are the consultant's written reports and permit applications. Hence, technical information required in site characterization studies must be presented in a lucid manner. Our experience shows that confused report organization is a sign of confused thinking. Poor English and sloppy presentation often cast questions on the technical validity of the report.

- Not backing the client during litigation. When asked to defend technical issues in court, consultants often get cold feet and refuse to stand on conclusions reached in their reports. Nothing corrodes confidence so quickly as requiring additional work to back-up report conclusions.

- The use of weasel wording. We believe that reports should be written in the active voice and the consultant should avoid using non-committal language.

- Out of sight, out of mind. Weeks or months may elapse while a project is dormant. Even if you're busy with other work, build and nurture the relationship with your client.

- Providing only service. Many professionals do the job well but forsake the relationship. Do provide the extra measure of service when you can, and give your project manager the time and budget to handhold the client. But also remember the personal side. Encourage social contact; cultivate a friendship as well as a professional relationship.

- "When's your next job?" Staying in touch after the job is finished only to inquire about future work is poor strategy. Make contact quarterly after completion, and provide a thorough project evaluation one year later -- at no cost to the client.

This sounds like a fairly extensive list of to-dos that have basically nothing to do with the activity of defining sub-surface profiles. However, the above "people problem" points often decides whether a project or client relationship is successful and that a project is completed correctly.

REFERENCES

Anderson, M.P. 1989, C.V. Theis & Heterogeneity, Recent Advances in Ground Water Hydrology, American Institute of Hydrology, Minneapolis, MN 55414.

Kazman, R.G. 1981, C.V. Theis' Contribution to Ground Water ,Advances in Ground Water Hydrology, American Institute of Hydrology, Minneapolis, MN 55414.

Kent. M. 1989. Bulletin of the Association of Engineering Geologists.

Legget R. F., & A. W. Hatheway 1988. Geology and Engineering, McGraw-Hill , New York NY, 613 p.

Office of Emergency & Remedial Response, 1989, Evaluation of Ground Water Extraction Remedies, Volume 1 EPA/540/2-89/054.

CHAPTER 2

PHASE I INVESTIGATIONS

2.1 INTRODUCTION TO PHASE I STUDIES

Site assessment Phase I preliminary investigations are designed to provide a comprehensive overview of available information concerning a site. The need for this preliminary phase is often forgotten in the rush to go and drill some holes on site to "see what is there." This urge should be resisted since many errors and misinterpretations in geologic data can occur when poorly placed exploration boreholes contact stratigraphic materials that are structurally complex. Any site assessment project where significant field data gathering is required, should be conducted in a phased approach. Phase I would consist of primarily a literature and field reconnaissance that would set the stage for later intensive Phase II field data gathering.

Although this text primarily concentrates on waste disposal or remediation projects, many if not all concepts presented below would be applicable for any site investigation where the facility may have some impact on the environment. Specific to waste disposal these preliminary Phase I investigations can be conducted for new sites, either greenfield or acquisitions, expansions to currently operating sites, evaluations of associative risk of a site or as a first step in a Superfund Remedial Investigation/ Feasibility Study (RI/FS) (Table 2-1). The scope of work for Phase I studies is some what different for each of the above; however, the basic elements remain the same for these preliminary review efforts.

A preliminary investigation should always summarize the available literature and provide as complete a picture as

Table 2-1 Types of Phase I Investigations

Phase I	Goal	Scope Summary Description
Fatal Flaw Analysis	Greenfield Siting of new facilities	Rapid screening of alternative sites by comparison of site features or ranking
Existing Monitoring	Definition of reliability and effectiveness of monitoring program	Information Review, System Evaluation Well-Specific Data, Well Performance Evaluation, Well Development
Environmental Audits	Property assessments & transfers	Facility background review
Acquisition Reviews	Interviews	Site reconnaissance Limited surface & subsurface characterizations Laboratory Indicators Parameter scans Report of findings
Monitoring Well Installations	Conceptual models	Accurate location of detection monitoring wells
Site Investigation Plan	Define Phase II RI Investigation	Establishment of field conditions to allow accurate location of field borings. Site visit. Definition of procedures to be used at boring locations.

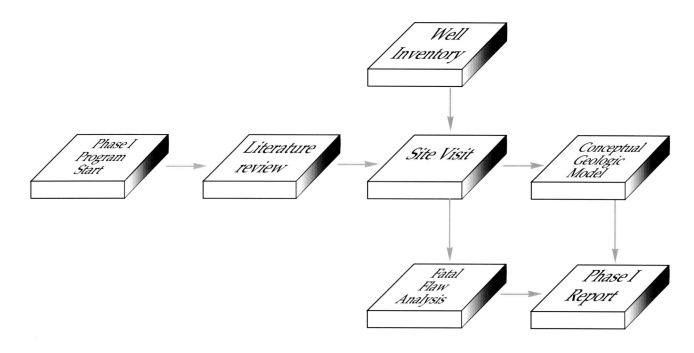

Figure 2-1 Phase I Program Summary

possible of the site, including facility operations and the basic geologic and hydrogeologic environment. Phase I studies for RI/FS investigations may require more additional programmatic components not fully covered by this manual, however, technical components for these assessment monitoring programs tend to have many similar goals and requirements. Table 2-1 illustrates the wide range of Phase I type investigations covered by this chapter. Site assessment requirements for landfill operating permits also differ from state to state, and review of state regulations should be an early component of the preliminary work. The Phase I scope of work, as shown in Figure 2-1, should be tailored to fit individual siting analysis needs. Suggested components of individual Phase I studies are provided in Table 2-2.

The preliminary investigation is based on a thorough review and analysis of existing information on the site or facility This review is supplemented by a site reconnaissance to substantiate concepts about the area developed as a result of this review and to provide additional information on problems requiring resolution. The final step in the preliminary investigation is the development of a conceptual model of the site and the completion of a Phase I site assessment report. The Phase I report may be called a fatal flaw analysis; monitoring well system evaluation; environmental audit or acquisition review; hazard identification or an RI work plan. Each of these Phase I documents may have somewhat different data analysis or reporting content, however, the primary goal is to set the stage for decisions on further actions associated

with the site. These actions may be to acquire the property, describe the field investigation necessary to define detection or assessment monitoring programs, or to assess the true or potential risk to human health and the environment.

The following discussion of a general Phase I investigation provides a basic template that would apply for most initial project, where decisions are required for later field studies. In addition the Phase I scope can be used for technical review of property acquisitions or of existing monitoring programs. These projects are described in more detail following the general Phase I description. The general Phase I preliminary investigation is divided into three main tasks:

• Existing Data Review,

• Site Reconnaissance; and

• Report of the results.

These initial Phase I projects are similar in the ways the data are reviewed, however, the specific reporting formats may be very different. This is especially true for regulatory programs planning documents as required under Superfund, RCRA, state remedial, or property transfer regulations. These reporting formats for regulatory programs must be fully addressed as the site assessment proceeds through the various phases.

A review of existing information is always required in Phase I project execution. This review may reveal fatal

Table 2-2 Phase I Program Summaries

Literature Review:- A review of existing information is always required in Phase I project execution.

Well Census:-An accurate well census of potable ground-and surface-water users represents one of the most important Phase I tasks associated with an assessment monitoring program

Site Visit:-After completion of Phase I literature search and review of the basic data checklist, one must visit the site to substantiate preliminary conclusions based on the information compiled during the literature search, and to obtain necessary additional site information.

Fatal Flaw:-Special care must be employed to define any "fatal flaws" of the sites in question. Refer to section 2.3 for examples of such flaws.

Conceptual Model:-Information obtained from the site reconnaissance, along with data secured during the literature search, provide the basis for the development of the conceptual model of the hydrogeologic conditions at the site. The use of conceptual models should be employed without fail for every type of Phase I project.

Phase I Report:-The purpose of the Preliminary Site Assessment (Phase I) Report is to summarize the literature review, findings, and recommendations of the initial field reconnaissance, and to provide the planning basis for the Phase II field investigations.

flaws in a potential Greenfield development, thus eliminating the site from further consideration. For detailed site assessment field studies, the review may put limits on the extent, depth, and location of required borings, piezometers, and monitoring wells; define the type and number of soil and water quality analyses; and identify potential hydrogeologic complexities and contingencies that may require further analysis. Such concerns may include the presence of faults, multiple and preexisting land uses, potential for reversible ground-water gradients, likelihood of variability of background water quality, uncertain stratigraphy, potential for ground-water flow oblique to the hydraulic gradient, variable background water quality, and initial evaluation of the complexity of the stratigraphy.

The planning, field assessment, design, and construction of any waste disposal facility require both regional information for planning and detailed site specific data. The large body of available regional data, however, has

made it increasingly difficult to obtain and review all the existing information pertaining to a given site or area. This Chapter is designed to expedite the literature search process by presenting a comprehensive description of the most widely used sources of geologic and related data for use by consultants and engineers. General site information can be gathered by using the sources listed in Table 2-3. The types of data required for site assessment investigations can be categorized as:

1. Site compliance & owner/operator data (Table 2-3)
2. Topographic data (Table 2-4)
3. Geologic data (stratigraphy, structure, geomorphology, geologic processes, and potential hazards) (Table 2-5)
4. Geophysical data (Table 2-6)
5. Remote sensing data (Table 2-7)
6. Ground and surface water data (Table 2-8)
7. Climatic data (Table 2-9)
8. Ecological data
9. Paleontological/historical data
10. Land uses
11. Holocene faulting

This section includes the most important sources of these categories of data, but additional data sources may be important on a state/local basis. A literature search should be made to gather as much data as possible at the start of the Phase I study. Much of this information can be gathered from both traditional information searches and computer interrogation of government and other data bases Many of these tradition sources and computer based systems are described within the following tabular material. Acquiring and assessing available data in any site investigation provides the most timely and cost effective beginning for any major waste disposal project.

2.2.1 Site Compliance & Owner / Operator Data

One of the first sources of available data for a site assessment project is from the owner/operator. Even greenfield siting projects require the gathering of specific information related to the proposed facility and evaluation of previous reports on the subject sites. Often project proponents will have performed preliminary evaluations of the property before going ahead with outside consulting services to further develop the facility. Commonly this owner/operator data will include air photos and feasibility reports. These data represents a good first step toward a complete Phase I investigation.

Operating or closed facilities often have long histories of investigations and regulatory compliance documentation. This is especially true for facilities that have historically handled hazardous waste. In order to have a complete

Table 2-3 Basic Data Sources

Source	Information Obtainable	Comments
Libraries	Earth science bibliographic indices	Many of the types of information discussed below can be obtained from libraries. Excellent library facilities are available at the U.S. Geological Survey offices (USGS) in Reston, VA; Denver, CO; and Menlo Park, CA. Local university libraries can contain good collections of earth science and related information and typically are depositories for federal documents. In addition, local public libraries normally have information on the physical and historical characteristics of the surrounding area.
Computer literature searches	Bibliographic indices	Perhaps one of the most useful and cost effective develop ments in bibliographic indexes has been the increased availability of computerized reference searches. Online computer searches save significant time and money by giving rapid retrieval of citations of all listed articles on a given subject and eliminate manual searching of annual cumulated indexes. A search is done by use of keywords, author names, or title words, and can be delimited by ranges of dates or a given number of the most recent or oldest references. The average search requires about 15 minutes of online searching and costs about $50 for computer time and offline printing of citations and abstracts.
SDI Services	Stored bibliographic searches	An extremely useful feature of computerized information retrieval systems is the Selective Dissemination of Information (SDI) Service. SDI is a procedure for storing search profiles and providing updates of an original computer search. The engineer should consult SDI Services for topics of relevance to the proposed project.
State and federal projects	Site specific assessment data for dams, harbors, river basin impoundments, and federal highways	Project reports contain data on all eight categories considered, as well as analysis, construction drawings, and references. These reports can generally be obtained most easily by contacting the responsible agency. Surface water and geological foundation conditions such as fracture orientation, permeability, faulting, rippability, and weathered profiles are particularly well covered in these investigations.
University sources	Engineering and geology theses	College and university geology theses, in most instances, are well-documented studies dealing with specific areas, generally prepared under the guidance of faculty members having expertise in the subject under investigation. Most theses are not published.
• Comprehensive dissertation index	Doctoral dissertations	Citations began in 1861 and include almost every doctoral dissertation accepted in North America thereafter. The index is available at larger library reference desks and is organized into 32 subject volumes and 5 author volumes. Specific titles are located through title keywords or author names. Ph.D. dissertations from all U.S. universities.
• AGI Directory of Geoscience Department	Faculty Members	Regular updates of faculty, specialties, and telephone date.

Table 2-3 Basic Data Sources

Source	Information Obtainable	Comments
• DATRIX II University Microfilms International 300 North Zeeb Road Ann Arbor, MI 48106 Tel: 800-521-3042 313-761-4700 (in Alaska, Hawaii, and Michigan)	Dissertations and Masters thesis	Using title keywords, a bibliography of relevant theses can be compiled and mailed to the user within one week. In addition, the DATRIX Alert system can automatically provide new bibliographic citations as they become available.
• United States Geology: A Dissertation Bibliography by State	Ph.D. dissertation or Masters thesis	Free index from University Microfilms International. Some universities do not submit dissertations to University Microfilms for reproduction or abstracting, however, and the dissertations from these schools do not appear in the United States Geology index. Citations for dissertations not abstracted must be located through DATRIX II or Comprehensive Dissertation Index.
• Dissertation Abstracts International, Volume B - Science and Engineering, a monthly publication of University Microfilms International	Extended abstracts of dissertations from more than 400 U.S. and Canadian universities	Once the citation for a specific dissertation has been obtained from the Comprehensive Dissertation Index or from DATRIX II, the abstract can be scanned to determine whether it is relevant to the project at hand. Since some universities do not participate, some theses indexed in the two sources listed above must be obtained directly from the author or the university at which the research was completed. .Abstracts of Masters theses available from University Microfilms are summarized in 150-word abstracts in Masters Abstracts and are also indexed by author and title keywords. Both Dissertation Abstracts International and Masters. Abstracts are available at many university libraries. A hard (paper) or microform (microfilm or microfiche) copy of any dissertation or thesis abstracted can be purchased from University Microfilms.
Local, state, and regional agencies	Soils, land use, flood plains, groundwater, aerial photographs	Local, state, and regional agencies. Local county, town, and city planning boards commonly provide data on general physical characteristics of areas within their jurisdiction. Many states maintain a department of the environment or natural resources. While the primary function of these agencies is not geological in nature, these organizations commonly have extensive information related to geology, remote sensing, and water.
Knowledgeable individuals	Historic information, past site owners and practices	Time can be saved in the initial stages of a data search by contacting knowledgeable individuals for references and an overview of an area, as well as for specific problems and details that may be unpublished. People to contact include university professors; and persons from relevant state and federal agencies such as state geological surveys, the USGS, or the Army Corps of Engineers; Corporate, Regional, and District Operations staff that have experience within the area. Local well drillers, consulting engineers, and architects.

Phase I evaluation the investigator should review these data available from both regulatory or compliance agencies and owner/operators:

- Preliminary reports, borings, siting studies and air photos;

- Local data on the location and characteristics of adjacent or nearby landfills, and other waste generators, transporters, and disposers (TSD survey) of hazardous waste;

- The listing of federal Superfund sites, both those on the the NPL and those included in the Comprehensive Environmental Response, Compensation and Liability Information System (CERLIS) database;

- Location data on the presence of underground storage tanks (UST's) which can be obtained by reviewing the required state notification form called for under published rules under CERLA;

- The location of above-ground storage tanks, available from the local fire marshal;

- Discharge permits such as NPDES, or Part A and Part B permits for the facility or nearby areas available from the owner/operator, state environmental agency and the regional EPA office;

- Information required by community right-to-know legislation and reported on Form R (SARA Title 111, Sec. 3 13);

- Compliance and inspection records for the facility or nearby facilities, available at both state and EPA regional levels;

- Notices of violations served; existing information or demand letters; citations and fines levied by federal, state and local agencies, and information on whether the site is under enforcement actions or administrative orders (actual or pending);

- Information from the Securities and Exchange Commission (SEC), which requires the reporting of any environmental liabilities and environmental lawsuits on Form 10-K; and

- Owner/operator records of environmental audits for the facility.

A Phase I review process may not require this level of data gathering for all sites, however, many times both owner/operator federal, state and local records become an integral, supporting part of the early parts of environmental site assessments.

2.2.2 Regional Base Map

Probably the simplest and potentially the most informative first step for an initial site, is the generation of a base map of the project area. These maps are used for many purposes, from locating the site in a regional context, to using the map as a base upon which project information can be drawn. These regional base map productions can be generated from road maps, air flight maps, naval shoreline maps or by use of the more popular USGS topographic maps.

Topographic maps are available from a number of sources and in a variety of scales. U.S. Geologic Survey (USGS) topographic maps are referred to as quadrangle maps. USGS provides maps covering a quadrangle area bounded by lines of latitude and longitude available in 7.5' series (1:24,000), 15' series (1:62,500), 30' series (1:125,000), and 2^O series (1:250,000) for most of the United States; however, many of the larger scales are out of print. The Defense Mapping Agency (DMA; formerly U.S. Army Map Service) provides maps based on the Universal Transverse Mercator (UTM) grid at scales of 1:25,000, 1:50,000, and 1:100,000. Other countries use scales ranging from 1:10,000 to 1:1,000,000, but coverage is often incomplete. A common scale available for many areas is 1:50,000. Additional sources for topographic data are given in Table 2-4.

All base maps should be photographically screened during the production of the base map so that map data appear as gray tones and, as such, will not compete with solid, darker lines added to the map as it is converted to specialty use. A 60 percent, dot-pattern density reduction is herein recommended as a convenient background gray.

The first step in obtaining a topographic map of the site area is to determine which map(s) are appropriate, i.e., what are the names of the specific quadrangles covering this area. This may be accomplished in any of several ways:

1. The USGS may be contacted in Arlington, Virginia, for areas east of the Mississippi River; and in Denver, Colorado, for areas west of the Mississippi River. They would provide the map name over the phone or send out an index map for the state in question. The engineer could then locate the area on the index map and obtain the name(s) of the needed quadrangle(s) in this manner.

2. Index map sets may be available from a local college or university. These could be consulted to determine which quadrangle is desired.

Table 2-4 Topographic Data

Source	Information Obtainable	Comments
Earth Science Information Center U.S. Geological Survey 507 National Center Reston, VA 22092 Tel: 703/860-6045 FTS 928-6045	Topographic quadrangle maps are available in several scales, most commonly 1:24,000, 1:62,500, and 1:250,000. 1:25,000 and 1:100,000-scale maps are available in limited numbers. Other scales are available for some areas.	The simplest method of selecting topographic maps is to refer to a state index map, which shows all the currently available topographic maps
Branch of Distribution U.S. Geological Survey Map Sales Box 25286, Federal Center Denver, CO 80225 Tel: 303/236-7477	Index and quadrangle maps for the eastern U.S. and for states west of the Mississippi River, including Alaska, Hawaii, and Louisiana.	A map should be ordered by name, series, and state. Mapping of an area is commonly available at two different scales. The quadrangle name is, in some instances, the same for both maps; where this occurs, it is especially important that the requestor specify the series designation, such as 7.5 minute (1:24,000), 15 minute (1:62,500), or two-degree (1:250,000).
Commercial map supply houses	Topographic and geologic maps.	Commercial map supply houses often have full state topographic inventories that may be out of print through national distribution centers.

3. Local town offices may be aware of which quadrangle covers their area (town engineer, planners, etc.).

4. Other nearby institutions or firms that deal with land holdings or have extensive properties in the area are likely to have USGS quadrangles for the area.

5. Maps may be purchased over-the-counter from local USGS offices, commercial map distributors, and other suppliers.

Most state solid waste permits require a topographic map presentation showing a distance of 1,000 to 5,000 feet around the proposed facility. This requirement will determine how many quadrangles are necessary to include and order from the supplier.

Map of Topographic Features

The USGS will provide, free on request, a one-page sheet which shows the topographic map symbols normally found on their maps. These features should also be noted on maps formed through photogrammetry, compiled by surveying, or obtained through town offices. Several of these symbols are described below:

- Contours — Contours are imaginary lines representing the ground surface at a constant elevation above or below sea level. Contour lines are normally shown in brown on USGS topographic maps. USGS maps vary in contour interval; they are either 5, 10, or 20 feet, depending on the relief involved. The contour interval is reported beneath the scale at the bottom of the map or may be determined by simple inspection of the map. Contour intervals depend on ground slope and map scale. Small contour intervals are used for flat areas; larger contour intervals for mountainous terrain.

- Scale — The scale shows the relationship between distance on a map and the corresponding distance on the ground. The scale of USGS topographic maps is provided at the bottom of the map in the form of a numerical ratio and graphically using bar scales marked in feet, miles, and kilometers. For example, the numerical ratio 1:24,000 indicates 1 inch on the map equals 24,000 inches on the earth; or 1 cm on the map equals 24,000 cm on the earth, etc. Any map reproduced for site assessment purposes should clearly indicate its scale using scale bars that are reduced or enlarged along with the map figure.

- Date — The date on which the map was compiled (year of photos used, etc.) should be inserted on the map where new maps or town maps are used. USGS quadrangles provide a date (year) in which aerial photographs were taken (from which the map was compiled) on the bottom left hand corner of the map. In addition, the date of any field check is provided. The most recent revisions to the map are normally added in purple, along with the year of any aerial photographs. Dates are also included in the bottom right hand corner of the map, under the quadrangle name. USGS attempts to update their maps every 5 years. These data should be included with portions of such maps used in site assessments.

• Surface Waters — Surface waters, including intermittent streams, must be shown on the topographic map supplied in the monitoring plan. For USGS topographic maps, USGS symbols should be consulted in determining the various types of surface waters shown. Perennial streams are normally shown as solid blue lines whose thickness relates to stream flow (size). Intermittent streams are usually shown as solid blue lines interrupted by dots at frequent intervals and no exaggeration of width. These surface water types may be contrasted with ephemeral streams which consist of a dry channel throughout most of the year, bearing water only during and after rainfall. Ephemeral streams need not be shown on the topographic site map. The author has often observed different opinions on the type of stream present on sites. Care must be taken to correctly establish the surface water type based on actual definitions of site stream flow conditions rather than using estimated stream types taken from old USGS maps. A correct definition for a particular stream is important to determine the applicability of State or Federal requirements. All topographic maps should also depict ponds and lakes. The USGS maps show lakes (usually blue), dams, large rapids, marshes, waterfalls, etc.

• Land Forms/Land Use — It's recommended that the site topographic map depict surrounding land uses such as residential, commercial, agricultural, or recreational. This information may be ascertained from sources in local town offices, by first hand observations, or by using the USGS quadrangle.

The USGS quadrangle indicates buildings (dwellings, places of employment, etc.) usually as solid hatch-marked squares. Schools, churches, and cemeteries are also indicated, as are barns or warehouses. Golf courses are often indicated, as are railroad tracks. Various boundary line types are used to delineate city limits, national or state reservations, small parks, land grants, etc. The breakdown of residential commercial /recreational uses may thus be set to a great degree from observing a USGS quadrangle map and updating the information as required.

Examples of Regional Base Maps

Figure 2-2 shows a regional base map generated by cutting down a USGS 7.5' topographic quadrangle to fit on an 8 inch by 11 inch sheet, and maintaining the original map scale. The solid waste fill area is shaded to enhance boundaries, and the required key information is contained on the figure. This type of simply produced location map can provide an important basic data source for further analysis and interpretation. Figure 2-3 shows a regional base map made by fitting two 7.5' USGS topographic quadrangle maps. As with the first regional base map, the fill areas and expansions are clearly marked on the drawing. Since the regional base map contains topography, rivers and streams, and cultural features, these data should provide a basic understanding of the local placement of the site within the regional picture.

List of Major Points

1. Is a regional location map provided in the Scope of Work?

2. Does the regional location map have the site boundaries clearly marked?

3. Is a suitable site specific topographic map available for the project area on a minimum scale of approximately 1 inch = 200 feet or at a scale consistent with the project area size?

4. Is the typical state requirement for a 2 foot or 5 foot contour interval established for the project area?

2.2.3 Geology/Geophysics/Ground Water

The typical Phase I scope of work requires that all available data describing the known regional geology and hydrogeology for the site area be reviewed, early in the project. In states such as California, where particular concerns of faulted and landslide areas have to be addressed throughout in a site investigation, the geologist must obtain all pertinent data on geologic hazards. The location and logs of existing exploratory pits, borings, or local water wells should be included within the review process. The literature search should be directed toward complete documentation of structural geology, local quarries, rock outcrops, and any bedrock or overburden well lithologies and well performance information. This Phase I task is the first step in the conceptual process leading toward a full understanding of the site processes that control ground-

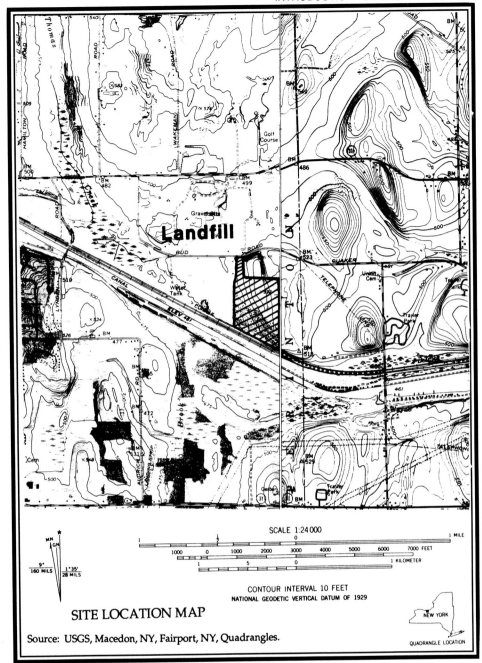

Figure 2-2 Regional Base Map Production

water movement at the subject site. The primary purposes of the Phase I geologic data review are to:

- Define and describe the specific regional geology (see Table 2-5) and the overall ground-water flow systems (occurrence, recharge, discharge, direction, and relative rates of movement) within the site (to 1 mi) and the vicinity (5 to 10 mi) of the site;

- Define the ambient or background ground-water quality in the areas of concern;

- Estimate the maximum elevation of the ground-water potentiometric surface;

- Evaluate the importance of the ground-water resource that might be affected by the operation of the facility; and

- Define potential sources of geophysical data as referenced in Table 2-6.

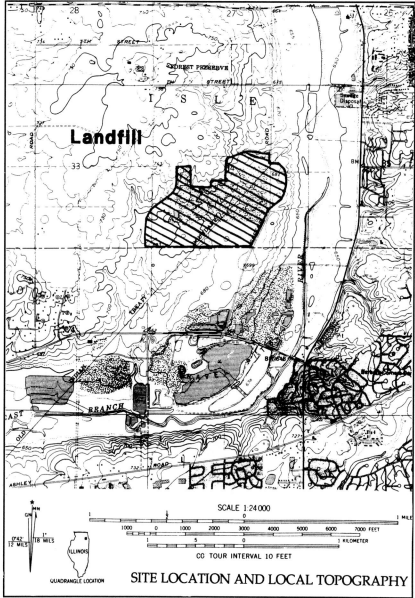

Figure 2-3 Site Location and Topography

Discussions with well drillers and other locally knowledgeable individuals often provide valuable information on bedrock aquifer drilling targets. The targets may delimit zones of high hydraulic conductivity that will provide directional flow-paths for ground-water movement. Particular attention should be given to recharge areas of significant ground-water aquifers. Basic sources of information include USGS publications and personnel, state agencies, national organizations, and universities.

Regional studies ranging from large (hundreds of square miles) to small areas (hundreds of acres) can be based on stereoscopic interpretation of aerial photographs, preferably obtained at two scale ranges — 1:60,000 to 1:40,000 and 1:20,000 to 1:8,000 — with the smaller scales providing an overview and the larger scales providing details. False-color infrared photographs may be obtained to locate vegetation vigor (the brighter the red, the healthier the vegetation), man-made features, high ground-water surfaces, and water bodies. Table 2-7 provides sources for remote-image data, from large scale images from space to more traditional aerial photography. Whenever possible, aerial photos, in stereo pairs, should be obtained for as many flight dates as possible. Aerial photos can also provide an excellent base for plotting of site features. There are many forms of remote imagery that show the site features evident on the topographic maps, as well as vegetation types, density, and image tone, and are useful for environmental as well as geologic studies. The selection of remote imagery depends upon availability, the purpose of the study, and the land area involved. Stereoscopic

Table 2 - 5   Geologic Data

Source	Information Obtainable	Comments
Geologic Indexes		
Geological Reference Sources: A Subject and Regional Bibliography of Publications and Maps in the Geological Sciences, Ward and others (1981)	Bibliographies of geologic information for each state in the U.S. and references general maps and ground water information for many sites.	Provides a useful starting place for many site assessments. A general section outlines various bibliographic and abstracting services, indexes and cataloges, and other sources of geologic references.
• A Guide to Information Sources in Mining, Minerals, and Geosciences, Kaplan (1965)	Describes more than 1,000 organizations in 142 countries. Its listings include name, address, telephone number, cable address, purpose and function, year organized, organizational structure, membership categories, and publication format. Federal and state agencies are listed for the U.S., as well as private scientific organizations, institutes, and associations.	An older useful guide. Part II lists more than 600 world–wide publications and periodicals including indexing and abstracting services, bibliographies, dictionaries, handbooks, journals, source directories, and yearbooks in most fields of geosciences.
• Bibliography and Index of Geology	Includes world–wide references and contains listings by author and subject.	This publication is issued monthly and cumulated annually by the American Geological Institute (AGI), and replaces separate indexes published by the U.S. Geological Survey through 1970 (North American references only) and the Geological Society of America until 1969 (references exclusive of North America only). Both publications merged in 1970 and were published by the Geological Society of America through 1978, when AGI continued its publication.
• GEOREF base	Bibliographic citations from 1961 to present.	The Bibliography and Index of Geology is also part of the GEOREF computerized data maintained by AGI.
• KWIC (Keyword–in–Contents) Index of Rock Mechanics Literature	Engineering geologic and geotechnical references.	The KWIC index is available in two volumes at many earth science libraries (Hoek, 1969; Jenkins and Brown, 1979).
• GEODEX Retrieval System with Matching Geotechnical Abstracts GEODEX International, Inc. P.O. Box 279 Sonoma, CA 95476	Engineering geologic and geotechnical references.	The GEODEX is a hierarchically organized system providing easy access to the geotechnical literature and can be used at many university libraries. The GEODEX system can be purchased on a subscription basis.
U.S. Federal Agencies		
• U.S. Geological Survey Branch of Distribution 604 South Pickett Street Alexandria, VA 22304	The U.S. Geological Survey (USGS) produces annually a large volume of information in many formats, including maps, reports, circulars, open–file reports, professional papers, bulletins, and many others.	To simplify the dissemination of this information, the USGS has issued a Circular (No. 777) entitled A Guide to Obtaining Information from the USGS (Clarke, et al., 1981). Circular 777 is updated annually and contains valuable information for anyone searching for earth sciences information.

Table 2 - 6 Geophysical Data

Source	Information Obtainable	Comments
U.S. Geological Survey Water Supply Papers	The most common types of data are obtained from seismic and resistivity surveys, but other types may also appear or be referenced. Water Supply Papers for an area can be by any of the computer searches or published indexes described in the first section of this paper.	The USGS also published geophysical maps of various types at relatively small scales for many areas of the U.S. Aeromagnetic maps have been completed for much of the U.S., although the flight altitude of several thousand meters and scale of 1:24,000 make these maps too general for most site specific work.
Well Log Libraries Electric Log Services P.O. Box 3150 Midland, TX 79702 Tel: 915/682-0591	Sample and electric logs for many petroleum wells can be obtained from one of several well log libraries in the U.S.	Sample logs generally extend to the surface, but geophysical logs start where the well casing ends — commonly at a depth of about 100 m. The type of geophysical logs available may include sonic velocity, radioactivity, caliper (borehole diameter), and others. The logs are indexed by survey section, and to obtain information on wells in a given area, it is necessary to compile a list of the townships, ranges, and section numbers covering the area.
Geophysical Survey Firms	Specific geophysical logs	Proprietary geophysical data can sometimes be obtained from private survey firms. In general, the original client must approve the exchange of information, and preference is given for academic purposes. If the information cannot be released, firms may be willing to provide references to published information they obtained before the survey, or information published as a result of the survey.
National Geophysical and Solar-Terrestrial Data Center Chief, Solid Earth Geophysics 325 Broadway Boulder, CO 80303 Tel: 303/497-6521 FTS 320-6521	NGSDC maintains a computer file of more than 136,000 earthquakes, known or suspected explosions and associated collapse phenomena coal bumps, rockbursts, quarry blasts, and other earth disturbances recorded worldwide by seismographs starting in January 1900. Historic U.S. earthquakes are included for the period starting in 1638. For $25, a search can be made for one of the following parameters: 1. Geographic area (circular or rectangular area) 2. Time period (starting 1638 for U.S.) 3. Magnitude range 4. Date 5. Time 6. Depth 7. Intensity (Modified Mercalli)	Site studies for many projects now require information regarding the seismicity of the region surrounding the site. The National Geophysical and Solar-Terrestrial Data Center (NGSDC) of the National Oceanic and Atmospheric Administration (NOAA) is a focal point for dissemination of earthquake data and information for both technical and general users, except for information on recent earthquakes. (Information about recent earthquakes can be obtained by contacting the USGS.)

examination and interpretation of aerial photographs is a basic and most productive analytical method.

Regional Geotechnical Data (from 1 to 10 mi radius)

This subsection discusses the types of information that should be included in a Phase I facility report to adequately characterize the regional geology at and near a site. Site specific geology should extend for approximately a 1 mile area around the facility, where as regional geology could extend out to a 1 to 10 mile radius, dependent on persistence of the geologic structure. The information obtained from literature sources includes:

Table 2 - 7 Remote Sensing

Source	Information Obtainable	Comments
USGS EROS Data Center • Additional information can be found in the publications The EROS Data Center and The Landsat Data User's Handbook. To obtain these publications, request further information, place an order, contact: User Service EROS Data Center U.S. Geological Survey Sioux Falls, SD 57198 Tel: 605/594-6151 FTS 784-7151	The Earth Resources Observation Systems (EROS) Program of the U.S. Department of the Interior, administered by the USGS, was established in 1966 to apply remote-sensing techniques to the inventory, monitoring, and management of natural resources. As part of this program, the EROS Program provides remotely-sensed data at nominal cost to anyone in the world.	The EROS Data Center, near Sioux Falls, South Dakota, is operated by the EROS program to provide access primarily to NASA's Landsat imagery, aerial photography acquired by the U.S. Department of the Interior, and photography and multi-spectral imagery acquired by NASA from research aircraft, Skylab, Apollo, and Gemini spacecraft. The primary functions of the Data Center are data storage and reproduction, user assistance, and training. The Data Center can provide a computer listing of all imagery on file for three geographic options: 1. Point search — all images or photographs with any portion falling over the specific point of longitude and latitude are included. 2. Area quadrilateral — any area of interest defined by four coordinates of longitude and latitude. All images or photographs with any coverage of the area are included. 3. Map specification — a point or area may be indicated on a map. (Options 1 and 2 are preferred by EROS.)
• Landsat Data	Landsat satellites sensor images in four spectral bands: Band 4 (emphasizes sediment-laden and shallow water) Band 5 (emphasizes cultural features) Band 6 (emphasizes vegetation, land/ water boundaries, and landforms) Band 7 (as above, with best penetration of haze) Band 5 gives the best general-purpose view of the earth's surface. Available are black and white images and false-color composites.	The Landsat satellites were designed to orbit the earth about 14 times each day at an altitude of 920 km, obtaining repetitive coverage every 18 days. The primary sensor aboard the satellites is a multi-spectral scanner that acquires parallelogram images 185 km per side in four spectral bands.
• NASA Aerial Photography	Photography is available in a wide variety of formats from flights at altitudes ranging from one to 18 km. Photographs generally come as 230 mm by 230 mm prints at scales of 1:60,000 or 1:120,000, and are available as black and white, color, or false-color infrared prints.	NASA aerial photography is directed at testing a variety of remote-sensing instruments and techniques in aerial flights over certain preselected test sites over the continental U.S.
• Aerial Mapping Photography	Aerial photographic coverage obtained by the USGS and other Federal agencies (other than the Soil Conservation Service) for mapping of the U.S. is available as 230 mm by 230 mm black and white prints which are taken at altitudes of 600 m to 12 km.	Because of the large number of individual photographs needed to show a region on the ground, photomosaic indexes are used to identify photographic coverage of a specific area. The Data Center has more than 50,000 such mosaics available for photographic selection.

Table 2 - 7 Remote Sensing (con)

Source	Information Obtainable	Comments
U.S. Department of Agriculture		
• Aerial Photography Field Office U.S. Department of Agriculture P.O. Box 30010 Salt Lake City, UT 84130 Tel: 801/524-5856 FTS 588-5857	Conventional aerial photography scales of 1:20,000 to 1:40,000.	Aerial photographs by the various agencies of the U.S. Department of Agriculture (Agricultural Stabilization and Conservation Service [ASCS], Soil Conservation Service [SCS], and Forest Service [USFS]) cover much of the U.S.
Other Sources		
• Coastal Mapping Division of NOAA An index for the collection can be obtained for free by contacting: Coastal Mapping Division National Oceanic and Atmospheric Administration 6001 Executive Boulevard Rockville, MD 20852 Tel: 301/443-8601 FTS 443-8601	The Coastal Mapping Division of NOAA maintains a file of color and black and white photographs of the tidal zone of the Atlantic, Gulf, and Pacific coasts. The scales of the photographs range from 1:20,000 to 1:60,000.	
• Bureau of Land Management For an index of the entire collection contact: U.S. Bureau of Land Management Larry Cunningham (SC-675) P.O. Box 25047 Denver, CO 80225-0047 Tel: 303/236-7991	The Bureau of Land Management has aerial photographic coverage of approximately 50 percent of its lands in 11 western states.	
• National Archives and Records Service Cartographic Archives Division General Services Administration 8 Pennsylvania Avenue, N.W. Washington, D.C. 20408 Tel: 703/756-6700	Airphoto coverage obtained before 1942 for portions of the U.S.	This service may be important for early documentation of site activities.
• Canadian airphoto coverage can be obtained from: National Aerial Photograph Library 615 Booth Street Ottawa, Ontario K1A 0E9 Canada Tel: 613/995-4560		

• Regional geologic tabular summaries, columns and cross-sections, used to identify pertinent geologic units and structure in the area; and,

• Regional geologic maps, used to depict the area wide locations of geologic units.

Sources used to collect the information are provided in the tables contained in this chapter.

Regional Geologic Cross-Sections

Available generalized geologic cross-sections that show subsurface geologic units and structures in the area of the site are typically presented in the Phase I report. These data

Table 2 - 7 Remote Sensing (con)

Source	Information Obtainable	Comments
• Canadian satellite imagery can be obtained from: Canadian Centre for Remote Sensing 2464 Sheffield Road Ottawa, Ontario K1A 0Y7 Canada Tel: 613/952-0202		
• Commercial Aerial Photo Firms For a listing of nearby firms specializing in these services, consult the yellow pages or contact: American Society of Photogrammetry 210 Little Falls Street Falls Church, VA 22046 Tel: 703/534-6617		In many instances, these firms retain the negatives for photographs flown for a variety of clients and readily sell prints to any interested users.

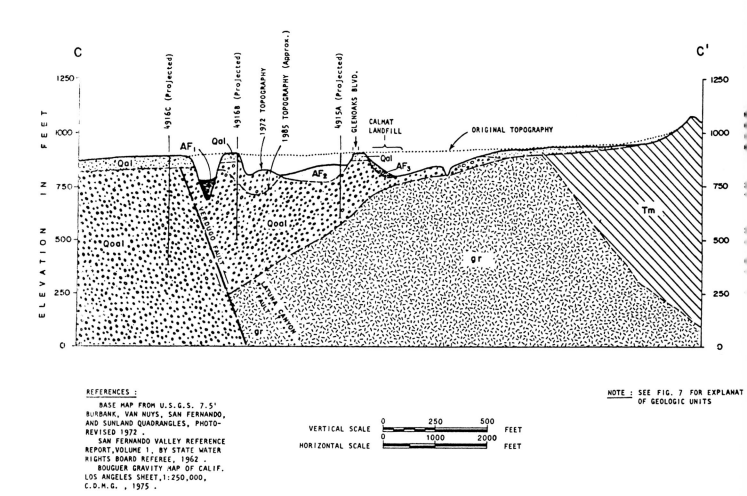

Figure 2-4   Regional Cross section

Time Stratig.			Rock Stratigraphy		GRAPHIC COLUMN	Thickness (Feet)	KINDS OF ROCK	
SYSTEM	SERIES	STAGE	MEGA-GROUP	GROUP	FORMATION			
QUAT.	PLEIS.				(See fig. 15)	0-350	Till, sand, gravel, silt, clay, peat, marl, loess	
PENN.	DESM.			Kewanee	Carbondale		0-125	Shale, sandstone, thin limestone, coal	
					Spoon		50-75	As above, but below No. 2 Coal	
MISS.	VAL.				Burl-Keokuk		0-700	Limestone	Only in Des Plaines
	KIND.				Hannibal			Shale, siltstone	Disturbance
DEV.	UP.				Grassy Creek		0-5	Shale in solution cavities in Silurian	
SILURIAN	NIAGARAN		Hunton		Racine		0-300	Dolomite, pure in reefs; mostly silty, argillaceous, cherty between reefs	
					Waukesha		0-30	Dolomite, even bedded, slightly silty	
					Joliet		40-60	Dolomite, shaly and red at base; white, silty, cherty above; pure at top	
	ALEX.				Kankakee		20-45	Dolomite; thin beds; green shale partings	
					Edgewood		0-100	Dolomite, cherty, shaly at base where thick	
ORDOVICIAN	CIN.	RICH.		Maquoketa	Neda		0-15	Oolite and shale, red	
					Brainard		0-100	Shale, dolomitic, greenish gray	
		MAY. ED.			Ft. Atkinson		5-50	Dolomite, green shale, coarse limestone	
					Scales		90-120	Shale, dolomitic, gray, brown, black	
	CHAMPLAINIAN	TRENT.	Ottawa	Galena	Wise Lake		170-210	Dolomite, buff, pure	
					Dunleith			Dolomite, pure to slightly shaly; locally limestone	
					Guttenberg		0-15	Dolomite, red specks and shale partings	
		BLACKRIVERAN		Platteville	Nachusa		0-50	Dolomite and limestone, pure, massive	
					Grand Detour		20-40	Dolomite and limestone; medium beds	
					Mifflin		20-50	Dolomite and limestone, shaly, thin beds	
					Pecatonica		20-50	Dolomite, pure, thick beds	
				Ancell	Glenwood		0-80	Sandstone and dolomite, silty; green shale	
					St. Peter		100-600	Sandstone, medium and fine grained; well rounded grains; chert rubble at base	
	CANADIAN		Knox	Prairie du Chien	Shakopee		0-70	Dolomite, sandy, oolitic cherty; algal mounds	
					New Richmond		0-35	Sandstone, fine to coarse	
					Oneota		190-250	Dolomite, pure, coarse grained; oolitic chert	
					Gunter		0-15	Sandstone, dolomitic	
CAMBRIAN	CROIXAN	TREMP.			Eminence		50-150	Dolomite, sandy	
					Potosi		90-220	Dolomite; drusy quartz in vugs	
		FRAN.			Franconia		50-200	Sandstone, glauconitic; dolomite; shale	
		DRESBACHIAN			Ironton		80-130	Sandstone, partly dolomitic, medium grained	
					Galesville		10-100	Sandstone, fine grained	
					Eau Claire		370-570	Siltstone, dolomite, sandstone and shale, glauconitic	
			Potsdam		Mt. Simon		1200-2900	Sandstone, fine to coarse; quartz pebbles in some beds	
PRE-CAM.								Granite	

Figure 2-5 Regional Geology

Source: Willman, W. H.

are intended to illustrate structural and stratigraphic characteristics of the region; when not available in the literature, such cross-sections can be constructed from regional geologic maps and their relationships to the site. Often, regional characteristics may be known elsewhere in the region but may sometimes be anticipated (though presently unknown) in the site area. In general, such relationships would be first identified during the Phase I investigation and further developed after Phase II drilling programs. Available regional cross-sections should be used to plan the drilling program so that sufficient data is gathered for each of the stratigraphic units present on site. As such, these maps should be used as part of the Phase II boring plan. As site specific soil and bedrock information becomes available, geologic cross-sections should be updated to more accurately represent relationships between the geology of the site and the region at large.

The investigator should have an overview understanding of the geologic setting of the facility, including the identification of the geologic provinces (e.g., Basin and Range Province), if applicable, and pertinent geologic history. A large-scale geologic map, available from published sources such as the U.S. Geological Survey, could be included with the text to illustrate geological units and structural features of the region.

Examples

A regional geologic section is presented in Figure 2-4. Borings were used to better define on-site relationships between the Verdugo and Latuna Canyon Faults at a site in California. The landfill and present topography is provided for comparison of soil and rock types expected in the excavation.

A second type of regional geologic information is the stratigraphic column, an example is shown on Figure 2-5. The figure presents a composite of bedrock stratigraphy in the Chicago area (Willman, 1971). Stratigraphic columns, such as Figure 2-5, can be used to illustrate relationships between geologic time (time-stratigraphy) and rock units (litho-stratigraphy). The column includes proper names of formations, members, etc.; a graphic presentation of important rock types, and location of unconformities; a range in thickness of each unit; and a column for inclusion information, such as important hydrogeologic properties of the units. Stratigraphic columns serve to summarize large quantities of information in a presentation, through which a reader can rapidly comprehend regional geologic relationships.

The tabular presentation of regional nomenclature and rock classification has been modified in Figure 2-6, "Geologic Units of Grayson County, Texas," located in the

System	Series and group	Formation	Thickness (feet)	Character of rocks	Water-bearing properties
Quaternary	Recent and Pleistocene series	Alluvium	0–60	Sand, gravel, clay, and silt.	Yields small to moderate quantities of water to domestic and industrial wells.
Cretaceous	Gulf series	Austin chalk	0–550	White to buff chalk, marl, and limestone.	Yields small quantities of hard water to shallow domestic wells.
		Eagle Ford shale	0–480	Bluish-black gypsiferous shale; thin beds of shale and limestone.	Yields small quantities of water to domestic wells.
		Woodbine formation	0–600	Crossbedded ferruginous sand, laminated shaly clay, lignite, and gypsiferous clay.	Principal aquifer in Grayson County. Furnishes large supplies for municipal, industrial, irrigation, and domestic use. Water is typically high in iron.
	Comanche series — Washita group	Grayson marl	0–50	Yellowish-brown fossiliferous clay and thin limestone.	Not known to yield water to wells in Grayson County.
		Main Street limestone	0–25	Hard white to brownish-white crystalline limestone alternating with layers of marl.	
		Pawpaw formation	0–80	Reddish-brown calcareous clay and yellowish-brown ferruginous sand.	Yields small to moderate quantities of water to wells in outcrop area.
		Weno clay	0–135	Dark-gray to tan shaly clay and thin beds of sand and limestone, fossiliferous.	
		Denton clay	0–60	Brownish-yellow clay; some hard sandstone beds; shell agglomerate in upper part.	
		Fort Worth limestone	0–70	Alternating limestone and marl, fossiliferous.	
		Duck Creek formation	0–130	Alternating limestone and marl; limestone predominating in lower part and marl in upper part; fossiliferous in lower part.	Not known to yield water to wells in Grayson County.
	Comanche series — Fredericksburg group	Kiamichi formation	0–35	Greenish clay and thin limestone beds; fossils abundant in upper part.	
		Goodland limestone	0–40	White fossiliferous limestone.	
		Walnut clay	0–22	Black gypsiferous fissile shale; ledges of shell breccia.	
	Trinity group	Undifferentiated rocks	600–1,200	Fine to medium sand and beds of red, purple, and gray clay.	Yields large supplies of water for municipal, industrial, and domestic uses. Water is saline in northern part of county.
Pennsylvanian	Undifferentiated		15,000±	Gray to red sandy shale, sandstone, and limestone.	Does not contain fresh water in Grayson County.

Figure 2-6 Regional Geology Grayson Co. Tx.

Source: Dames and Moore

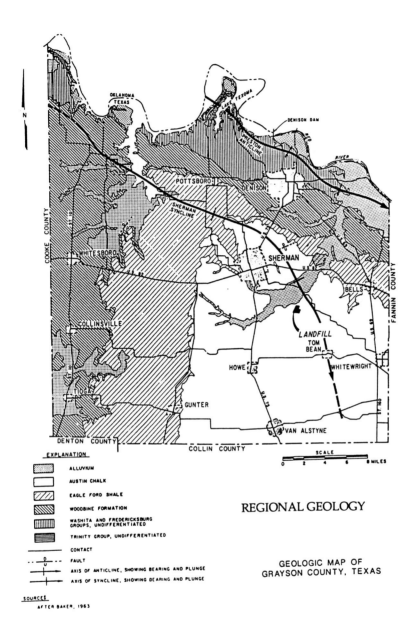

Figure 2-7 Regional Geologic Map, Grayson Co. Tx.

Source: Dames and Moore

north-central portion of the state. Figure 2-6 includes an interpretation of potential aquifers and aquicludes as they relate to specific rock units. The geology of Grayson County is further illustrated by a geologic map, shown in Figure 2-7. These data are extremely valuable for definition of regional aquifers and aquicludes for project siting expansion or aquifer remediation reports.

Regional Geologic Mapping- List of Major Points

1. Are geologic maps and cross-sections available in the literature that covers the site and adjacent areas.?

2. Is the scale of geologic information sufficient that the expected geologic units can be established before Phase II drilling. ?

3. Will the geologic structure shown on the above have any effect on the potential placement and depth of Phase II boreholes ?

4. Will the regional geologic information direct special sampling depths or areas due to specific geologic units or structures ?

5. Does the regional geologic data suggest potential high hydraulic conductivity zones to target for installation of piezometers?

6. Does the regional geologic data suggest special conceptual model components that will have important effects on waste disposal?

Well Census

The locations of all wells (differentiated as to public water, domestic, industrial, agricultural, and other) and springs, should be shown on a base map of the property and within 2,000 feet of the property boundaries. The base map should also include locations of wetlands and all surface water found in the area of the proposed site within 1/2 mile of the boundary. As part of the assessment, stream and river flows — where monitored by state and federal agencies — should be documented. This information is included to determine whether the location of a landfill could potentially cause degradation of a classified body of surface water.

Supporting information on the proximity and withdrawal rates of users and the availability of alternative drinking water supplies must be included in the documentation. The well inventory should include available data, such as: location, owner, surface elevation, aquifer tapped, water quality, well depth, casing size and depth, depth to water, estimated rate of pumpage, and date of inventory.

2.2.4 Soils

Agricultural maps and reports supply information on soil characteristics and topography that influence runoff and infiltration. Sources vary widely with respect to the quantity and quality of the information they contain. Soil survey maps, produced by the Soil Conservation Service (SCS) of the U.S. Department of Agriculture (the largest source of soils data), usually are plotted as overlays on aerial photographs at relatively large scales. They are prepared on a county-wide basis and illustrate the soil cover to a depth of about 6 feet (2 meters). They are based on pedological soil classifications and provide significant

information relating to residual soil and bedrock parent units. These classifications are often combined with symbols describing slopes, shallow ground water, and soil drainage. Recent SCS maps contain engineering-oriented data prepared by the Federal Highway Administration (FHWA; formerly the Bureau of Public Roads) in conjunction with the SCS. "Soils," as defined herein, includes all unconsolidated material above bedrock, and may include individually distinguishable hydrogeologic units.

Soil Conservation Service

For 50 years, the Soil Conservation Service (SCS), part of the U.S. Department of Agriculture, has been conducting periodic inventories of the nation's soil, water, and related natural resources. The primary environmental statistics program within the SCS is the National Resources Inventory, which provides data on the status, condition, and trends of these resources on non-federal land in the United States.

The majority of data collected by the NRI program are organized into eight general categories: soil characteristics and interpretations (including agricultural land capability); land cover; land use (including cropland, forestland, and prime farmland); erosion; land treatment (such as irrigation); conservation treatment needs; vegetative conditions (such as wetlands and pasture management); and potential for conversion to cropland. Data collection is primarily through field investigation, every five years.

Example of a Soils Map

Figure 2-8 is a published soil survey for an area of northern Illinois. The map illustrates the distribution of the various soil types present in the area. Information from such maps can be augmented through drilling programs to define the engineering and chemical properties of the soils. It is recommended to follow the Unified Soil Classification System (USCS) as the standard method for description of soils obtained in drilling activities. These USCS engineering descriptions must be compared with the SCS Soil Science descriptive system to develop the optimum understanding of the surface mapped and subsurface logged information.

A second type of presentation is a map of the surficial geology of unconsolidated material. In glacial environments, these presentations are extremely informative for planning of drilling activities and for investigations of borrow material. Figure 2-9 presents the surficial extent of bog, till, glaciofluvial, and ice contact deposits. Each of these materials would be expected to have significantly different hydraulic conductivity, shear strength, and

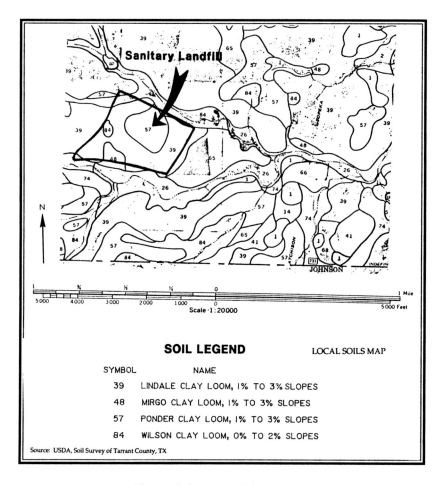

Figure 2-8 Local Soils Map

Source: USDA, Soil Survey of Tarrant Co. Tx.

physical attributes for landfill construction. Surficial geology of unconsolidated materials can be substituted for more traditional regional bedrock maps in areas of deep soils or glacial deposits. List of Main Points follows for soils maps:

1. Are soils maps available for the project area?

2. Are data sufficient for planning of Phase II drilling programs, or is additional surficial soils work necessary?

3. Has the site area been glaciated or are only residual soils present?

4. Are regional quaternary geologic maps available for the area?

2.2.5 Surface Water

The baseline data for surface-water hydrology should be first developed from the literature. The Phase I report should describe drainage systems, flow characteristics, and water quality of the site area. An important use of this information is to assist in determining ground water/surface water relationships at the site.

The 100-year flood plain is normally an exclusion zone for the disposal of solid waste. Equally important is the protection of wetlands and control of runoff from waste disposal areas into surface water bodies. The literature search should include data on the 100-year flood plain, wetlands, stream flow and runoff, water quality, and water use. These data may be obtained by review of the most recent U.S. Geological Survey, Army Corps of Engineers, or the Federal Insurance Administration (FIA) 100-Year Frequency Flood Plain Map for the area (if available), and by contacting cognizant State and local agencies. Table 2-8

Table 2 - 8 Hydrologic Data

Source	Information Obtainable	Comments
Water Publications of State Agencies, Giefer and Todd (1972, 1976)	This book lists state agencies involved with research related to water and also lists all publications of these agencies. In general, hydrologic data can be classified into four primary categories: stream discharge, stream water quality, ground water level, and ground water quality.	The trend for the past decade has been to compile such basic data in computerized data banks, and a number of such information systems are now available for private and public users. Many data now collected by federal and state water-related agencies are available through computer files, but most data collected by private consultants, local and county agencies, and well drilling contractors remain with the organization that gathered them.
Local Assistance Center of the National Water Data Exchange • NAWDEX U.S. Geological Survey 421 National Center Reston, VA 22092 Tel: 703/648-4000	NAWDEX identifies organizations that collect water data, offices within these obtained, alternate sources from which an organization's data may be obtained, the geographic areas in which an organization collects data, and the types of data collected. Information has been compiled for more than 600 organizations, and information on other organizations is added continually. More than 300,000 data collection sites are indexed.	NAWDEX, which began operation in 1976 is administered by the U.S. Geological Survey organizations from which the data may be consists of a computer directory system which locates sources of needed water data. The system helps to link data users to data collectors. For example, the NAWDEX Master Water Data Index can identify the sites at which water data are available in a geographic area, and the Water Data Sources Directory can then identify the names and addresses of organizations from which the data may be obtained. In addition, listings and summary counts of data, references to other water data systems, and bibliographic data services are available.
Published Water-Supply Studies and Data	Stream discharge, ground water level, and water quality data have been obtained during short-term, site-specific studies, and these data are typically available only in published or unpublished site reports. Data related to lakes, reservoirs, and wetlands are commonly found only in such reports.	Although significant progress has been made in computerizing surface- and ground-water data, the majority remains available only through published and unpublished reports.
Catalog of Information on Water Data	The reference consists of four parts: Part A: Stream flow and stage Part B: Quality of surface water Part C: Quality of ground water Part D: Aerial investigations and miscellaneous activities	Bibliographic publication indexes USGS sampling and measurement sites throughout the U.S.
Geologic and Water-Supply Reports and Maps (available for each state)		This publication lists references for each USGS division for each state or district; the listing, however, is by report number, requiring a scan of the entire list for information on a particular area.

Table 2 - 8 Hydrologic Data (con.)

Source	Information Obtainable	Comments

Water Resources Investigations, by State
 	
• Office of Water Data
  U.S. Geological Survey
  417 National Center
  12201 Sunrise Valley Drive
  Reston, VA 22092
 	
• Additional assistance can be obtained by contacting:
  Hydrologic Information Unit
  U.S. Geological Survey
  420 National Center
  12201 Sunrise Valley Drive
  Reston, VA 22092
	Listed are all agencies cooperating with the USGS in collecting water data, information on obtaining further information, and a selected list of references by both the USGS and cooperating agencies.	This booklet describes the projects and related publications for all current USGS work in a state or group of states. Also available is a useful summary folder with the same title that depicts hydrologic-data stations and hydrologic investigations in a district as of the date of publication.

Federal Flood Insurance Studies	To meet the provisions of the National Flood Insurance Act of 1968, the USGS, with funding by the Federal Insurance Administration, has mapped the 100-year floodplain of most municipal areas at a scale of 1:24,000.	Floodplain maps can be obtained free from the nearest district office of the USGS and commonly from other agencies, such as the relevant city, town, or county planning office, or the Federal Insurance Administration. In some areas, more detailed "Flood Insurance Studies" have been completed for the Federal Emergency Management Agency; these maps include 100-year and 500-year floodplain maps. The complete studies are available at the nearest USGS office, the relevant city, town, or county planning office, or the Federal Emergency Management Agency.

and 2-9 provides data sources that can be used for source information.

As a part of the National Flood Insurance Program, Flood Hazard Boundary Maps FHBM have been prepared for virtually all (20,238) communities that have been identified as "flood prone." The FHBM delimits the boundaries of the 100-year flood plain, although they do not provide elevations. Nearly 90 percent of the Federal Emergency Management Agency maps are of this type. In addition, some of the communities that have been accepted into the National Flood Insurance program have Flood Insurance Rate Maps which delimited the 100-year flood plain and also provide flood elevations. Where these maps are available, determination of the 100-year flood plain is straight-forward (see Figure 2-10). Such maps are most often included as part of a Flood Insurance Study for a particular political jurisdiction along a waterway. The U.S. Department of Housing and Urban Development, Federal Insurance Administration, which publishes such studies,

includes a fold-out map delimiting 100-year flood boundaries for various river reaches in the area covered. An original or clear copy of this map (or another suitable map) should be included in the Phase I report for all cases in which sites are located near a river. Hydraulic analyses used to determine protection measures are also provided in the flood insurance studies and should be included in the report.

If a 100-year flood level is not available from another source, the investigator may have to determine the level using a Flood Hazard Boundary Map. The investigator should use a qualified hydrologist for such determinations or may contact the FIA to assist in determining the 100-year flood elevation at the location in question. Investigators may also wish to undertake independent computations in those instances in which published flood levels are not believed sufficiently accurate or well supported by field evidence or historic hydrographs.

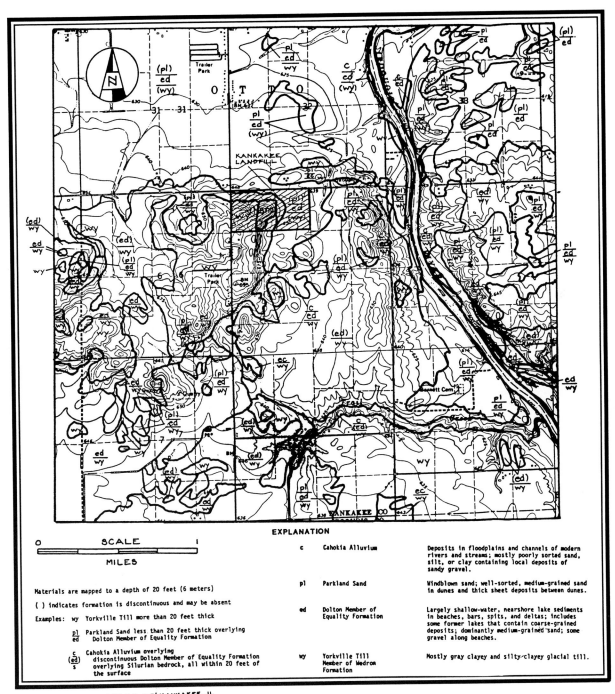

EXPLANATION

c	Cahokia Alluvium	Deposits in floodplains and channels of modern rivers and streams; mostly poorly sorted sand, silt, or clay containing local deposits of sandy gravel.
pl	Parkland Sand	Windblown sand; well-sorted, medium-grained sand in dunes and thick sheet deposits between dunes.
ed	Dolton Member of Equality Formation	Largely shallow-water, nearshore lake sediments in beaches, bars, spits, and deltas; includes some former lakes that contain coarse-grained deposits; dominantly medium-grained sand; some gravel along beaches.
wy	Yorkville Till Member of Wedron Formation	Mostly gray clayey and silty-clayey glacial till.

Materials are mapped to a depth of 20 feet (6 meters)

( ) indicates formation is discontinuous and may be absent

Examples: wy Yorkville Till more than 20 feet thick

 pl
 ── Parkland Sand less than 20 feet thick overlying
 ed Dolton Member of Equality Formation

  c
 (ed) Cahokia Alluvium overlying
 ───   discontinuous Dolton Member of Equality Formation
  s   overlying Silurian bedrock, all within 20 feet of
       the surface

SCALE

0   1

MILES

BASE MAP FROM U.S.G.S. 7.5' KANKAKEE, IL
QUADRANGLE. GEOLOGIC MAP FROM GROSS
(1981), AND KEMPTON (1987).

Figure 2-9 Regional Quaternary Geology

Source: Hydro-Search, Inc

FIA does not usually map flood plains that are less than 200 feet wide. In such situations, the engineer will have to make a determination of the extent of the 100-year flood plain. FIA mapping procedures should be used to make this determination, or the investigator can assume the flood plain is 200 feet wide, to be conservative. The investigator might also consult the U.S. Geological Survey (for availability of detailed flood-hazard maps of the area), the U.S. Army Corps of Engineers, the U.S. Soil Conservation Service, the State Office of Coastal Zone Management, or others for flood plain information. All data sources should be identified in the Phase I report. In addition, any special flooding conditions (e.g., wave action), which might be considered in designing, constructing, operating, or maintaining a facility to withstand washout from a 100-year flood, should be included on the map.

Surface Water Quality

As part of the regulatory program for the Federal Clean Water Act, many state environmental agencies, may have evaluated the existing quality of surface waters in the area and estimated current and future uses. The U.S.G.S. may have significant surface water quality data available through the computer based STORET system. Many state environmental agencies have assigned "designated uses" to surface water bodies in their state, as part of the process of developing water quality standards under Section 303 of the Clean Water Act. "Designated uses" of surface water bodies include protection and propagation of fish, shellfish, and wildlife; recreational uses; agricultural and industrial use; use as a public water supply; and navigational uses. A water body may have multiple designated uses. In assigning designated uses to a surface water body, regulatory agencies generally consider the highest attainable use to which a surface water body could be put, taking into account existing water quality and the potential for improving water quality. Therefore, designated uses may include both current and potential future uses. The biological, physical, and chemical data, upon which designated uses are based, are often contained in specific documents referred to as "water body surveys and assessments." These documents can be useful sources of information. An important first step in evaluating quality and uses of surface waters is to contact the appropriate state environmental agency or regional USEPA office,

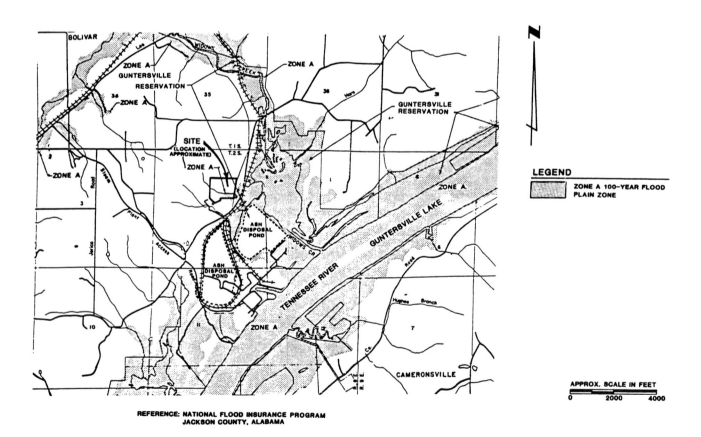

REFERENCE: NATIONAL FLOOD INSURANCE PROGRAM
JACKSON COUNTY, ALABAMA

Figure 2-10 100 Year Flood Plain Map

Source: Golder Associates, Inc.

Table 2 - 9 Climatic Data

Source	Information Obtainable	Comments
National Climatic Center (NCC) Federal Building Asheville, NC 28801 Tel: 704/259-0682	Readily available are data from the monthly publication Climatological Data, which reports temperature and precipitation statistics for all monitoring stations in a given state or region. An annual summary is also available.	The National Climatic Center (NCC) collects and catalogs nearly all U.S. weather records. Climatic data (which are essential for construction planning, environmental assessments, and conducting surface and ground–water modeling) can be obtained from the NCC.
	In addition to collecting basic data, NCC provides the following services: 1. Supply of publications, reference manuals, catalog of holdings, and data report atlases. 2. Data and map reproduction in various forms. 3. Analysis and preparation of statistical summaries. 4. Evaluation of data records for specific analytical requirements. 5. Library search for bibliographic references, abstracts, and documents. 6. Referral to organizations holding requested information. 7. Provision of general atmospheric sciences information.	NCC can provide data on file in hard (paper) copy, in microfiche, or on magnetic tape. For general summary statistics and maps, the publi–cation Climates of the States — NOAA Narrative Summaries, Tables, and Maps for Each State, by Gale Research Company (1980) is helpful.

responsible for water programs, to identify and obtain any available information on surface waters in the surrounding area.

U.S. Geological Survey Water Resources Data

The United States Geological Survey (USGS) boasts a staff of experts in the fields of geological and hydrological sciences. The USGS is primarily responsible for monitoring surface water systems and for collecting data on land surfaces, geology, and geography. Located within the U.S. Department of the Interior, the USGS provides important data assessments through its seven major statistics branches.

The Water Resources Assessment Program is primarily responsible for developing summary level statistics on water resources at the state and national levels appropriate for the preparation of USGA's biennial report, National Water Summary. Each of these reports focuses on a specific water resource theme, and statistics for major subjects covered at the national and state levels include: water availability (e.g., surface and ground water potential, use, and development); water quality (e.g., bottom sediment contamination, saline-water intrusion, hazardous wastes, radioactive wastes, and acidic precipitation); environmental hazards and land use (e.g., flooding, land subsidence, sinkholes, erosion, sedimentation, wetlands, and management activities. Other environmental activity, such as precipitation, streamflow, floods, and storms, is also studied by this program. Data summarized in the National Water Summary are compiled from existing USGS and other-agency data files, and are collected periodically from areas across the entire United States and Puerto Rico.

Both the National Hydrolic Benchmark Network program and the National Stream Quality Accounting Network provide a nationally uniform basis for assessing long-term trends in the physical and chemical characteristics of surface waters which are largely unaffected by land use activities. Water quality is assessed through the use of storage basins, where there are generally no man-made contaminants involved in data collection. The programs focus specifically on monitoring trends of pH, alkalinity, sulfate, nitrate, phosphorus, calcium, magnesium, sodium, potassium, chloride, suspended sediment,fecal coliform and streptococcal bacteria, dissolved oxygen, and dissolved oxygen deficit. Accurate data collection in insured by the use of various automated instruments, designed to measure and record the water conditions and trends. The frequency of data collection varies, depending upon site location and the type of water condition studied.

The goal of the National Land Use and Land Cover Mapping Program is to produce and distribute both land use and land cover maps, as well as the accompanying digitized data. Land use maps refer to the human activities that directly affect the land, while land cover maps are designed to describe the vegetation, water, natural surface, and artificial constructions at the land surface. The areas located on the maps are divided into nine major classes: urban or built-up land, agricultural land, rangeland, forest land, water areas, wetland, barren land, tundra, and perennial snow or ice. Each major class is then subdivided further, revealing a total of 37 minor classes. Data is collected throughout the continental United States by sophisticated remote sensing methods, including satellite imagery, high-altitude imagery, medium-altitude remote sensing, and low-altitude imagery.

The National Trends Network (NTN) is a program designed to monitor the chemistry of the precipitation that falls within United States borders. Formally established in 1983, NTN is the primary agency for deposition monitoring

2.2.6 Climatology

In site investigations, records of precipitation, temperature, wind speed and direction, evaporation, and humidity may be essential or useful supplemental data to provide information on the ground-water conditions. Climatic data are used primarily for estimating the seasonal variations and amounts of precipitation that may be available for ground-water recharge. Precipitation (10 year average) and other pertinent climatological data are needed to calculate water budgets. This information is available from NOAA, the National Weather Bureau, the State climatologist, local weather bureaus, and airport weather services. Table 2-10 describes data available from the National Climatic Center. On occasion, installation of a meteorological station in remote areas may be necessary to obtain more reliable local data.

2.2.7 Ecological Studies

The Federal RCRA regulations (Subtitle D) require that solid waste facilities, among other government sponsored or permitted activities or management practices, do not harm or threaten any plants, fish, or wildlife in danger of extinction. A number of states also have developed regulations preventing the siting of landfills in areas that would cause or contribute to the endangerment of threatened species. The literature search should identify known occurrences of threatened or endangered species or of critical habitat. Sources of this information include the U.S. Fish and Wildlife Service, U.S. Forest Service, State and local agencies, University departments, national and local wildlife and conservation organizations.

U.S. Department of Commerce

Within the United States Department of Commerce, the office responsible for the vast majority of environmental statistics gathering and processing is the National Oceanic and Atmospheric Administration (NOAA). NOAA's primary objective is monitoring research and assessment programs dealing with the environmental quality of marine ecosystems. In addition, NOAA is responsible for gathering all climate data which has been recorded by other agencies, and for re-distributing that information to the public in a more usable form. NOAA is divided into six major branches, and each is responsible for a slightly different facet of environmental statistics.

The Classified Shellfishing Waters branch of NOAA monitors commercial shellfishing waters (which provide oysters, clams, and mussels for nationwide consumption) for the presence of pollutants and coliform bacteria levels. Much of the program budget is used to monitor the nearly 2,000 classified shellfishing areas that have been defined within the East, West, and Gulf coasts of the United States. The data, which are obtained primarily through the use of questionnaires and interviews, are collected from individual state program administrators. Data were first compiled in 1966, and continue to and assessment programs collected approximately once every five years.

The Fisheries Statistics Program (FSP) has several appropriations, but its primary goal is to compile and assess statistics on commercial and recreational fishing locations across the country. In addition, the program is responsible for monitoring the packaging and storage of fish products, as well as keeping monthly data on the importing and exporting of those products. By utilizing this information, the FSP also helps develop policy guidelines for the coordinated collection and distribution of fishery statistics. Commercial data are obtained by several methods, including census of seafood buyers, logbook reviews, telephone surveys, and observer reporting. Data collection varies from daily to yearly, depending upon the subject and the area covered.

The Living Marine Resources (LMR) branch of NOAA gathers data from published sources on the distribution of marine species within the United States. The data gathered include information on distributions by life stage, statistics on commercial harvest, and the status of marine life quality and health. LMR provides extensive coverage of marine life in the Gulf of Mexico, the West Coast, and various seas of North America. Data are collected both from published literature and the databases of other environmental agencies.

The National Climatic Data Center (NCDC) is responsible for the collection, processing, and storage of meteorological and climatological data gathered from a network of stations around the world. NCDC boasts one of the largest and most extensive sources of environmental statistics within NOAA, with records dating back as far as the mid-19th century. NCDC collects data from several foreign and domestic organizations, including the National Weather Service, the Federal Aviation Administration, and the United States Air Force, Navy, and Coast Guard. Variations in climate are examined for both short-term and long-term periods, and statistical summaries of air pollution and oceanic elements are conducted on a regular basis. Some observations are made as frequently as every 15 minutes, but the majority of collections are made daily or monthly, depending upon the type and source of the information.

The National Coastal Pollutant Discharge Inventory Program contains a series of databases within NOAA's Strategic Assessment Program of U.S. coastal areas. This comprehensive system contains information on the amount of pollution being added to the major water bodies and tributaries within the United States. A large portion of these funds are appropriated obtaining pollution level estimates for nine major source categories and 17 pollutants, including those incurred from marine transportation operations and accidental oil spills. Estimates are taken seasonally, and based on a combination of computed methodologies and actual monitored observations.

Finally, the National Status and Trends Program, headed by 15 staff members and approximately 50 data gathering specialists located across the country, is responsible for providing information on the status and trends of environmental quality in estaurine and coastal areas. The program, which began in 1984, deals specifically with examining the distribution and intensity levels of contamination within the tissues of marine animals by geographic region. This program is devoted primarily to maintaining approximately 50 Benthic Surveillance sites and, since 1986, 150 Mussel Watch sites located throughout all U.S. coastal areas. At Benthic Surveillance sites, benthic fishes are collected for the chemical analysis of their livers. At Mussel Watch sites, bivalve mollusks are collected for analysis.

2.2 8 Historical and Archaeological Resources

Many states require mapping of archaeological and historic sites as part of the solid waste disposal permit application. Information on known, existing sites may be secured from literature surveys, contact with state historic preservation offices, university archaeology departments, and the National or State Registries of Historic Sites.

2.2.9 Land Use

State regulatory agencies have occasionally restricted solid waste landfill operations that overlie areas of potential coal, oil, and gas reserves. Previous land use, such as undermining, is extremely important to identify early in the Phase I investigation. Due to increased geologic hazards, undermined sites and location of oil or gas wells (abandoned or active) must be fully identified for the Phase I report. Land use can include facilities, ground-water supply wells, and other features important to understanding the regional setting. Figure 2-11 illustrates the municipal ground-water facilities within a proposed facility area in Iowa.

U.S. Fish and Wildlife Service

The U.S. Fish and Wildlife Service (FWS), a branch of the Department of the Interior, is devoted to monitoring environmental trends of marine ecosystems. Within the FWS, there are six programs dealing with environmental statistics and trends among various aspects of fish and wildlife health.

The National Contaminant Biomonitoring Program was designed in order to monitor and document specific trends in environmental contaminants that may threaten the health of fish and wildlife. At a network of 112 stations, located along major rivers across the country, fish and marine wildlife are chemically examined in order to assess the levels of harmful contaminants to which they are exposed. Since 1965, the monitoring program has continued at two-to four-year intervals.

Another program within the FWS, the National Survey of Fishing. Hunting. and Wildlife-Associated Recreation, represents one of the oldest and most comprehensive ongoing recreation surveys. The survey's purpose is to estimate the number of fishermen, hunters, and other sport-based wildlife recreation participants in the United States, as well as how often they participate and how much money they spend on these activities. Data are obtained through two-part surveys, which are conducted approximately every five years. The first phase of the survey is a telephone screen of over 100,000 households nationwide, to determine who in the household participated in wildlife activities for the sake of sport. The second part of the survey is a follow-up of the phone calls, including extensive interviews with those who confirmed regular participation in hunting, fishing, or camping.

In 1975, the FWS established the National Wetlands Inventory to compile technically sound data on the status of wetland resources in the United States. Data are compiled continuously and updated every ten years, as required by

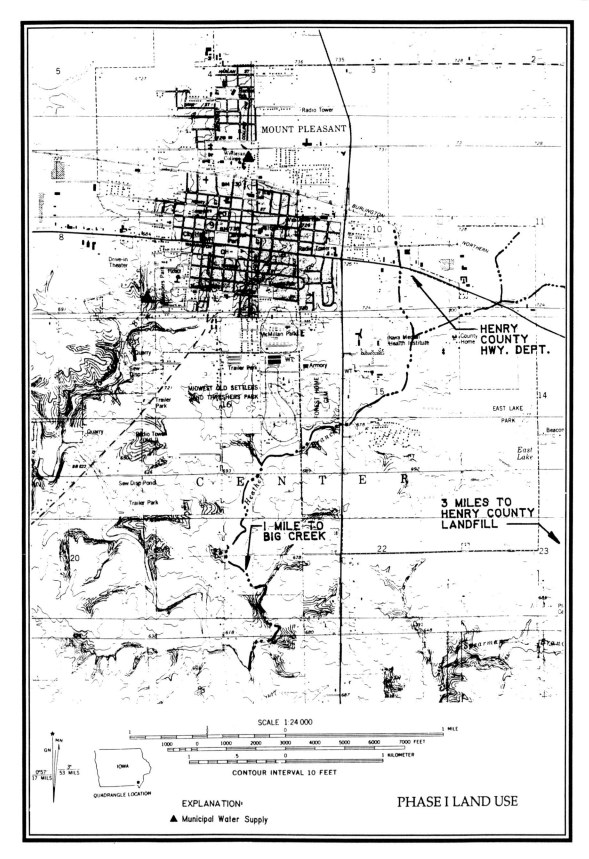

Figure 2-11   Phase I Land Use

Source:USGS, Mount Pleasant, IA, Quadrangle

the Emergency Wetlands Resources Act of 1986. Data are compiled by a series of intricate measurements. After being carefully divided into a four square mile area, each wetland region is surveyed by means of aerial photography. Changes over time are subjected to computer analysis, and assessed in terms of severity and probable cause(s).

The North American Breeding Bird Survey, which began in 1966, is responsible for providing a uniform basis for assessing long-term trends of bird populations throughout North America. Annual data collection is obtained primarily by census observation, taken within specified survey areas during the breeding season (usually June). The data is then processed and analyzed to assure validity.

Finally, the U.S. Fish and Wildlife Service Lands program is responsible for compiling the two primary data series that are reported to the general public: "Annual Report of Lands Under Control of the U.S. Fish and Wildlife Service" and the "Migratory Bird Conservation Commission Annual Report." Various data are collected for all FWS properties across the country, including National Wildlife Refuges, Waterfowl Production Areas, and National Fish Hatcheries, coordination areas, and administrative sites.

Environmental Protection Agency

The United States Environmental Protection Agency (EPA) spends in excess of a half a billion dollars annually collecting a wide variety of environmental information, most of which is collected by the states in response to specific media requirements. The programs described below are exemplary of these various data collection efforts.

The National Air Pollution Control Program of the Office of Air Quality Planning and Standards, is responsible for the collection and analysis of air quality and air pollution levels, which are then compared to national regulations, mandated by the National Ambient Air Quality Standards (NAAQS). Specific air quality criteria, examined daily and monitored for ten-year trends, include levels of sulfur dioxide, nitrogen dioxide, carbon monoxide, ozone, and lead. Data are then broken down into emission sources, as follows: transportation (e.g., motor vehicles and aircraft); stationary fuel combustion (e.g., coal, natural gas); industrial processes (e.g., mills and factories); solid waste disposal; and miscellaneous sources (e.g., forest fires and agricultural burning). Local trends in data are then compared with the national level, and estimates are made as to specific sources of environmental damage.

The Office of Ecological Processes and Effects Research is responsible for two major EPA data activities. The Environmental Monitoring and Assessment Program, Long-Term Monitoring Project (LTMP) was initiated in

1983. As one of the EPA's most recent and important programs, the Environmental Monitoring and Assessment Program (EMAP) was begun in 1988 with the intention of evaluating various environmental trends and assessing potential environmental problems. Its primary goal, essentially, is to anticipate environmental hazards and develop a means of prevention before they become insurmountable problems. The Long-Term Monitoring Project is now a specialized program within EMAP, specifically interested in monitoring trends in surface water variables, such as: pH, acid neutralizing capacity (ANC), calcium, magnesium, potassium, sodium, sulfate, chloride, nitrate, and aluminum. LTMP obtains data seasonally, with the aid of federally-affiliated agencies and universities located within six geographic regions across the United States.

The National Surface Water Survey is broken down into two parts: the National Lake Survey and the National Stream Survey. The purpose of the National Lake Survey is to monitor trends in the chemical and biological characteristics of lakes located in regions with traditionally high levels of acid deposition. The purpose of the National Stream Survey is to identify streams with high levels of acidity, as well as those with low acid-neutralizing capacity. Both programs collect data from a random selection of sites across the country, monitoring variables such as: acid neutralizing capacity (ANC), aluminum, ammonium, conductance, major ions, metals, nitrate, organics, pH, and sulfate.

The Comprehensive Environmental Response. Compensation and Liability Information System (CERCLIS) maintained by the Office of Emergency and Remedial Response, is a database that contains information on over 30,000 abandoned or uncontrolled hazardous waste sites across the country. Also included within the program are data on actions such as preliminary assessment, site inspection, and the date of final hazardous ranking determination. In addition, CERCLIS contains information such as: description of site (predominant land uses, waste treatment storage and disposal, and distance to nearest population); owner/generator information; regulatory and response history; waste description (physical state, type and quantity of waste); environmental information; and the planned events that occur at these sites, including starting and completing dates, prior year and current year obligations, and financial expenditures to date. Data are collected during inventory, assessment, and cleanup of uncontrolled hazardous waste sites across the United States.

The Environmental Radiation Ambient Monitoring System (ERAMS) in the Office of Radiation, was implemented in 1973 to monitor radioactivity associated with air, drinking water, surface water, and milk. Prior to 1973, environmental radiation data were collected by several national networks of sampling stations, which were operated by the U.S. Public Health Service. Once EPA was established, these networks were consolidated and modified into the single national network of ERAMS.

The ERAMS program subjects each liquid to different tests in order to assess specific chemical concentrations. Data for pasteurized milk include concentrations of: Iodine, Barium, Cesium, Potassium, Strontium, and Carbon. For air: Geiger-Mueller field estimates, gross beta, gamma scans, Plutonium, Uranium, and Krypton concentrations are noted. Precipitation is subjected to studies of Hydrogen, gross B, gamma scans, Plutonium and Uranium. Elements tested for concentration in drinking water include: Hydrogen, gamma scans, gross alpha, gross beta, Radium, Strontium, Plutonium, Uranium, and Iodine. Surface water is tested for Hydrogen and gamma scans. The ERAMS program includes 332 sampling stations across the country, which are located in areas where high levels of radiation are anticipated, such as downstream from nuclear power plants. Collection frequency differs among the various analyses.

The Office of Solid Waste Hazardous and Non-Hazardous Waste Surveys program collects data on the generation and management of hazardous and nonhazardous waste materials by site, state and across the entire United States. Hazardous waste data are available from the Resource Conservation and Recovery Information System (RCRIS) and the Biennial Report and two national surveys: the National Survey of Hazardous Waste Generators, and the 1986 National Survey of Treatment, Storage, Disposal and Recycling (TSDR) Facilities.

The Resource Conservation and Recovery Information System (RCRIS) is a database program used to store and assess information about hazardous waste sites located across the United States. Data from hazardous waste site inspections are used to identify trends in management procedures, including the adherence to governmental regulations. Data are collected monthly, and include specific elements such as the type and frequency of wastes emitted from a particular location.

The Biennial Report includes information such as: the number of generators and the amount of wastes generated; the number of waste management facilities; the amount of wastes managed (both by EPA region and by state); interstate waste flow; amounts of waste generated by waste type; and amounts of waste managed by waste management method. The National Survey of Hazardous Waste Generators and the 1986 National Survey of TSDR Facilities are both very similar to the Biennial Report, but go into greater detail on waste production and waste management processes. The program also keeps records on the generation and maintenance of nonhazardous wastes. Extensive data, obtained through a series of surveys, are compiled on wastes produced within industry and municipal landfill sites.

The Toxics Release Inventory (TRI), maintained by the Office of Toxic Substances, is an inventory of the 328 toxic chemicals released into air, water, and land disposal sites by more than 17,000 manufacturing facilities across the country. Data is mandated by the emergency Planning and Community Right-to-Know Act, and information includes emissions information for chemical disposal, off-site transfers of wastes, treatment processes and efficiencies, and waste reduction data. Data are collected yearly from sites across the United States .

The Water Quality and Biological Monitoring System (STORET, ODES, and BIOS) maintained by the Office of Water, contains an estimated by $1.7 billion dollars worth of water quality data used by scientists, researchers, and analysts to determine the quality of rivers, lakes, streams, coasts and ground water. STORET is EPA's oldest data management system, dating back to the early 1960's. It contains surface and ground water information from over 800,000 sampling sites. BIOS, a primary component of STORET, contains biological data collected from surveys conducted throughout the United States. The Ocean Data Evaluation System (ODES) was established in 1984 to address the specific needs of the marine and ocean programs.

2.2.10 Holocene Faulting Analysis

In those states where seismic conditions may affect the design of waste disposal facilities, the investigator should conduct a literature review of seismic potential within the Phase I scope of work. The U.S. Geological Survey and the geological surveys of the individual states are good sources of information concerning geology, capable (earthquake generating) faults, and historic earthquakes. The USGS operates nine Public Inquiries Offices to provide federal and state agencies and the general public with convenient access to information about Survey activities and products. Over-the-counter and mail order sales include topographic, geologic, and hydrologic maps and book reports relating to the geographic area of each office. Other sources include local offices of the Soil Conservation Service and the U.S. Department of Agriculture, local colleges and universities, county and city building departments or public works departments, and consulting geologists.

2.2.11 Basic Data Checklist

After the literature survey has been completed, the data should be reviewed with regard to potential effects of disposal of the site. This review includes identification of fatal flaws and establishes the need for additional information to be secured during site reconnaissance. Prior to conducting the site reconnaissance, the information

Table 2 - 10 Phase I Basic Data Checklist

A. MAPS AND CROSS SECTIONS
1. Planimetric
2. Topographic
3. Geologic
   (a) Structure
   (b) Stratigraphy
   (c) Lithology
4. Hydrologic
   (a) Location of wells, observation holes, and springs
   (b) Ground water table and piezometric contours
   (c) Depth to water
   (d) Quality of water
   (e) Recharge, discharge, and contributing areas
5. Vegetative Cover
6. Soils
7. Aerial Photographs

B. DATA ON WELLS, OBSERVATION HOLES, AND SPRINGS
1. Location, Depth, Diameter, Types of Wells, and Logs
2. Static and Pumping Water Level, Hydrographs, Yield, Specific Capacity, Quality of Water
3. Present and Projected Ground Water Development and Use
4. Corrosion, Incrustation, Well Interference, and Similar Operation and Maintenance Problems
5. Location, Type, Geologic Setting, and Hydrographs of Springs
6. Observation Well Networks
7. Water Sampling Sites

C. AQUIFER DATA
1. Type, Such as Unconfined, Artesian, or Perched
2. Thickness, Depths, and Formational Designation
3. Boundaries
4. Transmissivity, Storativity, and Permeability
5. Specific Retention
6. Discharge and Recharge
7. Ground and Surface Water Relationships
8. Aquifer Models

D. CLIMATIC DATA
1. Precipitation
2. Temperature
3. Evapotranspiration
4. Wind Velocities, Directions, and Intensities

E. SURFACE WATER
1. Use
2. Quality and Standards
3. Runoff Distribution, Reservoir Capacities, Inflow and Outflow Data
4. Return Flows, Section Gain or Loss
5. Recording Stations

F. ECOLOGICAL STUDIES
1. Endangered Species
2. Threatened Species
3. Critical Habitat

Source: Dept of the Interior

presented on the basic data checklist (Table 2-10; modified from USDI, 1981) should be reviewed to assure that the data compiled are as complete as possible.

2.3 SITE RECONNAISSANCE

After completion of Phase I literature search and review of the basic data checklist, the responsible engineer or geologist should visit the site to substantiate preliminary conclusions based on the information compiled during the literature search, and to obtain necessary additional site information. Each of the major types of Phase I studies has somewhat different site reconnaissance components. These components are discussed in the individual subsections.

2.4 PRELIMINARY CONCEPTUAL MODEL

Of prime importance to the use, (or past use), of a site for waste disposal are the hydrogeologic conditions, which must be incorporated into a conceptual model. Information obtained from the site reconnaissance, along with data secured during the literature search, provide the basis for the development of the conceptual model of the hydrogeologic conditions at the site. This model is an essential element for planning field investigations in Phase II, and is an ever-changing entity that can be revised as additional information is gathered and new interpretations are formulated. The use of conceptual models should be employed without fail for every type of Phase I project.

The initial conceptual model must incorporate all the essential known features of the physical system under study. With this constraint, conceptualization is tailored to an appropriate level of detail or sophistication for the problem under study. The degree of accuracy, required for various sites, differs. For simple installation of a few upgradient and four or five downgradient wells, the conceptual model presented in the Phase I report can define the geology, uppermost aquifer, and ground-water flow directions in relatively simple terms. Difficult geology, three-dimensional complex ground-water flow directions with vertical and horizontal components, or the presence of geologic pathways, require more involved descriptions of the physical system. Figures 2-12 and 2-13 illustrate examples of the effects of geology on the movement of leachate from a waste disposal site in cross-section views. The conceptual model must also be tailored to the amount, quality, and type of data available. The conceptual model must be consistent with the depositional history, hydraulic gradients, topographic features, and geophysical logs. Therefore, these parameters can be utilized to 'st the validity of the conceptual model. The number of these

parameters analyzed depends on the level of sophistication required.

After the conceptual model is formulated, the need for additional data may be apparent. Erroneous data, due to the sparsity of sampling points, (or sampling errors that may occur during the field investigations), can lead to an inappropriate conceptual model. Incorporating this data into the model should reveal these errors, usually through the inconsistencies of form. The erroneous data can be ignored or replaced if other information is available.

Conceptual models are revised and parameter estimates updated in order to reflect observations while maintaining physical plausibility. The final result of the Phase I conceptual model points toward additional Phase II field evaluations, which then leads to the desired locations for placement of monitoring wells, after full completion of the

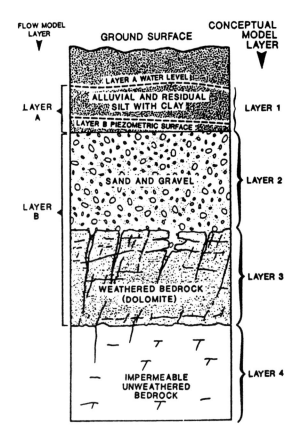

Figure 2-12 Phase I Conceptual Model

Source: U.S. EPA

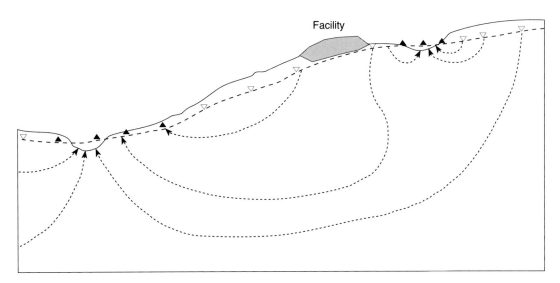

Figure 2-13 Simple Conceptual Model - Cross Section Flow

Phase II scope of work. Further Phase II refinement and linkage of conceptual models with flow net constructions are provided in Chapter 8.0.

2.5 PHASE I SITE ASSESSMENT REPORT

The purpose of the Preliminary Site Assessment (Phase I) Report is to summarize the literature review, findings, and recommendations of the initial field reconnaissance, and to provide the planning basis for the Phase II field investigations. The report should provide useful information, draw conclusions that can be substantiated, and make recommendations. The introduction to the main body of text will vary greatly, according to the major emphasis of the report. It should comprise a statement of the purpose and scope of the investigation, the conditions under which the Phase I study was conducted, acknowledgments, and a summary of previous work at the site. Pertinent information secured as a result of the literature search, contacts with knowledgeable individuals, and the site reconnaissance should be included within the report.

Conclusions about local or regional concerns, and indications of any noteworthy differences between project technical conclusions and those expressed in earlier publications, should be included within the text. Emphasis should be placed on those subject areas that are most relevant to the conduct of future work. Greenfield investigations for new facilities must establish fatal flaws or potential, serious environmental considerations, these should be described in detail. If there are no fatal flaws to preclude use of the site as a disposal area, then the major

emphasis of the report should be on the geology and ground-water conditions so that the Phase II field investigations can be conducted in a thorough and efficient manner.

An assessment of the potential impact of a proposed greenfield facility on endangered species should be included. This review of available data must be performed by a professionally trained ecologist. The Phase I botanical work may include a review of aerial photos to survey botanical zones. These zones should include established wetland areas associated with the site area. This Phase I environmental review is based on documented ranges of endangered or threatened species supplemented by a field visit. The results of the review of archaeological and historic sites, assisted by review of literature surveys, aerial photos, and a site visit, should also be documented.

The most important environmental consideration for siting a landfill is the regional and site specific hydrogeology. The factors controlling the potential migration pathways to the groundwater resources should be fully documented.to the extent of the available literature Sections on regional geology and hydrogeology should contain pertinent information on federal, state, and university studies of the area. Maps should support the text with appropriate, regional cross-sections. The location and logs of existing test pits, borings, or local water wells should be included. Particular attention should be given to recharge areas of significant ground-water aquifers.

If the area has been used previously as a landfill or is an expansion to an existing facility, the report should include descriptions of the materials disposed, if known, and the age of the waste, description of liners and cover material, and the extent of fill boundaries on site, all recorded cell-by-cell (sequential aerial photographs are generally useful for this purpose).

A site plan should be prepared, using the most recent aerial photography, to delimit areas subject to flooding during 100-year flood events. This information is necessary to determine if the site is located within a flood plain.

The final section of the report should provide recommendations concerning the placement of Phase II borings and piezometers, information about the sampling plan, and specifics of the field investigation activities. These field boring and sampling plans should be short, generally no more than 3 to 4 pages, and consisting primarily of tables providing locations, depths and, most importantly, the technical reasons for selecting the specific data points. These documents can be kept short by referencing ASTM or other appropriate "standards" documents. This may include company standard operating procedures (SOP'S), Quality Programs Procedures, or general standards for site investigations. All information should be clearly presented and use of terminology, units, and abbreviations should be consistent. The report should be reviewed thoroughly by the Project Manager and Senior Hydrogeologist assigned to the project.

The above sections 2.2 - 2.5 basic data sources represents a "standard" Phase I investigation that provides the main points of the majority of such investigations Additional project components special to greenfield, environmental audits, or evaluations of existing monitoring systems or project planning for RI/FS projects are provided below.

2.6 GREENFIELD SITING

Landfill siting studies are historically controversial because of potential conflicts with diverse public values. Typically, the greatest concern is voiced by residents, property owners and developers in the vicinity of potential sites who perceive that a landfill will have adverse effects on existing land uses, future growth patterns, or quality of life. People may recognize the need for solid waste management, but are frequently not willing to bear the burden of waste disposal, particularly for "other people's trash". Other public interests may be concerned about increased user costs if remote locations are considered in order to minimize local impacts, while still others support waste reduction, processing, or recycling technologies as an alternative to land disposal. Because of high public sensitivity to landfill siting and the potential for conflicting public interests, a responsive public involvement program is critical to a successful landfill siting process. One must recognize that land disposal is a sociopolitical process that must accommodate local and regional concerns. An environmentally unacceptable site has little chance for acceptance, however, even the technically best location for waste disposal can meet strong opposition.

Successful siting of a sanitary landfill requires sound technical studies (see Table 2-11) combined with the ability to obtain enough public support for the process to build consensus around an acceptable site. The public involvement program must strive to meet the following objectives:

- To establish and maintain the credibility of the overall siting process, including the environmental, engineering, and economic studies;

- To inform and educate the public as to the need for the land disposal facility, possible impacts on the environment, benefits to the public, and potential for reuse, recycling, and energy conversion (i.e., how land disposal fits into the overall solid waste management plan);

- To accurately identify and consider the values and concerns of the public, agencies and political entities; and

- To integrate public views and agency policy with technical data into overall decision making processes.

Greenfield siting of land disposal facilities are similar to other controversial facilities, can be divided into three major tasks:

- Site identification
- Site evaluation
- Site selection

The basic intent is to conduct a systematic study that is based on accurate data and uses this data in a defensible decision making process. A public information and involvement plan is typically prepared at the beginning of the study and a program is implemented throughout, as an integral part of the siting process. Public information and involvement activities are not the standard "public relations" format, but rather a community relations process that evolves throughout the siting process (see Figure 2-14). In addition to local and regional community (county) acceptance, state approval of the site must follow predefined state and federal siting criteria.

2.6.1 Siting Criteria

The criteria for siting future landfills vary widely among individual states and the federal government. While the first goal of landfill regulations is always to protect public

Table 2-11 Greenfield Siting Criteria

Criterion	Phase I: Site Identification	Phase II: Site Evaluation
1. Estimated Service Life	Try to delineate a site of areas sufficient to provide a long service life and adequate buffer area.	Compare acreage of alternative sites
2. Traffic	Locate site as close to a highway as reasonable given other criteria	Compare potential traffic impacts on existing roads (landfill traffic as % of existing ADT)
3. Ground Water Protection a. depth b. quality c. domestic users down-gradient	Choose areas with deeper ground water Choose areas with poorer quality water NA at this phase; land use avoidance will minimize potential impact downgradient.	Compare average depth Compare quality indices Determine and compare the number of domestic users within 1 mile.
4. Noise	Choose areas far from existing residences in quiet areas, or in already noisier areas.	Measure ambient noise levels, calculate and compare impact indices.
5. Visual	Choose area that is visually screened, can readily be screened, or is not near sensitive features.	Calculate and compare indices of potential visual impact; evaluate effectiveness screening or other alternatives.
6. Existing Land Use	Choose sites as far from existing residences Review Subtitle D Airport Restrictions.	Calculate and compare the number residences within 1 mile (or a greater distance if there is no difference at that one).
7. Future Land Use	Identify boundaries of proposed developments on the site or (if nearby; available); choose sites as far as possible from proposed developments.	Discuss prospects for other developments and general plan categories determine and calculate indices of future land use impacts ; compare sites.
8. Timeliness of Site Acquisition	Identify locations of private, state and federal lands; choose some sites of each land type to ensure site availability if acquisition problems arise with any given area.	Map land parcel boundaries; based on discussions with land management agencies, estimate months necessary to acquire; compare sites; Calculate, for each site, annualized.
9. Costs	NA in this phase;	The variety of siting areas identified will ensure variation in costs of land acquisition, site develop-haul costs, the most significant cost development, off-site development, variable. General debt service and haul costs.supplied in Phase II to specific sites. Studies during this phase will establish volumes, locations, and unit costs.
10. Cultural Resources	Avoid known historic or archaeological districts (required for projects sites on or near on state or federal lands).	NA, assuming avoidance of known sites or candidate landfill sites. If some cultural sites are nearby, however, may want to rate probabilities of cultural resources by landfill site. No matter what the results of this phase, a detailed reconnaissance of the selected site will be required for acquisition of state or federal lands.
11. Biological Resources	Avoid rare, endangered or sensitive biological species.	NA, assuming avoidance of sensitive biological resources in Phase I.
12. Economic Impacts	NA in this phase;	Avoidance of existing or proposed land uses will tend to minimize potential impacts. Based on existing, proposed and projected land uses, and on recent land values, characterize sites by relative property value impacts.
13. Geologic Hazards	Review Subtitle D Restrictions on locations.	Detailed geologic review of the potential site(s).
14. Meteorology	NA in this phase.	Identify prevailing wind direction(s) and frequencies; identify sites with existing or future residential developments within 2 miles.

Figure 2-14 Landfill Sitings Can Cause Strong Reactions in Local Land Owners

health and minimize environmental pollution, specific standards are unique to each state's needs. Factors considered by almost every state include the protection of ground water, wetland, and floodplains, as well as the avoidance of siting landfills in areas that are seismically or geologically unstable. These criteria in addition to other locational aspects are given in Table 2-11 for both Phase I and Phase II evaluations. The main functions of such standards are similar: protecting ground-water resources, surface-water quality, aesthetics and public health. U. S. EPA regulations, which revise Subtitle D of the Resources Conservation and Recovery Act, 40 CFR 25, (1991), also address location restrictions within floodplains, wetlands, fault areas, seismic impact zones, and unstable areas on a Federal level.

Ground-water Protection

Potentially, the most adverse environmental impact from landfills is contamination of the surrounding ground water. Based on EPA and municipal landfill reports from various states, ground-water contamination at waste disposal sites has been the most significant environmental problem and should be the primary concern in siting and permitting new landfills. Despite the importance of ground-water protection, many states at this date (1991) do not specify criteria prohibiting the location of landfill facilities over primary water supply and sole-source aquifers. However,

some states such as New York do set restrictions as to primary and principal aquifers.

A number of states regulate ground-water quality at the landfill boundary (or at some distance from the waste boundary) in a similar manner as the federal EPA's methodology called the "point of compliance". The point of compliance is the boundary at which the contaminant concentrations are negligible due to natural degradation, dispersion, and attention mechanisms. The U.S. EPA recommends (RCRA, Subtitle D regulations 1991) a maximum point of compliance 500 feet (150 meters) from the active face of the landfill or within the property of the landfill owner, which ever is less. Other methods for protection of ground water can be based on:

• Ground-water seepage velocity within the geologic formation on which the landfill is located;

• Horizontal setbacks to public water supply wellheads and aquifers can be determined by the rate and direction of ground-water flow, landfill design criteria, and provisions for corrective action in the event of failure of the leachate containment system; and,

• Exclusions to landfill location in areas where ground-water monitoring and site remediation cannot be conducted effectively.

Floodplain Protection

The EPA estimates that 14% of all existing municipal solid waste landfills are located within 100 year floodplains. The new restrictions outlined in 1988 (EPA, 40 CFR 257.257) require that landfills located in 100 year floodplains do not restrict the flow or reduce the water-storage capacity of the floodplain. In addition, measures must be taken to prevent the washout of waste.

Many states restrict landfills within floodplains using regulations similar to EPA's. Other states evaluated prohibit siting landfills within 100 year floodplains. The EPA has not required under drafts of Subtitle D a total ban on siting facilities in floodplains, however, local state acceptance of such siting locations should be fully evaluated.

Wetlands and Surface Water Protection

In siting studies one can expect setbacks from surface waters and wetlands to range from a minimum of 50 feet to a maximum of 1,000 feet. There are typically variable recommend waste setback distances for perennial streams than for wetland, ephemeral streams, lakes, or ponds. Typically, stream setbacks are considerably less than those for the surface water bodies such as lakes.

Within recent years, EPA has identified wetlands protection as a top priority, specifying aggressive implementation of the Clean Water Act (CWA). New facilities may not be located in wetland areas unless it is demonstrated that no significant degradation will result, or that adjacent areas can be developed as wetlands to obtain a no net loss of wetland areas. Therefore, new facilities are not generally located within wetlands as defined either by specific state regulatory agencies or by the U.S. Army Corps of Engineers.

The interactions of wetlands and other surface waters to the regional ground-water flow system requires technical evaluation when siting landfills, since many wetlands recharge ground water, and other discharge water into larger surface water systems.

Fault Areas and Seismic Impact Zones

A seismic impact zone is defined by EPA (40 CFR-258. 1991) as an area which has a 10% or greater probability that the maximum expected horizontal acceleration in bedrock will exceed 10% of gravity in 250 years. The new EPA regulations require that all structures in a landfill located within a seismic impact zone be designed to withstand the "maximum horizontal acceleration" as defined. This performance standard is designed to minimize the risk of slope and liner failure due to seismic motion. The regulations for fault areas require protection of facilities from deformation and displacement in the event of fault movement. Three western states prohibit siting landfills within 60 meters (about 200 feet) of faults that have moved during the past 10,000 to 12,000 years, termed as Holocene or active faulting.

Outside of U. S. western states Holocene faulting and the resultant landforms are rare. Therefore, simple surface inspections for Holocene faulting and associated landforms would not yield reliable information about recent events. As a result, general reference to seismic epicenter plots, historic records, and analysis of possible fault-related features expressed in Pleistocene and older deposits may be necessary. Holocene faulting is one of the most difficult technical issues to resolve in greenfield siting projects. In those seismically active areas care must be taken to examine for such recorded features in and around the project area.

Unstable Areas

An unstable area, as defined by the EPA, is a location which is susceptible to natural or human-induced events or forces capable of impairing the integrity of the structural components of a landfill that are designed for preventing contaminant releases. Such components include liners, leachate collection and removal systems, final cover and run-on and run-off collection systems. Unstable areas could include:

• Landslided materials that may still potentially move further;

• Weak and unstable soils that cannot support foundation structures as a result of expansion, shrinkage, and differential settlement; and

• Karst terrains (limestone and dolomitic bedrock) which have significant solution features.

States generally require that if a landfill is open or near an unstable area, it must be demonstrated that integrity can be maintained of the completed site's structural components. EPA defines the following soil conditions as weak or unstable: organic soils, expansive clays, sensitive clays, loess, and quick (high hydraulic head) conditions. There are currently (1992) no regulations regarding acceptable levels of differential settlement established by any state. Therefore, the geotechnical analysis of a landfill is up to the discretion of developers and state regulatory agencies on the individual design limits of allowable stability and settlement.

The siting of a landfill over permeable bedrock such as limestone or dolomite that has developed into a Karst terrain is technically very difficult. Fissures, cracks, and cavernous openings typical in a Karst environment can provide a pathway for the flow of leachate from the landfill. They may also provide uncertain foundation conditions due to differential settlement.

2.6.2 Site Identification

The primary goal of the initial site identification stage is to establish a set of potential locations which have a wide range of characteristics in terms of parcel size, distance from waste generators, ownership proximity to roads and highways, etc. The Phase I investigation is the vehicle to define these site attributes in a relatively quick and efficient manner. Sites with similar characteristics are grouped and screened down to the one or two best sites of each group through Phase I studies. The waste disposal goals and needs of the regional community must be brought into the evaluation is early as possible to establish the acceptability of the various classes of sites. This allows for comparisons among different classes of sites to prevent selection of similar candidate sites, all of which may have the same technical or sociopolitical problem. These difficulties should be discovered early in the study before extensive time and cost are expended on unacceptable sites..

Existing data are reviewed as part of a Phase I project to determine additional information needed to identify candidate sites. The "classical" site comparison project would collect appropriate information and map these data by criterion on a uniform base map so that overlays for each criterion can be used. This data are interpreted as offering either a siting "opportunity" or "constraint," and one or more composite maps of the siting areas under study are compiled to identify areas with the most opportunities or fewest constraints. Within such areas, potential candidate sites of the desired size are then delineated, and their advantages and disadvantages compiled. Special care must be employed to define any "fatal flaws" of the sites in question. Sites with these flaws should be eliminated from further study so that more effort can be concentrated on the best candidate sites. The typical siting study however, sometimes does not have the luxury of many varied site locations that would address any constraints. The more typical case would be three or four alternative site locations (or even a single site) that must be compared to a series of criteria that will make site rankings possible. These criteria should always include environmental geologic hydrogeologic and other potential "fatal" flaws that would make the site difficult or impossible to permit.

2.6.3 Site Evaluation

Phase I data concerning each specific evaluation criteria are collected and mapped on candidate site base maps. Conceptual-level designs for each site are then prepared concurrently with data evaluation studies, in sufficient detail to permit cost comparisons. Items such as run-on and run-off control, lining systems, access roads, environmental monitoring programs, utilities and fencing are evaluated for each candidate site. Conceptual site plans can be prepared for each candidate site for presentation at public meetings and/or for consideration by the project client. The conceptual design as described above would be separate from a Phase I "fatal flaw" analysis which should have been performed prior to this point.

Once studies are available to indicate the range of the site performance by criterion, the relative importance of each criterion to the siting decision can be rated. These sites are then rated numerically by the importance of each criterion in relationship to all criteria. These ratings are then combined with technical and environmental "performance" ratings of each site by each criterion to result in overall site desirability "scores." Interpretation of these scores, in conjunction with other non-quantifiable considerations and the level of public acceptability of each site, serve as the basis for further Phase II investigations on a limited number of sites.

The project proponent typically will use his own decision criteria in determining a preferred site. These criteria might include overall cost, ease of acquisition, capacity, public sentiment, environmental impacts, likelihood of receiving permits, etc. Throughout the siting process, many factors must be considered in the final decision, only one of which is public acceptability.

Phase I investigations of proposed new facilities require a significant level of effort, both to initially review competitive or alternative sites and then to select the optimum location(s) before proceeding with detailed field investigations. Greenfield siting, by definition, "to locate a new facility in some area not previously used for land disposal", can be an emotionally charged issue. The recent discussions in the popular press on the NIMBY (Not In My Back Yard) illustrated the difficulties with community opposition to landfill siting. For these reasons greenfield siting activities must be quite technically rigorous in both the Phase I technical reviews and all following technical field, design, construction and operations activities.

2.6.4 Greenfield Field Reconnaissance

Field reconnaissance for greenfield projects typically includes the Phase I project components. Since the goal of greenfield siting is to eliminate sites with fatal flaws quickly. The field reconnaissance should begin with a review of recent air photography of the proposed areas to define obvious adjacent site attributes, such as, population densities and flood plains. These data are combined with the factors shown in Table 2-9 in building the various siting criterion.

Verification of Data Adequacy

This investigation associated with greenfield siting activities should include a thorough inspection of all accessible outcrops, cut banks, and other surface exposures; and an examination of conditions, such as, erosion, landslides, seeps, springs, and other pertinent geologic conditions in areas in, and adjacent to, the site. All basic information (i.e., potential boring locations, fracture orientations, conceptual model formation) needed for the initial evaluation of the geologic and environmental characteristics of the site should be obtained during the Phase I site reconnaissance.

Additional Field Studies

As a result of the literature review and/or site reconnaissance surveys, it may also be advisable to conduct additional Phase II field studies prior to the initiation of an monitoring well drilling program. Such studies could also include, for example, an historical/archaeological survey by an archaeologist and one or more ecological surveys.

2.6.5 Greenfield Site Conceptualization

Phase I conceptual models of the greenfield areas should include as many features as possible for interpretation of ground-water flow conditions. Since new land disposal sites will likely have some type of liner and leachate collection systems employed, considerable effort should be expended in defining geotechnical and hydrogeologic conditions that could affect site design. Potential uplift pressures may need to be assessed in later Phase II field activities if the conceptual models indicate such conditions may exist. Since the conceptual model forms the basis for boring plans and field sampling, care must be exercised to identify conceptually the potential for such conditions, so they may be addressed in Phase II field programs.

2.6.6 Fatal Flaw/Preliminary Siting Report

The project components that may effect the successful siting of the facility should be fully described in the Phase I fatal flaw/preliminary siting report. Special care should be exercised in evaluation of down gradient potable water

users established by a well census of the proposed area. The Phase I report should provide an objective evaluation of the proposed sites so that Phase II field work can be directed toward specific siting issues as defined in Table 2-11.

2.7 MONITORING SYSTEM EVALUATIONS

Investigators are often called upon to review operations with some level of ground-water monitoring in-place for sometimes an extended period of time. Facilities with installed ground-water monitoring systems can be evaluated within a Phase I framework. The question should be asked; are monitoring wells located to fully determine potential risk of the site to the environment. The tasks of an evaluation would include:

- A review of all available information, including well construction logs and previous reports;

- An inventory of the location and condition of each facility well;

- Collection of geophysical logs of appropriate boreholes;

- Assessment of the performance (recovery rates) of each well;

- Evaluation sample collection techniques, and;

- Interpretation of historical water quality.

2.7.1 Reviewing the Information

One of the most critical tasks in evaluating an existing monitoring system is to assemble all available information on the system under review. Well construction logs can be found at a number of locations, for example, in a public repository (as with many Superfund sites), filed with the state water agency or geologic survey or alternatively the logs may be nonexisting. Documentation associated with a "designed" ground water monitoring system can typically be retrieved from the consulting firm originally responsible for the installation. Consulting houses typically maintain raw logging forms and paper copies of reports for up to ten years and, on occasion, maintain such documents permanently. Even when the original reports are destroyed consultants often maintain microfetch copies; these sources should always be checked. Consultants and clients share copyright on such data and under a review of a third party, the original client or owner must often direct the consulting firm to provide copies of the borehole and construction details.

Once the regional geologic data is established as part of a first task the available details of well construction should be then compared to the hydraulic properties (i.e.,recovery rates) of the target aquifer. The reasons for the selected locations and depths of the monitoring wells, if contained in the historic data, should be included in the review of the system. The evaluation of well construction details could include a review of the following items as available:

- Well identification number, permit number, and location coordinates;

- Installation dates, drilling methods, and contractors used to install the monitoring wells. Logs of boreholes drilled for stratigraphic interpretation including those holes converted to monitoring wells should be reviewed for consistency of geologic units;

- Depth to bedrock; if cores were not recovered to define the colluvium-alluvium interface with bedrock. Auger refusal, blow counts, and drilling response changes can be used as an estimate of this interface;

- Borehole depth (usually reported in feet above mean sea level) and borehole diameter,

- Casing (if any) type and diameter;

- Screen type, length of screen, and position of screen;

- Top-to-bottom elevation of screen;

- Top-to-bottom elevation of well filter pack;

- Top-to-bottom elevation of annular sealant (cement, bentonite, and grout types);

- Collar elevation; the top of the central well user should also be measured;

- Depth of individual well points or piezometers;

- Geologic interpretation of boring logs (very important)

- Development history of the well; and

- Type of dedicated sampling equipment installed in the well.

Older well systems often do not have all the above data, however, wells and boreholes installed by consultants, or under state and federal ground-water monitoring programs,

can provide much of this data. The verification of filter packed and screened intervals combined with cross-sectional information results in a certain level of confidence that a particular target geological zone is, in fact, being monitored. Development of a level of confidence in the monitoring sampling depth and locations provides one of the most important evaluation tools in initial acceptance of the water quality and water level data for the system. These data represent the first step in a typical Phase I investigation for a currently installed monitoring system. Additional evaluation steps that provide confidence that the monitoring system is continuing to work as designed and installed is the next step in the procedure.

2.7.2 Phase I Well Inventory and Field Conditions

Phase I field reconnaissance for evaluation of a current ground-water monitoring system relies on the inspection and testing of the in-place wells. One of the most important tasks within a Phase I site reconnaissance is to develop a site inventory of all well and piezometers included in the monitoring system.

If an inventory of all site monitoring wells and piezometers and their physical conditions is not part of the historic documentation, it should be generated as the first task in a field evaluation program. The well inventory should include:

- A map of surface and ground water sampling points on local and site-specific scale;

- A figure showing all monitoring wells, piezometers and known borings, where logs are available. Nested wells or piezometers should be referenced as to respective depth of the installations;

- The observed physical condition of the surface completions of all installations;

- Measurement of the diameter of inner and outer casing (if applicable).

- Measurement of depth (to bottom of casing) of wells. In-place pumps (if dedicated) should be pulled to measure the open depth of the hole. If the hole has silted-up, the well should be redeveloped to clean out fine-grained materials. It may be necessary to inspect the hole with a down-the-hole TV camera to define screened depths and conditions of the screens and risers. If a pump is to be rededicated to the hole, the appropriate lengths of hose or tubing should be attached to re-establish the pumps at the appropriate level; and

• Document the physical condition of the pump including any corrosion, wear and blockage of the system.

In developing a well inventory, investigators often find well vaults located in traffic areas (eg. a parking lot or the roadway of an industrial area). These vaults are constructed in the ground with the cover flush to the roadway. However, traffic over the vault cover can typically cause deterioration and cracking of the vault structure. In addition, roadway surfaces often are repaved without building-up the perimeter of the vault. These conditions can result in non-water tight seals leaking storm water into the depressed vault. If one finds surface water inside of a vault containing a capped monitoring well or piezometer, the cap should not be taken out to open the submerged well riser, since this will allow the pooled water to run down the well. The pooled water should be removed and the vault repaired before attempting to open the well casing. Rubber packers are also commercially available that can seal well casing risers and also provide a locking system. These vault structures should be inspected directly after wet weather to establish if the system is weather-tight to surface water intrusion. Down-hole TV cameras (see Figure 2-15) are extremely helpful for documenting inplace conditions of well screens and risers. These examinations of the well integrity often observes unusual well conditions down-hole.

2.7.3 Geophysical Data

In reviewing existing ground-water monitoring systems, one rarely finds complete documentation of the physical features of each monitoring well as discussed above. Some levels of baseline information concerning each monitoring well and piezometer point must be available to further use the existing monitoring network for additional ground-water sampling. Those wells and piezometer points with sketchy construction documentation may be geophysically logged to reconstruct some unknown portions of the well design. Monitoring wells are typically cased with PVC, (the most common casing and screen material), steel (for water supply or older monitoring wells), or may even be constructed of stainless steel components. Geologic cross-section estimates of confining units and aquifer zones, may be generally discernable by geophysical methods to roughly reconstruct general boring logs of material originally observed during drilling. Two of the most effective tools for logging within cased holes are gamma and neutron geophysics (Benson 1988). Gamma logs most often are used to differentiate between clay or non-clay stratigraphic zones. Neutron logs are used to define water-bearing zones (Benson 1988). These data must be supplemented by reasonably reliable boring logs collected from similar or nearby areas.

Figure 2-15 Down-hole TV Viewing Can Establish the True Condition of the Borehole

The cement bond log geophysical technique is commonly used to confirm or develop basic well construction information. This log is run-in well casings installed in holes having filter-packed intervals. This geophysical techniques can be used in conjunction with down-the-hole TV inspection to document the screened interval, and general conditions of the well casing and screens filter-pack zones, grouted or bentonite sealed sections of a monitoring well.

2.7.4 Testing for Hydraulic Connection

Monitoring well and piezometers systems often consist of closely nested or multiple wells in a single borehole. A common question arises as to the hydraulic isolation of the closely spaced monitoring points. These monitoring wells or piezometers be tested for hydraulic annular seal interconnections by the use of a limited pump test. Pump tests normally are conducted in aquifer characteristic testing programs to determine transmissivities, storage coefficients, and a wide range of aquifer parameters. However, short pump tests can also evaluate well or piezometer construction interconnections.

One procedure is to pump one well or piezometer point (the deepest) and plot water levels in adjacent measuring points. If a rapid draw down is observed in clustered or nested wells, a hydraulic connection is likely. In the case of nested piezometer points, if the response is sudden and parallels the draw down of the pumping well, the hydraulic connection is likely within the borehole and probably not a result of a vertical transmissive formation.

A hydraulic connection between nested piezometer points in the same borehole usually results from a faulty cement grout or bentonite annular seal. Whether the drawdown response results from a connection inside or outside the borehole can be assessed by calculating a formation drawdown response from known, previously determined, hydraulic conductivities. A comparison of the calculated formation drawdown response can be made to water level responses seen in hydraulic connection testing.

2.7.5 Well Performance Tests

Well performance can be evaluated through water level recovery rates after development for each well point that will be used for further monitoring. Wells with long recovery rates might indicate faulty well construction (where well seals cover the screened zones), screen closure (incrustation or corrosion) or relative low hydraulic conductivity of the formation. Evaluation of target aquifer materials as compared to the observed recovery rates provides keys to how these wells will perform during monitoring events. Recoveries should be known for each well so that sampling crews can judge the length of time necessary for ground-water collection activities. In general, if wells will not recover sufficiently for sampling within 24 hours, they should not be used for regular ground-water monitoring systems (ASTM D-5092). Water level measurements to determine hydraulic gradient should always be made before performance tests or water quality collection activities.

2.7.6 Existing Water Quality Data

The evaluation of existing water quality data is perhaps the most important, but difficult task included in a well evaluation program. Interpretation of previously collected data is usually included as one of the later tasks in the evaluation of the validity or usefulness of the historic water quality data.

A general approach to the evaluation of previously collected water quality data (once the sampling systems and wells are found acceptable), is to use parameter analysis by

trends in both time and space. Time based changes can include time series analysis, (such as analysis of quarterly data or seasonal changes) or review of the entire record period. Differentiation of geologic zones is almost always necessary in this analysis, and graphical plots such as histograms, stiff, or trilinear piper diagrams often are used to clarify vertical and horizontal trends in data. Simple statistics and analysis by inspection (i.e., you look at the data!) must be used to evaluate the variability of both the time series data and period of record data sets. Historic records of water quality data will typically fall into two general categories:

• Indicator parameters used for detection or assessment monitoring of site performance; and

• Broad based constituents selected for geochemical and background water quality analysis and regulatory comparisons.

The broad based chemical constituents selected for analysis are the water quality parameters indicative of the site conditions under study. They can provide insight into variations in water quality observed during the historic long-term sampling programs. Such comparisons are described further in Chapter 11.0. Indicator parameters may also be used to point out physical problems with particular well points; for example, a pH of 11 or 12 might be the result of alkaline grout effects reaching the filter pack. Unexpected variability in water quality results may be caused by a whole series of sampling, construction and environmental interferences:

• Vadose zone gases entering the screened portions of the well;

• Faulty installation of casing string, well screen, filter pack, or bentonite cement ;

• Confused well numbering;

• Inconsistent evacuation or sampling of collection points;

• Cross-contamination of undedicated sampling equipment (bailers);

• Careless or inconsistent sample handling or transport resulting in constituent contamination, precipitation, or volatilization;

• Extensive laboratory holding time;

• Poor surveying techniques; and

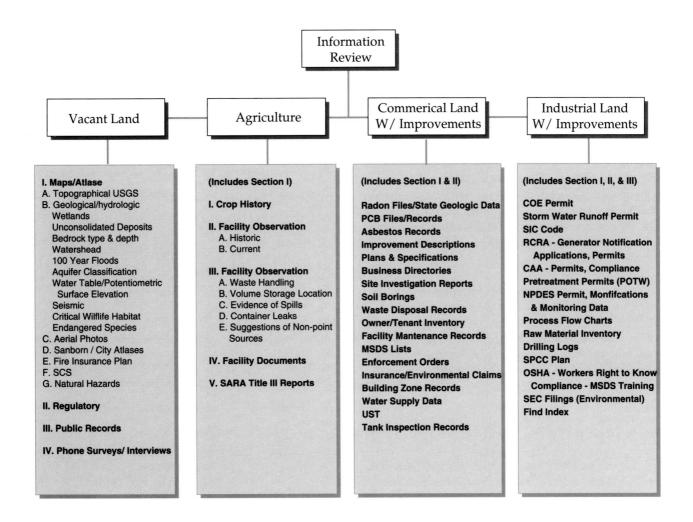

Figure 2-16 Information Review for Various Classes of Properties

• Improper documentation or transcription errors.

The tabular representation of water quality analytical results are the culmination of a long series of technical actions. Any of these actions can cause significant interferences in the results obtained from the analytical laboratory. A multitude of such water quality interferences, (or perhaps only one), might totally invalidate a sampling program and the further use of the historic data. Hence, the overall evaluation of such data before further use must be an important Phase I task for these evaluations of existing monitoring systems.

The report that documents the results of the existing monitoring system should also include a conceptual model/flownet construction. This construction provides for evaluating the flow path target monitoring zone. Chapter 7.0 describes the process of building these models. Some site characterizations are extensive enough to complete such construction, however, more likely than not, only after a full

Phase II field investigation can sufficient data be available for this task. Conceptual models must be attempted in every Phase I scope to organize the evaluation of where the facility flow paths are located. Background and down gradient well locations can be described based on the known geologic and hydrogeologic conditions.

2.8 ENVIRONMENTAL AUDITS AND PROPERTY ACQUISITION REVIEWS

Federal and state legislation over the last ten years, have spotlighted the risk of lending institutions, developers, and insurance firms of becoming an "owner or operator" of real estate that has been impacted by leakage or disposal of hazardous materials. Environmental site assessments are becoming common prior to the acquisition or sale of commercial real estate. These assessments establish the overall risk or liability the property represents to the purchaser. A facility assessment or environmental audit would conform to many of the components of a traditional Phase I investigation with particular emphasis on prior and current site use compared to waste disposal regulations (see Figure 2-15).

The assessment or environmental audit is developed through an evaluation of a facility's past and current operating, maintenance, and hazardous materials management practices. Key elements of an environmental assessment include site history, past disposal practices, soil and ground-water conditions, and the proximity of the site to drinking water supplies and surface water bodies. Environmental investigations of facilities have a number of levels of effort (from low to high) in site environmental reviews:

• A facility environmental *inventory* identifies the locations, type and quantities of hazardous materials,

• An *inspection* is a facility walk-through to identify hazardous materials to evaluate hazardous material management; this may include an initial sampling and analysis of any unknown materials.

• A compliance *audit* provides a broad review of site or facility liabilities.

• A facility *assessment* goes even further than the previous levels, and may incorporate aspects of health and safety risk assessments.

The design of the assessment must be directed to the needs of the purchaser and to the nature of the property. There are a number of degrees of project scope keyed into the previous bullets.

2.8.1 Compliance Audit

A compliance audit is defined as a one-to-one comparison between facility performance and environmental regulations. A compliance audit can evaluate the potential for a facility's operations liability to a state or federal environmental regulations, notices of violations (NOVs) of permit conditions, or third-party claims from nearby residents. The audit procedures formalizes an independent third-party verification of a corporation's compliance with statutory and regulatory requirements. Compliance audits can also evaluate the extent of a facilities adherence to corporate policies and guidelines on environmental protection, employee safety, handling of hazardous materials, and safe disposal of toxic

Table 2 - 12   Potential Site Environmental Liabilities

abandoned water wells	may provide pathways for aquifer contamination
underground storage tanks	costly to remove, remediation expensive if leaking
corroded barrels and drums	contents may have leaked into soil and ground water
unlabeled, filled drums	costly to analyze and dispose
asbestos	found in a variety of building products; expensive to remove
electrical transformers	may be a source of PCBs
batteries	lead electrodes can potentially contaminate soils
agricultural chemicals/containers	may have leached into soil and ground water
spilled paint and paint chips	may contain high concentrations of lead
leachate seeps	may indicate buried wastes at the site
ash or burned surface soil	wastes containing hazardous substances may have been burned

and hazardous chemical wastes. Objectives of an environmental audit can include any or all of:

• Document site operations and conditions;

• Compare site conditions to facility regulatory and operational requirements;

• Review procedures, policies, and guidelines related to hazardous waste management, and evaluate environmental monitoring plans;

• Identify potentially reusable toxic or hazardous materials, site exposure pathways to human health and the environment; and,

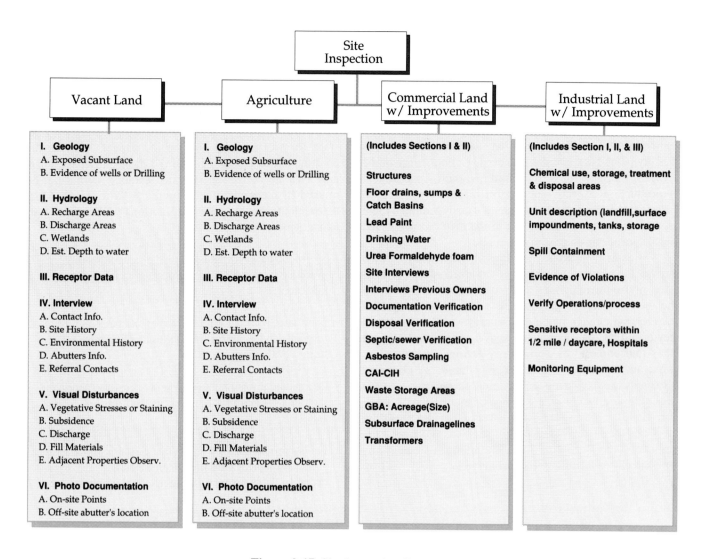

Figure 2-17 Site Inspection Components

• Document any potential risk for exposures to hazardous materials for past and current site activities and make recommendations to reduce potential risks.

These objectives are achieved through the above expressed review of pertinent regulatory and facility technical records, on-site inspections, interviews with site workers, corporate and plant management, and reporting the results of the audit.

Any environmental audit requires a review of the site's history, a review of federal, state and local government records, and an environmental site investigation. The first step of Phase I for site audits, is to conduct a title search that may reach back forty years or more. This, as a minimum, will establish the chain of ownership and identify some historical site users. Sites that may have been historically used for coal gas production or wood treatment may require title searches back to before the turn of the

century. The title information can be obtained through local title companies or county records, and can provide a first level of information on previous ownership and property use. Although county deeds office records may not indicate previous site use or the potential disposal of waste materials and toxic substances, they can provide some insight to the type of historic site use. This historical data use review provides the basis for later field interviews and for scoping the site Phase II field investigation. Past business directories (e.g., Sanborne Insurance Atlas) can be used to determine former industrial activities (Peterson, 1987). This search of the available data (literature) should also include the likely geologic conditions present on site. These data are established in a regional context through review of sources listed in Table 2-4 (Section 2.2).

Historic Site Use

Former site use can fit a number of general categories including rural farm land, vacant urban sites formerly used only for agricultural purposes, and sites which have experienced active industrial use for a variety of purposes. While many undeveloped areas are fully "greenfield" sites, other potential sites can have a long and complex industrial history going back into the 19th century. Former use of the property must be fully established early in the assessment procedure. Prospective purchasers of property or typical acquisition procedures of companies usually require a rapid appraisal of site conditions due to the involved negotiated legal agreements on property transfers.

The assessment scope should be organized and presented so that the potential buyer will be informed of the potential site environmental risks and any further costs of more detailed assessments (for example a Phase II investigation) can be paid through funding of set-asides of money. When reviewing the site the potential buyer should look beyond the "four corners" of the property and include other less obvious liabilities the seller may have. A common problems occurs when a seller closed operations years ago that still pose liabilities, a common example being a small site landfill. The buyer should have a clear idea about the scope of the site investigation. In addition, the buyer must gain access to all records, including information about off-site and closed facilities owned by the seller.

The seller's objectives in a property transaction will be to obtain a clear assignment of environmental liabilities and to account for the lender's legitimate need to be protected from unknown liabilities. Lenders are typically wary of transactions where the collateral (the property) may lose its value because of site contamination, or where the lender's lien on the property may be subordinate to one held by the state for site cleanup. Important aspects such as the construction of the facility and did the facility have liners and leachate collection systems installed should not be missed in the review(see Figures 2-15 & 2-16). Most assessments for property acquisitions will require four major phases:

- Site reconnaissance and review based upon readily available materials from appropriate sources outlined in this chapter;

- Interviews with present and former site personnel or local residents and a careful site tour to identify areas for further evaluation;

- Design and implementation where necessary, a targeted sampling program for both surface and subsurface areas; and,

- Timely production of a report that contains all the results of the assessment presented in an accurate, understandable manner for the potential purchaser.

Figure 2-18 Landfill Liner Construction Limits Discharge of Leachate to the Environment

Figure 2-18 Leachate Collection Systems Remove the Head Driving Forces

A Phase I site assessment generally involves bulleted tasks 1, 2, and 4 only, and each of these three tasks would be applicable to a greenfield site where little or no development activities have taken place. Sites with a potential for past dumping on the property, or where there may be some migration of ground-water contamination from adjacent property, must be carefully evaluated through review of historical airphoto data and an extensive site reconnaissance. The Phase II field assessments, when necessary for project acquisitions, as a minimum, should generally include some form of ground-water sampling, a geophysical survey of suspicious areas, and shallow soil borings to evaluate any previous disposal areas. Fitzsimmons and Sherwood (1987) suggest that all site assessments, including rural properties, should have this minimum amount of geotechnical work performed. However, property acquisitions may not permit sufficient time to locate, install, and sample monitoring wells and obtain from the laboratory the analytical results of the water quality.

Air Photo Analysis of Historic Data

Investigators must commonly reconstruct historic land use for environmental audits and acquisitions, greenfield/expansions and RI/FS projects. One of the most useful tools for gathering information about historic land use is sequential aerial photography (see Figure 2-19). Interpretation of aerial photography involves use of a stereoscope to aid interpretation of photographs acquired for sequential years. In essence, all years where stereoscopic coverage is available (including flights completed in the same year) would be reviewed for the site in question.

The major objective of aerial photographic interpretation is to identify changes in land use and site conditions over the period of photographic coverage, with primary emphasis placed on determining effects of man-made changes in topography and drainage, including potential effects on the hydrogeologic regime. The interpretation should, when appropriate, involve the search for features such as sinkholes and related dissolution depressions, evidence of excessive soil moisture, seeps, erosion and gullying, and lineaments/fracture trends.

Investigation of sequential aerial photography makes it possible to obtain a better understanding of natural conditions and man-made changes over the years. This enables investigators to identify historic conditions at the site and determine if changing conditions could cause adverse effects on surface- and ground-water hydrology. Typically, such an investigation would first address an overview of general site conditions during the periods covered by the flights. The majority of the historical air photo analysis should consist of an assessment of the chronologic changes in site conditions and the potential effects of such changes on the disposal of waste. The investigation should identify potential problem areas that may effect monitoring of surface or ground waters. The text of the investigation should include actual aerial photographs with pertinent annotations made on mylar overlays, which are registered to individual photographic frames. This is not a job for amateur air-photo interpreters!

In those cases where questions arise about depths of excavation, topographic contours (down to several foot intervals) can be generated by commercial air photo laboratories, even on sites flown 30-40 years ago. The technique can be much more cost effective in comparison with drilling or surface geophysics.

If review of historic airphoto data locates questionable disposal areas, they can be confirmed by further Phase II field assessment techniques such as geophysics, hand augering, and if necessary, borehole drilling, sampling, and chemical analysis of ground-water samples. Some labs offer 24-48 hour turn-around times for indicator parameters, hazardous metals and volatile organic chemicals (VOC's) that can quickly establish if significant ground-water contamination is present at the site.

2.8.2 Phase I Field Reconnaissance for Environmental Audits & Acquisitions

Environmental audits and acquisitions can have a major field component (reconnaissance) in the conduct of interviews with current and past employees. For commercial and industrial properties, a review of site records and interviews should be conducted with long-time facility employees. Maintenance engineers, and plant managers can provide a wealth of information on plant practices. A tour of the site should also be conducted with a facility or site representative. Table 2-12 provides areas that should receive special reconnaissance effort that may pose potential environmental liabilities to the prospective purchaser and Figure 2-16 site inspection issues. Inspection of the facility may include careful examination of physical plant structures, piping, material and liquid handling procedures (Bernath, 1988). Key containment features of product storage or use areas such as the condition of flooring in chemical storage or handling areas (i.e., cracked floors) should be examined for possible release to the underlying soils. The areas near sumps, drains, and trenches should also be examined for evidence of leakage through cracks or exposed soil areas. Underground piping can represent major leakage points that are very difficult to evaluate fully in a simple site reconnaissance.

Although one can expect that former and current "owner/operators" may be somewhat hesitant to discuss

Figure 2-20 Air Photo of a Mining Site

important environmental information about the site, interviews with these individuals may be used to determine past processes, raw and manufactured materials handled, locations of underground storage tanks, and the availability of site borings and water well logs. The local fire chief, town engineer, public works department, and governing officials can also provide valuable historical information on the site and should be contacted during the interview process.

During site acquisition activities, the prospective purchaser may also try to conceal adverse environmental or historical information which is important to the overall environmental assessment. This may occur because the purchaser wants to hide negative environmental facts from the financial lender or the insurance carrier. For this reason, it is recommended that the assessment or audit team obtain access to all parts of the facility and perform the assessment as independently as possible from the purchaser.

2.8.3 Supplemental Field Characterization for Property Acquisitions

Site assessments for property acquisitions may include a number of Phase I-II field data collection components. The extent of field collection activities is dependent on the time available for the acquisition review, the extent of previous site activities, and the overall size of the facility. Field characterization activities may also be necessary to confirm suspected contaminated areas or impacted ground water. This Phase I supplemental field characterization depends upon surface geophysical techniques and direct sampling of soils and ground water for analytical testing.

Surface Characterizations

Some field data collection activities, such as surface geophysical surveys, can be performed rapidly. Other field collection activities, such as installing monitoring wells, take a number of weeks to complete, with additional weeks to obtain and analyze ground-water samples. However, if reliable wells are available, these wells can be sampled and be analyzed (at a premium cost) within a week. These existing wells must be adequately documented as to the zone being sampled and the well constructing details. Following the review of air photography, a magnetometer or electro-magnetic (EM) ground conductivity survey can be used to delineate disturbed areas that may contain buried ferromagnetic debris, such as drums and abandoned USTs. False color infrared photography can provide excellent coverage for disturbed areas, springs and distressed vegetation. These points should be checked in the field with a geophysical survey and, if possible, backhoe reconnaissance trenches. The position of drains leaving the site buildings can often be located through air photo and geophysical techniques. These surface profile methods can also locate abandoned well casings that may need later costly decommissioning.

A variety of different types of geophysical equipment are available for evaluation of sites. They can range from simple metal detectors to sophisticated depth-penetrating radar. For most surveys a portable battery-operated magnetometer or metal detector can be used to locate shallow buried iron and steel objects. These devices are light weight and respond audibly when a difference in the magnetic field (or the actual metal object) is detected between the receiving and transmitting sensors.

EM ground conductivity meters measure the apparent conductivity of shallow earth materials. The instrument has a self-contained dipole transmitter and receiver with a spacing that can be varied for different depth penetration. Some of the meters provide output in a single survey, both ground conductivity and the location of buried ferro-metallic debris. The inductive EM method allows rapid survey of subsurface materials without the time-consuming ground contact probes necessary in electrical resistivity surveys. As with any technique reliant on conductivity, man-made surface features such as power lines, fences and railroad tracks can greatly affect performance of the technique.

Ground penetrating radar (GPR) provides detailed reflection records of the subsurface which can be used to identify the size, shape, and depth of buried debris. This equipment is most useful in areas of sandy soil and where shallow, detailed stratigraphy is needed in the assessment. Interpretation of the features observed in GPR can be particularly difficult in soils containing predominant clays.

Subsurface Characterizations

Environmental assessments for real estate transactions may require some level of borings or backhoe trenches to gather detailed site information in a relatively quick operation. Conducting subsurface exploration is important for a number of reasons:

• To characterize the general geologic conditions of the site,

• To evaluate the nature and thickness of fill materials, and

• To provide soil samples profiles for chemical screening.

Although shallow soil samples can be obtained using hand soil samplers and coring devices, truck or trailer-mounted drilling rigs are usually more efficient in the short time periods typically available for real estate transactions. Such boring and backhoe trenches will always require some level of negotiations for access rights.

If the stratigraphy of the site is poorly documented from previous investigation, some number of soil borings may be required to establish if geologic or man-made pathways provide hydrological connection with underlying aquifers, or discharge to surface water. Before any borings are drilled, a full literature search Phase I investigation should be completed to define regional geology and target drilling areas. A common practice is to target likely areas of contamination by increasing the density of borings or backhoe pits in areas near USTs, near drain systems which have contacted hazardous materials, and in areas of obvious surficial impact (stained soils, stressed vegetation, surface refuse, etc.), that were observed in historic air photo review and surface reconnaissance. Depth of borings and soil sampling should be based on site conditions and the product stored on site historically. This would consist of drilling down to confining units for tank areas handling dense non-aqueous phased liquids (DNAPL's) or down to the water table for tanks handling less dense (than water LNAPLE's) gasoline products. Special care is needed when drilling through areas where DNAPL's have been disposed as a product spill. Considerations to not cross-contaminate deeper aquifers by inadvertently drilling through perched DNAPL product zones must be applied during the drilling and sampling project phase.

Ground-Water Sampling

Representative "grab" samples of ground water can be quickly collected at sites with a shallow water table and permeable soils, by using any one of a series of "real time" sampling methods. A screened auger method (while drilling borings) or using of a direct push hydro-punch system, (see Chapter 5) can offer access to saturated permeable subsurface materials. Installation of new monitoring wells in borings may be essential to provide sufficient information on lithology and ground water contained in areas with low or highly variable hydraulic conductivity soils.

Installation of the more traditional monitoring of wells may not be possible with the typical time constraints imposed during acquisitions. As a general rule in areas where there exists a potential for LNAPLE's hydrocarbon contamination, the well screens should extend above the water table to allow entry of floating hydrocarbons. Remember, however, the thickness of product in the well will be greater than the product thickness in the aquifer, (see section 4.0. for details). In low hydraulic conductivity environments such as clays or glacial tills, sufficient time must be allowed for the ground water to enter the open hole and reach reasonably static water table levels. Estimates of this time can be made through Hvorslev's 90% hydrostatic time lag calculations (see Chapter 5.0). Special care must be exercised in establishment of the potential for these low hydraulic conductivity units to be fractured. Secondary porosity fractures in normally confining units can cause unexpected vertical hydraulic conductivity between shallow and deep units. The vertical component of flow can be determined from small diameter, nested piezometers installed in soil borings. Adequate care must always be exercised to seal the piezometers in permeable zones that will hydraulically react quickly.

As with the previous examples of Phase I investigations the results of the project should be fully described in the Phase I report. The report format will be somewhat different dependent upon the requirements of the purchaser or the state regulatory agency. In any case the conclusions of the investigation should be expressed in less than 30 pages of text with appropriate supporting documentation.

2.9 ASSESSMENT MONITORING PHASE I PROGRAMS

Assessment monitoring programs represent both a difficult technical, and potentially costly site evaluation

process. At least two federal programs (Superfund and RCRA), and many state level assessment monitoring regulations currently exist for hazardous and solid waste land disposal facilities. Although procedural and format differences exist between these various programs, the technical site evaluation process is essentially equivalent for any type of assessment monitoring program. The following section will first briefly describe both Superfund and RCRA assessment programs and then describe similar Phase I evaluations that would address this important planning phase of work. Further Phase II scopes of work for these assessment monitoring programs are addressed in Chapter 10 of this text and continue the goals of direct evaluation of site conditions in order to design an appropriate clean-up of the site to protect human health and the environment.

2.9.1 General Superfund Procedure

Superfund represents both a legal and procedural process that is tied to a series of federal laws the Superfund Amendments and Reauthorization Act (SARA, no relation!) and the National Contingency Plan (NCP). When looking at the process of Superfund, many organizational and planning activities are directly related to the phased investigation approach. The first step of the Superfund process is the identification of potentially hazardous sites which may require remedial action and their entry in a data base known as CERCLIS. At this point, or at any time thereafter, an emergency removal action may be conducted by the EPA at a site due to environmental conditions requiring rapid response actions, or because the situation at the site may significantly worsen before a full-scale remedial action can be implemented.

In the pre-remedial process, sites undergo a preliminary assessment (PA) and a site inspection (SI) which usually culminates in a scoring by the hazard ranking system (HRS). Currently, if a site scores over 28.5 on the HRS, it is placed on the National Priority List (NPL) where it becomes eligible for funding of investigative programs (Remedial Investigation /Feasibility Study [RI/FS]) and possible remedial action. Approximately 10% of all sites which are initially identified are finally listed on the NPL. Concomitant with this, the Agency for Toxic Substances Disease Registry (ATSDR) conducts a health assessment to determine if an imminent health threat exists or if further community public health studies (e.g., epidemiology and biological monitoring) are necessary for the site.

Once a site or in some cases, regional areas, are listed on the NPL, it eventually undergoes an RI/FS to determine the nature and extent of contamination and to evaluate alternatives for remedial action. The RI/FS program is shown in Figure 2-20. This program is divided into a series of tasks and phases. The RI and FS usually overlap in time

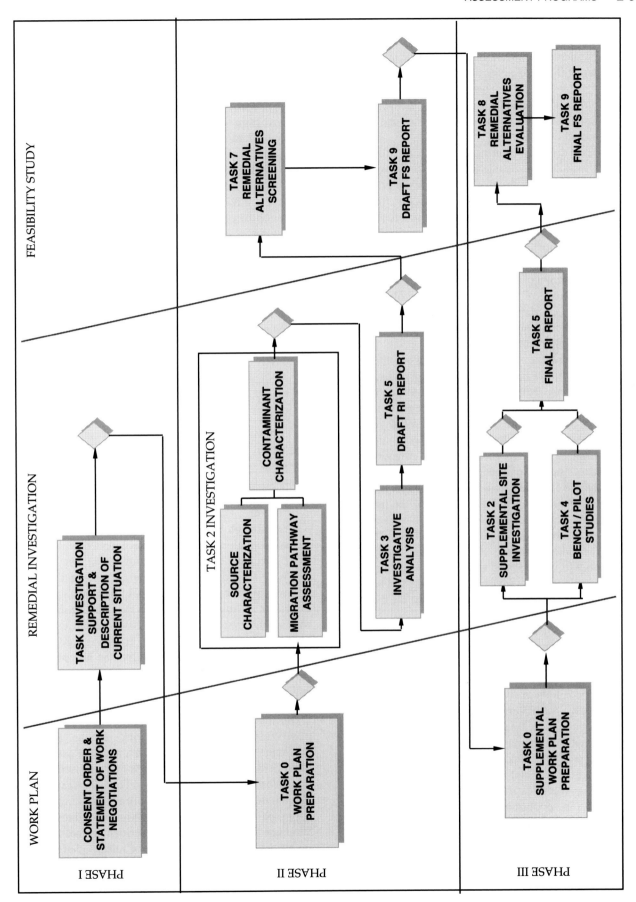

Figure 2-20 Fast Track Superfund RI/FS Program

as shown in Figure 2-20; for example, there can be initial scoping of alternatives while field data are being collected. The RI starts off with the preparation of a series of QA/QC, sampling, and work plans. This process is an evaluation of all data previously collected (e.g., during the SI/PA or by other investigation) and an in-depth cost and time proposal for the conduct of the RI/FS.

A Phase I investigation as described in this manual would be extended to include a number of these planning documents. The Phase I scope is similar to the preliminary assessment (PA) site inspection (SI) tasks of the formal Superfund investigation. The PA/SI also sets the stage for development of planning documents. A preliminary risk assessment, identification of applicable or relevant and appropriate requirements (ARAR's) boring and sampling plans, determination of data quality objectives (DQOs), and an initial screening of remedial alternatives often support the work plan. Once the components of the documents are approved by federal and state regulatory agencies, actual investigative work commences at the site.

Unfortunately, the majority of this work at the typical Superfund RI project involves the collection of samples for chemical analysis rather than directing the investigations toward fully understanding the site geologic conditions. Chapter 10 contains a full discussion into the goals and methods of the RI/FS investigation with suggested streamlining assessment techniques to concentrate on evaluation of site geologic and hydrogeologic conditions. These results are directly used to determine or estimate the nature and extent of contamination. During the time when field data is being collected, the initial screening of alternatives is performed by the investigation staff. After the analytical data leave the laboratory, they go through a process of data validation to insure that the data meet the U.S. EPA's QA/QC requirements. The field and analytical data are then used in the RI report to describe the nature and extent of contamination.

Another use of analytical results obtained from site samples is in the human health risk assessment or the public health evaluation performed as part of the RI/FS. The stated objective of the risk assessment is to assist the U.S. EPA in remedial alternative decisions which have a public health basis. Additionally, during this time, the FS progresses through its final evaluation of alternatives, with the result that one alternative is recommended to the U.S. EPA, Two additional risk assessment activities accompany the FS. The first assessment is a determination of preliminary remediation goals (cleanup levels) for contaminants in various media at the site; this determination takes health effects and ARARs into account. The second assessment is a health based screening of remedial alternatives which accompanies evaluations of long and short-term effectiveness and reduction of toxicity as required by SARA.

Following the completion of the RI/FS, the U.S. EPA issues a Record of Decision (ROD) which states the chosen remedy, justifies its choice and responds to comments received from the public on the RI/FS. The ROD may also decide on a no-action alternative, Additionally, a ROD may be issued for a portion or single operable unit at a site.

After the issuance of a ROD, the site proceeds to the remedial design (RD) stage which sets out the details of construction for remediation. This step may be preceded by a conceptual design and experience shows that most Superfund investigations require additional sampling and analysis over what was performed for the RI/FS. Once the RD is approved, the remedy is implemented as a remedial action (RA). When an effective cleanup has been completed, the site is removed from the NPL. If hazardous materials are left on site in a form where they are still toxic and potentially mobile, the site may be revisited every five years by the EPA to insure that the clean-up remains effective.

2.9.2 RCRA Site Remediation Program

The RCRA Facility Investigation (RFI) is generally equivalent in technical scope to the Superfund remedial investigation (RI/FS). Units are areas of concern that are determined in the RFI to be a likely source of significant continuing releases of hazardous wastes or hazardous constituents. The regulatory means of requiring the RFI is either through RCRA permit conditions (operating or closure/post-closure) or via enforcement orders [e.g., 3008 (h)]. Because of the Hazardous Substance Waste Act (HSWA) statutory language, the agencies must focus the RFI requirements on specific solid waste management units or known or suspected releases that are considered to be routine and systematic. The HSWA permit conditions or enforcement orders can range from very general (e.g., "evaluate the soil at.") to highly specific instruction (e.g., a specified number, depth, location of monitoring wells with samples analyzed for a given set of constituents at a set frequency).

Since the U.S. EPA in the RFI process is not required to positively confirm a continuing release, but merely determine that the "likelihood" of a release exists, the scope of the RFI can range from a single limited specified activity to a complex multi-media study. The investigation may be phased, initially allowing for confirmation or rebuttal of the suspected continuing release(s) through continual sampling. If release to the environment is verified, the second phase of investigation typically consists of release characterization.

This second phase of a RCRA RFI is much like a Superfund RI, includes:

- The type and quantity of hazardous wastes or constituents within and released from the unit;

- The media affected by the release(s);

- The current extent of the release, and,

- The rate and direction at which the releases are migrating.

Inter-media transfer of releases where applicable (e.g., Sediment or surface water releases) is also addressed during the RFI.

The responsible regulatory agency will typically interpret the RFI program and release findings. The agency will also evaluate the data quality of the RFI (i.e., due location criteria, sampling and analytical data quality objectives for the completed project agree with planning documents?) The analytical results of the RFI are then compared against established human health and environmental criteria. These criteria or "action" levels can be established for each environmental medium and exposure pathway, using the toxicological properties of the waste constituent and standardized exposure model assumptions. At this stage, in the RCRA facility investigation, if the continuing release of hazardous wastes is determined as a potential short-term or long-term threat to human health or the environment, the regulatory agency may require either interim corrective measures or a corrective measures study. This evaluation of human health and the environment risk factors is a crucial stage in the RCRA corrective action process.

Identifying and implementing interim corrective measures for the facility may be conducted at any time during the RFI. This procedure may be used in a case where, site conditions are identified that exposure to hazardous constituents is presently occurring or is imminent. In those sites where interim corrective measures may be needed, both the owner/operators and the regulator agency have a responsibility to identify and respond to these interim corrective action situations. The owner/operators and the regulatory agency must work together to assure the adequacy and acceptance of the data collected during the RFI and the final evaluation of those data.

2.9.3 Phase I Assessment Programs

There are many common Phase I goals that comprise important tasks in any assessment monitoring program. These "desktop" activities represent the early steps that permit full scale planning of the field activities that follow the Phase I scope of work. Both Superfund and RCRA have a number of required planning, documentation and safety documents that describe the overall process that will be followed during the assessment program. Although planning documents (such as generation of the QAPP) are specific to the individual remedial program, a number of basic Phase I tasks can greatly enhance the planning process to target drilling location of boreholes and the sampling process during later parts of the second phase field program. Although most RCRA and Superfund sites will have some historic ground-water monitoring, these monitoring systems may be poorly constructed, documented, or screened in the wrong locations and depths to fully evaluate the "rate and extent" of ground-water contamination at the facility. Often these sites or facilities have a minimum "detection" monitoring system that must be first evaluated and then supplemented by a significantly expanded assessment program beyond the traditional detection monitoring program. Unless the facility has had a comprehensive Phase I investigation completed as part of a detection monitoring or SI/PA program, it is recommended that a full Phase I scope of work be completed for every assessment monitoring project within the early planning process.

An investigation of releases from unregulated hazardous disposal areas or affected RCRA disposal units requires various types of technical information. This information is specific to the waste disposed or managed, unit type, design and operation the environment surrounding the unit or facility and the environmental medium to which contamination is being released. Although each medium (i.e., ground water, surface water, sediments) will require specific data and methodologies to investigate a release in the format of the Phase II field data collection, the following represents general guidance for Phase I investigation of several important elements: one is "desk top" in nature and the other focuses on field reconnaissance data collection:

The technical approach of full assessment monitoring program requires the investigator to examine extensive data on the site area and specific disposal practices. These data generally can be divided into the following categories:

- Regulatory history
- Facility and unit design;
- Waste characteristics;
- Environmental setting;
- Pollution migration pathways;
- Evidence of release;
- Environmental receptors, and
- Previous release events.

Investigative efforts will vary depending on which environmental pathway medium is considered most vulnerable. For example, unlined RCRA units are more likely to have soil and ground-water releases than lined units. A site's environmental setting will determine which media are of concern (e.g., shallow ground water or fractured subsoils). As such, the phase investigation approach must be based on regional understanding of the subsurface pointing toward the more direct site based Phase II geologic and hydrogeologic field data gathering effects. The scope of work for Phase I investigations, where an existing monitoring system is in place, should begin with:

• Review of the facility disposal practices
• Historic airphoto review
• When present on site, an evaluation of the current environmental monitoring system and the data acquired as per section 2.7
• Well Census
• Initial conceptual geologic or site model
• Phase I report
  - Surface Investigative Plan
  - Subsurface Investigative Plan

Formulation and implementation of field investigations, sampling and analysis, and/or monitoring procedures are designed to verify or rebut suspected releases (Phase I) and to evaluate the nature, extent and rate of migration of verified releases (Phase II). The latter phase can in turn be divided into logical technical steps.

Assessment monitoring programs typically have goals to evaluate the environment, establish the extent of contamination present at site, propose a technically correct and cost effective remedial solution. The above six tasks provide the raw material to adequately plan virtually any site remedial investigation for state or federal compliance. Additional documentation will be necessary for meeting program format submittals, such as, the quality assurance project plan (QAPP) or the project sampling plan. Each state and federal program has slightly different reporting requirements and the relevant programs must be rigorously followed. However, the six point Phase I program should be followed as the basic scope and included as part of the assessment planning process.

Documents such as the Superfund sampling plan should logically follow completion of the Phase I report or be composed essentially from the Phase I data. No matter what program is followed, at least, the above six points should be performed for the project area before selection of the relevant sampling points in the various planning documents required by Superfund. A strong case can be made to "hold off" the specific locating of sampling points for ground-water monitoring until the full Phase II study has been completed. Many of these projects spend most of the investigative dollars taking ground-water quality samples and then trying to figure out what the parameters values really mean. If the project establishes the geology and hydrogeology first selection of appropriate monitoring locations is relatively easy. The following sub-sections describe or reference important tasks within the six points.

Review of the Facility Disposal Practice

Phase I review of the facility disposal practices should consist of development of a series of data sets based on the characterization of potential leachate. Much of this data should be available for review if the facility represented an operational facility with some form of leachate collection and treatment. The mode of disposal or codisposal of waste materials can be particularly relevant for construction of migration pathway conceptual models. Closed disposal cells within more modern waste disposal facilities were often filled in previously excavated gravel, or construction material source areas. If source areas were stopped at the water table, at clay confining units, (for gravel or sand pits), or at saturated sandy soils (for a clay brick manufacturing site), contact of the waste with the ground-water surface can have a significant impact on the potential for leachate generation and migration. Much of this information can be established through the site interview process described in this chapter. Particular care should be taken to ask the specific questions dealing with base grade conditions present at the facility when the disposal practice began. If this is before the current or available staff, records of pre-existing conditions should be evaluated based on later regional geologic/hydrogeologic conditions established in the Phase I literature review.

Historic Air Photo Analysis

Historic air photo analysis can be the most efficient and cost effective technique available for Phase I assessment monitoring evaluation of post disposal practices, base grade elevations, saturated or unsaturated conditions, fill rates, and virtually any visually relevant information. Chapter 3.0 describes further air photo review activities which can span the range of structural interpretations of geology (fracture trace analysis) to direct topographic mapping of surface features with 2 foot contours of historic pit and excavation depths.

Evaluation of Current Monitoring Programs

During assessment monitoring programs one is often required to review data gathered from previous environmental monitoring. These data may be the primary cause for initiation of the assessment monitoring project. Section 2.7 provides the methods to review data generated from earlier environmental monitoring programs.

Well Census

An accurate well census of potable ground-and surface-water users represents one of the most important Phase I tasks associated with an assessment monitoring program, (or any environmental assessment or monitoring where ground water is involved). Well census data gathering can take a number of alternative directions. In those states where good well drilling data is available, (typically required by law), the location, logs and depths of local potable water wells are found at state geological surveys, department of health or departments of the environments, depend on state regulations. Unfortunately these data often include ground-water wells drilled only over the last 40 years or less, and older water well logs may not be available. Recent air photo review should form part of the well census work to evaluate the location of residences with potential to be locally affected by the facility under investigation. Discussions should also be held with local drillers on areas using ground-water sources and information on target aquifers for the area. These discussions are extremely important to be completed in the Phase I project work. They point the direction for Phase II field assessment work and provide important interpretations for conceptual modeling, (i.e., discharge areas, first water contacts, target monitoring zones etc.), for both the Phase I and Phase II investigations.

Although the raw data gathered from drillers logs filed at state offices are typically very rough, they often provide relative depths to bedrock, clay/sand interfaces, soil depths and types of bedrock contacted. These data can be quickly evaluated for usefulness and how pertinent the data will be to the assessment monitoring program. These data should be always be arranged into at least two formats:

- A map showing the location of each well relative to the facility; and

- Tabular presentations of the available driller log information

If sufficient regional geologic information is available for the area, (and the geology merits such presentations), cross-sectional presentations should be produced showing the relationship between perceived upgradient and downgradient potable water wells and the facility of interest. This initial cross-section should assist the Phase II field investigation scoping for site specific drilling and sampling and evaluation work. All adjacent potable water wells should be evaluated by a site visit to evaluate the current conditions and use of the wells. Particular care should be taken to evaluate near by wells for potential for contamination through facility leachate migration. If any potential exists it is highly recommended that potentially affected downgradient ground-water users be provided with bottled water until adequate testing of the potable well can be completed.

Phase I Assessment Report

Phase I reports for assessment monitoring programs should cover many of the same aspects as other Phase I reports described in section 2.5. However, additional format requirements specific to the federal or state program may require a significant increase in the planning efforts of later Phase II field investigations and the proposed interpretative and analytical programs. One should clearly define the deliverables necessary for the particular assessment monitoring program so that timely and complete documentation is available for the Phase I report. Although these documents, such as, sampling plans, Quality Assurance Project Plans, ARAR's and risk assessments are necessary relative to assessment project deliverables, the raw data that feeds into much of these documents is developed within the classical Phase I project scope of work. A properly conducted Phase I investigation with the resultant report will greatly assist the production of the required state and federal deliverables. Without the focusing of the Phase I data interpretations later expensive field data gathering efforts can and do collect irrelevant or redundant data. Often the somewhat random analytical results obtained from the unfocused sample collection, results in point values of water or soil chemical data without the knowledge on exactly what the value means in context to the whole site environment. In addition, reliance on sample collection specifically for analytical laboratory evaluation places the gathering of important site characterizations necessary for remediation of the facility as secondary and incomplete.

No matter what assessment monitoring program is applied to a facility the goal of the investigation must be to fully evaluate the facility geologic and hydrogeologic environment before selection of location for ground water or soil monitoring. The application of the various recommendations provided in this chapter and in later chapters on Phase II field programs can provide significant savings in both the assessment monitoring program design and implementation of the various remedial activities required for the facility.

REFERENCES

Benson R., 1988, Hazardous Material Control, Vol 1 No.4

Bernath, T., 1988, Environmental audit and property liability assessment: Pollution Engineering, Vol. XX, No. 9, pp. 110-11 5.

Fed. Regist 1982, 47(143, July 26), 32291.

Fed. Regist 1982, 47(143, July 26), 32295.

Fed. ,Regist 1982, 47(143, July 26), 32299.

Fitzsimmons, M. P. and Sherwood, J. K., 1987, The real estate lawyer's primer (and more) to Superfund: the environmental hazards of real estate transactions: Real Property Probate and Trust Journal, Vol. 22, No. 4, pp. 765-790.

Peterson, S., 1987, Historical risk assessment of environmental liabilities at former industrial properties, In Superfund 87' Proceedings, 8th National Conference: Hazardous Materials Control Research Institute, Washington, DC, pp. 45- 47.

U.S. EPA Procedures Manual for Groundwater Monitoring at Solid Waste Disposal facilities; EPA: Washington, D.C.,

U. S. Dept of the Interior 1981, Ground Water Manual, U.S. Government Printing Office, Washington DC 480 p.

Willman, H. B. ,1975, Handbook of Illinois Stratigraphy, Bulletin 95, Illinois State Geological Survey, 260p.

CHAPTER 3

PHASE II SURFICIAL FIELD INVESTIGATIONS

3.1 PHASE II COMPONENTS

A thorough understanding of the geologic and hydrologic regime is essential in a site assessment. For new sites, it is important to define existing hydrogeologic conditions so that the proposed facility may be properly constructed to minimize possible contamination of the ground water. For acquisitions or expansions to existing facilities, it is also important to understand the hydrogeology of the region and the site, so that the extent of any prior releases to the environment can be determined and proper corrective action implemented. Assessment monitoring programs under Superfund or RCRA also require full evaluation of hydrogeologic conditions to quantify rate and extent of contamination. Many of the aquifer characteristics established in the Phase II field assessment program will be directly applicable in design of the remediations for the facility.

Phase II field investigations are the first third of the site characterization and design of the ground–water monitoring system, as illustrated in Figure 3–1. Chapter 3.0 describes surface evaluation techniques. Chapter 4.0 discusses subsurface evaluation methods and Chapter 5.0 looks at assessment testing methods for the field and laboratory. These three chapters serve as the basis for Phase II field investigations in primary porosity geologic environments. Chapter 6.0 extends site assessment techniques to fractured rock. Chapters 7.0 to 9.0 describe data analysis, development of the conceptual models, and design of the monitoring system. The field investigation always provides data used for both design of the ground–water monitoring system, and engineering design of the site or the remediation.

Data from existing sources and on–site investigations serve as the basis for determining locations, numbers, and depths of monitoring wells that are most appropriate to a particular facility. In general, a Phase II description of the area should include a discussion of the following factors:

- Important climatic aspects of the area (e.g., precipitation and infiltration);

- Structural attitude, fracturing, and distribution of bedrock and overlying strata;

- Natural fractures in bedrock or over–consolidated site soils;

- Chemical and physical properties of underlying strata (soil and rock), including lithology, mineralogy, and hydraulic properties;

- Soil characteristics, including soil type, distribution, and

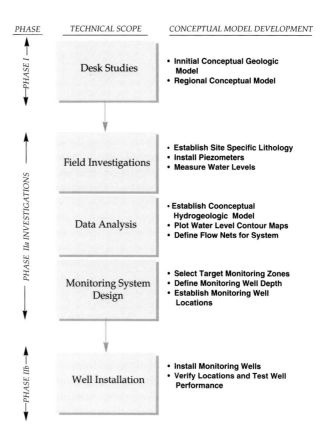

Figure 3-1 Phase II Characterization

attenuation properties;

• Ground-water regime, including depth to the potentiometric surface, aquifer types, flow paths, and flow rates;

• Anomalous geomorphic or structural geologic features that could represent potential contaminant flow paths; and,

• Environmental and human resources in the area.

This section identifies the many levels of technical information typically necessary to characterize a site for both assessment and detection ground–water monitoring and landfill design, beginning with an understanding of the surface of the site.

The complete Phase II characterization will be required for siting and development of most major solid or hazardous waste facilities. However, if the quality of ground water beneath the site is relatively poor (i.e., is not a current or potential future source of drinking water), or if a qualified engineer or geologist can show that there are likely to be no pathways for contamination of ground water, investigative and data gathering efforts can be appropriately reduced from those described in this chapter.

This section summarizes the investigative techniques used to collect data for site assessments. The purpose of this section is to provide the project investigator with an overview of these assessment techniques. Much of the information identified for characterizing the hydrogeology, associated with a site, will probably be obtained through extensive laboratory and field investigations, including hydrogeologic, geologic, soil, and water budget analyses, conducted by qualified professionals thoroughly familiar with such methods. Environmental information, if so required by the individual state regulation, also should be included to complete a single comprehensive investigation.

3.1.1 Field Parameters

Prior to specifying or initiating field work, all existing geologic and hydrologic data should be collected, compiled, and interpreted. This scope of work is described in the Phase I study (Chapter 2). After compiling and thoroughly reviewing the collected data, the investigator can properly begin planning a field investigation for Phase II. Some on-site investigation methods may be appropriate in one geologic setting but not in another. A combination of methods will likely be needed in most cases. The Phase I investigation "sets the stage", or scope through development of an evolving conceptual understanding of the site conditions. This conceptual understanding will direct the investigator to evaluate the respective components of the hydrogeologic regime. These pertinent assessment components may include both saturated and unsaturated evaluations. Vadose zone evaluations are especially important for land disposal facilities where thick unsaturated deposits separate the facility base grades from aquifer units.

The vadose zone extends from land surface to the water table. It has also been referred to as the zone of aeration, variably saturated zone, and unsaturated zone. Use of the latter term should be discouraged since the vadose zone contains moisture up to 100% saturation and, therefore, the term unsaturated could lead to misunderstanding of overall ground-water flow. In the humid areas, the normal vadose zone can be only a few feet thick, disappearing during high recharge periods when the water table is high. In arid areas, the vadose zone can be several hundred feet or more thick dependent on subsurface geologic units. One should not assume deep water tables in all desert environments, however, without adequate Phase I regional data.

Because the vadose zone overlies the saturated zone, facility releases at or near the land surface must pass through the vadose zone before reaching the ground-water surface. Both detection and assessment monitoring programs, on occasion require, both the vadose zone and the saturated zone be characterized for saturated and unsaturated hydrogeologic conditions. As will be discussed later, the vadose zone can be more difficult to characterize due to complex localized flow conditions, than is found in the saturated zone. On the other hand, because it is nearer to land surface, remedial actions may not require complete characterization of the vadose-zone flow system for certain site conditions and contaminants if the majority of the affected soils will be removed or treated inplace.

The vadose zone can be divided (from the top) into three layers: (1) zone of soil water, (2) intermediate zone, and (3) capillary fringe (Davis and DeWiest, 1966). The zone of soil water (see Fig. 3-2) extends from the land surface down to where soil moisture changes are minimal. Since it contains the root zone of plants, it represents the zones in which evapotranspiration is the major active process affecting recharge events. The amount of water and air in the saturated and vadose zones varies both spatially and temporally. This is one reason for the complex nature of the vadose-zone flow system. In general terms when precipitation, falls to the land surface portions of the flow runs-off via overland flow and some percentage infiltrates into the ground. Evaporation and transpiration reduce the quantity of infiltrating water. Evaporation can be defined as the process that converts the

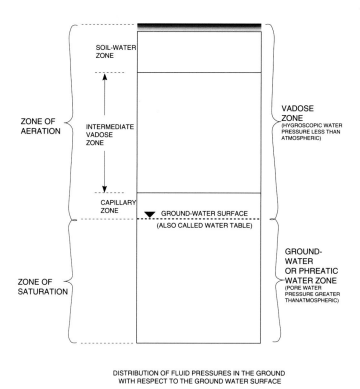

DISTRIBUTION OF FLUID PRESSURES IN THE GROUND
WITH RESPECT TO THE GROUND WATER SURFACE

Figure 3-2 Vadose and Saturated Zones

water at or near land surface to vapor. Transpiration is the process by which plant roots absorb water and release water vapor back to the atmosphere through their leaves and stems. These two processes are combined into the term evapotranspiration. In a typical hydrologic system a significant part of the infiltrating water is consumed by evapotranspiration. The water that is not incorporated in evapotranspiration will eventually reach the ground-water surface, as recharge. However, as the vadose zone geologic profile gets finer, or less permeable for example in areas of glacial clays and tills, the downward movement of recharge slows down to where it may take thousands of years for recharge to penetrate through overburden units to the bedrock. This observation has been confirmed through carbon 14 age dating of ground water contained in sand lenses that were incapsulated within glacial till materials. Vadose zone characterization may be important in those instances where there are requirements to (1) help understand recharge events and how release events may move through the vadose zone and (2) to use in design of land disposal unit caps for limiting infiltration and recharge to an upper most aquifer.

The capillary fringe (at the base of the vadose zone)

by definition extends upward from the water table until there is a decrease in soil moisture. Portions of this zone can be at 100% saturation. This zone can be expected to change in elevation as recharge/discharge causes the water table to fluctuate. The capillary fringe is formed due to a capillary rise effect caused by the surface tension between air and water. Therefore capillary zone thickness is dependent on the grain size of the geologic units. Fine grained materials such as clays may have a capillary rise on the order of 10 feet (3 meters) while course sands may have a capillary rise of only a few inches.

An important field parameter in any site assessment is the hydraulic head present at any particular point in the vadose and saturated zones. As part of the conceptual process of understanding ground-water flow, hydraulic head relationships must be known in the context of the three dimensional geologic environment.

Hydraulic head is made up of two components; an elevation head and a pressure head (see Fig. 3-2). At the water table, the pressure head is zero.(i.e., equal to atmospheric pressure) It increases below the water table and decreases above the water table. That is, pressure head is negative in the vadose zone, which is sometimes referred to as soil tension or suction, this negative pressure head is also characterized by a term known as relative permeability. Relative permeability varies between zero to one and represents the nonlinear function of saturation. Thus, for the vadose zone, in addition to determination of the saturated hydraulic conductivity, the relative permeability function must also be known to fully characterize shallow ground-water flow. The following sections describe applicable assessment techniques and test methods that can be used for both vadose and saturated zone evaluations. As such, care should be exercised to use appropriate methods for the hydraulic system under study. However, for most investigations one should concentrate on the saturated zone since these fully saturated units represent the most likely area for significant lateral flow offsite to potential down gradient ground-water users.

3.1.2 Field Task Components

Figure 3–3a is a flow–diagram illustrating the types of activities and analyses used in selecting locations for monitoring wells at the conclusion of the Phase II investigations. Each illustration is referenced to a Site Assessment Manual chapter or section number for easy referral. The important concept presented in the diagram is the linkage between individual project producibles (such as a geologic map or boring log) and collection of field data. All data should be collected for a purpose, and should be displayed in easily understood formats to aid interpretation.

Not all presentations illustrated in Figure 3–3a may be required for a specific site characterization; however, the logical progression of collecting field data, data analysis, and conceptualization should be followed, in order to design the monitoring system or to design an aquifer remediation. This conceptual scoping of a particular site area is illustrated in Figure 3-3b.

Subsequent sections provide a general discussion of the types of data necessary for characterizing the geology and hydrogeology of a site, and the types of investigative techniques that will likely be used to collect such data. Additional sources of information on investigative techniques can be found in the references included in this document. The investigator should realize that the amount of information and field efforts needed to fully characterize a site may be extensive. However, such assessments are critical to evaluating both the design of the ground–water monitoring program and other environmental issues related to location, operation and ultimate remediation of both hazardous and solid–waste landfill facilities.

A variety of investigative techniques are available to collect data for characterization of hydrogeology. The site–specific investigative program should include direct methods (e.g., borings, piezometers, geochemical analysis of soil samples) for determining the hydrogeology. Indirect methods (e.g., aerial photography, ground penetrating radar, earth resistivity borehole geophysical studies), also may provide valuable sources of additional information (such as porosity). Thus, the investigator should combine the use of direct and indirect techniques in the investigative program to produce an efficient and complete characterization of the facility. Different types of

DIAGRAM OF DELIVERABLES FOR A PHASE II
MONITORING WELL DESIGN PROJECT

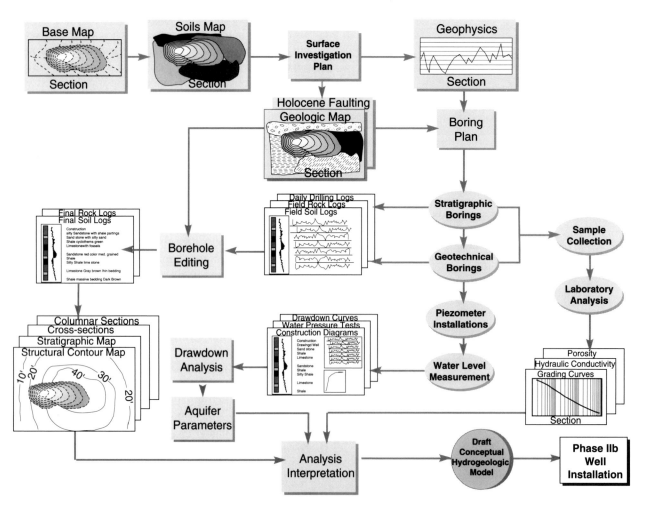

Figure 3-3a Phase II Flow Diagram

investigations, such as greenfield siting or expansions of existing facilities, may require somewhat different Phase II work–components. Table 3–1 shows the potential components of a Phase II investigation for greenfield, site expansions, remedial investigation/feasibility studies for Superfund or RCRA remediation projects, and investigations where fractured rock aquifers are present. The list only provides a general reference for comparison

and for collection of field data.

Phase II investigations must be of sufficient intensity to determine the conditions that may influence the design and construction of both the landfill and the ground–water monitoring–system. The extent of geologic investigation required for a particular site depends on: (1) complexity of the site conditions; (2) size of the landfill; and (3) potential damage, if there is functional failure in the containment..

EVALUATION OF GLACIAL TILL PERMEABILITY FEATURES

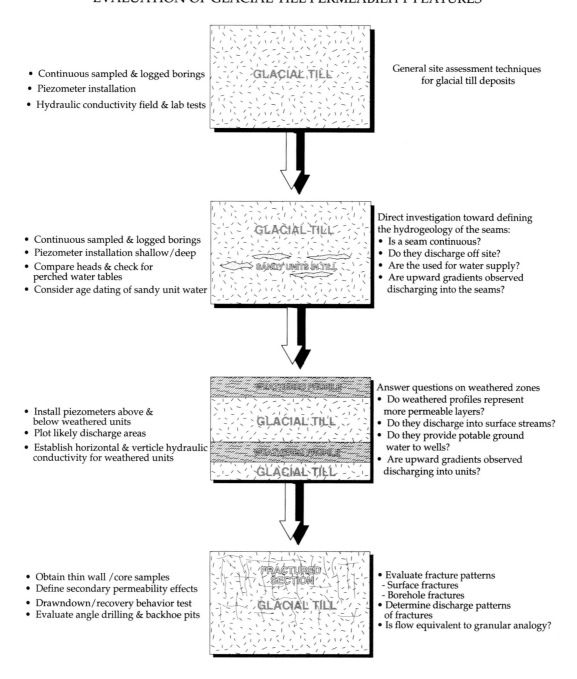

Figure 3-3b Example Phase II Characterization

PHASE II WORK COMPONENTS

Work Component	Expansions	New Sites**	RI/FS	Fractured–Rock Sites
Geologic Mapping	*	*	*	*
Geophysical Survey	+	+	+	+
EM Survey	+	+	+	+
Seismic Survey	+	+	+	+
Exploratory Drilling	*	*	*	*
Cable Tool	+	+	+	+
Rotary Tricone	+	+	+	+
Auger	*	*	*	+
Core	+	+	+	*
Percussion	+	+	+	*
Inclined Holes	+	+	+	*
Logging Test Holes	*	*	*	*
Borehole Geophysics	+	+	*	*
Laboratory Tests	*	*	*	*
Particle–size	*	*	*	*
Liquid and Plastic Limit	*	*	+	+
Engineering Classification	*	*	*	*
Moisture Density	*	*	*	*
Hydraulic Conductivity	*	*	*	*
Triaxial				
Strength/Permeability	+	+	+	+
Consolidation	+	+	+	+
Cation Exchange	*	*	*	*
Soil pH	*	*	*	*
Organic Content	+	+	+	+
Piezometers	*	*	*	*
Multipoint Piezometers	+	+	*	*
Monitoring Wells	*	*	*	*
Performance Tests	*	*	*	*

Table 3-1 Phase II Work Components

PHASE II WORK COMPONENTS (CON)

Work Component	Expansions	New Sites**	RI/FS	Fractured–Rock Sites
In–Situ Hydraulic Conductivity	–	–	+	*
Water Level Measurement	*	*	*	*
Conceptual Model	*	*	*	*
Fracture Analysis	+	+	*	*
Lithological Analysis	*	*	*	*
Hydraulic Conductivity Analysis	*	*	*	*
Gradient Analysis	*	*	*	*
Flow Net	*	*	*	*
Cross–Sections	*	*	*	*
Water Quality Analysis	+	+	*	+
Bedrock Contour Maps	*	*	*	*
Contour Maps – Potentiometric Surface	*	*	*	*
Field Ecological Survey	+	+	+	+
Historical/Archaeological	+	+	+	+
Phase II Report	*	*	*	*

**New sites that will not handle puceable materials (i.e., only construction waste) throughout its life, the facility can meet less stringent criteria.

* = Generally required.

+ = As necessary to meet requirements.

– = Generally not required.

Table 3-1 Phase II Work Components (con.)

This chapter has been divided into eleven subsections. Section 3.2 summarizes investigative techniques that are basic to data gathering efforts, the development of base maps upon which site data can be illustrated. Section 3.3 describes the types of soil classification systems used to describe both surface and subsurface materials. Sections 3.4 and 3.5 discuss regional and site–specific data required to characterize surficial geology and geophysics. Section 3.6 discusses climatic and hydrologic information needed to assess the effects of surface water patterns. Section 3.7 describes aspects of the surface and ground–water quality water quality and land use that should be assessed during the investigation. Section 3.8 summarizes field ecology surveys to evaluate endangered species in and around site areas. Section 3.9 provides general procedures for historical and archaeological surveys. Methods and tools of Holocene faulting investigations important for documentation of site conditions are provided in section 3.10

While this chapter is intended to give the investigator an idea of, and a guide to, the extent of potentially necessary data–gathering efforts, it should not be viewed as a "checklist" for the completeness of a site characterization. The specific types and amounts of information that must be collected and presented will vary from site to site, e.g., states with more stringent assessment requirements would mandate more information to characterize the facility.

Evaluation of any site requires a consideration of the geologic or hydrogeologic features that may affect ground-water flow. Figure 3-3b illustrates the building process used to characterize a glacial till for hydraulic conductivity features. Those features that would cause directional or enhanced flow must be evaluated for the particular site in question. In the case of the glacial till environment, sand layers, weathered zones, and potential fractures, if present, must be fully evaluated for effects on ground-water flow. The individual components of a Phase II investigation are linked to the overall complexity of the site. Figure 1-7 (in Chapter 1) provides a conceptual basis for planning site assessments referenced to the overall complexity of the geologic and hydrogeologic environment. It is highly recommended that the investigator work out the Phase II components on the basis of the Phase I regional information. In other words, link the extent of the investigation on the complexity of the system.

3.2 TOPOGRAPHIC MAP

The intent of requiring a topographic map, as part of a Phase II investigation, is to provide a base upon which information collected in the Phase II field investigation can be placed; and it also provides some of the required site–assessment parameters, such as contours, scale, date, etc. Two types of topographic maps are usually prepared for Phase II investigations:

- Regional Base Map

- Detailed Base Map of the Site

The regional base map, used for site location and general area features, was described in Chapter 2.0. These base maps are easily obtained from published sources and can be used directly for illustrative purposes. These maps, however, have limited use for the engineering design and location of environmental sampling points of a facility;

hence, a detailed base map of the site is normally required for Phase II investigations.

3.2.1 SITE TOPOGRAPHIC MAP

The topographic map of the site should contain 2 foot or 5 foot contour intervals. USGS maps (7.5 minute quadrangles) generally have contour intervals of 10 feet or more; therefore, a more detailed contour map may generally have to be produced for each site. Methods, which are applicable in producing or obtaining the necessary topographic maps, include:

- Obtaining a map from local town offices;

- On–site surveying to gather exact elevations and preparation of a contour map;

- The use of a photogrammetric survey company to fly the site and develop a map with a specified contour interval; and,

- The use of a USGS map, where it will meet the required qualifications.

Often, local town offices, such as the Building Department or Board of Assessors, have compiled large–scale maps which might be helpful in meeting the topographic map requirements. Municipal tax–maps, available from Town Clerks at local Town/City Halls, may provide useful information for a base map (e.g., boundary lines, rights–of–way, structures, pipelines). Such a map could be used as a base map if the contour intervals are 2 or 5 feet. If no suitable topographic map of the proposed site location exists, it may be necessary to measure and plot land elevations by conducting a stadia survey. This technique will provide the information necessary for plotting any desired contour interval.

The base map may be prepared, in part, from information from stereoscopic aerial photos (photogrammetry) yielding pertinent information on location of surface features. Departments of Transportation, Departments of Environmental Protection, and County Planning Departments generally catalogue such state or regional aerial photos. Agricultural, landscape design, and other related academic departments at colleges and universities throughout the country also maintain aerial photographs. Section 2.1 on the Phase I literature source can provide information on sources of air-photo base maps.

Another and more typical method which the investigator might choose to meet the topographic map requirements entails the use of a photogrammetric survey company to target the site and then photograph the area and produce the required maps. Such companies are available throughout the U.S. Their abilities include planning and scheduling the flight to collect data for the map, performing the flight, analyzing the results, and compiling a final topographic map. These companies can provide the topographic data in the form of mylars or digital format for later use in CAD/CAM systems. In addition (see Figure 3-4), the investigator could request that this company take the aerial photos needed for location of faults or other geologic information, thus fulfilling several site–assessment requirements at one time.

The project staff should carefully consider use of 2 or 5 foot contours in areas that have significant relief. If 2–foot contours are required (as in California) for submittals to state agencies, costs for photogrammetric surveys escalate quickly to $20,000 (1990 costs). This would provide flight services and production of a 1 inch to 200 feet (1:2400 scale) topographic map for a 1,200 acre site.

3.2.2 Site Base Map

In order to verify that the requirements for location of upgradient and downgradient wells are met, the preparation of a site base map is mandatory. The base map can be used throughout the process of locating wells to summarize and

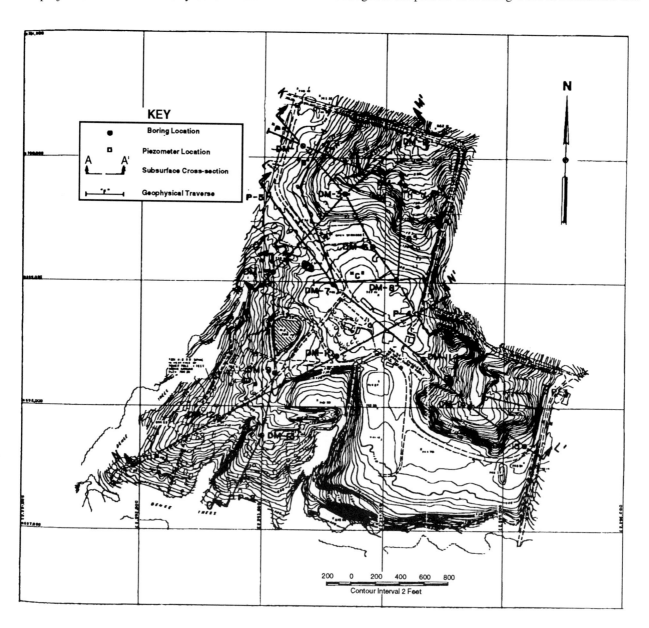

Figure 3- 4 Example Site Base Map

display information. In addition to its use for contour maps of the ground-water surface, a base map should also be used during planning and executing the on–site investigation, as well as, to serve for facility design. The area, to be covered by the base map, should be selected to best represent the significant features at the facility. The base map should extend beyond the facility boundaries to cover other areas that may be affected by disposal activities, i.e., off–site surface water supply wells, references to thematic maps, and water shed runoff drainage paths onto the site. USGS topographic quadrangles and aerial photographs may help in deciding how far to extend the base map.

Dependent on the size of the facility, an appropriate

scale for a base map is 1 inch equal to 200 feet (1:2400; i.e., approaching 1:2000, but not 1:3000), unless other factors override. (NOTE: The USGS, in converting to metric representations, commonly employs a scale of 1:2500, which is also a useful scale.). Although available aerial photographs do not usually provide this level of detail (i.e., 1 inch equals 2,000 feet is a commonly available scale for aerial photographs), information obtained on land surface features and man–made structures is very useful in base map preparation. The obvious selection of scale should be based on placement of the site features on a single sheet of 24" x 36" or 11" x 17" dimensions. Very large sites may require smaller scales to properly fit within the base map sheet. Important features

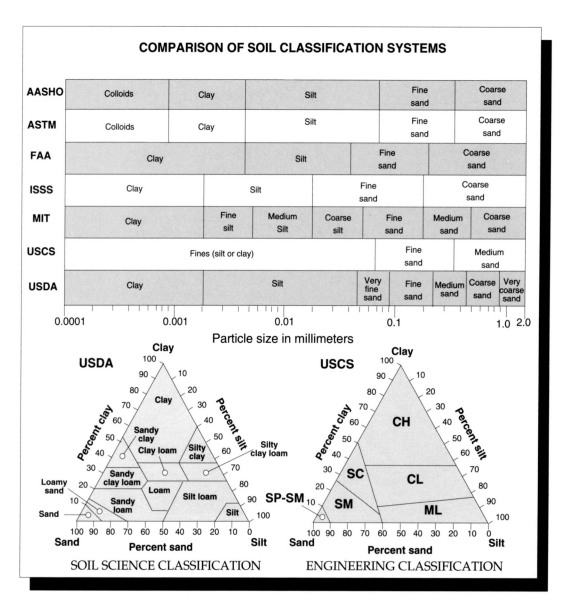

Figure 3-5 Soil Classification Schemes

which should be identified and located on the base map include:

- Facility–related structures (e.g., buildings, roads, parking lots, existing wells, pipelines, bench marks, soil and water sampling areas, borrow pits and quarried, stockpiled soil or rock, and foundation exploratory borings);

- Potential, adjacent sources of contamination (e.g., impoundments, other landfills, storage areas, septic tanks and drain fields) and adjacent waste management facilities;

- Direction of surface water bodies, drainage, and discharge points (e.g., streams, ponds, drainage patterns and divides);

- Withdrawals (e.g., wells, springs); and

- Vegetation.

An essential component of site mapping is to compare existing site configurations with the predevelopment topography and man–made features. This may be critical for interpretation of effects of prior land use on the ground–water system. Scale keys should always appear on the base map so that later reductions or enlargements are reflected in the size of the key.

3.3 SOILS

Soils data provides a basic source of information on the extent, thickness, and physical properties of suitable geotechnical construction materials for liners, drains, and daily and final cover. This data also provides a fundamental basis for decisions on the viability of soils at the facility base grade to contain or restrict the ground–water movement in or out of the site. Although published soils maps and aerial photographs of an area may provide useful information, detailed exploration of on–site soils may be necessary. This information is typically obtained through the use of the same investigative measures discussed later in Chapter 4.0 under subsurface geotechnical investigative methods, and by laboratory analyses of the samples collected. These surveys are especially important in areas of heterogeneous material that show significant variations in physical properties and/or stratigraphy. At least some of the soils information will be acquired through borings, which are conducted to investigate the subsurface stratigraphy. This section includes discussion of soil characteristics that should be investigated at each site. These characteristics include:

- Soil types and area extent as illustrated on soil maps; and

- Physical and chemical properties of each soil, such as organic carbon content, percentage clay, and moisture content.

In analyzing the surficial soils at homogeneous sites, published or otherwise available maps may be helpful if soils are undisturbed. While less likely to be available in adequate detail, published sources may also provide information about subsurface soils. If soils at the site surface have been disturbed, a soil survey may be conducted to determine the distribution of each type of soil. The first step in a soil investigation is identifying and classifying soil types. The U.S. Department of Agriculture's soil classification scheme (see Figure 3–5), based on soil grain–size distribution, is the typical presentation format. Most of the National Cooperative Soil Survey (NCSS) maps, or in common language, "agricultural soil survey" maps, in the USA have been published since 1899 by the Soil Conservation Service of the U.S. Department of Agriculture in cooperation with several Federal and State agencies, particularly in cooperation with State Agricultural Experiment Stations. Some soil surveys are also published by certain state organizations. The list of federally published soil surveys for the entire USA, together with a brief introduction, are periodically issued by the Soil Conservation Service of the Federal Department of Agriculture (for example, Anonymous, 1978). More up-to-date listings are frequently available from the local State Soil Conservation Surveys and related offices (for example, SCS, 1972). The state listing frequently contains important information on completed but still unpublished soil surveys and surveys in progress. Mailing addresses of the state offices are included in the Federal List of Published Soil Surveys (SCS, 1972). All soil survey maps are accompanied with a more or less explicit text describing topography, climate, agriculture, and soil series in the study area.

Some initial general soil surveys are of a reconnaissance character and covered a large area but had a relatively brief descriptive text. Most of the surveys published prior to 1957 were in a scale of 1:62,500 and 1:125,000 (1 in. equaling 1 or 2 mi), and their descriptive text is also rather brief. Most of the soil maps after 1957 are to scale 1:24,000; 1:20,000; and 1:15,840. In general, these maps are much more detailed than many available published geologic maps.

Many modern soil maps are printed on aerial photographs (ortho quads) base and the associated text contains detailed data on climate, soil formation, topography, agriculture, irrigation, drainage, conservation, etc. Some surveys have useful diagrammatic profiles and tabulations showing the relationship between soil series and topographic units.

3.3.1 Classification System

Main units shown on soil survey maps and identified by letter-number indices or colors are the so-called soil series or "soil series association" in which mapping units combine several soil series. These series are frequently subdivided into smaller units or phases.

Different systems of pedologic classification have been developed and are being employed at the present time. Most useful for the engineer and others dealing with shallow foundations in the U.S. and cooperating countries is the Thorp and Smith (1949) formulation of the original Marbut system (USDA, 1938). Marbut lists the following feature as essential for the definition of a soil unit:

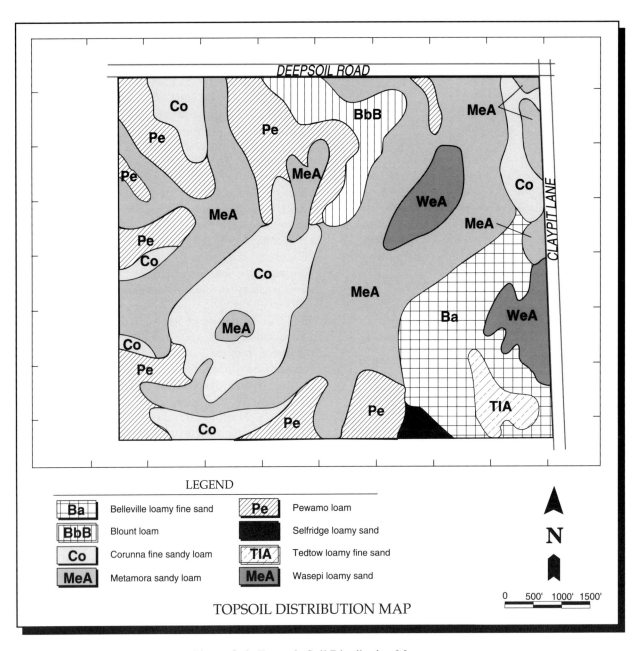

LEGEND

Ba	Belleville loamy fine sand
BbB	Blount loam
Co	Corunna fine sandy loam
MeA	Metamora sandy loam

Pe	Pewamo loam
	Selfridge loamy sand
TIA	Tedtow loamy fine sand
MeA	Wasepi loamy sand

N

0   500' 1000' 1500'

TOPSOIL DISTRIBUTION MAP

Figure 3-6 Example Soil Distribution Map

number, color, texture, structure, thickness, chemical and mineral composition, relative arrangement of the various horizons, and the geology of the parent material. An individual soil unit, the soil type has at least two names, a series or family name and a class (texture) name; for example, Sassafras loam. A soil series comprises all soils that have the same:

1. Parent material: (a) solid rock (igneous, sedimentary, metamorphic), (b) loose rock (gravels, sands, clays, other sediments);

2. Special features of parent geologic material (residual or transported by wind, water, ice, or combinations, or parent geologic formation);

3. Topographic position (rugged to depressed);

4. Natural drainage (excessive to poor); and

5. Profile characteristics.

The different series usually have geographic names indicative of the location where they were first recognized and described (e.g., Sassafras, Putnam, Cecil). It is the soil series which are most important to the geotechnical engineer and are described in detail and mapped by the U.S. Soil Service. For a list of published soil surveys of the U.S. and its territories, see SCS (1972). Identification of a soil in the field as belonging to a certain series makes automatically available the information already gathered regarding this soil.

By conducting a soil survey, the area extent of soil types at the site can be identified. Results should be presented on a plan–view soil map (see Figure 3-6). To supplement the surface analysis, a cross sectional analysis of the soils underlying a site can also be conducted. Results should be presented via maps showing soil thicknesses, types, and extents (lateral). The information needed to prepare these cross sections should be available from soil borings or test pits compiled in the field program. The location of each borehole or backhoe pit should be identified on each soil survey map.

3.3.2 Soil Properties and Geology

Soil properties that are important to measure or determine for understanding the engineering, physical and chemical properties include: porosity (total and effective), hydraulic conductivity, moisture content, organic carbon content, cation and anion exchange capacity, and grain–size distribution. These parameters should be available for each significant soil zone (i.e., several soil

types having similar characteristics) underlying a site, and should be presented in the site assessment report as a table summarizing soil information.

The geologic interpretation of soil maps typically used by investigators are based essentially on the: (1) composition of source (or parent) material for the different soil series; (2) age of the soil series, which controls the degree of development of the soil profile; and (3) topographic-geomorphic location of the different soil series. The first, and probably the most difficult, step in geologic interpretation of soil maps is to establish the sequence of geomorphic-geologic units in the area and their engineering characteristics. Many soil survey publications have a generalized profile showing the major geomorphic units and their relationship to the soil series. The geologic interpretation of some soil surveys, however, is sometimes difficult due to occasional differences in terminology used in pedology and geology.

The engineer must make the translation from soil survey nomenclature to engineering nomenclature for soil surveys to provide the optimum use (see Figure 3-5). Generally, these analysis efforts can provide keys to underlying geologic materials, especially in residual soil areas. Engineering data available from an interpretation of soil maps may sometimes be of greater interpretative value than data obtained from a review of common geologic maps. Particularly important are data on the presence and composition of the B horizon and on distribution of soluble salts illustrated on soil maps, but missing on geologic maps. The presence of the B horizon may cause serious operation-maintenance problems in facility excavations. For example, perched water bodies developed on B horizons can cause both numerous slides and slumps in cut banks These perched water tables also confuse ground-water monitoring system design.

3.4 GEOLOGIC MAPPING

Mapping the surficial geology of the site and adjacent areas represents an important early Phase II task. On the basis of the information obtained during the Phase I regional geology investigation, appropriate field mapping should be performed to define the general geology of the site, including soil conditions. The goal of this investigation is to identify the geologic conditions that may affect location of the facility monitoring system and those geologic conditions that may influence ground–water occurrence, movement, and recharge. The primary purpose of the mapping is to aid in selection of locations for exploration borings, piezometers, and later location of monitoring wells. This geologic mapping must be performed early in the Phase II work activity since geologic conditions have an important influence on the ultimate location and success of the monitoring well system. Sufficient time should be allowed to evaluate the geologic conditions before selecting subsurface investigation methods. The resultant geologic maps generated from this task should be refined after the Phase II drilling program to address three–dimensional distribution of the geologic and earth materials exposed or inferred within the area. This work is best based upon both geologic and geophysical interpretations of the site area.

In areas underlain by overconsolidated soil or fractured bedrock, secondary porosity may provide the major source of hydraulic conductivity. Here the geologic mapping program should include a detailed analysis of the orientation and distribution of fracture joints and joint sets. Fractures are characterized on the basis of information obtained from mapping of surface outcrops, tunnels, excavations, mines, adits, core drilling and logging. Among the parameters most commonly recorded are the locations, orientation, size, opening, and nature of infilling of individual fractures, and parameters describing the spatial relationships between fractures such as spacing and density. The data are then given either a geometrical representation in the form of schematic diagrams, fracture maps, depth profiles, parametric stereograms and polar or cyclographic stereograms, or a statistical representation in the form of histograms and statistical stereograms (also known as rose diagrams or rosettes). Chapter 6.0 specifically reviews the fractured rock environment where secondary porosity is the primary mode for ground-water movement.

3.4.1 Geological Map of the Site

Production of site geologic maps require that some level of standardization be employed so that subsequent evaluations can build on the field interpretation toward final conceptualization of ground-water flow.

A standard for engineering geologic mapping was issued by the Geologic Society of London (U.K.) in it's Quarterly Journal of Engineering Geology "The preparation of maps and plans in terms of engineering geology" was developed in the United Kingdom by the Geological Society Engineering Group Working Party (1972). Although somewhat modified later, this 89-page report was the first comprehensive work in English on the whole spectrum of engineering geology methodology, including what to do in the field, how to do it, and cartography of maps at scales of 1:10,000 or smaller and of engineering geologic plans at larger scales. Further comment about the Working Party Report symbols for lithology was made by Dearman (1974).

In 1968, the (IAEG) International Association of

Engineering Geologists established as its first commission the Working Group on Engineering Geological Mapping. The commission, composed of seven internationally known experts, produced its first major work— Engineering Geological Maps, A Guide to Their Preparation (Commission on Engineering Geological Mapping of the International Association of Engineering Geology, 1976). This is a comprehensive volume and is well illustrated by examples from actual maps, in a variety of scales. These maps are reproduced in color and in black and white.

Although not specifically designed for surface geologic mapping, the classifications of rock developed by the International Society for Rock Mechanics (Commission on Classification of Rocks and Rock Masses of the International Society for Rock Mechanics, and of soil by the American Society for Testing and Materials are pertinent sources of basic site mapping information. Various rock classification schemes were reviewed by Bieniawski (1988), who concluded that there was a need for limited standardization but that there are advantages to having several systems available for comparison on a given project and for application to various engineering requirements. A simplified system that does appear applicable to field classification, has been presented by Williamson (1984).

3.4.2 Mapping Procedure

First objective of a site assessment where geology is visible from the surface is the production of a general geological map for the site area, describing the rock type distribution, faults and tectonic structure. A site geological map is developed by using existing regional geological information (from phase I work), photogeology, obtained from historic recent flights and by onsite geological mapping. Available geologic maps obtained during the Phase I investigation can greatly help in planning a site specific geologic survey. Butler and Bell (1988) provide advice in reviewing geologic maps that apply to planning an additional site geologic survey:

1. Determine the regional setting by referring to a smaller-scale map that shows the relationships of the area in question to surrounding previously mapped areas.

2. Read the legend and other marginal information of the map to be studied, noting in particular the conventional symbols that are used.

3. Base your conclusions of detailed study on the evidence of more than one area or part of a map - it is quite common for a particular outcrop or exposure to

TABLE 3-2 GENERAL LITHOLOGICAL DESCRIPTIONS

1. Name of unit and/or brief rock name.

2. Specific locality or area to which description applies.

3. Thickness and overall structure or shape of unit in this area.

4. Main rock types and their disposition within unit.

5. Gross characteristics of area underlain by unit (topographic expression, color and type of soil, vegetation, nature of outcrops).

6. Characteristic structures of unit.

   a. Range of thicknesses and average thickness of beds or other layered structures.

   b. Shapes of beds or other structures (tabular, lenticular, lineate, etc.)

   c. Primary features within beds or other structures (grading, laminations, cross-bedding, channeling, distorted flow banding, inclusions, etc.).

   d. Characteristic secondary structures, especially cleavage and prominent weathering effects.

7. Fossils (especially if a lithologic characteristic of unit) .

   a. Distribution of fossils.

   b. Special characteristics of fossiliferous rocks.

   c. Position and condition of fossils (growth position, fragmental, rounded, pitted or fluted by solution, external or internal molds, etc.).

8. Description of rocks, with most abundant variety described first.

   a. Color, fresh and weathered (of wet or dry rock?).

   b. Induration (of weathered or completely fresh rock?).

   c. Grain sizes (range of sizes and principal or median size).

   d. Degree of sorting or equigranularity.

   e. Shapes of grains.

   f. Orientations or fabric of shaped grains, especially in relation to rock structures.

   g. Nature and amount of cement, matrix, or groundmass, if any.

   h. Nature and amount of pores (porosity), and any indications of permeability (is this of truly fresh rock?).

   i. Constitution of grains (mineral, lithic, fossil, glass) and their approximate percent by volume.

9. Nature of contacts.

   a. Sharp or gradational, with descriptions and dimensions of gradations.

   b. All evidence regarding possible unconformable relations.

   c. Criterion or criteria used in tracing contact in field.

Source : COMPTION 1962

represent a local anomaly or deviation from the general pattern.

4. Study the regional map as a whole, for example, by viewing it from a distance of a few yards, in order to recognize the broad pattern of rock relations that are shown in the area.

5. Relate your interpretation of the map to the historic environments and processes that created the rock/sand/clay structures that you are studying. In this way the geology of the area shown by the regional map will fit in the context of the specific site mapping area.

Approaches to geological mapping vary according to the amount of natural outcrops, complexity of geological structure, time constraints, and goals of the project. Topographical control of geological observations must be developed from topographical maps, aerial photographs, or for detailed work with the aid of a plane table and survey data. Investigations of these various geologic features must be as systematic and quantitative as possible. In general, the following requirements must be satisfied (Compton, 1962):

1. The stratigraphic frame of reference for measurements must be well understood.

2. Rock names and descriptions should be quantitative and based as much as possible on characteristics that have genetic meaning.

3. Small structures must be observed to determine conditions of sedimentation as well as stratigraphic sequence.

4. Methods of sampling and measuring must be precise enough to permit accurate correlation of data.

A single comprehensive study is generally preferred to several partial ones, for some data can be interpreted only in the light of others. Many projects, however, must be limited in scope, and must therefore be organized carefully in order to meet their special purposes. Lithologic

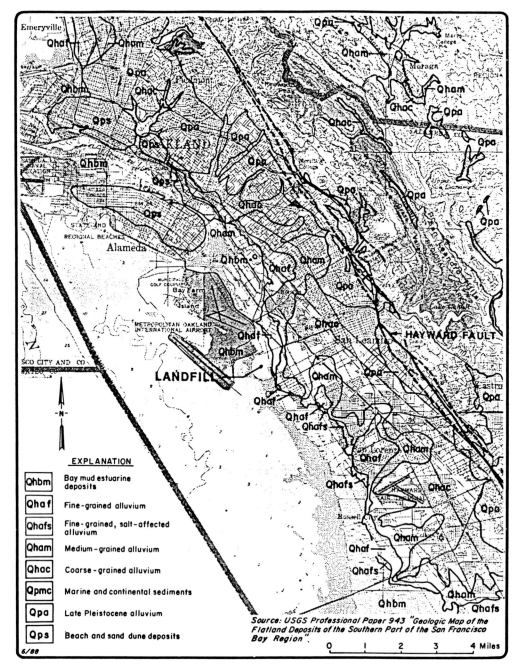

EXPLANATION

Qhbm	Bay mud estuarine deposits
Qhaf	Fine-grained alluvium
Qhafs	Fine-grained, salt-affected alluvium
Qham	Medium-grained alluvium
Qhac	Coarse-grained alluvium
Qpmc	Marine and continental sediments
Qpa	Late Pleistocene alluvium
Qps	Beach and sand dune deposits

6/88

Source: USGS Professional Paper 943 "Geologic Map of the Flatland Deposits of the Southern Part of the San Francisco Bay Region".

0   1   2   3   4 Miles

Source: EMCON Assoc.

Figure 3-7 Geologic Map

descriptions are more usable if recorded in a fairly systematic way, as by the outline (Compton, 1962) in Table 3-2.

Care must be used in determining the colors, induration, and mineralogy of units that are weathered almost everywhere, otherwise their "typical" recorded lithology can be totally unlike descriptions of the same unit in drill cores or mine samples. This does not mean that weathered materials should not be examined, weathering may make it possible to see structures and minerals that cannot be seen readily in fresh rock.

Mapping approaches can be divided into traverse mapping or area mapping dependent on the level of exposure. In addition, significant discussion is provided in Chapter 6.0 on fracture or discontinuity mapping. Such data is necessary for hydraulic conductivity assessments and conceptual modeling which fully requires knowledge on orientation components of the fractures present in the facilities' bedrock.

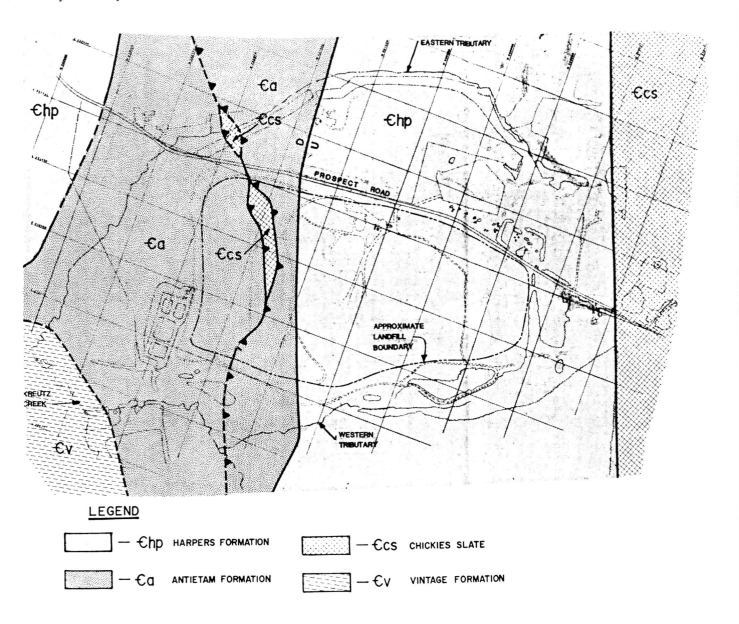

LEGEND

☐ — €hp HARPERS FORMATION   ☐ — €cs CHICKIES SLATE

☐ — €a ANTIETAM FORMATION   ☐ — €v VINTAGE FORMATION

Source: Golder Assoc. Inc.   **Figure 3-8** Surface Geologic Map

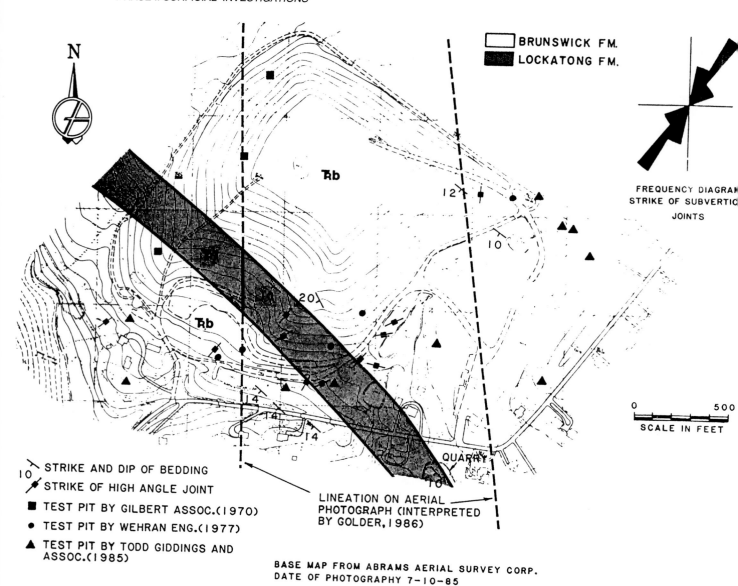

Figure 3-9 Facility Geology

Area Geologic Mapping

Area geologic mapping is recommended if the site possesses less than about 10% of natural outcrop. These mapping programs require every available outcrop delineated on the area base map, with geological observations plotted within these outlines. The end product is a drawing on which the geology and orientation measurements is plotted on a series of outcrop islands. As typical for these types of presentations, geological features between outcrops must be interpolated as inferred conditions with dashed lines One should provide clear explanations in the accompanying map text what geologic data was inferred from observed conditions. The "Manual of Field Geology' by R.R. Compton or "Field Geology" by Lahee should be consulted for further advice on general geological mapping.

Traverse Geologic Mapping

Sites with thin overburden and considerable rock exposure should have systematic mapping of geology to provide representative coverage of the area early in the project work tasks. These geologic site assessments are performed by laying out a system of parallel traverses across the area, about 100 ft. to (30 m) to 1000 ft (300 m) apart, dependent on outcrop conditions. The traverses are located on the project base map and surveyed in the field.

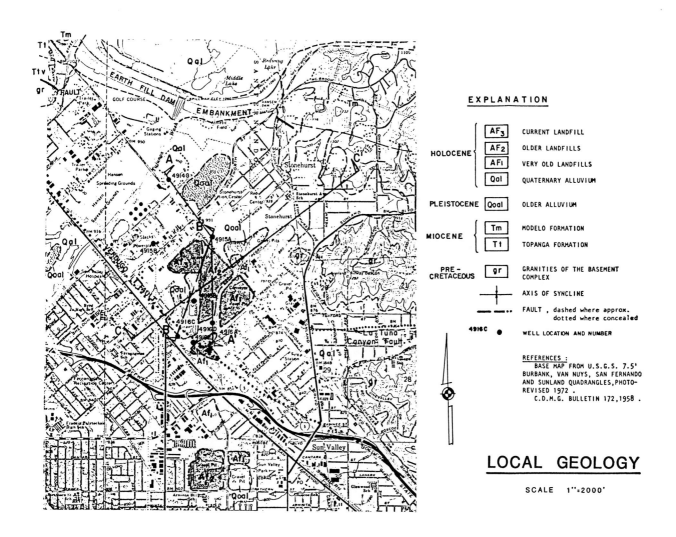

EXPLANATION

HOLOCENE
- AF₃ CURRENT LANDFILL
- AF2 OLDER LANDFILLS
- AFı VERY OLD LANDFILLS
- Qal QUATERNARY ALLUVIUM

PLEISTOCENE Qoal OLDER ALLUVIUM

MIOCENE
- Tm MODELO FORMATION
- Tı TOPANGA FORMATION

PRE-CRETACEOUS gr GRANITES OF THE BASEMENT COMPLEX

AXIS OF SYNCLINE

FAULT , dashed where approx. dotted where concealed

4916C ● WELL LOCATION AND NUMBER

REFERENCES :
BASE MAP FROM U.S.G.S. 7.5'
BURBANK, VAN NUYS, SAN FERNANDO
AND SUNLAND QUADRANGLES,PHOTO-
REVISED 1972 .
C.D.M.G. BULLETIN 172,1958 .

LOCAL GEOLOGY

SCALE 1"=2000'

Source: LeRoy Crandall and Associates   **Figure 3-10** Local Geology

The required observations along these traverses, are documented by the geologists and the observation locations are either paced out or measured by more accurate survey activities. Connections between traverses are typically made by interpolation and are clearly shown as dashed lines on the final geologic map. Due to the effects on ground-water flow at waste disposal sites, areas of complex geology, rock type boundaries, and faults have to be clearly mapped and described for later potential field drilling work.

The general geological map of a facility or proposed waste disposal site provides essential background information for decision making on general facility conditions but is insufficient for developing either detection or assessment ground-water monitoring programs. Rock and overburden materials should not only be known by geological names, but also provide an estimate of potential primary porosity flow paths and hydraulic conductivity. Fractured rock and, in some cases, partially fractured confining units require that the discontinuities must be described in more detail than is the case in general geological mapping. This discontinuity or fracture-set mapping procedure is described in Chapter 6.0 sections. The geologic map should clearly delineate the facility location and should show structural attitude, distribution, and lithology of surficial engineering geologic (soil) units, as well as bedrock types. Faults on or near the site should be located on the map and, (along with geologic

Figure 3-11 Example Air Photo

information), discussed in the geologic narrative of the reports, keeping in mind the potential of such features to act as pathways or barriers for the movement of ground water or the migration of leachate. While published sources will be of use (generally at scales of 1:24,000 or smaller), the level of detail needed (usually 1:2,400 or larger) for the inclusion of this information will probably require limited field and mapping (outcrop) surveys. Trenching and backhoe trenching may be needed to accurately document the location of Holocene (<13,000 yr. BP) faults. Professional judgment should be used in determining the area of map coverage. Figure 3–7 is a

regional geologic map of a California facility. The map includes recent unconsolidated alluvium and pre–Holocene faults.

Figure 3–8 is a map of the subsurface geology at a site in Pennsylvania. Details of pre–Holocene faulting are given along with surface features of the site. Relating surface and facility features to the underlying geology is a powerful analytical tool in site characterization studies. Additional examples of geologic presentations are shown in Figures 3-9 and 3-10.

3.4.3 Photography as Used in Mapping

An aerial photograph of the site should also be considered by the investigator for geologic mapping support and later inclusion in the report. A good quality photograph of the site area will allow identification of important features, such as surface water bodies, municipalities, industrial facilities, and residences. An aerial photo can be enlarged to the approximate scale of base maps for direct correlation of current site environmental conditions.

Important geologic features, identified during the Phase II site investigation process are studied and logged (for example, backhoe trench logs, borehole logs, and maps) in detail by the project geologists. Because of their graphic nature, these logs provide the most technically accurate descriptive data. Bedding, fault, and joint orientations, as well as, lithologic contacts with descriptions and precise measurements, must be clearly presented in the logs. While the technical descriptive data may be adequate for scientific review, the final decision of site suitability or remedial design often, if not always, must be made by non-technical personnel or public agency representatives, untrained in geology. It is in the decision-making process that the detailed photograph of important site geologic or structural features can be of greatest value to the project. Legal hearings can be extended for days while the specifics of a particular geologic - phenomenon are described and explained in technical jargon that few decision-makers (or judges) understand. It can not be over emphasized that these geologic documentation photographs should be of professional quality with proper lighting to best show the features under discussion. Such a photograph of the feature in question, with a simple, drafted acetate overlay delineating the primary geologic data, may save many hours of discussion and indecision. These photographs may consist of outcrops, trench or pit walls or large excavations and should include details of important geologic features.

In presenting the photogeologic data, either color or black and white prints can be used. Even though many copies of reports and documentary evidence are usually required during legal hearings, the reproduction costs of color prints are minor compared to the legal and consultant billings to prepare for the case. In either case, the presentation is similar. First, a size format is chosen. Typically 8 x 10 print size, black and white or color photographs. Enlarged prints can be taken into the field with overlay material (velum or acetate) and simplified logs drawn. The most readily identifiable features at each outcrop or exposure (e.g., lithologic units, faults, etc.) must be carefully delineated on the overlays.

In the final presentation, field log data should be transferred using pen and ink to clear acetate overlays. These overlays can be attached to the corresponding photographic prints, and overlay/photograph "packages" and incorporated into the final report. In reproducing the acetate overlays, copies can be made by a photographic lab using a lithographic-type film or by a photo-copier using Xerox-type transparency film.

3.4.4 Geologic Mapping Examples

Combining topographic map features with geology also offers a useful tool to understand the surface outcrop pattern and potential recharge/discharge patterns.

Figure 3–8 shows the regional bedrock geology of an area containing fractured rock in Pennsylvania. The figure includes a detailed geologic description of the site and a frequency diagram of the strike of subvertical joints observed in exposures and quarries.

Figure 3–9 is a map of the local surficial geology near a site in California. Holocene, Pleistocene, Miocene, and Pre–Cretaceous geologic units are mapped along with urban features adjacent to the site. Figure 3-10 shows an air photo example for a site facility area.

3.5 GEOPHYSICAL TECHNIQUES

Geophysical techniques for hydrogeologic investigations began to be commonly used in the 1950's, however, these methods were generally limited to siting locations for boreholes in small areas close to towns or industrial plants. The need to evaluate the extent of the aquifers and confining units in the context of subsurface waste disposal has considerably enlarged the role of geophysical technique in hydrogeology. Appropriate geophysical techniques often have scale factor considerations.

Figure 3-12 shows a regional site be considered at various dimensional scales. Although a specific point source waste disposal site may be only a part of an acre in size, leachable contaminants may be spread in long thin plumes. The extent of the contamination depends on the regional setting, including geology and hydrogeology, vegetation, receptor populations, water supply, discharges to surface water, and seasonal effects. It is necessary in any site assessment to understand the regional setting to gain a complete understanding of the local setting of a site. The regional picture is then gradually refined until the individual site heterogeneity can be conceptually understood to relate ground-water flow conditions.

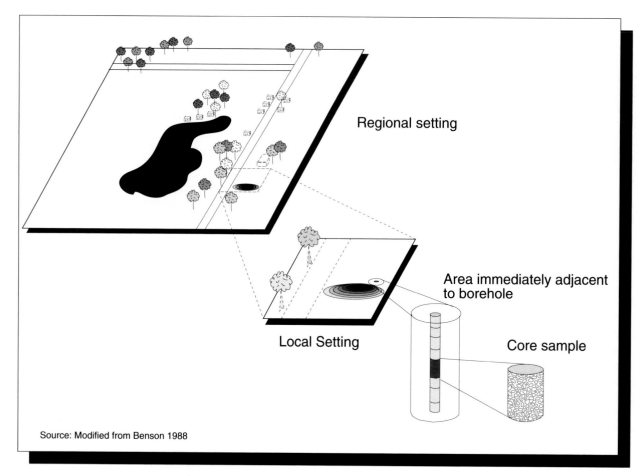

Source: Modified from Benson 1988

Figure 3-12 Regional to Core Sample Scales

Geophysics can be used to add the interpretative difference necessary for a successful site assessment. There are four primary objectives in the use of engineering geophysical methods in site assessments:

- Geological investigation: geophysical methods can play a role in mapping stratigraphy, determining the thickness of superficial deposits and the depth to bedrock, measuring weathered profiles, and are particularly important in establishment of erosional and structural features (e.g.) location of faults, dikes and buried channels, etc.)

- Resources assessment: aquifer location and assessment of ground-water quality, exploration of sand and gravel deposits, and rock for aggregate; identification of clay deposits.

- Detection of voids and buried waste material: e.g. mine-shafts, natural cavities, old waste disposal areas ,

pipelines, etc.

- Evaluation of site engineering parameters such as dynamic elastic moduli which can be used to solve many soil-structure interaction problems: soil corrosively for pipeline protection studies; rock rippability and rock quality evaluations.

Geophysical techniques are intended to supplement direct evaluation methods; they can not be a substitute for direct methods of site assessment, such as, drilling, trenching, etc. Geophysical methods should be considered as means of interpolating between, and extrapolating from, borehole data. By careful planning the number of Phase II borings required for definition of subsurface conditions can be greatly reduced if the proper geophysical methods are chosen to supplement the direct investigation program. The interpretation of geological conditions from borehole data alone in difficult or complex geology could lead to very misleading conceptual understanding. In cases such as faulted strata, or in the search for geological hazards like buried channels the use of the appropriate geophysical method is important to aid the correlation of geologic features between boreholes.

Geophysics can provide assurances that geologic conditions are consistent between direct measurement points (i.e., boreholes). As such, they can provide a certain level of risk reduction that unexpected subsurface conditions are not present. Geophysical techniques measure physical and electrical properties of subsurface materials. Table 3-3 shows applications of selected field investigation techniques for waste disposal sites and Table 3-4 shows more wide ranging uses for the most popular geophysical techniques.

Geophysical Survey Program Goals

Geophysics should be employed early in a phase II site assessment to solve specific problems or merely to provide supportive information cheaply and rapidly. The cost of conducting a geophysical survey in an area may be considerably lower than the cost of drilling even one deep exploratory borehole (Worthington 1975). In addition, a much larger area can be covered at a much greater speed. Geophysical investigations will never replace stratigraphic drilling, but they are relatively less expensive and can provide useful information on significant features of subsurface stratigraphy and structure. This potentially

This table is intended as a general guide. The application ratings given are based upon actual experience at a large number of sites. The rating system is based upon the ability of each method to produce results under general field conditions when compared to other methods applied to the same task. One must consider site-specific conditions before recommending an optimum approach.

In some cases a method rated 3 or NA may in fact solve the problem due to unique circumstances. For example, seismic refraction is rated NA for evaluating organic contaminants. However, in some cases where the contaminant flow is controlled by bedrock, the seismic method may provide an effective evaluation by mapping bedrock depth.

Application	GPR	EM	Res.	Seis.	MD	Mag.	OVA
Evaluation of natural geologic and hydrologic conditions							
Depth and thickness of soil and rock layers and vertical variations	1[a]	2	1	1	NA	NA	NA
Mapping lateral variations in soil and rock (fractures, karst features, etc.)	1[a]	1	2	2 (Refr.) 1 (Refl.)	NA	NA	NA
Depth of water table	3	2	1		NA	NA	NA
Evaluation of subsurface contamination and post-closure monitoring							
Inorganics (high TDS and electrically conducive)							
Early warning contaminant detection	3	1	2	NA	NA	NA	NA
Detailed lateral mapping	3	1	2	NA	NA	NA	NA
Vertical extent	3	2	1	NA	NA	NA	NA
Changes of plume with time (flow direction and rate)	3	1	2	NA	NA	NA	NA
Post cleanup/closure monitoring	3	1	2	NA	NA	NA	NA
Organics (typically nonconducive)							
Early warning contaminant detection	3	3	3	NA	NA	NA	1
Detailed lateral mapping	2[a]	2	3	NA	NA	NA	1
Vertical extent	2[a]	3	2	NA	NA	NA	2
Changes of plume with time (flow direction and rate)	3	3	3	NA	NA	NA	1
Post cleanup/closure monitoring	3	3	3	NA	NA	NA	1
Location of buried wastes and delineation of trench boundaries							
Bulk waste trenches—without metal	1	1	2	3	NA	NA	NA[d]
Bulk waste trenches—with metal	1	1	2	3	1[a]	1[b]	NA[d]
Depth of trenches and landfills	2	3	2	2	NA	NA	NA[d]
Detection of 55-gal steel drums	2[a]	2	NA	NA	1[a]	1	NA[d]
Estimates of depth and quantity of 55-gal steel drums	2[a]	3	3	NA	2	1	NA[d]
Location of utilities							
Buried pipes and tanks	1	1[c]	NA	NA	1[c]	1[b]	NA[c]
Potential pathways of contaminant migration via conduits and permeable trench backfill	1	2	NA	NA	2	2	NA[d]
Abandoned wells with metal casing	3	NA	NA	NA	2	1[b]	NA[d]

1 = Primary choice under most field conditions.
2 = Secondary choice under most field conditions.
3 = Limited field application under most field conditions.
NA = Not applicable.
[a] Shallow.
[b] Assumes ferrous metals to be present.
[c] Assumes metals to be present.
[d] Assumes no vapors present.
Note: Many site-specific conditions may dictate the choice of a method rated 2 or 3 in preference to a 1.

Source: Benson. 1988

Table 3-3 Applications of Selected Field Investigation Techniques for Waste Disposal Sites

APPLICATIONS	SEISMIC	SEISMIC MONITORING	SONAR	GRAVITY	MAGNETIC	RESISTIVITY	ELECTROMAGNETIC	RADAR	TIME-DOMAIN REFLECTOMETRY	RADIOMETRICS	BOREHOLE LOGGING
DEPTH TO BEDROCK	●					○	○	●			
FAULT DETECTION	●				●		●				●
FRACTURES IN ROCK		○						○		○	●
BURIED CHANNELS	●			○		●	●	●			
GROUND WATER SURFACE	○					○	○	●	○		●
SOIL WATER CONTENT	○					○	○	●	●	●	●
WATER DEPTH			●					●			
SUB-BOTTOM STRATIGRAPHY	●		●					●			
SEA BED SCOUR			●								
ICE THICKNESS							○	●			
PERMAFROST MAPPING	●					●	●	●	○		
PEAT THICKNESS	○					●	●	●			
SOIL STRATIGRAPHY	●					○	○	●		○	●
SAND & GRAVEL MAPPING	○					●	●	●			
LEACHATE PLUMES						●	●	○		●	●
SALT WATER INTRUSION						●	●	○	○		○
BURIED DRUMS					●	○	●	○			
BURIED PIPES & CABLES					○		●	○			
BURIED CAVITIES & TUNNELS	●			○	○		●	●			
VOIDS AROUND PIPES		○						○			
SUBSIDENCE (SLOPES & TUNNELS)	●	●		○			○	●	○		
PHYSICAL PROPERTIES	●	○		○	○	○	○	○	●	●	●
ELECTRICAL GROUNDING						●	●				
RIPPABILITY	●										
RADIOACTIVE HAZARDS											

SOURCE: Modified From MultiVIEW Geoservices Inc.

● **OFTEN APPLICABLE**
○ **SOMETIMES APPLICABLE**

Figure 3-12 Summary geophysical techniques

reduces the number of stratigraphic and geotechnical boreholes and can directly help in predicting their ultimate location and completion depths relative to site lithology and structure.

Geophysical surveys should be employed in the initial reconnaissance stage of the Phase II investigation when few, if any, boreholes have been drilled. On the basis of the geophysical surveys a minimum number of stratigraphic borings could be sited as confirmation of those surveys and to provide important control in their interpretation.

The design of a geophysical survey to support a site assessment for either ground-water detection or assessment monitoring programs should be prepared by a hydrogeologist and a geophysicist working together. The hydrogeologist sets out the goals of the survey and supplies the geophysicist with all the available geological and hydrological data. Geophysical methods are used to evaluate subsurface conditions for site assessments in three ways:

- Determination of geometry and properties of hydrogeologic units;

- Detection and mapping of ground water and soil contaminants; and

- Location of buried wastes, utility lines, underground tanks, etc.

Geophysical information can be acquired with a minimum of site disturbance, without causing further environmental damage to the site or pose risks to workers through direct intrusive drilling of waste areas. Site assessments that use both geologic and geophysical techniques can produce high quality data by focusing borings and piezometers into areas that represent the dominant flow paths. The use of these geophysical techniques in conjunction with other Phase II field assessment procedures locates highly targeted potential ground water monitoring sites that can produce more representative environmental sampling. Even if monitoring wells are already available on site, geophysical techniques can evaluate the condition and effectiveness of the well locations. For example, a geophysicist can assist in the evaluation of the location and performance of existing wells relative to the flow paths from disposal areas.

In addition to geophysical measurements taken from the surface, there are a variety of geophysical measurements that can be made down cased or uncased boreholes or monitor wells, as small as, two inches in diameter. These techniques are called down-hole geophysics or geophysical logging. These methods are reviewed within Chapter 6 relative to fractured rock analysis.

3.5.1 Planning a Geophysical Survey

The prime goals in geophysical survey planning is to ensure that they are an integral part of the overall Phase II investigation, and that they complement both surface geology mapping and subsurface assessment activities. The following steps are recommended by Darracott & McCann (1986), in designing a geophysical investigation.

Preliminary Meeting

The geophysicist should meet with project staff to decide on the following points:

1. To establish the exact nature of the geotechnical or hydrogeologic problem, for example: is the specific depth to bedrock really required for the project or is data required on conditions down to only a certain limited depth, such as the depth down to the facility base-grades. This sets direct guidance within the project scope as to the depth limits of the investigation.

2. Define if the problem may be solved by geophysical methods and which technique would be most appropriate in the particular geologic environment. At this early stage some agreement should be reached about the basic scope of work and proposed geophysical technique.

3. Confirm whether the site is suitable for the proposed geophysical survey. The selected geophysical survey must be performed to the sufficient depth and horizontal extent for use in the assessment program or in design efforts. Existing man-made interferences such as buried pipes, electrical power cables (buried and overhead) and fences should be considered in planning the survey, so that inappropriate geophysical methods are not used on a constrained site.

Nothing can erode client confidence more quickly than misuse of an incorrect geophysical technique after

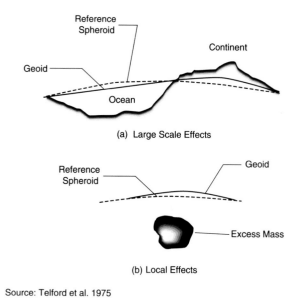

(a) Large Scale Effects

(b) Local Effects

Source: Telford et al. 1975

Figure 3-13 Gravity Geoid/Reference Spheriod

recommending the task in the project scope. If site conditions are questionable for successful use of a particular geophysical method, a phased trial area should be recommended before completion of the full scale survey. An honest appraisal of the results of the test survey will likely gain additional geophysical projects under more favorable site conditions. Whatever geophysical program is proposed the overall cost effectiveness of the geophysical program must be an important consideration for any field program. The geophysical survey must be planned to yield the maximum amount of reliable information consistent with the overall objectives of the site assessment.

The following subsection first describes the surface geophysical techniques that have been found to be most appropriate for site investigations. The following discussion is in the form of an introduction to geophysical techniques. No matter what geophysical techniques are applied on a project, they should only be conducted by experienced practitioners. The details of operation and field use are discussed by Telford & Others (1976), Griffiths & King (1981), Dobrin & Savit (1988), and Robinson & Coruh (1988). Geophysical applications, as used in site assessment projects will be then developed later in this subsection. These techniques are regularly used and are have proven effective for hydrogeologic and waste site assessments when used by experienced practitioners.

3.5.2 Geophysical Methods

The Gravity Method

The gravitational maps were the first geophysical maps which were applied for hydrogeological and petroleum-geological purposes as early as 1925 (Erdélyi & Gálfi 1988). Gravity surveys are to detect lateral changes in density within the subsurface through measurement of the changes produced in the gravitational field at the earth's surface. The gravity method is generally employed for preliminary reconnaissance purposes, being of most use for locating faulted and other structural boundaries and for the mapping of alluvium-filled valleys (Lennox and Carlson 1967). It is often used in conjunction with seismic refraction, resistivity, and other techniques which provide the necessary detail and control in the absence of borehole data.

Instruments that measure gravity can provide data for mapping major geologic features over hundreds of square miles or for detecting local fractures and cavities in rock. Gravity measurements respond to changes in the density of soil and rock.

The geophysical technique of measurement of the earth's gravity is based on the Law of Universal Gravitation, which states that every (point-like) particle of matter exerts a force of attraction on every other (point-like) particle. Garland (1965) defines the universal gravitational constant as : $G = 6.67 \times 10^{-8}$ degree - cm^2/g^2 This force is directly proportional to the product of the particle masses and is inversely proportional to the square of distance between the masses. This factor of proportionality is called the gravitational constant in physics and can be measured with very high precision by gravimeters.

The relationship between the constant of gravitation (G) and the acceleration due to gravity (g) with respect to a mass m_2 on the surface of the earth of mass M_e is as follows:

$$g = F/m_2 = GM_e/R^2 \qquad \text{Equation 3-2}$$

where R is the radius of the earth. This acceleration, g, termed the acceleration of gravity, is about 980 cm/sec2. For gravity prospecting investigations of subsurface structures, it is necessary to measures variations of 0.001 cm/sec^2 or less. Therefore, a smaller unit was introduced, the gal (after Galileo), which equals 1 cm/sec^2 and 1 milligal (mgal) equals 1 x 10-3 cm/sec^2. The magnitude of gravity on the earth's surface depends on five factors: latitude, elevation, topography of the surrounding terrain, earth tides, and variations in density in the subsurface.

This last factor is the only one of significance in gravity exploration and its effect is generally very much smaller than that of the other four combined.

In geophysical gravity surveys, we are primarily interested in the gravitational attraction of the geologic mass under assessment, therefore, the other large-scale gravitational effects from other Earth masses must be eliminated or filtered from the data. This filtering or correction process is called the "reduction" of gravity data (see Slaine and Leask 1988 for this process). That part of the Earth's gravity field, which remains after the field data is reduced, is known as the gravity anomaly, and provides the assessment tool of the overall gravitational effect of the body to be surveyed. This field is defined to an equipotential earth's surface at mean sea level, where the ocean's have been infilled and the land masses eroded. This equipotential surface is termed the reference spheroid and the gravitational acceleration since the earth is not a perfect sphere and undulations occur in the mean sea level, this surface is called the geoid. Figure 3-13 illustrates the comparison between the two terms.

Different kinds of anomaly maps may be obtained during the data reduction manipulation process. Bouguer-anomaly maps are the interpretative presentation that represents the gravity acceleration on a non-rotating Earth, if all masses would be reduced to sea level. The measured gravitational acceleration values used in these presentations must be corrected to eliminate non-geologic effects such as those caused by variations in elevation and latitude, topographic irregularities and by the stratum lying between the sea level surface and the place of the measuring station. This measurement is made up of deep-seated effects (the original Bouguer anomaly) and shallow effects (the local Bouguer anomaly). The local Bouguer anomaly is most often used in shallow ground water investigations, however, fractured bedrock underlying thick overburden materials may require the use of both anomaly maps.

The primary technique used in gravity survey data interpretation is field curve matching as shown in Figure 3-14. A conceptual model is constructed and its gravity field calculated. A comparison is then made of the results with the observed anomaly. The conceptual model is typically adjusted and its anomaly recalculated until the desired fit between the observed and calculated anomalies is achieved. There are sophisticated computer programs available which can interpret the effect of three-dimensional irregular bodies having spatially continuous or discontinuous density distributions.

There are two basic types of gravity surveys: a standard gravity survey and a microgravity survey. A standard survey uses widely spaced stations (100-1,000 foot intervals) and a standard gravity meter to cover many miles. Micro-gravity survey stations may be spaced anywhere from at 1 to 50-foot intervals. These measurements are made with a very sensitive micro-gravimeter, than can measure extremely small differences in the gravitational field down to 0.05 milligal. Micro-gravity surveys are used to detect and map local anomalies such as bedrock channels, fractures, and cavities. The main differences of micro-gravimetry methods as opposed to electrical and electromagnetic methods are as follows:

1. Conductive soil or overburden, which can mask high frequency electromagnetic surveys do not affect micro gravimetry;

2. It is not affected by buried metal pipes and power line currents; and

3. Vibration (from wind and traffic) can affect the microgravimetry measurements.

The micro-gravimeter is a very delicate instrument,

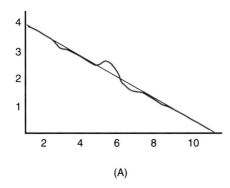

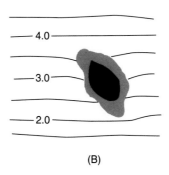

Figure 3-14 Gravity Curve Matching

and considerable care must be taken in its shipment, transport, and use. Field conditions such as ground noise, wind, humidity, and temperature will affect accuracy, so meteorological measurements must be made throughout the day at a base station to compensate for these effects. The results of a gravity survey can be a profile line or contour presentation, if data is obtained over a site area.

Seismic

Seismic surveys have, as their basis, the timing of artificially (e.g., by blasting or striking the ground surface) generated pulses of elastic energy propagated through the ground in the form of seismic waves and observed by electro-mechanical transducers, called geophones (see Figure 3-15). The seismic waves used in site assessments are of compressional, dilatational elastic waves, as of the same type as very low frequency sound waves. The electric signals of the transducers are recorded by a seismograph on film, magnetic tape, or directly to a digital format on which timing marks are also placed as part of the data.

Seismic velocity can be directly related to density and hardness of the soil and rock. Seismic techniques measure the travel time of seismic waves transmitted into the subsurface by an acoustic source. Because these waves travel at different velocities in various types of rock, this technique discerns the physical properties of bulk modules and shear modules of the soil and rock. Geophones arrayed along a surface line measures the travel times from the source and provides information on the thickness of subsurface layers, number of layers, and seismic velocity of each layer. Seismic waves coming from the interior of the ground arrive at the surface nearly perpendicularly, exciting the surface of the ground to move vertically. The vertical

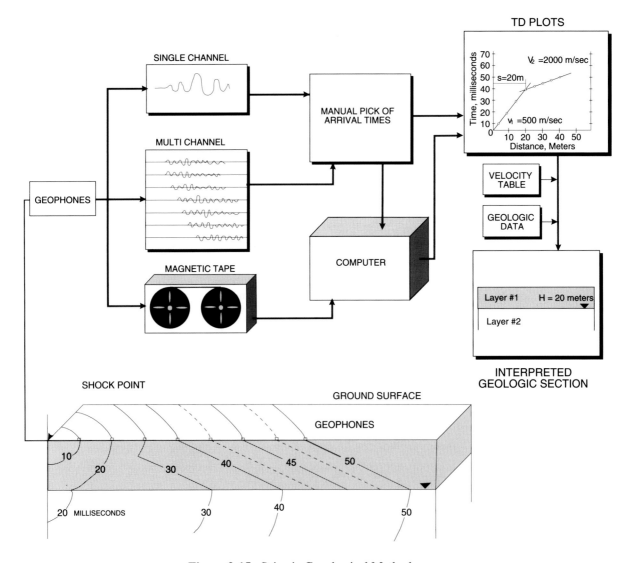

Figure 3-15 Seismic Geophysical Method

movement of the surface is sensed by the individual geophones and recorded by the seismometer on paper or film strips moving with constant rate correspond to the elapsed time during the motion of the ground. The resultant deviations are perpendicular to the time axis correspond to the amplitude of the ground motion. Each arrival of a seismic wave can be constructed on the seismogram as a "movement" of the corresponding trace. In Figure 3-15 seven trace readouts are shown which represent seven (out of typically twelve to twenty four traces) geophones placed along a survey line. The recorder instrument (Figure 3-16) can shadow the upper half of the sinusoid waves to make easy the following of arrivals from trace to trace.

Two geophysical methods described as <u>reflection</u> and <u>refraction</u> techniques are employed in seismic surveys. In both methods, travel times of seismic waves are measured, spreading out from the source to selected points of observation on the surface. The basic difference between the two methods lies in the character of the recorded arrival waves whose times of arrival are utilized. In the reflection method observations are made of seismic waves which have been reflected from elastic boundaries in the ground. In the refraction method travel time data are obtained for

seismic waves which have been critically refracted on boundaries separating two strata of different elastic properties.

Reflecting boundaries are interfaces which have different values of seismic wave velocity on their two sides. The refracting boundaries are elastic boundaries where critical refraction may occur are of the same character, but the wave velocity in the lower stratum must be higher than that in the upper one.

Refraction seismic methods are typically used in shallow investigations (up to a few hundred feet). Seismic refraction is the predominant seismic method sued for engineering and hazardous waste studies. The refraction method requires consideration in the spacing of the geophones which determines the resolution and depth of seismic measurements. Typically an array length four or five times the depth to be surveyed is required in such near surface programs. A need for data at greater depths requires a significantly greater amount of energy expended at the source. The quality of seismic refraction interpretations depends on both the type of modeling technique used and on the quality of the initial data. The amount and quality of refraction data is partially dependent on the number and placement of shots. Continuous reversed profiling (Telford et. al. 1976) provides optimal shot coverage, and is the predominant profiling methods currently in use. Obtaining quality refraction data at waste disposal sites will probably require signal filtering, to reduce the interference caused by extraneous seismic noise. Simple refraction surveys can detect only a limited number of layers and cannot detect lower velocity layers beneath higher velocity layers. A more detailed refraction survey can be carried out so that strata boundary depths are obtained under every geophone. Time and cost of these detailed refraction surveys may be twice that of a simple refraction survey because travel times must be measured from five locations and the resultant data require significantly more processing efforts. Many techniques have been proposed for modeling refraction data. the simplest techniques in common use have been described by, Mota (1954), and assume either gently dipping planar refractors or irregular horizontal refractors with continuous seismic velocities. These interpretation routines generally fall into two categories-intercept-time (ITM) and delay time methods (DTM). The main drawback to these methods is the assumptions used in these models, and their inability to accommodate lateral velocity changes, can limit their utility for modeling complex geologic environments. The generalized reciprocal method (GRM) as described by Palmer (1980) requires significantly more seismic data, but is superior to the ITM or DTM for modeling irregular dipping refractors and lateral velocity changes. Figure 3-17 illustrates an example of refraction

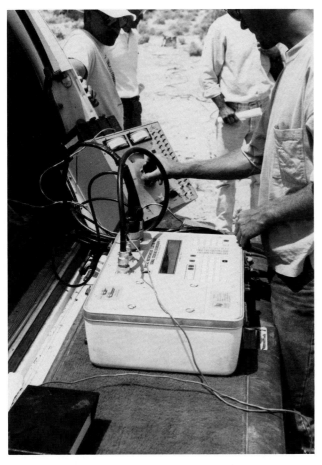

Figure 3- 16 Bison Seismograph Collecting Data

data sets.

Reflection seismic surveys can used on much deeper investigations using less energy than deeper refraction studies as shown on Figure 3-18. The reflection technique has been reported (Benson 1988) effective from depths of fifty feet to a few thousand feet and can provide a relatively detailed cross sectional representation of the subsurface features. The application of the seismic reflection technique to engineering and geological investigation is well documented in the literature (Hunter and Hobson, 1977; Hunter et al. 1982 (a) (b), 1984; Gagne et al., 1985). Haeni (1986) describes the successful application of marine seismic profiling to hydrogeological studies. Signal frequency for seismic work is dependent upon several factors, including the source signal and site stratigraphy. However the amount of high-frequency data obtained can be optimized by using high-frequency(> 80 Hz) geophones and by filtering the low-frequency end of the signal.

The cross section obtained from seismic reflection program look very similar to Ground Penetrating Radar (GPR) data, in that they provide a picture-like cross section. Reflection is very useful for mapping the top of bedrock, bedrock channels, alluvial and fluvial deposits, and for identification of important secondary hydraulic conductivity features such as fractures and channels. Seismic reflection data can also display phantom horizons in a similar manner to GPR traces and care should be used in their interpretation as lithologic boundaries.

Shallow reflection require special evaluation techniques be used to design the field testing program. The details of the "optimum window" and "optimum offset" shallow seismic reflection techniques utilizing a 12 channel seismograph are presented in a series of papers by Hunter et al. 1982 (a) (b) and 1984. The "optimum window" refers to the space in time where the subsurface reflector can be observed with minimum interference. The near side of the "optimum window" is located beyond the zone of "ground-roll" interference. "Ground-roll" is a term related to low-frequency vibrations traveling along the ground surface (Slaine 1989). The far side of the "optimum window" is usually the point where other events such as shallow overburden reflections start to interfere with the deeper or bedrock reflections.

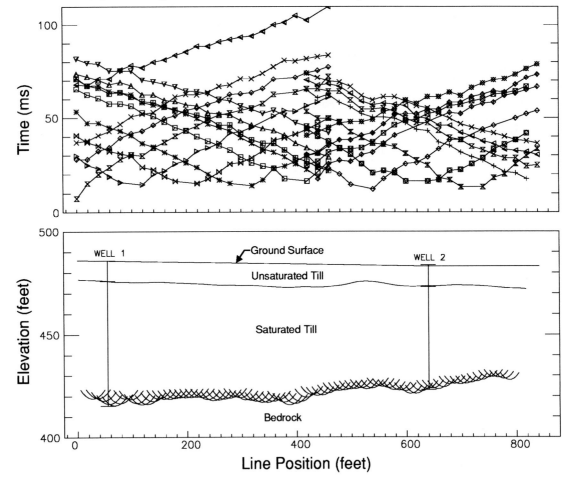

Figure 3-17 Example Refraction Data Sets

Within the "optimum window" a "common offset" is chosen meaning that the seismic section is shot one trace at a time with the same source to geophone spacing. Individual channels on a 12-channel seismography are then shot one at a time while maintaining the chosen offset. Data are stored on tape or in the memory of the seismograph for subsequent analysis. Computer processing of seismic records involves:

• Data enhancement with band pass filtering, trace gain normalization, automatic gain control, gain tapering and muting;
• Alignment using static corrections for variations in the first arrival refraction; and,
• Velocity analysis using normal move out corrections as described by Gagne et al (1985) to convert velocity data to depth measurements.

Figure 4-18 shows a result of a reflection study (Slaine 1989)

Metal Detection

Metal detectors have been extensively used in site investigations since they are important tools for detecting and delineating boundaries of filled areas or trenches containing metallic drums or other metallic waste, as shown on Figure 3-19. A metal detector is a continuously sensing instrument displaying metered results as the investigator walks along a traverse line. The range of these instruments is relatively short; their response is proportional to the cross section of the target (thickness of the target is relatively unimportant) and inversely proportional to the sixth power of the distance to the target (Benson et.al. [1984]).

Although metal detectors can be affected by pipes, fences, cars, etc., they have been used on many site assessments to effectively define trench boundaries and locate buried drums, utility lines, and tanks. They also can be used to select drilling locations clear of metal

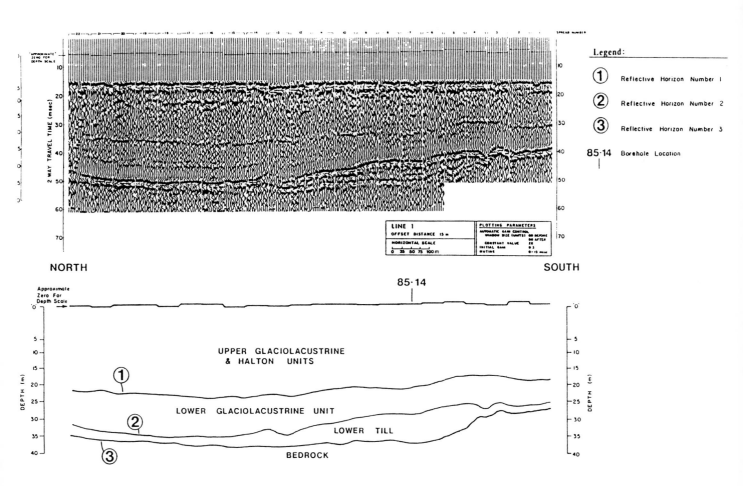

Figure 3-18 Example Reflection Results

obstructions or hazardous drum disposal areas.

Magnetometry

Mapping of the intensity of the Earth's magnetic field and the interpretation of variations in its intensity over a study area is an important geophysical technique used primarily in mineral and petroleum exploration in many types of geologic environments. These variations, called magnetic anomalies are the distortions of the regional magnetic field of the Earth, produced by materials of magnetic properties in the subsurface. Magnetic anomalies important to site assessments are due to the magnetization induced in a rock by the Earth's magnetic field call "induced magnetization." Magnetometers use a measurement of the intensity of the Earth's magnetic field to detect the presence at waste disposal sites of buried ferrous metals by magnetic variations. The magnetometer's response is proportional to the mass of the ferrous target and inversely proportional to the cube of the distance to the target. A single drum has been reported (Benson 1988) detected as deep as ten feet with large masses of drums detectable down to thirty feet.

The use of magnetic surveys in hard rock terrain for geologic assessments is limited practically to the regions of igneous and metamorphic rocks. The sedimentary rocks are non-magnetic, hence most common hard rock aquifers such as sandstone and carbonate rock show little benefit from magnetic surveys. Because of the extreme sensitivity of the earth's magnetic field to micro-scale anomalies, magnetometry works best in rural areas away from urban magnetic perturbations (Fences, power lines, underground pipes).

Two types of electrical magnetic instruments use different measurement techniques to obtain magnetic mapping data. The flux gate magnetometers use the frequency of the magnetic momentum of hydrogen ions after relaxing a strong magnetic field. The proton procession magnetometer directly measures the total intensity of the magnetic field present at a survey point. The proton procession magnetometer is based on a transducer that converts the Earth's magnetic field strength into an alternating voltage, which has a frequency proportional to the field strength. Proton procession magnetometers require the operator to stop for each measurement; fluxgate gradiometer magnetometers permit continuous data acquisition as the instrument is moved across the site (Figure 3-20). Continuous coverage is typically more suitable for detailed surveys and for mapping areas where complex anomalies are expected. Using an appropriate instrument is essential because the effectiveness of some magnetometers can be reduced or totally inhibited by noise or interference from time-

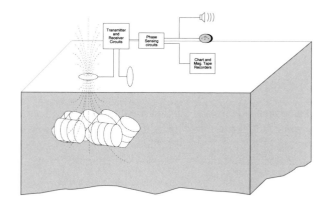

Figure 3-19 Metal Detection

variable changes in the earth's magnetic field, magnetic minerals in the soil, or iron and steel debris, pipes, buildings, passing vehicles, etc.

Electrical Geophysical Methods

The electrical conductivity of the Earth's materials can be studied by measuring the electrical potential distribution produced at the surface by an electric current supplied into the ground by means of electrodes, this is called the resistivity method, or by detecting the low frequency electromagnetic field produced by an alternating current introduced into the ground, called the electromagnetic method. The study of the decaying potential difference,

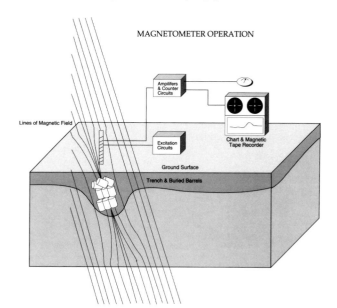

Figure 3-20 Magnetometry

when the injected electrical current was turned off, is known as the induced polarization method. Very short time duration electromagnetic pulses measured as reflected energy is used in the ground penetrating radar geophysical method. Electrical methods use a number of physical properties of rocks:

1. The specific electrical conductivity of rocks; i.e., their ability to conduct direct and low-frequency electrical currents;

2. Induced polarization which occurs when an electrical current is passed through them and then interrupted;

3. The permittivity of rocks, i.e. their ability to conduct high-frequency electrical current as applied by the ground-penetrating radar mapping system (Figure 3-21).

Figure 3-22 shows the various frequencies used by electrical geophysical methods both for laboratory evaluations and field site assessments.

Ground-Penetrating Radar (GPR)

Late in the 19th century scientists discovered that "Hertzian Waves" could penetrate matter. An early patent on what we now call radar, RAadio Detection And Ranging, was obtained in 1904. During atmospheric

nuclear testing in the 1950s researchers found that electromagnetic pulses (EMP) generated by the blasts in addition to disrupting power grids could also penetrate the Earth's surface.

Ground-Penetrating Radar, or GPR, is an active remote-sensing technique developed in the 1960s as a result of research on EMP. The technology allows researchers to see below, through, or into otherwise impenetrable solid objects

Unlike X-rays which are hazardous and must pass through the material to be studied and be recorded on a screen or film, GPR is a reflection technique that uses non-ionizing microwaves and radio waves to safely "see" into solid material with very low power requirements

Commercial and military frequency domain continuous wave (CW) radars generally transmit short wavelengths making up a narrow bandwidth often encompassing only one frequency. Frequency-domain return echoes can be analyzed for features such as Doppler shifts to determine velocity of a moving target.

Ground-penetrating radar on the other hand is a time-domain impulse radar, and transmits broad bandwidth pulses into geologic media such as ice or granite. Time-domain radars act as sounding devices similar to depth finders in boats

GPR provides the greatest resolution of all the surface geophysical methods. The high-frequency, very-short-time duration electromagnetic impulses in the 100 MHz-GHz range, are transmitted through an antenna system into the

Figure 3-21 GPR Represents one of the Most Common Geophysical Method used in Shallow Soil Investigations

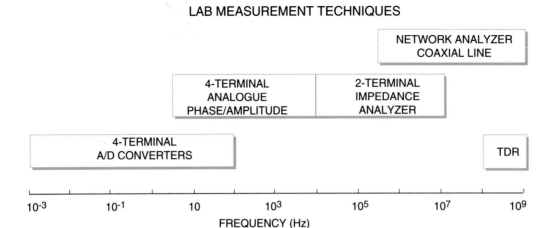

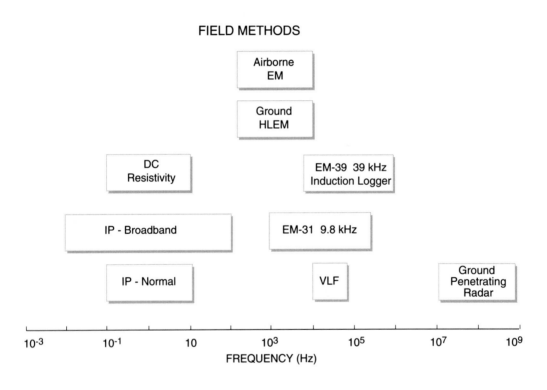

Figure 3-22   Frequencies Used By Electrical Geophysical Methods

earth. GPR impulses are reflected at interfaces where a dielectric constant change exists in the geologic units. The GPR reflected impulses are received at the surface and reflection depths are defined by the travel time from the surface to the reflecting surface and back again. The final graphic presentation is a cross section displaying the reflecting surfaces. Variations in the return signal are

continuously recorded to produce a cross-sectional profile of shallow subsurface conditions. Many different types of earth-radar instruments are available under different trade names but the basic principle is the same as is represented in Fig. 3-22.

An interpretable profile is produced by moving the GPR antenna over a ground surface and a cross section

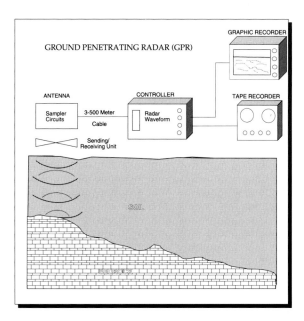

Figure 3-23 GPR Geophysical Methods

representation is made up from a collation of many pulses transmitted and received along the profile. An interface between soil and rock units having sufficiently different electrical properties (dielectric and conductive) will show up in the radar profile. The profile will also show differences caused by buried waste materials, drums, or buried pipes and cables.

Radar performance and depth of penetration are highly site-specific because of attenuation due to the higher electrical conductivity of subsurface materials or scattering. Generally, better overall penetration is achieved in dry, sandy, or rocky areas; poorer results are obtained in clay or conductive soils. Radar penetration in soil and rock to 30 ft is typical, with special GPR systems reading to 70 feet under optimum conditions, but penetration may be considerably less in some cases. Where saturated silts and clays are present, GPR penetration may be less than 3 ft, under dry conditions, with computer enhancement of the data sets, radar penetration depth may reach 45 feet. The ground-water surface will be detected in coarse-grained materials, though probably not in finer grained materials.

The continuous profile produced by GPR permits data to be gathered rapidly, thereby providing a large amount of data with substantial detail. This detail can, however, be difficult to interpret and only experienced staff should attempt to analyze the results from these surveys (see Figure 3-24). Since the profiles can be generated quickly in GPR, many lines can be run across a site to cross check the profiles. Lower frequencies of GPR (80 MHz) provide greater depths of penetration. Higher frequencies can provide better resolution. However, care must be exercised in the interpretation, since even strong reflectance features are often not discernible in samples collected by drilling or backhoe trenching programs. Since this technology was developed, the U.S. Geological Survey staff members have

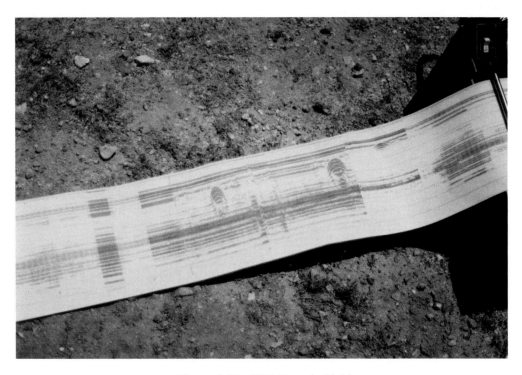

Figure 3-24: GPR Trace in Field

Figure 3-25   Geonics EM 34 Electromagnetic Induction Geophysical Method

compiled a bibliography on the subject that contains more than 4,000 citations Manufacturers of radar equipment include ERA Technology in the United Kingdom; Geophysical Survey Systems Inc., North Salem, N.H.; Sensors and Software, Mississauga, Ontario; and OYO Geospace Corp., Tokyo.

Electromagnetic (EM)

The electromagnetic geophysical methods have become one of the more popular techniques used in site assessment, since it provides a rapid measurement the electrical conductivity of subsurface soil, rock, and ground water. The EM method induces a current in the ground without ground-contact by placing an alternating current in a coil placed over the ground surface (Figure 3-25). Changes in phase and magnitude of the individual currents are measured by a receiver coil. By varying the frequency,

"depth sounding" can be produced without changing electrode distances. For EM resistivity methods, the exploration depth depends on the transmitter- receiver spacing and the coil configuration. Electrical conductivity is a function of the type of soil or rock, the porosity and permeability of the rock, and the fluids filling the pore spaces. Since the EM fields are attenuated strongly in conductive media, EM fields have a certain (effective) depth of penetration or skin depth. The effective exploration depth depends on the frequency of the EM field used and the resistivity of the ground. The conductivity (specific conductance) of the pore fluids contained in the soil or rock may dominate the EM field measurement. EM field programs are, therefore, useful for assessing natural hydrogeologic conditions and migration of higher conductivity facility leachate. Waste trench boundaries, drums, and metallic utility lines can also be located with EM techniques (see Figure 3-26). The primary limitations of EM methods are the narrow range of sensitivity, susceptibility of the electromagnetic signal to interference, and sensitivity of the instrument to misalignment. Detection and correction of alignment errors are part of the standard operating procedure for those EM devices that are sensitive to misalignment.

In general, standard EM mapping lacks the resolution and depth penetration of resistivity mapping, but has the advantage of being rapid and less expensive. Recent advances in high resolution EM techniques now allow the collection of massive numbers of quadrature and in-phase

Table 3-5 Depth Penetration EM 34-3

Intercoil Spacing Meters (m)	Feet (ft.)	Exploration Depth Horizontal Dipoles	Vertical Dipoles
10 m	(33 ft.)	7.5 (25 ft.)	15.0 m (49 ft.)
20 m	(66 ft.)	15.0 (49 ft.)	30.0 m (100 ft.)
40 m	(131 ft.)	30.0 (100 ft.)	60.0 m (195 ft.)

readings for site assessment mapping. The Geonics EM-31-DL device is used to simultaneously map the quadrature (terrain conductivity) and in-phase (metallic) response of the shallow subsurface. The EM-31 operates on the principle of electromagnetic induction and has an approximate depth of exploration of 4 m (13 feet). The unit creates an oscillating primary magnetic field in the earth through a transmitter coil. The primary magnetic field (H_p) creates a secondary magnetic field (H_s) that is detected by the receiver coil.

Geomics Limited recently attached a data logger system of the EM-31 model, which permits data to be digitally recorded and stored on-site. The data logger transmits information to a lap top computer where the data are initially analyzed on-site and stored on floppy disk. A data logger system tied to the geophysical-device in creased the amount of data one is able to measure during a survey along with increasing the overall efficiency of the system. The absolute apparent conductivity of the geologic materials and contained ground water are not necessarily diagnostic in themselves, however, variations in conductivity, observed both laterally and with depth, is the interpretation tool. EM anomalies in the data arrays (typically shown as contour maps of apparent resistivity) enable the investigator to define potential problem areas.

Lateral variations in conductivity can be mapped by a field technique called EM profiling. Profiling measurements have been reported by Benson (1988) being effective to depths ranging from 2.5 to 200 feet. These profile methods use frequency domain EM systems by applying a continuous alternating current of a fixed frequency. Continuous profiling measurement by EM can be obtained for shallow surveys (down to 50 feet). This method allows increased rates of data acquisition to improve resolution of small hydrogeologic features.

Resistivity

Resistivity geophysical methods measure the electrical resistivity of subsurface soil, rock, and ground water by applying a direct current (DC) or alternating current (AC) of very low frequency. Resistivity is the reciprocal of conductivity, the parameter measured by the EM technique. The resistivity method is based in principle on the contrast in resistivities of rocks, rather than on absolute resistivity values. In most rocks, electricity is conducted electrolytically by the interstitial fluid, and resistivity is controlled more by porosity, water content and water quality than by the resistivities of the solid matrix. Clay minerals, however, are capable of conducting electricity and the flow of current in a clay layer is both electronic and electrolytic. The conductive water film is very thin in coarse-grained sands and pebbles, thus the effect of saturation is very large on their resistivities. In fine-grained clays, however, where the water film fills the pore volume almost completely, resistivity changes little. These clay resistivity lows (or conductivity highs) also confound EM measurements. Resistivity contrasts, however, are sharp enough for practical site assessment purposes, and they exist locally in most cases in spite of a

Figure 3-26 EM 31 Geophysical Method in Field

wide range of regional variations.

The resistivity method pumps electrical current into the ground using a pair of surface electrodes (A and B) in Figure 3-27. The resulting ground potential (voltage) is then measured at the surface between a second pair of electrodes (N and M in Figure 3-27). The potential distribution within a homogeneous and isotropic rock containing the current electrodes at the surface is represented in Fig. 3-27. The subsurface resistivity is calculated from the separation and geometry of the electrode positions, the applied current, and the measured voltage.

The typical difference in the electric resistivity of aquifers and confining unit rocks (Fig. 3-27) points toward the applicability of resistivity methods in site assessments. Resistivity methods have been used successfully in stratigraphic, as well as, in structural ground-water assessments. Horizontal profiling is used to determine the variations of the apparent resistivity in the horizontal direction within a pre-selected range. This method uses a fixed electrode spacing set on the basis of studying the results of electrical soundings. The whole resistivity electrode array is moved (Figure 3-27) along a profile after each measurement is made. The value of apparent resistivity is plotted, generally, at the geometric center of the electrode array (Fig. 3-27). The method of horizontal profiling is very suitable to the study of hard rock terrain.

The technique of vertical electrical sounding (VES) is the process by which depth investigations are made by a series of resistivity measurements. Electrical sounding has its basis on the distance between the source electrodes and measurement of the potential difference between receiver and ejector points thus allowing the penetration of the resistivity technique; this is irrespective of the many alternative electrode arrays that may be used. The depth of probing depends also on the distance between the two current electrodes when linear arrays are used.

As expressed previously, specific resistivity of soil units and rocks (Fig. 3-27) varies within wide ranges. It is a very sensitive parameter, but there is no general correlation between the specific rock types and resultant resistivities. Thus, as the current electrode separation is increased during VES, the electrical potential distribution on the surface should be affected relatively more by the deep strata.

The first step in the interpretation of sounding field data is to prepare a graph in which the calculated apparent resistivities are plotted on the ordinates, variable parameters (dependent on the array used) are plotted on the graph abscissa. A significant difficulty in interpretation of resistivity sounding curves is the complexity of the curves. VES curve interpretation is one of the most intricate techniques in applied geophysics and should only be attempted by professionals with considerable experience in

HORIZONTAL RESISTIVITY MAPPING AND VERTICAL
ELECTRICAL SOUNDING (VES) GEOPHYSICAL SYSTEMS

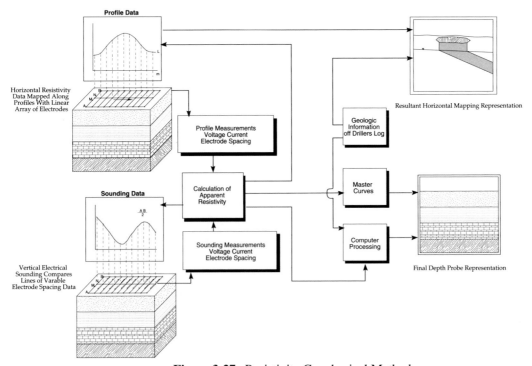

Figure 3-27 Resistivity Geophysical Method

the technique. There are some rules, however, which can serve as a basis for most cases in site assessment to define geologic conditions:

- Sediments are less resistive than igneous rocks;

- Basic igneous types are better conducting than the acidic ones;

- Clayey rocks possess lower resistivity than sandy types; and

- Stratified and schistose rocks display electric anisotropy, i.e., their specific resistivity is higher in the direction normal to the plane of schistosity than parallel to it.

Although time consuming, resistivity profiling has been a useful technique for mapping lateral changes in the subsurface. This field technique has been used to delineate leachate plumes and detecting changes in natural hydrogeologic conditions. Resistivity sounding requires considerable space; for example, to obtain data from a depth of 100 feet could require an overall array length (distance between the outermost pair of electrodes) of 900 to 1,200 feet. This method often constrained due to property fence lines and power lines in built-up areas.

Although measuring similar physical properties, when comparing EM and resistivity survey data from the same site, you can expect some differences in the results due to variance in the geometry of the measurement methods. The two techniques can be used for much the same purposes, and though both are affected by buried metal pipes, metal buildings, and fences, etc., it is believed that (Benson et.al. 1984) the resistivity technique may be less sensitive to these sources of error.

Induced Polarization (IP) Method

This less familiar method of electrical survey uses the passage of a direct current into the ground through two current electrodes for a period of several seconds, after which time the current is interrupted. The potential difference measured on this technique across the two potential electrodes does not drop to zero instantaneously, but after falling to a value, a V_0 that is a small fraction of V, the charging voltage, decays slowly over a period of several seconds. This phenomenon is known as induced polarization. Its magnitude is most commonly expressed by a parameter termed the chargeability which is the ratio V/V (millivolts per volt). The ratio of these voltages characterizes the decay of the "over voltage." The ground becomes polarized through the primary electric current. Upon turning off the polarizing current, the ground will gradually be discharged and returned to equilibrium.

The procedure for measuring induced polarization response in the field is very similar to that used in the direct current resistivity prospecting. Polarization is excited by galvanic current of a square-wave-form driven into the ground by two current electrodes. The effect of polarization is measured by noting the potential decay measured between two potential electrodes following the end of the exciting pulses. Induced polarization may be applied over a broad band from 100 to 10^{-2} Hz or within a "normal" band width of 10 to 10^{-1} Hz.

Although the greatest use of IP is in mineral exploration, this technique finds an application in hydrogeological surveys where it may be employed to distinguish between pure sands of high hydraulic conductivity (no IP effect) and sands with disseminated clays of lower hydraulic conductivity (high IP effect, e.g., Vacquier, Holmes, Kintzinger, and Lavergne 1957).

Organic Vapor Analysis

Although site assessments geophysical techniques do not normally consider measurement of organic vapors, the various field methods used in reconnaissance mapping of organic vapors in samples of soil rock or fluids are becoming important field techniques for assessment monitoring programs. These organic vapor analysis (OVA) detectors produce a measured current that is proportional to the concentration of vapor in the sample. Some portable OVA units have the capability to perform a gas chromatography scan that can also determine the type of organic compound present. OVA's have found use during site assessment sampling and drilling operations including:

- Establishing organic contamination (both vadose and ground water) at specific depths using a hollow Dutch cone type probe; (the Dutch Cone penetrometer method is discussed in the following chapter)

- Screening the head space of soil and water samples for organics;

- Evaluating cuttings for sampling while drilling to prescribe proper disposal; and

- Determine which samples will require further organic and physical laboratory analysis.

These methods provide alternative ways for defining "Hot Spots" for later detailed evaluation.

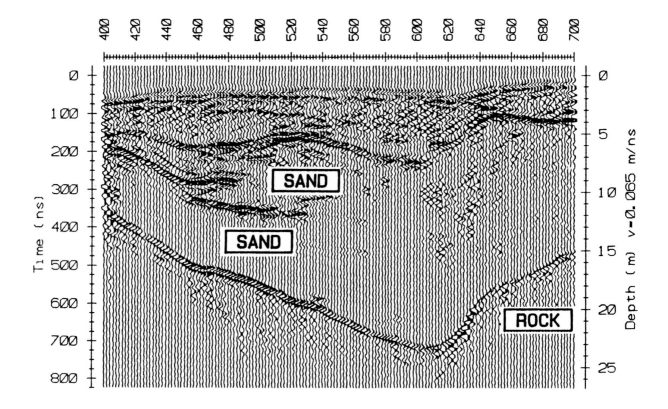

Figure 3-28   GPR Results   1

Figure 3-29 GSSI 120 Mhz GPR Antenna Being Pulled Along Road

3.5.3 Field Application of Geophysics

The most widely used geophysical technique to map lateral variations in soil and rock (such as fractures, Karst features, sand and clay lenses, and buried channels) are continuous profile GPR and EM. The EM technique has been applied in almost every environment and can often provide deeper information than GPR. Recently EM survey methods have included data loggers to allow more rapid data collection during field surveys. Increased quantities of data collected in a more efficient manner greatly increased the mapping resolution of the technique. An example of a resultant High Resolution EM survey (Slaine, 1989) was provided as a color contoured figure with contaminated soils, metallic objects and undisturbed soils all showing both different colors and overall saturation of colors. Thousands of individual EM readings were incorporated by Slaine to produce the color contour output. The purpose of the geophysical survey at this 16 acre site was to map buried wastes on the property so the site could be remediated prior to a property transfer. The data were continuously recorded along lines that were spaced 10 feet apart. Seismic reflection (although somewhat expensive) can provide good lateral and vertical information at much greater depths than GPR or the frequency domain EM. Slaine et.al. (1990) reported the results of a seismic reflection survey run to determine the integrity of a clay aquitard beneath a proposed hazardous waste site in southern Ontario. Continuity of the confining unit was demonstrated down to 35 meters along a 10 km transit line. A stratified gravel aquifer was defined for depth, thickness and lateral extent using high frequency seismic reflection techniques. The geophones were spaced at three meter intervals and shot gun shells were used as the energy source. The resultant processed common-offset seismic reflection section is shown back in Figure 3-18. The four 1,000 foot long transects defined the vertical and horizontal extent of the stratified gravel aquifer within the bedrock valley. High hydraulic conductivity units, such as the gravel aquifer in this case, are extremely important to fully establish in assessment monitoring programs since these features can have the potential to discharge some distance away from the facility. They can greatly effect the local movement of leachate plumes preferentially toward and into the high hydraulic conductivity units. This preferred flow path will typically confound attempts to

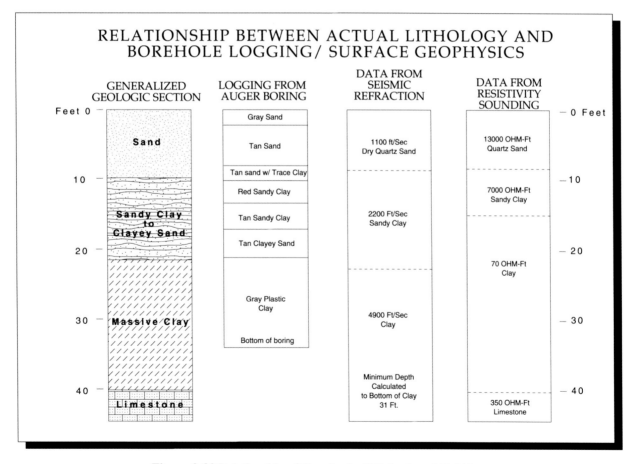

Figure 3-30 Relationship of Geophysical Methods and Field Logs

establish rate and extent of leachate migration, unless the location of these features are fully established within the Phase II field investigation.

Various geophysical methods provide different subsurface "pictures" of lithology. Figure 3-28 illustrates a GPR survey using 100 MHZ antennas on a 10 m spaced lines. The sequence of aeolian and fluvial sands overlying the bedrock is well visualized by this method. This exceptional GPR processed radar section shows the sands to have a maximum thickness of over 20 meters. Reflecting horizons in the sand indicate erosional unconformities which have a direct effect on the hydraulic conductivity . A complete ground-water flow model was successfully developed from the radar stratigraphy defined by the survey. Figure 3-30 shows a seismic and resistivity profile compared to a generalized geologic section (geologic section, edited borehole log) and actual logging data from an auger borehole. Each representation of subsurface condition provides reasonable correlation and also some significant differences in both the geophysics and logging profiles. Under optimum conditions GPR profiles can define complete stratigraphic sequences.

Downhole geophysical techniques have found wide use to improve the interpolation of geologic conditions between borings or to complement surface geophysical techniques. Geophysical logging between borings (hole-to-hole techniques) increases the volume of material sampled and may reduce the need for additional borings. These downhole techniques are covered in Chapter 6.0. The following subsections provide a selection of examples of the use of geophysics in site assessments. They were selected as much for a clear presentation of geophysical data as for demonstration of the results of the actual technique. One of the greatest difficulties in the full acceptance of geophysical surveys to support field investigations is the typical poor illustrative quality of the results of the geophysical survey. Often the inexperienced investigator will produce geophysical traces that require more than visual squinting at the figure to see what was interpreted from the data. Even relatively "standard" seismic refraction surveys can be presented in an easily understood format by preparing informative presentations. Seismic refraction survey data combined with selective drilling were used to map bedrock surface elevation. This data was used to define bedrock topography prior to shaft sinking and first level drifting in order to assure sufficient rock cover above a proposed mine workings Scaife, (1990). Bedrock topography data as illustrated in Figure 3-31 provide mining engineers with a basis for mine development design. A contour map and surface plot of the bedrock elevation data along with an isopach map of

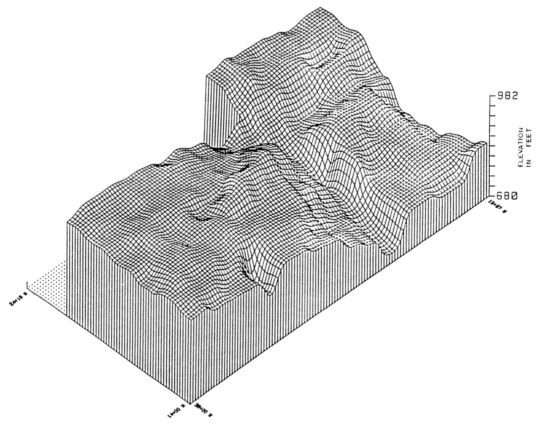

Figure 3-31 Seismic Reflection for Mining Application

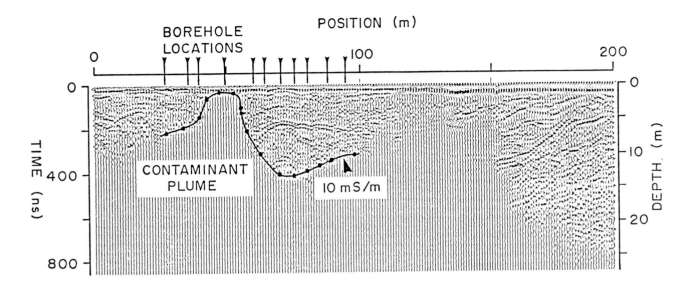

Figure 3-32 GPR Survey Results Survey

overburden thickness was delivered to the client within 30 days of project commencement.

The resultant geophysical survey should be presented in a clear format that a normally technically conversant individual from the sciences or engineering profession could understand by looking at the figurative information.

Detection of Inorganics

Inorganic parameters are often the key to evaluation of organics contained in ground water. This is especially relevant in landfill leachate migration projects. These inorganic indicator parameters such as chlorides, iron, manganese and many additional leachate parameters can

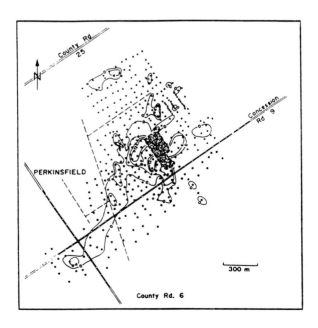

Figure 3-33 Location of EM Stations

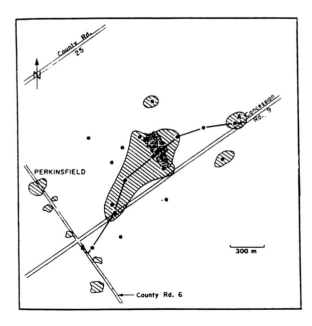

Figure 3-34 EM Results for Leachate plume

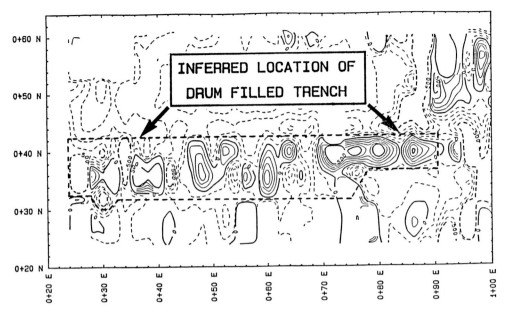

Figure 3-35 Vertical Gradient Magnetometer Results

raise the specific conductance of pore fluids within the area of the plume. Geophysical techniques that observe differences in subsurface conductivity often point toward potential target ground-water monitoring well locations. EM and resistivity methods have been the traditional geophysical methods for detecting and mapping inorganic plumes. EM however, has been the method of choice for site assessments because the rapid, non-contact nature of the method and the continuous data record. Both EM and resistivity methods can be greatly effected by noise interference from nearby fence lines, railroad tracks, buried pipes, or power lines. In these cases, if the geologic conditions are favorable (sandy soil), GPR methods may be applied successfully to shallow investigations.

A GPR survey Figure 3-32 was conducted to establish the orientation, lateral extent and depth of a landfill leachate plume. The radar survey clearly indicated the horizontal extent of the leachate plume along a survey line perpendicular to the regional ground-water flow direction. The high conductivity leachate plume absorbs the radar energy. Zones where radar reflections disappear indicate the presence of the signal absorbing leachate plume. Zones where pore water conductivity as obtained from existing monitoring wells, exceeded a value of 10mS/m, correlated with the areas of high radar attention. The stratigraphy of unaffected areas of the site were mapped down to 20 meters using the GPR technique. Geophysical survey equipment in the hands of experienced geophysics can provide many keys

to site conditions by what is <u>not observed</u> as by what <u>is observed</u> in the survey. Assessment monitoring situations typically require the "rate and extent" of contamination be determined. Rather than punching in numerous monitoring points, EM geophysical data can be obtained very rapidly in the field to aid in locating edge of leachate contamination. The results of 385 EM measurements are illustrated on figure 3-33. The contoured EM data show a plume-like anomaly migrating to the southwest from the landfill (Slaine, 1989). The results are shown on Figure 3-33 as decibels (Greenhouse and Saine 1983 and 1986) A decibel can be defined as a unit that describes a logarithmic ratio. The ratio for this contouring is a conductivity (resistivity) ratio normalized to a background conductivity (resistivity) value. The contours presented in Figure 3-33 are values only above background values. The method also, through the log scale, reduces congestion of contour lines in the immediate vicinity of the landfill or contaminant source. The EM survey was then used to locate assessment monitoring wells as shown on Figure 3-34. Ground-water chloride concentrations above the background level of 5 mg/l are shown on the Figure. Resistivity geophysical methods cannot produce continuous profile measurements as the EM method, due to the necessity of electrode movement. Resistivity methods may also be severely limited due to concrete or asphalt surfaces or due to highly resistive surface materials such as dry sand.

Assessment monitoring programs should include a geophysical component that looks at depth conductivity

/resistivity relationships. Both vertical sounding resistivity and EM methods can be used to assist in determining the depth of an inorganic leachate plume that may be generated by an unlined disposal facility. Geophysical applications used in the detection of contaminated ground water should be reliable down to several hundred feet; a depth frequently involved in ground-water assessment. The frequency domain EM method is limited to about 200 feet, hence, can cover depth requirements necessary for many of the assessment projects. Transient EM and resistivity depth criteria is virtually unlimited in the context of contamination site assessments, since, these methods can be accurate to few hundred to a few thousand feet (Benson 1984).

The physical length of the resistivity arrays or the EM coil separation necessary to measure the required depth can, however, be a limiting factor in these methods. The overall length of a resistivity array will typically be nine to twelve times the depth of interest so to measure to a depth of 200 feet, your array may need to be 2000 feet long. Longer arrays are also more subject to noise interference from electrical lines and such lengths can present access problems. EM coil separations only require about two times the depth of interest, therefore, EM methods can be used in much tighter sites than possible with resistivity methods.

Organic Detection

Geophysical surveys have been used to map organic contamination on many projects reported in the literature. However, direct measurement of the organics present in the subsurface by direct geophysical techniques has not been fully demonstrated as of early 1992. Indirect geophysical techniques for measurement of organics, however, is well described in the literature. Since a thin layer of hydrocarbons has the effect of compressing the capillary zone of the water table to a level that it can be detected by GPR, Olhoeft (1986) described the resultant GPR investigation over a pipeline in Minnesota. He reported a marked contrast in the reflective character of the radar records between undisturbed areas and zones where outwash was saturated with hydrocarbons. If the oily layer is thick enough, GPR may allow the hydrocarbon thickness to be estimated. Surface and downhole electrical methods may also be used to detect and map thicker floating

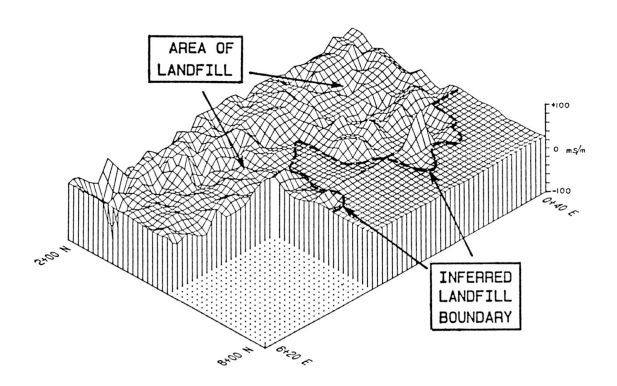

Figure 3-36 EM Conductivity Survey for Landfill Wastes

product spills. Cosgrave et. al. (1987) reported on a survey of a sandy aquifer near North Bay Ontario. A landfill leachate plume was reported imaged by the attenuation pattern of the radar-record. Vertical resolution was 15 cm and the 100 MHz radar record illustrated two areas of leachate movement adjacent to the facility.

Although direct evidence of detection of organic (not inorganic related) plumes is sparse in the literature, Saunders & Cox (1987) described the use of EM 31 conductivity surveys across a suspected leak of JP-4 aircraft fuel from an underground airport pipeline. The free product was believed sitting on a shallow water table in a sandy aquifer. The resultant EM in-phase and quadrature phase response were sited as direct evidence of detection of the fuel. The authors, however, acknowledged that this may be the result of volatile-related chemical changes in the vadose zone above the fuel.

Location of Buried Wastes

Site assessments commonly require the location of waste areas on and off of property boundaries. Buried wastes, disposal trenches, tanks, and utilities can be located using many different geophysical methods. GPR, EM, metal detection, and magnetometer methods can be used to quickly locate utilities and underground storage tanks before drilling on a property.

When metal is present in the form of trash, drums, tanks, pipes, or cables, the application of both metal detectors and magnetics becomes the primary approach to locate shallow drum disposal areas. Figure 3-35 illustrates the use of a vertical gradient magnetometer collected on 3 meter centers across a property. The data clearly shows a lineal series of isolated fluctuating magnetic anomalies which inferred the location of a buried drum filled trench. Upon test excavation, the small isolated magnetic anomalies were found to be generated by small targets such as individual pieces of rebar and a roll of metal strapping located 1.5 meters below surface. The site engineers were comfortable that no larger features such as buried drums or tanks were located at the site. The final report including processed magnetic data was delivered 7 working days after project commencement. These methods are unaffected by most soil types or the presence of other contaminants. All geophysical methods that rely on resistivity or conductance of subsurface materials can be made ineffective if buried non-hazardous wastes are present; for example, in a co-disposal facility that contains large volumes of municipal waste. Although GPR can be used to find drums, the method has some limitations, such as shallow depth of penetration, inability to detect drums due to orientation or geometry problems, and confusion due to other objects that give a similar response. Where

GPR can penetrate, a good evaluation of drain depths can be made.

Depths of trenches and landfills can be determined by resistivity, seismic refraction, gravity, and magnetics. If borings or wells exist, downhole geophysical logs may be used. In limited cases, radar may be useful for determining the depth of a shallow trench, if relatively non-conductive materials are present. Pit or trench slopes can also be estimated with radar; this can be used to extrapolate depth.

Metal detectors and EM will respond to ferrous and non-ferrous metals. As shown on Figure 3-36 EM conductivity surveys can effectively delineate the area of buried landfill debris.   The area of fluctuating electromagnetic conductivity values infer the presence of buried refuse, while the flat area of electromagnetic conductivity data indicates natural, undisturbed soils. A magnetometer will respond only to ferrous metals. It is helpful, therefore, to review the site history to evaluate

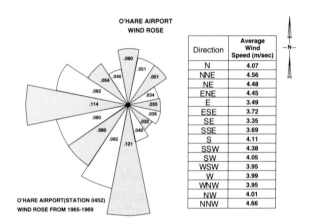

O'HARE AIRPORT WIND ROSE

Direction	Average Wind Speed (m/sec)
N	4.07
NNE	4.56
NE	4.48
ENE	4.45
E	3.49
ESE	3.72
SE	3.35
SSE	3.69
S	4.11
SSW	4.38
SW	4.05
WSW	3.95
W	3.99
WNW	3.95
NW	4.01
NNW	4.66

O'HARE AIRPORT(STATION 0452)
WIND ROSE FROM 1965-1969

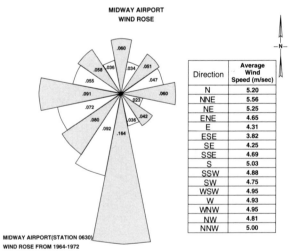

MIDWAY AIRPORT WIND ROSE

Direction	Average Wind Speed (m/sec)
N	5.20
NNE	5.56
NE	5.25
ENE	4.65
E	4.31
ESE	3.82
SE	4.25
SSE	4.69
S	5.03
SSW	4.88
SW	4.75
WSW	4.95
W	4.93
WNW	4.95
NW	4.81
NNW	5.00

MIDWAY AIRPORT(STATION 0630)
WIND ROSE FROM 1964-1972

NOTE:WIND ROSES SHOW PERCENTAGE OF TIME WIND BLEW FROM 16 COMPASS POINTS
SOURCE: USAPE PCGEMS, 1989

Figure 3-37 Annual Wind Rose Results

which metals to expect so that the proper instruments may be used.

3.6 HYDROLOGY

3.6.1 Meteorology And Climatology

This section describes the types of meteorological and climatic information that is typically collected during Phase II investigation. Meteorological stations are commonly erected on site to measure wind direction, wind speed, and temperature at a height of approximately 10 meters. An example of a wind rose format for direction and wind speed is shown on Figure 3-37.

Impacts of rainfall patterns can be divided into two categories: (1) surface water dilution potential, and (2) runoff potential. Surface water dilution potential affects the ability of nearby surface water to assimilate discharges and consequently reduce parameter concentrations. Runoff potential affects the potential transport of landfill materials to surface waters and the surrounding land. Some of the elements that define the climatic characteristics and hydrology of a region include precipitation (i.e., rainfall), temperature, evaporation, runoff, and infiltration. The type of information necessary to assess these phenomena are described below.

Precipitation

Monthly and annual precipitation and snowfall (expressed as the equivalent rainfall) can be obtained from the National Oceanic and Atmospheric Administration (NOAA), or the National Weather Service. Daily records of rainfall and snow are published in "Climatological Data" and "Hourly Precipitation" by the U.S. Environ-mental Data Service.

Regional precipitation data may be presented and used if they were generated within a reasonably close distance to the site (approximately 15 km) and are representative of rainfall and/or snowmelt conditions at the site. Regional data collected at greater distances from the site should be correlated with available on–site data. The monthly mean and range of these data, the specific time period from which the data came, and the location of the rain gauge(s) in relation to the facility should be provided. Precipitation data can be presented in tables showing monthly and yearly averages over a period of time, or as precipitation events as are shown on Figure 3-38. Precipitation is measured as a depth of water and is defined as the total amount of water that reaches land surface. It is measured with various rain

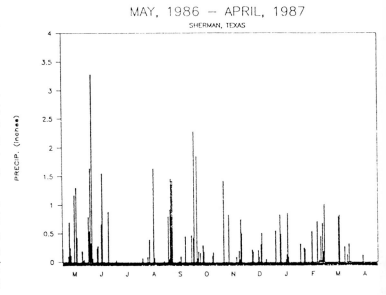

MAY, 1986 — APRIL, 1987
SHERMAN, TEXAS

Figure 3-38 Precipitation at Sherman, Texas

gauges (see Pictographic Figure 3-39). These data can also be used to relate rainfall and elevation of ground water in wells and piezometers as is shown on Figure 3-40.

This information should be available from the Federal Insurance Administration (FIA) in the form of maps or other data. If the facility design includes any special flood prevention devices (e.g., dikes, berms), these devices could also be shown on a site map. Any special site conditions that affect infiltration and runoff should also be discussed.

Temperature

Ambient air temperature (degree) data can be useful in the general assessment of the climatic setting of a site and may be useful in the assessment of potential volatilization of gases. This regional information should be available from similar sources used to obtain precipitation information. Temperature is generally reported as monthly and annual averages over the period of record, and is important in assessing evaporation.

Evaporation

Evaporation and transpiration (evapotranspiration) rates (depth of water per unit time) reflect the amount of precipitation returned to the air. Evaporation rates are measured by NOAA, and evapotranspiration rates can be estimated from these data. Evapotranspiration rates can also be obtained through site studies (preferably using tensiometers), or through published sources if the nearest data set (collected at a gauging station) is representative of the site conditions. Figure 3-31 shows the two major methods for defining evapotranspiration, the water balance methods and micro-meteorological methods. Standard pan evaporation equipment can be a component of the site weather station if required, however, these devices require daily attention.

Runoff

Surface runoff is of interest in assessing the transport of sediments over the land surface. Specifically, overland flow (i.e., the part of surface runoff that flows over the land surface toward channels) is of interest. The project engineer should identify the potential of overland flow to transport landfill wastes to land areas of particular interest (e.g., agricultural land) and surface water. This potential depends on the surrounding surface characteristics (e.g., slope, soil type, vegetation, paved areas). If the potential for the transport of materials is significant, then the engineer will likely have to conduct a detailed analysis of overland flow (i.e., the identification of the quantity and quality of overland flow). Such an analysis would likely require the use of a runoff simulation computer model such as the Storm Water Management Model (SWMM).

Infiltration

The maximum rate at which precipitation can enter the soil is the infiltration rate (depth of water per unit time). During precipitation events, all the water will infiltrate if the rainfall intensity is less than the infiltration capacity. If this capacity is exceeded, the excess rain cannot infiltrate and will produce surface runoff. The infiltration capacity can vary greatly due to the soil surface cover. If a soil is completely covered by a crop canopy, evaporation losses are negligible, and transpiration is the principal process by which water is lost from the root zone. The same environmental factors that control evaporation also control the potential transpiration.

Infiltration rates (average annual) can be important in the determination of the velocity of ground water moving downward through soil, as recharge to ground water and hence, in the modeling and determination of potential leachate transport rates. Records of estimated infiltration rates for an area may be available from sources such as the U.S. Department of Agriculture (Soil Conservation Service). However, it will probably be necessary to estimate this value by taking the average precipitation rate

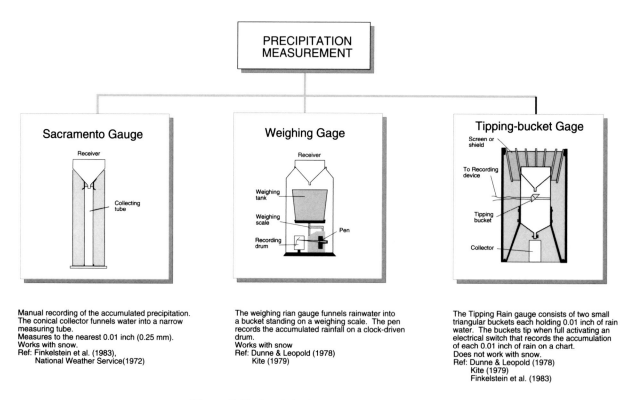

Figure 3-39 Precipitation Measurement

(average annual) and subtracting evapotranspiration and runoff rates (average annual). Actual field measurements of infiltration, through tensiometers in the vadose zone, are rarely performed. They are very expensive, difficult to perform correctly, and even more difficult to interpret.

3.6.2 Surface Water Hydrology

The purpose of surface water hydrology studies is to describe the drainage systems, flow characteristics, water quality of the streams and water bodies, and to aid in determining the ground water/surface water relationships. This information documents baseline conditions and forms the basis of assessing any future environmental impacts of the facility on surface water bodies.

The typical Phase II scope of work for surface water studies includes site reconnaissance by a combined hydrology/geology team, a stream gaging program, and various engineering analyses of hydrologic and stream

flow characteristics for a 12–month minimum time period with careful long–term extrapolation.

A review of pertinent literature published by federal, state, and local agencies (such as the U.S. Geological Survey, the U.S. Soil Conservation Service, and the various state Departments of Natural Resources) will be normally included in the Phase I report. Phase II tasks typically supplement the information from literature sources with more site specific information.

Periodic discharge measurements normally are made at selected stream sites using a standardized Price or pygmy current meter and standard U.S. Geological Survey techniques, as described by U.S.G.S. publication. These data are correlated with any continuous gaging stations operated by the U.S. Geological Survey along major surface water streams. All staff gages should be leveled to the U.S. Geological Survey NGVD, and nearby reference points should be established as permanent site monitoring stations. Lateral and longitudinal stream profiles are

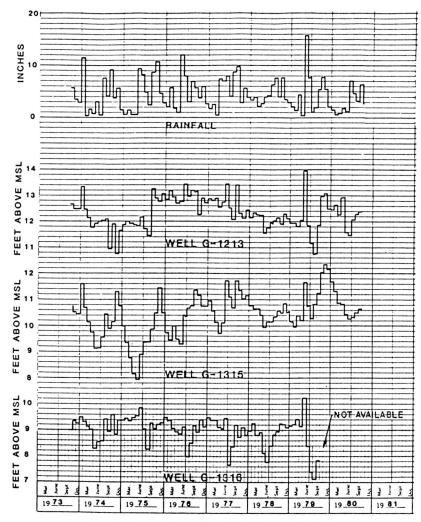

Figure 3-40 Rainfall Water Level Relationships

typically constructed, where appropriate to the investigation.

For a comprehensive investigation when surface water features and flow play an important role in the site assessment, staff gages in streams are normally monitored weekly by the field personnel, with staff gages in lakes read on a monthly basis. Precipitation data must be correlated with lake levels, and with stream flow data. Stage–discharge relationships should be determined for each stream gaging station. Drainage areas and basin boundaries can typically be determined from USGS 1:24,000 scale topographic maps. In order to calculate potential surface water flows in and around the site, a map of the drainage area should be prepared, based on site development topography (1:2,400) and stream flows. Figure 3-41 shows a drainage area map for an arid region in south–central Arizona. Each drainage area boundary is shown, along with soils of each drainage area. These data are used for evaluation of erosion potential, design–storm runoff–volumes, and peak rates of runoff. The major goals of such an analysis is to provide design criteria to: (1) manage off–site drainage from up–slope areas for subsequent down–slope release under non–damaging conditions; and (2) to detain on–site, for a specific period, a specific volume of runoff produced within the property boundaries following a design basis storm. The flow characteristics and flooding potential of streams should be established by means of the field gaging program, by correlation to any nearby long–term station operated by government agencies, and by use of standard engineering techniques. The cognizant state regulatory agencies should be contacted to determine the current design requirements for flood frequency or probable maximum precipitation criteria for design of required site flood control structures. The 10, 25, 50, and 100 year–frequency floods and the flooding potential should be addressed as appropriate requirements for design. A generalized water budget should be prepared so that the average annual rate of water movement into and out of the project area can be evaluated.

Sampling points may be established on streams in the area to monitor total suspended solids (TSS) as a function of stream flow. Selection of the number and specific location of each sampling point may be affected by the activities at the project. The sampling stations should be located at or near staff gages. Factors that should be considered in final selection of sample points include size of the drainage basin, slope of the land surface, topography, relief, height of vegetative canopy, and flow channel characteristics. Turbidity should be measured, using a turbidimeter, at the time sediment samples are collected from each station. Sediment samples should be collected using a suspended sediment sampler. For

comprehensive investigations, sampling may be required during periods of high precipitation or high runoff from snow melt. The frequency of turbidity measurements and collection of sediment samples should be increased to as often as daily, or even hourly, as appropriate for selected stations.

Measurement of total suspended solids within a given drainage basin provides a basis for calculating the erosion rates. As land is disturbed by project activities, monitoring of total suspended solids will provide information from which subsequent changes in the erosion rate can be evaluated.

Monitoring points established to evaluate the rate and volume of erosion, during the baseline period, will provide information to assist in design of effective site or cell–closure reclamation plans. In addition, these points can be used for monitoring, during and after completion of the project. During and following the field data collection, baseline information should be compiled and evaluated. A section on surface water will be prepared, to document existing conditions, and will be included within the site assessment report.

3.7 SURROUNDING LAND USE, WATER USE, AND WATER QUALITY CHARACTERISTICS

This section presents guidance on identifying and obtaining relevant information about water quality, water use, and land use. The Phase II program may require an extension of the preliminary Phase I studies, such as is often necessary for environmental impact analysis or risk assessments.

These factors can be grouped into three general categories: ground–water use and ground–water quality, surface water use and surface water quality, and characteristics of land use in the vicinity of the site. Accordingly, this section is organized into three basic sections. Section 3.7.1 discusses use of ground water and its quality characteristics, Section 3.7.2 discusses surface water use and quality characteristics, and Section 3.7.3 discusses land use of the surrounding area. Within each section, the type of recommended information is presented first, followed by a brief discussion of potential sources of information, suggestions for presenting the information, and how the information should be used in the assessment process.

This section presents separate discussions of the information needed to characterize ground water and surface water, both for use and quality, as well as land use of the area; recognizing, of course, that use of ground water, surface water, and surrounding land are all interrelated. Therefore, although this section provides a sequential discussion of quality and use of the ground

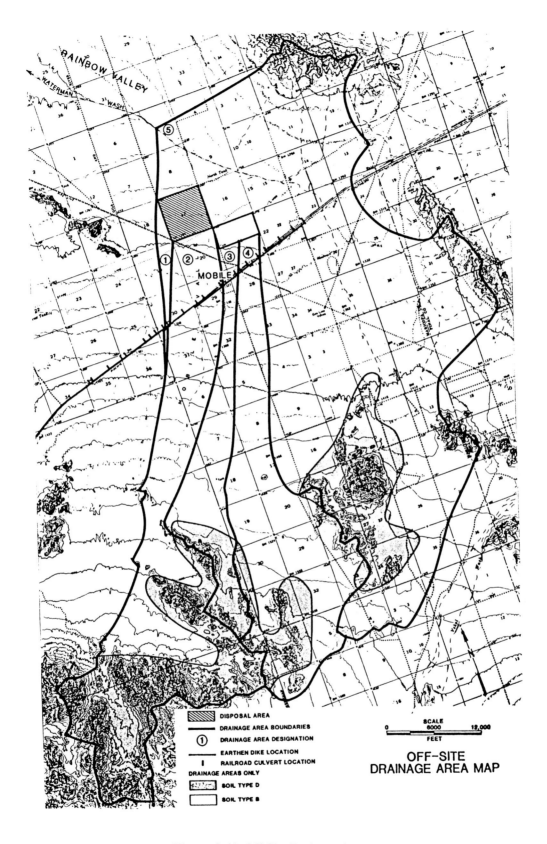

Figure 3-41 Off-Site Drainage Areas

water, surface water, and surrounding land, in practice it will probably be necessary for the investigator to obtain some information in all three of these areas before completing an evaluation of any one particular area. It is usually desirable to present data and findings in matrix form, for clarity and assimilation by the reader.

Not all of the water quality, water use, and land use parameters discussed in this chapter will require characterization at all sites. The investigator should exercise professional judgment, given the general guidelines presented in this section, in determining which water use and water quality parameters are relevant for the facility, and in determining the level of detail necessary to adequately support the EIR or risk–based assessments.

3.7.1 Ground-Water Use And Quality

Existing quality of ground water and current and potential future uses of ground water must be considered

by the engineer in evaluating the present and potential hazard to human health and the environment posed by a potential release of leachate from a waste management facility. Characterizing existing quality of ground water is an important task for two reasons. First, existing ground–water quality normally establishes the baseline conditions for evaluating risks to human health and the environment throughout the facility operational and post–closure periods. Second, existing ground–water quality, in part, determines current uses and affects future uses. In addition, determining ground–water uses is an important initial step in identifying potential pathways of exposure for risk assessment studies.

Proximity and Withdrawal Rates of Ground Water Users

The proximity and withdrawal rates of ground water by users within the site area are important factors in

WELL ID	LOCATION	STREET ADDRESS	ORIGINAL OWNER	DATE DRILLED	YIELD (GPM)	DEPTH (ft.)	GEOLOGIC UNIT
1	NE ¼ NE ¼ NW ¼ T3S R8E SECTION 1	3943 Van Top	Agey	11-5-74	DRY	115	Basal till
2	NE ¼ NW ¼ NW ¼ T3S R8E SECTION 1	3943 Van Top	Agey	11-8-74	8	125	Basal till Sand lense
3	W ¼ SW ¼ T3S R9E SECTION 7	8943 Handy Rd	Hollies	15-5-71	15	75	Quar. sand
4	SE ¼ SW ¼ NW ¼ T3S R8E SECTION 11	4357 Tyler Rd.	Roch	10-5-67	18	95	Quar. sand
5	NE ¼ sE ¼ sW ¼ T3S R8E SECTION 12	4578 Tyler	Pope	7-5-75	23	135	Quar. sand
6	NE ¼ NW ¼ NE ¼ T3S R8E SECTION 12	4547 Tyler	Actee	8-5-78	7	123	Basal till
7	SE ¼ NE ¼ NW ¼ T3S R9E SECTION 12	4589 Tyler	Upstart	7-5-71	12	85	Quar. sand
8	NW2 ¼ NE ¼ NW ¼ T3S R8E SECTION 1	4534 Tyler	Buber	9-5-68	---	73	Basal till
9	NE ¼ SE ¼ NE ¼ T3S R8E SECTION 1	4536 Tyler	Slopp	4-5-56	---	81	Basal till
10	SE ¼ NW ¼ NW ¼ T3S R8E SECTION 1	4783 Tyler	Crudd	2-5-71	20	65	Quar. sand

NOTE: Information obtained from Water Well Records provided by the
State Dept. of Natural Resources and the Geologic Survey
Geologic Units not listed on Well Records were estimated from drillers logs.

Table 3-5 Example Well Information

determining the potential adverse effects of a release on human health and the environment. The proximity of ground–water users will affect both the time it takes a leachate to reach the user and the concentration of parameters in the ground water. The withdrawal rate (i.e., the daily or annual volume of water pumped from an aquifer) is used to assess the total amount of contaminants to which the user is exposed and, in some cases, any influence the rate has on the direction and magnitude of ground–water flow.

Figure 3–42 shows a map of well locations for a region. Well notations are keyed into tables (example Table 3-6) providing information about individual wells. For each ground–water user identified in the surrounding area, the following information should be provided:

- Well location and distance from potential point of release;

- Well construction;

- Well depth;

- Type of user:
  – Potable (municipal and residential),
  – Domestic non–potable (e.g., lawn watering),
  – Industrial,
  – Agricultural,
  – Artificial recharge; and

- Estimated withdrawal rates (daily peak, annual, and seasonal).

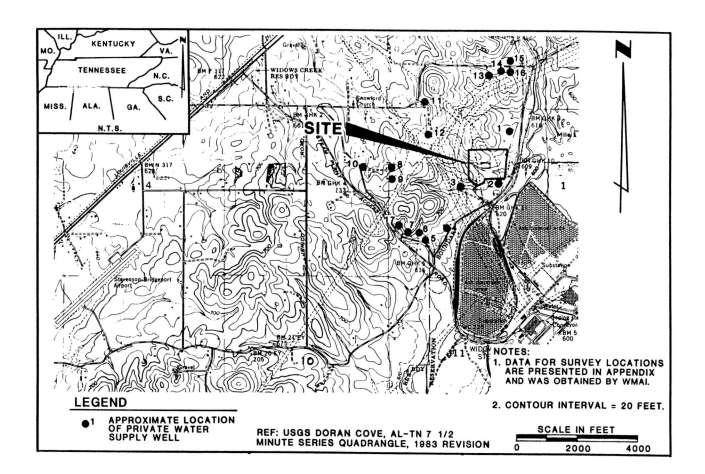

Figure 3-42 Well Location Map

The above information can be presented using a map and an accompanying summary sheet. Possible sources for information relating to proximity and withdrawal rates of ground water include local and regional water districts or companies; state agencies; Federal agencies (USEPA, U.S. Geological Survey, U.S. Department of Agriculture); and various data bases of state, federal, and private organizations. A detailed list of these sources is contained in Phase I sections of this document.

Existing Background Quality of Site–Area Ground Water

The existing quality of ground-water affects current and potential uses of ground water and also provides a baseline for evaluating the incremental potential for risk to human health and the environment due to a possible release of leachate from the facility. The quality of ground water can be affected by both naturally occurring sources (such as leaching of minerals from the aquifer medium) and human sources (such as leaking of petroleum or chemical products from underground storage tanks or the downward migration of pesticides and fertilizers from agricultural areas). The primary purpose of this subsection is to define guidelines for characterization of existing background quality of ground water in the site area (i.e., the ambient quality of the ground water prior to project implementation).

In evaluating background water quality of the site area, the investigator must consider not only possible background concentrations of the selected indicator chemicals, but also the background concentrations of other potential hazardous constituents from leachate. Existing contamination associated with indicator chemicals or other RCRA hazardous constituents may be due to:

- Natural conditions in the area;
- Prior or present releases from the unlined old landfill areas; or
- Prior or present releases from other upgradient sources in the surrounding area.

Assessing background concentrations of constituents is necessary to establish an existing baseline of ground–water quality to which the incremental effects of potential release can be added.

In cases where sufficient historical monitoring data are unavailable, the investigator may need to install an upgradient ground–water monitoring system or add to an existing system in order to adequately assess background quality of ground water. At a minimum, background water–quality should be determined based upon at least two separate confirmatory sampling of existing or newly installed monitoring wells.

3.7.2 Surface Water Use And Quality

Existing quality and current uses of surface water must be considered in evaluating the design of surface–water control–systems and discharges to surface water. Characterizing existing quality and current and future uses of surface water may be required in site assessments, directed toward evaluation of risk to: (1) establish the baseline conditions for evaluating risks to human health and the environment, and (2) determine probable water uses for identifying potential exposure pathways in surface water.

In assessing surface water use and quality characteristics, the investigator should evaluate those surface waters that might possibly become contaminated by a release, based on the hydrogeologic characterization discussed in Chapter 3. For example, it is unlikely that surface waters upgradient and distant from the facility would be contaminated by a release from the facility to be permitted. The investigator should consider the possibility of surface–water contamination via both surface runoff and ground–water flow via springs, seeps, or into gaining streams.

The following subsections explain what types of information relating to surface waters may be required and how this information may be used in evaluating the quality of surface waters, and in identifying current uses and drainage, and in predicting future water uses.

Existing Quality of Surface Water

Surface–water quality parameters are both physical and chemical in nature. Physical parameters of surface water quality, such as temperature and turbidity, may already have been measured in evaluating the hydrologic characteristics of surrounding surface water.

If available, the investigator should provide data of conventional surface water quality parameters, such as suspended solids, nutrients (e.g., nitrogen and phosphorous), total suspended sediment, oxygen–demand, salinity, hardness, alkalinity, pH, fecal coliform, and dissolved solids. Data on these parameters may be helpful in evaluating potential surface water uses and providing the base line for future impacts on surface water quality. For example, highly saline or turbid water may be unsuitable for drinking without extensive treatment. In many cases, state or local environmental agencies may already have obtained current data for these parameters for specific surface water bodies. If such data are unavailable for the area, water quality testing may be required. The results of water quality testing should be submitted in a clear, concise format along with necessary supporting documentation relating to conditions of sampling and analyses.

In both traditional site assessments and risk assessments, current users of water, state and local government agencies (such as water authorities), and natural resource management agencies should be contacted. In remote areas, nearby residents may also be able to supply useful information. If a surface water body has been assigned a "designated use" or "sole source"

status, the investigator should identify the designated use or uses and determine which of the designated uses are considered to be currently active.

3.7.3 Current and Future Uses of Surrounding Land

To some extent, current and future uses of surrounding land will have already been determined in identifying current and future uses of ground water and surface water. The purpose of obtaining information on the current and future uses of surrounding land in this section is to characterize surrounding agricultural, commercial, and residential land use (see Figure 3–43), and to identify any ecologically sensitive areas that could be adversely affected by operation of a proposed facility. Such ecologically sensitive areas may include:

• State, federal, and local parks;
• Wildlife refuges;
• Wilderness areas;
• Critical habitats for endangered and threatened species; and
• Wetlands.

Agricultural, commercial, and industrial land uses can usually be identified by contacting local land use regulatory authorities, such as planning and/or zoning boards, and reviewing appropriate land use plans and maps. Identifying ecologically sensitive areas may be more difficult. While some of these ecologically sensitive areas may be marked on U.S. Geological Survey topographic maps or other maps, many may not be marked. In particular, the location and boundaries of some critical habitats are not published or made readily available to the public.

To verify the existence of ecologically sensitive areas and to identify areas under consideration for protection, relevant state and federal government agencies (such as state and federal park agencies, fish and wildlife agencies, and private conservation groups (such as The Nature Conservancy) should be contacted. A list of potential agencies and organizations that may be able to provide information on surrounding land use is included in Chapter 2 of this document.

The investigator should include a brief narrative description of current and projected future uses of surrounding land in the Phase II report. The use of all land in the immediate surrounding area of the site should be carefully described. The engineer should devote special attention to mapping any ecologically sensitive habitats in the surrounding area. Two maps of the surrounding area, one identifying current land use and another identifying

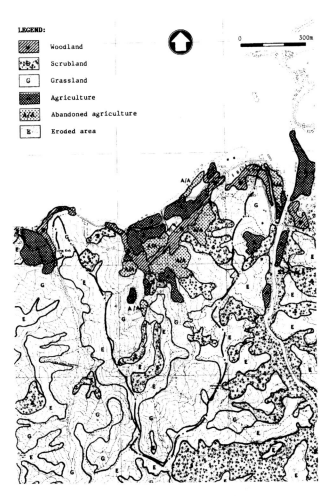

LEGEND:

Woodland	
Scrubland	
Grassland	
Agriculture	
Abandoned agriculture	
Eroded area	

0   300m

Figure 3-43 Current Land Use Map

known or reasonably projected future land uses, should be provided as part of the project documentation.

3.8 FIELD ECOLOGY SURVEYS

Field ecological surveys are very specialized technical investigations that must be scoped directly to the facility area. These surveys supplement the land use data collected in the previous investigations. Field ecology surveys can make the difference between project success or failure depending on the adequacy of the program. Endangered species are typically defined in Phase I activities; however,

Phase II comprehensive ecological surveys are long–term, (at least a year) and expensive investigations that should be handled by specialists in the field. These investigations are normally divided into two areas: aquatic and terrestrial ecology.

3.8.1 Aquatic Ecology

Studies of aquatic ecology are designed to define, in detail sufficient for assessment purposes, the aquatic ecosystems on and adjacent to the proposed facility areas. The work normally required to assess impacts to aquatic systems could include:

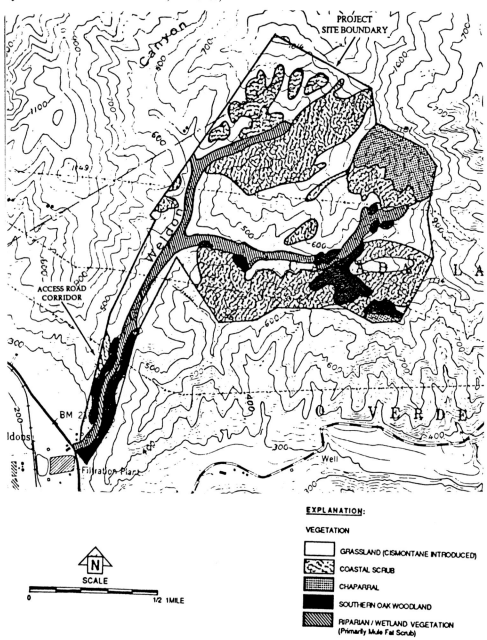

EXPLANATION:

VEGETATION

☐ GRASSLAND (CISMONTANE INTRODUCED)

COASTAL SCRUB

CHAPARRAL

■ SOUTHERN OAK WOODLAND

RIPARIAN / WETLAND VEGETATION
(Primarily Mule Fat Scrub)

Source: Dames and Moore

Figure 3-44 Wetland Mitigation Map

- A description of the biological baseline conditions in surface water bodies near the proposed site;

- The biological baseline conditions for surface water bodies in the immediate vicinity of the candidate disposal areas;

- Establishment of water quality monitoring programs in areas potentially affected by development of the disposal areas;

- An assessment of potential impacts on aquatic resources caused by construction and operation of the disposal facility; and,

- An evaluation of potential measures for mitigating or eliminating deleterious effects on the aquatic resources

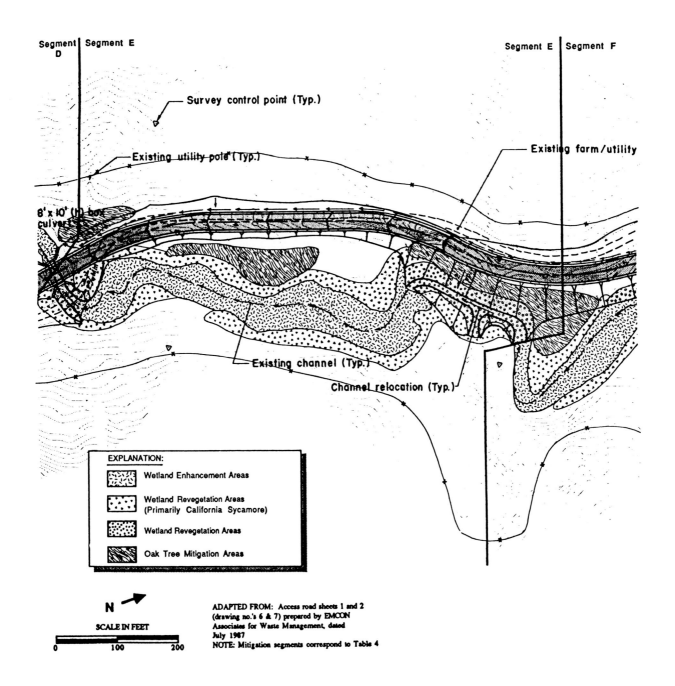

EXPLANATION:

Wetland Enhancement Areas

Wetland Revegetation Areas (Primarily California Sycamore)

Wetland Revegetation Areas

Oak Tree Mitigation Areas

N

SCALE IN FEET

0   100   200

ADAPTED FROM: Access road sheets 1 and 2 (drawing no.'s 6 & 7) prepared by EMCON Associates for Waste Management, dated July 1987
NOTE: Mitigation segments correspond to Table 4

Source: Dames and Moore

Figure 3-45 Principal Plant Communities

in the vicinity of the disposal site.

The study scope of work should be flexible so as to allow such modifications in the program as may be apparent as on–site experience is gained and design concepts become formalized.

3.8.2 Terrestrial Ecology

The terrestrial ecology program should be designed to identify and describe the major terrestrial communities within the proposed facility areas and to provide a baseline for analyzing potential impacts resulting from site development. Information developed within this program should include:

- Identification of important floral and fauna species including rare and endangered species, if present;

- Distribution and relative abundance of important biota;

- A map showing the distribution of the principal plant communities and the habitats of important fauna species;

- Site usage and tropic interrelationships of important biota, including habitat requirements; and,

- The regional importance of the site's fauna and flora populations based on a comparison of site data with a literature search describing the regional ecology.

Field studies of vegetation, mammals, birds, reptiles, and amphibians can all be part of a terrestrial ecology investigation. The exact nature and scope of the investigation should be based on the Phase I literature search and site reconnaissance. An example illustration of principal plant communities is provided in Figure 3–44. These data can then be used to establish impacts and mitigation analyses required by individual state environmental programs. Figure 3–45 shows proposed wetland enhancement, revegetation, and plant community mitigation areas. Such detailed plans for site environmental care can significantly reduce project development impacts.

3.9 HISTORICAL AND ARCHAEOLOGICAL SURVEYS

A regional overview of the area is typically developed, based primarily on literature review in Phase I. In addition, a reconnaissance survey may be required to confirm the presence or absence of significant historical or archaeological sites on or adjacent to the facility.

Reconnaissance surveys typically include the deployment of multiple transects which involve systematic examination of samples of subsurface conditions at regular intervals (for example at 15 meter intervals). Standard survey procedures also include visual examination of all disturbed earth areas such as road cuts, farmers' fields, spoil from rodent burrows, etc. These procedures are typically necessary in heavily forested areas with significant ground cover conditions.

These procedures result in the field verification of known or suspected locations of archaeological and historical sites. They may also result in the discovery of previously unknown sites. The precise location (on site topographic scale of about 1:2,400 or larger) of these new sites, together with maps which indicate areas which failed to produce evidence of remains, should be prepared for inclusion in the Phase II report. All recovered archaeological and historical debris should be washed, marked, cataloged, and deposited with an appropriate museum department of anthropology.

All newly discovered archaeological and historical remains should be recorded with the State Archaeologist and, if required, appropriate archaeological site codification numbers should be applied for. A period of laboratory analysis typically is included to follow the field work in order to process materials, inventory and evaluate cultural remains, and to prepare maps and photographs of these data.

3.10 METHODS AND TOOLS OF HOLOCENE FAULT INVESTIGATIONS

The methods and tools for conducting a comprehensive geologic investigation for Holocene faulting (within the last 11,000 to 13,000 years) include a review of published data, aerial reconnaissance, interpretation of aerial photographs, ground reconnaissance, and subsurface investigations. These tools are used to identify faults and lineations (linear features that suggest the presence of a fault) and to verify whether or not there has been fault displacement in Holocene time, and if so, such faults would be classified as "active" or "capable" (identical meanings). Federal regulations*, and many states, require consideration of faulting as a potential "fatal flaw" in a site considered for waste disposal (see Table 3-6). Proving a particular site does not contain Holocene faulting can represent a significant commitment in time and expense that requires a carefully planned field evaluation techniques.

This section contains a detailed description of these

---

*In general, these fault investigations will be required only in regions identified by the National Oceanographic and Atmospheric Administration (NOAA) as Zone 3 (Algemissan and Perkins, 1976)

TABLE 3-6 Comparison of U.S. Fault Criteria for Critical Structures

Agency*	Facility	Terminology	Activity criteria
NRC	Nuclear power plants	Capable fault	• 35,000-yr singular movement • 500,000-yr multiple movement • Macroseismicity • Structural relationship of faults
COE	Dams	Capable fault	• 35,000-yr movement • Macroseismicity (magnitude 3.5) • Structural arrangement of faults
DOT	Liquid natural gas facilities	Surface faulting (active fault)	• 35,000-yr singular movement • 500,000-yr multiple movement • Structural arrangement of faults
VA	Hospitals	Active fault	• IO,OOO-yr movement
BR	Dams	Active fault Inactive fault Indeterminate fault	• IOO,OOO-yr movement • Distribution of seismicity
EPA	Hazardous waste facilities	Active fault	• Movement in lifetime of waste toxicity
EPA	Solid waste facilities	Active fault	• Movement in during Holocene

*Agencies:
NRC Nuclear Regulatory Commission
COE Corps of Engineers
DOT Department of Transportation
VA Veterans Administration
BR Bureau of Reclamation
EPA Environmental Protection Agency

Source G. A. Robbins, L. A. White, and T. J. Bennett, "Seismic and Geologic Siting Regulations and Guidelines Formulated for Critical Structures by U.S. Federal Agencies," Report of the GSA Committee on Geology and Public Policy, Geological Society of America, Boulder, Colo., August 1979, pp. 8-11.

field evaluation techniques due to the complexity and difficulty in demonstrating that no Holocene faults are present within areas of waste disposal as modified from EPA (1982).

3.10.1 Review of Existing Data

A review of existing data is directed toward obtaining information on the seismicity of the site area (to a radius of 5 miles). Preliminary data on geology has been gathered in the Phase I program. Any available information on the geology, degree of how recent is the fault activity, and earthquakes, for which recorded information exists, should be reviewed to determine if Holocene faults have been identified in the project area. During this effort, it is important to further identify any structural geologic trends or geomorphic features that may be related to faulting. Seismicity data should include information about epicenter concentrations and depths of "felt" and instrumented earthquake activity, historical accounts of major earthquakes, and any related surface effects of faulting in the area. Each of these may indicate the presence and location of faults.

Information on the location of faults near the site and a record of their activity may exist as published data. Because fault–related studies of a particular site will not exist in many cases, geological information from a wider area than the specific local facility must be reviewed. The Phase I investigation establishes the potential for faulting in the facility area. After this assessment, one can determine if a Phase II field investigation (e.g., aerial reconnaissance, ground reconnaissance, borings) must be conducted to demonstrate compliance with a seismic location standard.

3.10.2 Analysis of Aerial Photographs

Studies of aerial photographs and other remote images

are an essential part of an investigation for Holocene faults. Experience has shown that aerial photographs may be of limited use in forested areas or in areas that were developed before the photographs were taken. However, in many cases, features of faults can be observed on aerial photographs that could not be readily identified during land–based studies. The traces of some faults can be easily identified through preliminary review of aerial photography. In some cases, the surface features of some faults, seen on the aerial photographs, may be so small and relatively insignificant from surface inspection that they can only be recognized as faults by experienced geologists with considerable background in studying active faults and in interpretation of aerial photographs. There is a large element of experience and judgment in the interpretation of aerial photographs, this is particularly true in the interpretation of Holocene fault activity. This is not a job for amateur photo-interpreters!

Much of the United States has been photographed in black and white at scales of 1:30,000 to 1:60,000. These photographs are especially useful for preliminary regional investigations. More detailed investigations require scales of 1:12,000 to 1:20,000, stereoscopic coverage is absolutely necessary for interpretation of fault traces.

Aerial photographs, not taken specifically for fault studies, are usually taken during midday. However, it is recommended that aerial photography flown specifically for detection of active faults should utilize low–sun–angle photography taken at different times of the day, and sometimes in different seasons of the year depending on the trend of the fault and the azimuth of the sun. The purpose of the photography is to record shadows cast from subtle surface irregularities representing fault scarps of sufficient vertical relief and/or youth as to yet remain detectable. Optimum lighting for fault analysis is obtained when the sun angle is nearly perpendicular to the trend of the fault, in the direction from the uphill side of the fault scarp toward the downhill side, at an angle slightly lower than the average slope of the scarp (usually 10 to 25 degrees). The most useful scale is about 1:12,000, with scales of 1:6,000 and 1:3,000 applicable for detailed analyses. While it may be beneficial to repeat the photography in both the early morning and late afternoon, experience shows that photographs taken either in the early morning or late afternoon provide most of the detail that can be obtained by flying at both times. Other types of remote images that may be helpful include infrared and color images, photographs from satellites, and high altitude aircraft photographs. The following points summarize fault analysis through aerial photography:

- The most useful type of photograph is the black–and–white, low–sun–angle, vertical aerial photograph.

- When photographs are taken for the purpose of a fault study, low–sun–angle black–and–white aerial photographs are superior to color photographs under the same conditions, which, in turn, are superior to black–and–white photographs taken at midday.

- Infrared photography can occasionally give a better indication of differences in near–surface ground–water level or contrasts in vegetation on the opposite sides of a fault with minimal surface expression, but low–sun–angle photographs are better for analyzing historic fault activity that has resulted in geomorphic evidence of displacement.

- Satellite and high–altitude aircraft images (both types at scales >1:60,000) can be useful for regional identification of geologic structures that may need additional evaluation using low–sun–angle photography or field studies.

Documentation of analysis of aerial photographs should include sources of photographs, photograph numbers, dates photographs were taken, type of photographs, and either copies of photographs or overlays on which analysis is made. Superfund image interpretation portfolios produced by the USEPA Environmental Monitoring Systems Laboratory (Las Vegas, Nevada) are of state–of–the–art quality.

3.10.3 Surface Geologic Reconnaissance

A surface geologic reconnaissance is made to observe geomorphic and geological features that have been noted in the literature or detected on aerial photographs and that may be associated with Holocene or older faults. If a fault has experienced Holocene activity, geomorphic evidence of faulting can be expected to occur at a number of locations along its trace. Confidence in conclusions drawn from a ground reconnaissance can be enhanced by:

- Extending the reconnaissance into the site area (5+ mile radius).

- Examining locations of known faults in the area to determine type of faulting and evidence of Holocene activity.

- Examining lineations or suspicious features identified during the aerial reconnaissance or as such appear on aerial photographs.

• Describing any evidence of stratigraphic continuity that would demonstrate a lack of Holocene fault activity in the vicinity of the site.

Ground reconnaissance should be documented by plotting geotechnical observations on a topographic map or photo overlay, and by providing 35 mm slides of observations, with photo location (station) and axis–of–view noted on an original geologic map prepared for the evaluation.

3.10.4 Subsurface Mapping

The purpose of mapping the subsurface is to demonstrate (1) the presence or absence of active faults within 200 feet (lateral distance) of portions of the facility where disposal of waste will be conducted; and (2) to determine if these faults have had displacement during Holocene time. Subsurface exploration may consist of:

• Geophysical investigations of subsurface conditions along traverses selected so as to cross suspected fault traces at nearly perpendicular orientations.

• Exploratory trenching and other extensive excavations to permit detailed and direct subsurface observations.

• Borings and backhoe pits to permit collection of data at specific locations and depths.

Geophysical methods are indirect methods to detect anomalies and variations in subsurface strata. These anomalies and variations may represent faults. Therefore, they require specific knowledge of subsurface conditions for reliable interpretation. Geophysical methods alone do not prove the absence of a fault, nor the age of fault activity.

Geophysical methods used for fault studies may include seismic refraction, seismic reflection, ground–penetrating radar, magnetism, resistivity, or gravity. These methods are described in standard textbooks on applied geophysics. The purpose of any geophysical investigation is to detect and locate subsurface geological structures or masses. During the investigation, measurements of variations in subsurface properties are obtained. These data are analyzed to surmise the causes of each variation.

Borings have traditionally been much less informative than exploratory trenching for locations of faults. Faulted zones are often rubbery and do not provide intact core well. Conversely, some faults are represented by only a single rupture plane or off–set that cannot be easily observed from cores.

Exploratory excavations are the most positive and definitive method to establish the location and potential activity of faults at a given location. The excavation is commonly made by a backhoe, a dozer, or some other means of exposing subsurface materials. Selection of the excavation equipment depends upon accessibility, depth of excavation, difficulty of excavation, cost, length of time the excavation is to remain open, and the potential environmental impact of the excavation. Guidelines as to scale, manner of recording detail, and logistics are presented by Hatheway and Leighton (1979).

The purpose of the excavation is to expose the subsurface materials below weathered soils at a depth sufficient for geologists to make a detailed evaluation of the excavation walls for inspection of potential age–determinable fault features. Excavations are expensive; hence, to ensure that the maximum amount of information is obtained from each excavation, the following procedures are recommended:

• Although each site is different, the excavation may be at least 10 to 20 feet deep. It will probably be necessary to excavate the trench below the depth influenced by man–made activities and weathering and soil forming processes in order to expose materials that will show age–determinable fault offset.

• The excavation should be perpendicular to the suspected trend of known faults, lineations, or suspicious features since additional faults generally occur parallel to other faults. Thus, maximum coverage can be obtained by such trenches.

• The excavation must be constructed with worker safety in mind and comply with local, State, and Federal requirements for safety. All permits for such excavations must be secured before excavation. Most excavations require shoring or walls laid back to reduce the potential for collapse.

• The trench should be inspected and logged by experienced individuals as significant fault features may be difficult to recognize.

• Sufficient time should be allowed to inspect, log, and photograph the excavation. After the excavation is backfilled, the excavation and photographic log are the only evidence of the investigation. It is very important that the log reflect the significant geotechnical details observed in the excavation. The graphic log should show scale, the length and depth of the trench, the contacts between various subsurface materials, representative strike and dip of bedding and/or joints,

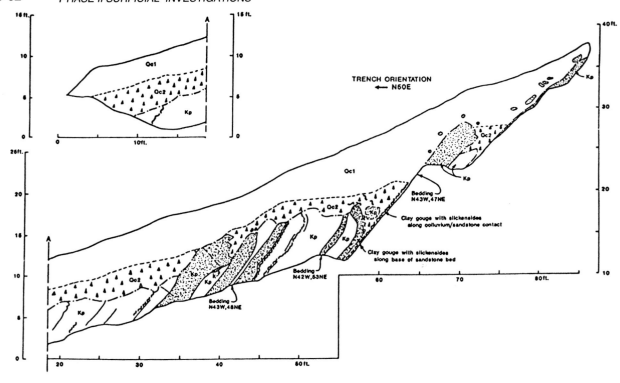

NOTE: - Elevation datum arbitary
- See figure A-2 for unit and symbol explanations
- See figure A-1 for trench locations

Figure 3-46 Exploration Trench Mapping

shear planes and faults, lithology, and continuous unfaulted material within the length of the trench. Figure 3–46 shows an example of a trench log prepared for a Holocene fault evaluation.

- The excavations should extend beyond the limits of the fault and the proposed facility boundary to identify parallel fault planes. In California, the required width of consideration is a 1/4–mile band for all potentially active faults identified by the State Division of Mines and Geology, and so depicted on their Open–file quadrangle maps (1:24,000 scale).

Generally, more than one excavation should be made across a fault or suspected fault to determine conclusively its location and relative activity. The number of trenches will depend on size of the facility (small–large), tectonic environment (active–inactive), style of faulting (strike–slip, normal, or reverse), type of surface materials (rock–Holocene materials), and results of initial trenching (simple–complex). One trench might be sufficient for a small site with a simple, narrow, well–defined fault in the subsurface; well recognized surface features; and continuous Holocene deposits. At least two trenches should

be used at large sites; sites with no evidence of subsurface faulting, but surface lineations or Holocene faults within 300 feet; a wide or complex fault pattern; a fault that changes width, direction, or pattern; or thrust or normal faults. Three, four, or more trenches may be needed to adequately locate a fault zone across a site; to demonstrate that the site is unfaulted; and/or to show that any faults, which cross the site, have not had displacement in Holocene time. The logging of exploratory excavations and the observations of "active" faults in exploratory trenches is discussed by Taylor and Cluff (1973).

3.10.5 Age Determinations

Age determination of faults is a relatively new field that presents many challenges for the investigator. Beyond simple geological mapping of trenches and excavations for fault features, the investigator should seek to identify and collect direct evidence of age of movement through carbon 14 (C^{14}) analysis of deposited wood chips and roots located directly above the offset strata. This data collection requires that, at least, small pieces of carbon based material be present in an undisturbed (by faulting) layer overlying the faulted strata. Typically, small chips can be found in boundary areas between rock canyons and stream valleys. C^{14} can now be used successfully for ages of up to 55,000 years.

3.10.6 Results

The results of the geologic study should be presented in a report and be of a technical quality acceptable to a geologist experienced in identifying and evaluating faults. The report should represent and describe the observations made during the investigation and be based on defensible geologic evidence. Contents of the report should include, but not be limited to, the following:

- Location and map of study area;

- Sources of available data;

- List of key scientists/engineers conducting study;

- Description of geologic setting and site conditions;

- Investigative methods and approaches;

- Aerial photographs or remote images interpreted (type, scale, source, index number);

- Trench logs (showing details, not diagrammatic), if appropriate;

- Photographs of the classified fault features;

- Geologic map with exploration locations (1 inch = 200 feet, for example), if appropriate;

- Site area geologic map (1 inch = 1,000 feet, for example);

- Regional geologic map (1 inch = 2,000 feet, for example);

- Results, conclusions, and basis of findings; and

- References cited.

Often, the data gathered must draw conclusions as to probability of the site being located above a Holocene fault. Unfortunately, much of the data gathered in these investigations is inconclusive in defining the absence of Holocene faults. Hence, the report must try and draw conclusions by using supporting evidence of regional geologic–tectonic settings and the local characteristics of regional Holocene faults. To accomplish this, the investigator must first define the regional setting, and then define the physical characteristics of the Holocene faulting as to the following:

- Orientation (example: NW–trending, right–lateral, strike slip)

- Length (example: 5 to 10 miles long)

- Rock displacements (example: Miocene rocks offset 1,000 ft)

- Loci of earthquakes (example: 1889 MMI 6.0 Antioch, CA, earthquake)

- Results of field investigation (example: results of trenching study)

- Surface geomorphic expression (example: recent faulting indicated by disturbed beds)

- Regulatory classification (example: identified as Holocene–aged fault by CDMG)

The data should be placed on convenient base maps showing the physical characteristics of the Holocene faulting activity. The characteristics of the Holocene faults should then be compared with features observed from mapping the surface and trenches. Tables of comparisons can be developed using the site characteristics established during the Holocene fault investigation.

3.11 ADDITIONAL FIELD SURVEYS

3.11.1 Acoustics

Equipment used in waste disposal operations make noise; as such, assessment of the effects of noise can play an important part in defining acceptance of the overall project in a community. For comprehensive environmental impact analysis, it is important to document the pre–operation ambient sound levels at the proposed site. Through assessment, by a comparison of the pre–operation and estimated post–operation data, the impact of a proposed facility on noise sensitive land use areas is established. Baseline sound levels are normally obtained by sampling the ambient sound as a function of time–of–day, season, and as a function of geographic location. To optimize the assessment, data are obtained at noise sensitive (human, domestic animals, and wildlife) land use locations.

The following baseline studies should be conducted during a comprehensive acoustic study:

• The proposed site and surrounding area should be reviewed through use of aerial photographs, USGS quadrangle maps, and other available maps;

• Potential noise–sensitive land–use areas should be identified by size, distance from the site, and other important parameters;

• Locations for measurement should be selected;

• Ambient sound levels should be recorded on tape; and

• The recorded data must be analyzed.

Ambient sounds are recorded on magnetic tape during representative sampling periods at each selected location. Field equipment typically consists of a sound–level meter and octave–band analyzer with accessories and a magnetic–recorder.

Tape–recorded data are then returned to an appropriate acoustics laboratory for analysis, using a real–time analyzer and a mini–computer. Each recording is used to construct an A–weighted sound level histogram and cumulative distribution for a typical measurement period. Octave–band statistical–distributions are normally obtained as well.

To evaluate the effect of the proposed facility on the sound–climate of the area, octave–band sound pressure–levels of principal noise–producing operations and equipment should be obtained from an existing facility, architect/engineer data, equipment manufacturers, and engineer files. These data would be extrapolated to the measurement positions initially used and added to the baseline sound–level data.

The estimated sound–levels resulting from construction and operation of the proposed facility are then compared with applicable State, Federal, and local noise regulations. They can also be evaluated using the current Federal (Environmental Protection Agency) ambient sound–level descriptor, day/night equivalent sound–level decibel number (Ldn), to assess the impact on the neighboring residents and workers. The assessment of impacts should consist of:

• Analysis of construction methodology and equipment;

• Analysis of operation noise, producing operations, and equipment;

• Estimation of community sound–level contribution by facility construction and operation;

• Comparison of estimated levels with current or proposed Federal, State, or local noise regulations; and

• Assessment of the proposed facility on public health and welfare.

When appropriate, general noise–abatement procedures should be suggested in the report to mitigate any adverse impact.

REFERENCES

Algemissan, S. T., and D. M. Perkins, 1976, A Probablistic Estimate of Maximum Acceleration in Rock in the Contiguous United States, U.S.G.S. Open File Report 76-416, U.S. Government Printing Office, Wasington D.C. 45p.

Benson, R. B., R. A. Glaccum and M. R. Noel, 1984, Geophysical Techniques for Sensing Buried Wastes and Waste Migration, EPA-600/7-84-064, 255 p.

Benson, R.C., Glaccum, R.A. and Noel, M.R. 1982, Geophysical techniques for sensing buried waste and waste migration. U.S.EPA Enviromental Monitoring Systems Laboratory, Office of Research and Development, Las Vegas, 236p.

Bieniawski, Z. T., 1988, Engineering Rock Mass Classifactions: John Wiley and Sons, New York, N. Y., 311 p.

Commission on Classification of Rocks and Rock Masses of the International Society for Rock Mechanics, 1981, Recommened Symbols for Engineering Geological Mapping: International Association of Engineering Geology Bulletin, No. 24, pp 227-234.

Commission on Engineering Geological Mapping of the International Association of Engineering Geology, 1976, Enineering Geological Maps, A Guide to Their Preparation: The UNESCO Press, Paris, France, 79 p.

Compton, R. R., 1962, Manual of Field Geology, John Wiley and Sons, New York, NY, 378 p.

Cosgrave, T. M., 1987 An investigation of shallow stratigraphic reflections from ground penetrating radar. Unpublished M.Sc. thesis. Department of Earth Science, University of Waterloo.

Darracott B. W. & D. M. McCann, 1986, Planning Engineering Geophysical Surveys, in Site Investigation Practice: Assesssing BS 5930, Geological Society Engineering Geology Special

Publications, No. 2, pp. 85-90.

Davis and DeWiest, 1966, Hydrogeology,Wiley, New York.

Dearman, 1974, Presentation of Information on Engineering Geological Maps and Plans: Quarterly Journal of Engineering Geology, Vol. 7, No. 3, pp 317-320.

Dobrin, M. B. & C. H. Savit, 1988, Introduction to Geophysical Prospecting: McGraw-Hill Book Co., New York, N. Y. 867p.

Erdélyi M.,& J. Gálfi, 1988, Surface and Subsurface Mapping in Hydrogeology, Wiley-Interscience Publications, 383 p.

Lahee , 1952. Field Geology

Gagne, R. M., S. E. Pullan, J. A. Hunter, 1985, A Shallow Seismic Reflection Method for use in Mapping Overburden Stratigraphy: Proc. 2nd NWWA Conference on Surface and Borehole Geophysical Methods in Ground Water Investigations, 132-145.

Garland , G. D., 1965, The Earth's Shape and Gravity, Pergamon Press, New York.

Greenhouse, J. P., and D. D. Saine, 1983, The Use of Reconnaissance Electromagnetic Methods to Map Contaminant Migration. Ground Water Monitoring Review, Vol. 3(2), pp. 47-59.

Greenhouse, J. P., and D. D. Saine, 1986, Geophysical Modeling and Mapping of Contaminated Ground Water Around Three Waste Disposal Sites in Southern Ontario, Canadian Geotechnical Journal, Vol. 23(3), pp. 372-384.

Griffiths D. H.,& R.F.King, 1981, Applied Geophysics for Geologists and Engineers: Elements of Geophysical Prospecting, Pergamon Press, New York, N.Y., 230p.

Haeni, F. P. 1985, Applications of Continuous Seismic Reflection Methods in Hydrologic Studies, Ground Water 24, 23-31.

Hammer, S. 1939. Terrain corrections for gravimeter stations. Geophysics, V. 4, p 184-194.

Hatheway, A, and B. Leighton, 1979, Exploratory Trenching, In: Hathaway, A. W. and C. R. McClure Jr. (editors),.Geology in the Siting of Nuclear Power Plants, Reviews in Engineering Geology, Vol. 4, The Geologic Society of America, Bolder, CO.

Hunter and Hobson, 1977, Reflections on Shallow Seismic Reflection Records; Geoexploration, V. 15, p. 183-193.

Hunter, J. A., R. A. Burnes, R. L. Good, H. A. MacAulay, and R. M. Gagne, 1982a, Optimum Field Techniques for Bedrock Reflection Mapping with the Multichannel Engineering Seismograph: in Current Research. Part B, Geol. Surv. of Canada, Paper 82-1B, 131-138.

Hunter, J. A., R. A. Burnes, R. M. Gagne, R. L. Good, and H. A. MacAulay, 1982b, Mating the Digital Engineering Seismograph with the Small Computer-some Useful Techniques: in Current Research. Part B, Geol. Surv. of Canada, Paper 82-1B, 125-129.

Hunter, J. A., S. E. Pullan, R. A. Burnes, R. M. Gagne, and R. L. Good, 1984, Shallow Seismic Reflection Mapping of the Overburden-bedrock Interface with the Engineering Seismograph-some Simple Techniques, Geophysics, 49, 1381-1385.

Lennox, D. H., and V. Carlson, 1967, Geophysical Exploration for Buried Valleys in an Area North of Two Hills, Alberta, Geophysics, Vol. 32, p. 331-362.

Mota, L 1954, Determination of dips and depths of geological layers by the seismic refraction method, Geophysics, v. 19, p 242-254.

Olhoeft, G.R., 1986. Direct detection of hydrocarbon and organic chemicals with ground penetrating radar and complex resistivity: in Proc. of the NWWA~API Conf. on Petroleum Hydrocarbons and Organic Chemicals in Ground Water, 1986, Houston, p. 284-305.

Palmer, D., 1980, The Generalized Reciprocal Method of Seismic Refraction Interpretation, K. B. S. Burke, Ed. Dept of Geology, University of New Brunswick, Fredericton, N. B., Canada, 104 pp.

Robinson, E. S., & C. Coruh, 1988, Basic Exploration Geophysics, John Wiley and Sons Inc., New York, N.Y., 562p.

Saunders, Wayne R. and Cox, S.A., 1987. Use of an electromagnetic induction technique in subsurface hydrocarbon investigations. in Proceedings of the First National Outdoor Action Conference, The National Water Well Assoiciation, Las Vegas, Nevada, May 18-21.

Scaife, J. 1990, Personal Communication

Slaine 1989, Applying Seismic Reflection Techniques to Hydrogeological Investigations,

Slaine, D. D., and , D. G. Leask, 1988, Gravity—Chapter in Guide to the Use of Geophysics in Engineering Geology, International Association of Engineering Geologists Commission on Site Investigations, Delft, Netherlands.

Slaine, D. D., P. E. Pehme, J. A. Hunter, S. E. Pullan, and J. P. Greenhouse, 1990, Mapping Overburden Stratigraphy at a Proposed Hazardous Waste Facility Using Shallow Seismic Reflection Methods, Society of Exploration Geophysicists Three Volume Special Publication on Environmental Geophysics, Salt Lake City, Utah, Volume II: Environmental and Groundwater, pp.273-280.

Soil Conservation Service, 1972, List of Published Soil Surveys, Washington, D. C.

Soil Conservation Service. 1972. List of published soil

surveys. U.S. Department of Agriculture, Washington, D.C.

Soil Conservation Service. 1972. List of published soil surveys. U.S. Department of Agriculture, Washington, D.C.

Taylor, C. L., and L. S. Cluff, 1973, Fault Activity and its Significance Assessed by Exploration, Conference on Tectonic Problems on the San Andreas Fault System Proceedings, Geologic Sciences, v. 13, School of Earth Sciences, Stanford Univ., Stanford, CA.

Telford, W. M., L. P. Geldart, R. E. Sheriff, and D. Keys, 1976, Applied Geophysics, Cambridge University Press, New York, N.Y., 860p.

Thorp, J., and G. D. Smith, 1949, Higher Categories of Soil Classification, Order, Suborder, and Great Soil Groups, Soil Science 67, pp. 117-126.

Thorp, J.; and G.D. Smith. 1949. Higher categories of soil classification, order, suborder, and great soil groups. Soil Science, 67, pp. 117–126.

U.S. Department of Agriculture. 1938. Soils and man. Yearbook of Agriculture, U.S. Department of Agriculture, Washington, D.C.

USDA, 1938, Marbut system, Soils and Man, Yearbook of Agriculture, Washington, D. C.

Vacquier, V.,C., R. Holmes, P. R. Kintzinger, and M.Lavergne, 1957, Prospecting for Ground-water by Induction Electrical Polarization, Geophysics 12, No. 3, 660p.

Williamson, D. A., 1984, Unified Rock Classification System, Bulletin of the Association of Engineering Geologists, Vol. XXI, No. 3 pp. 354-354.

CHAPTER 4

SUBSURFACE INVESTIGATIONS

4.1 TASK CORRELATION WITH GEOLOGY

Determinations of the location, depth, and total number of borings, piezometers and wells are typically important decisions to make within any site assessment project. Beyond pure technical needs for subsurface information there is often regulatory mandated guidance for coverage or extent of subsurface data. Some states require minimal borings or layouts on a grid pattern, using specific criteria for spacing and depth; other states impose minimum numbers based on acreage. State and local facility permit requirements should be reviewed before determining coverage of soil/rock borings to be installed during the site investigation.

Regardless of the state requirements, the number and location of borings should be sufficient to adequately characterize the geologic and ground-water conditions beneath the site, with special reference to the types of material, uniformity, potential leachate pathways, hydraulic conductivity, porosity, and depth to ground water.

The extent of the investigation for site assessments is determined by the character and variability of the subsurface and ground water, the type or importance of the project, and the amount of existing information. It is important that the general character and variability of the site area be established before deciding the basic principles of the design of the waste disposal facility. Investigations may include a range of 'methods', e.g. excavations, boreholes, and insitu testing. The factors determining the selections of a particular method are discussed below. The recommendations apply irrespective of the method adopted, and the term 'exploration point or borehole'; is used to describe a position where the subsurface is explored by any particular 'method.' Each combination of project and site is likely to be unique, and the following general points are guidance in planning the subsurface investigation and not as a set of rules to be applied rigidly in every case. Flexibility in approaching site investigation work can often provide for a successful result. The technical development of the project should be under continuous review since decisions on the design often influence the extent of the investigation.

Subsurface investigation work must always be supplementary to, and conditioned by previous studies of the local geological structure. The greater the natural variability of the ground or subsurface, the greater will be the extent of the investigation required to obtain an

EXAMPLES OF POTENTALLY INCORRECT GEOLOGIC INTERPRETATIONS OF BOREHOLE DATA

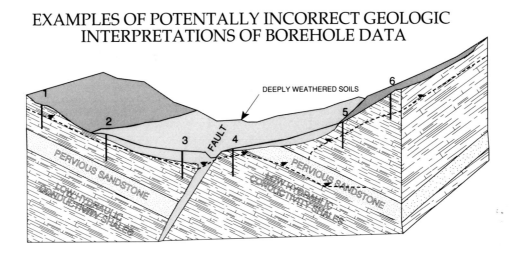

Figure 4-1 Borehole Geologic Interpretations A

indication of the character of the site. The depth of exploration is determined by the nature of the facility projected, but may be necessary to explore to greater depths at a limited number of points (called in this text "stratigraphic" boring) to establish the overall geological structure. Brief considerations of the possibilities of complex structures existing beneath relatively simple-looking ground surfaces will confirm the validity of the following recommendations. It is an especially important consideration in areas previously subjected to glacial action, where glacial drift now covers an original rock surface that might be totally unrelated to present-day topography. Even without the existence of glacial conditions, however, disastrous results have been known to occur when reliance was placed on the results of test boreholes that had not been correlated with the local and regional geology (Legget, 1962). Consider as an example the conditions shown in Figure 4-1. Cursory surface examination of exposed rock in the immediate vicinity of the proposed facility would show outcrops of shale only. Stratigraphic borings drilled, as shown to confirm what these rock outcrops appear to suggest, would give an entirely false picture of the actual subsurface conditions across the valley. A geological survey made of this site should include, at least, a general surface reconnaissance of the area geologic conditions. The fault and more permeable sandstone are the important target features in

this investigation, since they have significant influence over area ground-water flow. Observation of the outcrops along the river bed or through evaluation of changes in local topography, may have detected the fault. Even if the fault was not observed directly, a detailed examination of shale outcrops on the two sides of the valley would potentially show differences between the two deposits, through variations in color or bedding thickness, or perhaps fossil contents. The geologic mapping should be sufficient to show that they were not the same formation and thus to demonstrate some change in structure between the two sides of the valley. Regional data may point toward the potential for a major fault, but the absence of previously documented structural features should not discount adequate surface mapping of the geologic units. This mapping can direct the location and depths required for Phase II drilling programs. Figure 4-2 illustrates a case where the strata involved dip steeply across the site. Correlating the strata found in one drill hole (3) with those pierced by the adjacent hole (2) shows only shale overlying limestone. Often shale and limestone look quite similar in core, hence, it would be possible to confuse Shale A/Limestone A with Shale B/Limestone B. If borehole and surface mapping are not correlated and hole 3, for example, drilled only as far as the point shown, the existence of the fault and highly permeable sandstone would have been undetected, unless it was discovered

EXAMPLES OF POTENTALLY INCORRECT GEOLOGIC INTERPRETATIONS OF BOREHOLE DATA

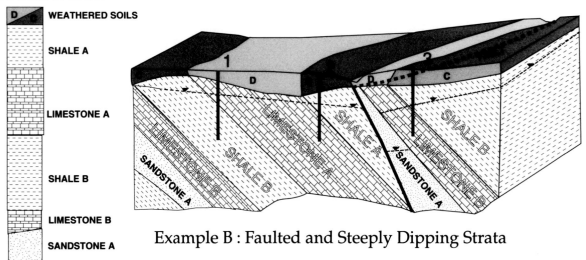

Example B : Faulted and Steeply Dipping Strata

Figure 4-2 Borehole Geologic Interpretations B

EXAMPLES OF POTENTALLY INCORRECT GEOLOGIC INTERPRETATIONS OF BOREHOLE DATA

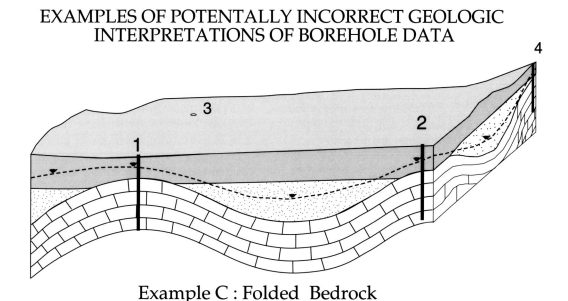

Example C : Folded Bedrock

Figure 4-3 Borehole Geologic Interpretations C

through surface geological investigations. In the case of surficial residual sails, (unit C and D), geologic mapping would be difficult without some prior knowledge of local faulting. As shown in Figure 4-2 relatively minor differences in surface exposure can develop into major structural complexity in the subsurface. Experience and knowledge of the region provide the best keys to being able to evaluate these types of dipping geologic units. Figure 4-3 shows a case in which casual surface examinations and the use of the subsurface information provided by the boreholes shown would be misleading because of the existence of folds in the strata, are covered by glacial sediments. The permeable sand channel deposits would go undetected from the pattern of borings shown on Figure 4-3.

The examples illustrated in the three cross sections are not given to show that stratigraphic borings and similar investigations are often faulty. Such a conclusion would, in fact, be incorrect, because in the cases cited the results from the boreholes may have been correct, but their interpretation might not have been, if the results were not correlated with the local geology. The structural complexities shown by the three examples have extreme implications to ground-water issues for both detection and assessment monitoring programs. From study of the cases described and many other similar instances, the following general guides from Legget (1962) were prepared for subsurface exploratory work. (For convenience, the word boreholes is used to describe all such subsurface work.)

1. No boreholes should be put down before at least a general geological survey of the area has been made.

2. Boreholes should always be located in relation to the local geological structure.

3. Boreholes should, whenever necessary, be carried to sufficient depths that they will definitely correlate the strata observed in adjacent holes by an "overlap" into at least one hydrostratigraphic unit.

4. In exploring superficial and glacial deposits, at least one borehole should whenever possible, be carried to bedrock.

5. In all cases of superficial deposits extending to no great depth, test borings should be taken into the rock for some specified distance, never less than 2 m (6 ft), but more than this if the nature of the work warrants the extra cost, or if large boulders are liable to be contacted.

6. Unusual care must be exercised in putting down test borings in areas known to have been subjected to glacial action, especially with regard to checking all rock found during drilling (as in 5) and for the possible existence of buried river valleys.

7. The three-dimensional nature of the work must always be remembered; for example, three boreholes properly located will define exactly the thickness, dip, and strike of any continuous buried stratum having a uniform dip.

8. Planning and contractual arrangements for the conduct of all test boring and drilling must be kept flexible, so that changes can be made immediately at the site, as the picture of the underground conditions gradually unfolds.

9. Special attention must be paid to ground water observed in test holes, careful observations being made of levels before work starts each day in holes that are in the process of being put down.

10. At least one test hole on every site should be cased and fitted with the necessary screen and filter arrangements, so that it may serve as an observation well for ground-water for the longest possible duration.

Legget (1962) made these recommendations some 30 years ago and they are as true today as the day he organized these guidelines. These 10 points should form the basis of both stratigraphic and geotechnical borings used in site investigations. Methods adopted for penetrating unconsolidated materials and solid rocks are naturally different, even though employed in the same hole. Detailed information concerning subsurface conditions can best be obtained through an adequate drilling program where, at least, one continuously logs the first few holes. The adequacy of a drilling program can become subjective in any site assessment. One should, however, use Legget's guidance to evaluate depth of drilling. Borehole location guidance is addressed in the following paragraphs. The objective of this subsurface program is to characterize the soil and bedrock beneath the site to depths sufficient to establish geology and ground-water conditions, and to plan the effective location, design, and installation of piezometers and monitoring wells. Furthermore, It is important to realize that you can only infer the detailed geology of a site from aerial photography, surface outcrops and subsurface information at the positions of the exploration points. The above examples illustrate possibilities that significant undetected variations or discontinuities in the subsurface can exist, including lateral or vertical variations within a given stratum. You can reduce uncertainties in any site investigation, but, except by complete excavation, can never be wholly eliminated by a more intensive investigation. This is why one should obtain a composite understanding of the variability of site subsurface conditions through comparative assessment techniques as illustrated in Figure 4-4

The overall goal is to understand the site conditions sufficiently to meet the project requirements, but not to overkill the technical components of the investigation. One may require X number of explorations to fully evaluate site conditions. 2X or 3X exploration may not necessarily provide significantly more information and may also affect the project negatively. This is especially true for holes drilled through confining units where later borehole completions and decommissioning work may leave the vertical containment of the unit in question.

The investigator will need to present a thorough characterization of surficial and subsurface geology at a site for both engineering design and ground-water monitoring. In order to describe in detail the geology beneath and adjacent to the site and, therefore, be able to identify potential pathways for migration of leachate, the investigator, (using qualified professionals), must collect and analyze individual strata beneath and adjacent to a site. To assess the geologic properties of strata beneath a site that are likely to influence the migration of ground water, the investigator will require results of a multi-faceted subsurface stratigraphic investigation including, as necessary:

- Exploratory borings;
- Backhoe pit excavations;
- Rock coring;
- Static probing devices;
- Borehole visual and geophysical logging;
- Sample collection;
- Geophysical surveys; and,
- Laboratory analyses.

In addition to the types of information discussed in Section 4.1, these investigations will also provide information on site features such as soil and aquifer characteristics.

4.2 BORING PLAN

The geologist or geotechnical engineer in charge of an investigation should prepare a boring plan to evaluate subsurface conditions. The plan should include proposed locations and anticipated depths of boreholes and the type and frequency of required soil and rock sampling. Determination of when to terminate boreholes and/or when additional boreholes are needed is dependent upon additional data that may require consideration of site-specific hydrogeologic factors. For example, a shallow (<50 ft) bedrock surface of low hydraulic conductivity may indicate a reasonable depth limit for investigation of an area. However, if the bedrock is highly fractured, penetration of the unit should be conducted so that sufficient understanding is developed on the potential for the unit to discharge ground water.

Two types of borings are typically employed in site assessment work. Stratigraphic borings are holes drilled

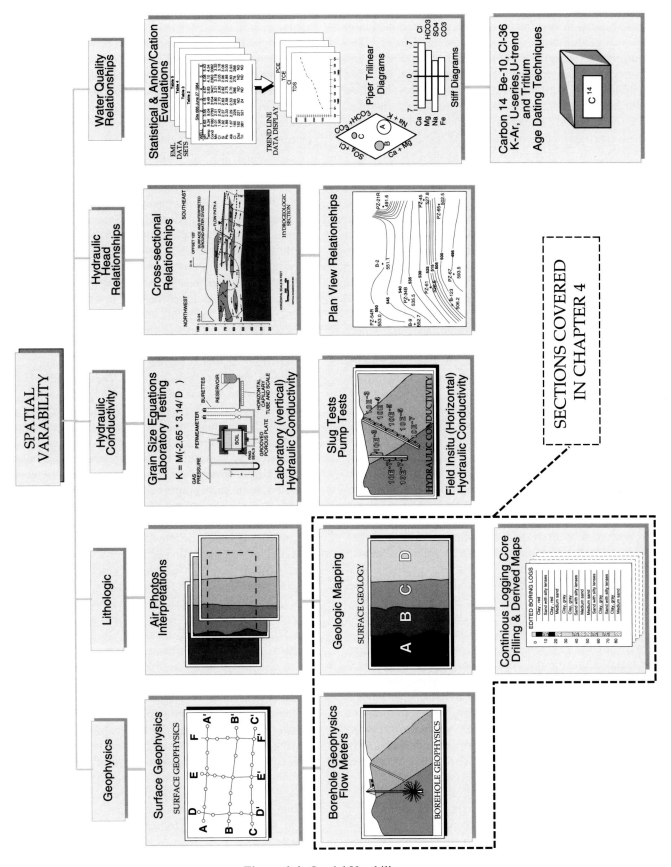

Figure 4-4   Spatial Varability

primarily to define the site stratigraphy down into bedrock. These borings drilled first to develop an overall picture of the site geologic and ground-water conditions. The stratigraphic borings should be continuously logged and soil/rock samples collected in order to develop the basic hydraulic and aquifer characteristics of the subsurface materials. Once completing the deep stratigraphic borings, as a rule of thumb, shallower geotechnical borings should be advanced down to approximately 25 feet below the base grades of the proposed or actual in-place facility, unless specific geologic conditions indicate that deeper borings are required. Geotechnical borings are used to define specific site condition necessary to design, operate or remediate the disposal facility. The points of exploration, (e.g., boreholes, insitu testing, backhoe pits), should be so located that a general geological view of the whole site can be obtained with adequate details of the engineering properties of the soils and rocks and of ground-water conditions. More detailed information should be obtained at positions of important facility structures, at points of special engineering difficulty or importance and where ground conditions are complicated, (e.g. suspected buried valleys and land slide areas). Rigid, preconceived patterns of backhoe pits, boreholes or nsitu testing should be avoided. In some cases, it will not be possible to locate structure until much of the ground investigation data has been obtained. In such cases, the program of investigations should be modified accordingly.

Planning of field (and office) activities have become extremely popular with the arrival of Superfund regulations and regulatory guidance that require Federal or State agency approval of plans before beginning of a major project. These mandatory planning documents have become massive efforts and, at times, too restrictive when field conditions turn out to be significantly different to those estimated a number of months previous to the field work. It is the author's opinion that these documents have become ends to themselves rather than leading to good technical field execution of investigative techniques.The borehole plan for most site assessment projects should be a maximum of 2 to 5 text pages in length, and should serve to briefly describe the depths, locations, and technical reasons for each boring. Table 4-1 contains the partial tabular information of a borehole plan that recommended a series of boreholes, piezometers and replacement monitoring wells. Where special sample collection activities are necessary for the project one should only reference the source of the technique and a brief description why the technique is necessary for the project.

As general guidance the boring plot plan should also allocate a minimum of one boring for each major geomorphic feature. The pattern (layout; array) of borings should facilitate development of detailed cross-sections

through the site area in order to sufficiently define the geology. It is recommended that the geotechnical borings be performed in a phased approach to permit modifications to the boring program, and that the number of borings in the active area of land filling be minimized. The important concept is to have an organized approach to the target depths to be drilled based on prior site information. Guidance on the location and depths of sampling, and piezometer location is provided in Figure 4-5.

Collect samples continuously down to 25 feet below the lowest elevation (base grades) of the facility and collected at either 5 foot intervals thereafter, or at depths at which there are observed changes in lithology. At times, complex subsurface conditions, or State regulations may require continuous sampling for all stratigraphic borings. Drilling deeper than 25 feet below base grade may be required in some areas depending upon the specific, hydrogeologic conditions beneath the site.

Continuously sampled borings in borrow areas (for liner construction material) are the basis for developing accurate estimates of usable volumes and characteristic of materials.

Drilling depths for deep stratigraphic borings are determined on a case-by-case basis. The depth should be, at a minimum, equal to or greater than the highly fractured/weathered bedrock/soil interface, as shown on Figure 4-5. In addition, at least the boring program should define the uppermost aquifer. Base the location of piezometers on the results of the deep stratigraphic boring program and, where appropriate, place piezometers in selected deep stratigraphic and shallow geotechnical borings.

During the Phase II investigations, drill exploratory borings in the landfill's foundation and potential borrow areas of a new or expanding facility, or around the margins of an existing facility. Only in rare cases, should borings be placed through refuse for piezometer installation, and never for monitoring wells. Base selection of drilling locations using geologic and geophysical interpretations (see Sections 3.4 and 3.5) in addition to results of Phase I studies. Plan borings necessary for expansions or greenfield waste disposal sites in accordance with the guidance provided in Table 4-2. If the borings recommended in Table 4-2 are insufficient to describe the geologic formations and ground-water flow patterns below the site, additional borings may be necessary.

For exploration efforts conducted in areas that have been contaminated by adjacent or previous land uses, the plan should also include methods for chemical analysis, storage, and/or disposal of the drilling materials, including drilling mud, fluids, cuttings, and water, as necessary.

In uniform geologic formations, you may reduce the number of borings if other techniques such as geophysics

GUIDANCE ON SEQUENCE OF DRILLING, SAMPLING & PIEZOMETER INSTALLATION

DRILLING & SAMPLING

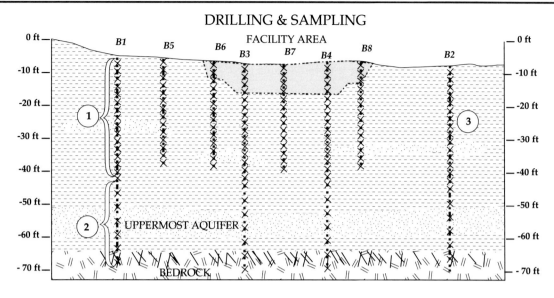

1. *Continuous sampling should be completed down to 25 feet below facility base grades*

2. *Sampling should be continued every five feet or at changes of lithology down to bedrock. Sampling/logging should continue down well into uppermost aquifer if regional aquifer system is deep or very thick.*

3. *Deeper stratigraphic borings should be completed first, then shallower geotechnical borings next.*

4. *Install small diameter piezometers in a selection of geotechnical and stratigraphic holes. Each major Hydrogeologic Unit should have at least 3 piezometers completed in the unit in order to construct potentiometric surfaces & flow nets.*

5. *Develop piezometers and test for horizontal hydraulic conductivity for each hydrogeologic unit.*

PIEZOMETER INSTALLATION

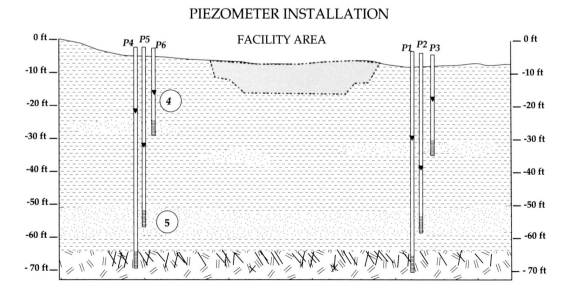

Figure 4-5 Borehohe Guidance for Drilling and Sampling

Table 4-1 Example Borehole Plan With Justification of Drilling and Piezometer Installation

Justificat.on: The lithology of the screened: interval at well M-3 is unknown, and the edge of the bedrock valley is nearby. This borehole will be conducted to determine the lithology of the screened interval in existing well M-3, to determine the depth to bedrock, and to determine the stratigraphy of the overlying sediments. If existing well M-3 is deemed to be completed in bedrock, then a new piezometer will not be installed.

G-7   1 borehole plus   1 piezo

Justification: The available data indicates that the bedrock is within 50 feet of the surface and that well M-4 is probably completed in bedrock (the original boring was not logged). The deep borehole will be conducted to determine the lithology at the screened interval, and the boring will then be plugged back. A shallower bedrock piezometer and the piezometer in the overlying sediments will be installed to measure vertical gradients within the bedrock and between units.

G-10:-- Justification: To determine depth to bedrock and lithology of overlying sediments. The bedrock piezometer will provide a third point (in addition to G-6 and G-7) for monitoring water levels in order to determine horizontal flaw direction and gradient within the bedrock. Vertical gradients between the overlying sediments and the bedrock will also be measured.

G-9:-- Justification: Additional measuring points are needed in the "Dense Sand" in order to determine the horizontal rate and direction of flaw within the sand. The "Dense Sand" piezometer will provide one such point. The shallow piezometer will provide a point for determination of vertical gradient at the G-9 location and will provide information on the presence of c,ground water in the shallow sediments.

<u>**TABLE XX**</u>

LOCATION ID	APPROXIMATE LOCATION	PIEZOMETER INSTALLED	DEPTH	GEOLOGIC UNIT OF BOREHOLE / PIEZOMETER
G-9	SOUTHEAST +39,100N -32,850E	YES	130'	DENSE SAND HSA and continuous sample to top of Dense Sand at estimated elevation 510 MSL. Enlarge boring and install surface casing. Continue borehole with cable tool Ql' rotary sampling to depth equivalent to M-l well screen. Install Piezometer in Dense Sand . (Assume surface elevation of 560')
		YES	50'	SEDIMENTS HSA through sediments overlying Dense Sand . Install Piezometer in overlying sediments.
G-10	SITE ENTRANCE +37,950N -32,800E	YES	90'	DENSE SAND Advance borehole with cable tool or rotary sampling to depth equivalent to M-l well screen. Install piezometer in Dense Sand. (Assume ground elevation of 520', near top of sand).
G-11	NEAR M-2 +37,700N -31,200E	YES	115'	DENSE SAND HSA and continuous sample to top of 'Dense Sand at estimated elevation 510' MSL. Enlarge boring and install surface casing. Continue borehole with cable tool cr rotary sampling to depth equivalent to M-l well screen. Install piezometer in Dense Sand. (Assume surface elevation of 545').
G-12	SITE CENTER +39,300N -32,000E	YES	185'?	DENSE SAND HSA and continuous sample to top of 'Dense Sand at estimated elevation SlD' MSL. Enlarge boring and install surface casing. Continue borehole with cable tool or rotary sampling, Possibly to depth equivalent to M-l well screen. Install Piezometer in Dense Sand . (Assume surface elevation of 615'). NOTE: 185' depth may be excessive.

Table 4-2 Recommened Rational for Minimal Allocation of Borings

Acreage	Number of Geotechnical Borings	Number of Deep Stratigraphic borings Recommended
Less than 10	3	1
10–49	6	2
50–99	10	4
100–200	15	5
More than 200	18 +1 boring/each additional 10 acres	6 + 1 boring/each additional 40 acres

is employed to correlate geologic units between boreholes. One or more stratigraphic test holes should be drilled prior to the selection of locations for near–surface (geotechnical borings generally to ≤50 feet) borings and piezometers for collection of geotechnical data and water levels. These holes, in any case, must be sufficiently deep to penetrate all pertinent materials. Boreholes are a cost–effective

Figure 4-6 Fractured Clays may only be Observable Through Back-hoe Pits and Excavations

means of providing physical documentation of subsurface conditions so that locations for monitoring wells and piezometers can be selected. If these "calibration" stratigraphic borings are properly documented and located, they may later be converted to piezometer installations. Care must also be taken to assure that the diameter of the borehole is of sufficient size to accommodate the instruments necessary to conduct borehole geophysical logging (usually ≥4 inches) if so required by the scope of the project. The information gathered from the stratigraphic and geotechnical borehole drilling programs also provides details about the engineering characteristics of the subsurface materials and observed ground-water conditions. From the analysis and evaluation of the information gathered from deeper stratigraphic boreholes, the geologist and the engineer confirm the geotechnical unit to be sampled and what laboratory analyses are needed during the shallow geotechnical drilling program. This determines the kind, number, and size of samples needed in the program. The data obtained from the stratigraphic boring program may cause a modification in the proposed drilling and sampling locations outlined in the Boring Plan. Care must be taken to keep the goals of the project in mind when changing borehole locations or sampling points. The reasons for changing borehole locations should be documented and discussed with the client before conducting the altered program.

Because weathered and fractured bedrock may represent a significant hydrologic component of ground–water flow, borings in areas of near–surface bedrock should be advanced to at least a depth representing reasonably unweathered bedrock, as determined by drilling penetration rates, RQD, or percent recovery (see Chapter 6.0). These boreholes should be cored with at least an NX–sized core barrel. Unconsolidated materials should be sampled to obtain an unambiguous picture of the lithology of the overburden. Detailed geologic logs of test holes must be obtained and retained (see section on Logging Test

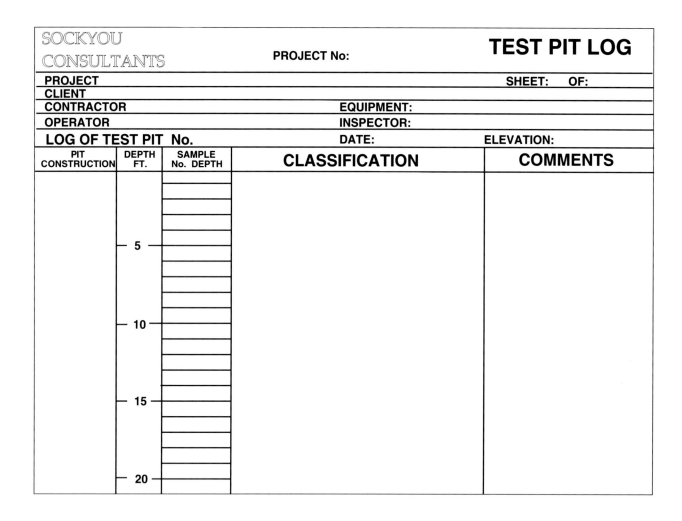

Figure 4-7 Trial Pit Log

Holes). It is essential that the location and ground levels for all exploration points be established, if necessary by survey.

4.3 SUBSURFACE METHODS

4.3.1 Backhoe Trial Pits

Shallow trial pits are usually dug using backhoe or bulldozer excavation methods. They are used in ground that will stand unsupported during excavation, and commonly extend to a depth of about 6 to 17 ft (2 to 5 m)(see Figure 4-6 and 4-7). It is essential for any such work that the pit sides are guarded against sudden collapse before allowing entrance of personnel into the pit. Many states have safety standards and shoring requirements for excavations that begin at depths of just over 1 meter. The spacing of shoring used to secure the excavation should be sufficiently wide to allow inspection of the pit sides, but

completely protective of the staff performing the inspection pit work.

Care must also be taken that gas or noxious vapors have not entered the trench or pit. Self contained breathing units should be available if needed for staff working in confined or deep pits where gases or insufficient oxygen may be present.

Shallow trial pits or trenches are excellent for rapid site investigations work since they permit the insitu condition of the ground to be examined in detail both laterally and vertically, and allow mass properties of the soil and rock to be assessed. They provide access for taking good quality (block) samples and allow for carrying out insitu field tests. Trial pits are particularly useful for investigating and sampling soil derived from insitu rock weathering and colluvium, both of which often exhibit a high degree of variability.

The field record of a shallow trial pit should include a plan giving the location and orientation of the pit, and a

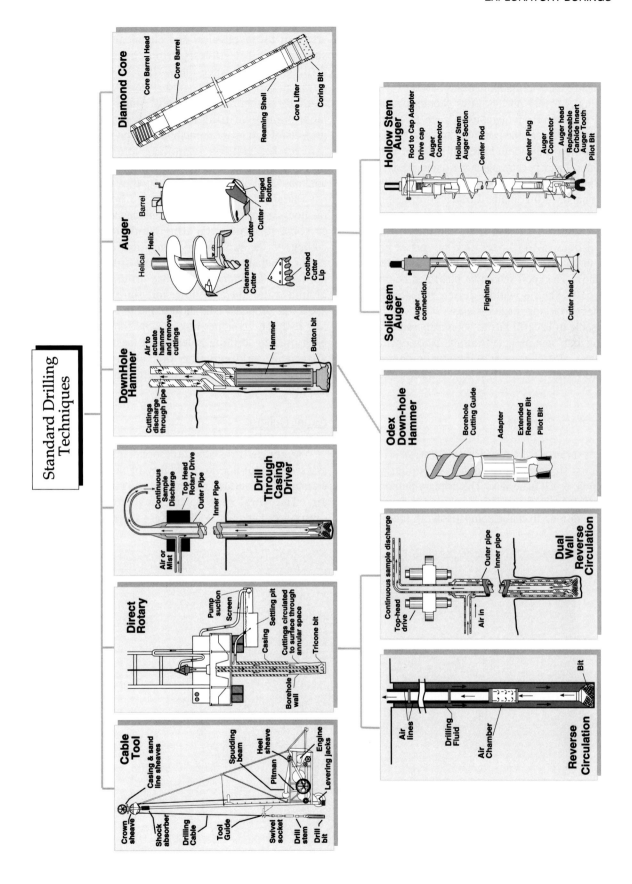

Figure 4-8 Drilling Techniques

dimensioned section showing the sides and floor. Fully describe the ground condition in accordance with ASTM Standards and when necessary, fully document all samples taken should using lithostratigraphic descriptions. Fig. 4-7 shows an example of trial pit log. Logs should always be supplemented with color photographs of each face and of the base of the pit. Record the positions and results of any field testing, which may include insitu density tests or Schmidt hammer, hand penetrometer and hand shear vane index tests. The subsection on Holocene faulting studies in the previous chapter 3.0 contains further information on shallow trenching techniques.

Stockpile material excavated from trial pits in such a manner that it does not fall back into the pit or cause instability of the pit excavation, (e.g., by surcharging the adjacent ground). The spoil should be placed and covered so as not to be washed downhill during rainstorms or allowed to enter surface drainage systems.

It is advisable to backfill pits as soon as possible after completing logging, sampling and testing, since open pits can be a hazard. Pits left open temporarily should be covered securely and fenced off, if readily accessible by the public.

4.3.2 Exploratory Borings

Site assessments that require evaluation of the subsurface must at some point penetrate underlying geologic soil rock or fill materials. The primary purpose of this action is to obtain information relative subsurface materials or conditions to make comparisons to other site areas, or to the physical and chemical properties of the subsurface. Site investigations have used numerous drilling methods since the early 1900's for recovering physical samples of the subsurface. A number of these methods shown on pictograph Figure 4-8 represent the most popular drilling techniques for site assessment projects. Each of these methods has positive and negative aspects for drilling, sampling and installations of piezometers and monitoring wells. U.S. EPA (1989) contains detailed descriptions of various drilling methods for installation of monitoring wells. These drilling methods can be divided into techniques that require the use of some form of circulating fluids or air to transmit cuttings to the surface and augers that do not require the use of circulating fluids. Cable tool drilling methods may not require additional fluid once drilling has reached the ground-water surface, however, fluid may be required above the ground-water surface and in geologic units with low hydraulic conductivity. Cable tool techniques have been used for drilling holes for at least several thousand years and, although time consuming, provide many positive aspects relative to obtaining representative samples.

Auger Drilling

Auger drilling has developed into a popular site assessment tool due to the ability of the technique to lift cuttings to the surface without the need of drilling fluids or air. The spiral shape of the auger tools conveys the subsurface materials to the surface for examination of the

Figure 4-9 Rotary Drilling Rig

Figure 4-10 Solid Stem Auger Drilling

cuttings. Four main types of augers are in common use for site assessment projects. Bucket and helical augers can provide large diameter (typically greater than 16 in) access to the subsurface for examination (using casing to keep the hole open), for drilling in landfills for gas collection system installations, special large volume sampling above the ground-water surface and in semi-consolidated cohesive materials. Hollow stem augers have generally replaced solid stem or continuous flight augers in site assessment work. Hollow stem augers allow sample collection through the tubular center stem, and access to the surface for construction of piezometers and monitoring

Figure 4-11 Hollow Stem Drilling is the Most Common Method used in Environmental Projects

wells. The most commonly used sizes of hollow stem augers (inside diameters (id)) are 3.25 inches and 4.25 inches for installation of two inches and smaller diameter piezometers and monitoring wells. Larger diameter installations up to 4 inches can be installed through 6.625 inch (id) hollow stem augers. Variations on hollow stem augers design even allow continuous cores cut while advancing the augers into the subsurface. Hollow stem augers would be used more extensively in site assessment work if it were not for material and depth limitations of the drilling system. As the drilling depth increases or the diameter of the auger increases; friction on the drill rods limits the ability of the tool to proceed further. Exact depths obtainable depends on equipment, auger size, geologic and ground-water conditions. When augering, the hollow stem is closed at its lower end by a plug, which may be removed so that the sampler can be lowered down through the stem and driven into the soil below the auger bit. The use of hollow-stemmed augers in cohesionless (sandy) soils often presents practical problems because it is difficult to prevent material from flowing into the hollow stem on removal of the plug. When contacting rock, the borehole can be extended by core-drilling through the hollow stem. Continuous-flight augering requires considerable mechanical power and weight so that the machine is generally mounted on a heavy vehicle (see Figure 4-9 to Figure 4-11). The drilling debris brought to the surface by auger flights gives only a very rough indication of the levels and character of the strata. Although drive samples may be taken through the hollow stem of the auger to identify geologic units, continuous sampling equipment are available from most drill rig and auger fabricators that allows continuous cores be cut during augering. When field conditions limit use of hollow stem equipment, drilling methods using circulating fluids or air must be used to obtain the desired results.

Rotary Open Hole Drilling and Rotary Core Drilling

Rotary drilling, in which the drill bit or casing shoe is rotated on the bottom of the borehole, has been a common method of subsurface exploration over the years. The drilling fluid, pumped down to the bit through hollow drill rods, lubricates the bit and flushes the drill cuttings up the borehole. The drilling fluid is commonly water or air, but drilling mud or air foam are often used in site assessments, when drilling conditions require additional support to maintain an open hole and more effective removal of cuttings. There are two basic types of rotary drilling:

• open hole (or full hole) drilling, in which the drill bit cuts all the material within the diameter of the borehole; and

Figure 4- 12 Rotary Drilling With Plain Water

• core drilling, in which an annular bit fixed to the outer rotating tube of a core-barrel cuts a core that is returned within the inner stationary tube of the core-barrel and brought to the surface for examination and testing.

Drill casing is normally used to support unstable ground or to seal off open fissures that cause a loss of drilling fluid. Alternatively, drilling mud or cement grout can be used to seal open fissures. Site assessment investigations are often performed in consolidated materials; in these conditions rotary core drilling has the important advantage over rotary open drilling of providing a core sample while the hole is being advanced. Core drilling is recommended for most assessment projects where secondary permeability features are important to the investigation. In rotary open hole drilling, drill cuttings brought to the surface in the flushing medium can only provide an indication of the lithology of the geologic units being contacted. However, rotary open hole drilling is useful for rapid advancement of a borehole required for field testing or instrument installation, and cored or driven samples may be obtained between drill runs even when the open hole technique is used.

Drilling is in part an art, and its success is dependent upon good practice and the skill of the operator, particularly when coring partially cemented, fractured, weathered and weak rocks, where considerable expertise is necessary to obtain full recovery of core of satisfactory quality. This is greatly influenced by the choice of core barrel and cutting bit type and configuration, and by the method of extruding, handling and preservation of the core.

Flushing Medium

Site investigations for solid or hazardous waste disposal facilities often require careful selection of drill bit flushing medium compatible with the desired project goals, equipment employed, and suitable for the ground to be drilled. Filtered air is the simplest flushing medium and perhaps the best when considering all environmental factors. Filtered air should be used, whenever possible, to advance core or rotary tricone drilling. Unfortunately even high pressure air may not have adequate uphole fluid velocity to remove drill cuttings sufficient to advance the borehole. In these cases one must gradually use more effective fluids to remove cuttings from the borehole. Other fluids used as flushing media are clean water, (Figure 4-12), drilling muds (which consist of water with

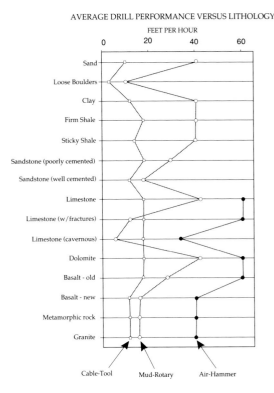

AVERAGE DRILL PERFORMANCE VERSUS LITHOLOGY

Figure 4-14   Drilling Performance Comparison

Source: Robert V. Colangelo

bentonite), water with an additive such as air foams (Figure 4-13) and polymer mixtures. The main advantage of these other flushing media is that drill cuttings may be removed at a lower flushing velocity and with less disturbance to the ground. The use of drilling mud can also minimize the need for temporary casing of the hole, as it helps to stabilize the sides and bottom of the hole in caving soils. However, drilling mud is not recommended if hydraulic conductivity tests are to be carried out in the borehole, or if piezometers are to be installed. Further guidance on the use of drilling mud is given by EPA (1989). The use of air foam as a flushing medium has become more popular as a preliminary before going to bentonite muds. This technique involves the injection of foaming concentrate and water into the air stream produced by a low volume, high pressure air compressor. A polymer stabilizer is added when drilling below the ground-water surface. The foam is forced down the drill rods in the conventional manner, and a slow-moving column of foam with the consistency of aerosol shaving cream carries the suspended cuttings to the surface. Compared with water, the air foam has a greater ability to maintain the cuttings in suspension, and the low uphole velocity and low volume of water used serve to reduce disturbance of the core and surrounding ground. The air foam also resists percolation into open fissures, and it stabilizes the borehole walls. The

Figure 4-13   Drilling With Foam may be Necessary

polymer stabilizer, however, has the disadvantage of coating the walls of the hole such that insitu hydraulic conductivity testing may not be representative of actual subsurface conditions.

Drilling Rates

In general terms the ability of each drilling method to penetrate subsurface materials is based on the drilling method and the individual unconsolidated material or bedrock being drilled. Figure 4-14 presents a comparison of various methods and sudsurface materials in respect of the individual advancement rates. Air hammer drilling rates can be as high as 60 feet/hour in hard rocks as compared to less than 20 feet/hour for cable tool drilling.

Inclined Drilling

Inclined boreholes can often be used to great advantage in ground investigations where some reliance for waste and leachate containment is being placed in the integrity of confining units. While inclined boreholes are generally more costly than similar vertical holes, they often allow additional geological data to be obtained on vertical discontinuities. Land disposal of waste material often relies on geologic units making up the base grades of the

Figure 4-15   Angle Auger Drilling

facility having very low hydraulic conductivity in all three (x, y, and z) directions. Inclined drilling can be performed with coring equipment or even hollow stem augers that cut core while drilling, (see Figure 4-15). Inclined holes may be found to deviate in both dip and direction from the intended orientation (Craig & Gray, 1985). This is particularly common in the vertical plane due to the weight of the drill rods causing the drill bit to dive downward. Some of the factors that contribute to deviation are:

- difficult ground conditions, such as bouldery soils or chert nodules which cause deflections in any direction; and

- alternatively hard/soft dipping geologic units which can cause the drill bit to reflect away from the harder materials;

- worn and undersized drill rods, rods much smaller than the borehole, and excessively flexible rods;

- drilling rig rotation and vibration, which tends to produce spiraling in the hole; and,

- excessive down bit thrusting pressure caused by the driller trying to increase drilling rates beyond the normal advance rate for the method.

Better directional control may be obtained by careful attention and adjustments to rotational speeds, thrust pressures, and the location of drill rod centralizers. In some difficult ground conditions, such as colluvium containing very hard boulders or chert nodules within a loose soil matrix, it may be very difficult to control the deviation of inclined holes.

Several instruments are available for measuring borehole orientation or deviation, such as the Eastmen-Whipstock single-shot or multi-shot photographic survey tools (see Figure 4-9), the Pajari mechanical single-shot surveying instrument and the ABEM Fotobor multi-shot photographic probe. The function of the Eastman and Pajari tools is to record the orientation of a gimbaled magnetic sphere, by photographic and mechanical means respectively. This precludes the use of these instruments within steel casings or attached to steel rods, or in areas where other magnetic disturbances may be anticipated. The ABEM Fotobor records cumulative deflection measurements photographically within the borehole and it can be used within steel casings. Multi-shot tools such as the ABEM Fotobor, or the multi-shot version of the Eastman-Whipstock tool, is useful for undertaking

Figure 4-16　Mass Samples Taken From Bucket Augers

Figure 4-17　Landfill Gas Coming From Augered Hole

complete borehole surveys, but may be cumbersome for taking repeated deviation checks during drilling, in which case single-shot tools may be preferable.

The drilling of boreholes to collect geotechnical data and to serve for installation of piezometers and wells, and their design, construction, and maintenance, are a specialized technology that rests only, in part, on scientific and engineering principles. There are many books (Gibson and Singer, 1971; Campbell and Lehr, 1973; U.S. Environmental Protection Agency, 1973a, 1976) that provide a comprehensive treatment of water well technology. In addition, Walton (1970) presents material on the technical aspects of ground–water hydrology, and his text includes many case histories of water well installation and evaluation of water wells. Hvorslev (1951), Kruseman and de Ridder (1970), and Campbell and Lehr (1973) discuss methods of piezometer construction, testing, and installation. Application of each technique is site–specific and each technique should be selected by the investigator in coordination with knowledgeable drilling contractors.

4.4 TEST HOLE SAMPLING

The main purposes of sampling are to establish the subsurface geological profile in detail, and to supply both disturbed and undisturbed materials for laboratory testing. The selection of a sampling technique depends on the quality of the sample that is required and the geologic and hydrogeologic character of the subsurface.

Direct Classification Of Soils

Sampling of soils and consolidated materials are typically performed through borings conducted during the Phase II scope of work. Although surface collection methods such as hand augers and backhoe excavations are used extensively, they are limited to defining approximately the top 6 to 17 ft (2m to 5m) of surficial materials. Standard penetration tests and continuous sampling methods should be used during drilling in unconsolidated overburden materials. These "disturbed" samples should be logged in the field by an experienced geotechnical engineer or geologist and sealed in glass containers for future examination and testing (if necessary).

As general guidance the overall behavior of the subsurface is often dictated by planes or zones of weakness that may be present (e.g, discontinuities). Therefore, it is possible to obtain a good sample of material that may not be fully representative of the mass. Because of this, and the frequent need to modify the sampling technique to suit the ground conditions, very close supervision of sampling is warranted in most site assessments. In choosing a sampling method, the goals of the project should be made clear whether mass properties or intact-material properties are to be determined within the site investigations. There are four main techniques for obtaining samples (Hvorslev, 1948):

- Taking disturbed samples from the drill tools (see Figure 4-16) or from excavating equipment in the course of boring or excavation;

- Drive sampling, in which a tube or split tube sampler having a sharp cutting edge at its lower end is forced into the ground either by a static thrust or by dynamic impact;

- Rotary sampling, in which a tube with a cutter at its lower end is rotated into the ground, thereby producing a core sample; and,

- Taking block samples specially by hand from a trial pit, shaft or adit.

Samples obtained by the last three techniques will often be sufficiently intact to enable the ground structure within the sample to be examined. Care must be take when drilling into any materials that may produce dangerous gas or liquids. Figure 4-17 shows augering of soils next to a landfill that was producing methane to the vadose zone. The figure show the gas being discharged around the auger flight. Such gas discharge may not be visible and can be very dangerous to the drilling crew (no smoking!).

The quality of such samples can vary considerably, depending on the technique and the ground conditions, and most will exhibit some degree of disturbance. Intact samples obtained by these last three techniques are usually taken in a vertical direction, but specially oriented samples may be required to investigate particular features that can effect directional hydraulic conductivity. The most common methods of soil and rock sampling used in site assessments are provided on Figure 4-18a. These methods have many additional equipment refinements not shown on

the flow chart; additional refinements to these basic methods are provided in Winterkorn and Fang (1974).

The sampling procedure used for site investigations is selected on the basis of the required quality of the sample, and on the suitability of the sample for appropriate laboratory tests. Laboratory classification tests are relevant primarily to soil, but the quality classification may be applied to rock samples in respect of other properties.

The necessary sample size to obtain a representative test result is determined largely by the secondary structure of the geologic units, often referred to as 'the ground fabric'. Where the fabric of the geologic units contains discontinuities of random orientation, the sample diameter, or width, should be as large as possible in relation to the spacing of discontinuities. Alternatively, where the ground contains strongly oriented discontinuities, e.g, in jointed rock, it may be necessary to take samples that have been specially oriented. For fine soils that are homogeneous and isotropic, samples as small as 35 mm in diameter may be used (as obtained from split spoon sampling). However, for general use, samples 100 mm in diameter are preferred, since the results of laboratory tests may be then more representative of the ground mass. In special cases, very large samples of the geologic units may be necessary to fully evaluate the fabric.

Frequency Of Sampling And Testing In Boreholes

A common question arising throughout scoping and execution of site assessments for waste disposal projects is the intensity of site sampling and testing. The frequency and extent of sampling and testing in a borehole ultimately depend on the information that is already available about the subsurface conditions from previous site work and the technical objectives of the investigation. Figure 4-18b provides additional summary guidance for common sampling tools for soil and rock.

The Phase II field work will cover three aspects, each of which may require a different sampling and testing program and may also require separate phasing of field work. However, the reality of site investigation work dictates that a single sampling program be conducted in most cases where the drilling has the flexibility to use alternative methods. These aspects are as follows:

1) The evaluation and classification of the lithologic character and geological structure of the subsurface, (borehole logging);

2) The determination of the properties of the various geologic units or materials whose locations have been determined in the previous step using routine techniques

COMMON SAMPLING TOOLS FOR SOIL AND ROCK

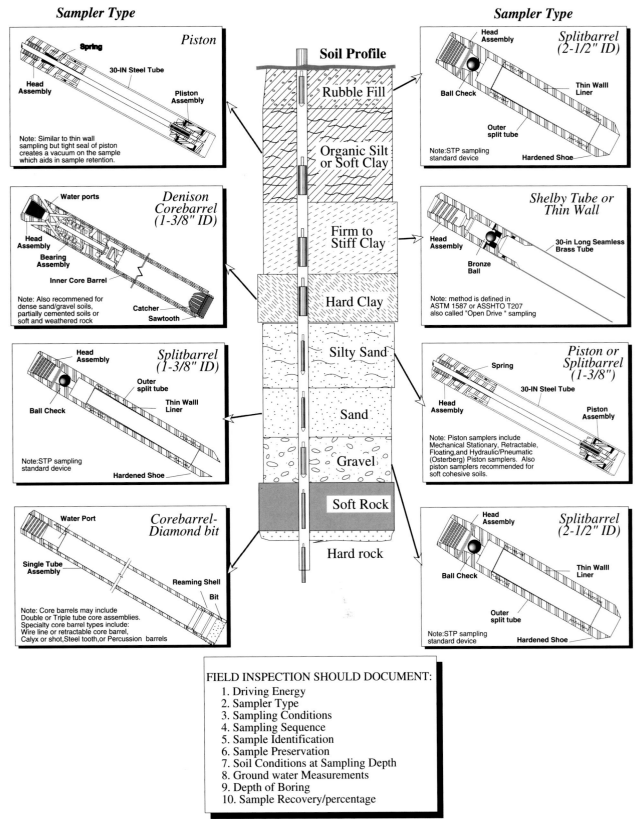

Sampler Type

Piston

Spring

30-IN Steel Tube

Head Assembly

Pliston Assembly

Note: Similar to thin wall sampling but tight seal of piston creates a vacuum on the sample which aids in sample retention.

Denison Corebarrel (1-3/8" ID)

Water ports

Head Assembly

Bearing Assembly

Inner Core Barrel

Catcher

Sawtooth

Note: Also recommened for dense sand/gravel soils, partially cemented soils or soft and weathered rock

Splitbarrel (1-3/8" ID)

Head Assembly

Outer split tube

Thin Walll Liner

Ball Check

Note:STP sampling standard device

Hardened Shoe

Corebarrel-Diamond bit

Water Port

Single Tube Assembly

Reaming Shell

Bit

Note: Core barrels may include Double or Triple tube core assemblies. Specialty core barrel types include: Wire line or retractable core barrel, Calyx or shot,Steel tooth,or Percussion barrels

Soil Profile

Rubble Fill

Organic Silt or Soft Clay

Firm to Stiff Clay

Hard Clay

Silty Sand

Sand

Gravel

Soft Rock

Hard rock

Sampler Type

Splitbarrel (2-1/2" ID)

Head Assembly

Ball Check

Thin Walll Liner

Outer split tube

Hardened Shoe

Note:STP sampling standard device

Shelby Tube or Thin Wall

Head Assembly

30-in Long Seamless Brass Tube

Bronze Ball

Note: method is defined in ASTM 1587 or ASSHTO T207 also called "Open Drive " sampling

Piston or Splitbarrel (1-3/8")

Spring

30-IN Steel Tube

Head Assembly

Piston Assembly

Note: Piston samplers include Mechanical Stationary, Retractable, Floating,and Hydraulic/Pneumatic (Osterberg) Piston samplers. Also piston samplers recommended for soft cohesive soils.

Splitbarrel (2-1/2" ID)

Head Assembly

Ball Check

Thin Walll Liner

Outer split tube

Hardened Shoe

Note:STP sampling standard device

FIELD INSPECTION SHOULD DOCUMENT:
1. Driving Energy
2. Sampler Type
3. Sampling Conditions
4. Sampling Sequence
5. Sample Identification
6. Sample Preservation
7. Soil Conditions at Sampling Depth
8. Ground water Measurements
9. Depth of Boring
10. Sample Recovery/percentage

Modified form Hunt(1984)

Figure 4-18a   Soil Sampling Methods

COMMON METHODS FOR SAMPLING SOIL & ROCK

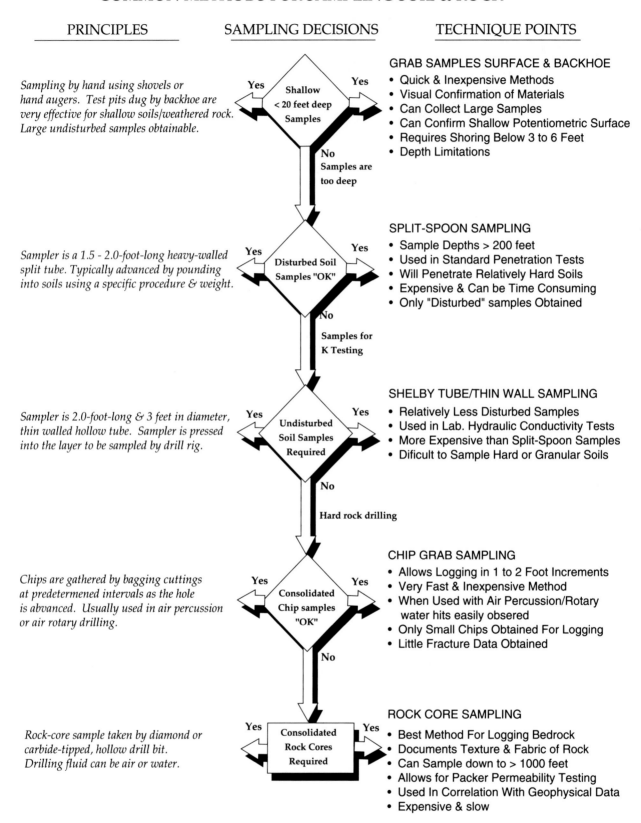

Figure 4- 18b   Common Methods of Soil Sampling

Figure 4-19 Standard Penetration Test (SPT) Sampler

for sampling and testing, (i.e. split spoon or Shelby tubes); and,

3) The use of special techniques of sampling for which routine techniques may give unsatisfactory results, (special sampling) or in situ testing such as with cone penetrometers.

Direct methods (e.g., lithologic analysis) should be employed to identify and to analyze the lithology and structural characteristics of the subsurface from data gathered in the Phase II field drilling program. Open-tube drive samplers consist essentially of a tube open at one end and fitted at the other end with means for attachment to the drill rods. Drive samples include all the methods shown on pictographic Figure 4-10 with the exception of the core sampler. A non return valve permits the escape of air or water as the sample enters the tube, and assists in retaining the sample when the tool is withdrawn from the ground. Before a drive sample is taken, the bottom of the borehole

or surface of the excavation should be cleared of loose or disturbed material as far as possible.

Even simple tests such as the Standard Penetration Test (STP) ASTM D-1586, (AASHTO T-206, BS-test 19 of 1377), should be considered for how valuable the actual "blow counts" will be in reaching project goals. Although split spoon (SPT) tests provide a sample for classification purposes, and the relative blow counts suggest a general comparison of the insitu subsurface conditions, the ultimate use of the disturbed samples and penetration data should be evaluated before using the test.

The sampler can be driven into the ground by dynamic means, using a drop weight or sliding hammer, or by a continuous static thrust, using a hydraulic jack or pulley block and tackle.

Thin-walled sampler The thin wall or Shelby tube sampling was introduced by H.A. Mohr in 1937 (Hathaway, 1988). Thin-walled drive samplers are used for soils that are particularly sensitive to sampling disturbance, and consists of a thin-walled steel tube whose lower end is shaped to form a cutting edge with a small inside clearance. The area ratio is about 10%. These samplers are suitable only for fine soils with a firm consistency, and free from larger rocks. Standard methods for thin-wall sampling have been established by ASTM D-1587 (AASHTO T-207, or BS 1377). They generally give reasonably undisturbed samples in all fine cohesive soils, including sensitive clays, provided that the soil has not been disturbed during the drilling of the borehole. Samples between 76 mm and 100 mm in diameter are normally obtained; samples up to 250 mm in diameter are sometimes required for special purposes.

Thin-walled stationary piston sampler. The thin-walled stationary piston sampler consists of a thin-walled sample tube containing a close-fitting sliding piston, which is slightly coned at its lower face. The first of the piston samples was developed by John Olsson of Sweden in 1922 and later modified by Dr. M. Juhl Hvorslev of the U.S. Army Corps of Engineers, (Hathaway, 1988). The sample tube is fitted to the drive head, which is connected to hollow drill rods. The piston is fixed to separate rods that pass through a sliding joint in the drive head and up inside the hollow rods. Clamping devices, operated at ground surface, enable the piston and sample tube to be locked together or the piston to be held stationary while the sample tube is driven down. The sample diameter is normally 75 mm or 100 mm, but samplers up to 250 mm diameter may be used for special soil conditions. Alternatively designed piston samplers include mechanical stationary, floating, retractable and hydraulic and pneumatic (Osterberg) piston samplers.

Initially, the piston is locked to the lower end of the sample tube to prevent water or cuttings from entering the sampler. In soft clay, with the piston in this position, the sampler can be pushed below the bottom of the borehole. When the sample depth is reached, the piston is held stationary and the sample tube is driven down by a static thrust until the drive head contacts the upper face of the piston. An automatic clamp in the drive head prevents the piston from dropping down and extruding the sample while the sampler is withdrawn.

The sampler is normally used in low strength fine soils and gives "undisturbed" samples in silt and clay, including sensitive clay. Its ability to take samples below the disturbed zone and to hold them during recovery gives an advantage over the thin-walled sampler. Although normally used in soft clays, special piston samplers have been designed for use in stiff clays.

4.4.1 Soil Classification

The classification of soils should be extended beyond the strict adherence to only using the United Soil Classification (USC) description. In addition to the USC system, lithostratigraphic descriptions should also be used in areas where glacial materials, primary porosity bedrock or unconsolidated sediments are present, and where heterogeneity shows hydraulic conductivity ranges of two orders of magnitude. Jennings, et al, 1973, provided additional guidance as to a systematic soils description procedure. The stratum (soil profile) should be described in terms of its moisture condition, color, consistency, structure, soil type and origin (MCCSSO). Soil described using this classification scheme (in addition to the USC classification) will provide basic information for the approximate quantitative assessment of the properties of the material and its depositional history. The basic skill required to use such classification schemes should be easily acquired by practicing civil, geotechnical engineers, geologists, and engineering geologists.

Moisture Conditions

The moisture conditions of each layer of geologic material should be described as a necessary start toward the description of soil consistency. Soil consistency is directly dependent on the moisture content at the time of sampling. The moisture condition would be recorded as: dry, slightly moist, moist, very moist, wet. As Jennings et al (1973) point out, "It will be appreciated that the interpretation of moisture condition in terms of approximate soil moisture content will depend on the grain size of the soil e.g. a sand with a moisture content of 5% to 10% will be observed to be wet, while a clay at the same moisture content may be dry or slightly moist." As such, an important task during logging is the provision of an

evaluation of the sample moisture conditions at the time of sampling.

Color

The observed color is important in the logging procedure to describe the soil and to correlate the particular layer in other site boreholes. Several methods may be used for obtaining a standard for comparison of soil color:

- Describe the soil color after mixing a small amount of soil with water to obtain a creamy paste. This method provides for a consistent color across the site to compare in different trial holes (with variable moisture contents); and,

- Describe the soil color as obtained from the fresh sample as the natural color.

In either case the resultant soil color should be compared to "standard" color references such as the Munsell Soil Color Charts. Many types of soil especially glacial and alluvial clays, have variable colors in section, described as mottled color. The color of the mottling should be described using the predominate color first with secondary colors following in the description.

Consistency

Consistency of a soil is the measure of toughness or hardness based on the effort necessary to dig into the soil or to remold the soil with the fingers. Research such as reported by Terzaghi and Peck (1967) and the Institution of Civil Engineers, London (1954) has related compressive strength and dry density to the consistency of the soil. Since soil classes fall generally into granular-free draining (non-cohesive) and cohesive due to the difference in drainage (hydraulic conductivity) is directly related to shear strength. During the logging process, the engineer or geologist must take particular note of whether the cohesive soil is dry, fissured or jointed. If joints exist, cracks may develop in tension zones and provide pathways through the otherwise low hydraulic conductivity soil matrix.

Structure

The term "structure" or in some cases fabric references the presence (or absence) of joints in the soil and the condition of these joints. The description of soil structure is one of the most often forgotten important criteria in the logging process. Since, non-cohesive soils

	Group Symbols[a]	Typical Names	Information Required for Describing Soils	Laboratory Classification Criteria
Coarse-grained soils — More than half of material is *larger* than No. 200 sieve size[b] (The No. 200 sieve size is about the smallest particle visible to naked eye) — **Gravels** More than half of coarse fraction is larger than No. 7 sieve size — **Clean gravels** (little or no fines)	GW	Well graded gravels, gravel-sand mixtures, little or no fines	Give typical name; indicate approximate percentages of sand and gravel; maximum size; angularity, surface condition, and hardness of the coarse grains; local or geologic name and other pertinent descriptive information; and symbols in parentheses. For undisturbed soils add information on stratification, degree of compactness, cementation, moisture conditions and drainage characteristics. *Example:* Silty sand, gravelly: about 20% hard, angular gravel particles ½-in. maximum size; rounded and subangular sand grains coarse to fine, about 15% non-plastic fines with low dry strength; well compacted and moist in place; alluvial sand; (SM).	$C_U = \dfrac{D_{60}}{D_{10}}$ Greater than 4 $\quad C_C = \dfrac{(D_{30})^2}{D_{10} \times D_{60}}$ Between 1 and 3
(For visual classification, the ¼ in. size may be used as equivalent to the No. 7 sieve size)	GP	Poorly graded gravels, gravel-sand mixtures, little or no fines		Not meeting all gradation requirements for GW
Gravels with fines (appreciable amount of fines)	GM	Silty gravels, poorly graded gravel-sand-silt mixtures		Atterberg limits below "A" line, or PI less than 4 — Above "A" line with PI between 4 and 7 are borderline cases requiring use of dual symbols
	GC	Clayey gravels, poorly graded gravel-sand-clay mixtures		Atterberg limits above "A" line, with PI greater than 7
Sands More than half of coarse fraction is smaller than No. 7 sieve size — **Clean sands** (little or no fines)	SW	Well graded sands, gravelly sands, little or no fines		$C_U = \dfrac{D_{60}}{D_{10}}$ Greater than 6 $\quad C_C = \dfrac{(D_{30})^2}{D_{10} \times D_{60}}$ Between 1 and 3
	SP	Poorly graded sands, gravelly sands, little or no fines		Not meeting all gradation requirements for SW
Sands with fines (appreciable amount of fines)	SM	Silty sands, poorly graded sand-silt mixtures		Atterberg limits below "A" line or PI less than 4 — Above "A" line with PI between 4 and 7 are borderline cases requiring use of dual symbols
	SC	Clayey sands, poorly graded sand-clay mixtures		Atterberg limits below "A" line with PI greater than 7

Determine percentages of gravel and sand from grain size curve. Depending on percentage of fines (fraction smaller than No. 200 sieve size) coarse grained soils are classified as follows: Less than 5% — GW, GP, SW, SP; More than 12% — GM, GC, SM, SC; 5% to 12% — Borderline cases requiring use of dual symbols

Use grain size curve in identifying the fractions as given under field identification

	Group Symbols[a]	Typical Names	Information Required for Describing Soils
Fine-grained soils — More than half of material is *smaller* than No. 200 sieve size[b] — **Silts and clays** liquid limit less than 50	ML	Inorganic silts and very fine sands, rock flour, silty or clayey fine sands with slight plasticity	Give typical name; indicate degree and character of plasticity, amount and maximum size of coarse grains; colour in wet condition, odour if any, local or geologic name, and other pertinent descriptive information, and symbol in parentheses. For undisturbed soils add information on structure, stratification, consistency in undisturbed and remoulded states, moisture and drainage conditions. *Example:* Clayey silt, brown; slightly plastic; small percentage of fine sand; numerous vertical root holes; firm and dry in place; loess; (ML).
	CL	Inorganic clays of low to medium plasticity, gravelly clays, sandy clays, silty clays, lean clays	
	OL	Organic silts and organic silt-clays of low plasticity	
Silts and clays liquid limit greater than 50	MH	Inorganic silts, micaceous or diatomaceous fine sandy or silty soils, elastic silts	
	CH	Inorganic clays of high plasticity, fat clays	
	OH	Organic clays of medium to high plasticity	
Highly Organic Soils	Pt	Peat and other highly organic soils	Readily identified by colour, odour, spongy feel and frequently by fibrous texture

Identification Procedures on Fraction Smaller than No. 40 Sieve Size

Group Symbols	Dry Strength (crushing characteristics)	Dilatancy (reaction to shaking)	Toughness (consistency near plastic limit)
ML	None to slight	Quick to slow	None
CL	Medium to high	None to very slow	Medium
OL	Slight to medium	Slow	Slight
MH	Slight to medium	Slow to none	Slight to medium
CH	High to very high	None	High
OH	Medium to high	None to very slow	Slight to medium

From Wagner, 1957.
[a] *Boundary classifications.* Soils possessing characteristics of two groups are designated by combinations of group symbols. For example GW-GC, well graded gravel-sand mixture with clay binder.
[b] All sieve sizes on this chart are U.S. standard.

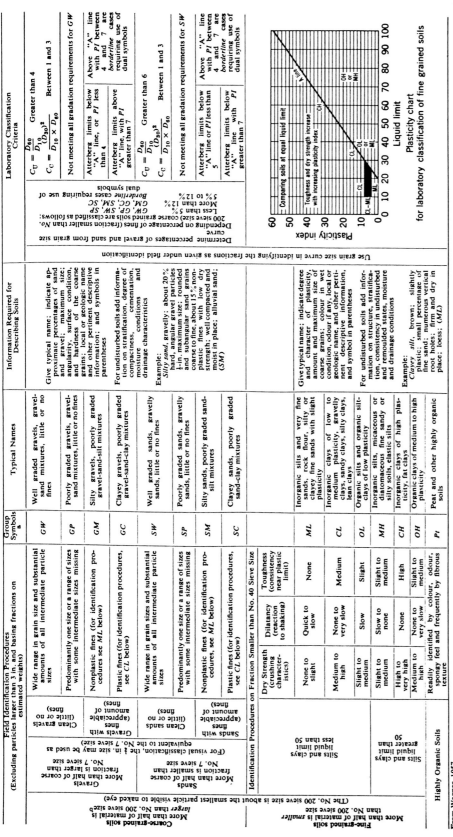

Plasticity chart for laboratory classification of fine grained soils

Field Identification Procedure for Fine Grained Soils or Fractions

These procedures are to be performed on the minus No. 40 sieve size particles, approximately 1/64 in. For field classification purposes, screening is not intended, simply remove by hand the coarse particles that interfere with the tests.

Dilatancy (Reaction to shaking):
After removing particles larger than No. 40 sieve size, prepare a pat of moist soil with a volume of about one-half cubic inch. Add enough water if necessary to make the soil soft but not sticky. Place the pat in the open palm of one hand and shake horizontally, striking vigorously against the other hand several times. A positive reaction consists of the appearance of water on the surface of the pat which changes to a livery consistency and becomes glossy. When the sample is squeezed between the fingers, the water and gloss disappear from the surface, the pat stiffens, and finally it cracks or crumbles. The rapidity of appearance of water during shaking and of its disappearance during squeezing assist in identifying the character of the fines in a soil. Very fine clean sands give the quickest and most distinct reaction whereas a plastic clay has no reaction. Inorganic silts, such as a typical rock flour, show a moderately quick reaction.

Dry Strength (Crushing characteristics):
After removing particles larger than No. 40 sieve size, mould a pat of soil to the consistency of putty, adding water if necessary. Allow the pat to dry completely by oven, sun or air drying, and then test its strength by breaking and crumbling between the fingers. This strength is a measure of the character and quantity of the colloidal fraction contained in the soil. The dry strength increases with increasing plasticity. High dry strength is characteristic for clays of the CH group. A typical inorganic silt possesses only very slight dry strength. Silty fine sands and silts have about the same slight dry strength, but can be distinguished by the feel when powdering the dried specimen. Fine sand feels gritty whereas a typical silt has the smooth feel of flour.

Toughness (Consistency near plastic limit):
After removing particles larger than the No. 40 sieve size, a specimen of soil about one-half inch cube in size, is moulded to the consistency of putty. If too dry, water must be added and if sticky, the specimen should be spread out in a thin layer and allowed to lose some moisture by evaporation. Then the specimen is rolled out by hand on a smooth surface or between the palms into a thread about one-eighth inch in diameter. The thread is then folded and re-rolled repeatedly. During this manipulation the moisture content is gradually reduced and the specimen stiffens, finally loses its plasticity, and crumbles when the plastic limit is reached.

After the thread crumbles, the pieces should be lumped together and a slight kneading action continued until the lump crumbles.
The tougher the thread near the plastic limit and the stiffer the lump when it finally crumbles, the more potent is the colloidal clay fraction in the soil. Weakness of the thread at the plastic limit and quick loss of coherence of the lump below the plastic limit indicate either inorganic clay of low plasticity, or materials such as kaolin-type clays and organic clays which occur below the A-line.
Highly organic clays have a very weak and spongy feel at the plastic limit.

Table 4-3 Unified Soil Classification System

(sand and silt) exhibit a granular structure, soil structure is typically not recorded, however, many structural features can be observed if reasonably undisturbed samples are collected during the drilling process. Cohesive soils also exhibit several types of structural characteristic. As a general practice any soil structure observable should be recorded in field logs. Terms used in describing soil structure are typically more specific to cohesive soils.

The term Intact indicates an absence of fissures or joints. If soft the soil may be plastic like butter, but if firm, it may exhibit tension breaks when cut.

The descriptive term Fissured describes the presence of closed joints. Joint surfaces are frequently observed stained with iron and manganese oxides. When cut, the soil tends to break along the joints. Residual soil fissures may coincide with joints pre-existing in the parent rock (Jennings 1973), or they may represent planes on which shear failure or due to previous tension cracking. Other causes of fissuring in soil may be evident if sufficient data is gathered on the soils historic geologic stress conditions.

Slickensided structures indicate the presence of soil fissures which are highly polished or glossy and frequently striated. Slickensides may be a sign of fairly recent near surface shearing movements in the soil, but similar shiny surfaces can also be developed on joint planes on which there has been no displacement.

Shattered describe the presence of fissures in which joints have opened up and permitted the entry of air. The soil fragments are usually stiff or very stiff and cubical or granular fragments are broken out when the soil is cut.

Shattering is usually a sign of past drying out. It is therefore often associated with heaving soils.

Micro-shattering is small scale shattering with the resultant fragments the size of sand grains. When micro-shattering is well developed, the soil may appear granular, but these grains can break-down into a clay or silt, (or some combination of clay and silt) when rubbed with water on the palm of the hand. Micro-shattering is also a sign of potential heaving conditions.

Stratified, laminated, foliated, etc., Many residual soils show the structure of the derived parent rock. Observation of these residual structures often provides identification of the parent rock material, e.g., types of bedding may provide a clue to the origin of residual soils. In some transported soil stratified materials consist of alternating layers of varying color or texture. Granular soils can show stratified layering of sands and silt that can cause heterogeneity in vertical and horizontal hydraulic conductivity. If the layers are less than about 6 mm thick the structure may be described as laminated (varved if the soil is silty or clayey).

If one observes strong layering in soils with geologic materials of differing hydraulic conductivity, this can help in deciding what field and laboratory tests are necessary to describe hydraulic conductivity of the subsurface materials.

4.4.2 Facies Codes

In addition to the physical descriptions presented above and the Unified Soil Classification System shown in

Table 4-4 Faces Codes

Diamict, D:	Sands, S:	Fine-grained (mud), F:
Dm: matrix supported	Sr: rippled	Fl: laminated
Dc: clast supported	St: trough cross-bedded	Fm: massive
D-m: massive	Sh: horizontal lamination	F-d: with dropstones
D-s: stratified	Sm: massive	
D-g: graded	Sg: graded	
Genetic interpretation ( ):	Sd: soft sediment deformation	
D—(r): resedimented		
D—(c): current reworked		
D—(s): sheared		

Modified from N. Eyles, C.H. Eyles and A.D. Miall, 1983

Table 4-5 LITHOFACIES AND SEDIMENTARY STRUCTURES OF LOW SINUOSITY AND GLACIOFLUVIAL DEPOSITS

Facies Code	Lithofacies	Sedimentary Structures	Interpretation
Gms	massive, matrix-supported gravel	none	debris flow deposits
Gm	massive or crudely bedded gravel	horizontal bedding, imbrication	longitudinal bars, lag deposits. sieve deposits
Gt	gravel, stratified	trough crossbeds	minor channel fills
Gp	gravel, stratified	planar crossbeds	linguoid bars or deltaic growths from old bar remnants
St	sand, medium to v. coarse, may be pebbly	solitary (theta) or grouped (pi) trough crossbeds	dunes (lower flow regime)
Sp	sand, medium to v. coarse, may be pebbly	solitary (alpha) or grouped (omikron) planar crossbeds	linguoid, transverse bars, sand waves (lower flow regime)
Sr	sand, very fine to coarse	ripple marks of all types	ripples (lower flow regime)
Sh	sand, very fine to very coarse, may be pebbly	horizontal lamination, parting or streaming lineation	planar bed now (1. and u. flow regime)
Sl	sand, fine	low angle (10 degree) crossbeds	scour tills, crevasse splays, antidunes
Se	erosional scours with intraclasts	crude crossbedding	scour fills
Ss	sand, fine to coarse, may be pebbly	broad, shallow scours including eta cross-stratification	scour fills
Sse, She, Spe	sand	analogous to Ss. Sh, Sp	eolian deposits
Fl	sand, silt, mud	fine lamination, very small ripples	overbank or waning flood deposits
Fsc	silt, mud	laminated to massive	backswamp deposits
Fcf	mud	massive with freshwater molluscs	backswamp pond deposits
Fm	mud, silt	massive, desiccation cracks	overbank or drape deposits
Fr	silt, mud	rootlet traces	peat-earth
C	coal, carbonaceous	plants, mud films	swamp deposits
P	carbonate	pedogenic features	soil

Modified from N. Eyles, C.H. Eyles and A.D. Miall, 1983

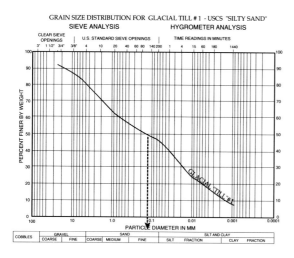

Figure 4-20 Grading Curve for Till (Diamict) #1

Table 4-3 an additional consideration should be made as to the visible structure in the sample or outcrop as represented in a classification system called "Facies Codes". These codes can be used for those sites where the ground-water flow system is affected by primary porosity heterogeneous depositional features. Examples include sand channel deposits contained in deltaic or fluvial depositional environments. The key is to use the Facies Codes when it can make a difference in the conceptual ground-water flow path. A simple Facies Code system (modified from Egles and Egles and Miall, 1983) is presented in Table 4-4 and additional tabular classifications in Tables 4-5 and 4-6 where sediments (both recent and lithified) are divided into three major groups:

- Diamicts: sediments composed of mixtures of many size sediments, such as glacial tills.

- Sands: sediments composed of predominantly sand sized particles, for example sandstone or outwash deposits, and,

- Fine grained: sediments composed or primarily fine sill and clay sized particles, such mudstone or shale.

Facies Codes provide a significant advancement to reconstructing sedimentary or depositional history at a site (Hughes, 1991). Facies Codes require careful examination of the small scale structural features, such as, resedimentation or reworking, and can be used to relate important permeability features of the rock or soil mass. Within the Diamict classification a basic separation of matrix or clast supported structures is important for direct application to hydrogeological investigations. A clast supported sediment is where individual grains are contacting each other leaving interstitial voids reasonably open to transmit ground water. A clast supported sediment does not specially require all one sized sand particles but rather, limited amounts of clay or very fine silt sized particles filling the voids between the larger particles. As the percentage of fine grained materials rises from 10% to above 20% the individual clasts are lifted or lose contact with other clasts. The sediment then becomes matrix supported and observed hydraulic conductivity dramatically fall down to below 10^{-6} cm/sec. Typically glacial tills or "Diamicts" often shows hydraulic conductivity below 10^{-7} cm/sec that then become effectively confining units, (in the absence of later secondary fracturing), so sought after for waste disposal sites. Many additional depositional environments also show similar clast/matrix supported relationships of lowering of hydraulic conductivity as the sediment changes from clast to matrix supported.

One should remember that Diamicts may be classified under many different USGS codes due to the requirement that the 50% passing grading curve line may be just one side of the other of the particle size passing the number 200 sieve. Figures 4-20 and 4-21 illustrate glacial Diamicts (Tills) where one soil is classed as a silty sand (SM) and the other a silty clay (CL). In actual practice the same soil could have been classed as a silty gravel (GM), silty sand (SM), silty or gravelly clay (CL). Figure 4-22 shows very different glacial sediments one a Diamict (called a "till"), the other an wellgraded outwash sand; although both samples were classified as silty sands (SM).

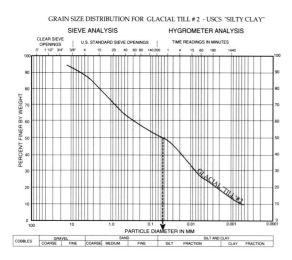

Figure 4-21 Grading Curve for Till (Diamict) #2

EXAMPLE GRAIN SIZE DISTRIBUTIONS FOR OUTWASH SANDS AND TILLS

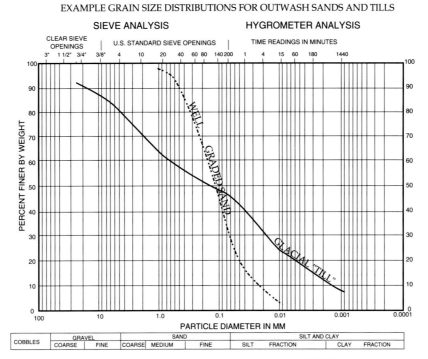

Figure 4-22 Comparison of Outwash and Till (Diamicts)

They recieve this classification under the Unified Soil Classification System (USCS) be cause thier 50% passing curve points are at the same place. Facies codes would have classified the Diamict as a DM (Diamict matrix supported) and the sand as SM (massive sand).   The true depositional environment for a particular facility may require more than simple grading curves, however, in these cases using the Facies Codes may provide much more insight into potential ground-water pathways.

Figure 4-23 shows an example of the classification applied to glacial materials.

4.4.3 The Unified Soil Classification System

Whatever drilling and sampling or trenching procedures is used in a stratigraphic investigation, the soil should be fully identified in accordance with the Unified Soil Classification System (USCS), as developed by Casagrande (1948) for the U.S. Army Corps of Engineers. A schematic representation of the system and its nomenclature is shown in Table 4-3 (from Lambe and Whitman, 1969). This description process is also fully described in ASTM Standard D-2487-85.

The system divides soil into coarse–grained (having more than 50 percent retained on the No. 200 sieve) and fine–grained (more than 50 percent passing through the No. 200 sieve). The coarse–grained materials are called gravels (G) if more than 50 percent of the coarse fraction is retained on the No. 7 sieve, and sands (S) if more than 50 percent of this fraction lies between the No. 40 and No.

200 sieve. The designations G and S are supplemented by the letters W, P, M, and C for well graded, poorly graded, silt– and clay–containing materials, respectively.

The soils that have more than 50 percent of silt–clay component (–200 sieve fraction) are subdivided into silts (M) and clays (C), depending on their liquid limit/plasticity index relationships, as shown in the diagram at the bottom of Table 4–3. Organic silts and clays are designated as O. Except for organic clays, the liquid limit (LL) and plasticity index (PI) values of the silts plot below the A–line and of the clays above, the M, C, and O groups are further differentiated by their liquid limits. A last group, designated Pt, contains organic peat, humus, and swamp soil possessed of great compressibility and low shear strength; for marginal soil double designations are used.

Details of the field and laboratory procedures, employed in the use of this classification and on the engineering characteristics of the various soil classes, are found in the Earth Manual of the USBR (1973). The USCS has been incorporated herein, along with the ASTM Visual–Manual Procedure for description and identification of soils (ASTM D–2488–84) and Classification of Soils for Engineering Purposes (ASTM D–2487–85). As such, these descriptive classification standards must be used in all site assessment projects. One should also check with local state requirements for soil classification systems. On occasion these classification schemes may differ from the Unified Soil Classification System. Even though other classification standards may be required (such as Facies Codes) the Unified Soil

Table 4-6 Diagnostic Criteria for Recognition of Common Matrix-Supported Diamict Lithofacies

Facies Code	lithofacies	Description
Dmm	Matrix-supported, massive	Structureless mud/sand/pebble admixture.
Dmm(r)	Dmm with evidence of resedimentation.	Initially appears structureless but careful cleaning, macro sectioning, or X-ray photography reveals subtle textural variability and fine structure (eg. silt or clay stringers with small flow noses.) Stratification less than 10 percent of unit thickness.
Dmm(c)	Dmm with evidence of current reworking.	Initially appears structureless but careful cleaning, macro sectioning or textural analysis reveals fine structures and textural variability produced by traction current activity (eg. isolated ripples or ripple trains). Stratification less than 10 percent of unit thickness.
Dmm(s)	Matrix-supported, massive, sheared.	Shear lamination caused by the shearing out of soft incompetent bedrock lithologies ("smudges") or slickensided bedding-plane shears resulting from subglacial shear.
Dms	Matrix-supported, stratified diamict.	Obvious textural differentiation or structure with diamict. Stratification more than 10 percent of unit thickness.
Dms(r)	Dms with evidence of resedimentation.	Flow noses frequently present; diamict may contain rafts of deformed silt/clay laminae and abundant silt/clay stringers and rip-up clasts. May show slight grading. Dms(r) units often have higher clast content than massive units clast clusters common. Clast fabric random or parallel to bedding. Erosion and incorporation of underlying material may be evident.
Dms(c)	Dms with evidence of current reworking.	Diamict often coarse (winnowed), interbedded with sandy, silty and gravelly beds showing evidence of traction current activity (eg. ripples, trough or planar cross-bedding). May be recorded as Dmm, St, Dms, Sr etc. according to scale of logging. Abundant sandy stringers in diamict. Units may have channelized bases.
Dmg	Matrix-supported, graded.	Diamict exhibits variable vertical grading in either matrix of clast content; may grade into Dcg.
Dmg(r)	Dmg—with evidence of resedimentation.	Clast imbrication common.

Modified from N. Eyles, C.H. Eyles and A.D. Miall, 1983

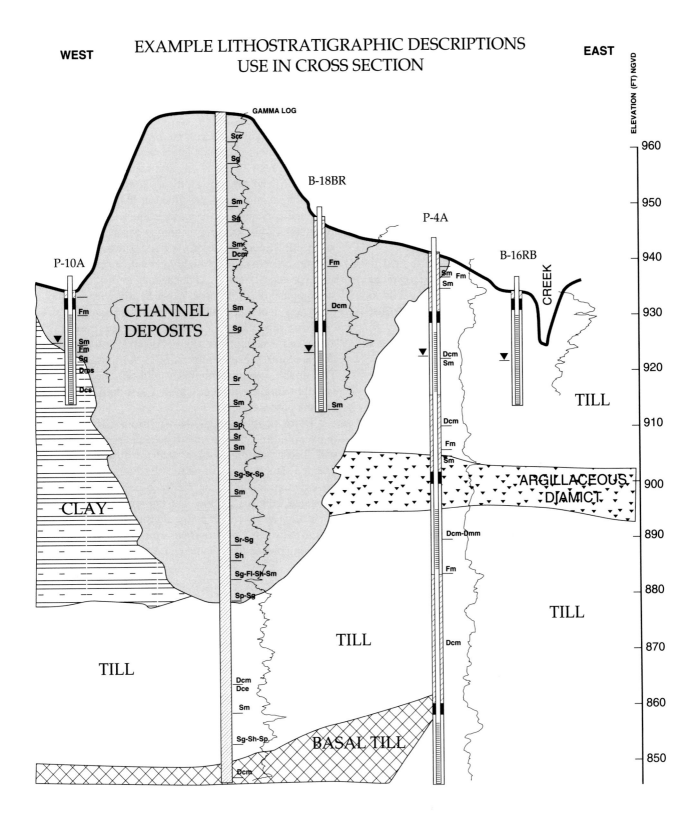

Figure 4-23   Use of Lithostratigraphic Descriptions

Classification System should always be employed even if used in conjunction with other systems.

At sites where shallow soils overlie bedrock, data from exploratory pits (shallow excavations) may be use to improve quality of collected data. These investigations serve to establish site–specific geologic conditions beneath the facility and to correlate it with geology of the region. It is important to correlate data obtained in exploratory pits with indirect (remote image and geophysical) studies for establishment of nature, extent, and depth of various geologic materials.

4.5 DETERMINATION OF THE SUBSURFACE PROFILE

In areas where acceptable information about the subsurface profile is available from previous investigations obtained from Phase I review work, it may be possible to reduce the need for the full spectrum of subsurface sampling and testing. Otherwise, it is necessary to determine, as far as, possible the location, character and structure of each geologic unit in the soil or (and) rock mass. Some of these heterogeneous zones may be quite thin; but ultimately significant to the investigation, and continuous sampling of the entire borehole may be required in order to obtain the required information. No part of the site investigation process exceeds the importance of collection of good subsurface data in an organized targeted approach.

Samples should be taken continuously or at close intervals supplemented by standard penetration tests (SPT's) dependent on project requirements. Continuous rotary coring should be performed in fresh to moderately decomposed rock. Where a run with the rotary core-barrel results in poor core recovery from semi-consolidated rock, one may switch an SPT sampler to try to recover a small drive sample. However, rotary coring equipment and techniques should be adjusted in order to obtain the best core recovery possible. Very little useful data is obtained from zones of poor core recovery, so special care must be exercised in these sections. Additional guidance is provided on core drilling and sampling in Chapter 6.0 specific to fractured rock.

In loose noncohesive soil derived from insitu rock weathering, colluvium and some fill materials, the ground profile can be defined by taking samples using a triple-tube core-barrel. In coarse granular soil, such as gravel, it is advisable to take disturbed samples from the drill tools, together with larger diameter (2.25 inch) split barrel standard penetration test samples at about 1 m intervals.

In fine cohesive soil, and some types of silty sands, consecutive drive samples can be typically obtained using the 100 mm diameter thin wall (Shelby tube) sampler. In soft clays, it is generally good practice to obtain at least one complete profile for the site using the continuous piston sampling technique to support non-continuous sampling efforts. Special sampling equipment may be required for taking long continuous samples in soft clay, loose silt and loose silty sand. Refer to Winterkorn & Fang (1976) & Hvorslev (1949) for additional guidance in soil sample collection in difficult ground conditions.

If a project does not have needs for 'undisturbed' laboratory tests, a particular soil sample, at the time of collection, can be split along its longitudinal axis and carefully examined and described in their fresh condition using the appropriate logging techniques and classification system(s). This splitting technique can be repeated later when soil is in a semi-dried state and, since the fabric may be more evident. Where highly variable ground conditions are expected, it may be advantageous to drill one or more stratigraphic boreholes first, either by rotary core sampling or by auger boring with continuous tube sampling. The cores or tube samples can then be examined to give guidance for sampling at selected depths in other boreholes that are sunk subsequently close to the initial boreholes. This double logging/sampling would only be necessary if special sampling methods hinder standard drilling and logging procedures.

Block samples are cut by hand from material exposed in excavations and are normally taken in rock and cohesive soil. The procedure is often used for obtaining specially oriented samples, and, in such cases, both the location and the orientation should be recorded before the sample is separated from the ground. The cutting of a block sample often takes an appreciable time during which there may be a tendency for the moisture content to change. The following precautions should be taken (BS 1981):

(a) No extraneous water should be allowed to come into contact with the sample.

(b) The sample should be protected from the wind and the direct rays of the sun.

(c) Immediately after the sample has been cut, the orientation should be marked and then it should be coated with paraffin wax and securely packed.

4.5.1 Handling and labelling of samples

Samples may have cost a considerable sum of money to obtain and should be treated with great care. The usefulness of the results of laboratory tests depends on the quality of the samples at the time they are tested. It is therefore important to establish a satisfactory procedure for handling and labelling the samples, and for their storage and transport. Samples should not be allowed to deteriorate, be readily identified and if required, easily drawn from the sample store age area. The samples should be protected from damaging frost, and from excessive heat and temperature variation, which may lead to deterioration in the sealing of the sample containers and subsequent damage to the samples.

All samples should be labeled immediately after being taken from a borehole or excavation. If the samples must be preserved at their natural moisture content, they should at the time of collection, be sealed in an air-tight container or coated in wax. The label should be marked with indelible ink securely affixed to the sample and duplicate sample information separating recorded on daily logging records. Since these samples are obtained at a great expense, identification of these important data sets should never be in question.

Samples taken with tube samplers (such as thin wall, Shelby samples) should be preserved using the following procedure that has been used successfully for many years. Immediately after the sample has been taken from the boring or excavation, the ends of the sample should be removed to a depth of about 25 mm and disturbed soil in the top of the sampler should also be removed. Several layers of molten wax, should then be applied to each end of the sample, forming a plug about 25 mm in thickness. The molten wax should be as cool as possible. It is essential that the sides of the tube be clean and free from extraneous soil. If the sample is very porous, a layer of waxed paper should first be placed over the end of the sample before waxing. Remaining space between the end of the tube or liner and the wax should be tightly packed with a nonabsorbent material and a close-fitting lid or screw-cap placed on each end of the tube or liner.

A label bearing the number of the sample, should be placed inside the container just under the lid. The label should be placed at the top of the sample. In addition, the number of the sample should be placed on the outside of the container, and the top or bottom of the sample should be indicated. The duplication in labeling may seem excessive, however, experience with sample handling points toward unquestionable identification as one of the most important issues to be addressed during the sampling process. The maintenance of a chain of custody may also be important relative to the entire sampling process. Standard Operations and Procedures (SOP's) may be required so that all necessary staff will have both procedural and documentation standards available for use during the project. One should, however, remember that staff involved with field sample collection work should be properly trained by experienced individuals before being allowed to work alone. The liners or containers should be packed in a way that will minimize damage by vibration and shock during transit.

4.5.2 Rock Coring Methods

As general guidance stratigraphic borings should be advanced into bedrock using air rotary techniques, if at all possible. This method permits relatively fast, economical drilling and provides for a direct observation of zones that yield ground water. Cores provide valuable information about stratigraphy, structure, and lithologic characteristics of rock contacted by the borehole. In homogeneous rocks, cuttings from air rotary drilling may be adequate for logging purposes, in conjunction with use of a portable flume or orifice meter to measure the volume of formation waters recovered during airlift production, if supported with geophysical logging. A field conductivity meter should also be used to measure and monitor any changes in the conductivity of any recovered borehole water.

Cored boreholes should be cased from the ground surface with appropriately sized casing cemented in place through overburden materials. The hole should be drilled by air rotary methods and potable water mist, if required. The air compressor must have a filter or oil trap. Summary guidance for obtaining optimum core recovery is as follows:

- In most rocks, satisfactory core will be obtained by using double-tube swivel type core barrels with core size not less than about 70 mm diameter (H size or equivalent).

- core size to 55 mm diameter (NX size or equivalent) for massive rocks; and,

- core size to between 100 mm and 150 mm diameter will usually give better results in weak, weathered or fractured rock.

In weak, weathered or fractured rock, better core recovery can be obtained using triple-tube core barrels in which the inner tube contains a removable sample tube or liner.

It is recommended, for important investigations, that core fractures observed during diamond core drilling should be oriented relative to the dip direction of any observable sedimentary bedding, or foliation planes of igneous or metamorphic rock. The fracture orientations can then be rotated relative to true north based on the results of field measurements of the strike and dip of the reference bedding or foliation planes. This method is a common structural geologic technique and produces reasonably accurate orientation where relatively steeply dipping bedding or foliation planes exist. Also, this approach is much less expensive than fully oriented core and will produce results well within acceptable accuracy and within the observed scatter in outcrop–measured joint sets. Additional orientation methods are described in Chapter 6.0 for fractured rock assessments.

It may be required that one of the coreholes be drilled on an angle from the vertical, and perpendicular to the reference bedding or foliation planes in the bedrock. Field conditions may dictate that the angle and orientation of the inclined borehole be modified based on local field conditions. Normally, the inclined boreholes should be advanced and cored as the last drilling task, accomplished after the discontinuities observed in the vertical boreholes are studied in combination with orientation measurements taken during the geological mapping. The direction of the inclined boreholes should then be chosen in the direction most likely to observe typical high–angle discontinuities (rather than the bedding planes) or the particular geologic unit (i.e., vertical joints, dikes, or faults). The borehole should be drilled at the angle most likely to be orthogonal to the discontinuities, within the limits of the drilling equipment.

Preservation and Documentation of Cores

The geologic characteristics and the intended use of the rock core determine the extent and type of preservation required. If engineering properties are to be determined from the core, it must be handled and preserved in such a way that the measured properties are not significantly influenced by mechanical damage and changes in environmental conditions during storage. The core is the sample of record for the subsurface geology at each borehole location and, as such, must be preserved for some period of time, in some cases indefinitely, for future geologic reference or study.

For any project a qualified person must be assigned to have curatorial responsibility for the core. This person must be technically competent in management of rock core samples and has a knowledge of the various uses for the core and the associated requirements for storage and preservation.

Requirements for use of the core range from simple measures, for which the only need is to identify and locate the various lithologic units, to complex and critical uses, in which detailed testing of the core are required for engineering design of hydraulic characterization. Priorities for multiple uses or different types of tests must sometimes be established when available cores are limited and when one test precludes another. For example, splitting a core for detailed geologic study prevents later strength testing, which requires an intact core.

Tests of mechanical properties, for structural design purposes, should be performed on a core in its natural moisture state, particularly if the rocks are argillaceous. Irreversible changes occur when such rocks are allowed to dry, often resulting in invalid data. The initial moisture content of such a core should, therefore, be preserved. Freezing of pore water in the core may reduce the strength of the rock. The high temperature associated with summertime unventilated storage, and temperatures alternating between hot and cold, may cause moisture to migrate from the core and weaken the rock due to differential thermal expansion and contraction between individual grains or crystals. Temperature extremes

Figure 4- 24 Core Logging Must be Done in the Field

should, therefore, be avoided, particularly for weak sedimentary or volcanoclastic rock types.

A weak rock core may become broken or further weakened by careless handling, such as dropping a core box, or by mechanical vibration and shock during transportation. Breaking the core reduces the sample lengths available for testing. Weakening caused by such mechanical stresses may lower measured strength parameters and may affect other properties. As noted below, core photography is a prime mean of recording rock condition before any subsequent events may alter its appearance.

The required preservation time may vary from, as short as, the end of a project to several years, and sometimes cores must be stored indefinitely. A core collected solely to identify bedrock lithology beneath a small facility may be needed only for a few months. For large facilities, such as RCRA Subtitle C facilities, it may be necessary to retain representative samples of the core for the operating life and post–closure monitoring period of the facility, because re–examination and testing may be required at some later time. Some states have regulations governing the storage and disposition of core. The cores should be stored in wooden core boxes even if short storage requirement is used on a project. Figure 4-24 shows an example core logging using NX sized cores. Cardboard core boxes may be usable for small diameter rock cores as are used in mining exploration, however, larger sized cores soon destroy cardboard core boxes.

The investigator should photograph each core with a 35 mm (or larger) camera format, using either slide or color print film to permanently record the unaltered appearance of the rock. The core should be cleaned by wiping with a cloth or, preferably, rinsing with water prior to obtaining photographs.

For rock placed in core boxes, one photograph should be made of each box once it is filled to capacity. Include the inside of the box lid, with appropriate data for identification and an engineer's scale along one edge of the box. When long, intact cores are preserved in single plastic tubes, make detail–revealing closeups of each core interval, in addition to a single photograph of the complete core. Take photographs before the core are obscured by protective seals and wraps, and before any deterioration begins in particularly fragile or sensitive rock types.

For a boxed core that is not particularly sensitive and for which maintenance of in–situ moisture content is not important, two photographs should be made: one with the core surface in a dry condition, and one in which the core is wet, to bring out visual details that would not otherwise be apparent.

Immediately upon recovery of the core, and preferably while such is still held in the driller's receiving rack, an inspection is made of all rock fractures, for the purpose of distinguishing between natural discontinuities and "mechanical" or drilling–induced fractures. All discontinuities should be physically marked by one stroke of a visible color in indelible ink.

This procedure may require photography both in the field and later at the storage facility, but it must be completed before removal of any test core, and before damage can occur from mishandling. Where it is impossible for a photo to show identification data marked directly on the sample or its container, then mount appropriately marked placards in the frame. Organize the photographs and mount in a folder for easy access and preservation.

Samples removed must be replaced by wood blocks, dates, and name of person responsible for the sampling action. Additional guidance on core logging and sample controls are described in Chapter 6.0 Fractured Rock.

4.6 LOGGING DATA FROM BOREHOLES

Logging is the recording of data concerning the materials and conditions observed in individual boreholes. It is imperative that logging provide an accurate description of subsurface conditions. It is equally imperative that records be concise, complete, and presented in standard descriptive terms that are readily understood and evaluated in the field, laboratory, and office activities related to analysis and design. Errors made in the field be inexperenced staff often have a way of preserving these same errors throughout the project, often with potentially expensive redrilling and reevaluation costs, if the errors are caught at all. Figure 4-25 provides an example of field staff logging samples as "sand" when the sample was actually lost out of the bottom of the sampler during collection. This single error was compounded by using the field logging data in the final logs and cross section constructions without catching the loss of sample substution for a sand zone as being incorrect. In this case the questionable holes were redrilled using grading curves to establish the type of materials present in the section. Carefull review of field data is important, however, experenced field staffing is essential.

The basic element of logging is a geologic description of the material between specified depths or elevations of geologic interest. This description includes such items as texture, structure, color, mineral content, moisture content, apparent relative hydraulic conductivity, and soil or rock type. To this must be added any information that indicates the engineering properties of the material. Examples are gradation, plasticity, and both the Unified Soil Classification System (Table 4-3), and Facies Code symbols, as determined by field identification. In addition, the results of any field test, such as the Standard Penetration Test (SPT), must be recorded along with the

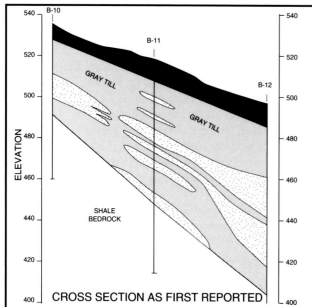

LITHOLOGY AND FIELD LOGGING

The cross section shows numerous sandy units within the `Gray Till. This portrated a very difficult field condition for installation of a waste disposal facility. Later field investigations however did not observe the thick sequences of outwashsand deposits within the fine grained tills. This was due to field description errors, where samples were described as sandy without collection of labortory samples for confirmation of soil type. The borehole log below with laboratory grading curves represents the actual lithology observed when borehole 12 was redrilled.

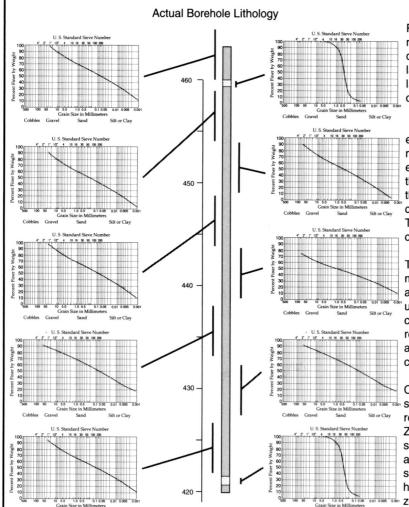

Review of the field logs showed that much of borehole B-12 was not recovered during the sampling process. Using a larger drill rig allowed collection and laboratory grading curves to be completed for the area in question.

The resultant borehole log between elevation 465' and 420' showed only two minor sand outwash deposits at elevations 460' and 421' one to two feet thick. The grading curves generated for the samples show the characteristic till configuration of a more or less 45⁰ slope. The outwash sand curves have the characteristic S shape.

The geologic processes that formed the minor sand units in the redrilled borehole are much of a concern that these sandy units will have area extent and be connected to major aquifer units. The revised subsurface cross section presents a very different picture than the initial cross secction presented above.

Care must always be taken to obtain samples from the subsurface that represents the actual site conditions. Zones of lost samples or lengths not sampled should not be automatically assumed to be the same as the last sample taken. This is especally true for higher (or lower) hydraulic conductivity zones.

Figure 4-25 Use of Grading Curves to Evaluate Site Conditions

specific borehole interval that was tested. Table 4-7 provides recommended boring log information.

Two types of borehole logs are normally prepared in connection with subsurface exploratory work. The first, called a field log, is commonly a detailed chronological record of drilling and sampling operations. The second, called a final log, is a graphic log of the lithology or soil type on which the pertinent subsurface information and the results of field and laboratory testing are reported. The boring log is the simplest form by which to illustrate results of a drilling program. An example Field Log – Soil Borehole (standard form) is provided in Figure 4-26. An edited Soil Borehole Log is given in Figure 4-27, with the completed Monitor Well Construction Summary as Figure 4-28. Appendix B provides detailed instructions for logging of geotechnical information on standard forms. The engineer should use similarly complete geotechnical forms on all site assessment projects. Field logs should be incorporated into the project files but not into final reports or documents.

The responsible site geologist or engineer must complete at least a preliminary field log of the core before it is packed to be transported. The preliminary log must include all data for identification of the borehole, personnel, and equipment involved; notations of depth intervals of core run; percentages of recovery; depth of lithologic contacts; types and locations of protection applied to samples; and any facts that would otherwise be unknown to whoever may complete a more detailed log, at a later time. Additional guidance for logging information is provided on Figure 4-29. It is desirable that detailed logs be completed by the same professional person who compiles the field log. It is advisable for the investigator immediately to make notations on the depths at which, in his/her judgment, any core losses occurred.

The investigator should complete a detailed field log, before leaving the drill site, in cases where the core is anticipated to deteriorate or otherwise change before being examined again.

Driller's logs are not acceptable for site assessment projects. A professional geologist or geotechnical engineer should be present, at the drill rig, at all times during drilling. He or she should fully log the borehole document the drilling process and supervising drilling operations. There is nothing as distressful as poorly documented borehole logs as the only record of thousands of dollars and hundreds of man hours spent on a site investigation. The responsible geologist or engineer has the professional duty to record in full detail the important aspects of the lithology as the hole is being drilled. In continuous soil sampling operations this will probably require a single professional per drill rig.

4.7 LOGGING ROCK CORE FOR RQD

RQD is a simple observational index, designed to be formulated at the time rock core is recovered from boreholes, yet to remain as a single semi-quantitative representation of the worth (quality) of that rock. For RQD to be reliable, rock core index number must be computed in a standard and unvarying manner (Hathaway 1990).

Rock Quality Designation was the invention of Don U. Deere, then Professor of Civil Engineering at the University of Illinois in 1963. Prior to introduction of RQD, engineering geologists could offer only stick logs of boreholes and overall assessments of what each engineering geologic unit represented in terms of a construction medium (Hathaway, 1990). Better-quality engineering geologic reports also contained reference to "percent recovery" of core.

Rock Quality Designation brought into the rock core description process a number that reflects a whole array of geological factors influencing rock mass character. RQD is defined as the ratio of the sum-length of all of the single pieces of naturally unfractured rock core of lengths equal to core longer than 10 cm (4 in.) to the total length of rock core attempted in the particular core run being assessed. Highest quality of all rock results in unbroken core of the entire length of the core barrel (generally 1.5 m or 5 ft) or of core segments that are all longer than 10 cm, as end-bounded by any form of natural discontinuity. By definition, RQD is assessed on the basis of detecting and summing the length of rock core separated by discontinuities (see Figure 4-30). Underlying the purpose of RQD is that higher "quality" rock will have higher RQD numbers; that is, higher "quality" rock will be stronger (compressive and shear strength), more dense, and have relatively fewer open discontinuities.

Note that the proper determination of RQD primarily depends on identification of which of the core "fractures" are discontinuities (natural; rock breaks) and which have been produced by the drilling process or by handling the core for placement in the core box. A common error in logging is ascribing drilling induced breaks in core as part of the natural discontinuities. As guidance, rotational scratches in the core probably indicate mechanical breaks caused by drilling.

RQD is meant to be an index property (see Table 4-8). This index property was defined by Deere as applying universally to NX rock core (Diameter of 54.7 mm or 2.16 in.). Values of RQD determined for core of other sizes can be useful, but the larger the core, the greater the potential for maintaining the presence of unopened or marginally healed discontinuities. Deere and Deere (1989), have

Table 4-7 Field Boring Log Information

General
- Project Name
- Hole Name/Number
- Bit Size/Auger Size
- Geologist's Name
- Driller's Firm and Name
- Hole Location; Map and Elevation
- Rig Type/Manufacturer/Model
- Date Started and Finished
- Petrologic/Lithologic Classification Scheme Used (Wentworth, Unified Soil Classification System)

Information Columns
- Depth
- Sample Location/Number
- Blow Counts and Advance Rate
- Percent Sample Recovery
- Narrative Description
- Depth to Saturation

Narrative Description
- Geologic Observations:
  - Soil/rock type
  - Color and stain
  - Gross petrology
  - Friability
  - Moisture content
  - Degree of weathering
  - Presence of carbonate
  - Fractures
  - Solution cavities
    - Bedding
    - Discontinuities; e.g., joints, foliation, shear planes
    - Water–bearing zones
    - Strike and dip
    - Fossils
  - Depositional structures
    - Organic content
    - Odor
    - Suspected contaminant (visually or by detector)
- Drilling Observations:
  - Loss of circulation
  - Advance rates
  - Rig chatter
  - Water levels/dates
  - Amount of air used,
  - air pressure
  - Drilling difficulties
  - Changes in drillin method or equipment
  - Readings from detective equipment, if any
  - Amount of water yield or loss during drilling at different depths
  - Amounts and types of any drilling liquids used
  - Running sands
  - Caving/hole instability
- Other Remarks:
  - Equipment failures
  - Possible contamination
  - Deviations fromdrilling plan
  - Weather

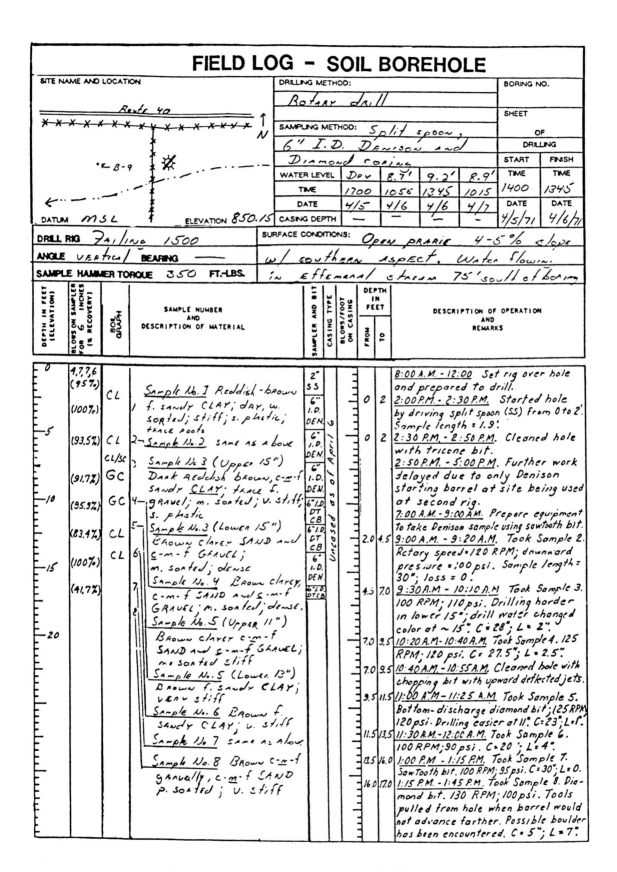

Figure 4-26   Example Borehole Field Log

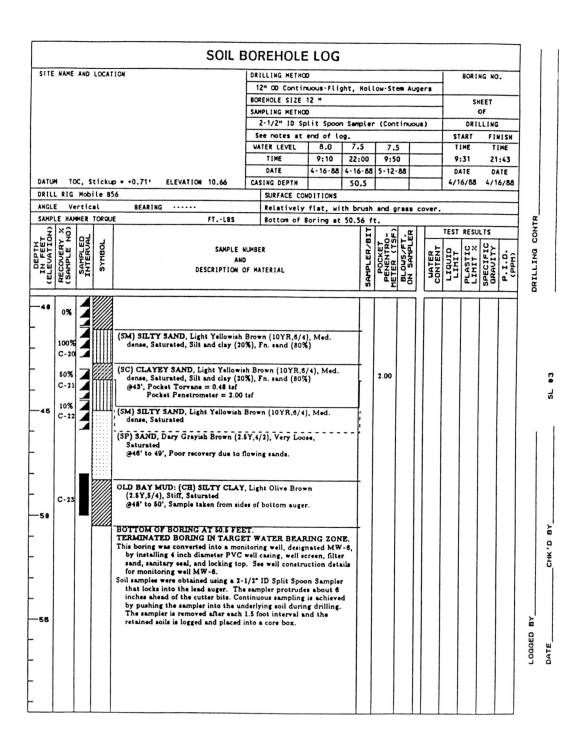

Figure 4-27 Example Borehole Edited Log

Well No. _____

Boring No. x-Ref: _____

MONITORING WELL CONSTRUCTION SUMMARY

Survey Coords: _____   Elevation Ground Level: _____

Top of Casing _____

Drilling Summary

Total Depth _____
Borehole Diameter _____
Casing Stick-up Height _____
Driller _____
_____
Rig _____
Bit (s) _____
Drilling Fluid _____
Protective Casing _____

Well Design Specifications

Basis: Geologic Log _____ Geophysical Log _____
Casing String (s): C = Casing S = Screen.

Depth	String (s)	Elevation
____ –		____ –
____ –		____ –
____ –		____ –
____ –		____ –
____ –		____ –
____ –		____ –

Casing: C1 _____
C2 _____
Screen S1 _____
S2 _____
Filter Pack _____
_____
Grout Seal: _____
_____
Bentonite Seal: _____
_____

Comments _____
_____
_____

Construction Time Log

	Start		Finish	
Task	Date	Time	Date	Time
Drilling	____	____	____	____
	____	____	____	____
	____	____	____	____
Geophys. Logging	____	____	____	____
Casing	____	____	____	____
	____	____	____	____
	____	____	____	____
Filter Placement	____	____	____	____
Cementing	____	____	____	____
Development	____	____	____	____
	____	____	____	____

Well Development:

_____
_____
_____
_____

Stabilization Test Data

Time	pH	Spec. Cond.	Temp (C)

Recovery Data

Q = S_0 =

% 100
R
E 80
C
O 60
V
E 40
R
Y 20
0

TIME   (   )

SITE NAME _____
LOCATION _____

WC

SUPERVISED BY _____
DATE _____

Figure 4-28 Example Monitoring Well Construction Diagram

Field Identification and Description of Soils

Basic Soil Type / Particle Size (mm) / Visual identification

Soil group	Basic Soil Type	Particle Size, mm	Visual identification
Very Coarse Soils	BOULDERS	200	Only seen complete in pits or exposures.
	COBBLES	60	Often difficult to recover from boreholes.
Coarse Soils (over 65% sand and Gravel sizes)	GRAVELS — coarse	20	Easily visible to naked eye; particle shape can be described; grading can be described.
	GRAVELS — medium	6	Well graded: wide range of grain sizes, well distributed. Poorly graded: not well graded. (May be uniform: size of most particles lies between narrow limits; or gap graded: an intermediate size of particle is markedly under-represented.)
	GRAVELS — fine	2	
	SANDS — coarse	0.6	Visible to naked eye; very little or no cohesion when dry; grading can be described.
	SANDS — medium	0.2	Well graded: wide range of grain sizes, well distributed. Poorly graded: not well graded. (May be uniform: size of most particles lies between narrow limits; or gap graded: an intermediate size of particle is markedly under-represented.)
	SANDS — fine	0.06	
Fine Soils (over 35% Silt and Clay sizes)	SILTS — coarse	0.02	Only coarse silt barely visible to naked eye; exhibits little plasticity and marked dilatancy; slightly granular or silky to the touch. Disintegrates in water; lumps dry quickly; possess cohesion but can be powdered easily between fingers.
	SILTS — medium	0.006	
	SILTS — fine	0.002	
	CLAYS		Dry lumps can be broken but not powdered between the fingers; they also disintegrate under water but more slowly than silt; smooth to the touch; exhibits plasticity but no dilatancy; sticks to the fingers and dries slowly; shrinks appreciably on drying usually showing cracks. Intermediate and high plasticity clays show these properties to a moderate and high degree, respectively.
Organic Soils	ORGANIC CLAY, SILT or SAND	Varies	Contains substantial amounts of organic vegetable matter.
	PEATS	Varies	Predominantly plant remains usually dark brown or black in colour, often with distinctive smell; low bulk density.

Particle nature and Plasticity

- Particle shape: Angular, Subangular, Subrounded, Rounded, Flat, Elongated
- Texture: Rough, Smooth, Polished
- Non-plastic or low plasticity
- Intermediate plasticity (Lean clay)
- High plasticity (Fat clay)

Composite Soil Types (mixtures of Basic Soil Types)

Scale of secondary constituents with coarse soils:

Term	% of clay or silt
slightly clayey / slightly silty GRAVEL or SAND	Under 5
clayey / silty GRAVEL or SAND	5 to 15
very clayey / very silty GRAVEL or SAND	15 to 35
Sandy GRAVEL / Gravelly SAND	Sand or gravel an important second constituent of the coarse fraction

For composite types describe as: clayey: fines are plastic, cohesive; silty: fines non-plastic or of low plasticity.

Scales of secondary constituents with fine soils:

Term	% of sand or gravel
Sandy / Gravelly CLAY or SILT	35 to 65
CLAY : SILT	under 35

Examples of Composite Types (Indicating preferred order for description)

Loose, brown, subangular very sandy, fine to coarse GRAVEL with small pockets of soft gray clay.
Medium dense, light brown, clayey, fine and medium SAND.
Stiff, orange brown, fissured sandy CLAY.
Firm, brown, thinly laminated SILT and CLAY.
Plastic, brown, amorphous PEAT.

Compactness/strength

Term	Field Test
Loose	By inspection of voids and particle packing
Dense	
Loose	Can be excavated with spade; 50 mm wooden peg can be driven
Dense	Requires pick for excavation; 50 mm wooden peg hard to drive
Slightly Cemented	Visual examination; pick removes soil in lumps which can be abraded
Soft or loose	Easily moulded or crushed by the fingers
Firm or dense	Can be moulded or crushed by strong pressure in the fingers
Very soft	Extrudes between fingers when squeezed in hand
Soft	Moulded by light finger pressure
Firm	Can be moulded by strong finger pressure
Stiff	Cannot be moulded by fingers. Can be indented by thumb
Very stiff	Can be indented by thumb nail
Firm	Fibers already compressed together
Spongy	Very compressible and open structure
Plastic	Can be moulded in hand and smears fingers

Structure

Term	Field Identification
Homogeneous	Deposit consists essentially of one type
Inter-stratified	Alternating layers of varying types or with bands or lenses of other materials
Hetero-geneous	A mixture of types
Weathered	Particles may be weathered and may show concentric layering
Fissured	Break into polyhedral fragments along fissures. Interval scale for spacing of discontinuities may be used.
Intact	No fissures
Homogeneous	Deposit consists essentially of one type
Inter-stratified	Alternating layers of varying types. Interval scale for thickness of layers may be used.
Weathered	Usually has crumb or columnar structure
Fibrous	Plant remains recognizable and retain some strength.
Amorphous	Recognizable plant remains absent

Interval Scales — Scale of bedding spacing

Term	Mean spacing mm
Very thickly bedded	over 2000
Thickly Bedded	2000 to 600
Medium Bedded	600 to 200
Thinly Bedded	200 to 60
Very Thinly Bedded	60 to 20
Thickly Laminated	20 to 6
Thinly Laminated	under 6

Scale of Spacing of Other Discontinuities

Term	Mean spacing mm
Very widely spaced	over 2000
Widely spaced	2000 to 600
Medium spaced	600 to 200
Thinly spaced	200 to 60
Very closely spaced	60 to 20
Extremely closely spaced	under 20

Color

Red, Pink, Yellow, Brown, Olive, Green, Blue, White, Gray, Black, etc. Supplemental as necessary with: Light, Dark, Mottled, etc. and Pinkish, Reddish, Yellowish, Brownish, etc.

Modified from BS 5930 Table 6

Figure 4-29 Example Soil Field Classification System

found that slightly smaller core (such as NQ, 47.6 mm or

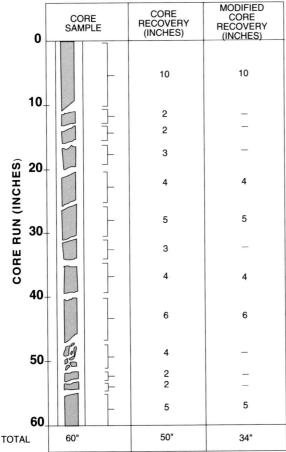

**MODIFIED CORE RECOVERY
AS AN INDEX OF ROCK QUALITY**

CORE SAMPLE	CORE RECOVERY (INCHES)	MODIFIED CORE RECOVERY (INCHES)
	10	10
	2	—
	2	—
	3	—
	4	4
	5	5
	3	—
	4	4
	6	6
	4	—
	2	—
	2	—
	5	5
TOTAL 60"	50"	34"

Core Recovery = 50"/60" = 84%   RQD = 34"/50" = 57%

RQD(Rock Quality Designation)	Description of Rock Quality
0 - 25	v. Poor
25 - 50	Poor
50 - 75	Fair
75 - 90	Good
90 - 100	Excellent

Figure 4-30   Example RQD calculation

1.88 in. diameter) as well as larger-diameter wire-line core is acceptable for computation of RQD. Notation should be made on the core logs when computing RQD for core other than NX size so that Subsequent investigators and reviewers are aware of the possible size effects on RQD.

The physical condition of the core is the best indication of the rock-mass characteristics in terms of all aspects of engineered construction, hence RQD provides a basic comparison value. It is fundamental to RQD that the host rock will have been "fractured" by the sum of physical

and chemical factors throughout its history (Table 4-8, Hathaway 1990). The evaluation of the nature and presence of the rock "fracturing" that controls RQD, must consider all types of discontinuities: non-systematic microfractures, bedding planes, foliation planes, shear planes, and faults. Any of these natural rock breaks are assumed equal in their affect on RQD.

Field determination of RQD requires that the logger proceed with a good intent (i.e. he, or she will only count natural breaks), proper training, and considerable experience in core logging. As simple as this ratio of two sums of core lengths appears, unskilled, untrained, and unsupervised core-logging personnel often defeat the simple intent of Deere's RQD ratio. Adherence to quality control (Table 4-9, Hathaway 1990) can produce a valuable record of rock quality that will stand the test of design and performance, throughout the investigation as well as inquiries during later potential legal depositions and testimony.

Core Recovery: Core recovery is calculated by dividing the total length of a core run into the total recovered length of core for that run:

$$R = L * 100 / T$$

where:
    R = core recovery in percent (dimensionless)
    L = total recovered length of drill core in feet
    T = total length of drill core run in feet.

Rock Quality Designation : RQD is calculated by dividing the total length of a drill core run into the summation of the lengths of all recovered pieces in a core run that equal or exceed twice the core diameter:

$$RQD = SUM(S1, .. SN) * 100 / T$$

where:
RQD = rock quality designation in percent
    (dimensionless)
S1 = length of drill core piece i whose length equals or
    exceeds twice the core diameter (feet)
T = total length of drill core run in feet
N = total number of pieces whose length equals or
    exceeds twice the core diameter.

Fracture-Frequency: Fracture frequency is calculated by dividing the total number of fractures recovered in a drill run by the length equal to that which the fractures are observed in the drill core:

$$F = N / (L - B)$$

where:

F = fracture frequency (l/feet)
N = total number of fractures observed in the core run
L = total recovered length of drill core in feet
B = total length of broken zone in feet.

Broken Zone: Broken zone is defined as sections of core where pieces of core are smaller than the core diameter and determination of individual fractures are not practical:

$$BZ = B * 100 / T$$

where:
BZ = broken zone in percent (dimensionless)
B = total length of broken zone in feet
T = total length of drill core run in feet.

Average Core Length: The average core length is calculated by dividing the number of whole pieces into the adjusted recovered length:

$$AL = (L - B) / X$$

where:
AL = average core piece length in feet
L = total recovered length of drill core in feet
B = total length of broken zone in feet
X = number of whole pieces recovered in the drill run.

*** *Note: fracture frequency is the reciprocal of average core length*

Longest Piece: The length of the longest piece is measured in feet for each drill run, and plotted against the other calculated parameters.

4.8 BOREHOLE GEOPHYSICAL LOGS

Borehole geophysical logs should be obtained from available vertical holes when appropriate to assist in determining site lithology. These logs are normally used in both fractured rock and rock containing secondary porosity. Uses of each are discussed below.

• Natural Gamma — of primary value in lithologic description; allows distinction between shale and sandstone as well as clay mineral content of shale.

• Spontaneous Potential — correlated with the natural gamma logs to pinpoint interfaces between clay–rich

shale and sandstone, by measuring the electric potential that naturally occurs at these boundaries.

• Resistivity — correlated with SP and natural gamma; helpful in determining zones of interconnected, high–porosity rock; these zones may be higher in hydraulic conductivity or conductivity of formation waters.

• Caliper (3–prong for maximum resolution) — run to allow for hole diameter corrections, if needed; also serve as an indication of rock quality (i.e., enlarged areas of the corehole usually indicate weak rock washed out during coring).

• Deviation Survey — for checking borehole alignment (vertically) and correcting elevation of geologic contacts.

• Temperature — To define thermal anomalies and associated contact zones or discontinuities.

Additional descriptions of downhole geophysical techniques are presented in Chapter 6.0 fractured rock assessments.

4.9 FIELD INSITU TESTS FOR PHYSICAL PARAMETERS

Field testing of geologic materials, both soil and rock provides a wide variety of physical properties. Although the test methods have many limitations the data acquired throughout these tests have found wide applications in site assessment evaluations. Applicability of these methods is provided on Figure 4-31. These methods can also be arranged into a pictographic Figure 4-32. Common in-situ tests used in waste disposal projects can be summarized:

• Field Density tests	Bulk density
• Vane Shear	Shear strength
• Direct Shear	Shear strength
• Pressure Meter	Shear strength
• Plate Load Test	Bearing Capacity/Settlement
• Standard Penetration Test (STP)	Empirical relationships
• Dynamic Cone Penetration	Qualitative consistency of soils
• Static Cone Penetration	Bearing capacity, settlement
• Permeability Tests	Horizontal hydraulic conductivity
• Piezometer	Head levels-pore water pressure

The various properties of soil or rock units that are necessary for site assessment geotechnical evaluations can be grouped into three classifications: shear strength,

Table 4-8 Causes of Low RQD

1. **Pervasive bedding-plane and foliation** discontinuities representing the origin of the rock.

2. **Metamorphic influences** caused by temperature and pressures of igneous intrusions, batholith to dike in size.

3. **Tectonic forces** that have brought the rocxk to and past its elastic limit.

4. **Mineral alteration** from temperature, pressure, and by cation exchange related to diagenesis and to ground water movement.

5. **Ground water dissolution** of soluble minerals and/or interstitial cements.

6. **Near-surface weathering** effects.

Source: A. Hathaway

deformation characteristics and hydraulic conductivity. In the early days of site assessments (i.e., before 1950) samples were traditionally gathered in the field and sent off to the laboratory for evaluation. Hvorslev and others pointed out in many technical papers that samples collected in the field were often disturbed and did not represent large enough volumes to be fully representative of the in-place or insitu conditions. It was observed that in certain cases it may be technically more feasible to determine the required insitu soil properties by means of insitu borehole testing. Improvements in apparatus, instrumentation, measurement techniques and analysis procedures have led to the increased use and acceptance in since the mid 1950's of insitu measurement techniques. The primary advantages of in-situ borehole testing (Hathaway, 1988) are the ability to:

- Determine properties of soil that cannot easily be sampled;
- Avoid sample disturbance, improper stress states, and changes in physical and biological environment that may influence laboratory testing; and,
- Test a volume of soil or rock that, in some cases, is larger than that which can be tested in the laboratory.

Even with the above advantages insitu testing can represent a more costly procedure and can have a number of test limitations:

- Uncertain empirical correlation between measured quantities and actual properties may exist;

Table 4-9 Quality Control Guidelines for Determination of RQD

1. Never let the drillers handle the core beyond removal from the core barrel. This avoids unnecessary breakage.

2. The logging geologist should personally place thew core into the core box.

3. Whenever possible (that is, when core is not wate-degradable, or there are other compelling site characterization resons such as are often present in environmental assessment work), wash the core first inspection, as the core is placed into the core box.

4. Always place a strike mark, in indelible ink, perpendicularly across all fractures judged, on first inspection, to be "natural" and therefore, representing true discontinuities.

5. Aligh the core in each core-box trough so that the string of core from each run fits together as closely as possible.

6. Then, with the core aligned uniformly for the core run, measure the length of individual segments of core, determining which are greater than or less than 10 cm (the standard RQD length determinant).

7. Always divide the total length of rock core fragments greater than 10 cm each by the total length of the attempted core run.

8. When an interval of poor-quality rock ("bad ground") is encountered as a successive string, in two adjacent core runs, consider recomputing the RQD for only their combined lengths. This additional RQD determination is then written in the log with a special annotation, thereby representing a realistic determination of the actual presence and location of poorer-quality rock.

Source: A. Hathaway

- Flow (in hydraulic conductivity tests) and stress direction cannot be independently varied; and
- Applied principal stress directions in the field test may differ from those in real problems.

The above tests are briefly described below, additional information concerning the tests is contained in the references. The methods that have important applications to site assessments for solid hazardous waste disposal, such as, piezometers, permeability tests, various Dutch cone penetrometer tests are described in more detail. These tests are summarized in Figure 4-31. Field tests for hydraulic conductivity are extensively reviewed for both vadose and below ground-water surface evaluations in Chapter 5.0.

Some field tests are relatively inexpensive and can be performed on a routine basis (e.g., field density tests, the various borehole and penetration tests, and the index tests). Other field tests are expensive and must be designed specifically to account for the nature of the works and the character of the ground mass. These later tests should not be undertaken before a comprehensive understanding is obtained of the geology and nature of the subsurface.

4.9.1 Field Density Tests

Field testing of soil bulk density is a common and useful procedure. When coupled with moisture content determinations, the test results can be used to obtain the dry density of the soil. A major use of such testing is for the control of compaction of embankments. Where it forms the 'field' portion of a relative compaction test, (the other portion of the test being carried out in the laboratory), field density testing may also be used in evaluation of insitu materials and old fills, where it provides a direct determination of density independent of the sampling disturbance normally present in laboratory tests.

In essence, most of the available methods of field density testing depend on the removal of a representative sample of soil, followed by determinations of the mass of the sample and the volume it occupied prior to removal. However, nuclear density methods are an exception to this need to remove a sample of soil for testing. Accurate measurements of sample volume are more difficult and may lead to significant variations in test results, depending on the technique used, which is in turn dependent on the nature of the soil being tested.

All the test methods described below require physical access to the soil insitu. Therefore, they are normally restricted to soil within 2 to 3 m of the surface, although they can also be used equally well within backhoe pits, caissons or shafts. Use of the neutron probe technique is

an exception to this depth limitation, since the neutron geophysical probe can be lowered down aluminum or seamless steel access tubes.

The methods described generally measure bulk density, and representative moisture contents are required if the dry density is to be calculated. Ideally, the weight of the moisture content sample should be determined on site, then the sample should be transported to the laboratory for oven drying in accordance with ASTM D-2216 or BSI (1957b), Test 1A. Otherwise, the entire sample has to be preserved in an airtight container until it can be weighed. Alternatively, a rapid determination of moisture content can be made using a microwave oven, the 'Speedy' moisture tester, or one of the rapid methods described in BSI (1957b), Test 1. However, all such rapid determinations should be thoroughly correlated with the standard oven-drying technique for the particular soil type being tested. In any case, moisture content samples should be as representative and as large as practical, or several determinations should be made in order to obtain a reliable mean value.

Nuclear Methods

Nuclear methods of density measurement at shallow depth are described in ASTM (1985e). They do not measure density directly, and calibration curves have to be confirmed for each soil type. Calibration involves measuring the densities of representative samples of the soils determined by one of the direct methods for moisture density. However, once this nuclear calibration curve has been generated, and assuming there are no significant changes in soil, the method is very much faster than the others. It is therefore most suited to situations where there is a continuous need for many density determinations over a period of time, and the soils do not vary to any significant extent. It should be noted that the density determined by nuclear methods is not necessarily the average density within the volume involved in the measurement but may represent only a segment of the sample being tested.

The measurement of moisture content at shallow depth by the nuclear technique is described in ASTM (1985h). In many modern nuclear instruments, measurements of both density and moisture content are made simultaneously.

Nuclear measurements of density and moisture content can also be made at depth by employing a geophysical nuclear probe within a borehole (Brown, 1981; Meigh and Skipp, 1960). This technique may be particularly useful when undistorted samples cannot be obtained readily, such as in some fine granular soils. As only indirect measurements are obtained, the limitations mentioned

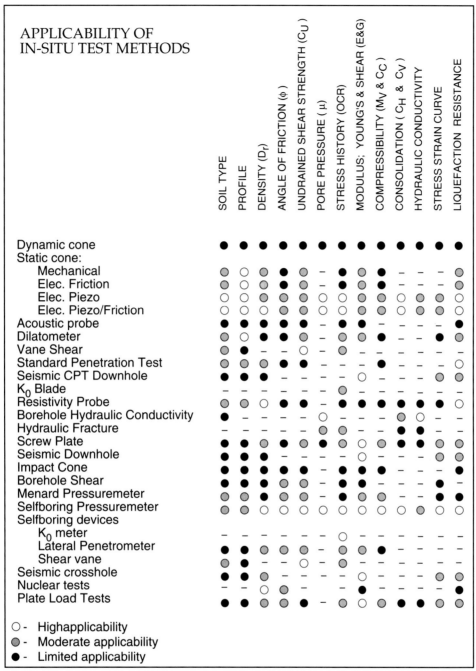

After Campanella & Robertson, (1982)

Figure 4-31 Comparison of Insitu Field Methods

above for shallow or surface nuclear techniques apply equally well to the nuclear geophysical probe.

All nuclear techniques use radioactive materials, and appropriate safety precautions must be followed. The use and handling of nuclear instruments should be fully in accordance with the manufacturer's recommendations and applicable regulations.

4.9.2 Stress Measurement In Soils

The analysis of the response of soil masses to applied loads requires reliable data on their strength and deformation characteristics, and, as these are stress dependent, a knowledge of the insitu state of stress assists in their evaluation by laboratory testing. Direct insitu

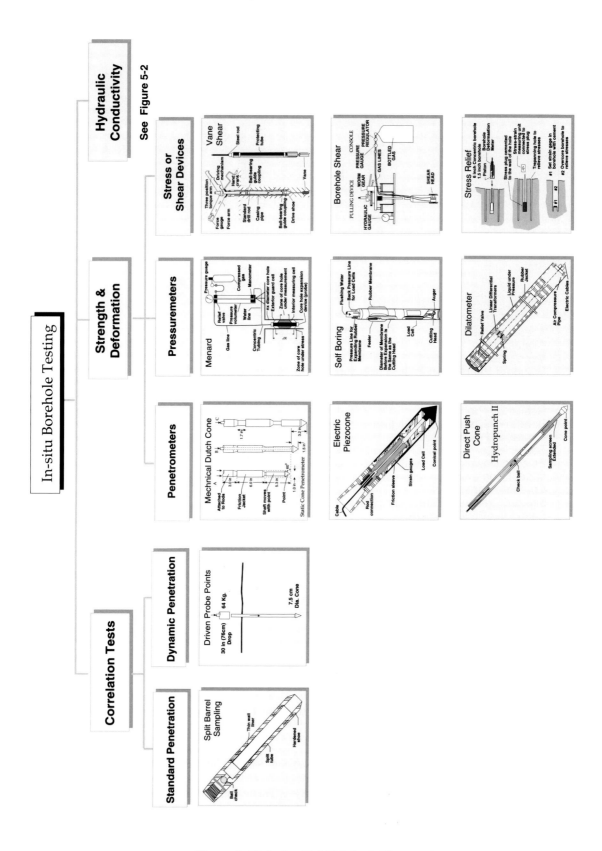

Figure 4-32 Insitu Field Methods Pictograph

measurement of the initial state of stress in soils is difficult because the disturbance created by gaining access to the soil or rock mass is generally non-reversible, and several times that produced by a stress-relieving technique. The accuracy of most instruments that have been developed suffers because of the disturbance created in the ground by drilling.

It is usual to measure only horizontal stress, and to make assumptions concerning the level of vertical stress based on the overburden depth. Only total stress may be measured; therefore, to determine the effective stress conditions, the pore water pressure at the test depth has to be measured or assumed. Methods of determining pore water pressure in the field is discussed in Section 4.9.

In soft clays, hydraulic pressure cells have been carefully jacked into the ground, or installed in a pre-bored hole (Kenney, 1967). The "Camkometer", a self boring pressuremeter, reduces disturbance to a minimum by fully supporting the ground it penetrates (Windle and Wroth, 1977). The total horizontal insitu stress may then be obtained by measuring the contact pressure. Facilities to measure pore pressures are available in the same instrument. Hydraulic fracturing has also been used to estimate minimum horizontal stresses in soft clay (Bjerrum and Anderson, 1972).

In large excavations, pressure cells are sometimes used to measure the contact pressure between the soil and a retaining structure. The type and position of a cell should be chosen with great care, because the introduction of the cell into the soil causes a redistribution of the stresses around it, and the errors depend on the geometry of the instrument. Details of the types of cells available and the problems that may be found when using them are given by Brown (1981) and Hanna (1985). Some of the factors that affect the accuracy of contact pressure cells are discussed by Pang (1986).

Insitu Direct Shear Tests

In this test, a sample of soil or rock is prepared and subjected to direct shearing insitu. The applied stresses and boundary conditions are similar to those in the laboratory direct shear test. Vertical stress is imposed by a hydraulic jack and shear forces are from a second jack to the point of failure. The test is generally designed to measure the peak shear strength of the intact material, or of a discontinuity (including a relict joint in soil), as a function of the normal stress acting on the shear plane. More than one test is generally required to obtain representative design parameters. Deere (1976) recommends at least five direct shear strength tests for each geologic feature to be tested. Each of these tests would be run with a different level of normal stress to

define points for the construction of a Mohr's envelope. The measurement of residual shear strength can present major practical problems in arranging for a sufficiently large length of travel of the shear box, but a useful indication of residual strength of the interface between concrete and rock or soil.

Insitu shear tests on soil may be carried out either within boreholes (Bauer and Demartinecourt, 1982; Handy and Fox, 1967) or near the ground surface (Brand et al, 1983b). Equipment for testing close to the ground surface in trial pits may be adopted to enable testing to be carried out within deep excavations, large diameter shafts or caissons. Insitu shear tests on specific discontinuities in rock may also be conducted using similar equipment; the results may be used to confirm the strength of discontinuities derived from laboratory tests and field roughness surveys.

Vane Shear

The field vane shear test is widely used in the determination of undrained insitu shear strength for soft sensitive clays. There are many types available varying from small hand vanes (Pilcon and Geonor) to vanes operated from large mechanical torque heads. Specific detail of a standardized vane shear test is described in ASTM D-2573 (18 of BS 1377-1975). The vane shear test consists of advancing a four-bladed vane to the required depth; then measuring the applied torque as the vane is turned at a constant rate. The undrained shear strength is then calculated from the measurement of the applied torque.

The test is normally restricted to fairly uniform cohesive fully saturated soils, and is used mainly for clay having an undrained shear strength up to about 100 kN/m2. The results are questionable in stronger clays, or if the soil tends to dilate on shearing or is fissured.

The vane shear test methods have many limitations relative to the test equipment, failure surfaces being predetermined and planes of failure being tested by the device. In clays the vane shear can only measure undrained shear strength. In free draining soil care must be taken to ensure that either a fully drained or a completely undrained shear strength is measured. Sand environments should not be tested by this method.

Even with these limitations the vane shear methods provide reliable strength index test of the relative measure of the variation of soil strength for subsurface materials at various depths and locations. With calibration the test also provides a convenient method to obtain undrained shear strength values for site soils.

Pressure Meters

Although several types of pressure meters exist, a fairly robust pressure meter design was first developed in 1954 by Menard in France. This device is first lowered into a pre-drilled borehole. The pressure of guard cells is adjusted to an estimated in-situ lateral stress at the depth of measurement. The center cell is then pressurized by water and the volume of water vs. pressure is plotted. The pressure meter is used to best advantage in soft silts, sensitive clays, brittle soils derived from glacial tills or weathered rock, interbedded or layered sand, silts and clays. Each of these materials is difficult to obtain in an "undisturbed" sample.

Two types of pressure meters are in use today: the Menard type and the Self-Boring type. The self boring pressure meter drills its own hole and overcomes some test limitations or problems with soil disturbance. The self boring pressure meter may also be called a Dilatometer or Camkometer. Hathaway (1988) discusses the use of pressure meters and additional test procedures can be found in , Menard (1965), Dixon (1970) and CGS (1978). Although the use of the pressure meter has been criticized by many, it has been shown to give very useful results in testing the degree of consolidation (i.e., dynamic consolidation) and is likely to continue to be used experimentally for other purposes. Menard claimed its use in giving a value for undrained shear strength. As such it shares many of the problems listed for the shear vane.

Standard Penetration Test

The Standard Penetration Test (STP) is probably the most widely used field test in subsurface exploration programs and has been defined by ASTM Method 1586 titled the "Penetration Test and Split-Barrel Sampling of Soils". The test consists of driving a split-barrel sampler (split spoon) into a soil deposit with a hammer of defined weight and recording the number of blows, (blow count, N), required to drive the sampler 30 cm (12 in.). Almost every geotechnical investigation contains evaluations of SPT's found in the field. It is important that the test is carried out precisely as described in ASTM D-1586 (test 19 of BS 1377: 1975), since even minor variations from the specified procedure can seriously affect the results. There is much published information linking the results of the test with other soil parameters and with the performance of structures. The main purpose of the test is to obtain an indication of the relative density of sands and gravels, but it has also been used to obtain an indication of the consistency of other soils (silt and clays) and of weak rocks (e.g., chalk). As such a number of correlations has been proposed for comparison of SPT values with relative density and bearing capacity. These are shown in Table 4-10 for both cohesive and noncohesive soils.

The great merit of the test, and the main reason for its widespread use is that it is simple and inexpensive. The soil strength parameters that can be inferred are approximate, but give a useful guide in-ground conditions where it may not be possible to obtain borehole samples of adequate quality, e.g., gravels, sands, silts, clay containing sand or gravel and weak rock. In conditions where the quality of the 'undisturbed' sample is suspect, e.g., very silty or very sandy clays, or hard clays, it is often advantageous to alternate the sampling with standard penetration tests, thereby obtaining a check on the strength.

Although a good case can be made to continue to use STP methods to provide a standard for comparison with other sites, unfortunately STP's do not provide sufficiently undisturbed samples for most site investigations for waste disposal projects. The money spent in hammering down a split spoon may be better spent in collecting continuous core using thin wall tubes or through internal hollow stem auger core cutting equipment. If samples are obtainable using the continuous sampling methods available for many of the hollow stem auger drilling systems, then continuous soil or semi-consolidated rock samples may be quickly obtained for detailed stratigraphic logging.

Table 4-10 Standard Penetration vs Relative Density and Consistency

Cohesionless Soils

No. of Blows (N)	Relative Density
0 - 4	Very Loose
4 - 10	Loose
10 - 30	Medium
30 - 50	Dense
Over 50	Very Dense

Cohesive Soils

Consistency	No. of Blows (N)	Unconfined Compressive Strength tons/ft.	kPa
Very soft	2	>0.35	<23.9
Soft	2 - 4	0.25-0.50	23.9-47.9
Firm	4 - 8	0.5-1.0	47.9-95.8
Stiff	8 - 15	1.0-2.0	95.8-191.5
Very stiff	15-30	2.0-4.0	191.5-383.0
Hard	>30	>4.0	>383.0

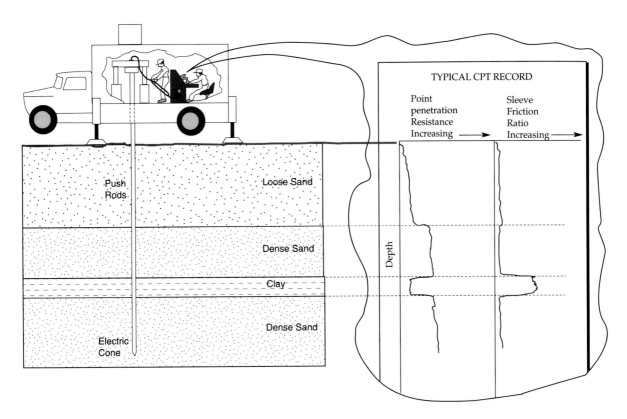

Figure 4-33 Direct Push Testing Procedures

4.9.3 Direct Push Cone Penetration Testing

Cone penetrometer test systems have been routinely used in Europe since 1917 and in the U.S. since about 1965. Several types of static probing equipment have been developed and are in use throughout the world (De Ruiter, 1982; Sanglerat, 1972). The basic principles of all systems are similar, in that a rod is pushed into the ground and the resistance on the tip (cone resistance) is measured by a mechanical, electrical or hydraulic system as shown in Figure 4-33. The resistance on a segment of the rod shaft (friction sleeve resistance) may also be measured. The cone penetrometer system has been recently expanded to include the ability to sample for chemical parameters as the probe is advanced. This is called the "direct push" method in sampling. Static probing, or cone penetration testing, is also known by a number of other descriptive terms, depending on the manufacturer or operator of the particular device being used.

There are several standards for cone penetration testing, given by the ISSMFE (1977) and the ASTM (1985). A number of others are under further development. Both of these test standards recognize a number of traditional types of penetrometers, and it is imperative that the actual type of instrument used in the test program is fully documented since the interpretation of the results depends on the equipment used. These types of penetrometers: mechanical, electrical, and direct push sampling are described further below.

Mechanical Cone Penetrometers

Two of the most common mechanical cones, the Dutch mantle cone and the Dutch friction sleeve cone were developed primarily at the Delft Soil Mechanics Laboratory, Holland, in the 1930's. With either type, the cone is pushed into the ground by a series of hollow push rods. With the mantle cone, the force on the cone is then measured as the cone is pushed downward by means of inner rods inside the push rods. This force is generally measured at the ground surface by a hydraulic load cell. With the friction sleeve cone, the same initial measurement is made, and then a second measurement is taken while the cone and friction sleeve are together pushed downward a further increment. The friction is calculated by deducting the former reading from the latter. This procedure is normally repeated at regular depth intervals of 0.2 or 0.25 m.

An alternative, quick continuous method of penetration is sometimes used with the mantle cone. In this method, the cone and push rods are pushed into the ground with the cone permanently extended and connected to the load cell. Accuracy of the system is reduced in this method, however, and care should be taken to check the free movement of the cone at frequent intervals.

For accurate work, the weight of the inner rods should be taken into account in calculations. In very soft soils when soundings are carried to a significant depth, the weight of the inner rods may exceed the force on the cone or cone plus jacket; in these circumstances, it is impossible to obtain readings.

Electrical Cone Penetrometers

There are a number of types of electrically-operated cone penetrometers in use in the United States which incorporate vibrating wire or impedance strain gauges for measuring the force on the cone and friction jacket (see Figure 4-34 for an example direct push chart of correlations for parameters measured during the test). In addition, immediately behind the cone tip are stainless steel filters for measuring the induced dynamic pore pressure generated through the test. With this set-up, penetrometer tip resistance (gc), side friction (fs) and pore pressures (u) can be measured as the probe is pushed into the soil at rates of 4 ft/min. (2 cm/sec). Figure 4-35 shows areas in the United States where cone penetration equipment may be usefull. Before using such equipment one should first confirm that the method is useable in the project area. The values of tip resistance, pore pressure, and side friction as well as depth, in duration and total pushing force are recorded continuously and stored on a computer tape for subsequent evaluation. Electrical cone penetrometers provide for on-line data correlation to produce soil behavior types, equivalent SPT "N" values, corrected equivalent SPT "N" values, drained friction angle (phi) for sand, and relative density. These interpreted parameters can provide significant enhancement of directly collected and logged stratigraphic data. Figure 4-36 illustrates results of a electrical cone penetrometer test in glacial sands and clays. Direct access to the data during or immediately after sounding is recommended to allow field evaluation of conditions. Provision should be made for calibration of the force-measuring system at regular intervals, preferably on site. An inclinometer built into the cone is available with some equipment.

Direct Push Dutch Cone Sampling

A third type of direct push penetrometer consists of a traditionally shaped tip Dutch cone with special ports that allow sampling of ground water at specific depths while the cone is being advanced into the subsurface. Such devices provide a rapid sampling system that can be of significant assistance in assessment monitoring programs. Both vertical and horizontal profiles can be developed for indicator parameters such as chloride, conductivity or VOC's. The direct push sampler can be combined with an infield Gas Chromatography lab to obtain reliable ground-water organic parameter values for the geologic system.

Uses and Limitations of the Test

The cone penetrometer test is relatively quick to carry out, and inexpensive in comparison with boring, sampling and laboratory testing. It has traditionally been used to predict driving resistance, skin friction, and the end bearing capacity of driven piles in granular soils. In recent years, the cone penetrometer test has also been used to give an indication of the continuous soil profile by interpretation of the ratio of friction sleeve and cone resistance. In addition, there is substantial published information relating cone resistance value with other soil parameters. More recent combinations of cone penetrometer and pore water sampling ability allows a rapid three-dimensional system

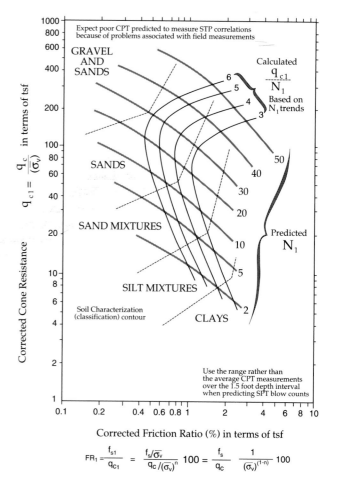

Figure 4-34 Direct Push Chart of Correlations

for evaluation of lithology, ground-water head levels and ground-water quality.

Combining of penetrometers and sampling methods can provide a significant tool for the rapid evaluation of three-dimensional water quality. It should be noted, however, that even with rapid collection of depth-location (x-y-z) and water quality, this method should not be used as an excuse for punching hundreds of holes around a site while looking for a leachate plume. Lithologic, hydraulic conductivity and gradient data should be known before sending out the penetrometer rig. One can then home-in on the potential target flow paths to evaluate in-situ water quality.

The cone penetrometer test is also the preferred substitute for the standard penetration test in soil conditions where results of the latter test are suspect, and where hard driving is not anticipated. The test is also commonly used as a rapid and economical means of interpolating between boreholes. Although it may be possible to estimate the type of soil through which the cone is passing as described above, it is preferable to carry out the test in conjunction with some other means of determining the general nature of the soil present such a direct sampling through augers. These known lithologic profiles are then calibrated to cone penetration results.

Extreme care must be taken in ground conditions of alternating confining units and overlying soft unconsolidated sands. While withdrawing the penetrometer the sand may enter the open hole before completion grouting of the boring. This may result in interconnecting upper sand units with lower units by the formation of inadvertent vertical drains. This could potentially result in environmental impairment of the site. In addition, extreme care should also be used if potential DNAPL's are perched on top of lower hydraulic conductivity geologic units. Punching holes through the perching units will effectively drain the heavier than water phased liquids down into lower aquifers and potentially cross-contaminate the entire system. Cone penetrometer manufacturers are currently working on methods of pumping sealants into the penetrometer hole as the device is withdrawn from the hole.

Cone penetration is limited by both the safe load that can be carried by the cone, and the thrust available for pushing it into the ground. It is also limited by the compressive strength of the inner rods. Because of limited cone capacity, penetration for any direct push device normally has to be terminated where dense sand or gravel, highly to moderately decomposed rock, or if cobbles are contacted.

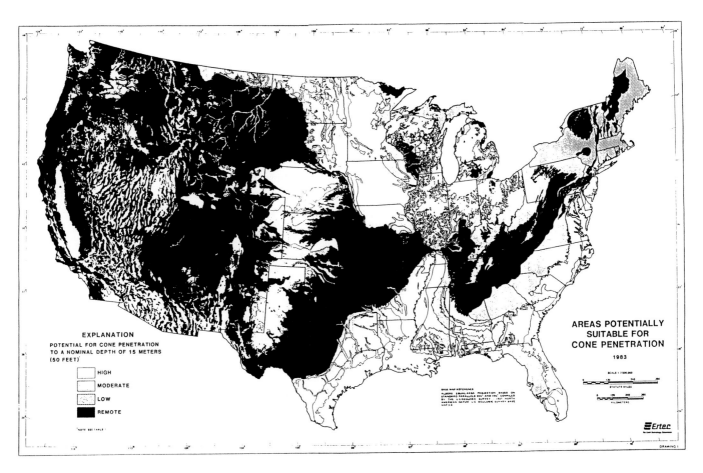

Figure 4-35 Potential Areas for Use of Piezocone Equipment

Source: ERTEC

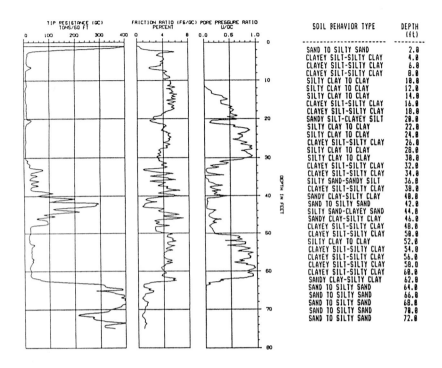

Figure 4-36 Example Data Set From Piezocone Equipment

4.10 PIEZOMETERS

Piezometers provide access to saturated units for measurement of ground-water head levels. Piezometers have come into wide use to provide a reliable method for the determination of hydrostatic pressures within strata, below the free ground-water level. Since piezometers screens are located below the free ground-water level, (or water "table"); these installations require a cased hole that is sealed above the open area of the device. This is to keep leakage to, or from the over lying strata. Hvorslev, (1949) recognized the importance of tight annular seals in piezometers and provided an example piezometer (pg. 80) that looks very similar to those installed today, more than 40 years later. Piezometers were initially used because soil mechanics problems required the measurement of the distribution of pore pressures within natural materials. One of the ways of obtaining this information is to carefully seal piezometers, containing permeable tips at their lower end, into the ground so that the tips are located at the positions from which pore pressure measurements are required. The level to which water will rise in these tubes is called a piezometric level. This terminology is applied to all such installations, which means that piezometric levels are being recorded from aquifers and confining units (clays, etc.) and from confined horizons and unconfined horizons. Hence the term 'piezometric level' has come to have two meanings, (i) any water level measured by a piezometer, and (ii) the water level above the top of a confined aquifer.

Unlike the water levels in piezometers the water levels in uncased holes or widely screened monitoring wells are a function of all the water pressures contacted by the hole, (integrated hydraulic head over the full screened area), and not just those at their tip. This means that the water levels in uncased holes need not always represent the true elevation of the water-table. Similarly, the water levels in piezometers may not represent the true water-table.

4.10.1 Piezometer Placement Decisions

After the preliminary Phase II field investigations are completed, the data must be reviewed. Maps and cross–sections may need revision as a result of drilling and borehole geophysical logging. A determination must be made concerning the use of the existing borings early in the Phase II program. A selected number of soil borings and deep stratigraphic borings should be converted to piezometers for measurement of ground-water levels. A considerable amount of hydrogeologic data will have been obtained during the drilling and sampling operation. These data should include:

- Depth and geologic unit in which ground water is found (i.e., throughout a water–bearing unit or only within joints and bedding planes);

- Water levels (at several intervals of hours/days after the borehole has recovered from effects of air–lift pumping);

- Specific conductance of the water, which should correlate with TDS;

- Rough estimates of aquifer parameters (from short–term air–lift recovery tests); and

- Penetration rates, rod drops, and zones of air loss.

A review of this information, and consideration of the information obtained during the Phase I study, should provide adequate information for locating borings that will be converted to small–diameter piezometers. The number of piezometers to be installed will depend upon several factors including:

- Number and location of borings;
- Geologic complexity of the site;
- Size of site;
- Number of aquifers or hydrostratigraphic units; and,
- Extent of existing or potential contamination.

Normally, the number of piezometers constructed for water level measurements will exceed the number of ground–water monitoring wells. As with drilling, sampling, and monitoring well installation, the need for professional control over installation of piezometers cannot be over–emphasized. In areas where near–surface studies are adequate to establish and monitor the ground–water regime, the program discussed in the following paragraphs may be implemented for the construction and installation of simple stand pipe (or Casagrande) piezometers in areas of future land disposal. The primary type of piezometer (summarized) consists of installation of a small diameter, 1 inch to 3/4–inch, PVC tube. The borings selected should generally be completed in accordance with the ASTM D5092 specification for monitoring well installation, with the following acceptable modifications:

- The piezometers should be of 3/4 to 1 inch PVC with machine–cut slots to obtain the open screened area. The screened length should be based on drilling observations of permeable zones and should adequately monitor pressure heads within individual zones.

- Small–diameter piezometers may be installed in areas where future land filling may require removal of the piezometer at some point during the life of the facility. These installations should be carefully considered so as not to represent pathways into underlying aquifer units, in the advent of future land filling operations. In such instances, upon decommissioning, the boring must be carefully survey–located then drilled–out and back–grouted before being covered by the landfill. Deep vadose zones should not have cross–connecting holes drilled beneath the landfill.

- Generally, only one piezometer should be completed in each hole in unconsolidated sediments; however, cored holes or very stable open or cored holes may have multiple completions, if such are properly designed.

Water level recovery in the piezometer should be monitored for hours to days, in order to estimate hydraulic conductivity of the screened interval and to determine static water level at the time of installation.

The manner in which boreholes drilled for lithology and sampling should be completed is the subject of considerable current controversy. Experience on projects over the past several years has shown that wells, regardless of their initial purpose, eventually become construed as "monitoring wells" and that the data from these wells are subject to criticism, often used improperly (leading to erroneous results), and frequently create difficulties in dealing with regulatory agencies. Other regulatory agencies do not recognize piezometer as primarily head measurement devices but rather larger diameter (i.e., New York State 360 regulations that require two inch diameter piezometers completed as a single monitoring well) analogies to monitoring wells. In Wisconsin piezometers are traditional monitoring wells completed below the ground-water surface.

Measurement of hydraulic head values requires specific design components that set piezometers apart from monitoring wells. Monitoring wells for detection programs typically have longer screened sections ranging from 10 to greater than 30 feet long. These monitoring wells need diameters around 2 inch nominal sizes to permit sampling devices to fit in the well. Conversely, piezometer installations are designed for quick response to hydraulic head changes, hence, they are of small diameter with short screened zones. For this reason, the use of small–diameter piezometers (3/4 to 1 inch), pneumatic transducer, or multiport systems have come into favor where measurement of water levels is the primary data collected. In general, those boreholes penetrating the ground–water surface would have at least a single port (porous tube) piezometer installed. This installation can be rapid since only filter sand covers the screened zone (plus 3 feet above the screened zone). Then the piezometer can be grouted to the surface with an expansive, high–solids bentonite–grout. The surface completion should be similar to the ASTM D 5092 standard for ground–water monitoring wells to protect the riser–pipe "stick–up" of the piezometer.

Geotechnical and stratigraphic boreholes provide the basic site geologic lithology in field investigations. These data are, however, only a part of the total picture necessary for definition of appropriate monitoring locations within the geologic environment. Potentiometric head (and gravity) is the driving force in ground-water movement. This driving force should be established in any detailed hydrogeologic investigation. Such investigations involving determination of the free ground-water level or levels, and hydrostatic pressures in various geologic units often require the installation of short screened small diameter standpipe piezometers, or in lower hydraulic conductivity units, hydraulic, pneumatic or electrical piezometers.

The installation of multiple piezometer points in a single hole has received significant controversy over the years due to the difficulty of hydraulically separating the various screened zones. The effectiveness of multiple point completions (say 5 to 20 points) was clearly demonstrated in research efforts (Cherry 1982), and in projects associated with waste disposal sites. The Borden test site has illustrated the technical benefits of detailed cross-sectional hydraulic and water quality data developed from multiple completion piezometers. Through these and other research efforts, a number of multiple point piezometer devices have come into popular use in recent years. Table 4-13 provides a summary of some of these systems. Simple multiple open tube piezometers in a single borehole were often found to be difficult to install properly, since intercommunications of pressure heads between the screened zones were occasionally observed between the various piezometer tips. This construction difficulty was due to the multiple risers restricting placement of sand (for filter packing screened zones) and annular sealant bridging within the restricted borehole diameter between the screened zones. This type of multiple completion left little room for tremmie pipes to assure accurate backfill placement of grouts and sand packs.

4.10.2 Multiple Piezometers in a Single Borehole

There are several design approaches that can be used for the completion of multiple piezometers in a single borehole. Several examples are illustrated in Figure 4-37. If the formation is sufficiently cohesive so that sand pack and seals can be placed in discrete zones without collapse, several stand-pipe piezometers can be placed in the

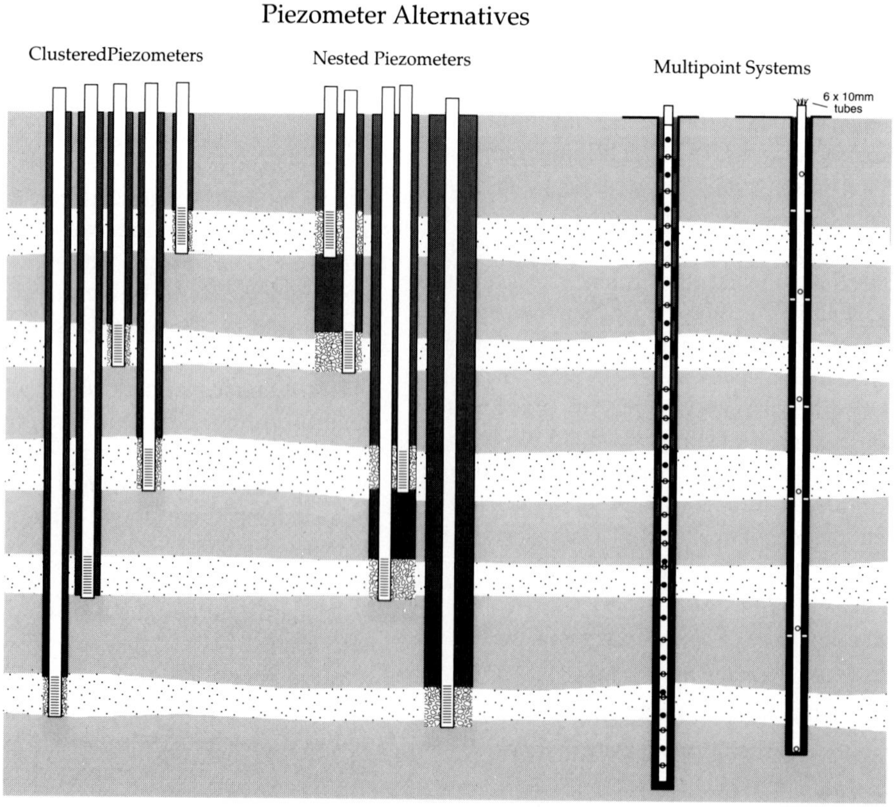

Piezometer Alternatives

Figure 4-37 Well Completion Classifications

borehole. The construction difficulties in single borehole multiple stand pipe piezometers can be addressed through use of spacer devices that keep the piezometer risers separated and allow for use of one of the risers as a tremmie pipe. This device allows placement of up to four one-inch piezometers to depths typical in hydrogeological investigations through a four-inch ID (6-1/2" OD) hollow stem auger. Multiple point piezometer construction requires an adequate seal between the piezometer tips. If a poor seal is constructed, vertical movement of water may occur along the borehole. This can cause the head levels from the piezometers to be an average reading of the pressure levels in the fully screened formation as a result of preferential vertical movement of ground water between the interconnected piezometer tips. Unfortunately, the existence of leaky seals in a multiple piezometer installation can, in some cases be difficult to detect. Mcalf et. al., (1981) reported a technique of measuring the static water levels in the surrounding piezometer tips of an installation, while one is being stressed by pumping. If the head levels do not change, leaky seals are probably not a problem.

A piezometer nest (i.e., piezometer cluster) as shown in Figure 4-37 is composed of individual piezometers in separate boreholes, at different depths, is generally considered to have a lower probability of problems due to potentially leaky seals. However, the increased cost of drilling may outweigh the benefit derived from separate installations. In the design of a ground-water head measurement network, the choice of piezometer nests versus multiple piezometers in a single borehole is a balance between the higher reliability associated with the more expensive nest compared to a potentially lower reliability associated with the less expensive multiple piezometer installation. With the relatively short term requirements for most site assessment projects the multiple point piezometer may represent the most cost effective technique for obtaining a number of head level readings at a single map location. These multiple head measurements provide important vertical head data used in construction of flow nets, conceptual and computer models.

Multiple-Port Casings

Multi-level installations contained in a single borehole make use of a single casing (pipe) fitted with openings (ports) leading into the formation at several depths. The design of the ports, and the ways in which they are operated and accessed from the surface are dependent on the specific design of the installation.

In relatively cohesive formations casings can be installed with sand packs around the ports, and with seals or packers between the ports. In cohesionless formations, the seals may be unnecessary, since the formation

materials will collapse around the casing when it is installed.

One type of multiple-port casing installation suitable for shallow, unconfined aquifers was described by Hansen and Harris (1974). The device consists of fiberglass probes imbedded in a sand matrix retained in a well point. The portions of the sand matrix containing each fiberglass probe is isolated from adjacent ones by caulking material. Each probe is connected to the surface by a 0.6 cm (1/4 in) tube contained in a casing that extends from the well point to the surface. The apparatus is implanted in the formation at a specified depth and sampled by suction-lift.

A multi-point installation described in Patton (1979) is marketed by Westbay Instruments Ltd. (see Figure 6-38). It consists of a single string of coupled PVC pipe with a variable number of ports along its length. Packers are installed around the PVC pipe to isolate the individual

Figure 4-38 Installation of Multi-point

ports. There are two types of ports that can be coupled into the system, a dual pressure-measurement and water-sampling port, and a port with a much larger screened opening for pumping large volumes of water from the formation. The ports used for pressure measurements and water sampling is operated by means of control mechanisms mounted on down-hole monitoring tools that are moved up and down the casing. The tool used for hydraulic head measurement contains a pressure transducer with a surface electronic readout. The sampling tool contains a pneumatic bailer. Vispi (1980) reported that the pressure measuring device gave comparable results with head levels obtained from traditional stand pipe piezometers, but that the pneumatic bailer tool was reported to operate very slowly. The pumping ports consist of a larger screen-fitted opening with a sleeve-type window, opened and closed by control mechanisms mounted on a special tool. A packer-fitted submersible pump used to pump water from the borehole zone.

A specialized head measurement and water sampling system was developed at the University of Waterloo and is marketed by the Solinst Company. The system uses a PVC casing string to isolate and protect multiple sampling tubes. The casing used in the installation is of a modular construction, using flush threaded joints, fitted with O-ring seals. For use in consolidated rock boreholes, a feature of the system is the self-inflating packer module. Packers are used above and below each sampling port module, to isolate the strata to be sampled.

Up to 6 sampling tubes can be installed in a 3"(N) borehole and 10 tubes in a 4" (H) sized installation. (See Figure 4-38). Samples can be obtained using dedicated samplers in each port, or a portable sampling pump. For boreholes in overburden or as an alternate method in rock, packers can be omitted from the casing string. Sampling points are then isolated in the conventional manner, using bentonite and sand. A major advantage of the Waterloo System, when using this method, is that the sampling tubes are confined and protected within the central PVC casing string. Installation of the Waterloo Multi-level Sampling System is achieved quickly and easily, as it is assembled as it is lowered into the borehole. As each port is put into position, a new sampling tube is connected to the system. Each successive module or section of casing is then threaded over these sampling tubes.

The packer chemical is activated when water is poured down the central PVC string, during installation. Over the following 24 hours the chemical swells and pushes the rubber against the side of the borehole wall, and forms a tight seal.

The rubber sleeve ensures the seal against the borehole wall and prevents chemical reactions between the compound and the formation water outside the casing.

Woven Kevlar cuffs provide the strength required to withstand high differential pressures across the packer, which may be encountered in some installations. When a portable pump is to be used in conjunction with the Waterloo System, water level measurements can be taken, with a dipmeter, directly in the sampling tubes.

4.10.3 Piezometer Devices

Installation of piezometers during a field investigation should be regarded as an integral part of the overall assessment system. The typical field program commences with drilling and includes documentation of ground water observed during drilling, detailed examination of soil and rock cores, standing borehole water level measurement and some form of in-situ hydraulic conductivity measurement. The next step, (which may occur before hydraulic conductivity testing), consists of the selection and emplacement of the appropriate piezometer potentiometric head measuring system. The previous section described a number of multiple completions. These multiple (and single) completions also require placement of some type or 'class' of piezometer device in order that hydraulic head level can be measured from the installation. The basis for selection of the class of piezometer is dependent on the rate at which the piezometer device will react to changes in hydraulic head, the need for the piezometer to serve as a reconnaissance tool for water quality assessments, and the need for in-situ hydraulic conductivity testing.

The installation of a piezometer system requires full consideration of the location and depth within the specific geologic unit. The general location must be chosen so that it provides ground-water potentiometric head information relevant to facility or remedial design, foundation stability and ground-water monitoring system design. The position of the piezometer installation in a 'depth and location' context should provide for full hydraulic connection with the soil or rock mass. Field conditions such as hydraulic conductivity and water quality will affect both the types of instrument chosen and the installation procedure. Piezometer selection and installation process are described in the following subsections.

Piezometer Classifications

A general classification of piezometers with important system characteristics, is given in Table 4-11 and illustrated in Figure 4-39. The appropriate piezometer installation is almost always based on the need to obtain hydraulic data in sufficient time for project requirements. Open tube or Casagrande piezometers, (the most common design), can be limited due to the requirement for sufficient water entering the screened zone to fill the open tube as

Piezometer Evaluations

Gauge Pressure	Piezometer Type	Pressure Range	Response Time	De-airing Capability	Remote Reading Capability	Long-Term Reliability	Other — Advantages	Other — Disadvantages	Recommendations
Positive	Open-hydraulic (Casagrande)	Atmospheric to top of standpipe	Slow	Self de-airing	Not normally, but possible with bubbler system	Very good	Cheap, simple to read & maintain; insitu permeability measurement possible.	Vandal damage often irreparable.	First choice for measurement within positive pressure range unless rapid response or remote readingd required; response peaks can be detected by use of Halcrow buckets system.
	Closed-Hydraulic (Low air entry pressure)	Any positive pressure	Moderate	Can be de-aired	Yes	Depends on pressure measuring system 1) Mercury manometer- very good 2) Bourdon gauge - poor in humid atmosphere 3) Pressure Transducer- moderate but easily replaced.	Fairly cheap; insitu permeability measurement possible; can be made vandal proof if required.	Gauge house usually required; regular de-airing necessary; uncovered tubing liable to rodent damage if left exposed.	Usefull when remote reading, and for artesian pressure.
	Closed-Hydraulic (High air entry pressure)	-1 atmosphere to any positive pressure	Moderate	Can be de-aired	Yes	As above	Fairly cheap; insitu permeability measurements in low permeability soil are possible	As above; very regular de-airing required when measuring suctions.	Useful for measuring small suctions
	Pneumatic	Any positive pressure	Rapid	Cannot be de-aired; only partially self de-airing	Yes some head loss over long distances	Moderate to poor, but very little long term experence available	Fairly cheap; no gauge house required	No method of checking if pore water or pore air pressure is measured	Only suitable when tip almost always below ground water level and no large suction occur.
	Electric vibrating wire type	Any positive pressure	Rapid	As above	Yes but special cable required	Signal quality degenerates with time; instrument life about ten years, but reliability of instrument that cannot be checked is always a question.	--	As above; expensive zero reading liable to drift and cannot be checked.	Not generally recommended
	Electric resistance type	Any positive pressure	Rapid	As above	Yes, but with care because of transmission losses	Poor	--	As above	Not recommended
Negative (suction)	Tensiometer	-1 Atmosphere to positive pressure	Moderate to rapid	Can be de-aired	Yes	Good	Cheap, simple to read and maintain.	Vandal damage often irreparable; regular de-airing required.	First choice for measuring pore suction.
	Psychrometer	Below -1 atmosphere	Variable	Not relevant	Short distances only	Instrument life one to two years; little long term experence available.	--	Not accurate between 0 and -1 atmospheres.	Research stage.

Table 4-11 Piezometer Selection Matrix

pressure changes occur in the soil or rock matrix. Hvorslev (1942) described resultant hydraulic conductivity and time delays necessary to register 90% of the pressure change within piezometer devices. Hvorslev's work fifty years ago forms the basis for piezometer selection in site assessment projects.

4.10.4 Response Characteristics

Type A - Type A piezometers consist of an open tube or stand pipe installation called the Casagrande piezometer. The type A piezometer could simply consist of small diameter (3/4") PVC plastic pipe with factory slotted screens. These simple open-tube designs have been augmented for cored hole installations into multiple-packer ported systems. These types of piezometer completions have been successfully installed to depths in excess of 100m (300 feet). Stand pipe completions however, can suffer from very slow response times for installations completed in fine-grained materials (see Figure 4-40). The slow response times for simple open tube piezometers can be significantly improved by placement of a solid PVC or aluminum "dummy" within the riser. The use of this device reduces the volume of water necessary to change the hydraulic head within the open tube. Since the plastic dummy reduces the well bore storage effects by lowering the amount of water required to fill the open tube from the full volume of the hole (V = 3/4 d • H) to V = (3/4 dt • H) - (3/4 dd • H) hydraulic response time improves dramatically.

Type B - The B class of closed tube hydraulics (both low and high air entries) are used primarily for

experimental use in laboratories typically for negative pore water pressures. As such, they will not be discussed further in the context of field installations.

Type C - These type C mechanical diaphragm piezometers were divided by Canmet (1982) into two subdivisions. Piezometer Type C1 requires a pore water volume change each time a reading is taken by the device. Piezometer Type C2 uses water pressure against a diaphragm to compare with a measured air pressure. As such, they have negligible pore water volume demands.

Type D - These so-called vibrating wire strain gauge transducer piezometers are among the fastest responding piezometers. These are negligible volume change devices, and therefore, have excellent pressure response characteristics under most operating conditions.

Type E. Electrical resistance type piezometer are generally not recommended for site assessments due to questionable long-term reliability. They will not be discussed further.

4.10.5 Readout Methods

Type A - -The water level in the stand pipe is read by means of a dip meter or electronic tape — a probe consisting of a graduated insulated cable, with a lower (weighted) end, terminating in two electrical contacts. The contacts are connected to a resistance meter or to a lamp or buzzer. The cable is lowered down the stand pipe until indication is observed that the water level has been reached.

Type C - These "bubblier" type piezometers employ supply and return air or oil tubes for pressure readout, and require specialized readout equipment.

Type D - The operating method for this installation consists of impulsing a tensioned wire and measuring the natural frequency of the resulting vibrations. The electrical output employed is ideal for remote monitoring of head data since it can be electronically amplified for transmission over any desired distance without affecting signal accuracy. Remote readouts of the head data are possible for later down-loading to computers for interpretation of time/head level information. There are many commercial sources of D Class transducers that provide a full system that can measure rapidly changing pressure head data in wells for multiple points. Since these systems are often used in open tube wells and piezometers, they can only respond as rapidly as the open tube can respond (see Figure 4-40), which is inherently a slow

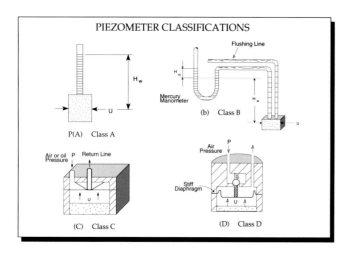

Figure 4-39 Piezometer Classifications

reacting design. These type piezometers can, however, show improve performance in open hole piezometers by using "dummies" as illustrated in Figure 4-41, to reduce the well bore storage effects.

4.10.6 Piezometer Selection

The important physical properties required of a piezometer installation are ruggedness, adequate accuracy and long term reliability for those systems dedicated for a piezometer installation. As general guidance, the simpler the installation, the more likely the device will perform satisfactorily for the required assessment period. Piezometer devices must be selected on the basis of three criteria:

• The expected hydraulic conductivity of the geologic material;
• the need for later testing of the installation for in-situ hydraulic conductivity; and,
• the need for later reconnaissance water quality sampling.

Accurate hydraulic response to changes in ground-water pressure requires that the volume of water necessary to operate the piezometer device must be readily available from the geologic unit without a significant pressure change. High volume demand piezometers are, therefore, unsuitable for measuring rapid pressure changes in low hydraulic conductivity materials. If sufficient response times are available for the geologic unit to yield water to the piezometer device, relatively large volume demand piezometers may be used. For example, in systems with little monthly potentiometric surface variations, large volume demand devices are appropriation; however, in geologic environments that have more rapid fluctuations or where stress tests (i.e., pump tests) require more accurate responsive head measurements, lower volume piezometer devices should be installed.

The relative response characteristics of various piezometers are illustrated by the "time lag", (Hvorslev, 1947), the time taken for the piezometer to reach equilibrium with ground-water conditions next to the piezometer. The relationship between hydraulic conductivity and the time taken to reach 90% of the true pressure value are shown for various piezometers and wells in Figure 4-40. In the many cases where aquifer pressure

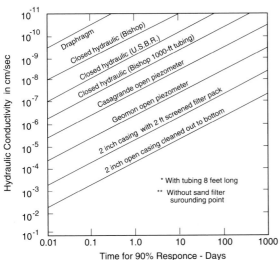

Approximate Hydrostatic Time Lags

Modified from Hvorslev, U.S. Corps of Engineers, W.E.S.

Figure 4-40 Piezometer Response Rates

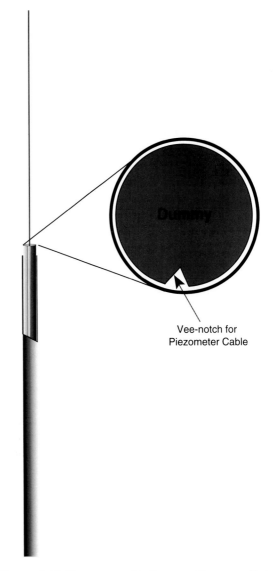

Vee-notch for
Piezometer Cable

Figure 4-41 Dummy used to Improve Response Rates

changes are expected over a period of weeks rather than of days or hours, open tube designs may be appropriate for the project. Stand pipe piezometers have been successfully used in fine grained geologic materials with hydraulic conductivity as low as 10^{-7} cm/sec. The time to achieve 95% recovery, however, took several months, even with the use of "dummy" blocks to reduce volume demands.

Problems of Slow Recovery in Piezometers

Type A piezometers give satisfactory response times where water table variation is small, or where the hydraulic conductivity is relatively high. For example, a typical 95% response time for a 3/8 inch (10 mm) internal diameter stand pipe piezometer having a collector zone 10 feet (3 m) long by 1.75 inches (44 mm) in diameter and located in a material with permeability $k = 10^{-5}$ cm/sec would be 10 minutes; if $k = 10^{-8}$ cm/sec the time would be seven days. A two-inch piezometer would require 900 days with a similar material. To measure piezometric heads that vary significantly over short time periods in low permeability environments, it is necessary to use piezometers with low volume demand; i.e., Class D or those of low volume demand in Class C.

Limitations of depth must also be considered in the design of piezometer systems. Failure of connectors under the weight of the riser pipe may limit the use of open tube piezometers. PVC open tube piezometers have performed satisfactorily for most ground-water investigations to depths as great as 300-400 feet, however, they may require some form of additional support. Since classes C, and D readout cables or tubes weigh less than open tube designs they can be placed under difficult soil conditions with higher hydraulic heads where type A piezometers would be very difficult to install.

4.10.7 Piezometer Installation Methods

Piezometers used in site assessments for measurement of potentiometric heads must be correctly installed to obtain reliable results and meaningful predictions of hydrogeologic conditions. Installation details can vary with the instrument and ground conditions; however, the introduction of filter pack or annular sealant materials and piezometer equipment should be, if possible, through a cased hole. Working through a hollow stem auger, temporary rotary, or cable tool casing, can greatly improve the reliability and control of the piezometer installation by reduction of caving or bridging of materials in the hole. Several factors are important in piezometer construction:

• If the piezometer is to measure a representative value of the ground water pressure heads at a given depth, the

borehole must not act as a potential flow path. Therefore, the hole must be sealed along its length with grout of about the same, or lower hydraulic conductivity as the adjacent soil/bedrock. The piezometer screen or tip must also have full access or contact to the ground water in the adjacent geologic material. There must be sufficient ground-water flow into the piezometer tip to satisfy the volume demands of the instrument within project specific response times. The higher the volume demand of the piezometer, the more care is necessary to provide a sufficiently permeable hydraulic collector zone. In fractured rock, this can be accomplished by use of three-prong caliper geophysical tools to locate portions of the borehole with significant fracture zones. These data may also be supplemented by stressed flow meter geophysical tests to establish likely areas of ground-water inflow into the boring.

• In unconsolidated, fine-grained materials, continuous logging of boreholes should be used to define potentially more permeable zones that can yield water sufficient to a piezometer tip, so that reasonable 90% response times are maintained for the project.

Piezometer installation may take place in specially drilled holes or in geotechnical or stratigraphic holes put down during the early stages of site assessment. The borehole diameter required for installation of piezometers is generally dependent on the depths necessary to obtain the specific ground-water information and the number of piezometer tips that will be completed in the borehole.

Piezometers in unconsolidated materials are typically constructed through 6.5 inch OD hollow stem augers or rotary tricone holes. In geologic materials with sufficient hydraulic conductivity, up to four open tube piezometers may be completed (if care is used!) through a 6.5" OD hollow stem auger using riser spacers. Hollow stem augers of this diameter can typically drill to depths of 75 to 150 feet depending on the drilling equipment, the geologic materials, and hydrogeologic conditions of the site. Conventional diamond drilling is used for obtaining rock cores this leaves a core hole available for installation of single or special multiple completion piezometers. Single-cored holes with special multiple point completions, as discussed in the previous subsection, can provide the most cost effective system for measurement of pressure heads in the fractured rock environment.

Deeper installations of piezometers may require staged drilling. This is especially true in areas of potential or confirmed subsurface contamination. Since multiple piezometer installations require more riser lines, larger diameter core holes such as the H series may be necessary to obtain the proper clearance for the down-the-hole device.

Installation Methods for Piezometers

Piezometer installations require careful construction procedures that are only slightly less rigorous than the installation of monitoring wells used for extraction of water quality samples. Installation of piezometers can be much more difficult than detection monitoring wells due to adverse geologic and hydrogeologic ground conditions. Piezometer installation methods can be divided into the filter/seal and fully grouted methods. The type of piezometer installation technique used on a particular project are dependent on the hydraulic conductivity of the bedrock or overburden materials. Low permeability environments and project requirements may allow the fully grouted procedure, which greatly eases construction time and costs for the installation.

Stand Pipe Piezometers

Normal sizes of risers used in piezometers range from approximately 1/2 inch to 1 inch (12 mm to 25 mm) inside diameter. If deep installations (>200 feet) are necessary for the project, the strength of riser screw connections should be reviewed to assure that the screw connections will be able to take the weight stress at full length of the riser string. Flush coupled tubing should be used to minimize installation problems with the piezometers.

The piezometer completion must have a permeable graded filter around the tip screen, or in the case of multiple installations, a graded filter on top of the previously grouted stage. It is essential that filter sands used in construction of piezometers are clean and contain virtually no silt or clay particles. Procedures used in construction of an open tube piezometer through a hollow stem auger hole would be as follows:

- After reaching placement depth, a one foot (0.3 m) minimum layer of filter material is placed initially in the hole bottom to avoid problems of any residual fines left in the hole after drilling;

- The riser pipe and screen are lowered into position. Lengths of tube are progressively added at the top of the hole in the case of rigid PVC pipes. Continuous rolled semi-flexible PVC pipes are lowered directly into the hole.

- The hollow stem drill string is gradually raised to the top of the proposed filter. If caving ground conditions are suspected at the depth of the piezometer tip, this operation can be delayed until after the filter pack has

been placed. However, extreme care should be used not to lock in the riser in the hollow stem by over sanding filter material inside the casing. Under these conditions pulling the hollow stem will pull the entire assembly. Careful tamping of the filter may later be required after pulling the hollow stem casing in this case. A designed filter pack sufficient to cover the perforated zone of the tip plus three feet (1 m) is poured or washed through a tremmie pipe to avoid blockage. The position of the top of the filter should be carefully checked after placement and should be compared with the measured and calculated volume.

- The hollow stem string is then raised and a layer of fine filter sand is poured, or preferably pumped through the tremmie pipe, on top of the coarse filter pack.

- The piezometer is then sealed in the borehole by using appropriate annular sealant. Pumping through side discharge tremmie is necessary to ensure that the annular sealant is introduced at the correct level and not injected downward. If the piezometer will be used to collect "grab" samples for water quality, a three foot (1 m) bentonite seal should be placed above the fine filter sand before placement of the annular sealant. This is to help reduce alkali grout interferences in the piezometer tip.

Annular Sealant Mixes

Annular sealants of the following mixes have been used successfully for installation of piezometers in both fractured rocks and unconsolidated geologic units: Careful consideration should be made of the long term use of the piezometer. If reconnaissance water quality samples will be recovered from the installation, any accelerators or metallic powders used in the annular seals may influence samples collected from the screened portion of the installation. This is especially true for fine-grained, low hydraulic conductivity geologic materials where vertical gradients can move residual grout liquids downward. The following mixes have been used successfully for placement of annular seals:

A. Cement-bentonite-accelerator grout at a ratio of approximately: 5 lb. bentonite : 95 lb. cement, to 6.5 gals. of water and selecting accelerator quantities so that adequate strength and suitable setting times are obtained depending on requirements imposed by the particular site. Bentonite reduces shrinkage and improves pumpability of the grout. The accelerator is used to decrease grout setting time, which is particularly important for multiple installations in a single hole.

B. Cement-accelerator-expander grout similar to the above but using an expanding agent such as powdered aluminum to obtain a hard, strong seal, with slight expansion.

C. Bentonite can be used either in a pellet chip or granular form in holes less than 100 feet deep, or pumped through tremmies as a bentonite slurry to moderate depths. This method using pellets may be very difficult if the hole has any appreciable depth of water.

The quantity of grout required to adequately seal the blank (unmonitored) portion of the installation should be determined from the borehole volume. Sufficient time must be allowed for the grout to be pumped into position, while allowing for progressive withdrawal of the tremmie and drilling casing. The grouting operation must be carried out slowly with sufficient control to insure thorough sealing of zones between piezometer screened and filter packed areas. If grouting is stopped between stages and allowed to set overnight, fluid at the top of the grout column must be flushed out prior to grouting the next stage.

After placing, a typical grout will take several hours to settle. The position of the grout upper surface should be carefully checked if multiple installations are required. Additional grouting steps may be necessary to achieve the required level.

The three above steps are then repeated for each stage of a multi-stage installation after withdrawing sections of the casing or drill string and grouting between each tip filter pack section. Some common problems that have been observed in installation of piezometers are listed in Table 4-13 along with some possible remedies.

Type C, and D Piezometers

The method of installing these types of piezometers, including the placing of a graded filter pack in the collector zone and sealing with grout is very similar to that for stand pipe completions, as described in previous subsection. Certain differences in procedure are necessary as described below:

• Piezometer installations in deep holes require that all tubes, connectors and cables must be fully checked to ensure adequate strength in tension;

• Readout tubes or cables should be handled carefully through the use of cable reels. These reels can help to prevent mistreatment of the cables by stretching or kinking the readout cables;

• In cases of strong artesian (upward) flow conditions in the borehole, it may be necessary to attach long cylindrical weights below the piezometer tip to allow positioning the device to the desired depth;

• Each piezometer tip should be checked for correct operation immediately after lowering the device into position, before placement of sand filter or annular seals; and,

• Extreme care should be taken with multiple cable installations to properly mark each completion as to reference depth to avoid later errors or confusion in elevations.

Fully Grouted Procedure

Site investigations in fractured and fine-grained materials may not require highly permeable filter packs be placed around piezometer points. If low volume response piezometers must be used for monitoring steady-state ground-water conditions (Class C2 or Class D types), installation procedures may be greatly simplified by using a fully grouted or continuous placement procedure.

In this technique, the piezometer tips are normally taped or strapped together in a single string and are suspended at the required depths together with a tremmie pipe that is initially lowered to the bottom of the hole.

The grout mix used in the annular seal should be carefully controlled and metered during the placement. Special mixing and watering equipment is typically required. The grout is placed as the side discharge tremmie pipe, casing or drill rods, are withdrawn. The grout mix is varied according to downhole depth so as to place a relatively permeable mix around the piezometer tips. Obviously accurate depths must be known for all piezometer points so that the more permeable sandy grout mix can be placed next to the tips.

Pure sand containing a minimum of cement should be used at the depths around the piezometer tips. A pure cement grout with a suitable expanding agent should be used between piezometers. These devices do not allow for later collection of ground-water "grab" samples due to the class of piezometer used in this installation.

Natural Filter Pack Seal

In geologic environments that consist of medium sand to gravels, a natural filter pack may be formed by the collapse of the geologic material around the piezometer tip as the auger or temporary casing is withdrawn from the hole. These installations may be very difficult to grout due to the collapsing sands closing up the hole. In such environments where very high vertical hydraulic conductivity is present, installation of annular seals may not be necessary.

Care must be taken however, that no confining units will be breached by such "natural" back filing construction, and that the vertical hydraulic conductivity of the completed installation does not greatly exceed the overall hydraulic conductivity of the aquifer. This technique is only acceptable in thick sand and gravel deposits. Surface seals would still be used under these procedures.

4.10.8 Verification of Piezometer Operation

Piezometers often come into question as to long-term performance and the accuracy of the resultant potentiometric levels. This is especially true for installations that were completed in previous investigations by other consulting. The following guidance provides a number of procedures for assessing the operation of piezometers.

Checking Installation Response Times

Readings should be taken of the potentiometric levels, as soon as possible, after installation of the instrument. The measured potentiometric heads should be expected to change for some period after installation, dependent on the drilling method used, and the hydraulic conductivity of the geologic media. A time reference to this increase or decrease in heads can be compared to Hvorslev 90% response times for the size and type of piezometer.

Seal Effectiveness- Multiple-point piezometers have been historically held in suspect due to the belief in the ineffectiveness of annular seals. The effectiveness of sealing piezometer tips can be evaluated by measuring any change in piezometric pressure resulting from the introduction of water to the borehole above the annular grout seals. There should be no short-term response in the piezometer tip unless the piezometer is located in an uncased hole in geologic materials of high hydraulic conductivity.

Piezometer Performance Tests

Stand-pipe or Casagrande piezometer performance can be tested through a number of procedures based on falling head tests. For stand-pipe installations, the water level is increased by the rapid addition of a known quantity of water. The water level should be measured until a reasonable assurance is developed that the piezometer tip reacts as expected for the geologic material. Care must be taken in very low hydraulic conductivity environments that

the falling head test does not affect readings for long periods of time inconsistent with project requirements. In reasonably permeable environments, the head levels should return to the pre-test reading according to calculated rates. If the imposed water level remains stationary, the stand-pipe tip is probably blocked. The drop in head with time should be recorded and retained as an initial calibration to compare with future checks of the installation.

Some piezometer tips in the Type C categories have a by-pass system that can be used to apply a measured pressure on the piezometer tip. Normal functioning of the installation is demonstrated if the pressure recorded by the tip is equal to the pressure applied through the by-pass system.

Long-Term Piezometer Evaluations - The best demonstration of proper functioning of piezometer installations is long-term consistent records of potentiometric levels recorded from the device. Comparisons can also be made with other area piezometers located in a similar geologic environment. Potentiometric level pressures that become erratic and depart from general behavior are usually the first indications of malfunctions. Pneumatic or electric type piezometers, most frequently indicate zero or maximum readings; typically indicative of blocked or broken pipes and valves or short circuits in the electrical or hydraulic systems.

The designation of a piezometer layout for investigation or monitoring purposes should address the likelihood of instrument failure. The number and frequency of piezometer tips in the hydrogeologic units should depend on the importance of the site and the length of time for which potentiometric head level monitoring is required. In assessment or detection monitoring programs, where the resultant piezometers are only required for short-time periods, a reasonable percentage (10-15%) of ineffective piezometer tips should be accommodated in the design. These investigations are typically too short in duration to allow replacement of piezometer installations.

4.10.9 Rehabilitation Methods

Some pneumatic and electric piezometer tips have provision for flushing the tip with water under pressure to counteract silting. Stand pipes can similarly be cleaned by injecting water or air under pressure to wash out the accumulated fines. Generally, however, flushing is of only temporary effectiveness once silt and clay has entered the collector zone, and must be repeated frequently. Attempts to clear partially blocked stand pipe tips are rarely successful.

In some cases, it may be possible to convert a stand pipe piezometer with a blocked screened zone into a Type C2 or D piezometer. Several versions of miniature piezometer-packers have recently been introduced, having diameters down to about 0.5 inches (1 cm). These instruments can be lowered down a stand pipe and sealed in place at the bottom. They will then accurately measure the pressure at the tip since they do not require large volumes of water to measure the resultant pressure heads.

4.11 WATER LEVEL MEASUREMENT

Installation of piezometers and monitoring wells should offer the opportunity to provide precise water level measurements in a sufficient number of points and at a sufficient frequency to gauge both seasonal and average flow directions, changes in elevation or water levels, and to account for seasonal or temporal fluctuation of water levels. There are numerous ways to measure hydraulic head in the saturated or vadose zones. For convenience, both zones are discussed here and defined on Figure 4-42. The accuracy of depth-to-water measurements is discussed in a recent Superfund Ground Water Issue paper by Jerry Thornhill. The steel tape (wetted tape) method is generally considered the most precise method however, electronic tapes are now the defacto standard for water level measurements, due to the convenience of the equipment. Pressure transducers can be used in either the saturated or vadose zones. They are useful for making frequent measurements, such as during a slug or pump test.

In a saturated medium, the hydraulic head, H, is measured at a point using a piezometer (see Figure 4-37) and is defined as the elevation (pressure head) at which the water surface stands in an open piezometer tube terminated at a given point in the geologic medium. Hydraulic head is a combination of pressure head and elevation head (distance of the measuring point above a reference level datum). The reference level chosen for measurement of H is arbitrarily chosen as sea level. The hydraulic head is a potential function, the potential energy per unit weight of the ground water.

These concepts of hydraulic head, pressure head, and gravitational (or elevation) head may be applied to the vadose zone. A common device used to measure the hydraulic head in the vadose zone is a tensiometer. Tensiometers are completed in the soil by a porous cup, permeable to water, where the pores of the cup are filled with water; but impermeable to air. The porous cup is used to establish hydraulic contact between the water in the tensiometer and the soil water. For the vadose zone, the pressure head is inherently negative, i.e., the free water surface in the open arm of the manometer will stand below the point of termination in the soil. The effective pressure range of a standard tensiometer, 0 to about -0.08 MPa, is limited by the fact that negative pressures are measured with reference to atmospheric pressure. The reader is referred to Chapter 2 of Freeze and Cherry's "Groundwater" (1979) for an extensive review of the terms associated with ground-water and vadose zone issues.

In those situations where the relationship between ground water and surface water is critical, the use of continuous water level and stage level recording instruments should be considered. Continuous water level measurements, known as a hydrograph gives a picture of both short and long-term ground-water level fluctuations at a given site. Experience has shown that the hydrographic curve shape is often influenced by the frequency of the measurements (Sweet et al, 1990). The examples of hydrographs shown below were produced by continuous recording equipment. Less frequent readings taken manually at less frequent intervals would give far less detailed and possible a different picture from these continuous measurements. As a general statement, the water level in the ground is not static and can indeed vary by tens of feet from season to season, so the value of a single measurement only represents a single point in time. This is especially true for fractured rocks with low porosity (1-2%). Relatively small quantities of recharge can cause large changes in observed ground-water levels. The extent and rate of ground-water level change can require monthly or even continuous recording to obtain of short term fluctuations. Tidal effects on an aquifer system are shown in Fig 4-43. Tidal effects are caused by loading the aquifer

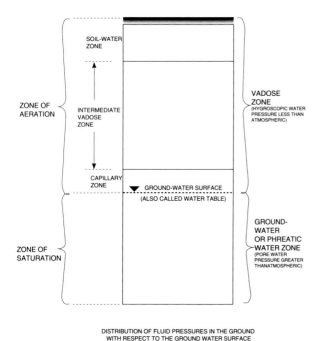

DISTRIBUTION OF FLUID PRESSURES IN THE GROUND WITH RESPECT TO THE GROUND WATER SURFACE

Figure 4-42 Terms used in Pressure-Water Level Relationships

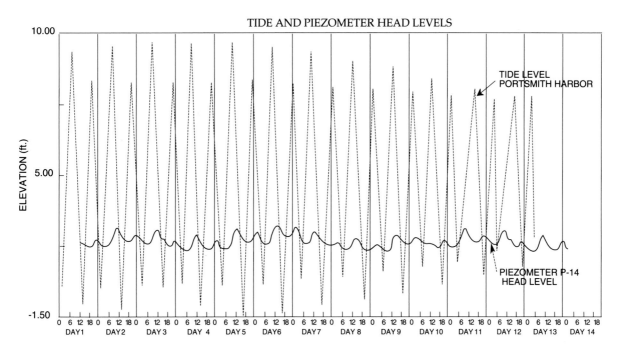

Figure 4-43 Hydrograph of Tides and Water Level Relationships

so that water is hydraulically squeezed out of the aquifer, and not physically pushing water into the aquifer.

Fig 4-44 shows changes in ground-water levels due to pumping and seasonal effects.

4.11.1 Barometric Efficiency

Barometric pressure can effect changes in ground-water level in an observation borehole as a result of the change in loading of an aquifer. A borehole located in an aquifer will show the full effect of barometric changes on

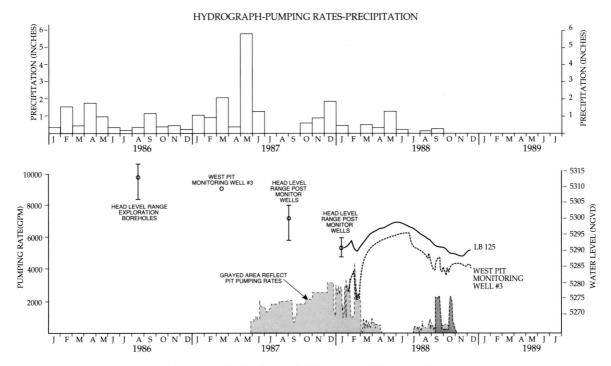

Figure 4-44 Hydrograph-Water Level Relationships

the water surface in the open borehole (due to direct connection with the atmosphere). However, the water in the aquifer itself will only feel a partial barometric effect since the structure of the aquifer absorbs some of the loading change. High barometric pressure will tend to push down first on the open water in the borehole, since the full barometric effect does not reach the rest of the water in the aquifer. Thus ground water is pushed back into the aquifer from the borehole, thus lowering the water level in the borehole. One would expect in unconfined aquifers, changes in atmospheric pressure would transmit equally to the ground-water surface in the aquifer and in a well. You would expect no pressure differences to occur. However, air trapped in pores below the ground-water surface, will respond to barometric pressure changes and cause a similar behavior as observed in confined aquifers; but with smaller fluctuations in water level. Hydraulic conductivity often varies vertically in an aquifer under the influence of variable lithology. This may produce

confining effects, particularly in deep aquifers. In these cases effects of barometric pressure changes will be reflected in observation borehole water levels.

To interpret ground-water level data affected by barometric changes, the data should be corrected to enable other head level influences to be clearly seen. One must first calculate the barometric efficiency of the aquifer. This evaluation must be done for each piezometer or well, since local lithological differences can cause barometric efficiency to change with location. Barometric efficiency can be defined as the change in water pressure, divided by the change in barometric pressure expressed as either a percentage or a decimal. Both barometric pressure and piezometric head data must be expressed in the same units, so one must convert the barometric readings to centimeters of water. Figure 4-45 illustrates one method of calculating barometric efficiency, by plotting the two data sets so that they can be compared visually for head level-time relationships. You should always plot the increasing atmospheric pressure in a downward direction. This example illustrated the change in water level and its corresponding change in atmospheric pressure as estimated from the graph. An alternative method is to plot a graph of barometric pressure on the x - axis and water levels on the y - axis. Figure 4-45c was generated using the data from Figure 4-45b. If one obtains a poor fit of data plotted this can be attributed to the well being located some distance from where the barometric measurements are taken by the recording meteorological station. This separation distance will cause a time lag between the two data sets, which will vary depending upon prevailing weather.

Pumping tests conducted to determine aquifer properties always require correction of ground-water level data for barometric changes. Important pump testing requires that a recording barometer should be situated close to the test site. Experience with these measurements has shown that the majority of barometric efficiency values fall in the range of twenty to eighty per cent. To correct your ground-water level for barometric efficiency effects, follow the steps shown on Figure 4-45. However, Weeks reported (1979) that barometric effects can occur in deep, apparently unconfined aquifers. Multiple piezometers, sealed at various levels in an aquifer can show that barometric efficiency changes with depth. Also Weeks (1979) reported that if sealed water pressure measuring devices are used in the site investigation differences between these and open standpipes may be observed.

A consideration of the hydrographs above shows the average water pressures of all the different layers open to the monitoring point or response length. As the aquifer is not uniform it has a significant number of different layers. Each of these stratigraphic units represents different hydraulic conductivity and provides different piezometric

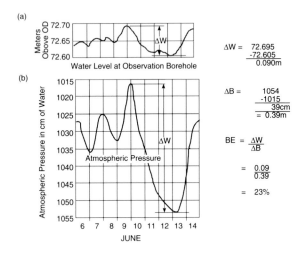

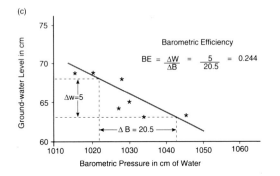

Source: Modified from Field Hydrogeology, Brassington, 1988.

Figure 4-45 Barometric Efficiency Determinations

heads to the observation point. The interpretation of the data therefore must consider three-dimensional flow.

4.11.2 Head Level Data and Aquifer Flow

It is essential during the investigation of a facility area to identify the individual elements of the aquifer and to observe the response of each element. Permeable sands and gravels, or rock aquifers are readily identified during the initial boring and instrumentation can be installed in the appropriate hydrostratigraphic unit(s). This has traditionally taken the form of standpipe piezometers or monitoring wells completed in the sands and gravels (with widely screened zones), or open completion boreholes in the rock aquifers. The problem with the unreliability of water level readings in open boreholes in fissured aquifers has been discussed by Rushton & Howard (1982) where the different layers within a sandstone aquifer were only identified after the installation of multiple piezometers. The presence of different piezometric heads within, what at first sight appears to be a uniform aquifer, (or confining unit) has also been demonstrated by the use of borehole packers (Brassington 1985) and by a combined multiple packer/piezometer installation.

Head level effects are important in the interpretation of the response of the aquifer to remediation (aquifer restoration), and in particular to interpreting whether the aquifer will show an unconfined or confined response. The presence of low hydraulic conductivity geologic (confining) units, across which there is a hydraulic gradient, will also have direct effects on potential local flow adjacent to a facility. In particular, bedrock aquifers having direct connection through fissures or very high hydraulic conductivity zones may give higher drawdown at more remote locations during pumping tests. It is therefore very important that the hydrogeology of the site area is fully understood before placing great emphasis on the interpretation of any ground-water level measurements.

It is important, as with any physical measurement, that the individual ground-water level to be measured is understood and that the measuring device is appropriate for the project in question. For aquifers this means that the construction of the water level observation piezometers must be carefully designed to monitor the correct hydrostratigraphic horizons. To observe the true effects of climate, both seasonal and daily, and to detect man-made effects, it is often necessary to have continuous readings at a given site as is shown on Figure 4-46. However, this is not necessary for all the boreholes within a group, provided that frequent readings are made on the remaining boreholes so that local pattern to emerge. In practice, once-a-month head level reading is adequate for most applications and may even be stretched to quarterly readings for evaluation

of long-term effects within an aquifer. On a small restricted area, such as a site investigation, weekly readings may be required to establish the site relationships before increasing the interval to once a month. In site assessments these data can provide the numerical values necessary for verifying maximum and minimum ground-water levels, together with the likely effects and response time to climatic and other water level changes.

Ground-water levels are normally read in the field with an electric dipper or a calibrated chalked tape. These devices are capable of an accuracy of 10 mm (0.4 inches), however, the frequency of head level reading is considered even more important for site assessments than the relative accuracy of an individual head level reading. Whereas, once it would have been almost impossible to collect and process, for example, hourly readings from twenty points for six months, now with the advent of data loggers and desk top computers, such intensive data collection and processing is possible. It can be seen from the hydrographs presented above that the reasons for ground-water level changes can be many and varied, and that these must be identified, as far as possible, for each site. The best way to achieve this goal is by the use of continuous recording equipment on one or more observation points at a site, coupled with a series of readings at less frequent intervals on all the other observation points. This technique has been successfully used both in regional aquifers and on site assessment projects.

For more routine conditions, it is generally recommended that monthly readings of water levels be performed for the first year for new waste disposal sites, expansions, and for those wells/piezometers used in RI/FS investigations. After the initial year, a quarterly measurement of water levels can be used if more frequent measurements are not required by permits or specific site issues.

4.11.3 Accuracy in Water Level Reading

The major factors affecting water level accuracy are (1) measurement of water levels (operator and sounding equipment precision), (2) deviation of the well from vertical, (3) surveying precision of the well head, and (4) natural water level fluctuations.

The measurement of water levels involves both human operators and sounding equipment. Depths to water within each well are typically measured manually using electric line sounders. The length of cable in the well (below the top of the well casing or sounder tube) can be measured to the nearest 0.1 foot by reading numbers marked on the cable. The remaining fraction of a foot is determined with a tape measure. Water depth measurements are recorded to the nearest 0.01 foot. Although one can write down depth measurements to this level, the actual accuracy of the measurement may be significantly greater than this hundredth of a foot value. The following two sections address the potential sources of error that affect accuracy of water level measurements due to operators and equipment (see Table 4-12, Source : Sweet et al.).

Operator Errors:- Operator care and consistency are always important when obtaining water level data at any site. Poor weather can also compound operator errors. An untrained or unaware operator is liable to make several types of measurement errors;

• Misreading the measuring tape or sounder markings;

• Poor recording of data such as number transposition or illegible entries;

• Not recognizing equipment failures or faulty readings due to equipment sensitivity ; and

• Not measuring to the surveyed (or V) mark on the casing

Trained operators can avoid mistakes by using consistent methods, being aware of equipment inadequacies, and being alert to the significance of each reading. Review of previous water level readings will often prevent gross misreading or number transposition Part of the data collection protocol should include direct comparison of each measurement with previously recorded water level readings

A source of operator errors that cannot be resolved is the consistency between two different operators. For example, differences occur depending on whether the reading is taken while lowering the sounder or raising the sounder at the water air interface. Using the same sounding equipment on the same day and time at the same well, but with two different operators, can obtain significantly different head level readings.

Equipment: Some equipment errors are unavoidable Others simply require attention to procedure to reduce or eliminate the error Large water level discrepancies can be observed from:

• Condensation in the sounding tube can cause sounding equipment to be activated prematurely;

• Sounder wire stretching due to the force needed to break a vacuum bond between the sounder and casing or simply due to the weight of the sounder line in deep holes;

• Obscure cable markings or loose incremental marking rings;

• Kinking and wrapping of sounder wire within the well; and

• Sounder malfunctions due to weak batteries or breaking connections.

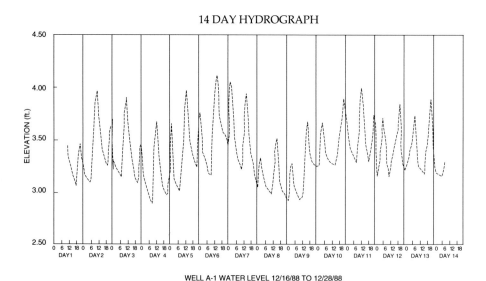

Figure 4-46 Hydrograph of Over 14 Day Period

Table 4-12 Summary of Methods for Manual Measurement of Well Water Levels in Non-flowing and Flowing Wells.

Measurement Method	Measurement Accuracy in Feet	Major Interference or Disadvantage
Nonflowing Wells		
Wetted-tape	0.01	Cascading water or casing wall water
Air-line	0.25	Air line or fitting leaks; gage inaccuracies
Electrical	0.02 to 0.1	Cable wear; hydrocarbons on water surface
Transducer	0.01 to 0.1	Temperature changes; electronic drift; blocked capillary
Float	0.02 to 0.5	Float or cable drag; float size and lag
Popper	0.1	Well noise; well pipes and pumps; well depth
Acoustic probe	0.02	Cascading water; hydrocarbon on well water surface
Ultrasonics	0.02 to 0.1	Temperature changes; well pipes and pumps; casing joints
Flowing Wells		
Casing extensions	0.1	Limited range; awkward to implement
Manometer/ pressure gage	0.1 to 0.5	Gage inaccuracies; calibration required
Transducers	0.02	Temperature changes; electronic drift

Electronic sounders should be frequently calibrated to check for stretching.

Well Deviation: Accuracy of water levels is also affected by the deviation of the well from the vertical. This well deviation error can escalate greatly as piezometers and wells go down to depths past 200 to 300 feet. For a well, which is not exactly vertical, actual vertical depths are shorter than those measured with an electric line sounder. Over the length of these deep installations, the error introduced by deviation can be significant. Fortunately this type of error can be corrected through the use of a correction factor.

To obtain a correction factor to account for deviation, a geophysical survey must be conducted using a downhole drift tube, magnetic multishot or gyroscope probe This downhole probe contains a transducer for sensing inclination and direction of inclination of the probe The magnetic-multishot surveys use a tool which consists of a 1.75-inch-diameter probe which records a compass and inclinometer reading on film as the probe is lowered down the hole A gyroscopic survey used in wells with steel casing material utilizes a 1.75-inch probe with a spin motor that is aligned to a surface reference of known orientation, readings are recorded on file as the probe is lowered down the hole.

After a deviation survey is taken the water level data is adjusted to record the actual water level thus compensating for the deviation in the well. The deviation survey corrects for the well's variation from vertical; however, the probes and equipment used to measure the deviations of the well have a limited accuracy. The survey companies have reported the accuracy of the equipment and survey technique to be 0.30 foot This accuracy limit cannot be eliminated when calculating the water level elevation

Elevation Survey: An accurate topographic survey is essential to obtain reliable water level measurements. Despite the precision equipment, surveying mistakes can cause large errors. Surveyed elevations for both ground and top of the casing should be obtained. However, inconsistent placement of top of casing measurements can create a large source of error. For most wells the top portion of the casing is not flat. For consistency, the surveyors should be requested to mark the top of casing with a Vee notch exactly where the elevation measurement has been made to verify that it is at the same point from which water level measurements are taken. Each time the well head configuration is changed a new survey is required. A sudden drop in water level at a well may be caused when a well head configuration was changed and a new survey had not been performed

Natural Water Level Fluctuations:— Water level fluctuations can result from natural forces such as barometric pressure, earth tides, or a combination of such forces. Other outside influences such as pumping, sampling, and recharge will also change water levels from static. Usually these changes can be noted in the field or in the office. However, these readings often must be disregarded or re-evaluated when incorporating these anomalous values into larger data base.

Large changes in water level can be caused by barometric pressure changes that reflect passing weather fronts. Responses at water levels in site wells to barometric change are often synchronized in each well; however, the magnitude of fluctuation is different in each well due to both location and depth variations.

Because these larger fluctuations occur over longer periods of time their effect on water level measurements can be eliminated by comparison between wells by taking site water levels within a 24 hour period during stable weather. When comparing water levels at two wells the measured difference in elevation may be greater or less than the true elevation by up to 0.2 foot. This error can significantly bias gradient computations in a system with a flat or shallow hydraulic gradient system.

Elevation surveys can achieve an accuracy of 0.01 foot However, duplicate elevation surveys illustrate that the accuracy of a survey without precise control can be 0.10 foot. Well deviation surveys can eliminate the error introduced by a non-vertical well but the equipment used for the survey has a resolution limit of 0.30 foot. Table 4-14 shows the various accuracy of water level measuring techniques (Nielsen 1991). For hydrogeologlc analysis that involve more than one well, such as the determination of the hydraulic gradient or preparation of a ground-water contour map, all the factors are totaled to determine the overall accuracy of the ground-water level measurements. In the analysis of a single well record, the accuracy of the well deviation survey and the well elevation survey do not affect the quality of the record since these factors do not vary with time. The accuracy of the measurement proceeds, operator, and equipment and the natural water level fluctuations, however need to be considered within the project

Phased Product:—Phased liquids can also affect liquid readings in boreholes. Light non aqueous phased liquids(LNAPL's) often will not set off electronic tape measures and the floating product can coat the water sensor tip so the point where the actual water surface occurs will be misread.

4.11.4 General Guidance for Water Level Measurements

In order for the investigator to initially determine the elevation of the piezometric (unconfined) or potentiometric (confined) ground–water surface in any monitoring well or piezometer, several criteria should be considered during the investigation:

- The casing height should be measured by a licensed surveyor to an accuracy of +0.01 feet, relative to the local USGS benchmark or the National Geodetic Vertical Datum (NGVD), 1929. This may require the establishment of a topographic benchmark on the facility property.

- Generally, water level measurements from boreholes, piezometers, or monitoring wells employed to provide data to construct a single–date potentiometric surface should be collected within a 24–hour period. This practice is adequate if the magnitude of change is small (say, <2 ft) over a single climatic season. There are other situations, however, that necessitate that all measurements be taken within a shorter time interval:

–Tidally influenced aquifers;

–Aquifers affected by recharge from or discharge into rivers, swamps, impoundments, and/or unlined ditches;

–Aquifers stressed by intermittent pumping of production wells; and

–Aquifers being actively recharged due to a precipitation event.

• Water levels are measured by lowering an electronic probe into the piezometer boring. Water levels should be measured and recorded to +0.01 feet in shallow (<50 ft deep) boreholes. Deeper boreholes will probably have accuracy levels = .05 to .2 feet.

• A surveyed mark should be placed on the casing for use as a measuring point. Many times the lip of the

riser pipe is not horizontal, and a small "v" notch should be cut in the riser to indicate a surveyed measurement point.

• Water levels in piezometers should have been allowed to stabilize for a minimum of 24 hours after construction and development, prior to measurement. In low yield (gallons per day or less) situations, recovery may take longer.

4.12 FLOWMETER BOREHOLE TESTING

One of the best methodologies for finding vertical distribution of hydraulic conductivity is through the borehole impeller meter test. This test can be performed in any consolidated unit(s) where there exists a variable flow rate distribution along the borehole or well screen.

During an impeller stress test of a well the well is pumped at a measured rate and the distribution of flow throughout the screened segment is also recorded.

Various types of flowmeters based on impeller or heat-pulse technology have been devised for measuring flow or some directional component of flow (Q) (see Figure 5-47 for examples). Spinner type impeller flowmeters measure a minimum velocity ranging from 3 to 10 ft/min (1 to 3m/min), where this range is useful for large volume pump testing in water supply projects, it is less useful for site evaluations for the establishment of

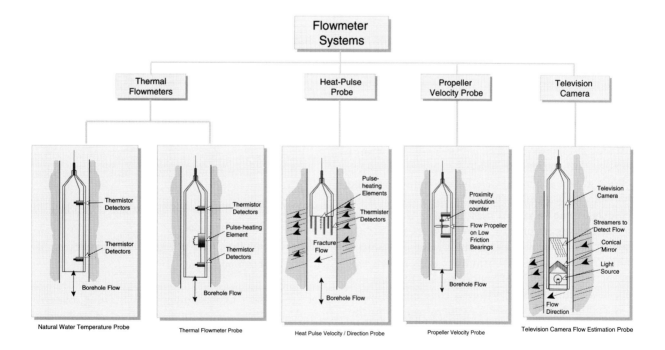

Figure 4-47 Flowmeter Systems

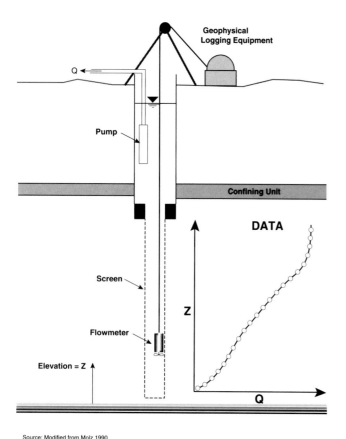

Source: Modified from Molz 1990

Figure 4-48 Flowmeter Test Set-up

vertical changes in hydraulic conductivity. This limitation of impeller systems have forced the use of more sensitive methods for the measurement of vertical variations of flow into the well. Low-flow sensitive meters are based on heat-pulse, electromagnetic or tracer-release technology as described by Keys and MacCary, 1971: Hess,1982, 1986: Hada, 1977: Keys and Sullivan, 1978; Morin et al., 1988 and Molz et al., 1989.

The concept behind the impeller flowmeter stress test is illustrated in Figure 4-48. The procedure is to first run a caliper log to establish the hole or screen diameter. Variations in hole diameter must be accounted for since up-hole velocity is greatly affected by the diameter of the borehole and these variations must be taken into account when calculating discharge. The flow meter is first lowered into the borehole and a small pump is placed a limited distance into the ground water above the flowmeter. The pump should be operated at a constant rate without dewatering the well. Once a constant rate/water level is obtained, the flowmeter is then lowered to the bottom of the well and a measurement of discharge rate is obtained in terms of impeller-generated electrical pulses

over a selected period of time. The meter is then raised a few feet and another reading is taken. This procedure is repeated until the entire section of interest is measured. The entire procedure may be repeated several times to compare stability of the readings.

An alternative methodology (called the trolling method) is to raise the flowmeter from the hole bottom continuously at a slow rate, and then lower the flowmeter back down at the same rate. This procedure is repeated at an additional two rates both up and down the borehole. Although both methods can produce good results, Molz (1991) reported that for fine scale ground-water applications, the stationary mode was believed to be better suited for obtaining higher quality data.

4.13 DOCUMENTING OF BOREHOLES, PIEZOMETERS, AND WELLS

The docimentation of all installation activities should be handeled through individual piezometer completion sheets and summary figures that describe the project "as built" completions. Figure 4-49 shows an example document that illustrates the important features of such presentations. These documents must always be linked to individual field borehole logs, edited logs and well completion diagrams as show in section 4.6

4.14 DECOMMISSIONING OF BOREHOLES, PIEZOMETERS, AND WELLS

Field personnel are commonly called upon to decommission open boreholes, piezometers, and wells. These decommissioning activities are typically required by state regulations. Figure 4-50 provides guidance for closure and decommissioning of a variety of installations ranging from open geotechnical boreholes to ground–water monitoring wells completed to ASTM standards. The guidance is based on three different procedures:

- Simple cement/bentonite ground backfill of an open hole.
- Auger overdrilling an installation and backfilling with cement/bentonite grout.
- Backfilling the casing of a known construction well with cement/bentonite grout.

The most extensive decommissioning, overdrilling, and grouting procedure is directed toward those installations that may contribute to movement of contaminated water down to aquifers. Before deciding on

GROUND-WATER INSTALLATIONS

WELL NO.	DIA.	A	B	C	D	E	F	G
MW1	2"	1.80	117.0	122.0	121.46	162.0	162.0	163.30
MW2	2"	2.14	118.0	122.0	122.94	162.0	162.0	163.64
MW3	2"	1.15	65.0	70.0	83.26	100.0	100.0	103.32
MW4	2"	1.71	74.0	79.0	87.77	119.0	119.0	121.15
MW5	2"	2.05	75.0	80.0	79.14	120.0	120.0	123.21
MW6	2"	2.03	72.0	77.0	72.04	117.0	117.0	119.05
MW7	4"	1.89	60.0	66.0	61.94	201.0	201.0	203.53
P8	4"	1.87	177.0	182.0	194.86	222.0	232.0	234.89
P9	2"	1.63	145.0	152.0	163.80	192.0	192.0	193.87
P10	2"	1.63	95.0	100.0	112.85	140.0	140.0	141.63
P11	2"	2.05	75.0	81.0	79.14	120.0	120.0	123.21
P12	2"	2.03	72.0	76.0	72.04	117.0	117.0	119.05
P13	3/4"	1.85	61.0	65.0	62.84	201.0	201.0	203.53
P14	3/4"	1.87	178.0	183.0	194.86	222.0	232.0	234.89
P15	3/4"	1.53	145.0	152.0	163.81	192.0	192.0	193.87
P16	3/4"	1.63	94.0	101.0	112.65	140.0	140.0	141.63
P17	3/4"	1.71	74.0	79.0	87.77	119.0	119.0	121.15
P18	3/4"	2.05	75.0	80.0	79.14	120.0	120.0	123.21
P19	3/4"	2.13	72.0	77.9	72.04	117.0	117.0	119.05
P20	3/4"	1.91	61.0	67.0	61.94	201.0	201.0	203.53
P21	3/4"	1.87	177.0	182.0	193.86	222.0	232.0	234.89
P22	3/4"	1.63	145.0	152.0	163.80	192.0	192.0	193.87
P23	3/4"	1.73	95.0	100.0	113.85	140.0	140.0	141.63
P24	3/4"	2.05	76.0	82.0	79.14	120.0	120.0	123.21
P25	3/4"	2.04	71.0	76.1	72.04	117.0	117.0	119.05
P26	3/4"	1.85	61.0	66.0	62.94	201.0	201.0	203.53
P27	3/4"	1.87	178.0	183.3	194.76	222.0	232.0	234.89
P28	3/4"	1.63	143.0	152.0	163.71	192.0	192.0	193.87
P29	3/4"	1.62	95.0	102.1	112.55	140.0	140.0	141.63

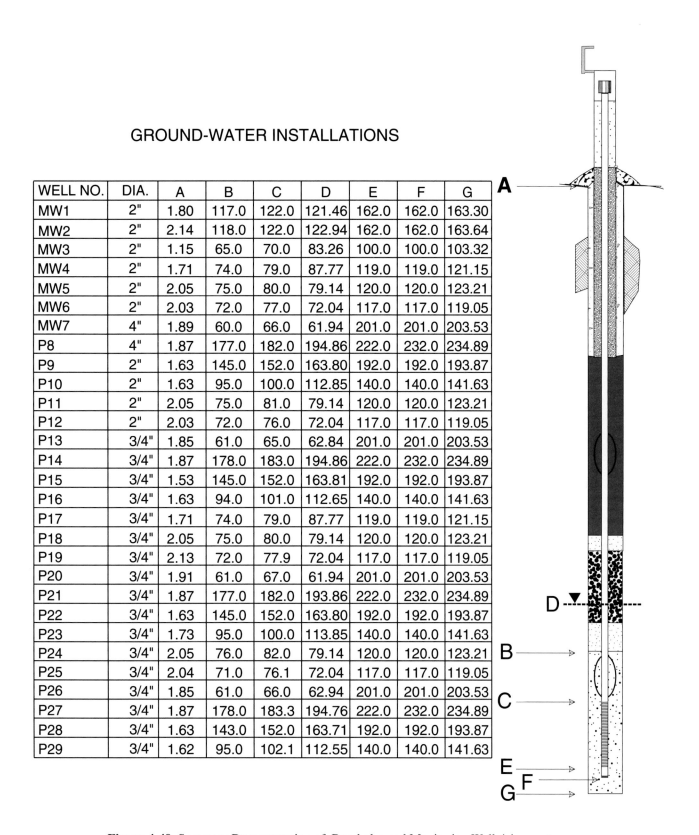

Figure 4-49 Summary Documentation of Boreholes and Monitoring Wells/piezometers

BOREHOLE, PIEZOMETER & MONITORING WELL DECOMMISSIONING GUIDANCE

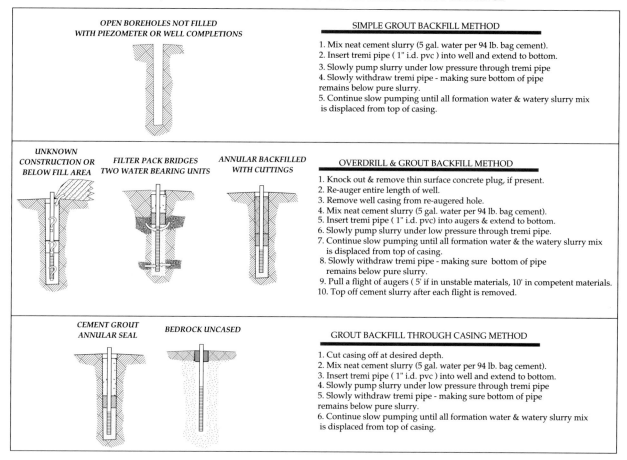

Figure 4-50 Decomissioning of Boreholes and Monitoring Wells/piezometers

the appropriate decommissioning technique, state closure or abandonment regulations should be reviewed for applicability to the site.

REFERENCS

ASTM (1985). Standard test method for penetration test and split-barrel sampling of soil. Test Designation D1586-67. 1985 Annual Book of ASTM Standards. American Society for Testing and Materials, Philadelphia, vol. 04.08, pp 298-303.

ASTM (1985). Standard test method for deep, quasi-static, cone and friction cone penetration tests of soil. Test Designation D3441-79. 1985 Annual Book of ASTM Standards, American Society for Testing and Materials,Philadelphia, vol. 04.08, pp 550-557.

ASTM D-1587, Practice for thin-wall sampling of Soils, 1991 Annual Book of ASTM Standards. American Society for Testing and Materials, Philadelphia, vol. 04.08,

ASTM D-2216-80, Method for Laboratory Determination of Water (moisture) Content of Soil, Rock, and Soil-Aggregate Mixtures, Annual Book of ASTM Standards, American Society for Testing and Materials, Philadelphia, vol. 04.08

ASTM D-2573 Vane shear Test Designation D1586-84. 1991 Annual Book of ASTM Standards. American Society for Testing and Materials, Philadelphia, vol. 04.08.

ASTM D5092 Design and Installation of Ground water monitoring well in Aquifers, Test Designation D5092-91. 1991 Annual Book of ASTM Standards. American Society for Testing and Materials, Philadelphia, vol. 04.08.

.ASTM D–2488–84, Practice for Description and Identification of Soils (Visual - Manual Procedure), Annual Book of ASTM Standards, American Society for Testing and Materials, Philadelphia, vol. 04.08.

ASTM, 1991, Classification of Soils for Engineering Purposes (ASTM D2487-69) Annual Book of ASTM Standards, American Society for Testing and Materials, Philadelphia, vol. 04.08, pp

ASTM, 1991, Standard test method for penetration test and split-barrel sampling of soil. Test Designation D1586-84. 1991 Annual Book of ASTM Standards. American Society for Testing and Materials, Philadelphia, vol. 04.08, pp 298-303.

Bauer, G.E. & Demartinecourt, J.P. (1982). The modified borehole shear device. Geotechnical Testing Journal, vol. 6, pp 24-29.

Bjerrum , L. and K. H. Anderson, 1972, In situ Measurement of Lateral Pressures in Clay, Proc. 5th European Conf. on Soil Mech. and Found. Eng. Madrid.

Brand, E.W., Hencher, S.R. & Youdan, D.G. (1983a). Rock slope engineering in Hong Kong. Proceedings of the Fifth International Rock Mechanics Congress, Melbourne, vol. 1, pp C17-C24.

Brassington, F. C., S. Walthall, 1985, Field Techniques Using Borehole Packers in Hydrogeological Investigation, Quarterly Journal of Engineering Geology, 28, 181-195.

Brown, E.T. (Editor), 1981, Rock Characterization Testing and Monitoring: ISRM Suggested Methods, Pergamon Press, Oxford, 211 p.

Brown, E.T. (Editor), 1981,. Rock Characterization Testing and Monitoring: ISRM Suggested Methods, Pergamon Press, Oxford, 211 p.

Campbell, M. D.and J. H. Lehr, 1973, Water Well Technology, McGraw-Hill, New York, N.Y., 681 pp.

CANMET, 1977, Ground-water, Pit Slope Manual, Report No. 77-01.

Casagrande, A., 1948, Classification and Identification of Soils, Trans. ASCE p. 901-992.

Cherry, J. A., and P. E. Johnson, 1982, A Multilevel Device for Hydraulic Head Monitoring and Groundwater Sampling in Fractured Rock, Ground Water Monitoring Review

Coates, D. F., Classification of rocks for rock mechanics , International Journal Rock Mech. and Mining Science, 1, 1964.

Craig, D.J. & Gray, I. (1985). Groundwater Lowering by Horizontal Drains. GCO Publication No. 2/85, Geotechnical Control Office, Hong Kong, 123 p.

Deere, D. U., "Technical Descriptions of rock cores for engineering purposes", Rock Mechanics. *Geol.*,l, 1963.

DeRuiter, J. (1982). The static cone penetration test, state of the art report. Proceedings of the Second European Symposium on Penetration Testing, Amsterdam, vol. 2, pp 389-405.

Eyles, C. H., and H. Eyles , 1983, A Glaciomarine Model for Late Precambrian Diamictites of the Port Askaig Formation, Scotland, Geology, V. 11 p. 692-696.

Gibson, U. P., and R. D. Singer, 1971, Water Well Manual, Premier Press, Berkeley CA., 156 pp.

Hada, S. 1977. Utilization and Interpretation of Micro Flowmeter, Engineering Geology (Japan), 18 26-37.

Handy, R.L. & Fox, N.S. (1967). A soil borehole direct-shear test device. Highway Research News, USA no. 27, pp 42-52.

Hanna, T.H.,1985, Field Instrumentation in Geotechnical Engineering. Trans Tech Publications, Clausthal, Germany, 843 p.

Hansen, E. A. and Harris, 1974, A groundwater Profile Sampler, Water Resources Research, 10(2):375.

Hathaway A. 1990. Bulletin of the Association of Engineering Geologists Vol 45.

Hathaway, A. W., 1988, Manual on Subsurface Investigations, American Association of State Highway and Transportation Officials, Inc. Washington D.C.

Hess, A. E. 1982. A heat-pulse Flowmeter for Measuring low Velocities in Boreholes, U.S. Geological Survey Open File Report 82-699, 40 pp.

Hess, A. E. 1986. Identifying Hydraulically Conductive Fractures with a Slow-velocity Borehole Flowmeter, Canadian Geotechnical Journal, 23, 69-78.

Hughes, 1991, Personal Communication

Hvorslev, 1951, Time Lag in the Observation of Ground Water Levels and Pressures, Bulletin No. 36, USA/COE Waterways Experimental Station, Vicksburg, MS.

Hvorslev, M.J.,1949,. Subsurface Exploration and Sampling of Soils for Civil Engineering Purposes. V S Army Waterways Experiment Station, Vicksburg, Mississippi, 521 p.

ISSMFE (1977). Report of the subcommittee on standardization of penetration testing in Europe, Appendix 5. Proceedinqs of the Ninth International Conference on Soil Mechanics and Foundation Enqineerinq, Tokyo, vol. 3, PP 95-117.

Jennings,J. E., A. B. A. Brink, and A. A. B. Williams, 1973,Soil Profiling for Civil Engineering Purposes in South Africa, Transactions of the SAICE, p. 3-11.

Kenney, T. C., 1967, Field Measurements of In Situ Stressess in Quick Clay, Proc. Geotechnical Conference, Oslo, p. 49-55.

Keys, W. S. 1971. Application of borehole Geophysics to Water Resources Investigations, Techniques of water-resources investigations of the United States Geological Survey, NTIS, Washington DC, Book 2, Chapt. 109-114.

Keys, W. S., and J. K. Sullivan. 1978, Role of Borehole Geophysics in Defining the Physical Characteristics of the Raft River Geothermal Reservoir, Idaho Geophysics, 44, 1116-1141.

Kruseman, G. P., and N. A. DeRidder, 1976, Analysis and Evaluation of Pumping Test Data. International Institue for Land Reclamation and Improvement, Wageningen, The Netherlands, 200 pp.

Lambe and Whitman, 1969, Soil Mechanics, John Wiley and Sons, Inc. Lambe, W. T.,1951, Soil Testing for Engineers, John Wiley and Sons, London

Legget, R. F., 1962 " Geology and Engineering", McGraw-Hill, New York, 613 p.

Meigh, A.C., Skipp, B.O. 1960, Gamma-ray and neutron methods of measuring soil density and moisture. Ge'otechniaue, vol. 10, pp 110-126.

Menard, L. (1965). Regles pour le calcul de la force portante et du tassement des foundations en fonction des resultats pressiometriques. (Rules for the calculation of bearing capacity and foundation settlement based on pressuremeter tests). Proceedings of the Sixth International Conference on Soil Mechanics and Foundations Engineering, Montreal, vol. 2, pp 295-299.

Molz, F. J., et al., The Impeller Meter for Measuring Aquifer Permeability Variations: Evaluation and Comparison with Other Tests, Water Resources Research, 25, 1677-1683.

Molz F. J., et al., 1990. A New Approach and Methodologies for Characterizing the Hydrogeologic Properties of Aquifers, EPA/600/2-90/002, 204 pp.

Morin, R. H. et. al., 1988a. Determining the distribution of Hydraulic Conductivity in a Fractured Limestone Aquifer by Simultaneous Injection and Geophysical Logging, Ground Water, 26, 587-595.

Nielsen, D. M., 1991, (editor) Practical Handbook of Ground-water Monitoring, Lewis Publishers, Chelsea, MI., 717p

.Pang, P.L.R. (1986). A new boundary stress transducer for small soil models in the centrifuge. Geotechnical Testing Journal, vol. 9, pp 72-79.

Patton, F. D., 1979, Groundwater Instrumentation for Mining Project, In: Proc. First International Mine Drainage Symposium, Denver CO., Miller Freeman Publ.

Peck, A. J., and R. M. Rabbidge, 1966, Soil Water Potential: Direct Measurements by a New Technique, Science, Vol. 151, pp. 1385-1386.

Peck, A. J., and R. M. Rabbidge, 1969, Design and Performance of an Osmotic Tensiometer for Measureing Capillary Potential, Soil Science Society Am. Proc., Vol. 33, pp. 430-436.

Rushton, R. R. & K. W. F. Howard, 1982, The Unreliability of Open Observational Boreholes in Unconfined Aquifer pumping Tests, Ground Water 20, 346-350.

Sanglerat, G. , 1972, The penetrometer and Soil Explorations, Elsevier Publishing Company.

Sweet et al, 1990, Water Level Monitoring-Achieval Accuracy and Precision, in ASTM STP No. 1053, Ed. Nielson and Johnson , Ground Water and Vadose Zone Monitoring, 1916 Race st., Philadelphia Pa. 19103, pp 178-192.

Terzaghi and Peck, 1949, Soil Mechanics in Engineering Practice, J. Wiley, Second Edition, New York.

Thornhill. J , 1989, Accuracy of Depth to Water Measurements, Superfund Ground Water Issue USEPA, Office of Solid Waste and Emergency Response

U.S. Environmental Protection Agency, 1973, Manual of Individual Water Supply Systems, Office of Water Porgrams, Water Supply Division, EPS 430/9-74-007, Washington, D.C.

U.S. Environmental Protection Agency, 1976, Manual of Water Well Construction Practices, Peport USEPA 570/9-75-001, Washington, D.C. 156 pp.

U.S. EPA, 1989, Handbook of Suggested Practices for the Design and Installation of Ground-water Monitoring Wells, EPA 600/4-89//034.

Vispi, M. A., 1980, Multiple-port Piezometer Installation at Rocky Mountain Arsenal, Report No. WES/MP/GL-80-12.

Walton, W. C., 1970, Groundwater Resource Evaluation, McGraw-Hill, New York.

Weeks, E. P., 1979, Determining the Ratio of Horizontal to Vertical Permeability by Aquifer-test Analysis, Water Resources Research, Vol. 5, No. 1, pp. 196-214.

Windle, D., Wroth, C.P., 1977, The use of a self-boring pressuremeter to determine the undrained properties of clays. Ground Engineering, vol. 10, no. 6, pp 37-46.

Winterkorn and Fang, 1975,. Foundation Engineering Handbook, Van Nostrand Reinhold Company, New York.

CHAPTER 5

ENVIRONMENTAL TESTING

5.1 INTRODUCTION

Beyond the insitu geotechnical tests performed to evaluate strength and deformation, (discussed in the last chapter), lie field and laboratory tests to evaluate aquifer properties and the environmental factors that affect the hydrogeologic system. Aquifer properties may include hydraulic conductivity and porosity of individual samples, hydraulic conductivity of the area immediately adjacent to boreholes or larger scale pump testing to evaluate large volumes of aquifer or confining unit materials for a wide variety of indices. These tests can also include chemical and recharge evaluation of subsurface units. A number of factors that must be considered during any environmental sampling and testing program are as follows:

1. Are the tests relevant to the particular problems being considered? For example, when hydraulic conductivity parameters are required, the soil or rock aquifer stressed by the test should be subjected to pumping stresses which are as close as possible to both the initial, and resultant changes of stress which will occur in the full scale situation. In other words, vertical hydraulic conductivity determined from laboratory permeameter tests may have little relevance to the horizontal hydraulic conductivity values necessary to design and operate a full scale pump and treat system.

2. Is the volume of soil or rock affected by the test, as representative as possible, of the geologic unit. Variations in both the subsurface composition and fabric need to be considered. At any one location it is necessary that the volume of geologic material undergoing testing should contain a representative number of the fabric features likely to be critical in the full scale situation. Major changes in stratification can often be found from agricultural soil survey data, geological studies including the geomorphology and geophysical methods. Data on the spacing and orientations of laminations and other discontinuities can only be obtained from exposures in excavations or from detailed inspection of samples of adequate size and known orientation. In some geologic units indications of the variability can be obtained by continuous profile measurements using borehole geophysics or cone penetration tests. It should however be realized that changes or lack of change in the profile measurements of these devices may not reflect actual changes in the mass properties of interest. Test data recovered from both field and laboratory environmental testing may be interpreted through a number of alternative methods.

3. Consideration should be given to the methods available for interpreting the test data. Where interpretation relies on calibrations with other tests, or the results are incorporated into empirical relationships with full scale behavior, it is essential to have full details of both the geologic materials (including fabric features) and the environmental test procedures in order to establish their validity in the assessment questions at hand. The relative merits of making simple tests which are adequately calibrated against other tests, or full-scale behavior in a wide range of soils, (such as STP's), or the carrying out of limited numbers of more complex tests, should be carefully considered with in the investigation. A common occurrence within site assessments is including relatively few very complex tests, such as a single large pump test or a long term infiltration study. Complex laboratory tests are also generally only warranted when good representative samples are available and the test specimens themselves are fully representative of the soil or rock in the mass. Similarly, complex field tests are only justified when adequate precautions can be taken during the installation of the test equipment to reduce disturbance of the soil around the test device to a reasonable level. As a guide to the interpretation of the tests, empirical relationships showing a wide range of scatter are usually indicative of either unsatisfactory test methods or that the correlations have been obtained over too wide a range of materials.

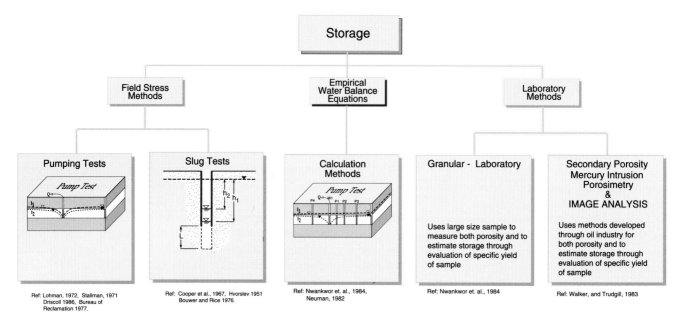

Figure 5-1 Pictograph Storage

4. Major site assessment projects often have sufficient scope and budget to allow the collection of large volumes of site specific technical data. These larger projects provide the investigator with the means to make much more accurate assessments of geologic parameters through larger scale field tests. These field tests and/or full-scale trials should be considered at an early stage of the project. These tests can often lead to substantial overall economies and increased confidence in the designs. On extensive sites careful consideration is necessary in the location of large scale tests and full scale trials, as well as, in the choice of comparative tests required to extend the use of the data to other locations. Data gathered can range from aquifer test data to individual soil characteristics, the key in such programs is to organize a comprehensive and cost effective test program that will answer real technical issues of the site. These issues may include such items as location of monitoring wells, foundation stability or aquifer remediation. The conduct of borehole drilling, sampling and testing at the site provides an opportunity to evaluate the hydrogeologic environment, in which the following conditions and characteristics may be determined:

• Rates of ground–water flow;
• Elevation of seasonal or cyclical maximum high ground–water surface (uppermost aquifer);
• Ground–water recharge and discharge areas and their interrelationships; and,
• Base level information on existing ground-water quality.

In order to develop these data, laboratory and field tests must be performed on the installations completed during the Phase II field investigation.

The aquifer properties typically important for site assessments include both storage properties and hydraulic conductivity. The methods used to measure or estimate aquifer storage properties are listed in Pictograph Figure 5-1; methods to measure or estimate hydraulic conductivity are listed in Pictograph Figure 5-2. The determination of aquifer properties (i.e., Transmissivity and Storage), begins with applying a known stress to the ground-water system, followed by measuring the response to that stress over space and/or time. The observed aquifer response is matched to a computed response relative to boundary conditions, and the corresponding parameters determined.

Bradbury and Muldoon (1990) evaluated many types of laboratory and field hydraulic conductivity test methods. They observed that hydraulic conductivity must be viewed in terms of the operational scale of the measurement, based on the scale of the problem at hand and the volume of the material of interest. The hydraulic conductivity of a given geologic unit was observed by Bradbury and Muldoon (1990) to increase as the operational scale of measurement increases. Figure 5-2 shows the relationships between laboratory permeameter tests, slug tests, single well tests and multiple well pump tests. The relationship shows that large volume stress tests generally showed larger values of hydraulic conductivities. More importantly, the field tests (slug, single well and multi well pump tests) were reasonably close to each other. This is an important

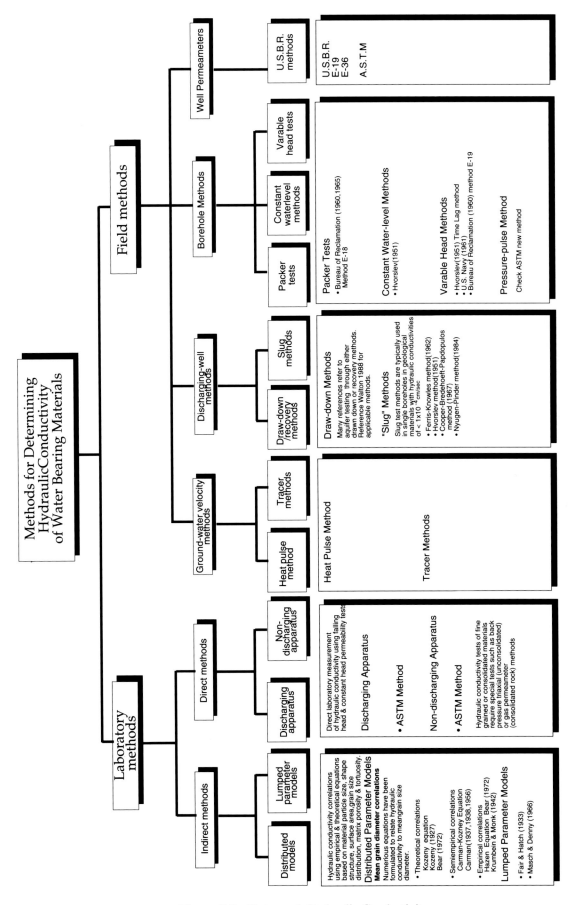

Figure 5-2   Pictograph Hydraulic Conductivity

finding relative to the field evaluation techniques used in site investigation projects. Although the volume of soil or rock tested differs through the various test methods residual local effects are probably still evident throughout the test methods. We therefore see slug tests and multi well pump tests with, at least, order of magnitude similar hydraulic conductivities. This may be due to reasonably homogeneous aquifers used in the testing, or a result of the local hydraulic conductivity of the aquifer controlling the overall pump testing results. One should consider the potential overall spatial variability of the system under investigation before taking point values, (i.e. slug tests, packer tests and short term single well pump tests), as representative of an entire area. Many point values, however, can simulate understanding of area wide values.

In general, one finds the more detailed the site investigation, the more heterogeneous the geologic unit are observed to be. Methods used to determine spatial variability as depicted in Pictograph Figure 4-3 were developed in Chapter 4.0 from information in Waldrop et al. (1989) and Taylor et al. (1990). These methods can help to understand the spatial variability of a site or regional area that can direct or •target' point value measurements to evaluate site wide aquifer characteristics. In other words, if one understands spatial variability of a site one can better target field and laboratory tests.

Site assessments have traditionally had saturated hydraulic conductivity tests performed at piezometers or wells, by use of appropriate hydrogeologic methods. Testing of field hydraulic conductivity of geologic materials through simple, (at least in concept), variable head or constant head techniques represented a relatively quick method that established a reasonably defensible hydraulic conductivity value for each of the geologic units tested. These traditional borehole variable or constant head tests only represent a portion of the range of methods applicable for evaluating both saturated and vadose zone hydraulic conductivities for porous geologic materials. The primary use of these additional methods is to increase the opportunity to establish the heterogeneity of the subsurface.

Hydraulic conductivity (K) of a porous medium, is the volume of water that will move in a unit time under a unit hydraulic gradient through a unit area measured at right angles to the direction of flow (Lohman, et al., 1972). Irregularities in ground–water flow, which result from variations in hydraulic conductivity of subsurface materials, should be identified within the site assessment. Soils and sedimentary rocks occurring in alternating layers or zones of varying K should be identified, if they may influence ground–water flow. Because the pattern of

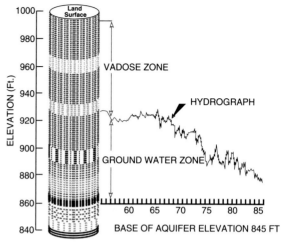

Cross Section through The Vadose Zone and Ground-water Zone
(Modified from Ayers and Branson, 1973)

Figure 5-3 Vadose-Saturated Zone

ground–water flow and distribution of K are major influences on the process of identifying final locations of wells, it is important to establish both average and worst–case K values of geologic units underlying and adjacent to the facility. Methods for determining K values of subsurface material samples are discussed in the following section.

The interrelationship and classification of the various methods for determination of hydraulic conductivity for water–bearing earth materials is shown by Figure 5-2. The methods can be divided into two major categories: laboratory methods, and field methods. The field methods are then divided into four sections: ground–water velocity methods, discharging–well methods, borehole tests, and shallow–well (unsaturated / vadose–zone) permeameter tests. The latter three methods all depend on pumping or changing the head in borings.

In reviewing these methods we will first describe determination of saturated hydraulic conductivity above the ground-water surface (see Figure 5-3). Later in this chapter, methods for evaluating unsaturated hydraulic conductivity will be outlined and finally, fully saturated hydraulic conductivity measurements below the water table will be described using a full complement of evaluation methods.

The investigator must be fully conversant with these techniques and strive to adequately document the performance of each test. Many of these aquifer stress tests can be performed quickly; however, frequent head

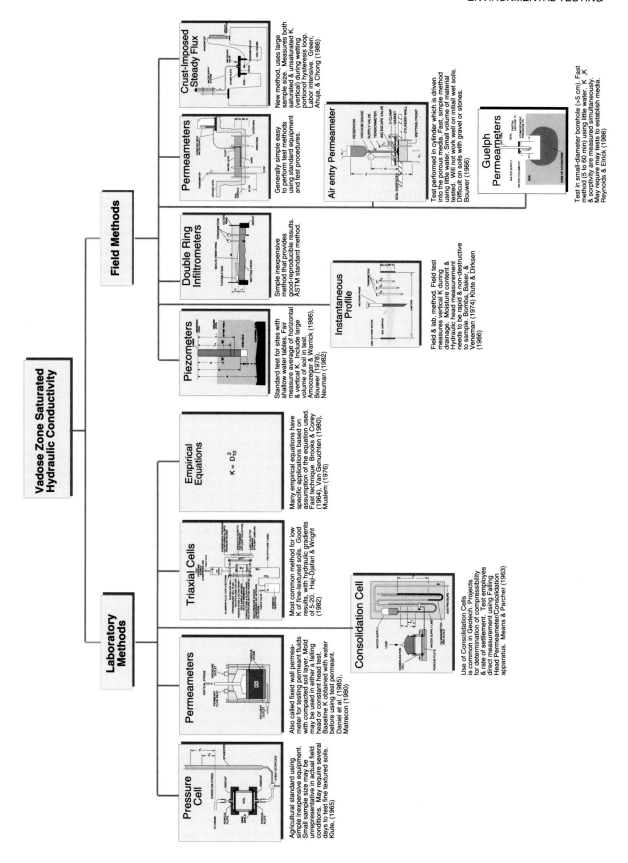

Figure 5-4 Pictograph Saturated Hydraulic Conductivite of Vadose Zone

level/time data must be taken, generally in the first 10 minutes, to establish reliable values for hydraulic conductivity for each monitoring well and piezometer. These tests usually only provide horizontal hydraulic conductivity values for the area immediately adjacent to the piezometer, however, if the geologic unit under investigation is reasonably homogeneous the near field results can often be extrapolated to other portions of the same hydrostratigraphic unit.

5.2 METHODS FOR MEASURING SATURATED HYDRAULIC CONDUCTIVITY ABOVE THE WATER TABLE

An important type of liquid movement relevant in all types of land disposal facilities is flow in the unsaturated zone extending down to the ground water table (see Figure 5-3). This flow is controlled by the unsaturated hydraulic conductivity of the vadose zone materials and various preferential flow paths. Unsaturated hydraulic conductivity is more difficult to measure than saturated hydraulic conductivity, since, unsaturated conductivity varies with both moisture content and pressure head. While saturated hydraulic conductivity is a constant value, unsaturated hydraulic conductivity must be determined over a range of moisture contents and pressure heads.

The test methods make certain assumptions to enable the evaluation of saturated hydraulic conductivity. Simpler methods such as infiltration tests rely on many more frequently violated flow assumptions to arrive at a resultant value. These potential sources of testing analytical error and bias should be understood, and care must be taken to avoid these errors to the extent possible, while conducting field tests. Most test methods for hydraulic conductivity assume that soils are vertically uniform or isotropic, a condition which seldom applies for natural soils. The appropriate methods used for a particular field test should be chosen based on a number of points: the cost of testing; the accuracy and precision required; the depth at which a layer is to be tested; the characteristics of the soil profile (uniform or layered); and, the approximate hydraulic conductivity range expected.

The overall accuracy of hydraulic conductivity tests is known to be highly dependent on the spatial variability of soils or rocks. Nielsen et. al. (1973) examined the highly variable nature of that hydraulic properties of field soils. They recommend taking numerous readings (at closely spaced locations) to characterize a "site" or field-sized area. This study reported that field-saturated hydraulic conductivity values tend to be log normally distributed rather than normally distributed, hence a majority of the net flux of water flow was in a few more permeable zones.

There are a number of methods (see pictograph Figure 5-4) available for determining the field saturated hydraulic conductivity of unsaturated materials above the water table. Most of these procedures require measurement of the infiltration or seepage rate of water into soil or geologic materials from an infiltrometer or permeameter device. Infiltrometers are used to measure hydraulic conductivity at the soil surface, and permeameters determine conductivity at different depths within the soil profile. These methods can be listed:

Field	Laboratory
• Borehole Permeameter	1. Pressure Cells
• Constant Head Methods	2. Compaction Molds
• Variable Head Methods	3. Consolidation Cells
• Guelph Permeameter	4. Tri-axial Apparatus
Infiltrometers	5. Emperical Equations
Single-ring Infiltrometers	
Double-ring Infiltrometers	
• Air-entry Permeameter	
• Double-tube Method	
• Crust-Imposed Steady State (Cube Method)	

5.2.1 Borehole Permeameter

Methods to determine hydraulic conductivity of near surface soils through borehole permeameter methods encompass a wide range of test procedures and analytical solution methods. The common feature among the the many borehole permeameter tests is that the rate of water infiltration into a cylindrical borehole is used to determine saturated field hydraulic conductivity. There are numerous varieties of borehole permeability tests depending on the geometry and materials located at the point of measurement. Besides the cylindrical constant head method described below as U.S.B.R. test designation E-19, there are spherical cavities, sand-filled cavities, piezometers completed in sand filled zones, or even piezometers driven directly into the soil. Jezequel and Mieussems (1975) reported that the driven piezometer point was promising for low hydraulic conductivity clays.

One of the most used borehole permeameter tests is the constant-head borehole infiltration test, as described by the Bureau of Reclamation Earth Manual (1968) as Test Designation E-19. This method consists of measuring the rate at which water flows outward from an encased hole under constant head (see Figure 5-5). The borehole geometry, borehole radius, steady state flow rate, and height of water within the borehole, and along with test duration consideration, are typically used in the solution. Variations of this test are available by conducting multiple

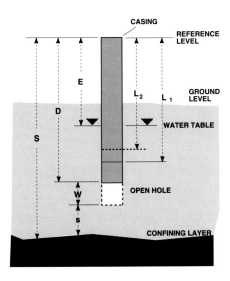

Figure 5-5 Borehole Permeameter

constant head borehole infiltration tests within the same borehole. This alternative results in different water levels within the borehole for each individual test. Then results from one or more trials at different heights of water are solved simultaneously to independently find hydraulic conductivity and capillarity effects.

Borehole permeameter test methods try to roughly simulate three-dimensional flow components due to lateral, as well as downward, saturated and unsaturated flow components. The configuration of the flow adjacent to the borehole is dependent on the geometry of the borehole, the hydraulic properties of the soil and the level of capillary suction of the soil. Since many of the earlier analytical solutions for falling-head and constant-head type borehole tests do not account for the effects of unsaturated flow away from the borehole. Glover (1953) and the U.S. Bureau of Reclamation (1978) have proposed borehole methods which are entirely dependent on "free surface" solutions to ground water flow. Stephens, et al., (1987); Philip (1985); and Reynolds and Elrick (1987), reported unsaturated flow in fine texture soils such as clays can greatly affect the infiltration rate from a borehole and must be considered in the solution for hydraulic conductivity. These investigators propose testing methods and/or solutions which consider unsaturated flow away from a wetted bulb around the augered borehole.

5.2.2 Infiltrometer Method

Infiltrometer methods are designed to measure the rate of infiltration at a soil surface. Hence, they do not directly measure field saturated hydraulic conductivity. This measurement point can also be a surface in a trench or excavation. Infiltration rates are known to be influenced both by saturated hydraulic conductivity, as well as, *soil capillary* effects. Capillary soil effect can be defined as the ability of dry soil to pull water away from a zone of saturation more rapidly than if the soil were saturated uniformly. The magnitude of the soil capillary effect can be estimated by using an initial moisture content at the time of testing, soil pore size, soil physical characteristics (texture, structure), and additional factors dependent on the particular method used. Capillary effects are reduced by continuing the test until obtaining steady-state infiltration.

Infiltrometers simply consist of a metal cylinder set at shallow depths into the soil. Variations on the infiltrometer techniques include the single-ring infiltrometer, the double-ring infiltrometer, and the infiltration gradient method. The literature describes many alternatives to the design and procedural use of these methods to determine the field-saturated hydraulic conductivity of material within the unsaturated zone. These methods are, however, similar since they all employ a steady volumetric flux of water infiltrating into the soil enclosed within the infiltrometer ring. Darcy's Equation for saturated flow is the basis for the determining the saturated hydraulic conductivity of soils from the test design and water flow rate. Infiltrometer tests require assumptions that the volume of soil being tested is field-saturated and that the saturated hydraulic conductivity of the soil mass is a function of both the flow rate and the applied hydraulic gradient.

Infiltrometer methods as described, should be used cautiously to determine the saturated hydraulic conductivity since infiltration is affected by both the materials' hydraulic conductivity, as well as, by capillary effects. Infiltration measurements can be sensitive to: 1) disruption of the infiltration surface (e.g., compaction, sealing by rain splash), 2) presence of textural stratification, 3) chemistry of the water used and, 4) water temperature. Water which is low in salts or high in sodium is dispersive and may result in lower calculated values of K.

Infiltrometer tests have been found useful for measuring rates of infiltration but these tests cannot directly provide a measure of field-saturated hydraulic conductivity. This is due to entrapped air within the wetting front, causing unsaturated soil conditions during infiltration tests. Experience has shown (Bouwer, 1966, Reynolds and Elrick, 1986) that "field saturated", (with intraped air), hydraulic conductivity is approximately 50 to 75 per cent less than K the true saturated hydraulic conductivity of the material.

Single Ring Infiltrometer

The single ring infiltrometer test typically consists of a cylindrical ring 30 cm or larger in diameter which is driven several centimeters into the soil. The test is begun when water is added within the ring on the surface of the soil. Covers are often used on the upper surface of the ring apparatus to prevent evaporation. The volumetric rate of water added to the ring must be sufficient to maintain a constant head within the ring. If the head of water within the ring is relatively large, an alternative falling head type test may be used to calculate the flow rate by measurement of the rate of decline of the water level within the ring, and the hydraulic head for a later time portion of the test. Infiltration rate is calculated after the water flow has sufficiently stabilized in the test method. The infiltrometer is removed directly after completion of the infiltration test and the depth to the wetting front is measured either visually, with a penetrometer-type probe, or by collection of soil samples using ASTM test method D4643 "Moisture Content Determinations".

As an attempt to deal with local geologic heterogeneities, large-scale, single-ring infiltrometer tests can be designed by construction of a ponded infiltration basin. These rectangular basins may be as large as several, to many meters on a side. A measured flow rate of water is

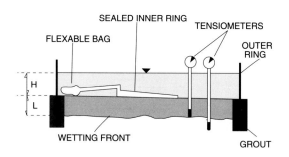

Figure 5-6 Double Ring Infiltrometer

added to the basin to maintain a constant head of water within the pond. If the depth of ponding is negligible compared to the depth of the wetting front, the saturated hydraulic conductivity of the soil is calculated to be equal to the steady state flux of water across the soil surface within the basin.

Since with most applications, the wetting front is allowed to propagate below the bottom of the ring, the single ring infiltrometer is subject to divergent flow conditions, due to the effects of unsaturated flow heterogeneities and anisotropy. These effects may lead to inaccuracies in the determination of saturated hydraulic conductivity.

Figure 5-7 Large Area Double Ring Infiltrometer

Double Ring Infiltrometer

The double ring infiltrometer's underlying principles and method of operation are similar to the single ring infiltrometer. The one exception in this test method is that an outer ring is included to ensure that one-dimensional downward flow exists within the test horizon of the inner ring. Infiltrating water through the outer ring acts as a hydraulic barrier to lateral movement of water from the inner ring (see Figure 5-6 and 5-7). Double ring infiltrometers may be either open to the atmosphere, or most common to the method, the inner ring is covered to reduce evaporation. The flow rate is measured for open double ring infiltrometers directly from the rate of decline of the water level within the inner ring as a falling head test. For the constant head method; the rate of water input necessary to maintain a stable head within the inner ring is used for calculation of flow rates. The sealed double ring infiltrometers defines a flow rate measurement by weighing a sealed flexible bag supple reservoir for the inner ring.

Test method ASTM D3385-88 is applicable for infiltration rates in the range of 10^{-2} cm/s to 10^{-5} cm/s. A modified double-ring infiltrometer method recommended for infiltration rates from 10^{-5} cm/s to 10^{-8} cm/s is currently under development (Tratwiew, 1988).

5.2.3 Double-Tube Method

The double tube method is described as a means of measuring the horizontal, as well as the vertical, field-saturated hydraulic conductivity of material in the vadose zone (Boersma 1965). The double tube method (Bouwer 1961, 1962, 1964), uses two coaxial cylinders placed in an auger hole as shown in Figure 5-8. The difference between the rate of flow in the inner cylinder and the simultaneous rate of combined flow from in the inner and outer cylinders is used to calculate K.

Depending on the hydraulic conductivity of the soil, the double-tube method may require over 200 liters of water and two to six hours for completion. The method is generally not suitable for rocky soils because of the difficulty in driving tubes into the ground. Due to soil disturbance around the inner tube, the diameter of the outer tube should be at least two times that of the inner tube. The K value obtained utilizing this method is affected by both the horizontal and vertical hydraulic conductivities of soil.

5.2.4 Air-Entry Permeameter

The air-entry permeameter technique is a method were a 25 cm diameter cylindrical unit of undisturbed soil is isolated within an infiltration cylinder driven into the soil approximately 15 to 25 cm. A head of water is then applied to the upper end of the soil unit, causing water to penetrate the soil with a wetting front. The infiltration rate is measured until the wetting front is close to the bottom of the cylinder which is determined by use of an implanted tensiometer Figure 5-9 provides an illustration of an air entry-permeameter.

The air-entry permeameter is similar in design and operation to a single ring infiltrometer. For example, the field saturated hydraulic conductivity is used to calculate the volumetric flux of water into the soil within a single permeameter ring. The air-entry permeameter measures the air-entry pressure of the soil. Air-entry pressure is used in the test as an approximation of the wetting front pressure head. This pressure head then allows determination of the hydraulic gradient, and consequently saturated hydraulic conductivity.

Flow rates are measured by observing the decline of the water level within the reservoir. After a predetermined

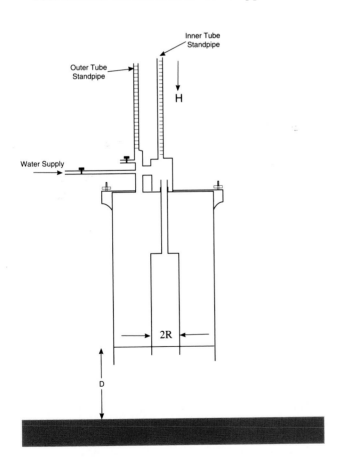

Source: Klute, 1986

Figure 5-8 Double Tube Method

volume of water has infiltrated, (this volume must be based upon the estimated available storage of the soil interval contained within the ring), and the flow rate is relatively stable, infiltration is terminated and the wetted profile is allowed to drain. The minimum air pressure measured inside of the permeameter ring attained during drainage over the standing water is the air-entry valve. Once the minimum pressure is achieved, the permeameter is removed, and the depth to the wetting front is measured (Amoozegar and Warrick, 1986).

Many of the same restrictions and assumptions apply for the air-entry permeameter tests as for the infiltrometer methods. However, one-dimensional vertically downward flow is ensured, since the wetting front is not allowed to advance below the bottom of the permeameter ring. In addition, since the hydraulic gradient is measured during the test, the infiltration rate need not necessarily reach steady state during the first portion of the test. The depth of the wetting front after completion of the test can be a problem, since visual determination is especially difficult in soils with higher initial moisture content.

As soon as the minimum pressure is reached, air begins to bubble up through the wetting front. Field-saturated K can be calculated from the critical "air-entry value" or minimum pressure. Field-saturated K is approximately equal to 1/2 of K, in most soils or 1/4 of K in heavy soils.

5.2.5 Gypsum Crust-Cube Method

The cube technique is a field method for measuring

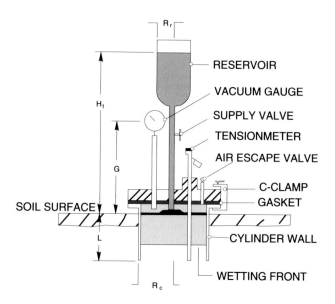

Figure 5-9 Air-Entry Permeameter Method

both the vertical and horizontal saturated hydraulic conductivity of a cube of soil 25 cm on a side. The cube is cut out in a field sampling mass and covered with gypsum. The vertical saturated hydraulic conductivity is measured by infiltrating water into the exposed upper surface of the cube. The horizontal hydraulic conductivity of the cube is obtained by turning the cube and repeating the vertical test. The field set-up to obtain the cube is illustrated on Figure 5-10. Bouwa and Dekkert, (1981), provide additional guidance to the cube method. This relatively inexpensive test may require relatively long periods of time to complete on clay soils. Some inaccuracy may develop due to questionable saturation levels and potential swelling of the sample materials. This test does, however, offer the benefit of large sample size and can include minor discontinuities that may effect soil hydraulic conductivity.

5.3 METHODS FOR MEASURING UNSATURATED HYDRAULIC CONDUCTIVITY

In the vadose zone, the void space is partly filled by air and partly by water. The moisture content or volumetric water content represents the quantity of water present at a certain time at a point in the porous media. The maximum value of volumetric water content occurs when all voids are filled with water; the minimum value occurs when all voids are filled with air. Thus, it varies between zero to the value of the soil porosity.

Measuring vadose-zone hydraulic conductivity values are very difficult because head gradients, flow rates, and moisture content or pressure head also must be measured.

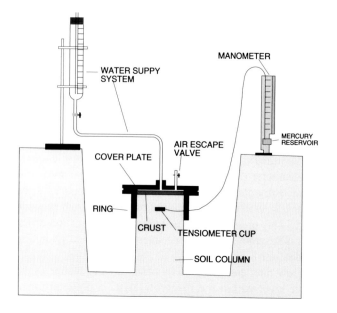

Figure 5-10 Gypsum Crust-Cube Set-up

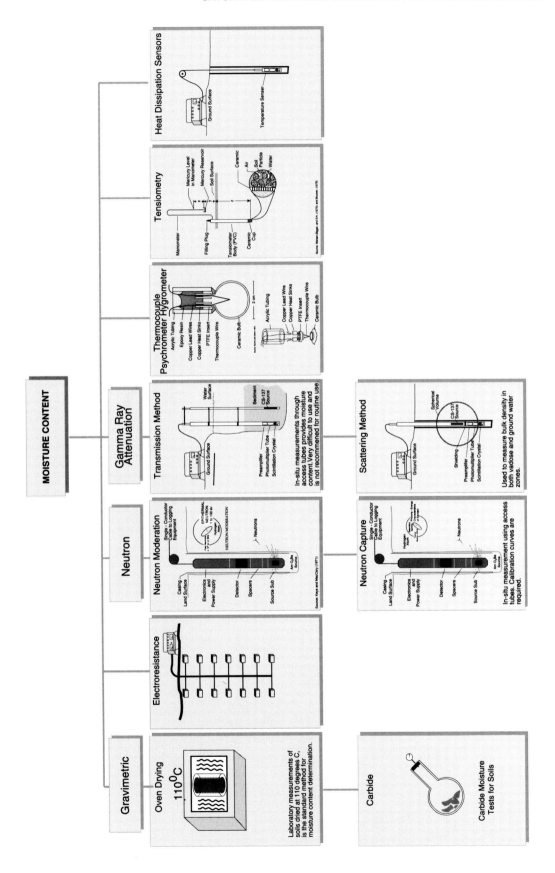

Figure 5-11 Moisture Content

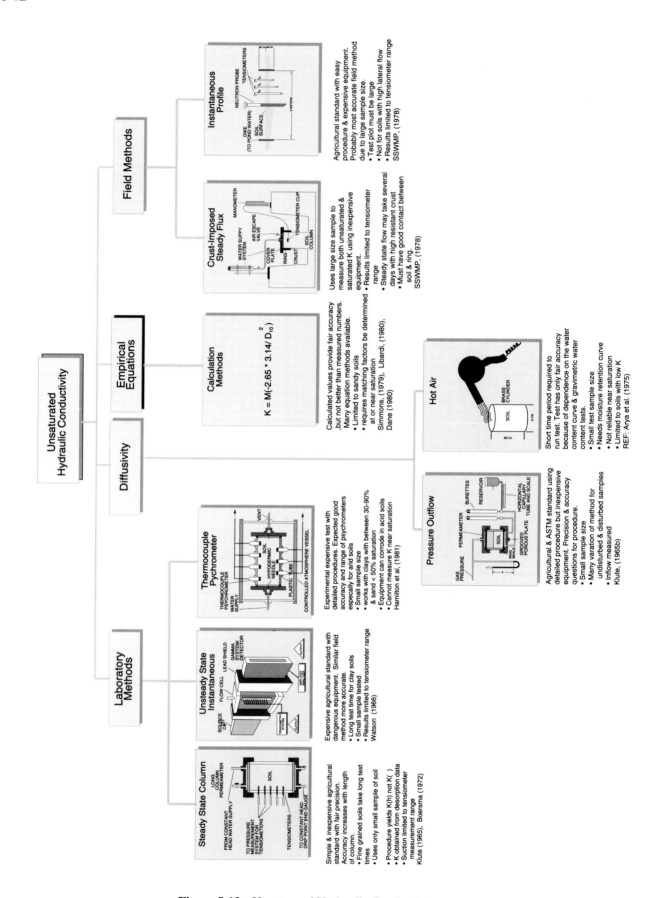

Figure 5-12   Unsaturated Hydraulic Conductivity

Factors that influence these measurements include: soil texture, soil structure, initial water content, a shallow ground-water level, water temperature, entrapped air, biological activity, entrained sediment in the applied water, and chemistry of the applied water (Wilson, 1982).

Changes in moisture content are important to detect in vadose zone testing. For example, under a waste disposal facility leachate collection sump, monitoring for changes in moisture content could be used to indicate leaks in the containment. Determining moisture content with depth can also be used to locate potential or seasonal perched water zones   Various methods used to measure moisture content are shown in Pictographic Figure 5-11. The recommended techniques for most field projects are gravimetric and neutron scattering methods. The neutron moisture meter containing both a source of fast neutrons and a slow neutron detector is lowered into the soil through an access tube. Neutrons are emitted by the source (for example, radium or americium-beryllium) at a very high speed. When these neutrons collide with a small atom such as hydrogen contained in soil water, their direction of movement is changed and they lose part of their energy. These "slowed" neutrons are measured by a detector tube and a scalar. The reading is related to the soil moisture content in the vadose zone and porosity in the saturated zone.

Gravimetric moisture content measurements are made by weighing soils before and after drying. If the moisture characteristic curve is known (Figure 5-13), then pressure

head can be measured using a tensiometer. Similarly, relative permeability can be determined in the vadose zone. The relative permeability is a normalized coefficient, which when multiplied by the saturated hydraulic conductivity, yields the vadose-zone or unsaturated hydraulic conductivity. It is typically presented as either a function of capillary pressure or saturation, and varies from one to zero. A number of empirical equations have been developed for approximating the vadose-zone permeability of isotropic porous media. Three commonly used equations for estimating the vadose-zone hydraulic conductivity are by Brooks-Corey (1964), Mualem (1976), and van Genuchten (1980). Methods to determine the vadose-zone hydraulic conductivity are listed below and shown on Figure 5-12. They represent field and laboratory tests that can be used within site assessment programs for evaluation of unsaturated hydraulic conductivity.

- Field Tests

- Instantaneous Profile
- Crust

- Laboratory Tests

- Steady-state Column
- Unsteady State Instantaneous Profile
- Thermocouple Psychrometer

- Empirical Equation

- Diffusivity

- Pressure Outflow
- Hot Air Method

The above laboratory tests are described in more detail in Section 5.11.2.

5.3.1 Instantaneous Profile Method (Ip)

The traditional instantaneous profile in-situ technique uses a level plot of soil, diked to pond 2-3 cm of water (see Figure 5-14). After addition of water, the plot is covered to prevent evaporation during drainage into the soil. At frequent intervals, both pressure heads and water content values are measured by any one of a number of alternative techniques. A nest of tensiometers and neutron access tubing can be, for example, installed at varying depths in the center of the basin. Movement of the wetting front is detected with a neutron probe. The test is continued until the wetting front passes the bottom of the instrumented

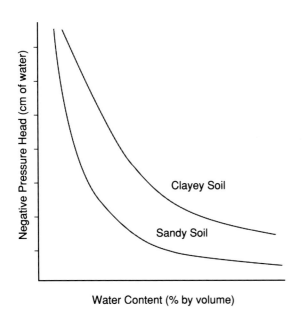

Source: Hillel, 1971

Figure 5-13 Soil Characteristic Curve

horizon.

A detailed description of the procedure for calculating unsaturated hydraulic conductivity (or diffusivity) with different depth increments is provided in Green and others (1986). Graphical plots of tensiometric data, and soil water content data through time are used within the procedure to estimate instantaneous water flux at known levels of water content and water potential.

The IP test method is reported (Green et.al., 1986) to be a relatively accurate test of unsaturated hydraulic conductivity. However, it is costly and time consuming to perform. General errors in measurement of water content or water potential can affect the accuracy of calculated unsaturated hydraulic conductivity. Rapid changes in barometric pressure may affect soil water potential readings. As with all methods, stratification within the soil profile being measured will also affect accuracy. If there is a ground-water surface within about three to four feet of the base of the zone of measurement, the test should not be performed (Watson 1966).

5.3.2 Crust-Imposed - Steady Flux

The crust-imposed, steady flux method also called the gypsum crust technique, is a field adaptation of the laboratory procedure for flow through an impeding layer. The gypsum crust method is similar to infiltrometer methods, since both use the measured rate of water flux across an infiltrative surface. The test method consists of a crust composed of varying mixtures of gypsum and coarse sand poured over the surface of an exposed excavated cylinder of soil. After the crust cures, water is ponded on the crust. The presence of the gypsum crust causes unsaturated soil conditions to form beneath the crust layer.

The method can also be used for saturated hydraulic conductivity measurements. The extension to saturated conditions requires the addition of a barrier to flow (after wetting the sides of the pedestal is plastered).

The crust method includes instrumentation of a series of tensiometers to measure water potential below the gypsum crust. Data recorded includes the rate of flux of water necessary to maintain a constant head over the gypsum crust and the diameter of the cylinder.

The gypsum crust method provides a single measurement of unsaturated hydraulic conductivity as a function of measured water potential for each crust constructed. In the crust method, a steady unsaturated flux of water is reached with a unit hydraulic gradient (influenced only by gravity). Under these conditions, the measured water flux is equal to the hydraulic conductivity. For those conditions where a unit hydraulic gradient does not form due to presence of textural variable layers, or due to compacted zones, several tensiometers can be installed at the top and bottom of the zone of investigation.

The crust method provides a single measurement of unsaturated hydraulic conductivity at a specific water potential which is read off of a tensiometer installed below the crust. Tensiometer readings must also be made accurately. Crusts of different resistances yield different points on the K(h) curve. A series of crusts ranging from greater to lower resistance assures that the data points fall on the wetting curve. The water potentials which evolve below the crust are a function of the crust material used (e.g., specific gypsum/sand mixture). A measurement of the steady-state flow rate can take hours to achieve in this test especially under crusts of high resistance. The geometry of the excavated block of soil is also critical to the analytical solution. Therefore, soil cylinders with a consistent diameter must be accurately obtained. This procedure is difficult in highly-structured or rocky soils (Bouma et. al. 1971).

5.4 BOREHOLE TESTS OF FIELD HYDRAULIC CONDUCTIVITY

The determination of in-situ hydraulic conductivity in boreholes is one of the most important field evaluation methods used in site assessments for waste disposal projects. Numerous test methods have been designed by governmental agencies in the United States for the determination of estimated hydraulic conductivities through exploration boreholes. These procedures, (see Figure 5-15), are widely used due to the relative economy of measuring hydraulic conductivities of moderate volumes of soils and rocks in existing site characterization boreholes. Valuable references are found in publications

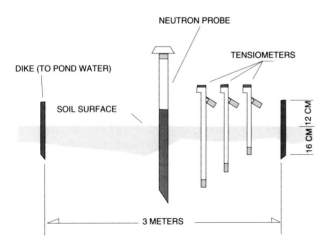

Figure 5-14   Instantaneous Profile

TYPICAL METHODS USED TO PREDICT HYDROGEOLOGIC SYSTEM PARAMETERS

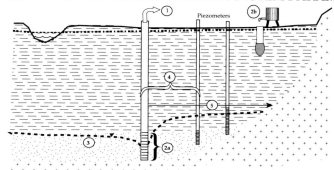

⓵ Sample
⓶ₐ Slug & Packer Test
⓶♭ Well Permeameter Test
③ Pump Test
④ Ground-water Velocity
⑤ Pump Test With Piezometers

OBJECTIVES	*SCALE*	*SCOPE*

OBJECTIVES

1) SAMPLE POINT EVALUATION
Can obtain test values for verticle K's and many inportant soil characteristics such as:
• *Porosity*
• *Hydraulic Conductivity*
• *Grain size distributions*

2) NEAR BOREHOLE EVALUATION
2a) PACKER & SLUG TESTING
These tests are relatively simple & less costly than aquifer tests. Usually conducted in conjunction with exploratory drilling & installed piezometers.
2b) WELL PERMEAMETER METHOD
Used to determine the K values of a soil in place. Consists of measuring the rate at which water flows from an uncased hole

3) LOCAL AQUIFER EVALUATION
Single well tests are commonly used to establish yield potential, well efficiency & aquifer types. Site Assessments use these tests to define:
• *Transmissivity*
• *Hydrogeologic Boundaries (recognized)*
• *Hydrogeologic yield*

4) DIRECT VELOCITY FIELD EVALUATION
Used when actual measurements are required of the flow direction & rates. Can provide for calibration of computer models when using indicator parameters.

5) HYDROGEOLOGIC SYSTEM EVALUATIONS
Methods provide the most comprehensive assessment of:
• *Transmissivity*
• *Storage & yield*
• *Hydrogeologic Boundaries (defined)*
• *Leakage*

SCALE

Yes — **Only Soil Samples** — Yes
No Insitu Tests Required

Yes — **K Value Adjacent to Borehole** — Yes
No, Need Local Aquifer Parameters

Yes — **Local Aquifer Values "OK"** — Yes
No Require Directional Velocity

Yes — **Aquifer Flow Velocity** — Yes
No Must Define Boundries & Storage

Full Aquifer Characterization

SCOPE

LAB TESTS OF BOREHOLE SAMPLES
• *Grain Size (ASTM D422-63)*
• *Direct Testing Methods*
 - *Permeameter-Falling Head (ASTM D2434-68)*
 - *Tri-axial/back pressure Tests(Corps. of ENG. EM1110-2-1906 App. VIII 7*

HYDRAULIC CONDUCTIVITY IN SINGLE HOLES
SLUG TESTS
• *Ferris-Knowles Method*
• *Hvorslev Method*
• *Bouwer-Rice Method*
• *Cooper-Bredehoeft-Papadopulos Method*
PACKER TESTS
• *Single Packer - USBR E-18*
• *Double Packer - USBR E-18*
WELL PERMEAMETER METHODS
• *USBR E-19*
• *USBR E-39*

SINGLE WELL AQUIFER TEST METHODS
MULTIPLE RATE
• *See text for references for multiple rate testing*
CONSTANT RATE
• *See text for references for constant rate tests*

GROUND-WATER VELOCITY METHODS
TRACER TESTS
• *Chloride*
• *Natural/release Related Indicator Parameters*
DIRECT MEASUREMENT
• *Thermal Probes*

PUMP TESTS WITH OBSERVATION PIEZOMETERS
CONSTANT RATE TESTS
• *Test Method Dependent on Aquifer Conditions, references in text*

Figure 5-15   Test Methods

of the U.S. Bureau of Reclamation (1960, 1965, 1966), the U.S. Navy Facilities Engineering Command (1971), the U.S. Geological Survey (1942), and many others, some of which will be quoted in the following pages on borehole testing. Perhaps the most useful compendium is that of Cedergren (1967).

Unfortunately, practicing hydrogeologists in the U.S. (including those in the regulatory agencies) utilize a mixture of traditional U.S. Geological Survey or water well industry units (e.g. transmissivity in gpd/ft; dimensions in feet) and SI units (e.g., hydraulic conductivity in cm/s). This document has continued this questionable practice, but the author has made every attempt to clearly define the units used and to simplify "mixed" calculations.

These important classifications of field tests in boreholes involve the application of an hydraulic pressure in the borehole different from that in the adjacent ground, and the measurement of the flow due to this difference. The pressure in the borehole may be increased through the introduction of water into the hole called a falling-head or inflow test, or it may be decreased by pumping water out of the hole, in a rising-head recovery or outflow test. The pressure may be held constant during a test (a constant-head test), or it may be allowed to equalize to its original value (a variable-head test). The technique is strictly applicable only to the measurement of hydraulic conductivity of soils or other geologic units below the ground-water surface, although an approximate assessment may be made in the vadose zone. However, this approximate value will reflect the infiltration capacity of the subsurface material rather than its unsaturated hydraulic conductivity. A great variety of tests are included under this heading, varying from the very crude, where simple problems can be solved by simple means, to the very sophisticated when the nature of the problem at hand demands more refined data.

For most types of subsurface investigations, field tests yield more reliable horizontal hydraulic conductivity data than those carried out in the laboratory. This difference are due to the larger volume of material tested in the field that will contact more permeable layered strata and because the soil is tested in-situ, thereby avoiding the typical disturbance associated with sampling. Laboratory tests for hydraulic conductivity yield vertical K values due to the testing procedure being applied on the top (or bottom) of the sample. The appropriate choice of drilling method and application of careful drilling technique are necessary minimize disturbance of the soil for single borehole hydraulic conductivity testing. In granular soils, the borehole bottom may be loosened during drilling or sampling thus increasing hydraulic conductivity. In layered deposits of varying hydraulic conductivity, a skin of remoulded or smeared material may be formed on the walls of the borehole, (called skin effects) thus reducing the horizontal hydraulic conductivity. Alternatively, layered sediments (especially silty layers in sandy deposits) will show much higher field hydraulic conductivity, than laboratory hydraulic conductivity. This is a result of the predominant horizontal K values obtained from short-term field testing, compared to the predominant vertical K results obtained from laboratory testing. In very soft marine clays, it is very difficult to carry out a successful hydraulic conductivity test because of the low hydraulic conductivity of the soil, its compressibility, and the possibility of hydraulic fracture arising from the relatively large head require for a falling-head or constant-head test.

One must make active decisions on the type of field hydraulic conductivity test to be used for each situation based on project requirements. Constant-head tests are likely to give more accurate results than variable-head tests, but, on the other hand, variable-head tests are simpler to perform. The water pressure used during the test should be less than that which will disrupt or fracture the geologic units by hydraulic fracturing. In general, it is recommended that the total increase in water pressure should not exceed one half the effective overburden pressure. This rule of thumb applies to both single borehole testing as well as packer permeability testing. With soils of high hydraulic conductivity, greater than about 10 cm/s, flow rates are likely to be very large and head losses at entry or exit points and in the borehole may be high. With these exceptionally permeable materials one should strongly consider use of field pumping tests. These pump tests permit the pressure head distribution, as it moves through the aquifer, to be measured by piezometers located radially away from the borehole. If a hydraulic conductivity test is carried out within a borehole using the drill casing, the lower limit of hydraulic conductivity that can be measured reliably is determined by the water-tightness of the casing joints and by the ability to seal the casing into the ground at the required test depth. In lower hydraulic conductivity soils and unweathered rock, it is advisable to carry out the test using a standpipe or piezometer which is sealed within the test length using annular sealant grout. In soil of low hydraulic conductivity, the flow rate may be very small, and measurements subject to significant error due to simple changes in temperature of the measuring apparatus over the course of the long test.

The hydraulic conductivity of a compressible soil is influenced by the effective stress at which it is measured, and there may be significant differences between the results of inflow tests, in which effective stress is reduced, and the results of outflow tests, in which it is increased. One should, during site assessments, try to use a test method that models the actual field conditions, as closely as possible. Where the conditions indicate increasing effective stress, such in aquifer recharge projects, a rising-head test should be used. In the case of decreasing effective stress, such as when assessing the quantity of inflow into an excavation, or a pump and treat collection system, a falling-head test would be considered most appropriate. The hydraulic conductivity of soil around the borehole may also be influenced by local recent changes in its stress history due to the installation of the borehole and any previous tests performed on it.

Execution of the borehole hydraulic conductivity test requires much expertise, and small errors in technique lead large errors in results. Even with considerable care, an individual test result is often accurate to <u>one significant</u> figure only (i.e. 3 x 10^{-4} cm/sec). Accuracy will usually be improved by analyzing the results of a series of field tests. However, in many types of geologic units, particularly stratified soil or jointed rock, there may be a very wide variation in hydraulic conductivity, and the calculated mass hydraulic conductivity may be greatly influenced by a relatively thin layer of high hydraulic conductivity; or in fractured rock, a major open joint. Considerable care and geologic insight is needed in interpreting field data. In cases where a highly reliable result is required, the program of borehole hydraulic conductivity tests such as slug testing or single borehole testing, is generally followed by a full-scale pump test as described below.

5.4.1 Preparations For A Test For Hydraulic Conductivity

In the simplest form of field hydraulic conductivity test, preparation consists of cleaning out the bottom of an uncased borehole. The test is then conducted by measuring the rate of flow of water out of the borehole into the soil, or vice versa, through the open hole. Alternatives to this test design would have a cased hole with a bottom flush with the bottom of the hole or the borehole may be extended some distance beyond the bottom of the casing, thus increasing the surface area through which water can flow. If necessary, the uncased part of the borehole may be held open by a filter pack. Misleading results can arise in this test configuration if any return flow occurs up the outside of the casing (thus miscalculating the area of the test). Also the casing must have tight joints, especially for tests in soils with moderate to low hydraulic conductivity. These testing alternatives are illustrated in Figure 5-15.

For more accurate measurements, a perforated tube or a suitable piezometer tip is installed, which is then surrounded by a granular filter pack to prevent caving or erosion of the ground. It is essential that the filter material used has a hydraulic conductivity significantly greater than that of the soil being tested. Hydraulic conductivity tests can also be carried out at various depths in the borehole as drilling progresses. Figures 5-16 through 5-20 show suitable test arrangements for single well tests.

Before a hydraulic conductivity test is performed, it is essential to determine the level of the adjacent natural ground-water surface by one of the methods described in Chapter 4.0. Care must be used in evaluation of these water levels, since, measurements taken soon after cessation of drilling usually do not represent equilibrium values, and a series of head measurements may be necessary. If piezometers are installed in exploration boreholes, the piezometric data obtained from these installations may provide a check on the measurements taken directly before and during the test.

The period required for constant-head tests is decreased and the interpretation simplified if short lengths of borehole are used for the test. Pore pressures should be in equilibrium before the test is performed, and with clays (or claystones) of low hydraulic conductivity it can take several months for the pore pressures set up by the drilling of the borehole to equalize. For soils derived from outwash glacial materials, insitu rock weathering and colluvium, equalization of the heads typically occurs very much faster. The following points should be used as a guideline in selection of the appropriate single borehole test method:

1. Either the rising or the falling head methods (variable head) should be used if the hydraulic conductivity is low enough to permit accurate determination of the water level;

2. Only clean water at a similar temperature to ground water should be used in the falling level test, since the flow is from the hole to the surrounding soil and clogging of the soil pores may occur if using sediment–latent water;

3. If rising–head methods are used, a sounding should be made in the boring after completion of the test to check for a loosened or quick condition has occurred in the soil due to too great a gradient imposed during the test; and,

4. Constant head testing is used in those cases in which hydraulic conductivity is so high as to preclude accurate measurement of the rising or falling water level.

Data must be recorded for each test, regardless of the type of test performed, to include:

- Depth from ground surface to the ground–water surface, both before and upon completion of the test;

- Inside diameter of the casing;

- Height of the casing above ground surface;

- Length of the casing during the test;

- Diameter of the borehole below the casing;

- Depth to the bottom of the boring, from the top of the casing;

- Depth to the standing water level from the top of the casing; and,

- Description of the material tested (in the intake or outflow interval).

5.4.2 Constant Water Level Methods

The U.S. Bureau of Reclamation (1960, 1965) has developed test Designation E–18, in which clean water is pumped into a cased hole in the configuration of Figure 5-

16. A constant-head test is normally conducted as an inflow test in which arrangements are made for water to flow into the ground under a sensibly constant head (+/- 10%). It is essential to use clean water at a temperature similar to the ground water. It will not be possible to achieve a constant head if the ground-water level is not constant or the head lost by friction in the pipes is significant. Where a high flow rate is anticipated and where the installation comprises a piezometer tip surrounded by a filter material, two standpipes should be installed, one to supply the water and the other to measure the head in the filter material surrounding the piezometer tip. The rate of flow of water is adjusted until a constant head is achieved and, in the simplest form of test, flow is allowed to continue until a steady rate of flow is achieved. In some ground, this may take a long period of time, and, in such cases the method suggested by Gibson (1963) may be used, in which the actual rate of flow is measured and recorded at intervals from the commencement of the test.

A constant head is maintained (with or without pressure) by adding water to the borehole through a water measuring device. The tests may be made either below the ground–water level (a) or above the water level (b). If necessary, pressure can be increased to add to the available hydrostatic head, as is shown in (c) and (d). In addition to the data listed in the above general discussion, the data recorded should consist of the amount of water added to the casing at 1, 2, and 5 minutes after the start of the test and at 5 minute intervals thereafter until an adequate determination of the hydraulic conductivity has been made. The value of hydraulic conductivity K is calculated for the test by the equation:

$$K = \frac{q}{5.5rh} \qquad \text{Equation 5-1}$$

q = the constant rate of flow into the hole,
r = the inside radius of the casing, and
h = the differential head of water used in maintaining the study rate.

Head is calculated by finding the difference between the ground–water level and the elevation of the sustained water level in the casing. In those tests above the ground–water level, h is the depth of water in the hole. In those cases where pressure tests are made, the applied pressure in feet of water, where 1 psi = 2.3 ft., is added to the hydrostatic head, to compute the total head h. Cedergren (1967) gives the most frequent causes of error for borehole tests, which apply particularly to constant head tests:

1. Leakage along casing and around packers;

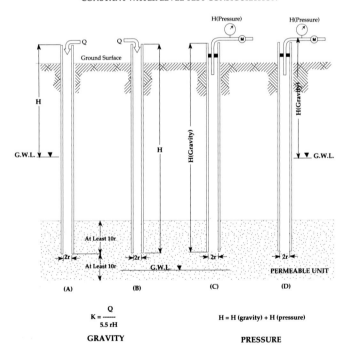

Figure 5-16   Arrangement Constant Head Test

2. Clogging due to sloughing of fines or sediment in the test water;

3. Air locking due to gas bubbles in soil or water; and,

4. Flow of water into fractures in soft rocks that are opened by excessive (dz) head in test holes.

Typical layouts for constant head pump–in and pump–out tests are shown in Figures 5-17 and 5-18 which were modified from CASECO (1964).

5.4.3 Rising or Falling Water Level Tests — Variable Head Tests

The variable head test is generally carried out in soils of relatively low hydraulic conductivity. Two methods have been in wide use for the determination of hydraulic conductivity by the measurement of the rate of flow of water either into or out of a piezometer or boring (Dames & Moore, 1974). The U.S. Navy Facilities Engineering Command (1971) has standard methods for performing variable head tests to estimate the hydraulic conductivity of soils and rocks. The borehole water level is raised or lowered (hence, rising head level test or falling head test) from its equilibrium position and readings are taken of water levels at periodic intervals as it returns to equilibrium. Figure 5-19 and 5-20 illustrate typical layouts for rising and falling level tests modified from CASECO (1961).

The second method, which is usually referred to as the time lag method (U.S. Army Corps of Engineers, 1951), consists of bailing the water out, (or pumping in), of the borehole and observing the rate of rise (or fall) of the water level in the borehole at intervals until the rise (or fall) in water level becomes undetectable. The rate is observed by measuring the elapsed time and the depth of the water surface below the top of the casing. This flow either into or out of the borehole is affected by a "time lag" since whenever a piezometer or other monitoring device is installed, the initial hydrostatic pressure recorded by the device is seldom equal to that in the surrounding soil. Flow also occurs with a corresponding time lag whenever the surrounding pore pressure increases or decreases. The time lag theory has been used extensively as a practical device for determining the hydraulic conductivity of soils in the field. Interpretation of time lag (Hvorslev method)

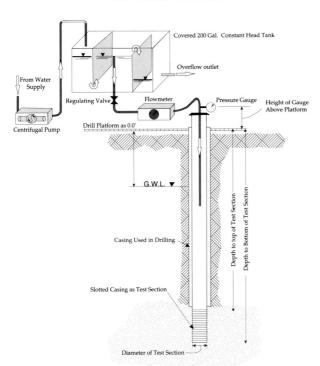

Figure 5-17   Pump In Test

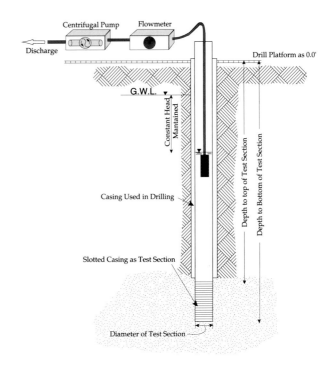

Figure 5-18   Pump Out Test

is described in Chapter 7.0 for some simple piezometer point configurations.

5.4.4 Slug Tests

Hydraulic conductivity can be determined in the field using either single or multiple well tests. For units having reasonably moderate to low hydraulic conductivity (say, below 10^{-4} cm/s), single well tests are generally used (i.e., a slug test). In evaluating the accuracy or completeness of hydraulic conductivity data, the investigator should be aware that:

1. Hydraulic conductivity determinations, based upon multiple well tests, (pump tests with observation piezometers), are preferred due to the large portion of the aquifer tested. However, given the significant cost of multiple well tests, these tests should be selected on the importance or project needs of determining more accurate assessments of the aquifer characteristics.

2. Multiple well tests provide more complete information because they characterize a greater portion of the aquifer.

3. The use of single–well tests will require that more

individual tests be conducted at different locations in order to adequately estimate hydrostratigraphic unit heterogeneity.

4. Water quality of the pumped water will have an important factor in which type of test is performed. High–volume tests make disposal of contaminated water expensive (for example, in a Superfund RI) where residual waste issues must be addressed on the project.

Single–well tests, more commonly referred to as slug tests, are performed by suddenly adding or removing a slug (known volume) of water from a well and observing the recovery of the water surface to its original level. Solid PVC slugs are commonly used to displace water levels within the piezometer. This procedure produces a more "instantaneous" head change than addition or subtraction of a known volume of water. Similar results can be achieved by pressurizing the well casing, depressing the water level, and suddenly releasing the pressure to simulate removal of water from the well. Slug tests, however, offer a quick and inexpensive field method of obtaining in situ hydraulic conductivity values. These generally give a good approximation of horizontal hydraulic conductivity values for the localized zone surrounding a well with hydraulic conductivities less than about 10^{-2} cm/sec. Analysis of slug test data can be performed in the

TYPICAL LAYOUT FOR FALLING HEAD TEST

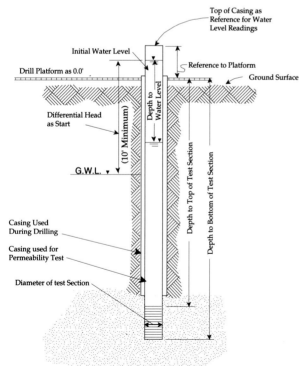

Figure 5-19 Falling Head Tests

TYPICAL LAYOUT FOR RISING HEAD TEST

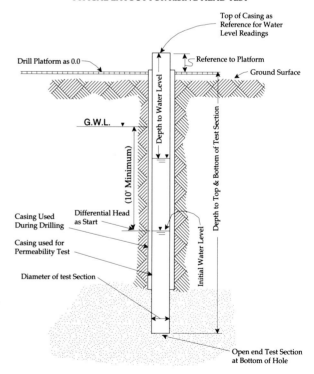

Figure 5-20 Rising Head Tests

traditional analytical techniques described in Chapter 7.0. The vertical extent of well screening must be considered in the test evaluation since this will control the part of the geologic formation that is being tested. The following general sequence of actions is required to conduct a slug test:

• The static ground–water level is first determined; and,

• A slug is injected into or withdrawn from the ground water.

The first operation is either to fill the piezometer tube with water, (falling-head test), or to force the water level down through a header device and placement of positive air pressure,.in the piezometer Alternatively the head levels in the borehole can be changed by insertion of a solid PVC "dummy" into the borehole, allowing the head to equilibrate then remove the dummy and measure the recovery of the water level (rising-head test). The head in the borehole is then allowed to equalize with that in the ground, the actual head being measured at intervals of time from the commencement of the test. The depth of the borehole should be checked to determine whether any sediment has come out of suspension or whether the bottom of the borehole has heaved during the test period. Note that the analysis assumes an instantaneous change in volume with this event recorded at elapsed time equal to zero. The test procedure then continues:

• Ground–water levels (depths) are measured and recorded with corresponding elapsed times. A number of measurements are required over time to adequately represent the test. Typically, a high density of measurements is necessary during the early stages of the test with the number of measurements decreasing over time; and,

• Measurements continue until the ground–water level approaches equilibrium.

The time required for a slug test to provide sufficient data is related to the volume of the slug, the hydraulic conductivity of the subsurface strata being tested, and the manner in which the well is constructed. These factors must be such that several incremental changes in ground–water level can be practically measured during the test interval. Hand–held tapes or automatic recording methods can be employed during the test. Care must be taken to record sufficient numbers of incremental water levels during the test period. Automatic methods for measurement of head changes in small diameter piezometers have greatly simplified the application of

hydraulic conductivity testing in site assessment projects. These devices consist of pressure transducers that record water level and time for later downloading to small computers for data analysis. Sufficient numbers of tests should be run to provide a representative measurement of hydraulic conductivity and to document lateral variations of hydraulic conductivity at various depths in the subsurface.

For hydraulic units having high (say $>10^{-3}$ cm/s) hydraulic conductivity where evaluation of storage and boundary conditions are important, multi–well pumping tests are preferred. Multiple well tests, more commonly referred to as pumping tests, are performed by pumping water from one well and recording the resulting drawdown in nearby observation wells. Tests conducted with wells screened in the same water–bearing formation provide hydraulic conductivity data. Tests conducted with wells screened in different water–bearing zones furnish information concerning hydraulic communication between units.

Heterogeneity in aquifer materials will cause variations in hydraulic conductivity that should be evaluated and, if possible, quantified. Additionally, hydraulic conductivity may show variations with the direction of measurement. It is important that measurements define hydraulic conductivity, both vertically and horizontally, as components of the vector(s) of ground–water flow across the site. In assessing the completeness of hydraulic conductivity measurements, the investigator should also consider geologic characterization information from the boring program. Zones of high hydraulic conductivity or discontinuities identified from drilling logs should also be considered in the determination of hydraulic conductivity. Depending on the requirements, the designed well performance test will allow determination of some or all of the following points:

1. Yield characteristics and flow potential of the well or piezometer;

2. Efficiency of the installation performance as an indication of its hydraulic condition;

3. Confirmation of the aquifer type;

4. Determination of the hydraulic properties of the aquifer system; and,

5. Prediction of the effect(s) of present and/or future ground-water withdrawal from the piezometer or well based on hydrogeologic conditions in the aquifer.

The first two items are best derived from a test

procedure in which pumping is increased incrementally, while the remaining items require a constant withdrawal rate of reasonable duration, preferably using one or more observation wells. To gather information on all the items specified, it is necessary to combine into a single test schedule two types of pumping: a multiple–rate performance test, and a constant–rate aquifer test. Figure 5-21 illustrates the various pumping schedules for typical field testing for hydraulic conductivity.

Typically, the piezometer to be tested may be located in an individual exploratory borehole. In this case, it will probably be a small–diameter hole drilled primarily to evaluate the lithological sequence, but can also be used with certain limitations for an aquifer test. In the case of the aquifer test designed for establishing area–wide properties, the hole is likely to be of medium to large diameter, capable of accommodating a pump. Although a number of submersible pumps can fit down a 2 inch well, at least a 4 inch and preferably larger diameter casing is required to conduct a full scale pump test. If concerned solely with maximizing yield, the test may be restricted to a performance test with steps or stages, each of several hours or more duration. With this length of duration, there may be no need to conduct a separate aquifer test for determination of aquifer properties.

For individual pumping wells without observation piezometers, only the transmissivity and not the storativity

of the aquifer can be calculated. Where the test well is an investigation type with accompanying observation point(s) or a production well associated with others in a well field, then it is possible to identify and determine a much wider range of parameters than was possible with the single–well situation. The full pumping test schedule can be used and the tabulation listed in Table 5-1, which summarizes the type of test and resulting parameters.

There are numerous published formulae for calculating hydraulic conductivity or permeability from these tests, many of them partly empirical. Those given by Hvorslev (1951), which are reproduced in in Section 7.0 are much used and cover a large number of conditions. They are based on the assumption that the effect of soil compressibility is negligible. The method given in Gibson (1963) for the constant-head test is also indicated. This gives a more accurate result with compressible soils.

It must be emphasized that the formula given in Chapter 7.0 are steady-state equations suitable for calculation of hydraulic conductivity when the test is carried out below the ground-water surface. In site assessments for waste disposal projects it is often necessary to measure hydraulic conductivity above the water table. In this case, the steady-state equations can only be used if the time over which the test is conducted becomes very long.

ALTERNATIVE WELL & PIEZOMETER TESTING SCHEDULES

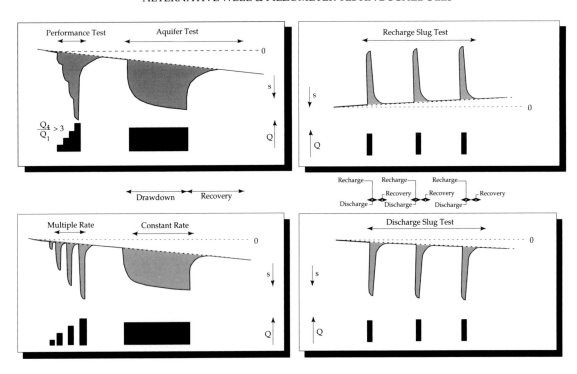

Figure 5-21 Well & Piezometer Testing Schedules

5.4.5 Aquifer Pump Test Procedures

Site assessments commonly require application of larger scale pump testing to evaluate potentially complex problems of ground-water extraction and replenishment. Examples of these larger scale problems include optimum location of extraction wells, reliability of confining units to control vertical leakage, artificial recharge, determination of available storage, boundary conditions and directional hydraulic conductivity. All of these field questions can be resolved using mathematical methods developed over the last 120 years. The results obtained from these mathematical methods are highly dependent on field derived values of the hydraulic characteristics with initial and boundary conditions. Wrong assumptions, or poor field techniques will lead to incorrect evaluations of the parameters of interest. The pump test, where water is extracted at a constant rate and drawdown are observed in nearby observation points was first evaluated by Thiem (1870) through formula to derive aquifer properties from the resultant aquifer stresses (drawdown). Since the work of Thiem many pump test formula and procedures have been derived to evaluate aquifer properties from pump testing. General discussion is provided below on test design and execution not covered by previous publications more specific to pump testing, such as Kruseman and DeRidder (1970) Rosco Moss Company (1990) and Walton (1987). After this discussion a specific application of these guidelines will be provided applied to testing of a confining unit to derive values of leakage, an important concept for waste disposal sites.

Test Design And Execution

Any aquifer stress or well performance test must be planned with an appreciation of the prevailing hydrogeological conditions, by utilizing all existing information to develop a conceptual model for the aquifer system. Then, by adopting a reasonable range of values for the variable parameters in the Theis (1935) equations for non–steady state solutions for drawdown, one can derive a set of pretest data that will provide a basis for selection of pumping well discharge, duration of the test, and location of observation piezometers. This pretest design is important to establish the eventual water level drawdown and, hence, the required location distances of the observation piezometers. This pretest analysis is required to:

• Avoid drilling observation piezometers in areas where drawdown will be less than a few inches; and,

• To establish pumping rates and test durations sufficient to obtain significant drawdown in existing piezometers.

An example of pretest analysis is provided for a special case of evaluation of a confining unit. The pretest procedures are similar (but may be less rigorous) for the typical full scale pump test. The importance of generation of a conceptual model and pretest evaluation should not be ignored in any aquifer evaluation.

Duration of testing should theoretically continue until sufficient data have been gathered for the purpose of the test. In practice, the test is run for periods of time specified in advance, such as one or two days, to as much as one or two weeks for large scale or difficult situations. Tests carried out on confined aquifers will probably require a shorter time duration than those representative of unconfined (watertable) conditions. The distance to observation wells can be estimated through a number of key aquifer criteria:

• The greater the hydraulic conductivity the greater the

Table 5-1 Hydrogeologic Well Tests and Measurement Parameters

Type of Test	Pumping Well Multiple Rate	Well Constant Rate	Piezometers Constant Rate	"Slug" Tests
Yield Potential	√	–	–	–
Well Efficiency	√	–	–	–
Aquifer Type	–	√	√	–
Aquifer Limits	–	-	√	–
Transmissivity	–	√	√	√
Hydraulic Conductivity	–	√	√	√
Storativity	–	–	√	–
Leakage	–	-	√	–
Drawdown Prediction	√	√	√	√

distance to the observation well (R);
- For the same hydraulic conductivity R (unconfined) < R (confined);
- To avoid the effect of partial penetration R > 1.5 x m K_H/K_V; where m is the aquifer thickness and,
- If distance-drawdown method will be used, then three observation wells should provide at least one logarithmic cycle of the distance-drawdown data.

Typical spacing is 100, 400, and 1,000 feet. (Walton 1967) Table 5-2 illustrates some estimates of distances to observation points for various types of geologic/hydrogeologic systems. The distance values provided in the table are applicable if no other recommendation or supporting information are available for the particular aquifer system. After selecting the distance (with particular reference to distance II and III observation points), check whether the radius of influence is >> than the distance concerned. Several equations have been developed to estimate the radius of influence, however, none is reliable under every field circumstance. For estimation of radius of influence for evaluation of

drawdown at observation points before the execution of the pump test, the "Kozeny" formula may provide acceptable values.

$$R = \sqrt{\frac{12t}{Sy}}\sqrt{\frac{QK}{\pi}} \quad \text{For a unconfined System (Eq. 5-2)}$$

$$R = \sqrt{\frac{12t}{S}}\sqrt{\frac{QK}{\pi}} \quad \text{For a Confined System (Eq. 5-3)}$$

R = Radius of influence [m]
t = Time of projected pumping (unsteady state) [s]
Q = Anticipated discharge rate [m^3/S]
Sy = Specific yield [percentage]
S = Storativity [specific storage * thickness]

Table 5-3 provides representative value ranges of specific storage for a number of geologic materials.

Recommended Distances to Observation Points for Pump Tests

GEOLOGIC UN IT	K cm/sec.	AQUIFER	Distance to the Observation Point (m)		
			I	II	III
HIGHLY FRACTURED	> 10^{-3}	Confined	50 - 100	100 - 200 *	200 - 500 *
		W.T.	20 - 25	30 - 60	100 -200
GRAVEL AND UNIFORM COURSE SAND	10^{-3} 10^{-4}	Confined	20 - 50	50 - 100 *	150 - 200 *
		W.T.	10 - 20	30 - 40	50 - 100
MEDIUM SAND NON-UNIFORM SAND SLIGHTLY FRACTURED CONSOLIDATED ROCK SLIGHTLY CEMENTED SANDSTONE	10^{-4} 10^{-5}	Confined	15 - 30	40 - 50	50 - 100
		W.T.	8 - 15	20 - 30	40 - 50
FINE SAND SANDSTONE	10^{-5} 10^{-6}	Confined	10 - 20	30 - 40	40 - 60
		W.T.	8 - 10 **	10 - 20	30 - 40
SILTY SAND	< 10^{-6}	Confined	6 - 15 **	15 - 20	20 - 30
		W.T.	4 - 8 **	8 -15	15 - 20

* = Rather Farther   ** = Rather Closer

Source : Pazdro Z., Hyrrogeologica Ogolna, Wydawnictwa Geologiczne, Warszawa, 1983

Table 5-2   Estimates of Distances for Radius of Influences

Table 5-3 Estimates of Specific Storage

Specific Storage 1/m	Materials
6×10^{-3} - 2×10^{-4}	Clay
3×10^{-4} - 1×10^{-5}	Sand and gravel
2×10^{-5} - 1×10^{-6}	Rock, fractured

This simple computation to evaluate drawdown-radius of influence is developed further in following sections on pre pump test evaluations.

Water levels should be measured in each piezometer or well from a distinctly–marked datum point by means of manual or automatic devices that should have been calibrated beforehand in order that readings can be made to the nearest 0.1 inch. Where measurements are to be made within the pumping well, it is desirable that a narrow–diameter (say, <0.50 inch) tube, (called a stilling tube) with perforations in the lowest section be introduced into the annulus between rising main and well lining. This allows readings taken inside during the test are unaffected by surface turbulence within the pumping well. Observation of water levels should be made for as long as possible, prior to, and after the period of pumping, in order to enable determination of the nature, frequency, and magnitude of any water level changes not directly attributable to withdrawal from the well.

Conditions other than pumped well drawdowns can cause a lowering (or rise) in head levels observed in piezometers during a pump test. The causes of these water level changes may be detected in both confined and unconfined aquifers, but a major category of loading / unloading head changes (stresses) are restricted to confined aquifers. They may be produced by a variety of natural or artificial aquifer stresses, but the most common effect is that resulting from changes in atmospheric pressure. The inverse relationship between barometric pressure and water level changes may be quantified in terms of the barometric efficiency of the well from measurements made in advance, and the test data thus corrected for any such changes identified from a recording barometer operated at or near the well site.

The frequency of water–level measurements made during the test period should be sufficient to provide a reasonable spacing of points on the logarithmic scale normally used for data plotting. Where autographic recording is available, the data points can be selected as desired; but for manual measurement, it is preferable to adopt some such sequence as:

- Every 0.5 minute from start to 5 minutes;
- Every 1 minute from 6 to 15 minutes;
- Every 5 minutes from 20 to 60 minutes;
- Every 15 minutes from 75 to 300 minutes;
- Every 30 minutes from 5–1/2 to 10 hours; and,
- Every 1 hour from 11 hours to end of test.

Such field data are best recorded directly onto prepared forms with any computations, adjustments, or corrections left until less demanding conditions are present.

Discharge from the pumped well should be accurately measured and recorded either continuously or periodically throughout the duration of pumping. Measurement can be made by means of weirs (V–notch or rectangular), orifice plates, or free–pipe discharge used in conjunction with readily available tabulations, or simply by noting the time taken to fill receptacles of known volume.

Withdrawal (pump) rate should not be allowed to vary more than +/-5 percent from the specified discharge, otherwise data analysis may be made so complicated as to be meaningless. It is very important that the discharged water be carried sufficiently far away from the test well for recirculation to be minimized, and, on occasions, this may involve distances in excess of 1,000 feet.

In concluding the remarks on general pump test procedures, it is essential to emphasize that a flexible approach is adopted in order to deal with the inevitable equipment failure or human error. The test specification may anticipate some of the more obvious failures, but the presence on site of the responsible hydrogeologist is to be preferred to delegation of decision making to the driller or pump attendant.

Most of the discussion on borehole and aquifer testing has focused on determining hydraulic conductivity of the geologic unit, however, many of the methods for determining hydraulic conductivity also give an estimate of storage properties as shown on (Figure 5-15 and Table 5-1). Hydraulic conductivity is needed to calculate ground-water velocities and dissolved inoganic and organic parameter travel times. Storage properties are also important for the following reasons: (l) porosity is used in rate of travel time calculations, (2) porosity also is used to estimate the ground-water mass in place, and (3) the storage properties determine how rapidly the flow system will respond to pumpage. This latter factor is important for pump-and-treat systems where pulsed pumping is used. The storage properties can be used to help determine the cycle duration of pumping. Various analytical evaluation methods for pump test data are described in Chapter 7.0.

Since the analytical methods used, and therefore the ultimate results, are to a very large extent dependent on the accuracy of the available data, it is essential not only that care is undertaken in data collection, but also that considerable thought is given in advance to the design of the test.

Assessment of Confining Unit

One of the important uses of full scale aquifer testing within site assessment projects is the field testing of confining layers to demonstrate a lack of interconnection between designed containment units (such as liners and leachate collection systems) and deeper aquifers. Section 1.4 of the RCRA Ground-Water Monitoring Technical Enforcement Guidance Document (EPA 1986) qualitatively discusses the concept of a lower boundary or confining layer to an uppermost aquifer. Section 9.3.3.1 of the Permit Applicant's Guidance Manual for Hazardous Waste Land Treatment, Storage, and Disposal Facilities (EPA 1984) provides guidance that the lack of interconnection can be assumed if the following conditions are met:

1. The unit is laterally continuous and has a hydraulic conductivity of 1×10^{-6} cm/sec or less;

2. There are measurably different piezometric levels in water bearing zones above or below the aquitard;

3. Only minor leakage (occurs across the unit) during a pump test; and,

4. The unit prohibits response in a lower (or upper) water-bearing unit during a pump test within the overlying (underlying) water bearing unit.

The assessment of confining units through pump testing is perhaps the most demanding of the field tests for insitu hydraulic conductivity. As such, the various procedures used in such an investigation will be extensively reviewed in the following subsections. The evaluation methods as described represent somewhat of an extreme for the typical site assessment, but the procedure may be the only recourse to demonstrating hydraulic isolation of waste disposal areas. This technique is also extremely important for evaluation of secondary porosity features in confining units, (such as vertical fractures), since larger volumes of the hydraulic system can be tested during an aquifer pump test.

The federal guidance for evaluation of interconnection of geologic units evolve around hydraulic conductivity measurement through various techniques. Four alternative techniques exist for measuring the hydraulic conductivity of a confining unit:

- laboratory hydraulic conductivity tests on core samples;
- slug tests (either rising or falling head) in individual boreholes;
- installing piezometers in the unit and measuring their hydraulic pressure response to a load (e.g., a stockpile of soil); or,
- a pumping test of an adjacent aquifer with multiple observation piezometers in the confining unit.

Laboratory hydraulic conductivity tests provide individual measurements of small volumes of collected samples of the confining unit. Although these laboratory tests can provide reasonably reliable measurements of hydraulic conductivity, questions often arise on how representative the actual sample really is as compared to the confining unit as a whole. These questions also arise due to larger scale secondary hydraulic conductivity features that are not represented in the laboratory sample tested. To some extent the second method, the slug tests also evaluate hydraulic conductivity directly adjacent to the tested well. The applicability and potential cost effectiveness of the third method, which is based on established soil mechanics concepts, has been demonstrated by van der Kamp and Maathuis (1985). Unfortunately, this technique has not been applied on sufficient projects to find fully regulatory acceptance. Of the four methods, the pumping test can generate the most representative value of the average or effective vertical hydraulic conductivity of a confining unit because a much larger, (relative to either the laboratory or slug tests), volume of material is involved in the test. This scale factor has the important effect of including the contribution of both the fine-grain mass matrix (primary porosity) and any possible joints or fractures (secondary porosity) to the measured bulk hydraulic conductivity of the unit.

The four regulatory conditions for demonstrating lack of hydraulic interconnection obviously can only be met with data from a organized site assessment, where accurate subsurface data has been collected in stratigraphic and geotechnical boreholes. Installation of piezometers during the Phase II investigation then provides measurement points to evaluate head level changes during the performance of a pumping test. Consequently, the purpose of the test method and procedures described in the following subsections is to obtain data to reinforce the second regulatory condition (differences in measured head levels) and to provide data to determine if the first, third, and fourth conditions outlined in the preceding subsection are met by the confining unit under study. The following subsections detail a set of procedures that can be used in conducting a complex pump test, where accurate placement of observational wells or piezometers are important for evaluations of drawdowns. Misplacement of the piezometers or insufficient length of pumping can ultimately result in little or no useable drawdown data.

The initial key to obtaining reliable data from a pump test, is pretest analysis through first establishment of conceptual models and then testing the potential drawdowns through estimated aquifer parameters. The parameters are evaluated by plugging representative values into appropriate analytical equations developed for the analysis of pump tests. The following section steps through the evaluation of a confining unit. All the steps evaluate 'potential responses' as an important part of the aquifer test design. This design process can also be used on less complex situations, however, it is recommended never to attempt a pump test without some form of pretest

design. This pretest procedures may include some form of step testing where pump rates are gradually increased to evaluate well performance. This is especially important in fractured rock investigation described in Chapter 6.0.

Pre-Test Analysis Of Conceptual Hydrogeologic Model

The purpose of conducting a pre-test analysis using a conceptual hydrogeologic model of a multi-layer aquifer system, is to size the pumping well, design the piezometer network, and estimate the necessary duration of the proposed pumping test. An example conceptual hydrogeologic model including both known thicknesses and assumed hydraulic parameters is shown on Figure 5-22. As indicated on this figure, the hydraulic parameters of the confining units and aquifers are not relevant to this pre-

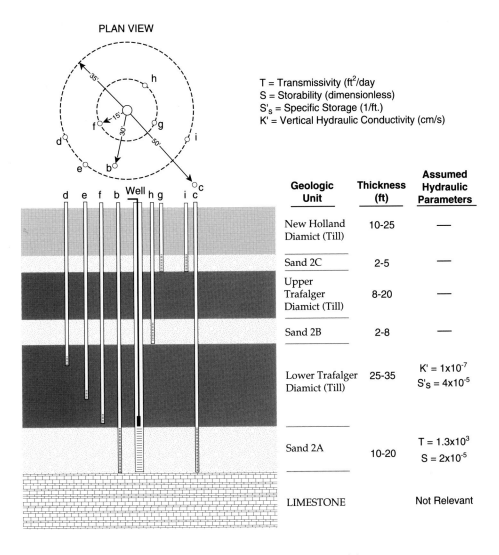

PLAN VIEW

T = Transmissivity (ft^2/day
S = Storability (dimensionless)
S'_s = Specific Storage (1/ft.)
K' = Vertical Hydraulic Conductivity (cm/s)

Geologic Unit	Thickness (ft)	Assumed Hydraulic Parameters
New Holland Diamict (Till)	10-25	—
Sand 2C	2-5	—
Upper Trafalger Diamict (Till)	8-20	—
Sand 2B	2-8	—
Lower Trafalger Diamict (Till)	25-35	$K' = 1x10^{-7}$ $S'_s = 4x10^{-5}$
Sand 2A	10-20	$T = 1.3x10^3$ $S = 2x10^{-5}$
LIMESTONE	Not Relevant	

Figure 5-22 Pre-test Conceptual Model

test analysis.

Unfortunately, practicing hydrogeologists in the U.S. (including those in the regulatory agencies) utilize a mixture of traditional U.S. Geological Survey or water well industry units (e.g. transmissivity in gpd/ft; dimensions in feet) and SI units (e.g., hydraulic conductivity in cm/s). This document has continued this questionable practice, but the author has made every attempt to clearly define the units used and to simplify "mixed" calculations

Potential Yield of the Aquifer

The potential maximum yield of a well completed into the sand aquifer as shown in the conceptual model Figure 5-16 can be estimated from the Theis non-equilibrium equation which can be written as:

$$Q = \frac{4\pi Ts}{192.5W(u)} \qquad \text{Equation 5-4}$$

Q = discharge rate (gpm),
s = drawdown (ft),
T = transmissivity (ft^2/day)

and W(u) is the Theis well function that is available from published tables for the dimensionless parameter

$$u = \frac{rw^2 S}{4Tt} \qquad \text{Equation 5-5}$$

where:
rw = radius of well (ft),
S = storativity (dimensionless), and
t = time (days).

In order to calculate drawdown a well diameter, pumping period estimated and transmissivity is required. Assuming a nominal 8-inch diameter well for purposes of the pretest evaluation, a pumping period of 10 days, and the estimated values of:

T = 1.3 x 10^3 ft^2/day
S = 2 x 10^{-5} (confined value)

as given on Figure 5-22, it can be shown that

u = 4.3 x 10^{-11}

and from published tables of W(u) vs u,

W(u) = 23.3

Field measured water levels, as shown on Figure 5-22 show a confined potentiometric surface of the sand aquifer approximately 55 feet above the top of the aquifer. In order to evaluate a pumped well, assumptions must be made as to the available drawdown. The available drawdown consists of both the well efficiency and the maximum drawdown until the aquifer goes unconfined. Maintaining at least 5 feet of "free-board" to prevent the aquifer from becoming unconfined (which would significantly complicate the analysis), and assuming 85 percent efficiency (i.e., 15 percent of the drawdown in the conceptual model would be due to well loss), the maximum available draw-down would be 42.5 feet. Inserting these values into Equation 5-4, the potential maximum yield of a well completed into this unit is estimated to be on the order of 150 gpm. Although a nominal 6-inch well would house a pump capable of producing 150 gpm at the given lift. One must be conservative, in case the sand aquifer is found to be more transmissive. This minimizes well losses and assures adequate flow around the pump to keep it from overheating. In this case a nominal 8-inch well should be constructed for an adequately sized pumping well.

The next step is the prediction of the effects of the stress (pumping) on the aquifer or in this specific case, the predicted response in the confining (aquitard) unit.

Predicted Confining Unit (Aquitard) Response

The theory of flow through aquitards as developed by Neuman and Witherspoon (1972), states there is both a distance and a time limitation for data to be used in the ratio method of analysis. They define a dimensionless parameter, β, and its limit as

$$\beta = \frac{r}{4}\sqrt{\frac{S's K'}{TS}} < 1.0 \qquad \text{Equation 5-6}$$

where:
S's = specific storage of the aquitard (ft^{-1})
K' = vertical hydraulic conductivity of the aquitard, (ft/day)
r = radial distance from the pumping well (-ft),
and T and S are as previously defined.

Re-arranging Equation 5-6, you can obtain the equation:

$$r < \frac{4}{\sqrt{\dfrac{S'_s K'}{TS}}} \qquad \text{Equation 5-7}$$

Values for the pre-test conceptual hydrogeologic model shown on Figure 5-22 provide input to calculate that data from any wells within 6,100 feet of the pumping well should provide valid data.

The Newman and Witherspoon theory behind the ratio method assumes that the aquitard is of infinite thickness. However, in an aquitard of finite thickness, there is a specific time before the pressure transient through the aquifer is significantly affected by an overlying aquifer. Neuman and Witherspoon (1972) indicate that valid data can be collected from an aquitard of finite thickness providing

$$t < 0.1 b'^2 \frac{S'_s}{K'} \qquad \text{Equation 5-8}$$

Where b' is the thickness (feet) of the aquitard and the other terms were previously defined..

For the values given on Figure 5-22, (assuming b' = 30 feet), the critical time, t_c, (i.e., the time beyond which data become invalid) can be calculated from Equation 5-8 to be 12.5 days. A wide range of values can be calculated from Equation 5-8 depending on the conceptual model values of b', K' and Ss. If various assumed values based on the conceptual models are combined, one can estimate critical times ranging from 1 to 20 days.

The drawdown that will be propagated into the aquitard must also be quantified whether the measurable values will be sufficiently above the background "noise" level in the potentiometric surface of the aquitard. The conceptual model defines that the aquitard is approximately 30 feet thick, and that there are three piezometers at elevations of 7.5, 15.0 and 22.5 feet, respectively above the top of the sand aquifer. All piezometers are located at a radial distance of 30 feet from the pumping well (Figure 5-22). Using Equations 5-4 and 5-5, one can calculate that for an assumed pumping rate of 150 gpm and the aquifer characteristics given on Figure 5-22, the predicted drawdown in the aquifer at a radial distance of 30 feet and at a time of 10 days will be approximately 25 feet. There are two dimensionless time parameters, one for flow in the aquifer:

$$t_D = \frac{tT}{r^2 s} \qquad \text{Equation 5-9}$$

and one for flow in the aquitard:

$$t'_D = \frac{K't}{S'_s z^2} \qquad \text{Equation 5-10}$$

where z is the distance (feet) above the top of the aquifer and all other variables are as previously defined. Using the conceptual model aquifer characteristics (Figure 5-16), one can establish that at 10 days and at the given radial distance of 30 feet, $t_D = 7.2 \times 10^5$. Using Equation 5-10 and the conceptual model aquitard values, it can also be shown that:

$$t'_D = \frac{70}{z^2} \qquad \text{Equation 5-11}$$

Now combining a) the above relationship, b) the type curve for $t_D = 7.2 \times 10^5$ (which can be interpolated between the type curves for $t_D = 10^2$ and $t_D = 10^{10}$) shown on Figure 5-23 (generated by Neuman and Witherspoon, 1969), and c, the calculated aquifer drawdown (s = 25 ft), Table 5-4 can be computed to predict the drawdown, s'(ft), in each of the aquitard piezometers:

The resultant predictions of drawdowns should be easily differentially from "noise" (establish from pretest observations in piezometers and pumping well estimated on the order of 0.5 ft for this case) in the aquitard piezometers.

An important consideration in the design of any piezometers located in confining units is their responsiveness to pressure transients across the aquitard. This has important effects on both the length of time it will take newly installed piezometers to come to hydrostatic equilibrium with the formation pressure and on their ability to measure the pressure changes in the aquitard. This time-response has direct bearing on the type of piezometer installation used in the confining unit. One measure of this responsiveness is the "basic time lag" as defined by

Table 5-4 Predicted Drawdowns in Aquitard Piezometers

piezometer	(ft)	tn	s'/s	(ft)
f	7.5	1.2	0.5	12
e	15	0.31	0.18	4.5
d	22.5	0.14	0.06	1.5

Hvorslev (1951). For a well point of finite length in an isotropic, porous medium, the basic time lag is:

$$T_0 = \frac{r_s{}^2 \ln\left(\dfrac{L}{R}\right)}{2LK}$$

Equation 5-12

T_0 = basic time lag (days),
$r_s{}^2$ = radius of piezometer standpipe (ft),
L = intake length (ft),
R = intake radius (ft), and
K = hydraulic conductivity of medium (ft/day).

Hvorslev (1951) defines, in simple physical terms, the basic time lag as the elapsed time before a piezometer recovers to 37 percent of the head difference between its static level and the level induced by a sudden removal or addition of water. Hvorslev (1951) suggests that the value of $2.3T_0$ which corresponds to 90 percent recovery, is adequate for most field situations.

Table 5-5 , arrays the value of $2.3T_0$ for various values of hydraulic conductivity and the dimensions of the piezometers that will be designed for use in the project. Included in the calculations is the critical time, t_c, as defined above. As shown in this table, the resultant 90 percent time lag in the response of the piezometers and the duration during which such test data would be theoretically valid could severely limit the period over which usable data can be collected. Such slow response times for typical open tube piezometer completions can be improved through a number of methods. Using type C and D piezometers, (pneumatic or vibrating wire), installations can greatly improve response characteristics. This, however, does not mean that one would simply lower a vibrating transducer down a standpipe piezometer. In this case, the response time remains the same as without the transducer. The window of 90% response time would not improve, since the transducer was not sealed into the standpipe to reduce the volume requirements of the device necessary to cause a rise in the head level.

Table 5-5 : Time Lag in Aquitard Piezometers "Modified"

K (cm/s)	K (ft/day)	2.3 T_0 (days)	t_c (days)	Modified 2.3 T_0 (days)
1×10^{-8}	2.8×10^{-5}	131	130	23
1×10^{-7}	2.8×10^{-4}	13.1	13	2.3
1×10^{-6}	2.8×10^{-3}	1.31	1.3	0.23
1×10^{-5}	2.8×10^{-2}	0.131	0.13	0.023

A technique used to increase this "window" of time, is to essentially fill the piezometer tube with a reasonably close fitting solid aluminum rod that contains a groove for the transducer cable. Simple geometric calculations for the volumetric storage of the standpipe, shows that this will reduce the time lag by a factor of 5.6. Table 5-5 also includes this modified value, indicating that this simple field modification will allow a much longer period of time for obtaining usable test data.

As described above, the time lag in piezometers completed into such "tight" confining units can be significant and limit the ability to obtain usable data. In order to minimize the time it takes open tube piezometers to recover to hydrostatic conditions upon completion of drilling, (assuming the hole was drilled with air) water can be added to immediately reach the predicted piezometric level before the pump test. This assumes that one knows the unstressed head levels sufficiently to adjust the piezometer artificially and additional time is available to obtain natural readjustment to actual hydrostratigraphic values.

Data Acquisition System

The conduct of pump tests using observation piezometers is a demanding task that should be fully planned and conceptually modeled before the field execution. Data collection activities can be greatly simplified through the use of calibrated data acquisition systems. A data acquisition system typically consists of a network of vibrating wire transducers suspended into each open tube piezometer in conjunction with a programmable data logger. Before execution of the actual test one should fully understand the idiosyncrasies of your data logger.

Each data logger transducer system will have individual inconsistencies that are often only seen after completion of the pump test through anomalies in the final data evaluation. Power cables for the pump can cause interference to transducers, hence, they should not be put directly into the pumping well (i.e., use a stilling tube). Other than during the step-drawdown test and to insure the water level does not go below a pre-determined depth, data from the pumping well is typically the least important of all of the monitoring points. Consequently, water level data from the pumping well can be collected less frequently using a standard electronic sounder.

Pre-test Monitoring

Prior to beginning the actual pump test, head level data should be taken for a sufficient period of time in all of the piezometers that will be used during the evaluation in order to:

- determine the nominal variations (or "noise") in the potentiometric levels of the aquifer and aquitard;

- obtain information on the barometric response of the aquifer system; and,

- ensure the confining unit piezometers have essentially come to hydrostatic equilibrium.

Pre-analysis data calculations provided by these head level observations can provide the various time/head level trends of the aquifer and confining unit. If recovery times will be too long for the investigation, the water levels in the piezometers may be artificially set (by either adding or removing water) to their estimated hydrostatic level to minimize the time lag. Generally, you should perform pretest head level measurements for twice as long as calculated to perform the actual pump test.

Monitoring During Test

During the pumping portion of test, water level data should be collected from all aquifer and confining unit piezometers on a preprogrammed logarithmic time scale (using standard pumping test intervals) for the first 4 hours and every 0.5 of an hour thereafter.

After the pumping portion of the test is completed addition recovery head level data should be collected; based on the recoveries water levels may be required to be measured in the confining unit piezometers for a period of as long as 30 days. The actual recovery in the aquifer will be much more rapid and the length of post test monitoring should also be based on the unit response with time.

Barometric data should be obtained during the entire pretest, test and post testing period. This data can be conveniently recorded using an absolute pressure transducer connected to an automatic data logger.

Another point one should bear in mind is the expectancy to see evidence of the "Noordbergum Effect" in the confining unit piezometers (see Keller et al, 1986). It typically shows up as a temporary increase of water level when pumping starts and a temporary decrease when pumping stops. This effect is due to radial compression of the aquifer during pumping, caused by the lowered pore pressure near the pumping well. Exact analysis and prediction of these reverse water-level fluctuations would be very complex, but they can be expected to be relatively small and to dissipate within a time much less than the critical time t_c. The effect should not significantly interfere with the value of the leakance test in the confining unit.

5.5 METHOD OF ANALYSIS OF LEAKAGE THROUGH CONFINING UNITS

Introduction:-- A common problem in evaluation of confining unit associated with waste disposal facilities is the evaluation of the vertical hydraulic conductivity of these materials. One obvious method for obtaining vertical K values is to perform laboratory permeability tests on cores or soil plugs as are described later in this chapter (Freeze and Cherry, 1979). This technique is well developed and applied often, but suffers from the usual problems of sample disturbance and the question of non-representative sampling of the in place units. In this chapter techniques will be presented that, when successful, will result in a vertical K value averaged over a volume of subsurface material much larger than that contained in a core. Hydraulic conductivity is the constant of proportionality in Darcy's law,

$$V = -K \, dh/dx \qquad \text{Equation 5-13}$$

where:
K = hydraulic conductivity.
V = Darcy's velocity.
x = distance.
h = hydraulic head.
dh/dx = hydraulic gradient.

Hydraulic conductivity, which is sometimes called the coefficient of permeability, has been shown to be related to the fluid properties and the permeability of the porous medium by the following formula (Hubbert, 1940):

$$K = k\rho g/\mu \qquad \text{Equation 5-14}$$

where:
k = specific or intrinsic permeability of the porous medium.

the fluid properties and the permeability of the porous medium by the following formula (Hubbert, 1940):

$$K = k\rho g/\mu \qquad \text{Equation 5-14}$$

where:
k = specific or intrinsic permeability of the porous medium.
ρ = density of fluid.
μ = dynamic viscosity of fluid.
g = gravitational acceleration.

Intrinsic permeability k, which is a function of mean grain diameter, grain size distribution, sphericity, and roundness of the grains, is a measure of the ability of the medium to transfer fluids, independent of the density and viscosity of any particular fluid. Values of hydraulic conductivity of a geological formation can vary in three dimensional subsurface space. This property of the medium is called heterogeneity. They can also show variations with the direction of measurement at any given point. This property is called anisotropy and is quite common in sedimentary rocks. In such rocks, hydraulic conductivity along the layers is sometimes several orders of magnitude larger than across the layers. This property becomes especially important in layered formations where some thin layers of very low permeability appear within highly permeable sediments. Anisotropy is also quite common in fractured rocks where aperture and spacing of joints varies with direction.

Bredehoeft-Papadopulos Single-Well Test:-- Bredehoeft and Papadopulos (1980) have proposed a method of measuring hydraulic conductivity which is a modification of the conventional slug test. The purpose of this test is to measure in-situ horizontal hydraulic conductivity of what would be normally be considered as very low hydraulic conductivity units such as clays, or rock units with either "tight" fractures, (closed or filled, if they exist at all!), or matrix rock between fractures.

The test set-up for (a) an unconsolidated formation and (b) a consolidated formation is shown on Figure 5-23. A common condition observed in low hydraulic conductivity units (depending on the time elapsed since the hole was drilled), the water level in the hole may not have stabilized to the ambient hydraulic head at the interval to be tested. Before starting the test, the test system is filled with water and, after a period of observing the water level for ambient conditions, the test interval is suddenly pressurized by injecting an additional amount of water with a high pressure pump. The test interval is then shut-in, and the head change Ho caused by the pressurization is allowed

to decay. As water slowly penetrates into the formation, Ho will drop. The variation of Ho with time is recorded.

In a conventional slug test the water flow into the formation comes directly from the volume of stored water in the system under normal hydrostatic pressure. Conversely, the driving force governing the movement of water during this test from the well into the formation is the expansion that the water stored within the pressurized system undergoes as the head, or the pressure within the system, declines. Thus, the rate at which water flows from the well is equal to the rate of expansion. A conventional slug test in a formation with hydraulic conductivity of K = 10^{-10} cm/s may last more than one year whereas the modified slug test method as discussed here may take only a few hours. The solution for the modified slug test has been presented in the form

$$H/Ho = F(\alpha, \beta) \qquad \text{Equation 5-15}$$

where Ho and H are values of head measurement in the hole at the time of shut-in and following that with respect to the background head, respectively. α and β are given by

$$\alpha = \frac{\pi r_s^2 S}{V_w C_w \rho_w g} \qquad \text{Equation 5-16}$$

$$\beta = \frac{\pi T t}{V_w C_w \rho_w g} \qquad \text{Equation 5-17}$$

where:
r_s = radius of well in the tested interval.
t = time.
S = storage coefficient of the tested interval.
Vw = volume of water within the pressurized section of the system.
Cw = compressibility of water.
ρw = density of water.
T = transmissivity of the tested interval.
g = gravitational acceleration.

Tables of the function F(α, β) for a large range of variation of α and β are given by the above authors as well as Cooper et al. [1967] and Papadopulos et al. [1973].

Major assumptions applied in development of this method are as follows:

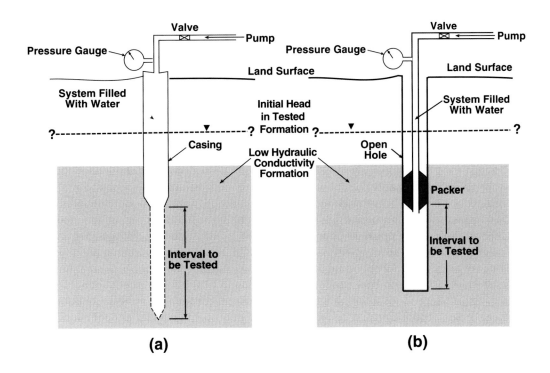

Source: Bredehoeft and Papadopulas, 1980

Figure 5-23 Pressure Test Arrangement

• Flow in the tested interval is radial, which will also imply that the flow at any distance from the well is limited to the radial zone defined by the tested interval.

• Hydraulic properties of the formation remain constant throughout the test.

• The casing and the formation on the side of the borehole containing the water are rigid and do not expand or contract during the test.

• Before the system is pressurized, the water level in the well has come to a near equilibrium condition with the aquifer.

Analysis of Field Data

Bredehoeft and Papadopulos have proposed two different techniques, one for $\alpha < 0.1$ and the other for $\alpha > 0.1$. If $\alpha < 0.1$ the following steps should be taken.

• Prepare a family of type curves, one for each α, of $F(\alpha, \beta)$ against, β on semilogarithmic paper. A table giving

the value of $F(\alpha, \beta)$ as function of α and, β is presented by Bredehoeft and Papadopulos (1980).

• Plot observed values of H/Ho versus time t on another semilog paper of the same scale as the type curves.

• Match the observed curve with one of the type curves keeping the β and taxes coincident and moving the plots horizontally.

• Note the value of α of the matched type curve, and the values of, β and t from the match point.

• Calculate values of S and T from the definitions of α and β given by equations (5-16) and (5-17).

The above method is not suitable for $\alpha > 0.1$. In this range of α, this method can only give the product of transmissivity and storage coefficient, TS. This product may be calculated by matching the field curve of H/Ho versus time t with a type curve family of $F(\alpha, \beta)$ versus the product $\alpha\beta$ (Fig. 5-24).

The major assumption employed in this method is that "volumetric changes due to expansion and contraction of other components of the system are negligible." In other

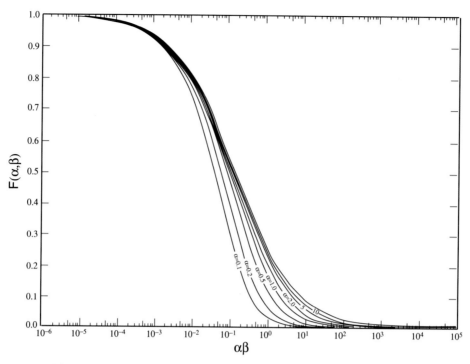

Figure 5-24 Type Curve of the Function F(α,β) vs parameter $\alpha\beta$

words, expansion of the pipes and contraction of the rock in the test zone is negligible relative to that of water. This assumption may introduce large errors into the calculation of hydraulic conductivity. Neuzil (1982) has referred to a test in which the compressibility in the shut-in well was approximately six times larger than the compressibility of water.

The other major assumption which was employed in this method was that before the system was pressurized, either the water level in the well had come to a near equilibrium condition with the aquifer or that the observed trend could be extended throughout the test. Neuzil (1982) has pointed out that this assumption may also lead to erroneous results. He argues that the pressure changes due to nonequilibrium conditions before shut-in become much more rapid after the well is pressurized. Neuzil (1982) has proposed the following modifications in the setup and procedure for performing the test.

• Modify the test equipment to that shown on Figure 5-25.

• Fill the borehole with water and set two packers near each other.

• Set up two pressure transducers as shown in the Figure 5-25.

• Close the valve, shutting in the test section, and monitor the pressures in both sections until they are changing very slowly.

• Open the valve, pressurize the test section by pumping in a known volume of water, and reclose the valve.

• Measure the net pressure decay (slug) by subtracting the decline due to transient flow prior to the test from the measured total pressure.

• Analyze data using the technique prepared by Bredehoeft and Papadopulos (1980) as was mentioned before, except that the term for the compressibility of water Cw is replaced by the ratio c, defined as

$$c = (\Delta V/v)/\Delta P \qquad \text{Equation 5-18}$$

where:
v = the volume of the shut-in section; and,
ΔV = the volume of water added to generate a pressure change of ΔP.

Neuzil (1982) indicates that a rise in pressure measured by the transducer between the two packers may indicate leakage upward from the test section. However, two other phenomena may cause some rise of pressure in the middle section. One is increase of pressure inside the formation

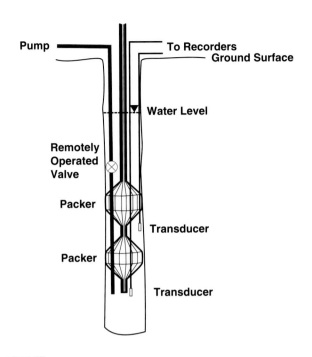

Source: Neuzil, 1982

Figure 5-25 Borehole Instrumentation

- *The hydraulic conductivity measured by these tests is only representative of a small zone around the testing interval. A thin lens of very small permeability located between injection and measuring zones could lead to an erroneously low vertical hydraulic conductivity, even if it is only locally present. This problem may be overcome by conducting several tests within the total thickness of a given formation. however, the lateral variation of vertical hydraulic conductivity could be another problem which requires either other types of testing or performance of a number of single-well tests.*

- *Because the horizontal hydraulic conductivity of sedimentary materials is usually much larger than the vertical hydraulic conductivity, flow lines generated by either injection or pumping in these tests are predominantly horizontal. Therefore, a long time may be required to have significant pressure disturbances in measuring intervals located vertically above or below the flow zone. A small pressure change together with the possibility of leakage behind the casing due to poor cementing will result in an increased degree of uncertainty in the credibility of these tests in tight formations.*

- *Measurement of change of pressure due to pumping or injection in single-well tests is another source of uncertainty. This is because the test may often start before the pressure at the measuring interval has stabilized. One way to handle this problem is to minimize the volume of the measurement cavity in the well with the help of extra packers. This will shorten the time required for pressure stabilization.*

- *In a single-well test, injection is preferred over pumping unless the well will flow without artificial lift (Earlougher, 1980). In a tight formation, indeed, injection is the only feasible way to test.*

- *The injection or pumping zone should be packed off to minimize well bore storage."*

Tests with Two or More Wells

Those tests involving two or more wells can measure the response of a much larger volume of rocks or soils than tests from a single well. Therefore, the value of hydraulic conductivity obtained from multiple well tests is usually more representative of the large scale behavior of the formation. Unfortunately, these tests often cannot be directly used within the formation of interest, once the hydraulic conductivity of that formation becomes very low. Wells completed in very low hydraulic conductivity materials are unable to produce fluid for the required test period and pressure heads take long periods of time to be transmitted through the unit. Even by injecting fluide into these wells it could take years before any useful response can be measured in observation wells at a distance of 5 to 10 m. One would then have an interesting time seperating out natural varations in head level from stress induced head changes.

Neuman and Witherspoon's "ratio method":-- Data from a pumping test to determine the effects of leakage through aquitards (confining units) can be analyzed using what is commonly referred to as the ratio method. Although the theory behind the ratio method is somewhat complex (see Neuman and Witherspoon, 1972), its use with applicable field problems is actually very easy. In fact, it is simpler and much less subjective than many of the aquifer test methods which require curve matching. Neuman and Witherspoon's (1972) "ratio method" is based on the assumption that flow in the aquitard is vertical. The horizontal hydraulic gradients in a permeable aquifer in areas near the pumped well, will be relatively large and, therefore, flow in the aquitard will deviate somewhat from the vertical. Fortunately, the "ratio" method is "robust" in that it is not sensitive to such small deviations from the

horizontal hydraulic gradients in a permeable aquifer in areas near the pumped well, will be relatively large and, therefore, flow in the aquitard will deviate somewhat from the vertical. Fortunately, the "ratio" method is "robust" in that it is not sensitive to such small deviations from the assumed conditions. If non-vertical flow were really important, and could effect the value of hydraulic conductivity in the confining unit, a more exact analysis using numerical modelling could be carried out. In practice, the most critical parameter in application of the "ratio method" is often the value of Z, the height of the confining unit piezometer above the top of the aquifer. If the interface between the aquifer and the aquitard is irregular or takes the form of a gradual transition, there may be considerable uncertainty as to the value of Z that should be used in the ratio method calculations. In any case, this means that careful attention should be paid to the elevation and nature of the interface between the aquifer and confining unit. A brief summary of the application of the ratio method to the test of a confining unit is as follows.

Using values of transmissivity, T (ft^2/day), and storativity, S (dimensionless), for the aquifer derivable with data from observation wells b and c (Figure 5-22), the aquifer dimensionless time parameter is calculated, by Equation 5-19

$$t'_D = \frac{K't}{S'_s z^2} \qquad \text{Equation 5-19}$$

where:
t = time (days), and
r = radial distance from the pumping well (ft)

The time parameter establishes which of the type curves is to be used (see Figure 5-26). For each of the confining units piezometers included in the test, the ratio between their measured drawdown, s' (ft), and the drawdown in the aquifer, s (ft), at the same radial distance at the same time:

$$s' / s \qquad \text{Equation 5-20}$$

is provided by the ratio. Drawdown in the aquifer at an equivalent distance is either interpolated or extrapolated from the drawdown vs distance (logarithmic) plot provide by the drawdowns in observation piezometers b and c. Although the assumed linear drawdown vs distance (logarithmic) relationship is not absolutely valid for a leaky aquifer (Neuman and Witherspoon, 1972), the error over relatively these closely spaced piezometers would be

considered minor. The value of the aquitard dimensionless time parameter can be evaluated by using the following equation:

$$t_D = \frac{tT}{r^2 s} \qquad \text{Equation 5-21}$$

where:

K' = vertical hydraulic conductivity of the aquitard (ft2/day),
S's = specific storage of the aquitard (ft and
z = vertical distance of the piezometer above the top of the aquifer (ft)

t_D the aquifer time parameter is then simply read off the horizontal axis of the graph on Figure 5-26 for the point at which the value of the drawdown ratio (S'/S) intersects the selected type curve. Inserting the appropriate values of t and z into Equation 5-21 the value of hydraulic diffusivity for the aquitard can be calculated by the derived equation:

$$K' / S's \qquad \text{Equation 5-22}$$

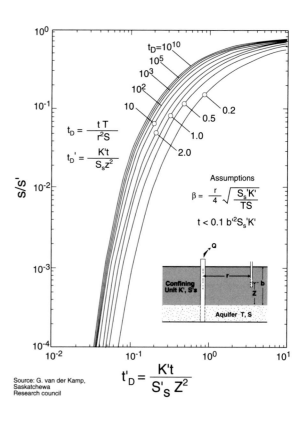

Figure 5-26   Ratio Method Type Curves

S's = 0.305 [$\rho g\,(\alpha + n\beta)$]   Equation 5-23

where:
ρ = density of the water (kg/m^3)
g = gravitational constant, m/s),
α = compressibility of the aquitard, (m s^2/kg),
n = porosity of the aquitard (dimensionless), and
β = compressibility of the water (m s^2/kg).

Assuming the constant values for ρ, g, and β given in Section 8.0, Equation 5-23 can be re-written as

S's = $3.0 \times 10^3 \alpha + 1.3 \times 10^{-10} n$ Equation 5-24

Values of compressibility for clays are in the range of 10^{-6} to 10^{-8} m s^2/kg and the porosity of clays is typically in the range of 0.4 to 0.7 (Freeze and Cherry, 1979). If the tested properties of the confining unit fall into such typical ranges, the second term in Equation 5-24 becomes insignificant when compared to the first term. Consequently, the specific storage of the confining unit would be able to be calculated simply from

S's = 3.0×10^3 Equation 5-25

The remaining unknown term, the effective vertical hydraulic conductivity K' of the confining unit can be obtained by substituting the value derived in Equation 5-25 (or 5-24, if necessary) into Equation 5-22.

Another version of leakance analysis is given in Walton (1987). However, Walton's method is not very explanatory, uses very cumbersome units, and relies on less than precise interpolation between values given in a table to determine t_D, however as a check, one should use this or other methods to confirm the results of the test. Relevant case histories applying the ratio method to field problems involving confining units (glacial till) are presented in Keller et al. (1986) and Grisak and Cherry (1975).

Packer Hydraulic Conductivity Tests

Packer tests provide a means of assessing permeability of earth materials surrounding a definite, preselected test interval. The technique is particularly useful in rock exhibiting only secondary hydraulic conductivity, due to the presence of discontinuities. The procedure used for packer hydraulic conductivity tests depends upon the condition of the rock. Figure 5-27 shows the arrangement for performing packer tests in consolidated material where the tests are completed in stages (A) and (B) with a single packer arrangement, and (C) and (D) where a double packer system is used. In rock that is not subject to cave–in, the following double packer method is used. After the borehole has been completed, it is filled with clear water, surged, and washed out. The double packer test apparatus is then inserted into the hole until the top packer is at the top of the rock. Both packers are then

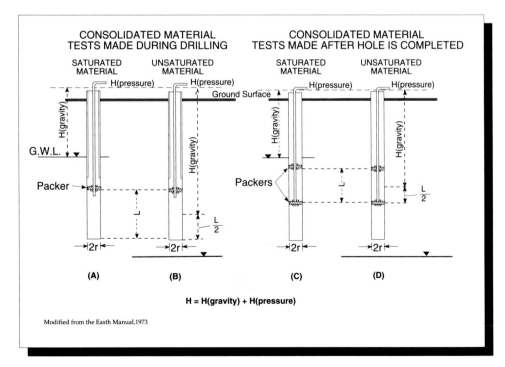

Figure 5-27 Packer Permeability Tests

expanded and water under pressure is introduced into the hole, first between the packers and then below the lower packer. Observations of the elapsed time and the volume of water pumped at different pressures are recorded as detailed in the section on pumping. Upon completion of the test, the packer apparatus is lowered a distance equal to the space between the packers and the test is repeated. This procedure is continued until the entire length of the hole has been tested or until there is no measurable loss of water in the hole below the lower packer. If the rock in which the hole is being drilled is subject to cave–in, the pressure test is conducted after each advance of the hole for a length equal to the maximum permissible unsupported length of hole or the distance between the packers, whichever is less. In this case, the test is limited, of course, to the zone between the packers.

In jointed and/or bedded or foliated rock, packer intervals should be selected to incorporate individual discontinuities or clusters of such fractures, so that an accurate representation of secondary hydraulic conductivity may be obtained.

Regardless of which procedure is used, a minimum of three pressure increments should be used for each section tested. The magnitude of these pressures are commonly 15, 30, and 45 psi above the natural piezometric level. However, in no case should the excess pressure above the natural piezometric level be greater than 1 psi per foot of soil and rock overburden above the upper packer. This limitation is imposed to insure against possible heaving and hydrofracture damage to the foundation rock. In general, each of the above pressures should be maintained for 10 minutes or until a uniform rate of flow is attained, whichever is longer. If a uniform rate of flow is not reached in a reasonable time, the investigator must use personal discretion in terminating the test. The quantity of flow for each pressure increment should be recorded at 1, 2, and 5 minutes after the start of the test, and for each 5 minute interval thereafter. Upon completion of the tests at 15, 30, and 45 psi, the pressure should be reduced to 30 and 15 psi, and the rate of flow and elapsed time should once more be recorded in a similar manner for each of these pressure increments.

Observation of the water inflow quantities, taken with increasing and decreasing pressure, permits evaluation of the nature of open discontinuities in the rock. For example, a linear variation of flow with pressure indicates openings that neither increase or decrease in size. If the curve of flow versus pressure is concave upward, it indicates the openings are enlarging; if convex, the openings are becoming plugged. Additional data required for each test are as follows: (1) depth of hole at time of each test, (2) depth to bottom of top packer, (3) depth to top of bottom packer, (4) depth to water level in borehole

at frequent intervals, (5) elevation of piezometric level, (6) length of test section, (7) borehole radius, (8) length of packer, (9) height of pressure gauge above ground surface, (10) height of water swivel above ground surface, and (11) description of geologic host material being tested in each interval. Item (4) is important since a rise in water level in the borehole may indicate leakage around the packers. Methods for evaluating data from packer tests are included in Chapter 7.

5.6 LABORATORY ANALYSES

5.6.1 Soils

Laboratory tests of soils collected during the field drilling task define four basic properties of soils and their suitability for use on waste disposal sites. These soil characteristics can be summarized as:

- Index and Mechanical Properties;
- Strength and Compressibility;
- Hydraulic Conductivity; and
- Chemistry.

Laboratory test procedures for these groupings are provided in Figure 5-28 for a number of standard reference documents or standards developed by various "standards" organizations. The use of standard technical methods in the development or measurement of site assessment data should never be overlooked in any project.

Determination of soil physical and chemical properties in site characterizations is a major part of the RI/FS process at hazardous wastes sites, where it is essential for evaluating the fate and transport of contaminants in the soil system. Fate and transport studies of soil contaminants are directly used for developing exposure assessments, risk assessments and remedial design strategies.

The purpose if soil sample collection with the RI/FS may be to evaluate a wide variety of geochemical reactions such as, chemical leaching to ground water, retardation within the soil column, leaching/retardation within caps, resuspension of contaminated soil as dust, and chemical retardation within the aquifer matrix.

The interrelation of soil chemical and physical processes in the soil system is complex. Sorption of chemicals can vary with concentration, pH, moisture content, and sorbent. This is further complicated by the contact of the chemical contaminant on specific soil properties and interactions with other chemical contaminants. The movement of solutes in the unsaturated zone is dependent on soil physical properties and infiltration which must be evaluated as a total system.

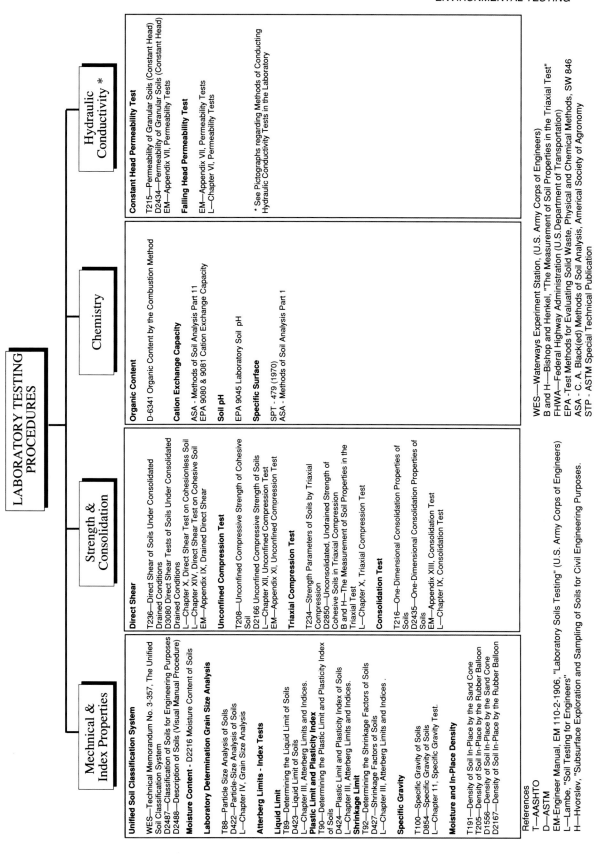

Figure 5-28 Laboratory Testing Procedures

Soil systems can also extremely heterogeneous and anisotropic. A chemical or physical property measured or determined in one location may not adequately characterize the entire site. Care must be taken that values such as hydraulic conductivity, fraction of organic carbon (foc), pH, soil moisture, sorption coefficients, and bulk density are representative of field conditions and adequately describe the total volume of soil under consideration within the assessment. Statistical techniques may be required to establish this correlation due to the typical heterogeneity of soil existing on a site.

The following laboratory tests are designed to provide information on the quantity and physical characteristics of soil located on project sites. For specific waste disposal projects individual tests must be selected to evaluate both borrow and landfill cut areas for greenfield and facility expansion projects. Guidelines for the numbers of tests are given below for typical waste disposal projects; however, good engineering judgment must be applied to adequately describe the quality and quantity of the individual soils located on the property. RI/FS investigations are too specific to site disposal history for general guidelines on the quantity of tests necessary to evaluate the remediations. Chapter 10 includes general descriptions of geostatistical methods for the evaluation of quantifies of soil tests for RI/FS investigations under assessment monitoring programs.

The Phase II investigations for disposal projects will generally consist of: (1) either drilling and sampling soil borings or excavating shallow exploratory pits, (2) obtaining bulk samples of representative soils, and (3) evaluating these samples for use as borrow material. Establishment of borrow material is an important part of many Phase II programs for waste disposal. Although not currently available (1992), it is anticipated that both the American Society for Testing Materials (ASTM) and the American Society of Agronomy will be compiling specific methods directly relevant for contaminated soils in the near future. The specific references for the analytical methods discussed are cited. The purpose of the methods, with a brief description, and limitations of the methods are briefly discussed in this section. It is not the intent of this section to give a detailed description or procedure for each method, since that information is available in the references cited. It is the intent rather to list various methods that are generally the most applicable for selection within approiate laboratory testing programs.

The Representative Sample

The difficulties of obtaining a representative soil

sample for laboratory evaluation has been recognized for at least forty years. Hvorslev (1949) classified five basic types of potential disturbances to soil that range from slight disturbance to severe:

- Changes in stress conditions;
- Changes in water content and void ratio;
- Disturbance of the soil structure;
- Chemical changes; and,
- Mixing and segregation of the soil constituents.

These five classes or levels of disturbances have continued importance today, since site assessments results may be greatly affected by any one of the above points. The simple (or complex) act of sampling and transporting soil samples to the laboratory can effect any or all of the above classes of soil disturbance. Recent reviews of ground-water sampling procedures (Nielsen 1991) have looked at sampling and transportation difficulties of pulling water quality samples from monitoring wells and moving them to the analytical laboratory. Obtaining representative soil samples has conceptually similar, but very different real operational problems beginning with the act of sampling, through to the soil laboratory tests and concluding with the data evaluations.

It is expected that by the time the sample has arrived in the laboratory it has been fully logged according to descriptive standards, USCS, and Facies Codes as approiate for the project. Additional classification may be conducted at the laboratory based on the results of grading and index tests, but the basic classification work must be completed in the field.

Mechanical Index Properties

Evaluation of waste disposal sites must incorporate an organized sampling and testing program of the facility subsoils to fully understand site geologic conditions. Grading curves, plasticity index properties (called Attenberg limits) and many additional physical properties of the soil and bedrock are normally evaluated as the first step of the design of new facilities or expansions to waste disposal sites. The evaluation of physical properties are also important for remedial projects, where long term stability of the design must be assured for predictable time periods into the future.

Soil Moisture Content

The gravimetric-oven-drying procedure is the standard by which all methods of measuring soil moisture content

Soil Moisture Content

The gravimetric-oven-drying procedure is the standard by which all methods of measuring soil moisture content are compared. Oven-dry water content is expressed on a weight or volume basis, producing gravimetric (Og) and volumetric (Ov) content values.

The water contents of soils commonly range from 5% to 35%. However, for soil types that contain a high percentage of clay or organic material, very high water content may be observed; sometimes unconsolidated muds can show moisture contents in excess of 100%. In general, the water content depends on void ratio, particle size, clay minerals, organic content, and ground-water conditions.

Soil moisture content determinations are essential for fate and transport estimations. The soil moisture content (% pore water) is one test used in vadose transport equations to modified water input from infiltration. Soil moisture has a direct influence on the sorption processes and should be considered in sorption studies or estimations. Soil moisture determinations also allow conversion of soil masses and chemical concentrations on a wet weight basis to dry weight equivalent masses and concentrations.

ASTM Method D-2216 determines the free water or pore water content as a percentage based on moist and oven dried soil weight differences. Method D-2216 may generate erroneous information for soils with a high organic material content, high clay content (halloysite, smectite, and illite/smectite), high gypsum content, or a high salt content in the pore water. Additional problems may be observed with the oven drying method as oven temperatures may vary significantly, and convectional temperature differences may introduce a systmatic error.

Relative Density

The unit weight or density of a material is defined as the weight of material divided by its volume, including solids, liquids and voids. The in-place density of soils is considered a function of deposition, gradation, and loading history. For example a fine grained soil deposited by wind or water and not subsequently subjected to loading will be relatively loose; if subsequently consolidated by the weight of overlying soil deposits or glacial action the soil density will be materially increased.

In general, when measuring the degree of compaction of a soil, the dry density-weight of solids divided by total volume is used. The relative density value offers a convenient measure of the degree of compactness of a soil in a fill or embankment, and also provides a significant indication of the susceptibility of the soil to liquefaction.

The relative density of a soil is the density relative to the limiting values described as its loosest state and its densest state. A soil in its loosest state has a relative density of zero, and a soil in its densest state has a relative density of 100%. Because the density of a soil is directly related to the void ratio, the relative density, Dr, can be expressed in terms of the void ratios:

$$Dr = \frac{e_{max} - e}{e_{min}}$$

Equation 5-26

where:
e max is the void ratio of the loosest state;
e min is the void ratio of the densest state; and,
e is the void ratio of the soil as it exists.

The specific gravity of a rock or soil may be defined as the ratio between the unit weight of the substance to the unit weight of pure water at 4^0C.

Classification of Soils for Engineering Purposes (ASTM D2487-69)

Test method ASTM D2487 classifies soils from any geographic location into categories representing the results of prescribed laboratory tests to determine the particle-size characteristics, the liquid limit, and the plasticity index. The assigning of a group name and symbol(s) along with the descriptive information required in ASTM practice D2488 can be used to describe a soil to aid in the evaluation of its significant properties for engineering use. The various groupings of this classification system have been devised to correlate in a general way with the engineering behavior of soils. This test method provides a useful first step in any field or laboratory investigation for geotechnical engineering purposes. Additional classification methods to further describe soil samples are included in Chapter 4.0 may be more important in hydrogeologic investigations than pure engineering classifications as represented by D2487.

Particle–Size Analysis of Soils (ASTM D422).

Particle size information is used to generate soil classifications, size distribution curves, and textural classifications (e.g. USDA and USCS classification) for both engineering and descriptive purposes. Particle size and distribution determinations aid in characterizing the physical soil environment where migration can occur. This information can also aid in lithologic correlations. To some extent, the grain size curve for sands can be related to engineering behavior such as soil hydraulic conductivity, frost susceptibility, angle of internal friction, bearing capacity and liquefaction potential

Particle or grain size determinations are either visually estimated or determined by sieving in the field. In the laboratory, grain or particle size is determined by sieving (>75 μm) and hydrometer techniques for the fine fraction (<75 μm). A sieve analysis consists of passing a sample through a set of standard sieves and weighing the amount retained on each sieve. The results are plotted on a grain size distribution curve in the form of percent fines by weight versus particle size to a log scale. The shape of the grain size curve is indicative of the grading. A "uniformly" graded (or poorly sorted) soil has a grain size curve that is nearly vertical and a "well-graded" (also poorly sorted) soil has a relatively flat curve that extends across several log cycles of particle size.

The hydrometer (sedimentation) analysis for the fine soil fraction is based on Stokes law, which relates the velocity at which a spherical particle falls through a fluid medium to the diameter and specific gravity of the particle and the viscosity of the fluid. The particle size is obtained by measuring the density of the soil-water suspension using a hydrometer. For soils with both coarse and fine constituents, a combined analysis should be performed. The sieve analysis is performed on soil retained on the No. 200 sieve and the hydrometer analysis is performed on soil passing the No. 200 sieve. Example grain size distributions for various soil types are illustrated on Figure 5-29.

Various different organizations (USDA, ISSS, ASTM, BS, and CSSC) have developed different textural classification systems based on particle size. Care must be taken when correlating a descriptive term from one system to another that the size ranges are consistent. Additionally, special soil treatments not addressed by the ASTM D422-63 test procedure including; removal of carbonates by acid NaOAC treatments, iron oxides by sodium dithionite-citrate treatments, salts by washing, and organic material by hydrogen peroxide pre-treatments are often required to accurately determine the particle size for textural classifications.

5.6.2 Index Tests

The original soil consistency tests of Attenberg (1911) have been further developed and standardized to the present methods described by ASTM and AASHTO. The Atterburg limits are used to determine the plasticity of soils. The liquid limit defines the boundary between the liquid and plastic states, and the plastic limit the boundary between the plastic and solid states. The difference between these two water content values is the range over which the soil remains plastic and is called the plasticity index. This parameter is used to classify soils for estimating other physical properties.

Water content is one of the most important index properties for a given soil. It is defined as the ratio of

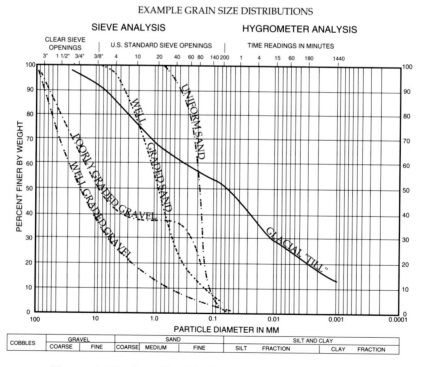

Figure 5-29 Grain Size Distribution for Example Soils

is a measure of the difference between the water content and the plastic limit, provides a qualitative assessment of the soil compressibility, sensitivity and preconsolidation. These index property relationships are shown in Figure 5-29.

Liquid and Plastic Limits (ASTM D4318–83)

The plasticity characteristics of cohesive material are described by the liquid limit, plastic limit and shrinkage limit, or Atterberg limits   Soils may be classified according to their plasticity, using the difference between liquid limit and plastic limit, or the plasticity index (Fig 5-30). Plasticity characteristics of various soils are tabulated in Table 5-7.

The liquid limit (LL) is defined as the water content at which a standard groove closes after 25 blows in a liquid limit percussion device first developed by Casagrande. Although both ASTM-D4318 and BS 1377 (1975) use this device a specification difference between the two standards produces different results for the liquid limit. The plastic limit (PL) attempts to measure the water content at which the soil begins to crumble when rolled into 3.2 mm (0.125 in.) diameter threads. The thread should break into numerous pieces between 3.2 mm (0.125 in.) and 9.5 mm (0.375 in.) long. The purpose of the limits is to aid in the classification of fine-grained soils (silts and clays) to evaluate the uniformity of a deposit and to provide some general correlations with engineering properties. The liquid and plastic limits are, however, not well correlated with engineering properties that are a function of soil structure or its undisturbed state (Hathaway, 1988). However, some general empirical correlations for fine-grained soils have been developed based on index properties, natural water content and Atterberg limits. It is recommended that limits be determined on fine grained soil starting with the soil at or near the natural water content (Lambe, 1951). Soil that has been dried (air or oven) should be thoroughly mixed with water and allowed to equilibrate for several days before testing. Soils with significant organic content should not be dried prior to testing.

Soil Bulk Density

Estimates of soil bulk density for soils are a primary tool in waste disposal projects to control installation of liners, leachate collection blankets and final covers of facilities. In addition bulk density values are needed in assessment monitoring programs to convert contaminant concentrations in soil from a weight to weight basis to a weight to volume basis. If particle densities are known, soil porosities can also be calculated for the soil. Bulk density measurements are also extensively used in both analytical and mathematical equations in assessment monitoring programs for estimating subsurface fate and transport effects. Soil bulk density is the ratio of the dry mass of solid to the bulk volume of solid. This volume includes both solids and pore space. Bulk density determination ref: Methods of Soil Analysis Part 1, (Klute, 1986) (core, clod, and balloon methods); Annual Book of ASTM Standards: D2167-84 (balloon), D1556-82 (sand-cone), and D2937-83 (core), consists of weighing a dried soil sample and comparing that value to the known sample volume (using the core method) or a determined volume (clod or in-situ excavation volume balloon or sand-cone methods). Core and clod soil bulk density methods are the most widely used and recommended methodologies.

The difficulty of accurately estimating soil volume is probably the major limiting factor in bulk density determinations. Because bulk density determinations are on a dry weight basis, variations in drying can generate errors. Bulk density may vary with the dregree of soil packing, and with total moisture content in soils containing appreciable amounts of swelling clays. For soil particles, the void ratio can be determined from the following

Table 5-7: Typical plasticity characteristics

Soil	Liquid limit	Plastic limit	Plastic index	Specific gravity
Lean Clay	32	18	14	2.75
Fat Clay	80	28	52	2.75
Sandy Clay	40	20	20	2.75
Sand	Non-plastic			2.65
Gravel-sand -silt	Non-plastic			2.68
Rock Fill	Non-plastic			2.70

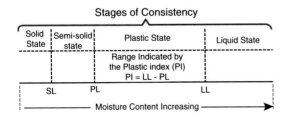

Figure 5-30 Index property Tests

errors. Bulk density may vary with the dregree of soil packing, and with total moisture content in soils containing appreciable amounts of swelling clays. For soil particles, the void ratio can be determined from the following relationship:

$$e = \gamma w Gs/\gamma d - 1 \qquad \text{Equation 5-27}$$

where:
γw = unit weight of water,
Gs = specific gravity of the soil particles,
γd = dry density, and
e = void ratio (the ratio of the volume of voids to the volume of solids in a soil).

The relative density of cohesionless soils is obtained by determining the void ratio of the soil as provided in Equation 5-27. Relative density can also be given by the relationship between dry density and void ratios for a given specific gravity as:

$$Dr = \frac{\lambda d_{max}(\lambda d - \lambda d_{min})}{\lambda d(\lambda d_{max} - \lambda d_{min})} \qquad \text{Equation 5-28}$$

Where:
Dr = relative density expressed as a percentage,
γ_{dmax} = dry density of the soil in its densest state,
γ_{dmin} = dry density of the soil in its loosest state,
γ_d = dry density of the soil being tested,

Approximate estimates of the relative density of cohesionless soils may be obtained in the field from the standard penetration test or STP as described by ASTM D1586-84. Empirical relationships which allow correlation of penetration resistance obtained by standard penetration tests (SPT'S) to relative density have been developed.

This relationship is illustrated in Chapter 4 under STP tests. Soundings with cone penetrometers can also be used for an approximate determination of relative density of the soil in question. Approximate correlations have also been made between standard penetration test blow count, N, and the consistency of clay. This relationship (given in Chapter 4) should be used only as a guide and general comparison.

Soil Particle Specific Gravity and Density

The specific gravity of a soil is defined as the ratio of the weight in air of a given volume of soil particles to the weight in air of an equal volume of distilled water at a temperature of four degrees Celsius. The specific gravity of a soil is a basic value that is used in computations for most laboratory tests. The specific gravity of soil samples is usually measured by a pycnometer and converted to mean particle density [ref: Annual Book of ASTM Standards D854-83 (pycnometer) (1987); Methods of Soil Analysis, Part 1 (pycnometer and submersion) (Klute, 1986)]. Particle density for soils is defined as the weighted mean density of particles in a soil sample volume. Particle density is expressed as the ratio of mass of soil particles to the sample volume, excluding pore spaces. In addition, the specific gravity is often used to relate the weight of a soil to its volume of solids for use in phase relationships, such as unit weight, void ratio, moisture content, and degree of saturation. The specific gravity is, however, of only limited value for identification or classification of most soils because the specific gravities of most soils fall within a narrow range.

Estimates of particle density and bulk density can be used to estimate void space as an approximation of porosity, (shown in Table 5-8), which may be required for evaluating subsurface migration of contaminants.

The pycnometer method can yield very precise specific gravity and density determinations when properly used. However, soils with significant soluble salts or carbonates can result in test errors. the concept of particle generally refers to solid mineral particles only. Soil specific gravity determinations by this method assume that water can fill all the void space. Soils with excessive organic material may cause interference effects which would require pretreatments prior to analysis.

Shear Strength

The shear strength of a soil is a commonly requested test procedure to evaluate in a soil mass the resistance to the displacement of adjacent soil elements along a plane or curved surface. The laboratory tests traditionally used to determine shear strength, the characteristics value at which the material fails in rupture or shear, are the direct shear, unconfined compression and triaxial compression tests

Shear resistance of soil is due to both cohesion and friction. The physics of shear strength determinations are well defined by many researchers especially, good references are Winterkorn and Fang (1975), Terzaghi and Peck (1949), and Lambe and Whitman (1969): see for further details. A soil mass may be considered to be a compressible skeleton of solid particles. In saturated soils the void spaces are filled with both water and air. Shear stresses are carried only by the skeleton of solid particles, whereas the normal stresses on any plane is carried by both the solid particles and the pore water.

Coarse-grained soils generally exhibit little or no

Table 5-8   Soil Density Relationships

Description	Porosity	Void ratio	Water Content	Unit Weight			
				lb/ft^3	kg/m^3	lb/ft^3	kg/m^3
Uniform Loose Sand	46	0.85	32	90	1441	118	1890
Uniform dense sand	34	0.51	19	109	1746	130	2082
Mixed-grained sand, loose	40	0.67	25	99	1585	124	1986
Mixed-grained sand, dense	30	0.43	16	116	1858	135	2162
Glacial till, very mixed-grained	20	0.25	9	132	2114	145	2322
Soft glacial clay	55	1.2	45	76	1217	110	1762
Stiff glacial clay	37	0.6	22	105	1681	129	2066
Soft slight organic clay	66	1.9	70	58	929	98	1570
Soft very organic clay	75	3.0	110	42	672	89	1425
Soft bentonite	84	5.2	194	27	432	80	1281
Rock fill	41	0.7	25	100	1601	125	2002

w = water content when saturated, in per cent dry weight

yd = unit weight in dry state

y = unit weight in saturated state

cohesion (i.e., cohesionless) and therefore the shear strength depends primarily on the frictional resistance of the soil. An estimate of the shear strength of the cohesionless soil in situ can be difficult to determine back in the laboratory because the strength can vary with density or critical void ratio, composition of the soil (particle size, gradation and angularity of soil particles), non-homogeneous of the deposit and the loading conditions (Hathaway 1988). Therefore, the soil should be tested in the laboratory under conditions which simulate the most critical condition in the field.

Estimating strength from the results of laboratory tests ideally calls for performance of tests that will duplicate in situ conditions as closely as possible. As was expressed at the beginning of this section, it is very difficult to achieve this field/lab correlation for many reasons; such as, sample disturbance, unknown in situ stresses and equipment and testing limitations that impose non-uniform stresses or the wrong stress system on the soil sample in the laboratory. This difficulty of achieving representative values for shear strength in laboratory tests have prompted the conduct of a large-scale standard shear box test in the field. A block sample of soil or rock is enclosed within a box and vertical and horizontal loads are applied. The load-displacement curve is then obtained and analyzed, as in the laboratory test.

Soil shear strength parameters are affected by many test factors including rate of loading, method of loading, principal stress ratios, rate of specimen strain, total specimen strain, degree of saturation and drainage conditions. The data obtained from shear strength tests are normally presented in terms of effective stresses. During testing, both total stresses and pore water pressures are measured. The effective stresses are determined by subtracting pore water pressure from total stress.

Triaxial Compression Test

The triaxial test is the most common and versatile test available to determine the stress-strain properties of soil. In the triaxial compression test, a cylindrical specimen is sealed in a rubber membrane and placed in a cell and subjected to fluid pressure. Triaxial tests are generally classified as to the condition of drainage during application of the cell pressure and loading, respectively, as follows:

Unconsolidated-Undrained (UU). Test or quick (Q) Test - No drainage is permitted during application of the cell pressure or confining stress and no drainage of the soil sample is allowed during application of the deviator stress. This test is generally performed on undisturbed saturated samples of fine grained soils (clay, silt and peat) The unconfined compression test is a special case of the UU test with the confining stress equal to zero. The principal stress difference at failure is called the unconfined compressive strength.

Consolidated-Undrained (CU) or Consolidated Quick (R) Test - Drainage is permitted during application of the confining stress so that the specimen is fully consolidated during the test. No drainage of the sample is allowed during application of the deviator stress. The CU test is performed on undisturbed samples of cohesive soil, on reconstituted specimens of cohesionless soil and, in some instances, on undisturbed samples of cohesionless soils which have developed some apparent cohesion resulting from partial drainage.

Consolidated-Drained.(CD) or Slow (S) Test - Drainage of the sample is permitted both during application of the confining stress and the deviator stress, such that the specimen is fully consolidated under the confining stress. The rate of applied stress is slow enough that essentially no change in the initial pore pressure occurs. The principal application of the results of CD tests on cohesive soils is for the case where either construction will occur at a sufficiently slow rate that no excess pore pressures will develop or sufficient time will have elapsed that all excess pore pressures will have dissipated. The principal application to cohesionless soils is to determine the effective friction angle.

Consolidation Tests

Consolidation tests determine the compressibility of a soil and the rate at which it will consolidate when loaded. Consolidation test procedures are described in ASTM D2435. The two soil properties usually obtained from a consolidation test are Cc, the compression index, which indicates the compressibility of a soil, and Cv, the coefficient of consolidation, which indicates the rate of compression under load.

Consolidation can be defined as the volume change at a "constant" load caused by the transfer of total stress from excess pore pressure to effective stress as drainage occurs. When load is applied to a saturated soil mass, the load is carried partly by the mineral skeleton and partly by the pore fluid. With time, the water will be squeezed out of the soil consistent with the rate of pore pressure dissipation and the soil mass will consolidate. When a load is applied to a saturated deposit of clay, there will be three types of settlement:

- *Initial settlement:* related to the undrained shear deformation of clay;
- *Consolidation settlement:* volume changes related to the dissipation of excess pore pressure; and,
- *Secondary Compression* (consolidation): volume

changes related to essentially constant effective stress, after complete dissipation of excess pore pressure.

In general, the magnitude of consolidation settlement will be of greatest concern for most cases. The laboratory test most commonly used to evaluate consolidation settlement is the oedometer test or one dimensional consolidation test. To perform an oedometer (consolidation) test, the soil is first placed in an oedometer ring and stress is applied to the soil specimen along the vertical axis. The test is conducted by loading the soil sample in stages equal to twice the previous load. This doubling of the applied load on the soil specimen causes a vertical deformation in the sample which is recorded versus time for each loading increment. Because strain in the horizontal direction is prevented by the ring, the vertical strain is considered equal to the volumetric strain.

The test is generally performed on a specimen of clay that is 19 or 25 mm (0.75 or 1.0 in.) in thickness and 64 mm (2.50 in.) in diameter. The 64 mm ring is the most common size ring used with the test equipment because the soil specimen can be obtained from a 76 mm (3-in.) thinwall tube sample. The information that may be obtained from the test include (Hathaway, 1988):

- *Compressibility of the soil,* for one-dimensional loading as defined by the compression curve, (vertical strain, Ev, or void ratio, e, plotted versus log consolidation stress, avc)

- *Maximum previous stress,* EVm~ as determined by empirical procedures from the compression curve.

- *Coefficient of consolidation, Cv,* using curve fitting techniques, developed from the Terzaghi theory of consolidation, applied to the deformation versus time curves.

- *Rate of secondary compression ,* as defined by the slope of the deformation versus log time plot after primary consolidation is completed.

The stress-strain or compressibility characteristics of clays as obtained from the consolidation test are highly dependent upon stress history of the materials. The stress history of a clay deposit (for example) refers to the existing stresses and the degree of over-consolidation. If the vertical consolidation stress currently acting on the clay is the greatest that has ever been applied to the clay, the clay is called normally consolidated. If the soil exists at a degree of prestress less than existing overburden pressures, such as would result from hydrostatic pressures reducing overburden load, these soils can be considered as under

to previous glacial ice loads, removal of existing overburden soils by erosion, desiccation or lowering of the ground-water level. The stress history of a sample can provide insight into the geologic conditions that caused the over-consolidation. For example, using the level of consolidation as the basis for relating glacial stages of sediments deposited by various ice advances or retreats.

Consolidation test data are typically presented in the form of a plot of void ratio e, or vertical strain ε_v, versus, consolidation stress σ_{vc}, graphics should be presented on a log scale.

5.6.3 CHEMICAL SOIL TESTS

Cation Exchange Capacity Tests

The USEPA Test Methods 9080 and 9081 from "Test Methods for Evaluating Solid Waste" (1991) were adopted from standard methods for soil analysis used by the American Society of Agronomy .

Soil cation exchange capacity (CEC) results can be used to approximate the maximum sorptive capability of a soil for positively charged polar organic compounds and inorganic cationic species. These tests are used for both solid waste disposal and remedial assessment projects. Soil CEC data does not generate a soil specific sorption coefficient; however, it provides a screening estimation for potential parameter retardation.

The preferred method is SW 846 Method 9081 (sodium acetate) which works equally well for calcareous or non-calcareous soils. In this method, a soil sample is mixed with an alkaline sodium acetate solution, resulting in sodium ion exchange with matrix cations. Ammonium acetate is then used to exchange the sorbed sodium; and the supernate is analyzed for sodium by atomic adsorption spectroscopy. The method of cation-exchange capacity by summation (Chapman, 1965; in Methods of Soil Analysis, Part 11, Page, et al, 1982) is recommended for highly acid soils. In this method, barium is exchanged to determine exchangeable hydrogen

Soil pH: Tests for soil pH values are by electrometric method referenced in Chapman (1965) or by the U. S. Environmental Protection Agency, Test Methods for Evaluating Solid Waste, Physical and Chemical Methods, SW 846 Method 9045 (1986).

Soil pH determinations are performed on a sample/distilled water mixture or slurry using a pH meter. Appropriate instrument analytical protocols are followed and temperature corrections incorporated. For soils high in salts alternative buffering solutions should be employed.

Soil pH determination are important for fate and transport assessments since the sorption process is often pH dependent for both organic and inorganic compounds. A determination of pH is normally required for soil systems where the sorption process is expected to be pH dependent, such as, mobility of metallic ions or for pH controlled bacteria bioremediation projects. This method may give erroneous determinations for soils with very high or very low pH. Improperly cleaned and maintained electrodes can also generate misrepresentative values. No method of soil preparation is mentioned in SW 846-1985; therefore, a natural unprepared and untreated sample is assumed in the testing process.

5.6.4 General Soil Sampling Criteria

Selection of the spatial locations and depths of soil samples represents an important task during the phase II scope of work. Sufficient balance must be maintained between the numbers of test samples and maintaining reasonable control on soils laboratory costs and later professional interpretation of the test results. Only general recommendations are provided, since only professional judgement and experience can fully evaluate the site specific project data requirements.

Care must be taken to specify sufficient test samples from those zones being selected for piezometer head level monitoring. The classification of soils for engineering purposes must be made initially during drilling, and reviewed after laboratory testing to confirm the field classification. While specific projects may require more or less of a particular test, general guidance for greenfield or expansion solid waste facilities are presented below. Moisture–density tests should be performed at the rate of one test for each material type found during drilling of each borehole, with more tests performed on material suitable for use as a bottom liner. Both organic content and hydraulic conductivity tests should be run for each type of material, unless sound engineering judgment suggests that additional tests are required. Consolidation tests should also be performed on each type of material. Cation exchange tests, which determine the potential capacity of the soil to adsorb leachate constituents, should be performed on every different material found at the proposed site.

Early in Phase II tasks (typically in the project proposal costing effort) a general quantification of the proposed field and laboratory soil sampling program is known, however, field subsurface conditions observed during the backhoe pitting and borehole drilling work will necessitate later revising of these early estimates. Laboratory evaluations of soils supports field observations and provides data beyond that obtainable from simple borehole evaluation tasks. Do not skimp on these tests in hopes that

later revising of these early estimates. Laboratory evaluations of soils supports field observations and provides data beyond that obtainable from simple borehole evaluation tasks. Do not skimp on these tests in hopes that the site will have simple, uncomplicated subsurface conditions.

5.6.5 Laboratory Permeability Tests

Darcy, the 19th century French hydrologist, showed experimentally that the discharge of water, Q, flowing through soil of cross–sectional area A was proportional to the imposed gradient i or:

$$Q = K\,i\,A \qquad \text{Equation 5-22}$$

The coefficient of proportionality, K, has been called "Darcy's coefficient of hydraulic conductivity" or "coefficient of hydraulic conductivity" or "hydraulic conductivity." The three terms are often used in an interchangeable manner, although the term hydraulic conductivity is most commonly used in ground–water literature. It is now generally accepted that a strict interpretation of meaning associates any particular pore fluid with permeability, and that ground water is associated with hydraulic conductivity. Soil science literature and rock mechanics have used lower case k to describe hydraulic conductivity and upper case K to represent permeability. The ground-water sciences, however, use exactly the opposite; hydraulic conductivity is represented in equations as 'K' (upper case) and 'k' (lower case) is reserved for intrinsic permeability. As will be described, below these term are not interchangeable in the case of ground-water sciences..

The two most common laboratory methods used for determining hydraulic conductivity are the variable head method and the constant head method. Both have test methods defined by Lambe (1951) and by a more recent ASTM standard test methods (D2434–68 for constant head tests of granular soils). Both types of tests, however, do not give reliable test results for fine–grained materials, in which more than 10% of the soil passes a No. 200 sieve. These finer grained materials require the use of back–pressure triaxial test methods for cohesive soils, a complicated procedure requiring carefully sampled soil and costing in the range of $250–$500 each (1992). Laboratory hydraulic conductivity test specimens may be either undisturbed or representative disturbed samples, depending on the purpose for which they are to be tested. The method of testing for hydraulic conductivity depends on the hydraulic conductivity range of the soil to be tested. The constant-head permeameter is used to determine the hydraulic conductivity of granular soils, whereas the

falling-head type is more suitable for soils of medium hydraulic conductivity such as more silty materials. For soils of very low hydraulic conductivity < 10-5 cm/sec, consolidation or triaxial test data may provide more representative data.

Examples of permeability (hydraulic conductivity) test devices and the methods used for testing are described briefly below. The emphasis of the discussion is on features that distinguish the various devices and the advantages and disadvantages. The relative merits of fixed-and flexible-wall permeameters are discussed by Daniel et al., 1985. Laboratory tests for hydraulic conductivity are organized on pictograph Figure 5-24.

Pressure Cell

The accepted standard test for laboratory determination of saturated permeability in the agricultural sciences (but not the geotechnical sciences), is called the pressure cell test. In this procedure a soil sample, which may be an 'undisturbed' core, a sample compacted in a mold, or a volume of soil, is placed in a metal pressure cell. After the soil is initially saturated, it is connected to a standpipe. The permeant fluid is introduced through the standpipe and forced through the pressure cell under a falling head. The standpipe may or may not be connected to a source of air pressure to superimpose a pressure head over the fluid column. The pressure cell apparatus for a falling head test

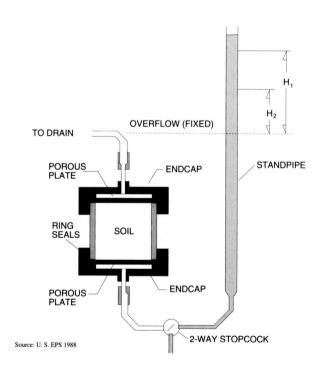

Figure 5-31 Pressure Cell Method

accomplished by submerging one end of the core in a pan of water for 16 hours with the other end open to the air. This procedure however, may not be adequate to saturate the sample completely prior to the test. Vacuum wetting and fluctuating external gas pressure are alternate saturation techniques that may be used with the pressure cell.

Compaction Permeameter

The compaction permeameter (also called a fixed-wall permeameter) has been developed for testing permeant fluids with a compacted soil layer. One advantage of this method over the pressure cell is that the sample is compacted directly in the permeability test device. As a result, a better seal is obtained between the sample and the walls of the test vessel. To ensure a good sidewall seal, some workers apply a bentonite slurry to the inside of the chamber before the sample is loaded into the device. Silicon grease has also been tried to insure a tight fit within the walls of the test vessel. As with the pressure cell, the compaction mold may be used in either a falling head test or a constant head test. The compaction permeameter may be modified so that elevated pressures can be superimposed to reduce the testing time. A modified compaction permeameter is illustrated in Pictograph Figure 5-32. Not shown on the diagram is the source of compressed air with a water trap, regulator, and pressure meter. Also not shown is a fraction collector with automatic timer that handles the collection and isolation of effluent samples.

The sample is leveled and the fluid chamber is slowly filled to begin the test so as not to disturb the sample surface. The filled permeameter is allowed to stand for a period of time (typically overnight or longer) to allow hydration and swelling of the clay. Pressure is applied to the sample only after it has completely hydrated. In compatibility testing, baseline permeability determinations are often obtained with the standard permeant fluid before the test permeant is introduced. Alternatively, baseline values are obtained in separate tests with other samples of the same soil.

Triaxial Cells

Triaxial compression testing was introduced in the previous section to determine the stress-strain properties of soil. These tests can also be used to evaluate the hydraulic conductivity of cohesive and non-cohesive fine grained soils.

Samples prepared for triaxial testing consist of cylindrical columns of compacted soil encased laterally with a flexible membrane (often latex rubber) and enclosed at the ends with porous stones. The enclosed soil sample is placed in a water-filled cell that can be pressurized to provide a confining pressure on the sides of the sample (Figure 5-33). The sample to be tested may be prepared in a compaction mold and extruded for testing in the triaxial cell. Samples are typically 2 to 4 inches in diameter. Relatively undisturbed Shelby tube samples may also be tested.

Triaxial permeability tests are usually performed by passing permeant liquids (typically water) upwards into the sample under pressure while maintaining a lower pressure

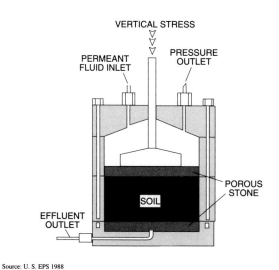

Source: U. S. EPS 1988

Figure 5-32 Consolidation Permeameter

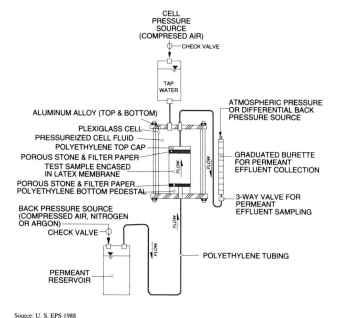

Source: U. S. EPS 1988

Figure 5-33 Tri-axial Test Arrangement

tested.

Triaxial permeability tests are usually performed by passing permeant liquids (typically water) upwards into the sample under pressure while maintaining a lower pressure at the exit port on the top of the sample. A confining pressure somewhat greater than the pressure under which the liquids enter the soil is imposed to press the flexible membrane firmly against the soil sample, preventing flow along the sidewall. In liner studies the confining pressure on the soil sample should be selected to simulate the lateral pressures a material will experience in the clay liner.

Sample saturation is accomplished by forcing a standard permeant liquid upward through the sample. Backpressure may be applied to hasten the dissolution of trapped air bubbles or gases generated by reactions between permeant liquid and the sample. The use of backpressure ensures virtually complete saturation of the sample, a factor that probably contributes to the good reproducibility of triaxial tests. Pressure regulators and electronic pressure transducers are used to control and monitor sample stress conditions and to assess the saturation state within the sample during testing.

In a typical test for leachate compatibility with liner or base-grade soils, sufficient standard liquid is passed through the sample to establish a stable baseline hydraulic conductivity before the test permeant fluid is introduced. For some tests the test permeant is introduced directly with baseline being determined with another soil sample. The test may be conducted as a constant head or falling head test. Hydraulic conductivity, leachate volume, and time increment are reported for each volume of test fluid passed through the sample. These data are used to plot hydraulic conductivity versus pore volumes of test fluid passed.

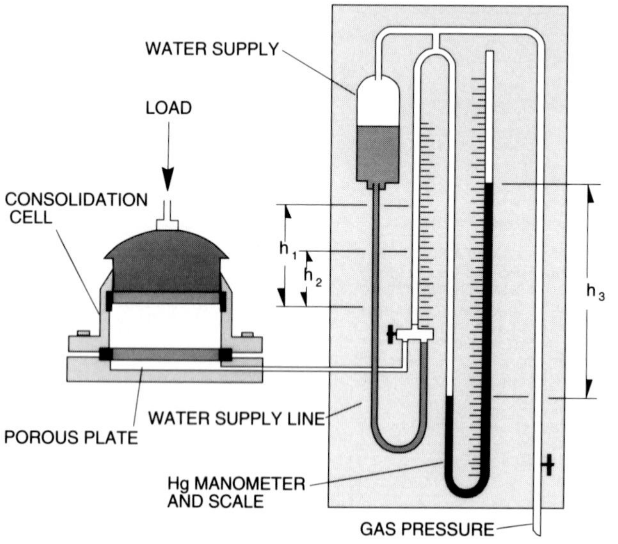

Consolidation Cells

Consolidation cells (consolidometers) are commonly used in the field of geotechnical engineering to determine the compressibility and rate of settlement of soils. Consolidation occurs when water is squeezed out of the soil and is therefore a function of permeability. A fixed-ring consolidation cell can also be used to measure hydraulic conductivity (Figure 5-34).

The consolidation cell method is routinely used in testing hydraulic conductivity for applications such as earth dams, retaining walls, and slurry trenches. The method however, has not been widely used in the evaluation of chemical compatibility with clay liner material.

5.6.6 Laboratory Evaluations Of Hydraulic Conductivity

Laboratory hydraulic conductivity tests are most commonly performed on a small, "undisturbed" sample of the rock or sediment collected during the field investigation. Measurements are made on samples commonly obtained by Shelby tube or thin wall samples. Because of the ways in which the various test are performed, the results are generally reported for vertical hydraulic conductivity only. Since only a small volume of material is tested in typical laboratory hydraulic conductivity procedures, heterogeneity in the geologic unit(s) under investigation is not directly determined, except through use of an extensive sampling program. However, in granular materials, particle size grading curves (a simple and inexpensive test procedure) can be used to approximate the hydraulic conductivity of granular materials, using empirical correlations. These values can be used to supplement and support laboratory hydraulic conductivity testing programs.

Hydraulic Conductivity And Permeability

The coefficient of proportionality, K, which appears in the various forms of the Darcy equation is known as hydraulic conductivity. Hydraulic conductivity may be defined as in an isotropic porous medium as the specific discharge per unit hydraulic gradient. 'I' is a measure of the ease with which a fluid is transported through a porous

WATER SUPPLY

LOAD

CONSOLIDATION CELL

h_1

h_2

h_3

WATER SUPPLY LINE

POROUS PLATE

Hg MANOMETER AND SCALE

GAS PRESSURE

Source: U. S. EPS 1988

Figure 5-34 Consolidation Cell Test Arrangement

material with the dimensions length/time. Hydraulic conductivity is a result of both the fluid and matrix properties of the materials tested. The relevant matrix properties are primarily grain (or pore) size distribution, shape of grains (or pores), specific surface, tortuously and porosity. The relevant fluid properties are density, p, and viscosity. Nutting (1930) expressed hydraulic conductivity as:

$$K = k\rho g/\mu \qquad \text{Equation 5-29}$$

where k (lowercase k), having the dimensions (L^2) is called the permeability (or intrinsic permeability) of the porous matrix. The coefficient k depends solely upon the properties of the solid matrix and the ratio $\rho g/\mu$ represents the influence of the fluid properties on hydraulic conductivity. Generally, throughout this text we deal with the flow of water through a porous medium, and as such, reference will be to maintain hydraulic conductivity K (upper case K). Permeability, k (lower case), is measured in the metric system in cm^2 or in m^2. For water at 20°C, one can define following conversion:

$$K = 1 \text{ cm/sec is equivalent to } k = 1.02 \times 10^{-5} \text{ cm}^2$$

Permeability (k) has traditionally been used by the engineering profession interchangeable with hydraulic conductivity (K), however, the terms are not equivalent and should not be mixed incorrectly in discussions.

Permeability Correlations

The literature contains numerous equations and predictive techniques which relate permeability, k, to a series of properties of the solid matrix. The simplest approach is to seek a correlation between hydraulic conductivity and total porosity. However, it can be concluded that such an approach is generally futile (except for comparison of otherwise identical media) owing to the strong dependence of flow rate upon width continuity, shape and tortuosity of the conducting pores. This is the reason why coarse-textured soils (with less total porosity and fewer individual pores, but lager and more uniformly sized pores) will have greater saturated hydraulic conductivity than fine-textured soils (which have greater total porosity, but smaller more irregularly-sized, tortuous pores).

A method listed in Figure 5-4 is grain size analysis for the determine hydraulic conductivity. Since Hazen (1892), a number of formulas have been proposed that relate some measure of grain size to hydraulic conductivity (for example, Fair and Hatch, 1933; Krumbein and Monk, 1942; Masch and Denny, 1966; and Er-Hui, 1989). These

formulas are empirical with hydraulic conductivity proportional to a function of representative grain diameters. These formulas are considered reasonably accurate for sandy soils, with the accuracy significantly decreasing when the samples are predominantly silt or clay. They may be very useful, however, in the early stages of a field investigation when large numbers of samples can be taken for grain size analysis. They also may be helpful in estimating hydraulic conductivity in the vadose zone, which can be a difficult task (see Chapter Section 5.3). Permeability correlations can be placed into three basic groups:

• Those purely empirically developed through experimentation;

• Purely theoretical obtained from the theoretical derivation of Darcy's equation using conceptual models; and,

• Those which are semi-empirical where permeability is related to various matrix properties through a conceptual model with experimentally determined numerical coefficients.

Most equations which relate permeability to the porous matrix properties are of the general form Bear (1972):

$$K = f_1(s)f_2(n)d^2 \qquad \text{Equation 5-30}$$

where:

s = a dimensionless parameter (or parameters) that express the effect of the shape of the grains (or pores);
$f_1(s)$ = a shape factor;
$f_2(n)$ = the porosity shape factor; and,
d = the effective (or mean) diameter of the grains.

The effective diameter is usually taken as the harmonic or geometric mean of the grains, or as d_{10} i.e., a diameter such that 10% (by weight) of the porous medium consists of grains smaller than it (ASTM 1990). The product $f_1(s)f_2(n)$ often appears as a single coefficient, C, in the relationship between k and s and results in:

$$k = Cd^2 ; \quad K = Cd^2\rho g/\mu \qquad \text{Equation 5-31}$$

Note that the relationship $\rho g/\mu$ is added to convert between permeability and hydraulic conductivity. This conversion between k and K is often hidden within equations as constants and coefficients. Several investigators have studied the relationships between permeability and pore size distribution (Marshall, 1958).

Correlations Using Mean Grain Diameter

Numerous equations have been formulated which relate permeability to the mean grain diameter. These equations date back to 1892 with some of the original research of Hazen. The equations have progressed through many derivations, however, the form is generally that of Equation 5-31 and have been arrived at from one of three techniques: purely theoretical, semi-empirical, or purely empirical studies.

Theoretical Correlations

Expressions for permeability using theoretical correlations have been based from derivations of the darcy equation using the conceptual model approach. Fundamental conceptual models forming the basis for the development of these theoretical permeability equations are derived from capillary tube models, fissure models, hydraulic radius models and resistance to flow models.

Kozeny (1927) proposed a widely accepted derivations of permeability and its relationship to porous medium properties. This derivation was modified by Carman (1937, 1938, 1956). The porous medium is treated as a bundle of capillary tubes of equal length but not necessarily of circular cross-sections in the theory proposed by Kozeny. The Navier-Stokes Equation was solved simultaneously by Kozeny for all channels passing through a cross-section normal to the flow in the porous medium, to obtain the equation for permeability in the form:

$$k = \frac{C_o(n)^3}{M^2}$$
Equation 5-32

where C_0 is a Kozeny's numerical coefficient that varies according to the geometrical form of the individual channels in the model. Numerous other expressions which relate permeability to the mean grain diameter have been derived from conceptual models. For a description, the reader is referred to Bear (1972).

Semi-empirical Correlations

Semi-empirical relations have been found between permeability and matrix properties using numerical coefficients developed from experimentally determined theoretical treatments. The numerical coefficients must be evaluated for each group of similar porous matrices. These semi-emperical correlations are the most popular equations used in assessment projects.

One of the most widely accepted theories on the relation of saturated hydraulic conductivity to the geometric properties of porous media is the Kozeny-Carmen theory which is based on the concept of hydraulic radius. The measure of hydraulic radius is the ratio of the volume to the surface of the pore space, or the average ratio of the cross-sectional area of the pores to their circumferences. The Kozenny-Carmen Equation is shown below.

Carmen (1937, 1938, 1956) modified the Kozeny Equation to obtain Equations (5-33, 5-34) and evaluated $C_0 T = 1/5$ experimentally. The resulting equation is known widely as the Carmen-Kozeny Equation:

$$k = \frac{\dfrac{n^3}{(1-n)^2}}{5M^2}$$
Equation 5-33

$$K = \frac{1}{k_0 T^2 S^2 [\dfrac{n^3}{(1-u)^2}]^2}$$
Equation 5-34

K = saturated hydraulic conductivity
k = pore shape factor (approx. 2.5)
T = tortuosity factor (approx. square root of 2)
S = specific surface area per unit volume of particles
n = porosity

or:

$$k = \frac{n^3}{180(1-n)^2} d_m^2$$
Equation 5-35

If one is lucky enough have ground water in the system at 20^0C then the equation reduces (Harleman et al 1963) to

$$K \text{ (cm/sec)} = 64.1 \, d_{10}^2$$

where d_{10} (in cm) is the effective grain diameter.

Although the Kozeny-Carmen Equation works well for the description of saturated hydraulic conductivity in uniformly graded sands and some silts, serious discrepancies are found in clays (Michaels and Lin, 1954, Lambe, 1955, and Olsen, 1962). The failure of this theory stems from the original assumptions which are:

- no pores are sealed off;
- pores are distributed at random;
- pores are reasonably uniform in size;
- porosity is not too high;
- diffusion phenomena are absent; and,
- fluid motion occurs like motion through a batch of capillaries (Scheidegger, 1974).

Soil systems which can satisfy every assumption are rare with the exception of uniformly graded sands upon which the Kozeny-Carmen equation was founded.

Empirical Correlations

The Hazen relationship is the most widely used method to estimate the permeability of sands through grading curves. Early investigations into permeability relationship used experimentally determine k from direct permeability laboratory measurements and to calculate C from d^2/k. Hazen (1892) adopted this approach on filter sands and arrived at the following expression:

$$K = 100 \, d^2_{10} \ (cm/s) \qquad \text{Equation 5-36}$$

where, K (hydraulic conductivity) is in cm/s and d_{10} as previously defined, is in cm and the uniformity coefficient is between 2-5. In this case, the coefficient equals $C\rho g/\mu$. Experimental data suggest that the finer particles in the porous media have the most influence on permeability (or hydraulic conductivity), and hence the use of d_{10} in Hazen's Equation. If the uniformity coefficient = 1, then

$$K = 150 \, d^2_{10} \ cm/sec \qquad \text{Equation 5-37}$$

Correlations Using Grain (or pore) Size Distributions

Since permeability is a measure of the ease with which a fluid moves through the porous medium, certain relationships must exist between the grain (or pore) size distributions of the porous medium and its permeability. Fair and Hatch (1933) formulated the following empirical expression involving porosity and several other parameters using dimensional considerations:

$$K = \frac{\rho g}{\mu} \left[\frac{n^3}{(1-n)^2} \right] \left[m \left(\frac{\theta}{100} \sum \frac{P}{d_m} \right)^2 \right]^{-1} \qquad \text{Equation 5-38}$$

where:

K = the hydraulic conductivity in units consistent with ρ, g, d_m,

m = viscosity of the fluid in units of mass per time-length

ρ = fluid density in units of mass per unit volume,

n = the porosity

m = a packing factor ($\pm$ about 5)

θ = sand grain shape factor; 6.0 for spherical grains and 7.7 for angular grains,

P = percentage of sand held between adjacent sieves,

d_m = geometric mean of rated sizes of adjacented sieves in units of length, and

g = acceleration of gravity, (32 feet/sec^2)

In the phi scale system the negative of the logarithm to the base 2 of the particle diameter in mm is used. Masch and Denny (1966) determined experimental relationships between statistical relationship of grain size distribution and hydraulic conductivity. The parameters included measures of the average size, dispersion, skewness, peakedness, and modality of the sample distributions. Masch and Denny produced several graphs that related hydraulic conductivity to the various statistical parameters. On the basis of these plots, they defined a group of curves to predict laboratory hydraulic conductivity values for samples with random statistical distribution. This set of prediction curves, given in Figure 5-26, incorporate only average size and dispersion since these parameters were found to best describe the relationship between hydraulic conductivity and the grain size properties of the porous mediums investigated.

The basic procedure one follows to determine hydraulic conductivity using the Masch and Denny method of statistical correlation is to obtain the median diameter MD_{50}, calculate and read the hydraulic conductivity in units of US gallons/day/ft^2 directly from the predictive curves given in Figure 5-35. The value obtained from the predictive curves is for a temperature of 15.6^{o}C. Other temperatures require adjustments for variations of density from the standard test temperature. This useful range is equivalent to about 10^{-2} cm/s to 10^{-4} cm/s respectively.

Powers (1981) provides a graphical solution for hydraulic conductivity using d_{50} (mm) sieve analysis with uniformity coefficient, and soil density. Values of K are read directly off the various density hydraulic conductivity graphs (see Figure 5-36). The uniformity coefficient (Cu) is defined as the ratio of the D_{60} size of the soil (the size than which 60% of the soil is finer) to the D_{10}.

The above can be used where appropriate, to support field tests of hydraulic conductivity. Building an adequate data base of both field and laboratory hydraulic conductivity tests then provides the basis for definition of vertical and horizontal components of hydraulic conductivity in later flow net constructions.

5.6.7 Laboratory Tests For Effective Porosity

An important aquifer characteristic of geologic environments used in time of travel calculations, is effective porosity. Fetter (1980) defines effective porosity as follows: "The amount of interconnected pore space through which fluids can pass, expressed as a percent of bulk volume. Part of the total porosity will be occupied by static fluid being held to the mineral surface by surface tension, so effective porosity will be less than total porosity." The relative value of effective porosity for fine grained sediments such as clays has been the point of significant disagreements between regulatory agencies and

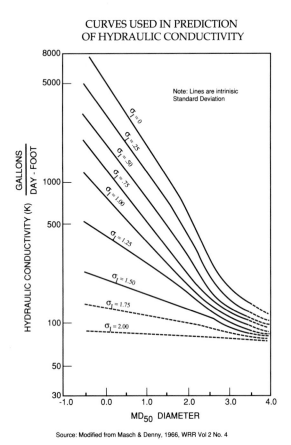

CURVES USED IN PREDICTION
OF HYDRAULIC CONDUCTIVITY

Source: Modified from Masch & Denny, 1966, WRR Vol 2 No. 4

Figure 5-35 Masch and Denny Curves

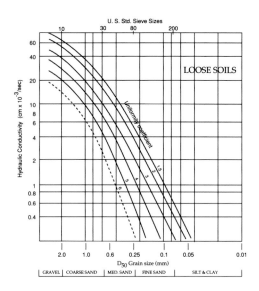

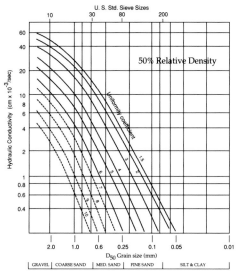

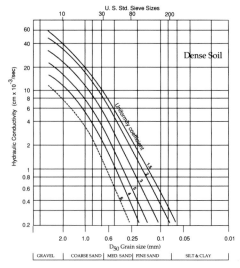

Source: Modified from Powers(1981)

Figure 5-36 Powers Curves for Density and K

the regulated community for a number of years since effective porosity is used in the following equation of seepage velocity:

$$V = \frac{Ki}{n_e}$$ Equation 5-39

V is the average lineal velocity;
n_e is the effective porosity.
i is the gradient, and K is hydraulic conductivity.

There is a current controversy centering around using very low values of ne for fine grained clay sediments. A U.S. EPA guidance document (EPA 1986) proposed using values of effective porosity on the level of 1% for clays. The basis of this low figure is from specific yield determinations of sediments derived from water supply literature (see Walton 1970, p. 34). However, the Glossary of Geology (1987) defines effective porosity as follows: "The percentage of the total volume of a given mass of soil or rock that consists of interconnected interstices. The use of this term as a syn. of specific yield is to be "discouraged." The definition of effective porosity is recognized as not synonymous with specific yield in the technical literature. This departure of n_e from specific yield is especially important in fine grained sediments, since the value of lineal velocity V is so dependent on effective porosity. For example, using a low effective porosity of 1% (as compared to 30%) for a clay confining unit, will change the calculations of lineal velocity as follows:

Assuming K = 1×10^{-7} cm/sec
i = dh/dl = 0.5
n_e = 0.01, 0.30
1.57 m/year (at 0.01 n_e)
0.053 m/year (at 0.30 n_e)

Considering the average facility can have thirty years of active life the difference in the two calculations can represent the difference between breaking through confining units some thirty times faster than with the larger value of n_e. The controversy surrounding n_e is difficult, in some ways to understand on a scientific basis, considering effective porosity can be calculated from breakthrough column tests. The following test is described for quantification of n_e for a series of glacial till (confining unit) samples. This testing was performed to: 1) determine the effective transport porosity; and, 2) to allow evaluation of the effect of leachate on the hydraulic conductivity of these potential clay liner materials. These results can ultimately be used in design of recompacted clay liners for future hazardous waste disposal cells at a potential site, and can be used to estimate effective porosity of unfractured confining unit materials.

A testing program should include both hydraulic conductivity testing and saline solution breakthrough evaluation of a number of confining unit samples. In addition, one should also conduct several reverse breakthrough tests using fresh water as the permeant.

Breakthrough and Leachate Solutions

The initial stages of laboratory tests for specific porosity, including saturation and hydraulic conductivity determination, should be conducted using actual site water obtained during the site assessment investigation.

An example tracer solution used for the column breakthrough portion of the tests consisted of a lithium chloride salt in water. The concentrations of lithium and chloride in the breakthrough tracer solution for the example below were:

• Lithium: 4,800 mg/l
• Chloride: 25,000 mg/l

Experience with the initial portion of column breakthrough testing shows that one will observe that the chloride ions will pass through soil samples at a much faster rate than the lithium ions. The likely reason is that due to cation exchange effects within the clay particles slowed down the passage of the lithium ions. The variation of the observed chloride concentration with time is considered to be the more accurate indicator of breakthrough time and, hence, pore fluid velocity. Therefore, only chloride ion concentrations is measured for the reverse breakthrough tests to maintain a conservative approach to the testing program.

The tracer solution is placed into a tri-axial cell upgradient accumulator (see Figure 5-33) and the concentration of lithium and chloride is then measured in the fluid that permeates out of the sample and is collected in the downgradient accumulator. During the initial portion of the test a gradient of 5 pounds per square inch (psi) is imposed on the sample. One may find with low hydraulic conductivity geologic test materials there is a resistant low flow rate, therefore, the gradient may be increased later in the test program to 10 psi. During the breakthrough portion of the test, at least 2 pore volumes of fluid should be passed through the sample.

Determination of Effective Porosity

Based on the principles of Darcian flow, the velocity

(v) of a soluble constituent flowing through a porous medium can be estimated using the Equation 5-39:

$$v = Ki/n_e$$

where:
K is the hydraulic conductivity);
i = the hydraulic gradient, and,
n_e = the effective porosity.

For the purposes of calculating breakthrough velocity, breakthrough is defined as the point where the constituent concentration in the permeant leaving the sample is fifty percent of that entering the sample. The effective porosity is the porosity through which flow is actually occurring. As such, it is always less than or equal to the total porosity. The incorporation of effective porosity into the Darcy equation, then, takes into account bound water and closed voids within the sample.

Since it is typical for effective porosity samples under going testing have different values for hydraulic conductivity, it would be considered appropriate to examine the change in concentration of the constituent being measured in the permeant (in this case, chloride) leaving the sample in terms of the number of pore volumes of fluid passing through the sample. Since these small samples are relatively homogeneous, the ratio of the effective porosity to the total porosity is equal to the number of pore volumes passing through the sample at breakthrough. Values of effective porosity based on this method of analysis are shown in Table 5-9. The effective porosity obtained in the test series for glacial tills ranged from 0.26 to 0.34. (Golder 1987)

These values can be compared with work performed by Rawls, Brakensiek & Saxton (1982) on 1,323 soils with 5,350 horizons from 32 states that provided the total porosity, residual saturation, and effective porosity (among other soil properties) for this large spectrum of soil tests. Table 5-10 shows effective porosity for clay at 38.5%. This value is consistent with column tests performed in site investigations for clays.

5.6.8 Laboratory Hydraulic Properties Of Rocks

Samples for laboratory testing should be selected from available cores, and should be representative of site bedrock geologic units. Where possible, horizontal and vertical plugs of 25 mm diameter should be cut from drilled rock cores of 50–100 mm diameter. However, if the sample is friable and subject to breakup in handling, testing it may not be possible to measure hydraulic conductivity. The engineer should keep in mind that such laboratory

Table 5-9 Summary of Effective Porosity Results

Sample	Porosity	Pore Volumes at Breakthrough	Effective Porosity
TP-1	0.38	1.05	0.29
TP-3	0.38	0.68	0.26
TP-5	0.39	0.88	0.34
TP-9	0.38	0.88	0.33

procedures represent tests on core that is essentially so disturbed in terms of preservation of open discontinuities as to be of limited actual value.

Hydraulic Conductivity

The hydraulic conductivity of rock plugs can be measured in a standard gas permeameter, with nitrogen as the test fluid (API, 1966), with results empirically corrected for gas slippage and expressed for water at 20°C (approximate ground–water temperature in the field).

Effective Porosity, Bulk, and Matrix Density

The rock plugs or sample fragments may then be tested by the liquid resaturation method, employing propanol as the test fluid (API, 1966) in order to avoid a swelling reaction with any altered clayey materials, a situation that would lead to erroneous permeability determinations.

Pore–Size Distribution and Specific Yield

If the rock has exceptional porosity, further laboratory measurements can be undertaken on a smaller number of selected samples to investigate the distribution of pore–sizes and gravity–drainage characteristics.

When dealing with friable rock, the preferred method is by mercury injection porosimeter (Fatt, 1956). This technique involves forcing mercury into the pore spaces of the sample, progressively smaller pore necks being invaded at successively higher pressures, generating a curve of invaded pore–volume against injection–pressure.

The curves should be examined in detail to determine if an incomplete invasion of mercury has occurred. This could result from one of two reasons: either there exist many pores that are too small to be invaded, or some of the

TABLE 5-10 Porosity, Residual Saturation, and Effective Porosity of Common Soils

Texture Class	Sample Size	Total Porosity (ϕ). cm³/cm³	Residual Saturation (ϕ_r) cm³/cm³	Effective Porosity (ϕ_c). cm³/cm³
Sand	762	0.437 (0.374: 0.500)	0.020 (0.001: 0.039)	0.417 (0.354: 0.480)
Loamy Sand	338	0.437 (0.368: 0.506)	0.035 (0.003: 0.067)	0.401 (0.329: 0.473)
Sandy Loam	666	0.453 (0.351: 0.555)	0.041 (0.0: 0.106)	0.412 (0.283: 0.541)
Loam	383	0.463 (0.375: 0.551)	0.027 (0.0: 0.074)	0.434 (0.334: 0.534)
Silt Loam	1206	0.501 (0.420: 0.582)	0.015 (0.0: 0.058)	0.486 (0.394: 0.578)
Sandy Clay Loam	498	0.398 (0.332: 0.464)	0.068 (0.0: 0.137)	0.330 (0.235: 0.425)
Clay Loam	366	0.464 (0.409: 0.519)	0.076 (0.0: 0.174)	0.390 (0.279: 0.501)
Silty Clay Loam	689	0.471 (0.428: 0.524)	0.040 (0.0: 0.118)	0.432 (0.347: 0.517)
Sandy Clay	45	0.430 (0.370: 0.490)	0.109 (0.0: 0.205)	0.321 (0.207: 0.435)
Silty Clay	127	0.479 (0.425: 0.533)	0.056 (0.0: 0.136)	0.423 (0.334: 0.512)
Clay	291	0.475 (0.427: 0.523)	0.090 (0.0: 0.195)	0.385 (0.269: 0.501)

**First line is the mean value
Second line is + one standard deviation about the mean

Reference: Rawls, W.J., D.C. Brakensiek, K.E. Saxton 1982
Transactions of the ASCE Soil & Water Division Paper 81-2510, pp. 1318

pores are so large that they were invaded before the instrument could measure them. To test such possibilities, one can perform direct tests of specific–yield by centrifugation (Price, 1977) on representative samples of the friable rock.

REFERENCES

American Petroleum Institute, 1966, Recommened Practice for Core Analysis Procedure, API Manual API-RP40, Dallas.

Amoozegar, A.and A. W. Warrick, 1986 Hydraulic Conductivity of Saturated Soils - Field Methods, pp. 735-70. Methods of Soil Analysis Part 1 : Physical and Mineralogical Methods. Agronomy Monograph 9. American Society of Agronomy, Methods, WI.

ASTM, 1991, Annual Book of ASTM Standards D854-83 (pycnometer) (1987); Annual Book of ASTM Standards, American Society for Testing and Materials, Philadelphia, vol. 04.08, pp

ASTM, 1991, Annual Book of ASTM Standards: D2167-84 (balloon), D1556-82 (sand-cone), and D2937-83 (core)Annual Book of ASTM Standards, American Society for Testing and Materials, Philadelphia, vol. 04.08, pp

ASTM, 1991, D2435 Consolidation test procedures

ASTM, 1991, D3385-88,Annual Book of ASTM Standards, American Society for Testing and Materials, Philadelphia, vol. 04.08, pp

ASTM, 1991, D4318–83, Liquid and Plastic Limits Annual Book of ASTM Standards, American Society for Testing and Materials, Philadelphia, vol. 04.08, pp

ASTM, 1991, Method D-2216, Annual Book of ASTM Standards, American Society for Testing and Materials, Philadelphia, vol. 04.08, pp

ASTM, 1991, Classification of Soils for Engineering Purposes (ASTM D2487-69) Annual Book of ASTM Standards, American Society for Testing and Materials, Philadelphia, vol. 04.08, pp

ASTM, 1991, standard test methods (D2434–68 for constant head tests of granular soils, Annual Book of ASTM Standards, American Society for Testing and Materials, Philadelphia, vol. 04.08, pp

ASTM, 1991, test method D4643 "Moisture Content Determinations' Annual Book of ASTM Standards, American Society for Testing and Materials, Philadelphia, vol. 04.08, pp

ASTM, 1991, Standard test method for penetration test and split-barrel sampling of soil. Test Designation D1586-84. 1991 Annual Book of ASTM Standards. American Society for Testing and Materials,

Philadelphia, vol. 04.08, pp 298-303.

ASTM, 1991, Standard test method for density and unit weight of soil inplace by the rubber balloon method. Test Designation D2167-84. 1991 Annual Book of ASTM Standards, American Society for Testing and Materials, Philadelphia, vol. 04.08, pp 342-347.

ASTM, 1991, Standard test method for triaxial compressive strength of undrained rock core specimens without pore pressure measurements. Test Designation D2664-80. 1991 Annual Book of ASTM Standards, American Society for Testing and Materials, Philadelphia, vol. 04.08, pp 429-434.

ASTM, 1991, Standard test method for laboratory determination of pulse velocities and ultrasonic elastic constants of rock. Test Designation D2845-83. 1991 Annual Book of ASTM Standards, American Society for Testing and Materials, Philadelphia, vol. 04.08, pp 445-452.

ASTM, 1991, Standard test methods for density of soil and soil aggregate in place by nuclear method (shallow depth). Test Designation D2922-81). 1991 Annual Book of ASTM Standards, American Society for Testing and Materials, Philadelphia, vol. 04.08, pp 463-472.

ASTM, 1991, Standard test method for direct tensile strength of intact rock core specimens. Test Designation D2936-84. 1991 Annual Book of ASTM Standards, American Society for Testing and Materials, Philadelphia, vol. 04.08, pp 473-477.

ASTM, 1991, Standard test method for unconfined compressive strength of intact rock core specimens. Test Designation D2938-79. 1991 Annual Book of ASTM Standards, American Society for Testing and Materials, Philadelphia, vol. 04.08, pp 484-487.

ASTM 1991. Standard test method for moisture content of soil and soil aggregate in place by nuclear methods (shallow depth). Test Designation D3017-78. 1991 Annual Book of ASTM Standards. American Society for Testing and Materials, Philadelphia, vol. 04.08, pp 508-513.

ASTM, 1991, Standard test method for direct shear test of soils under consolidated drained conditions. Test Designation D3080-72 (Reapproved 1979). 1991 Annual Book of ASTM Standards, American Society for Testing and Materials, Philadelphia, vol. 04.08, pp 514-518.

ASTM, 1991, Standard test method for elastic moduli of intact rock core specimens in uniaxial compression. Test Designation D3148-80. 1991 Annual Book of ASTM Standards, American Society for Testing and Materials, Philadelphia, vol. 04.08, pp 519-525.

ASTM, 1991, Standard test method for deep, quasi-static, cone and friction cone penetration tests of soil. Test

Materials, Philadelphia, vol. 04.08, pp 519-525.

ASTM, 1991, Standard test method for deep, quasi-static, cone and friction cone penetration tests of soil. Test Designation D3441-79. 1991 Annual Book of ASTM Standards, American Society for Testing and Materials,Philadelphia, vol. 04.08, pp 550-557.

Attenberg A., 1911, Die Plastizitat der Tone, Int. Mitt. fur Bodenkunde, I; 10-43.

Bear J., 1972, Dynamics of Fluids in Porous Media, American Elsevier Publishing Co., New York

Boersma, L., 1965, Field Measurement of Hydraulic Conductivity Below a Water Table, Methods of Soil Analysis Part 1: Physical and Mineralogical Methods Agronomy Monograph No. 9, American Society of Agronomy, Madison WI. pp 22-252.

Bouma, j. D. I. Hillel, F. D. Hole, and C. R. Amerman, 1971, Field Measurement of Hydraulic Conductivity by Infiltration Through Artifical Crusts, Soil Sci. Soc. Amer. Proc. 35: 362-364.

Bouwa, J. and L. W. Dekkert, 1981, A Method of Measuring the Vertical and Horizontal Saturated Hydraulic Conductivity of Clay Soil With Macropores, Soil Sci. Soc. Amer. J. 45: 662-663.

Bouwer, H., 1961, A Double Tube Method for Field Measureing Hydraulic Conductivity of Soil in Sites Above a Water Table, Soil Science, Soc. Amer. Proc. 25:334-342.

Bouwer, H., 1962, Field Determination of Hydraulic Conductivity Above a Water Table With a Double-tube Method, Soil Science, Soc. Amer. Proc. 26:330-342.

Bouwer, H., 1964, Measuring Horizontal and Vertical Hydraulic Conductivity of Soil with the Double-tube Method, Soil Science, Soc. Amer. Proc. 28:19-23.

Bouwer, H., 1966, Rapid Field Measurement Air Entry Valve and Hydraulic Conductivity of Soil as Significant Parameters in Flow Syster Analysis, Water Resources Research, Vol. 2, No. 4, pp. 729-738

Bradbury, K. R., andM. A. Muldoon, 1990. Hydraulic Conductivity Determinations in Unlithified Glacial and Fluvial Materials, in Ground Water and Vadise Zone Monitoring Nielsen and Johnson, Ed. ASTM STP 1053pp138-152

Brooks, R. H. and A. T. Corey, 1964, Hydraulic Properties of Porous Media, Hydrology Paper No. 3. Colorado State University, Fort Collins, CO.

BSI, 1974, Specification for Core Drilling Equipment (BS 4019:1974). Part 1 - Basic Equipment. British Standards Institution, London, 152 p.

BSI, 1975a, Methods for Sampling and Testing of Mineral Aggregates, Sand and Filters (BS 812:1975). Part 1 - Sampling. Size, Shape and Classification. British Standards Institution, London, 24 p.

BSI, 1975b, Methods of Test for Soil for Civil Engineering Purposes (BS 1377:1975). British Standards Institution, London, 144 p.

BSI, 1981a, Code of Practice for Site Investigations (BS 5930:1981). British Standards Institution, London, 148 p.

BSI, 1981b, Code of Practice for Earthworks (BS 6031:1981). British Standards Institution. London, 88 p.

BSI, 1986, British Standard Code of Practice for Foundations (BS 8004: 1986). British Standards Institution, London, 150 p.

Carman, P. C.,1937, Fluid Flow Through a Granular Bed, Trans., Inst., Inst., Chem. Eng., London UK, 150-156.

Carman, P. C., 1938, Determination of the Specific Surface of Powders, J. Soc. Chem. Indus., 57, 224-234.

Carman, P. C., 1956, Flow of Gases Through Porous Media, Butterworths, London.

CASECO, 1964, Mica Project Field Permeability Tests in Soil, July.

Cedergren,H. R.,1967, Seepage, Drainage, and Flow Nets, John Wiley &Sons, New York.

Chapman, H.D., 1965, Cation–Exchange Capacity in C.A. Black [ed.] Methods of Soil Analysis, Part 2, American Society of Agronomy, Madison, WI.

Dames & Moore, 1974, Manual of Ground-Water Practices, Los Angeles Ca.

Daniel et al., 1985, Fixed Wall Vs. Flexible-wall Permeameters, Geotechnical Testing Journal, Vol. 7, No. 3, pp 113-122.

Er-Hui, Z., 1989, Experimental Research on Permeability of Granular Media, Ground Water, Vol. 27, No. 6, pp. 848-854.

Fair, G. M.and Hatch, L. P., 1933, Fundamental Factors Governing the Streamine Flow of Water Through Sand, J. Am. Water Works Assoc. 25, 1531-1565.

Fatt, I, 1956, The Network Model of Porous Media, Trans. Amer. Inst. Min. Metal. Petrol. Engrs.,207, pp. 144-159.

Fetter, C. W.,1980, Applied Hydrology, C. E. Merril Publishing Co., Columbus, Oh.

Freeze and Cherry, 1979, Groundwater, Englewood Cliffs, N.J. Prentice-Hall, 604 p.

Gibson, R.E. ,1963,. An analysis of system flexibility and its effect on timelag in pore-water pressure measurements. Geotechnique, vol. 13, pp 1-11.

Glover, 1953, Flow for a Test Hole Located Above Groundwater Level, in Theory and Problems of water Percolation, by C. N. Zangar, U.S. Department of Interior, Bureau, Bureau of Reclaimation, Engineering Monograph No. 8, pp 69-71.

Green, R. E., L. R. Ahuja, and S. K. Chong,.1986, Hydraulic Conductivity, Difusivity, and Sorptivity of Unsaturated Soils - field Methods, In:Methods of Soil Analysis, Part 1, A. Klute, Editor, Agronomy

Monograph No. 9, 2nd Edition. Soil Science Society of America, Madison, WI.

Grisak, G. E. and J. A. Cherry, 1975, Hydrologic Characteristics and Responses of Fractured Till and Clay Confining a Shallow Aquifer, Canadian Geotechnical Journal, Vol. 12, pp 23-43.

Harleman, D.R.H., Mehthorn, P. F., and Rumer, R. R.,1963, Dispersion-Permeability Correlation in Porous Media, J. Hydraul. Div. Amer. Soc. Civil Eng. No. HY2, 89, 67-85.

Hathaway, A. W., 1988, Manual on Subsurface Investigations, American Association of State Highway and Transportation Officials, Inc. Washington D.C.

Hazen, A., 1892, Experments Upon the Purification of Sewage and Water at the Lawrence Experiment Station, In: 23rd Annual Report, Massachusetts State Board of Health.

Hvorslev, 1951, Time Lag in the Observation of Ground Water Levels and Pressures, Bulletin No. 36, USA/COE Waterways Experimental Station, Vicksburg, MS.

Hvorslev, 1949, Subsurface Exploration and Sampling of Soils for Civil Engineering Purposes,USA/COE Waterways Experimental Station, Vicksburg, 521 p.

Jezequel, J. F. and C. Mieussems, 1975, In Situ Measurement of Coefficients of Pearmeability and Consolidation in Fine Soils, In: Proceedings of the Conference on In Situ Measurement of Soil Properties, A.S.C.E., New York, Vol. I, pp 208-224.

Keller, C. K., G. van der Kamp, and J. A. Cherry, 1986, Fracture Permeability and Ground water Flow in Clayey Till Near Saskatoon, Saskatchewan, Canadian Geotechnical Journal, vol. 23, no. 2, pp. 229-240.

Klute, 1986, (core, clod, and balloon methods); In:Methods of Soil Analysis, Part 1, A. Klute, Editor, Agronomy Monograph No. 9, 2nd Edition. Soil Science Society of America, Madison, WI.

Kozeny, J., 1927, Uber Kapillore Leintung des Wossers im Boden, Sitzungsber, Akad. Wiss. Wien 136, 271-306.

Krumbein, W. C. and G. D. Monk, 1942, Permeability as a Function of the Size Parameters of Unconsolidated Sand, Petroleum Technology, July.

Kruseman, G. P.and N. A. DeRidder, 1976, Analysis and Evaluation of Pumping Test Data. International Institue for Land Reclamation and Improvement, Wageningen, The Netherlands, 200 pp.

Lambe and Whitman,1969, Soil Mechanics, John Wiley and Sons, Inc. Lambe, W. T.,1951, Soil Testing for Engineers, John Wiley and Sons, London

Lambe, W. T.,1955, The Permeability of Fine Grained Soils, ASTM 163, American Society for Testing and Materials, Philadelphia, Pennsylvania. pp. 55-67.

Masch F. D. and Denny, K. J., 1966, Grain Size Distribution and its Effects on the Permeability of Unconsolidated Sands, Water Res. Res., Vol. 2 No. 4, 665-667.

Marshall, T. J., 1958, A Relationship Between Permeability and Size Distribution of Pores, J. Soil Sci. Oxford Univ. Press, 9, pp. 1-8.

Michaels,A. S. and C. S. Lin, 1954, The Permeability of Kaolinite, Industrial and Eng. Chem. 46: 1239-1246.

Mualem, Y.,1976, A New Model for Predicting the Hydraulic Conductivity of Unsaturated Porous Media, Water Resources Res., 12:513-522.

Neuman, S. P.,and P. A. Witherspoon, 1972, Field Determination of the Hydraulic Properties of Leaky Multiple Aquifer Systems, Water Resources Research, Vol. 8, No. 5, p. 1284-1298.

Neuman, S. P., and P. A.Witherspoon, 1969, Applicability of Current Theories of Flow in leaky Aquifers, Resources Research, Vol. 5, No. 4, p. 817-829.

Nielsen, D. M., 1991, (editor) Practical Handbook of Ground-water Monitoring, Lewis Publishers, Chelsea, MI., 717p.

Nielsen , D. R. , J. W. Biggar, and K. T. Erh, 1973, Spatial Variability of Field-Measured Soil-Water Properties, Hilgardia 42: 215-259.

Nutting, 1930, Physical Analysis of Oil Sands, Bull. Amer. Ass. Petr. Geol. 14, 1337-1349.

Olsen, H. W., 1962, Hydraulic Flow Through Saturated Clays, Proc. of the Ninth National Conf. on Clays and clay Minerals, pp. 131-161.

Page, A. L.,R. H. Miller, and D. R. Keeney (ED), 1982, Agronomy, No. 9, Part 2, 2nd Ed.; Methods of Soil Analysis; Part 2 - Chemical and Microbiological Properties. American Society of Agronomy, Madison, WI.

Philip, R. J., 1985, Approximate Analysis of the Borehole Permeameter in Unsaturated Soil, Water Resources Res. 21(7):1025-1033.

Powers, J. P. 1981. Construction Dewatering, JohnWiley ans Sons., New York, N.Y.

Price, M., 1977, Specific Yield Determinations From a Consilidated Sandstone Aquifer, J. Hydrol., 3, pp. 147-156.

Rawls, W.J., D.C. Brakensiek, K.E. Saxton, 1982, Transactions of the ASCE Soil & Water Division, Paper 81-2510, pp. 1318.

Reynolds and Elrick, 1987, A laboratory and Numerical Assessment of the Guelph Permeameter Method, Soil Science, Vol. 144, pp.282-299.

Reynolds and Elrick, 1986, A method for Simultaneous in Situ Measurement in the Vadose Zone for Field-Saturated Hydraulic Conductivity, Sorptivity and the Conductivity-Pressure Head Relationships, Ground

Water Monitoring Review, Vol. 6, No. 1, pp. 84-95.

Rosco Moss Company, 1990, Handbook of Ground Water Development, John Wiley and sons Inc. New York, N.Y. 493p.

Stephens, D. B., K. Lambet, and D. Watson, 1987, Regression Models for Hydraulic Conductivity and Field Test of the Borehole Permeameter, Water Resources Res. 23:2207-2214.

Taylor , K., S. Wheatcraft, J. Hess, J. Hayworth, and F. Molz, 1990, Evaluation of Methods for Determining the Vertical Distribution of Hydraulic Conductivity, Ground Water, Vol. 28, No. 1, pp 88-98.

Terzaghi and Peck, 1949, Soil Mechanics in Engineering Practice, J. Wiley, Second Edition, New York.

Theis, C. V., 1935, The Relation Between the Lowering of the Piezometric Surface and the Rate and Duration of Discharge of a Well using Ground-Water Storage, American Geophysical Union Transactions, Vol. 16, Pt 2, p. 519-524.

Tratwiew, S., 1988, Draft Test Method for Field Measurement of Infiltration Rate Using A Double Ring Infiltrometer with a Sealed Inner Ring, Transactions of the ASCE Soil & Water Division Paper 81-2510, pp. 1318.

U. S. Environmental Protection Agency, 1986, Test Methods for Evaluating Solid Waste, Physical and Chemical Methods, SW 846, Method 9045, U.S. Government Printing Office Washington D.C.

U.S. Environmental Protection Agency, 1986, RCRA Ground-Water Monitoring Technical Enforcement Guidance Document, U.S. Government Printing Office Washington D.C.

U.S. Bureau of Reclamation, 1960, Earth Manual, Test Designation, E–18 pp. 541-546,Test Designation E–19, U.S. Government Printing Office Washington D.C., pp.546-562.

U.S. Bureau of Reclamation, 1965, Design of Small Dams,U.S. Government Printing Office Washington D.C. pp 193-196.

U.S. Bureau of Reclamation, 1978, Drainage Manual, U.S. Govt. Printing Office, Washington D.C.

USBR, 1974,. Earth Manual. (Second edition) - United States Bureau of Reclamation, US Government Printer, Washington D.C., 810 p.

U.S. Geological Survey, 1942, Methods For Determining Permeability of Water-boring Materials, Water Supply Paper No. 887, U.S. Government Printing Office Washington D.C.

U.S. Navy Facilities Engineering Command, 1971, Design Manual, Soil Mechnics, Foundations and Earth Structures, NAV-FAC DM-&, Washington D.C.: U.S. Government Printing Office.

USEPA, 1984, Permit Applicant's Guidance Manual for Hazardous Waste Land Treatment, Storage, and Disposal Facilities, Washington D.C.

USEPA, 1991, Test Methods for Evaluating Solid Waste, SW 846, Method 9081 (sodium acetate) and 9080, Office of Solid Waste and Emergency Response, Washington D.C.

van der Kamp and Maathuis , 1985, Excess Hydraulic Head in Aquitards Under Solid Waste Emplacements In Hydrogeology of Rocks of Low Permeability, Momoires of the Tucson Congress of the International Association of Hydrologists, vol. XVII, part 1, pp. 118-126.

van Genuchten, M. T., 1980, A Closed Form Equation for Predicting the Hydraulic Conductivity of Unsaturated Soils, Soil Science Society of America Journal, Vol. 44, pp. 892-898.

Waldrop, W. R., K. R. Rehfeldt, L. W. Gelhar, J. B. Southard, and A. M. Dasinger, 1989, Estimates of Macrodispersivity Based on Analysis of Hydraulic Conductivity Varability at the MADE Site, Electric Power Research Institute Report, EPRI EN-6405, Palo Alto, CA.

Walton , W. C., 1987, Groundwater Pumping Test Design and Analysis, Lewis Publishers, Chelsea, MI. 201 pp.

Walton, W. C., 1967, Groundwater Pumping Tests Design and Analysis, Lewis Publishers, Chelsea, MI 201 pp.

Walton, W. C., 1970, Groundwater Resource Evaluation, McGraw-Hill, New York.

Walton, W.C., 1962, Selected analytical methods for well and aquifer evaluation, Illinois State Water Survey, Urbana, Illinois, Bulletin no. 49, 81 p.

Watson, K. K., 1966, An Instantaneous Profile Method for Determinating the Hydraulic Conductivity of Unsuraded Porous Materials, Water Resources Res. 2: 709-715.

Wilson, L. G., 1982, Monitoring in the Vadose Zone, Part II, Ground Water Monitoring Review, Vol.2, No. 4, p. 31.

Winterkorn and Fang, 1975,. Foundation Engineering Handbook, Van Nostrand Reinhold Company, New York

CHAPTER 6

FRACTURED ROCK ASSESSMENTS

6.1 INTRODUCTION

Site assessments in fractured rock present many unique problems that require somewhat different field and laboratory methods than would be used in granular materials. A critical element in the successful evaluation of fractured geologic materials is the ability to identify and evaluate migration pathways through the bedrock. There are a variety of analytical procedures that investigators can employ to assist in identifying these principal migration pathways. These methods will be addressed within the content of the Phase I and Phase II site assessment scope of work. The rocks discussed under this heading include indurated rocks, such as igneous, intrusive (plutonic) and extrusive (eruptive) types, metamorphic rocks and indurated sedimentary rocks (sandstone, quartzite, shale, etc.). The fractures and fissures in these rocks remain relatively unchanged as the water circulates, unlike limestone and dolomites in which solution and erosion cause the fissures and fractures to grow larger through geologic time.

Witherspoon (et al. 1979) characterized fractured systems as "rock blocks bound by discrete discontinuities comprised of fractures, joints and shear zones, usually occurring in sets with similar geometry.' These fractured rock systems are shown in Figure 6-1 and are referred to as 'dual-porosity' systems. Scale of the investigation also is pertinent to the assessment method used in the rock mass evaluation. (Benson 1988). Figure 6-2 provides a matrix comparison of fracture analysis methods to the relative size of area evaluated. Each of these methods have direct application to determining regional discontinuities (such as faults) local-site discontinuities (as in shear zones and fracture sets) and smaller scale features observable in rock cores. Scale of observation is also pertinent in fractured rock assessments (see Figure 6-2). Large scale or megascopic features such as faults represent major discontinuities that can greatly modify ground-water flow conditions. Mesoscopic features such as joints, fractures and sheer zones can modify local ground-water flow so that with a sufficient diversity of discontinuities, an equivalent primary porosity condition can be considered

present in the rock mass. Microscopic fractures can also contribute a small but significant increase in rock mass porosity. Each of these levels of discontinuities should be evaluated within the site assessment.

The study of the hydrogeology of fractured and fissured rocks differs greatly from the study of friable rocks with interstitial porosity. Hence, it has been the subject of relatively moderate scientific work, and it occupies only a very small place in current textbooks. However, recent high-level, nuclear repository investigations have highlighted the difficulties of fractured rock assessment. Unfortunately many investigators lack a clear understanding of how to tackle hydrogeological problems in fissured and fractured rock terrain.

This chapter will discuss the field and office assessment techniques for evaluation of fractured rock systems for all phases of waste disposal investigations. The study of the hydrogeology of friable or primary porosity rock aquifers may begin with the measurement of porosity and hydraulic conductivity on a small volume of rock, the results of which are then extrapolated to the whole aquifer. The

Fractured Rock Dual Porosity Model

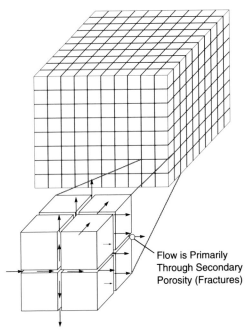

Flow is Primarily Through Secondary Porosity (Fractures)

Figure 6-1 Dual Porosity Systems

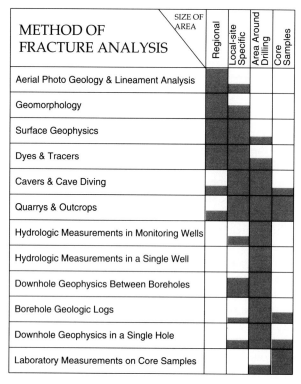

METHOD OF FRACTURE ANALYSIS	Regional	Local-site Specific	Area Around Drilling	Core Samples
Aerial Photo Geology & Lineament Analysis	■	□		
Geomorphology	■			
Surface Geophysics	■		■	
Dyes & Tracers			■	
Cavers & Cave Diving	□			
	■			
Quarrys & Outcrops	□			
			■	
Hydrologic Measurements in Monitoring Wells	□		■	
Hydrologic Measurements in a Single Well		□		
Downhole Geophysics Between Boreholes		■	■	
Borehole Geologic Logs			■	■
Downhole Geophysics in a Single Hole		■	■	
Laboratory Measurements on Core Samples			■	■

Source: Modified from Benson (1988)

Figure 6-2 Scale Factors in Fractured Rock

study of the hydrogeology of fractured and fissured rocks, where the secondary porosity openings are (by comparison) widely separated, requires a conceptual change of scale from a laboratory sized volume of rock with potentially very low primary porosity, to hundreds or thousands of cubic meters of rock with highly specific secondary porosity features or discontinuities. Thus, broad scale field work must often take the place of detailed laboratory test assessments of hydraulic conductivity and porosity. Fractured rock evaluation requires that geological field study has priority over sample collection and laboratory testing. Just as the study of pores and the water circulation within them is essential for the understanding of unconsolidated, primary porosity geologic materials; so the study of water circulation in secondary porosity fractures is essential for an understanding of ground-water movement within a mass of indurated rock.

6.1.1 Fractures and Joints

According to their most common definition, and following the International Society of Rock Mechanics (ISRM, 1978) convention, fractures are discontinuities in rock along which there is little or no displacement parallel to the discontinuity surface. These fractures range from <0.5m to >1,000m in length and microns to 5 cm in width. In a geomechanical sense, a fracture is a surface at which a loss of cohesion has taken place as a result of rupture.

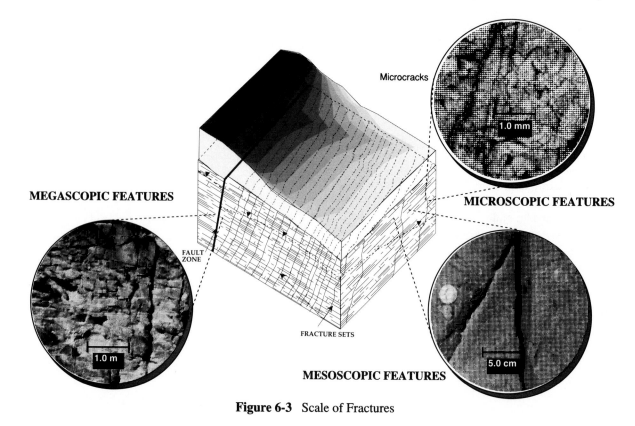

MEGASCOPIC FEATURES

FAULT ZONE

FRACTURE SETS

Microcracks

1.0 mm

MICROSCOPIC FEATURES

1.0 m

5.0 cm

MESOSCOPIC FEATURES

Figure 6-3 Scale of Fractures

APPROACH FOR INVESTIGATING THE GROUND WATER
FLOW CHARACTERISTICS OF FRACTURED ROCK

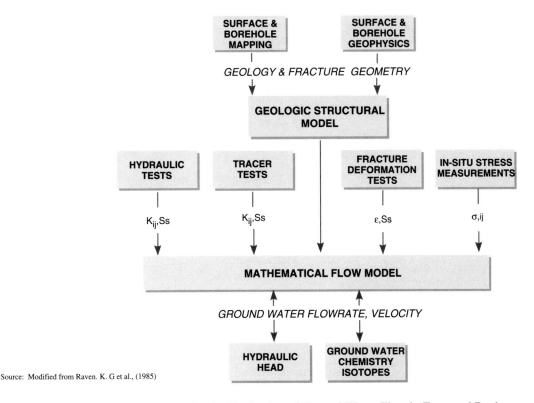

Source: Modified from Raven. K. G et al., (1985)

Figure 6-4   Design for Evaluation of Ground Water Flow in Fractured Rocks

According to this definition, the term fracture is synonymous with the term joint. Fractures and joints filled with secondary minerals are often referred to as veins. Microfractures or microcracks are fractures of the order of millimeters or less in length (100-1,000 microns long) which often appear in clusters and may form networks on the scale of a porous medium (see Figure 6-3). They are penny shaped, may be healed or open and typically are uniformly distributed or related to other structures. While some researches refer to microcracks as fissures, most investigators imply the term fissure to mean the same as fracture (not necessarily microfractures). The term microfracture is subjective designating a fracture of large aperture and/or extent, or simply any fracture larger than a microcrack. On this microscopic scale, porosity is represented by microcracks in crystalline rock. An example of porosity due to microcracks of granites ranges from 0.10-0.30% (.001-.003).

Many sub-parallel fractures concentrated in a relatively narrow band form a fracture zone (Figure 6-3). Shear zones are fracture zones that have undergone, or have been created by, shear (stresses causing displacement parallel to the discontinuities). Shear zones are the typical form of

secondary porosity in igneous/metamorphic terrain and may represent a fracture zone along which displacement has occurred. Joints may be systematic (occur in sets of parallel or sub-parallel planes or curved surfaces) or nonsystematic (more or less random, often curved and non-intersecting). First order fractures cross several layers, while second order or strata bound fractures are confined to a single layer. Extension fractures are due to displacement in a normal direction only.

Fractures can be induced by a man's activities such as drilling, coring, excavating, and blasting. Fractures form and propagate due to large concentrations of stress (primarily tension and shear), their orientation being controlled by the acting stresses. Induced fractures tend to have fresh, clean surfaces: in cores, they often develop parallel or normal to the axis. Natural fractures can be recognized by slickensides (due to repeated shearing), mineralization, and their tendency to appear in sub-parallel sets.

Where fractures appear in folded beds or in association with faults, they are often of a tectonic origin formed by diastrophism related to the folding and faulting events. Other stress-related forming mechanisms include differential stresses due to erosion of overburden, shrinkage of shale and shaly sands due to the loss of water, and the shrinkage of igneous rocks due to temperature variations. Landslides and rock falls may also create shallow fractures near the soil surface. Structure-related fractures usually show a consistency of orientation in the field and a certain degree of regularity over given areas and volumes of rock. Highly irregular and curved discontinuities that do not show a consistency of orientations are more typical to surficial phenomena such as landslides and desiccation.

When a rock is very heavily fractured into pseudo prismatic blocks that appear to fit together (sometimes categorized as breccia), mapping of individual fractures is typically not possible. When the rock is finally broken to such an extent that the individual pieces do not fit together, the rock is referred to as rubble and there are no available criteria to evaluate fracture density for such fractured rock units.

Stylolites are irregular planes of discontinuity in limestone that look like sutures and often form roughly parallel to bedding. Their seams are formed by an insoluble residue and tend to either terminate laterally or converge into residual clay seams. As such, they can act as small scale barriers to ground water flow. Stylolites that intersect pre-existing fractures also cause a reduction in their degree of hydraulic connectivity between fracture sets. They form in rocks of varying facies and geological age, and are thought to form by a pressure-soluble process.

6.1.2 Faults

Faults are long (up to many km in length), deep fractures or fracture zones, with a vertical or oblique attitude, separating two rock units that have undergone relative, large-scale displacement in the shear direction. They are commonly some distance apart and parallel to each other, thus forming wide or long, vertical or sub-vertical planar discontinuities. They are at places intersected by (or associated with) networks of other fractures forming vertical or sub-vertical blocks. In neither case are these units and blocks common to another fault system either horizontally or obliquely. Faults may be associated with pinnate fractures that generally parallel the main fault orientation. Even faults that run in the same direction may intersect each other at depth by reason of a difference in inclination to the vertical. Thus, faults may act as drains for intersected aquifers; as sub-surface dams to the aquifers by juxtaposing less permeable barriers to the flow of water through aquifers; or cause little discernible change in the local ground-water flow conditions. Water circulation through faults or fault zones

is not uniform and is frequently localized, because faults are not necessarily open at all places or in all directions. Site assessments of ground-water condition adjacent to faults rely heavily on direct measurements of hydraulic head in piezometers to sort-out the local effects of the fault on ground-water flow. Rarely can drilling programs directly assess the hydraulic conductivity of the fault zone: even when using angle core drilling techniques to evaluate the fault plane. If head measurements next to the fault zone show higher heads on the upgradient side, the fault is probably acting in some fashion, as a barrier to ground-water flow.

6.1.3 Hydraulic Conductivity

In geologic environments where fractured bedrock is covered by thick, overburdened deposits, only rough estimates and comparisons of hydraulic conductivity are usually possible. However, these comparisons should be based on the results of in-field testing and core drilling. In those environments where the bedrock can be examined directly, for example, in quarries, outcrops and road cuts, much better assessment is possible. One rock mass may be more permeable than another because the fractures are more open and more numerous, as a result of its being more affected by tectonic movements.

The hydraulic conductivity of the fracture system of the rock mass as a whole is almost always of more interest than the ability of an individual fracture to transmit water,

Figure 6-5 Example Fractured Rock Design Set

for the typical scale of a facility assessment. The hydraulic conductivity cannot be estimated, of course, unless the mass of rock is sufficiently large. The hydraulic conductivity of the mass as a whole depends on the collective hydraulic conductivity of each of the fractures of an interconnecting system: in other words, on the extent to which they are open, on the number of fractures per unit of volume and, especially, on the degree of inter-communication between fractures. An opinion on the relative fracture orientations should be based on direct field observations of the relative fracture frequency and directional components. When fractures are numerous and well interconnected, traditional pumping tests with observation piezometers can provide approximate values of hydraulic conductivity and directional flow components. As fractures become less numerous or have relatively smaller fracture aperture, traditional packer permeability methods become the method of choice for defining hydraulic conductivity of a rock mass.

A generalized approach for fractured rock assessments is presented in Figure 6-4 and illustrates the components of such investigations. The basis for developing potential migration pathways to define many of the above fracture systems, lie in the conduct of well-structured field investigation programs: including application of specific geologic mapping and geophysical techniques. The information gathered during the various surveys are then employed to supplement and direct the boring, piezometer, and finally, the monitoring well installation. This scale of the fracturing can affect the characterization techniques of individual quantification or mass averaging technique, (Raven et. al., 1985). Detailed hydrogeological investigations of large structural discontinuities, such as fracture zones, faults, and shear zones (see Figure 6-5), may require an individual discrete approach. However, most typically fractured systems, where smaller fractures or joints predominate, would be impractical to attempt individual fracture assessments. These bedrock environments would require averaging methods or the continuum approach.

6.2 PHASE I INVESTIGATION - FRACTURED ROCK

The Phase I desk-based study specifically for fractured rock would follow a similar scope as the Phase I study for a granular environment described in chapter 2.0. However, the investigator should concentrate on completely documenting, through literature and field reconnaissance, of structural geology, local quarries, rock outcrops, and existing bedrock well performance information. Fracture characterizations often begin with reviews of large-scale structural features using air-photo analysis. Fracture trace analysis provides readily gathered information on the occurrence and directional component of fractures. Since the probability of striking higher well yields increases in areas with zones of fractures, field assessments should concentrate on location of these surface expressions of these fracture traces. Figure 6-6 shows the relationship between fracture traces and zones of fractures. Recent publications (NWWA 1988) provide guidelines as to selection of drill sites based on examining aerial

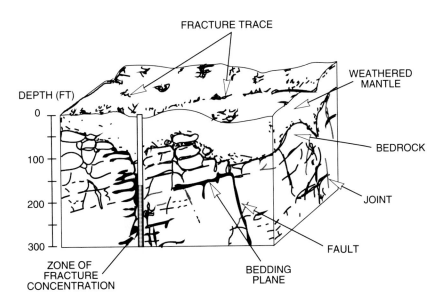

Figure 6-6 Fracture Trace Surface Expressions

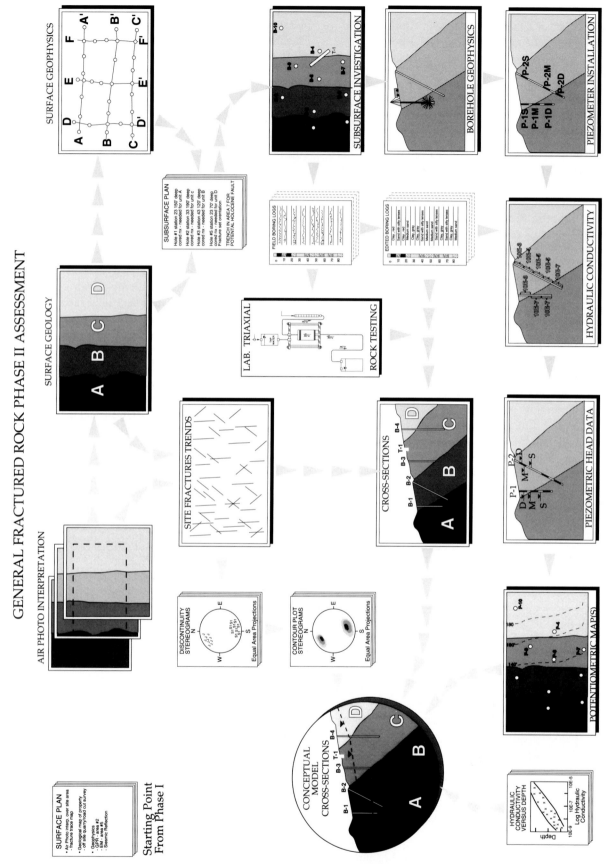

Figure 6-7 General Fractured Rock Phase II Investigation

photographs stereo scopically for surficial expressions of lineal zones of subsurface fractures. Conclusions as to zones of higher yields can be expected in these sometimes faint topographic expressions; because, the less resistant "fracture zones" are more prone to weathering, concentrate flow in interconnected more numerous fracture sets, and provide better connectivity to the screens of piezometers and wells. The success of an assessment can often be related to determination of locations of individual fracture zones and the conduct of representation hydraulic testing and sampling of these features in the Phase II field investigations.

Fracture trace analysis also serves an important function in greenfield siting studies where potential Holocene faulting is important to both locate and, if at all possible, age date. Fracture trace analysis is used in this function to locate trenching areas for detailed assessment actions.

Discussions with well drillers and other locally knowledgeable individuals on bedrock aquifer drilling targets that often providing zones of high hydraulic conductivity or yield, can often give important insight into bedrock ground-water directional flow paths. Discussions with local knowledgeable individuals are a critical aspect of Phase I studies.

6.3 PHASE II INVESTIGATION FRACTURED ROCK

Fractured rock investigations have many similar components or tasks that will have variable levels of intensity or technical effort dependent on the specific site conditions. The components can be divided into ten tasks within a typical fractured rock Phase II scope of work (see Figure 6-7) including: geologic mapping, surface geophysics, drilling and logging core holes, down-the-hole geophysics, packer permeability or pump tests, construction of single multi-level piezometers, laboratory tests, completion of geologic / hydrogeologic model selection, and installation of monitoring wells and report the results of the investigation. For complex or significant assessment monitoring investigations, additional tasks may include tracer tests, isotope chemistry, fracture deformation tests, in-situ stress measurements, and mathematical flow models to define the basic mass flow system of the bedrock.

The main goal of the Phase II fractured rock investigation for the design of a detection monitoring system is the placement of monitoring wells in the most likely or rapid flow path from the facility of interest (see Figure 6-8 for example of local discharge from fracture

Figure 6-8   Local Discharge From Fracture Set

sets)). In many cases, the monitoring wells would be used only to confirm the data gathered with small diameter piezometers installed during the Phase II field investigation. However, the monitoring wells located by Phase II data assessment provide the "official" water quality results upon which many technical and economic decisions will be based. Assessment monitoring techniques in fractured rocks also would rely heavily on piezometers to define flow conditions present at site. These flow systems would be later assessed for verifying monitoring well installations to confirm in-place water quality. The assessment monitoring Phase II scope of work also includes full characterization of the aquifer properties (or confirming the nature) of the rock mass. These data may be used later in design of ground-water remediation programs.

To summarize, the combined geology / hydrogeology / geophysical data gathered in phase II investigations define those zones to be monitored by the official monitoring wells for both assessment and detection monitoring programs.

6.4 GEOLOGIC MAPPING

The investigator should conduct a geologic mapping program on the surficial material present in and around the project area. If sufficiently detailed maps are available for the assessment, this task may not be required under the Phase II scope. The primary purpose of the mapping is to aid in selection of locations of piezometers for head level measurement and later location of monitoring wells. These initial surficial geologic maps developed early in the Phase II process are normally expected to be extended or refined on the basis of the later drilling and field analysis program.

Fractures are characterized on the basis of information obtained from mapping of surface outcrops, excavations, quarries, tunnels, mines, adits, and core drilling, coring logging, and testing, both in the laboratory and the in the field. Among the rock mass discontinuity parameters most commonly recorded during geological mapping, are the locations, orientation, size, opening, and nature of infilling of individual fractures. Additionally, parameters describing the spatial relationships between fractures such as spacing and densities are important to fully document. Surface fracture mapping typically determines the orientation trends and fracture spacing. Only estimates of fracture hydraulic conductivity anisotropy is obtainable during surface geological mapping. Fracture characterization methodology is relatively straight forward, but can be subject to a number of biases. Terzaghi, 1972 described the relative biases due to orientation of fractures.

For example, blasting induced fracture biases can occur when conducting surface mapping of road cut and quarry exposures. These induced fractures can be found as not equivalent in form and orientation to pre-blasting fracture patterns. A typical field fracture mapping exercise would include the layout and orientation of a grid of scan lines on the rock surface. Fractures would be measured for orientation, trace length and spacing. Particular care must be taken to define fracture width and fillings. CANMET (1977) can be used to correct for biases in the data collection. The data are then given either a geometrical representation in the form of schematic diagrams, fracture maps, depth profiles, parametric stereograms and polar or cyclographic stereograms, or a statistical representation in the form of histograms and statistical stereograms (also known as rose diagrams or rosettes).

Several types of geologic environments require that rock discontinuities must be described in more detail than is generally the case in general geological mapping. These environments include all types of fractured rock, and in some cases may include, low hydraulic conductivity confining units with unexpected secondary porosity fracture systems. This discontinuity or fracture-set mapping procedure is described in the following subsections. As a first step in the geologic mapping of a site is understanding the basic rock types and structure of the area. The Phase I investigation typically goes a long way toward defining the local and regional geology; this then forms the jumping off point for detailed site geologic and discontinuity mapping.

6.4.1 Direct Observations Used in Mapping

Direct mapping of road cuts, excavations, outcrops, and drill cores provides the most reliable and easy to gather data preferred in site assessment for waste disposal facilities. Indirect observations by air photography, geophysical borehole logging and surface geophysical explorations provide important data and help to map areas with difficult access: such observations also allow for faster area wide surveys. Indirect methods provide important data on regional trends and help to build the conceptual model of the rock mass. These data also provide the key criteria upon which the overall assessment approach is selected to direct the field evaluation; whether the discrete or continuum approach will be most appropriate for later hydraulic conductivity testing. At times both approaches may be required to fully evaluate the rock mass properties.

Rock Types: The basis for the general geologic mapping of a facility is directed toward the classification and locations of various rock types present on, and adjacent to the site. Rock types are classified according to geological origin into three major groups:

sedimentary   e.g.,   sandstone
metamorphic   e.g.   slate
igneous   e.g.   granite

Weakness planes such as bedding and schistosity on the scale of a hand specimen can be suggested by the rock type name. Further classification into rock types is based mainly on mineral composition, grain size and texture. The level of weathering of the rock mass can be very important to properly quantify during surface geologic mapping. Weathering of the rock mass can significantly increase (or decrease) the hydraulic conductivity. The geologist responsible for the mapping of the geologic conditions should be sufficiently experienced to evaluate the site for definite expectations of rock type distribution and fabric. For geotechnical characterization, the rock types are not always specific enough and some determinations of hardness, point load, or compressive strength should be considered for further quantification of the rock mass strength. Hydrogeologic investigations are interested in zones of highly fractured permeable, or weathered rock that can enhance rates and volumes of ground-water flow.

Several types of lithological mapping are possible in bedrock environments. The first system represents the traditional engineering description where the primary logging work is directed toward descriptions of discontinuities (i.e., fractures). Alternatively the traditional engineering descriptions should be supplemented by lithologic descriptions of geologic units that hold important keys to potential ground-water pathways. These logging techniques use descriptions of the form of deposition (Faces Codes) as a key into resolving the third dimension in the subsurface environment. Lithostratigraphic logging using these Faces Codes can provide the keys to sorting out site heterogeneity due to primary depositional environments and facies changes.

6.4.2 Natural Outcrops

Natural rock outcrops and roadcuts provide the most cost effective source of information and should be fully documented on area maps and field forms in the vicinity of the site. In addition, the local surface topography, creeks and springs must be fully documented because later excavations, building activities and disposal at the site may cover the original ground surface. Historic airphoto analysis is extremely valuable for definition of pre-existing

Figure 6-9 Spring Produced From Fracture Set

Table 6-1 General Fracture Key Parameters

Location : For sites with sufficient exposure the traverse line system allows the accurate position of a minor discontinuity specified by the distance from the origin of the traverse line. The origin and direction of the traverse line are typically given in the site coordinate system.

Orientation : Discontinuity orientation is described by dip vector, e.g., 180^0 (dip direction) 60^0 (dip angle) = 180/60, or by strike and dip. Analytical results should be presented in reports as mean and standard deviation of orientation.

Length : The length or size of a minor discontinuity indicates the portion of a rock mass where rock mass material continuity is interrupted by features such as fractures. Limited exposure makes the full determination of the exact length in the field difficult. In many cases it can be judged from geological experience, e.g., bedding joints that are 100% continuous in road cut or excavation walls or by estimating the length from observed fracture trace exposures through air photo analysis.

Spacing : Normally measured either as a true spacing (perpendicular to fracture orientation), or an apparent spacing (along a specific direction). Spacing is used in defining fracture domains and may be further used in rock mechanics failure analysis. The result of spacing measurements is a frequency diagram of the fracture site.

Fillings and Fracture Surfaces: During field mapping, fillings are characterized by mineral or geologic name and thickness. If potential enhanced flow or removal of the fillings is recognized, special sampling and testing should be conducted.

Ground water: Ground-water occurrence in minor discontinuities can also have an important effect on ground-water flow and resultant head conditions in the rock mass. It is recommended that a similar description process of ground-water occurrence is used for minor discontinuities as described above with major discontinuities.

Aperture: Aperture is the perpendicular distance between walls of an open discontinuity. Aperture width may be important, on occasion, to record in connection with potential increases in hydraulic conductivity and rock mass deformation parameters. This rock character is not typically measured on a routine basis for projects.

and developing site conditions. These photos can even be used to provide topographic contour maps of the pre-existing site area, fracture traces and the development of historic disposal areas. Knowledge of preconstruction topography is essential because pre-existing water courses, even when surface flow has been diverted, can still concentrate water seepage areas along specific pathways.

6.4.3 Trenches

Sites located in areas with reasonably thin overburden, and where natural outcrops are practically nonexistent, may require artificial exposures excavated to obtain any chance for surface mapping. Trenches can be excavated through shallow overburden or weathered materials to define near surface conditions. As a general rule once opened, trenches should be mapped in detail. Trenches can be dug by bulldozers with a ripper blade or by a backhoe. Light blasting may sometimes be necessary in areas of tough rock. Trenches can be practical, in some cases to a depth of 25 ft. (8 m), with proper shoring of the excavation to insure worker safety. Depths of overburden should be known before trenching from previous drilling or geophysical work. To obtain the maximal benefits from trenches, they should be perpendicular to the strike of lithological sequences, alteration zones, or major structural discontinuities. Refer to Section 3.0 for additional procedures related to trenching of unconsolidated materials. Special care must be exercised when entering trenched holes near waste disposal areas. Migrating methane, carbon dioxide or other dangerous gases may collect in the trench bottoms and endanger worker safety.

Information about fracture orientation is obtained from surface or subsurface outcrops by means of a Brunton (or similar design) compass. If magnetic interference due to the presence of metallic equipment or hematite and magnetite is a problem, one can use a direct-reading azimuth protractor (DRAP) of the kind designed by the U.S. Bureau of Mines.

6.4.4 Fracture-Set Mapping

Geologic maps as a general tool provides the rock type boundaries and major discontinuities as a basis for more detailed fracture-set mapping. At outcrop locations the prevailing fracture sets are documented, showing orientation, spacing, extent, and other fracture details such as aperture. In such geologic mapping efforts the geologist is often required to make judgments on the level of features to map, since it would be technically infeasible to map all minor fractures or discontinuities, even on a well-exposed site. The geologist should evaluate the fracture areas or domains as follows:

- Before detailed line mapping is begun, a reconnaissance geologic survey should be completed to define site geology and structure as the basis for discontinuity mapping. An important definition of fracture set mapping refers to the relative density or homogeneity of the fracture sets; called structural domains. Areas or structural domains are typically delineated where minor discontinuities are statistically homogeneous. Statistical homogeneity exists in an area if a number of joints can be "taken" from a subarea and "exchanged" for those of another subarea without resulting in a significant change in the rock mass properties.

- These areas or domains are then documented for discontinuity orientation, length, spacing, fracture filling, aperture and ground-water conditions.

One of the most important characteristic of discontinuities in waste disposal site assessments is orientation of the fractures. Domains are therefore recommended to be predominantly defined on the basis of discontinuity orientation (see Table 6-1). However, in the strict sense of fracture set mapping, domains or areas of statistical homogeneity can be defined for any characteristic, spacing, fracture filling, ground-water fracture flow conditions, etc.

Major Discontinuities

A major discontinuity can be defined as a geologic feature that is unique and mappable on the scale used for the project. Typically this would be faults, shear zones or major fracture sets, unconformities, geologic boundaries or contacts. Major discontinuities should be documented by the properties described below.

Location: Since accurate locations of discontinuities are important for the evaluation of recharge potential of the fractured system, surveyed traverse line system should be laid-out for mapping, and the position of a discontinuity is then specified by the distance from the origin of the traverse line. Major discontinuities are typically plotted directly on geologic maps or scale controlled air photos. The physical location of the discontinuity is also documented in field log books.

Orientation: Discontinuities are mapped as idealized planes and their orientations are described by strike and dip or by their dip vectors.

Fillings and Fracture Surfaces: Fracture filling can control the rate and direction of ground-water flow by their presence or absence. In addition, fracture fillings can be affected by waste leachates that may have the potential to remove portions of the fracture blockage. Fillings may be unconsolidated materials such as clay, sand, silt, altered rock, or soluble substances such as calcite, rock salt, and gypsum. During field mapping, fillings are characterized by a descriptive name and observed thickness. If the fracture filling may be potentially soluble in the waste leachates, one should consider special sampling and testing.

Ground Water: Geological discontinuities, if open to almost any extent can allow ground-water flow, or may form important barriers to flow if filled with low permeability fillings or gouge. Fundamental changes in flow direction can occur if open water-bearing fractures build higher upgradient pore pressures and result in a change the rock mass ground-water flow. For mapping purposes a six-point scale of ground-water flow from a discontinuity, based on CANMET (1977) ranging from apparently tight to free-flowing, is given below:

- The discontinuity is visually tight water flow along it does not appear possible.

- The discontinuity is dry with no evidence of water flow.

- The discontinuity is dry with no evidence of past water flow, e.g., iron staining of discontinuity surface.

- The discontinuity is damp but no free water present.

- The discontinuity shows seepage, occasional drops of water, but no continuous flow.

- The discontinuity shows a continuous flow of water (estimate gallons or liters/minute and describe pressure, i.e., low, medium, high).

The majority of flow from unweathered fractured rock is through open fractures. Often springs and weep lines can vary in flow significantly from season to season and year to year. Careful inspection of discontinuities for historic ground-water flow can form an important part of the geologic mapping of the rock mass. If at all possible, high ground-water level seasonal information can provide significantly better insight into shallow ground-water movement in fractured rocks. Figure 6-7 shows an active area of discharge form a source area some 2 Km away. The spring discharge increased many orders of magnitude due to the recharge area of the fractures receiving impoundment surface water.

Minor Discontinuities

Field properties of minor discontinuities should also be evaluated carefully to decide:

• If the rock mass minor discontinuity property is essential;
• Has some value to ground-water flow determination;
• Has specific limited value; or,
• Is not relevant to the assessment of a particular site.

Omissions from the listed properties should, however, be made only after careful consideration of the effects of the field properties on ground-water flow conditions.

Design Fracture Sets

Documentation of the above major and minor discontinuities for field attributes, provide several alternative ways to use the data developed during the field mapping work. Geologic fracture sets, or groups of fractures sharing similar attributes and are formed under similar historic stress and strain conditions. Stereograms of conjugate joint sets and stress ellipse analysis provides the basis for defining the direction of maximum principal stresses or the compressional stresses historically acting on the rock mass. The concept of the "design fracture set" is introduced as an extension to the conceptual model. The design fracture set is the result of the historical stresses acting on the rock mass. In addition, design fracture sets are groups of fractures possessing a specific important attribute, as shown on Figure 6-10, usable for the evaluation of hydrogeologic or engineering problems. Normally the classification of geologic fracture sets is made primarily on the basis of orientation, whereas design fracture sets are normally identified on the basis of a key parameter or aspect of fracture character/attributes. Orientation data are generally presented with the use of stereographic projections; a method to display three dimensional orientations of fractures in a projection of two dimensions. The following subsection describes construction of stereographic projections. Design fracture sets can also be described as stereographic projections and as with other visual presentations described below

6.5 SURFACE GEOPHYSICS

Geophysics can often be employed during the hydrogeological investigation to solve specific problems supplement surface or subsurface geologic data or merely to provide information cheaply and rapidly.

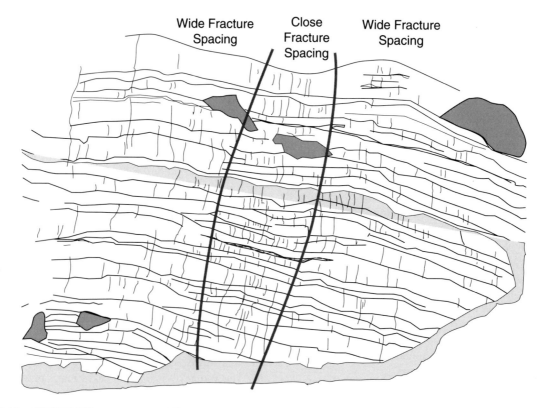

Figure 6-10 Design Fracture Set

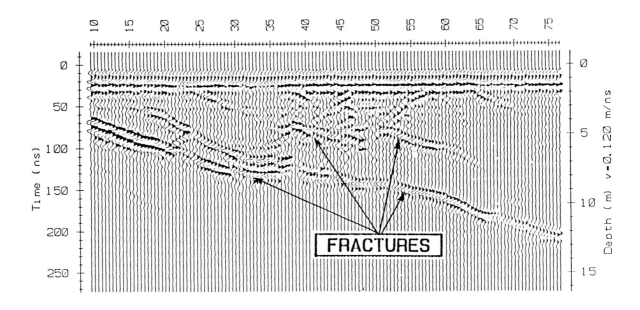

Figure 6-11 GPR Trace Showing Fracture Set

On the basis of surface geophysical surveys, depth to bedrock, water tables, and anomalous geologic conditions can be assessed for later drilling. Seismic surveys also provide useful information in fracture surveys. The seismic survey should tie into existing borings for control of depth and for correlation of lithologies with velocities. Seismic velocities can also be used for later ripability analyses. Geophysics for large scale bedrock structural features often employ electromagnetic methods; EM 31 and EM 34 are commonly used, however, EM 16R is most widely used for fracture assessments (Novakowski, 1991). Additional geophysical methods for fractured rock environments also include galvanic resistivity, radar, magnetic and other remote sensing methods. In the proper geologic environment even techniques such as ground penetrating radar (GPR) can be used to evaluate fractures. Figure 6-11 illustrates subhorizontal fracture sets that could serve as near surface pathways for ground-water flow.

6.6 BORING PLAN

An extremely important predrilling task is to plan the lay out the location of each borehole to be drilled within the Phase II scope of work. The knowledge gained from the regional Phase I review, fracture trace analysis, surface geology, and geophysics must be used to select the location for each of the borings proposed during the phase II project. Exact borehole locations are extremely important for fractured rock investigations. The mislocation of a borehole by drilling on the surface expression of a dipping fracture can result in a piezometer or well that has very poor hydraulic-head level recovery. Poor well performance due to sparse fracture intercepts can confuse and complicate site evaluations. Surface bedrock exposure of fractures can add the third dimension to fracture trace air photo evaluations. Borehole field drilling sites must accommodate dipping fracture orientations so that the fracture sets are intercepted below the ground-water surface. Each borehole should be described in a brief borehole plan as to location and reason for selecting the particular drilling site. This plan should be short (three to four pages) of tabular material providing, (among other data), proposed borehole number, drilling/sampling method, estimated depth, down-hole geophysical method proposed, piezometer numbers/types and finally, rational for the location of the boring. The goal of this effort should be to provide a reasoned approach for every penetration into the subsurface.

This effort is especially important for sites where DNAPL product spills are present, or suspected, so that proper precautions can be taken to reduce the potential for breaching local confining units or inadvertent interconnection of fracture sets.

6.7 DRILLING AND LOGGING CORE HOLES

The objective of the field drilling program is to characterize the bedrock beneath the site to depths sufficient to establish ground-water flow from the facility, to take samples and log observed lithology, to install piezometers and in later field drilling work, target locations for monitoring wells. Borehole data is also used to trace regional features and locate fractures at depth.

Recognizing that core drilling provides the greatest amount of information in a fractured rock environment, these methods will be discussed first. Core drilling in fractured rock provides the investigator with a visual basis in which to collect important site data. Diamond core drilling is the primary tool to determine the orientation of major discontinuities not mappable from the surface. In general, at least three core holes are required in any fracture bedrock investigation. In practice this minimum number is probably rarely used since site conditions often require significantly more borings to evaluate major discontinuities. This method is only suitable to detect major, planar, continuous discontinuities. The collection of core data is significantly enhanced by observing the orientation of discontinuities in the core. To evaluate fracture orientation data from borehole cores where natural reference planes (such as regular foliation or bedding) are

unavailable the investigators must first obtain a reference that will be parallel to the core axis. A reference line may be obtained for cored holes in a variety of ways including core scribing, core indenting, taking an impression of the core stub and paint marks on the core (Goodman, 1976). To determine the orientation of discontinuities in a single borehole, the core obtained in the drilling program must be oriented in some manner. This may be difficult in practice, but the following means are available:

- By the use of naturally occurring structural features or lithologic markers such as bedding or schistosity which are of known orientation for the site;

- By the use of marking devices in the core barrel such as Christensen-Huegel or BHP Systems;

- By devices obtaining imprints of the bottom or sides of the borehole, as in the Craelius method, clay pot impression, impression packers;

- By downhole viewing with T.V. cameras, acoustic televiewers or periscopes; and,

- By the integral sampling method, where the recovered core has been reinforced with a bar of known orientation.

All of these methods require an inclined hole for the marking techniques to allow determination of the lower side of the hole. This is where the reference line is drawn. One of the most effective techniques for core obtained from inclined holes uses a simple core indentor (Figure 6-12). The indentor device is lowered on a wireline to the bottom of the hole at the end of each core run. When the hole bottom is reached, the indentor is raised several feet then released. This allows the indentor to slide along the lower side of the borehole and indent the lower side of the face of the core stub. As each core run is removed from the hole the following procedure is used (Gale 1984) for split triple tube core drilling.

1. The split triple-tube is opened and a preliminary reference line is drawn on the core parallel to the PVC pipe using a felt pen. This reference line permanently shows the relative position of each core section as it came from the barrel. A split PVC pipe is then placed on top of the core and securely taped to the lower split tube.

2. The core is then rotated so that it sits in the split PVC pipe and the split triple-tube is removed.

3. The top piece of core is rotated in the pipe so that the

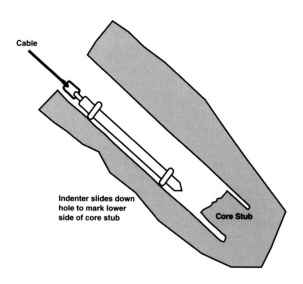

Cable

Indenter slides down hole to mark lower side of core stub

Core Stub

Figure 6-12 Core Indentor

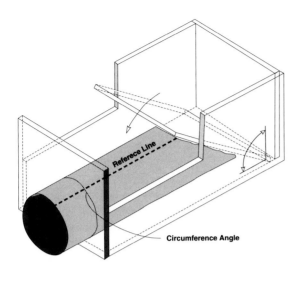

Figure 6-13a Simple Goniometer

indentation on the core is adjacent to the side of the pipe. Each subsequent piece of core is then rotated so that the surfaces of each core section mate properly.

4. Once the core is considered to be properly reconstructed a final reference line (representing the lower side of the hole) is drawn on the core. The core is now ready for determining the values of α and β for each fracture. Angles are measured with a simple goniometer as shown in Figure 6-13a. Angles are measured using a strip of mylar with 0-360^0 markings also shown in Figure 6-29. Core orientation can also be provided by borehole periscope techniques described in the following section.

Fractures are planar features whose orientation in space can be fully described by two angular measurements. As a borehole penetrates a fracture plane the resulting trace of the fracture is left on the borehole wall. If the borehole is perpendicular to the fracture the core surface will be circular If the borehole is not perpendicular to the fracture the trace will be elliptical on the borehole walls. Figure 6-13b shows a section of core with an elliptical trace generated by intersection of the borehole with the fracture plane (β plane). Also shown in Figure 6-13b are the major and minor axes of the elliptical trace. Measurement of angles α (angle between the core axis and major axis of ellipse) and β (angle measured clockwise around core in the B plane from reference line to the dip side of the major axis of the ellipse) can be used to describe the orientation of a fracture with respect to a reference line drawn parallel to the core axis. Knowing the bearing and plunge of the

borehole (thus the orientation of the reference line) one can determine the strike and dip of the fracture plane from α and β data using techniques described by Goodman (1976). Values of α and β may be determined from the core by direct measurement or by direct observation of the borehole walls using acoustic and televiewer borehole television or borehole periscope equipment.

The cost of core orientation ranges from about 1-1/2 to 2 times the cost of conventional diamond drilling, depending on the nature of the technique and geology. Core orientation has been used primarily in NX sized and larger cores.

Other drilling methods such as rotary wash, down the hole percussion, drilling and double tube "Becker" drillings provide little information on fractures and as such, should be used after sufficient knowledge is obtained from core drilling stratigraphic holes. To this end, the typical scope of work in this field phase includes:

1. Setting a PVC surface casing through the overburden weathered bedrock to depths sufficient to penetrate into solid rock and cement in place with an appropriate bentonite/cement grout. The overburden material should be sampled as appropriate so that sufficient information is gathered to establish the overburden lithology and evaluate ground-water head conditions.

2. Core the rock using diamond bit air rotary methods to the full required depth. Structurally log all fractures and note indications of water movement. Samples of rock

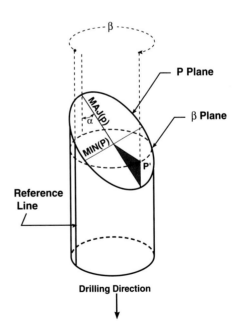

Figure 6-13b Core Section Elements

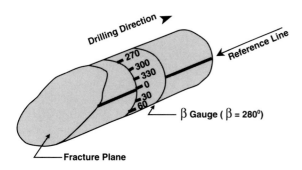

β Gauge (β = 280⁰)

Fracture Plane

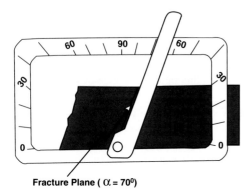

Fracture Plane (α = 70⁰)

Figure 6-14 Oriented Core Section

should be evaluated for laboratory testing.

3. The orientation of the regional fracture pattern may require that one of the core holes be drilled at some angle from vertical in a direction perpendicular to the bedding planes in the bedrock. The core log prepared by experienced investigators, using oriented core is one of the most important data sources for field investigation of fractured rock (see Figure 6-14). This text based representations of the geology will survive the actual rock cores in most cases and provide an easily handled description of the results of the core drilling and logging process. Care must be taken to fully document the important aspects of the core recovered and the drilling observation as described in Table 6-2.

4. Standard geophysical logs are useful methods for geologic correlation in the vertical holes and can include three-prong caliper, natural gamma, gamma-gamma, spontaneous potential, and resistivity. Other useful geophysical techniques could include borehole acoustic televiewer, and borehole TV. Borehole temperature logs and low-velocity flow meters are extremely useful for identifying large fractures that yield water to the borehole.

5. Aquifer characteristics of the rock mass should be obtained through either pump test procedures in larger yielding boreholes or packer hydraulic conductivity tests in the cored holes using appropriate equipment and test methods. The key to the hydraulic conductivity method used is dependent on the mass conductivity of the rock and the need for individual quantification of the secondary porosity, or defining if mass averaging techniques would be more appropriate.

6. A selection of vertical and angle (if drilled) holes should have piezometers, either single or multiple point, completed at various permeable zones in the rock. Piezometer classifications are addressed later in this chapter and the investigators should carefully select piezometer tips that can react sufficiently fast for the bedrock environment.

7. Water level recovery in open tube piezometers should be monitored to allow estimates of hydraulic conductivity of the screened interval and to establish static water-levels for the fractured rock mass.

8. Long-term pump tests may be necessary to verify directional components of hydraulic conductivity to evaluate the fracture flow system.

9. The field data must be assimilated into a composite whole bringing in each component of the air photo, geologic and fracture mapping into focus and agreement with field drilling data. Appropriate graphical, statistical and conceptual models should be prepared for describing the geologic/hydrogeologic environment.

10. After completion of the field assessment and interpretation component of the Phase II program, monitoring wells should be drilled and installed in the preselected locations based on the ground-water flow directions. Care must be taken to locate monitoring wells along fracture traces so that wells will yield sufficient water to monitoring point. Care must be taken to understand the fractured rock environment sufficiently <u>before</u> selecting locations for monitoring wells.

Each of the above tasks will be described in detail in the following chapter sub-sections. Particular attention to field assessment techniques is provided to guide the evaluation of flow conditions in fractured rocks, specifically for design of detection monitoring and assessment of facilities for potential discharges into the environment. As such, the high end of hydraulic conductivity of fractured rocks is more important in such

Table 6-2 Information Typically Recorded on Fractured Rock Logs

1. Reference information comprising the project number, title, and location shown on a small drawing with reference features; the borehole exploration number or letter designation; the borehole exploration location by survey coordinates, station and offset, or referred to permanent nearby structures; the inclination of the boring and, if inclined, the bearing or azimuth of the dip of the hole; the reference elevation, that is, the elevation from which all depth measurements are made; and the datum of the reference elevations, e.g., Mean Sea Level, etc.

2. Personnel information including the names of the drilling contractor, the driller, and the inspecting engineer or geologist.

3. Equipment data consisting of the manufacturer's name and model designation for the drill rig, motorized equipment used to excavate test pits and trenches, and seepage and pressure testing equipment.

4. Sampling and coring information that can consist of the following:

a. All coring and sampling operations document the following on field logs necessary for the project requirements: The sample type and number, the inside diameter, outside diameter, and length of the sample tube; the depths at the start and at the completion of the sampling drive or push or coring run; the length of sample or core recovered; the recovery defined as the ratio, expressed in percentage, of the length of sample or core recovered to the length of the sampling drive or push or coring run. A complete visual description should be recorded on the field log of each sample or core including color both wet and dry, type of material, density, or consistency of soil, hardness of rocks, stratification rock structure, moisture conditions, etc. The description should be made by the responsible engineer or geologist immediately following the retrieval of the sample or core so that it represents the "as retrieved" classification. This documentation becomes very important when sampling materials that tend to break down on exposure. Among these are air slaking shale and rocks containing expansive clay minerals.

b. For soil or rock coring: downward hydraulic pressure used in the drilling rate of penetration

c. For drive samplers: The weight and height of drop of the drive hammer and the number of blows required for each six inches of penetration of the sampler as per ASTM Standard Penetration Tests (SPT's).

d. For pressed or pushed samplers: The hydraulic pressure required to push the sampler into the ground and the rate of penetration.

e. For rock coring: The Rock Quality Designation (RQD) developed by Deere (1964). The RQD is the ratio, expressed as a percent, of the aggregate length of core pieces over four inches long in a run divided by the length of the run, S. Description of material penetrated but not sampled as determined from drilling or chopping action or action or changes in the color of the drill water.

f. Drilling methods that return chips to the surface allow the investigator to collect and log portions of the rock mass. These chips should be collected as returned to the surface in piles representing 1 to 3 foot samples of the rock.

6. Casing information consisting of the size of the casing; the depth at which casing was added; the length of casing added; the final depth of the bottom of the casing; the weight and height of drop of the hammer and the number of blows for each 12 inches of penetration for driven casing; and, for drilled casing, the average rotational speed and downward pressure on the casing and the average rate of penetration.

7. Packer permeability or falling head hydraulic conductivity test information comprising the depths at which tests were performed and the time required for each test. The actual test data is recorded on specific forms designed for that purpose.

8. Ground water information consisting of the depth to the water surface as first contacted and water level in the hole, recorded daily, at the start and close of work. These readings should be continued for some period of time after completion of the hole until the water level in the hole has stabilized.

9. Boreholes located in rocks and overburden masses of low hydraulic conductivity may require significant periods of time to recover to static conditions. A schedule of recovery measurements should be established for these environments. Artesian pressure information including the depths at which artesian pressures were encountered, the measured heads, and the time at which each measurement was made.

10. Elevation of the top and bottom of the hole and the top of the rock.

11. The date and time of all drilling, testing and sampling operations and delays including, but not limited to, equipment cleaning seepage and pressure testing, additionally, machine breakdown, and injuries.

12. Miscellaneous information which may aid in the interpretation of subsurface conditions. Drilling observations such as rod drops can have a significant impact on conceptual models and these data should include the depth at which drill water or air is lost or regained, the amount and color of the return water, and the depth at which a change in drilling action occurs. The latter would include the depth at which rod vibration starts or stops, the depth at which the rate of penetration or ease of penetration, changes in penetration rates, etc.

13. Any additional information which the driller, engineer, or geologist considers pertinent to the interpretation of subsurface conditions.

monitoring and assessment activities. Very low hydraulic conductivity, indurated or igneous rocks can act more like confining units to movement of leachates. The discussion will concentrate on techniques usable for establishing flow paths and hydraulic conductivity above 10^{-7} cm/sec.

6.7.1 Logging Of Subsurface Explorations

Subsurface explorations require an organized documentation scheme where the responsible geologist or engineer assigned to the drilling rig must fully describe the subsurface explorations on a series of logs. Two types of logs normally are prepared in connection with subsurface exploratory work. The first, called a field log, should be an extremely detailed chronological record of drilling and sampling operations. The second, called a final or edited log, is primarily a graphical log of the lithology upon which the pertinent subsurface information and the results of field and laboratory testing are reported. Field log data should be recorded on individually numbered log sheets that can be referenced to edited logs and piezometer/well construction diagrams. Details of both types of logs are described below.

Field Logs Of Subsurface Explorations

A legible, concise, and complete record of all significant information pertaining to the drilling and sampling operations within each borehole must be maintained concurrent with the advancement of the borehole. Typical drilling specifications require the maintenance of such records, called field logs, by the investigator and upon completion of the hole, the submittal of two copies of the log (for QA purposes) to the owner. These records are kept on individually numbered forms prepared specifically for the purpose of documenting the results of the drilling process. Quality assurance programs typically require such traceability and dual record keeping actions.

The format for field logs can vary from one organization to another but, in general, the basic field information recorded is the same. An example of one type of field log is presented in Figure 6-15 with an edited borehole example Figure 6-16. Logs such as these are used in the selection of representative samples for laboratory testing, in the preparation of geological profiles, and in the preparation of the final logs that are referenced or contained in engineering reports and contract drawings. Therefore, they must include all information necessary to completely define the subsurface profile and ground-water conditions. The log should be a complete chronological record of the drilling and sampling operations within the

hole, including delays. The recommended information to be recorded is the following:

Fracture Character

Both surface and internal characteristics of fractures should be logged These characteristics include:

- Evaluation of natural fracture and fractures induced during drilling;
- Evaluation of both open and closed fractures with a possible estimation of aperture;
- Evaluation of the form of fracture surfaces;
- Evaluation of the roughness of the fracture surfaces;
- Evaluation of any weathering present on fracture surfaces;
- Identification and description of fracture fillings; and,
- Description of the rock containing the fracture.

The most complete data is obtained from rock cores where fractures can be opened to see details of their surface and internal characteristics. The following is an outline of features to be logged by core logging techniques. More detailed information can be found in ISRM Commission on Standardization of Field and Lab Tests (1978) and Piteau (1971).

Natural fractures can usually be readily distinguished from breaks induced by drilling by inspection of the core. Induced breaks will appear fresh, generally at 90^0 to core axis and irregular in surface character. Natural fractures often will be weathered, slickensided or contain gouge.

Closed fractures are fractures which are visible in the core and have sufficient cohesion not to have been opened during drilling. Open fractures can have surfaces which mate very closely or they may have very poorly mating surfaces. One can make the observation that the poorer the degree of fracture mating, the greater the effective aperture of the fracture and the greater the opportunity for fluid flow.

Fracture surface form can be described as planar, curved (undulating) or irregular (stepped) (Figure 6-17). Fracture form distinctions assist in determining the origin and continuity of each fracture and can be used in conjunction with a fracture surface roughness classification.

The roughness classification provided below is based on that outlined by Piteau (1971). Fracture surfaces can be considered to range from rough to smooth or slickensided where "slickensided" indicates clear evidence of previous shear displacement along the fracture (Table 6-3).

Fracture surfaces may appear fresh to highly weathered depending on their geologic history, depth below ground

ROCK BOREHOLE LOG

SITE NAME AND LOCATION:	DRILLING METHOD:		BORING NO.
Proposed Landfill	Wireline NX core		**P-109C**
	Through 6" casing		SHEET
	SAMPLING METHOD: NX-Coring		**3 of 5**

		DRILLING	
		START	FINISH
WATER LEVEL		TIME	TIME
TIME		14:30	16:30
DATE		DATE	DATE
DATUM: MSL ELEVATION: 487.3	CASING DEPTH	01/18/89	01/25/89

DRILL RIG: Acker AD-2   SURFACE CONDITIONS: Gravelly field

ANGLE: vertical   BEARING:

SAMPLE HAMMER TORQUE:

DRILLING CONTR Empire Soils, Inc.
Mark Bravch

DEPTH IN FEET ELEVATION	BLOWS/6IN ON SAMPLER (RECOVERY)	CORES				SOIL DESCRIPTION OR ROCK LITHOLOGY	SYMBOL	ROCK STRUCTURE	SAMPLER / BIT	CASING TYPE	BLOWS/FOOT ON CASING	TESTS		
		RUN NO.	NO. / SIZE OF CORE PIECES	% RECOVERY	RQD							DEPTH (ft)		PERM.-CM/SEC
												FROM	TO	

		8	95	17.5		Greenish gray SHALE		A couple low angle fractures w/silickensides				80.5	90.5	No Take
90						Red SHALE, tr. grn. mottl. red Clay seams (<2") @ 91'& 93.8'		Scattered low angle fractures w/silickensides						
95		9	111	17.5		Red & grn. mottled SHALE, little gypsum		Several low angle fractures, some w/silickensides				90.5	100	8.1x10⁻⁷
100						Increased gypsum content (25-40%) gypsum appears to have grn. shale halos								
		10	100	5		Pale grn./red mottled SHALE slightly pitted, laminated w/ small nodules		Gypsum filled veins, tiny fractures				100	110.4	4.1x10⁻⁷
						Red SHALE, little gypsum, possible tiny red Clay seams from 105-106		Scattered low angle fractures w/slickensides, a couple high angle fractures w/slickensides						

— NX Size Core Barrel —

SL.00525

LOGGED BY Mike Watkins
DATE 01/25/89 CHK'D BY G.L.C.

Figure 6-15 Example Borehole Log

ROCK BOREHOLE LOG

SITE NAME AND LOCATION: Proposed Landfill	DRILLING METHOD: Wireline NX core; Through 6" casing	BORING NO. P-109C
	SAMPLING METHOD: NX-Coring	SHEET 4 of 5

WATER LEVEL					DRILLING START	DRILLING FINISH
TIME					TIME 14:30	TIME 16:30
DATE					DATE 01/18/89	DATE 01/25/89
CASING DEPTH						

DATUM: MSL ELEVATION: 487.3

DRILL RIG: Acker AD-2 SURFACE CONDITIONS: Gravelly field

ANGLE: vertical BEARING:

SAMPLE HAMMER TORQUE:

DEPTH IN FEET *ELEVATION*	BLOWS/6IN ON SAMPLER (RECOVERY)	CORES: RUN NO.	CORES: NO./SIZE OF CORE PIECES	CORES: % RECOVERY	CORES: RQD	SOIL DESCRIPTION OR ROCK LITHOLOGY	SYMBOL	ROCK STRUCTURE	SAMPLER / BIT	CASING TYPE	BLOWS/FOOT ON CASING	TESTS: DEPTH (ft) FROM	TESTS: DEPTH (ft) TO	TESTS: PERM.-CM/SEC
		10		100	5				NX Size Core Barrel					
		11		97	55	Grn. SHALE, lit. red mottl. Red/grn. mottled SHALE						100	110.4	4.1×10^{-7}
110						White-Red gypsum, some Red/grn. Shale Red SHALE & CLAY seams		Couple low angle fractures w/slickensides						
		12		100	23	Red SHALE, some Grn. mottl. scattered gypsum filled nodules and veins Green SHALE Lt. Gray SHALE, slightly pitted, lit. grn. Shale, abundunt gypsum								
115						Grn. SHALE, some gray mottling, some gypsum as nodules and vein filling Red SHALE, tr. grn. mottling		Couple high angle fractures filled w/gypsum, couple low angle fractures coated w/gypsum						
120		13		96	37	Grn. SHALE, some red & gray Shale, little gypsum Alternating layers of Red grn, gray, SHALE Grn. SHALE, tr. red & gray mottling		Abundant bedding fractures filled w/gypsum				120	130	1.43×10^{-5}
						Gray SHALE pitted at places								

DRILLING CONTR Empire Soils, Inc.

LOGGED BY Mike Watkins

SL.00526

Figure 6-16 Example Edited Borehole Log

surface and degree of ground-water movement. Iron stains and a slightly worn appearance suggests moderate to high weathering of the fracture surface.

Table 6-3 Piteau (1971)

Category	Degree of Roughness	Sketch
1	Slickensided Surface	
2	Smooth	
3	Defined Steps	
4	Small Steps	
5	Very Rough	

Fracture fillings or gouge may be caused by weathering of materials derived from breakage of the country rock, alteration products, precipitation from fluids or intrusions of igneous materials. The type, color, thickness and hardness should be included in the logging process. The gouge or filling type can be recorded with appropriate symbols for calcite, chlorite etc. Rouge thickness should be measured and described on the logging forms. A set of simple mechanical tests based on the physical properties of rock was set up by Piteau (1971) to establish classification

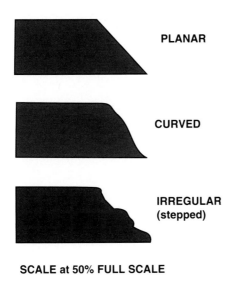

PLANAR

CURVED

IRREGULAR (stepped)

SCALE at 50% FULL SCALE

Figure 6-17 Fracture Surface Form

of their hardness. The appropriate symbols are recorded on the logging sheets. The complete range of hardness classification in the field can be established as described on Table 6-4.

6.7.2 Edited Logs

The edited log is an 'edited' reduced version of the field log, refined on the basis of a technical review of the lithology contacted and the results of laboratory tests. The edited final log will probably have a limited amount of laboratory test data incorporated into it. The final log must present a clear, concise, accurate picture of subsurface conditions, and represents the beginning of the conceptualization process. It provides the data in a form in which subsurface information is utilized by the design engineer. Final logs also are used in project reports and can be provided as a part of the contract documents or request, or for a proposal for construction of the project. Information which should be shown on final logs is illustrated by Figure 6-16 shows a convenient format for use when a boring is totally or primarily in rock. The major advantages of a separate log format for rock holes is that the lithologic description as presented in a soils log required differing description from the structure

---

Table 6-4 Hardness Classification for Rocks

S1 : Very soft - easily moulded in fingers; shows distinct heelmarks.

S2 : Soft - moulds in fingers with strong pressure; faint heelmarks

S3 : Firm - very difficult to mould in fingers; difficult to cut with a hand spade.

S4 : Stiff - cannot be moulded in fingers; cannot be cut with hand spade and requires hand-picking for excavation.

S5 : Very stiff - very tough and difficult to move with handpick; requires pneumatic spade for excavation

R1 : Very soft rock - material crumbles under firm blows with a sharp end of a geological pick and can be peeled off with a knife; it is too hard to cut a triaxial sample by hand.

R2 : Soft rock - can just be scraped and peeled with a knife; indentations 1/16 in. to 1/8 in. show in the specimen with blows of the pick point.

R3 .: Hard rock - cannot be scraped or peeled with a hammer end of a geological pick with a single firm blow.

R4 : Very hard rock - hand-held specimen breaks with hammer end of pick under more than one blow.

R5 : Very, very hard rock - specimen requires many blows with geological pick to break through intact material.

description of the rock borehole log. This allows for a more rapid visualization of the stratification and provides space for a more convenient and adequate presentation of structural features.

6.7.3 Storage Of Core Samples

Rock cores are stored in partitioned boxes of the type and size shown in Figure 6-18. The cores are boxed in the same sequence in which they were taken from the drill hole. Their arrangement in the boxes is as follows. With the core box opened so that the hinged cover is away from the viewer and the partitioned section is adjacent to him, the core is arranged in order of decreasing elevation starting at the left end of the partition nearest the hinges, proceeding to the right and continuing from left to right in succeeding partitioned areas. Core boxes are numbered in sequence with Box No. 1 containing the core of highest

elevation. The cores from each drilling run are separated from the core from adjacent runs by wooden blocks on which the depths of the beginning and end of the run are clearly and permanently marked as indicated in Figure 6-18. Blocks, marked as illustrated in the figure, also are used to indicate core loss. If the loss can be pinpointed, the block is placed at the depth of the loss; otherwise, it is placed at the end of the run in which the loss occurred. Labeling of the box is discussed in a subsequent paragraph. It is a good practice to obtain color photographs of the cores in each box as soon as practical after completion of a hole. This provides a record of the "as retrieved" condition of the cores, which permits the investigator to review the nature of the rock, as required, at subsequent times. Also, the photographs provide a record of the correct sequence of the core pieces in case the core box is spilled accidentally or cores are not returned to their proper place by persons examining them. The photographs should be taken from

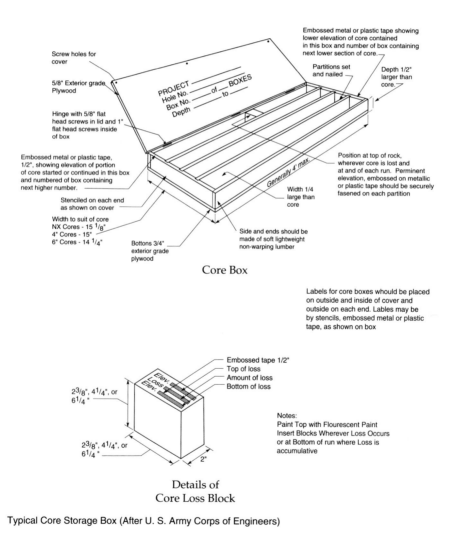

Core Box

Details of
Core Loss Block

Typical Core Storage Box (After U. S. Army Corps of Engineers)

Figure 6-18 Core Storage Box Design

directly above the box and should include the inside of the cover of the core box which contains the project name, boring number, box number, and depth covered by the box. A maximum of two core boxes, and preferably one box, should be included in each photograph. If the cores have dried out prior to taking the photographs, the cores should be wet with a light water spray or a damp cloth to accentuate the color of the cores.

6.7.4 Core Sample Labeling

Labeling of samples is an important quality control activity that should be a part of the overall site documentation system. All samples should be clearly marked so as to leave no doubt as to their exact source. In general, the information shown should include the project name; exploration identifying number, such as the boring or test pit number; sample number; top elevation of the hole or pit; depth of the sample below ground surface; description of the material; and, where applicable, the number of blows to drive the sampler. In addition, sampling tubes or liners should be marked to indicate the top of the liner and the level of the top and bottom of the sample within the liner.

All sample markings should be weather and wear-proof. Jar samples are commonly marked with gummed labels. These should be overlaid with clear tape. Similar labels are often used on sampling tubes and liners. However, it is preferable to identify such samples by painting directly on the tubes. Block samples should have an internal and external level. The internal label may be enclosed in a plastic envelope and sealed in with the last

Figure 6-20 Core Box Storage Must be Organized

layer of wax. The outer label may be pasted or stapled to the shipping box or, preferably, the information should be identified with a plastic enclosed label as shown in Figure 6-18. Note that the project name and the box number should be stenciled on both ends of the box as well as both sides of the cover. This makes stacked core box identification (Figure 6-20), more efficient without shuffling boxes.

6.8 DOWN-THE-HOLE GEOPHYSICS

Geophysical logs should be run in the available vertical holes and should assist in determining lithology and areas of weak and possibly more permeable rock. Geophysical logs are not normally run in angle holes since lowering of the instruments by gravity encounters major problems at

Figure 6-19 Example Core Box Documentation

Figure 6-22 Downhole Geophysical Probe

angles flatter than 10 degrees from the vertical. The logs normally used in fractured rock and uses of each are discussed below.

A wide range of downhole geophysical logging techniques are available for determining characteristics of soil, rock, or fluid along the length of a boring or monitoring well. These techniques provide continuous in-situ measurements that are often more representative of actual conditions than measurements made on samples

Figure 6-21 Geophysical Logging in the Field

removed from the hole. A number of logs are available; to adequately assess subsurface conditions one often must run a suite of logs. Some of these logs will provide measurements from inside plastic or steel casings and some can be used in both the unsaturated and saturated zones.

Downhole geophysical measurements can be used to identify and correlate geologic strata. Since thin layers and subtleties not readily detected in core samples can be resolved by down-hole logging, logs can significantly improve the interpolation of strata between borings. This is an important advantage of downhole techniques for use at waste sites because it provides information on the uniformity of subsurface conditions. For example, one can identify permeable zones such as sand lenses, fractures, and rock cavities that may provide pathways for contaminant migration. The continuity and integrity of confining zones or aquicludes can also be evaluated.

In addition to providing valuable information on subsurface geology, downhole logging techniques can be used to evaluate conditions within boreholes or monitoring wells and to detect contamination. For example, electromagnetic induction logs can detect contamination through polyvinyl chloride (PVC) casings. Downhole television cameras can be used to observe fractures in open boreholes and to evaluate conditions within a cased well (see Figure 6-20).

6.8.1 Applications

To determine which borehole geophysical methods would be most appropriate at a given site, the investigator needs to know the general soil and geologic conditions, the area of the site, site access, topography, and the approximate depth to ground-water surface and bedrock. The approximate background-specific conductance of ground water and the potential contaminants present at site also useful for determining the appropriate borehole geophysical techniques.

Even at sites for which limited site-specific data exists, a great deal of general information is usually available from the U.S. Geological Survey and various state agencies, which can be used to guide the investigation and develop a specific list of questions. For instance, at almost every site three questions must be answered (Benson, 1988) before remedial work can begin.

- What are the natural hydrogeologic conditions, both regional and local?

- Where does contamination exist and where is it likely to migrate?

- What are the buried hazards (i.e., wastes, tanks, utilities) to consider?

The geophysical techniques described can help provide answers to each of these questions; however, no one method will provide all the answers. When geophysical methods are improperly used or applied to the wrong problem they will be ineffective. It is extremely important that the site investigation be directed by someone with extensive field experience and a thorough understanding of the capabilities, advantages, and limitations of each technique. Table 6-5 give applications of each method (Benson 1988), but they are intended only as a guide.

Site-specific conditions will always be the dominant factor in planning a borehole geophysical investigation and, as yet, there is no simple way to include these factors in an application table. Furthermore, scheduling and budget constraints often affect the method choice. Almost all downhole geophysical techniques have application for assessing vertical variations in the subsurface. The most appropriate techniques will depend on type of hole (open, steel, PVC casing, etc.) and whether the surrounding soil is saturated. Nuclear logs can be run in open holes or holes cased with steel or PVC to obtain data on geologic strata outside the casing. Induction logs can be run on open or PVC-cased holes. A downhole television camera can be used to inspect open boreholes to locate fractures or observe the condition of subsurface materials. The major advantage of downhole geophysical techniques is that resolution of geologic units is independent of depth.

6.8.2 Logging for Strata Properties

Many aquifer properties have been quoted as being obtainable from geophysical logs (Scott-Keys 1989). These include strata geometry, resistivity, formation-resistivity factor, fluid resistivity (or conductivity), porosity, bulk density, hydraulic conductivity, specific yield, and grain size. Although the last three parameters are of the greatest interest to the hydrogeologist; no log measures these parameters directly; and only in certain strictly limited environments can estimates of these parameters be made from a combination of geophysical logs. Of the remaining parameters porosity is of great interest. This is, however, only relevant in granular aquifers where the inter-granular porosity is the major component in the hydraulic conductivity. In 'brittle' or hard-rock aquifers where fissure or secondary permeability provides the yield to boreholes and wells, the porosity of the rock matrix may be largely irrelevant. Clean, granular formations on the other hand are very common throughout the world and their porosity can be measured in a number of ways, using resistivity, gamma-gamma or neutron logs. The sonic log can also give porosity but is a very expensive tool and is rarely used in ground-water studies. Porosity can be obtained directly from nuclear logs. The gamma-gamma tool utilizes a gamma source of cobalt-60 or cesium-137 shielded from a detector. Gamma photons penetrate the borehole fluid, casing and rocks. Compton scattering takes place and therefore the gamma radiation absorbed is proportional to the electron density of the material penetrated and this is approximately proportional to the bulk density of most rocks. Properly calibrated gamma-gamma tools are designed to produce a log which when corrected can be calibrated in bulk density unit.

Corrections are required for casing type, the kind of fluid in the borehole column and for borehole rugosity. The various neutron tools are very useful in water well logging as the tool response is due mainly to hydrogen and therefore to water. Different detectors are used to measure the effects of a neutron source in the surrounding rocks, and the fluid contained therein. The resulting logs of detector count rate can be related to the American Petroleum Institute (API) units and porosity if the tool has been properly calibrated. As with gamma-gamma tools, neutron tools can be used in cased or uncased holes in air, mud or formation water. The tools are expensive and are not generally calibrated when purchased. Considerable scientific resourcefulness is therefore required by the operators if these tools are to be used to their full advantage.

TABLE 6- 5   Downhole Measurement and Sampling Techniques

Technique/application	Water-filled hole, saturated zone			Dry hole, unsaturated zone			Hole to hole
	Uncased	PVC	Steel	Uncased	PVC	Steel	hole
Induction Soil and rock identification, geologic correlation, soil and rock porosity, pore fluid conductivity	yes	yes	no	yes	yes	no	no
Resistivity Soil and rock identification, geologic correlation, soil and rock porosity, pore tluid resistivity	yes	no	no	no	no	no	yes
Natural gamma Soil and rock identification, geologic correlation and rock porosity, pore fluid resistivity	yes	yes	yes	yes	yes	yes	no
Gamma-gamma Bulk density determination	yes	yes	yes	yes	yes	yes	no
Neutron Above water table moisture content total below water table porosity	yes	yes	yes	yes	yes	yes	no
Caliper Hole-diameter log corrections, cavity and large fracture location and assessment	yes	yes[a]	yes[a]	yes	yes[a]	yes[a]	no
Single pointresistance Soil and rock identification, geologic correlation, resistive rock fracture detection	yes	no	no	no	no	no	no
Spontaneous potential (SP) Soil and rock identification, geologic correlation	yes	no	no	no	no	no	no
Temperature Fluids and flow characterization, casing leak location	yes	yes[a]	yes[a]	yes[a]	no	no	no
Fluid conductivity Fluids and flow characterization, casing leak location, water quality determination	yes	yes[a]	yes[a]	no	no	no	no
Acoustic log Soil and rock identification, soil and rock porosity, fracture detection, elastic moduli determination	yes	no	no	no	no	no	no
Seismic wave velocity Soil and rock identification, compressive strength of materials, elastic moduli determination	yes	yes	yes	yes	yes	no	yes
Television Visual image of conditions, fracture and cavity evaluation, structure inspection	yes	yes[a]	yes[a]	yes	yes[a]	yes[a]	yes
Radar Soil and rock identification, void/fracture location	yes	yes	no	yes	yes	no	yes
Scanning sonar Size and shape determination of cavities in water-filled holes	yes	no	no	no	no	no	no
Organic vapor analyzer Total organic vapor measurement in soil and water, quality and quantity identification of specific organic vapors	yes	yes[b]	yes[b]	yes	yes[b]	yes[b]	no
Packers Isolation of sections of a hole for hydrologic testing and water sampling	yes	yes[a]	yes[a]	no	no	no	no
Pump/slug tests Hydrologic properties determination	yes	yes	yes	no	no	no	yes
Magnetometer (Mag.) Location of steel casing or drilling hazards	yes	yes	yes[a]	yes	yes	yes	yes[a]
Metal detector (MD Location of drilling hazards	yes	yes	no	yes	yes	no	no
In-situ chemical sensors Selected chemical parameters measurements	yes	yes[b]	yes[b]	yes[a]	no	no	no

a = Limited applications. b = Samples through screen.   Note: The ratings are based on most common applications; exceptions exist.

Source: Benson 1988

6.8.3 Fluid logging and hydraulic properties

Geophysical logs that measure fluid parameters in boreholes filled with formation water provide data on hydraulic properties of the borehole column and adjacent strata. The most important use of these tools is the detection of levels of fluid movement into and out of the borehole and the determination of the magnitude and direction of downhole fluid movement. The main tools for these investigations include: Fluid temperature; Differential fluid temperature; Fluid resistivity; Differential fluid resistivity; Flowmeters, both electronic and electro-mechanical; Tracer injector; Fluid sample collectors; Closed circuit television-axial and radial; Neutron with various detectors; and Caliper, single and multi.

Zones of water movement into and out of a borehole column are characterized by anomalies in the normal fluid resistivity and fluid temperature profiles. Using differential fluid temperature and fluid resistivity logs very small zones of water movement into and out of the borehole can be detected. Fluid resistivity or conductivity is normally measured by conductivity cells incorporated into the probe with thermistors which measure temperature.

Vertical fluid movement in a borehole column can be caused naturally by artesian flow, interchange of water between aquifers in the same borehole due to differing piezometric head, and by pumping. The presence of such flow indicated by fluid logging should be verified by flowmeter. Inexpensive mechanical impeller flow meters are most useful where flow velocities are high or moderate. However, below velocities of about 3 cm/sec, thermal flow meters are now available which have no moving parts and

Figure 6-23   CCTV Probe

depend on the response of a transducer to the flow of water. The heat pulse flow meter is one example of this type. Very low velocities can be measured and therefore these tools are useful in both pumped and unpumped boreholes. Such meters are more expensive than mechanical flow meters and like many other geophysical tools need to be accurately calibrated to be of maximum use.

Normally the flow meter is used by stratigraphy positioning the tool at various levels in the borehole. Impeller rotations at each point are converted to electrical pulses and are counted and plotted as fluid velocity. Increased sensitivity can be achieved if a continuous impeller flow meter log can be run in the borehole. This involves running the meter through the borehole at a constant line speed. The pulse rate output is used by a standard rate meter similar to that used for, a gamma ray log to provide a continuous log. The tool is run both up and down the borehole at several different rates to obtain variable hole indices. Caliper logs must be run in conjunction with flow meters, since variations in hole diameter can affect velocities. Flow meters can be run both unstressed, as described above, or stressed where the flow meter is first lowered in a borehole and a small pump is installed above the tool. While pumping the well the flow meter is lowered and raised at various rates. The boundaries of zones of differing fluid velocities show up as deflections on the log while levels of inflow frequently produce major peaks.

Alternative methods for investigating low flows include the following of an injected saline tracer with a fluid resistivity tool, the following of visible tracers with closed-circuit underwater television equipment. Lightweight streamers can also be used on probes within the field of view of underwater television cameras to detect flow. To summarize, in fissured and granular aquifers geophysical fluid column logging techniques are currently being used to provide information on the following:

- The number of aquifers in a given well or area;

- The frequency of aquifers in a vertical section;

- The exact location of productive fissures and joints where these provide the main hydraulic conductivity features;

- The thickness of productive zones;

- Whether a particular aquifer unit is giving water to, or taking water from, the borehole;

- The lateral extent and continuity of aquifers;

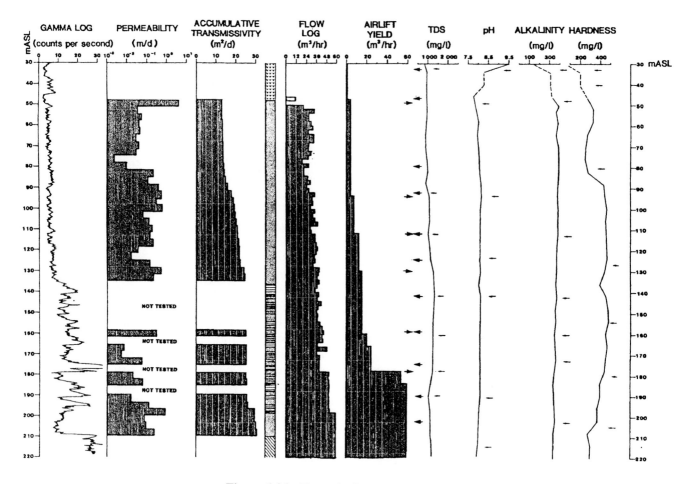

Figure 6-24   Example Composite Flow Zone Evaluation

- Water quality interpretation where mixing of water from different aquifers is taking place;

- The quantifying of flow from different aquifers in the same borehole.

The key to water quality determination from geophysical logs is the accurate measurement of fluid resistivity Rw at a known temperature. In boreholes drilled with formation water it is easy to run fluid resistivity and fluid temperature logs but the fluid in the borehole column may no longer be similar to the formation water remaining in the various aquifers. Mixing may have taken place during drilling, the fluid column may have natural water movement in it or it may have been contaminated by surface water. All these factors need assessing before evaluating the resultant geophysical logs

6.9 INVESTIGATION OF BOREHOLE CONDITION AND GEOMETRY

In production boreholes, monitoring wells, or any open borehole filled with clear water there are considerable potential uses of closed-circuit under-water television (CCTV) in site assessments. The uses of CCTV include the investigation of casing condition, such as, the location of fractures, breaks and holes, the investigation of corrosion or bacterial growths and similar problems associated with borehole screens and gravel packs. In particular the clogging of screens is a common problem investigated by CCTV. The method allows the location and inspection of debris in a borehole and sidewall collapses, caving and slumping. At sites where the original constructional detail of a borehole are unknown CCTV can be used to locate casing, diameter changes and even mining adits which were unknown or forgotten. The effect of firing explosives to increase borehole yield can be investigated as can the effects of hydrochloric acidization

Example Rose Diagram Format

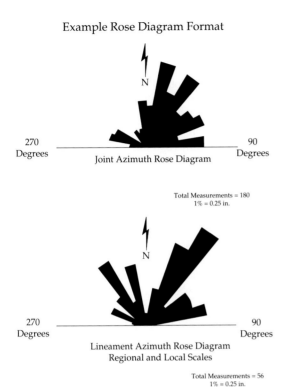

270
Degrees

Joint Azimuth Rose Diagram

90
Degrees

Total Measurements = 180
1% = 0.25 in.

270
Degrees

Lineament Azimuth Rose Diagram
Regional and Local Scales

90
Degrees

Total Measurements = 56
1% = 0.25 in.

Figure 6-25 Jcount Azimuth 'rose' Diagram

of boreholes in limestone.

CCTV is used in conjunction with fluid resistivity logs, fluid temperature logs, and caliper logs to investigate fissuring and fissure geometry in hard-rock aquifers. Vertical or high angular fissuring in the borehole wall can be evaluated with CCTV. Caliper log often miss such these fracture features and fluid logs are best suited to the investigation of horizontal fractures. Generally underwater television cameras are in cylindrical watertight housings which can be submerged to depths of up to 1000 m. (see Figure 6-23). Normal one inch or half-inch vidicon tubes are used, allowing construction of cameras with diameters of less than 50 mm Standard vidicon cameras with a 50 to 150 W quartz-halogen light source are generally used. In murky water a large increase in lighting power rarely improves definition and even in clear water opalescence caused by even a low density of suspended particles can cause problems. One must confirm turbidity for the wells since water clarity is essential for CCTV logging.

Borehole caliper logging provides a useful mechanical log and is a commonly available tool in most commercial equipment. Frequently a caliper log will point to the need for further investigation with CCTV and should be run as routine with the electrical or gamma ray logs. Furthermore it is essential for the correction of nearly every other log where quantitative interpretation is required.

Downhole measurements can be used to improve the interpolation of geologic conditions between borings or to complement surface geophysical techniques. Geophysical logging between boring (hole-to-hole techniques) increases the volume of material sampled and may eliminate the need for additional borings. Natural gamma logs are of primary value in defining lithology penetrated by the boring and allow distinction between shale and sandstone as well as the variable clay content of shale.

Spontaneous potential logs can be used to correlate with natural gamma logs to evaluate a geophysical interface between clay rich shale and sandstone by measuring the electric potential that naturally occurs at these boundaries. Resistivity logs can be used to correlate log data with the SP and natural gamma logging for evaluation of zones of interconnected, high porosity rock.

Caliper logs (three prong should be used for maximum resolution) are valuable since they allow for hole diameter corrections, and can also serve as a rough indication of rock quality; i.e., enlarged areas of the corehole usually indicate weaker rock zones washed out during coring.

Down-hole geophysics should be supplemented by observations made while drilling, since high hydraulic conductivity zones are often indicated by increased drilling rates, a loss of drilling fluid, and poor core recovery. However, the only sure way to evaluate individual fractures in a borehole is to fully examine the cores recovered in drilling or use an optical or acoustic televiewer. Some borehole geophysical logs may, in an otherwise homogeneous formation and in a borehole of constant diameter, register an anomaly in the normal sensor response due to the presence of a fracture or, more commonly, a fracture zone. The anomaly may be caused by the presence of water in the fractures or by the infilling material. The anomaly may be distinct if the fracture is widely open, and imperceptible if the fracture is filled or has a small aperture. Even a combination of geophysical logs will point only to selected fractures and will seldom indicate whether the fracture is natural or induced. While quantitative estimates (mainly of porosity) are sometimes possible, geophysical logs remain useful primarily as qualitative tools. The greatest benefit of geophysical logs is the composite presentation of the geophysical logs and borehole observations. These presentations (see Figure 6-24) provide a powerful tool for evaluation of potential flow zones that would be included in some form of well testing.

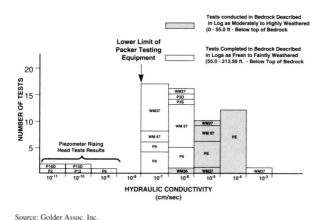

Source: Golder Assoc. Inc.

Figure 6-26 Depth/Hydraulic Conductivity Diagram

Direction Orientation Diagrams

For certain geological purposes only the direction of a planar structure is recorded. Examples include direction of cross-bed inclination. Joints visible on air photos often can be measured only according to their trend (strike direction) of lineaments; strike of joints, rather than dip, is plotted in this situation.

Lineation directions also can be plotted according to orientation in plan view. Figure 6-25 is a hypothetical *histogram* of joint azimuth orientation data. In a histogram the area of a given class is proportional to the percentage

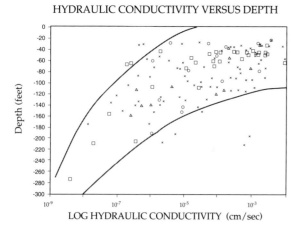

Note: Center point of test plotted

Source: Golder Assoc. Inc.

Figure 6-27 Depth/Hydraulic Conductivity Diagram

of total values in that class. For a histogram plotted on linear axes, plotted height of a class is proportional to percentage abundance in the class.

A *circular histogram* is a plot of class frequency on a compass rose. Direction is plotted on a compass rose; relative abundance is plotted with radius equal to height of class in Figure 6-26. This procedure is used by most geologists to portray distribution of directional data.

These comparisons joint and lineament azimuth rose diagrams provide both regional and local aspects to the structural trends of the area under study. The procedure of Figure 6-26 violates one concept of histogram theory. The area plotted for a class is not proportional to class abundance. Hydraulic conductivity tests can also be illustrated in histograms of depth or geologic unit relative to the hydraulic conductivity tests. These relationships are illustrated on Figure 6-27. The relative geologic unit tests and their hydraulic conductivity are distributed over the range of conductivity observed in the field. The histogram presentation allows a number of relationships to be shown concurrently. As such these presentations can be extremely helpful in describing the conceptual geology and hydraulic conductivity relationships for the area under investigation.

Hydraulic Conductivity-Depth Diagrams

It is commonly observed that there is a general decrease in hydraulic conductivity with depth in hard and consolidated bedrock (see Davis and DeWiest, 1966). Unfortunately this observation is often neglected in site assessments. This field derived observation has extreme importance in establishing zones of very low hydraulic conductivity; in addition, some basis can be established for practical confining unit depths that may limit the extent of bedrock requiring ground-water remediation. Figure 6-27 shows hydraulic conductivity versus depth ranges for a site in central Pennsylvania: 95% of the hydraulic conductivity tests are within the heavy lines shown. The tests show the hydraulic conductivity values can be expected with a 95% reliability to be below 10^{-7} cm/sec below a depth of 230 feet. On sites where reliance must be placed on some cut-off point of low hydraulic conductivity, these diagrams should be provided in the project analysis.

Structural Domains and Design Sectors

After the results of the geological investigation have been plotted and analyzed, it is necessary to carry out a check on the adequacy of the initial conceptual structural domain boundaries chosen during reconnaissance mapping. There are several different methods based on graphical or numerical techniques, beginning with the orientation of discontinuities. In most cases the plotting of mean discontinuity orientations on a survey plan will show whether the orientation of discontinuity sets:

• stay the same
• varies continuously, or whether there are;
• abrupt changes which require further division of a structural domain.

Hydrogeologic assessments are rarely affected by minimal changes of mean orientation and would have little consequence on overall design of a monitoring system or ground-water recovery system. They may become important, however, if the orientations of discontinuities influence ground-water flow sufficiently to effect risk assessment results.

Orientation of Discontinuities

Results of general geological data gathering are commonly displayed on maps and sections covering such aspects as:

• Rock type boundaries, age relationships, and bedding plane orientation;

• Major faults and shears;

• Orientation of joints; and,

• Ground-water occurrences.

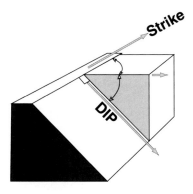

Source: Modified from Hoek and Bray (1981)

Figure 6-28   Planar Elements

Data compilations of this kind provide essential information to develop a geological conceptual model for a site. Such a model is valuable when considering the extrapolation of a geological data base to general ground-water flow directions; however, site assessments for past waste disposal require additional information to allow the location and later interpretation of Phase II field hydraulic conductivity measurements. This may require the following additional data be collected:

• structural geological maps may be required to show, in addition to major structures, the orientation of minor discontinuities on equal area nets.

• Separate listings supplied on;
  - mean and dispersion of discontinuity orientation;
  - mean and dispersion of discontinuity spacing;
  - mean and dispersion of discontinuity aperture

• Finally, fracture domain boundaries to evaluate the directional components of potential flow for later domain specific hydraulic conductivity testing.

Bullet one and two require special plotting procedures or analytical efforts, (which are explained below), with graphical or numerical examples. In some cases, alternatives are suggested which provide quick and often sufficiently accurate estimates. Bullet three is carried out in the final phase of the geological data analysis and should provide the qualitative guidance for locating boreholes for further assessment of the subsurface fracture system.

6.10 STEREOGRAPHIC PROJECTION IN SITE ASSESSMENTS

The geologic information obtained from the investigation of a site for geotechnical assessment purposes is generally recorded in the form of notes, maps, cross sections, drill logs, logs of investigation boreholes and trenches, supplemented by photographs, block diagrams, and conceptual models. The recorded quantitative data in general relate to the location, dimensions and orientation of the various geologic features observed during the field investigations. Of fundamental importance from the assessment and monitoring point of view are the attitudes of the fabric* elements. The term "fabric" is used generally in the sense defined by ISRM (1978) A fabric can be considered as a three-dimensionally ordered array of structural discontinuities. It may be noted that the terms homogeneous, heterogeneous, isotropic, and anisotropic applied to a fabric are statistically defined phenomena and therefore depend on the scale or relative dimensions of the

GRAPHICAL REPRESENTATIONS OF STRUCTURAL DATA

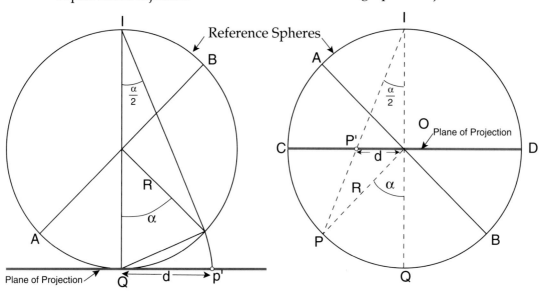

Equal-area Projection   Stereographic Projection

Equal-area Schmidt net   Stereographic Wulff net

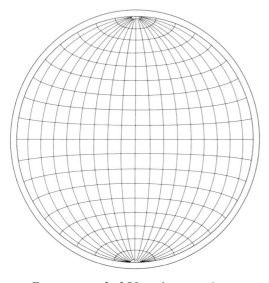
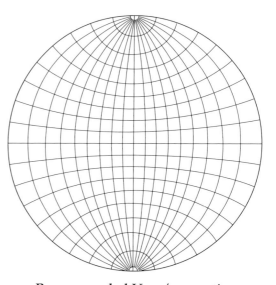

Recommended Uses/properties

Statistical studies of orentation of joints
or other structural elements.Equal-area
is representative of patterns of distribution
of datawhere a large number of readings
of poles, planes, lineations, or crystallographic directions have been assembled

Projection preserves areas but not angles.
Planes project as 4th order quadrics, lines
project as points.

Recommended Uses/properties

Solution of problems of crystallography
and structural geology, certain problems
in structural petrolog. Sometimes non-
statistical presentation of joint orentation
data is best illustrated on stereographic
projections.

Projection preserves angles but not areas.
Planes project as circles, lines project as points.

Figure 6-29   Stereographic Projection

domain under consideration. These fabric elements may provide the majority of flow in fine grained sediments (with secondary fractures) and in hard rocks with secondary porosity elements such as joints, bedding planes, fractures, shear zones and faults. These secondary porosity elements all have the geometric properties of either a plane or a line and their attitudes or orientations are independent of scale or dimension.

Traditionally in geologic investigations, the orientation of a planar element, such as a bedding surface or a joint surface, is given by reference to geographic axes, for example, "strike' which is the geographic azimuth (i.e., bearing from the North) of the line of intersection between the planar surface and a horizontal plane; and "dip' which is the angle between the horizontal and the surface measured normal to the strike. In a similar fashion, the orientation of a linear element is given in terms of the "trend', and "plunge', which is the angle between the horizontal and the line in this vertical plane (see Figure 6-28).

It is of interest that geologic "structural analysis', which is concerned especially with orientation data relating to planes and lines (fabric elements) and their intersection, including the preferred orientation of fabric elements, provides basic data that is essential for the ground-water flow analysis of a fractured consolidated rock mass.

The stereographic projection is a tool that is in common use for the solution of problems involving the orientation of planes and lines and their geometric relationships in crystal morphology and structural geology, and can be used in a similar way for ground water problems in fractured rock.

Mean Orientation of Discontinuity Sets

At most sites, bedrock joints will occur with systematic orientations due to stresses during the geologic history of the rock mass. Joints parallel to each other are called a joint system. Joint sets are defined by the mean orientation and a value of the standard deviation about this mean. To obtain mean and the standard deviation for joint orientations the data are treated as follows:

- plotting of observations on equal-area nets

- determination of joint sets (clusters of joint poles) from contouring

- determination of mean orientations in each cluster

- determination of the standard deviation about the mean

6.10.1 Stereographic Projections

Although the attitude or orientation of fabric elements can be conveniently described by points on or intersections with the surface of a reference sphere it is convenient to have means of showing this on a two dimensional paper. One of the most satisfactory methods of doing this is the stereographic projection originated by Hipparchus in second century BC.

In many instances during the analysis of complicated geological structures it is convenient to be able to represent data in a simple two-dimensional manner. The stereographic and equal area projections are powerful graphical tools of use in portraying three-dimensional orientation data in two dimensions or in graphically solving complicated three dimensional problems. The problems capable of being solved using these projections are those involving the angles between lines and planes rather than those concerned entirely with the relative positions of lines and planes in space.

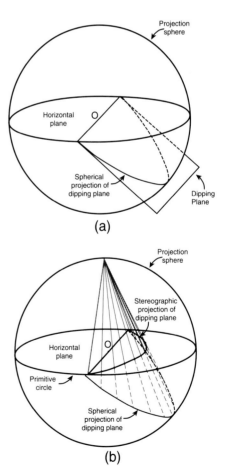

Figure 6-30 Equiangle Stereographic Projection

The Stereographic or Equiangular Projection

The essential features of the stereographic projection are illustrated in Figure 6-29. The orientation of a plane such as a bedding, cleavage, schistosity or fault surface may be represented by imagining that plane to pass through the center, O of a sphere of radius R. This sphere is known as the projection sphere. The dipping plane of interest intersects the projection sphere in a circle that has the same. radius as the sphere itself and is accordingly called a Great Circle. This great circle is the spherical projection of the dipping plane (Fig. 6-30). For each orientation of a plane there is a unique great circle

To obtain a two-dimensional projection of the dipping plane a convenient projection plane is selected through the center O of the sphere. In structural geology this is commonly the horizontal plane although in some applications it may be more convenient to select a plane with some other orientation as the projection plane

The stereographic projection differs from the equal-area projection, (which will be fully described in the following subsection) in several important respects. One of the most important qualities is that an area of a given size on the surface of the reference sphere is represented in the stereographic projection by an area of which the size increases significantly with the increasing distance from the center of the projection circle. This property of the stereographic projection makes the equal area projection (Schmidt net) more convenient for statistical investigations of the orientation of joints or other structural elements.

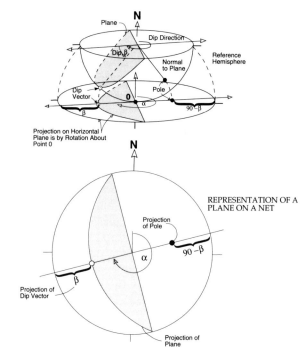

LOWER HEMISPHERE EQUAL AREA PROJECTION

Figure 6-32   Equal Area Projection

Source: Modified from CANMET (1977)

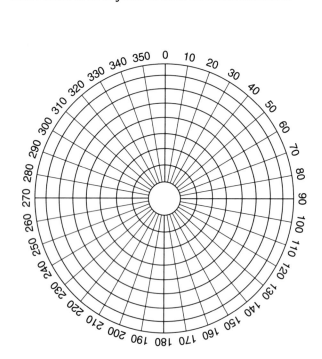

Figure 6-31   Polar Stereographic Projection

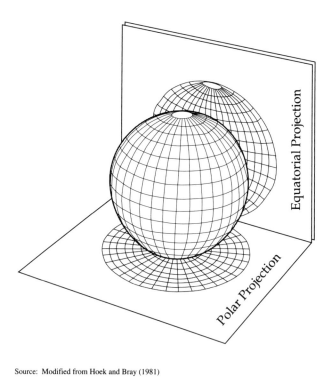

Source: Modified from Hoek and Bray (1981)

Figure 6-33   Stereographic Polar Projection

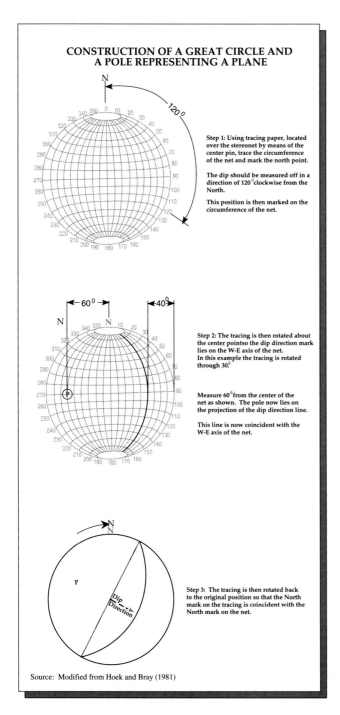

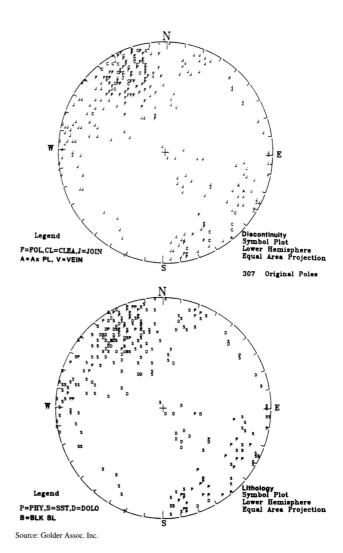

Figure 6-34 Equal Area Pole Projection

this purpose two types are in common use:

(1) Polar Stereonet. In this net the projection plane is the equatorial plane of the latitude-longitude grid and the projection point is one of the poles of the grid. The construction is illustrated in Fig. 6-31. The pole of the latitude-longitude grid opposite the projection point forms the center of the projection. An example of the polar stereonet is shown in Fig. 6-31. The meridians appear as straight lines radiating from the center of the projection and the parallels of latitude as circles.

(2) Meridional or Equatorial Stereonet. In this net a meridional plane of the latitude-longitude grid is taken as the projection plane and the projection point then lie on the equator of the grid at 90^0 from this meridional plane. An example of the meridional stereonet is shown in Figure 6-33.

6.10.2 Stereographic Nets

If we lay out, on the reference sphere, a grid of latitude and longitude coordinates and construct a stereographic projection of this grid, the resulting "net' is a very useful tool for graphic solution of many of the problems arising in connection with fabric elements and rock mechanics. For

Figure 6-35 Discontinuities Projection

Both types are used, for engineering work; they are complementary to each other. It is frequently found that the cardinal geographic points N, S. E, and W, and full circle azimuth bearings in degrees, are marked around the primitive circle.

In plotting field data these calibrations are very useful because relative to the center of the projection the stereographic projection maintains true direction. For other purposes, such as in crystallography, they are simple convenient calibrations relative to an arbitrary reference direction. Furthermore, it should be realized that the latitude-longitude grid on the stereonet has no relation to the latitudes and longitudes used as geographical coordinates on the earth's surface even though it is identical to that used for map making. In geological and engineering use it is just a convenient grid for measuring angles in three dimensions. The meridional net is often called a "Wulff" net after the Russian crystallographer who introduced its use in crystallographic work in 1802.

6.10.3 Lambert Equal Area Nets

The meridional stereographic nets have the disadvantage that areas on the surface of the reference

STEREOGRAMS OF CLEAVAGE (S$_2$) DATA

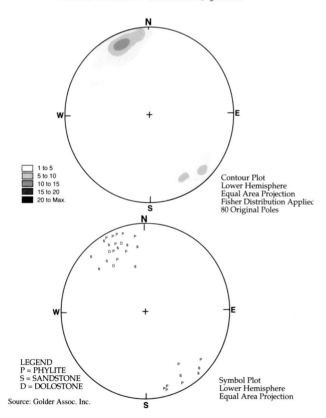

Figure 6-36   Stereogram Cleavage

STEREOGRAMS OF FOLIATION (S$_1$) DATA

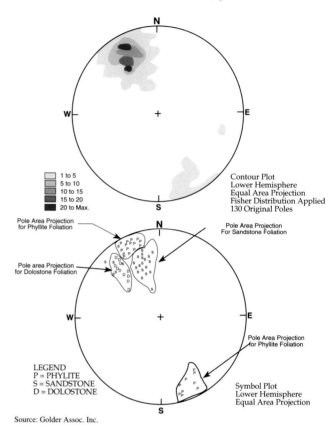

Figure 6-37   Stereogram Foliation

sphere are distorted in the projection. This is due to the fact that although the scale of the projection is the same in all directions at any given point it decreases from the primitive circle towards the center. The property of equality of area throughout a projection (as it is on the surface of the sphere) is incompatible with the property of conformability. Lambert (1728-1777) developed an equal area projection that has become the standard tool for field assessment of fractured rock. As general guidance, in any work requiring statistical assessment of the attitudes of fabric elements the use of equal area nets is essential.

This equal area net represents the projection of coordinates of a hemisphere onto a plane so that the relationship of relative areas is maintained (Fig. 6-32). This projection is available as an equatorial and as a polar equal-area net (see Figure 6-33). For plotting strike and dip measurements all discontinuities are considered to pass through the center of the hemisphere (Figure 6-32). By convention, only the lower hemisphere is used.

The orientation of a plane can be represented by a great circle or trace which is the intersection of the plane with the hemisphere (Figure 6-32). This requires counting the strike clockwise from north on the periphery. The dip is

STEREOGRAMS OF JOINT DATA

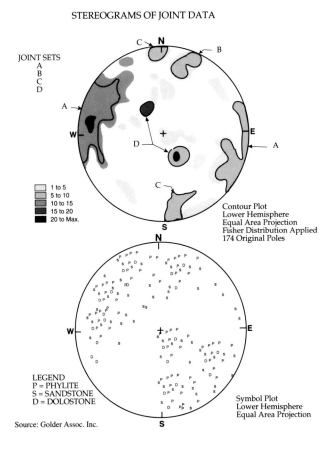

Figure 6-38   Stereogram Joints

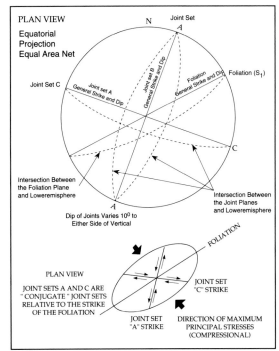

Source: Golder Assoc. Inc.

Figure 6-39 Composite Stereogram

6.10.4 Cluster Delineation and Shape

The first step in obtaining mean orientations of joint sets requires that clusters of joints can be visually delineated on the equal-area net (Fig 6-33). In cases where too many poles are plotted, the clusters might not be clearly recognizable, and the Schmidt equal area contouring method, (Fig 6-36 to 6-38) can be used to clarify and enhance the pattern. In general, the contouring

plotted at right angles to the strike, counting the angles of dip from the periphery to the center

One can achieve significant space saving for representation of many measurements by plotting of only the poles of each plane. This is obtained at the intersection of the normal to the plane with the lower hemisphere. To plot the pole, the dip angle is counted from the center of the net at right angles to the strike towards the periphery (Figures 6 -34a, 6-34b, 6-34c).

The plotting of poles is usually done on an overlay of the polar equal-area net to avoid reorientation of the overlay. Both type nets will yield the same geometric distribution of poles but if graphical manipulations are necessary, such as rotation of clusters or construction of great circles, the equatorial net has to be used. Figure 6-35 represent 307 measurements of discontinuities classified first in 6-35a as joints (J) cleavage (CL) and foliation (F) and by lithology in 6-35b as an example. Stereograms of cleavage, foliation, and joint data can be individually plotted and contoured as shown on Figures 6-36, 6-37 and 6-38. Many computer based systems are now available to automatically plot poles and construct equal area diagrams.

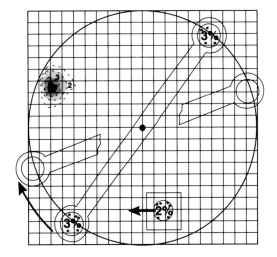

Contouring a Pole with a Point Counter

Figure 6-40 Peripheral Counter

involves superimposing a square grid on the equal-area net. A 1% circle is applied at each grid intersection and the number of poles in the circle is counted and noted on the grid intersection.

In order to produce such a diagram the poles of all observed joints are plotted in either equal-area or in stereographic projection. The resulting diagram is then contoured on the basis of a specific procedure known as counting out. A circle, known as a counting circle, typically consists of 1% of the large plotting net. The center counting circle is moved systematically from grid intersection to grid intersection while labeling each intersection with the number of points visible inside the center counter as shown in Figure 6-40. In counting out diagrams plotted in stereographic projection, special templates must be used that have increasing counting areas as one moves away from the center of the projection. Either projection method requires the use of peripheral counter (see Figure 6-40). The peripheral counter has two counting circles at opposite ends of a bar. At an initial position along the peripheral cells, the total number of points that lie in either counting circle is recorded along the circumference of the projection circle. The peripheral counter is then moved to the next area along the projection circle and the procedure repeated. This procedure continues until the points are counted in the remaining peripheral cells. The values recorded on the diagram are then used as the basis for contouring of the data.

Basic text on structural geology such as Billings (1942) and CANMET (1977) provide instructions on these procedures and a number of computer based packages are available to perform the majority of the work.

Statistical tests such as the Fisher distribution, can be used for evaluation of the significance of preferred orientations. As general guidance if clusters of joints or fractures are not clearly recognizable on the contoured plots, the orientation data should be assumed to be random or uniform.

6.10.5 Fitting a Distribution

The evaluation of a confidence range surrounding the calculated mean orientation of the discontinuity requires that some true frequency distribution function must be fitted to the observed data.

Selection of a suitable statistical distribution function depends mainly on the shape of the cluster that can vary from a point maximum over an elongated (elliptical) distribution to a girdle around the perimeter of the equal-area net. Due to the properties of the equal-area net, shapes of clusters are distorted if not plotted close to the center. Statistics are available for point maximal on a sphere (Fisher distribution) and are best dealt with by computer calculation. Figure 6-36 to 6-38 show contour plots using the Fisher distribution for Foliation (S) cleavage (S) and joint data sets.

6.10.6 Errors in Frequency Contours of Orientation Data

With the wide use of equal-area net presentations it is often taken for granted by investigators that the contoured data not only describes the existence of pole concentrations, but also shows the relative importance of individual joint sets by the number of observations shown. If many measurements have been taken over the site area, using different data collection techniques, such as surface mapping and borehole data, this is generally true. If measurements are however taken only along a limited length and single orientation core hole or a line traverse in a single direction, some joint maximal can be severely under represented. Due to the number of observations changing with direction; e.g., a strong bias can develop about horizontal fractures if sampling is done only along a

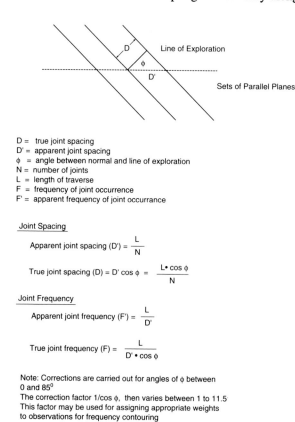

D = true joint spacing
D' = apparent joint spacing
ϕ = angle between normal and line of exploration
N = number of joints
L = length of traverse
F = frequency of joint occurrence
F' = apparent frequency of joint occurrance

Joint Spacing

Apparent joint spacing (D') = $\dfrac{L}{N}$

True joint spacing (D) = D' cos ϕ = $\dfrac{L \cdot \cos \phi}{N}$

Joint Frequency

Apparent joint frequency (F') = $\dfrac{L}{D'}$

True joint frequency (F) = $\dfrac{L}{D' \cdot \cos \phi}$

Note: Corrections are carried out for angles of ϕ between 0 and 85°
The correction factor 1/cos ϕ, then varies between 1 to 11.5
This factor may be used for assigning appropriate weights to observations for frequency contouring

Source: Modified from CANMET (1977)

Figure 6-41 Bias in Borehole Drilling Data

horizontal line traverse. In backhoe pits, or due to excavation wall irregularities there will often be sufficient three-dimensional exposure available so that joint sets can be adequately mapped. However, core hole data sampling bias can be severe, especially for vertical boreholes drilled to investigate vertical or subvertical joint or fracture sets. If there is any single investigative technique requiring rigorous attention, it is the evaluation of potential vertical fractures in both consolidated rocks, and fine grained clays

Statistical Analysis of Dip Angles

DIP OBSERVATIONS: 33, 38, 40, 40, 40,
41, 45, 47, 47, 48, 48, 48, 50, 50, 50, 51, 51, 51,
51, 51, 51, 51, 52, 52, 54, 55, 55, 55, 56, 58, 58,
58, 60, 60, 65, 65, 66 (n = 39)
Mean = 50.67
st. dev. = 7.64

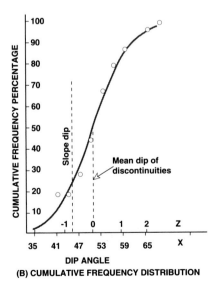

(A) FREQUENCY DISTRIBUTION

(B) CUMULATIVE FREQUENCY DISTRIBUTION

Source: Modified from CANMET

Figure 6-42 Frequency Distribution for Rock Dip Data

or glacial tills. If a comparison of relative frequency is required, then each observation is weighted by a factor C.

$$C = 1/COS\ \phi \qquad \text{Equation 6-4}$$

Where ϕ = angle between normal of a plane and direction of sampling (Figure 6-41). Cos ϕ varies between 0 to 1 and C thus varies between ∞ and 1. It is usual however to truncate ϕ at 85, which then reduces the maximum weight of a measurement to a value of 11.5. Correction can overcompensate and before applying it, it is desirable to compare results from different mapping directions within the same domain. If frequency patterns are similar, a directional bias correction may not be warranted.

6.10.7 Spacing Of Discontinuities

Spacing is used as a qualitative measure to indicate the importance of discontinuities occurring in a set and provides direct input along with fracture aperture into computer models that can define vertical migration potential. The spacing of fractures of the same orientation is described by the distance perpendicular to adjacent fractures. From the length of traverse, L, and the number of fractures, N, the spacing is:

$$D = L/N \qquad \text{Equation 6-5}$$

If the fractures are not intersected at right angles by the line traverse or drill hole direction, a correction has to be applied to obtain the true spacing:

$$D = D'\ cos.\ \phi \qquad \text{Equation 6-6}$$

D = true spacing
D' = apparent spacing
f = angle between traverse direction and pole to fracture plane

Frequency distributions of discontinuity spacing follow a negative exponential distribution as shown in Fig 6-42. The negative exponential distribution is:

$$f\ (x) = \lambda\ \rho^{-\lambda x} \qquad \text{Equation 6-7}$$

Where:
f (x) = frequency of a discontinuity spacing x
λ = average number of discontinuities per meter

The near fracture spacing and the standard deviation are

both equal to $1/\lambda$. The probability density function of sample data in Figure 6-42a is plotted on a logarithmic scale in Figure 6-42b.

6.11 WELL TESTING METHODS

Conceptually the application of methods to test the hydraulic conductivity of open boreholes can be divided into either a fixed interval (mass averaging) approach or a discrete zone approach. These methods can provide different conceptual understanding of the mode of hydraulic conductivity present within a rock mass. As such, the investigator should select the appropriate methodology or the basis of the criteria presented below. In addition, further test methods must be selected to perform the actual assessment of hydraulic conductivity.

6.11.1 General Conceptual Approach

Fractured rock environments where small evenly spaced fractures or joints provide the predominant flow conducts, individual quantification of these discontinuities would be impractical. In these cases mass averaging methods (described as the continuum approach) would provide more reliable assessments of the rock mass hydraulic conductivity. Since these average mass quantification can apply to fractured rock systems of many orders of magnitude different hydraulic conductivity, a number of alternative assessment techniques can be used to evaluate bedrock hydraulic conductivity. Where fractures are numerous and well interconnected, traditional aquifer pump testing can provide appropriate values of hydraulic conductivity and directional ground-water flow components. Where fractures are less numerous or poorly interconnected, traditional packer permeability tests can be used to evaluate the mass average hydraulic conductivity. In bedrock systems where fracture systems are widely

spaced special packer tests may be required to evaluate the very low resultant hydraulic conductivity. The mass averaging or continuum approach has its root in civil engineering practice, where packer tests are commonly run in this manner. This approach has been called the Fixed Interval Length (FIL) approach. The continuum approach standard method (for a rock mass of "average" hydraulic conductivity) uses a preset packer interval, and the hole is tested over its entire length (Carlsson, et al., 1983). Transitivity data are then used to develop hydraulic conductivity-depth curves, as well as, cumulative density curves on log probability plots. Once a transmissivity profile is developed with the coarse packer spacing, additional tests may be run using smaller zone lengths, often as small as 10% to 50% of the original spacing, to better define specific fractures in the borehole. In rocks having more highly conductive fractures pump tests are inherently of a mass averaging approach since pump tests should continue until a reasonably radial drawdown is obtained to simulate primary porosity conditions. Later subsections evaluate these larger scale pump tests in fractured rock.

The mass averaging approach appears to have its best application in rocks where the hydraulic conductivity of the rock mass is distributed among a large number of fractures whose flow properties cannot be distinguished from core observations or geophysical logs. This approach (as illustrated in Figure 6-43) is also well suited to supporting stochastic analyses of the flow system, as the data are gathered systematically and explicit distributions are developed of hydraulic parameters.

Discrete-Zone (Dz) Approach

Hydraulic testing of fractured rock presents many practical difficulties to the investigator. First of all, the ability of the borehole to yield water through

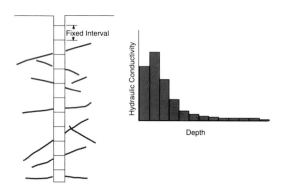

Figure 6-43a Mass Averaging Approach Conceptual Model

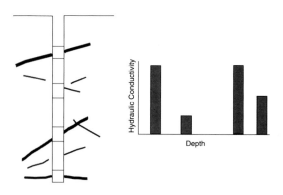

Figure 6-43b Discrete Approach Conceptual Model

interconnected fractures can dictate the methods used to calculate the hydraulic conductivity of the rock mass. In addition, the scale of fracturing can affect the field characterization techniques. In cases where large structural features, such as fracture zones, faults, and shear zones are present in the rock mass; a discrete zone (DZ) approach may be more applicable. This discrete technique would rely on testing an "individual" secondary permeability feature. If the feature is sufficiently permeable, aquifer pump testing may also provide localized storage and boundary conditions. Features with lower hydraulic conductivity may require traditional packer permeability testing. The discrete approach requires one to select the zones of the hole to be tested based on core and geophysical logs. These zones are generally individual conduits, (i.e. solutioned fractures) or individual fracture zones, or clean unweathered faulted systems. The focus of the majority of the effort in the discrete zone approach is to define hydraulic characteristics of a small number of discrete conductive fractures in the borehole. The remaining portions of the borehole are either not tested or they are tested in very long sections to assure that no significant fractures have been overlooked in the discrete zone selection process.

The discrete methodology appears to be best suited to bedrock sites which have few individual fractures. The conductivity of the rock at these sites should be truly concentrated in a few well-defined zones. The discrete approach is conceptually more effective at sites where good correlation have been established between transmissivity, observed core geology, and geophysical logs. Figure 6-43b is a schematic illustration of the discrete approach. This approach may also be important in detection and assessment monitoring evaluations where individual structural features can truly affect the localized ground-water flow conditions. In order to relate aquifer characteristics to the values obtained in discrete zone testing the discrete approach requires the use of computers and extensive knowledge of the fractures and their properties, e.g. see Wilson (1970); Wilson and Witherspoon (1970, 1974); Witherspoon et al. (1979 a,b); Gringarten and Witherspoon (1972); Mathews and Russel (1967). Other aspects of the discrete approach are treated by Freeman and Natanson (1959), Zheltov (1961), Parsons (1966), Gringarten and Ramey (1974), and Boulton and Streltsova (1977), Streltsova-Adams (1978), and Gringarten (1982).

Comparison Of The Sampling Approaches

The primary difference between the mass averaging or continuum approach and the discrete zone approach to fractured rock lies in the degree of the conceptual understanding of the site in question. Many types of low hydraulic conductivity, crystalline, consolidated sedimentary and metamorphic rock sites appear to have a relatively frequent fractures. One can expect to find that few of these fractures are major conduits, and they cannot be identified from the rest of the fracture population except by a targeted flow test.

The discrete zone approach was based on work by Snow (1965) that consists of using discrete fracture data to describe the rock mass in the form of a hydraulic conductivity tensor, that is, an anisotropic equivalent continuous porous medium. This approach requires fundamental geotechnical components of careful surface mapping of the fracture system, drilling of boreholes approximately perpendicular to the principal fracture sets determined from the field mapping, oriented core drilling, detailed logging of the core to define both the characteristics and geometry of the fracture planes, detailed fracture investigation through packer injection tests to determine effective fracture apertures, and mathematical integration of these data into the form of a hydraulic conductivity tensor. Although this method requires the evaluation of large amounts of data, Gale (1982), felt that significant benefits can be gained from development of a conceptual model based on the flow characteristics of the equivalent porous medium that are developed at the level of the individual flow conduits, or even single fractures

The main application of the mass continuum approach should be to greenfield or remedial investigation sites where the bedrock conditions are not well understood. Hence, the sampling strategy should be employed in the initial stages of site exploration. Once a good understanding of a site has been developed, and conceptual models of the flow system have been validated, the discrete zone approach may be the more cost effective of the assessment techniques.

Fracture trace techniques often relate directly to the discrete zone evaluation method since these surface expressions of highly permeable fractured systems can produce yields of ground water far in excess of the randomly contacted fracture system.

Categories of Fractured Rock Hydraulic Conductivity

Hydraulic conductivity testing of core and open boreholes can be divided into several categories relative to the overall ability of the bedrock to accept or produce water. Pump testing of fractured bedrock that is highly transmissive represents interpretative complexities rather than operational difficulties that would be important in lower hydraulic conductivity fractured rock.

Packer permeability tests represent the "standard" methodology for determining the hydraulic conductivity of reasonably permeable bedrock. Reasonably "permeable" can be defined as between 10^{-3} to 10^{-6} cm/sec. Since packer permeability tests are limited in upward and lower ranges by equipment (i.e., water supply lines, flow meters, low flow limitations), highly permeable rocks or very low hydraulic conductivity rock masses must be tested by alternative methods specific for high volume or low volume test conditions. The following sections describe test methods for:

1. $>10^{-4}$ cm/sec - "high" hydraulic conductivity pump test methods for highly integrated and permeable fracture systems;

2. 10^{-3} to 10^{-6} cm/sec - "normal" hydraulic conductivity test methods for fractured rock;

3. 10^{-6} to 10^{-9} cm/sec - "low" hydraulic conductivity test methods

The appropriate methods must be selected by the investigator to fit the expected hydraulic conductivity and connectivity of the fractured geologic medium.

6.11.2 PUMP TEST EVALUATION OF FRACTURED ROCK

The literature has addressed the inadequacy of analytical methods to describe flow characteristics of fractured and carbonate rock aquifers, (See Freeze and Cherry, 1979). One of the basic assumptions of the radial flow formulas is that flow takes place through a homogeneous medium having the same properties in all directions. Even casual inspection of fractured and carbonate rocks reveals little or no intra-granular porosity. The void space is in the form of joints, fractures, and solution channels. It should be apparent in the typical fractured rock environment that the various combinations of fracture location, fracture width, fracture content infilling, (with fine grained or crystalline material), and amount of ground water contained in storage adjacent to the well, can result in almost any type of drawdown curve. A number of reviews of the hydraulics of wells in fractured, and jointed, rocks have appeared in the literature (Lewis and Burgy, 1963, 1964; Moore, 1973; Zdankus, 1975).

All the traditional equations and graphs for the evaluation of pumping test data have been formulated on the assumption that the ground water is flowing through a porous medium. Later, supplementary equations were developed by others such as Hantush and Papadopoulus to describe the movement of ground water through fractured media. While the latter equations are sometimes applicable to specific types of fractured systems, their accuracy would not always be significantly higher than the conventional equations, (see Smith and Vaughan , 1985) Although the basic assumptions applicable to all pump tests are not met precisely, radial flow analytical concepts can still be applicable for fractured and carbonate rock. Zeizel et. al. (1962) inferred that the dolomite in Illinois contains numerous fractures and crevices which are connected on an area basis. They stated, "Such a network of openings can give a resultant regional effect equivalent to a radially homogeneous aquifer." Because the water flows in fractures and crevices that bring the water directly or by complex interconnection into wells, the flow system assumes, at least, some of the characteristics of a linear channel in the immediate vicinity of a pumped well where the flow departs from laminar. Thus, the leaky artesian formula may describe drawdown on an area basis with reasonable accuracy, but does not completely describe the drawdown in a pumped well.

The following rule of thumb should be considered when making a decision as to the type of equation to be used in the evaluation of the pumping test. If the area of influence during a pumping test is large in comparison to the scale of the fractures and joints in that area, providing that these structures do not have a single orientation, then the conventional porous flow equations can be used for the calculation of the transmissive and storage characteristics of the aquifer. The same other constraints (see Lorman 1972) as in the case of porous flow media, will still have to be considered when using these equations in the evaluation of fractured media.

One of the difficulties in working with both fractured and carbonate aquifers is the seeming inconsistency in the hydraulic characteristics of wells within a small area. To a great extent this is caused by conditions in the vicinity of the borehole. These irregularities may be of great consequence initially, but dwindle to small importance as the cone of depression becomes very large. As with the fractured rock, the larger the area considered, the more nearly some carbonate aquifers assume the hydraulic characteristics of a homogeneous medium.

Figure 6-44a to 6-44d show the gradual evaluation of a drawdown cone as water is pumped from a single borehole at 1000 m^3/day. Although some anisotropic conditions are evident in the first few days, the overall cone of depression develops into a reasonably radial condition as measured by the 29 piezometers used in the 30 day pump test. With known values of discharge and the measurement of the

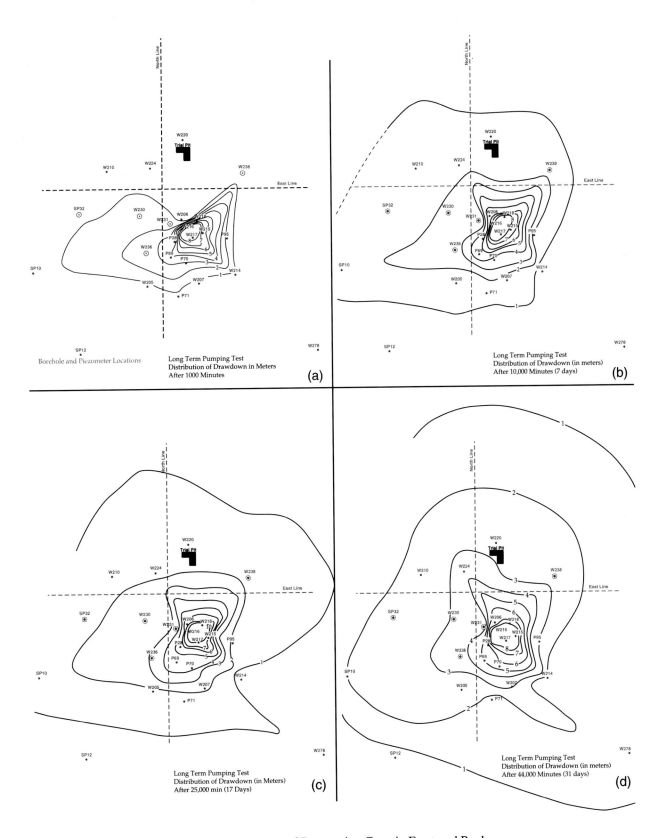

Figure 6-44a-d Shape of Dewatering Cone in Fractured Rock

volume of rock dewatered during the pump test an estimate of rock mass porosity can also be made for the test area.

Carbonate aquifers in well-developed Karst terrain or areas of considerable relief, may however, never react analogous to a radial flow homogeneous medium. Large solution cavities may be developed along fracture zones leaving the adjacent rock relatively impervious. Under these conditions, the resultant storage effect might resemble the emptying of a tank rather than reflect the instantaneous pressure release caused by the expansion of water and the elasticity of the aquifer skeleton represented by an artesian storage coefficient.

Typical Pumping Test Curves For Fractured Media

Pumping test curves for large scale fractured media should, in concept, be no different from curves for porous media. Confined (Theis, Cooper-Jacob), semi-confined and leaky (Walton), unconfined (Theis, Cooper-Jacob) and delayed drainage (Boulton), environments can be found in both porous and fractured media. In order to evaluate pumping test data, for fractured rock environment it is therefore essential that the conceptual physical conditions under which the ground water occurs in the rock mass be determined first, where after the appropriate solution technique can be selected.

The purpose of pumping tests and subsequent analysis is to both establish aquifer characteristics and to enable predictions of well yields. Aquifer characteristic and local isotropy are important in defining facility target monitoring zones and potential flow directions. Computed hydraulic characteristics can also be used to mathematically simulate various schemes of development or remedial alternatives. The magnitude of interference between wells can also be computed to determine the practical design of ground-water recovery well fields. Recharge must be considered in determining the potential yield of wells located in fractured rock due to the typically low storativity and resultant high rates of recharge. Unfortunately, in fractured and carbonate aquifers these techniques yield only estimates, at best, when done prior to drilling and testing the actual wells. In areas where data is abundant these estimates of rock mass aquifer characteristics may be very good. In poorly understood fractured rock it is probably impossible to predict with any degree of accuracy the yield of a well before drilling.

Adequate pump-test data (from a well-executed aquifer stress test) however, makes it possible to predict average well yields for a particular rock mass with a reasonable degree of accuracy. In each well having drilled through a single or multiple water-bearing zone, there exists a pumping rate (called the critical pumping rate), at which drawdown in the well will stabilize at a specific discharge rate. With local water levels and recharge remaining stable this well will be perhaps best characterized by its long-term calculated specific capacity on the basis of the •critical pumping rate'. Individual values for well loss, dewatering, critical pumping level and the rate of drawdown with time, must be fully considered in determining the yields of individual wells in fractured rock

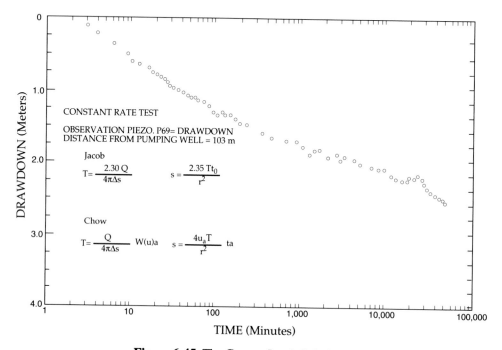

Figure 6-45 The Cooper-Jacob Solution

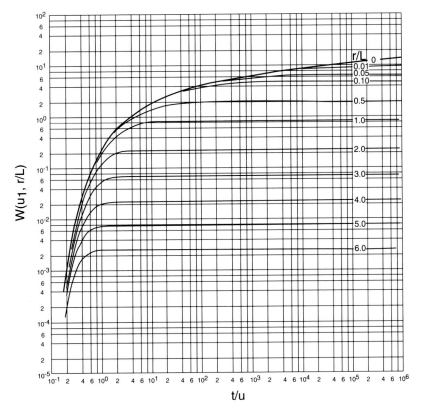

Figure 6-46 The Walton Solution Type Curves

environments. If critical pumping levels are above the top of the aquifer, a straight forward solution is possible:

- Theoretical drawdown is directly proportional to the pumping rate for a given time, so test drawdown must be corrected for well loss;

- The theoretical drawdown for any rate can then be computed and the actual drawdown for a given time can be predicted by adding the well loss for the new pumping rate;

- The rate of drawdown can be determined by the modified non-leaky artesian formula, using the transmissivity (T) determined from the test and extending the results; and,

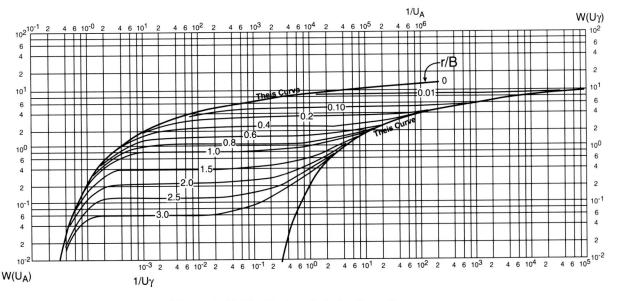

Figure 6-47 The Boulton Solution Type Curves

- When recharge characteristics are not known, an arbitrary period of no recharge is selected;

If dewatering of some fracture zones supplying ground water to the well is expected, the above analysis would not apply. If a critical pumping level is selected for production or constant rate tests, which is below pumping levels produced during the initial step test, the hydrologist must base his judgment on other data and should be very cautious in his use of the data. Disregarding of the relationship of fractured rock well pumping levels and fracture zones can cause erroneously optimistic prediction of yields. Walton (1962) described the empirical projection of yield-drawdown relationships under such dewatering conditions based on step test data. There are a number of alternative plotting methods to display the results of the step test. Drawdown at the end of each step can be plotted against the corresponding pumping rate and a best fit curve drawn through the points. Alternatively water levels can be plotted instead of drawdown so a comparison with the well log can be made. One should always try to match the results of step test data (as well as, constant rate test results) and the physical or geophysical logging of the borehole. Cross-checking data gathered through different methods is always an important procedure in fractured rock investigations.

The Theis Or Cooper-Jacob Solution :—The Theis and the Cooper-Jacob solutions for the evaluation of pumping test data are the most used analytical techniques for fractured rock studies. The Cooper-Jacob solution was derived from the Theis approach and the Cooper-Jacob approach is generally more useful for fractured rock

evaluation due to the resultant straight line data plot on semi-log paper (Figure 6-46). Deviations from a straight line are relatively easy to evaluate and provides for visual assessment of reasons for these deviations.

The Walton Solution Leaky conditions may result where the stressed aquifer is overlain by one or more hydrostratigraphic units with lower transmissive characteristics than the aquifer. The Walton Solution type curves are shown in Figure 6-46. The curves represent increasing degrees of leakage into the target aquifer. The Walton method provides a straight forward procedure to determine the leakage coefficient. An important key to evaluation of leaky aquifer systems is that all the curves flatten off after a while.

A second important characteristic of leaky aquifers is that the ground-water surface will recover (or rebound) faster after the pump has been shut down, than is normally seen in non-leaky aquifer environments. One should expect that both the drawdown and recovery data traces result in smooth curves, without abrupt changes. This can make the curve matching more difficult with the flatten and smooth data curves present in leaky systems.

The Boulton Solution:— Unconfined conditions require the use of the Boulton solution, however, this method should only be used where the fractured aquifer consists of a whole range of fracture aperture sizes. The dewatering curve for the pumping well should look like that in Figure 6-47. The flattening of the curve, normally some minutes after pumping commenced, is due to the continuous drainage of the smaller pores in the aquifer (Hodgson, 1983). After these pores are drained, the curve can be expected to steepens. These systems can be easily

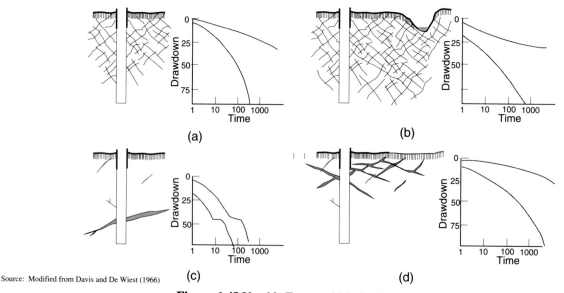

Source: Modified from Davis and De Wiest (1966)

Figure 6-48 Varable Fractured Media Drawdowns

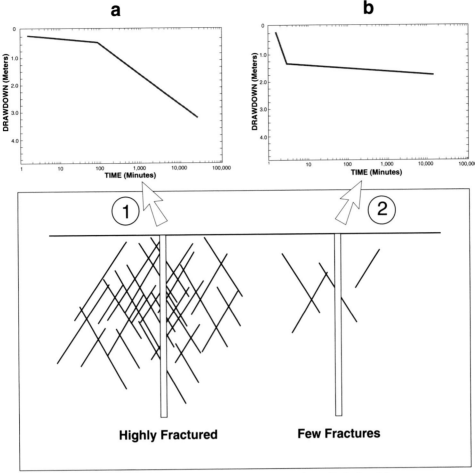

Figure 6-49 Fractured Rock Drawdown Relationships

Source: Modified from Hodgson (1984)

mistaken for leaky conditions when tests are not sufficiently extended to reach the later portions of the curve.

Differential Fractured Media Curves:—In variably fractured media, pump test evaluations are often complicated by boundary conditions which are not represented in the above analytical curves. Figure 6-49 illustrates a few cases of typical drawdown curves for selected geological conditions.

A feature that is very common in wells located in fracture rock is an initially high or moderate initial yield that decreases rapidly with time. The cause is usually insufficient storage of ground water or poor communication of fractures in the vicinity of the well.

Figure 6-48 shows an additional typical drawdown curve for fractured rock environments as modified from Davis and DeWiess (1966) Figure 6-48b shows the response related to pumping in fractured rock that may have several levels of open, saturated fractures. Figure 6-48c illustrates the typical case where fracture filling material is released during pumping causing minor rises in

pumping levels and turbidity problem. If this well had been pumped initially at a high rate for the development purpose of removing fine materials loosely held within the fractures, the well's yield might have been increased significantly, and the turbidity problem reduced significantly before using the well for monitoring purposes. These storage and communication problems with fractured rock can be viewed as representing boundary conditions that are not accounted through the typical pump test curves.

Two of these conditions are illustrated (Hodgson, 1983) in Figure 6-49. The assumption is made in this conceptual model that the density of the joints and fractures is not the same throughout the model area. Conducting a pump test in hole A, that is located in an area of less than average hydraulic conductivity will result in a drawdown curve (straight line Cooper-Jacob) as in Figure 6-49a. The first part of this curve represents the aquifer characteristics in the vicinity of the well, with the latter part of the curve would represent regional characteristics. Since these curves are the result of both local and regional fracture patterns yielding water to the test well, the actual transmissivity of the rock mass probably lies between the calculated values that would be obtained from using either portion of the curve (Hodgson 1983).

A pumping hole at position B in Figure 6-49 will produce a Cooper-Jacob drawdown curve. The local and regional fracture patterns although sufficient for conducting a full scale pump test will produce two (or more) different slopes in a similar manner as the previous example. Especially in the case of hole A, the drawdown curve can be mistaken for the curve for leaky conditions. These two curves on first look are very similar, however, there are a number of distinct differences between the curves that can provide clues to the fractured aquifer environment. These two major differences being the following:

• True leaky conditions will result in a flat drawdown curve; fractured rock pump test data will never go completely flat; and,

• Fractured rock environments have very rapid initial recovery after stopping the discharge phase of test. Figure 6-50b illustrates this initial quick recovery with a later marked decrease in the rate of recovery. True leaky conditions will not show the initial very rapid recovery data.

These concepts can be extended to include typical boundary conditions present in both granular and fractured systems Figures 6-50 and 6-51 illustrate the typical effects of an impermeable boundary and a recharge boundary, respectively. In the first example the drawdown in the well increases due to the "impermeable" fault causing a reduction of flow to the well. In the second example the drawdown stabilizes as the cone of depression (or pressure relief) reaches a surface water body, or in some cases, an overburden unit made up of a primary porosity aquifers,

and begins to draw from it. It should be also observed that this type of stabilized drawdown curve may be caused by the return of water discharged from too near the tested well, and thus reentered the fracture system from the ground surface. From these curves it is evident there may be many other drawdown curve shapes and unique interpretations of the drawdown curves than those presented. For this reason it should be stressed that caution is required in conducting and evaluating pumping tests. A short-term test should never be the basis for determining the long-range yield potential of a fractured rock mass ground-water potential, especially for long term pump and treat systems.

Much more can be said on the topic of the interpretation of drawdown curves under fracture flow conditions, that goes beyond the simple fracture density-borehole location. For the purpose of this evaluation, the following near-field and far-field consideration will explain a number of restrictive conditions under which ground-water movement takes place in fractured and carbonate rocks.

6.11.3 Evaluation of Near-field Hydraulic Parameters

The traditional methods for evaluation in observation and pumped wells were described as the various solution curves for the conceptual fractured rock conditions. There is, however, early time (or "near-field") and boundary (or "far-field") conditions that modify the traditional analytical curves such as the Theis or Boulton solutions. The relationships of these near field and far field effects are shown on Figure 6-52. These near field considerations on the traditional pump test curves include skin effects and partial penetration.

Outer, or far field boundary conditions consist of constant pressure, no flow and water table boundary

Equivalent Fractured Rock Barrier Boundary

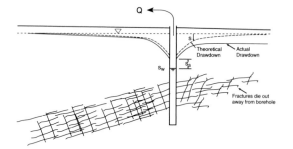

Equivalent Fractured Rock Recharge Boundary

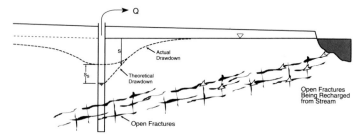

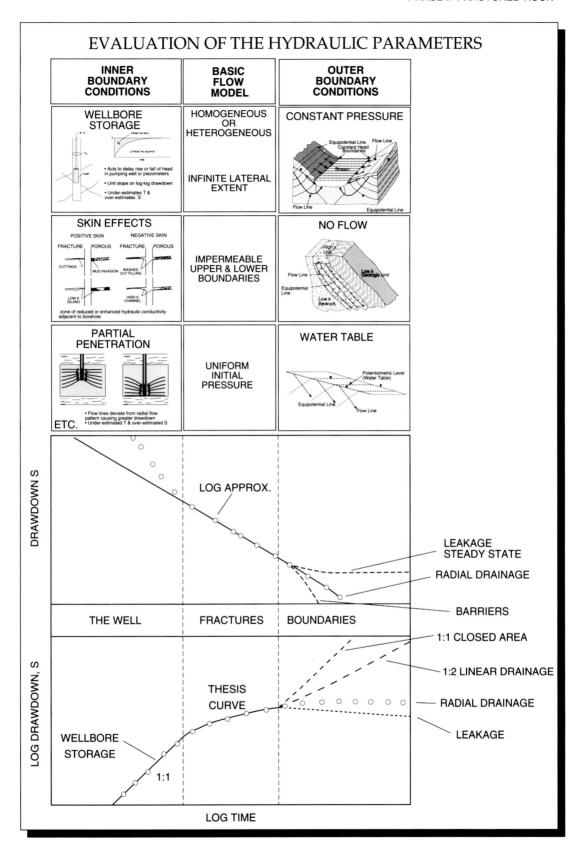

Source: Modified from Unesco (1984)

Figure 6-52 Main Portions of Data Plots

conditions. These near-field and far-field boundary conditions bracket those projections of the drawdown curves that represent analytical pump test curves. In general terms analysis of fractured rock pump test results require many assumptions on orientation of fractures, and both near-field and far-field influences on the resultant drawdown data. Kruseman and de Ridder (1990) provide summaries of the test methods for uniformly-fractured aquifers based on the double-porosity concept. These methods are complex and may not provide significant additional assisstance beyond traditional single porosity methods. One should always graph the data in both log-log and semi-log plots, then try to decompose the resultant curves into nead-field and far-field effects to select appropriate single porosity or double porosity methods to use on the mid-field data.

Skin Effects

The Theis well function is based on small diameter well data and line source solutions, that encompasses all water bearing zones in the formation. In reality, however, many of these zones may be totally or partially plugged. In primary porosity environments this plugging is primarily due to smearing of fine-grained material along the borehole wall during drilling. In some cases in fractured bedrock, the fractures supplying water to the well may be partially blocked or may have a considerably larger hydraulic conductivity than the gross transmissivity of the conductive structure of the rock. These two near field effects may be evaluated by employing a skin factor (see Figure 6-53).

If the near well borehole zone is plugged with cuttings or fracture filling clays or minerals, skin effects may be present. A set of assumptions must be specified for the evaluation. This is true for all the various methods for the evaluation of flow in fractured rocks where the actual orentations of the fractures and the various early nearfield effects are unknown. Ramey (1982) has presented a review of the well-loss function and related skin effects that provide additional insight into this effect.

As discussed previously, with the addition of the skin effect factor a constant additional drawdown, is produced. This does not change the slope of the linear part of the curve and its application nevertheless results in the following familiar equation:

$$T = \frac{0.183Q}{\Delta s}$$ Equation 6-8

If the additional drawdown were known, the skin factor could be calculated. However, if a reasonable estimate of the storage coefficient is available, the skin factor can be established by calculating the intersection point between the draw-down line without skin effects and the zero drawdown axis.

Well Bore Storage

The second near field effect that may be evident during early times is the well bore storage. Since some water is stored in the well, and when pumping starts most of this early volume of water is taken from the well and not from the aquifer, hence the drawdown well be less than would be predicted from pump test theory.

For very short time drawdown data plots as a 1:1 straight line on a logarithmic graph. Since this near-field effect only applies until the well bore storage is expended, the drawdown data may soon follow the Theis curve.

6.11.4 Evaluation Procedure

Fractured rock environments require somewhat different evaluation procedures due to the heterogeneous nature of the fracture-borehole interface. In order to determine the hydraulic properties from a set of drawdown data, one should:

• Plot data in both logarithmic and semi-logarithmic diagrams .

• Estimate the storage coefficient - This value should be based on experience in the specific geographic area;

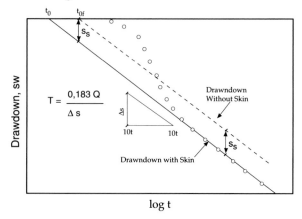

Figure 6-53 Skin Factor

• From the semilog plot, the transmissivity and skin factor are evaluated according to the procedures illustrated in Figure 6-53;

• Evaluate the data curves for both near-field and far-field influences. If the test data can be evaluated by single porosity methods proceed in the traditional curve matching techniques. If the data fits the dual porosity type curves proceed with methods listed in Kruseman and de Ridder (1990);

One should always evaluate the single porosity methods first since experience has shown that approximations of the hydraulic parameters of the fracture system tapped by the well can be obtained through use of these models if consideration is made in the relative pumping levels obtained during the pump testing.

6.11.5 Field Measurements

A successful pump test in fractured rocks requires more than just a functional knowledge of the use of type curves and equations. Fractured rocks require great care in both the design and performance of the pump test. For example, barometric changes can have significant effects on the observed head levels, due to typical low confined storage values obtained from fractured rocks. As with any pump test procedure constant and steady pumping of the borehole are required because interruptions and variable production rates from the borehole may make a proper evaluation of the pump test impossible.

Before beginning the pump test, measurements should be made for a sufficient period in order to determine the static water level and trends in the water level. If anisotropy is expected, investigators should plan to locate observation piezometers at several orientations to the structural fabric of the rock mass. One way to plan for anisotropy and heterogeneity is to include more than just the minimum number of observational piezometers. The exact location aspects of some of the observational piezometers to hydraulic discontinuities may cause a few of the piezometer points to respond somewhat anomalously, but taken as a whole, the remaining observational piezometers can help to provide reasonably useful drawdown data.

A typical field program for field hydraulic properties begin by evaluating the data from pumping tests by first attempting to employ the standard analytical methods. Although this evaluation typically achieves varying degrees of success, it is generally possible to obtain reasonably realistic values for rock mass hydraulic properties. As work progresses on the evaluation, one technique is to use apparent anomalies in the various pump test analyses as aids in the interpretation. A description of these analyses along with specific examples illustrating the procedures follows.

It may be especially difficult to select the rate and duration of pumping in fractured environments with lower hydraulic conductivity (say $< 10^{-3}$ cm/sec). Extensive step testing may be needed to find a pumping rate that is high enough to maintain constant discharge without adversely dewatering the well.

The time schedule for water-level measurements in fractured rock should follow a similar geometric series as with primary porosity materials', since, the drawdown functions approach logarithmic functions. The following pump schedule is recommended: First at rest level; then every 1/2, 1, 2, 3, 4, 5, 6, 7, 8, 9, 10 minutes, and then; every 12, 14, 16,18, 20, 25, 30, 45, 60, 90, 100, 150, 180 minutes, and then; every 4, 6, 8, 10, 12, 16, 20, 24, 30, 36, 48 hours, and then the following days; morning, noon and evening, for the pre-estimated pump test period. During the test, records should be kept of the water level, time of the measurement, pumping rate and any other factors that may affect the test data. When the test is stopped, the same program of measurements should be implemented in order to obtain reliable recovery data.

Well Yield

As was expressed earlier in this section it is normally not possible to determine the final yield of a well in fractured rock by a short term pumping test. Long-term capacity, as necessary for the typical evaluation of ground-water recovery system is affected by climatic variations, interference from other wells, etc. This suggests that the yield from a well or a group of wells must be calculated on the basis of long term observations of drawdown. This long term yield of a well becomes particularly important when balancing the production of both individual and groups of wells.

The short-term pump test can, however, provide data for the calculation of the preliminary yield of the well. Together with data from other wells, these data provide information for use in site-wide or near regional analysis. In evaluating the qualitative features of the data curves of Figure 6-52, these curves can be subdivided into three main parts (UNESCO 1984) which are influenced by:

• The well (near field effects);

• The fractures surrounding the well; and,

• The boundary conditions (far field effects).

The straight line of the log approximation is established in the Cooper-Jacob semilog plot during the mid-portion of the curve where the main influence on the drawdown is due fractures. If the curve turns upward due to boundary influence it indicates leakage to the aquifer and that the drawdown approaches a steady state. For the radial drainage and the steady state cases, a preliminary yield may be derived by the Thiem equation. The influence radius may be determined by the test results, but an estimated value based on experience may suffice, since the log function grows very slowly in the range $Ro/r_w > 1000$. The drawdown, s_w, applied in the equation as a general guideline should not exceed 50 percent of the possible drawdown in the well, and in any case, should be based on the step drawdown determined critical pumping levels of the well.

Hydraulic barriers, if present in the system, should show up clearly in the resultant semilog plot. The log-log plot indicates, the type of barrier system which is evaluated by the slope of the later part of the data curve when far field conditions apply. A 1:2 slope usually indicates a linear drainage and a 1:1 slope a drainage area that is closed on all sides. In both these cases, continued pumping would be advisable until recharge or leakage effects produce a steady state condition, which provides the basis to calculate the yield.

When conducting field tests in any specific project, one must consider that local geologic conditions play important roles in the design of the aquifer test. A long term pump test in fractured rock can provide a lot of data regarding the distribution of transmissivity and storage values. The investigator must be very careful to select the analytical technique which best describes the physical conditions in the field. In addition to the near field well effects and the far field boundary effects, there are other complicating factors in interpretation of fractured rock aquifer test results. These factors can have bearing on the validity of the derived aquifer parameters, such as, the calculated storage of the system. For example, storage values are typically dependent on three factors:

• The transmissivity value, which may be different from point to point;

• The t_0 value of the dewatering curve, which is likely not the true value for areas of differential fracturing; and,

• The distance of the observation hole from the pumping well, the latter being an important consideration in fractured material.

Equations used to calculate the storage coefficient are based on assumptions that the distance between the observation and abstraction boreholes is the shortest distance between the two holes. This may often not be the case for fractured rock, since the flow of the ground water may in fact be in a zig-zag fashion, along the numerous secondary structures, until it reaches the abstraction point. This will have the effect that the distance of travel is longer, resulting in calculation of erroneous high values for storage.

The analysis of constant rate tests to determine values for transmissivity and storage coefficient for fractured rock systems is based primarily on time-drawdown data. The shape of the cone of depression, for a given pumping rate, is determined by the hydraulic characteristics throughout the entire area encompassed by the cone. In considering the variable nature of fractured and carbonate rocks it is unrealistic to assume that distance drawdown relationships in given directions are representative of the entire area. However, after the cone becomes very large, local variations become relatively insignificant. The rate of decline of the cone becomes an integration of the hydraulic characteristics in the area of the cone.

Time-drawdown derived values of transmissivity (T) obtained from shallow well data (partially penetrating) are usually found too high to be consistent with values obtained from the pumping well or deep observation piezometers. This can be attributed to the use of Q (test pumping rate) in the artesian (Theis) formulas. Since the zones penetrated by the individual observation points contribute only part of the water being pumped from the pumping well, the discharge which caused drawdown in the shallow partially penetrating well can be expected to be some value less than the recorded test Q. In effect the actual Q cannot realistically be determined with any precision. In such cases it would be just as unrealistic to attempt standard partial penetration corrections in these variably fractured environments, since, the degree of hydraulic connection may be as variable as the rock itself. The same uncertainty is possible in fully penetrating observation wells, but the degree of error is probably not as significant as with partially penetrating observation piezometers.

Time drawdown derived values of T computed from test well data are considered valid, as long as production pumping levels are maintained above the water-yielding zones. If dewatering occurs in these producing zones, the hydrologist must deal with the resultant erratic drawdown data. In analyzing pumping-well data, the graphical solution of the modified non-leaky artesian formula (Cooper and Jacob, 1946) is commonly employed for the determination of transmissivity; where:

$$T = \frac{264Q}{\Delta s} \qquad \text{Equation 6-9}$$

T = transmissivity,
Q = discharge, in gpm
Δs = drawdown difference per log cycle, in ft.

The Cooper and Jacob semi-log method is used more often than any other analytical solution due to the convenience of analyzing various portions of the straight line curves. Figure 6-54 and 6-55b shows a time-drawdown curves where significant dewatering has occurred. Dewatering is evident at the abrupt break in slope. The increased slope at later time periods represents the lower T local to the pump well rather than the full volume of the fractured aquifer. In order to obtain valid values of T the hydrogeologist must use pumping data obtained prior to dewatering of the local fractures. Figure 6-57 also has recovery data plotted for use in calculating T values for the test. Recovery data should be calculated for comparison to predewatering draw-down data. However, computations must use residual drawdown (the difference between the projected non-pumping levels and recovery levels) in order to eliminate effects of fracture dewatering. Fracture rock investigations often require many pump tests to fully evaluate a rock mass. In these cases observation well data are often not available for use in evaluation of

rock mass aquifer properties. Recovery data is essential for evaluation of pumping well tests in fractured rock.

Although transmissivity results can be reasonably reliable for fractured rock assessments, vertical hydraulic conductivity estimates based solely on pumping test analysis are usually inadequate to accurately describe field conditions. More realistic conductivity values can be determined from detailed studies of major pumping centers where some level of hydraulic equilibrium has been established in the rock mass.

Specific Capacity Determinations

The traditional method of expressing the yield of a well is in terms of specific capacity. Specific capacity (Q/s) is expressed as the yield of a well in gallons per minute per foot of drawdown (gpm/ft) for a specified time and rate. In porous media it is assumed that specific capacity is constant for a given well regardless of the time period and pumping rate. Fractured rock systems can be expected to deviate significantly from these assumptions typically based on primary porosity-granular media. These deviations from expected specific capacity can greatly affect the design and ultimate performance of ground-water control, and pump and treat systems. Specific capacities for fractured rock can only be compared by considering, a number of important physical factors of the bedrock system.

These hydraulic properties can be computed theoretically by traditional aquifer assessment techniques. The general difficulty with calculated specific capacity is that one considers only head loss (drawdown) due to laminar flow. The specific capacity calculation does not consider head loss caused by turbulent flow, in the well, and in the immediate vicinity of the well. Downhole observations made in fractured aquifers indicate that turbulent flow would probably dominate as water enters the well and flows through the borehole. The actual specific capacity can be represented by the following equation:

$$\frac{Q}{S_w} = \frac{Q}{(s + s_t)} \qquad \text{Equation 6-10}$$

Where:
Q = pumping rate, in gpm
S_w = drawdown in the pumped well, in feet.
s = drawdown due to laminar movement of water through the aquifer, in feet.
S_t = drawdown due to turbulent flow in and near the well bore, in feet.

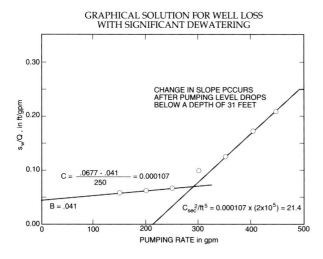

GRAPHICAL SOLUTION FOR WELL LOSS WITH SIGNIFICANT DEWATERING

Figure 6-54 Data Plots With Significant Dewatering

The total drawdown in the pumped well, considering both laminar and turbulent flow, may be expressed by the following equation (Jacob 1946):

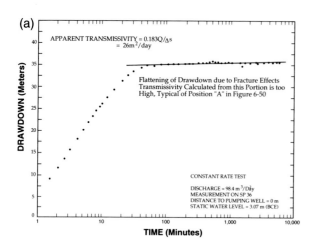

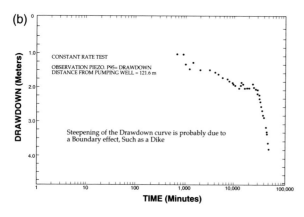

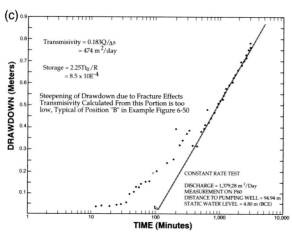

Figure 6-55 Fractured Rock Data Plots

$$S_w = BQ + CQ^2 \qquad\qquad \text{Equation 6-11}$$

where:
Sw = drawdown in the pumped well, in feet.
B = aquifer constant 25
C = well-loss constant, sec/ft
Q = pumping rate, in cubic feet per second (cfs).

The BQ term in the above equation is called the "formation loss" and the CQ term is the "well loss". C can be calculated by a formula derived by Jacob (1946) that utilizes data obtained from a step test. The solution for C requires computations involving increments of drawdown and pumping rate for successive steps.

Rorabaugh (1953) derived a more exact method for evaluation of well loss that may be also applicable to fractured rocks. Rorabaugh's method uses the exponent for turbulent flow expressed as an unknown constant "n". The "well loss" term in the equation is then CQ_n. An empirical solution is then obtained from step-test data by a graphical solution of the equation on logarithmic coordinate paper. A graphical solution has the advantage of averaging field data and eliminating or adjusting poor data. The Rorabaugh method is fairly straight forward, but does require some trial and error computations to reach a final solution.

A third graphical solution for well was described by Bruin and Hudson (1955) which solves the original equation where well loss is represented by the term CQ_n. Bruin and Hudson provide a simple solution for well loss that does not require the conversion of the pumping rate to cubic feet per second. Values for C are in units of ft/gpm. Multiplying by a conversion factor of 2×10^5 gives C in the units of the other two solutions.

Graphical or iterative solutions, as described above. have a definite interpretative advantage in fractured rock environments over averaged results obtained from successive groups of data, as in the Jacob method. The inherent heterogeneous nature of fractured geologic systems should be approached by assessment techniques that can average data obtained over larger brackets of the stressed system.

Well loss solutions for fractured and carbonate appear to provide valid field numbers, as long as, the test pumping levels (and production levels) are above significant water-bearing zones. Once dewatering of these specific zones occurs, the hydraulic performance of a given well can be expected to change drastically. As the pumping level falls below a water-yielding zone, the maximum contribution of water is obtained from that zone as it drains into the hole. The contribution to the well cannot be increased by additional drawdown, in fact, the rate of flow from that zone typically will decrease with time as this area is dewatered. If the pumping rate from the well is increased, the additional water must be supplied by the lower zones. This effectively increases turbulence and well loss as the well is pumped down. The well loss constant in this case is related to the number of fractures and hydraulic properties of the fractures contributing water to a given well. The result of this dewatering is an additional compound drawdown due to a combination of dewatering and increasing well loss. Since the two variables of dewatering and increased well loss cannot be separated, reliable values for C can no longer be computed. In some fractured rock wells this change may be gradual. However, in many instances, these factors increase abruptly upon dewatering of a significant water yielding zone. Once this dewatering (or rapid drawdown) effect is reached a well is said to have reached the 'critical' pumping level for that well. Because there are many pitfalls in the determination of well loss under fractured rock conditions, a simple field verification of the of value C can be made for the test well. The procedure is as follows:

• Values of s (total incremental drawdown) versus Q (total pumping rate) for each step is plotted on rectangular coordinate paper as shown in Figure 6-55;

• A straight line is extended through the plotted point down to the zero pumping rate; and

• C is the slope of the straight line, and the value for B is

equivalent to the line intercept with the Sw/Q axis.

• Since the theoretical drawdown is directly proportional to the pumping rate, the corrected drawdown should fall on a straight line drawn through the point of origin.

If necessary, values of C can be adjusted to give a better straight line approximation. Some estimates of dewatering may be calculated as shown by Figure 6-55, for wells where significant dewatering are observed in the drawdown data. The break in slope points to a dewatered fracture zone. Although the calculated values of B and C obtained from the steeper line are meaningless, the first three steps provide a representative value for C, so long as, the critical pumping level is not exceeded..

Step test data can be empirically used to predict yields considering overall pumping time. Figure 6-56 illustrates this technique. Procedures to apply the correction are as follows:

• Values of S versus Q for each step is plotted on rectangular coordinate paper as shown in Figure 6-56;

• Values of S represent total incremental drawdown and Q represents total pumping rate of each step; and,

• Each value of drawdown must be corrected for well loss and re-plotted against the corresponding pumping rate.

Since the theoretical drawdown is directly proportional to the pumping rate, the corrected drawdown should plot along a straight line through the origin. In this case where a pump test was carried out for twenty-four hours drawdown as plotted against the pumping rate falls beneath the corresponding step-test curve, adjustments to the curves can be applied. By using reasonable technical judgments a second curve can be approximated that provides a yield-drawdown relationship for the test period. Time-drawdown data is often projected through extension of slopes for predicting yields of wells. The yield-drawdown curve of step test data can be used to check the validity of these extensions. For example, a predicted yield and drawdown which falls to the right of the step-test curve cannot possibly be correct. This method does not yield an exact solution for these parameters, but if used properly they can give practical results.

Mathematical computations to predict drawdown with the non-leaky artesian formula (Theis) should be used with extreme caution for a pumped well in fractured or carbonate aquifers. These methods require an assumption that effective radius is equal to the nominal radius of the well. Such an assumption can be questionable in fractured rock and is seldom justified for carbonate rock wells.

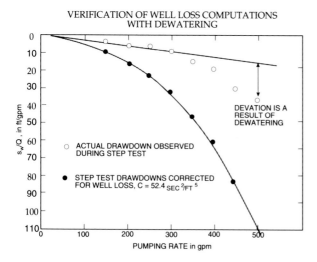

Source: Modified from State of OHIO (1982)

Figure 6-56 Dewatering Estimates Using Step Tests

Corrected test drawdown can be compared to distance drawdown graphs based on the hydraulic characteristics determined from each field test analysis. The nominal radius is assumed for the prediction of general relationships as specific capacity versus T in Figure 6-56. However, it is not recommended to use this assumption when empirical data from pumping tests is available.

It is often necessary to use fracture trace and surface geologic mapping to specifically locate production wells for aquifer restoration and pump and treat systems in variably fractured rock environments. These field data should help to adjust the locating of the drilling point. Rarely should pre-specified spacing be used for any monitoring or ground-water recovery design. Each point should be selected based on the variable observed field condition present in these rock mass fracture systems.

6.11.6 "Normal" Hydraulic Conductivity Test Methods - Packer Permeability Tests

"Normally" permeable fractured rocks is a subjective decision to be made by the investigative staff. However, if the hydraulic conductivity of the bedrock fracture system is expected to range between 10^{-3} and 10^{-6} cm/sec the "normal" packer permeability tests can provide reliable values for in-situ hydraulic conductivity. Hydraulic conductivity above and below these values may cause system limitations that necessitate evaluations using aquifer pump tests (highly permeable environments) or special low volume shut-in packer testing (sparingly permeable environments).

The U.S. Bureau of Reclamation (1960) developed one of the most useful methods (U.S.B.R. Method E-18) for the determination of insitu hydraulic conductivity through the use of packers to segregate portions of a drilled hole. CASECO (1964), Dames & Moore (1974) and Harza (1972) have made use of the packer technique for many large scale geotechnical site investigations in both consolidated soil and rock. Figure 6-57 gives the arrangement for the performance of hydraulic conductivity tests in holes that will remain open without casing. Hydraulic conductivity are calculated by the formula:

$$K = \frac{q}{2Lh}\log_e \frac{L}{r} \qquad \text{where } L \geq 10r$$

$$K = \frac{q}{2Lh}Sinh^{-1}\frac{L}{2r} \qquad \text{where } 10r > L \geq r$$

Equation 6-12a,b

r = radius of the hole tested
L = length of the section of the hole being tested
q = the constant rate of flow into the hole
h = differential head
$\log_e$ = natural logarithm
$sin\ h^{-1}$ = arc hyperbolic sine

In the above equation any consistent units may be used in calculating the hydraulic conductivity. Figure 6-58 taken from Lambe and Whitman (1969) can be used to convert numbers into the various hydraulic conductivity units. These formulas have best validity when the thickness of the stratum tested is as least 5L, and they are considered by Dames & Moore (1974) to be more accurate for tests below ground-water table than above it.

The following sections will review in some detail important considerations in borehole pressure testing that were not considered or quantified by the original Earth Manual U.S.B.R. (1960) test description. In addition, a general procedure for borehole pressure testing common to several references is given in the following text.

Net Pressure

An important consideration in borehole pressure tests is the water pressure applied between the packers (or under one packer) which is referred to as the net pressure. The net pressure applied to a section of borehole is the algebraic sum of the following three pressure terms as shown:

Net Pressure = gauge pressure + column pressure - friction losses.

All pressure terms are expressed in units of feet of water. Column pressure is measured directly in feet. Friction losses due to the flow of water through the test pipe are calculated in feet. However, gauge pressure is usually measured in pounds per square inch (psi) and must be converted to feet of water by the following factor:

Gauge Pressure (psi)/0.433 = feet of water.   Equ. 6-13

Gauge Pressure

The gauge pressure is the water pressure measured on a gauge at the ground surface (or other reference point such as the top of casing). Since a pressure difference of a few psi is equivalent to several feet of water, the gauge pressure must be measured very precisely. A high quality gauge that is accurate to +1 psi should be used for packer permeability tests is recommended by both Harza (1972) and Dames & Moore (1976).

A common problem noted by Dames & Moore (1974) was the type of pump used in the test. If a reciprocating pump is supplied by the drilling contractor, the pressure gauges may exhibit so much vibration that they become unreadable. To eliminate this problem, it was recommended that the contractor should install one or more surge chambers in series between the pump and the pressure gauge to dampen this vibration. If this does not eliminate the pressure change problem, a new system is required using a different pumping setup.

Column Pressure

The column pressure is the static pressure due to the weight of water in the test pipe above the water table. Column pressure is therefore equal to either the depth to the upper packer or the depth to ground water, whichever is smaller.

Above the water table, the column pressure is equal to the distance between the bottom of the upper packer and the surface reference point (where the pressure gauge is situated). This distance is measured to the nearest foot. Sometimes the lower reference point is chosen as the center of the tested interval, rather than the depth of the

upper packer. However, the difference has a negligible effect on the calculations for a packer spacing of less than 10 feet.

Below the water table, the column pressure is always equal to the depth of the water table with respect to the surface reference point. The depth to ground water should be measured to the nearest foot immediately before lowering the packer string into the hole.

Friction Loss

The head loss from friction of the pipe causes a back pressure that is recorded on the pressure gauge, but is not applied to the formation. Friction loss is a function of the flow rate and the length of pipe between the pressure gauge and the packer. Both Harza (1972) and Dames & Moore (1974) note the importance of calculation of friction losses that were not considered by Lowe & Zaccheo (1975), Cedergren (1967) and U.S.B.R. (1960).

Two methods may be used to determine friction loss during a particular test. The first method is to use handbook values for head loss for a given pipe diameter and roughness condition. Figure 6-59 shows an example of a theoretical flow rate Vs head loss relationship for 3/4-inch rough iron pipe. For example, if a flow rate of 5 gpm

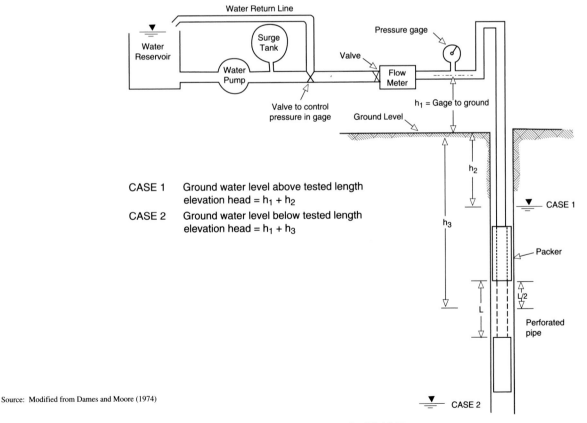

CASE 1   Ground water level above tested length
          elevation head = $h_1 + h_2$

CASE 2   Ground water level below tested length
          elevation head = $h_1 + h_3$

Source: Modified from Dames and Moore (1974)

Figure 6-57 Test Arrangement for Field Test

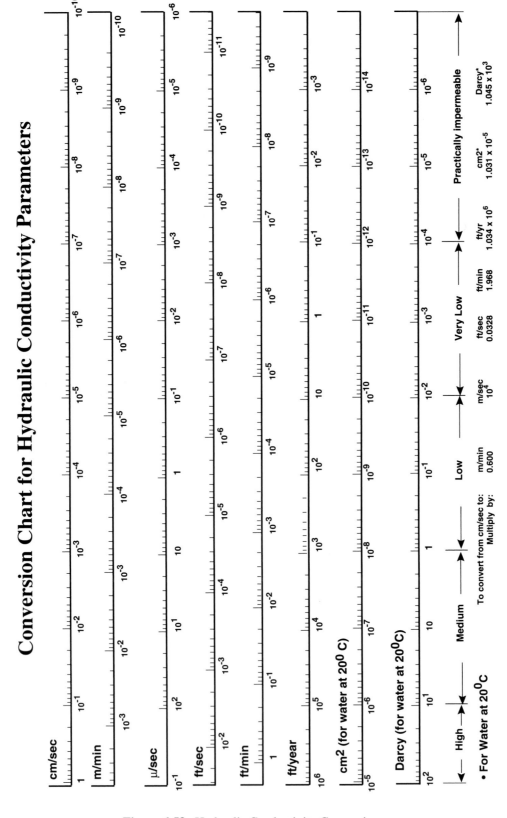

Figure 6-58 Hydraulic Conductivity Conversions

and 125 feet of pipe are given for a particular test, the friction loss is estimated as follows:

Friction loss = 0.43 feet/foot (at 5 gpm on graph) X 125 feet of pipe = 54 feet (of water)   Equ. 6-13

Friction losses at high flow rates can have a dominating effect in the calculation of net pressure. Conversely, friction losses at very low flow rates (less than 1 gpm) are insignificant and can be disregarded.

The second method of determining friction loss is to perform a calibration test in the field. Flow rate versus head loss is plotted on log-log paper, and the resulting relationship is used in the same manner as the handbook curve. Figure 6-60 shows a layout for this purpose.

A record of values of $P = P_1 - P_2$ for a given flow Q is made. The graph shows the head loss per 100 feet of pipe for different discharges. For different lengths of pipe, the head loss is assumed to be linear.

Test Water Quality

The water used for pressure testing must meet nominal specifications to maintain a valid test result. Generally, fine grained, low permeability rocks are more sensitive to water quality than are rocks of high hydraulic conductivity (Dames & Moore 1974).

Most studies of field measurement of hydraulic conductivity recognize one of the first considerations of the test is turbidity in the test water. Water containing visible suspended solids will tend to plug any formation fine-grained enough to filter out the solid material. This includes many soils and rocks that have hydraulic conductivity < 10^{-3} cm/sec. Therefore, clean water must always be used for pressure testing.

The second consideration brought out by CASECO (1964) and Dames & Moore (1974) is test water temperature. Test water temperature should be as close to natural ground-water temperature, as possible. Water supplies for pressure testing may range from near freezing to temperatures over 80^0F. The kinematic viscosity of water ranges from about 0.009 to 0.018 Stokes between these limits. Since hydraulic conductivity values obtained from pressure tests are controlled partially by viscosity of the fluid, an error is possible if "warm" or "cold" water is used. The maximum likely range of this error is a factor of about 2 for waters between 32^0 and 80^0F. For test water within 10 degrees of natural formation temperature, this error is very small. If one must use test water of differing temperature than formation water, then it is better to use test water slightly warmer than the formation temperature. Cold water may contain more dissolved gas than warm water. If gas-saturated cold water is pumped into the ground and becomes warmed, gas will come out of solution which will reduce the apparent hydraulic conductivity of the formation. This effect is most significant in fine-grained rocks where porosity can be

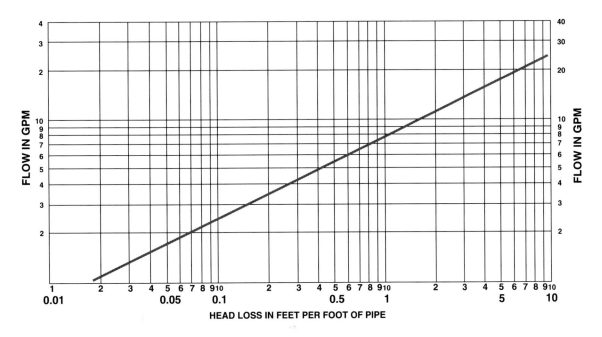

Source: Modified from Dames and Moore (1974)

Figure 6-59 Friction Loss Estimates

blocked by the presence of gas bubbles.

Measurement Procedures - Preliminary Setup

The following steps are conducted prior to testing each interval in a boring:

1. The depth to ground water is measured with water level indicator;

2. The distance between the pressure gauge and ground is measured and recorded at the location of top of the hole;

3. Equipment is set up as shown in Figure 6-57. The exact distance between packers is measured and recorded; and,

4. All tests should be conducted from top to bottom of hole unless advised otherwise. The top of the bottom packer is used as reference point for depth measurements.

Testing

The following steps recommended by both Harza (1972) and Dames & Moore (1974) are used during the testing of each interval:

1. The packer is lowered to interval to be tested.

2. Air pressure is increased slowly to seal the packers.

3. The water-return valve is opened fully. Then, the operator is instructed to start the water pump <u>at low speed.</u>

4. The maximum gage pressure is recommended as 1 psi per foot of rock depth except as noted below. This pressure should not exceeded, since hydrofracturing of the rock is likely to occur during test. The minimum recommended pressure for testing is 20 psi.

Pressure of 1 psi per foot of rock depth can open joints of bedding planes and increase hydraulic conductivity in the following cases:

• Shallow rock up to a depth of approximately 50 feet. This depth varies with the quality of the rock;

• The rock is low strength, highly jointed or thin bedded, or

• The test is done near a valley as in the case of many boreholes along a dam axis.

In the above cases the maximum gage pressure should be 3/4 psi per foot of rock depth or less. The maximum pressure used is a judgment decision based on the quality of rock.

5. The gauge pressure is built-up by partially closing the return valve until the desired pressure is obtained. If the return valve is fully closed and the desired pressure is not obtained, the valve is opened and the operator instructed to increase the pump speed.

6. When the desired pressure is reached, the flow meter reading is recorded and timing with the stopwatch begins. The flow meter readings are then recorded.

7. Flow meter readings are taken at the end of the first, second and third minutes. After the third minute, readings every 2,3 or 5 minutes are considered adequate.

8. During step no. 7, a constant watch is kept on the gauge pressure which is adjusted if necessary. In most instances, it starts decreasing or increasing due to variations in pump speed, or loosening of material in the rock.

9. The tests are finished when a constant rate of loss is obtained. In most instances, 10 minutes is an adequate time to obtain a constant flow and adequate data.

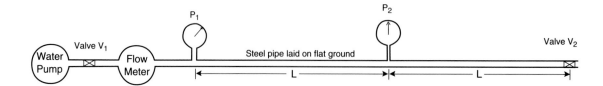

Source: Modified from Dames and Moore (1974)

Figure 6-60 Test Arrangement for Friction Loss Test

10. Subsequently, the same section is tested using a somewhat higher allowable pressure.(but not exceeding maximum recommended). Typically test pressures of 1/2 maximum allowed, full pressure allowed, and again 1/2 full pressure is used in testing rock formations. Steps 5 to 9 are repeated. However, for shallow depths only one test may be possible, due to the 1 psi per foot of rock depth maximum, and 20 psi minimum.

11. When all tests are completed at an interval, the pump speed is reduced or shut off, the return valve opened fully, the flow meter valve closed and the air valve to release the air from the packer is opened. These operations must be done in the sequence explained above.

12. Steps 5 through 11 are repeated for each interval tested.

13. The last 10 feet of depth of the borehole are not tested. Loose rock and sediment commonly accumulate at the bottom of the hole and the packer systems can get stuck down the hole.

Data Requirements

The importance of detailed data recording was stressed by CASECO (1964), Harza (1972) and Dames & Moore (1974), which includes the following information:

1. Depth of hole at time of each test;

2. depth to bottom of top packer;

3. depth to top of bottom packer;

4. depth to water level in borehole at frequent intervals;

5. elevation of piezometric level;

6. length of test section;

7. radius of hole;

8. length of packer;

9. height of pressure gauge above ground surface;

10. height of water swivel above ground surface; and

11. description of material tested.

The formulas for calculation of hydraulic conductivity with packer tests give only an approximate value of K since they are based on several simplifying assumptions. They do, however, give values of the correct magnitude and as suitable for most geotechnical and hydrogeologic investigations. A graphical solution of Equation 6-12 is given in Figure 6-61 from Davis & Sorensen (1969).

The test procedure used depends upon the condition of the rock. In rock which is not subject to cave-in, the following method is in general use. After the borehole has been completed, it is filled with clear water, surged, and washed out. The test apparatus is then inserted into the hole until the top packer is at the top of the rock. Both packers are then expanded and water under pressure is introduced into the hole, first between the packers and then below the lower packer. Observations of the elapsed time and the volume of water pumped at different pressures are recorded as detailed in the section on pumping below. Upon completion of the test, the apparatus is lowered a distance equal to the space between the packers and the test is repeated. This procedure is continued until the entire length of the hole has been tested or until there is no measurable loss of water in the hole below the lower packer. If the rock in which the hole is being drilled is subject to cave-in, the pressure test is conducted after each advance of the hole for a length equal to the maximum permissible unsupported length of hole or the distance between the packers, whichever is less. In this case, the test is limited, of course, to the zone between the packers.

Regardless of which procedure is used, a minimum of three pressures should be used for each section tested. The magnitude of these pressures are commonly 15, 30, and 45 psi above the natural piezometric level. However, in no case should the excess pressure above the natural piezometric level be greater than 1 psi per foot of soil and rock overburden above the upper packer. This limitation is imposed to insure against possible heaving and damage to the base grade foundation rock and obtaining falsely high values for hydraulic conductivity. In general, each of the above pressures should be maintained for 10 minutes or until a uniform rate of flow is attained, whichever is longer. If a uniform rate of flow is not reached in a reasonable time, the engineer must use his discretion in terminating the test. The quantity of flow for each pressure should be recorded at 1, 2, and 5 minutes and for each 5 minute interval thereafter. Upon completion of the tests at 15, 30, and 45 psi, the pressure should be reduced to 30 and 15 psi, respectively, and the rate of flow and elapsed time should once more be recorded in a similar manner.

Observation of the water take with increasing and decreasing pressure permits evaluation of the nature of the openings in the rock. For example, a linear variation of flow with pressure indicates an opening which neither

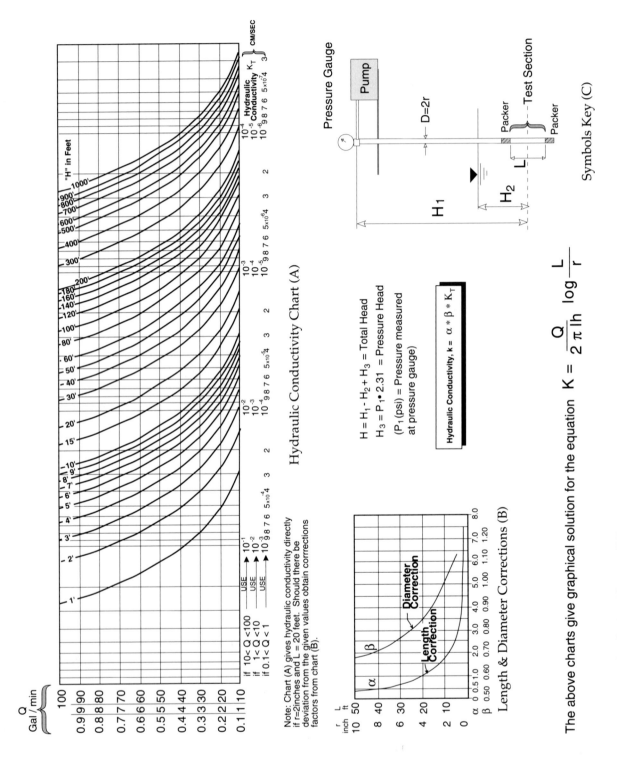

Figure 6-61 Graphical Solution for Packer Tests

increases or decreases in size. If the curve of flow versus pressure is concave upward, it indicates the openings are enlarging; if convex, the openings are becoming plugged. Item (4) is important since a rise in water level in the borehole may indicate leakage from the test section.

6.11.7 "Low" Hydraulic Conductivity Test Methods

Low hydraulic conductivity bedrock provides the site assessment professional with many obstacles beginning with difficult conceptual movement of ground water to complex equipment requirements necessary to perform the evaluation.

The equipment used in testing low hydraulic conductivity fractured rock environments require sensitive and high accurate instruments to evaluate the insitu hydraulic conductivity (see Figure 6-62). Equipment calibrations and the procedures used in the various field testing methods become extremely important in obtaining accurate data. In general these 'low' hydraulic conductivity environments would range between 10^{-6} to below 10^{-9} cm/sec, and in most hydrostratigraphic environments these units would be considered as confining units or aquitards. Evaluation of these low hydraulic conductivity environments can be extremely important for confirming unit containment of hazardous waste or the evaluation of separation of aquifers from contaminated zones. The equipment for testing low hydraulic conductivity fractured rocks can be divided into, surface equipment, down-the-hole system and the data acquisition system.

Surface Equipment

Surface equipment includes the pumps, flow meters, and control systems necessarily to conduct the test program. Due to the difficulties of monitoring constant pump pressures, pressure tanks are used to run the constant pressure tests. Tanks should be insulated or otherwise protected from temperature fluctuations. Additional flow meters (e.g., turbine flow meters) may be used in combination with bubble tube flow meters in parallel with higher volume meters. If a pressure transducer is used in the surface instrumentation to monitor injection pressure, it should be located downstream of the flow meter and any other parts of fittings that may have high pressure losses

The Down-hole System

A down-hole testing system for low hydraulic conductivity rock includes the packers, down-hole valving, down-hole instrumentation, tubing from the surface to the test zone, and pressurization lines for down-hole equipment in a field set up as shown in Figure 6-63. Packers used in fractured rock assessments should be inflatable and capable of withstanding a differential pressure of about 5 MPa. The packers can be either water or nitrogen inflated.

An essential for conduct of the tests in these fractured rock environments is the down-hole valving of the main flow line. The down-hole valve must be easily operable from the surface in a period of a few seconds. The speed of opening the valve controls the early build-up rate of pressure for instantaneous tests and also affects the duration of the pressure pulse in pulse tests.

Down-hole measuring instruments used in the test include pressure transducers and temperature sensors. Pressure transducers are used to measure hydraulic or pressure heads in the test zone, the packers, and, if possible, above and below the packers in the hole. A thermal sensor should be placed to provide an accurate reading of the test zone temperature conditions. The pressure transducers should be set for rapid data acquisition rates through computer based data storage.

Data Acquisition

Down-hole hydraulic testing in fractured rock is an extremely costly and exacting technique. Thus, efforts should be directed toward insuring that data necessary for interpretation and analysis of the results are not lost. Therefore, the digital system should be backed by an analog recording system, such as a set of strip chart recorders.

Data acquisition systems for low hydraulic conductivity fractured rock environments should contain both digital and analog filtering and recording capabilities for multiple data channels. The data acquisition system should be capable to recording at high rates, as fast as 1-2 readings per second during the early time of the test.

General

Injection water temperature must be maintained as closely as possible to the down-hole conditions since temperature can affect the accuracy of well testing. The injection water should also be chemically similar to ground-water down-hole. Water quality measurements of the borehole water can be used to identify incompatibility with fresh injection water, or if significantly different density conditions exist between the formation water and the injection fluids.

Reviews of well testing methods for low permeability rock have identified three major test methods for assessing hydrologic properties from single boreholes (Almen, et al., 1986; Golder Associates, 1987, Table 6-6). The methods are the: 1) constant-pressure injection test: 2) constant-rate injection test: and 3) pressure-pulse test (see Figure 6-64). The following section describes these well test methods, and discusses, in general, how the tests can be performed and analyzed for low hydraulic conductivity rocks.

Constant Pressure Injection Test

Constant pressure injection tests are one of the basic procedures typically used in low hydraulic conductivity rocks assessment programs. The test is performed using the following general steps:

1. Inflatable packers are lowered to the desired test interval.

2. The down-hole packers are inflated and the system is charged with water. The down-hole packers are inflated and the system is charged with water. If a down-hole valve is present in the system above the packers, the valve is shut in and the pressure is monitored until the system reaches stability.

3. The injection line is then pressurized to the predetermined constant head, and the valve opened to begin the test.

4. Constant pressure is maintained and the flow rate is monitored for a typical period of fifteen minutes to several hours.

5. The pressure is then released and the test section is allowed to recover and the pressure recovery data is recorded until reasonable levels of recovery are obtained for the system.

There are a number of methods available to analyze data from constant pressure injection tests to evaluate hydrologic properties of the test interval. It's important to obtain a complete picture of the hydraulic character of the

INJECTION TANK SYSTEM

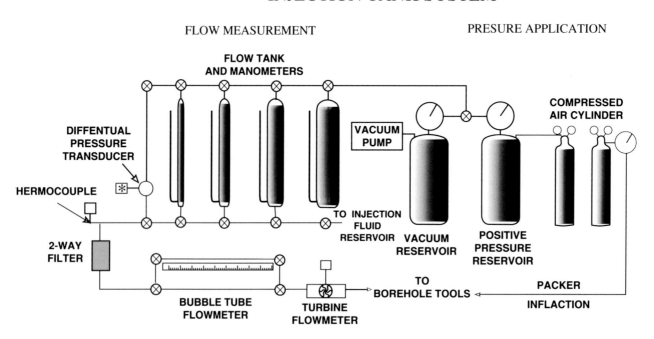

Source: Modified from Novakowskie (1991)　　**Figure 6-62** Configuration of Test Equipment

rock mass by the evaluation of the data both in the flowing period and the pressure recovery period as with any pump test procedure. Analyses for the flowing period would include a form of steady flow analysis using a selection of the following alternative plots of flow, Q, and time, t, for transient analysis:

- Log flow-log time for traditional type curve analysis;

- Semi-log analysis using 1/Q versus log t;

- Plot of 1/Q versus square root of t' and,

- Plot of 1/Q versus fourth root of t.

The pressure recovery data can also be analyzed using conventional pump test type curves methods for pressure-recovery analysis. Standard methods used for interpreting recovery after constant-flow testing (Uraiet, 1979, Ehlig-Economides, 1979) are summarized below:

- Log-pressure, log time analysis using constant rate type curves (the constant rate used is the last rate in the flowing period);

- Pressure recovery versus log-time;

- Horner plot of pressure recovery versus t+dt/dt (where dt is the time since pumping ceased and t is the time since pumping began);

- Plot of pressure recovery versus I/t (to check for spherical flow); and,

- Plot of pressure versus fourth root of t (bi-linear flow).

In addition to the above test methods transient test analysis can also provide data on the transmissivity and storativity of the test interval, as well as, well-bore storage, skin effects, hydraulic boundaries, and flow geometry, in a similar manner to near-field and far-field analysis used in aquifer pump tests for highly transmissive fractured

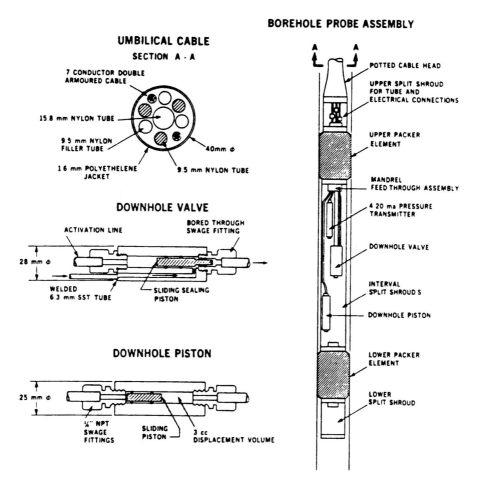

Figure 6-63 Down-hole Test Equipment

aquifers. Table 6-6 summarizes the parameters that may be obtained from each of the listed analyses as discussed by Almen et al. (1980) and Golder Associates (1987).

Since these tests are performed in a single stressed borehole the storativity of low hydraulic conductivity rock is much more difficult or impossible to fully quantify. The accuracy of both skin and boundary calculations is also directly tied to the storativity value obtained from well tests. Measurement of storativity is, an important aquifer parameter, hence, other methods of determination of storativity can be used in a site assessment. The available methods for radial flow tests in fractured low hydraulic conductivity environments include:

- type curve matching;

- calculation from semi-log plots of 1/Q and pressure versus time; and,

- calculations from well-bore storage duration.

Type curve matching is a standard method for calculation of storativity used since the 1930s. However, the determination of storativity requires a good match of the test data to type curves. In stiff rock, near-steady conditions may be established quickly and the near horizontal curve makes a good transient flow match difficult.

Storativity may also be estimated from the plot of 1/Q versus log time using Jacob's method (Jacob and Lohman, 1952). Calculation of S is obtained by the intersection of the straight line with the time axis, t_c using:

$$S = \frac{2.25 T t_c}{r_w^2}$$

Equation 6-14

This method also may be applied to semi-log measured recovery plots.

The duration of well-bore storage in the pressure recovery test can be used as an alternative method. Earlougher (1977) shows that well-bore storage ends when t_D is about 60. If transmissivity is known from other methods, such as the semi-log straight line or a type curve match, the storativity can be calculated from a rearranged definition of t_D, letting $t_D = 60$:

$$S = \frac{T t_w s}{60 r_w^2}$$

Equation 6-15

$t_w s$ = duration of well-bore storage in actual time. The stepped constant-pressure test is an important variation of the constant-pressure test. A stepped pressure test can be used to check validity of linear laminar flow conditions. An example of a pressure flow rate plot from a stepped constant rate test is shown in Figure 6-65. Stepped pressure tests are performed in a similar manner as aquifer step tests (with increasing Q) by increasing the constant pressure in the test by regular increments. The pressure increments should last ten minutes or until the flow rate is steady.

Constant Rate Injection Test

The second major test method that finds wide application in hydraulic conductivity testing of low permeability fractured rock is the constant rate injection test. The constant rate test is the basic test of both water and petroleum reservoir exploration, and there are many type curve solutions available for analysis of the data under a wide range of test conditions, configurations, flow geometry, and boundary conditions.

Constant rate injection tests are performed by the injection of water into a packed off section of a borehole at a constant flow rate, similar to the traditional packer permeability tests. Hydraulic properties of the rock are

Plot Type / Hydrologic Parameter	Transmissivity	Storativity	Skin	Wellbore Storage	Boundaries	Spherical Flow	Linear Flow	Radial Flow	Bilinear Flow	Deformation	Turbulence
Constant Pressure (Disturbance Period)											
P - Q (steady flow)	O									O	O
log t - log Q	O	O		O	O		O				
log t - 1/Q	O	O	O		O			O			
1/Q - $\sqrt{t}$	O					O					
1/Q - $\sqrt[3]{t}$	O								O		
Constant Rate (Disturbance Period)											
log t - log p	O	O		O	O		O				
log t - p	O	O	O		O			O			
p - $\sqrt{t}$	O					O					
p - $\sqrt[3]{t}$	O								O		
Recovery											
Horner plot: Δp - log (t+Δt) / Δt	O		O	O	O						
MDH, log t - Δp	O			O	O						
Δp - $\sqrt{t}$	O					O					
Δp - $\sqrt[4]{t}$	O								O		
log Δp - log t	O							O			
Slug /Pulse											
|p - p| vs. log t	O	O									

Source: Modified from Battelle (1987)

Table 6-6 Rock Parameters Obtainable from Tests

then determined from the observed transient pressure response to the injection. The test results may also be analyzed using the pressure recovery response following the injection.

A major limitation of constant rate tests (Golder, 1987) is the effect of near-field well bore storage. Well bore storage includes volume associated with the changes of water level in the hole, hydraulic performance of the test equipment, and actual deformation of the borehole wall. Constant rate injection tests can be performed and analyzed using procedures similar to those for the constant pressure injection:

• Packer sets are located to isolate a test zone, and then the down-hole valve is shut-in to stabilize the pressure;

• When the zone pressure stables the valve is opened and water is injected at a constant rate for a period of fifteen minutes to three hours, dependent on local well-bore storage effects; and,

• The valve is then shut and the pressure is monitored for a time equivalent to the duration of the injection period.

The data for constant rate tests can be analyzed using a similar approach as constant pressure tests using the following plots of pressure, p, and time, t:

• Type curve matching of log p and log t;

• Plot of pressure and log t;

• Plot of pressure and square root t; and,

• Plot of pressure and fourth root of t.

Pressure recovery analyses use similar plots as the constant pressure tests. The uses of the data analyses are tabulated in Table 6-6.

Pressure-Pulse Test

The third major method for determination of aquifer properties of low hydraulic conductivity rocks is the pressure pulse test. The pulse test method evaluates hydraulic properties from the pressure decay after introduction of a sudden change of pressure response in a test zone. The pulse test can be considered as analogous to a slug test when the pulse is a change of water level in a borehole. Alternatively, the pulse may be a pressure pulse to a test interval where the main pressure control valve has been shut. The pressure-pulse test is included in Table 6-6.

The pressure pulse method provides results more rapidly than slug tests since the volume change required to dissipate the pressure pulse is very small. Due to the small volumes of water involved, and the relatively low compressibility of the water, the pressure-pulse test has been reported very sensitive to both thermal and well history pressure effects, (Grisak et. al., 1985). Pressure-pulse tests are typically performed as follows:

• The packers are set at the required interval and the down-hole valve is shut off to stabilize pressure;

• When the pressure stabilizes, the water level in the injection line is raised or lowered to an elevation above or below the head in the test interval;

• The test is begun by opening the down-hole valve;

• For a pressure-pulse test the valve must be closed immediately, as soon as the test interval pressure has equilibrated with the pressure in the tubing;

• For a slug test the valve would remain open, and the pressure decay is measured through the change in water level; and,

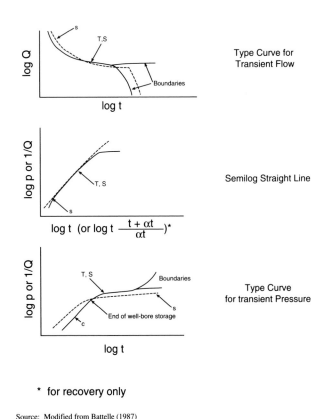

* for recovery only

Source: Modified from Battelle (1987)

Figure 6-64 Type Curve Analysis Alternatives

• The test should continue for three hours or until the pressure in the test zone has recovered to at least 70 percent of the stable pretest value.

Pressure pulse test data can be analyzed using the type curves of Bredehoeft and Papadapoulos (1980) for pulse tests and Cooper et al. (1967) for slug tests. Alternative methods for fracture analysis are available such as Wang, et al. (1977) for pulse tests or Barker and Black (1983) for slug tests.

Calibration Of Pressure Tests

The results of any hydraulic testing of fractured rocks are highly dependent on proper calibration of the pump test equipment used in the test program. Injection/measurement systems for fractured rock assessments can range from complex mechanical/hydraulic/electrical systems to simple flow meter/packer systems. Since the quantity of water injected in low hydraulic conductivity testing is so small these systems require full calibration to obtain reliable test results.

Calibration of instruments used in hydraulic conductivity testing is typically performed in parallel with the set-up of the equipment. All flow meters used in hydraulic conductivity testing should be fully calibrated over the expected range of application. In case of flow devices calibrated at the factory, turbine meters or other direct flow measuring devices should be checked by measuring the volume of flow over a specified time under steady flow conditions. The flow calibration may be checked either by a graduated container or by water weight. The temperature of the fluid should be noted for use in determining the proper water density if weight calculations are used in the calibration procedure. Temperature sensors should also be calibrated against a reliable standard, such as a high accuracy laboratory thermometer. Pressure transducers can be calibrated against a dead weight tester. If the range of application of the transducer in the test is less than half of the transducer's rated range, the calibration should be performed in ten steps over the expected range of application.

Closed flow tanks are also commonly used for constant pressure flow measurement tests. In a flow tank set-up, the volume of injected fluid is measured by observing the change of water level in the tank with time. The water in such a system level is established using a differential pressure transducer. The tank and the transducer may be calibrated together by draining the tank into a graduated container and noting volume of water discharged for a

given change in water level (or differential pressure reading). Tank calibrations are typically performed at three pressures: atmospheric pressure, half the expected maximum injection pressure, and the maximum expected injection pressure.

Downhole packer systems should be tested in a length of equivalent diameter casing to check for packer leakage and for packer system performance. Performance of the packer system is tested as follows:

• The packers are placed in the casing and inflated. Care will be taken to assure that the zone between the packers is water-filled and completely free of air;

• A known volume of water is then injected into the straddle zone, and the resulting increase in pressure is measured; and,

• This test should be performed in at least five pressure steps up to the expected injection pressure in the field.

The packer system should be bench scale tested before use in field testing situations. The straddle packer assembly is placed in a well casing then the portion of the casing lying between the packers contains a valve that is opened to simulate the flow of water into the rock. A simulated calibration test can be run with the data acquisition system fully installed to check the performance of instruments and the collection of the data.

Electronic recording devices must also be included in the calibrator procedure. A strip-chart recorder is tested by generating a range of known calibrated voltages and

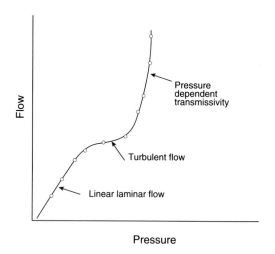

Source: Modified from Battelle (1987)

Figure 6-65 Stepped Constant Rate Field Test

measuring the displacement of the chart pens using at least five points over the channel's voltage range.

6.12 MULTI LEVEL HEAD MEASUREMENTS

How test holes drilled for lithology and sampling should be completed has been the subject of considerable controversy. Experience on projects over the past number years has shown that wells, regardless of their initial purpose, become construed as "monitoring wells" and that the data from these wells are subject to criticism, are often used improperly (which leads to erroneous results), and frequently create difficulties with regulatory agencies.

One should always locate well screens and piezometer tips based on fractures observed in surface mapping, core holes, geophysical and hydraulic conductivity testing. Precise vertical placement of these screens or open areas in zones that can yield water is also extremely important for the overall success of the monitoring and assessment program.

The need to obtain detailed information on the vertical distribution of hydraulic head, combined with the escalating cost of drilling, has led to the development of multi-level piezometer installations. This installation may also offer the possibility of small quantity sampling of the ground water at more than one depth in a single borehole.

Short screens such as those used in most multilevel piezometer installations and in many single-level installations provide hydraulic head information pertinent to relatively small zones of the aquifer. This kind of information is necessary for detailed mapping of target monitoring zones in a vertical sense. Multi-level hydraulic head data when combined with hydraulic conductivity and lithology allow the evaluation of flow net data with conceptual models. Generally this approach requires that sufficient head measurement points be installed to give consistent trends in both horizontal and vertical gradients. Multiple installations contained within a single borehole can offer a cost effective method for these important fractured rock head measurements

Chapter 5.0 provides information on the selection and installation of head level measuring instruments that are based on the ability of the geologic material to yield water to a monitoring zone. Fractured rocks can cause extreme difficulties in representative measurement of heads, since some zones with little or no fracturing have limited ability to yield water to the measuring device. As general guidance one should fit the ability of the borehole zone to yield water to the device to be used in the zone. Hvorslev (1949) provides guidance in selection of the instrumentation by its 90% response time. Although various types of piezometers can react more rapidly due to

low water requirements for the devices, these fast reacting installations cannot supply a water sample for water quality evaluation. As such, one should evaluate the need for the system recovery rates in the fractured rock environment. If the fracture system has the ability to yield water at a rate sufficient for the environmental assessment, then piezometer completions (such as open tube design) that have relatively large water needs can be used in the system. However, if the rock mass only yields water slowly and accurate readouts of head levels are necessary, then piezometer completions may be required that are not amenable to later water quality sampling.

One should fully review the need for such devices and evaluate the project and system requirements before selecting the multi-point system. A recent comparison of several types of multi-point systems was described by Ridgeway and Larssen (1990). They regarded selection of multi-point installation should be considered in the context of the overall technical requirements and objectives of a project. A number of commercial multi-point systems are described below that represent only several of the many systems currently available.

A multi-port installation suitable for fractured rock systems is marketed by Westbay Instruments, Ltd. The Westbay system was described in Patton (1989) and Ridgeway and Larssen (1990). Figure 6-66 shows a comparison between Westbay and Solist multi-point design. It consists of a single string of coupled PVC pipe with a series of ports along its length. Packers are installed around the PVC pipe to isolate ports from each other. The two types of ports coupled into the system consist of pressure-measurement and water-sampling port, and a port with a screened opening for pumping larger quantities of water from the formation. The ports that are used for pressure measurements and water sampling are operated by means of control mechanisms mounted on monitoring tools. These tools are moved up and down the casing to open selected ports. The tool used for pressure measurement contains a surface readout pressure transducer. The tool for sampling contains a pneumatic bailer. Vispi (1980) reported that the pressure measuring device was found to be satisfactory (giving results that were comparable with those obtained from traditional standpipe piezometers), but that the pneumatic bailer tool was very slow to operate.

The system has been reported expensive, but becomes economically viable in situations where many monitoring points are required at great depths (Vispi, 1980). The majority of the expense is tied to the Westbay probe that can be used in many monitoring locations and thus spread costs over the entire system. The Westbay system is also particularly useful in fractured rock where many sampling levels are necessary because of the complexity of the

WESTBAY MP MONITORING WELL   WATERLOO MONITORING WELL

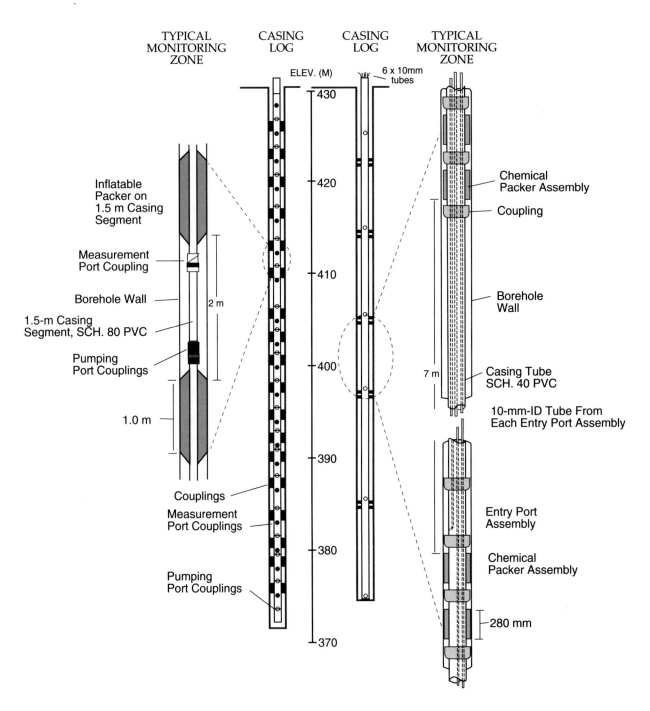

Source: Modified from Ridgway and Larssen (1990)

Figure 6-66 Two Types of Multi-point Piezometers

permeability network and the uncertainty with regard to the specific zones through which potential contaminants will travel.

The Solist Waterloo multiport system uses a PVC casing string to isolate and protect multiple sampling tubes. The casing is of a modular construction, using flush threaded joints, fitted with O-ring seals. For boreholes in rock a unique feature of the system is the specially designed self-inflating packer module. Packers are used above and below each sampling port module, to isolate the strata to be sampled. From each port module a small diameter sampling tube passes up the center of the casing string to the surface (see Figure 6-67).

Up to six sampling tubes can be installed in a 3"(N) borehole and 10 tubes in a 4"(H sized installation. Samples can be obtained using dedicated samplers in each port, or a portable sampling pump. For boreholes in overburden or as an alternate method in rock, packers can be omitted from the casing string. Sampling points are then isolated in the conventional manner, using bentonite and sand. Installation would normally be in larger diameter boreholes.

Installation of the Waterloo multilevel sampling system is achieved quickly (see Figures 6-67 and 6-68) as it is assembled as it is lowered into the borehole. As each port is put into position a new sampling tube is connected. Each successive module or section of casing is then threaded over these sampling tubes. Water is added to the inside of the PVC casing string, as needed, to counteract buoyancy. The process continues until the entire string is installed. The result is a set of small diameter tubes, each connected to an individually isolated sampling port, which terminate in a manifold.

The port module, as shown, has a stainless steel stem that connects the small diameter sampling tube to the port. The port is isolated at the required location with packer units both immediately above and immediately below it. For dedicated use of Triple Tube or Double Valve pumps, these may be fitted within the small diameter tubing directly above the port. Pneumatic or Vibrating Wire Pressure Transducers may also be dedicated within separate port modules.

The packer modules consist of a length of slotted PVC casing onto which is mounted a sleeve of Dowell chemical sealant, held in place by a gum rubber sheath. The packers are strengthened at each end with woven Kevlar cuffs and securely fixed in place by stainless steel clamps at either end. The packer chemical is activated when water is poured down the central PVC string, during installation. Over the following 24 hours the chemical swells and pushes the rubber against the side of the borehole wall, to form a tight seal. The process of expansion continues for some time, ensuring a permanent seal between sampling locations.

Figure 6-67 Multi-point Equipment Before Installation

When a portable pump is to be used in conjunction with the Waterloo System, water level measurements can be taken, with a dipmeter, directly in the sampling tubes. When dedicated pumps are to be used, direct water level measurements cannot be taken. Water pressure may be estimated by calculation of the volume of water extracted. Alternatively, additional ports may be installed at each location. These would allow direct water level measurement, or could be fitted with dedicated pneumatic or vibrating wire pressure transducers.

The author favors the use of small diameter piezometers (1 - 3/4 inch) or multi-port systems where measurements of water heads are the primary data collected. In general, those holes penetrating the water table would have at least a single port (open tube) piezometer installed. This installation can be rapid since only filter sand covering the screened zone (plus three feet over the screened zone), then the piezometer can be grouted to the surface with an expansive grout. The

surface completion should be similar to the standard for monitoring wells to protect the surface stick-up of the piezometer.

6.13 LABORATORY TESTS

Laboratory tests of overburden soils collected during the field drilling task defines four basic properties of soils and rock and their suitability for use on landfill sites. These soil characteristics can be summarized:

• Strength
• Compressibility
• Hydraulic conductivity
• Chemistry

Section 5.0 describes laboratory tests are designed to provide information on the quantity and physical characteristics of both borrow cut areas and landfill foundation areas. Guidelines for the numbers of tests are given below; however, good engineering judgment must apply to adequately describe the quality and quantity of the individual soils and rocks located on the property.

The investigations will generally consist of (1) either drilling and sampling soil borings or excavating shallow test pits, (2) obtaining bulk samples of representative soils, and (3) evaluating these samples for use as borrow material. Overburden soil samples should be evaluated by performing the following laboratory tests as appropriate:

A. Particle-size analysis of soils (ASTM D422-63).

B. Liquid and plastic limits (ASTM D4318-83, cohesive soils).

C. Classification of soils for engineering purposes (ASTM D2487 - 69).

D. Moisture/density relationship using five compaction points (standard proctor) to determine the maximum dry density and ultimate moisture content (ASTM D698-78).

E. Permeability at 90% of standard proctor density within +2 percentage points of optimum moisture content. A falling head test method using back pressure can be done in general accordance with Corps of Engineers' Manual EM 1110 2-1906, "Laboratory Soils Testing," Appendix VII.7.

Figure 6-68 Multi-point Equipment Installation

The particle-size, liquid and plastic limit tests should (cohesive soils) be performed so that approximately 10 tests are completed for each soil type. Care must also be taken to select particle-size test samples for those soil zones being selected for piezometer monitoring. The classification of soils for engineering purposes must be made initially during drilling and reviewed during laboratory testing to confirm the classification. Moisture density tests should be made at the rate of five per each material type found in the drilling. Hydraulic conductivity tests should be run for each material type, unless sound engineering judgment suggests that additional tests are required.

6.14 GEOLOGIC/HYDROGEOLOGIC CONCEPTUAL MODEL

After collection of the geologic, hydraulic, and chemical data, an assessment must be made as to the ground-water heads, quality and the pattern of fractures in the bedrock. This assessment should include several analyses, as detailed below.

6.14.1 Data Analysis

Fracture analysis must be performed to refine the structural geologic information previously collected during the Phase I and Phase II field rock coring program. The investigator should analyze the data by stereographic projection plots rather than by simple rose diagrams. Stereographic projection graphically presents the strike and dip angles of fractures rather than only the strike as with rose diagrams. This method graphically group fractures by similar strike and dip and the dominant joint sets can easily be identified. Fracture orientation data are most conveniently represented on equal-area projection diagrams in which the contours define areas of equal pole density (percentage poles per unit area) on the lower hemisphere. Fracture analyses should include both core fractures and surface fractures measured during geologic mapping. Surface data are typically biased toward vertical and sub-vertical fractures, and data from vertical boreholes are biased toward horizontal and sub-horizontal fractures. The bias is reduced by combining the surface and the borehole data in a single diagram. In addition, core fractures should be reviewed for indications of water flow of the existence of infilling. Results from down-hole packer tests should be compared with the observed fractures. Data analysis can be grouped into: lithologic, hydraulic conductivity and gradient analysis.

Lithologic Analysis must be performed and should include interpretation of the geophysical logs, drill cuttings or cores, outcrop investigation, and review of the regional geologic literature. The results of this analysis should make comparison of the bedrock at the site to the regional geologic formations and members. This type of comparison is important if other regional information such as fracture patterns and ambient geochemistry are to be utilized in the investigation.

Hydraulic conductivity analysis should include a qualitative review of the geophysical logs, packer permeability tests, and analysis of the rising head tests performed in each well or piezometer. This analysis will result in identification of zones of high and low hydraulic conductivity in the bedrock that can be correlated with lithology and fracture patterns. Relationships between depth and hydraulic conductivity should be diagrammed for each hydrostratigraphic unit and composite for all units on site.

Gradient analysis must include both vertical and horizontal gradients and should identify the directions and magnitudes of the driving forces acting on the ground-water system. Multiple completion piezometers, if installed, should provide ground-water head data within discrete strata in the bedrock and should allow determination of vertical gradients. By plotting the ground-water levels on the site, the investigator should prepare a water table or potentiometric head contour map. This data, when coupled with hydraulic conductivity and fracture orientation information, should allow estimation of ground-water flow direction and velocity.

6.14.2 Conceptual Models For Flow Geometry

The important result of a Phase II investigation analysis is the construction of the conceptual model that must incorporate all the essential features of the fractured physical system under study. With this constraint, the conceptualization is tailored to an appropriate level of detail or sophistication for the problem under study. The degree of accuracy required for various problems differs. For simple installation of a few upgradient and three downgradient wells, the conceptual model presented in the phase I report may sufficiently define the geology, uppermost aquifer, and ground-water flow directions in relatively simple terms. Difficult geology, complex ground-water flow directions with vertical and horizontal components would require more involved descriptions of the physical system. The conceptual model must also be tailored to the amount, quality, and type of data available.

For fractured rock systems, a full phase II conceptual model formulation must be completed before locating monitoring wells. This model is extremely important due to the extreme ground-water directional variability normally observed in the fractured rock geologic environment.

The sophistication of the conceptual model thus depends on project objectives, data availability, aquifer geometry and features, project facilities and resources, project budget, project schedule, and site access.

Feasibility level studies involve less time, money, and data, and require only a fair "accuracy" in their predictions. Consequently, a much simpler conceptual model may be called for. Final designs and/or large-scale analysis involving significant financial investments or risks require complete information, more time and money, and a more sophisticated conceptualization of the problem. Once the conceptual model is proposed in Phase I, we may recognize the need for additional data. Erroneous data due to sampling errors, or the scarcity of sampling points can lead to an inappropriate conceptual model. Incorporating this data into the model should reveal these errors, usually through inconsistencies. The erroneous data can be ignored or replaced if other information is available. The final Phase II conceptual model, however, should directly point to the appropriate locations for the monitoring wells.

Fractured rock ground-water flow systems can be interpreted by aquifer stress tests. These interpretation methods for well tests date back to the 1930's generally consider homogeneous aquifers (or equivalent porosity models for fractures) and very simple geometry. The most common geometry for most aquifer system analyses is that of radial flow model dictated by the application of analytical solutions to Darcy's equation. However, as described a number of times in this chapter, fractured rock systems may not behave as an equivalent porous system.

One of the important steps in conceptualization of a fractured rock aquifer system is the definition of the flow geometry. The three major conceptual flow geometry are radial flow, linear flow, and spherical flow (Figure 6-69). The actual geometry of the flow relative to a fractured rock environment may be deduced from analysis of the flow or pressure recovery data during or after a stress (pumping) test of the system. The results may exhibit one or all of these forms during a test, the typical progression from lower dimension (linear) to higher dimension (spherical) as the pressure effect spreads to larger volumes of rock. This spreading to larger volumes of rock mass during a pump test allows the use of radial equations for calculations of aquifer properties.

Linear flow conditions are represented by flow lines that do not diverge away from the hole; therefore, the cross-sectional area of flow does not increase with distance from the well. Linear flow conditions are likely to occur with vertical fractures or if the fracture flow is strongly channeled. Channeling is caused by flow in one or more independent stream tubes as may be represented in Karst environments.. Linear flow conditions can be evaluated using straight line plots of pump test data with log 1/Q and log time having a slope between 0.25 and 0.5. Alternatively, linear flow appears as a straight line in a plot of 1/Q against the square root or fourth root of time (for linear and bi-linear flow, respectively).

Radial flow condition in fractured rock is typified by flow lines that radiate in all directions perpendicular to the hole in fractures primarily normal to the borehole. The fractures should have more-or-less uniform mass transmissivity to properly assess such flow systems. Alternatively the pump test should continue for long time periods to incorporate significant volumes pf bedrock into the test. Radial flow can be characterized in pump test data by a straight line relationship between 1/Q and log time.

Spherical flow may be observed if the flow lines diverge in all directions from the test zone. Spherical flow conditions may be expected in permeable rock where the flow diverges homogeneously in all directions from the test zone. A straight line on a plot of 1/Q versus the inverse square root of time indicates spherical flow conditions.

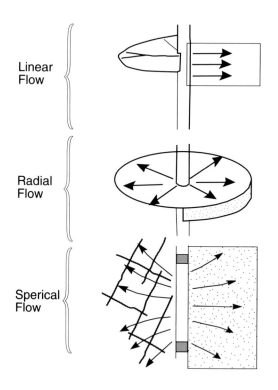

Source: Modified from Battelle (1987)

Figure 6-69 Conceptual Models for Fractured Flow

Interpretations of flow geometric should be tempered with the knowledge that even simple conceptual fractured rock models, can have a composite fracture pattern that confuse interpretations as to flow conditions. Channeling of flow in discrete tubes, may lead to a linear flow model, even in fractures that macroscopically have a radial geometry with respect to the borehole. Fracture flow test results may also produce evidence of mixed flow geometry, such as early time linear flow followed by radial flow in later time. This flow behavior would allow use of equivalent porosity models during the evaluation of remedial pumping alternatives.

Evaluation of fractured rock environments can be very difficult and require a combination of techniques to fully understand flow conditions. Examination of borehole test records from zones with fractures having a variety of orientations (varying from vertical to horizontal with respect to the borehole axis) should be made to assess flow in fracture patterns, since they may provide indications that flow geometry and resultant leachate migration controlled by fracture orientation. The combination of borehole core logs, geophysical caliper logging and results of hydraulic testing of the borehole provides insight into flow conditions present in the bedrock.

The hydraulic conductivity of the fracture system of the rock formation as a whole is generally of more interest than the ability of an individual fracture to transmit water. The hydraulic conductivity cannot be estimated, of course, unless the mass of rock is sufficiently large. The hydraulic conductivity of the rock mass as a whole depends on the collective conductivity of each of the fractures of an interconnecting system, in other words, on the extent to which they are open, on the number of fractures per unit of volume and, especially, on the degree of inter-communication between fractures. When fractures are numerous and well interconnected, pumping tests can provide approximate values of hydraulic conductivity directional components. Long term pump tests, or series of individual packer permeability tests that evaluates sufficient portions of the rock mass can provide insight into the flow conditions present at site. One should not lose sight of the "big picture" in such evaluations. Head levels obtained from piezometers combined with field drilling observations of lithology, hydraulic conductivity, and a major share of professional experience with such environments are the main focus of the conceptual process. Only rough estimates and comparisons may be possible. One rock mass may be more permeable than another because the fractures are more open and more numerous as

a result of its being more affected by tectonic movements. An opinion may be finally based only on field observations.

This opinion requires a large dose of experience that should be backed-up by quantifiable testing of the inplace rock mass. In general, the larger the volume of rock tested, the more likely the field observations and evaluations with approach to the real value or directional component of flow.

6.15 LOCATION AND INSTALLATION OF MONITORING WELLS

The evaluation of a fractured rock flow system often requires technically difficult field testing and assessments tied together with a large measure of interpretation. As such the above chapter provided many of the technical "tools" of fractured rock investigations. Some fractured rock investigations may be relatively simple with ground water moving as in a equivalent primary porosity geologic material. Other fractured rock systems may befuddle the most experienced hydrogeologist using all the technical tools of his trade. Although, one may be able to evaluate a fractured rock system with a small number of well placed boreholes and a lot of luck, to arrive at a correct conceptualization, many of the tools provided in this chapter will ease the conceptualization process. Once the field and office data has been completely analyzed and a full conceptual model completed, the responsible investigator can move toward locating the detection or assessment monitoring wells. Locations and conceptual depths of each monitoring point should be located in the field for the subsequent drilling and well installations.

A considerable amount of hydrogeologic data should also be obtainable during the air-rotary drilling of the monitoring well holes. These data should include:

- Where ground water is found (i.e., throughout a water-bearing unit or only within joints and bedding plane partings);

- Water levels (after the borehole has recovered from air-lift pumping effects);

- Conductivity of the water (an indicator of the gross parameter, TDS;

- Rough estimates of transmissivity (from short-term air-lift recovery tests); and,

• Penetration rates, rod drops, and air loss zones.

All monitoring wells should be completed as per the ASTM monitoring well recommended standards D5092. These standards provide location, installation, design and completion and documentation guidance for the monitoring program.

6.16 REFERENCES

Almen,K-E., J. E. Andersson, L. Carlsson, K. Hansson, and N-A. Larsson, 1986. Hydraulic Testing in Crystalline Rock, A Comparative Study of a Single Hole Test Methods, Technical Report 86-27, Swedish Geological Company, Uppsala, Sweden.

ASTM, 1991. D4318–83, Liquid and Plastic Limits Annual Book of ASTM Standards, American Society for Testing and Materials, Philadelphia, vol. 04.08, pp

Barker, J. A., and J. H. Black, 1983. Slug Tests in Fishered Aquifers, Water Resources Research, Vol. 19, No. 6, pp. 1558-1564.

Battelle Laboratories, Office of Waste Technology Development Report No. 833-1017-105. 1987, Evaluation of Hydrologic Test Parameters. 49 p.

Benson, R.C. 1988. Surface and downhole geophysical techniques for hazardous waste site investigation. Hazardous Waste Control, Vol. 1, No. 2.

Billings M. P. 1954. Structural Geology, 2nd. Ed., Prentice-Hall, New York.

Boulton, N.S., and T. D. Streltsova , 1977. Unsteady Flow to a Pumped Well in a Fissured Waterbearing Formation, Journal of Hydrology, Vol. 3, p124-130.

Bredehoeft, J. D., and S. S. Papadopulos, 1980. A Method for Determining the Hydraulic Properties of Tight Formations, Water Resources Research. Vol. 16, No. 1, pp. 233-238.

Bruin, J. and H. E. Hudson, 1955. Selected Methods for Pumping Test Analysis, Illinois State Water Survey, Report of Investigation 25.

CANMET, 1977. Ground-water, Pit Slope Manual, Report No. 77-01.

Carlsson, L., A. Winberg, and B. Grundfelt, 1983. Model Calculations of the Groundwater Flow at Finnsjon. Fjallveden. Gidea and Kamlunoe. KBS Technical Report TR 83-45.

CASECO Consultants, Ltd. 1964. Field permeability tests in soil and rock for the Mica Creek Project.

Cedergren, H.R. 1967. Seepage, drainage, and flow nets. John Wiley & Sons, New York, pp. 184–204.

Cooper, H.H., Jr.; J.D. Bredehoeft; and I.S. Papadopulos. 1967. Response of a finite–diameter well to an instantaneous charge of water. Water Resources Research, Vol. 3, pp. 263–269.

Cooper,H. H., and C. E. Jacob, 1946. A Generalized Graphical Method of Evaluating Formation Constants and Summarizing Well Field History, Am.Geophys. Union Trans. 27(4).

Corps of Engineers' Manual EM 1110 2-1906, "Laboratory Soils Testing," Appendix VII.7.

Dames & Moore, 1974. Manual of ground–water practices.

Davis, C. V. & K. E. Sorensen, 1969. Handbook of Applied Hydraulics, 3rd Ed., McGraw-Hill, New York, N.Y.

Davis, S. N. and R. J. M. DeWiess,1966. Hydrogeology, John Wiley, New York, N.Y.

Davison, C. C., and G. R. Simmons, 1983. "The Research Program at the Canadian Underground Research Laboratory," Proceeding of the NEA Workshop on Geological Disposal of Radioactive Waste and In Situ Experiments in Granite. (October 25-27, 1982, Stockholm, Sweden) pp. 197-219, Nuclear Energy Agency, Organization for Economic Cooperation and Development, Paris, France.

Deere, D. U., 1968. Geological Considerations, Rock Mechanics in Engineering Practice, K. G. Stagg and O. C. Zienkiewcz, Eds., Wiley, New York, pp. 1-20.

Doe, T. W., and J. D. Osnes, 1985. Interpretation of Fracture Geometry from Well Tests, International Symposium on Fundamentals of Rock Joints. (September 15-20, 1985), Bjorkliden, Sweden.

Earlougher, R. C., Jr., 1977. Advances in Well Test Analysis. Monograph Volume 5, Society of Petroleum Engineers of AIME.

Ehlig-Economides, C. A., 1979. Well Test Analysis for Wells Produced at Constant Pressure, Ph.D. Thesis, Stanford University.

Elsworth, D., and T. W. Doe, 1986. Application of Non-Linear Flow Laws in Determining Rock Fissure Geometry From Single Borehole Pumping Tests," International Journal of Rock Mechanics. Mineral Science. and Geomechanics Abstracts. Vol. 23, No. 3, pp. 245-254.

Freeman, H. A., and S. G. Natanson, 1959. Recovery Problems in a Fractured-pore system, Kinkuk Field, Proc. 5th World Pet. Congr., Sec. II pp. 297-317.

Freeze, R.A., and J.A. Cherry. 1979. Ground water. Prentice–Hall, Inc., NJ, 604 pp.

Gale J. E., 1984. Class notes on Fractured Rock Hydrology.

Gale J.E., 1982, Assessing the Permeability Characteristics of Fractured Rock, Recent Trends in Hydrogeology, G.S.A. Special Paper 189 Boulder Co. 446p.

Gale J.E., 1982. Assessing the Permeability Characteristics

of Fractured Rock, Recent Trends in Hydrogeology, G.S.A. Special Paper 189 Boulder Co. 446p.

Golder Associates, 1987. Scoping Evaluation of Hydrologic Methods for Testing Single Holes in Low Permeability Rock, Report to Battelle Office of Waste Technology Development, 10 p.

Goodman, R. E. 1976. Methods of Geological Engineering in Discontinuous Rocks, West Publishing, N. Y. 422 p.

Gringarten and Ramey, 1974. Unsteady-state pressure distributions created by a Well With a Single Horizontal Fracture, Partial Penetration, or Restricted Entry

Gringarten, A. C., and P. A. Witherspoon , 1972. A Method of Analyzing pumping Test Data from Fractured Aquifers, Proc. Symp. Percolation in Fractured Rock, Int. Soc. Rock Mech., Stuttgart T3, pp.131-139

Gringarten, A. C., 1982. Flow Test Evaluation of Fractured Reservoirs, in: Proc. Symp. Recent Trends in Hydrogeology, Feb. 8-9, 1979, Lawrence Berkeley Laboratory, Calif., Geol. Soc. of America, Special paper, No. 189, pp.237-264.

Grisak, G. E., J. F. Pickens, D. W. Belanger, and J. D. Avis, 1985. Hydrogeological Testing of Crystalline Rocks During the NAGRA Deep Drilling Program. Technisher Bericht 85-08, GTC Geologic Testing Consultants, Ottawa, Ontario, and Intera Technologies, Houston, Texas.

Harza, Engineering Co. 1972. Memorandum on Water Pressure Tests, Chicago Il.

Hodgson F. D. I., 1983. Evaluation of the Pering Ground-water Investigation, Cape Provence, Institute for Ground-water Studies, University of the O.F.S., BLOEMFONTEIN 9300, RSA.

Hvorslev, M. J., 1949. Subsurface Exploration and Sampling of Soils for Civil Engineering Purposes, U.S. Army Corps of Engineers, Waterways Experimental Station, Vicksburg, Mississippi. International Association of Hydrogeologists, 1985. Hydrogeology of Rocks of Low Permeability, Vol XVII

ISRM, 1978. Commission on Standardization of Field and Lab Tests, Suggested Methods for the Quantitative Description of Discontinuities in Rock Masses, Rock Mech. Min. Sci. and Geomech. Abstr. Vol 15, Pergammon Press Ltd. Great Britian, pp 319-368.

Jacob, C. E., 1946. Drawdown Test to Determine Effective Radius of Artesian Well, Proc. Am. Soc. Civil Engrs., 72 (5).Jacob, C. E., and S. Lohman, 1952. Non-steady Flow to a Well of Constant Drawdown in an Extensive Aquifer, Transactions of the American Geophysical Union, Vol. 33, No. 4, pp. 559-569.

Keys, W. S., 1989. Borehole Geophysics Applied to Ground water Investigations, National Water Well Association, Dublin, OH, 313 p.

Kruseman G. P. and N. A. de Ridder, 1990. Analysis and Evaluation of Pumping Test Data, Second Edition, ILRI Publication 47, P. O. Box 45, 6700 AA Wageningen, The Netherlands, 377 p.

Lewis, C. and R. H. Burgy, 1963. Hydraulic Characteristics of Fractured and Jointed Rock, National Water Well Exposition, September 1963, San Francisco, Cal. Ground Water, Vol. 2, No. 3, pp. 4-9

Lewis, C. and R. H. Burgy, 1964. The Relationship Between Oak Tree Roots and Ground Water in Fractured Rock as Determined by Tritum Tracing, J. Geophys. Res., Vol. 69, No. 12, pp. 2579-2588.Lohman, S.W. 1972. Ground–water hydraulics. U.S.G.S. Prof. Paper 708, 70 pp.

Louis, C., and T. Maini, 1970. Determination of In-Situ Hydraulic Properties in Jointed Rock, Proceedings of the 2nd Congress International Society of Rock Mechanics. Belgrade, Yugoslavia, Vol. 1, pp. 235-245.

Mathews, C. S., and D. G. Russel, 1967. Pressure Build-up and Flow Tests in Wells, Soc. Pet. Eng. Monogr. Ser., Vol. 1.

Moore, G. K., 1973, Hydraulic of Sheetlike Solution Cavities, Ground Water, Vol. 11, No. 4, pp. 4-11.

Noorishad, J., and T. Doe, 1982. Numerical Simulation of Fluid Injection into Deformable Fractures, Proceeding of the 23rd U.S. Rock Mechanics Symposium." Issues in Rock Mechanics," Society of Mining Engineers, AIME, pp. 645-663.

Novakowski, K. S. 1991. Class notes on DNAPL contamination, Univ of Waterloo short course.

Papadopulos, I. S., and H. H. Cooper, 1967. Drawdown in a Well of Large Diameter. Water Resources Research. Vol. 3, No. 1, pp. 241-244.

Patton F. D. 1989. Ground Water Instrumentation for Determining the Effect of Minor Geologic Details on Engineering Projects, The Art and Science of Geotechnical Engineering at the Dawn of the 21st. Century, a volume honoring Ralph B. Peck, Univ of Il. Urbana, pp 73-93

Piteau, D. R., 1971. Geological Factors Significant to the Stability of Slopes Cut in Rock, Symp. on Planning Open Pit Mines, Johannesburg, Balkema, Amsterdam pp. 33-53.

Raven, K. G. et. al., 1985. Field Investigations of a Small Ground Water Flow System in Fractured Monzonitic Gneiss, Symposium on the Hydrogeology of Low Permeability Rocks, pp 72-82.

Ramey, H. J. 1982. Well-loss Function and the Skin Effect, A Review. in Proc. Symp. Recent Trends in

Hydrogeology, Geol. Soc. of America, Special Paper No. 189, pp 265-272.

Ridgeway, W.R., D. Larssen , 1990. A Comparison of Two Multiple-level Ground-Water Monitoring Systems, In Ground Water and Vadose Zone Monitoring, Nielsen/Johnson editors, ASTM STP 1053.

Rorabaugh, M.I. 1953. Graphical and theoretical analysis of step–drawdown test of artesian wells. Am. Soc. Civil Engineers Trans., Vol. 79, Separate 362, 23 pp.

Smith, E. D. and N. D. Vaughan, 1985. Experence with Aquifer Testing and Analysis in Fractured Low-permeability Sedimentary Rocks Exhibiting Non-radial Pumping Response, Report DOE Contract No. DE-AC05-840R21400.

Snow D. T. 1965. A Parallel Plate Model of Fractured Permeable Media, Ph-D Dissertation, Univ. of Calif., Berkeley

Snow, D. T., 1970. The Frequency and Apertures of Fractures in Rock, Journal of Rock Mechanics and Mineral Science. Vol. 7, pp. 23-40.

Streltsova-Adams, T. D., 1978, Well Hydraulics in Heterogeneous Aquifer Formations, In: Advances in Hydroscience, Vol. II, New York, Academic Press, pp. 357-423.

U.S. Bureau of Reclamation. 1960. Earth manual. U.S.B.R. Method E-18, U.S. Government Printing Office, Washington, D.C.

UNESCO, 1984. Ground Water in Hard Rocks, Report 33, [II] SC.83/XX-33/A, 228 pp.

Uraiet, A. A. ,1979. Transient Pressure Behavior in a Cylindrical Reservoir Produced by a Well at Constant Bottom Hole Pressure, Ph.D. Thesis, Univ. of Tulsa. OK.

Vispi, M. A., 1980. Multiple-port Piezometer Installation at Rocky Mountain Arsenal, Report No. WES/MP/GL-80-12.

Walton, W. C., 1970. Groundwater Resource Evaluation, McGraw-Hill, New York.

Wang, J. S. Y., T. N. Narasimhan, C. F. Tsang, and P. A. Witherspoon, 1978. Transient Flow in Tight Formations," Proceedings of Invitational Well-Testing Symposium, (October 19-21, 1977, Berkeley, California), LBL-7027, Lawrence Berkeley Laboratory, Berkeley, California.

Wilson , R. C.and P. A. Witherspoon, 1970. An Investigation of Laminar Flow in Fractured Porous Rocks,Univ. of California, Berkeley, Civil Engineering Department, (Pub. No. 70-6).

Wilson , R. C.and P. A. Witherspoon, 1974. Steady State Flow in Rigid Networks of Fractures, Water Resour. Res., Vol. 10, No. 2, pp. 328-335.

Wilson, R. C., 1970. An Investigation of Laminar Flow on Fractured Porous Rock, Ph.D. Thesis, Civil Engineering Department, Institute of Transport and Traffic Engineering, Univ. of California, Berkeley.

Witherspoon, P. A., et al., 1979a. Observations of a Potential Size Effect in Experimentatal Determination of the Hyrdaulic Properties of Fractures, Water Resour. Res., Vol. 15, No.5, pp. 1142-1146.

Witherspoon, P. A., et al., 1979b. Validity of Cubic Law for Fluid Flow in a Deformable Rock Fracture, Special report from Dep. Materials, Science and Mineral Engineering, University of California, 27pp.

Zdankus, N. T., 1975. Analysis of Shallow Hard rock Well Pumping and Recovery Test Data, Ground Water, Vol. 12, No. 5, pp. 310-317.

Zeizel, A. J., et. al., 1962. Ground-Water Resources of DuPage County, Illinois, Illinois State Water Survey and Geological Survey, Coop. Ground-water Report No. 2.

Zheltov, S. P., 1961. Single-phase Liquid Flow Through Deformed Fractured Non-porous Rock, Appl. Math. Tech. Phys. (USSR), Vol. 6, pp. 187-189.

Ramey, H. J. 1982. Well-loss Function and the Skin Effect, A Review. in Proc. Symp. Recent Trends in Hydrogeology, Geol. Soc. of America, Special Paper No. 189, pp 265-272.

Ridgeway, W.R., D. Larssen , 1990. A Comparison of Two Multiple-level Ground-Water Monitoring Systems, In Ground Water and Vadose Zone Monitoring, Nielsen/Johnson editors, ASTM STP 1053.

Rorabaugh, M.I. 1953. Graphical and theoretical analysis of step–drawdown test of artesian wells. Am. Soc. Civil Engineers Trans., Vol. 79, Separate 362, 23 pp.

Smith, E. D. and N. D. Vaughan, 1985. Experence with Aquifer Testing and Analysis in Fractured Low-permeability Sedimentary Rocks Exhibiting Non-radial Pumping Response, Report DOE Contract No. DE-AC05-840R21400.

Snow D. T. 1965. A Parallel Plate Model of Fractured Permeable Media, Ph-D Dissertation, Univ. of Calif., Berkeley

Snow, D. T., 1970. The Frequency and Apertures of Fractures in Rock, Journal of Rock Mechanics and Mineral Science. Vol. 7, pp. 23-40.

Streltsova-Adams, T. D., 1978, Well Hydraulics in Heterogeneous Aquifer Formations, In: Advances in Hydroscience, Vol. II, New York, Academic Press, pp. 357-423.

U.S. Bureau of Reclamation. 1960. Earth manual. U.S.B.R. Method E-18, U.S. Government Printing Office, Washington, D.C.

UNESCO, 1984. Ground Water in Hard Rocks, Report 33, [II] SC.83/XX-33/A, 228 pp.

Uraiet, A. A. ,1979. Transient Pressure Behavior in a Cylindrical Reservoir Produced by a Well at Constant Bottom Hole Pressure, Ph.D. Thesis, Univ. of Tulsa. OK.

Vispi, M. A., 1980. Multiple-port Piezometer Installation at Rocky Mountain Arsenal, Report No. WES/MP/GL-80-12.

Walton, W. C., 1970. Groundwater Resource Evaluation, McGraw-Hill, New York.

Wang, J. S. Y., T. N. Narasimhan, C. F. Tsang, and P. A. Witherspoon, 1978. Transient Flow in Tight Formations," Proceedings of Invitational Well-Testing Symposium, (October 19-21, 1977, Berkeley, California), LBL-7027, Lawrence Berkeley Laboratory, Berkeley, California.

Wilson , R. C.and P. A. Witherspoon, 1970. An Investigation of Laminar Flow in Fractured Porous Rocks,Univ. of California, Berkeley, Civil Engineering Department, (Pub. No. 70-6).

Wilson , R. C.and P. A. Witherspoon, 1974. Steady State Flow in Rigid Networks of Fractures, Water Resour. Res., Vol. 10, No. 2, pp. 328-335.

Wilson, R. C., 1970. An Investigation of Laminar Flow on Fractured Porous Rock, Ph.D. Thesis, Civil Engineering Department, Institute of Transport and Traffic Engineering, Univ. of California, Berkeley.

Witherspoon, P. A., et al., 1979a. Observations of a Potential Size Effect in Experimentatal Determination of the Hyrdaulic Properties of Fractures, Water Resour. Res., Vol. 15, No.5, pp. 1142-1146.

Witherspoon, P. A., et al., 1979b. Validity of Cubic Law for Fluid Flow in a Deformable Rock Fracture, Special report from Dep. Materials, Science and Mineral Engineering, University of California, 27pp.

Zdankus, N. T., 1975. Analysis of Shallow Hard rock Well Pumping and Recovery Test Data, Ground Water, Vol. 12, No. 5, pp. 310-317.

Zeizel, A. J., et. al., 1962. Ground-Water Resources of DuPage County, Illinois, Illinois State Water Survey and Geological Survey, Coop. Ground-water Report No. 2.

Zheltov, S. P., 1961. Single-phase Liquid Flow Through Deformed Fractured Non-porous Rock, Appl. Math. Tech. Phys. (USSR), Vol. 6, pp. 187-189.

CHAPTER 7

DATA ANALYSIS AND INTERPRETATION

The basic goals of data analysis and interpretation are to increase one's knowledge and understanding of site conditions, both surficial and subsurface through evaluation of data collected during the Phase I and Phase II scopes of work. These analyses and interpretation efforts attempt to define the actual field conditions and hydrogeologic processes, based on the necessarily limited sampling and observation points used during the field program. Site assessment projects may have many data analysis aspects. They may include identification of conceptual physical models of the subsurface, definition of previously unknown conditions and establishment of recurring physical patterns and ultimately determination of the physical or chemical causes of the phenomenon under investigation.

Once the goals of a site assessment are established within a Phase I project scope, the data are recorded or collected during the Phase II field program (over a period of time or at different locations). These data, as a first step, could be compared with previous regional data to evaluate if the site data is in some way predictable. This is generally termed as finding if the data has recognizable structure. Data gathered in site assessment projects may have many components within the data structure. The technical challenge is finding the underlying significance in the data. The characteristics of a data set that has a structural component that includes a tendency to take a preferred value, a similarity to adjacent values (in time or space), and a dependence on the value of other variables observed at the same location (Green 1985).

The evaluation of a data sets structural component can imply that the data values are related in some way to other variable(s) that defines position of the elements of interest to the project. This can be results of logging of borehole data (i.e., SPT's or packer permeability tests) related into geologic or hydrostratigraphic units. Data may also be related by time (variations in hydraulic heads measured in piezometers), linear position (SPT measurements along a section line), or spatial position (soil sample analytical results). Once some level of structure is observed, the structural element can be removed from the data to produce a residual field. Handling of data sets showing residuals are commonly evaluated through looking for unnaturally high or low values called "anomalies'. The next step in the analysis may be to evaluate the anomalies for internal structure. Site assessment projects often rely on identification and sorting out anomalies, since they often point toward areas of further investigation.

Structure of data sets may also present if the data can be grouped into distinct classes based on the data values. This is the basis for much of the geologic or hydrostratigraphic classification used in relating specific site conditions to potential flow paths of ground-water movement. In trying to classify geologic units on the basis of observed lithology or determine values, (such as geophysical logging results); the investigator must first establish the general or specific properties of the groups. For example, background water quality results may be very different for different soil or bedrock formations and anomalies become difficult to evaluate before an understanding is developed on the spatial variability of the ambient water quality. If this is the case, an independent observational method(s) may be employed to sort out the apparent local variability. These independent methods for evaluation of spatial variability are illustrated in pictograph Figure 7-1. When these methods fail, the statistical behavior of the data may be investigated to evaluate the multivariate statistical relationships through complex methods such as factor analysis (Green 1985).

The problem of interpretation is then to relate the effects of variations in the proposed conceptual model to variations in the data. Several approaches can be used to develop this relation:

- Calculate the expected observations consistent with the data, given a known functional relationship, (called the "direct' approach); and

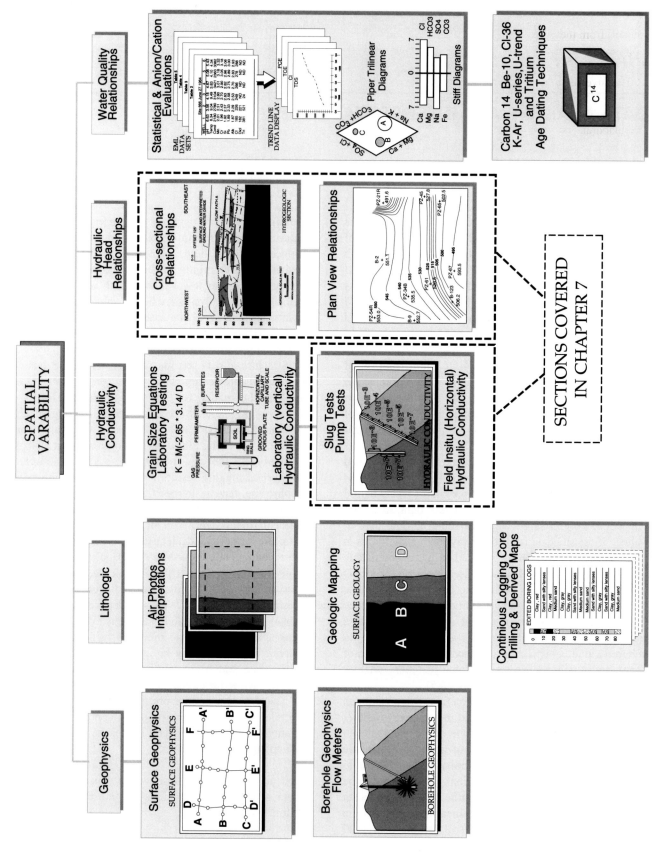

Figure 7-1 Spacial Varaibility Assessment Methods

• Define the nature of sources (i.e., the parameter of the model) from the observations (called the "inverse" solution approach)

The inverse solution approach is generally considered (Green, 1985) to be a more difficult, since it requires a more detailed knowledge of the physical processes and more elaborate measures to protect against the influences of spurious data components. Interpretation problems with data often crop-up due to the lack of uniqueness between a number of different models (i.e., the models all fit the data equally well). One can evaluate the similarity of the models in several ways:

• The number of parameters "i"; the model may be reduced to the point where only one model produces a best fit to the data; and,

• A more exacting methodology is to evaluate the range of model parameters which define the set of all acceptable models.

The result for either method may demonstrate that the data will provide little useful information on their own, if ranges are observed to be high. In effect the result of a particular solution is found for a very restricted model.

The way around this is to include other available sources of information at the same time. For example, in Pictograph Figure 7-1 spatial variability is established through a number of alternative evaluation techniques (i.e., logging lithology, hydraulic conductivity testing, hydrochemical testing, etc.). Each technique gathers independent data on the conceptual model representing the site. Interpretation of the data gathered during the Phase I and II programs, in terms of conceptual models, are only part of the process. The conceptualizations are further extended for use in decision making for a range of ground-water monitoring and remediation projects. The development of conceptual geologic and hydrogeologic models is further addressed in Chapter 8.0, where specific evaluation techniques are developed into the conceptual model process.

Site assessments require the analysis and interpretation of large volumes of data. Experience in data analysis techniques can greatly aid in the establishment of geologic and hydrogeologic criteria that will support design components of solid–waste landfill facilities and placement and design of the ground–water monitoring system. Alternatively, the data analysis segment of a Phase II assessment monitoring program (such as Superfund RI/FS

investigations or RCRA site remedial investigation) will take field data and evaluate the most likely target monitoring zones and locations to establish the assessment monitoring system.

Many of the deliverables of a Phase II study are developed and refined during data analysis and interpretation. Data analysis begins in the field during collection of the basic geologic and hydrogeologic data within the various surface and subsurface mapping tasks. This analysis and integration continue in the office where raw field data is reduced and combined with laboratory derived information. The resultant data are combined into various tabular, map or cross-sectional presentations relating observed field conditions with important site criteria that affect potential disposal actions at the site. The basic goals of data analysis and interpretation may require some or all of the following information:

Establish lithology

- Stratigraphic names
– Classification of soil/weak rock/rock units
– Extent of soil/weak rock/rock units
– Description of lithologic units
– Depositional history/environments of deposition
– Geometry of lithologic units
– Geologic structure/tectonic history

Engineering Properties

- Physical properties
- Moisture/density
- Compaction
- Consolidation
- Shear strength parameters (natural and compacted states)

Provide key hydrogeologic characteristic to the conceptual models

- Hydraulic conductivity
- Porosity
- Gradients
- Specific yield

Define aquifer characteristics

- Boundaries
- Type of aquifers

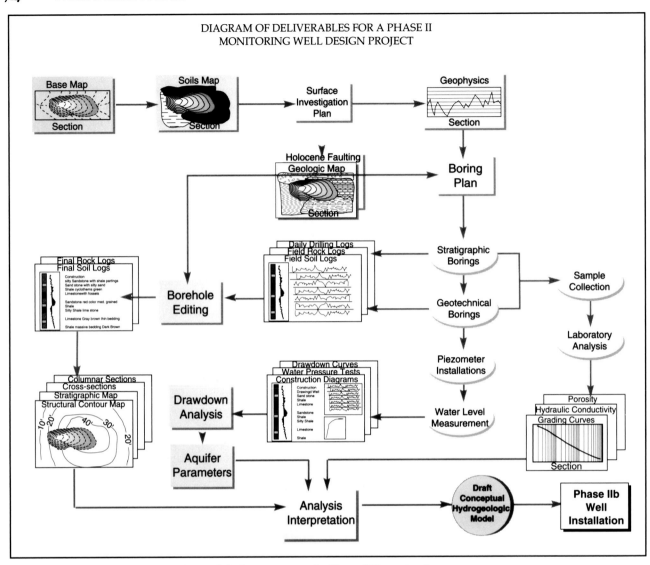

Figure 7-2 Components of a Phase II Investigation

- Saturated/unsaturated conditions

Define water quality baseline conditions

- Ambient water quality;
- Off-site influences; and,
- Diurnal and seasonal effects.

Figure 7-2 is a diagram of the deliverables specific to a Phase II project that illustrates interrelationships among these items. The data analysis and interpretation component are highlighted within the dashed line. Each item is discussed in the following chapter subsections. The geotechnical and hydrogeologic data, collected during the Phase II investigations, can also be used in landfill design activities, RI/FS projects, and risk assessment studies

because techniques for collection and assessment of basic data are basically similar.

Design of ground–water monitoring–systems depends, to a considerable extent, upon the local conditions. In any assessment, the overriding factor is the geology; without a proper understanding of the geological framework of a ground–water system, it is impossible to properly design the ground–water monitoring system or to evaluate site ground-water remediations.

The type of data is classified in Table 7–1. Each of these presentation formats may be required, (based on project needs), and may be used in analysis and presentation of hydrogeologic data. The requirements of a project dictate the individual work items necessary to finally define the locations for installation of facility monitoring wells. These wells may represent detection

Table 7-1 Components of a Phase II Investigation

Narrative Description of Geology
Topographic Maps
Boring Logs or Coring Logs
Geologic or Stratigraphic Columns or Maps
Geologic Cross–Sections
Geologic or Soil Maps
Raw Data and Interpretive Analysis of Geophysical Studies
Raw Data and Interpretive Analysis of Material Tests
Narrative Description of Ground Water with Flow Patterns
Ground–Water Surface or Potentiometric Maps (Plan–View) with Flow Lines
Structure Contour Maps of Aquifer and Confining Layers (Plan–View)
Raw Data and Interpretive Analysis of Slug Tests, Pump Tests, and Tracers Studies

monitoring as would be required under a RCRA subtitle D or C program, or assessment monitoring actions under other federal programs such as, Superfund or RCRA remedial actions. Remediation programs under state control follow similar evaluation procedures as Superfund. RCRA subtitle C and D also contain major remediation components that require a site characterization process similar to Superfund characterizations.

This chapter introduces the major hydrogeologic characteristics that serve to describe how aquifer systems work. Characteristics, such as, hydraulic conductivity, porosity, hydraulic gradients, aquifer or confining unit lithology and shape, are described in detail in the remainder of Subsections 7.1 through 7.3. These subsections describe the various components of a Phase II project report as shown in Figure 7–2.

7.1 CHARACTERIZATION OF LITHOLOGY

The first step in a Phase II study is to define the probable ground–water flow–systems, as observed in the field, and to describe relationships among the lithology, folding or structural elements, and fracturing of the rock or soil types. This requires the establishment of a form of hydrogeological mapping–units that will depend initially on both field estimates of the relative water–bearing potentials of the rocks or soils, and development of flow–components for the system. The classification of subsurface lithology represents one of the most challenging aspects of field investigations. Although junior staff are often assigned the task of supervising drilling efforts and logging of borehole samples, the importance of technically accurate sampling and logging should not be under

estimated. Junior staffs should be properly trained in logging and sampling skills along with prior briefing on the expected subsurface conditions. Entire investigations have gone wrong due to incorrect or incomplete field logging. This is an important aspect that should not be overlooked. As with any geologic investigation, the raw data, gathered in the field, must be organized into detailed descriptions of lithology and measured sections to form a basis for correlation between boreholes. After establishment of defined units, normal geological procedures may be followed to construct a geologic map. Normally, only the boundaries of defined units should be mapped; but if it is necessary to emphasize geologic relationships or different potential pathways for ground water within one unit, hydrogeological subdivisions of the units may be delimited. Over–simplification should be avoided, but a balance must be obtained between what is applicable to the regional ground–water flow and what is of local interest for defining the ground–water monitoring system at the facility.

The process of developing detailed description of lithology is based on first characterizing the lithology beneath the site with a consistent and organized approach. This classification is carried through the surface mapping, interpretation of borehole data incorporated into edited logs, and used to construct structural contour and isopach maps.

7.1.1 Stratigraphic Names

Aquifers are often known by their stratigraphic names (e.g., the Ogalalla in the high plains). If an aquifer beneath a site is formationally named (as evidenced by common usage or published reports), such a name should be included in the Phase II discussion and included on edited logs.

7.1.2 Classification of Units

The investigator should classify the hydrogeologic units within and below the uppermost aquifer on the basis of their lithology and hydrogeologic properties. The classification should generally extend from the surface down to the aquitard underlying the uppermost aquifer. The lithology or composition of the aquifer(s) materials can be determined during drilling (including stratigraphic borings). The classification of units should be graphically presented as a hydrogeologic column or cross–section, with an accompanying description. The graphics can be used in combination with the maps for delimiting the extent of the hydrogeologic units.

7.1.3 Extent of Units

The geologist should delimit the area extent, vertical extent, and geometry of significant hydrogeologic units. The determination of the vertical extent of the uppermost aquifer will be discussed below in "Aquifer Boundaries". The extent of these units can be presented in several ways. For complex settings, the most desirable presentation is a series of structural contour maps for the top and/or bottom of each unit. Vertical sections or columns and isopach maps are also frequently used. However, vertical sections and isopach maps may not contain all the information available. They are most useful as a supplement to the structure contour maps. Because the construction of any of these diagrams involves interpolation and extrapolation of limited data, the diagrams should also show the location of control points. For simple geologic settings, in which the hydrogeologic units are laterally extensive and flat lying, structure contour maps may not be necessary. For example, a table listing the elevations of the top or bottom of each unit may be adequate. These exhibits can be presented separately with a narrative discussion of the aquifer or aquifer system, or they may be included along with the stratigraphic cross–sections. The location of any local faults should also be indicated. Any stratigraphic units that behave as confining layers (aquitard/aquiclude) should be identified as such.

7.1.4 Description of Units

The interpretation and analysis of field data, presented in a form that can aid in the overall understanding of the hydrogeologic system, requires those narrative descriptions of the geology be prepared for inclusion in the conceptual hydrogeologic model. Regional setting, stratigraphy, structure, and geologic history may all form components in description of the hydrogeology of the site. These can further be defined as:

Regional Setting –– Geologic relationships and geologic history are required as a framework for site–specific studies. Typical examples are regional structure, regionally significant stratigraphic sequences, distribution of igneous bodies, and regional patterns of metamorphism.

Rock Units (Stratigraphy) — This section may be introduced by a general description of the rock sequence and its genetic interrelationships. The section consists of systematic descriptions of the units in the order of decreasing age, typically including: (a) discussion of the principal kinds of rocks, the thickness or shape of the unit, and perhaps its geographic distribution; (b) brief explanation of unit name, including reference to its original definition and any revisions; (c) complete description of rocks and their lateral variations; (d) description of contacts; (e) fossils or isotopic age analyses and probable geologic age of the unit; and (f) interpreted origin and environment of deposition.

Structural Geology — An introduction may clarify geographic and age relations in complex areas. Descriptions of folds, faults, joints, brecciated rock, intrusive relations, and melanges may be organized into a manner that clarifies their interrelationships.

Geologic History — A chronology of interpreted events, including the interrelationships among depositional history, igneous activity, metamorphic events, structural geology, and tectonic history.

The task of establishing geologic regional settings specific to a particular site, can present significant difficulties to an investigator. Although some geologic environments can provide clear-cut definition as to the subsurface geologic units (i.e., sandstone, shale, etc.), comparisons of relatively unconsolidated coastal sediments may be difficult in practice. Figure 7-3 shows variations in stratigraphic interpretations between 1980 and 1986 within a single facility area. Seven investigative reports on the specific New Jersey coastal plain site provided seven differing interpretations (Hathaway 1989) on subsurface geologic units down to the very moderate depth of thirty meters (100 feet). Although an exact stratigraphic classification of the subsurface geologic units may not be necessary on every project, a reasonable assurance should

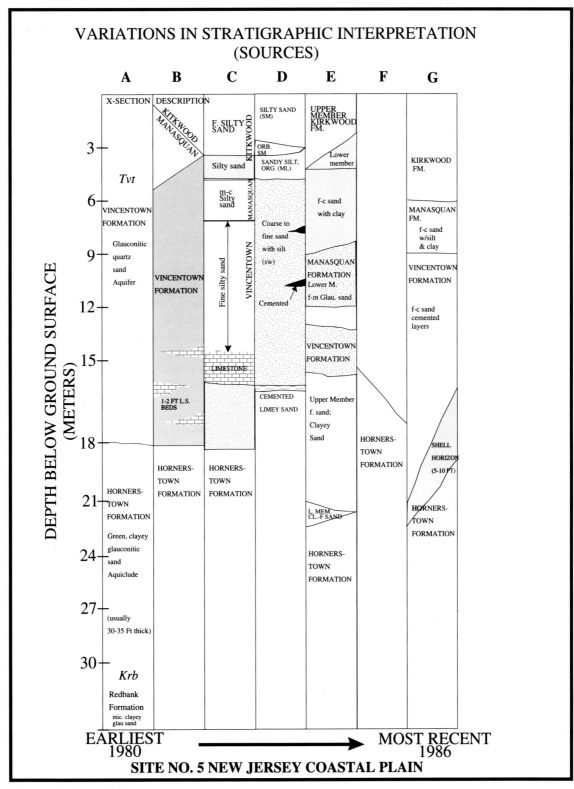

Source: Hathaway 1991

Figure 7-3 Alternative Interpretations of Geologic Conditions

be established as to the lithology expected in the subsurface. Remedial investigations and assessments monitoring evaluations require thorough understanding of subsurface stratigraphy so that geologic and hydrogeologic relationships can be correlated between boreholes. The stratigraphic interpretations must clearly represent the lithology observed in borehole drilling operations as an important part of the geologic conceptuation process.

Topographic maps serve as the format on which to summarize and display information collected throughout Phase II investigation. Chapters 2.0 through 6.0 provides the features that should be located on the base map, and which may be necessary for design of ground–water monitoring–systems. The required scales should be convenient for illustrating site geologic and site assessment features. Although larger–scale drawings (such as 1 inch equals 100 feet) are important for design work, accurate Phase II base map work can often fit onto an 11 inch by 14 inch foldout illustration that can be bound into the Phase II reports.

7.1.5 Edited Final Boring–Logs

The final log is essentially a polished refinement of the field log, based on subsequent examination of the samples and laboratory analysis. The final log often incorporates a limited amount of laboratory test data as an aid to further evaluations. The final logs should present a clear, concise, accurate picture of subsurface conditions, because it is commonly the basis from which site assessment interpretations and design parameters are derived.

The final logs are derived from the hand written raw field logs and should be included within the report of the investigations. The field logs, however, should be maintained as the primary data source at the consultant's office with copies sent to the project engineer. There is generally no reason to include unedited field logs within a Phase II report.

As with mapping procedures, the portrayal of the geological environment in hydrogeology relies upon standard geological methods and maps. Some of the more important maps used in the portrayal of geologic materials relative to ground–water measurement are as follows:

• Geologic columns;
• Bedrock geologic maps with cross–sections;
• Structure contour maps;
• Isopach maps;
• Fence diagrams; and,
• Contour map of the depth to the potentiometric surface or to aquifer, if confined.

Although it is important that a bedrock geologic map should be prepared, it may also be necessary to create a surficial geologic map depicting the more important soil or surface geologic units that may be of significance in drilling operations. The presence of thick (say, >50 feet) unconsolidated overburden deposits may overwhelmingly dictate location of wells. In other cases unconsolidated deposits can play a minor role in ground-water flow as compared to bedrock. Thick sections of coarse, clastic materials may also influence the type of drilling method used in the final installation of the monitoring wells. Each site must be addressed on its individual merits and geologic setting.

7.1.6 Geologic Columns

Individual boring logs represent the observed lithology at a particular point. Each boring can encounter anything from slightly, to radically different lithology and elevational contacts. Data should be incorporated into a typical stratigraphic sequence, columnar section, or as a schematic cross section of site stratigraphy. These illustrations are made to show the sequence and original stratigraphic relations of the formations or geologic units at a site. The height of each formation or unit in the column represents the approximate relative thickness of the formation. The column cannot be drawn absolutely because the units vary in thickness from place to place. If the strata are inclined or folded, the thickness of each member must be computed. In this respect, a columnar section differs from a well log, unless the borehole penetrates horizontal strata. Since columnar sections are used to summarize many boreholes they must represent a conceptuation of the actual geologic subsurface at any particular location.

Brief descriptions of each unit may be included, to the right of the column, as in Figure 7-4, or the column may be accompanied by an explanation consisting of a small box for each lithologic symbol and for other symbols. The following elements of a stratigraphic column are essential to the understanding of the figure:

1. Title — indicating topic; general location; and whether the section is single (measured in one coherent segment), composite (pieced from two or more sections), averaged, or generalized.
2. Name(s) of geologist(s), organization, and date of the survey.
3. Method of measurement.
4. Graphic scale.
5. Map location or description of locality.
6. Major chronostratigraphic units, if known.

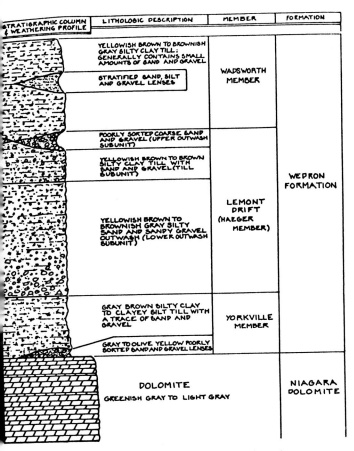

Figure 7-4 Schematic Site Stratigraphy

Source: Hydro-Search, Inc

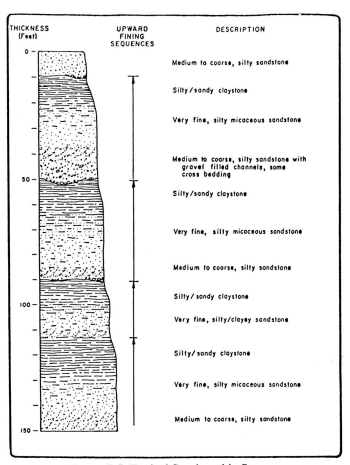

Figure 7-5 Typical Stratigraphic Sequence

Source: EMCON Associates

7. Lesser chronostratigraphic units, if known.
8. Names and boundaries of rock units.
9. Graphic column composed of standard soil or rock lithologic patterns.
10. Unconformities.
11. Faults, with thickness of tectonic gaps, if known.
12. Covered intervals, as measured.
13. Positions of index or marker beds.
14. Positions of important samples, with number, and supporting data.

Other kinds of information that may be included are:

15. Designations of formal or informal measured units.
16. An irregular edge indicating relative resistance of the rocks to weathering and erosion.
17. Summary descriptions of formations or other units (especially desirable if the section will not be accompanied by an explanatory text).

18. Thicknesses of units or range in thickness, if thickness varies significantly.
19. Intervals of deformed rocks.
20. Symbols or numbers indicating kinds of fossils, primary structures, porosity, cementing agents, shows of petroleum, and any other pertinent observations.

Some of the latter features may be added directly to the lithologic column. Columns are constructed from the stratigraphic base upward and should be plotted first in pencil in order to insure spaces for gaps at fault displacements and unconformities. Sections that are thicker than the height of the figure can be broken into two or more segments, with the stratigraphic base at the lower left and the top at the upper right. Bedding and unit boundaries are drawn horizontally except in detailed sections or generalized sections of distinctly non-tabular deposits, as some gravels and volcanic units, or when unconformities are present.

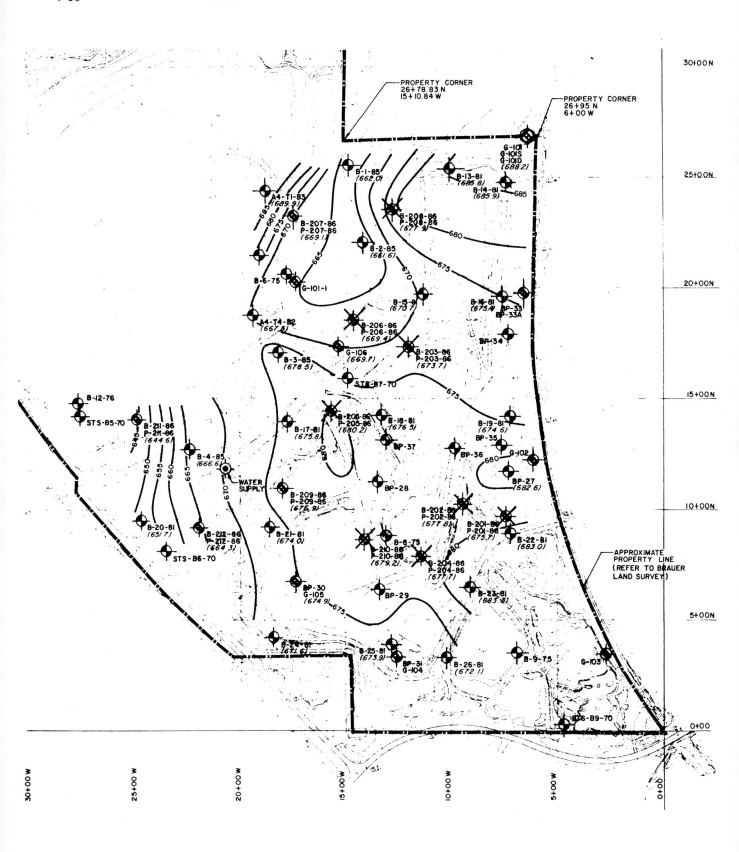

Figure 7-6 Geologic Structure Map Example 1

Source: Donohue

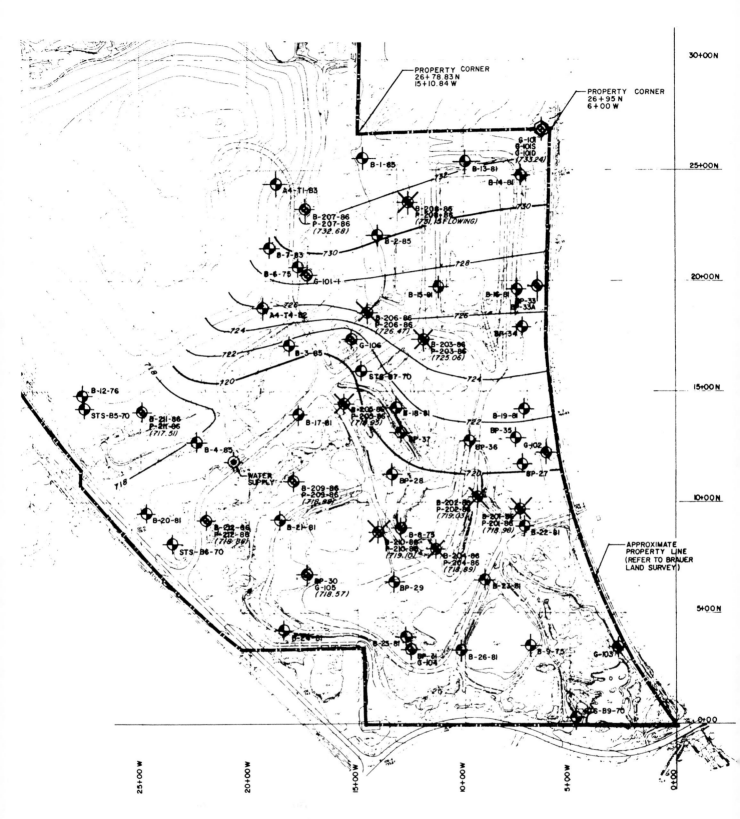

Figure 7-7 Structural Geologic Map

Figures 7-4 and 7-5 show typical columnar sections of geologic material at two facilities. The reader should take notice that the term "till," when used, should always be modified as to type (lodgment or basal versus ablation). Likewise, the glacial term "drift" is very ambiguous and should be replaced by more specific terminology. Lithostratigraphic facies codes should be used where appropriate to the geologic setting. These codes are particularly useful in glacial materials due to the wide range of sediments produced by glaciers and the important hydraulic relationships played by the various geologic units. These sections represent the simple, conceptual geology for a site and can provide a significantly increased understanding of the relationships between rock units.

The geologic columnar sections provide a visual statement that can be easily understood by non-technical staff. As such they provide a conceptual basis for understanding the physical occurrence of geological materials present at site.

7.1.7 Geologic Structural Maps

A knowledge of the general character and form of the geologic units in the region is necessary for interpretation of the subsurface structure, especially where stratigraphic control is sparse. Structural contour maps should be constructed so that the geologic features coincide with the regional trends or tendencies, but include site–specific data necessary to define ground–water movement. It is easy to construct a geologic structure contour map that is technically correct; but based on poor data. One should not forget the eventual analytical importance placed on these maps demands that they make good geological sense.

Maps should show locations of all measurements, samples, or observations, as well as, the type of data obtained. Derivative maps present information extrapolated over relevant areas, generally by colors, patterns, or contours. Structure contour maps can be derivative maps showing elevations on a contact or other surface of interest, and isopach maps by definition are contoured so as to show thicknesses of a unit or of overburden.

The investigator should use the data collected from the subsurface (borehole) investigations to prepare, as necessary, stratigraphic cross sections or isopach maps depicting the subsurface site stratigraphy. These data help greatly to visualize the most probable ground–water flow system. Several cross sections may be required to depict significant geologic or structural trends and reflect geologic/structural features in relation to local and regional ground–water flow. The area and vertical extent of the geologic units can be presented in several ways:

- For complex settings, the most desirable presentation is a series of structural contour maps for the top or bottom of each unit; and,

- Vertical sections and isopach maps can also be used since they are generally more graphic and are useful as supplements to the structural contour maps;

Structural contour or isopach maps are generated from boring logs using elevations of key stratigraphic–horizons which have been obtained during the drilling program. Structural contour maps can be effectively combined with piezometric ground–water surfaces for a particular formation or unit as is shown on Figures 7-6 and 7-7.

Key to presentation and interpretation of these data is the accurate and consistent field mapping of the information. If one wishes to generate a structural contour map of a particular geologic unit, or the piezometric levels observed in a hydrostratigraphic unit, the feature being mapped must be recognizable and be consistently mapped in the field. For example, if a presentation is prepared showing the piezometric levels of a formation (say the Robein fm shown in Figure 7-7), then the data used to generate the piezometric level contours must only come from piezometers completed within the target formation (or rather the target hydrostratigraphic unit within the formation or geologic unit). You must always compare apples with apples rather than apples with oranges in these presentations.

Contouring of Data

Much of the interpretive information gathered in the Phase II field investigation is subjected to some form of contour mapping. Geologists must always relate numerical values obtained from field investigation to the geographic location of the data points. One of the most useful ways in site assessments to study area trends of data are to construct isopleth maps, which may also be called isoline maps, isogram, or isocontour maps. Isopleths are groups of points that have equally numerical value and are interpolated from the plotted locations of the sample points and these respective numerical values. The most common example used in site assessments is the construction of contour lines of equal elevation for topographic maps. Flow nets are also commonly used to illustrate hydrogeologic conditions. Since flow nets are drawn using equipotentials that represent lines of equal head, they are a form of isopleth construction. Other types of isopleths include isopach maps and structure contour maps. An isopach map typically shows the thickness of a specified

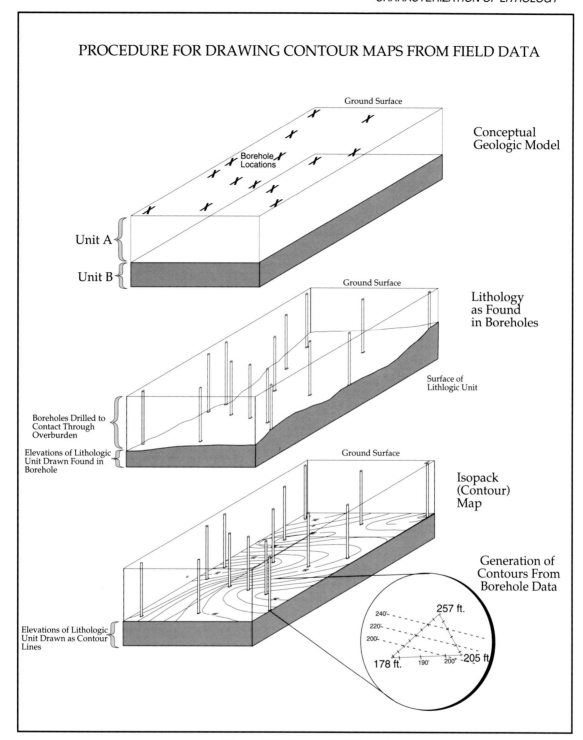

PROCEDURE FOR DRAWING CONTOUR MAPS FROM FIELD DATA

Figure 7-8 Contouring Method

stratigraphic unit throughout a geographic area. Structure contours represent connected points of equal altitude on a geologic surface such as a formational contact, a confining zone or a fault surface.

The objective of contour mapping (in a general sense) is to portray the form of a hydraulic or physical surface. A contour map is simply an illustration of a three–dimensional surface on a two–dimensional piece of paper. Vertical dimensions of the map may represent elevation above sea level, thickness, or some other quantity, as shown by means of isolines or lines of equal value. Contour lines are isolines of elevation, whether representing the ground surface or the top of a subsurface formation. Terminology of use of contours, however, usually calls any isoline a contour whether it depicts hydraulic head, elevation, thickness, porosity, concentration, composition, or some other physical or chemical property.

Since the contour lines on a map connect points of equal numerical value, the space between two successive contour lines contains only points whose values fall within the interval defined by the two contour lines. In most circumstances, the value of the surface cannot be determined at every possible location, nor can its value be measured at any specific point we might choose. Usually, only scattered measurements of the surface are made at a relatively few numbers of control points, such as monitor well locations, seismic shot points, or the sites where water–quality samples have been taken.

Procedure for Drawing Contour Maps

Construction principles used to plot contours on topographic maps are basically the same as those used to prepare contour diagrams for site assessments. Each known point numerical value is first plotted at its proper location on the base map or diagram. Trial sketching of the isopleths is then performed by interpolation between each pair of known points. The inclusion of the interpolation of a third measurement point is to establish lines of equal numerical values as shown on Figure 7-8. Isopleths should usually be sketched lightly first, starting either with the highest or lowest contour line that will appear on the map, and then finished by rounding out any sharp curves not based on actual data. Typically one would also begin in areas where more data sets are available for interpolation.

The contouring of data is often a difficult undertaking since there is no one best way for determining the position of specific isopleths on a map. Often one must consider that "correctness" in any constructed contour (isopleth) map must be based on initial conceptual models. This is especially true where data is sparse. Rettger (1929) described three distinct types of procedures for drawing isopleths: mechanical contouring, parallel contouring, and equi-dip contouring. Figures 7-9b through 7-9d illustrate widely different interpretation in derived contour maps with the same data or control pointsfrom Figure 7-9a. Mechanical contouring represents the majority of hand and computer based methods; as such this method will be described in some detail below.

Mechanical Contouring: Mechanical contouring technique assumes that the slope perpendicular to the contour is uniform along lines connecting control points.

Therefore contours are defined in arithmetic proportion. The data points of Figure 7-9a are connected by straight lines to form a series of irregular triangles. The intersection of a specific contour with such a connecting line is determined by proportional interpolation between each pair of control points. By using this method, contour points are obtained between each pair of control points, and those points with the same value are connected to form contour lines. Figure 7-9b is a contour map prepared from the data of Fig 7-9a using mechanical contouring technique. Geographers generally use the mechanical contouring technique to prepare a topographic map. In gathering field data for the map the surveyor deliberately places control points at significant topographic points where the slope (elevation change per unit horizontal distance) and contour spacing, is essentially uniform between topographic control points. Once these topographic controls are established, one can alter simple mechanical contouring by sketching minor topographic deviations from the regular slopes configurations.

In site assessment situations the investigator cannot evaluate the subsurface features to pick out control points at the exact positions of "highs" and "lows" of the population values. Instead we can use only those data that are obtained by whatever subsurface sampling process determines the location of the control points. It is expected there will be uncertainty about the actual direction of surface trends, or the positions of actual maxima or minima for contours in the map region. It is unlikely that sample points will be practically so closely spaced that true subsurface trends will be fully understood, given the cost and time constraints typical for site assessments. Figures 7-9c and 7-9d shows alternative contouring using the same data points as 7-9b. The varations between these representations of what could be the actual surface can point out the potential difficulties of over dependence on widely spaced control points.

The difficulty with interpretation of control points obtained from drilling programs is that they are not placed at significant localities with respect to subsurface isopleth highs or lows. Many geologic contour maps, particularly those based on only subsurface control, are based on some sampling locality scheme not related to the true configuration of the surface of interest. Sampling schemes such as, random sampling, sampling on a rectangular grid, or simply using all data available, even if it was not located by predetermined pattern, can induce misconceptions of the true surface of interest. One must remember that no matter how authoritarian the contours look on a structural map, the actual surface mapped may have locally a radically different configuration. A key to the interpretation of these presentations is the consistency of

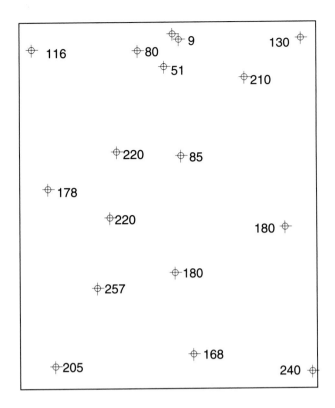

Figure 7-9a Contour Methods Basic Data

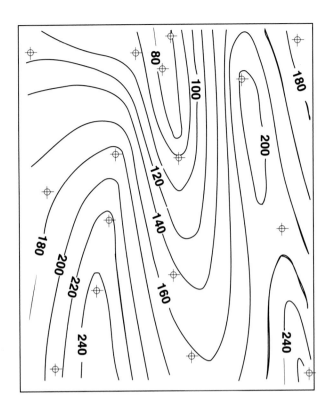

Figure 7-9c Contour Method Example 2

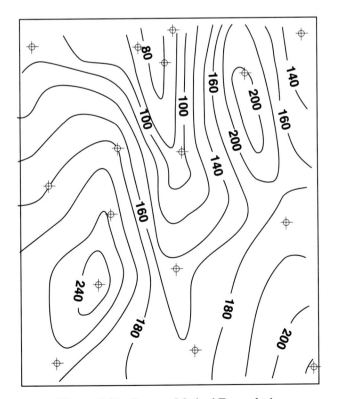

Figure 7-9b Contour Method Example 1

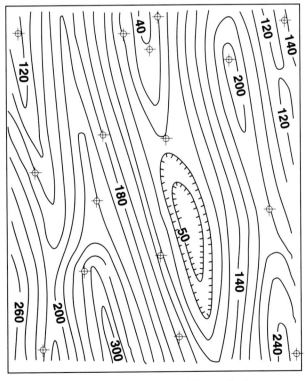

Figure 7-9d Contour Method Example 3

data used in the construction of the map. Such subsurface maps have inherent lack of reliability when compared to the more familiar topographic map that was based on control points carefully chosen from scanning the true configuration of the irregular surface.

Structure Contour Maps

Structure contours connecting points of equal altitude on a geologic surface are drawn as separate structure contour maps or are overprinted on geologic maps to show the area variation of the vertical distance of a geologic horizon above a datum surface (elevation of the geologic horizon, if the datum is sea level). Structure contours are sketched from points of known elevation on the key horizon or bed being contoured.

The name interpretative contouring has been applied by Bishop (1960, p.47-48) to the method she considers best for interpreting regions with few control points. These structure contours deviate slightly from the mechanical procedure described above to indicate surface patterns that are consistent with the observed subsurface stratigraphy obtained from the drilling program, as well as, the known, (or supposed), structural habit of the region obtained from surface mapping and Phase I work. The more sparse the data points, the more varied the possible interpretations.

If the geologist has results of surficial geologic field work on outcropping structures, he (or she) can obtain structural contour control approaching the quality of that used for topographic contours. A knowledge of stratigraphic thickness of geologic units allows the investigator to calculate the approximate elevation of a reference horizon.

The principles employed to plot elevation contours on topographic maps are similar to those used to prepare all kinds of isopleth presentations, including equipotentials or flow net diagrams. The numerical values for each known point is first plotted at its proper location on the map, cross section or diagram. Sketching of the isopleths is then performed by the interpolation of data between each pair of known points. Isolines are typically sketched lightly first, starting either with the highest or lowest line occurring on the map, and then adjusting the lines by rounding out sharp curves not based on actual data.

There is no one best way for determining the position of specific contour or isopleth on a map. While surface topography can be easily checked for true elevation changes, subsurface features or units of hydraulic heads, as illustrated by isopleths, are not so easily evaluated for correct values. A correct isopleth map is considered to be consistent with the assumptions and conceptual models used as guidelines during construction of the map. There

are a few basic rules for hand contouring a group of numbers on a map or diagram:

1. Each contour line, of given value, must pass between those points whose numerical values are higher and lower than the contour value.

2. No contour can cross itself or any other contour. There are two exceptions to this rule: overturned or recumbent anticlines and reverse faults. In practice, the underside of a recumbent anticline and that part of the datum lying below a thrust segment are ordinarily omitted on a contour map because of the confusion of lines. Occasionally, it is desirable to show the relationship by contouring the "hidden" segments with dotted or dashed lines.

3. Two or more contours may merge into a single line only where the datum is vertical or where faulting has displaced the datum along the strike by an amount equal to or exceeding the contour interval.

4. Where data are sparse, the contours should be dashed or terminated.

5. Contour intervals should be carefully chosen based on spacing, precision, and range of data.

A map can be contoured so all technical requirements are fully satisfied, yet fail to convey the probable real world conditions. Hand drawn and computer generated contours require careful review for validity. Drawing a contour map by computer usually involves an intermediate step — the construction of a mathematical model of the surface; this must be performed by the computer before the contour lines themselves can be generated.

A computer contouring program traces contour lines by a precise mathematical relationship based on the geometry of the control points. A geologist, however, contours not only the control points, but also his/her concepts and ideas about what the surface should look like. If these preconceived ideas are indeed correct, a competent geologist should be able to create a map superior to a computer–made product. On the other hand, if the geologist's preconceptions are erroneous, the finished map is likely to be seriously in error. Numerous computer based contouring packages are available, and they can all produce accurate representations of the surfaces. Care should, however, be exercised to subject the computer generated surface to real world comparisons and scrutiny, based on suitable experience.

Contouring is used extensively for interpretation of site assessment data and analysis for generation of derived maps. Isopach maps are used in conjunction with structural contour maps to determine drilling depths for monitoring wells and are as important as the latter in Phase II operations. Isopach maps showing the estimated depths from surface to a target horizon are particularly useful. Their representation of a geological unit can assist greatly in understanding the regional environment, and they are also of considerable importance in assessing the spatial distribution of water–bearing formations that may serve as potential ground–water flow paths.

The selection of a suitable horizon for representation by structure contours depends upon the complexity of the system being investigated. The upper and lower horizons of an aquifer or system is usually most significant for waste management facilities.

The most common contour map is a representation of bedrock topography (Figure 7-10). These contour surfaces are typically drawn on known elevations of the contact

between overburden materials and bedrock units. The drilling data should be reviewed carefully to confirm if refusal or actual bedrock was observed during sampling of the borehole. Confusion between refusal by a boulder (or glacial erratic) and refusal at the bedrock surface often leads to inaccurate estimates of depth to bedrock. This is especially a problem when designing ground–water cut–off walls. Cost estimates of cut–off walls can radically change if bedrock surfaces are deeper than expected.

7.1.8 Cross–Sections

The construction of subsurface cross–sections, using information from boring logs, geophysical surveys, and background information, is useful for presenting and evaluating this data, especially for complex hydro-geological conditions.

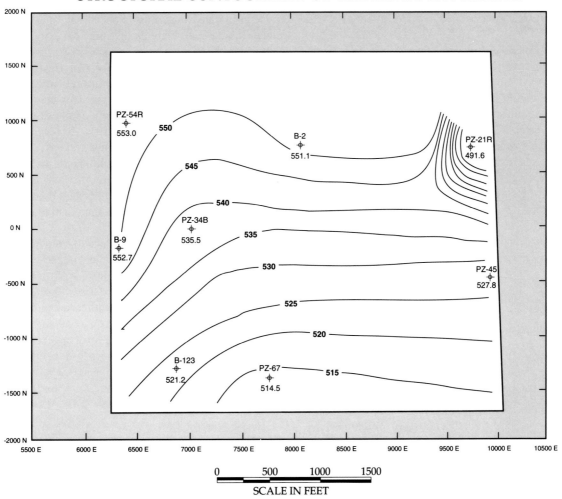

STRUCTURAL CONTOUR MAP OF THE BEDROCK SURFACE

Figure 7-10 Bedrock Contour Map

This section will describe the integration of geologic information obtained in Phase I and II investigations to produce an interpretation of the structure of a site area to use for cross-sectional, conceptual, geologic analysis and generation of accurate flow nets. These flow-net constructions are then used for determination of target flow zones and ground water monitoring locations.

The vertical and horizontal scales of the cross sections used for hydrogeologic flownet interpretations should be the same (i.e., 100 feet vertical to 100 feet horizontal scale). Although large area cross sections with relatively shallow boring are difficult to illustrate in reports, the initial interpretation should have equivalent vertical and horizontal scales. This allows observed or calculated dips of lithological boundaries and thicknesses of lithological units to be plotted directly on the section, and provides a typically completed section with a true scale representation of the geology and hydraulic conductivity relationships. When the geology is sufficiently understood for the site area, vertical scales may be exaggerated for explaining geologic features in the cross section in reports and drawings.

Cross-sectional constructions from edited boring logs and field mapping efforts are a progressive effort. As the construction of the cross section progresses, the investigation should obtain a clearer appreciation of the three-dimensional structure of the area. It is usually important to establish most of the near-surface geology through surface mapping before extending the construction to deeper levels by using borehole or geophysical data.

The following sequence of work tasks is generally employed for the construction of most sections, however, try to keep a flexible approach in development of the cross section. Figure 7-11 illustrates the process in an outline format for construction of cross sections for monitoring system design used in hydrogeologic projects.

Preliminary Borehole Data

Much of the data used in construction of cross sections are developed from borehole drilling programs. As was described earlier, accurate field logging of the subsurface combined with selected laboratory results provides the basis for all derived logs, cross-sections and maps that come later. Errors in logging can be wrongly preserved and lead to fully incorrect interpretations and conceptualizations. This is especially true for Pleistocene glacial areas where surficial unconsolidated deposits either cover bedrock or provide little indication to subsurface materials. A first step is to prepare borehole data collected during the field investigation to define major subsurface geologic lithologic boundaries. The scale of the borehole logs should be the same as the proposed cross sections.

Edited Borehole Data

Edited logs should represent both major and minor lithologic boundaries that will have an effect on the geologic and hydrogeologic interpretations. The flow net and conceptual models depend on accurate interpretation and editing of borehole data to represent the essential features lithologic and hydraulic conductivity of the subsurface.

Geologic/Geophysical Surficial Data

In those non-glacial areas where bedrock can be reasonably mapped from the surface, geologic maps should be prepared in conjunction with the borehole data to illustrate the surface exposures of both consolidated and unconsolidated materials. Special care must be taken to map lithographic features that may have some effects on the potential movement of ground water. Dipping, folded and faulted strata require special care in field assessment of primary porosity (intergranular) and secondary permeability features (fracturing). This may require field mapping of individual units such as claystone/sandstone that can represent significantly different hydraulic conductivity or secondary fracture systems. Considerable geologic technical field judgment should be employed to select important lithologic features that may affect ground-water movement. Mapping scale should be sufficiently selective to concentrate on these important features and not try to map every minor unit.

Those sites with deeply weathered profiles may require excavation of back-hoe pits to expose near surface geology and weathered profiles. Additional surficial geologic interpretations may be assisted through the use of SCS soil maps that relate weathered residual soils to bedrock geology. Road cuts, quarries and excavations should be used to improve knowledge of geology. Surficial geophysics can also provide indications of subsurface lithology in areas where surface geologic mapping is difficult.

Cross-Section Lines

Once surficial geologic maps and edited borehole logs are complete, a suitable cross-section line should be selected. The appropriate cross-section line should consider borehole locations, geologic features, ground-water flow components, and proposed or actual facility

CROSS-SECTION CONSTRUCTION PROCEDURE

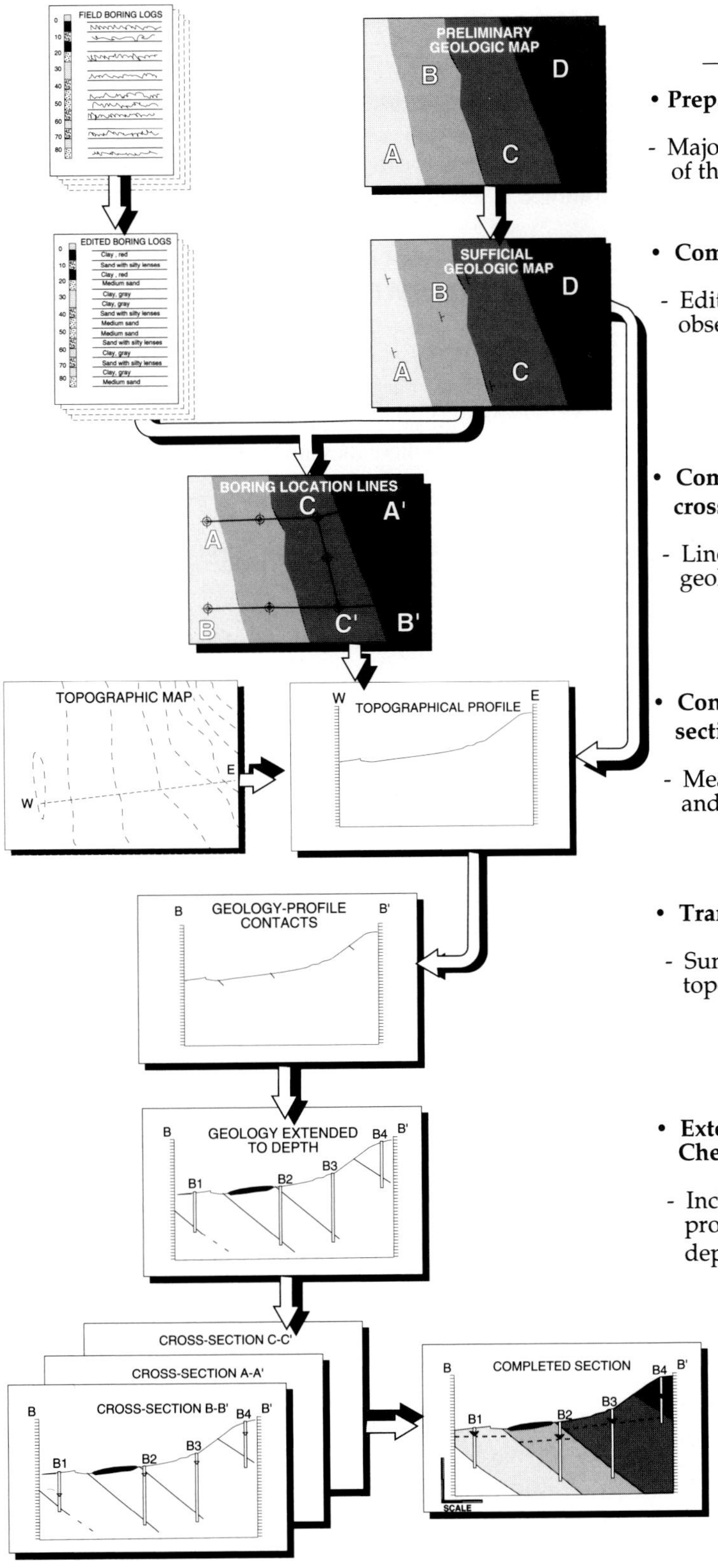

WORK ACTIVITIES

- **Prepare preliminary geolologic map & field boring logs**

    - Major geologic boundaries establish the basis for much of the later evaluation.

- **Complete edited boring logs & final geologic maps**

    - Edited logs must preserve essential geologic features observed in site investigations.

- **Combine geologic maps with boring locations & select cross-section lines**

    - Lines must consider field data collection, locations, geologic structure, recharge and discharge areas.

- **Construct profile from topographic map along cross-section lines**

    - Measured profiles are drawn from topographic data and facility features.

- **Transfer surface geology to topographic profile:**

    - Surface outcrops, if present, are used to tie-in topograpgy to geology.

- **Extend geology using borehole data and Check for plausibility:**

    - Inclusion of geophysical, borehole and surface data provides the criteria for completing geologic data to depth.

- **Include all hydrogeologic data & complete additional sections:**

    - Faults, springs, and all features that may affect local ground-water flow are then added to the cross-section.

Figure 7-11 Cross Section Construction Flow Diagram

water flow components, and proposed or actual facility boundaries.

Topographical Profiles

A measured topographical profile should be drawn for each cross-section line. These topographical profiles should illustrate all facility and topographic elevations, surface water, and discharge areas located within the profile area. These data can be obtained from topographic maps of the site area or surveyed along the selected line.

Topographical Profiles and Lithological Boundaries

Superficial deposits such as alluvium and colluvium can be shallow and inadvertently omitted from the cross-section construction. Inclusion of geology and hydrogeology of superficial deposits is extremely important for defining shallow ground-water flow conditions. Often shallow permeable deposits or weathered zones can transmit the majority of ground water from recharge to discharge areas.

Faults in Cross-Section

Faults are extremely important features both for Holocene (active or recent) faulting regulatory criteria and as high variable hydraulic conductivity features. The location of all known faults should be established on cross-section lines. Estimated or known strike and dip lines should also be illustrated on the section with appropriate annotations. Initially extend each fault line to only a reasonably shallow depth since the fault may later be found to die out, change direction at depth, or be intersected by another structural feature.

Unconformities in Cross-Section

Since unconformities typically represent significantly different lithologies and relative weathering profiles, unconformities should be fully represented within the cross-section construction. If the dip of the overlying lithological units is known or can be assumed to be the same as that of the unconformity, construct a short line at the angle of dip; in further constructions use the same procedure as for the overlying units. Use a distinctive line symbol to represent the unconformity in the section. Overburden materials such as glacial sediments require

borehole data to define contacts between bedrock, weathered bedrock, and unconsolidated materials. These unconformities can represent highly permeable zones.

Stratigraphic Boundaries

Cross-section construction activity should proceed as follows:

1. The observed bedrock dips, with their strike should be marked on the topographical profile.

2. Construct dip lines at the observed angle of dip and guidelines perpendicular to the dip. If faults or unconformities are present, guidelines must stop at the point where they intersect such structural features.

3. Define lithological boundaries on the topographical profile.

4. Establish locations of any fold axial traces on the profile.

5. Locate lines to establish the lithological boundaries parallel to the dip lines.

6. Inspect the thicknesses of lithological units drawn to this point, by comparison with other data on the geologic and topographic map.

7. Complete adjustment to the topographical cross section or to the locations of changes of dip relative to the positions of guidelines.

Map Inspection

Areas of the map adjacent to the line of section should be inspected to see whether there are overlying superficial or unconsolidated deposits, lithological units, or structures. Laterally discontinuous units (such as lenticular sedimentary deposits) should be checked since they may or may not be present at depth One should examine the regional area information (including any descriptive material on discontinuities) for background information about lithological units or structures. Data from boreholes, underground mines, quarries, or geophysical survey data is added to the cross section at this point.

Lithological units or structures can be indicated conceptually and marked as such. Recognition of these features may be essential to the interpretation of the cross section. The relative hydraulic conductivity relationships of the units are also extremely important to include within the cross-section interpretation. As the order of magnitude hydraulic conductivity contrasts increase, the greater effects the features will have on ground-water flows.

Borehole Data in Cross-Section

Borehole data collected during Phase II field investigations represents the second important data source for construction of the cross section. Often in glacial areas with thick unconsolidated overburden or in thick weathered rock units, these data are the only information available for construction of the sections. Field borehole data should be edited to illustrate the essential features of the lithologic section. The construction of the cross section should be continued in depth using similar vertical scale as the topographic surface profile. Where available, nearby borehole data is not located along the section line, lateral projection of the data is allowable, where supported by the regional data. The limit to which the construction should be carried is usually set by one of the following criteria:

- Conceptual model objectives have been reached (e.g., the physical configuration of geologic units that will affect ground-water flow conditions is sufficiently established);

- The cross-section construction adequately illustrates the lithology and structure of the site area;

- You have reached the limit of reasonable extrapolation of the surface information; and,

- Borehole data along the section is not adequate to extend the construction.

Reasonable extrapolation of surface information often limits cross-section construction before a full conceptualization is reached for the site geology and hydrogeology. The probable or possible extrapolation of the stratigraphy and structure to areas beyond available 'good' data should then be indicated by the use of dashed lines, question marks, or other symbols to indicate the lower certainty of these portions of the cross section. Due to poor site access, weathering, or overburden coverage in some parts of the section, there may be no immediate way of determining what lies below the surface materials. In these cases, information should be gathered from adjacent areas or consideration given to additional lithologic data gathering in order to make the interpretation of the section as complete as possible.

Plausibility of Geological Features

The interpretation of the cross section up to this stage has been based on geometrical construction from the data available from surficial, geophysical, and geologic information and borehole data. In some cases, however, rigorous geometrical principles developed from surface mapping and borehole data may conflict with geological realities established from literature review. A series of technical questions can be used to check construction points of the cross section, based in part on the conceptual geology established from literature review tasks, and based in part on the concept of reversing structural and sedimentary processes to restore geologic units to their original orientations as depositional layers:

1. Are the thicknesses of lithological units on the section matching unit thicknesses determined from both the geologic map and borehole derived stratigraphic sections?

2. Do the thicknesses of stratigraphic units remain constant across the section, or do they vary in a manner consistent with their depositional history, both in total thickness and in the angle between top and bottom surfaces?

3. In faulted areas, are the thicknesses of a rock unit each side of a fault equivalent? Do its geological boundaries maintain a similar angle with the fault plane? If the original positions of the rocks each side of the fault were restored by reversing the movement of the fault, would the lithology and structure be continuous from one side to the other?

4. Is there a topological similarity between the lithological relationships shown on the map and on the cross section? The shape of a fold (angular, gently curved, box-shaped, etc.) is usually comparable on the horizontal surface of the map. The vertical cross section; the map and the cross section of an unconformity are often directly comparable.

5. Are the unit relationships consistent with what is known of the geological history of the area (this is especially important in glacial terrain)?

6. Do geologic sections agree with residual soil patterns obtained from SCS maps?

7. Do spring lines agree with fracture, faulting or lithologic variability?

The section should be adjusted as necessary to accommodate inconsistencies observed in the system. The preparation of cross-sections represents an important part of the conceptual understanding of a geologic environment. Points within the mass of geologic material are only known with full certainty directly within the boreholes. Once extending beyond the borehole certainty decreases. This decrease in certainty may be relatively minor in 'layer cake' geology, or major in certain glacial deposits where heterogeneity of the deposits is the dominate characteristic. The investigator must weigh the geologic environment and select the basis for extending geologic or hydro-stratigraphic units between known data points. This work requires use of conceptual models to aid in evaluation of the project lithology.

Cross Section Completion

The final version of the section should be prepared by inking or through computer CAD/CAM constructions. Add an appropriate scale. Label all lithological units and structures, geographical locations, and the locations of the ends of the section. Include an explanatory key of all the symbols and abbreviations used. The appearance and intelligibility of the section are often improved if one or two significant lithological units are colored or shaded so that the interpretation of the geological structure and the relationships of the different parts of the section are made as clear as possible. For example, alternating sandstone and claystone can be represented by shading one of the two materials.

It may be necessary to construct more than one cross section in order to illustrate and understand the three-dimensional geometry of an area. These cross sections can be combined to generate fence diagrams that illustrate the geology and potential flow conditions of the facility. These constructions should include scales for each of the three directional components illustrated on the fence diagram. These cross sections can be used later in flow net constructions as part of the process of locating target monitoring zones for ground-water monitoring.

Schematic Site Stratigraphy

Cross sections to illustrate site stratigraphy can be further enhanced by the development of a schematic history of site sedimentation. Figure 7-12 illustrates the reconstructed sedimentary history of an over-bank deposit

in Louisiana. Each cross section shows the development of depositional layers with the associated sea level changes and weathering cycles. Such conceptual model cross sections are extremely valuable for addressing the results of field boring programs. Facies codes used within lithostratigraphic mapping structure can extend the conceptual model to link observed geologic features to depositional environments. Used in conjunction with a base map, these figures add a third dimension to the hydrogeologic framework of the facility that will be useful in evaluation of ground-water flow systems or for later meeting location requirements for monitoring wells. Each cross section should contain the following:

- The depth, thickness, and area extent of each stratigraphic unit;
- All stratigraphic zones and lenses within the near–surface zone of saturation;
- Lithology of significant formations/strata;
- Significant structural features;
- Stratigraphic contacts between significant formations/strata;
- Zones of high hydraulic conductivity or fractures;
- The location of each borehole and depth of termination;
- Depth to the zone of saturation; and,
- Depiction of any geophysical logs.

A scale of no greater than 1 inch: 200 feet is suggested for these cross–sections. With an adequate number of cross–sections, a very useful and illustrative fence diagram (3–dimensional) can be constructed. A table that summarizes the subsurface geologic information should also be submitted as part of the Phase II report.

Using the prepared cross–sections and elevations of the water surface, the depth to the ground–water surface can be determined for purposes of selecting depths for monitoring. Cross sections should be updated and improved as new data are obtained. Where complex subsurface conditions are indicated, additional boreholes, should be drilled for verification of data.

All cross–sections should be identified on a location map as is shown on Figure 7-13. Individual cross–sections are normally keyed by alpha and alpha prime titles. Sections of surficial deposits commonly are constructed with exaggerated vertical scales, in order that thicknesses and depths to units can be measured directly from the illustrations. The resulting exaggeration of surfaces and stratigraphic dip may require explanation to the reader if the nature of the study is sensitive to perceived thicknesses. Whenever possible, exaggerated scale cross-section figures

Schematic Development of Site Stratigraphy

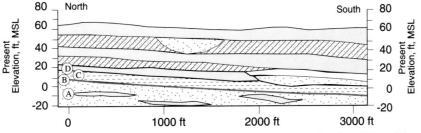

A. Soils A,B,C and D deposited during rising sea level, soil D oxidized

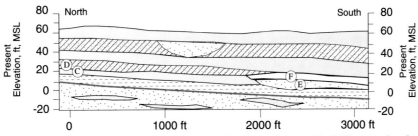

**B. Soils C & D incised by river during temporarily falling sea level.
Soils E & F deposited by river diring rising sea level & not oxidized**

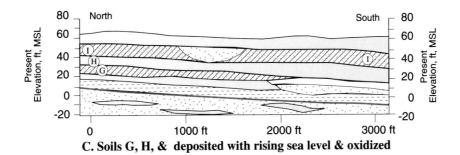

C. Soils G, H, & deposited with rising sea level & oxidized

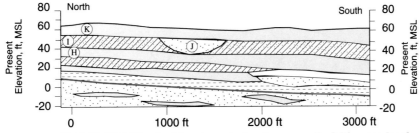

**D. Soils H & I incised by river during temporarily falling sea level.
Soils J & K deposited during rising sea level & not oxidized.**

LEGEND

A. Deltaic gray sands with clay interbeds
B. Organic backswamp silts & clays
C. Varved gray clays
D. Red clays with tan mottling

E. Gray sands
F. Gray clays & silts
G. Red or tan silts & fine sands
H. Red slickensided clay

I. Red fine sands, silts & clays with gray mottling
J. Gray sand
K. Gray clays

Modified from Woodward Clyde (1989)

Figure 7-12 Schematic Development of Site Geology

should be supported by an unexaggerated scale cross-section of the strata present at site. Both diagrammatic presentations can be illustrated on the same figure to allow for comparison purposes. Figure 7-14 shows a geologic computer generated cross–section from the cross–section location map (Figure 7-13). Computer drawn geologic cross–sections may contain a tremendous amount of data and should be reviewed carefully by an experienced geologist/geotechnical engineer to work-out common relationships between units. Figure 7-15 shows a hand

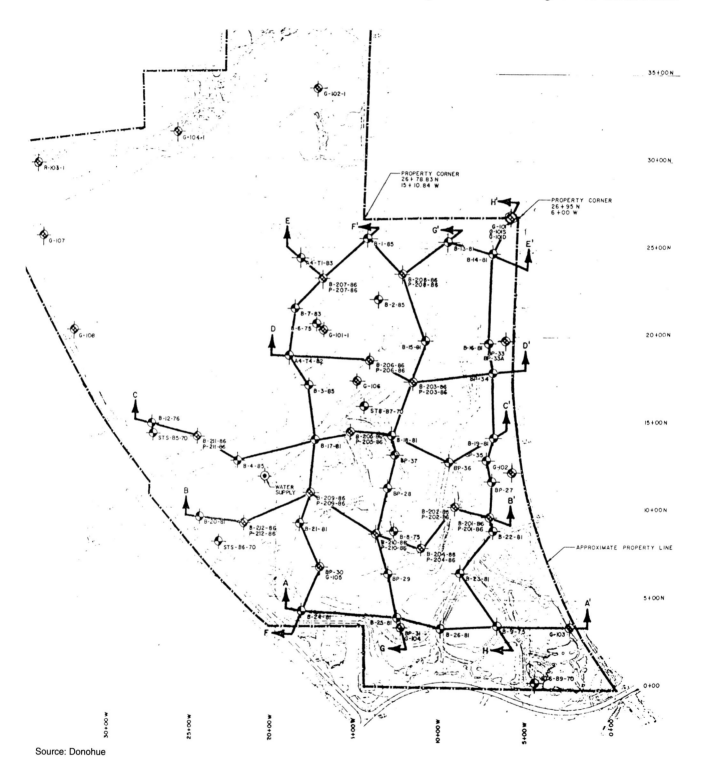

Source: Donohue

Figure 7-13 Cross Section Location Diagram

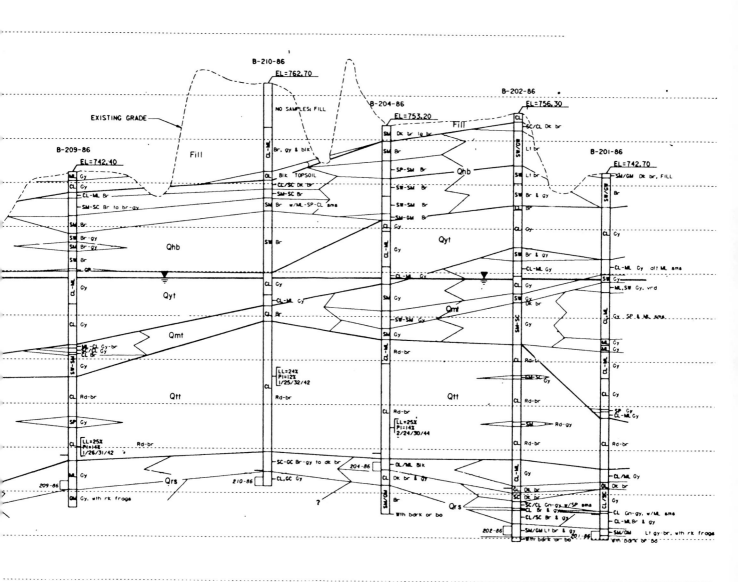

Source: Donohue

Figure 7-14 Computer Generated Cross Section Example

drawn, (and human interpreted), geological section for the same site illustrated in Figure 7-14. This presentation is much more valuable to the project than the too detailed computer generated representation of the section. Since cross-sections are part of the conceptual process, highly detailed data showing each and every layer, make the goal to understand the important geologic units present at site much more difficult. These geologic units shown in Figure 7-15 represent a conceptualization of the actual elevation where the strata would be contacted in a borehole. As with individual graphic columnar sections, the resultant elevation of strata in this conceptual presentation represents the average elevation of lithologic contacts.

Excellent hand-drawn cross-sections can be prepared by experienced geological professionals. Figure 7-16 shows the relationships between fine-grained clay lodgment till units and permeable sands within the Valparaiso lodgment till/fluvial units. This figure also suggests potential water bearing zones that represent

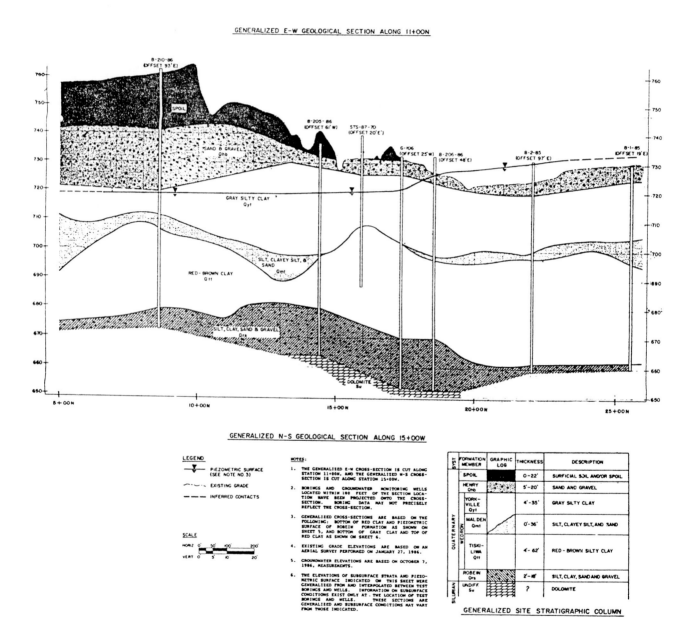

Source: Donohue

Figure 7-15 Hand Drawn Conceptual Model of Figure 7-14

candidate zones for screening of monitoring wells or piezometers.

Cross–sections should be constructed so as to be illustrative conceptual models of site geology and hydrogeology. Figure 7-17 shows a hydrogeologic cross–section of a site in New York State. In addition to the physical descriptions of the surficial materials, screened zones and in–situ hydraulic conductivity are given on the figure. These visual representations of multi-characteristics of the geology go a long way toward summarizing the likely flow systems for such complex sites.

7.2 AQUIFER CHARACTERISTICS

Using data from borings, detailed information should be presented in the Phase II investigation describing the aquifer system and the hydrogeologic properties of the saturated zone underlying the site. Methods used to obtain this information should also be discussed in the final report. The investigator should characterize the uppermost aquifer, as well as, any lower aquifers that are directly interconnected hydraulically with the aquifer of interest.

Described below are several characteristics that should be identified for aquifers that may be affected by the facility. The characteristics should be included as part of the interpretation of the conceptual model.

7.2.1 Aquifer Boundaries

The identification of a lower boundary (or confining layer) can delimit the vertical extent of the uppermost aquifer. Some hydrogeologic settings (e.g., alluvial depositional environments) do not contain clear-cut lower boundaries of aquifers. In such a case, it may be adequate to limit characterization to the expected depth of downward migration of a contaminant. Professional judgment should be used in assessing interconnection between upper and lower aquifers (and the resulting levels of efforts required to characterize lower aquifers, once committed to such an effort). A lack of interconnection may be indicated if the lower boundary of the upper aquifer consists of a thick confining layer (aquitard). Evidence of a significant degree of interconnection between aquifers may be indicated if:

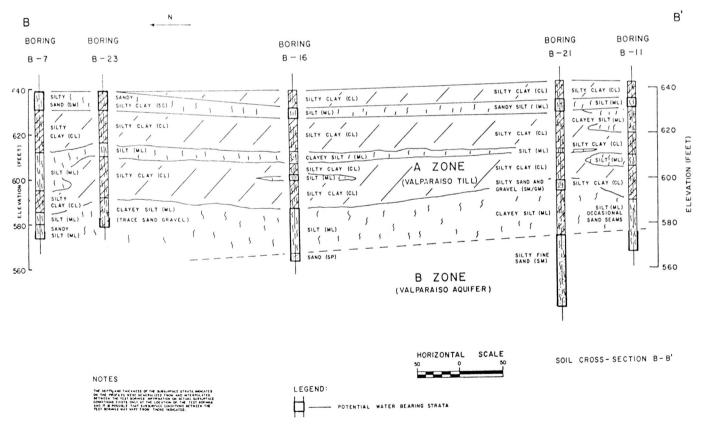

Source: Canonie Environmental

Figure 7-16 Hand Drawn Conceptual Model of Glacial Environment

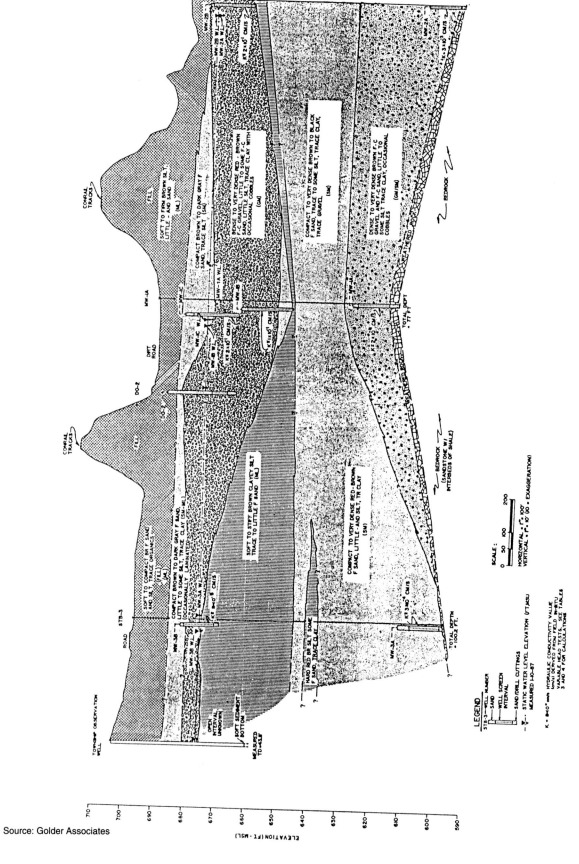

Figure 7-17 Hydrogeologic Cross Section of a New York Site

Source: Golder Associates

- The confining layer pinches out within the site boundary;

- The aquitard is fractured rock or in a Karst system;

- Hydraulic heads measured in the two aquifers show similar elevations;

- Numerous inadequately sealed wells penetrate the aquitard; or,

- Pumping (or injection) tests show a significant response in the aquifer on the other side of the aquitard.

It is also important to identify the presence and position of any hydraulic boundaries (such as fault gouge barriers or confining beds) that limit the aquifer system at a site.

7.2.2 Type of Aquifer

Following the identification of aquifer boundaries, the aquifer(s) beneath the site should be identified as being either unconfined (where the ground–water surface forms the upper boundary), or confined (as confined between two layers of low hydraulic conductivity beds in a stratigraphic sequence). Traditionally, confining zones have been termed aquitards (low hydraulic conductivity) and aquicludes ("impermeable'). In the present context of hydrogeologic nomenclature and related environmental protection imperatives, the latter term should no longer be used, without significant forethought and site–specific knowledge.

Real world aquifers include a full spectrum between the classical confined aquifers and unconfined aquifers. In general, interpretations of piezometer heads are important indicators of the type of aquifer present. Data gathered during drilling are important in deciding the type of aquifer system present. This is especially true for low hydraulic conductivity environments, where ground water will take a significant period of time to reach static ground–water levels in a borehole.

7.2.3 Saturated Zones

Saturated zones, above the uppermost aquifer (such as low hydraulic conductivity clay soil), can also act as pathways for migration of contaminants and, if present, should be identified and described. Low hydraulic conductivity geologic units located above the uppermost aquifer must often be evaluated for the ability to yield water to shallow boreholes and piezometers. A

reoccurring question arises in such evaluations; where is the ground-water level (saturated zone) located in the unit? Simply drilling a hole in the unit and observing where saturated material is found, often does not establish to a reasonable extent, the actual point of the saturated zone. This inability to establish fully saturated conditions are due to the low horizontal hydraulic conductivity of the material. The geologic unit yields ground water so slowly to the open hole, that leaving the hole open for 24 hours may not allow sufficient time for the borehole to fill (or even to show signs of any liquids). If the hydraulic conductivity is less than 10^{-6} cm/sec boreholes will fill very slowly. Probably the only reliable method available to investigators for these situations is to install in the target geologic units piezometer tips that require very little free water to establish a head level. The Hvorslev 90% recovery graphs provides the key relationships between piezometer completions and the relative hydraulic conductivity of the geologic unit. The perched water zones or perched aquifers present underneath a site should be identified as part of an overall site evaluation.

Similar questions can arise on the level of full saturation of confining units located below perched water systems and above fully saturated and confined aquifers. As with the shallow confining units, deeper confining units can be evaluated in a similar fashion by locating piezometer tips that require little free water to establish hydraulic heads within the unit. If sufficient time is available for long-term head measurements, conventional open tip piezometer completions can be used. 90% reaction times can be improved somewhat through the use of 'dummy' PVC or aluminum boreholes fillers that reduce the volume requirement of the installation.

Perched water zones are usually difficult to fully define, especially as to their area extent, hydraulic properties, and manner of recharge. Most shallow perched water bodies can successfully be dewatered for waste management purposes, if enough effort is expended in their characterization. Perched water is especially common in overburden areas of glacial outwash deposits.

One of the most important geologic criteria for evaluation of perched water systems is the confirming of the structural surface of the confining unit. Since by definition, a perched water table is located above the regional aquifer ground-water surface, (i.e., as in the unsaturated or vadose zone), gravity flow along the top of the confining unit represents the main driving force of ground-water or contaminant flow. Hence, sufficient data must be available to fully define the surface of the top of the confining unit. These data are then developed into isopach maps that depict the geologic surface as shown in Figure 7-18. The procedure to develop these maps are similar to that used in topographic map production. It is extremely important to fully understand the geologic surfaces that can direct ground-water flow or provide windows between aquifers. Isopact presentations can provide insight into orientation of confining beds and to thin areas of low hydraulic conductivity materials that may provide unexpected surprise in development or remediation of the respective facility.

7.3 CHARACTERISTICS OF GROUND–WATER FLOW

The amount of ground water at a site and the direction in which it flows are important factors to consider in waste disposal site assessment projects, because they are essential components in an analysis of potential leachate transport in ground water.

This section describes evaluation of the site characteristics with respect to recharge and discharge zones, ground–water flow directions, and ground–water flow rates, necessary to define the conceptual movement of ground water below and adjacent to a facility. In addition to the inclusion of maps and tables as discussed in this section, the Phase II report should present a thorough narrative discussion of investigative results identifying:

• All dominant ground–water flow–directions, including both horizontal and vertical components in upper and (significant) lower aquifers.

ISOPACH MAP OF THE BASAL TILL

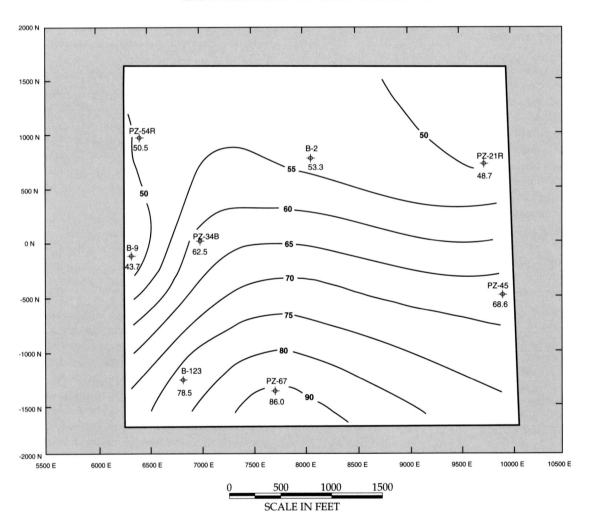

Figure 7-18 Isopach Maps That Depict the Geologic Surface

- Relationships of ground–water flow to discharge or from recharge areas.

- Temporal and seasonal variations in ground–water levels.

As with other data identified in this chapter, ground–water flow–characteristics are important in the understanding of the hydrogeologic setting, and in determination of the design for the monitoring – well – system requirements.

7.3.1 Recharge/Discharge Zones

To aid in the understanding of the ground–water flow–regime, and to aid in the identification of potential paths for leachate migration, the location of any proximate recharge or discharge zones should be identified in the report. This identification is determined in part by location of the site within the watershed. The investigator probably does not need to quantify this information; rather, a general indication of discharge or recharge characteristics could be presented.

For unconfined aquifers, recharge areas are usually topographic highs, while discharge occurs in topographic lows. Discharge/recharge areas also indicate relative depth to ground water. In discharge areas, the ground–water surface is found close to or at the land surface; while at recharge areas, there is often an unsaturated zone between the ground–water surface and the land surface. A ground–water level contour–map can be used to locate these areas.

Recharge and discharge in confined aquifers is more complex, and either may occur where the aquifer is exposed, or may be due to recharge in the form of upward leakage in areas of upward hydraulic gradient. Recharge can also occur by downward flow through relatively low hydraulic conductivity confining layer.

7.3.2 Ground–Water Flow Directions

The hydrogeologic field investigations should include a program for precise monitoring of the ground–water levels, including area variations on temporal and seasonal bases. This program will involve measurement of water levels in the observation piezometers, installed for the purpose of investigating the saturated zone. Water level data also are used to establish the depth to ground water at a site, as well as, in establishing hydraulic gradients and flow directions. The data gathered by measurement of piezometer heads can be illustrated in a variety of formats that show important aspects of site hydrogeology.

Ground–water contour–maps, piezometric surfaces, and depth–to–water maps all illustrate potential flow directions.

Care should be exercised in the preparation, use, and evaluation of ground–water level and depth maps. Initially, only a limited number of control points (observation wells or piezometers, etc.) can normally be used, and ground–water conditions between the points may deviate widely from the expected generalities. Furthermore, unless the piezometers or wells are constructed to reflect specific aquifer or confining unit site conditions, composite hydraulic head levels, such as would be reflected by a combined piezometric (unconfined) and potentiometric (confined) levels, may be reflected. Wide screened monitoring wells, piezometers cutting the ground-water surface and piezometers screened within at depth in saturated hydrostratigraphic zones can all display fundamentally different hydraulic data. This could yield erroneous and misleading data presented on potentiometric maps.

The basic principle of ground–water flow holds that water moves from a higher hydraulic level (or potential) toward the lower. The contours on ground– water – elevation contour–maps are those of equal potential, and direction of movement is, under isotropic conditions, at right angles to the contours. This is true whether the contours are of a free–water–surface (piezometric) or of a potentiometric surface. In an unconfined aquifer, the contours often tend to parallel the contours of the land surface. At many locations, however, there is little apparent relationship between surface and subsurface flow.

Mounding of ground water can result from downward seepage of surface water or upward leakage from deeper artesian (confined) aquifers. In an ideal unconfined aquifer, gradients from the center of a recharge mound will decrease radically, at a declining rate. An impermeable boundary or change in transmissivity will affect this pattern and may provide clues in determining locations of such changes.

7.3.3 Ground–Water Contour Maps

A ground–water surface contour–map is the most commonly constructed and most useful map for studies of unconfined ground water. It is a contour map of the potentiometric surface, and the contour lines are usually lines of equal elevation (see Figure 7-19). The map is constructed using water level elevations from observation wells, and piezometer that cut the ground-water surface, stream and lake surfaces, and spring discharge points for elevation control.

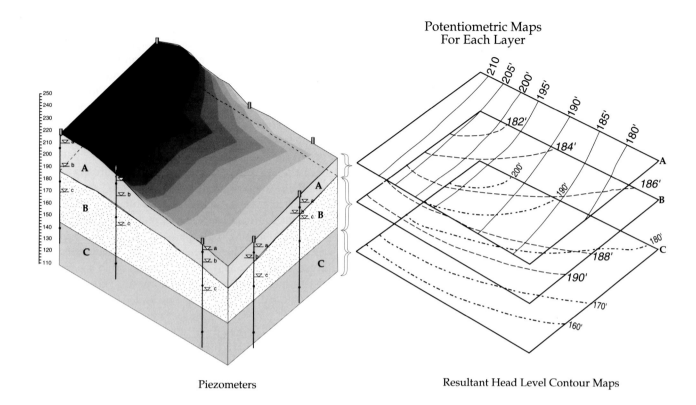

Figure 7-19 Ground Water Contour Surface

Ground-water surface maps are particularly important for the display and evaluation of directional flow. These maps are especially sensitive to increased vertical scale expansion as a means of enhancement to the illustration. Figure 7-20a shows the typical ground-water surface map (sometimes called ground-water table). The illustration although reasonably clear can still prove increased perspective once placed on a perspective grid shown on Figure 7- 20b. The vertical scale exaggeration at ten times is insufficient to fully illustrate flow directions for the site. However, with a fifty times exaggeration the three-dimensional diagram of the ground-water surface provides the necessary perspective to evaluate potential flow directions. One should however, not assume the three dimensional nature of the perspective fully represents the three-dimensional aspect (X-Y-Z) of ground-water flow. These perspectives only illustrate the unconfined surface of ground water. As such, the ground-water could be moving in a downward component not illustrated by these piezometers, which were selected to cut the unconfined ground-water surface. In actuality a series of these three-dimensional surfaces could be prepared to show hydraulic

head data for each hydrostratigraphic zone monitored by piezometers. One must be sure to only use the head data from piezometers completed within the same hydrostratigraphic unit. In other words you must compare "apples with apples" to obtain correct interpretations of ground-water directional flow.

Maps of the Potentiometric Surface

A map of the potentiometric surface is similar to a ground–water level contour–map, except that it is based on the potentiometric heads measured in piezometers or tightly sealed wells, which penetrate a single confined aquifer (see Figure 7-21).

Maps of the Depth–to–Ground Water

Depth–to–ground–water maps are of particular interest when considering drainage and dewatering projects. Such derived maps show the depth–to–water from the ground surface. They are most easily prepared by overlaying a ground–water contour–map on a surface topographic map. The points at which the two data sets contour intersect are a whole number of feet apart in elevation and represent the control points for drawing a contour map of depth to water. They can also be prepared by calculating the depth–to–water from the ground surface and placing this depth on a map, at the location of the observation well. Contours are then drawn by using these points using standard techniques.

Analysis of conditions defined by ground–water contours is made in accordance with Darcy's law, Q = KiA. Accordingly, the spacing of contours (the gradient) is dependent on the flow rate, on the aquifer thickness, and

hydraulic conductivity. If continuity of the flow rate is assumed, the spacing for contour lines depends only on aquifer thickness and hydraulic conductivity. Thus, area changes in contour spacing may be indicative of change in aquifer conditions. However, in view of the heterogeneity of most aquifers, apparent changes in gradients must be carefully interpreted with consideration of all possible combinations of factors. Parameters necessary to determine ground–water flow directions include:

• **Depth to Ground Water** — Water levels should be submitted with the Phase II report for each piezometer or well. Data can be presented in tabular form and should

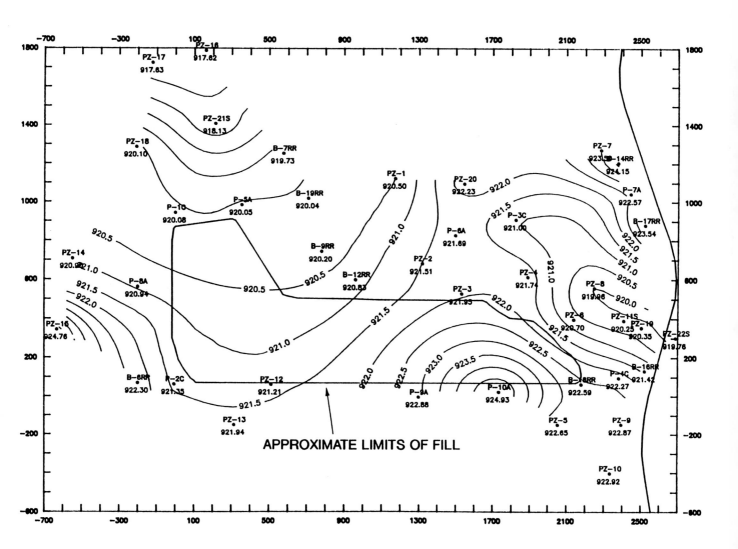

Source: P. E. LaMoreaux & Assoc.

Figure 7-20a Ground Water Contours Example - Plan View

include locations and identification of wells, depth of wells, screened intervals, ground–water elevations, and sampling date. The ground–water level can be indicated on the stratigraphic maps for illustration of the relationship of subsurface materials to ground water. Measurements are typically represented as the average depth to ground water at each piezometer. If significant variations in levels occur due to temporal or seasonal factors, it will be important to provide a time–series presentation of the data.

• **Potentiometric Surface** — Using water level (hydraulic head) distributions at the site, water–level contour maps, (Figure 7-22) should be prepared. These contour maps should also clearly include the facility(s) location, ground–water elevations, and isopachs of

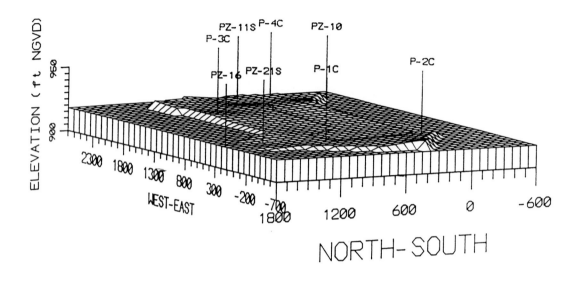

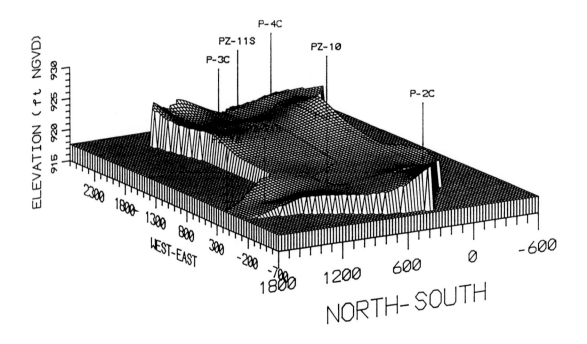

Figure 7-20b/c Ground Water Contours Example Expanded to 10 and 50 Times Scales

aquifers. For sites with unconfined aquifers and no underlying significant (interconnected) hydraulic units, just one piezometric contour map should be prepared for flow–net analysis and development of a conceptual model. For sites with more than one aquifer unit, contour maps of potentiometric surfaces, (as determined from piezometer measurements), should be prepared for each lower unit, which can serve as a flow path from the facility.

• **Hydraulic Gradients** — Using the contour maps of the potentiometric surface discussed above, one can establish hydraulic gradients (the change in elevation of the potentiometric surface over distance). This determination will represent the hydraulic gradient in the horizontal direction. Hydraulic gradients should also be evaluated for

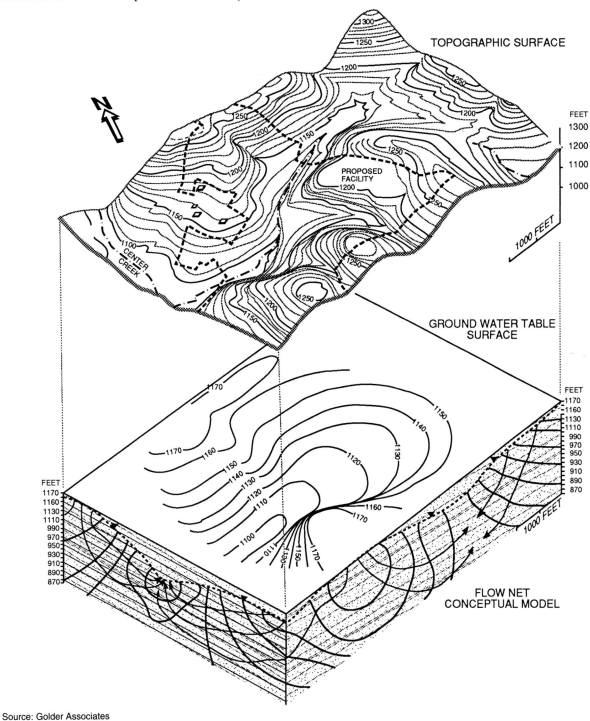

Source: Golder Associates

Figure 7-21 Potentiometric Surface Map

any significant lower units or confined aquifers that may serve as a flow path from the facility.

• **Vertical Components of Flow** — In addition to considering the components of flow in the horizontal direction, the investigator must assess vertical components of ground–water flow. Both vertical and horizontal flow components are somewhat misnomers; since, there is only one flow continuum present; water levels in piezometers establish heads within a three–dimensional flow system. However, to represent the data from nested piezometers, "vertical" components of flow will normally be considered as separate from "horizontal" gradients. This evaluation may require installation of piezometers in clusters. A cluster of piezometers, or piezometer nest, is a closely spaced group of wells screened at different depths to measure vertical variations in hydraulic head. Placement of vertically nested piezometers in closely–spaced, separate boreholes was, at one time, the preferred method to gather such data. This single hole-single point piezometer completion was the result of early attempts to place more than one piezometer in a small diameter borehole (typically NX core holes). These core holes were less than 3 inches in diameter and although allowing two or three 3/4" piezometers to be placed into the core hole, there was insufficient clearance to adequately place bentonite pellets in the correct position to seal the screened zones of the piezometers. Information obtained from

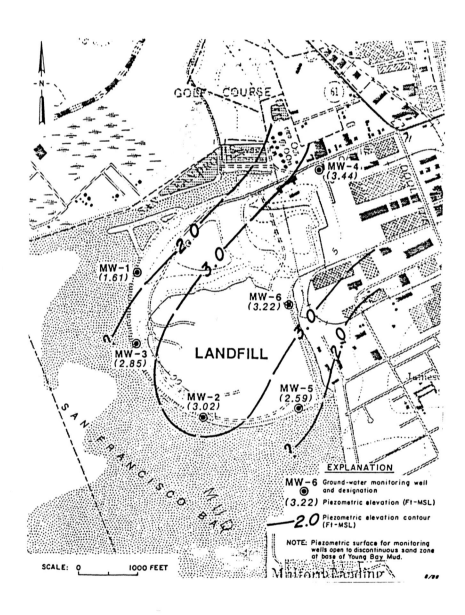

Figure 7-22 Ground Water Surface Map

poorly constructed (leaking) multiple piezometer placement in single boreholes confused head level readings and resulted in use of erroneous data. Now, however, multiple point piezometers in cored holes, or 3 to 4 small diameter "piezometers" can be completed with greater reliability using high–solids (>20 percent solids) bentonite slurry between the screened zones in a single hole.

Collected data should be organized in tabular form for each well nest. Data should include well locations, identification of wells, well depths, screened intervals, ground–water elevations, and date of sampling. Determination of vertical flow–gradients should be made from the piezometric measurements, made in nested or nearby wells. These data can be used to make determination of discharge and recharge zones, and confining unit (aquitard) characteristics. The measurement of hydraulic heads from nested piezometers is a basic data set to construct flow nets, (a vertical cross–section of the site illustrating a pattern [or flow lines] of hydraulic heads). Flow net constructions are fully described in Chapter 8.0.

The hydrologic fluctuations and other factors, discussed above, that may make the determination of flow patterns unreliable, can often be overcome by an expanded effort in water level monitoring. For seasonal variations in water levels, a higher frequency head level monitoring schedule may be necessary. For low horizontal gradients, the effects of short–term changes in water levels can be analyzed by installation of continuous recorders in selected wells. In aquifers having significant vertical gradients, piezometers, completed at various depths, may be required in order to provide a three–dimensional description of the ground-water flow field. For heterogeneous and anisotropic aquifers, more water level piezometers, wells, and more field tests for hydraulic properties typically required to full evaluate site conditions.

7.3.4 Other Considerations For Assessing Characteristics Of Ground–Water Flow

Factors such as hydrologic fluctuations, (e.g., seasonal variations, well pumping, and tidal processes), low or high gradients, and aquifer heterogeneity or anisotropy can result in variations in ground–water levels and resultant flow patterns. Accurate determination of ground–water flow under these circumstances can be difficult. Any program undertaken to investigate ground–water flow patterns at a site should identify and assess these external processes that contribute to, or affect ground–water flow patterns. If these processes are not evaluated, the uncertainty introduced by neglecting them should be estimated. Table 7-2 shows a number of potential influences on the level of ground water in a piezometer or well. Several of these considerations are discussed below:

- **Seasonal Variations** –– Seasonal variations in ground–water use and recharge can cause significant changes in ground–water flow directions. In extreme situations, a flow reversal can occur. For sites where this phenomenon may be important, maps of the piezometric surface should be submitted that represent yearly averages and the two seasonal extremes. In addition, the investigator should provide information describing the temporal changes in ground–water flow direction using records compiled over a period of no less than one year. Seasonal variations in flow direction are more likely to occur in unconfined systems, due to generally more locally prevalent recharge and discharge conditions. Seasonal variations in water table elevations can range from minimal change to 10's of feet. Especially affected are fractured rock and Karst geologic environments where porosity values may be very low,(i.e., ranging to below 2%). Available recharge waters moving into open fractures or solutioned joints can rapidly raise observed head levels with even little volumetric inputs.

- **Pumping** - - Off–site or on–site pumping from wells may affect the direction of ground–water flow. Municipal, industrial, or agricultural ground–water use may significantly change ground–water flow patterns and levels

	Unconfined	Confined	Natural	Man-induced	Short-lived	Diurnal	Seasonal	Long-term	Climatic influence
Ground water recharge (infiltration to the ground water surface)	✓		✓				✓		✓
Air entrapment during ground water recharge	✓		✓		✓				✓
Evapotranspiration and phreatophytic consumption	✓		✓			✓			✓
Bank-storage effects near streams	✓		✓				✓		✓
Tidal effects near oceans	✓	✓	✓			✓			✓
Atmospheric pressure effects	✓	✓	✓			✓			
External loading of confined aquifers		✓		✓	✓				
Earthquakes		✓	✓		✓				
Ground-water pumpage	✓	✓		✓				✓	
Deep-well injection		✓		✓				✓	
Artificial recharge: leakage from ponds lagoons, and landfills	✓			✓				✓	
Agricultural irrigation and drainage	✓			✓				✓	✓
Geotechnial drainage of open pit mines slopes, tunnels, etc.	✓			✓				✓	

Source: Modified from Freeze and Cherry, 1979

Table 7-2 Fluctuations in Ground Water Levels

over time. Pumpage may be seasonal or more dependent upon water consumption patterns. Frequent measurement of water levels in piezometers must be used to detect such water use patterns. For sites where such variations occur, the rate of ground–water withdrawal in the vicinity of the facility should be summarized in tabular form and include: well location, depth, type of user, and withdrawal rates. The zone of impact created by any major withdrawal from a well or well field should be identified on the site map. The analysis of these ground-water withdrawal patterns may include projection of drawdown curves through solving of analytical equations or computer numerical modeling. Long term observation of ground-water flow direction may be required for land disposal facilities. These facilities may require assessments of flow directions lasting 30 to 50 years when combining siting, operational and post-closure monitoring requirements. Changes in ground-water flow directions should be evaluated, at least, yearly to confirm that down gradient wells remain truly down gradient. Reverses in deep ground-water flow direction were observed during the late 1980's when the Chicago Deep Tunnel system was constructed in the Silurian dolomite bedrock. Dewatering or depressuration of the bedrock caused water levels in deep wells to drop sufficiently to reverse, (at least locally), the observed ground-water flow direction. One should not be surprised when man-induced changes in ground-water flow components are observed in long term monitoring programs.

• **Tidal Processes** — Natural processes (see Figure 7-23) such as riverine, estuarine, or marine tidal movement may result in variations of well water levels. The investigator should document the effects of such patterns where locally important.

• **Gradient Considerations** — In areas of low or almost flat horizontal gradients, small errors in water level measurements, or small transient changes in water levels can make determination of flow direction and rates unreliable. Determination of flow patterns is also difficult where high or steep vertical gradients exist (i.e., often observed in surficial units). Often, a near–surface, shallow, unconfined aquifer may overlie an aquifer of higher hydraulic conductivity, resulting in vertical head gradients through a "leaky" intervening confining unit. Flat horizontal gradients may also be indicative of very high hydraulic conductivity. Water table contour maps can be used to evaluate changes in lateral hydraulic conductivity. Relatively flat contours on these maps may represent good target zones for ground-water monitoring since they may represent areas of higher hydraulic conductivity.

7.4 HYDROGEOLOGIC PROPERTIES

Each of the significant stratigraphic units located in the zone of saturation adjacent to the facility should be characterized for hydrogeologic properties such as hydraulic conductivity (vertical and horizontal) and

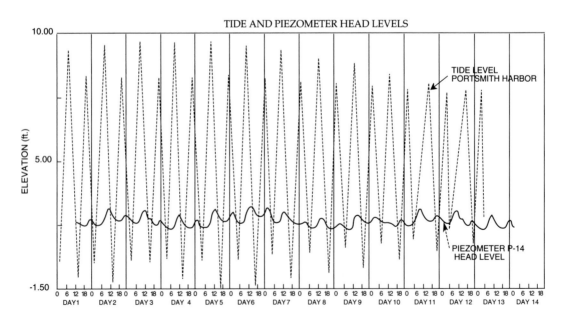

Figure 7-23 Tidal Processes Observed in Monitoring Wells

effective porosity. These parameters describe aquifer characteristics that control the movement of ground water and, hence, the ability of the aquifer to retain or pass potential contaminants. Both hydraulic conductivity and effective porosity are needed for a basis for a general understanding of the hydrogeologic setting at a site. Hydraulic conductivity and porosity of aquifer materials can be determined by using laboratory and/or field methods. Tests that are conducted to define the necessary hydrogeologic properties of stratigraphic units should be performed in the field. Laboratory tests may be used to substantiate results of field tests, but should not be the sole basis for determining aquifer characteristics. Because each of these parameters can vary within the same aquifer, any area variations must be identified. The amount of data needed to accurately determine hydrogeologic parameters increase with increasing heterogeneity of the aquifer. For example, an aquifer of extensive, homogeneous beach–sand (SP) will require less investigation than a glacial unit consisting of lenticular deposits of outwash sand and gravel (GW/SW), interbedded with clayey lodgment till (CL). Listed below are several hydrogeologic parameters that should be identified for the typical aquifer system.

7.4.1 Hydraulic Conductivity

Hydraulic conductivity (K) is a measure of the ease by which a medium transmits ground water. It is one of the few physical parameters that includes values that may span more than 13 orders of magnitude. Although this entire range will probably not be exhibited at a single site, Phase II hydrogeologic field investigations often reveal variations in K values that range over many orders of magnitude. Aquifer heterogeneity may be related to either a single stratum that had a geologically–complex depositional environment (i.e., glaciated terrain of ablation and lodgment till and outwash sands), or a system of layered geologic units such as a sand–clay–sand sequence. Depending on placement of piezometers, fractured bedrock can also exhibit extreme variability in computed hydraulic conductivity.

To accurately characterize complex flow domains, an appropriate number of piezometers are typically placed in the saturated zone, during Phase II field investigations. Nested piezometers are used to determine both K values for stratified deposits, and the three–dimensional hydraulic heads.

Hydraulic conductivity (K; distance/time) refers to the ability of aquifer materials to transmit water, which in turn controls the rate at which ground water will flow under a given hydraulic gradient. Hydraulic conductivity is controlled by the size, amount, and interconnection of void spaces within the aquifer, which may occur as a consequence of intergranular porosity, fracturing, bedding planes, etc. Ground–water contaminants have the potential to move more rapidly in an aquifer unit with a high hydraulic conductivity. For these reasons hydraulic conductivity should be determined for each aquifer unit that may be affected by the facility. In addition, hydraulic conductivity should be determined on any semipermeable or confining beds, present in the subsurface, through which water may transmit to or from the aquifer.

7.4.2 Hydraulic Gradient

Hydraulic gradient (i), the driving force of ground–water flow, is determined by dividing the change in head over the distance between the points of measurement along the flow length. Gradients can range from greater than 1.0 near a point of ground–water discharge to less than 0.0001, a value associated with extensive areas of flat terrain or very high hydraulic conductivity. Factors influencing hydraulic gradients include:

• Aquifer characteristics;

• Conditions at the boundaries of the flow domain under consideration (i.e., river elevation, ground–water discharge from the underlying bedrock into a sand and gravel aquifer, etc.;

• Inputs and outputs to the system (e.g., rainfall, evapotranspiration, and infiltration); and,

• Man–made influences.

Gradients can be established through use of ground–water contour maps and nested piezometers. Since gradients are related to a three dimensional flow, care must be taken in context of directional components. If piezometers are completed in permeable aquifer units where flow components are essentially horizontal, then the estimates of ground–water contours and flow directions should reflect gradients usable in rate calculations.

7.4.3 Porosity

The porosity of a rock is a measure of the interstitial space of the rock and is expressed quantitatively as the percentage or fraction of the total volume of rock occupied by interstices (Meinzer, 1959). In general, a porosity greater than 20 percent is considered large, a porosity between 5 and 20 percent is considered medium, and a porosity less than 5 percent is considered small.

The porosity of a sedimentary deposit depends chiefly on the shape and arrangement of its constituent particles; the degree of assortment of those particles; the cementation and compaction, to which it has been subjected since deposition; the removal of mineral matter through dissolution by percolating waters; and, the amount of fracturing of the rock. In deposits composed of well–sorted and well–rounded particles, grain size has no influence on porosity; thus, a boulder deposit may have the same porosity as a body of fine–sand. The porosity of many deposits can be increased by the irregular angular shapes of its constituent grains. Porosity decreases with increases in the variety of size of grains; such as smaller grains filling interstices between larger grains.

7.4.4 Specific Yield

The specific yield of an aquifer is a measure of the water–yielding capacity of the rock and is expressed quantitatively as the fraction of the total volume of rock occupied by the ultimate volume of water released from or added to storage, in an unconfined aquifer, per unit (horizontal) area of aquifer and per unit decline or rise of the ground–water surface.

7.4.5 Effective Porosity

If there are dead–end pores within the ground–water system or ground water is for some reason immobile, then the entire pore–space (the total porosity) is not effective in transmitting fluid and a correction to the porosity must be made. This adjusted value (dimensionless) of porosity is called the effective porosity. Effective porosity, therefore, refers to the amount of interconnected pore–space available for transmitting water. The specific retention of a rock is a measure of the water retaining capacity of the rock and is expressed quantitatively as the percentage of the total volume of rock occupied by ground water that will be retained interstitially against the force of gravity. The porosity of a saturated rock is equal to the sum of specific yield and the specific retention. Because effective porosity is used in the calculation of velocity of ground–water movement, there has been a movement to use very small effective porosity and, hence, significantly increase velocity.

Care must be taken when calculating velocity from very low effective porosity (such as 1 or 2 percent) estimated from specific yield values for non–granular (cohesive) materials such as clays. Column tracer testing shows that velocity measurements should be made with a more realistic 30 (plus) percent effective porosity values for clays.

Effective porosity (n_e [%]) is the amount of interconnected pore space in soils or sediments through which fluids can pass, and is expressed as a percentage of bulk volume. Although effective porosity is important in determining time of travel (TOT), its impact on V is limited to a much smaller range of potential values, unlike K and I. A thorough site characterization where measurement of TOT is important for permitting of a facility should provide measured values of ne. However, effective porosity data may not be available for some sites; it can be estimated with little influence on the validity of the TOT calculation. Chapter 6.0 describes the laboratory procedures that can be used to establish n_e for geologic units found on site that form part of the flow path from the facility to downgradient areas. This drive to use lower n_e values are a result of near-field molecular diffusion through clay units causing local apparent higher velocity than would be observed through simple transport phenomenon alone. Care must be taken to include field observations in selecting of values of ne for a particular geologic unit. Secondary hydraulic conductivity features must also be accounted for in selecting reliable ne values. If a particular low hydraulic conductivity geologic material is shown to be fractured, the fracture system would predominate thus causing a higher hydraulic conductivity and lower ne to be used in the velocity calculations.

Effective porosity should not be confused with total porosity, specific yield, or gravity drainage. Use of any of these parameters as an estimate or substitute for effective porosity can affect resulting estimates of TOT by several orders of magnitude. In some cases, such as for coarse-grained soils, the use of gravity-drainable porosity or specific yield may be acceptable. Effective porosity of fine-grained soils can be calculated using compacted soil columns where a tracer parameter (chloride) is forced through the column under a specific head or gradient. Knowing the hydraulic conductivity of the material from back-pressured triaxial tests allows back calculation of ne once breakthrough of the indicator parameter reaches the base of the soil column.

Ground–Water Flow Rates: The investigator is typically required to provide an assessment of ground–water flow rate or velocities beneath a site for both detector and assessment monitoring programs. Along with flow direction, this assessment is one of the most important pieces of hydrogeologic information supplied in the conceptualization of hydrogeology. Ground–water flow rates are commonly used to demonstrate projected travel times. They can be directly determined by tracer tests, or calculated by using aquifer parameters developed during the Phase II field investigation.

Field Determination of Flow Rates: Field techniques for the direct measurement of the rate of ground–water flow, such as tracer tests, are difficult to perform and are not often required in typical site investigations. Rather, information obtained from analysis of the hydrogeologic properties and flow directions (hydraulic gradients) will allow the calculation of a reasonable representation of ground–water flow–velocity by way of a simple modification to Darcy's law, as discussed below.

Calculations of Flow Rates: Darcy's law is based on empirical evidence that the flux water through an aquifer is proportional to the hydraulic gradient. The constant of proportionality is the hydraulic conductivity. The flux of ground water flowing through an aquifer can be calculated using the following equation for average linear velocity, V, derived from Darcy's equation for saturated flow:

$$V = \frac{K \cdot I}{n_e} \qquad \text{Equation 7-1}$$

Where:
K = saturated hydraulic conductivity,
I = the hydraulic gradient (equal to the change in head divided by the length of the flow path [dh/dl]), and
n_e = the effective porosity.

The linear velocity, V, represents the rate of ground–water flow through pore spaces only. As shown in the Equation 7-1, V is proportional to K and I, and inversely proportional to ne. Because this equation is a linear relationship, the effect of parameter variability on V can be demonstrated.

• If K or I is increased an order of magnitude, V will increase one order of magnitude.

• Conversely, a doubling of ne will decrease the calculated V to one–half its original value.

This sensitivity relationship demonstrates that accurate measurements of K, I, and ne are required to define average linear–velocities. Site–specific value for each of these parameters, especially K and I, are obtained from a thorough Phase II hydrogeologic investigation, ne is a more difficult parameter to evaluate in the field. Laboratory tests based on column break-through tests.

Section 5.0 provides the procedures for calculation of ne for samples collected from Phase II field drilling programs. The measurement or estimate of each variable is discussed in detail below.

7.5 AQUIFER TESTS

Analyses of results of systematic observations of water level changes during stress testing of the aquifer yield values of aquifer characteristics. The extent and reliability of these analyses are dependent on features of the test including duration, number of observation points, and method of analysis. Two general types of analyses are available for determination of aquifer characteristics:

• Steady state or equilibrium methods which yield values of transmissivity and hydraulic conductivity; and,

• Transient or nonequilibrium methods which also yield storativity and boundary conditions.

The principal difference between the two methods is that the transient method permits analysis of ground–water conditions, which change with time and involve storage, whereas the steady state method does not. Test analyses also require the inclusion of the hydrologic and geologic setting of the aquifer. Conditions that should be known include: location, character, and distance of nearby bodies of surface water; depth, thickness, and stratigraphic conditions of the aquifer; and construction details of the test well and of any observation wells.

The following subsections describe commonly used analytical models to solve for local hydraulic conductivity commonly called "slug testing". Following the discussion of slug tests additional guidance is provided on analysis of full scale aquifer pump tests obtained from larger scale programs. Detailed discussions of the numerous aquifer stress testing available in the literature will not be provided below, however, quantitative guidelines for appropriate analysis techniques and the resultant parameters obtainable through the techniques will be discussed. For further information on any of these analytical solutions, see the listed papers and the standard textbooks by Freeze and Cherry (1979), Fetter (1980), Driscoll (1986), and Walton (1962).

7.5.1 Techniques For Analysis Of Data From Slug Tests

There are four popular analytical techniques, used in the analysis of slug test data, these are discussed below.

These techniques constitute empirical relationships that are used to solve for the hydraulic conductivity, in the aquifer containing the test well or piezometer. Both confined and unconfined aquifer conditions are addressed, with two different approaches to solve for hydraulic conductivity under each set of conditions. The applicability of each analytical technique is contingent upon data that meet specified criteria for validity, as described below:

• **Ferris–Knowles Method (1954, 1963)** -- This method is applied to wells completed in confined aquifers with hydraulic conductivity less than 0.05 cm/sec. The analysis of the data involves an arithmetic plot of residual drawdown versus the reciprocal of time. Validity of the analysis is determined by fitting a straight line through the plot of the data points, in which the line intersects the origin (0, 0). The equation used to solve for hydraulic conductivity, K, using this method is as follows:

$$K = \frac{q\frac{1}{t}}{4\pi \cdot H(t)L} \qquad \text{Equation 7-2}$$

Where:
q = Volume discharged (slug volume)
H(t) = Residual drawdown at time = t
t = Time
L = Screen length

The values of $1/t$ and H(t) are obtained from a straight line fit through the plot of slug test data.

• **Hvorslev Method (1951):**-- The research associated with this method was originally performed in near–surface saturated soils, which is analogous to wells completed in unconfined aquifers. Analysis of this method involves a semilogarithmic plot of the residual drawdown divided by the instantaneous drawdown versus time. Validity of the analysis is ascertained by fitting a straight line through the plot of the data points so that the line intercepts the coordinate 1, 0. Basic equations are applied for different configurations of soil. Hydraulic conductivity is proportional to a "shape factor" and inversely proportional to a "time lag." The shape factor is determined from the characteristics and dimensions of the test well, while the time lag is determined from the semilogarithmic plot. In the field, the basic time lag is determined by either raising or lowering the head in a piezometer and recording the head at a number of time intervals. One then plots the head ratio (h/h_0) on a log and by time on an arithmetic scale to obtain the relationships necessary for time lag calculation. The basic time lag is the time at which the head ratio equals 0.37

To interpret a set of field recovery data, the data are plotted in the form of Figure 7-24. the value of T_0 is measured graphically, and K is determined from Equation 7-3(1). For a piezometer of length L and radius R Figure (7-24), with ratio of L / R > 8, the shape factor, F has a resultant equation for K as:

$$K = \frac{r^2 \ln \dfrac{L}{R_w}}{2LT_0} \qquad \text{Equation 7-3}$$

Where:
r = Well casing radius
L = Screen length
R_w = Borehole radius
T_0 = Time lag when residual drawdown/instantaneous
    drawdown = $10^{-0.4343}$

Or for the most common type of monitoring wells the equation relating the hydraulic conductivity to the well response is:

$$K_h = \frac{r_c^2 \ln \dfrac{mL}{r_w}}{2Lt} \cdot \ln \frac{H-h}{H-H_0} \qquad \text{Equation 7-4 (1)}$$

where
K_h = the horizontal hydraulic conductivity
r_c = radius of the well casing
r_w = radius of the wellbore
L = length of the sand pack
t = time since initial perturbation in head in well
h = head in well at any given time
H_0 = head in well at time 0
H = static elevation of the piezometric surface
m = $(K_h/K_v)0.5$
K_v = vertical hydraulic conductivity

(Fig. 7-24a). Equation 7-4 (l) is only an approximate equation, assumes that the specific storage of the formation, S_s, is equal to zero, and is applicable only if $[mL/(2r_w)] > 4$. If all of the assumptions used in the derivation of Eq. 7-4 (l) are true, then a plot of $\ln[(H-h)/(H-H_0)]$ versus t should yield a straight line with a slope of c. The horizontal hydraulic conductivity can then be calculated from

Formulas for Determination of Permeability

Case	Constant Head	Variable Head	Basic Time Lag
A LABORATORY PERMEAMETER	$k_v = \dfrac{4 \cdot q \cdot L}{\pi \cdot D^2 \cdot H_c}$	$k_v = \dfrac{d^2 \cdot L}{D^2 \cdot (t_2 - t_1)} \; ln \dfrac{H_1}{H_2}$ $k_v = \dfrac{d^2 \cdot L}{t_2 - t_1} \; ln \dfrac{H_1}{H_2} \;\; for\; d=D$	$k_v = \dfrac{d^2 \cdot L}{D^2 \cdot T}$ $k_v = \dfrac{L}{T} \qquad for\; d = D$
B FLUSH BOTTOM AT IMPERVIOUS BOUNDARY	$k_m = \dfrac{q}{2 \cdot D \cdot H_c}$	$k_m = \dfrac{\pi d^2}{8\, D \cdot (t_2 - t_1)} \; ln \dfrac{H_1}{H_2}$ $k_m = \dfrac{d^2 \cdot L}{8(t_2 - t_1)} \; ln \dfrac{H_1}{H_2} \;\; for\; d=D$	$k_m = \dfrac{\pi d^2}{8\, D \cdot T}$ $k_m = \dfrac{\pi \cdot D}{8 \cdot T} \qquad for\; d = D$
C FLUSH BOTTOM IN UNIFORM SOIL	$k_m = \dfrac{q}{2.75 \cdot D \cdot H_c}$	$k_m = \dfrac{\pi d^2}{11\, D \cdot (t_2 - t_1)} \; ln \dfrac{H_1}{H_2}$ $k_m = \dfrac{d^2 \cdot L}{11(t_2 - t_1)} \; ln \dfrac{H_1}{H_2} \;\; for\; d=D$	$k_m = \dfrac{\pi d^2}{11\, D \cdot T}$ $k_m = \dfrac{\pi\, D}{11 \cdot T} \qquad for\; d = D$
F WELL POINT FILTER AT IMPERVIOUS BOUNDARY	$k_h = \dfrac{q \cdot ln\left[\dfrac{2mL}{D} + \sqrt{1+\left(\dfrac{2mL}{D}\right)^2}\right]}{2 \cdot \pi \cdot L \cdot H_c}$	$k_h = \dfrac{d^2 \cdot ln\left[\dfrac{2mL}{D} + \sqrt{1+\left(\dfrac{2mL}{D}\right)^2}\right]}{8 \cdot L \cdot (t_2 - t_1)} ln\dfrac{H_1}{H_2}$ $k_h = \dfrac{d^2 \cdot ln\left(\dfrac{2mL}{D}\right)}{8 \cdot L \cdot (t_2 - t_1)} \; ln\dfrac{H_1}{H_2} \; for \; \dfrac{2mL}{D} > 4$	$k_h = \dfrac{d^2 \cdot ln\left[\dfrac{2mL}{D} + \sqrt{1+\left(\dfrac{2mL}{D}\right)^2}\right]}{8 \cdot L \cdot T}$ $k_h = \dfrac{d^2 \cdot ln\left(\dfrac{4mL}{D}\right)}{8 \cdot L \cdot T} \; for \; \dfrac{2mL}{D} > 4$
G WELL POINT FILTER IN UNIFORM SOIL	$k_h = \dfrac{q \cdot ln\left[\dfrac{mL}{D} + \sqrt{1+\left(\dfrac{mL}{D}\right)^2}\right]}{2 \cdot \pi \cdot L \cdot H_c}$	$k_h = \dfrac{d^2 \cdot ln\left[\dfrac{mL}{D} + \sqrt{1+\left(\dfrac{mL}{D}\right)^2}\right]}{8 \cdot L \cdot (t_2 - t_1)} ln\dfrac{H_1}{H_2}$ $k_h = \dfrac{d^2 \cdot ln\left(\dfrac{2mL}{D}\right)}{8 \cdot L \cdot (t_2 - t_1)} \; ln\dfrac{H_1}{H_2} \; for \; \dfrac{mL}{D} > 4$	$k_h = \dfrac{d^2 \cdot ln\left[\dfrac{mL}{D} + \sqrt{1+\left(\dfrac{mL}{D}\right)^2}\right]}{8 \cdot L \cdot T}$ $k_h = \dfrac{d^2 \cdot ln\left(\dfrac{2mL}{D}\right)}{8 \cdot L \cdot T} \; for \; \dfrac{mL}{D} > 4$

Notation

D = diam. intake, sample, cm
d = diameter, standpipe, cm
L = length, intake, sample, cm
H_c = constant piez. head, cm
H_1 = piez. head for $t = t_1$, cm
H_2 = piez. head for $t = t_2$, cm
q = flow of water, cm³/sec
t = time, sec
T = basic time lag, sec
k_v = vert. hydraulic cond. ground, cm/sec
k_v' = vert. hydraulic cond. casing, cm/sec
k_h = horz. hydraulic cond. ground, cm/sec
k_m = mean coeff. hydraulic cond. cm/sec
m = transformation ratio

$k_m = \sqrt{k_h \cdot k_v} \qquad m = \sqrt{k_h / k_v}$

$ln = log_e = 2.3 \; log_{10}$

PLOT OF OBSERVATIONS

HEAD RATIO H_t/H_0 (LOG SCALE)

$1.0\, h_0$
$0.37\, h_0$ — $t = T$
$0.1\, h_0$ — $t = 2.3 \cdot T$

TIME, t, (ARITHMETIC SCALE)
Determination basic time lag T

Assumptions

- Soil at intake
- Infinite depth and directional isotropy (k and k constant
- No disturbance, segregation, swelling or consolidation of soil
- No sedimentation or leakage
- No air or gas in soil, well point, or pipe
- Hydraulic losses in pipes, well point or filter negligible
  (After Hvorslev, U.S. Corps of Engineers W.E.S.)

Figure 7-24a Horslev Equations For Common Field Situations

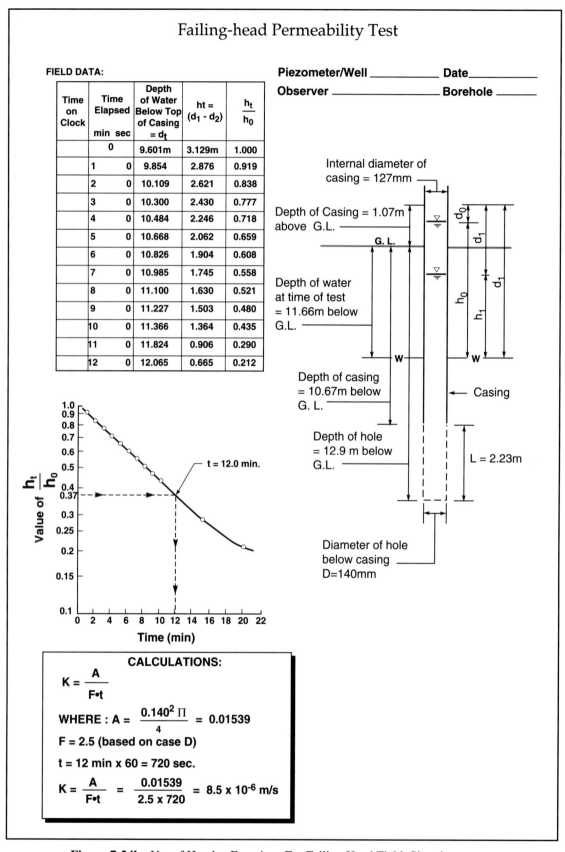

Failing-head Permeability Test

FIELD DATA:

Time on Clock	Time Elapsed min sec	Depth of Water Below Top of Casing = d_t	ht = $(d_1 - d_2)$	$\dfrac{h_t}{h_0}$
	0	9.601m	3.129m	1.000
1	0	9.854	2.876	0.919
2	0	10.109	2.621	0.838
3	0	10.300	2.430	0.777
4	0	10.484	2.246	0.718
5	0	10.668	2.062	0.659
6	0	10.826	1.904	0.608
7	0	10.985	1.745	0.558
8	0	11.100	1.630	0.521
9	0	11.227	1.503	0.480
10	0	11.366	1.364	0.435
11	0	11.824	0.906	0.290
12	0	12.065	0.665	0.212

Piezometer/Well _____ Date_____

Observer _____ Borehole _____

Internal diameter of casing = 127mm

Depth of Casing = 1.07m above G.L.

Depth of water at time of test = 11.66m below G.L.

Depth of casing = 10.67m below G. L.

Depth of hole = 12.9 m below G.L.

L = 2.23m

Casing

Diameter of hole below casing D=140mm

t = 12.0 min.

Value of $\dfrac{h_t}{h_0}$

Time (min)

CALCULATIONS:

$$K = \frac{A}{F \cdot t}$$

WHERE : $A = \dfrac{0.140^2\,\Pi}{4} = 0.01539$

F = 2.5 (based on case D)

t = 12 min x 60 = 720 sec.

$$K = \frac{A}{F \cdot t} = \frac{0.01539}{2.5 \times 720} = 8.5 \times 10^{-6} \text{ m/s}$$

Figure 7-24b Use of Horslev Equations For Falling Head Field Situation

$$K_h = \frac{r^2{}_c \ln \dfrac{mL}{r_w}}{2L_c}$$ Equation 7-4 (2)

The slope, c, is also called the basic time lag and can be determined directly from the semilog plot as the time when (H/H_0) is equal to 0.37 (Fig. 7-24b).

• **Bouwer–Rice Method (1976)**— The solution for hydraulic conductivity using this method is more involved than the other methods, resulting from empirical derivations. This method is applied to either partially or completely penetrating wells in unconfined aquifers. The procedure was developed using an electrical analog model that resolved combinations of flow for various well and aquifer configurations. Bouwer and Rice developed a series of curves for determining applicable coefficients for the field configuration tested. Analysis involves a semilogarithmic plot of residual drawdown versus time. A straight line is fitting to the early–time data for use in calculating conductivity. The equation used to solve for hydraulic conductivity, K, using this method is as follows:

$$K = r_c{}^2 \ln \frac{R_e}{R_w} \bullet \ln \frac{H^0}{2L(t) \bullet H(t)}$$

or,

$$K = \frac{r_c{}^2 \ln \dfrac{R_e}{r_w}}{2L_e} \bullet \frac{1}{t} \bullet \ln \frac{y_0}{y_t}$$ Equation 7-5

Where:
r_c = Well radius (inside blank casing)
R_e = Effective radius
R_w = Borehole radius, including filter pack

This equation is identical to Equation (7-3) except for the term $\ln(R_e/r_w)$ which is $\ln(L/r_w)$ in the Hvorslev (1951) solution (Equation 7-3). The parameter R_e is the effective horizontal radius over which the instantaneous slug, $(H-H_0)$, is dissipated. Bouwer and Rice (1976) determined the value of $\ln(R_e/r_w)$ for different values of L, r_w, D (saturated thickness of the aquifer), and H' (vertical distance from the water table to the bottom of the monitoring well) by the use of an electrical analog model. For a given L/r_w, parameters can be read from a graph and

used in empirical equations to calcuate $\ln(R_e/r_w)$. Like the Hvorslev (1951), Equation. (7-5) assumes the specific storage of the formation is equal to zero. A plot of $\ln[(H-h)/(H-H_0)]$ versus time should yield a straight line with a slope of $2LK_h/[r^2c \ln(R_e/r_w)]$ The evaluation of r_w and r_c requires the knowledge of the thickness of the filter pack or developed zone. We want to know the hydraulic conductivity of the geologic unit tested not that of the filter pack surrounding the screened area of the well. Figure 7-25a illustrates the geometry and symbols of a piezometer or well in an unconfined aquifer. The dimensional variables are defined as follows:

L = Screen length
H = Distance from bottom of well screen to water table
D = Saturated thickness of aquifer
y = Vertical distance between water level in the well and water table
y_0 = initial water level (maximum drawdown)
y_1 = water level at some time t.

The term $(\ln[R_e/R_w])$ is determined with one of two empirical relationships. These relationships use one or two empirical coefficients derived from curves based on the value of L_e/R_w.

In general terms the value of r_w is the sum of the radius of the well screen plus the thickness of the filter pack or developed zone. The value for r_c is dependent on the elevation y of water in the borehole at t = 0. If y is in the blank casing above the screen, the rc value is equal to the inside diameter of the casing. If y however, is in the well screen r_c should he corrected for the filter pack using estimates of the porosity of the pack. The Bouwer and Rice method requires that R_c is evaluated for a partially penetrating well by:

$$\ln \frac{r_c}{r_w} = \left\{ \frac{1.1}{\ln \dfrac{H}{r_w}} + \frac{A + B \bullet \ln \left[D - \dfrac{H}{r_w} \right]}{\dfrac{1}{r_w}} \right\}^{-1}$$ Equ. 7-6

Fully penetrating wells require the following relationships be used for evaluating R_c:

$$\ln \frac{R_c}{r_w} = \left\{ \frac{1.1}{\ln \dfrac{H}{r_w}} + \frac{C}{\dfrac{L}{r_w}} \right\}^{-1}$$ Equation 7-7

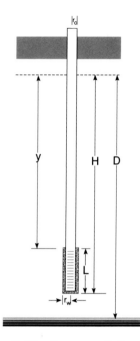

Figure 7-25a Geometry & Symbols

The terms A, B, C are dimensionless coefficients that are functions of L/rw evaluated from Figure 7-25b . The Bouwer and Rice method holds the terms K, r_c, r_w and L constants within Equation 7-6. Other variables are obtained as follows from plotted field data as log yt verses t:

- Variable y_o is the intercept of the best fit line with the y axis;

- Variable yt is any other value of y along the best fit line at some time t; and,

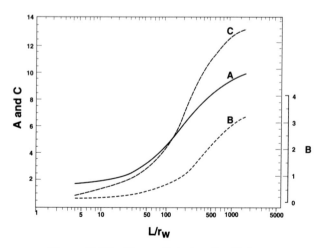

Figure 7-25b Curves Relating Coefficients A, B, & C to L/r_w

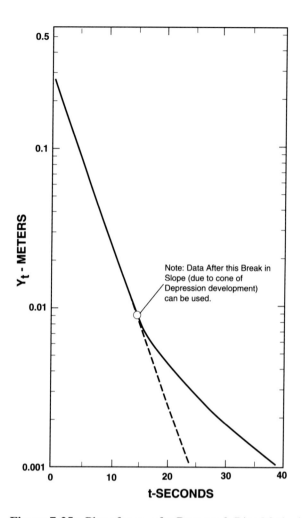

Figure 7-25c Plot of y vs. t for Bouwer & Rice Method

- Substitute values back to Equation 7-6 (or 7-7) to obtain hydraulic conductivity value.

Cooper–Bredehoeft–Papadopulos Method - This method applies to wells completed in confined aquifers with hydraulic conductivity less than 0.05 cm/sec. The analysis of the data involves a semilogarithmic plot of residual drawdown divided by the initial residual drawdown versus time. In other words, the water-level data in the test well, expressed as a fraction of H_o (the initial head rise in the well) that is H / H_o is plotted versus time, t, on semilog format of the same scale as that of the type-curve plot. The plot is then matched to a set of type curves shown on Figure 7-26. Resulting match points for B, t, and oc obtained while keeping the baselines the same (shift plot horizontally) are then substituted into an equation as time values for derivation of hydraulic conductivity. The equation used to solve for hydraulic conductivity, K, using this method is as follows:

$$K = \frac{r^2}{tL} \qquad \text{Equation 7-8}$$

Where:

r = Well casing radius

t = Time corresponding to a match point on a type curve

L = Screen length

The transmissivity is calculated from:

$$T = \frac{\beta r c^2}{t} \qquad \text{Equation 7-9}$$

Where the value β is determined from the match point you select when superimposing the plot of the test data over the best-fit type curve. The match point is selected so that the resultant point gives a value of one to many of the equation factors.

The storage coefficient is calculated from:

$$S = \alpha \frac{rc^2}{r\alpha^2} \qquad \text{Equation 7-10}$$

As with β, α is determined from the best-fit curve, using a specified value of α.

The determination of K and T was recognized by Cooper and Others (1967) as being more reliable than S due to the similar shape of the curves used in the matching procedure. Tables of the F (B, L) are given by Cooper and others (1967) for values of β from 10^{-3} to 2.15×10^2 and for values of L from 10^{-6} to 10^{-10} in order to apply the method to low storage coefficient geologic units. The Cooper et al. (1967) solution is based on the assumption that:

1) the medium is homogeneous, isotropic, and of infinite areal extent;
2) the formation is confined;
3) the confining layers are nonleaky; and,
4) the well is screened over the entire thickness of the formation.

A more recent method for the calculation of hydraulic parameters from slug tests has been presented by Nguyen and Pinder (1984). Their method is based on a three-dimensional, axisymmetric representation of the flow field with a non-zero specific storage. The three—dimensional nature of the formulation allows the partial penetration of the well screen to be taken into account. The specific

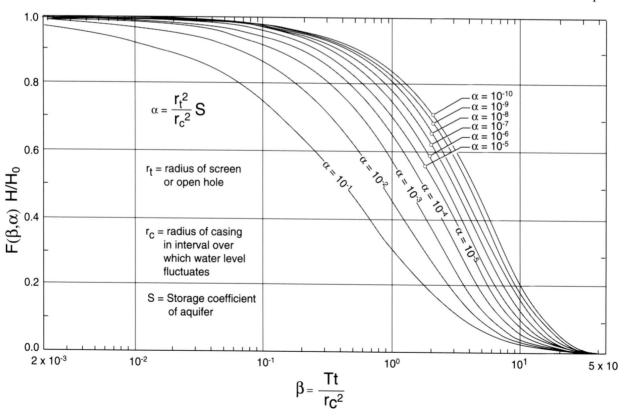

Figure 7-26 Match Point Curves Cooper-Bredehoeft Method

storage is determined by plotting $\log[(H-h)/(H-H_o)]$ versus Log(t). The slope of this line (C_1) is determined and the specific storage is calculated from:

$$Ss = \frac{(r_c^2 C_1)}{(r^2 w_b^2 L)}$$

Equation 7-11

where r_c is the radius of the standpipe, rw is the radius of the borehole, b is the thickness of the aquifer, and L is the length of the intake. The hydraulic conductivity, K, is found by plotting $\ln(dH/dt)$ versus $(1/t)$, determining the slope of the curve (C_2), solving:

$$K = \frac{r_c^2 C_l}{(4C_2 L)}$$

Equation 7-12

for K. The derivative, dH/dt, can be approximated from the data by finite difference techniques. This method is not commonly used at this time, however, the equations are a better representation of the physics than the other methods described above. Therefore, if sufficient data are available so that accurate estimations of the derivatives can be made, this method should provide the best estimates for the methods described here for hydraulic parameters from slug tests.

	Nonleaky Confined Aquifers					Leaky Confined Aquifers			
	Isotropic				Anisotropic		Unconfined Aquifers		
	Theis (1935) Nonequilibrium	Theis (1935) Recovery	Modified non-equilibrium Cooper and Jacob (1964)	Slug test Cooper and Others (1967)	Radial-Vertical Hantush (1966a & b) Weeks (1969)	Hantush & Jacob (1955) Hantush (1960)	Neuman (1975)	Theis (1935)
Stress on Aquifer:								
Constant discharge	●	●	●	○	●	●	●	●
Varable discharge	○		○	○	○	○	○	○
Instantaneous hydraulic-head change	○		○	●	○	○	○	○
Drawdown Measurement:								
Control well	[2]●	●	[2]●	●	○	○	○	[2]●
Obsrevation well	●	●	●	○	●	●	●	●
Aquifer penetration:								
Control well:								
Full	●	●	●	[3]●	○	●	[4]●	[5]●
Partial	○	○	○	[3]○	●	○	[4]●	[5]●
Observation well:								
Full	●	●	●	○	●	●	[4]●	[5]●
Partial	●	●	●	○	●	○	[4]●	[5]●
Single-well test								
Application	●	[2]●	●	○	[2]●	[2]●	[2]●	[2]●

Source: Modified from Bedinger & Reed, 1988   ● = Applicable evaluation method

[1] Variable discharge can be treated by several methods and by preparation of special type curves for the discharge function.

[2] Measurements in the control well may be used to obtain values of transmissivity, but values of storage coefficient are unreliable unless the effective radius of the well is known.

[3] Slug tests are commonly assumed to test the interval of the control well open to the aquifer. The assumption may be reasonable where the aquifer is stratified and horizontal hydraulic conductivity is much greater than vertical hydraulic conductivity.

[4] Neuman (1975) presenteds a general method for partially penetrating control and observation wells.

[5] The Theis (1935) solution is applicable to unconfined aquifers for specific conditions of fully and partially penetrating control and observation wells and for certain distances and time of pumping.

Table 7-3 Aquifer Testing Methods

7.5.2 Analytical Techniques for Pump Test Data

Table 7–3 summarizes several of the more commonly used techniques for analyzing pump test data. Most site assessments have single observation piezometers, previously installed public water wells, or monitoring wells available for testing. Since new well or piezometer installations are typically developed before use, they can at least be easily tested for recovery after the development pumping.

The recovery test, often used after an aquifer pumping test, applies to unsteady radial flow in a confined aquifer and assumes that the well fully penetrates the aquifer. Data required for this test include the average pumping rate (Q), well radius (r), time since pumping began (t), time since pumping ended (t'), and residual drawdown (s'). The residual drawdown is defined as the difference between the initial water level in the well and the water level at any later time, t'. The analysis requires that r be small and t' be large.

Analysis of the recovery test is similar to the Cooper–Jacob solution for unsteady radial flow and is described by Todd (1980). In this method, the residual drawdown (s') is plotted against the logarithm of the time ratio (t/t') and fitted with a line. The slope of this line is used to calculate the hydraulic conductivity. The transmissivity is calculated by:

$$T = 2.30 Q \frac{\Delta \log(t / t')}{4\pi} \Delta s' \qquad \text{Equation 7-13}$$

Where:
T = transmissivity (cm^2/s)
Q = Average discharge rate (cm^3/s)
t/t' = Ratio of time since pumping began to time since pumping ended
s' = Residual drawdown = initial water level – water level at time t' (cm)

The hydraulic conductivity, K, is determined by dividing the transmissivity by the aquifer thickness, which is assumed to be equal to the screen length of the installation.

The recovery test can be biased toward higher values because fine particles may be removed from the surrounding formation as the borehole is developed. A second problem with the recovery method is that it is based on the assumption that the aquifer is fully penetrated by the piezometer or well screens — an assumption that is not usually valid. Constructing the piezometers to meet this condition for relatively thin permeable zones or high hydraulic conductivity contact zones can still make the

analysis method valid. The test can be performed relatively quickly and can be performed as part of a well development program to give additional information on the geologic unit, in which the installation is screened.

Full–scale pumping tests with observation piezometers represent the most expensive (say, $15,000–$30,000 each or in special projects many times this value) and, typically, these are the most informative method for defining aquifer characteristics.

The referenced analytical techniques are generally adequate for estimating the performance of wells in confined aquifers, as well as, in unconfined aquifers where the drawdown is a small percentage of the aquifer thickness and the discharging well is fully penetrating. Corrections for partial penetration of the discharging well, large drawdown in unconfined aquifers, and anisotropy has been derived in the literature, but adequate data for application of the corrections are often not readily available. Details of these sources can be found in Bedinger and Read (1988), Kruseman and DeRidder (1976, 1990), and Walton (1987).

7.5.3 Pump Testing

Large scale pump testing or "stress" testing of an aquifer is becoming more common in site investigations for waste disposal or remediation projects. Many authors have provided papers and books on the subject of evaluation of pump test analysis. Notable among these are Kruseman and de Ridder (1990), Stallman (1971), Lohman (1972), Weeks (1977, 1980), Reed (1980), Walton (1987) and Bedinger and Reed 1988. Each of these documents provides detailed reviews on the pump test literature. Van der Kamp (1985) provided systematic guidelines on appropriate pump test data analysis techniques based on a few easily evaluated criteria. Van der Kamp extended the work of Kruseman and deRidder (1971) and Weeks (1977,

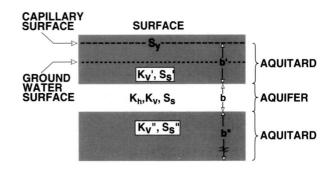

Figure 7-27 Pump Test Curve Model

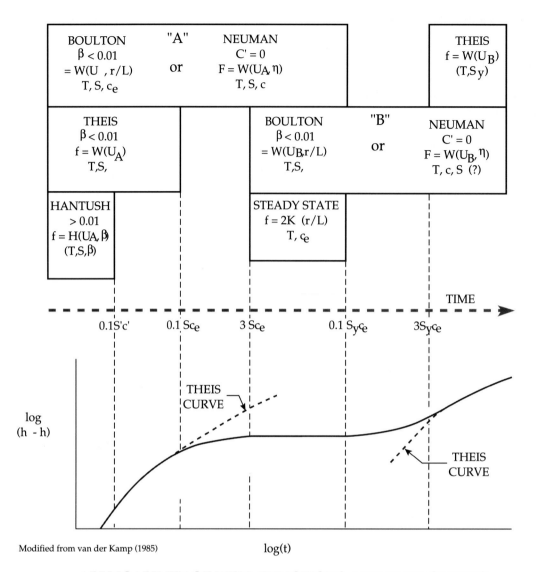

APPLICABILITY OF WELL FUNCTIONS AND TYPE CURVES

Source: van der Kamp
Saskatchewan Research Council

Figure 7-28 Applicability of Basic Type Curves

1980) to develop the quantitative criteria presented below in the evaluation of pump test data sets.

As with the entire site characterization process the conceptual model is the prime reference guide to the evaluation of pump test data. Figure 7-27 shows a hydrogeologic model representing a three layer system; a single aquifer overlain by a confining unit. As Van der Kamp (1985) notes, this model can represent a wide variety of field situations by varying the thickness of the overlying confining unit from zero upwards. Hence, the model can represent all cases from a water table (unconfined) aquifer to deep confined aquifers. The underlying confining unit is considered as a no flow boundary for this discussion. The model is bounded at the top by a capillary fringe or capillary surface above the ground-water surface. This saturated capillary must be considered during evaluation of pump tests under water table conditions (Nwamkor, 1984).

The following assumptions are made: (a) The formations are homogeneous, horizontally isotropic of constant thickness and of infinite horizontal extent. (b) Horizontal flow in the aquitard is negligible. (c) Drawdown of the capillary surface is small compared to the saturated thickness of the formation in which the capillary surface occurs. (d) Release of water from storage with change of hydraulic head is instantaneous both within the formations and at the capillary surface. (e) The pumping well is screened over the entire thickness of the aquifer, and storage in the pumping well and observation wells are negligible. The type curves discussed in this chapter are all based on this assumption. Analysis techniques for evaluating conceptual models where not all of these assumptions are valid, are reviewed in the pumping test literature referred to earlier. Bedinger and Reed (1988) especially deal with an extensive listing of aquifer test methods. The hydrogeological characteristics of the geologic units are defined in the usual manner, with the "primed' symbols referring to the aquitards:

b, b', b"	saturated thickness
K_h, K'_h, K''_h	horizontal hydraulic conductivity
K_v, K'_v, K''_v	vertical hydraulic conductivity
S_s, S'_s, S''_s	specific storage coefficient
S_y	specific yield at the capillary surface
$T = K_h b$	transmissivity of the aquifer
$S = S_s b$, $S' = S'_s b$, $S'' = S''_s b''$	storage coefficient
$c = b/K_v$, $c' = b'/K_v$, $c'' = b''/K''_v$	vertical hydraulic resistance
$c_e = c' + c/3$	effective hydraulic resistance
$L = (Tc_e)^{1/2}$	leakage factor

The concepts of hydraulic resistance (c) and leakage factor (L) is especially useful in describing pumping test phenomena. With regard to the various possible models for flow to a pumping well Van der Kamp introduced the following parameters:

$$u_A = \frac{r^2 S}{4Tt}$$ Equation 7-13

$$u_B = \frac{r^2 Sy}{4Tt}$$ Equation 7-14

$$B = \frac{r}{4b}\left(\frac{K'_v S'_s}{K_h S_s}\right)^{\frac{1}{2}} + \frac{r}{4b}\left(\frac{K''_v S''_s}{K_h S_s}\right)^{\frac{1}{2}}$$ Equ. 7-15

$$\eta = \frac{r}{Tc}$$ Equation 7-16

In these expressions r is the distance from the pumped well and t is time since start of pumping. The general equation for drawdown (h_o-h) due to pumping at a steady rate Q from single well is:

$$h - h_0 = \frac{Q}{4\pi T} f\left(u_A, u_B, B, \frac{r}{L}, \eta\right)$$ Equation 7-17

The function f here represents a general well function which for particular cases generally reduced to known functions such as the well-known Theis (1935) well function W(uA). The significance of the other parameters will become clear in the following discussion.

Applicability Of Well Functions And Type Curves

The pump test literature describes a large variety of well functions and associated type curves for the analysis of pumping tests in infinite homogeneous aquifers. The type curves for particular cases considered in the literature falls along a single curve as is shown in Figure 7-28.

The essential element of this diagram (Van der Kamp, 1985) is the time axis, progressing from left to right (small time to large time), through four points that mark the approximate limits of applicability of particular type curves. The various available pumping test conceptual models, their ranges of applicability and the parameters that can be determined are shown above the time axis. Below the time axis a schematic drawdown curve is shown illustrating at which ranges of the curve the various well functions may be applied. Figure 7-28 shows five different sets of basic type curves along with the associated time limits and other limiting conditions.

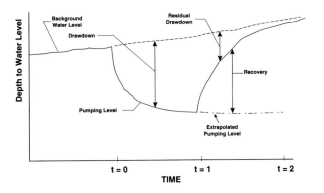

Figure 7-29 Corrected Depth to Water Verses Time

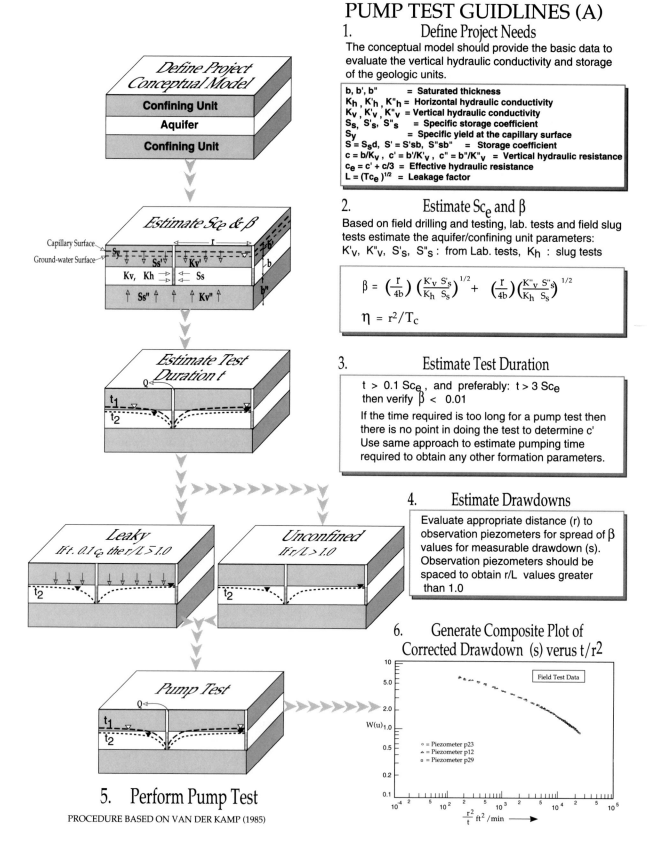

PUMP TEST GUIDLINES (A)

1.   Define Project Needs

The conceptual model should provide the basic data to evaluate the vertical hydraulic conductivity and storage of the geologic units.

b, b', b'' = Saturated thickness
K_h , K'_h , K''_h = Horizontal hydraulic conductivity
K_v , K'_v , K''_v = Vertical hydraulic conductivity
S_s, S'_s, S''_s = Specific storage coefficient
S_y = Specific yield at the capillary surface
$S = S_s d, S' = S'_s b, S'' = S''_s b$ = Storage coefficient
$c = b/K_v, c' = b'/K'_v, c'' = b''/K''_v$ = Vertical hydraulic resistance
$c_e = c' + c/3$ = Effective hydraulic resistance
$L = (Tc_e)^{1/2}$ = Leakage factor

2.   Estimate Sc_e and β

Based on field drilling and testing, lab. tests and field slug tests estimate the aquifer/confining unit parameters:
K'_v, K''_v, S'_s, S''_s : from Lab. tests, K_h : slug tests

$$\beta = \left(\frac{r}{4b}\right)\left(\frac{K'_v \ S'_s}{K_h \ S_s}\right)^{1/2} + \left(\frac{r}{4b}\right)\left(\frac{K''_v \ S''_s}{K_h \ S_s}\right)^{1/2}$$

$$\eta = r^2/T_c$$

3.   Estimate Test Duration

$t > 0.1 \ Sc_e$, and preferably: $t > 3 \ Sc_e$
then verify $\beta < 0.01$

If the time required is too long for a pump test then there is no point in doing the test to determine c'
Use same approach to estimate pumping time required to obtain any other formation parameters.

4.   Estimate Drawdowns

Evaluate appropriate distance (r) to observation piezometers for spread of β values for measurable drawdown (s). Observation piezometers should be spaced to obtain r/L values greater than 1.0

6.   Generate Composite Plot of Corrected Drawdown (s) verus t/r^2

Field Test Data

○ = Piezometer p23
▲ = Piezometer p12
□ = Piezometer p29

$W(u)$

$\frac{r^2}{t}$ ft^2/min

Define Project Conceptual Model
Confining Unit
Aquifer
Confining Unit

Estimate Sc_e & β
Capillary Surface
Ground-water Surface
S_y
S'_s K_v'
K_v, K_h S_s
S_s'' K_v''

Estimate Test Duration t

Leaky
If t. 0.1 c_e the r/L > 1.0

Unconfined
If r/L > 1.0

Pump Test

5.   Perform Pump Test

PROCEDURE BASED ON VAN DER KAMP (1985)

Figure 7- 29a   Data Plotting Method For Determining Pump Test Values

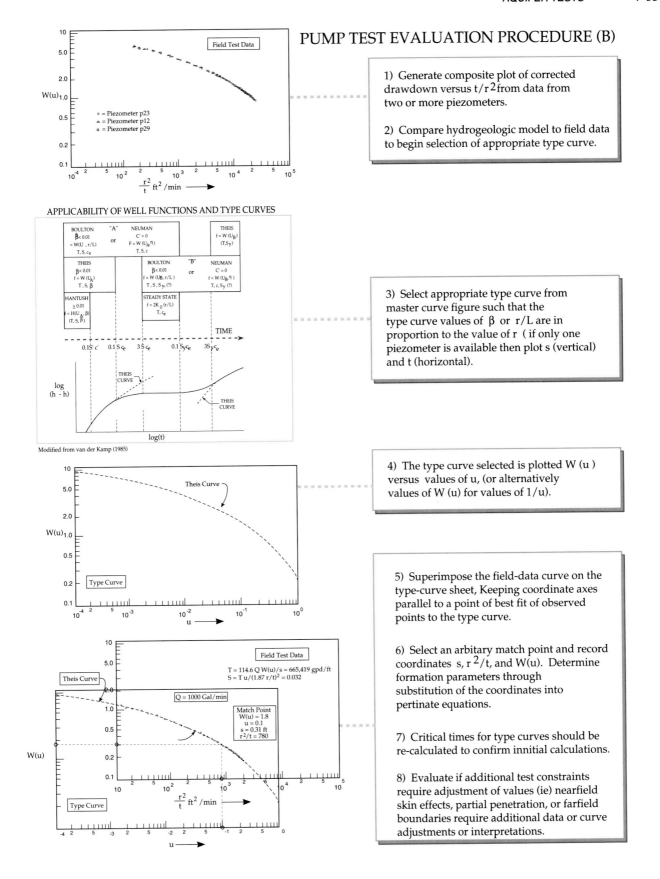

PUMP TEST EVALUATION PROCEDURE (B)

1) Generate composite plot of corrected drawdown versus t/r^2 from data from two or more piezometers.

2) Compare hydrogeologic model to field data to begin selection of appropriate type curve.

APPLICABILITY OF WELL FUNCTIONS AND TYPE CURVES

Modified from van der Kamp (1985)

3) Select appropriate type curve from master curve figure such that the type curve values of β or r/L are in proportion to the value of r (if only one piezometer is available then plot s (vertical) and t (horizontal).

4) The type curve selected is plotted $W(u)$ versus values of u, (or alternatively values of $W(u)$ for values of $1/u$.

5) Superimpose the field-data curve on the type-curve sheet, Keeping coordinate axes parallel to a point of best fit of observed points to the type curve.

6) Select an arbitary match point and record coordinates s, r^2/t, and $W(u)$. Determine formation parameters through substitution of the coordinates into pertinate equations.

7) Critical times for type curves should be re-calculated to confirm innitial calculations.

8) Evaluate if additional test constraints require adjustment of values (ie) nearfield skin effects, partial penetration, or farfield boundaries require additional data or curve adjustments or interpretations.

Figure 7- 29b Data Ploting Method For Determining Pump Test Values (con.)

After completion of the pump test a number of data manipulation steps are required to transform the raw field data into values usable for plotting and comparison to type curves. Weeks (1980) describes a series of steps beginning with an initial plot of the observed depths to water and discharge as a function of time for each of the piezometers used in the test program. Since these head level observation points are monitored for a period of time before beginning the pump test, (typically for twice the length of the subsequent pump test period) the resultant hydrographs should provide indications as to long term antecedent trends. Short term water level aberrations may be caused by barometric pressure changes, tide induced or near by man made pumping effects must be evaluated for extent of interference on the long term head level trends. It is of extreme importance to remove the effects of all short term aberrations from the data acquired during the pump test through applicable correlation techniques.

The "cleaned-up" data is then used to prepare hydrographs that show corrected depth to water verses time. The long term pre-test trend line is then extrapolated as if there was no pump test. These data as shown by Figure 7-29 provide actual drawdown by subtracting the trend value of depth to water from the corrected observed depth to water value.

Additional adjustments in drawdown may be necessary even at this point in the evaluation due to fundamental conditions such as dewatering of unconfined aquifers (see Jacob 1963 and Neuman 1975) or due to partial penetration effects of the pumping well. One can see from inspection of the many adjustments that may be required in evaluating pump test that essentially "clean data" should be used in the interpretation. This holds especially for computer based interpretations through auto-type curve fitting of data sets. Few, if any of these programs will evaluate pretest trends to the level necessary for reliable determination of aquifer parameters.

Type curve selection and application presents additional potential error factors when applying the data to a selection of available type curves. The responsible individual must always consider the conceptual model in the selection of the type curve to be used in the pump test evaluation. The mechanical manipulation of the type curve and the resultant test data to solve for aquifer properties follows a similar procedure in each instance. A recent text by Roscoe Moss (1990) provides examples of the matching procedure for many of the important type curve solutions, as such, this text will review the selection of the appropriate type curve, using the original Theis (1935) nonequilibrium drawdown method. Application of other important type curve solutions that represent the composite curve given in Figure 7-28, such as, Hantush (1960) for leaky confined aquifers, and the method of Bolton (1963)

and Neuman (1972) for unconfined aquifers, follows the same general procedure. The application of the correct type curve requires more than a good match of the corrected drawdown curve from a single piezometer to a selected type curve. In every case possible a composite plot should be constructed from corrected well test data. These data can then be matched with the appropriate type curve solutions as illustrated in Figure 7-29a, b.

These composite plots of piezometers (with variable r) should provide proportional type curve value of B or r/L to the value of r for each piezometer. One would expect if a complete match cannot be made then the resultant transmissivity, storage and other formation parameters calculated from the procedure would be suspect (Van der Kamp 1985). This composite curve matching provides a cross-check that the actual field hydrogeologic conditions are represented in the hydrogeologic model used to develop the original type curve solution. The formational parameters as shown in Figure 7-28 are then obtained by application of the selected well function and pump test data in the appropriate time ranges. The resultant formational parameters can then be used to back calculate the critical times to insure that the selected curves actually represent the applicable curve at the time interval used in the type curve matching procedure. Van der Kamp (1985) points out that the five major type curves shown on Figure 7-28 can be reduced to two basic sets of type curves, the Hantush $H(U_A,B)$ and Boulton $W(U_A,r/L)$ and $W(U_B,r/L)$ curves. This is due to the limiting cases of the Boulton curves bounded by the Theis curve and the steady state curve.

Additional useful aquifer test methods such as recovery methods, Cooper-Jacob's straight line distance-drawdown, step-drawdown and partial penetration methods are reviewed by Bedinger and Weeks (1988) and Roscoe Moss (1990). These methods, in addition to many other contained in the literature, provide the hydrogeologist with alternative analytical tools for evaluation of formation parameters. The Cooper-Jacob's straight line method also called the Modified Nonequilibrium Method, is probably used more than any other aquifer test analysis method. This method is used extensively with evaluation of boundary conditions and, as such, is very useful for fractured rock testing. Chapter 6.0 describes the use of the straight line method as applied to fractured rocks. The following section describes a step by step application of a Neuman type B analysis.

Example Analysis Procedures

Following manipulation and reduction procedures, the resultant pump test data can be analyzed by a number of methods for nonsteady-state flow for unconfined aquifers. One example that commonly must be addressed in site evaluations for waste disposal, is the potential leakage through confining units that are to contain the waste. Figure 7-30 illustrates a conceptual model of a confining unit overlying an aquifer that was tested according to the normal procedures in Section 5.0.

The Neuman (1974) analysis was considered suitable for the conditions at the facility, based on the geologic/ hydrogeologic characteristics and the trends displayed by time-drawdown graphs obtained during the pump test. Neuman analysis takes into account the delayed yield, that usually occurs in pumping tests completed in unconfined aquifers. Either the Boulton or Neuman procedure could have been used, however, though the Boulton analysis takes into account the delayed yield phenomena, the Neuman analysis is based on the same concept but has the advantage that more hydrogeological parameters can be calculated from the data set.

The delayed yield defines a general trend of the log-log drawdown curve graphs (drawdown versus time) and typically consists of three segments (Figure 7-31):

The first segment of the curve only, covers a short period after pumping has been initiated. This segment illustrates the response of the unconfined aquifer is similar in early time data to a confined aquifer. Gravity drainage has not yet started in the system. Water is released instantaneously from storage by the elastic deformation of the aquifer and by the expansion of the water itself.

The second segment of the time-drawdown data curve represents the decrease in the rate of drawdown due to the vertical replenishment of the aquifer by gravity drainage from the interstices above the cone of depression.

The third segment has a steeper slope (in other words, an increase in the rate of drawdown); the trend then approaches the Theis curve. In the third segment there is equilibrium reached between the gravity drainage and the rate of fall of the water table.

The full development of these drawdown segments depends on the pumping rate, distance from the pumping well and the possible boundaries (heterogeneity) contacted by the cone of depression during its expansion. It is an assumption that the pumping well and the observation well are perforated throughout the entire saturated thickness of the aquifer therefore; the drawdown in the observation well is given by:

$$s(r,t) = \frac{Q}{4\pi T} \cdot S_D \qquad \text{Equation 7-18}$$

s : drawdown;
r : radial distance from the pumping well:

t : time since pumping started;
Q : pumping rate;
T : transmissivity;
s_D: dimensionless drawdown represented numerically by an integral function.

Two sets of type curves were developed by Neuman, expressed in terms of three independent dimensionless parameters: σ, β, t_s (dimensionless time with respect to the specific elastic storage) or t_y (dimensionless time with respect to the specific yield). The dimensionless time parameters are related to each other by the equation:

$$t_y = \sigma \cdot t_s \qquad \text{Equation 7-19}$$

σ and β were defined by Neuman (1975) as follows:

$$\sigma = \frac{S}{S_y} \qquad \text{Equation 7-20}$$

$$\beta = \frac{K_D r^2}{b^2} \qquad \text{Equation 7-21}$$

The dimensionless time parameters are defined as follows:

$$t_s = \frac{Tt}{S_s r^2} \qquad \text{Equation 7-22}$$

$$t_y = \frac{Tt}{S_y r^2} \qquad \text{Equation 7-23}$$

Where:
S : storage coefficient = Ss b;
Ss : specific elastic storage (Ft-1)
Sy : specific yield;
b : initial saturated thickness of the aquifer;
K_D: degree of vertical anisotropy = Kz (vertical hydraulic conductivity)/Kr (horizontal hydraulic conductivity);
r : radial distance from the pumping well;

The key to this method is when σ approaches zero (S is much less than S_y) only two independent variables remain and s_D can be integrated for different values of β. The results are two asymptotic families of type curves: type A (s_D versus t_s) and type B (s_D versus ty). Both A & B type curves approach a set of horizontal asymptotes, the length of which depend on the value of σ. The procedure used to determine aquifer parameters are as follows:

- Plot the family of "Neuman type curves" s_D versus t_s and t_y for a practical range of β values on log-log paper;

- Plot the corrected drawdown (s) against the corresponding time for a given observation well at a distance r from the pumped well On another sheet of log-log paper, with the same scale as was used to plot the type curves;

- Superimpose the time-drawdown field data curve on type B curves. Keeping the coordinate axes parallel at all times, that the s_D axis parallel with the s axis and the t_s/v axis parallel with the t axis. The overlying data sets used and type curves must be adjusted until, as much as, possible late time-drawdown field data falls on one of the type B curves. Record the β value of the selected type B curve;

- Select an arbitrary point B and record the coordinates S_d, t_y and s, t; and,

- Calculate the transmissivity (T) and the specific yield (S_y) using the equations:

$$T = \frac{114.6 Q s_D}{s}$$
Equation 7-24

$$Sy = \frac{0.1337 T t}{t_y r^2}$$
Equation 7-25

Q is in gpm and the other parameters in gallons per day per foot system. The early time data should also be evaluated for elastic storage coefficient using the following procedures:

- As before, superimpose the field data on the type A curves, keeping the coordinate axes of both graphs parallel to each other and matching as much of the earliest time-drawdown data to a particular type curve as possible. The value β corresponding to this type curve must be the same as that obtained previously from the type B curves;

- Select an arbitrary point A on the superimposed curves and note its coordinates s_D, t_s and s, t;

- Calculate the transmissivity and the elastic storage coefficient;

$$S = \frac{0.1337 T t}{t_s r^2}$$
Equation 7-26

- In practice the transmissivity, is taken as the average of the early and late data, unless boundary effects are evident in the data the horizontal hydraulic conductivity (Kr) can be calculated with the formula;

$$K_r = \frac{T}{b}$$
Equation 7-27

- Vertical anisotropy (K_D) is evaluated from the value of β according to:

$$K_D = \frac{\beta \cdot b^2}{r^2}$$
Equation 7-28

- Values of K_r and K_D, obtained previously, can be used to determine the vertical hydraulic conductivity (Kz) with the following relationship;

$$K_z = K_D K_r$$
Equation 7-29

The parameter σ is calculated from:

$$\sigma = \frac{S}{S_y}$$
Equation 7-30

Specific storage (Ft-l) of the aquifer can be calculated from:

$$S_s = \frac{S}{b}$$
Equation 7-31

Unconfined aquifers are often thick and piezometers are compiled in the unit as partially penetrating. When either the pumping well or the observation well is screened only through a portion of the saturated thickness of the aquifer the extended Neuman model requires additional parameters be taken into account. The integral function for partially penetrating conditions that describes the dimensionless drawdown (s_D) is expressed in terms of six independent dimensionless parameters: σ, β, l_D, d_D, Z_{1D}, Z_{2D} and t_s or t_y where:

- σ, β, t_s and t_y have the same definition as the fully saturated model described above.

- l_D = ratio between distance from the initial water table to bottom of the screen in pumping well (l) and initial saturated aquifer thickness (b) = l/b.

- d_D = ratio between distance from the initial water table to top of the screens in pumping well (d) and initial saturated aquifer thickness (b) = d/b.

- Z_{1D} = ratio between the vertical distance from the bottom of the aquifer to bottom of the screens in the pumping well and initial saturated aquifer thickness.

- Z_{2D} = ratio between the vertical distance from the bottom of the aquifer to top of the screens in the pumping well and initial saturated aquifer thickness.

Fully penetrating condition allows the number of independent parameters to be reduced from three to two by letting σ approach zero. Partially penetrating conditions allow the same parameter reduction procedure provided that the geometric factors (l_D, d_D Z_{1D} Z_{2D}) are known. A special set of theoretical curves must be developed to use this procedure in accordance with the geometrical distribution and construction of the pumping well and the observation wells. Driscoll (1987) considers that a distance from the pumping well equal to twice the aquifer thickness, the effect of partial penetration is no longer observed in the observation piezometers. The effect of partial penetration on the drawdown in an unconfined aquifer decreases with radial distance from the pumping well and with the ratio $K_D = K_z / K_r$ Neuman (1974). At distances greater than $r = b / K_D^{1/2}$ this effect disappears completely when time exceeds $t = 10\ S_y r^2/T$ and the drawdown data follow the late Theis curve in terms of t_y. If the condition is not satisfied a set of theoretical type curves has to be used.

Distance - Drawdown Analysis

Distance-drawdown analysis can be performed using a number of wells at variable distances between the pumping well and observation point. Usually three wells are used to produce the graphs of drawdown versus distance. The formulae used to determine the transmissivity (T) and the storage coefficient (S) is given below:

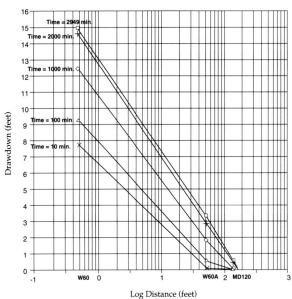

DISTANCE VERSUS DRAWDOWN
Wells W60, W60A, WD120

Time = 2000 min.
Q = 31.9 GPM
Δs = 12.9 - 7.0 = 5.9 feet
r_0 = 180 feet

Figure 7-30 Distance-Drawdown Graph

$$T = \frac{528Q}{s} \quad \text{(a)}$$

$$S = \frac{0.3Tt}{r_o^2} \quad \text{(b)}$$

Equation 7-32(a, b)

where:
T: Transmissivity in gpd/ft;
Q: Pumping rate in gpm;
s: slope of the distance-drawdown graph expressed as the change in drawdown, in feet, between any two values of distance on the log scale whose ratio is 10.
t: time since pumping started in days;
r_o^2: intercept of the extended straight line at 0 (zero) drawdown, in feet.

An example of the graphs of drawdown versus logarithm of distance are presented in the Figures 7-32. As it can be seen at 2000 minutes since the pumping started the corresponding drawdown for the considered wells fall along a straight line on the semi-logarithmic graphs. This moment of the pumping test was used in the distance - drawdown analysis.

Aquifer Radial Anisotropy

The Neuman model takes into account the vertical anisotropy of the aquifer. The distance - drawdown analysis can be used to assess radial anisotropy of an aquifer since the analysis generates the average hydraulic conductivity along a line of piezometers. By comparison of lines of observation points along perpendicular lines of wells one can calculate a value for anisotropy.

Distance-drawdown analysis performed along several sets of observation wells, set along the strike of geologic formations and perpendicular to the strike represented the major axis of hydraulic conductivity. The degree of radial anisotropy in the vicinity of a pumping well (KR) is typically calculated as the ratio between the hydraulic conductivity along the major resultant axis of anisotropy and the hydraulic conductivity along the minor axis of anisotropy

The shape of drawdown cones (measured from many piezometers) provide large scale views of the anisotropic hydraulic conductivity conditions. In general one should plot the shape of drawdown cones at a series of time steps. Anisotropic conditions should be evident by the shape of the drawdown cones. Often initial anisotropic shaped drawdown cones later develop fully radial shapes at long time periods. These shapes can be back calculated to define porosity values from the total volume of geologic material dewatered divided by the total discharge during the test period.

Recovery Analysis

To confirm results obtained during a pump test, recovery data can also be analyzed using the Theis Recovery Method. The results obtained using the Theis recovery analysis should be the same order of magnitude as those determined for the unsteady state condition during the pumping period. The formula used to determine the transmissivity (T) is presented below.

$$T = \frac{264Q}{s} \qquad \text{Equation 7-33}$$

where:
T: Transmissivity in gpd/ft;
Q: Pumping rate in gpm;
s: slope of the distance-drawdown graph expressed as the change in drawdown, in feet, between any two

values of distance on the log scale with a ratio is 10.

Summary of the Results

One should always include summary ranges of the average hydrogeological parameters obtained by the various analysis methods for the area influenced by the pumping test. Reporting of the data should include at least the following data:

- A drawing of the location of the test area, that includes the well and piezometers, and any likly recharging or barrier boundaries;

- The lithologic cross-section for the site based on the Phase II investigation that includes the depth of the well screens and all piezometers used in the test. Also include the location of local wells and piezometers that either did not respond or were not used in the pump testing. This data should include all pertinant information about the observation points, such as distance to the points, screened depths and hydrostratigraphic units.

- Tabular data sets of the pre-pumping, pumping and post-pumping water level information for the well and piezometers;

- Barometeric data for the entire water level data set;

- Hydrographs that show the corrections applied to the raw data sets;

- Time-drawdown and distance drawdown curves with calculations applied, if possible, directly to the curves;

- A discussion on the conduct of the test and the resultant models and calculations used in the evaluations. You should always include problems and evaluations of the data accuracy;

- Always include the entire data set as an appendix so that reviewers may also conduct inspection of the models and calculations used in the evaluations.

7.6 CONCLUSIONS

At the conclusions of the Phase II field and office studies, the physical, geologic, and hydrogeologic setting of the site should be well established. The following data should be available for inclusion into the conceptual model:

- Topographic maps;
- Boring logs;
- Interpretative cross–sections;
- Interpretative analysis of geophysical studies;
- Material test results and interpretations;
- Narrative description of ground water with flow patterns;
- Piezometric or potentiometric maps, with interpreted flow lines;
- Analysis of aquifer characterization tests; and,
- Structural contour maps of uppermost aquifer and confining layers in plan view

These data are then used to construct the unified conceptual hydrogeologic models and flow nets which are then used to design the monitoring system for the facility. The following chapter addresses the process of geologic conceptualization and flow net construction techniques specific to geologic models and ground water monitoring programs. These models and flow net construction, are important for both detection and assessment monitoring design, as well as, aquifer remediation design elements.

REFERENCES

Bedinger M. S. and J. E. Reed, 1988. Practical Guide to Aquifer Test Analysis, Environmental Monitoring Systems Laboratory, Office of Research and Development, U.S. EPA Las Vegas, Nevada 89193-3478, Interagency Agreement DW 14932431-01-1.

Boulton, N.S. 1963. Analysis of data from non–equilibrium pumping tests allowing for delayed yield from storage. Inst. Civil Engineers Proc., Vol. 26, pp. 469–482.

Bouwer, H. and R. C. Rice, 1976. A Slug Test for Determining Hydraulic Conductivity of Unconfined Aquifers with Completely or Partially Penetrating Wells, Water Resources Research. v. 12 pp. 423-428.

Cooper, H.H., Jr.; J.D. Bredehoeft; and I.S. Papadopulos. 1967. Response of a finite–diameter well to an instantaneous charge of water. Water Resources Research, Vol. 3, pp. 263–269.

Driscoll, F. G., 1987. Groundwater and Wells. Johnson Division, St. Paul Minn., 1089 p.

Ferris, J. G., and D. B. Knowles, 1954. The Slug Test for Estimating Transmissibility. U.S. Geological Survey Ground Water Note 26, 6 p.

Ferris, J. G., and D. B. Knowles, 1963. Slug-injection Test for Estimating the Coefficient of Transmissibility and Drawdown. U.S. Geological Survey Water-Supply Paper 1536-I, p 299-304

Fetter, C. W., 1980. Applied Hydrogeology. Charles E. Merrill Co. Columbus, OH.

Freeze, R.A.; and J.A. Cherry. 1979. Ground water. Prentice–Hall, Inc., NJ, 604 pp.

Green, W. R., 1985. Computer-aided Data Analysis, A practical Guide. John Wiley and Sons, New York, 268 pp.

Hantush, M.S. 1960. Modification of the theory of leaky aquifers. Jour. Geophys. Research, Vol. 65, No. 11, pp. 3713–3725.

Hathaway A., 1989. RQD in Geotechnical Investigations, AEG News

Hvorslev, M. J. 1951. Time lag and soil Permeability in Groundwater Observations. Bulletin No. 36, Waterways Experiment Station, Corps of Engineers, Vicksburg, MS.

Jacob, C. E., 1963. Determining the Permeability of Water-table Aquifers, In Bentall, Ray, compiler, Methods of Determining Permeability, Transmissibility, and Drawdown. U.S. Geological Survey, Water-Supply Paper 1536-1, p. 245-271.

Kruseman, G.P.; and N.A. DeRidder., 1970. Analysis and Evaluation of Pumping test Data. Bulletin 11, International Institute for Land Reclamation and Improvement, Waginengen, The Netherlands, 190 p.

Kruseman, G.P.; and N.A. DeRidder. 1976. Analysis and evaluation of pump test data. International Institute for Land Reclamation and Improvement, Wageningen, Netherlands Bulletin 11, 200 pp.

Lohman, S.W. 1972. Ground–water hydraulics. U.S.G.S. Prof. Paper 708, 70 pp.

Meinzer, O. E., 1939. Ground Water in the United States. U.S. Geological Survey Water-Supply Paper 836-D, pp 157-229.

Neuman, S.P. 1972. Theory of flow in unconfined aquifers considering delayed response of the water table. Water Resources Research, Vol. 8, No. 4, pp. 1031–1045.

Neuman, S.P. 1974. Effect of partial penetration on flow in unconfined aquifers considering delayed gravity response. Water Resources Research, Vol. 10, No. 2, pp. 303–312.

Neuman, S.P. 1975. Analysis of pumping test data from anisotropic unconfined aquifers considering delayed gravity response. Water Resources Research, Vol. 11, No. 2, pp. 329–342.

Reed, J. E. 1980. Type Curves for Selected Problems of Flow to Wells in Confined Aquifers. U.S. Geological Survey Techniques of Water-Resources Investigations, Book 3, Ch. B3, 106 p.

Roscoe Moss , 1990, Handbook of Ground Water Development. Wiley Interscience, 493 pp.

Stallman, R. W. 1971. Aquifer-test Design, Observation, and data analysis. U.S. Geological Survey Techniques of Water-Resources Investigations, Book 3, Chap. B1, 26 p.

Theis, C.V. 1935. The relationship between the lowering of piezometric surface and the rate and duration of discharge of a well using ground–water storage. Am. Geophys. Union Trans., Vol. 14, Pt. 2, pp. 519–524.

Todd, D.K. 1980. Ground water hydrology. Second Edition, John Wiley & Sons, NY, 535 pp.

Van der Kamp G. 1985. Brief Quantitative Guidelines for the Design and Analysis of Pumping Tests. Hydrogeology in the Service of Men, Memoires of the 18th Congress of the International Association of Hydrogeologists, Cambridge, p 197-206.

Walton, W. C. 1962. Selected Analytical Methods for Well and Aquifer Evaluation. Illinois Department of Registration and Education Bulletin 49. 81 p.

Weeks E. P. 1964. Field Methods for Determining Vertical Permeability and Aquifer Anisotropy. U.S. Geological Survey Professional Paper 501-D, p. D193-D198.

Weeks E. P. .1969. Determining the Ratio of Horizontal to Vertical Permeability by Aquifer-test analysis. Water Resources Research. Vol. 5, No. 1, p 196-214.

CHAPTER 8

THE CONCEPTUAL MODEL

There are two objectives in the use of conceptual models for the evaluation of ground–water flow at waste disposal sites. These are:

- Site Characterization — To develop a sound and informed understanding of the geology and geohydrology, the natural hydraulic character and chemical evolution of the ground–water system, ground–water flow directions and rates; and,

- Prediction — To forecast or predict the nature, rate, and volume of movement of leachate into ground–water monitoring zones.

The site characterization and prediction process are explained in this chapter, as it employs data gathered in the Phase I and Phase II programs. These data are formatted into additional visual presentations, such as the maps and cross–sections that were discussed in Chapter 7.0. Review of these derivative presentations naturally leads to completion of the conceptual hydrogeologic model for the facility of interest. Figure 8–1 illustrates an overview of the model conceptualization process using data developed during the Phase II investigation, and displayed as derivative maps, cross–sections, and fence diagrams.

The conceptual model of the physical geologic and hydrogeologic system can be represented by more than the typical two or three dimensional drawings common in the literature. These representations of the real world field conditions can also be enhanced by including within the conceptualization process ground-water flow net constructions. These composite evaluations provide the investigator with important insight into directional components and rates of flow of the system under investigation.

Review of data leads to a conceptual model of the ground–water system and its interaction with surface water bodies and human activities. The term "conceptual model' is a convenient designation for visualization of the physical

system, which is formed in the mind of a geohydrologist. It is always an idealization and simplification of the actual physical situation.

An aquifer is a complex, three–dimensional geologic–system containing water. The aquifer is bounded by other geologic units, the ground surface, and/or bodies of surface water. At many locations, for example, it is entirely appropriate to represent this concept as a single layer of finite horizontal extent. This is done by applying an averaging process to the third dimension, the vertical, resulting in what is sometimes known as the vertically integrated approach. Editing of borehole logs and

OVERVIEW OF SITE CONCEPTUAL MODEL PROCESS

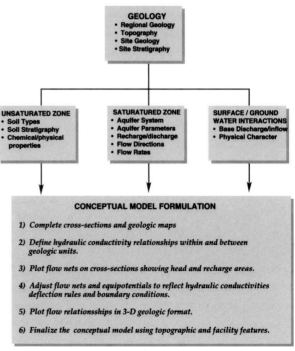

Figure 8-1 The Conceptual Process

conversion into a final geologic–column is a similar process; however, unified conceptual and flow net models should include both geologic and hydraulic conditions contributing to ground–water movement. Many of the major contributions to the field of ground–water hydrology — the Theim equation, Theis equation, Hantush–Jacob leaky equation, Dupuit–Forcheimer equation, the Dupuit approximation, etc., are based on the approach of integrating average conditions into an analytical "conceptual model".

The unified conceptual/flow model must incorporate all the essential features of the physical system under study. With this constraint, conceptualization is tailored to an appropriate level or sophistication to describe the environment under study. The degree of accuracy required for various geologic conditions differs. For example, an unconfined aquifer in a permeable, homogeneous, sandy lithology may require only simple cross–sections to illustrate the conceptual system. If there is a significant and identifiable three–dimensional gradient with more than a single permeable layer, more complex geologic models, including complex flow net construction, may be necessary to illustrate the system. The conceptual process may require consideration of facies or depositional models to fully understand the geologic system. These models are

supported in Phase II field lithostratigraphic mapping and logging procedures.

Phase I studies involve less time, less money, less data, and require only a fair "accuracy" in their predictions of the conceptual system. Consequently, a much simpler conceptual model is required. Design of the final conceptual model for important facilities and/or large–scale ground–water restoration projects, involving significant financial investments or risks, requires complete information, more time and money, and a more sophisticated manner of conceptualization.

The following subsections first reviews the "geologic" model where principal emphasis is placed on arriving at a correct interpretation of the geologic environment. A conceptual geologic model often is developed over time with many alternative configurations of the model are tried and discarded. This process will be developed to illustrate the common procedure typically employed in these cases.

The geologic conceptual model is then further supported by hydrogeologic data developed by field gathered hydraulic head levels. Flow nets are developed as pertinent to geologic systems. The integration of geologic data with flow data completes the process leading to selection of target zones for ground-water monitoring system design. Flow net construction may only be an

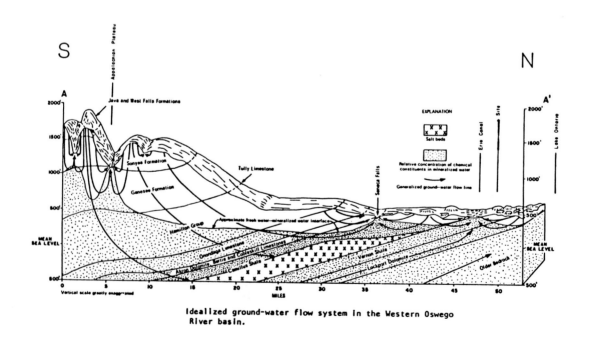

Idealized ground-water flow system in the Western Oswego River basin.

Figure 8-2 Cross Sectional Information on Regional Scale

Source: Wehran Engineering

initial step toward full computer modeling using numerical techniques. These flow net constructions, however, can be completed in a matter of hours and would even be recommended as a preliminary step before extensive computer modeling, to serve as a guide and cross check of the numerical work. These flow net constructions can provide steady state results that are often sufficiently accurate for many ground-water issues that arise during site assessment projects.

8.1 REGIONAL AND SITE GEOLOGY

In order to define the geologic environment of a specific area, the regional and site–specific geology must be thoroughly characterized. These data must establish the properties and features of individual geologic units beneath and near the site, through use of:

- Regional geologic maps and cross–sections that are used to characterize area–wide geologic units facies, depositional and structural features, to depths incorporating probable ground–water movement as it could impact or be impacted by facility construction and operation;

- Topographic maps used to characterize site–specific topographic relief;

- Stratigraphic maps and cross sections used to characterize detailed site–specific geologic conditions; and,

- Aerial photography, as used to illustrate information such as vegetation, springs, gaining or losing stream conditions, wetland areas, and important elements of geologic structure.

The above data will have been gathered within the Phase I and Phase II programs. Cross–sections and stratigraphic maps should be prepared so that a clear technical basis is used to derive the conceptual geologic and hydrogeologic models.

The entire process of geologic conceptualization is built-up by regional data to establish the geologic setting and there by gradually building an understanding of geologic conditions at the site from the drilling program. Interpretation of data through cross–sections and maps strengthens the conceptual model until a final picture of the geology is established. The final conceptual geologic model is a site–specific representation of the geologic system under and adjacent to the facility. Geologic

information used to construct conceptual geologic models typically consists of the following:

- The depth, thickness, and area extent of each stratigraphic unit, including weathering horizons;

- Stratigraphic zones and lenses within the near–surface zone of saturation;

- Stratigraphic contacts between significant formations/strata;

- Significant structural features such as discontinuity sets, faults, and folds; and,

- Zones of high hydraulic conductivity or shearing/faulting.

Several cross–sections may be required to depict significant geologic or structural trends and to reflect geologic/structural features in relation to local and regional ground–water flow. The area and vertical extent of the geologic units can be presented in several ways. For complex settings, the most desirable presentation is a series of structure contour maps representing the physical character of the tops and/or bottoms of each unit. Cross–sections and isopach maps can also be used because they are generally good graphic supplements to the structure contour maps. These cross–sections can be combined into a fence diagram (three–dimensional) that can serve as the basis for conceptualization of the geology.

Conceptualization is a way of achieving a graphic idealization of the actual geologic conditions; the investigator must, therefore, consider the geologic features that would affect ground–water flow and ground–water quality. Minor features that are not important to the overall picture should not be transferred to representations of the conceptual model.

The conceptualization process is illustrated through a series of Figures (8–2 through 8–6). Figure 8–2 provides a cross–section showing regional bedrock and ground–water flow for an area in upstate New York. The conceptualization was developed entirely from a Phase I literature search and geologic experience in the region.

The regional cross–section shows the site to be located over the dipping Vernon Shale and unconsolidated glacial deposits. The regional ground–water discharges into the Erie Canal. The deeper bedrock units are also shown discharging ground water from depths beneath the Appalachian Plateau, through the Vernon Shale, and into the Erie Canal.

This information allowed the investigators to plan a program to drill to the Vernon Shale and confirm the depth of glacial deposits; ice contact deposits, and weathering in the shale. Figure 8–3 is an example of a final boring log from the drilling program. The lithology of glacial till deposits, residual soil, and weathered/fractured bedrock were confirmed by the detailed description of split–spoon samples, taken during boring effort.

A series of borings is presented on a cross–section in Figure 8–4. Boring B–8 illustrates the geologic section down to competent shale. The conceptualization process was started in this figure since many of the minor sand lenses are grouped within the till body as undifferentiated till deposits. If the unit contained significant sand layers, as are represented in boring B–2 as glacial outwash, these units should be shown on the cross–section.

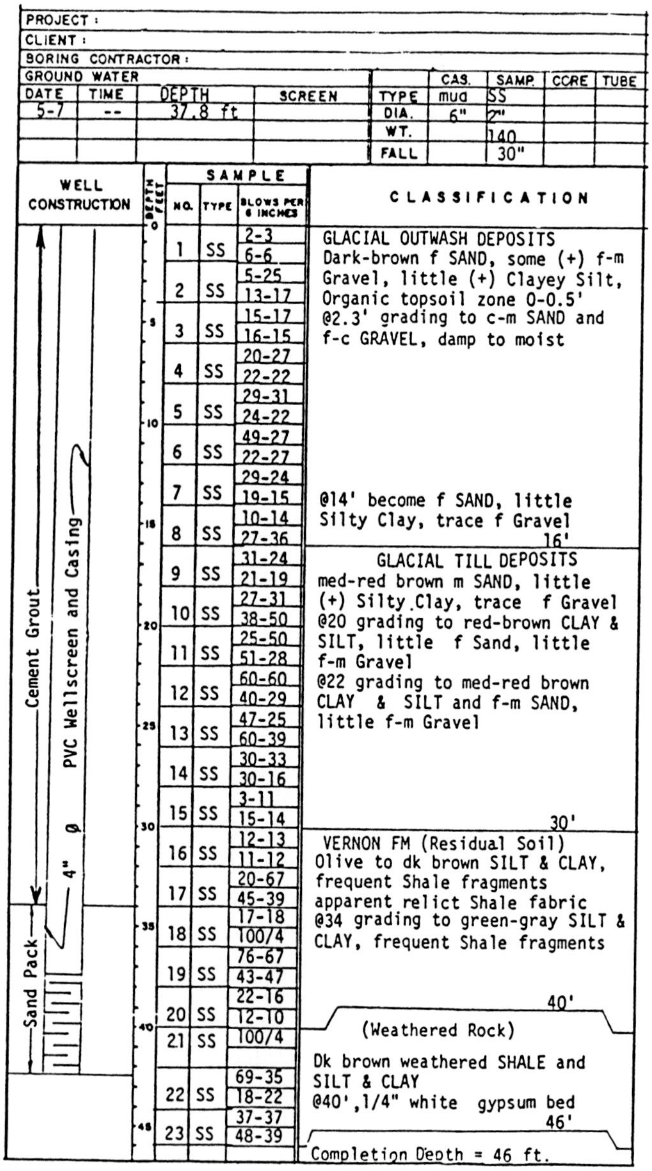

Figure 8-3 Final Boring Log Example

Source: Wehran Engineering

The conceptualization process goes further, as shown in the Figure 8–5 fence–diagram, which includes many cross section views. These diagrams are three–dimensional representations, constructed using several cross–sections that are helpful in presenting an area view of geologic and ground–water conditions. As with cross sections, they are based on the logs of the field borings, measurements of ground–water levels, and topography. However, the conceptual process required one additional step to aid in the evaluation of the site.

The final step in the conceptualization process is reduction of the information, on cross sections or fence diagrams, to the essential elements of geology. The picture should be a clear and concise conceptualization of the

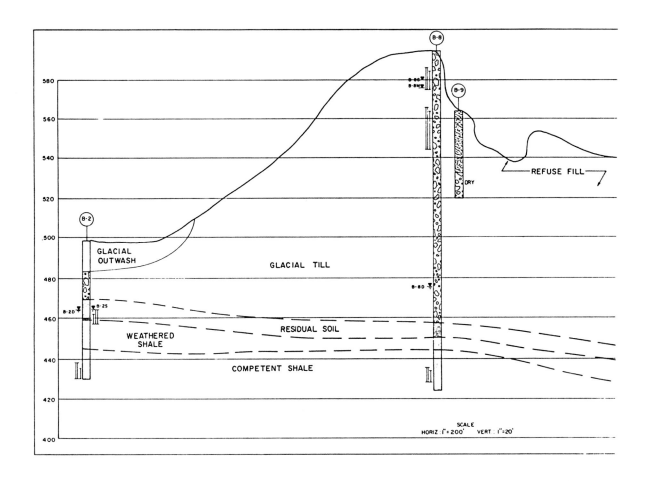

Figure 8-4 Cross Sectional View of Site Data

geology, presenting all important features necessary for understanding the interrelationships between geology and ground–water flow below the site.

Figure 8-6 presents the final conceptual model of the geology and the preliminary ground–water flow–directions for the facility. The arrows do not represent an actual flow net, however, the directional components shown provide

the target monitoring zone for the site. Weathered and unweathered basal tills are shown as separate conceptual elements, due to inherent differences in hydraulic conductivity of the two units. Residual–soil overlying the fractured and weathered contact with the Vernon Shale is not identified because it has a hydraulic conductivity similar to that of unweathered tills. In such an instance,

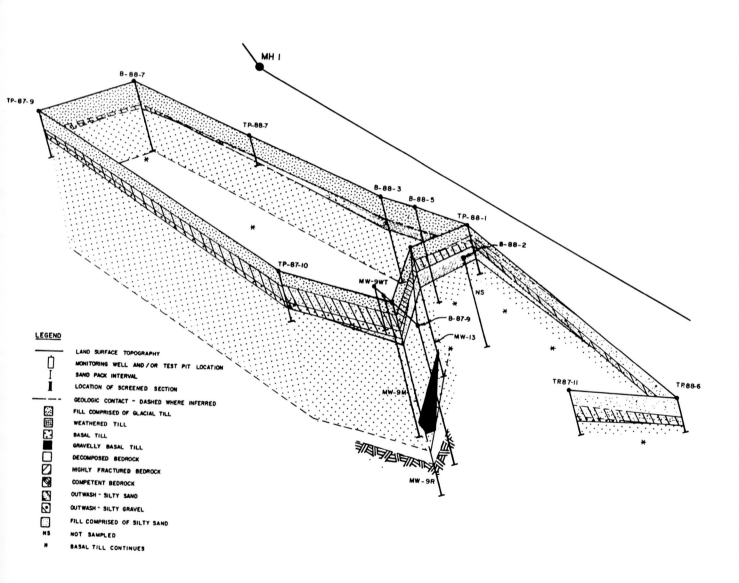

LEGEND

——	LAND SURFACE TOPOGRAPHY
	MONITORING WELL AND / OR TEST PIT LOCATION
	SAND PACK INTERVAL
	LOCATION OF SCREENED SECTION
– – –	GEOLOGIC CONTACT – DASHED WHERE INFERRED
	FILL COMPRISED OF GLACIAL TILL
	WEATHERED TILL
	BASAL TILL
	GRAVELLY BASAL TILL
	DECOMPOSED BEDROCK
	HIGHLY FRACTURED BEDROCK
	COMPETENT BEDROCK
	OUTWASH – SILTY SAND
	OUTWASH – SILTY GRAVEL
	FILL COMPRISED OF SILTY SAND
NS	NOT SAMPLED
✳	BASAL TILL CONTINUES

Figure 8-5 Fence Diagram of Site Geology

Source: Wehran Engineering

the geologic–unit nomenclature should show the combined names (lacking in this figure). Weathered and fractured bedrock is, however, distinguished from competent bedrock due to its differences in hydraulic conductivity. This system is later in this chapter integrated with a flow net to fully evaluate potential for alternative detection and assessment monitoring well location.

8.2 CHARACTERISTICS OF THE SATURATED ZONE

Each of the significant stratigraphic units in the zone of saturation should be characterized by determination of its hydrogeologic properties such as hydraulic conductivity (vertical and horizontal) and effective porosity. These parameters describe aquifer characteristics that control ground–water movement and, hence, the ability of the aquifer to retain or pass potential leachate. Both are needed for a general understanding of the hydrogeologic setting at a site, and for completing the conceptual hydrogeologic model for design of ground–water monitoring systems. Typically, the amount of data necessary to complete a conceptual hydrogeologic model will differ for each geologic environment in question.

For example, an aquifer in extensive, homogeneous beach–sand (SP under the Unified Soil Classification System) will require less investigation than a glacial unit consisting of lenticular deposits of outwash interbedded with clayey till (GW/CL). The USC system should be supplemented by lithostratigraphic descriptions (Faces Codes), for example; is the sediment in question clast supported or matrix supported? When fine grained sediments reach 20% within a particular sandy unit the sand sized clasts lose contact and the unit becomes matrix supported. Hydraulic conductivity drops significantly when this point is reached in a particular unit (Hughs, 1991). This drop in hydraulic conductivity can range as much as three to four orders of magnitude. There are two types of fundamental aquifer characteristics required to define a hydrogeologic conceptual model: characteristics of the aquifer and characteristics of ground–water flow. Aquifer characteristics, for each hydrogeologic unit, should include:

• Hydraulic conductivity
• Effective porosity
• Specific yield/storage (as required for the project)

Additional aquifer characteristics such as transmissivity, and chemical attenuation properties can be derived; and estimated dispersivity values may be necessary for ground–water transport modeling to estimate ultimately (long–term) concentrations at points of exposure. These aquifer characteristics support ground–water flow characteristics used to define the quantity and the direction of ground–water flow. Both aquifer and flow characteristics are necessary to define the conceptual movement of ground water away from a facility

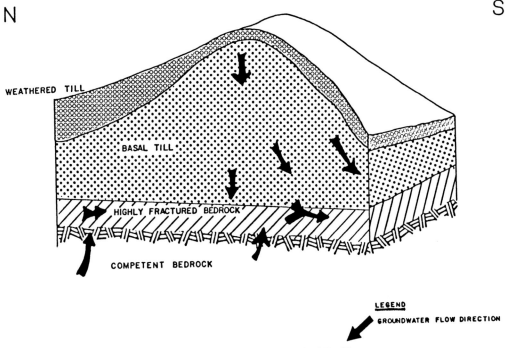

Figure 8-6 Final Conceptual Model of Facility

Source: Wehran Engineering

and toward a monitoring point. In order to define the general flow characteristics necessary to develop conceptual hydrogeologic models, the following data are required:

- Boundaries and zones of recharge/discharge;
- Flow directions; and,
- Flow rates.

Each of these flow characteristics provides components necessary to evaluate the geologic units likely discharging from the facility and to select target zones for monitoring.

8.2.1 Boundaries And Recharge/Discharge Zones

To aid in the understanding of the ground–water flow–regime and identification of potential pathways to target for monitoring, the location of any proximate zones of recharge or discharge should be identified within the hydrogeologic conceptual–model. This identification is determined in part by the general site location in the watershed. Figure 8–7 shows regional discharge/recharge areas from Hubbert (1940) for an unconfined aquifer. The investigator probably does not need to quantify such information in detail; rather, a general indication of discharge or recharge characteristics are used as part of the conceptualization process.

For unconfined aquifers, recharge areas are usually topographic highs. Discharge/recharge areas also indicate relative depth to unconfined ground water. In discharge areas, ground water is found close to or at the land surface; while at recharge areas, there is often a deep unsaturated zone between the unconfined ground–water surface and the land surface. Discharge areas may be represented by streams, springs, rivers, swamps or ultimately the ocean. Discharge areas may be also related to more permeable aquifers, dry stream beds (with a subsurface discharge to thalweg) or to areas of wetland type vegetation. Freeze and Witherspoon (1967) concluded in a computer based analysis that discharge areas within a particular basin are smaller than recharge areas. They estimated that studies with three dimensional models the actual percentage of discharge area represented approximately 7% of the total area of the basins. Further discussion of computer based flow net based discharge-recharge is contained in section 8.4 on flow net construction. A ground–water contour map can be used as an important tool to help locate these areas.

Recharge and discharge in confined aquifers are typically more complex than for unconfined aquifers. Discharge and recharge may occur where the aquifer is exposed. Some discharge may also occur in the form of upward leakage in areas of upward hydraulic gradient and

leaky confining units. Recharge can also occur by downward ground-water flow through leaks in the confining layers (aquitards). Topographic information can provide significant insight into ground–water movement; however, topography can also mislead an investigator, when aquifers are confined or perched. Ground water in such aquifers can often move in unexpected directions. The investigator must pay particular attention to potential mounding of ground water and directional complexities that may be covered by low hydraulic conductivity subgrade materials. Specific conditions can only be discerned through analysis of lithology, (perhaps through depositional models), and measurement of piezometric hydraulic heads, which should be gathered in the Phase II field program.

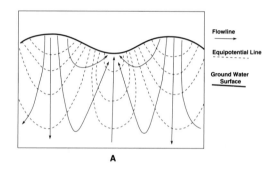

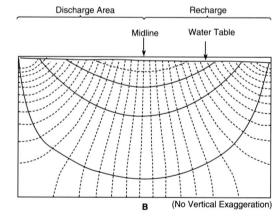

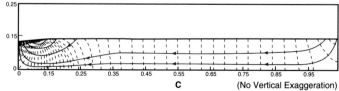

Figure 8-7 Regional Conceptual Flow Patterns

8.2.2 Ground-Water Flow-Directions

The Phase II investigations include a program for precise monitoring of the area and temporal variations in ground–water levels. This program involves the measurement of water levels or hydraulic head levels in piezometers installed for the purpose of investigating the saturated zone. These data are used to define ground–water flow–directions during development of the ground–water monitoring system. Parameters necessary to measure or ascertain for use in determination of ground–water flow directions include:

- Depth to ground water;
- Potentiometric surfaces;
- Hydraulic gradients; and
- Vertical components of flow.

Each of these parameters was defined within Chapters 3.0, 4.0 and 5.0 of this manual; hence, the following text will describe use of these data in development of the conceptual hydrogeologic model.

Figure 8–8 shows the potentiometric surface of a glacial outwash sand and gravel (SW) unit that has facility-wide extent. Piezometers are located on the map, which also provides water levels measured on a single day during

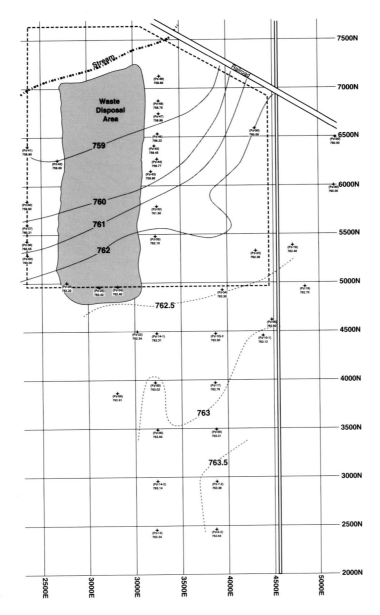

Source: Golder Associates

Figure 8-8 Regional Conceptual Flow Pattern in Glacial Environment

stable barometric conditions. Because the unit is confined within a fine–grained lodgment till (or diamict), water level readings are affected mainly by changes in barometric pressure. Interpretations of ground–water flow–direction (for the geologic unit measured) should always be based on water level measurements from a single, continuous, permeable unit. This unit can be classed as a "hydrostratigraphic" unit. Much of the inconsistencies observed in potentiometric contour maps are due to piezometers installed in either different hydrostratigraphic zones, or in thick, low hydraulic conductivity confining units that have strong downward gradients. Slight variations in piezometer tip depth can result in large inconsistencies in observed hydraulic head level for such confining units.

Cross–sectional stratigraphic data is provided in Figure 8–9. A cross–section without vertical exaggeration is provided in Figure 8–10. Cross–sections without exaggerated vertical scales are helpful for conceptualizing the zones that will serve as discharge points. Such a view provides a perspective on relative thickness and regional extent of the units.

The data provided in Figure 8–9 was used in conjunction with hydraulic conductivity data to sketch a flow net for the cross–section (Figure 8–11). The flow net provides the conceptual ground–water movement for the facility, from which the potential target monitoring zones can be determined. Two zones, above and below the till units, would probably represent the target zones for monitoring areas, as based on the nearly–horizontal flow regime at the site.

Constructing piezometric or potentiometric contour maps from raw data, then combining the data with stratigraphic cross–sections to develop flow nets, can be a tedious effort when geologic systems become complex; however, these derived illustrations of ground-water flow are essential for selection of target monitoring zones and design of the monitoring system.

8.3 CONCEPTUAL GEOLOGIC MODELS

Much of the conceptualization process for hydrogeologic systems was developed from the traditional engineering approach for water balance evaluations. This can be simply illustrated as a layered aquifer system with both recharge (from the surface) and flow horizontally through the geologic block diagram shown in Figure 8-12. Each pertinent layer of the site geologic units are

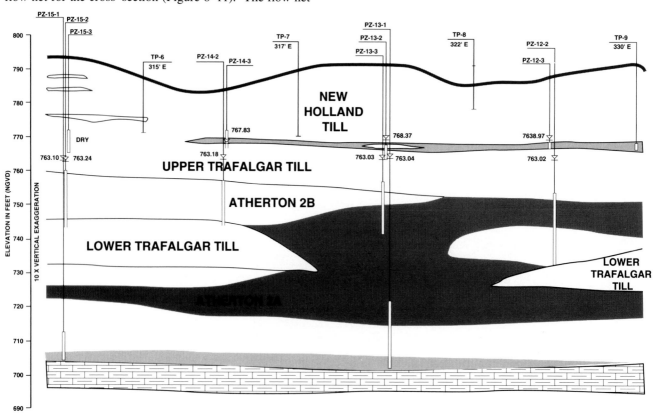

Figure 8-9 Cross Sectional Stratigraphy

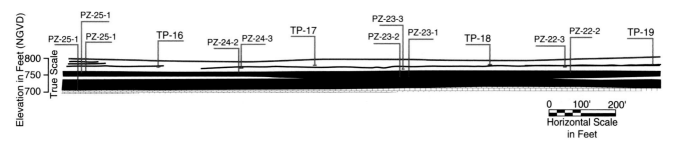

Source: Modified from Golder Associates

Figure 8-10 Cross Sectional View Without Vertical Exaggeration

represented in the diagram with "relative' thickness and includes values for hydraulic conductivity. The illustration represents an idealized system with little variability in geologic unit lateral thickness and unit hydraulic conductivity. Although actual field situations will normally vary somewhat (or can vary widely), the conceptualization process dictates that the model represents an idealization and simplification of the actual physical situation. If the actual field data showed significant differences in thickness of the lithology in the site cross-section from one end of the site to the other, the model should include some concept as to this spatial variability. The concept of unit thickness variability may

not have any relevance to the model unless the observed variability exceeds some acceptable limits. These limits depend on the overall size of the model and may only be important if the thickness has relevance to the overall model. One can expect that there will be considerable variability in the results of hydraulic conductivity testing due to both variability in the test procedure and to the natural variability of the geologic units. As with minor thickness variability in the field testing program; if the hydraulic conductivity of a single hydrostratigraphic unit can be represented as a normal curve of values with similar hydraulic conductivity across the unit, then the unit can be considered as a single hydrostratigraphic unit. Any

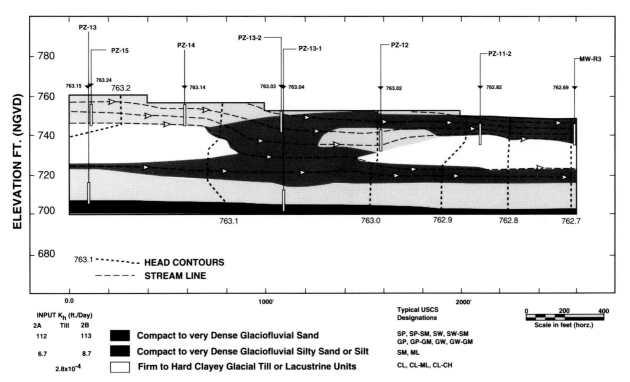

Source: Modified from Golder Associates   **Figure 8-11** Flow Net Construction

geologic unit can be expected to show both horizontal and vertical variability as is illustrated in Figure 8-13.

The illustrations represents an obvious layering of the geologic units , however, the bell shaped curves of horizontal hydraulic conductivity show the natural ranges expected in naturally occurring materials. Although unit B shows two orders of magnitude variations in average hydraulic conductivity from borehole L to bore N, an individual measurement of K (in layer B) could range from 10^{-2} cm/sec to 10^{-4} cm/sec in borehole L to 10^{-4} cm/sec to 10^{-6} cm/sec in borehole N. Since layer A averages around 5×10^{-6} cm/sec some confusion can result when one depends only on determinations of hydraulic conductivity as the deciding factor in the geologic conceptual model formulation. The decision process of developing a conceptual geologic model for a site must include the geologic information gathered from the literature, where typical type sections can point toward proper interpretation of lithological data gathered during the field boring program. However, specific site conditions can also change radically from conditions depicted in regional studies. Type sections are the starting point where further site work derives the specifics that develop the conceptual model.

This tuning in of the site to regional data represents one of the first tasks in developing an conceptual model. Previously defined hydrostratigraphic units may be already available for inclusion into the site specific conceptual model. These hydrostratigraphic units may be formations, units or individual stratigraphic zones that represent likely confining or, more highly, conductive zones. Probably the best guidance in this area is to consider the scale of the features. First evaluate which features or geologic units would have some effects on ground-water flow in the area of interest. These should be defined in the borehole drilling and logging process. The conceptualization process also must consider structure of the geologic system in the Phase I investigation so that areas that may require investigation can be targeted in the Phase II field program. Lateral variations, such as facies changes, should be fully included in the conceptualization process if these changes affect the flow of ground water through the geologic units present on site.

Examples of the building process of hydrogeologic conceptualization is shown in Figure 8-14a through 8-14e. These five illustrations represent a process of conceptualization from early miss-steps to later more acceptable realization of the relatively uncomplex hydrogeologic system. Illustration 8-14a represents the first attempt in construction a conceptual model for a landfill located on coastal deposits. The figure shows a landfill composed of typical solid waste with a significant level of leachate. The geologic units serving as the landfill base-grade consists of fine to medium, clean sand with a number of separated clay units. This type of representation is typically called a "cartoon' due to the unrealistic

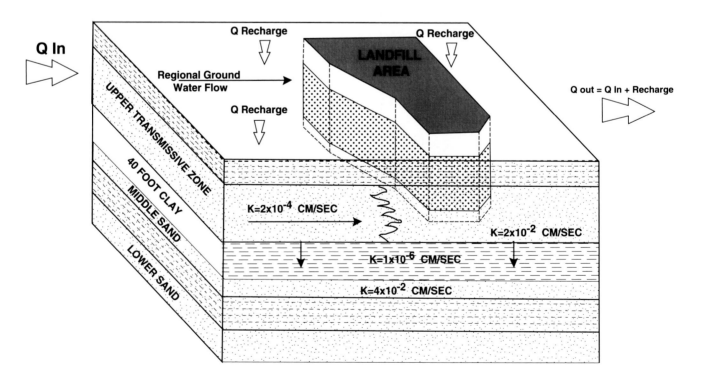

Figure 8-12 Horizontal Flow Through System

representations of real-world geologic and site features. The foundation sand and gravel of the facility is shown as somehow able to contain a significant height of leachate within a mass of solid waste (with a probable waste hydraulic conductivity somewhere between 10^{-4} to 10^{-5} cm/sec). It is unlikely that the highly permeable sand would contain, (like a bathtub) leachate within the landfill limit, given that the landfill is most likely unlined. The water level (or water table) shown on the figure is also too flat for the suggested leachate migration suggested in the cartoon. Lastly, the physical form of the clay layers shows little insight by the illustrator in likely geologic processes.

The same cross-section was modified at a later date (shown in 8-14b) to include bedrock and the formation names of the unconsolidated deposits (Potomac-Rariton Magothy formations) Three different water levels are shown draining presumably from point A toward the landfill. Although this representation is probably better than the first attempt, local lithologic variations in the

unconsolidated formations are provided and little insight into why we are viewing three separate water levels is provided by the model. As with the previous model, leachate levels in the landfill are shown above regional water levels. As with the previous cartoon, given that the regional aquifer consists of highly permeable sands it would be unlikely that the leachate levels shown could be maintained at the levels shown. The model is also static with little consideration into ground-water flow. Geologic model figure 8-14c has greatly improved cross-sectional information tied into the regional water surfaces. Although the various clay units are shown in a more realistic manner, the water "table' shown (with a date 10-27-83) does not show any deflection downward over the area without clay (between well 6 and well 5). It is unlikely that the ground-water level could maintain the heads shown over this area given the wide variations between the region water table and the local piezometric surface. One would expect a downward deflection in head levels in the ground-water

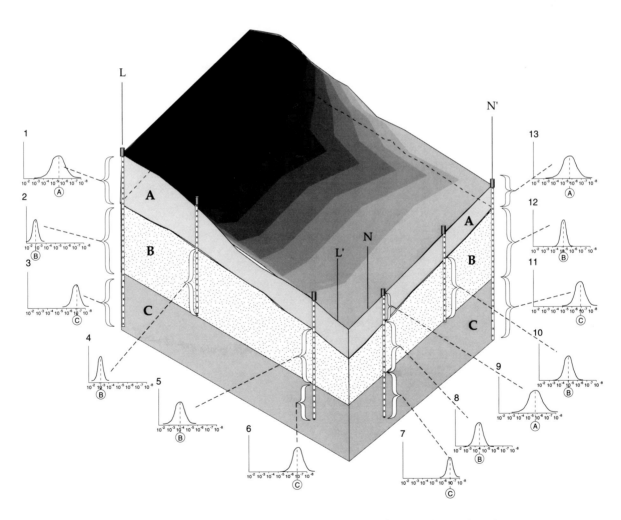

Figure 8-13 Horizontal and Vertical Variability of Geologic Units

level in the "window" area and an upward deflection (or interconnection) of the piezometric surface with the regional ground-water level. Conceptual models must always show logical solutions to the interconnections between geologic units and ground-water flow systems.

Occasionally, conceptualization of both geologic and ground-water flow systems requires more than one illustration to fully transmit an understanding of site conditions. Figures 8-14d and 8-14e show two conceptual models of the site previously described in 8-14a to 8-14c. Figure 8-14d shows the relationship between shallow clay units and perched shallow ground water discharging into the deeper aquifer. Figure 8-14e shows a different perspective of the same site area. Historic context of both previous water tables (dating back to 1956) show regional ground-water level lowering through 1990 due to water supply pumpage. As with any conceptual model one must consider the regional and local geologic and hydrogeologic conditions and express these relationships in a easily understood pictorial representation.

Conceptual models are, however, more that simple illustrations of the lithology. Conceptual models are constructed from detailed knowledge of the geologic and hydrogeologic conditions present on site. The following section provides examples of conceptual geologic models constructed from an understanding of the processes that formed the geologic units present on site. Each conceptual model is first introduced by a short description of the geologic/hydrogeologic conditions. Then the geologic conceptual model is described and a review of the evaluation of the model relative to ground-water conditions at site.

8.4 EXAMPLES OF GEOLOGIC CONCEPTUAL MODELS

8.4.1 Glacial Geologic Model

Site A is located in pre-Illinoian glacial drift and alluvium, and includes recent alluvial and aeolian deposits. Figure 8-15 shows a fence diagram of the site area Shallow pre-Illinoian aquifer materials consist of the well-sorted sand of the Sankoty Member of the Banner Formation. Recent alluvial and aeolian deposits include the Cahokia alluvium and Parkland Sands. The Cahokia alluvium consists of sands, silts, and gravels deposited by recent rivers and streams. Regionally, the Cahokia alluvium is present as recent sediment deposited in the Illinois River Basin. The thickness of these deposits in the Illinois River drainageways is unknown, but the deposits commonly are less than 50 feet thick (Willman, et.al., 1975). The Parkland Sands consist of wind-blown sands

also deposited along the Illinois River Basin. The thickness of these deposits along the Illinois River Basin. The thickness of these deposits along the Illinois River drainageway is unknown, but the deposits have been noted to average 20 to 40 feet thick (Willard, et. al., 1975). The Parkland Sands is noted here as a potential shallow aquifer

ALTERNATIVE CONCEPTUAL MODELS OF SAME SITE

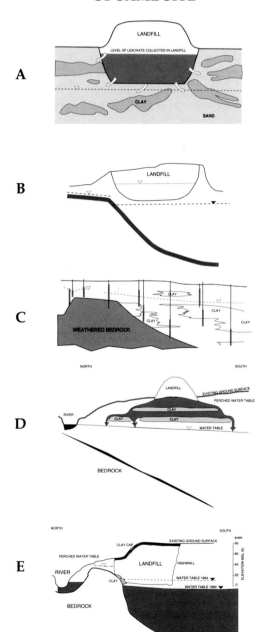

Figure 8-14a- d Conceptual "Cartoons" of Geology

material because of its permeable character, but may not be an actual water-bearing unit because of its physical elevation at or above existing water table conditions.

In the site area, the thicknesses of the glacial aquifer materials overlying bedrock are estimated to be 100 to 150 feet. The elevation of ground water at the site ranges from approximately 508 ft. msl at the eastern portion of the site to approximately 500 ft. msl at the western edge of the area. Therefore, ground water has been established to flow east to west across the site. Regionally, recharge to the Sankoty is expected to occur from upland areas east of the area. Local flow is expected to discharge to creeks approximately one-half mile to the West of the site. Intermediate components of flow may discharge to the Illinois River four to five miles to the West of the site where the stage of the river is approximately 440 ft. msl.

Ground water that occurs in Illinois and Wisconsin-aged glacial tills in the immediate area of the facility exists under perched conditions. Where glacial and post glacial erosion has dissected outwash units of these glacial tills, ground water within these units discharges to the surface as seeps along the dissected drainageways. Movement of residual water within these units, however, is also dependent upon the hydraulic conductivity of the outwash units. In areas of eastern and northeastern of the area where Wisconsin and Illinoian-aged tills have not been dissected by glacial and/or port-glacial erosion, water bearing outwash units of the till materials may be usable aquifers.

Bedrock

Pennsylvanian-aged sandstone and limestone are possible bedrock aquifer materials. Bedrock in the area is at an elevation of approximately 400 ft. msl. None of the water well records show completion within the Pennsylvanian-aged bedrock.

Conceptual Hydrogeologic Model

The geology of the site can be described as a layered system consisting of variable thickness of Wisconsin-aged loess overlying Illinoian-stage glacial tills and outwash units; which in turn overlie pre-Illinoian-aged till of predominantly outwash deposits. Subsequent to deposition, the glacial materials have been dissected by the processes of glacial and post glacial erosion creating the glacial plateau on which the facility is currently situated. Perched ground-water conditions have been found in outwash deposits of the Illinoian-aged glacial tills. The permanent ground-water surface is found within the Sankoty Sand Aquifer.

A conceptual hydrogeologic model for the site is presented on Figure 8-16. The following sections include discussion for the hydrogeologic system relating to historic development of the geology and hydrogeology to the present.

Evaluation of Geology and Hydrogeology

Deposition of glacial till and outwash units appear to have been in a layer cake fashion with younger units being deposited conformably on top of each older unit (Figure 8-15). Exceptions to this appear near edges of the existing bluff slopes where the glacial units appear to have been deposited on sloping surfaces. The phenomenon is best defined near the southwest corner of the facility on Figure 8-15 where the current topography appears to mimic the paleogeography. Geologically, glacial till units of the Radnor, and Vandalia Tills appear to have been deposited from advancing ice fronts. Silt, sand, and sand and gravel deposits within specific till units may have also been deposited from advancing ice fronts.

Outwash deposits, such as the unnamed sand and gravel, the sand unit above the Vandalia Till, and the Sankoty Sand appear to have been deposited by water processes. These three deposits appear to be laterally continuous from borehole information for the site. Silt deposits of the Wisconsin stage and the unnamed silt appears to have been deposited by aeolian or wind processes.

The present day geomorphology of the site is predominantly a result of glacial meltwaters dissecting the unconsolidated glacial materials of the Wisconsin, Illinoian, and pre-Illinoian stages of glaciation along the southern extent of the glacial bluff. In essence, the glacial and post glacial activities of the current creek drainage basin have removed the Wisconsin-stage loess, and the till and outwash units of the Ilinoian stage to expose the underlying Sankoty Sand Aquifer. The creek drainage basin appears to have been one of a number of glacial tributaries which eventually discharge to the major river system.

Subsequent to formation of the glacial bluff areas and subsidence of glacial meltwaters, it would appear that water within outwash units of unnamed sand and gravel, and the outwash unit overlying the Vandalia Till would have been allowed to freely drain along the dissected bluff areas. Figure 8-15 presents potential drainage pathways of the outwash units will be dependent upon the geometry and slope of the unit and whether or not the unit has been dissected by former erosion activities. Referencing Figure 8-16, areas of perched water or potentially perched waters occur in two different areas: 1) in silts, sands or sand and

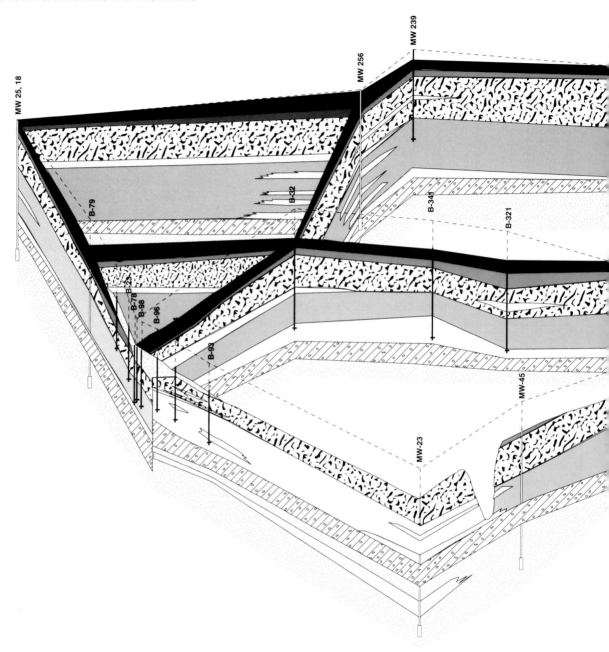

Figure 8-15a Fence Diagram Example

Source: Modified from Hydro-Search, Inc

gravel deposits included within major glacial till units, or 2) in more laterally contiguous outwash units overlying less permeable glacial till units. As previously noted, and as depicted in Figure 8-16, the presence or absence and flow direction of water within the units are a function of the slope and geometry of the confining till unit and whether or not the outwash unit is exposed along glacial bluff areas.

Recharge of perched water to the outwash units may occur in several ways. The primary source of recharge is anticipated to be the release of connate water from overlying glacial till units. Another source of perched water may result from infiltration due to precipitation. This type of recharge is greatest where perched outwash units are exposed to a continuous source of water, such as a stream.

Water table conditions occur within the Sankoty Sand Aquifer below the landfill site. Referencing Figure 8-15a, and 8-15b, the Sankoty Sand Aquifer is along the creek drainage basin where the Sankoty Sand is exposed near the

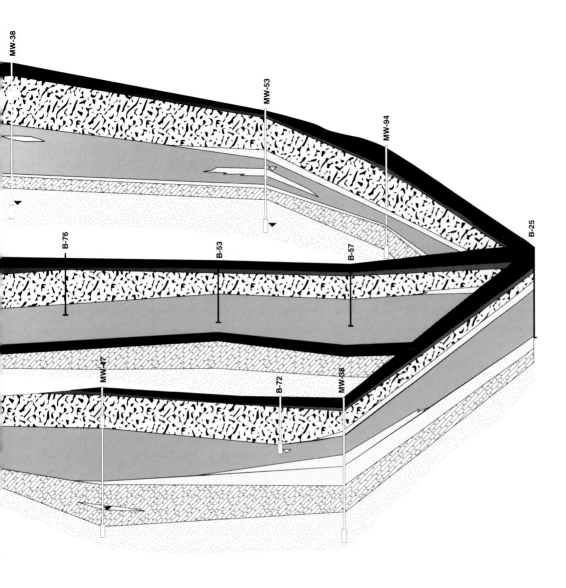

Figure 8-15b Fence Diagram Example

Source: Modified from Hydro-Search, Inc

surface. Primary recharge to the Sankoty is relatively near the surface. Secondary recharge to the Sankoty may occur where overland runoff from upland areas may be discharged to lowland areas where the release of perched water from overlying glacial tills. In this case local components of ground-water flow are expected to discharge to the creek. Intermediate and regional component ground-water flow is from east to west near the site.

Important points:—The description of the regional and site geologic and hydrogeologic conditions provide the basic data for making decisions on ground-water monitoring of the facility. The descriptive hydrogeologic model offers the kind of information that points toward target monitoring zones. This data that includes recharge areas and discharge areas for both perched and deeper ground-water level aquifers. Even though this is an extremely complete description of the local hydrogeologic conditions, additional evaluation of the need to monitor

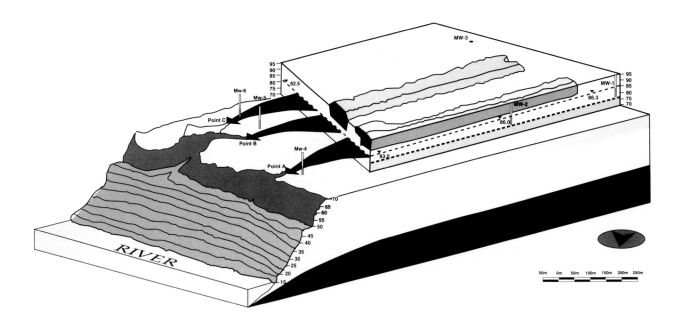

Figure 8-16 Conceptual Model Example

Source: Modified from ADS

perched ground-water systems (as the upper most aquifer) must be included in the design approach. One can appreciate through inspection of the conceptual model that this is geologically a complex site. However relatively simple decisions can be made on the background (or upgradient) locations of monitoring wells in the continuous Sankoty aquifer. Down gradient monitoring wells would be indicated in the Sankoty aquifer on the western side of the site. Evaluations as to screening depth would be necessary, beyond the data presented above, to further target the monitoring wells.

Figure 8-16 provides an example of a conceptual model of a sandy aquifer overlying a fine grained clay unit. Discharge from the facility flows toward the surface discharge points in the small streams. The use of such visual representations can greatly aid in understanding how surface and ground water systems interact.

8.4.2 Unconsolidated Deposits Conceptual Model

The purpose of the site B investigation was to review the regional geology to confirm the subsurface conditions, and to provide an interpretation of the hydrogeological conditions particularly as they affect the design of the facility.

The geological fence diagram shown in Figures 8-17a, and 8-17b confirmed that the site lies within the Prairie Formation (Late Quaternary). As such, the property generally is underlain by fine grained cohesive sediments laid down during the Mississippi River delta building process. In addition, the site is traversed by three channel sands that have been termed "Lower", "Middle", and "Upper". The depth to the three sands is typically 37 ft., 20 ft., and 6 ft., respectively. The piezometric levels in the channel sands indicate that the lower and middle sands are not hydraulically connected whereas the middle and upper units probably are connected. Superimposed on this pattern is a system of random local sand "stringers" generally located less than 20 ft. deep.

The site is underlain by thick beds of competent, plastic and relatively impervious subsoil. These fine grained cohesive materials are incised by local channel sands and sand stringers. The cohesive units display a broad range of plasticity and can be classified on that basis into organic clay, clay, silty clay and clayey silt. In general, the thickness of the individual beds for the latter three units varies from 10 to 15 ft. Ground surface is generally underlain by a silty clay unit that overlies a clay stratum. The organic clay constitutes a marker horizon and is located at depth. Clayey silt and silt deposits, where observed, generally flank the channel sands and represent overbank deposits of the channel units.

Regarding ground-water conditions, the potentiometric surface is typically 6 to 8 feet below ground surface with seasonal fluctuations of a few feet. The flow direction across the site is from northeast to southwest. The ground water level in the Middle Channel Sand decreases across the site at a gradient of about 0.0012. The flow velocity within this unit is estimated to range from 4 to 19 ft/year.

Interpreted Site Geology

The site location, based on the Phase I literature review, is landward of the estimated position of sea level during the Late Quaternary in which the Prairie Formation was being deposited. The sediments observed on site are believed predominantly non-marine to marginally marine and exhibit characteristics associated with deltaic aggradation (delta building) processes and not coastal or offshore marine processes. Typical features of this type of system initially consist of braided streams becoming meandering streams as the river gradient decreases. As different deltas were built into the Gulf of Mexico, broad meander belts developed. Also associated with this environment are organic-rich back-swamp deposits that lie adjacent to meander belts and natural levees.

Phase II site investigation soil borings and cone penetrometer testing to validate the model described above. There have been three significant sands observed at different elevations across the site. An organic clay (OH), probably indicative of a back-swamp deposit, can be found with reasonable consistency across the site at about elevations of 11 to 12 ft. This bed serves as an excellent time-stratigraphic marker. In addition, most borings adjacent to the three sands show a sand-silt mixture indicative of natural levee or near-river overbank deposits.

The three significant channel sands found on the site occur at different elevation with respect to their organic clay marker bed. These sands are designated the Lower, Middle, and Upper channel sands, in reference to their relative position in the stratigraphic column. The Lower Channel sand unit was found in the northern part of the site at about MSL elevation -10 to 3 ft. This unit trends generally east-west, apparently meandering across the northern edge of the site. The largest of the three sands, the Middle Channel sand, can be found in the central part of the site at about MCL elevation 7 to 22 ft. The sand trends east-west and apparently does not meander significantly within the borders of the site. It has incised into the OH bed, removing it altogether. The upper channel sand is found in the southwest corner of the site at about MCL elevation 20-34 ft. Little can be inferred about the trend or shape of this sand because of its relatively small area of intersection with the site borders.

Several other significant sands have been found at different elevation across the site. These are probably indicative of smaller tributary channels that are often quite numerous in the dendritic drainage pattern commonly associated with deltaic deposits.

The relationship of the Lower, Middle, and Upper channels to each other is not clearly understood at this time. A plausible explanation of their formation is that they represent three meander scars of the same river. However, ground water levels indicate that the hydrostatic head of the Lower and Middle channel sands differ by about 3 to 4 feet, the Middle sand having the higher hydrostatic head. If these sands were a result of a continuous meander pattern, one would expect that there would be hydraulic connection between the sands and the heads would be similar. The hydraulic difference may indicate that one or more of the meander channels have been "cut off", resulting in the formation of oxbow lakes.

The discussion of the geologic conceptual model provides lithologic and sedimentary relationships of unconsolidated Quaternary sediments. These silty to clayey deposits can be extremely difficult to evaluate without detailed continuous field logging of boreholes. The fence diagram Figures 8-17a and 8-17b is the result of data from over fifty boreholes. Additional borehole information would probably not have made any difference in the interpretation or in the conceptual model. In fact, the conceptual geologic model could have been adequately constructed with Phase I regional information and one half to one forth of the drilling completed on site. Once the basic conceptual model is established, (in this case deltaic non-marine with sand channel stream deposits), additional borehole examination of the subsurface should be used to validate the model. Excessive over drilling of a site to gather redundant subsurface data for waste disposal investigations can significantly cross connect hydrostratigraphic units. Although grouting-off of boreholes is commonly required, one cannot guarantee that there will be complete sealing on the borings. Once these penetrations are made through confining units, vertical hydraulic conductivity can be significantly affected by excessive over-drilling of a site to demonstrate the existence of a continuous confining unit.

8.4.3 Consolidated Bedrock Conceptual Model

Site C conceptualization began with a review of the geology of the site, including the major stratigraphic units, and must include ground-water occurrence and potentiometric heads. This information is then combined to develop a conceptual model depicting the general hydrogeology of the site. Such a model may assist the understanding of site conditions by providing a framework for placing the site into general hydrogeologic perspective.

The following represents a conceptual geologic model of a relatively simple layered consolidated rock site where fracturing does not greatly affect the interpretation of the

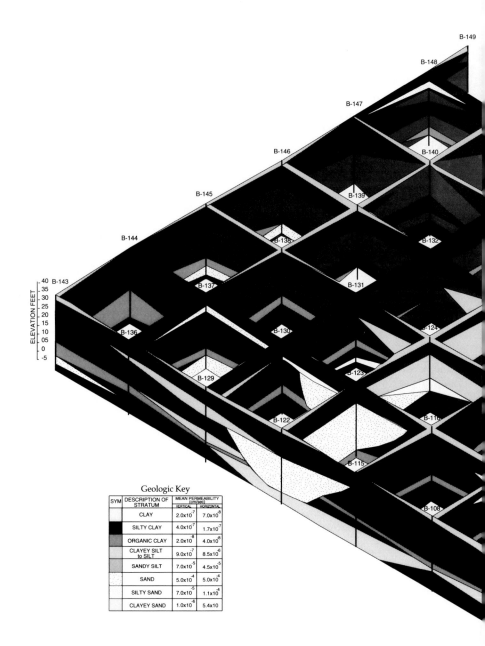

Geologic Key

SYM	DESCRIPTION OF STRATUM	MEAN PERMEABILITY (cm/sec)	
		VERTICAL	HORIZONTAL
	CLAY	2.0×10^{-7}	7.0×10^{-8}
	SILTY CLAY	4.0×10^{-7}	1.7×10^{-7}
	ORGANIC CLAY	2.0×10^{-8}	4.0×10^{-8}
	CLAYEY SILT to SILT	9.0×10^{-7}	8.5×10^{-6}
	SANDY SILT	7.0×10^{-5}	4.5×10^{-5}
	SAND	5.0×10^{-4}	5.0×10^{-4}
	SILTY SAND	7.0×10^{-5}	1.1×10^{-4}
	CLAYEY SAND	1.0×10^{-6}	5.4×10^{-6}

Figure 8-17a Fence Diagram Example B

Source: Modified from Soil Testing Engineers, Inc.

geologic model.

Figure 8-18 is an idealized hydrogeologic profile (i.e., a conceptual model) for the site. The above described hydrostratigraphic units are fundamentally a reflection of Appalachian plateau geology with alternating lithologies of varying hydraulic conductivity, resulting in multiple zones of saturation. Vertical and horizontal lithologic variations (or facies changes) make it difficult to correlate lithologic units between borings. The basis for identification of the hydrostratigraphic units described above is the general lithologic characteristics from borehole samples, albeit with variations, as well as, the tendency for piezometric heads to occur, or be missing from discreet elevation zones. Figure 8-18 also shows an inset enlargement of local conditions along the site margin where mine spoil has been cast beyond the limits of the 'No. 5 Clay'. In these

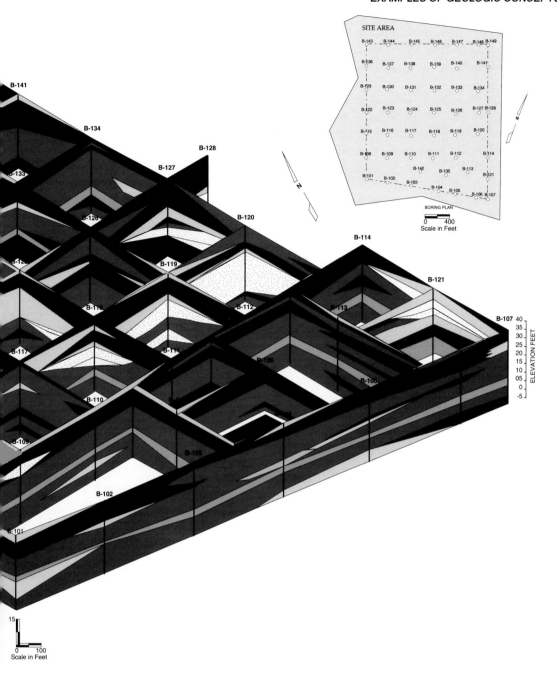

Figure 8-17b Fence Diagram Example B

areas, test pits and borings show either an absence of the 'No. 5 Clay' or its occurrence at elevation below the normal or expectable top surface elevation. The basic hydrostratigraphic units are indicated in descending order as follows:

• Mine Spoil; the mine spoil is readily distinguished by its physical appearance in borehole samples, test pits, and outcrops. The mine spoil has a thickness which varies

from 15 to 48 feet, and a hydraulic conductivity of 10^{-5} to 10^{-6} centimeters per second. This unit is predominantly unsaturated except for a thin, perched ground-water zone at its base. The predominant flow direction appears to be horizontal, as indicated by plan view head relations, seeps and springs along lateral margins, and a dry, crumbly texture for the bulk of the underlying 'No. 5 Clay'. Piezometric heads in the mine spoil and deeper saturated zones indicate a potential for downward flow, although whether or not such flow exists has not been established. If downward flow does occur, it most likely would be in the form of quantitatively small amounts of leakage.

- 'No, 5 Clay'; this unit represents the regionally described No. 5 Clay and the underlying weathered zone. Because these materials are difficult to distinguish by their appearance, they are combined (Note: conceptual models often combine similar materials). The contact between the true clay and the weathered bedrock is gradational. The 'No. 5 Clay' was identified in all borings and test pits except locally along the site margins where it was removed by erosion or mining. The thickness of the 'No. 5 Clay' generally ranges from about ten feet in the central portions of the site to in excess of 30 feet along the margins. As already noted, the "No. 5 Clay' is generally unsaturated, except for the upper few feet. Slight dampness or moisture was also found along small scale heterogeneity (stratification or textural/structural variations) in core samples.

- Shale/Mudstone Bedrock; the 'No. 5 Clay' grades downward into consolidated bedrock. Core samples show an upper zone that is predominantly shale or mudstone, but with interbedded layers or lenses of sandstone. This unit includes a saturated zone (or multiple zones) with head elevations in wells screened in this unit of about 1165 to 1175 feet (MSL). Lateral head differences in the few wells screened in this unit suggest southwesterly flow. Vertical head differences between this unit and deeper strata indicate a potential for limited downward flow.

- Unsaturated Sandstone; beneath the shale/mudstone strata is a sandstone unit, the upper portion of which trends to be unsaturated. Three wells screened in this unit indicate that the unsaturated zone extends over an elevation range of 1148 to 1172 feet.

- Saturated Sandstone; the lower most hydrostratigraphic unit is a saturated sandstone. Wells screened in this unit

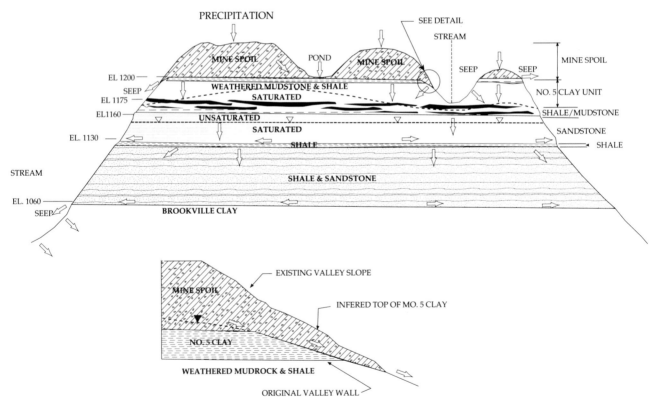

Source: Modified from Wehran Eng.

Figure 8-18 Conceptual Model of Claystone/Sandstone Site - Example C

have head elevations — on the order of 1135 to 1145 feet. Lateral head differences suggest a flow to the Southeast, although the degree of data control points is limited in this unit.

In summary, the conceptual model shown in Figure 8-18 and described above depicts the general conditions bearing on site suitability for landfilling. The mine spoil is relatively easily identified and is everywhere laterally continuous, extending out to the site margins. The 'No. 5 Clay' is likewise relatively easily identified and is similarly laterally continuous, except for local gaps along the site margin. Because this investigation focused on the saturated, more permeable zones, there is relatively little information on the permeability of this unit and any hydraulic heads. The top of the 'No. 5 Clay' occurs characteristically at an elevation of approximately 1200 feet (MCL). The bottom surface of the unit is variable in elevation due to greater thicknesses along the lateral margins. The bedrock stratigraphy is primarily established on the basis of general lithologies and especially elevation differences in piezometric heads for wells screened in the different units. Although leakage through the 'No. 5 Clay' has not been quantified, the upper saturated bedrock zone in the shale mudstone potentially represents the uppermost aquifer beneath the facility.

The above geologic conceptual model agrees with a general description of Appalachian claystone and siltstone. The vertical hydraulic conductivity is extremely limited in this environment, so long as, secondary fracture systems do not breach the shale/mudstone units. Indication of the limited vertical flows are the spring systems occurring at major lithologic boundaries such as the Number 5 clay. Field observation of the integrity of the clay or mudstone units should always form part of the Phase II geologic mapping tasks. Visual examination of fractures and lithological continuity must verify the presence of the confining units and provide reasonable assurance that secondary porosity does not provide vertical pathways for ground-water movements.

8.4.4 Consolidated Channel Deposits Conceptual Model

Site D is underlain by a sedimentary rock sequence more than 13,000 feet thick that exists within the Denver Basin. This structure, a north-south trending asymmetrical basin with a gently dipping east flank, was formed during late Cretaceous and early Tertiary time. During its formation, the basin was the site of fluvial deposition of sediments eroded from the mountains to the West.

The upper six formations beneath the site (with a combined thickness of over 8,000 feet) are the Dawson, Denver, Arapahoe, Laramie, Fox Hills Sandstone, and Pierre Shale. The regional dip of the sedimentary units beneath the site is towards the West at approximately 1 degree.

The two formations of primary interest to this investigation are the Dawson and Denver Formations. The lower approximately 100 feet of the Dawson Formation forms the near-surface bedrock at the site. The lithology of this formation is complex, consisting of interbedded claystone, siltstone, and lenticular sandstone. Some of these materials contain reworked sediments from the underlying Denver Formation. Tracing distinguishable lithologic units for lateral distances in excess of a few hundred feet is difficult within the Dawson.

A weathering profile has developed within the upper 25 to 50 feet of the bedrock that is identifiable by a dominance of mottling, brown and orange-brown colors, friability, and iron and manganese oxide stained fractures. In general, the depth to unweathered bedrock is greatest near the tops of ridges and is shallowest along stream drainages. The unweathered Dawson Formation tends to have a dark to blue-gray color, is without evident mottling, has an absence of fractures and oxide straining and is less friable.

The underlying Denver Formation is similar to the Dawson Formation in that it consists of interbedded claystone, siltstone, and sandstone. The main distinguishable difference between the two formations is color. The Dawson is generally gray beneath the weathered zone, whereas the Denver exhibits a darker greenish-gray color. This darker color reflects the presence of altered volcanic debris in the rock. The thickness of the Denver Formation is estimated to be about 800 feet beneath the site.

Regional Hydrogeology

The principle water-bearing formations within the Denver Basin are Dawson, Denver, Arapahoe, and Laramie-Fox Hills (the latter two are usually considered to be a single hydrostratigraphic unit). It should be noted that the USGS and Colorado Department of Natural Resources (which includes the office of the State Engineer) use the term "aquifer" synonymously with "formation" in referring to these geologic units. With the exception of the Laramie-Foxhills, the true aquifers (in a hydrogeologic sense) are individual beds or groups of beds of sandstone and conglomerate within the predominantly argillaceous sequences of sediments. Nonetheless, the term "aquifer" has been used for entire formations for regulatory (e.g., for the issuance of water rights) purposes and in much of the literature on the water resources of the Denver Basin. This conceptual description however, will use the term "aquifer" in the more limited and technically correct sense only when referring to primary water-yielding geologic units.

The water-bearing formations of primary interest are the Dawson and Denver sedimentary units. The units within the Dawson Formation that comprise the true aquifers beneath the site consist primarily of coarse-grained and poorly to moderately well consolidated sandstone and conglomerate. These represent the channels in the fluvial depositional environment that existed in this

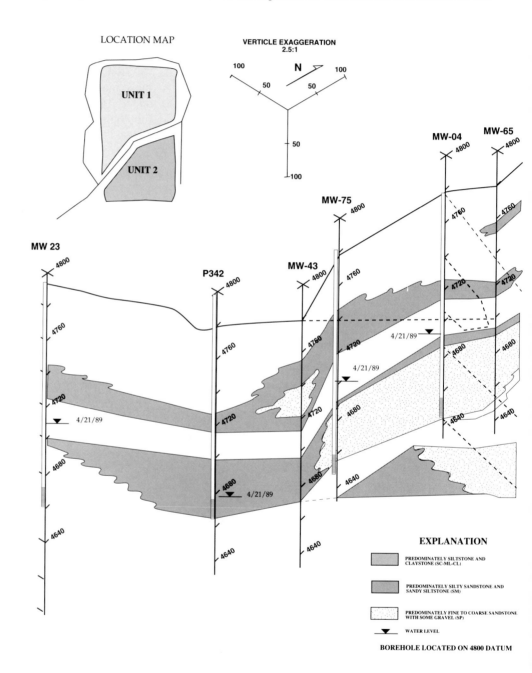

Source: Modified from Hydro-Search, Inc & Golder and Assoc. Inc.

Figure 8-19a Fence Diagram Example D

area in the late Cretaceous and early Tertiary time. Overall, the Dawson Formation on a regional basis is estimated to contain an average of 45 percent sandstone, conglomerate, and siltstone (Robson and Romero, 1981).

Site Geology

The stratigraphic conditions beneath the facility as defined by all coreholes and boreholes drilled to date, are shown in a fence diagram in figure 8-19. The complex fluvial stratigraphy has been depicted according to the

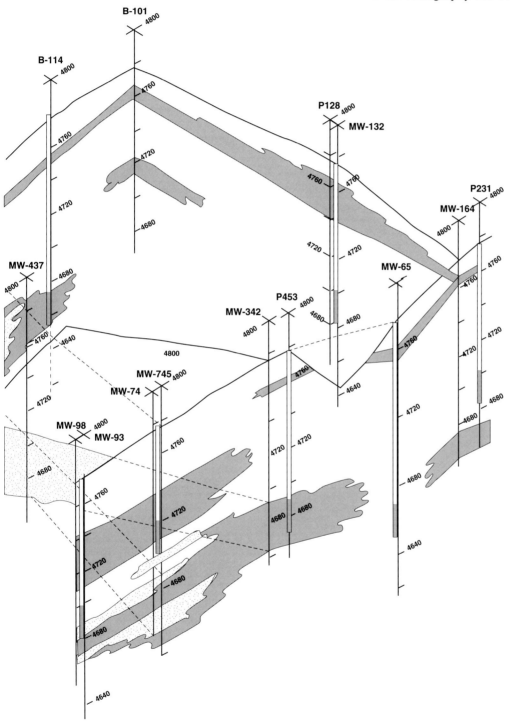

Figure 8-19b Fence Diagram Example D

predominant lithology in each portion of the sections. Therefore, a rock sequence with a generalized description of "claystone" may contain some relatively thin units of coarse-grained materials. Similarly, a sequence identified as "sandstone" may contain some claystone or siltstone layers. Interbedded sequences are shown in the cross sections as a mixture of sandstone and claystone.

As indicated in the fence diagrams, argillaceous sediments comprise the actual majority of the geologic materials beneath the facility. (Note: that regional geology may differ significantly from actual specific site conditions.) Within this fine-grained matrix, however, there are three distinct sandstone (with some gravel) units which, based on lateral variation in grain size, are most likely east-west trending buried stream channels. The uppermost sandstone occurs at a depth of 10 to 20 feet on the north side of the facility and pinches out o the south. The intermediate sandstone, which is as much as 40 to 80 feet, is primarily beneath the western side of the facility. The largest sandstone unit is the lower sandstone channel which is located at a depth between 80 to 140 feet. This gravels, representing the center of this channel deposit is the key hydrostratigraphic unit beneath the facility and should be the primary target for ground-water monitoring.

Hydrogeologic Framework

Ground-water levels have been monitored in eight piezometers and in nine wells. Together with the regional hydrogeology and the well-defined local stratigraphy, head data describe the following hydrogeologic framework (shown schematically in Figure 8-19) for the shallow ground-water system in the vicinity of the ponds.

1. The site is located within the regional recharge area of the Dawson-Denver aquifer system characterized by deeper stratigraphic units having lower potentiometric surfaces. This is evidenced by the 5690' ground-water elevation (MSL) in the lower sandstone and the 5682' ground-water elevation of the coal.

2. Based on all data collected to date, the intermediate sandstone is unsaturated. This indicates that it is above the ground-water level and that the limited vertical flux through the sandstone is not great enough to impound or perch ground water at the base of the unit.

3. The upper sandstone on the north side of the site does contain perched ground water. This is because the rate of vertical infiltration to the unit is greater than the rate at which ground water can leak out through the bottom of the sandstone.

4. It is not known whether the perched water in the upper sandstone results in a ground-water mound on the north that is connected to the lower sandstone by continuously saturated claystone or whether there is a wetting front below the perched water and a zone of unsaturated material between it and the lower sandstone. In either case, there is a hydraulic gradient from the up sandstone southward and downward into the lower sandstone. This downward infiltrating ground water will move primarily in the claystone and not laterally into the unsaturated intermediate sandstone. This is because the saturated hydraulic conductivity of the claystone most likely would be greater than the unsaturated hydraulic conductivity of the intermediate sandstone. This phenomenon, which has been well demonstrated by the laboratory models of Palquist and Johnson (1962), will probably keep the perched water to the north away from the intermediate sandstone.

5. The lower sandstone is the uppermost saturated zone beneath the ponds. Based on current water levels, this sandstone channel is an unconfined aquifer bordered by less permeable units with lower water levels. The gradient within the unit is from east to west. This suggests the channel could be cross-cutting hydrogeologic feature whose recharge zone may be along the creek at an elevation greater that 5691' MSL and whose discharge point could be along the creek at some lower elevation.

6. Although the predominant direction of ground-water flow through the unsaturated zone in this recharge area is downward, the lower full saturated sandstone (which has a significantly higher hydraulic conductivity than the surrounding claystone) acts as a "short circuit" in the flow system and permits lateral flow where sandstone's discharge to areas of lower head.

The previous subsections that describe heterogeneity in the geologic materials can have widely variable ground-water flow directions from the expected patterns. For example, the channel deposits described in the Denver-Dawson Formations can represent lateral drains to the low hydraulic conductivity claystone deposits. The localized flow would be similar in cross section as to Figures 5.8b and 5.9 (b) and (c) in Freeze and Cheery (1979, pp. 176 and 177). The discharge to the single point (the channel) would then discharge along the channel deposit to areas of lower hydraulic head.

The Denver-Dawson channel and mudstone deposits presents a series of problems for ground-water monitoring design. Locating a monitoring well in the claystone would provide a highly ineffective monitoring location. The well could take several months to recover after purging before water quality sampling efforts. Given the USEPA's general sampling guidance (SW 846) to purge and resample within twenty-four hours, the accurate evaluation of water quality from such low hydraulic conductivity environments would be highly problematical. In addition, locating specific monitoring wells in more highly permeable sand channel deposits can also be difficult due to the localized condition of these deposits. Depth-area and discharge-directional criteria for the sand channel deposits must be understood sufficiently to define background and down gradient locations for monitoring system design. Only with sufficient geologic understanding, supported by piezometer hydraulic head data and possibly geophysical data will provide sufficient information to locate ground water monitoring wells.

8.4.5 Ground-water Flow Oblique to Hydraulic Gradient

A number of the previous examples may have "complications" from the expected ground-water flow directions. These "complications" can be due to high hydraulic conductivity channel type deposits that may not be detected through the normal site investigation.

By definition, ground-water flow is perpendicular to the hydraulic gradient in aquifers that are homogeneous and isotropic. Since many aquifer conceptualizations are based on homogeneous and isotropic conditions, it has become common place to use the term "direction of ground-water flow" interchangeable with "direction of hydraulic gradient". In most cases, however, the aquifers exhibit heterogeneous and anisotropic conditions which result in ground-water flow being tangential or oblique to the prevailing hydraulic gradient. This condition must be considered when evaluating ground-water flow systems for the design of ground-water monitoring networks and/or the restoration of contaminated aquifers. In evaluating and developing conceptual models of site ground-water systems, several basic factors may be helpful in predicting the occurrence of flow oblique to gradient. Examples of the conceptualization of flow oblique to gradient are provided on Figures 8-20 and 8-21. The first factor is that the hydraulic gradient is primarily controlled by the position and geometry of the recharge and discharge areas. The flow, on the other hand, is controlled generally by geologic or subsurface features which serve to restrict, redirect, or concentrate the flow of ground water. It must be remembered that ground water, like electricity, will seek the path of least resistance (i.e., the highest hydraulic conductivity material in the ground-water flow path) in

response to a pressure differential. This phenomenon is most pronounced in geologic formations which are either fractured due to secondary weathering or orogenic processes; or in materials which, by their depositional nature, are graded and sorted.

In most environments, the hydraulic gradient of the uppermost aquifer is consistent with the gradient of the land surface, as evidenced by topographic contours. In these environments, the ground-water recharge areas generally coincide with areas of higher elevation; whereas the discharge areas coincide with the lower elevations in the watershed, generally identified by the presence of streams, rivers, and lakes. In the vertical dimension, the presence of upward or downward gradients can be predicted by identifying the position of the site within the ground-water flow system. Again, by definition, a recharge area is one in which hydraulic heads decrease with depth; a discharge area, conversely, is one in which heads increase with depth. Therefore, by determining the location of the site relative to the overall ground-water flow system, the hydrogeologist or geologist can estimate the general pattern of lateral and vertical gradients at a site.

The geologic factors that may result in a flow oblique to gradient can also be predicted by prior evaluation of the site geology and depositional history. Generally, continental depositional environments (fluvial, glacial, pluvial, colluvial, aeolian) tend to produce deposits that are graded and poorly sorted and which exhibit pronounced anisotropic conditions. These deposits are particularly suited to lithostratigraphic mapping techniques that can link depositional models to the observed anisotropic geology. These anisotropic conditions generally persist through the processes of rock consolidation and solidification. Therefore, by evaluating the depositional history of a site and the orientation of the geologic processes that resulted in the deposition, the geologist or hydrogeologist can often predict the general direction of anisotropy within a given deposit. In consolidated rock environments, the initial fabric or structure of the deposit may be altered by secondary physical and chemical processes. These processes result in apertures or openings in the rock matrix which may provide preferential pathways for ground-water migration. Prediction of the direction of these apertures or fractures can be extremely difficult, particularly in geologic units which have a complex orogenic history. In rock formations which have only been deformed by a single event, the fractures are generally most pronounced along the strike of the rock. Secondary fractures are also common at 60% angles from the strike. In formations that have undergone multiple orogenic events and are intensely fractured and folded, prediction of anisotropic orientation is generally not possible without extensive background drilling and testing data.

In developing the conceptual model at the end of Phase I, one should expect a reasonable prediction of the occurrence of flow oblique to gradient and, if possible, the direction of that flow. Hydrogeologic reports, should allow one to distinguish between upgradient/downgradient and upflow/downflow. It is the author's position that ground-water monitoring wells should be located in accordance with the prevailing direction of ground-water flow and not the prevailing direction of hydraulic gradient, particularly in those conceptual cases where flow can be oblique to gradient.

Confirmation of anisotropic and heterogeneous conditions can best be obtained through detailed analysis of borehole data and reconstruction of the depositional environment. Geophysical and piezometric data can often confirm or supplement the stratigraphic interpretations. However, qualification of anisotropic conditions can generally only be accomplished through intensive pump testing programs. Such programs are generally undertaken only when aquifer restoration is being considered as a remedial action. In fractured rock environments, the installation of angle boring and the acquisition of oriented core is necessary to determine fracture frequency and orientation. These data provides the geologic insight into the formulation of a conceptual model that addresses the

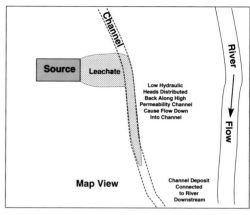

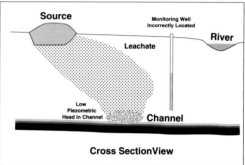

Figure 8-21 Conceptual Oblique Flow Conditions

heterogeneity and resultant ground-water flow directions.

8.4.6 Unconsolidated Deposits With Oblique Flow Conceptual Model

The following section describes a geologic environment when heterogeneity affect the general ground-water flow directions. These highly varied geologic units are described to illustrate the field conditions that lead to an unexpected ground-water flow direction.

Glacial deposits can range in hydraulic conductivity from greater than 10^2 (gravels) to $< 10^{-7}$ cm/sec. Such a great range of hydraulic conductivity offers potential for ground water flow oblique to the expected ground-water direction. The geologic history of the glacial deposits is first discussed, then the evaluation of ground-water flow.

Approximately 1.5 to 2 million years ago, thick ice sheets developed in northeast Canada and expanded into the United States. Four primary advances and recessions have been established from geologic mapping of deposits along the edge of the glacial advances. During the first three advances, the Nebraskan, Kansan, and Illinoian, hundreds of feet of sediments were stripped from the landscapes of Canada, Wisconsin, Minnesota and the

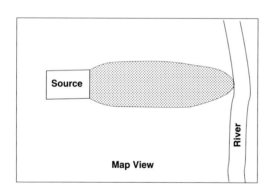

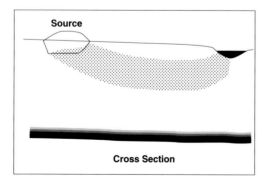

Figure 8-20 Traditional View of Discharge

Dakotas and redeposited in broad till plains stretching from Indiana to Colorado.

Beginning about 75,000 years ago, after a prolonged warm period, the Laurantide Ice Sheet expanded again from its center west of Hudson Bay and advanced southward across Minnesota. This signified the beginning of the last major glacial advance in Wisconsin. In Minnesota, the Wisconsin glaciation is marked by six major advance and recession cycles with the major landforms and deposits being attributable primarily to the last two of these cycles. Deposits from the three earlier glaciations were removed, buried, or reworked by the Wisconsin glaciation.

Approximately 30-35,000 years ago, renewed growth of the Rainey-Superior Lobes sent a thick sheet of ice from northeastern Minnesota and the Lake Superior Basin into central Minnesota to a point 35-40 miles southwest of the site. As this advance stalled, the prominent St. Croix moraine was formed at its margin. As the ice sheet melted, high velocity streams at the bottom of the ice cut tunnel valleys into the frozen, unconsolidated sediments beneath the glacier. The trace of these valleys can often be interpreted by mapping lineations on aerial photographs and topographic maps (Wright, 1973). These valleys are commonly associated with linear strings of lakes, wetland depressions, and eskers, the latter of which commonly have crests at the level of the adjacent till plain. The presence of these eskers, commonly bounded by marginal troughs as depicted on Figure 8-22 indicate that the subglacial streams responsible for the initial cutting of the tunnel valleys eventually lost energy and changed from erosional to depositional environments.

As the Rainey-Superior Lobe melted, a blanket of reddish, poorly graded material, the Superior Till was left as ground moraine over the entire area of inundation. Further melting as the lobe retreated caused an outwash plain of sands and gravels to form at the ice margin.

Before all of the ice remnants of the Rainey-Superior Lobe melted, the Grantsburg sublobe blocked southward drainage from the Rainey-Superior Lobe and Glacial Lake Grantsburg was formed. As the Grantsburg ice sheet melted a second till was deposited over the area. This till is characteristically a gray, fine grained calcareous material and is generally visually distinct from the red Superior Till appears to have been chemically reduced upon burial and has been converted to a dark gray color. In these locations the Grantsburg and Superior tills can be distinguished only by texture and gradation.

The Glacial Mississippi River maintained its position along the front margin of the Grantsburg Sublobe

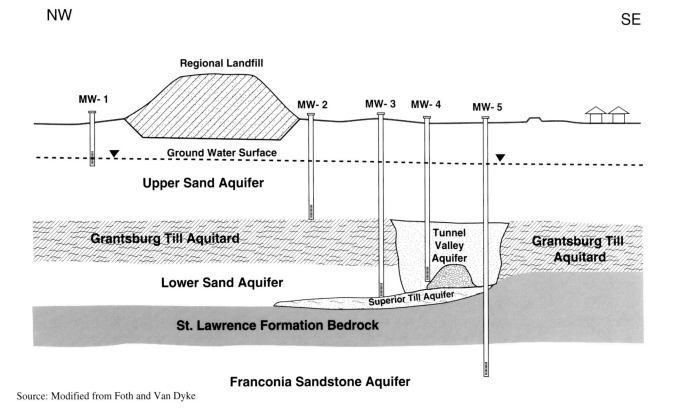

Figure 8-22 Example of Oblique Flow Conditions in Granular Environments

throughout its melting and retreat. As the river migrated southwestward to its present-day course it deposited a thick mantle of outwash sand known as the Sand Plain. This unit forms what is referred to as the Upper Sand unit in Figure 8-22. Occurrence of the Tunnel Valley as mentioned in the previous section, Wright (1973) traced several tunnel valleys in east central Minnesota including one immediately to the East and southeast of the site. Test drilling and photo interpretations have generally defined the presence and geometry of this valley. The Phase II drilling also established the presence within the valley of an esker that has been buried under the Anoka Sand Plain. The surface expression of the tunnel valley and esker is shown by the location and orientation of the lakes and surface depressions near the site. These depressions are the result of the melting of remnant ice blocks that were buried during the retreat of the Rainey-Superior Lobe.

In spite of the uncertainty regarding the exact time of origin, the presence of a tunnel valley, bounded vertically by till plains, partially filled with as esker, and covered by the Sand Plain, is an accurate conceptual model. The esker is believed to be continuous from north of the East Lake to where the esker and tunnel valley would likely be truncated by the Mississippi River valley train deposits.

Figure 8–22 illustrates the separation of permeable aquifers and much less permeable aquitard/confining units. The range in hydraulic conductivity for this conceptual model was over seven orders of magnitude; hence, a clear understanding of the extent of the aquifer/confining units was important before analysis of potential flow directions. In this case, the tunnel valley aquifer, with hydraulic conductivity of 10^{-1} cm/sec, provided a preferred pathway for ground–water movement from the site. Subsequent to the conceptual model formulation, it was believed that ground water moved directly in a west to east direction. These directional flow estimates were based on simple ground-water contour maps developed from numerous monitoring wells. The highly permeable tunnel valley deposit, however, represented a ground-water flow pathway that is discharging to areas of lower hydraulic heads. The actual ground-water flow from the facility was some 90 degrees different from what would be expected from interpretation of the ground-water contour map of the site area. Difficulties arise in the flattening of the contours within the high permeability feature. Placement of piezometers within the permeability feature would have shown the very low head levels in the tunnel gravels, however, difficulties in locating these features result in few actual measurement points being placed in these deposits.

Figure 8-23 shows an example of two conceptual

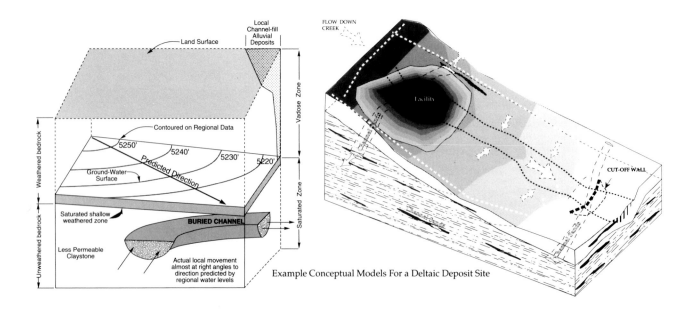

Example Conceptual Models For a Deltaic Deposit Site

Figure 8-23 Conceptual Models of Granular Geologic Conditions With Oblique Flow

models of the same deltaic deposits. These units are primarily claystone with channel sandstone deposits. These two conceptual models also illustrate the fence diagram Figure 8-18a-b, conceptually.

8.5 THE FLOW NET

Construction of flow nets offers a direct method for defining the most likely direction of ground–water movement. Flow nets are the basic tool used in site assessment and monitoring well system design to illustrate regional and local flow patterns that may require monitoring or later aquifer remediation. A flow net is a two–dimensional model of the ground–water system that identifies ground–water flow directions and head levels. These are a means of portraying a graphic solution to the Laplace equation that governs steady state flow. Flow nets can also be conveniently used to identify suitable locations for monitoring wells, as well as, screened intervals of the wells.

8.5.1 Introduction

Flow net analysis is a graphical method of solution for several different kinds of ground-water flow and seepage problems. This text will concentrate on the use of these flow net constructions for selection of the target monitoring zones for assessing the performance of land disposal facilities.

Flow nets are particularly useful because they illustrate the principles of fluid motion and the influence of various types of boundary conditions on flow patterns. They also provide a means, which does not require specialized mathematical procedures, for obtaining approximate solutions to many flow problems. For one variation of this method, the hydraulic head distribution in the geologic media is unknown in cross section and is to be determined to solve for ground-water flow paths. To determine the solution, a family of curves parallel to the flow direction called flow lines are drawn orthogonally to another family of curves called equipotential lines. Each flow line represents the path a fluid particle would follow through a porous medium, and successive flow lines are constructed so that an equal quantity of discharge is contained between each flow path. The first set is referred to as equipotential lines, which are locations of equal hydraulic head. Equipotentials represent the height of the potentiometric surface of a confined aquifer or the piezometric height of the unconfined water above a reference datum plane; or, alternatively, a discharge elevation. Since ground water moves in the direction of the highest hydraulic gradient,

the resultant flow line in isotropic geologic systems is perpendicular to equipotential lines. Hence, flow lines cross equipotential lines at right angles.

One general limitation on this method lies in the amount of time needed to obtain a solution because these flow net constructions require trial and error sketching of flow lines and equipotential lines. This limitation becomes especially severe for free-surface problems having complicated boundaries and involving more than one hydraulic conductivity. Flow lines represent an additional conceptualization of ground-water movement in a geologic system; as such, only a relatively few of these sets must be drawn during the analysis. A reasonable level of effort in many geologic environments can achieve sufficient conceptual understanding of ground-water flow to select relatively accurate monitoring well locations from these flow net constructions. This method can also be useful for calculation of flow rates and quantities for selected geologic sections

Another variant of the method uses a contour map of either a potentiometric surface or ground-water surface that has been constructed from known water-level data. Pairs of flow lines are drawn orthogonally to the contour lines only at selected points. Then discharge differences can be computed between several cross sections, each bounded by the pair of flow lines and oriented parallel to the contours. Differences between the discharges at two or more of the cross sections can be related to withdrawals by wells, recharge, changes in ground-water storage, or other phenomena. Knowing all but one of these variables, the remaining one can be computed. If all of these variables are known, it is also possible to use the discharge difference to estimate transmissivity.

Descriptive models of regional and local steady state ground–water flow in an unconfined aquifer was first presented by Hubbert (1940). He demonstrated that the hydraulic head at a point in a potential field represents the elevation to which water will rise in a piezometer that is open only at that point. At the point where an equipotential line intersects the potentiometric surface, water in the piezometer will rise to the ground–water surface. Elsewhere, water in a piezometer intersecting the equipotential line may be above or below the ground–water surface, depending upon the relative hydraulic potential (see figure 8-24).

Hubbert's model was for an unconfined aquifer of great depths. Additional models were later presented by Toth (1962 and 1963), and Freeze and Witherspoon (1967). These flow models extended the flow path concepts to more complex hydrologic systems and allowed direct analysis of small basins to define areas of recharge and discharge.

A number of popular texts review the basics of flow net construction for site–specific, small–scale flow systems. Cedergren (1967), Freeze and Cherry (1979), and Todd (1980) all describe the basic criteria for generation of flow nets and provide the mathematical derivations. A more recent publication of the U.S. Environmental Protection Agency (1986) provides additional information on ground–water flow nets and flow line construction related to the time–of–travel calculations. These methods of flow net construction are part of the traditional instruction in undergraduate hydrogeology and soil engineering courses, and must be used in establishing ground–water flow directions for monitoring well system design.

Sand tank constructions have been used since the 1950's to illustrate ground-water flow. C.V. Theis, H. Skibiski, and J. Lehr developed the use of sand tank models to represent a wide variety of subsurface flow and lithologic conditions. More recently these physical models were used by Swiss Researcher to represent Dense Non Aqueous Phased Liquids (DNAPL'S) flow in both vadose and fully saturated conditions. These physical models use dyed water to represent ground-water flow paths. They provide an unparalleled instructional technique to illustrate the interaction between lithology, and ground-water recharge and discharge points. The National Water Well Association markets a series of 35mm slides that show time-laps sequence photographs of various geologic flow situations.

Figure 8-25 illustrates ground-water discharge to a gaining stream. The boundary conditions of a constant head at the recharge, (ends of the model) and discharge at the stream, (in the center of the model) where water is removed and finally the no flow boundary at the bottom of the model causes the resultant flow lines as shown on the model. Ground water as shown in the dye traces moves from higher to lower hydraulic pressure. Especially important is the area of "no flow" directly below the discharge point in the sand tank model. The flowing dye lines discharge to the lowest hydraulic head levels, as shown, cannot cross over to the other side of the model, due to the higher head levels present across the stream area. The only way ground-water flow lines could cross the discharge area shown in the model, would be for a lower (than the models' constant head discharge point) discharge point to be present. Such lower head points could be due to a highly transmissive geologic unit discharging to a lower hydraulic head area. This lower head level is present within the physical model. This recharge/discharge relationship is the basis for many decisions that must be made on both detection and assessment monitoring design.

Estimating depths necessary to intercept target monitoring zones can also be illustrated by sand tank

models. Figure 8-26 shows discharge from a rectangular shaped recharge area. The flow line dye traces are discharging to the bottoms left and right in the model. One can observe that a shallow screening of a monitoring point next to the recharge area would miss deeper discharges from the facility. Sand tank models provide a graphic view of depth, location, and flow relationships.

Sand tank models can also provide insight into the localized effects of aquifer heterogeneity and ground-water flow. Figure 8-27 illustrates a downward ground-water flow through various fine and course grained layers. Although deflections, (according to the tangent law) are observed, these course sand units are not discharging in a horizontal direction. The lowest hydraulic head at the bottom of the model "attracts" the flow lines as represented by the dye traces. Questions regarding flow in minor sandy units are common during evaluation of ground water monitoring systems. If these small more permeable sandy

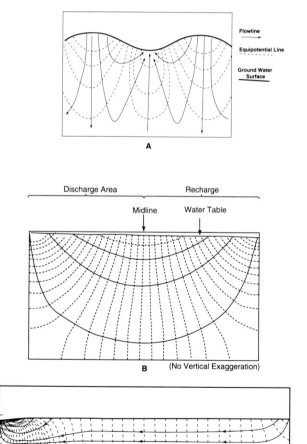

Figure 8-24 Hubert's Regional Flow Model

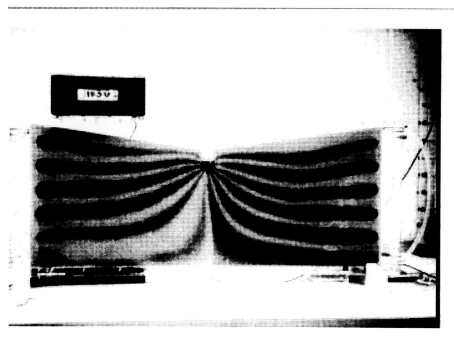

Source: J. Lehr

Figure 8-25 Sand Tank Flow Model Example

units are not connected to a discharging unit, these units will not have the low hydraulic heads necessary to "attract' the ground-water flow. Sand tank models can be extremely illustrative in representing local recharge and discharge relationships.

Regional ground-water flow in nonhomogeneous, anisotropic systems can be investigated by means of both analytical and numerical methods. Toth (1962 and 1963) used analytical solutions to evaluate models for basin wide flow patterns. Freeze and Witherspoon (1967) evaluated basins using variations of water table configurations and hydraulic conductivity variations. They developed a series of two dimensional cross sectional views. These series of illustrations, which were later reprinted in Back and

Source: J. Lehr

Figure 8-26 Reflection of Flow through Materials with Variable Hydraulic Conductivity

Freeze, Physical Hydrogeology (1985), provide excellent insight into ground-water basin flow conditions.

Only a few of these illustrations are reproduced below, however, they accurately portray many of the issues that must be addressed in development of flow net constructions. Computer based cross-sectional presentations extend the applicability of simple sand tank models to more complex geologic and hydraulic situations.

The effect of ground-water surface configuration on regional ground-water flow through homogeneous isotropic geologic units is illustrated in Figure 8-28a to 8-28c.

8.5.2 Darcy's Equation and the Flow Net

Flow nets have their basis as practical solutions to Laplace equations. As such, they require that Darcy's Law is valid and essential assumption of flow apply to the system. In order to introduce the concept of flow net constructions, the development and validity of Darcy's Law will be reviewed as fundamental to the understanding of ground-water flow. As described in Chapter 5, Darcy's Law provides the means for determining hydraulic conductivity of soils and rock both in the field and laboratory. The relationship has been defined in a series of forms, such as:

$$Q=kiA$$

$$Q=kiAt \qquad \text{Equation(s) 8-1}$$

$$q=kiA=V_dA$$

In these expressions, Q or q is the seepage quantity, t is time, k is Darcy's coefficient of permeability, i is the hydraulic gradient that is also equivalent to:

$$v = ki = q/A = V_d$$

$$V_s = ki/n_e$$

A is the total cross-sectional area normal to the direction of flow, including both void spaces and solids, V_d (or v) is the discharge velocity, and V_s is the seepage velocity. The effective porosity is defined as the ratio of the actual volume of pore spaces through which water is seeping to the total volume. Discussions as to effective porosity and velocity calculations are further developed in Chapter 5.0.

Darcy's Law has been the focus of ground-water flow analysis since the recognition of the difficulties of further development of Newton's Law of Friction together with the classical Navier-Stokes equations of hydrodynamics to describe the behavior of fluids under motion in porous media (Cedergren 1967). Darcy's development of the quantitative representation in 1859 of the simple relationships of fluids flowing in porous media provided the basic understanding that has been used in the intervening years. Muskat (1937), Taylor (1948), Leonard (1962) and Cedergren (1967) have presented discussions of permeability (or hydraulic conductivity) and Darcy's Law. Each of these references defines the valid ranges of Darcy's Law for a range of sediment sizes and gradients. The fundamental consideration of the nature of flow in porous media, as pointed out by Cedergren (1967) have led investigators to conclude that Darcy's Law of proportionality of macroscopic velocity and hydraulic gradient is an accurate representation of the "law of flow" as long as the velocity is low. Taylor (1948) has described the slow transition states from laminar flow (i.e., Darcian flow) to a slightly turbulent condition (where Darcy flow does not apply). He concludes that uniform sand with a grain size of 0.5 mm or less will always have laminar flow for gradients of 100%. Even with gradients as high as

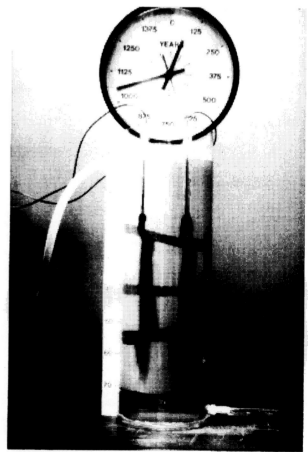

Source: J. Lehr

Figure 8-27 Downward Ground Water Flow

COMPUTER GENERATED FLOW NET CONSTRUCTIONS

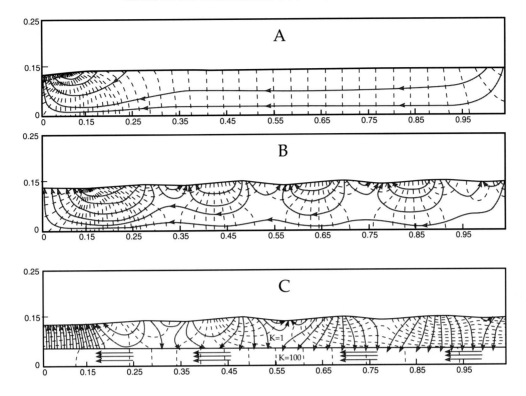

Source: Freeze and Witherspoon, 1967

Figure 8-28a-c Regional Flow Through Geologic Units

800%, the flow will remain laminar for sands with particle diameters below 0.25 mm.

Very low gradients have also been investigated by Fishel (1935) for validity to Darcy's Law. He reported that under heads as low as 2 to 3 inches to the mile in sands, Darcy's Law fully applied. Seepage solutions that require validity to Darcy's Law include:

• Flow nets for steady state seepage;
• Calculations for the velocity of water masses;
• Approximation solutions to non-steady seepage for moving saturation lines using flow nets; and,
• Seepage quantity determinations in saturated soil and rock.

The first of the above four bullets represent the focus of the practical solution of seepage through geologic cross sections of material with different hydraulic conductivity for both isotropic and anisotropic conditions. Flow nets for steady state seepage conditions provide investigators a tool for selection of target monitoring zones where seepage from the facility would move most rapidly and meet regulatory criteria for immediate detection of discharges

from the facility.

Flow Nets

Flow net construction is a graphical method for solution of forms of streamline flow that can be represented by the Laplace equation. The Laplace equation for flow of water through porous media requires a series of assumptions that the following conditions apply:

• Homogeneous soil;

• The media is fully saturated;

• Consolidation and expansion of the media is not a factor;

• The geologic media and water is incompressible; and,

• Ground-water flow is laminar (hence, Darcy's Law is valid).

Flow nets are very useful since they illustrate the principals of fluid motion and the influence of various types of boundary conditions on the fluid flow patterns. As water flows through porous geologic materials, ground water moves through all the interconnected pore spaces. The path of the water particles can be represented by individual "flow' lines. There is an infinite number of flow lines possible, however, only those needed provide sufficient details for satisfactory construction accuracy are

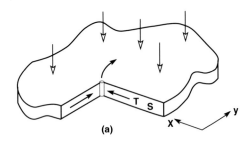

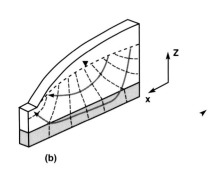

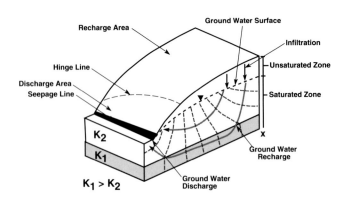

Figure 8-29 One, Two, & Three Dimensional Flow Models

Source: J. Cherry

drawn on the flow net.

Each flow line represents the path a fluid particle would follow through a porous geologic medium, and successive flow lines are constructed so that an equal quantity of discharge is contained between each flow line segment. A simple illustration of flow lines is shown on Figure 8-29 for one, two, and three dimensional models. These flow systems can be simply represented by a model containing saturated sand within a horizontal tube with two end reservoirs that illustrates the nature of flow lines. The driving force for the model consists of the differential head h between the two reservoirs. This flow from left to right is highlighted by dye streaks in the illustration. These streaks represent flow lines. All flow net constructions could be represented through similar sand tank models where dye traces could represent flow lines. However, construction of sand tank models is very tedious and flow nets can be drawn on paper that can represent the flow conditions (dye traces) in sand tank models. Flow through the model can be either horizontal as shown in Figure 8-30a or vertical as shown in 8-30b.

8.5.3 Flow Net Theory

A fundamental rule is that each flow path in a flow net must transmit the same quantity of water. Therefore, the total flow of ground water through the system; Q, is equal to the flow quantity in each flow path, q, multiplied by the total number of flow paths, F. Also, the total head loss, H, experienced in moving through one flow path of the entire flow net, is equivalent to the head loss experienced in passing through any equipotential space multiplied by the number of equipotential spaces, N. The total quantity of water that flows through a given mass of geologic material is equal to the sum of the quantities of water through each flow path in the flow net. Flow, q, through any path in a flow net is defined by Darcy's Law:

$$q = k \, i \, A \qquad\qquad \text{Equation 8-10}$$

where:

k = the coefficient of hydraulic conductivity (cm/sec or ft/day),
i = the hydraulic gradient (dimensionless), and
A = the cross sectional area through which flow occurs (Sq. meters or sq. ft.)

The value of q obtained from the Equation is for a unit width in the third dimension perpendicular to the cross section. Hence, the units of q are meters cubed/sec per meter of width. The total flow through a single flow path

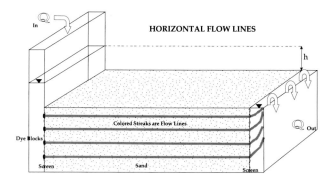

HORIZONTAL FLOW LINES

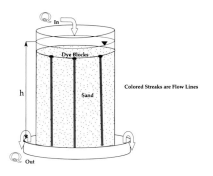

VERTICAL FLOW LINES

Figure 8-30 Flow Line Construction Example

is found by multiplying q by the width of interest. When one considers a flow net consisting of squares of dimension s x 1 with head loss, h, through a single equipotential space (U.S. EPA 1986), Darcy's Law reduces to:

$$q = khs / l \qquad \text{Equation 8-11}$$

The flow net is by definition composed of rectilinear spaces that approximate squares, hence s = 1 and:

$$q = k \cdot h \qquad \text{Equation 8-12}$$

As defined:

$$h = H/N \qquad \text{Equation 8-13}$$

so that:

$$q = H / K \qquad \text{Equation 8-14}$$

Since the flow, q, through any square is described by:

$$q = Q / F \qquad \text{Equation 8-15}$$

Where F is the total number of flow paths, the total flow, Q, can be calculated as follows:

$$Q = kH \, F / N \qquad \text{Equation 8-16}$$

Therefore, the total quantity of water that will pass through a unit width of a given subsurface geologic unit can be calculated by using a flow net for the cross section using the hydraulic conductivity multiplied by the total head difference and the ratio of the number of flow paths to the number of equipotential spaces.

Each equipotential line represents a line of equal hydraulic head, and these lines are drawn so that there is an equal head loss between each successive segment. The second type of lines used in flow net construction are the equipotential lines. These lines are shown in Figure 8-31. Each equipotential line represents a line of equal hydraulic head. These lines are drawn so that there is an equal head loss between each successive line. As represented in Figure 8-31, the gradual frictional loss of energy between the two reservoirs can be quantified through head levels in piezometers located with the sand mass. Piezometers located along an equipotential line will show equal head levels. Since the model has an impermeable top and bottom (no flow), the equipotential lines are perpendicular to the boundaries.

If the flow lines of Figure 8-30 are combined with the equipotential lines of Figure 8–31, a very simple flow net is obtained (Figure 8-32). Although this flow net is extremely simple, it possesses the properties of a true flow net. The flow lines intersect the equipotential lines at right angles, and the net is composed of squares.

Figures 8-30 and 8-31 illustrate the basic meaning of flow lines and equipotential lines, and Figure 8-32 shows a simple flow net. Most flow nets are composed of curves, and the "square' figures are not true squares but curvilinear squares. The strict requirement that a net must be composed of square figures is met only if the average width of an area is equal to its average length. As pointed out by Cedergren (1967), water will rise to the same level in piezometers installed anywhere along a given equipotential line. This is mandatory since the same energy level exists everywhere along a given equipotential line.

As with the flow lines, equipotentials are independent of the orientation of the geologic element vertical (Figure 8-32, or any other) equipotential lines can be illustrated by flow nets. Flow lines and equipotential lines are then combined to form the construction components of flow nets. All flow nets must meet a number of base

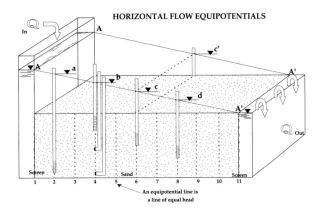

HORIZONTAL FLOW EQUIPOTENTIALS

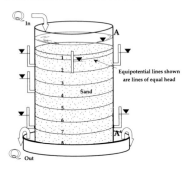

VERTICAL FLOW EQUIPOTENTIALS

Figure 8-31a, b   Horizontal and Vertical Flow Examples

requirements:

1. Flow lines and equipotentials should intersect at right angles and form geometric figures that are "squares". The requirement for squares can be met with any length to width ratio.

2. All boundary conditions must be satisfied:

   • Equipotential lines must meet impermeable boundaries at right angles; and,
   • Equipotential lines must be parallel to constant head boundaries.

3. Adjacent equipotential lines represent equal head losses. This would apply to the series of successive equipotential drops. The flow net should be drawn to contain a whole number of stream tubes. The number of equipotential drops depends on the physical shape of the cross section of interest. The last drop at each end or both ends of the net may have a fractional drop.

4. The number of stream tubes must remain the same throughout the constructions. An equivalent discharge

must flow between adjacent pairs of flow lines. As with the previous rule, a stream tube near a boundary may be fractional, but this fraction should be the same everywhere. This rule guarantees that the flow net will be composed of squares except at seepage boundaries where the net is cut-off.

5. At any point in the flow net, the spacing of adjacent lines is inversely proportional to the hydraulic gradient (i) and the seepage velocity (V_s).

The requirement for flow net constructions that represents "squares" can be easily understood for simple models presented so far, however, most flow nets will not have the impermeable boundaries represented in these models (that cause simple, parallel flow lines). Most flow consists of curved lines that are more difficult to picture as a series of "squares," but rather curvilinear squares. Casagrande (1937) provided a simple illustrative procedure (Figure 8-33) to demonstrate the actual square nature of curvilineal geometric figures.

The simple requirement that a net must be composed of square figures is met if the average width of an area equals the average length. For example, if the distance between points 9 and 10 equal the distance between points 11 and 12, then the area represents a square (a circle could be

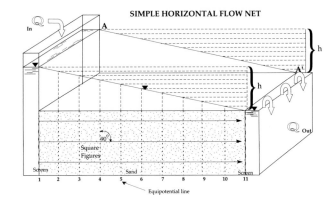

SIMPLE HORIZONTAL FLOW NET

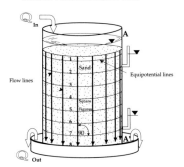

SIMPLE VERTICAL FLOW NET

Figure 8-32a, b Simple Flownet Construction

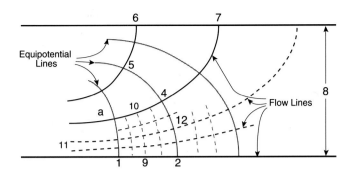

Figure 8-33 Curvilinear Net Construction

drawn within the Figure 8-33 points 1, 2, 3 and 4). This square nature of the figure becomes clear if subdividing lines (and squares) are drawn with the 1, 2, 3 and 4 points of the figure.

8.5.4 Application Of Flow Nets To Conceptual Models

A flow net is a two-dimensional model of a ground-water system that identifies ground-water flow directions and can be used to calculate ground water flow rates and quantity of flow. Flow nets can also be used to identify suitable locations for monitoring wells, as well as, the target monitoring zone, for the most appropriate depth for the screened interval of the wells. The conceptual hydrogeologic model of a site can be linked with a flow net construction. The combined model can then be tested, (similar to verifying a computer flow model), by installing additional piezometers at selected locations and comparing the actual head values at these locations with those predicted by the flow net construction.

Any ground-water system in the field can be represented by the three-dimensional set of equipotential surfaces and orthogonal flow lines. If a plan view or a two-dimensional cross section is drawn to represent this conceptual flow system, the resultant equipotential lines and flow lines constitute a flow net. A flow net can be later used to determine the distribution of heads, velocity distribution, flow paths and flow rates, and the general flow pattern in a ground water system (McWhorter and Sunada, 1977).

One should always remember that two dimensional representations of three dimensional flow only provides part of the picture. Several cross sections orthogonal to each other or true three dimensional representations of flow may be necessary to fully describe local flow conditions at site. Four basic types of ground water

systems exist (U.S. EPA, 1986) based on the distribution of hydraulic conductivity:

• homogeneous and isotropic;

• homogeneous and anisotropic;

• heterogeneous and isotropic; and,

• heterogeneous and anisotropic.

Figure 8-34 is a graphical representation of the four types of systems, where the hydraulic conductivity (horizontal and vertical) is represented in vector form and shown at two different locations within each aquifer. Although geologic (or any other) material can have variable hydraulic conductivity in any of three-dimensions, hydraulic conductivity measurement has dictated the horizontal and vertical conventions. Through laboratory measurements (vertical), and field in-situ hydraulic conductivity tests (horizontal measurements) two directional orientations are provided for flow net construction.

Geologic materials are homogeneous if the hydraulic conductivity does not vary spatially, whereas such materials are heterogeneous if hydraulic conductivity does vary spatially. If the hydraulic conductivity varies with the direction of measurement at a point (for example, when the vertical hydraulic conductivity is different from the horizontal conductivity), the geologic material is anisotropic at that measurement point.

Decisions On Use Of Flow Nets

Flow nets may not be always appropriate in every site assessment project. There are a number of geologic and hydrogeologic situations in which the accurate constructions and use of flow nets is difficult or impossible. These conditions occur when there are scaling difficulties within complex geologic settings under conditions of little three-dimensional hydrologic data for the ground-water system, and when ground-water flow conditions do not sufficiently conform to Darcy's Law. One has physical scaling difficulties when the aquifer and/or geologic hydrostragraphic layers associated with a particular ground-water system are thin in relation to the length of the flow net. Unless the scale is exaggerated the flow net for this condition will be made up of squares that are too small to accurately draw.

A common problem with flow net construction obtained from monitoring well data is the lack of three-dimensional hydrologic data or hydrologically equivalent data for a

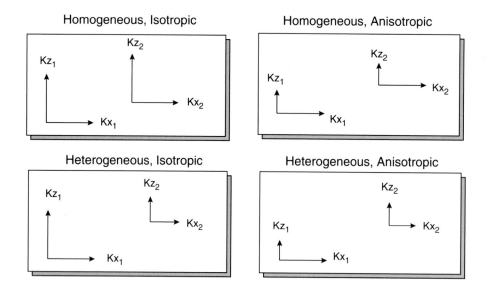

K = hydraulic conductivity; cm/sec
x and z represent horizontal and vertical directions respectively

Source: Freeze and Cherry, 1979

Figure 8-34   Four Types of Heterogeneous Conditions

ground-water flow system. Hydrologic testing must be available at various depths within an aquifer. Monitoring wells that screen wide portions of an aquifer can only provide integrated hydraulic head data rather than point values. If, however, flow is horizontal in the aquifer, (or hydrostratigraphic unit), then the almost vertical equipotential line could be fairly represented by the monitoring well water level. One should fully consider the reliability of using such data sources before including in the project data base. You must also obtain values for both the vertical and horizontal hydraulic conductivity and the vertical and horizontal gradient in the aquifer to provide reliable interpretive results. These data must be available before a flow net can be constructed.

Questions can arise with two types of ground-water systems on the applicability of Darcy's Law. The first is a system in which ground-water flows through materials with low hydraulic conductivity under extremely low gradients (Freeze and Cherry, 1979) and the second is a system in which a large amount of flow passes through materials with very high hydraulic conductivity. Since Darcy's Law expresses linear relationships and requires that flow is laminar, a system with high hydraulic conductivity, can have turbulent flow conditions. Turbulent flow can be characteristic of Karst limestone and dolomite, cavernous volcanics, and some fractured rock systems (U.S. EPA 1986). Construction of flow nets for areas of turbulent flow would not provide accurate results.

The concept of time of travel (TOT) has been used in federal guidance decision models for site location criteria The flow path of least resistance can usually be identified by inspection once a flow net is constructed for the site in question. This particular application of flow nets provides a very conservative approach to assessing ground-water vulnerability beneath a waste management site. However, these time of travel calculations are useful to define relative rates of ground-water flow under site specific constraints.

8.5.5 Hydrologic Considerations

In using flow nets for ground–water monitoring well design problems, it should be recognized that the solutions are no better than the idealized cross sections drawn. For a given section, however, the flow net can give an accurate solution for flow quantities. To enable proper construction of a flow net for use with a conceptual geologic model, certain hydrologic parameters of the ground–water system must be known, including:

• Distribution of vertical and horizontal head;
• Vertical and horizontal hydraulic conductivity of the saturated zone;
• Thickness of saturated layers; and,
• Boundary conditions.

Facility Head Distribution

Piezometers installed during the Phase II field tasks are used to determine the distribution of head throughout the site area. Head measurements made for the flow net construction must be time equivalent; that is, all piezometric measurements must be made coincidentally. Alternatively, all head measurements must be made under the same ground–water conditions. Piezometers should be spatially distributed and placed at varying depths to determine the existence and magnitude of vertical and horizontal gradients (actually the three-dimensional components of flow). When significant vertical flow components exist in the system, the flow direction cannot be derived simply based on inspection of the potentiometric surface in two dimensions. Three–dimensional views of the potentiometric surface would be required to fully interpret the flow direction.

Ground water will flow, however, from areas of high hydraulic head to areas of low hydraulic head. Figure 8-35 illustrates the hydraulic heads in a series of piezometers located in recharging and discharging geologic environments. In recharging zones, deeper piezometers show lower ground–water levels. This is due to the lowering of equipotentials with depth in recharge areas. In discharging environments, deeper piezometers show higher water levels or heads, as illustrated in Figure 8-35. Simple homogeneous, isotopic systems, as illustrated, require consideration of depth below the ground–water surface to correctly use the head data. Strong recharge/discharge systems require full understanding of both the vertical and horizontal gradients. Special care should be maintained to not mix piezometer readings from different hydrostratigraphic units or unrelated elevations in generating ground–water contour maps.

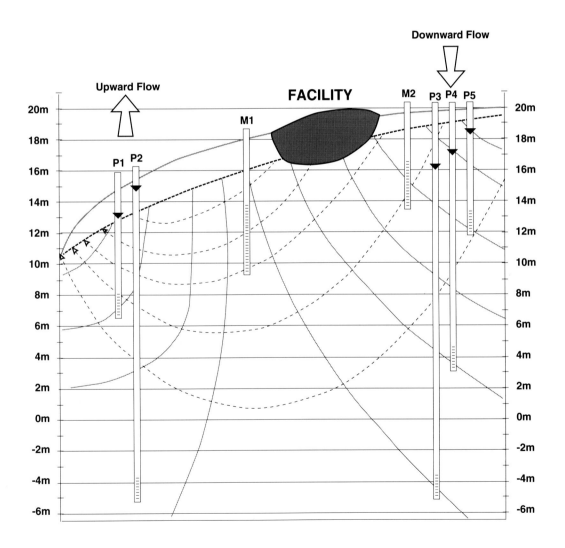

Figure 8-35 Recharge/Discharge Piezometer Head Relationships

Aquifer Thickness and Extent

The physical location and thickness of an aquifer (or any geologic strata) are determined during the Phase II field program by evaluation of geologic logs or by geophysical techniques. Geologic logs generated from boreholes show changes in lithology (the characteristics of the geologic material) indicating the relative hydraulic conductivity of materials. Various geophysical techniques, both downhole and surface, discussed in Chapter 3.0, can be used to assist in the evaluation of the thickness and extent of geologic units. The observed geologic materials establish the physical framework to establish flow nets and conceptual models for the facility.

Hydraulic Conductivity

Vertical and horizontal hydraulic conductivity is a primary component of Phase II site assessment tests. Several laboratory and field methods can be used to determine the saturated and unsaturated hydraulic conductivity of soils, including tracer tests, auger–hole tests, and pumping tests of wells. Most ground–water systems consist of mixtures of aquifer and aquitard units. Flow path analysis must consider ground–water movements through aquifers and across aquitards. The relative hydraulic conductivity of these units can vary many orders of magnitude; hence, aquifers offer the least resistance to flow. This results in a head loss per unit of distance along a flow line ten to many thousands of times less in aquifers than in aquitards. Therefore, lateral flow in aquitards usually is negligible, and the flow lines concentrate in aquifers and run parallel to aquifer boundaries.

Boundary Conditions

The boundary conditions of the area of investigation must be known to properly construct a flow net. The boundary conditions are used as the boundaries of the flow net. The three general types of boundaries are: (1) impermeable boundaries, (2) constant–head boundaries, and (3) unconfined aquifer boundaries (Freeze and Cherry, 1979), and (4) surface of seepage. Ground water will not flow across an impermeable boundary; it flows parallel to these boundaries. Unfractured granite is an example of an impermeable boundary (Figure 8-36a). A boundary where the hydraulic head is constant is termed a constant head boundary. Ground–water flow at a constant–head boundary is perpendicular to the boundary. Examples of constant–head boundaries are lakes, streams, and ponds (Figure 8-36b). The unconfined ground–water boundary is the upper surface of an unconfined aquifer, and is a surface of known and variable head. Flow can be at any angle in relation to the piezometric surface due to recharge and the regional ground–water gradient (Figure 8-36c). The surface of seepage boundary is where water leaves the porous medium as seepage and enters the open air. The pressure at this boundary is considered constant. Since h - z = o, constant, equipotential lines will intersect the boundary at constant intervals. This boundary cannot be considered either an equipotential or a flow line in the construction of the flow net. The boundary conditions of an aquifer can be determined after a review of the hydrogeologic data for a site and then considering the flow system within the conceptual model framework.

Boundary conditions of a particular cross section must

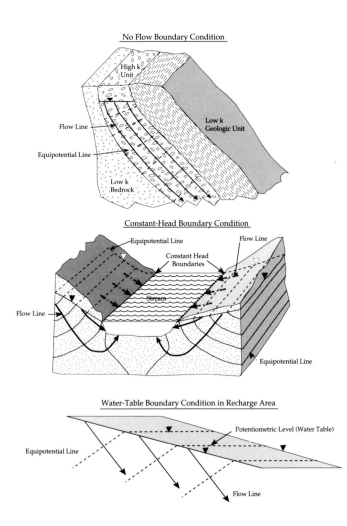

Figure 8-36 Boundary Conditions

be established through regional and specific investigation of site conditions. After assessing the hydrologic parameters of the ground-water system at the site of concern, the necessary data should be available to construct flow nets and resultant determination of ground water flow direction from the facility to potential down gradient receptor points.

8.5.6 Basic Guidance On Flow Net Construction

The most basic flow net construction consists of a homogeneous and isotropic system relative to hydraulic conductivity. The homogeneous and isotropic of geologic medium although rarely occurring in nature, can serve to describe the basic rules of flow net construction. The basic rules must be applied for all hydrogeologic media, with necessary modifications made to account for heterogeneity or anisotropic conditions of the system under examination. The fundamental rules (U.S. EPA 1986) and properties of flow nets are summarized below:

1. Equipotential lines and flow lines must intersect at 90 degree angles.
2. The geometric figures formed by the intersection of equipotential lines and flow lines must approximate squares.
3. Equipotential lines must be parallel to constant-head boundaries (constant-head boundaries are equipotential lines).
4. Equipotential lines must meet impermeable boundaries at right angles (impermeable boundaries are analogous to flow lines).
5. Each flow path in a flow net must transmit the same quantity of water (q).
6. The head difference (h) between any pair of equipotential lines is constant throughout the flow net.
7. At any point in the flow net, the spacing of adjacent lines is inversely proportional to the hydraulic gradient (i) and the seepage velocity (Vs).

Flow net sketching can be sufficiently accurate for assessing most ground-water flow condition relative to water quality monitoring issues. Patience and a certain degree of intuition will develop with practice in construction of these nets. Even with the difficulties in generation of accurate flow nets, the precision associated with these sketches is likely comparable to that associated with the measurement of field and laboratory values for hydraulic conductivity of the various site geologic units.

Flow Net Construction Steps

A relatively small number of flow lines are necessary to adequately characterize flow conditions at any particular site. The use of three to five flow lines will generally be sufficient for most investigations. With this in mind, the following steps should be used to construct a flow net:

1) Draw a geologic cross section, on a 1:1 scale, of the geologic units of concern in the direction of flow. At times expanded vertical scales may be useful, but with consideration that some changes will occur in flow lines and equipotential interception angles.

2) Establish all points of known hydraulic head from the Phase II data and draw tie lines between them by traversing the shortest possible distances and avoiding crossing of lines.

3) The tie lines constructed in Step 2 should be used to interpolate other hydraulic head values for the purpose of sketching equipotential contour lines. The accuracy of this interpolation procedure will depend upon the number and location of piezometer points for assessment of known hydraulic head.

4) With the known geologic data, establish two boundary flow lines.

5) By trial-and-error, one would sketch intermediate flow and equipotential lines, consistent right angles and squares should be formed in the sketching process.

6) These lines continued to be sketch until inconsistent shapes (i.e., angles other than right angles or rectangles that are not squares) start to develop in the flow net construction.

7) Successive trials should be made until the flow net is fully consistent Each inconsistency noted will indicate the direction and magnitude of change for the next trial.

One obtains the most effective results if only a few lines should be used in constructing the flow net. Any transitions that exist in the net should be smooth and the size of the spaces should change gradually. The following steps should be viewed as additional suggestions for drawing flow nets and not as fixed rules. Experience of each individual will determine what methods are most beneficial to the analysis technique.

a. The cross section to be studied should be drawn on one side of tracing paper, and the flow net construction should be drawn on the other side. It can then be traced onto the former side or onto another sheet of paper.

b. Just enough flow lines and equipotential lines should be used to bring out the essential features of the site. If detail needs to be emphasized in some parts of the flow net (i.e., facility features), flow lines and equipotential lines in those parts can be subdivided after the flow net is completed.

c. The scale of the drawing should be just large enough to draw essential details. If necessary a long flow net construction, (i.e., 1 to 1 cross section with an extensive horizontal dimension), can be subdivided into regions that will fit on single sheets of paper.

d. The boundary conditions, especially prefixed flow lines and equipotential lines, should be plotted before starting to draw the full flow net.

e. Either the number of streamtubes or the number of equipotential drops should be made a whole number to simplify flow net construction. However, if necessary, both streamtubes can be fractional on the ends of the net.

f. The overall shape of the flow net should be kept well in mind while working on details of the system. Because the shape of each section of a flow net affects the rest of the construction, a small portion should never be refined before the net is almost completed.

At most facilities, vertical and horizontal head data are obtained from well and piezometer measurements, and from free surfaces such as springs, lagoons, ponds and swamps. Data gathered from sources not related to the controlled area evaluated in the site assessment should be inspected carefully before use. Often, potable water supply wells (used for water level measurement) have long, open screened sections or have slotting at variable elevations below the ground-water surface for measurement of fluid pressure. These measurement devices provide very different head data from piezometers with short screened zones. As general guidance the open interval on a piezometer should be as short as possible with the midpoint of the interval being the measuring point for use in the flow net construction.

Preexisting wells with long screened sections may be used to obtain approximate metric levels if the midpoint of the open interval is used. The head measured in such a well is the integrated average of all the different heads over the entire length of the open interval. In this instance, it is important to note that if vertical gradients are present, the measured head can be a function of the screened length of the well and the variable depths of the screening. This must be considered when piezometric data are collected from such wells to be interpreted for the purpose of establishing hydraulic head conditions.

8.5.7 Example Flow Net Constructions

Homogeneous Isotropic Flow System

Figure 8-37a as an introduction to flow net analysis procedures shows a cross section of a homogeneous, isotropic system with no vertical hydraulic gradients. As a typical example the cross section has been drawn parallel to the direction of flow. The water level elevation is 122 m in Well A and 120 m in Well B. The aquifer consists of fine sand with a hydraulic conductivity of 10^{-4} cm/sec. The top and bottom of the aquifer are considered in these cases as impermeable (no-flow) boundaries and, as such, represent flow lines. The flow lines in this figure form a single flow path, which is sufficient in this flow net example. For this example, it is assumed that flow is the vertical equipotential lines are drawn at Wells A and B due to the horizontal flow. Intermediate equipotential lines are drawn by equally dividing the space between Wells A and B into squares (Figure 8-37b). Once the flow net has been constructed, the flow rate can be calculated using the equation:

$$Q = kFH / N \hspace{3cm} \text{Equation 8-17}$$

where:
 Q = flow rate
 k = coefficient of hydraulic conductivity
 = 10^{-4} cm/sec = 10^{-6} m/sec
 F = number of flow paths = 1
 H = total head drop = 2 m
 N = number of equipotential spaces = 4

Using the flow net constructed for this problem,

$$Q = (10^{-6} \text{ m/sec})(1)(2m) / 4 \hspace{1.5cm} \text{Equation 8-18}$$

$Q = 5 \times 10^{-7}$ m^3/sec per meter of width

N is the number of equipotential spaces in one flow path rather than the number of equipotential lines; therefore N is one less than the number of equipotential lines. The flow net in Figure 8-37b is the simplest representation that can be drawn for the system. A more detailed flow net (Figure 8-37c) can be constructed for this system, but the calculated flow rate is the same. From the flow net in Figure 8-37c, there are 5 flow paths and 20 equipotential spaces. Thus, the calculated flow rate is:

$$Q = (10^{-6} \text{ m/sec})(5)(2m) / 20 \quad \text{Equation } 8\text{-}19$$

$$Q = 5 \times 10^{-7} \text{ m}^3/\text{sec per meter of width;}$$

and is equal to the value calculated from the flow net in Figure 8-37b.

Heterogeneous, Isotropic Systems

Few ground water flow systems can be adequately studied as a section with a single hydraulic conductivity. Natural geologic materials can have over seven orders of magnitude variation in hydraulic conductivity within a few feet (for example, glacial out-wash gravels underlying till). As a general rule greater head loss should be expected to occur in materials with low hydraulic conductivity than in materials with high hydraulic conductivity. Flow lines tend to follow or parallel zones of contact between materials that have differences in hydraulic conductivity of 100, (two orders of magnitude), or more. Flow nets drawn for materials with a difference in hydraulic conductivity of a factor of 100 will look the same if the ratio of conductivity is 10^{-7} to 10^{-5} or 10^{-3} to 10^{-1}. However, variations will be evident in the quantity of flow and the calculated time of travel. Directional differences in hydraulic conductivity within the same geologic layer (i.e., anisotropy) are also of importance in flow net construction.

Construction of flow nets must consider not only hydraulic conductivity differences between layers, but also variations in horizontal and vertical hydraulic conductivity. Horizontal and vertical hydraulic conductivity are convenient representations for the test results from in-situ field tests (horizontal hydraulic conductivity) and laboratory permeability or triaxial tests (as vertical hydraulic conductivity). In reality, hydraulic conductivity of geologic material can be represented in any direction but test methods have formalized the concept of vertical and horizontal conductivity.

Beginning with a layered system with two different hydraulic conductivity; water flowing through the two materials will have flow lines that bend at the boundary. This phenomenon of bending of flow lines at changes in hydraulic conductivity is due to the conservation of energy, where the flow (or any natural energy such as light rays) is deflected when passing through boundaries.

Cedergren (1967) describes flow between materials of different hydraulic conductivity conceptually "all factors being equal, the higher the permeability, the smaller the area required to pass a given volume of water. Conversely, the lower the hydraulic conductivity, the greater the area required." This relationship is also applicable to the energy necessary for water to flow through porous media. Water level contour maps often show the effects of variations in hydraulic conductivity from point to point. Since the rate or loss in energy is reliable to the steepness of the hydraulic gradient, steep gradients are often observed in areas of lower hydraulic conductivity and flatter gradients in areas of higher hydraulic conductivity.

The deflection (and velocity change) of flow through geologic boundaries is derived through use of Darcy's Law and by geometry described by the Law of Tangents. This relationship is described by the ratio of the hydraulic conductivity of two different materials is equal to the ratio of the tangents of the two angles formed by the ground water flow lines so that:

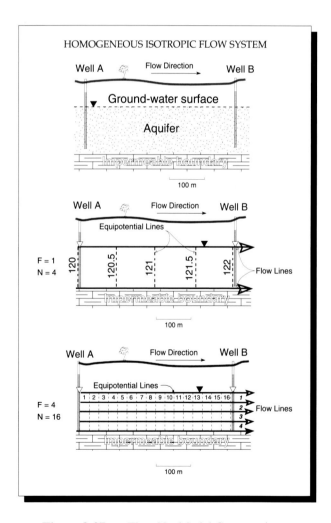

Figure 8-37a-c Flow Net Model Construction

$$k_1 / k_2 = \tan \theta_2 / \tan \theta_1 \qquad \text{Equation 8-20}$$

The relative shape of this flow line deflection is shown in Figure 8-38 Figure 8-38a shows flow from higher conductivity and back again to the original higher hydraulic conductivity. The second part of Figure 8-38b shows flow from low hydraulic conductivity to higher and again back to lower conductivity material. In these illustrations, both flow lines and equipotential lines are similarly deflected.

The areas formed by the intersecting flow and equipotential lines also can be described by ratios of the two hydraulic conductivities by the following equation:

$$c / d = k_2 / k_1 \qquad \text{Equation 8-21}$$

The relationship of longer or shorter rectangles based on the ratio of the two hydraulic conductivities, can be visualized by considering the energy necessary to push water through geologic materials. The cross-sectional area required for water to flow from low hydraulic conductivity units to high hydraulic conductivity units is less, hence, the rectangles will stretch out in the higher hydraulic conductivity material. This stretching effect will be demonstrated by lower resultant gradients. Flow from higher hydraulic conductivity to lower conductivity, results in a shortening of the geometric figures shorten, because steeper gradients are required and more cross sectional area is required to pass the flow. Since conservation of mass applies in the flow between the two materials, the ratios will hold according to Equation 8-21.

Heterogeneous, isotropic ground-water systems usually consist of two or more layers of materials with different lithologies and hydraulic conductivity. This heterogeneity may result from vertical layering, sloping strata, fault zones, igneous injection, or the existence of man-made structures, such as slurry walls to control rates of seepage. Ground-water flow in heterogeneous, isotropic systems is controlled by both the hydraulic conductivity of the layers, as well as by boundaries within the system.

The rules for construction of flow nets for heterogeneous, isotropic systems are the same as for homogeneous, isotropic systems, except that the "tangent law" (see above) must be satisfied at geologic boundaries of variable hydraulic conductivity. If squares are created in one portion of a formation, squares must be created throughout that formation and throughout other formations that have the same hydraulic conductivity. Rectangles will be created in associated formations that have different hydraulic conductivity (Freeze and Cherry, 1979). Flow lines tend to be parallel to the zone of contact between materials in the medium with higher hydraulic conductivity, and perpendicular to contacts between

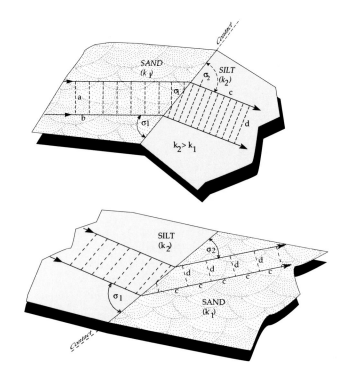

Figure 8-38 Flow Net Refraction

materials in the medium with lower hydraulic conductivity (Figure 8-38).

The illustrations presented above also show the shape of the rectangles that exist in the downstream material. Equipotential lines are also deflected when they cross conductivity boundaries because they are perpendicular to flow lines. It is impossible to construct a flow net for a heterogeneous, isotropic system in which only squares are created. However, the intersections of flow lines and equipotential lines must still form right angles (the flow net will consist of squares and rectangles).

Before beginning construction of the flow net for heterogeneous, isotropic systems, one should look for the dominating parts of the cross section and determine whether the hydraulic conductivities are in series or parallel. The system is in series when the cross section perpendicular to the flow direction is of one hydraulic conductivity, and different hydraulic conductivity regions occur sequentially in the flow direction (fig. 8-39). Conversely, hydraulic conductivities are in parallel when more than one region occurs perpendicular to the flow direction, and most flow lines remain in the same region throughout the net (fig. 8-40b).

For geologic systems where hydraulic conductivity is in series, draw a preliminary flow net choosing one region to have "squares" and making length-to-width ratios in the

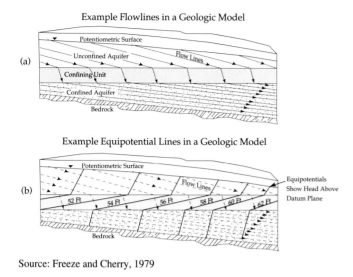

Example Flowlines in a Geologic Model

(a)

Example Equipotential Lines in a Geologic Model

(b)

Source: Freeze and Cherry, 1979

Figure 8-39 Flow Lines in Geologic Model

other region as nearly correct as possible. If the flow net has a free surface, the slope of the free surface will be greater in the region of lower conductivity (fig. 8-40b). Repeated adjustments will be necessary to yield the final flow net.

When seepage occurs through two hydraulic conductivity zones that are basically in parallel, flow through the more pervious zone usually dominates the flow pattern (fig 8-40a). A flow net can be constructed for the

more pervious part assuming temporarily that the other part is impermeable. The equipotential lines are then extended into the less permeable zone, and, by repeated adjustments, the flow net is completed. This process works best when the hydraulic conductivities to be compared are much different, for example at least one order of magnitude.

A check on accuracy for all composite sections can be made by sub-dividing the "rectangles' into a number of parts equal to the number of times the conductivity is higher (or lower than the hydraulic conductivity in the region where "squares' are drawn. Each subdivision should be a "square.' Whether the subdivisions are of the flow tubes, or the equipotential drops, depend on whether the conductivity is lower or higher, respectively, than in the region of I squares (figs. 3-40a and 3-40b). To obtain the quantity of seepage through any composite section, count the total number of equipotential drops and streamtubes and use Equation 8-17. The conductivity used should be the one for the region of "squares.'

Heterogeneous Isotropic Ground-Water Systems

The example waste facility (Figure 8-41) is located in a recharge area over a heterogeneous aquifer. The conceptual geology of the site is based on two geologic units. The upper unit is a 20 m thick layer of silt sand with a hydraulic conductivity of 5×10^{-7} m/sec ($k1$). The lower layer is a 10 m thick layer of sand with a hydraulic conductivity of 1×10^{-5} m/sec ($k2$). The flow net is constructed using the following procedures:

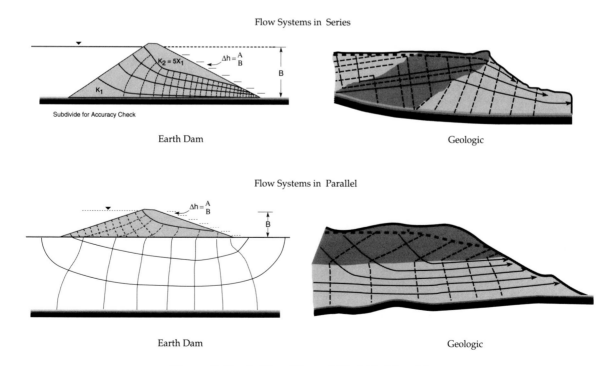

Flow Systems in Series

Earth Dam   Geologic

Flow Systems in Parallel

Earth Dam   Geologic

Figure 8-40a, b Flow Through Variable Materials

- Construct a conceptual model cross section on a 1:1 scale;

- Add the location of screens, head levels and a base line vertical scale;

- Draw tie lines between the measured head levels (as between 16m heads in p-17b and p-18a, 12m between p-19a and p-18b);

- Using the constructed lines, interpolate the equipotential head values for the cross section (18m, 14m and 10m equipotential);

- Construct the initial flow net lines perpendicular to the equipotentials as shown in Figure 8-42;

- Equipotential lines that do not originate at a head measurement point should be determined from the ratio 36m / 2m = x / 1m;

In the case of the example provided in Figure 8-43 is the distance from the intersection of the equipotential and the flow line to the measuring point. Figure 8-43 distance is calculated at 18 meters.

- Continue plotting equipotential lines along the tie lines until all points of equal head are connected;

- Construct flow lines at right angles to the equipotential lines to form squares;

- As ground water flows into the more permeable sand layers, an acute angle σ_1, is formed by the flow lines. This angle, is used in the tangent law calculations to determine the angle of deflection σ_2;

- Enter the hydraulic conductivity of each geologic unit and the tangent of σ_1, to calculate the tangent of σ_2;

- With calculation of σ_2 in degrees the angle of refraction is used to plot flow lines into the sand layer. (the example is 12.1°);

- The ratio of the adjacent units hydraulic conductivity are used to define the dimensions of the rectangles in the sand layer. Equation 8-21 is used to calculate the dimensions using C as the length of the rectangle and d as the width.

The example Figure 8-44 shows the width d, of the rectangles in the sand equal to 1 meter and the length c equal to 20 meters. In order to evaluate the flow path for the facility to be monitored one should plot flow lines from the edges of the disposal area to select the depth-location relationships. The resultant flow net will directly show the most direct target monitoring zone of the facility.

LITHOLOGIC CROSS-SECTION

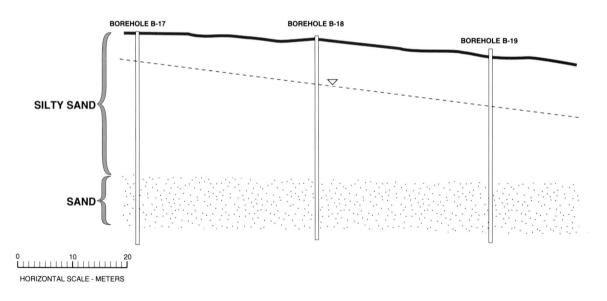

Figure 8-41 Lithologic Cross Section

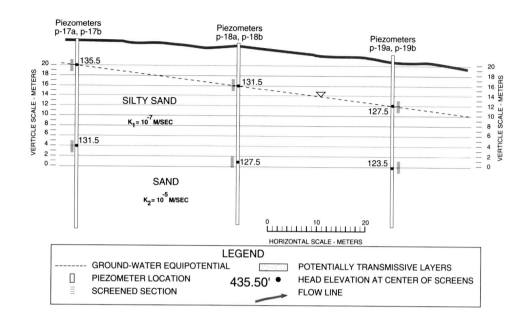

Figure 8-42 Lithologic and Head Data

Heterogeneous Isotropic Ground-Water Systems (Three Layers)

The previous example of a heterogeneous isotropic ground-water system with two units represents a simple layered case. Real world ground-water flow systems often consist of three layers, with overburden on top of a weathered or highly fractured bedrock overlying a much less fractured or unweathered deeper bedrock.

These three layer systems can be evaluated in a similar Phase II program as described in Chapters 3 and 4. Lithology, hydraulic conductivity, hydraulic heads along with physical boundary conditions provide the necessary tools to construct the flow net for a three layered system. Especially important, and often overlooked, is the evaluation of deeper bedrock conditions. If the deeper

INTERPOLATION & PLOTTING OF EQUIPOTENTIAL LINES

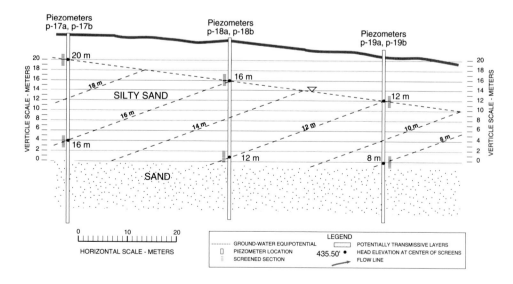

Figure 8-43 Plotting of Equipotential Lines

INNITAL FLOW NET

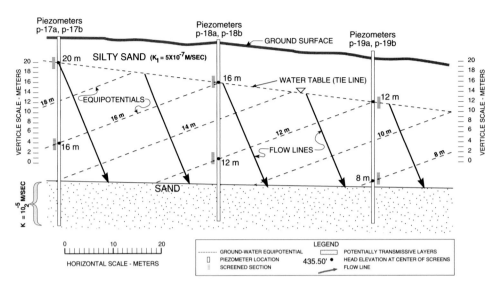

Figure 8-44 Initial Flow New Construction

bedrock has a much lower hydraulic conductivity than overlying units it will act effectively as a traditional confining unit, however, this must be first proven if reliance is to be placed on the bedrock to act as a very limited flow zone. Very deep bedrock having overlying units with much higher hydraulic conductivity (say, three orders of magnitude) may be considered as confining units. However, each site must be individually evaluated for potential deep bedrock ground-water movements. In addition, density flow conditions, such as, DNAPL product spills may cause even deep lying low hydraulic conductivity bedrock to be contaminated. These density controlled pure-product flow conditions must include both stratigraphic surface consideration (i.e., high density liquids flow stratigraphically down-hill on confining unit), as well as, traditional flow net constructions to evaluate the soluble water borne product flow. This diluted contaminant flow will generally move in a similar manner

FLOW NET FOR SILTY SAND & SAND UNITS

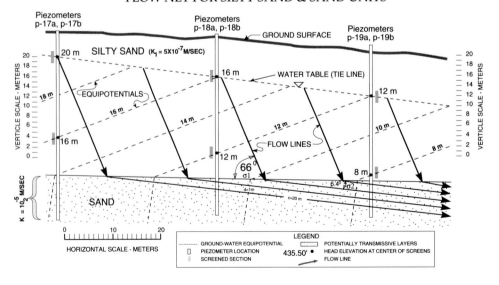

Figure 8-45 Silty Sand and Sand Layer Flow New Construction

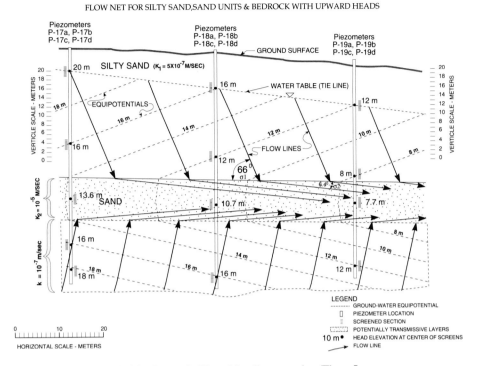

Figure 8-46 Example Flow Net Construction Three Layers

as normal ground-water.

The procedure described above for the two layer system must be extended to include flow components of the deeper unit. Hydraulic heads of the deeper unit provide the directional components necessary to construct the flow net. Figure 8-46 shows a three layered system with hydraulic conductivity of 5×10^{-7} m/sec, 1×10^{-5} m/sec and 1×10^{-7} m/sec respectively from top to bottom. Hydraulic heads are downward in all units. The flow net is extended through the various units down into the third layer with the appropriate reflection angles calculated from Equation 8-. In the case of Figure 8-46 the angle of reflection is calculated as :

$$1 \times 10^{-5} \text{ m/sec} / 5 \times 10^{-7} \text{ m/sec} = \text{Tan } 66^{\circ}/\text{Tan } x = 6.4^{\circ}$$

The refraction angle for the deep unit is 2.1°. Flow lines should be continued into the deeper layer using the calculated refraction angle. Scale considerations are important in these constructions. The 1:1 scale allows evaluation of facility boundary flow lines for monitoring system design. If the facility boundary flow lines enter the third geologic unit before reaching the facility edge (where it can be sampled with traditional monitoring wells) considerations must be made to include the deeper unit within the monitoring program. This decision must also

include evaluation of the relative hydraulic conductivity of the deepest unit; whether this unit is sufficiently permeable to represent a target monitoring zone. Low hydraulic conductivity geologic materials that represent confining units rarely provide "good' locations for ground-water monitoring (see ASTM D5092).

If the deeper units show higher hydraulic heads, as illustrated in Figure 8-46, then an upward flow must be accounted for in the flow net construction. The procedures for construction of these nets are as above based on use of the Tangent Law. The completed construction shown on Figure 8-46 represents a flow net for a silty-sand and sand and bedrock units with the bedrock having upward gradients. Since the middle sand unit is acting as a discharging zone, it represents a target monitoring zone. Deeper bedrock, (with upward gradients), probably represents a discharging environment from upgradient recharging areas, such as, bedrock upland areas. One must be sure to evaluate the recharge and discharge areas of the bedrock to be sure that the facility is not locally recharging the deeper bedrock unit in upland areas. This will probably require multi point piezometer evaluations to assure that consistent upward gradients exist throughout the bedrock surrounding the facility area. As with the consistent downward ground-water flow conditions, as shown in Figure 8-47, discharging units provide the target

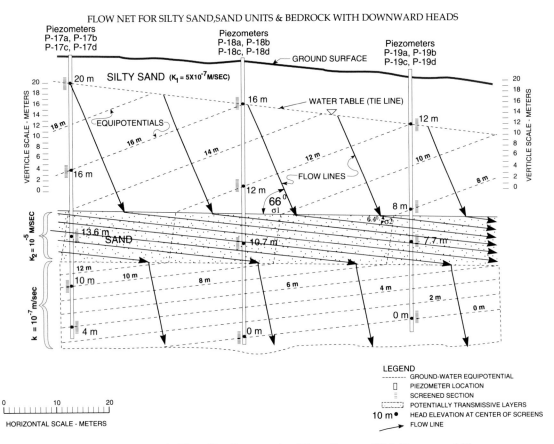

Figure 8-47   Example Flow Net Construction Three Layers With Downward Flow

monitoring zone for the facility. If the discharging unit is very thick the facility boundary flow lines may be contained within only a portion of the strata. In this case the detection monitoring wells can be designed with partial penetration of the discharging unit. However, local variations in hydraulic conductivity and piezometric heads may cause the flow to preferentially move into deeper portions of the discharging strata. With reasonably thick units (say 10 to 40 feet) fully penetrating monitoring well screen probably represents a prudent target monitoring strategy.

Homogeneous, Anisotropic Systems

The hydraulic conductivity within the flow system as illustrated in the previous example was the same in all directions; at any given point this system was both homogeneous and isotropic. In some sediments, such as

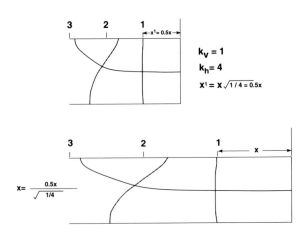

Figure 8-48b   Transformation of the Flow Net

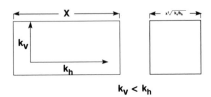

Figure 8-48a   Transformation of the Flow Net

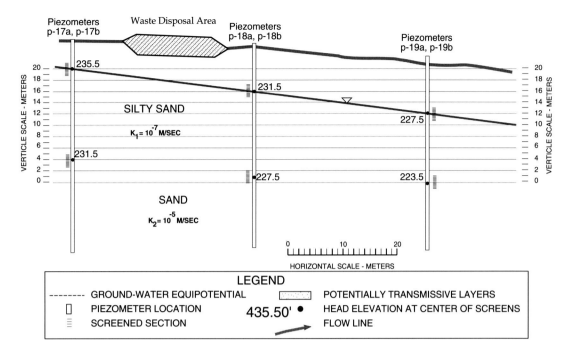

Figure 8-49 Original Dimensions of the Section for a Homogeneous Anisotropic System

INTERPOLATION & CONTOURING OF EQUIPOTENTIAL LINES

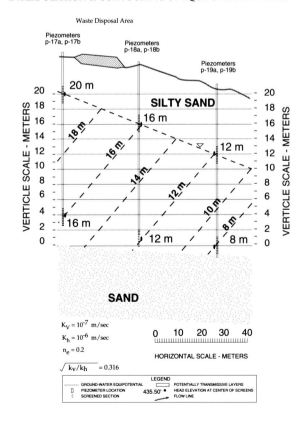

Figure 8-50a Transformed Section of Equipotential Lines

clays and silts, micro scale anisotropy is due to flat particles deposited in sheet-like beds. These tabular deposits decrease vertical hydraulic conductivity and result in relatively higher horizontal hydraulic conductivity. Anisotropy (large scale) can also result from lenses, channels or pockets of material of different hydraulic conductivity within a matrix. A medium that is predominantly clay, but includes sand stringers or varved deposits (alternative graded bedding) will have a higher horizontal conductivity than clay without the sand stringers.

When constructing flow nets for anisotropic media, you must decrease the dimensions of the cross section in the direction of the higher hydraulic conductivity, or increase the dimension of the cross section in the direction of the lower hydraulic conductivity. With a horizontal hydraulic conductivity, kh, greater than the vertical hydraulic conductivity, kv, one must then reduce the reconstructed section to a narrower horizontal dimension or expanded to increase the vertical dimension. If the reverse is true (kv > kh), the section could be stretched horizontally or reduced vertically (Cedergren, 1977). Referring to Figure 8-48a, the expression $\sqrt{k_v/k_h}$ is multiplied by the horizontal dimension (x) to transform the cross section (Figure 8-48b). To transform the cross section vertically, one would multiply the vertical dimension by the expression, k_h/k_v. Once the cross section is transformed, the flow net is then drawn as if it were in isotropic media. The dimension that

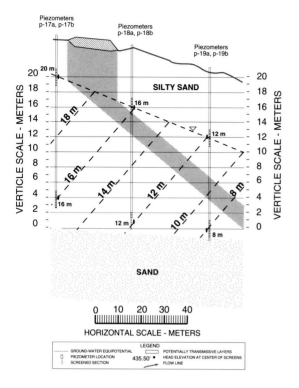

Flow net Construction in Silty Sand Layer

Figure 8-50b Flow net Construction in Silty Sand Layer

has not been transformed remains the same.

To obtain the original dimensions of the cross-section after construction of the flow net one divides the transformed dimension by the expression used in the transformation, $\sqrt{k_v/k_h}$ or $\sqrt{k_h/k_v}$. After returning the cross section has been returned to its original dimensions, the flow net is reconstructed using the same number of flow lines and equipotential lines. Transformation such as this does not change the ratio of F to N and will not change any of the calculations made on the transformed section. The new flow net is composed of rectangles elongated in the direction of high hydraulic conductivity. Due to the reconstruction of the net the intersections between flow lines and equipotential lines may not necessarily be at right angles. Figure 8-49 illustrates the procedure of returning a flow net to its original dimension, a value of horizontal hydraulic conductivity four times greater than the value of vertical hydraulic conductivity is used for the transformation. Figure 8-50a and 8-50b is the transformed flow net and Figure 8-51 is the flow net in the original scale.

After transformation the flow net must be rechecked to be sure that a reasonable construction is obtained. As a key to this reality check, the rectangles should be elongated in the direction of greatest hydraulic conductivity. As with the previous example the ratio of F to N will remain the same and $\sqrt{k_v\ k_h}$ is used as the effective hydraulic conductivity for construction with the transformed section.

Section Transformed Back to Original Dimensions

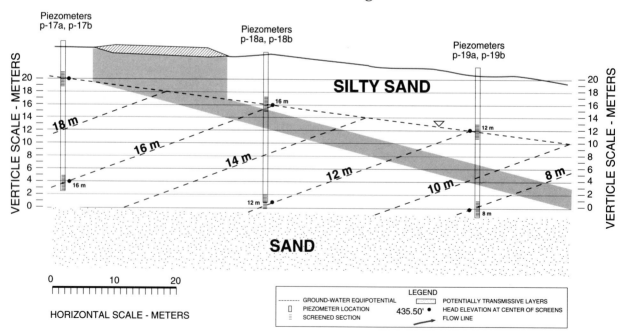

Figure 8-51 Flow net Returned to Original Dimensions

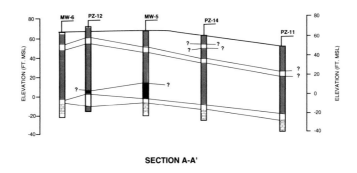

SECTION A-A'

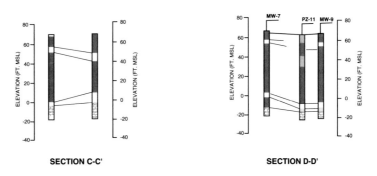

SECTION C-C'   **SECTION D-D'**

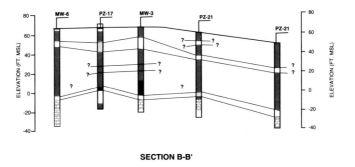

SECTION B-B'

Source: Modified from Golder Associates

Figure 8-52 Number of Geologic Cross Sections for Site

Example of A Homogeneous Anisotropic Flow System

An example of a homogeneous anisotropic flow system with the geology and system orientation of the site the same as that shown earlier in Figure 8-41 provides a simple case to evaluate. This example uses the same head distribution but has different hydraulic conductivity from that shown in Figure 8-42. Since this example only requires the additional consideration of the anisotropic silt layer we will only consider the flow construction in the silt layer. The silt layer had field hydraulic conductivity (slug tests) ten times the laboratory obtained vertical hydraulic conductivity.

The hydraulic properties of the silt layer are:

$k_v = 5 \times 10^{-7}$ m/sec

$k_h = 5 \times 10^{-6}$ m/sec

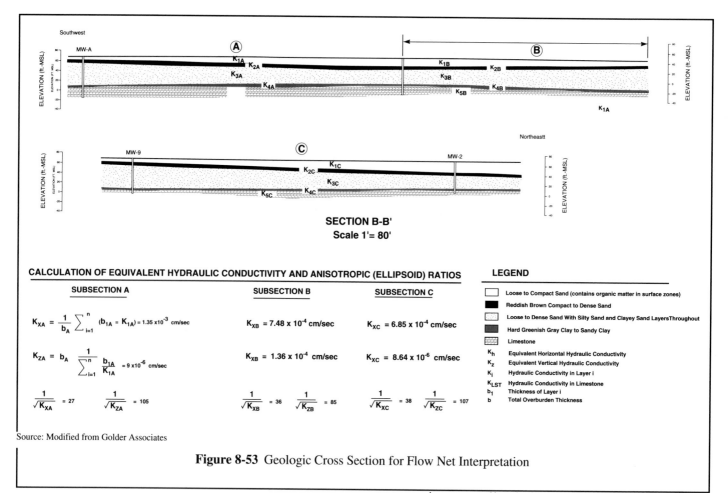

SECTION B-B'
Scale 1'= 80'

CALCULATION OF EQUIVALENT HYDRAULIC CONDUCTIVITY AND ANISOTROPIC (ELLIPSOID) RATIOS

SUBSECTION A	SUBSECTION B	SUBSECTION C
$K_{XA} = \dfrac{1}{b_A} \sum\limits_{i=1}^{n} (b_{1A} = K_{1A}) = 1.35 \times 10^{-3}$ cm/sec	$K_{XB} = 7.48 \times 10^{-4}$ cm/sec	$K_{XC} = 6.85 \times 10^{-4}$ cm/sec
$K_{ZA} = b_A \dfrac{1}{\sum\limits_{i=1}^{n} \dfrac{b_{1A}}{K_{1A}}} = 9 \times 10^{-6}$ cm/sec	$K_{XB} = 1.36 \times 10^{-4}$ cm/sec	$K_{XC} = 8.64 \times 10^{-6}$ cm/sec
$\dfrac{1}{\sqrt{K_{XA}}} = 27 \qquad \dfrac{1}{\sqrt{K_{ZA}}} = 105$	$\dfrac{1}{\sqrt{K_{XB}}} = 36 \qquad \dfrac{1}{\sqrt{K_{ZB}}} = 85$	$\dfrac{1}{\sqrt{K_{XC}}} = 38 \qquad \dfrac{1}{\sqrt{K_{ZC}}} = 107$

LEGEND

☐ Loose to Compact Sand (contains organic matter in surface zones)
■ Reddish Brown Compact to Dense Sand
☐ Loose to Dense Sand With Silty Sand and Clayey Sand Layers Throughout
▨ Hard Greenish Gray Clay to Sandy Clay
▨ Limestone
K_h Equivalent Horizontal Hydraulic Conductivity
K_z Equivalent Vertical Hydraulic Conductivity
K_i Hydraulic Conductivity in Layer i
K_{LST} Hydraulic Conductivity in Limestone
b_1 Thickness of Layer i
b Total Overburden Thickness

Source: Modified from Golder Associates

Figure 8-53 Geologic Cross Section for Flow Net Interpretation

$n_e = 0.25$

The value of the square root of k_v/k_h for the section is 0.316. Several alternative adjustments can be used to obtain the necessary scale changes. The horizontal scale can be reduced to 31.6 percent of the original, and/or both the horizontal and vertical scales could be increased or decreased so that the ratio of these dimensions is 0.316. In this example, the horizontal scale is 1 in. = 6 m and the vertical scale is 1 in. = 1.896 m (note that 1.896/6 = 0.316). The transformed geologic section is shown in Figure 8-50. The flow net is constructed as if the sediments are isotropic, as shown in Figure 8-49. Figure 8-50 illustrates the boundary flow lines that originate in the facility area and the flow net for the silty sand layer returned to the original dimensions.

Flow Net Construction in Heterogeneous, Anisotropic Systems

The progression from simple homogeneous systems to more complex flow net constructions follows the basic

progression as outlined above. Heterogeneous and anisotropic systems must include the additional factor of the ratio between vertical hydraulic conductivity and horizontal hydraulic conductivity in the uppermost layer of the system must be assumed to be representative of the directional ratio of conductivity in the lower layers. This assumption of $k_h > k_v$ is based on unconsolidated or semiconsolidated geologic material. This assumption would apply for most unconsolidated sedimentary deposits with one exception. This one unconsolidated strata exception would be for vertically fractured glacial tills. Consolidated bedrock must also be evaluated for directional hydraulic conductivity. To begin construction of the flow net with heterogeneous anisotropic conditions separate flow nets would have to be constructed for each hydrostratigraphic layer in the system by using information from the upper layer as the starting point for construction of a lower layer. Additionally, the dimensions of this new flow net would have to be adjusted to the conductivity ratio in the next layer and so on throughout the system. The following provides an example of the calculations and net constructions for a heterogeneous, anisotropic layered system.

Mass Balance Calculations • Step 1. A typical geological section line (Figures 8-52 and 8-53) was chosen as the representative hydrogeological section for flow net construction. The section was divided into three sub-sections (for construction convenience), based on the geological interpretation of lateral variations in the soils in the unconfined aquifer. The sub-sections A, B and C are shown in Figure 8-52 and are based on an apparent increase in silt, clay and "mudstone" content in the soils towards the north-east. The division between the sub-sections is somewhat arbitrary, since the geological log descriptions by various investigations were difficult to correlate. However, the divisions used correspond to descriptive changes in the lithology of the unconfined aquifer.

Step 2. The geologic units in each sub-section of the geological section are then assigned a hydraulic conductivity (Figure 8-52 and Table 8-l). In sub-section A, the values correspond approximately to those measured by field pump testing, except In sub-sections B and C, the hydraulic conductivities were decreased to correspond to the increasing silt, clay and "mudstone" content of the geologic units The values chosen are considered consistent with the geologic unit observed in field sampling programs.

Step 3. Hydraulic heads for the upper unconfined zone, the lower unconfined zone and the limestone were interpolated from the potentiometric data shown on Figures 8-53, 8-54, and 8-55, and are plotted on the geological section on Figure 8-52. The assumption is made that the limestone and unconfined zone are hydraulically connected, since there is no strong evidence at the facility for a continuous confining layer. Generally, the data indicate a slight ground-water mound near the facility, causing almost vertical flow and an increase in verticality of the equipotential lines toward the discharge zones to the Northeast, this results in an increase in localized horizontal flow.

Step 4. Each sub-section of the unconfined zone in the geologic cross section was resolved as a homogeneous, anisotropic system by calculating a horizontal (K_h) and vertical (K_v) equivalent hydraulic conductivity from:

$$Kx = \sum_{i=1}^{n} \frac{Ki \cdot bi}{b}$$
Equation 8-22

$$Kz = \frac{b}{\sum_{i=1}^{n} \frac{bi}{Ki}}$$
Equation 8-23

where:
b = total thickness of unconfined zone, ft;
bi = thickness of layer i in unconfined zone,ft;
Ki = hydraulic conductivity of layer i, ft/sec. (assumed homogeneous & isotropic)
n = number of layers

The calculations are shown on Figure 8-53 Figure 8-53 also shows the conceptual model resulting from this approach.

Step 5. Flow lines for each sub-section are resolved, taking account of non-orthogonality with equipotential lines using the method of hydraulic conductivity ellipse constructions as described by Bear and Bogan (1965), Liakopoulog (1965) and Maasland (1957),. The resultant ellipse is shown below:

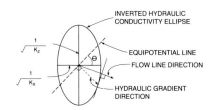

The hydrogeological flow net is shown on Figure 8-57.

Step 6. For each sub-section of the conceptual hydrogeologic section, the parameters required to calculate flow velocity, flow rate and travel times within flow tubes are computed. These parameters are;

$K_{s'}$ = the system hydraulic conductivity resolved from the hydraulic conductivity ellipse as shown below:

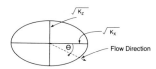

$$\frac{1}{K_S} = \frac{\cos^2 \theta}{K_X} + \frac{\sin^2 \theta}{K_Z}$$

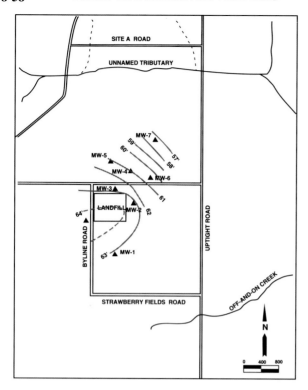

Figure 8-54 Potentiometrig Heads Shallow Unit

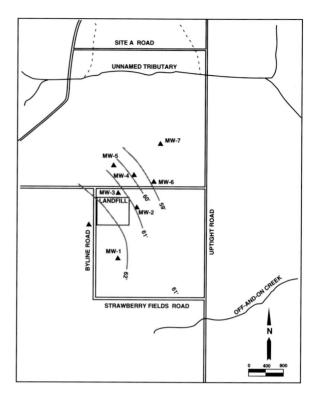

Figure 8-56 Potentiometrig Heads Deep Unit

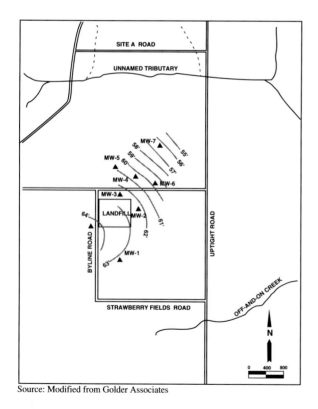

Source: Modified from Golder Associates

Figure 8-55 Potentiometrig Heads Medium Depth Unit

n = the material effective porosity, estimated as 0.20 in the water table zone and semi-confining unit.

i = the hydraulic gradient, computed as the change in head within the flow tube between the piezometric surfaces in the water table zone and in the limestone divided by the length of flow tube (LWT) above the limestone.

A = the area of the flow tube, taken as the thickness of the flow tube per unit width.

Thus:
Average flow velocity =

$$V_{wt} = \frac{K_s}{n} \bullet i \quad ft \, / \, sec$$

Equation 8-25

Travel time =

$$TWT = \frac{L_{WT}}{V_{WT}} \quad sec\, s$$

Equation 8-25

Flow rate = $\quad Q_{WT} = AK_s \bullet i \quad ft.^3/sec$ Equation 8-27

Step 7. Potential recharge rate into the flow tubes is calculated using the flow rate in the flow tubes from:

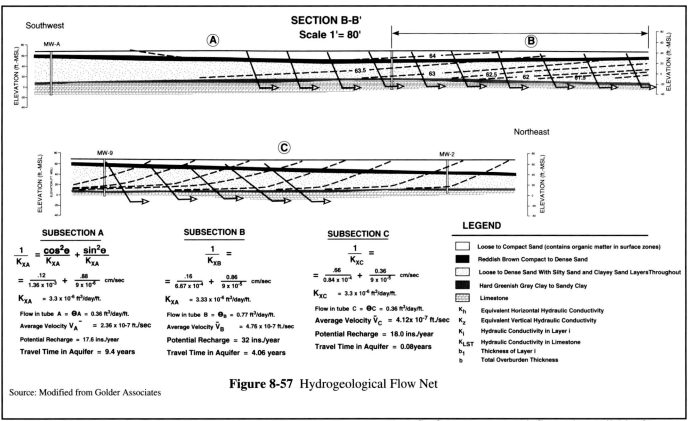

Figure 8-57 Hydrogeological Flow Net

Source: Modified from Golder Associates

$$r = \frac{Q_{WT}}{A_S} \text{ ft / year}$$ Equation 8-28

Where:
r = Potential recharge rate/flow tube in the saturated unconfined unit (ft./year)

As = Surface area over each flow tube available for recharge (ft.2) per unit width.

The potential recharge rate/flow tube is checked against the published values for the site area of 10 to 16 inches per year. If the r values indicate each sub-section is capable of taking this recharge, the calculation is considered acceptable.

Step 8. Flow velocity, flow rates and travel times in the

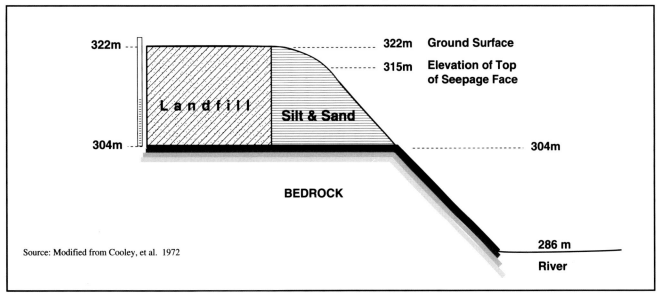

Figure 8-58 Seepage Face Conditions Example

limestone unit are then calculated from:

$$\text{Average Velocity, } T_{LST} = \frac{K_{LST} \cdot i}{n} \qquad \text{Equation 8-29}$$

where:

K_{LST} = hydraulic conductivity in the limestone
  $= 2 \times 10^{-2}$ cm/sec
n = effective porosity of the limestone (0.1)
i = hydraulic gradient in the limestone = 0.0027

Flow Rate, $Q_{LST} = AKi$

Where:
A = area of limestone per unit width = 25 ft²/ft.

$$\text{Travel Time, } T_{LST} = \frac{L_{LST}}{V_{LST}} \qquad \text{Equation 8-30}$$

where:
  L_{LST} = distance from north-east corner of
  the facility to borehole clusters EMW-3.

8.5.8 FLOW NETS IN SPECIAL SETTINGS

Several special cases of construction of flow nets for seepage face conditions and free surface/water table conditions are developed below. Because these conditions are important for interceptor trench design and discharge into excavations, the governing equations and construction procedures are important for many types of site assessments.

SEEPAGE FACE CONDITIONS

Seepage face conditions developed in a saturated-unsaturated flow system are typically considered a free outflow boundary, such as an excavation. Freeze and Cherry (1979) describe this phenomenon and analysis techniques for drawing seepage face flow nets. The intersection of the ground-water surface and the ground surface at the excavation face defines the upper boundary of the seepage face.

An example is a waste disposal site located adjacent to a riverbank (U.S. EPA 1986). A shallow upgradient ground-water surface introduces water that passes through the waste area and flows through a layer of sand and silt that forms the top portion of the river bank. The water table and river bank intersect at the top of the seepage face at an elevation of 315 m. A confining bedrock boundary is located below the sand and silt deposits.

A piezometer located upgradient from the waste disposal area facility in an adjacent upland area shows that the ground-water level is 10 feet (3 meters) below ground surface during the high rainfall summer months. A conceptual model based on the cross section shown in Figure 8-58, provides the basis for ground-water discharge calculation into the river. The flow net Figure 8-59 uses the hill top as a constant-head boundary. The underlying bedrock is considered as a confining unit (impermeable

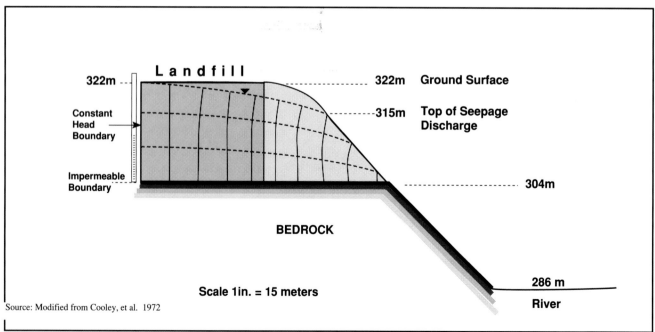

Source: Modified from Cooley, et al. 1972

Figure 8-58 Flow net for Seepage Face Example

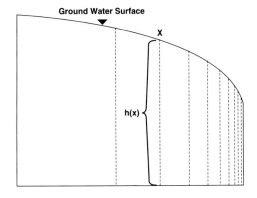

Figure 8-59 X & h(x) for Dupuit-Forchheimer Calculations

boundary), and the water flows out of the seepage face. The exact top of the seepage face (see Figure 8-59) is unknown, however, trial and error estimates or approximate location from observations in the field, provides the complicating factor in evaluating this type of field problem. The flow through a unit width represented by the cross section is determined from the basic equation for flow net quantity calculations:

$$Q = KFH/N \qquad \text{Equation 8-29}$$

where, in this example:

F = 3	(from flow net)
K = 1 x 10^{-5} m/sec	(from data)
H = 18 m	(from flow net)
N = 11	(from flow net)

Therefore:

$$Q = \frac{(1 \times 10^{-5}\text{m/sec}) \ (3) \ (18\text{m})}{11}$$

$$Q = 4.9 \times 10^5 \ \text{m}^3/\text{sec per meter of width}$$

8.5.8 Free Surface/Water 'Table' Flow Conditions

If the ground-water surface itself approximates a flow line, it represents a boundary and no vertical gradient exists is referred to as free-surface flow. There are two methods (USEPA, 1986) that can be used to calculate flow for free-surface flow conditions:

• Solve the flow problem using the Dupuit-Forchheiure theory to construct a flow net; and,
• Calculate the flow as in the previous sections.

With a simple flow net construction method, however, the position of the entire free surface may not be known and errors may develop in the construction. To solve the flow problem where free-surface conditions are important the Dupuit-Forchheimer theory of free-surface flow is used. This theory is based on two assumptions (U.S. EPA 1986):

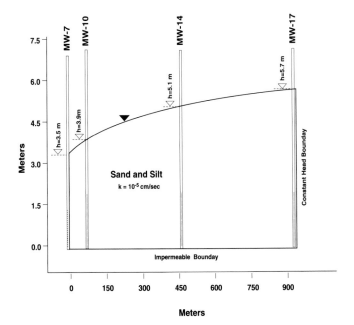

Figure 8-60a Cross Section for Free-Surface Flow

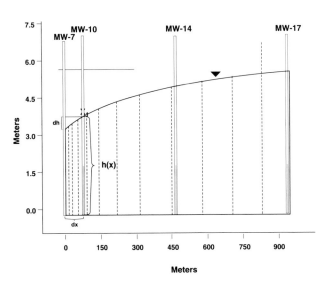

Figure 8-60b Solution for Free-Surface Flow

- Flow lines are assumed to be horizontal and equipotential lines are assumed to be vertical; and
- The hydraulic gradient is assumed to be equal to the slope of the free water surface and to be invariant with depth.

An empirical approximation under these assumptions can be used to calculate flow, as follows:

$$Q = k\, h(x)\, dv/dx \qquad \text{Equation 8-30}$$

Where:

   $h(x)$ is the elevation of the ground-water surface above the datum used for the flow system at x,

The gradient (d_h/d_x) is the slope of the free surface, $\Delta h/\Delta x$ at x (Figure 8-60a). This equation in theory is representative of a free surface that forms a parabola. The Dupuit-Forchheimer theory is believed to produce the most accurate calculated results when the slope of the free surface is small and when the depth of the unconfined aquifer is shallow (Freeze and Cherry, 1979).

Figure 8-61 shows a simple conceptual cross section in the direction of ground-water flow at a site where there is significant free-surface flow. To better illustrate the free-surface flow the figure is vertically exaggerated by a factor of 100. Dupuit-Forchheimer flow net is then constructed on the basis of the conceptual cross-section One flow path is used with each equipotential line representing the same drop in head (Figure 8-60b).

The flow rate is calculated using the Dupuit-Forschheimer solution at point X_1, where $h(x) = 3.9$ m. The quantity d_h/d_x, at X_1, is measured from the figure, resulting in $d_h = 0.6$ m and $d_x = 115$ m. Therefore, $d_h/d_x = 5.2 \times 10^{-3}$ m/m.

Using the equation:

$$Q = Kh(x)\, d_h/d_x \qquad \text{Equation 8-31}$$

$$Q = (1 \times 10^{-5}\ \text{m/sec})(3.9\ \text{m})(5.2 \times 10^{-3}\ \text{m/m})(1\ \text{m})$$
$$Q = 2 \times 10\ \text{m /sec}$$

8.5.9 AREA FLOW NETS

The flow net construction processes described up to this point have been of vertical cross sections, most useful for target monitoring zone selection. However, two-dimensional flow net analysis can also be conducted on water table or piezometric surfaces:

- If it is assumed that there is no variation in hydraulic head vertically in the aquifer (Ferris, et al., 1962, pp. 139-144), or,
- If it is assumed that each equipotential line represents the mean in the vertical.

It is also required that the saturated thickness of the aquifer be constant for each above point because any variation in thickness from point to point causes a change in velocity from point to point. As an example, in order for discharge for the section to remain constant along each flow tube, the flow net would be composed of rectangles of varying length-to-width ratios. This would be a very difficult flow net to draw. A constant aquifer thickness is rarely observed in site assessments. However, sufficiently accurate flow nets can be drawn where the variation in saturated thickness is small compared with the total saturated thickness. In these cases an average thickness should be used to compute the discharge through the flow net. The equation:

$$q = KH\, n_t / n_d \qquad \text{Equation 8-31}$$

can be rewritten for area ground-water flow analysis using the relation:

$$Q = qb \qquad \text{Equation 8-32}$$

Where:
   Q = total discharge through the flow net;
   q = discharge per unit thickness; and,
   b = average saturated thickness of the aquifer.

The resulting relationship is:

$$Q = KbH\frac{n_t}{n_d} = TH\frac{n_t}{n_d} \qquad \text{Equation 8-33}$$

Where:
   T = aquifer transmissivity
   n_t = the number of stream tubes
   H = the total head loss through the region
   n_d = the number of potential or head drops in the flow net

If the area transmissivity is found to be anisotropic, the flow net can be transformed as described earlier for homogeneous anisotropic systems. Area based nonhomogeneous regions can also be analyzed, but if more than two regions the flow nets are difficult to draw and numerical models may be required to fully evaluate the flow system. Fig. 8-61. represents a flow net of ground-water flow from a perennial stream to a discharging well (modified from Ferris, et al., 1962, p. 142)

An example of a flow net drawn for a discharging well located between a recharge boundary (a perennial stream) and a barrier boundary is shown in fig. 8-61. The well acts as an internal boundary of known head; therefore, the value of each equipotential drop is determined from the head difference between the perennial stream and the well.

Modified Area Flow Net Analysis

A modified method of aril flow net analysis can be developed using a map of either ground-water surface or piezometric contours. For each area being studied in a given region, two flow lines are drawn and used as limits for vertical cross sections along selected contours (fig. 8-65). As with the previous area flow net analysis the assumption must be made that hydraulic head and conductivity do not vary with depth, to compute flow through the aquifer at the selected cross sections. A flow net made of exact squares does not have to be drawn, so aquifer conductivity and saturated thickness can vary somewhat within each area being studied. Also, unsteady-state cases can be analyzed approximately. It is convenient to position each cross section at the average distance between two successive contours. In order to compute the flow across each cross section, a convenient form of Darcy's law was developed by Foley, Walton, and Drescher (1953) from the form $Q = TiW$ by defining:

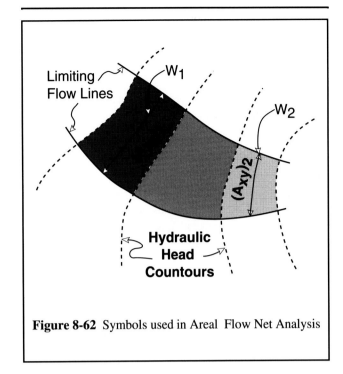

Figure 8-62 Symbols used in Areal Flow Net Analysis

$$i = c / L_a \qquad\qquad \text{Equation } 8\text{-}34$$

where:
$L_a = A_{xy} / W$
$c = $ contour interval for equipotential

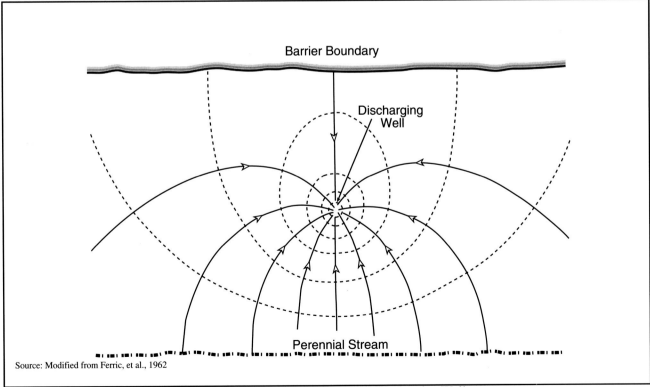

Figure 8-61 Flow net From Stream to Discharging Well

A_{xy} = surface area bounded by the two successive contours andthe limiting flow lines; and,

W = average width (i.e., parallel to contours) within A_{xy}.

Fig. 8-62 illustrates two cross sections bounded by two flow lines. In this case flow across both cross sections would be computed. The surface area of the total area being studied in Figure 8-62, AQ is represented by the vertical line pattern. The areas, $(A_{xy})_1$ and $(A_{xy})_2$ are represented by the diagonal line patterns. The subscripts, 1 and 2, refer to the two cross sections across which flow is being computed. If there is vertical leakage into or from an aquifer bounded by a confining bed above or below, it can be computed from Walton (1962, p. 22):

$$Q_c = K'A_c \, \Delta h / b' \qquad \text{Equation } 8\text{-}35$$

where:

Q_c = leakage through the leaky confining bed;

K' = vertical hydraulic conductivity of the leaky confining bed;

A_c = area of leaky confining bed within the study area;

Δh = difference between the head in the aquifer and the head at the distal surface of the confining bed; and,

b' = the thickness of the confining bed

As with cross-sectional net constructions, flow lines are refracted across a conductivity boundary. If the conductivity contrast is large enough, the 90° refraction implied by horizontal flow in the aquifer and vertical flow in the confining bed (or aquitard) is a good approximation. A simple construction where vertical leakage take place between units is shown in fig. 8-63 (from Cooley et. al., 1972). The conceptual model shows leakage from aquifer A to aquifer B through the confining bed. Leakage occurs because the water table is higher than the piezometric surface for aquifer B. A general water balance or continuity equation can be formulated to apply to the section of aquifer between the limiting flow lines and two cross sections. This equation is derived from Equation 8-36.

$$Q_1 - Q_2 = A_1 S_f(t)\Delta h/\Delta t + D - R \qquad \text{Equation 8-36}$$

It represents a modification of one presented in Walton (1962, p. 22) and is stated as:

$$Q_1 - Q2 =$$

$$2.1 \times 10^8 \, A_1 S \, \Delta h/\Delta t + D - R_s A_1 + Q_c \qquad \text{Equation 8-37}$$

where:

$Q_1 - Q_2$ = average difference in discharge in gpd between the two cross sections during the time interval t computed using Equation 8-34. The discharge Q_1 is upstream from Q_2

A_1 = the map areas between the two assumed flow cross section and the limited flow lines in miles^2

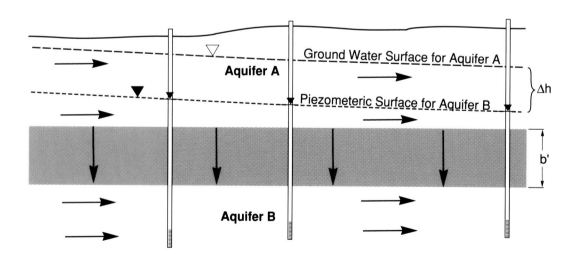

Figure 8-63 Vertical Leakage to a Confined Aquifer

(see Fig 8-62). If a leaky confining bed is present then $A_l = A_c$.

S = coefficient of storage (dimensionless). This will equal Sy for an unconfined case.

$\Delta h/\Delta t$ = an average rate of water table or piezometric surface rise in ft/day calculated as:
(average h in the area at time t_2 - average h in area at time t_1)/(t_2 - t_1), $t_2 > t_1$). The quantity will be negative for a decline.

D = discharge rate in the section in gpd.

R_s = net recharge rate in gpd/mi^2. This variable is assumed to represent the balance between the actual average recharge rate and the average ground-water evapotranspiration rate over the interval of time, Δt.

The last term in Equation 8-37, Qc, is positive for leakage from the aquifer. Equation 8-37 contains several assumptions. The most important ones are:

• Vertical flow exists through confining beds and horizontal flow exists through the aquifers (see Fig. 8-65).
• Dupuit's simplifying assumptions hold for analysis of water table aquifers. In this case, T = Kb where b = f(h)
• Conductivity at all elevations in the aquifer or confining beds are equal.
• No changes in storage in the confining bed(s) takes place during the time interval Dt.
• Darcy's law is valid.
• The coefficient of storage for the aquifer is constant.

The above assumptions are rarely fully met in practice, and average values are assumed to be constant numbers. As with any assumptions used for hydrogeologic evaluations, large deviations from the assumptions can produce large errors. This method is very useful where the assumptions are appropriately met and where all but one of the quantities in Equation 8-37 is either known are can be shown to be negligible. Unknown values can be computed for a number of areas in the region covered by the water table or piezometric maps and geological data. The method also can be used to check the consistency of data when all of the quantities can be estimated using other methods. The equation should approximately balance for all areas in the region.

8.6 FLOW NET/CONCEPTUAL MODEL FORMULATION

The process of conceptualization of a ground-water system requires the bringing together of much of the data

gathered during the Phase II study. Several examples of this data organization are provided below.

Written discussions of the drawings and calculations completed for a project are of great importance and should not be left to inexperenced staff. This interpretative discussion must include all major aspects of the evaluation than have revalence to the project. Subsurface information gathered during field drilling are typically edited into final boring logs and cross sections as shown on Figure 8-52 (please refer back to these previous figures). These cross sections, constructed through edited borehole logs, go through a second conceptual process of a generalized cross section.

8.6.1 Example Unconsolidated Materials

A generalized cross section for the example area the following subsection fully links conceptual hydrogeologic models and flow nets to provide targeted monitoring locations for site assessment project. This method provides the necessary location and depth information for both detection and assessment monitoring design. A descriptive conceptual model for an example site area will first introduce the conceptual process followed by the incorporation of flow net evaluations for the site area as was constructed to describe the mechanics of flow net construction. This series of figures from 8-52 to 8-57 provides the construction tools and techical back-up for the following discussions.

The stratigraphic section provided by the series of cross sections (Figure 8-52) contains five separate layers ranging from loose sand to clay overlying limestone. A series of water level measurements was taken for shallow and deep unconsolidated aquifers and the limestone bedrock aquifer illustrated in Figures 8-54 to 56.

The descriptive conceptual site hydrogeologic model developed from the geology, the potentiometric data, hydraulic characteristic test data, and a general water balance have the following characteristics:

1. The system includes an anisotropic unconfined aquifer (sand unit with silts and clays) separated from a limestone aquifer by a semi-confining clay/marl unit. There was strong evidence for a layer at shallow depth, within the unconfined aquifer.

2. The unconfined aquifer may be considered to closely mirror the ground surface, since the variability in hydraulic conductivity is not significantly greater than one order of magnitude vertically or horizontally. Values of hydraulic conductivity in the aquifer vary from

10^{-3} cm/s and greater in the sand portions, to 10^{-4} cm/s in the hard, calcified zone.

3. The semi–confining aquitard separating the unconfined aquifer and the limestone aquifer is primarily a clay with hydraulic conductivity in the range of 10^{-5} cm/s. This unit is hydraulically discontinuous (leaky).

4. The limestone aquifer is fractured and/or dissolutioned and, therefore, quite heterogeneous.

5. The overall hydraulic conductivity of the limestone aquifer is on the order of 10^{-2} cm/s, but varies locally depending on the secondary hydraulic conductivity of the discontinuities (dissolution–widened joints).

6. Recharge for the area is occurring across the plateau where the landfill is located, and discharge is in the valleys of the adjacent river system.

7. Ground–water flow in the site area is predominantly vertical down to the limestone, and then horizontal toward the stream valleys.

The conceptual hydrogeological model is based on the descriptions of site and vicinity conditions developed during Phase I and II investigations. Modifications in the flow net analysis were made to account for variability of soil conditions within the unconfined aquifer, particularly downgradient from the site, in the northeast and east directions, where the available data indicate the surficial material to contain a greater clay content. The conceptual model was tested using flow net analyses and water balance estimates in selected flow tubes.

Method of Analysis

Figure 8-53 shows the geological section and hydrogeological parameters for the flow net analysis. The method used to define the hydrogeological flow regime near the facility is the graphical construction technique to account for heterogeneity and anisotropy in the system, and using interpolated hydraulic heads based on field measurements. This construction was totally hand drawn without support of a numerical computer model of the system. The method used to develop the flow net shown on Figure 8-57 is described in Freeze and Cherry (1979, pp. 174) and follows the procedure described in section 8.0 of this text. Pertinent flow net calculations and the construction method for flow lines are shown on Figures 8-

53 and 8-57.

Discussion of Flow Net/Conceptual Model Results

Once the linked conceptual hydrogeologic model is combined with a flow net construction, an important next step is evaluation of the results of the conceptual/flow net drawings. The facility site described above in the previous subsection provides sufficient data to fully evaluate the area flow system. The flow net analysis described above has shown the following:

1. Flow lines in the vicinity of the landfill are relatively steep (vertical) in the unconfined aquifer and horizontal in the limestone, indicating short flow paths to the limestone from higher elevations in the aquifer. This scenario is consistent with a recharge area having no significant topographic variations and a high hydraulic conductivity zone underlying a lower hydraulic conductivity semi–confining zone. Close to the discharge zone, the flow lines became more horizontal in the unconfined aquifer.

2. Average travel times for water movement from the facility, down through the unconfined aquifer, into the limestone, and then to the vicinity of boreholes A-2 and A–3 is about 4.7 years through the unconfined aquifer and 2.5 years through the limestone, for a total of about 7 years. Travel times for other parts of the downgradient system, from surface to the limestone, are greater than 5 years, reflecting the lower hydraulic conductivity of the downgradient system.

3. The flow lines show that, under natural gradients, potential seepage from the facility would enter the limestone upgradient of wells A–2 or A–3 and should not be detected in the potentiometric surface zone at these well locations. A surface source of contaminants in the vicinity of wells A–2 and A–3 would take some 3 to 4 years to reach the 50–foot level in either borehole. Thus, contamination of wells A–2 and A–3 is more likely due to activities in the areas of the wells than due to the facility.

As described above the combination of conceptual models and flow nets provide a powerful tool for evaluation of site ground–water movements and predictions on potential leachate flow directions so that geologic units can be targeted for the monitoring well design described in Chapter 9.0.

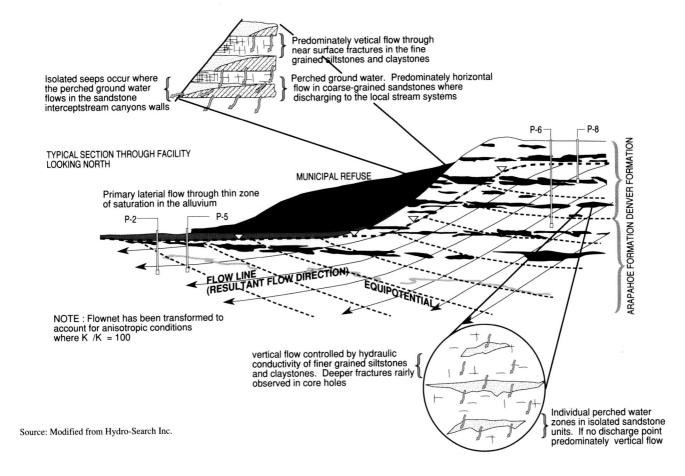

Figure 8-64 Low Hydraulic Conductivity Materials With Channel Deposits

8.6.2 Example-Consolidated Rock

A second example of the conceptual model building with linkage to flow net construction provides additional insight into the process. This example is based on semiconsolidated deltaic sandstone and claystone bedrock with well defined sand channel deposits.

The geologic units on site consist of unconsolidated deposits and bedrock of the Denver and Arapahoe Formations. The upland area along the north and east portions of the site area typically consists of surficial materials (alluvium, colluvium, residual soil, and weathered bedrock). These consist of silty clays and clayey sands with isolated cemented layers. Underlying the surficial materials, (exposed on the sites' cliff face), are interbedded claystone, siltstone, and sandstone of the Denver Formation.

Underlying the Denver Formation is the Arapahoe Formation, locally characterized by coarse sandstone and

conglomerate with interbedded siltstone and claystone. The contact between the Arapahoe and Denver Formations is believed to be near the base of the topographic expression of the site's cliff face.

In an earlier geologic study of the area, Lincoln-DeVore (1982) divided the bedrock exposed along the cliff face into the upper and lower units of the Dawson Group. Since the northern portion of the face has been covered with solid waste materials since the time of the Lincoln-DeVore study, their description of the two units is highly descriptive of the facility lithology repeated as follows:

"Upper Unit". The Upper Unit is well exposed on the site and is characterized by an interbedded series of sandstones, claystones, and siltstones. It is estimated that sandstone comprises 60% of this unit while the siltstone and claystone comprise about 40%. The sandstone within this unit is characteristically tan to light brown in color and vary from arkosic and medium to coarse grained to fine to medium grained and silty to micaeous. Scattered throughout the unit is petrified wood and carbonaceous debris. Within this unit are a few resistant, ledge-forming sandstones which range in thickness from only a few

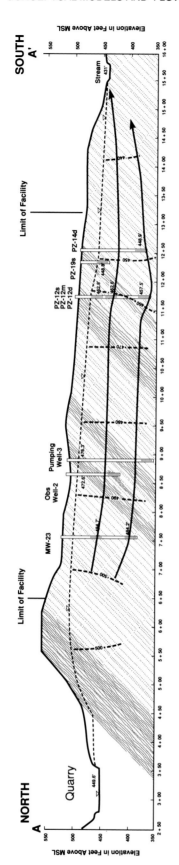

First Try Without Consideration of Bedrock Varability

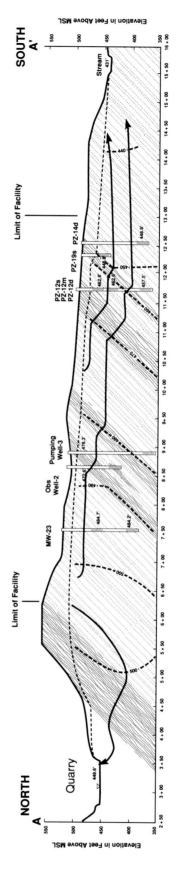

Second Try Using Bedrock Varability in Flow Net Construction

Source: Modified from Wehran Engineering

Figure 8-65 a-b Alternative Flow Net Construction - Layered Rock

inches to an estimated 20 feet. Although these resistant ledges and cliff formers dominate the topography by virtue of the ledges and cliffs that they form, it is estimated that they make up 10% or less of the exposed unit. The sandstone in this Upper Unit typically exhibit laminations from magnetite grains and small to large scale cross-bedding. A few of the thicker sandstones are lenticular in nature and can be seen to pinch out or are reduced from thicknesses of 20 feet to somewhat less than 2 feet within the perimeter of the site. Other sandstones are more or less tabular and can be traced for considerable distances across the site. The thicknesses of the various sandstone beds range from only a few inches to approximately 20 feet. The claystones and siltstones that are interbedded with the sandstones in the upper unit are typically gray to brown to tan in color and range from a few inches to about 8 feet in thickness.

Only about 20 to 30 feet of the Lower Unit is exposed on-site. The contact between the Upper and Lower Unit is drawn at the base of a distinctive white sandstone marker bed that is well exposed on the ridge crest in the south/central portion of the site. Although this white marker bed is not entirely continuous and dies out in the central portion of the site in the main canyon area, it also occurs west of the main drainage on the prominent ridge.

The lower unit described by Lincoln-Devore (1982) may correspond to the Arapahoe Formation as referred to in this report. Regardless, since the Denver Formation is reported to be about 200 feet in this area, it is probable that the Arapahoe Formation bedrock underlies the surficial deposits on the canyon floor area of the site. Surficial deposits at the site below the cliff face consist primarily of alluvium. The majority of the site topography at the base of the cliff topographic feature has been altered due to site activities. Prior to site regarding activities, alluvium occurred at the base of the cliff face and along the former surface drainage features.

Conceptual Hydrogeologic Model

A conceptual flow net of the site is presented in cross section on Figure 8-64. The site is characterized as a ground-water recharge area for the Denver Formation beneath the upland area of the site and the Arapahoe Formation on the canyon floor beneath the landfill. Figure 8-68a presents an approximation of the uppermost water table. Water level data indicate that the cliff face causes a localized alteration of the regional flow system in that a water table divide is present near the crest of the ridge. To the East and North of the ground-water divide, the primary direction of flow in the Denver Formation is laterally towards the Southeast through erratically distributed, more

permeable sandstone lenses.

To the South and West of the ground-water divide near the cliff face, the direction of ground-water flow in the Denver Formation is towards the site cliff face. An enlarged view of the flow paths is presented in Figure 8-64. Anisotropic geologic conditions cause ground water to flow towards the cliff face laterally within more permeable sandstone lenses. Vertical leakage occurs beneath the sandstone layers through subvertical fractures within less permeable siltstone and claystone units. As observed during cliff face reconnaissance, vertical leakage occurs at a greater rate near the cliff face because fracture frequency and width increase towards the cliff face due to stress relief from natural erosion. The results of in-situ testing during the field program indicated that the hydraulic conductivity of subgrade materials on the excavation bench is about 6×10^{-6} cm/sec. This value is greater than the average hydraulic conductivity value measured in the canyon floor 1×10^{-6} cm/sec.

As shown schematically in Figure 8-64, the resultant direction of flow along the cliff face roughly parallels the cliff face. The conceptual flow net has been constructed to reflect differences in hydraulic conductivity in the horizontal and vertical directions of flow. The ratio of 100 horizontal to 1 vertical is used to represent the anisotropic conditions within the stratified sedimentary rocks typical of those occurring at the Landfill (Freeze & Cherry, 1979). As a result of the transformation, streamlines and equipotentials are not orthogonal, but rather are deformed in the direction of the observed higher hydraulic conductivity.

8.7 CORRELATION OF FLOW NETS AND GEOLOGIC STRUCTURES

Structural effects on ground-water flow are often misinterpreted in site assessments. While the effect of layered strata of different hydraulic conductivity on the flow system is represented by the Tangent Law, dipping beds of variable hydraulic conductivity require considerable care in the linkage between the conceptual model and flow net construction.

Figures 8-65a and 8-65b illustrate alternative flow net constructions for dipping heterogeneous rock. Section B-B (Figure 8-65a was originally drawn using the head levels observed from a series of multiple point piezometers located along the cross-section line. This flow net shows almost vertical equipotentials and the resultant almost horizontal flow lines. If the site geology were composed of relatively homogeneous/isotropic units this interpretation would be acceptable, however, the real world geologic conditions are alternative shales and sandstones with significantly different hydraulic conductivities. Although the equipotentials drawn in 8-65a agree with the head levels observed in the section's piezometers, this construction is not unique and requires further consideration to obtain a more representative flow net.

Figure 8-65b shows a redrawn net; where consideration of the variable hydraulic conductivity has refracted the flow lines to represent the tangent rule. The equipotential lines have been drawn to reflect the likely head loss through the lower hydraulic conductivity geologic units. The redrawn net provides a more consistent example of flow conditions represented by the dipping alternative shales and sandstones. This is especially important for consideration of the depth needed to screen monitoring wells for both detection and assessment monitoring design. Alternatively layers of dipping bedrock must be evaluated for the effects of fracturing as a modifying feature to the geologic units variable hydraulic conductivity. In some cases the resultant effective hydraulic conductivity (which is a combination of primary and secondary porosity) can modify flow conditions so that relative recharge/discharge relationships can hold true. For example, although unfractured hydraulic conductivity of the sandstone/shale geologic units of the section would vary 3 to 4 orders of magnitude, the fractures interconnecting the two units have modified the effective hydraulic conductivity of the units to one order of magnitude. Hence, ground water can discharge through the various layers to down gradient discharge points along the section shown.

In those cases where fractures are not significant as a modifying factor for increasing the effective hydraulic conductivity then discharge along strike of the geologic units would be more likely flow directions for ground water. Figure 8-66 illustrates a regional conceptual model of an anticlinal structure source to depths of 2 kms deep that was based on oil and gas explorations. A facility based on a limb of the anticline could be expected to overlie dipping rocks as are shown in Figure 8-67. The individual beds consist of alternating claystone and sandstone layers that have at least 4 orders of magnitude different hydraulic conductivity. Fractures are not significant in this particular situation so flow conditions can be expected to be highly modified by the layered rocks. This condition is highlighted by site piezometer data that shows inconsistent head level data for each of the sandstone units monitored. Although the head levels have higher elevations in the "upgradient" piezometers, monitoring head levels in intermediate sandstone units illustrates the inconsistencies in the flow system due to the highly varied hydraulic conductivity. In effect, if one could show head elevations behind the two dimensional surfaces of the paper you would see lower hydraulic head

elevation in each of the sandstone units. Respectively, a similar view in front of the paper's surface would also show a lower hydraulic head value for each of the sandstone units monitored. We therefore have, at least locally, a mounding condition along strike of the sandstone units. Hence, ground-water flow is parallel to the strike of the bedding and flow nets must consider the individual sandstone units as bounded by the claystone confining units. Since mounding exists in this particular hydrogeologic environment, flow conditions should be represented by a localized flow cell discharging in either nearby adjacent areas or down toward the regional discharge flow system.

Additional flow considerations with layered dipping bedrock systems must also address down-dip movement of ground water, if the down-dip strata are discharging to a lower head aquifer system or to a lower surface water discharge point. The particular example used above was part of a major deep seated basin of highly saline brines that were not connected to potable water aquifers, nor discharging to surface waters. In this case, downward movement of ground water down dip was considered as not important since the overall majority of the ground-water flow was along strike of the sandstone units.

This conceptual model illustrates the importance of rigorous understanding of the site geology and hydrogeology before designing a monitoring program. Without adequate understanding of hydraulic head conditions, in a 3-D sense, it would have been highly unlikely that a resultant monitoring system with (wide screened zones) would have sorted out the overall components of flow. Dipping heterogeneous rocks must have piezometers installed in lithologic borings with very

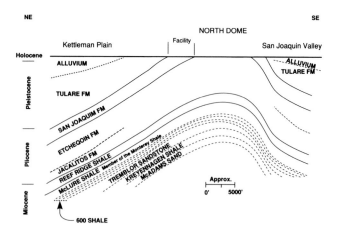

Figure 8-66 Anticline Regional Structure

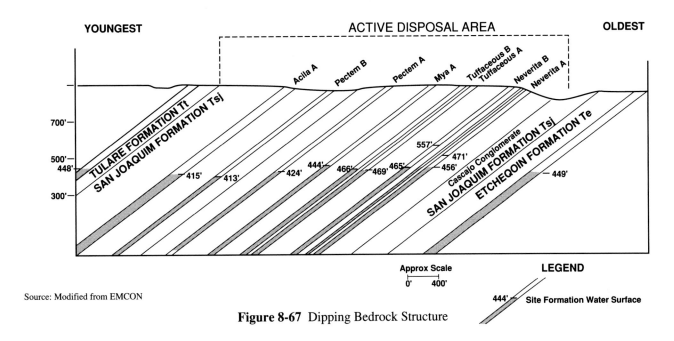

Source: Modified from EMCON

Figure 8-67 Dipping Bedrock Structure

accurate logging and testing data to evaluate the potential for ground-water flow parallel to the strike of the rock mass. Additional complications such as secondary porosity through fracturing must be also fully evaluated before any consideration is given as to the position of the facility monitoring well system.

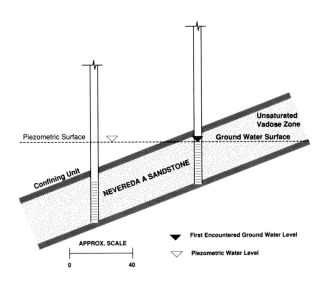

Source: Modified from EMCON

Figure 8-68 Vadose, Unconfined and Confined Conditions

8.7.1 Ground Water Discharge to Streams

The discharge of ground water aquifer to surface water bodies is an important situation that must be evaluated to define the potential limits for assessment monitoring investigations. As was illustrated in sand tank and computer numerical models, discharge to areas of low hydraulic head represents the area neutral density leachate plumes must move toward. Figure 8-69 illustrates a flow net of a system that fully discharges to the local stream system between the elevations of 1300 to 1400 foot msl. Keys to the interpretation of these flow systems can be listed as follows:

- Very low or neutral vertical gradients are observed in piezometer sites in upland areas;

- Strong horizontal gradients are typically found;

- Strong upward vertical gradients are observed in piezometer sets near surface water discharge points;

- Similar strong upward gradients are found on the opposite side of the surface water; and,

- Confining units may be found underlying the shallow flow cell which limit downward movement of ground water.

Figure 8-69 shows almost a classical case of almost

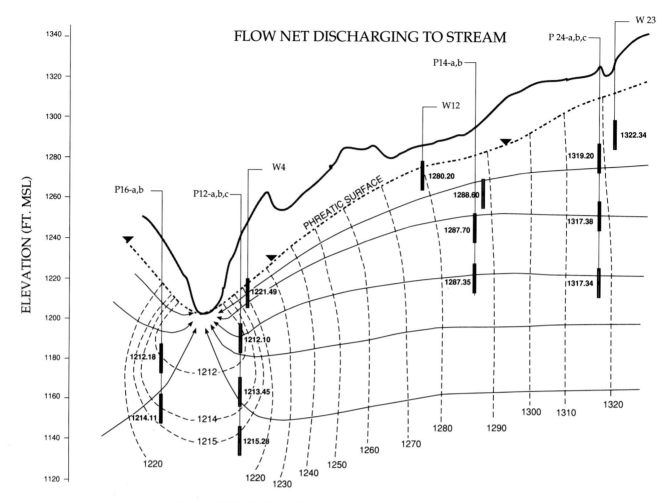

Figure 8-69 Shallow Flow System Discharge to A Large River

complete discharge of the cross-sections flow to the stream discharge zone. Considering this illustration represents a 10 to 1 scale from horizontal to vertical, the actual flow net would have been virtually flat within the area of the net construction. Deeper piezometers extending down to a regional flow system may have shown ground-water flow lines moving beyond the local discharge area of the stream. Keys to interpretations for these systems must be based on:

• Stronger downward gradients than observed in a fully discharging system;

• Downward gradients in deeper piezometers adjacent to local streams;

• High density leachates that moved to deeper systems; and,

• Unbounded systems where no confining units exist at depth.

The importance of establishing the limits of these discharging cells cannot be over emphasized. Extent of the investigative program, risk assessments, assessment monitoring system design remedial system design, and well census data collection activities are all directly related to the potential for affected ground water to move past a local discharge boundary.

8.8 CONCLUSIONS

This chapter developed the integration of flow net constructions and conceptual models. Just as some sites are not amenable to computer modeling, some of these sites are also difficult to draw reasonably accurate flow

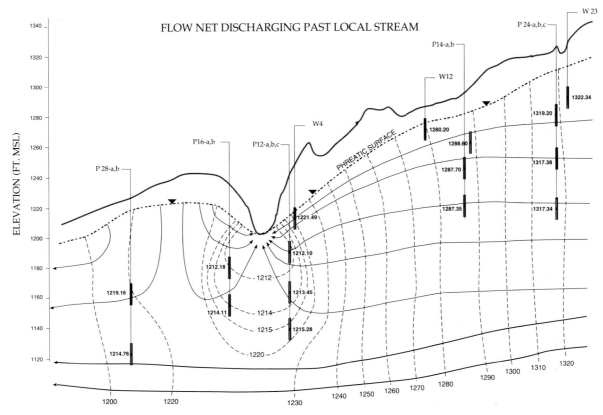

Figure 8-70 Shallow Flow System Discharge

nets. It is recommended that conceptual models be generated for all project sites even if an integrated flow net construction is difficult or impossible for the geologic environment. The conceptual model can provide the key to an effective ground-water monitoring program or aquifer remediation. As such, conceptual models should form part of the overall site interpretation, before designs of the facility monitoring or aquifer remediation.

REFERENCES

ASTM, 1991, Annual Book of ASTM Standards D5092 (Designing and Installation of Ground Water Monitoring Wells in Aquifers); Annual Book of ASTM Standards, American Society for Testing and Materials, Philadelphia, vol. 04.08, pp

Back, W., and R. A. Freeze, 1985. Physical Hydrogeology

Bear, J. and Dagan, G., The Relationship Between Solutions of Flow Problems in Isotropic and Anisotropic Soil. J. Hydrol.,3, 1965.

Casagrande, A. 1937. Seepage Through Dams, Harvard University Pub. 209, reprinted from Journal of the New England Water Works Assoc., June, 1937.

Cedergren, H.R. 1967. Seepage, drainage, and flow nets. John Wiley & Sons, New York, pp. 184–204.

Cooley, R. L. et. al., 1972. Hydrologic Engineering Methods for Water Resources Development, Vol.10; Principles of Ground-Water Hydrology, Corp. of Engineers, Davis California.

Ferris, J.G., D.B. Knowles, R.H. Brown, and R.W. Stallman. 1962. Theory of aquifer tests. U.S. Geol. Survey Water–Supply Paper 1536–E, pp. 69–174.

Fishel, V. C., 1935. Further Tests of Permeability with Low Hydraulic Gradients, Transactions, American Geophysics Union, p. 503.

Foley, F. C., W. C. Walton and W. J. Dreschler, 1953. Ground-water Conditions in the Milwaukee-Waukesha Area, Wisconsin, U. S. Geological Survey Water Supply Paper 1229.

Freeze, R.A., and J.A. Cherry. 1979. Ground water. Prentice–Hall, Inc., NJ, 604 pp.

Freeze, R.A., and P.A. Witherspoon. 1967. Theoretical analysis of regional ground water flow. American Geophysical Union, Water Resources Research, 3, pp. 623–634.

Hubbert, M.K. 1940. Theory of ground water motion. J. Geol., 48, pp. 785–944.

Hughs, T., 1991, Personal Communication.

Leonards, G. A. (1962), Foundation Engineering, McGraw-Hill Book Co., New York, pp. 107-39.

Liakopoulos, A.C., "Variation of the Permeability Tensor Ellipsoid in Homogeneous Anisotropic Soils." Water Resources Research,l, 1965.

Marcus, H. and Evenson, D.E., 196. Directional Permeability in Anisotropic Porous Media. Water Resources Center Contribution No. 31., Univ. of California, Berkeley.

Maasland, M., 1957. Soil Anisotropy and Land Drainage. Drainage of Agricultural Lands, ed. J.N. Luthin. American Society of Agronomy, Madison, Wis.

McWhorter and Sunada, 1977 Ground-Water Hydrology and Hydraulics, Water Resources Publications, Frot Collins, CO. 80522

Muskat, M. , 1937. Flow of Homogeneous Fluids Through Porous Media. McGraw-Hill Book Co., New York, (Litho-graphed by Edwards Bros., Ann Arbor, Michigan, 139 p.

Taylor, D. W. 1948. Fundamentals of Soil Mechanics, John Wiley and Sons, New York, pp. 97-123.

Todd, D.K. 1980. Groundwater Hydrology (2nd edition). John Wiley and Sons. 535 p.

Toth, J. 1963. A theoretical analysis of ground water flow in small drainage basins. J. Geophys. Res., 68, pp. 4796–4812. (Also published in Natl. Res. Council Canada, Proc. Hydrol. Symp. 3, Ground Water, pp. 75–96, 1963.)

Toth, J.A. 1962. Journal of Geophysical Research, 67:4375–4387.

U.S. Environmental Protection Agency, 1986.Criteria for Identifying Areas of Vulnerable Hydrogeology Under the Resource Conservation and Recovery Act, Appendix B, Interm Final. PB86-224979.

Walton, W.C., 1962. Selected analytical methods for well and aquifer evaluation. Illinois Department of Registration and Education Bulletin 49. 81 p.

Willman, et.al., 1975. Handbook of Illinois State Geologic Survey Bulletin 95, Urbana, Illinois 261 pages

CHAPTER 9

MONITORING SYSTEM DESIGN

In designing an "optimum or ideal" ground-water monitoring system, a wish list of system attributes can be formulated:

- Three dimensional array of monitoring points for discrete sampling and hydraulic testing;

- Continuous real time measurements of chemical parameters of chemical parameters and hydraulic head at each monitoring point,

- As few as possible holes penetrating the facility area;

- Sufficient monitoring points so that complex geology will not confound detection of potential releases from the facility;

- Significant releases observable by sufficiently frequent measurements of indicator parameters;

- Reliable installations to maintain reproducibility and representativeness of the sampling; and,

- Convenient maintenance and quality auditing.

These seven attributes of an ideal ground-water monitoring system are not obtainable by current technology, however, many aspects of these points can be approached through a well-designed monitoring system using a conceptual understanding of the geologic conditions present on the target site.

Since the majority of ground-water systems move slowly and in predictable pathways (if you understand the subsurface environment sufficiently!), the attributes of continuous monitoring at many points can be achieved if the target monitoring zone (ASTM, 1990) is carefully selected and a reasonable sampling period is established.

Techniques provided in previous chapters of this manual combined with those described in the ASTM D 5092 standard "Practice for the Design and Installation of Ground Water Monitoring Wells in Aquifers" will effectively address the remaining 5 points expressed above.

Selection of the proper locations for monitoring wells should be based on a holistic approach to the evaluation of a specific site. The placement of the wells in this process must weigh and balance data collected in the field, laboratory, and office.

The question "How much detection monitoring is enough?", when answered in the context of the number of monitoring wells required at a site, will be entirely site–specific. In general, the monitoring system designer should ensure that convincing evidence is established for each assumption, and for demonstrating the basic capability of the system to produce ground–water samples representative of both upgradient (background) and downgradient conditions. General rules of thumb are provided in this chapter, but the reader should bear in mind that "enough" is a subjective determination, both for the questions of "how much" monitoring is necessary to provide a monitoring system capable of detecting ground-water-contamination, and "how much" demonstration is required to convince a regulatory agency of that capability.

The key word in most regulatory programs that require ground-water monitoring is demonstration that the system is capable. The owner or operator of a facility required to monitor ground water must install and implement a monitoring system capable of determining the facility's impact on ground water; it must be capable of yielding representative ground–water samples for analyses. The number, locations, and depths of the detection monitoring wells must be such that the system is "capable" of the "prompt" detection of any statistically significant differences in indicator parameters.

The monitoring system designer must base decisions on sufficient numbers and locations of monitoring wells on performance-oriented criteria to describe a sufficient monitoring system since, it will be a very unusual monitoring situation in which as few as several downgradient wells would insure system capability. Some very simple geologic environments can be effectively monitored with the U.S. EPA–specified (RCRA 1982) minimum system of one upgradient and three downgradient wells, however, this level of monitoring may be representative of very few sites. It is not uncommon for monitoring systems to employ many sampling points in detection monitoring. This is especially true for sites located in heavily regulated states. This is also true for facilities in operation over long periods of time and consisting of multiple cells or expansions.

9.1 REGULATORY CONCEPTS IN FACILITY MONITORING

Many regulatory concepts surround monitoring of all types of facilities, but specifically hazardous and solid waste management facilities evolve around compliance with U.S. EPA regulations on ground–water monitoring. Specifically RCRA regulations (40 CFR 264.97) set requirements for ground-water monitoring at hazardous waste sites. For example, the owner or operator of a hazardous waste management facility must comply with the following requirements for any ground–water

monitoring program developed to satisfy §264.98, §264.99, or §264.100:

(a) The ground–water monitoring system must consist of a sufficient number of wells, installed at appropriate locations and depths to yield ground–water samples from the uppermost aquifer that:

(1) Represent the quality of background (upgradient) ground water that has not been affected by possible leakage from a facility; and

(2) Represent the quality of ground water downgradient of the facility.

(b) If a facility contains more than one regulated unit, separate ground–water monitoring systems are not required for each regulated unit provided that provisions for sampling the ground water in the uppermost aquifer will enable detection and measurement at the compliance point of hazardous constituents from the regulated units that have entered the ground water in the uppermost aquifer.

Many millions of dollars and millions of words in reports and meetings have been spent on defining exactly what these relatively few lines of text really mean in the context of actual monitoring of hazardous waste sites.

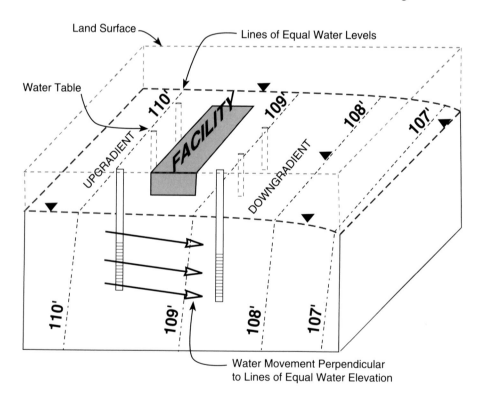

Figure 9-1 Regulatory Context of Detection Monitoring

Both RCRA Subtitle C (hazardous waste) and Subtitle D (solid municipal waste) facilities are required to meet these basic points of detection monitoring programs. This Federal rule can be depicted in a single figure that illustrates the concept of detection monitoring. Figure 9–1 shows a conceptual presentation of the §264.97 guidance on placement of detection monitoring wells.

The RCRA Ground–Water Monitoring Technical Enforcement Guidance Document (TEGD), (U.S. EPA, 1986) provided additional guidance on placement and number of upgradient or background wells by recommending that these wells are:

• Located beyond the upgradient extent of possible contamination from the hazardous waste management unit so that they reflect background water quality;

• Screened at the same stratigraphic horizon(s) as downgradient wells to ensure comparability of data; and,

• Of sufficient number to account for heterogeneity in background ground-water quality.

The conceptual homogeneous unconfined uppermost aquifer, Figure 9–1, would still meet the above three TEGD requirements; however, this conceptual hydrogeologic condition is seldom observed in the field as such a simple unconfined aquifer flow system. RCRA Subtitle D (Oct. 13, 1991) includes many components specific to ground-water monitoring.

Section 258.51 (ground-water monitoring systems) requires that:

(a) A ground-water monitoring system must be installed that consists of a sufficient number of wells, installed at appropriate locations and depths, to yield ground-water samples from the uppermost aquifer (as defined in Section 258.2) that:

(1) Represents the quality of background ground water that has not been affected by leakage from a unit. A determination of background quality may include sampling of wells that are not hydraulically upgradient of the waste management area where:

(i) Hydrogeologic conditions do not allow the owner or operator to determine what wells are hydraulically upgradient; or

(ii) Sampling at other wells will provide an indication of background ground-water quality that is as

representative or more representative than that provided by the upgradient wells; and

(2) Represent the quality of ground water passing the relevant points of compliance specified by Director of an approved State under Section 258.40(d) or at the waste management unit boundary in unapproved States. The downgradient monitoring system must be installed at the relevant point of compliance specified by the Director of an approved State under Section 258.40(d) or at the waste management unit boundary in unapproved States that ensures detection of ground-water contamination in the uppermost aquifer. When physical obstacles preclude installation of ground-water monitoring wells at the relevant point of compliance at existing units, the down-gradient monitoring system may be installed at the closest practicable distance hydraulically down-gradient from the relevant point of compliance specified by the Director of an approved State under Section 258.40 that ensures detection of ground-water contamination in the uppermost aquifer.

The Director of an approved State under Section 258.51 (d) may approve a multi-unit ground-water monitoring system instead of separate ground-water monitoring system meets the requirement of Section 258.51 (a) and will be as protective of human health and the environment as individual monitoring systems for each MSWLF unit, based on the following factors:

(1) Number, spacing, and orientation of the MSWLF unite;
(2) Hydrogeologic setting;
(3) Site history;
(4) Engineering design of the MSWLF units, and
(5) Type of waste accepted at the MSWLF units.

RCRA Subtitle D Section 258.51 in section (2) goes on to require that in subsection (d) The number, spacing, and depths of monitoring systems shall be:

(1) Determined based upon site-specific technical information that must include thorough characterization of:
(i) Aquifer thickness, ground-water flow rates, ground-water flow direction including seasonal and temporal fluctuations in ground-water flow; and,

(ii) Saturated and unsaturated geologic units and fill materials overlying the uppermost aquifer, materials comprising the uppermost aquifer, and materials comprising the confining units defining the lower boundary

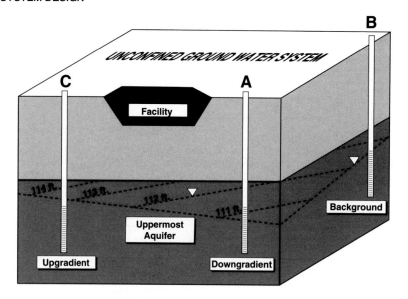

Figure 9-2a Unconfined Ground Water System

of the uppermost aquifer; including, but not limited to: thicknesses, stratigraphy, lithology, hydraulic conductivities, porosities and effective porosities.

The above technical Subtitle D requirements are comprehensive as to the full definition of site geological and hydrogeologic conditions. Many waste disposal facilities are located in complex geologic environments in which preliminary (or very extensive) site assessment investigations are required to properly locate the wells for detection ground–water monitoring systems as described above in the new RCRA Subtitle D regulation Layering of geologic units of significantly different hydraulic conductivity complicates the simple conceptual picture

described by the Federal rule. Figures 9–2a and 9–2b show a two–layer system with the uppermost aquifer consisting of homogeneous isotropic sand below a near surface silt/clay unit of lower hydraulic conductivity.

In Figure 9–2a, the uppermost aquifer is unconfined, in that water only partly fills an aquifer, (i.e., potentiometric surface) the upper surface of the saturated zone is free to rise and decline. Where water completely fills an aquifer that is overlain by a confining bed as shown in Figure 9-2b the aquifer is said to be confined by the lower hydraulic conductivity unit. Downgradient well positions are shown as point B in both figures. Both upgradient and background wells are also shown in these figures. The concept of background representing not hydraulically

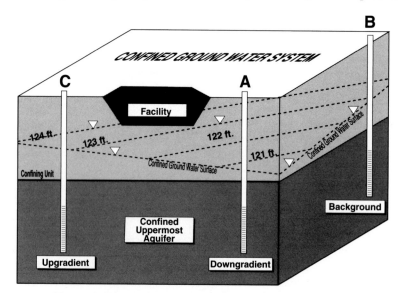

Figure 9-2b Confined Ground Water System

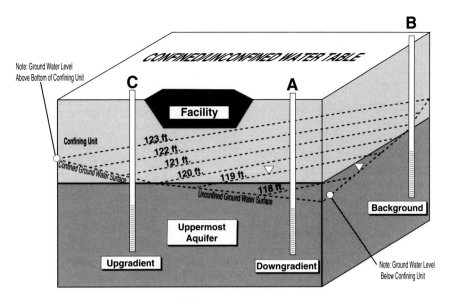

Figure 9-2c Unconfined/confined Ground Water System

"upgradient" locations but reflecting general water quality of the uppermost aquifer are represented by point C. In each case, the sand unit should be considered as the uppermost aquifer for the following reasons:

• The sand unit has regional extent and is saturated;

• The sand unit has sufficient hydraulic conductivity to produce usable quantities of water to springs or wells; and,

• The sand unit would be the zone in which leachate from the facility could migrate horizontally away from the site to potentially affect human health and the environment.

Much of the early concern of regulatory agencies in respect of subtitle C detection monitoring programs is with meeting Federal regulations in 40 CFR 265.91; which describe ground–water monitoring system requirements for interim status hazardous waste disposal facilities. These regulations state: "A ground–water monitoring system must be capable of yielding ground–water samples for analysis and must consist of:

(1) Monitoring wells (at least one) installed hydraulically upgradient (i.e., in the direction of increasing static head) from the limit of the waste management area. Their number, locations, and depths must be sufficient to yield ground–water samples that are:

(i) Representative of background ground–water quality in the uppermost aquifer near the facility; and

(ii) Not affected by the facility; and

(2) Monitoring wells (at least three) installed hydraulically downgradient (i.e., in the direction of decreasing static head) at the limit of the waste management area. Their number, locations, and depths must ensure that they immediately detect any statistically significant amounts of hazardous waste or hazardous waste constituents that migrate from the waste management area to the uppermost aquifer." This interim status rule on ground–water monitoring has several key features different from §264.97 rules that have been widely used in defining what a detection monitoring system should consist of, specifically:

• One monitoring well upgradient and three downgradient from a facility; and,

• The system must have "immediate" detection capabilities.

While "immediate" detection is open to widely variable interpretation, especially considering the slow movement of ground water, the TEGD (U.S. EPA, 1986) provides some additional guidance on how to meet the "immediate" criteria by placing detection monitoring wells immediately adjacent to the waste management unit. The Federal Subtitle D regulations proposed for non–hazardous solid waste sites, (early 1989), set 150-meter buffer zones (or property boundary, whichever is less) for placement of monitoring wells. This buffer zone was also included in the final rules which might include a compliance boundary set at the edge of the waste management unit (EPA Oct., 1991). Reducing these Federal regulations to a series of criteria based on concepts presented in this text would result in a series of technical points. The detection monitoring system should have:

• Sufficient wells, both upgradient (background) and downgradient, to detect discharges from the regulated facility; and,

• Wells located within a flow path from the regulated facility in the uppermost aquifer;

Furthermore, the uppermost aquifer should have sufficient hydraulic conductivity and extent so that sampling could be conducted within the waste unit boundary for both hazardous waste Subtitle C facilities and Subtitle D solid waste sites. An adequate detection monitoring program can be designed for any geologic/hydrogeologic environment using the above criteria. The following chapter sections present conceptual models for detection monitoring programs for a wide variety of hydrogeologic environments.

Prior to selecting the locations and depths for the screened intervals for ground-water monitoring wells, the ground–water monitoring system designer must have, at a minimum:

• Performed a complete site characterization;

• Established a conceptual hydrogeologic model for the site;

• Constructed a ground-water flow net ; and,

• Located facility boundaries and waste disposal areas.

Each of these tasks provides data that will be used to select the target monitoring zones for the monitoring system. The remaining sections describe the monitoring system process summarized in, Figure 9–3. Examples of the process are included to assist in the design conceptualization process.

9.2 DATA ANALYSIS REQUIRED FOR DESIGN

Geologic factors (related chiefly to geologic formations and their water–bearing properties) and hydrologic factors (related to the movement of water in the formations) must be known in some detail to properly design a ground-water

monitoring system. These data are normally developed in a field investigation as described elsewhere in this manual

The geologic framework of a site includes the lithology, texture, structure, mineralogy, and distribution of the unconsolidated and consolidated earth materials through which ground–water flows. The hydraulic properties of these earth materials depend upon the geologic framework. Thus, the geologic framework of the facility heavily influences the design of the ground-water monitoring system. Elements of the hydrogeologic framework and the site hydrogeology that should be considered in ground-water monitoring system design include:

• The spatial location and configuration of the uppermost aquifer and its hydraulic properties (e.g., horizontal and vertical hydraulic conductivity, depth and location of ground–water surface, seasonal fluctuations of ground–water surface elevation);

• Hydraulic gradient (vertical and horizontal) within the geologic materials underlying the facility;

• Discharge and recharge areas of the site; and,

• Facility operational considerations.

These data are used to establish the locations of both upgradient and downgradient wells in the uppermost aquifer. Both upgradient and downgradient wells should be located in the direction of ground-water flow along flow pathways most likely to transport ground water and the potential contaminants contained in ground water. These pathways should be identified from data gained from existing information and the phased site investigations. The objectives of the field site investigations and subsequent data analysis and interpretations to provide some or all of the following information:

• Lithologic characteristics of the subsurface, including;
  – Established stratigraphic names;
  – Classification of hydrogeologic units;
  – Extent of hydrogeologic units;

• Key hydrogeologic characteristics used to describe the site to the conceptual model including;
  – Hydraulic conductivity (vertical and horizontal);
  – Porosity;
  – Gradient (vertical and horizontal);
  – Specific yield;

• Aquifer characteristics including;

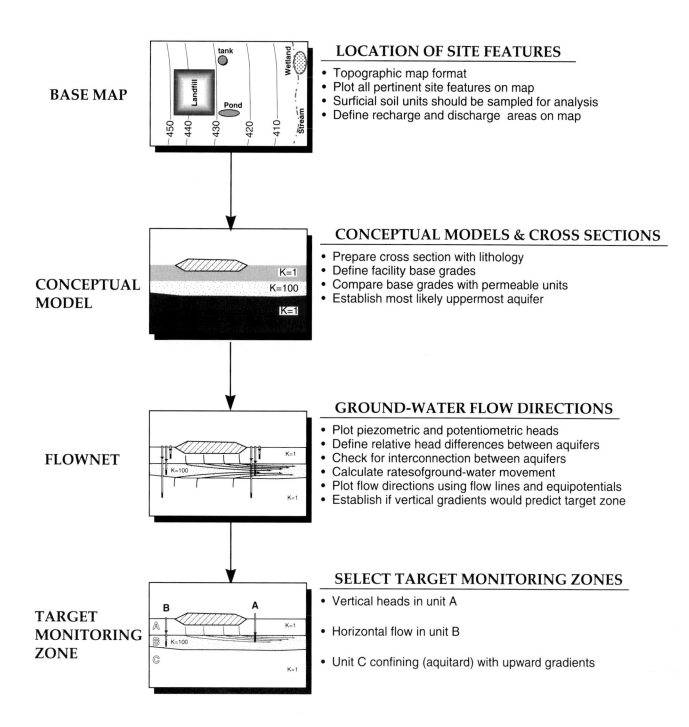

BASE MAP

LOCATION OF SITE FEATURES

- Topographic map format
- Plot all pertinent site features on map
- Surficial soil units should be sampled for analysis
- Define recharge and discharge areas on map

CONCEPTUAL MODEL

CONCEPTUAL MODELS & CROSS SECTIONS

- Prepare cross section with lithology
- Define facility base grades
- Compare base grades with permeable units
- Establish most likely uppermost aquifer

FLOWNET

GROUND-WATER FLOW DIRECTIONS

- Plot piezometric and potentiometric heads
- Define relative head differences between aquifers
- Check for interconnection between aquifers
- Calculate ratesofground-water movement
- Plot flow directions using flow lines and equipotentials
- Establish if vertical gradients would predict target zone

TARGET MONITORING ZONE

SELECT TARGET MONITORING ZONES

- Vertical heads in unit A

- Horizontal flow in unit B

- Unit C confining (aquitard) with upward gradients

FIGURE 9-3 Monitoring System Design Summary

– Boundaries;
– Type of aquifers; and,
– Saturated/unsaturated conditions.

Each piece of data is an important building block in establishing the conceptual hydrogeologic model and targeting zones to be monitored. These data are used in combination to define the uppermost aquifer and hydraulic gradients to allow the construction of flow net that will provide identification of aquifer flow pathways so that target monitoring zones can be selected.

9.3 SELECT TARGET MONITORING ZONES

The first task in the design of a detection ground–water monitoring system is the selection of the target monitoring zone (see ASTM 5092-90). The logic used in selection of the target monitoring zone is illustrated in Figure 9–4. A review of features of the facility to be monitored, used in combination with conceptual models and flow nets provide the system designer with the information to select those zones that will provide a high level of certainty that releases from the facility will be "immediately' detected. The concept of the target monitoring zone was developed as a means of directing the ground–water monitoring system designer toward placement of well screens in the uppermost aquifer at locations and depths that would have the highest likelihood of detecting leakage from a facility. Target monitoring zone is defined in ASTM standard D-5092 as the ground-water flow path from a particular area or facility in which monitoring wells will be installed. The target monitoring zone should be a stratum (strata) in which there is a reasonable expectation that a vertically placed well will intercept migrating contaminants. This target zone usually lies in the saturated geologic unit in which ground-water flow rates are the highest because it possesses the highest hydraulic conductivity of the material adjacent to or underlying the facility of interest. Figure 9-4 illustrates the process of selection of a target monitoring zone using information on facility features, geologic characteristics and hydraulic characteristics gathered during the preliminary field investigations. This selection process can be described as a series of steps:

- **Step 1: Locate Site Features on a Topographic Base Map Format:** site features should be compared to information on geologic and soil maps to define the location of important facility components in relation to the distribution of surficial materials. Any likely recharge/discharge areas (streams, wetlands or other surface-water) should be located.

- **Step 2: Cross Section Construction and Conceptual Model Development:** cross sections should be constructed, based on boring logs and/or geophysical traverses. These sections should be compared with the location of site features and facility components. The base grades of the facility should be plotted on cross–section of sensitive geologic units or ground-water flow pathways. A conceptual model should be constructed to establish the site geological framework and to illustrate distribution of geologic materials of differing hydraulic conductivity.

- **Step 3: Use Flow net to Define Likely Direction Ground–Water Flow:** Construction of flow nets will assist in defining the gradient and direction of ground-water flow in the uppermost aquifer. The rates of flow along flow paths can be calculated from the information provided by the flow net using equations from Chapter 7.0. Vertical gradients can be used to predict target zones by comparison of relative heads between units. Interconnections between aquifers can be predicted from relationships of relating hydraulic conductivity to hydraulic heads for the units defined in the conceptual models.

- **Step 4: Select Target Monitoring Zones:** The zone meeting the regulatory definition of the uppermost aquifer, which also shows primarily horizontal ground–water movement under or adjacent to the facility would therefore represent the target monitoring zone. This zone would probably represent a permeable unit that is discharging to other permeable units or to local discharge areas. The system designer should be aware of the flow paths within the uppermost aquifer that would represent the most likely zones of ground–water movement away from the facility. These zones, are typically those with the highest hydraulic conductivity would be the focus of the detection monitoring system. If interconnected aquifers are present, these units should be monitored as necessary to provide safeguards of downgradient ground-water users.

This four–step procedure for selecting the target monitoring zone must be flexible enough to accommodate environmental effects due to seasonal changes in gradient, or due to future plans to expand or alter the configuration of the facility. The target monitoring zone might include only a portion of a very thick aquifer (for example, the top 30 feet), or span several geologic units (as in the case of a thin, permeable, unconsolidated unit overlying weathered and/or fractured bedrock). These target zones represent the

SUMMARY OF GROUND-WATER MONITORING SYSTEM DESIGN PROCESS

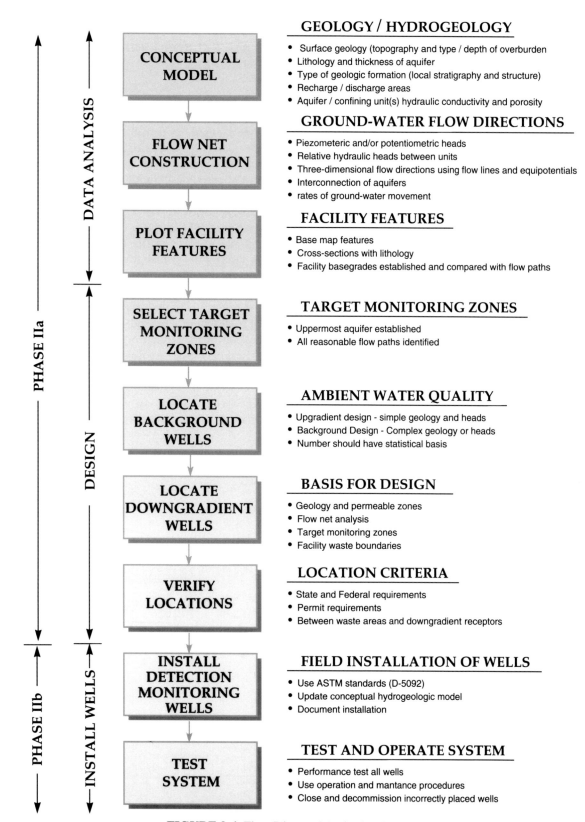

FIGURE 9-4 Flow DiagramMonitoring System Design

Plan View

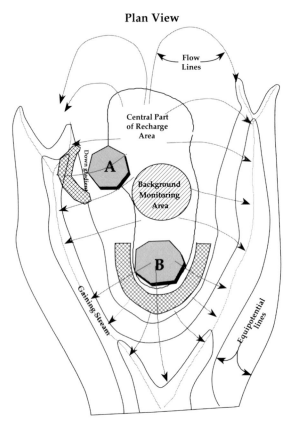

FIGURE 9-5 Potential Target Monitoring Areas

and/or fractured bedrock). These target zones represent the proper location for placement of monitoring wells.

9.4 LOCATE BACKGROUND AND DOWN GRADIENT WELLS

9.4.1 Gradients

The basis for detection monitoring programs is knowledge of the hydraulically upgradient and downgradient direction from the site to be monitored. Figure 9–1 illustrated a simple relationship of ground–water movement from higher potentiometric surface elevations (upgradient) to lower potentiometric elevations (downgradient). This simple conceptual model of a homogeneous aquifer is the basis for much of the regulatory thought on ground–water monitoring. It is also unfortunately rare to find such a simple flow configuration in the real world.

After the selection of the target monitoring zone(s), the next step in design of a ground–water monitoring system is the location of upgradient or background monitoring wells. The conceptual geologic model and flow net construction will have defined the uppermost aquifer and the relative direction of ground-water flow, both vertically and horizontally. Selection of upgradient wells for analysis should be based not only on this information, but

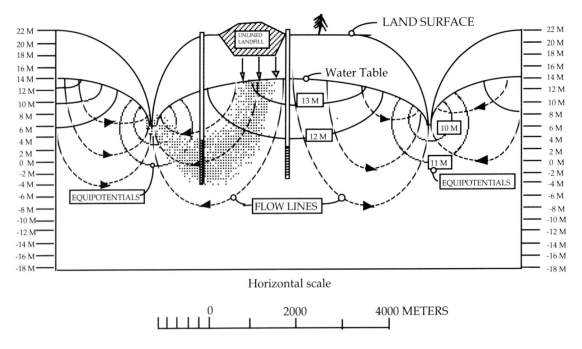

FIGURE 9-6 Cross Section of Target Monitoring Zones

also on other factors mainly relating to the physical presence of the facility. The numbers of upgradient or background wells installed at a site must be based on the size of the facility, the geologic/hydrogeologic environment, and the ability to satisfy statistical criteria for analysis of water quality data. As general guidance, it is very difficult to pass any type of statistical test (applicable under RCRA), unless more than one upgradient well is used in the monitoring system. This is due to the natural spatial variability observed in geologic environments. This spatial variability must be evaluated during the design process so that sufficient background water quality data is available for background to downgradient water quality statistical comparisons.

The TEGD (U.S. EPA, 1986) defines upgradient wells as "one or more wells that are placed hydraulically upgradient of the site and are capable of yielding ground–water samples that are representative of regional conditions and not affected by the regulated facility."

This usage of the term "upgradient" is consistent with 40 CFR 265.91, which links background and upgradient for interim RCRA sites. Background wells would meet the 40 CFR 264.97 test to "represent the quality of background water that has not been affected by leakage from a regulated, unit and represent the quality of ground water passing the point of compliance." The term "upgradient" can be a difficult concept to demonstrate in ground–water monitoring system design, because field conditions may not match the simple, regulatory models. As a closing statement on the relationship between upgradient and downgradient wells, a correctly designed (located) detection monitoring well will only have "upgradient" the facility to be monitored. This is due to the placement of the downgradient well screen within the flow path from the facility. The designer must closely consider site-specific hydraulic conditions to accurately locate upgradient monitoring wells because ground water does not always flow as expected in simple regulatory models; horizontally from upgradient to downgradient areas.

Simple single–aquifer flow systems are established by a clear understanding of the directional movements of ground water through evaluation of the ground-water gradients across a site. Figures 9–5 and 9–6 illustrate, in plan view and cross–section, the flow around a gaining stream, where discharging ground water provides the stream's base flow. Figure 9–7a and 9-7b illustrate conversely, flow around a losing stream, where surface water supports adjacent ground-water levels.

In each of these cases, this simple system provides directional components to allow the positioning of ground–water monitoring wells. Figures 9–5 and 9–6 illustrate a facility (B) located in a recharge area that

discharges to streams on either side of the facility. Ground-water flow lines are shown in plan and cross section. Because the facility is sitting directly on top of the recharge area, the down gradient flow zone is composed of a wide arc around the facility. This example provides perhaps the simplest gradient controlled system.

Potential target zones for a detection monitoring system are shown in Figure 9–5. Upgradient background water quality target zones should be sufficiently within the recharge area so as not to be affected by the facility. Several conclusions can be drawn from Figures 9–5 and 9–6:

• Facility A would have its downgradient monitoring wells located within the ground-water flow lines shown. This facility location would have background monitoring well(s), located in the central recharge area.

• Facility B would have an upgradient or background well in the area indicated. Because the facility is located directly within the local recharge area, this would not be considered an upgradient well, but rather a background well that represents water quality similar to a well that would be upgradient from the facility.

• Actual flow conditions would result in a water table significantly flatter than that shown in Figure 9–6, vertical exaggeration (of approximately 125 to 1) makes the flow lines appear to travel deeper than would be represented in real life. The vertical scale indicates that the monitoring wells installed at the site should be screened from 19 to 24 meters below ground surface to intercept the ground-water flow (and any contained contaminants) emanating from beneath the site.

Figure 9–7a and 9-7b illustrate conversely, a "losing stream" condition and the resultant monitoring target zones for Facility A. Because the stream in this illustration is recharging ground water, and thus represents the highest point of upgradient ground water, target monitoring zones are located along the flow lines shown on Figure 9–7. Depths of screen placement must be based on the projected vertical gradients in the area. One can observe from the example provided that the location of a detection monitoring system is particularly sensitive to whatever the stream is "gaining" or "losing". This relatively simple complication can lead to incorrect location of downgradient detection monitoring wells. Piezometers located perpendicular to the stream including the careful evaluation of stream flow can provide the basic data to define the recharge/discharge relationship of the surface and ground-water system.

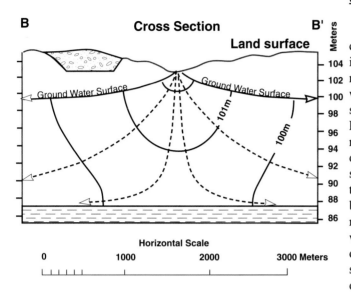

FIGURE 9-7 a Losing Stream Target Monitoring Zones

Steep/Flat Gradients

Even simple, single–aquifer systems require consideration of local gradients adjacent to the facility of interest. In an area with a relatively flat gradient it is necessary to consider possible ground-water flow in what would normally be an upgradient direction. In an area with steep gradient on the water-table surface, as shown in Figure 9–8 (typical of lower hydraulic conductivity materials), there is little potential for reversal of flow directions. The target monitoring zone in an area with such a steep gradient would normally be much narrower than in the flat gradient environment. The relationship between horizontal and downward gradients are still required to establish the depth of the detection monitoring well screens. Figure 9-8 shows placement of two downgradient piezometers (A & B). A monitoring well screened at B would meet the regulatory criteria of being downgradient from the facility (and from upgradient piezometer C), however, the flow path screened by B would be too deep to intercept flow from the unlined

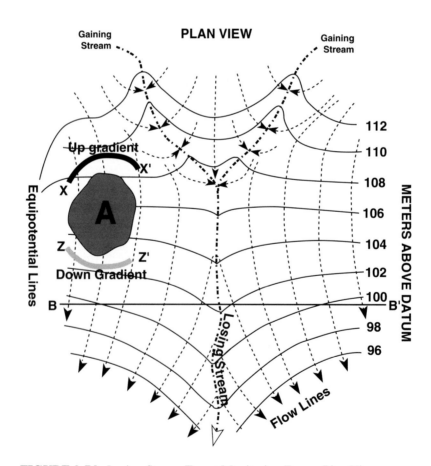

FIGURE 9-7 b Losing Stream Target Monitoring Zones- Plan View

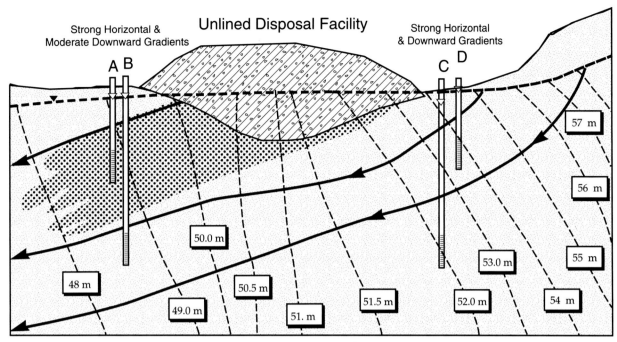

FIGURE 9-8 Steep Gradient Facilities Example

facility. Detection monitoring wells located at the depth of piezometer A would correctly monitor the facility.

Figures 9–9 and 9–10 show an unlined landfill within an area where the hydraulic gradient is low. The target monitoring zone is characteristically thicker than it would be in an area with high hydraulic gradient (as shown in Figure 9–8). The discharge directions shown in these figures probably represent a common hydraulic condition for unlined facilities. The cross-sectional view (Figure 9-9) shows both intermediate and shallow flow cells. The deeper intermediate flow system, at least in this case, is not affected by the facility. The local independent shallow flow cell is discharging in, what could be viewed as, both downgradient and a perceived upgradient direction. The upgradient component due to the higher heads observed at piezometer D (52.0 m) and lower heads in the other three piezometers (A, B, and C). These discharge recharge cells relationships require significant interpretative skills by the designer, as well as, sufficient field piezometric data. A plan view of this type system is shown in Figure 9-10. The local shallow ground-water system discharging from the facility caused a disturbance in the regional ground-water system. The actual disturbance may be difficult to establish in the field, so sufficient care should be exercised to locate background monitoring wells out of the local cell influence. The localized flow cells depicted in Figure 6-9 and 6-10 probably represent, more often than not, a typical flow net for a low gradient site. The local flow cells discharging around the topographically higher site, would

cause downgradient monitoring wells to be located in what would typically be called, an upgradient location. Deeper screening of the detection monitoring wells at locations B or C would place the wells in the deeper intermediate flow cell. As such, they would not represent truly downgradient monitoring of the facility.

Procedures for Gradient Controlled Sites

Even with simple-single hydraulic conductivity environments, care must be taken to fully understand the three-dimensional nature of ground-water flow. As general guidance the designer should:

1. Establish lithology and gradients as with single–aquifer systems;

2. Compare natural (base line) gradients across the site and hydraulic conductivity of aquifer; and,

3. Select position for upgradient monitoring well, as in position D of Figures 9-9 and 9–10.

Gradient Control/Flow nets

Unfortunately most real-world geologic systems are not composed of simple single layers. Once observed field conditions include stacked, variable hydraulic conductivity layers somewhat more difficult evaluations of how ground water is affected by the variable geologic materials, comes into play.

Figure 9–11 shows an unconfined aquifer separated from a confined aquifer by a low hydraulic conductivity confining bed. Ground–water movement through this system involves flow not only through the aquifers but also across the confining bed. The hydraulic conductivity of aquifers are tens to thousands of times greater than those of confining beds, the result since, for a given rate of flow, the head loss per unit of distance along a flow line is tens to thousands of times less in aquifers than it is in confining beds. Consequently, lateral flow in confining beds usually is negligible, and flow in aquifers tends to be parallel to aquifer boundaries, as shown in Figure 9–11.

Differences in the hydraulic conductivity of aquifers and confining beds cause refraction or bending of flow lines at their boundaries, as was described in Chapter 8.0 on conceptual models and flow nets. As flow lines move from aquifers into confining beds, they are refracted toward the direction perpendicular to the boundary. In other words, they are refracted in the direction that produces the shortest flow path in the confining bed. As the flow lines emerge from the confining bed, they are refracted toward the direction parallel to the boundary

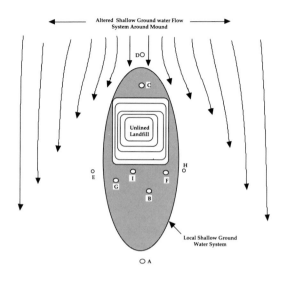

FIGURE 9-10 Map View of Local Flow Cells

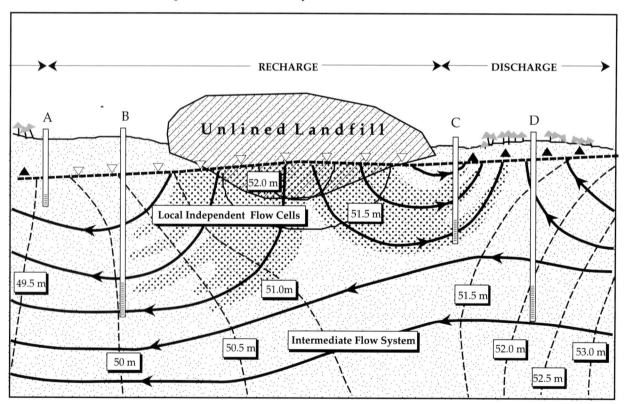

FIGURE 9-9 Conceptual Model of Local Flow Cells

Example Equipotential Lines in a Geologic Model

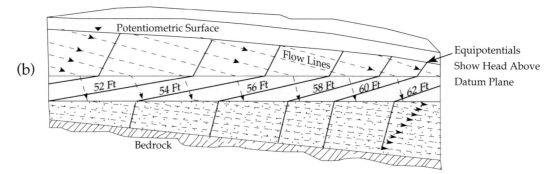

FIGURE 9-11 Unconfined/confined Flow Nets

(Figure 9–11). Hence, ground water tends to move horizontally in aquifers and vertically in confining beds or low hydraulic conductivity materials. This observation has a high level of importance in the design location and depths of the facility detection monitoring system.

Lateral flow components in aquifers have direct relevance to ground–water monitoring system design, since the physical location and depth of the wells must correspond to the overall three-dimensional components of flow typically at the edge of the facility. Most detection monitoring programs concentrate on establishing target monitoring zones in the uppermost aquifers beneath a site. These target monitoring zones are directly correlated with the hydrostratigraphic zone that has the highest rate of flow away from the facility so that "immediate" detection of leachate migration could be detected from the facility. Some assessment programs may involve monitoring the uppermost aquifer, deeper aquifers, and zones between the uppermost and deeper aquifers. As a general statement; the three-dimensional ground-water flow established by piezometer hydraulic-head relationships are necessary for either detection or assessment monitoring design.

The movement of water through aquifer/confining unit systems is controlled by the vertical and horizontal hydraulic conductivity, the thicknesses of the aquifers and confining beds, the recharge and discharge (boundary) areas, and hydraulic gradients. Because of the relatively large head loss that occurs as water moves across confining beds, the most vigorous circulation of ground water normally occurs through the shallowest aquifers.

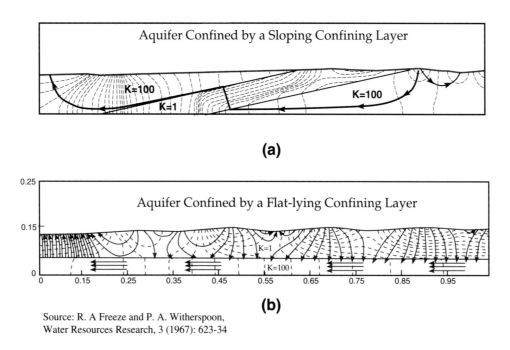

Source: R. A Freeze and P. A. Witherspoon,
Water Resources Research, 3 (1967): 623-34

FIGURE 9-12 a-b Regional Ground-Water Flow in Confined Aquifer

Movement generally becomes slower as depth increases (Heath, 1985). The uppermost aquifers will usually show contamination first (unless a direct conduit for downward movement exists into deeper aquifers) and thus must be served by monitoring efforts. The concentration of flow lines in aquifers is illustrated further by Figures 9–12a and 9–12b (Freeze and Witherspoon, 1967). Aquifers bounded by a sloping confining layer and a flat–lying confining unit, as may present for example, in glaciated regions where low hydraulic conductivity tills overlie higher hydraulic conductivity outwash sand and gravel aquifers. Nearly vertical flow occurs through the generally thick, low hydraulic conductivity materials, while nearly horizontal flow occurs within the underlying aquifer. The aquifer represents the only zone in which ground water moving away from a facility could be properly intercepted and monitored, and thus, should be considered the target monitoring zone.

This concept is further illustrated in Figures 9–13 where the piezometers installed at increasing depths in the confined aquifer and in the confining zone indicate that a strong downward gradient exists in the fine–grained overburden material. Monitoring wells located in Figure 9-13 at A-3 and B-3 would represent background and downgradient, respectively. The entire target zone should be screened in both these locations. Figure 9-14 illustrates an unconfined flow system in a recharge area. Recharge areas with strong downward gradients may require special consideration of localized shallow flow cells. Depth-location relationships are especially important in such situations. For example, downgradient monitoring wells in the unconfined aquifer, shown in Figure 9-14, should be

located in a target zone screened at or below the interval screened by piezometer B-2.

Figure 9–15 illustrates the potential ground-water flow paths to a discharging stream. Both upgradient wells (A and B) and downgradient wells (C) are shown in this simple conceptual illustration. However, even this relatively simple conceptual model can demonstrate how a shallow downgradient well (C) would not intercept potential leachate flow from the unlined waste disposal area. The downgradient ground–water monitoring point for facilities located in discharge areas must be designed on the basis of shallow, near-surface discharge to wetlands or streams. Upgradient wells should be screened in shallow flow paths, as illustrated by well B. Deeper upgradient wells (as illustrated by well A) would probably suffice, but may not represent ground-water flowing in the target monitoring zone.

Ground–water monitoring in complex alluvial deposits often presents difficult problems in respect to identification of target monitoring zones. These deposits often have shallow sandy zones encapsulated within low hydraulic conductivity sediments. Sand tank experiments have shown that these discontinuous sandy deposits do not affect the downward movement of ground water when strong downward gradients exist. Figure 9–16 shows such a conceptual situation. Shallow permeable zones contained within the low hydraulic conductivity materials do not have significant horizontal gradients; vertical gradients usually predominate in such environments. Monitoring points located adjacent to a facility located in these deposits, (such as well A), may not represent a target monitoring zone. Only where significant horizontal flow exists, as in the regional (uppermost) aquifer, would a

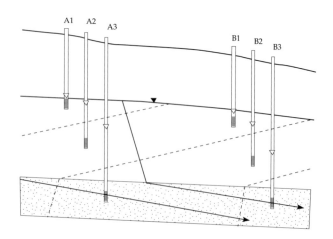

FIGURE 9-13 Confined Aquifer Piezometer Nest

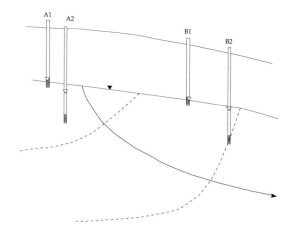

FIGURE 9-14 Unconfined Aquifer Piezometer Nest

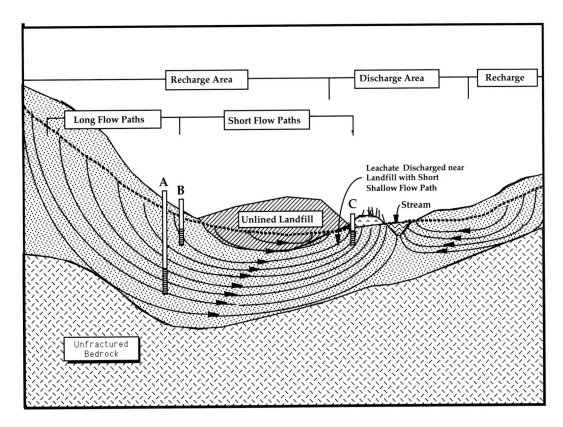

FIGURE 9-15 Shallow Discharging Ground-water System

horizontally downgradient target flow path be found. Well B represents a correct downgradient monitoring point for this situation. However, upper permeable units may represent uppermost aquifers if they have sufficient hydraulic conductivity, and are of sufficient extent to serve as a water source for off–site ground-water users. These more permeable, sandy lenses, channels, and tabular deposits have been observed in many types of geologic environments. These units can range from recent glacial deposits, such as, tills with interlayered outwash sands, to unconsolidated overbank deposits associated with alluvial river deposits, to consolidated claystone deposits with interbedded channel sandstone deposits. There are four important criteria for establishment of the need to monitor saturated sand lenses located within lower hydraulic conductivity units. These criteria are:

• Differential hydraulic conductivity;
• Directional hydraulic heads;
• Unit prevalence; and,
• Unit thickness.

Differential hydraulic conductivity refers to the variation in hydraulic conductivity observed between geologic units. Directional hydraulic head refers to the potential flow directions observed from piezometers located within individual units. Unit prevalence is a qualitative judgment based on the overall site stratigraphic characterization. Unit thickness is defined from the site field drilling program and is based on simple thickness of the observed sandy units.

Each of these criteria must be considered in order for a system designer to decide if a particular permeable unit would require monitoring as a target monitoring zone. Differential hydraulic conductivity represents an order of magnitude comparison of the sandy unit to the adjacent matrix materials. Freeze and Cherry (1979, pg. 173) state that "In aquifer–aquitard systems with permeability contrasts of two orders of magnitude or more, flow lines tend to become almost horizontal in the aquifers and almost vertical in the aquitards." This flow must, however, require that the aquifer discharges into other permeable units, to surface water, or is pumped from the system. Directional heads provide an indication as to the discharge potential of the sandy units. If vertical directional heads are discharging upward (from below the unit) and downwards (from above the unit) into the sandy layers, it is likely that the unit discharges into adjacent lower head areas. The unit prevalence criteria provide an indication as to how continuous the layer is in the field. These data are

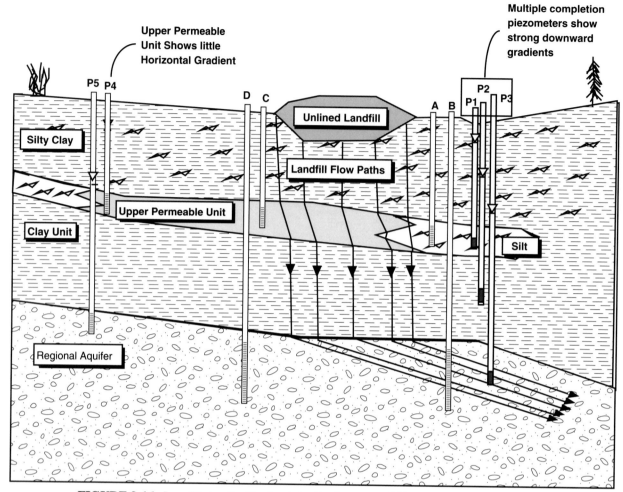

FIGURE 9-16 Low Hydraulic Conductivity Environments With Non-discharging Sand Lenses

gathered during field borehole drilling activity to demonstrate the continuity of the unit in the site area. As general guidance, if all the Phase II borings (100%) contacted the definable unit at equivalent elevations, it is likely that the geologic stratum is continuous. If this contact percentage falls to 50% or shows an elevation variability, the unit is much less likely to represent a continuous feature that should be monitored. An understanding of the depositional history of the geologic unit probably represents the best method for evaluating the continuity of more permeable deposits that could discharge ground water to down gradient, off site areas. Channel deposits may have been cut off by agrading streams during the geologic past. However, sufficient stratigraphic data should be established to confirm such assumptions before ruling out discharge through such linear features. The drilling program also establishes the relevant thickness of these units. If the thicknesses of the saturated permeable

units are very thick (say 100 feet), it would be likely that the unit would require monitoring. As the thickness lessens, the other factors or criteria become important in the overall decision to monitor or not monitor the unit, as the uppermost aquifer. The last criteria, water use of the units, can outweigh all the other factors, assuming that the unit is hydraulically connected between the facility and the downgradient water users. Each of these factors must be weighed in the decision process.

A system designer is often required to monitor "all potential pathways' by regulatory agencies. Rather than blindly installing in monitoring wells in every conceivable permeable unit, the author recommends using technical reasoning on flow path interception. Only monitor geologic units that have a reasonable chance for flow toward downgradient receptors. This may include human or biological receptors. Stick to detection monitoring in the classical "uppermost aquifer' that is discharging off-

LEVELS OF DISCHARGE FROM SANDY UNITS

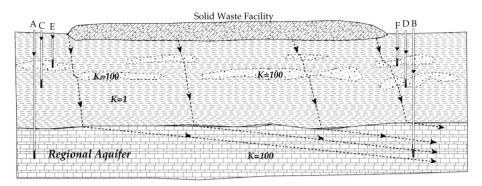

SAND LENSES WITH LITTLE HORIZONTAL DISCHARGE

KEYS IN UNDERSTANDING CONCEPTUAL MODEL
- Confirm sand lenses in drilling program
- Piezometers show verticle gradients in all cases
- Regional aquifer known as productive unit

FIGURE 9-17a Conceptual Model With Non-discharging Sand Lenses

site. In the majority of cases you will meet both the letter of the law and limit long term liability issues with this approach.

If a discontinuous sand unit with a differential hydraulic conductivity of one order of magnitude shows piezometric heads passing through the unit (i.e., heads continued downward through sandy unit) and few borings contacted the approximately 1-foot thick unit, we would not consider the unit as a target monitoring zone. If that unit is 10 feet thick and was contacted by only a few borings, it may be necessary to monitor the unit as the

uppermost aquifer, or it may not. However, if the saturated unit was 20 feet thick, was penetrated by all borings, and showed piezometric heads discharging into the unit from above and below, the unit would probably be monitored as the "uppermost aquifer."

Figures 9–17a through 9–17d illustrate the use of this concept with a series of conceptual models with various levels of discharge from sandy units. The levels of discharge range from almost none in Figure 9–17a, to significant discharge between the unconsolidated and bedrock systems. The interpretation of site hydrogeologic

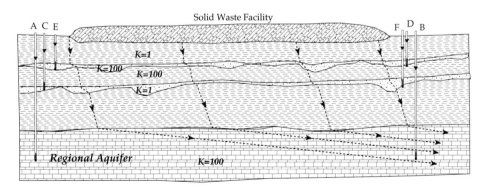

CONTINUOUS SAND LAYERS WITH LITTLE HORIZONTAL DISCHARGE

KEYS TO UNDERSTANDING CONCEPTUAL MODEL
- Well defined sand units present in most boreholes
- Sand layers do not have regional extent
- Downward gradients present in all cases
- Regional aquifer known as productive unit

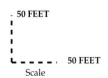

FIGURE 9-17b Conceptual Model With Non-discharging Sand Layers

LEVELS OF DISCHARGE FROM SANDY UNITS

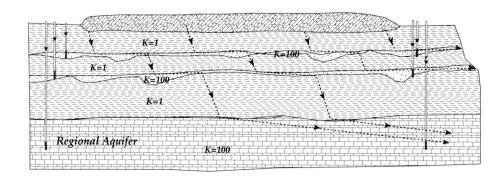

SAND LAYERS WITH SIGNIFICANT HORIZONTAL DISCHARGE

KEYS IN UNDERSTANDING CONCEPTUAL MODEL
• Confirmed sand layers in drilling program
• Verticle gradients in only fine grained units
• Regional aquifer known as productive unit
• Sand layers used locally or discharge to streams

FIGURE 9-17c Conceptual Model With Discharging Sand Layers

condition and thus the design of the monitoring system, in each case, would be based on the following key points:

• The lateral extent and thickness of the various geologic material present;
• The hydraulic conductivity of each of the individual lithologic units;
• The gradients obtained from piezometers placed in each of the permeable units;

• The discharge/recharge potentials of the geologic units present on site.

The conceptualizations and flow net constructions should be based on illustrations with 1 to 1 scales. Figures 9-17a through 9-17d provide some additional keys to interpretation of the appropriate monitoring locations. The conditions depicted in figures 9–17a and 9–17b would point toward the ground-water detection monitoring only be conducted in the regional (uppermost) aquifer. Ground-

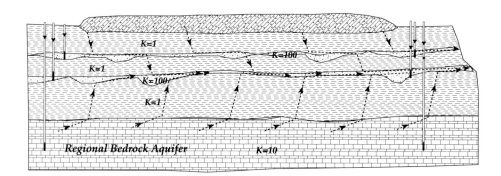

SAND LAYERS DISCHARGEING BOTH BEDROCK AND OVERBURDEN

KEYS TO UNDERSTANDING CONCEPTUAL MODEL
• Well defined sand units present in all boreholes
• Sand layers have regional use as water supply
• Gradients discharge into sand layers
• Regional aquifer discharges into sand layer

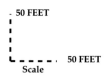

FIGURE 9-17d Conceptual Model With Discharging Sand Layers

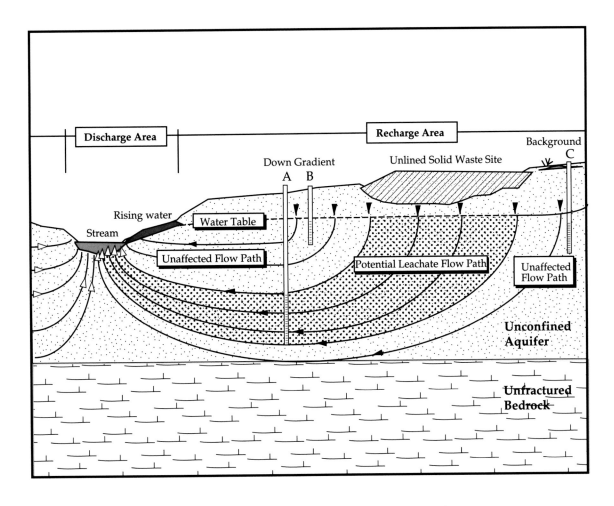

FIGURE 9-18 Conceptual Model of Discharging SandUnits

water discharges down through the two sandy layers (one discontinuous in 9-17b), into the regional bedrock aquifer. The conditions depicted in Figure 9–17d would indicate that sandy unit #2 would represent the better detection monitoring target as the uppermost aquifer. The first sandy unit in Figure 9–17d would not represent an effective monitoring location because of its thin limited discharging nature. Ground-water samples obtained from this unit only be representative of conditions along the edge of the facility within the flow path shown as Y — Y' in Figure 9-17d. While a case may be made that a monitoring well located at D may be necessary to evaluate the Y - Y' area along one side of the facility, virtually 95% of the area would be monitored if wells were placed at down gradient locations in sandy units.

Figure 9–17c represents a situation in which both sandy unit and the regional system should be monitored. This decision to monitor both sandy units should be weighed on the basis of additional site characterization work to determine the regional extent and current/future use of the

sandy units. If the second sandy unit represents a likely flow path and hence a target monitoring zone from the facility, it should be included in the monitoring program.

Detailed evaluations of layered geologic units can be used to define the specific discharging more permeable strata next to a waste disposal area. Figure 9-18 shows a relatively detailed evaluation of a cross sectional area forty feet deep and 200 feet wide. The waste disposal area is just to the left of piezometer pz-13 and a leachate collection line is shown between pz-13 and pz-1 about 10 feet below the ground surface. The flow net superimposed on the cross section was based on both piezometers and wells screened along the cross section line. Recognizing that widely screened zones provide an integrated hydraulic head value, more validity was placed on hydraulic data gathered from piezometers. The results of this linked cross sectional and flow net construction shows the discharging nature of the deeper continuous sandy layer for both deeper less permeable units, and the site area. Selection was made to monitor at a location within the relatively thin

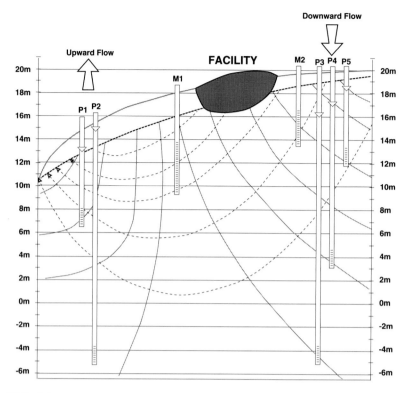

FIGURE 9-19 Gradient Comparisons for Recharging and DischargingAreas

discharging sandy zone. Although there may be some upward movement of ground water from deeper, less permeable units as shown by the upward flow arrows, a screening of the upper 10 foot portion of the 20 foot thick sandy unit at pz-7 (between elevation 417' to 407') would probably represent the optimum location and depth for detection monitoring.

Multiple Piezometers To Establish Flow Relationships

Heads established by multiple piezometers can identify the potential flow paths from a facility in homogeneous materials. Figure 9–19 illustrates an upgradient area of recharge and downgradient discharge point as defined by water levels measured in piezometers. The downgradient piezometers show an upward vertical gradient, while upgradient piezometers show a downward gradient.

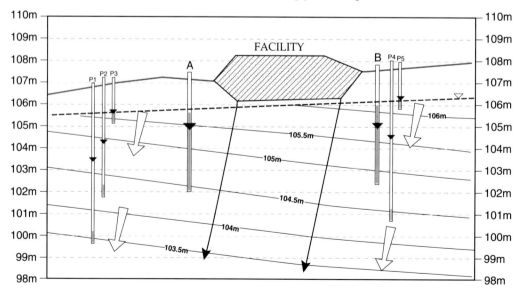

FIGURE 9-20 Conceptual Recharging Conditions

Figure 9–20 illustrates a recharge condition both in background and downgradient piezometers. The heads shown in monitoring wells A and B represent the average of the equipotential heads contacted by the screened area of the wells.

9.4.2 Geologic Control

Geologic controls over ground–water movement represent the most critical factors that should be considered in ground–water monitoring network design. The goal, in most cases, is to define the most likely zone in which ground water moves beneath a facility and, hence, the most likely zone for any possible contaminant movement to occur and be detected. The following discussion first addresses simple geologic systems where design of the monitoring system is relatively straight forward based on the geology and ground-water flow directions. The discussion then moves to more complex systems that require significant site assessment and conceptualization to design an appropriate monitoring system. The discussion also includes design for perched water conditions. Some of the following examples include unlined waste disposal sites where leachate is shown to dramatize the potential flow paths and target monitoring zones.

Single Aquifer / Homogeneous

The single homogeneous aquifer represents the simplest geologic environment in which to design a detection monitoring system. The single homogeneous/isotropic aquifer system only requires the following steps to define the target monitoring zone:

1. Evaluate aquifer geometry, thickness, and vertical and horizontal hydraulic conductivity variability by way of continuously sampled stratigraphic borings logged to confirm homogeneous and isotropic conditions within each layer.

2. Prepare a conceptual geologic/hydrogeologic model and plot potential target monitoring zones;

3. Construct flow nets using water level/piezometric head information from piezometers or observations wells; and confirming target monitoring zones

4. Install wells to monitor potential contaminant flow paths.

Figure 9–20 illustrates the subsurface movement of leachate from an unlined solid waste facility in a humid environment. Selection of appropriate screen depths for down gradient wells is relatively simple using the procedure above. Figure 9–21 (Freeze and Cherry, 1979) represents isoconcentrations of chloride next to an unlined solid waste landfill. The contours were based on water quality obtained from numerous, closely spaced sampling points screened at various depths. The location of the target monitoring zone here would be the center line of the chloride plume. The center point, with the highest chloride concentrations, represents the most direct flow path away from the landfill. Monitoring wells located in this zone (along the highest chloride contour) would provide the earliest detection of leachate excursion away from the facility. Figure 9-21 (field determined flowfields) and

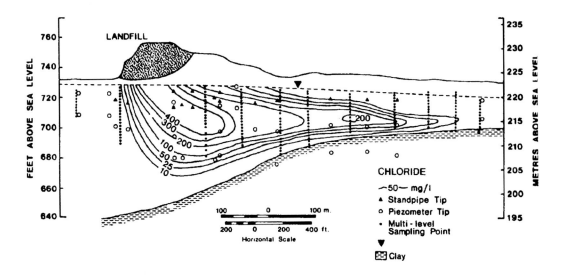

FIGURE 9-21 Example Leachate Water Quality Plume- Field Determined

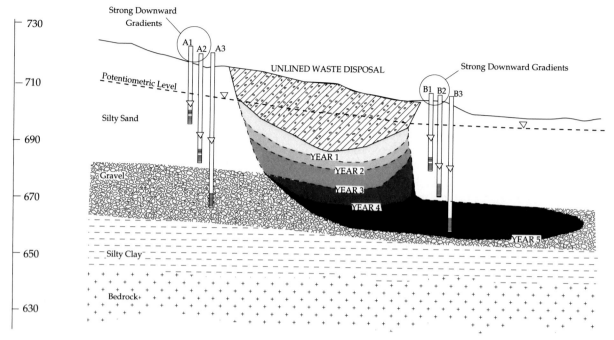

FIGURE 9-22 Time Sequence Leachate Plume

Figure 9-20 constructed from a flow net provide essentially the same solution to ground-water flow for this particular hydrogeologic site environment.

Single Aquifer of Variable Hydraulic Conductivity

Differences in hydraulic conductivity due to changes in stratigraphy with depth can point the way toward establishment of an effective monitoring system. The procedure for design would likely include the following steps:

1. Determine the horizontal extent and thickness of individual geologic units by evaluating geologic logs of continuously sampled stratigraphic borings; to a depth of least 25 feet below base of the facility. This suggested depth is used as a rule of thumb, and actual depths may vary based on site conditions.

2. Establish hydraulic conductivity for each unit from results of field and laboratory tests confirming isotropic conditions;

3. Construct a flow net based on observed hydraulic heads (from piezometers) and a conceptual

geologic hydrogeologic model to select target monitoring zone(s); and

4. Install monitoring wells based on defined target zones that represent primarily horizontal movement of ground water.

If anisotropic conditions are observed in the difference between laboratory and field hydraulic conductivity test results, the procedures for construction of anisotropic and heterogeneous conditions will be necessary to properly draw the flow net. As a general rule the difference between hydraulic conductivity measurements in the field, as compared to those evaluated in the laboratory, may not be a result of actual isotropic aquifers. Since laboratory measurements are made from relatively small volume samples, and field hydraulic conductivity measurements are made on screened sections of five feet or more, some natural variations in hydraulic conductivity should be expected. A comparison of the values obtained from the field and laboratory should be made in conjunction with both the samples collected and logs of the lithology. If the comparisons of the values show little reasons for a wide variation in hydraulic conductivity then conduct an inspection of the samples provided to the laboratory. Special care should be taken that the majority of the samples collected in the field represent the typical lithology, rather than, the exception to the typical field

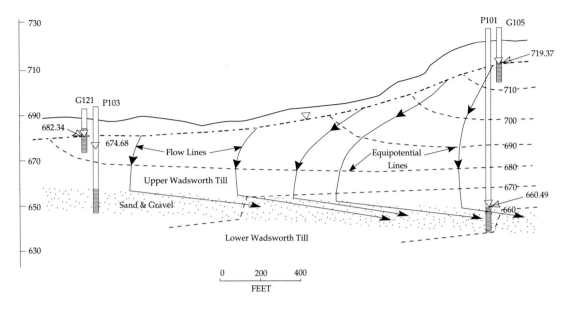

FIGURE 9-23 Conceptual Flow in Layered Deposits

conditions. In any case, it is probably rare to obtain exactly the same hydraulic conductivity from an individual field and laboratory test. Since we are sampling a range of hydraulic conductivities of the geologic strata, the smaller the sample tested, the more likely is the sample to test differently from an average value as obtained from field tests of screened sections of the units. For these reasons it is recommended that sufficiently large numbers of laboratory samples be collected from each hydrostratigraphic unit to provide a representation test value. As a rule of thumb, three laboratory samples for

each hydrostratigraphic unit should provide a minimum value for comparison to field obtained hydraulic conductivity. For large or complex sites many additional laboratory determinations may be necessary.

Figure 9–22 depicts a time sequential leachate plume from an unlined solid waste facility. Leachate movement in the system is represented by primarily vertical flow in the lower hydraulic conductivity units and the horizontal excursion in more permeable silty sands and gravel zones. The monitoring system for this facility would consist of wells screened either in the silty sand unit directly next to

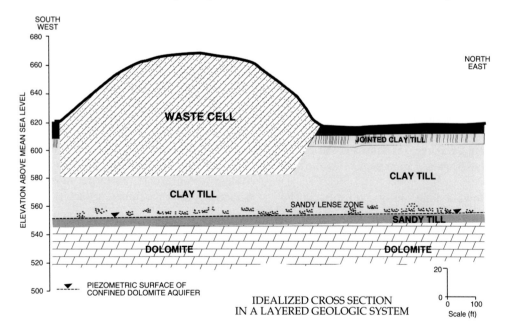

FIGURE 9-24 Conceptual Model in Layered Deposits

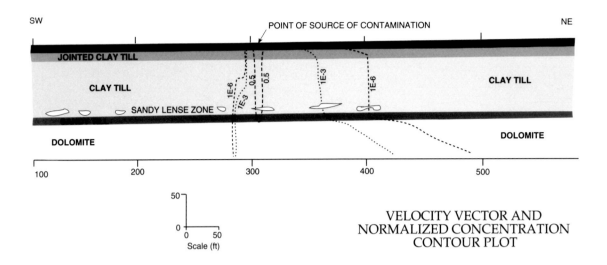

FIGURE 9-25 Computer Flow Results Model in Layered Deposits

the facility, in the gravel, or in both units. The extent of the geologic units, the potential for off-site migration and the current or potential use of the water contained in the units are some of the deciding factors in the actual system design. If the silty sand is discontinuous, the gravel would be the primary monitoring target zone. However, if the silty sand extended beyond the site boundaries, and sufficient horizontal flow exists to be monitored effectively at the edge of the facility, both the silty sand and the gravel would be targets for ground–water monitoring. The silty sands represent the probable first affected unit, and the gravel would most likely represent a water supply for off–site downgradient water users. An important key is the potential for horizontal migration in the silty sand unit. If flow nets show discharge of the silty sand unit to downgradient receptors then this unit would likely represent the "uppermost' aquifer. As such, detection monitoring should be conducted within this unit. Conversely if this silty unit shows strong downward gradients and does not discharge, (as shown in the Figure 9-22), then little would be gained from monitoring the silty sand unit.

Figure 9–23 shows a sand and gravel unit as the uppermost aquifer beneath two tills. Typical of near–surface, low hydraulic conductivity units, ground–water flow is nearly vertical in the tills. Ground water then flows horizontally in the much higher hydraulic conductivity sand and gravel aquifer. This sand and gravel unit is the only potential target monitoring zone for a facility located in this type of environment. The dominance of vertical flow in low hydraulic conductivity

deposits and horizontal flow in continuous, permeable zones is very typical. In glaciated regions, deeper sand and gravel, valley fill or outwash deposits are often in direct contact with underlying weathered or highly fractured bedrock. Such systems would represent a composite target monitoring zone. Small lenses of sand within a mass of low conductivity material, however, do not represent adequate targets for monitoring. Thin or discontinuous, sand lenses will not provide the low heads necessary for horizontal movement of ground water away from a facility. Figures 9–24 and 9–25 represent the idealized cross–section of a facility located in a clay till above a bedrock aquifer. A series of discontinuous sand seams was present within the clay till. Numerical modeling of the system provided the velocity vector and concentration contour plots shown on Figure 9–25. A point source of contamination was simulated in the modeling project. The point source produced a plume that moved horizontally in near surface material (the jointed till), vertically downward through the clay till and sand lenses; and finally, horizontally in the underlying dolomite bedrock. The dolomite represents the target monitoring zone in this situation, due to the following factors:

- The near–surface, jointed till is shallow and does not represent a flow path away from the base of the facility;

- The near–surface tills can be influenced by vertical recharge events that are not associated with ground water passing beneath the facility (i.e., not in the flow path);

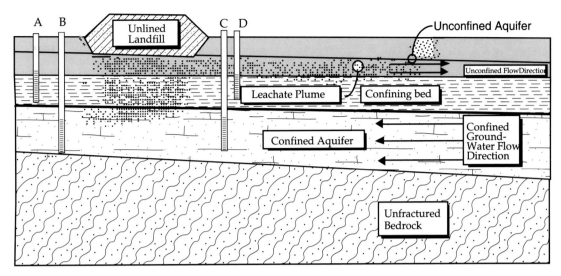

FIGURE 9-26 Two Layer Flow Model

- The thick clay tills and the minor sand lenses do not represent aquifers;

- The thick clay till and enclosed sand lenses, when considered as composite units, have primarily vertical ground-water flow components;

- The dolomites can yield water to monitoring wells and do represent a horizontal flow path from the facility.

Therefore, the dolomites would represent the target monitoring zone for the facility.

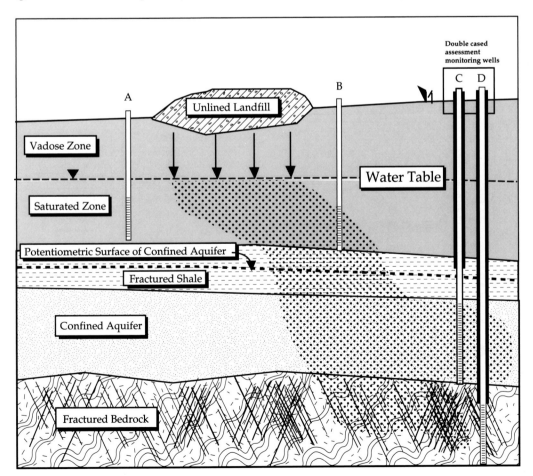

FIGURE 9-27 Three Layer Flow Model

Multiple Aquifers

Multiple aquifers represent a challenge to the ground–water monitoring system designer. Ground water in layered aquifers often moves in different directions. Thus multiple aquifers require a more complex understanding of the three–dimensional hydrogeologic data to accurately establish a capable monitoring system.

Figure 9–26 shows a two–aquifer system with ground-water flow in opposite directions. Such a geologic environment would require significant geologic and hydrogeologic characterization to establish target flow paths from the facility. The following procedure is recommended to establish a ground-water monitoring system for a two–aquifer system as is shown in Figure 9–26:

1. Install stratigraphic borings using continuous sampling techniques from the surface through all overburden units down to competent bedrock;

2. Complete piezometers in each aquifer so that vertical and horizontal gradients can be established for the each of the aquifers;

3. Establish hydraulic conductivity for each unit by conducting field insitu and laboratory hydraulic conductivity tests;

4. Construct flow net constructions for each aquifer; and

5. Establish the target monitoring zones.

If the goal of the monitoring system is to provide "immediate' detection of any contamination released from a facility (i.e., as in a detection monitoring program), the target monitoring zone should be the unconfined uppermost aquifer. If the goal of the system is assess the extent of contamination emanating from a site (i.e., as in an assessment monitoring program), defining the rate and extent of contaminant movement would require monitoring in both the upper and lower aquifers. If a nearby surface stream serves as a base flow discharge point for one of the aquifers, the stream would probably also require water quality monitoring. The monitoring program should also define if there is underflow beneath the stream.

Figure 9–27 shows a three-aquifer system including a deep, interconnected, fractured bedrock aquifer. As with the two–aquifer system, the assessment technique should be as follows:

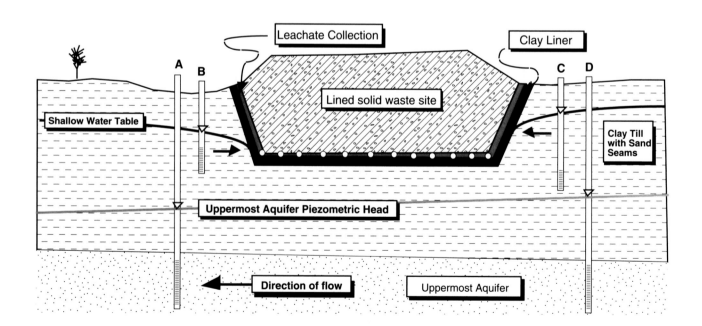

FIGURE 9-28 Effect of Drains on Low Hydraulic Conductivity Materials

1. Install borings to take soil samples sufficient to characterize the unconsolidated materials down to competent bedrock. Determine if continuous sampling and logging techniques is necessary for the geologic environment. The presence of fractured bedrock indicates that rock core drilling would be required to evaluate fractures and bedrock hydraulic conductivity;

2. Complete a series of piezometers in each geologic unit to establish hydraulic gradients;

3. Establish hydraulic conductivity (horizontal or vertical) for each geologic unit, including confining units;

4. Construct flow nets and piezometric contour maps for each aquifer;

5. Develop a geologic/hydrologic conceptual modes and establish target monitoring zones; and

6. Install monitoring wells. Assessment monitoring wells in deeper units should be double–cased through the overlying units as necessary to prevent cross-communications between units.

As with the two–aquifer system, a monitoring system installed for the purpose of detecting contamination would focus on the uppermost aquifer to provide immediate detection of leachate from the facility.

The shaded area in Figure 9-27 represents wide spread contamination that provides many challenges in assessment monitoring programs. Typically if assessment programs require full project planning at the project start (such as in the Superfund program), these deeper contamination zones are often not included in sampling programs. A phased program that includes full geologic conceptualization and flow net construction should be completed before generation of sampling plans for ground-water quality. Chapter 10.0 provides additional guidance for assessment monitoring evaluations.

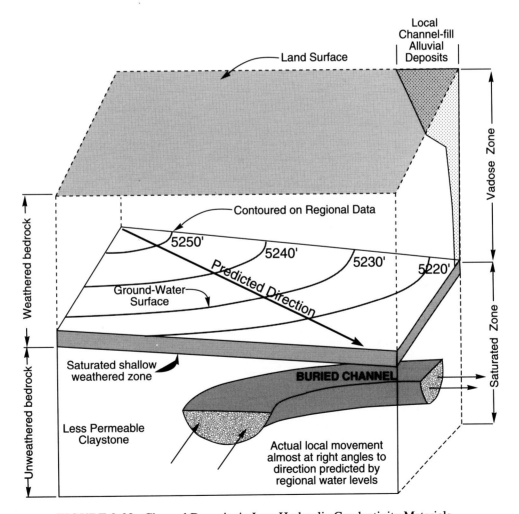

FIGURE 9-29   Channel Deposits in Low Hydraulic Conductivity Materials

Low Hydraulic Conductivity Environments

Probably the most difficult (and error prone) geologic environment in which to design a ground–water monitoring system is thick, low hydraulic conductivity materials overlying an aquifer at depth. Much of the controversy surrounding ground–water monitoring of hazardous waste sites is based on the difficulties in interpreting ground–water movement in low hydraulic conductivity environments. Figure 9–28 illustrates a facility located in a thick low hydraulic conductivity clay overlying a high hydraulic conductivity sand. The sand is confined and the clay contains minor sand lenses. The shallow, unconfined ground–water surface is affected by a facility leachate collection system that acts as a subdrain to the adjacent low hydraulic conductivity clay, and thus masking the directional components of the shallow ground water flow. Piezometers should be installed within the clay and the uppermost aquifer, (the lower sand), in order to define vertical gradients and to assist in selection of the target monitoring zones.

Geologic environments that consist of primarily low hydraulic conductivity units containing higher hydraulic conductivity deposits of significant lateral extent require comprehensive preliminary hydrogeologic investigations to define the target monitoring zone(s). An example of the kind of conceptual geologic descriptions necessary for evaluating generally low hydraulic conductivity

environments is provided below. Figure 9–29 illustrates a conceptual model of a buried channel located in much less permeable claystone. One example of this type of lithologic system is the Cretaceous Dawson Formation in the Denver, Colorado area, which was deposited in a fluvial, deltaic environment. The Dawson stratigraphic sequence consists of depth–uncorrelatable, vertically stacked sandstone channel deposits, each of which is isolated within a fine–grained claystone that originated as backswamp deposits in the Cretaceous delta. Thin, isolated sandstone lenses (as viewed in cross sections) are present in the sequence that are characteristic of levee splay deposits and minor overbank deposits. The majority of geologic materials in such a sequence are matrix supported diamicts that have very low hydraulic conductivity. The channel deposits represent clast supported units. These channel deposits can provide discharge pathways both to recent alluvial materials present in ephemeral stream channels and to adjacent claystone units.

On the basis of on evaluation of the depositional environment, through detail core analysis using Facies Codes, the Dawson Formation deposits were determined as laid down in a delta that gradually was uplifted by ancestral Rocky Mountain tectonics, in early Tertiary time. Different depositional characteristics of each of the sand sequences observed in cored boreholes emphasize that sands were deposited by separate and different stream

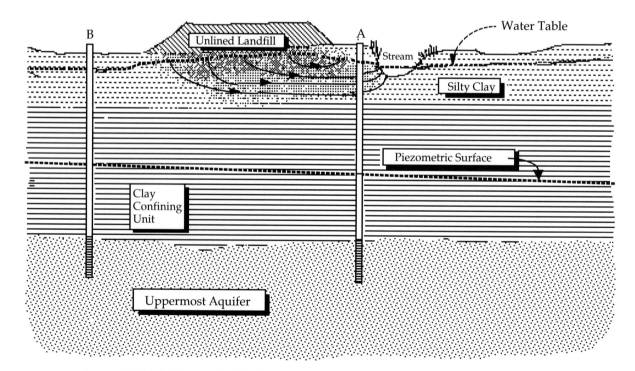

FIGURE 9-30 Low Hydraulic Conductivity Materials With Weathered Upper Units

systems and, therefore, not vertically interconnected. Minor sand lenses, such as the levee splay deposits or overbank matrix supported sands that were deposited in backswamps, also have limited area extent. They are connected horizontally over short distances and are vertically separated from other sandstone in the system by the intervening claystone.

Near–surface Dawson claystones are typically weathered and can become seasonally saturated as a perched ground water system with sufficient hydraulic conductivity to comprise a target monitoring zone. This weathered zone can be easily defined by shallow (<30 feet) borings and piezometers. The deeper sand channel deposits, however, present a more difficult directional flow analysis problem. In this environment, each sand channel deposit has its primary component of flow in the stratigraphically down dip direction. Structural warping of the geologic units tilted the channel deposits back toward the original source direction. As such these deposits are very difficult to evaluate for discharge direction. Monitoring such a heterogeneous geologic environment requires detailed field drilling sites and boring log information must be located through geophysical methods to define the depth and location of these channels. These channels may be secondary target monitoring zones since they likely serve as subdrain systems for the shallow, unweathered claystones.

Low hydraulic conductivity environments also may have a more permeable upper unit discharging locally to a stream or river. Figure 9–30 illustrates a thick clay unit confining a regional uppermost aquifer. The near surface weathered silty clay unit has relatively low hydraulic conductivity with minor sandy units, and the conceptual unlined landfill discharges leachate as seeps or springs near the landfill base and into the stream. These sites typically look bad due to the surface discharge of leachates, however, these surface discharges represent rejected leachate that did not move into subsurface pathways. Monitoring of such an environment would probably include alternative sampling of the stream. Visual inspection of local streams can provide insight into springs and small discharge areas.

Because the deep aquifer is confined by the thick clay unit and the unlined landfill discharges to the stream, monitoring wells for a detection monitoring program would probably be located between the stream and the landfill. However, the relatively low hydraulic conductivity of the near–surface materials will make monitoring difficult in practice due to long (days to weeks) rates of recovery time for the wells. Additional piezometers should be located across the stream to verify that ground–water discharge along both sides of the stream.

9.4.3 Geologic Structural Control

Geologic structures such as dipping beds, faults, cross-bedding, and facies changes can greatly affect the rate and direction of ground–water movement. The monitoring system designer must consider geologic structural controls throughout the entire preliminary site investigation to insure adequate site characterization. As with other field complexities the development of a conceptual model is the key to successful ground–water monitoring system design in structurally controlled environments. Geologic structures affect ground–water movement in several ways:

- Acting as more permeable flow paths, because of higher primary porosity (cross-bedded sands) or through secondary hydraulic conductivity enhancement (natural fractures); and

- Acting as either barriers to ground–water flow or as conduits for ground–water flow, as do many fault zones, depending upon the nature of the material in the fault zone.

If the fault zone consists of finely ground rock and clay (gouge), the material may have a very low (say, $<10^{-6}$ cm/s) hydraulic conductivity. Significant differences in ground–water levels can occur across such faults. The hydrogeologist should be alert to large (say, >20%), unexplained differences in water levels across a site in faulted environments. These differences may be due to fault gouge retarding ground–water flow across the fault. Impounding faults can occur in unconsolidated materials with clay present, as well as, in sedimentary and even in igneous rocks. For example interbedded shales, which normally would not hinder lateral ground–water flow, can have weathered clay products smeared along the fault. Fault–zone barriers are relatively common in the ground–water basins of southern California. However one should not lose sight that faults can also act as conduits to ground-water flow. These systems should be evaluated by careful observation of water levels next to the faulted units. If the potentiometric surface flattens over the faulted area then it is a likely higher hydraulic conductivity zone that is discharging to other aquifers or surface discharge points.

Definition of geologic structures as considerations in ground–water monitoring design should include the following points:

- Identification of major geologic structures specifically and generically, in regional studies under the preliminary site investigations;

- Identification of potential fault areas through literature and air photo review;

- Identification of springs, vegetation changes, and surface geology through site reconnaissance prior to drilling;

- Establishment of an initial conceptual hydrogeologic model;

- Installation of borings placed to define geologic structure, variable water levels, and gradients in each geologic unit;

- Reconciliation of logs of stratigraphic borings and piezometer water levels with the conceptual geologic models; and,

- Interpretation of structural contour, ground–water elevation, and piezometric level contour maps to develop a linked hydrogeologic conceptual model with flow nets to identify target monitoring zones.

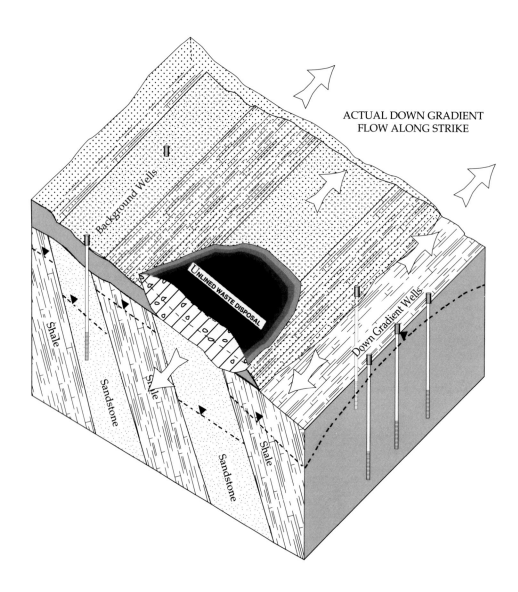

FIGURE 9-31 Structural Control of Ground Water Flow

The installation of the monitoring system should reconcile the full office and field data to complete the monitoring system.

The effect of geologic structures such as the simple dipping orientation of the bedrock on leachate movement is illustrated in Figure 9–31. The steeply dipping, alternating beds of sandstone and shale can have significantly different (say, 1×10^{-4} to 1×10^{-7} cm/s) hydraulic conductivity. The three–dimensional view (Figure 9–31) shows preferential movement of ground water along strike of the sandstone. The view illustrates the down-dip movement of contaminants. A detection monitoring system located in the shales would not establish the early leakage from the unlined site. One indication of the highly variable hydraulic conductivity of the rock mass is the overland

flow of leachate. Because the leachate cannot move rapidly into the sandstone (and less so in the shales), leachate is rejected to the surface over the shale outcrop. A leachate seep occurs at the contact between the sandstone and shales. Surface seeps (and springs) are excellent indicators of changes in formation hydraulic conductivity, and should always be considered in development of conceptual hydrogeologic models.

Perched Ground Water

Monitoring programs for perched ground–water environments present a number of complications to the monitoring system designer. Figure 9–32 presents potential leachate pathways from an unlined waste disposal

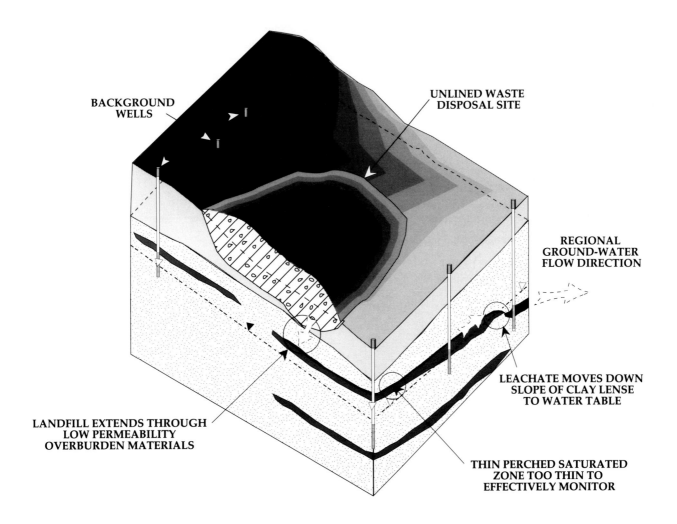

FIGURE 9-32 Perched Structural Control of Ground-Water Flow

site. Perched ground–water does not follow regional ground–water gradients, but rather, will flow along a hydraulic conductivity interface due to gravity, as shown in Figure 9–33. The approach used to designing a monitoring system for this type of perched water condition should begin early in the site assessment of the facility. The design procedures would include the following:

1. Evaluate the lateral extent and thickness of various geologic units down to at least 25 feet below the base of the facility through continuous sampling of soil borings. Particular attention must be paid to the presence of saturated zones above low hydraulic conductivity/fine grained layers. A rule–of–thumb is that a potential perched zone may occur at a hydraulic conductivity contrast of two orders of magnitude. A three order of magnitude hydraulic conductivity contract will almost always cause perched ground water. Contrasts between sand and silt or clay will likely show such a three-order-of-magnitude variation. Such contrasts in hydraulic conductivity will result in a thin perched water zone that will make interpretation of flow direction very difficult if not properly recognized. Geologic cross sections can help identify potential locations of perched water.

2. Carefully evaluate road cuts in the vicinity of the site for the presence of geologic units that could cause perched conditions. The units in the road cut, may already have been drained, however, the perched water

tables may be located by vegetation concentration above the perching unit or during winter, frozen ground water discharges.

3. Sufficient continuously sampled borings should be drilled to define the horizontal extent and elevations of potential low hydraulic conductivity zones above the regional ground–water surface.

4. Piezometers (at least three) should be completed in each geologic unit, including permeable units above potential perching units to establish the presence of thin layers of perched water. Care must be taken to complete the piezometers so that the screened zone begins directly at the interface between the perching unit and the overlying more permeable unit. If perching unit is very thin (say less than one foot) or if it has a very thin saturated area (2 to 4 inches), the need to monitor the surface of the unit should be re-evaluated. Very thin perching units are often discontinuous and it may be more important to understand the geologic history of the system to predict the orientation of the perching systems rather than one individual unit.

5. A structural contour map of the top surface of the low hydraulic conductivity unit should be constructed to define potential perched water flow directions. The contour map should be combined with cross sections

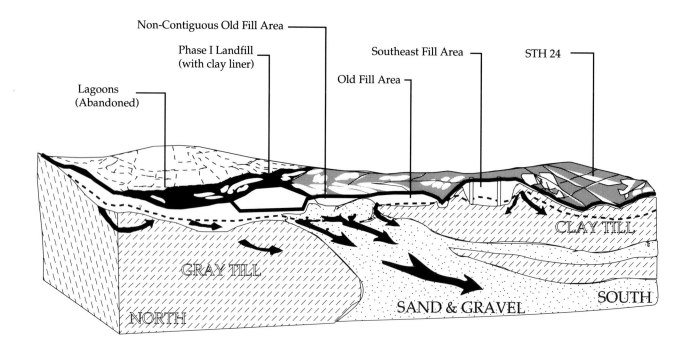

FIGURE 9-33 Conceptual Discharging to Unconsolidated Materials

showing water levels to establish perched water flow paths.

6. If the perched saturated zone is below the base of the facility and sufficiently thick to be characterized as an "uppermost aquifer," or if it is sufficiently thick to allow collection of adequate samples to serve as an early detection location, it should be considered a target monitoring zone.

7   If the perched saturated zone is too thin (say, <2 feet) to be saturated year around, the monitoring system should be installed in the first permanently saturated zone, (the uppermost aquifer), beneath the perched zone, as shown in Figure 9–32.

A detection monitoring system could be based in either the perched water body, if a sufficiently thick (say, >10 feet) saturated zone exists, or within the deeper uppermost aquifer, as based on piezometric gradients. A key point in determining if there is a potential for perched water bodies is the stratigraphy present at the site. If clay or other fine–grained materials are present near the surface, the potential for occurrence of perched water bodies is greatly reduced, due to limited recharge. If highly permeable (say, >1x10^{-3}) material exists near the surface with less permeable material below, then perched water bodies are more likely. The lower the amount of recharge, the less likely that the hydraulic conductivity contrast will act as a significant perching mechanism. Monitoring beneath the perched zone for "rate and extent" qualification would concentrate, in this case, on the first aquifer beneath the perched zone rather than on perched zone itself.

Although the potential may be present on a site for individual clay units to perch ground water, the limited extent or continuity of the perching unit may not require the definition of the individual low hydraulic conductivity units. Figure 9-33 illustrates a conceptual cross section/flow net of unconsolidated deposits overlying a regional confining unit. In this example the individual clay units were limited horizontally so ground water could flow through "windows" of the clay deposits. A single unconfined water surface was established with little or no perching conditions. Flow path A discharges from around piezometers D-15 toward the Southeast working through and around the various flat-lying clay units. Alternative flow paths could be constructed from the D-15 area to the Northwest, since this location represents a recharge-upland area. Discharge is toward streams (gaining streams) cutting into the Coopers formation to the Northwest and southeast of the facility area. The projected flow path A would be perhaps best monitored in the area of D-13 since the majority of flow for the system passes below the clay unit at this location. To the Northwest, monitoring of the zone between 50-60 feet above NGDV would provide a secondary flow path target from the site area.

Secondary Hydraulic Conductivity

There are three basic types of ground–water occurrence and movement, as shown in Figures 9–34a through 9–34c. Figure 9–34a shows primary porosity where ground water moves through the interstices (voids) between sand–sized grains. Figure 9–34b shows ground–water movement through fractures that represent secondary porosity. Figure 9–34c, shows ground-water movement through solution channels developed in a carbonate rock; another type of secondary porosity. A particular geologic environment could consist of any or all of these media.

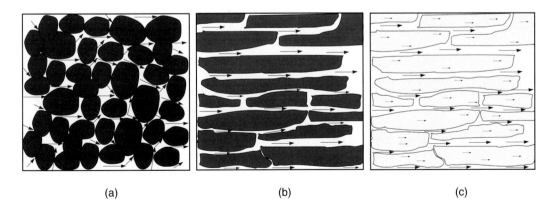

(a)   (b)   (c)

FIGURE 9-34 Various Forms of Ground-Water Flow

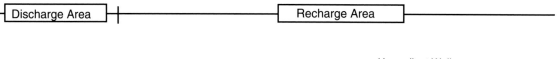

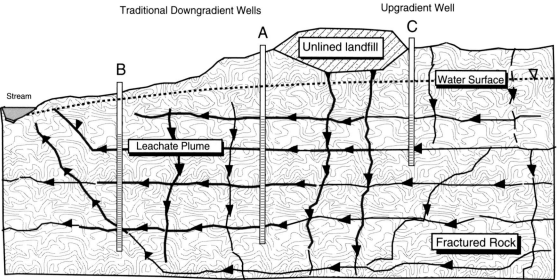

FIGURE 9-35 Fractured Rock Control of Ground-Water Flow

The preliminary field investigation should determine the dominant flow mechanism beneath the facility to be monitored so that the appropriate locations for monitoring wells can be selected.

Fractured or solution-channeled carbonate rocks provide special problems in ground–water monitoring system design. Often there will be highly directional ground–water movement along discontinuities or dissolution–widened joints. The success of any monitoring system in a fractured or solution-channeled environment requires knowledge of the joint or fracture patterns. In some instances, remote image interpretation and special field techniques (i.e., tracer tests) can point toward the target monitoring zone in a secondary porosity environment.

Most consolidated rocks (with the exception of some indurated sandstone and conglomerates) have few primary intergranular openings for ground–water flow and usually have much lower hydraulic conductivity value than their unconsolidated equivalents. Ground–water flow in bedrock aquifers often takes place through secondary openings such as fractures (joints, bedding planes), and/or solution-channels. The investigator designing a monitoring system should fully identify areas in which this factor is important, and should do so at an early stage in the preliminary site investigation. Although regional flow patterns should be well established in preliminary investigation, it is often very difficult to predict ground–water flow through a set of fractures or solution channels on a site–specific scale (e.g., in the vicinity of a

monitoring well). Thus, facilities located over bedrock aquifers should employ additional investigative techniques (e.g., fracture–trace analysis, geologic mapping and pumping tests specifically designed to evaluate anisotropy) to adequately determine likely ground–water flow pathways.

Fractured rock environments require consideration of specific flow paths to define the target zones monitoring system design. Chapter 4 discusses the Phase II field investigation tasks for a fractured rock environment. Figure 9–35 conceptually shows the individual ground–water flow paths in a fractured rock environment of a single rock type. Leachate is shown moving down from an unlined landfill toward a series of fracture sets that control local ground–water flow. Detection monitoring system design would place screened zones both upgradient and downgradient of the facility. Individual screen depths must be based on the results of the boring program and the observed fractures or weathered zones rather than on only observed head levels. In an assessment monitoring situation, long (say, >15 feet) screened zones should be avoided to reduce the potential for cross-contamination caused by leachate entering an upper zone and moving downward into uncontaminated zones. Fracture patterns can be highly localized and unpredictable, as shown in Figure 9–36, or more evenly distributed and predictable as illustrated in Figure 9–37. Often, both primary and secondary porosity are present in bedrock units, as illustrated by Figure 9–38, so the preliminary site assessment must include measurement of the hydraulic and

STRUCTURAL EFFECTS ON GROUND WATER MONITORING
FAULTING & PERMEABLE FRACTURE SETS

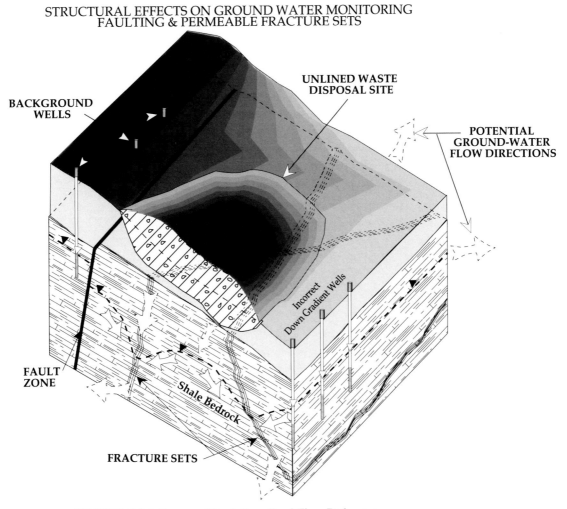

FIGURE 9-36 Fractured Rock Localized Flow Pathways

geologic parameters for each of the media present at the site. The approaches for design in a fractured geologic environment should follow the recommended procedures:

1. Evaluate fracture patterns using background information, aerial photographs (fracture trace analysis), and measurement of fractures at surface exposures.

2. Establish core drilling program at the site that may include use of rock quality designation (RQD), fracture orientation of core, borehole video logging, and detailed visual logging. Packer hydraulic conductivity tests should be considered for use in the assessment, in which packer intervals should be selected to test specific, observed discontinuities. Consideration should also be given to including angle core drilling in those areas where vertical fractures may be present.

3. Implement borehole geophysical surveys, such as caliper logs , flow logs, and temperature surveys.

4. Install multiple piezometers or multipoint completions for assessment of hydraulic conditions in individual fracture zones detected in the coring and geologic logging process.

5. Measure piezometric heads and gradients in relationship to joint patterns ("sets"). and

6. Establish a conceptual model defining the target flow zones in plan and cross section.

A detection monitoring system can be effective, in a fractured–rock environment if the wells are screened in highly–permeable fractures (those which flow into the borehole) downgradient from the facility. These systems

STRUCTURAL EFFECTS ON GROUND WATER MONITORING
FAULTING & HOMOGENEOUS FRACTURE SETS

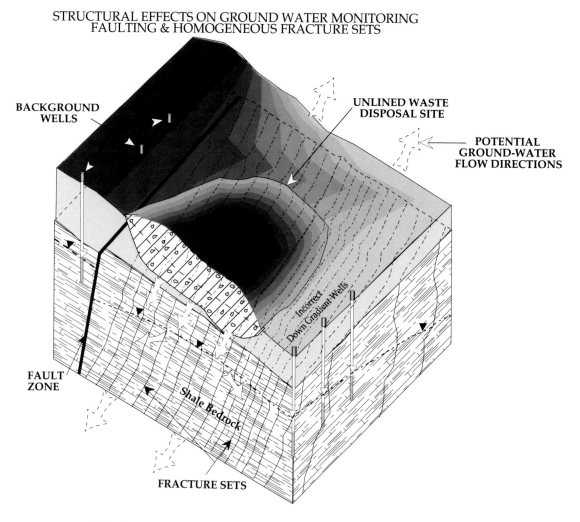

FIGURE 9-37 Fractured Rock - Evenly Distributed Flow Pathways

can react very quickly (say, days to weeks) to leachate releases from the facility.

Solution-channeled bedrock (Karst) terrain present additional challenges to the designer of a ground–water monitoring network because monitoring wells can miss permeable solution channel joints and may even end up as dry holes. Quinlan (1990) provided a full description of ground-water monitoring in Karst terrain. Karst environment not only requires consideration where to locate (i.e., a three-dimensional context) both background and downgradient wells, and springs but, also when to monitor the extremely fast reacting system. Quinlan (1990) recommends the following procedures for design of monitoring systems in Karst terrain:

• Review the regional and local geologic and hydrogeologic literature for the area in question;
• Evaluate topographic and geologic maps;

• Conduct a survey of springs;
• Map regional and local potentiometric surface;
• Organize a dye-tracing study based on references above;
• Perform the first dye-trace, preferably during moderate flow conditions;
• Evaluate the results of the first dye trace and modify if necessary the design of the tracing study;
• Determine whether local springs to the facility are characterized by conduit or diffuse flow;
• Perform additional dye-traces during moderate flow conditions, always modifying the tracing plan, as necessary, in the light of the results of the previous trace results. Quinlan (1989) noted that for most facilities it is necessary to perform only two dye-traces during moderate flow conditions;
• Repeat selected traces during base flow and flood flow conditions;

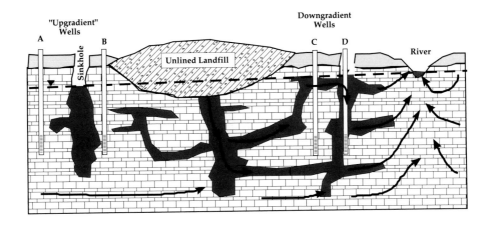

FIGURE 9-38 Karst Rock Localized Flow Pathways

- Integrate dye-tracing results, available potentiometric data, conductivity and turbidity data used to discriminate between conduit and diffuse flow into a monitoring plan; and,
- Have the entire project area reviewed.

General guidance for detection monitoring in Karst terrain must also include consideration of background well locations. In general terms background well locations must be based on:

- Negative results from dye tracing tests;
- Locations in similar rocks and geochemistry as downgradient sites; and,
- Locations selected in similar cultural environments.

Sampling for water quality in Karst terrain also does not meet the typical regulatory model for biannual or quarterly sampling periods. Quinlan (1990) recommends sampling based on storm and meltwater events. Figure 9–38 illustrates monitoring of an unlined facility, in a Karst area, where sinkholes are present beneath and next to the facility. Normally, such a setting is not easy to monitor, but a monitoring system can be developed to determine the facility's impact on the environment.

The assessment monitoring procedure would be similar to that used above with the possible addition of other surface geophysical surveys. Ground penetrating radar, electromagnetic conductivity, and seismic refraction surveys can help identify some zones of solution channeling and deep weathering within the rock mass.

Selection of target monitoring zones in the plan view and cross section should be prepared before any installation of any monitoring wells. In Karst systems, gradients are typically very low and require very accurate surveys for directional definition of ground-water flow.

9.5 SEPARATION OF ADJACENT MONITORING PROGRAMS

New landfills built to the state-of-art design criteria are commonly being constructed next to traditional disposal areas which many not have liners and leachate collection systems. Even with extensive double composite liners the new facilities must demonstrate the long term engineering performance of the new cells through ground-water detection monitoring programs. Selection of the proper locations for detection monitoring wells should be based on a holistic approach to the evaluation of a specific site. The placement of the ground-water wells in this process must weigh and balance data collected in the field, laboratory, and office. This is especially true for multiple facilities located on adjacent properties.

The target monitoring zones are further useful for the separation of detection ground-water monitoring systems from adjacent facilities that may have been unlined or are currently impacting water quality. The following three case histories illustrate how ground-water flow concepts can be effectively used to evaluate the optimum locations and depths of a detection monitoring system.

Simple Gradient Control:—Facility 'A' represents a relatively simple condition where the existing twenty (20) acre landfill is located in a downgradient position from a twenty five (25) acre expansion as shown on Figure 9-39. The expansion has a 60 mil HDPE liner and leachate collection. Ground-water surface contours, shown on Figure 9-39, when combined with the hydrogeological cross-section (Figure 9-40) show the target monitoring zone to be the Pleistocene Terrace deposits of interlayered sand and clayey sands. The underlying Choctawhatchee

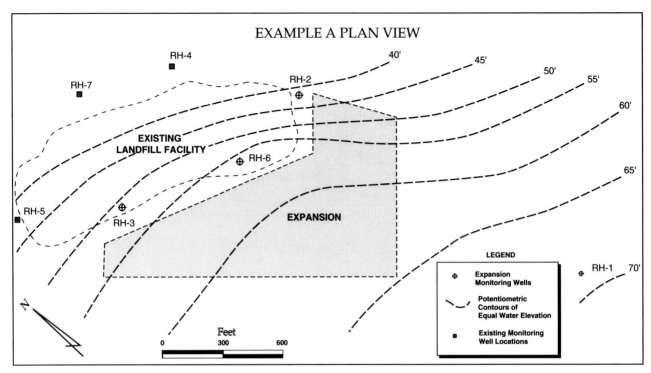

FIGURE 9-39 Gradient Controlled Ground Water Surface Countors

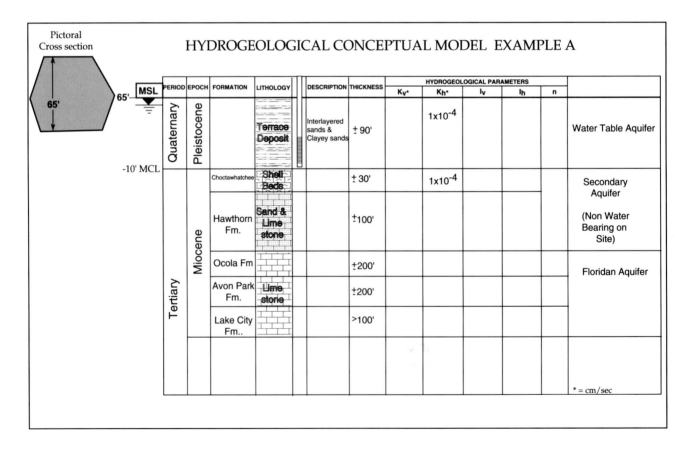

FIGURE 9-40 Hydrogeological Conceptual Model

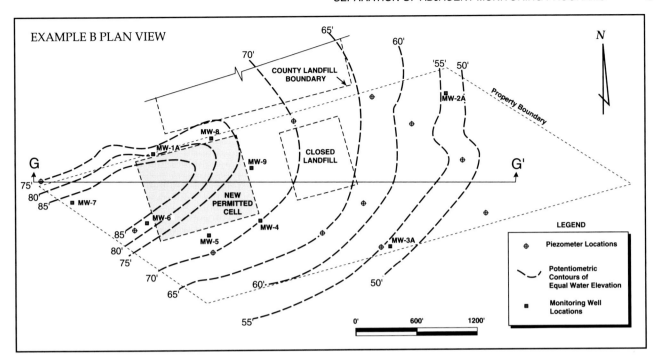

FIGURE 9-41 Plan View of Facility

formation was not judged to be significantly more permeable than the Terrace deposits and both formations are underlain by the locally non-water bearing Hawthorn formation. As such, ground-water would not have a significantly downward movement adjacent to the site and detection monitoring wells should be located between the two facilities. The flatness of the ground-water contours as they pass under the site shows that there is not significant mounding in the existing facility. This facility would not cause the area between the two facilities to be discharge point for the existing site. Wells located between the facilities with the observed ground-water flow conditions would monitor the expansion. At least two background wells are located between the 65' and 70' ground-water contour line. The well depths should be completed between the base grade of the expansion and the base of the Terrace deposits.

Gradient And Lithology Control:— Site `B' represents a slightly more complex example where a ground-water divide and lithology complicate ground-water monitoring conditions. Figure 9-41 shows a 16 acre lined cell with leachate collection next to a five (5) acre closed commercial, and large closed County landfill. The hydrogeologic cross-section, Figure 9-42 illustrates the lithology and base grade configurations for the site. The site's geology/hydrogeology is dominated by the thick (100 ft.) Cooper formation that acts as a regional confining unit. The shallow aquifer Pleistocene sands and clays have

a generally unsaturated upper sands unit with an intervening clay unit overlying a saturated lower sands unit.

The hydrogeologic cross-section, as presented, is insufficient to fully evaluate the localized flow conditions for selecting locations for detection monitoring wells. Cross-section G-G' (Figure 9-43) must be evaluated in combination with water table contours shown on Figure 9-41. The water table contours show a ground-water divide occurring along a natural ridge area. The new lined cell is located along the nose of the ridge and on top of the ground-water divide. Monitoring of sites located on ground-water divides require almost a three sided approach to the detection monitoring program. If the cell was located directly on top of a hill down gradient could be in all four directions and for monitoring background ground-water quality would be generally more complicated. Background water quality for the example `B' site would, however, be along the ridge line in the locations shown between MW-7 and MW-6. Downgradient locations would be located from MW-1A clock-wise around through MW-5. Since the site has sufficient ground-water gradients from the new cell toward the old closed county and commercial sites, local discharge between the old and new sites would not be considered a problem. The cross-section of G-G' illustrates a combined conceptual model of the geology site cross-section and the observed ground-water flow paths. Flow path A represents a potential flow line from below the lined facility in one direction toward a local creek

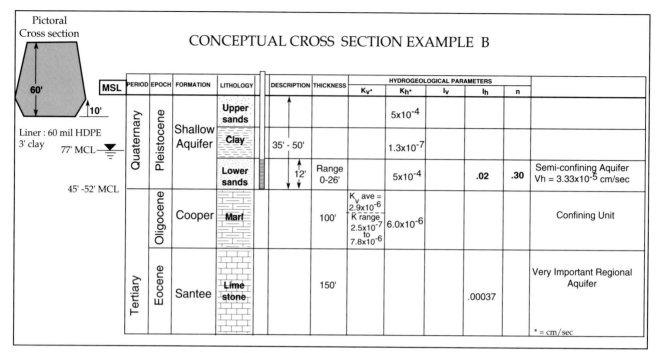

FIGURE 9-42 Conceptual of Facility B

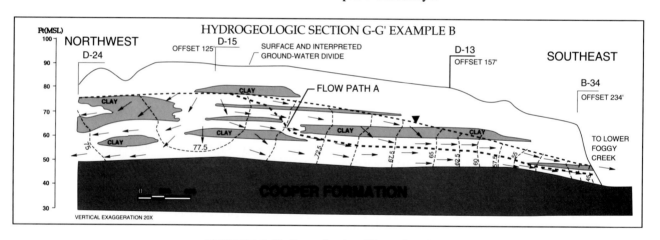

FIGURE 9-43 Cross Section View of Facility

discharge point. A flow path could also be drawn in the opposite direction (as represented by the ground-water divide shown in the Figure 9-43).

The well screen depths that would most effectively cover the facility in a detection monitoring system should be located in the sandy unit above the Coopers Formation, directly below the first saturated Pleistocene clay unit. Downgradient detection monitoring wells should be located in both cross-section directions as indicated by the ground-water flow arrows. Information on local lithology and ground-water flow nets make selection of potential depths for well screens a relatively straight forward task.

The previous two examples represent primarily horizontal ground-water flow conditions; the following example site "C', however, represents a more complex three-dimensional example requiring complete linkage of flow and lithology.

Complex Ground Water Flow Conditions:—
Site 'C' represents a new 70 acre lined site with leachate collection next to a closed 80 acre MSW landfill (Figure 9-44). The closed site is unlined without leachate collection facilities. The uppermost aquifer as shown in Figure 9-45 is the Pleistocene Alluvium (35 feet thick). The underlying

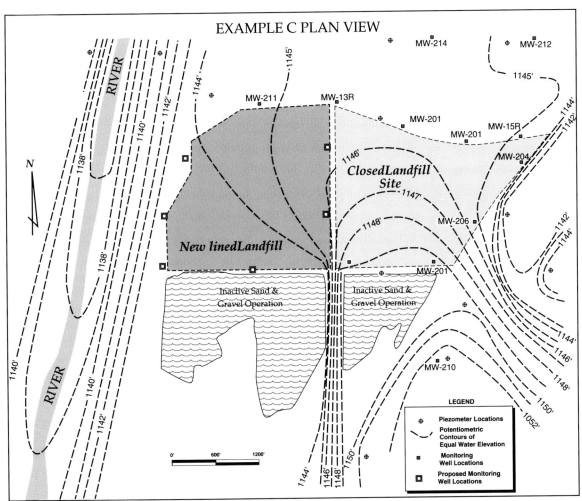

FIGURE 9-44 Ground Water Contour Map

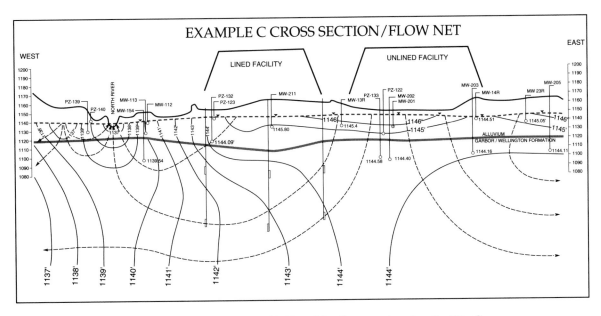

FIGURE 9-45 Flow Net Construction for Facility C

Garber-Wellington Formation (300 to 400 ft. thick) is the main regional aquifer. The Garber-Wellington Formation shows hydraulic conductivity similar to the overlying alluvium, so simple horizontal flow conditions cannot be assured for detection monitoring purposes. The ground-water contour plan (Figure 9-44) shows complex localized flow conditions exist at site due to the effects of the inactive sand and gravel operations (now full of water), and the river to the West of the site. Ground-water movement is generally from the Southwest to the West of the new facility, however, local discharges also occur to the East of the closed facility. The physical condition of having an unlined site generally upgradient of a new facility can make interpretation and comparison of background and downgradient water quality very difficult. The ground-water flow conditions are furthered complicated by having no confining units to separate the uppermost aquifer (alluvium) from deeper regional aquifers.

This site must be evaluated through use of a linked conceptual model and flow net construction cutting an East-West cross-section. Figure 9-45 illustrates site flow conditions through the uppermost aquifer into the Garber-Wellington and discharging in two directions (West & East). The lined site's potential flow path discharges to the river and as such, can be monitored with wells completed between 1080' and 1130' above MSL. This fifty foot zone represents the most likely flow path for potential discharges from the lined facility. Completion of deeper monitoring wells would intercept flow paths coming from the unlined facility and, as such, would not represent true downgradient conditions from the active disposal site.

Background wells must also be carefully chosen relative to adjacent facilities and the overall geology and ground-water flow conditions. The example 'C' monitoring wells are located between the new facility and the river in an arc down to the sand and gravel operation south of the lined cell.

Background wells are located between the closed site and the new facility. As with the downgradient detection monitoring wells the background water-quality monitoring points must be carefully selected relative to depth of the well screens. They should probably be no deeper than being screened between elevation 1140' and 1130' feet above MSL. As with any detection ground-water monitoring system located next to an unlined facility, the wells must be entirely screened below the seasonal low water table. This is to insure that landfill gas potentially moving in the vadose zone from the unlined site would not enter a monitoring well and cross-contaminate ground-water samples taken from the well. This type of cross-contamination can greatly confuse detection monitoring

analytical results. Gas movement in the vadose zone should be monitored through a separate gas monitoring network designed especially for the purpose.

9.6 DENSITY CONTROL

In this discussion, various monitoring system designs will be reviewed to illustrate some of the major implications of density and immiscibility in the design of ground water sampling networks.

In the preceding monitoring well system designs, it was assumed that the waste solutions were miscible in water and that the density of the water was not altered by the presence of the solutes. Although typical leachates from codisposal, hazardous waste and solid waste disposal facilities fit this profile, there are a number of monitoring situations in which these neutral density assumptions are not appropriate. In particular, waste such as brines or other water-based industrial effluents can be miscible in water but have a density significantly greater than water. Oily wastes and other organic-based fluids, may also contribute to the waste solutions that are immiscible in water. Physical properties of waste solutions can result in transport characteristics quite unlike those normally associated with such neutral density leachates.

The effect of density in miscible displacements is considered in Bear (1972), while several books including Schwille (1988), Corey (1977), and Collins (1961) consider the mechanics of immiscible fluids. These are represented schematically in Figure 9-46.

Figure 9-46a shows a contaminant plume developing as a result of seepage of a dense miscible fluid into the ground water zone. As shown, the contaminants tend to move vertically downward to the bottom of the aquifer. Once on the bottom, the movement of the plume is governed largely by the topography of the confining unit below the aquifer, and thus, the direction of flow will not necessarily be in the direction of regional ground water flow. Due to dispersion, diluted contaminants near the edge of the plume will be contributed to the local ground-water flow system. Thus, several areas of contamination may be present, the major plume of dense fluid and the adjacent ground-water zone contaminated by diluted levels of dense fluid as a result of dispersion at the perimeter of the plume. Establishing the area of contaminated ground water would require typical methods of investigations as for any assessment monitoring design. Locating the major pool of contamination, the dense plume would require knowledge of the surface of the lower permeability layer, and the installation of sampling points near the bottom of the aquifer. In the case of high density, miscible fluids over a period of time, the density would decrease to the

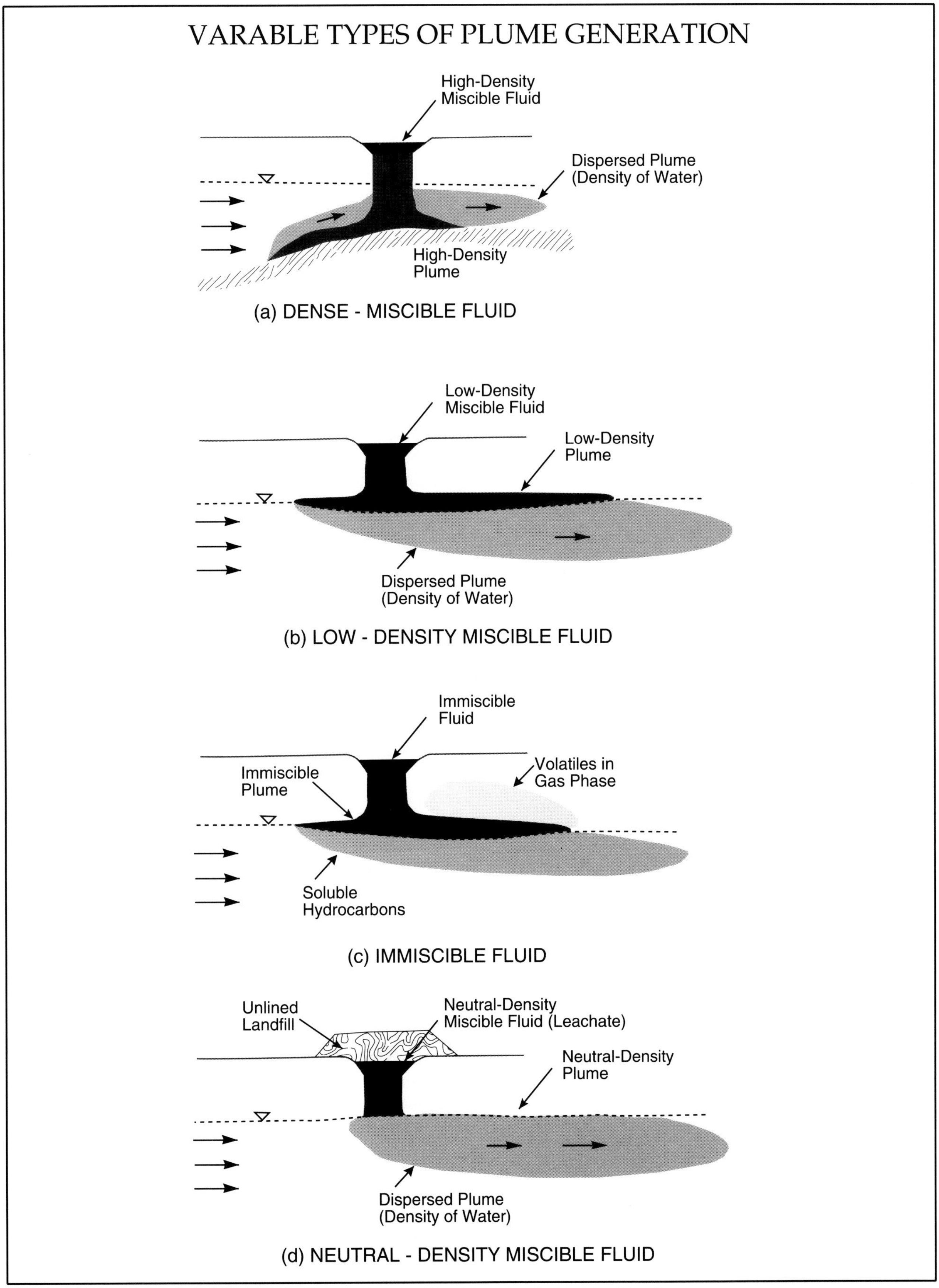

VARABLE TYPES OF PLUME GENERATION

High-Density
Miscible Fluid

Dispersed Plume
(Density of Water)

High-Density
Plume

(a) DENSE - MISCIBLE FLUID

Low-Density
Miscible Fluid

Low-Density
Plume

Dispersed Plume
(Density of Water)

(b) LOW - DENSITY MISCIBLE FLUID

Immiscible
Fluid

Immiscible
Plume

Volatiles in
Gas Phase

Soluble
Hydrocarbons

(c) IMMISCIBLE FLUID

Unlined
Landfill

Neutral-Density
Miscible Fluid (Leachate)

Neutral-Density
Plume

Dispersed Plume
(Density of Water)

(d) NEUTRAL - DENSITY MISCIBLE FLUID

FIGURE 9-46 a-d Density Consideration in Facility Monitoring

point where the plume would migrate according to the local ground water flow conditions (Cherry et al., 1982). In the case of a dense immiscible fluid, soluble constituents would be contributed to the local flow system, however, the denser immiscible plume would likely remain at the bottom of the aquifer for a prolonged period of time moving according to gravity.

Figure 9-46b represents the migration of a miscible phase having a density less than that of water. In this case, the major zone of contamination occurs near the top of the saturated zone and migration is controlled by the directional slope of the water table. As a result of dispersion, contaminants would be contributed to the regional flow system and the plume would gradually be dissipated over potentially long distances. In the case of miscible, less dense (than water) product spills, monitoring points should be concentrated in the upper part of the aquifer if vertical flow components are not significant.

Figure 9-46c represents the infiltration and migration of an immiscible fluid having a density less than or similar to that of water. In this case, the main area of contamination occurs near the top of the saturated zone and the plume is

dispersed in the direction of ground water flow. As the immiscible phase moves through a porous medium, a residual amount of fluid is retained in the medium in a relatively immobile state to slowly leak into ground water as it flows past the residual contaminated units. Immiscible fluids have soluble constituents that will tend to be leached from the fluid and migrate as part of the regional ground-water flow system. In addition, volatile constituents could be contributed to the gas phase in the zone above the water table.

The final figure 9-46d represents a neutral density miscible fluid moving from an unlined landfill into a local ground water aquifer. Since the leachate plume moves in a similar manner to that of the ground-water flow in the aquifer, monitoring would be based on target monitoring zones and three dimensional ground water flow components. Ground-water monitoring system design for the above contaminants should use both conceptual models and flow net constructions along with consideration of the leachate density.

This presentation is designed to share information with you regarding the conceptual design and implementation of a vadose zone monitoring system. Definitions and a brief discussion of the assumptions and properties governing vadose zone transport are presented. Step by step guidelines and procedures for designing a vadose zone monitoring system follow this discussion.

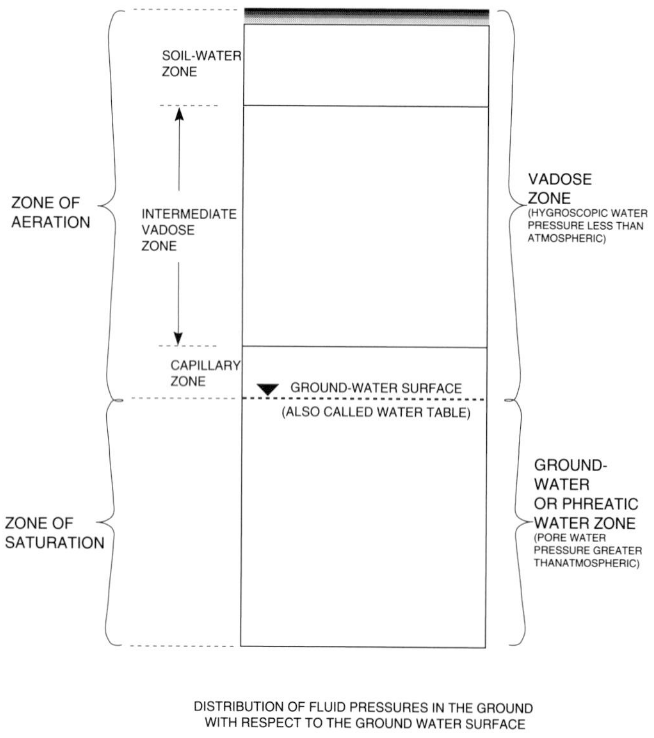

9.7 FUNDAMENTAL MONITORING CONCEPTS THE VADOSE ZONE

An important point for any type monitoring programs is the regulatory requirement to install the system. OMB currently is rewriting (1992) Part B of RCRA to require Vadose Zone Monitoring. The Revised Part B is anticipated to be published in the Federal Register as early as June 1991. The new law will require Vadose Zone Monitoring in any reauthorized Part B permit.

Subtitle D requires Vadose Zone Monitoring should be determined at the State level for MSW facilities. The California Legislature has introduced the Calderon Bill that is to require retrofitting all Subtitle C and Subtitle D sites with a Vadose Zone Monitoring System. California has promulgated under the California Code of Regulations (CCR) Title 23, Chapter 15, article 5 that requires vadose zone monitoring for solid waste disposal sites.

DISTRIBUTION OF FLUID PRESSURES IN THE GROUND
WITH RESPECT TO THE GROUND WATER SURFACE

FIGURE 9-47 Vadose and Ground-water Zones

9.7.1 Definitions and Units of Measurement

The vadose zone (also called the zone of aeration or unsaturated zone as shown in Figure 9-47) is defined as the geologic column that extends from the ground surface to the first major saturated formation. A soil or unconsolidated sediments can be conceptualized as a three phase system composed of liquids, solids and gases. The Solid phase consists of mineral and organic matter while the liquid phase occupies the pore space. The liquid phase is often referred to as soil pore water. The soil solution may partially fill the soil pores "unsaturated" case, the liquid exists as thin films along particle surfaces, as liquid wedges around the particle contacts, or as isolated bodies in small pore spaces.

The matric potential (also referred to as capillary potential. tension and soil suction) is the result of capillary and adsorptive forces acting on the soil matrix. The matric potential is a negative pressure and can be measured in unsaturated soils by tensiometers. As a soil becomes wet, for example, the matric potential increases from a very low (negative value) to zero value (i.e., saturated).

The soil suction can be measured with a tensiometer and is reported in units of bars, or centibars (l bar = 1 atmosphere). Soil suction can range from 0, which means the soil is saturated, to as high as 600 bars, as found in some extremely dry soils. This becomes important when discussing pore-liquid samplers or suction Iysimeters, because unsaturated pore-liquid movement takes place only in the 0 — 20 cbar range. A pore-liquid sample can be obtained only in the 0 - 60 cbar range. If the soil suction is greater than 60 cbars, then the suction lysimeter will not work.

The soil water potential is the observation that liquid flows in the vadose zone as a function of its' potential and kinetic energy. Since vadose water movement is slow, the kinetic energy is normally considered negligible. Therefore the primary energy source is that which determines the state and movement of soil water. Potential energy is determined by the liquid position and the drive of vadose soil water to move to a lower potential energy state. Soil water potentials are often expressed in terms of the height of a reference body of water that will develop an equivalent pressure or suction.

Units used in describing tension and energy of the soil water potential includes the bar (1 bar equals 0.987 atmospheres = 1017 cm of H_2O = 10^{-1} megapascal). A common unit conversion is that 1 cbar is roughly equal to 1 kPa that is roughly equal to 100 cm of H_2O. The soil moisture distribution dictates the unsaturated hydraulic conductivity. Because the soil moisture most always varies, the unsaturated hydraulic conductivity may be highly variable, even in a homogeneous lithology.

A common observation in vadose zone flow systems is the movement of pore liquids through fine grained soils rather than course grained materials.

The Richard's Principle says that in unsaturated flow (0 - 20 bars), pore liquids will not readily move from a fine-grained to a coarse-grained material. The contact will act as a boundary, and fluids will build up, or mound, before entering the coarse-grained material (if they enter at all). As the liquid enters the coarse-grained material, "fingers" of flow referred to as "Taylor Instabilities" develop. These "fingers" appear to form as the result of some preferential flow path force; however, the actual cause is unknown. If the soil is completely dried and then rewet, the "fingers" of flow develop at new locations. These "Taylor Instabilities" results in unpredictable flow patterns, that make the Vadose Zone modeling difficult.

9.7.2 Implementation Of A Vadose Zone Monitoring System

Designing a vadose zone monitoring system can be reduced to five major tasks as:

- project scoping
- site characterization
- equipment selection
- equipment placement
- documentation

Each of these steps is briefly discussed.

Project Scoping

Defining the purpose or objective of the monitoring program is required prior to designing the system. The objectives and expectations need to be realistically defined as part of a project scoping effort. Monitoring objectives must become part of the conceptual understanding of the site. Vadose zone monitoring programs require a clear understanding of site geology since, the ability of the proposed system to obtain a sample of pore liquids or measure changes in soil moisture, is directly related to subsurface conditions. These data are gathered during site characterization activities.

Site Characterization

Site characterization for a vadose zone detection monitoring system consists primarily of collecting soil/geologic and hydrogeologic information:

- Textural information of major soil groups, including; depth, stratification sequence, estimated porosity;

- Depth to ground water and seasonal ground-water flux; and,

- Saturated hydraulic conductivity of major soil groups.

These data are needed in designing a vadose zone monitoring system to assess the applicability of the various sampling and evaluation techniques to the specific site. This is accomplished best by obtaining continuous cores of the entire Vadose Zone, or to an appropriate depth, at a specified number of locations depending on size of the site. The matrix potential of these core samples can be measured top to bottom in the field by use of a quick-draw tensiometer. This tool is simply pushed into the sidewall of the core and takes approximately three minutes to equilibrate. The soil moisture may vary significantly at lithology changes.

Soil Vapor Sampling / Monitoring

Soil vapor hydraulic conductivity should not be measured in the lab by forcing the liquids out of the sample before running the test. The lab K (vapor) should be run on the sample at the same soil tension as was measured in the field. This will assure that the core is not wet or dried out relative to actual formation moisture content.

TDR (time domain reflectometry) can be used to measure the formation water content independent of lithology. This is accomplished by measuring the capacitance of the soil.

Soil psychrometers measure the relationship between negative soil-water potential and the relative humidity of soil water. The uses of psychrometers are commonly associated with erroneous data. This is because an insignificant change in relative humidity, which the instrument measures, are associated with a significant change in both water content and soil tension. This is illustrated by the following table. Typical scopes of work for vadose zone monitoring system design may include:

- Unsaturated hydraulic conductivity of major soil groups (field and laboratory-as expected the field measurement is more representative);

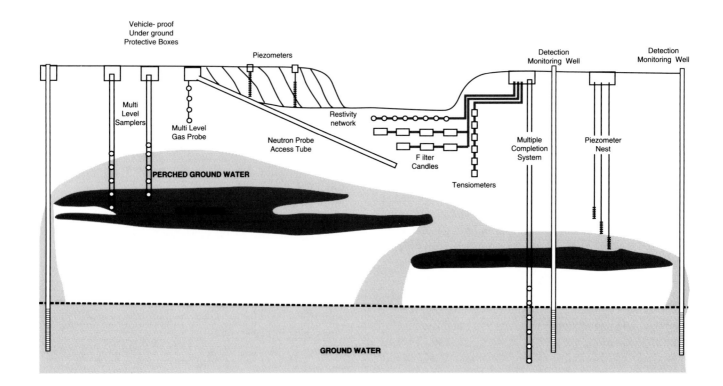

FIGURE 9-48 Example Vadose Monitoring Placement in Facility

- Drying branch of a soil moisture characteristic curve for each major soil type;

- Estimated or field measured unsaturated hydraulic conductivity (Bouma et al., 1974, 1983);

- Gas (air) permeability of major soil types as measured in field;

- Grain size distribution of major soil groups;

- Identification of major textural interfaces or gradations (mean grain size diameter >1.5 times greater than adjoining soils); and,

- Saturated hydraulic conductivity of all major soil types.

Once this subsurface information is collected, a conceptual or graphical model of the soils at the facility should be developed in a similar fashion as used for saturated monitoring programs. This model will begin to provide one with an intuitive sense of optimum equipment locations for placement.

Uncertainties associated with the soil moisture transport through the vadose zone must be considered to identify the most probable location to detect a facility release (i.e., optimum equipment placement location). For example, it might be qualitatively determined that placement of any device at a depth less than 5 feet below the ground surface may be impractical due to a preponderance of roots, earthworm activity, burrowing animals, etc.

At the conclusion of this step, the optimum locations for equipment placement for the purpose of leak detection can be identified and prioritized based on the purpose of the monitoring system.

Vadose Zone Equipment

The purpose of the vadose zone monitoring system, leachate properties, and characterization of the subsurface environment will dictate the most suitable technology(s) for the site. Many types of vadose monitoring equipment may be employed in conjunction with saturated monitoring equipment as are shown on Figure 9-48. For example, if pore water collection is desired, an active system (i.e., lysimeters) in combination with a passive (non-sampling system) such as tensiometers may be the most appropriate design. Both depth and location issues should be considered in the vadose zone detection monitoring program. In some cases new facilities or cells will have

sump areas for leachate collection systems. These sumps are at the lowest portion of the facility or cell and represent likely points of limited area that can concentrate vadose zone monitoring. Some base grades are often designed to be located 5 to 10 feet above seasonal ground water high. Large pans lysimeters located between sump base grades and the ground-water highs have a high likelihood of detecting leakage from these sumps. There will be cases that the limiting factors of the geologic and hydrogeologic systems will dictate a single system. In most cases, there will be a threshold at which more money does not buy an appreciable increase in detection probability for a given technology.

Various pore-liquid sampling equipment are appropriate for vadose systems including BAT samplers, suction lysimeters, pan lysimeters, gypsum blocks, resistivity measurements, magnetometry, and neutron moderation. Some of the important systems were:

- Soil psychrometers measure the relationship between negative soil-water potential and the relative humidity of soil water. The use of psychrometers is commonly associated with erroneous data. This is because an insignificant change in relative humidity, which the instrument measures, are associated with a significant change in both water content and soil tension. This is illustrated by the following chart.

Relative Bars	Humidity	Water Content (by weight)
0.1	99.9926	15 %
1.0	99.926	6 %
5.0	99.637	3 %
10.0	99.26	2 %
15.0	98.89	1.5%

- Neutron Moderation - At a site where the soil suction is greater than 2 bars, the neutron probe is the best monitoring choice. The probe is two inches in diameter and one-foot long. It has a 50 millicurie Americium Beryllium high-energy, neutron-emitting source. The probe emits the neutrons and measures the moderated, or slowed down, neutrons that return. The number of moderated neutrons correlates to the hydrogen ion concentration in a 32 cm radius around the probe. Because the mass of the hydrogen ion is very close to the mass of the neutrons, the collision of these results in the maximum energy loss. Therefore, the maximum amount of energy loss is a function of the formation's hydrogen concentration. Thus, the neutron probe measures the concentration of water. In the presence of water and methane, the gas will have little influence on the reading because the hydrogen concentration of the methane is very low compared to that of the water. Also, the neutron probe will work through PVC, HDPE, or stainless steel casing, and can penetrate through up to two feet of cement and/or bentonite grout.

The particular technology selected for the vadose system should be field tested in a few units at the site. Calibration techniques over anticipated range of conditions for example soil moisture content can provide assurances that the selected system does, in fact, look as designed. These trial tests allow refinements or modifications to be evaluated before actual installation of the final system. The final installation may also require considerable assembly and calibration efforts before final equipment check-out.

Vadose Equipment Location

Location of the equipment in the optimum or prioritized target monitoring zone is usually the least difficult of the various steps. Whenever possible the vadose monitoring system should be installed as part of a new cell or facility, rather than an afterthought. Since the majority of flow in the vadose zone is vertical, location of vadose zone monitoring technology has to be primarily under portions of the facility. External sumps are facility design features that allow for reasonably effective vadose system design. Retrofitting vadose monitoring systems may present the least optimum condition, since these installations probably require access to below the facility. A key task performed during this process is position of the drilling rig to access the target zone. For example, one may bulldoze a trench next to a surface impoundment so that the drill rig can drill horizontally under a surface impoundment rather than slant boring.

Borehole location and documentation of the angle of drilling (if used) should be recorded throughout the installation process. Extensive logging of the soil layers using continuous sampling techniques should be performed during the drilling. Continuous sampling and logging form the basis for adjusting the target zone during installation. If the actual target zone once reached consists of course grained rather than fine grained sands (as originally believed from the early site characterization phase), placement of the vadose monitoring system can be adjusted

so that it is installed in a material with the physical attributes closer to that required by the equipment selected.

Once the target zone is reached the device(s) are placed at the appropriate depth(s), the hole should be backfilled with the correct annular sealant and the unit fully tested prior to moving to the next installation. For example, if a porous cup sampler is installed, a vacuum should be drawn at the surface and sample collected. A gypsum block requires evaluation of an electrical signal be compared to the calibration curve prepared for that unit at the approximated moisture content of the backfill. If the device is shown to be fully functioning, the borehole can be completed and the above ground tubing/wires, etc., permanently labeled.

Documentation

The final but important step in the vadose zone monitoring process is the documentation of each installation. One should not lose sight that monitoring programs are built with proper documentation. Such information should include all installation data (field notes, survey information, types of units, depth of installation, etc.) maintenance problems encountered after installation; calibration data for each unit, and the name of manufacturer and model number of all equipment installed in the borehole.

As part of the documentation a maintenance guide should be created, specific to the vadose monitoring technology and equipment. This guide should be simply worded so that diagnostic tests of any malfunctioning equipment could be easily performed by technicians.

9.8 EXAMPLE MONITORING DESIGN

9.8.1 Interpreted Site Geology

The site is considered to be landward of the estimated position of sea level during the time in which the Prairie Formation was being deposited (Late Quaternary). Therefore, it is expected that the sediments found on site would be predominantly non-marine to marginally marine and would exhibit characteristics associated with deltaic aggradation (delta building) processes and not coastal or offshore marine processes. Typical features of this type of system initially consist of braided streams becoming meandering streams as the river gradient decreases. As different deltas were built into the Gulf of Mexico, broad meander belts developed. Also associated with this environment are organic-rich back-swamp deposits which lie adjacent to meander belts and natural levees.

Examination of an NHAP false color infrared aerial photograph revealed no surficial expressions of these features. Thus, is was necessary to rely on the results of subsurface investigations.

The results of a hydrogeological investigation for the proposed landfill are shown conceptually in Figures 9-49 through 9-51. The purpose of the investigation was to review the geology of the site, to confirm the subsurface conditions as determined previously at the site by others and to provide an interpretation of the hydrogeological conditions particularly as they affect the design of the Landfill.

The geological review confirmed that the site lies within the Prairie Formation. As such the property generally is under-lain by fine grained cohesive sediments laid down during the delta building process. In addition, the site is traversed by three channel sands which have been termed "Lower", "Middle", and "Upper". The depth to the three sands is typically 37 ft, 20 ft., and 6 ft., respectively. The piezometric levels in the channel sands indicate that the lower and middle sands are not hydraulically connected whereas the middle and upper units probably are connected. Superimposed on this pattern is a system of random local sand "stringers" generally located less than 20 ft. deep.

To augment previous geotechnical data, a field investigation was undertaken. This involved 5 sampled boreholes, 9 cone penetrometer tests, 1 test trench and the installation of 6 piezometers.

In general, the site is underlain by thick beds of competent, plastic and relatively impervious subsoil. These materials are incised by local channel sands and sand stringers. The cohesive units display a broad range of plasticity and have been classified on that basis into organic clay, clay, silty clay and clayey silt. In general, the thickness of the individual beds for the latter three units varies from 10 to 15 ft. Ground surface is generally underlain by a silty clay unit which overlies a clay stratum. The organic clay constitutes a marker horizon and is located at depth. Clayey silt and silt deposits, where encountered, generally flank the channel sands and are believed to be overbank deposits.

Regarding groundwater conditions, the potentiometric surface is typically 6 to 8 feet below ground surface with seasonal fluctuations of a few feet. The flow direction across the site is from northeast to southwest. The ground-water level in the Middle Channel Sand decreases across the site at a gradient of about 0.0012. The flow velocity within this unit is estimated to range from 4 to 19 ft./year.

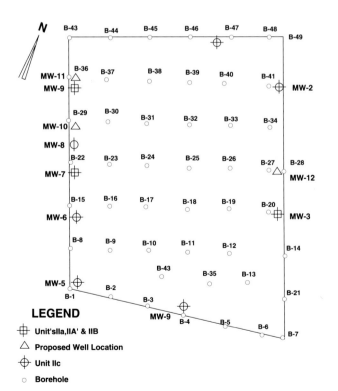

FIGURE 9-49 Layout of Boreholes at Facility

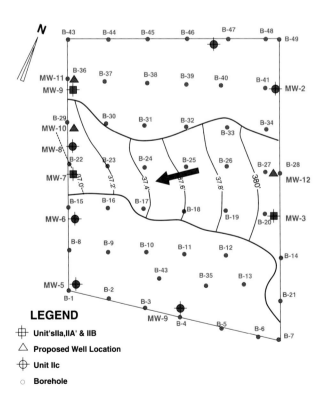

FIGURE 9-50 Piezometric Levels

HYDROGEOLOGIC CONCEPTUAL MODEL

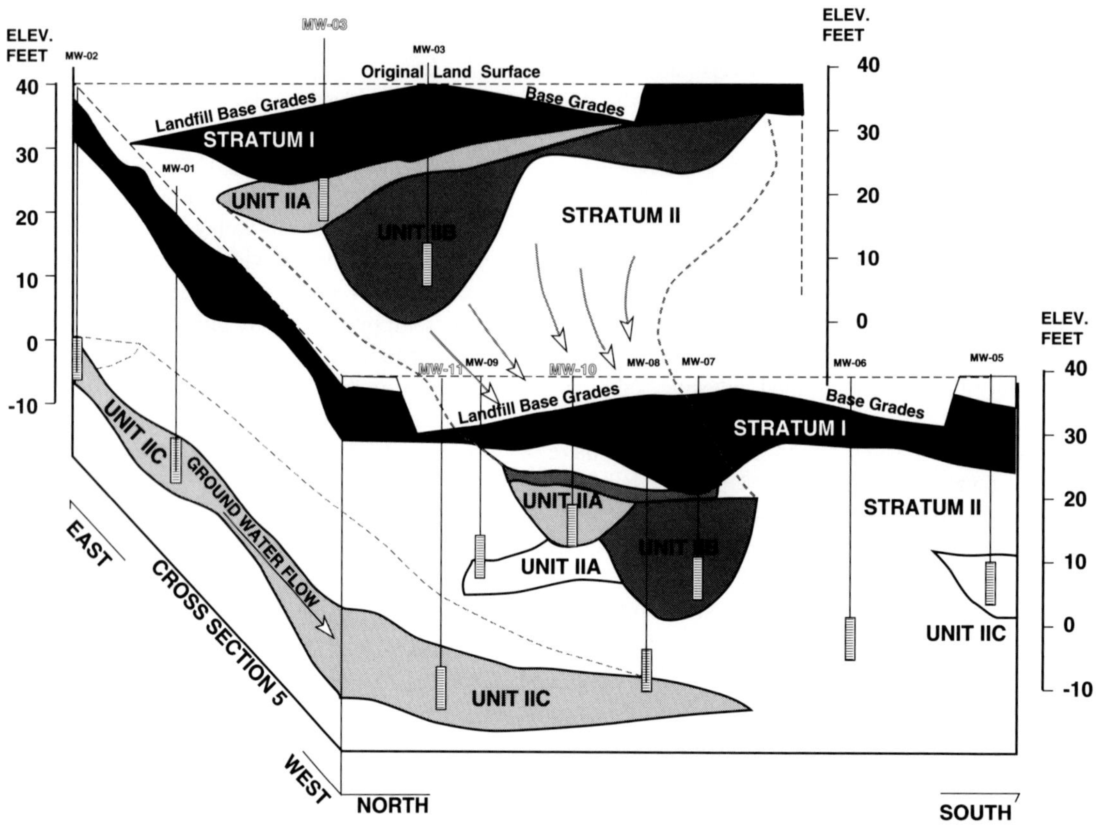

FIGURE 9-51 Conceptual Hydrogeologic Model

9.8.2 Monitoring System Design

Once the conceptual hydrogeologic model is constructed fully based on the data gathered during the Phase II investigation, design of the detection monitoring system is relatively simple. The flow into the sand channel units represents the most rapid and significant point that would should contamination if there should be a release from the facility. The target monitoring zone is represented by the sand channel that crosses the site and serves as a local discharge system for the site area.

REFERENCES

Arbhabhirama, A., and C. Kridakorn, 1968. Steady downward flow to a water table. Water Resources Research. Vol. 4.

Brooks, and Corey, 1975. Drainage characteristics of a soil. Soil Science Society of America Proceedings.

39:251-255.

Bouma, Baker, and Veneman, 1974. Measurement of water movement in soil pedons above the water table. University of WisconsinExtension. Geological and Natural History Survey, Information Circular Number 27.

Diment, G., and K. Watson, 1983. Stability analysis of water movement in unsaturated porous materials. 2. Numerical studies. Water Resources Research. 19(4)1002-1010.

Diment, and Watson, 1985. Stability analysis of water movement in unsaturated porous materials: Experimental studies. Water Resources Research. 21:979-984.

Gardner, W., and J. Chatelain, 1947. Thermodynamic potential and soil moisture. Soil Scie~ce Society of America Proceedings. 11:100-102.

Germann, P., and K. Beven, 1981a. Water flow in soil macropores I. An experimental approach. Journal of Soil Science. 31:1-13.

Germann, P., and K. Beven, 1981b. Water flow in soil macropores II. A combined flow model. Journal of Soil Science, 32:15-29.

Glass, R., T. Steenhuis, and J. Parlange, 1988. Wetting front instability as a rapid and far-reaching hydrologic process in the vadose zone. Journal of Contaminant Hydrology. 3:207-226.

Kirkham, D., and W. Powers, 1972. Advanced Soil Physics. WileyInterscience, New York.

Klute, A., 1972. The determination of the hydraulic conductivity and diffusivity of unsaturated soils. Soil Science. 113:264-276.

Miller, E., 1975. Physics of swelling and cracking soils. Journal of Colloid and Interface Science. 52t3)434-443.

Muralem, Y., 1976. A catalogue of the hydraulic properties of unsaturated soils. Haifa, Israel. Israel Institute of Technology.

Quinlan, J. F. 1990. Special Problems of Ground-water Monitoring in Karst Terranes in Ground Water and Vadose Zone Monitoring, Nielsen and Johnson ed. ASTM STP 1053, 1916 Race st. Philadelphia, PA 19103, p.275-307.

Richards, L., 1931. Capillary conduction of liquids through porous mediums. Physics. Vol. 1.

Scott, and Clothier, 1983. A transient method for measuring soil water diffusivity and unsaturated hydraulic conductivity. Soil Science Society of America Journal. 47:1068-1072.

Simpson T., and R. Cunningham, 1982. The occurrence of flow channels in soils. Journal of Environmental Quality. 1(1):2930.

Steenhuis, T., and J. Parlange, 1988. Simulating preferential flow of water and solutes on hillslopes. Conference on Validation of Flow and Transport Models for the Unsaturated Zone. Ruidoso, New Mexico. May 23-26. pg. 11.

Taylor, A. 1950. The instability of liquid surface when accelerated in a direction perpendicular to their planes. Proceedings of the Royal Society. 201:192-195.

Warner, G., and J. Nieber, 1988. CT scanning of macropores in soil columns. Winter Meeting of American Society of Agricultural Engineers. Paper No. 88-2632. Presented at the International Winter Meeting at the Hyatt Regency, Chicago, Dec. 13-16, 1988. pg. 13.

White R., 1985. The influence of macropores on the transport of dissolved and suspended matter through soil. Advances in Soil Science, Vol 3., Springer Verlig New York Inc., pg. 95-113.

CHAPTER 10

ASSESSMENT MONITORING DESIGN

Assessment monitoring programs typically are employed once the owner/operator of the facility has detected a statistically valid increase in indicator parameters in his detection monitoring program. Assessment monitoring activities may also be initiated due to previous hazardous waste disposal activities that may nominate the site through a State or Federal regulatory program as a significant risk to human health and the environment. Federal and State regulations have codified assessment monitoring programs throughout the Resource Conservation & Recovery Act (RCRA), Subtitles C and D, and Comprehensive Environmental Response, Compensation and Liability Act (CERCLA) "Superfund" regulations. Although it may seem like a daunting task to review assessment monitoring programs for such complex regulations, the actual scope of the site evaluation technical work components are essentially similar. These federal programs are compared in Figure 10-1 for major goals of assessment investigations. These goals can be summarized into the following points:

- Identify releases needing further investigations;

- Characterize nature, extent, and rate of release;

- Evaluate alternatives and identify remedy(s);

- Purpose selected remedy;

- Public participation;

- Authorize selected remedy; and,

- Design and implement chosen remedy.

These Federal programs are also followed by very similar State remedial action programs that could easily be categorized into the same set of goals. Hence, assessment programs go beyond pure evaluation of "rate and extent" of ground-water contamination to include definition of all contaminated media and the selected remedy.

These regulations place considerable reliance on collection of ground-water quality samples to determine both the rate and extent of contaminant movement. Once these data are established, appropriate corrective actions could range from just continued monitoring to an aggressive aquifer and site remediation program. Unfortunately, Federal and State mandated programs have often primarily relied on collection of extensive organic contaminant parameters lists, to the detriment of gathering sufficient geologic data to both define the extent of affected ground water, and establish the basic hydrogeologic characteristics sufficient to design a remediation for the facility.

The technical difficulties in design of assessment monitoring programs that follow the above seven points can be eased through application of a site assessment program based on a phased approach similar to the analysis used in detection monitoring. An investigation program that put understanding the geology and hydrogeology first allows the remaining five assessment goals be addressed in a more direct manner. The following chapter summarizes some of the common aspects of Federal and State assessment monitoring programs and details program guidance for both specific determination of the extent of affected aquifers or soils and acquisition of sufficient data to design appropriate aquifer restoration techniques.

The guidance provided below evaluates site complexity and general program components applicable for such evaluations. Special guidance is also provided for site evaluations specific to non-aqueous phased liquids (NAPL's) since density may have a significant affect on the scope of a facility assessment.

10.1 RCRA ASSESSMENT MONITORING PROGRAMS

Congress significantly expanded the federal role in controlling the management of waste materials in the United States with the passage of the RCRA of 1976. Subtitle C of RCRA required EPA to establish a

RCRA / CERCLA COMPARISON

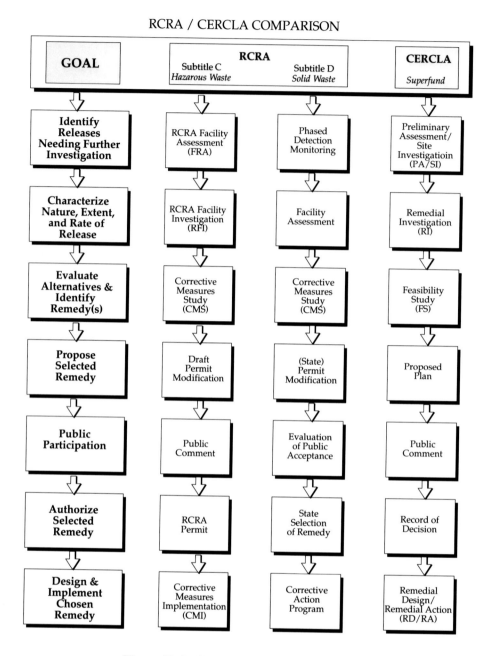

Figure 10-1 Comparisons of Remedial Programs

comprehensive regulatory program to ensure proper management of hazardous waste. Subtitle D of RCRA established a cooperative framework for involving federal, state and local governments in solid waste disposal. The federal role in the Subtitle D is to provide technical assistance to states and regions for planning and developing waste management practices and to furnish financial assistance to enable states to implement these

measures. The planning and implementation of solid waste programs under Subtitle D will remain, in the majority, a state and local function.

Much of the monitoring technology used to date has been based on a RCRA guidance document "Technical Enforcement Guidance Document", 1987 (TEGD). While this document is used as a basis for a RCRA monitoring program, a number of important issues regarding ground-water monitoring have made portions of the TEGD out-dated. The TEGD, which was issued in 1986, is far from the last word in the rapidly developing field of ground-water monitoring technology.— For example, the TEGD indicates that "fluorocarbon resins or stainless steel should be specified for use in the saturated zone when volatile organics are to be determined". Moreover, according to the TEGD, "PVC is appropriate only for trace metals and nonvolatile organics". There is no scientific data to suggest that PVC is not appropriate for use in ground-water monitoring wells and, in fact, there is substantial data to support such use. Indeed, the recently finalized ASTM Committee D-18 "Recommended Practice for Design and Installation of Ground Water Monitoring Wells in Aquifers" D 5092-90 specifies that well screens and risers can be composed of PVC, stainless steel, fiberglass, or fluoropolymer materials. Another example of the TEGD's obsolescence regards the use of statistical analysis of ground-water monitoring data to determine if a release has actually occurred.

The TEGD recommends that the Cochran's Approximation to the Behrans-Fisher (CABF) t-test be used as the statistical test for a detection monitoring program. Since issuing the TEGD, EPA itself expressed doubt regarding the validity of the CABF t-test for ground-water monitoring programs in the October 11, 1988 final rule, Statistical Methods for Evaluating Ground-Water Monitoring Data from Hazardous Waste Facilities. EPA concluded that the this statistical methodology was not appropriate for use because: (1) the replicate sampling method required under the then current Part 264 Subpart F regulations was not appropriate for the CABF procedure; (2) the CABF procedure did not adequately consider the number of comparisons that must be made under the regulations; and (3) the CABF did not control for seasonal variation. Moreover, the CABF test generates an unacceptable rate of false positives and has placed facilities in compliance monitoring programs when, in fact, no releases had actually occurred.

Ground-water remedial programs should always be based on using solutions that provide a reasonable assurance that the affected media will actually reach some acceptable clean-up level. This may sound like a given for conducting any remedial project, however, there are circumstances when, (because of widespread ground-water contamination stemming from numerous other sources), performing a corrective action remedy at a waste disposal facility will produce no measurable improvement in ground-water quality. Experience over the past few years has shown that most, if not all, contaminated aquifers cannot, at present, be restored to a condition compatible with national health-based water quality standards. Several recent studies support this conclusion. For example, EPA

(1990) recently released a study involving 19 Superfund sites where ground-water pump-and-treat systems have been in operation for up to 10 years. This study concluded that, in most cases, although the initial phases of remediation had achieved significant mass removal of contaminants, there had been little success in reducing concentrations to target levels. Typically, these systems achieved an initial drop in concentrations by factors ranging from 2 to 10, followed by a leveling off with no further decline in contaminant concentrations. Similarly, an independent analysis recently completed by the Oak Ridge National Laboratory concluded that restoration of contaminated aquifers to health-based levels may be impossible. In addition, leading environmental and ground-water scientists have predicted that continuous pumping for 100 to 200 years may be necessary to reduce contaminant concentrations by a factor of 100 (assuming ideal conditions), and that for sites with "Light or Dense Non-Aqueous Phase Liquids" (LNAPL or DNAPL) contaminants, restoration to drinking water standards may simply be unachievable. Although implementation of remedial programs are the ultimate goal for the Federal and State programs discussed in this chapter, one should not lose sight of the reality in the extreme difficulty of evaluating and actually successfully cleaning-up affected aquifers.

10.1.1 RCRA Subtitle C

The passage of the revised RCRA legislation in 1984, detailed assessment monitoring of waste disposal facilities within several sections of 10 CFR for hazardous waste sites. Sections 265.93(d)(4), 270.14(c)(4) and 270.14(c)(2) provide enforcement officials with the regulatory authority needed to review an owner/operator's assessment monitoring program. These sections describe the following criteria:

- 265.93(d)(4) An assessment monitoring plan must be capable of determining:

  (i) Whether hazardous waste or hazardous waste constituents have entered the ground water, and

  (ii) The rate and extent of migration of hazardous waste or hazardous waste constituents in the ground water.

- 270.14(c)(4) The Part B applicant must include in the submittal a "Description of any plume of contamination that has entered the ground water from a regulated unit at the time the application was submitted that:

(i) Delineates the extent of the plume, and

(ii) Identifies the concentration of each Appendix VIII ... constituent ... throughout the plume ..."

• 270.14(c)(2) The Part B applicant must submit, among other things, an "identification of the uppermost aquifer and aquifers hydraulically interconnected beneath the facility property, including ground water flow direction and rate, and the basis for such identification (i.e., the information obtained from hydrogeologic investigations of the facility area)."

Technical Approach RCRA Site Remediation Program

The RCRA Facility Investigation (RFI) is generally equivalent in technical scope to the CERCLA remedial investigation. Units are areas of concern that are determined in the RCRA Facility Assessment (RFA) to be a likely source of significant continuing releases of hazardous wastes or hazardous constituents. The regulatory means of requiring the RFA is either through RCRA permit conditions (operating or closure/post-closure) or via enforcement orders [e.g., 3008(h)]. Because of the Hazardous and Solid Waste Act (HSWA) statutory language, the agencies must focus the RFI requirements on specific solid waste management units, or known or suspected releases that are considered to be routine and systematic. The HSWA permit conditions or enforcement orders can range from very general (e.g., "characterize the ground water at . . . ") to very specific (e.g., a specified number, depth, location and frequency of samples analyzed for a given set of constituents). Since the regulatory Agency in the RFA, is not required to positively confirm a continuing release, but merely determine that the "likelihood" of a release exists, the scope of the RFI can range from a limited specified activity to a complex multi-media study.

The investigation may be phased, initially allowing for verification or rebuttal of the suspected continuing release(s). If release to the environment is verified, the second phase of investigation typically consists of release characterization. This second phase of a RCRA RFI is much like a Superfund RI, includes: (1) the type and quantity of hazardous wastes or constituents within and released from the unit, (2) the media affected by the release(s), (3) the current extent of the release, and (4) the rate and direction at which the releases are migrating. Inter-media transfer of releases (e.g., evaporation of organic compounds from contaminated soil to the atmosphere) may also be addressed where applicable

during the RFI. Investigation of potential releases from RCRA units requires establishment of various types of technical information early in the RFI. This information is specific to the RCRA waste managed, the unit type (for example lagoon or landfill), design and operation of the facility, the environment surrounding the unit or facility, and the environmental medium to which contamination may be released. Although each medium (i.e., ground water, surface water, sediments) will require specific data and methodologies to investigate a release, the following represents a general strategy for a RCRA RFI project.

The technical approach of the RFI requires the investigator to examine extensive data on the facility and specific units at the facility. These data generally can be divided into the following categories:

• Regulatory history;	(Typically estab-
• Facility and unit design;	lished in Phase I
• Waste characteristics;	work)...
• Previous release events.	
• Environmental setting;	
• Pollution migration pathways;	(Typically estab-
• Evidence of release;	lished in phase II
• Environmental receptors	field investigations)

Specific factors in each category that must be considered will vary depending on which environmental pathway medium is considered most vulnerable. For example, unlined in-ground units are more likely to have soil and ground-water releases than lined units. A facility's environmental setting will determine which media are of concern (e.g., shallow ground water or fractured subsoils). As such, the phase investigation approach must be based on first a regional understanding of the subsurface that points toward a more direct site based Phase II geologic and hydrogeologic field data gathering.

RCRA Facility Investigation

Assessment monitoring programs as detailed later in this chapter can be used to define the various site indices that provide for direct evaluation of the potential release and the remediations required by RCRA.

In reviewing the RFI phase I and II investigative effort, the regulatory agency will typically interpret the release findings. The first interpretive emphasis of the investigation is primarily on the data quality (i.e., were location criteria, sampling and analytical data quality objectives defined and accomplished according to planning documents?). The results of the RFI are then compared against established human health and environmental criteria.

It is important to remember that the direct assessment techniques described in this chapter will typically provide water quality values that are directly within the man part of the migration pathway. These targeted analytical values are generally much higher than the somewhat random values obtained from typical assessment programs. The more concentrated values produce higher criteria or "action" levels for each environmental medium and exposure pathway, using the toxicological properties of the waste constituent and standardized exposure model assumptions. At this stage, in the RCRA facility investigation, if the continuing release of hazardous wastes is determined as a potential short-term or long-term threat to human health and the environment, the regulatory agency may require either interim corrective measures or a corrective measures study. This evaluation of human health and the environment risk factors is a crucial stage in the RCRA corrective action process. Since the location of assessment monitoring points are placed directly within ground-water flow lines from the facility in question, these values allow better evaluation of migration pathway risk assessment, potential receptors and worst-case water quality parameter leads. With the data gathered within the targeted field assessment program, water quality values and remedial design criteria are available early in the program.

In addressing releases from Solid Waste Management Units (SWMUs) to the environment, the RFI is followed by corrective measures. That is, a release, or source of contamination, have been identified, and the owner/operator must initiate a remedial response. As in the Superfund program, remedial action objectives of a particular corrective action are site-specific, therefore, quantitative goals that define the level of cleanup are required to achieve the response objectives. These goals include any preliminary cleanup levels for environmental media affected by a release, the area of attainment and the remedial time-frame.

The above objectives are accomplished through the Corrective Measures Study (CMS) and the Corrective Measures Implementation (CMI) by identifying, designing and implementing the appropriate remedial strategy, all in accordance with published guidance. The CMS serves as a recommendation to the U.S. EPA or the State, while the CMI is the allowed time frame for the actual corrective measures.

Corrective Measures Study (CMS)

The first step in the CM phase is the development and implementation of the CMS to determine the most effective remedial option to correct potential or actual environmental impact and human exposure threats posed by releases of hazardous wastes or constituents. Regardless of whether the remedial response effort is conducted under CERCLA or RCRA authority, the objectives of the RCRA CMS, or Superfund feasibility study, are to utilize technical knowledge and propose actions to control the source of the contamination (by preventing or mitigating the continued migration of contamination by removing, stabilizing and/or containing the contaminants) and/or actions to abate problems posed by the migration of substances from their original source into the environment.

Through the CMS, the owner/operator must technically demonstrate that the response action proposed effectively abates the threats to human health and the environment posed by the release(s). This typically requires the analysis of several remedial technologies in detail sufficient to show that the recommended measures effectively remove the threats posed by the release. To do so, the owner/operators must assess these alternatives in terms of their technical feasibility (including reliability and requirements for long-term operation and maintenance), their ability to meet public health protection requirements and their ability to protect the environment and any adverse environmental effects of the measures. The owner/operator also should consider any institutional constraints to implementation of the measures, such as off-site capacity problems and potential public opposition.

The RCRA approach to assessing the level of remedial action required for environmental media is similar to that of CERCLA and generally is based on the following criteria:

- Overall protection of human health and the environment;

- Compliance with regulatory programs (e.g., CERCLA or RCRA);

- Short-term effectiveness;

- Long-term effectiveness and permanence;

- Reduction of toxicity, mobility or volume of hazardous wastes and/or waste constituents;

- Implementability;

- Cost;

- U.S. EPA and/or State acceptance, and;

- Community acceptance

The first two criteria are the basic regulatory

requirements, while the next five criteria are interactively used to analyze and compare the options. The final two criteria are considerations in the overall evaluation. In some cases, it is possible for owner/operators to analyze and present to the Agency or State only a single alternative that meets public health and environmental requirements. This situation is often the case at facilities that have taken "interim corrective measures" and thus have had an opportunity to evaluate the remedial strategy and the associated operations to determine their effectiveness.

RCRA final remedies will be required to meet applicable, possibly current, health and environmental standards promulgated under RCRA and other laws. For example, at regulated units, ground-water releases are subject to the ground-water protection standards, possibly consisting of the following:

• Constituent specific maximum concentration limits (MCLs)

• The background level of that constituent in ground water; and

• An approved alternate concentration limit (ACL) where approval would be based on criterion set forth in the RCRA regulatory framework.

For soil, soil gas, surface water, ground water and air emissions problems that cannot be addressed by existing health based or regulatory standards, the US EPA currently is assessing the appropriate technical approach. One possible alternative is to establish appropriate health-based standards on a case-by-case basis. Once the owner/ operator proposes the remedial strategy(s) for addressing releases to environmental media and the SWMU itself, the U.S. EPA or the State will evaluate the owner/operator recommendation and approve or disapprove it. During the review process, the owner/operator must be prepared to provide the technical support for his or her proposition and must be open to negotiations. The views of the public on the proposed measures and the financial assurance demonstration also will normally be considered by the State and U.S. EPA in making these technical and cost decisions.

Corrective Measures Implementation

Once the U.S. EPA, the State, and the owner/operator agree on the remedial approach, the owner/operators must design and construct the selected response action. After construction, the appropriate measures needed to operate, maintain and monitor the remedy will be taken by the owner/operators. These activities will be called out by

permit condition or a compliance order and these activities must be completed by the owner/operators with some level of oversight by the U.S. EPA or State. Since the actual remedial operations serve to provide data concerning the effectiveness of the corrective action, it is essential that these data are used as criterion in determining whether the operations should be modified over time to meet the project cleanup objectives.

10.1.2 RCRA Subtitle D

In developing the corrective action provision of RCRA Subtitle D, Subpart E, for solid waste facilities the U.S. EPA reviewed the processes, approaches, and historical experiences gained through CERCLA remedial actions and the corrective action program under the hazardous waste regulations of RCRA Subtitle C. The assessment process contain in 40CFR §258.56 is begun when the ground-water trigger levels have been verified exceeded during Phase II monitoring as is shown in Figure 10-2. Components of the 1991 RCRA Subtitle D assessment monitoring and remedial actions sections of these regulations are provided below in italic type face referenced section by section.

Assessment monitoring program [§258.55]

• *Within 90 days of triggering an assessment monitoring program, and annually thereafter, the owner or operator must sample and analyze the ground-water for all constituents identified in Appendix II. A minimum of one sample must be collected and analyzed during each sampling event. For any constituent detected in the downgradient wells as a result of the complete Appendix II analysis, a minimum of four independent samples from each well (background and downgradient) must be collected and analyzed to establish background for the constituents.*

The Director of an approved State may specify an appropriate subset of wells to be sampled and analyzed.

The Director of an approved State may delete any Appendix II monitoring parameters.

• *The Director of an approved State may establish an alternate frequency for repeated sampling and analysis considering the factors in §258.55(c)(1) through (6).*

{Note: Specifying a subset of wells, deleting Appendix II parameters, and establishing an alternate frequency for assessment monitoring are not available options in

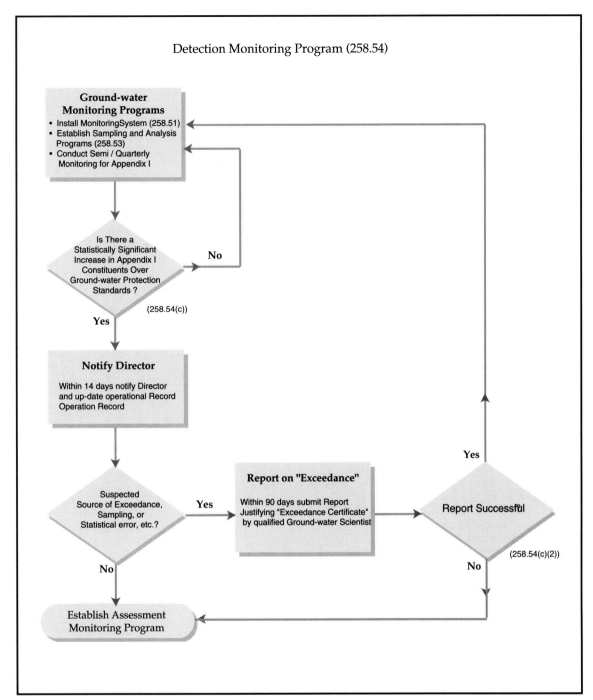

Figure 10-2a RCRA Sub-title D Detection Monitoring

unapproved states.}

• After obtaining the results from the initial or subsequent sampling events, the owner or operator must:

- Within 14 days, place a notice in the operating record identifying the constituents that have been detected and notify the State Director that the notice has been placed

in the operating record.

- Within 90 days, and on at least a semi-annual basis thereafter, resample all wells, conduct analyses for all constituents in Appendix I (or alternative list) and for the Appendix II constituents that are detected per §258.55(b). The Director of an approved State may specify an alternate monitoring frequency for the

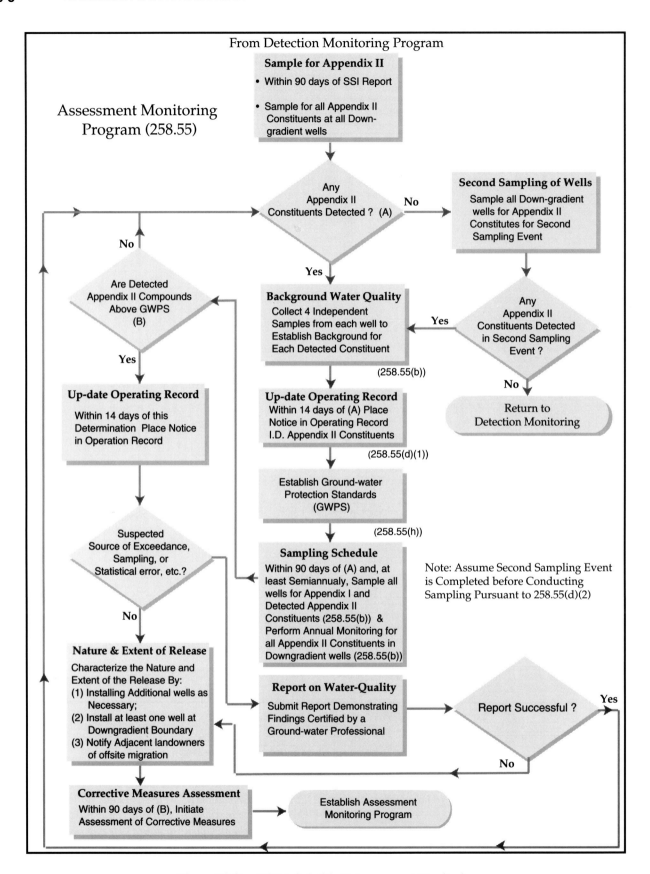

Figure 10-2b   RCRA Sub-title D Assessment Monitoring

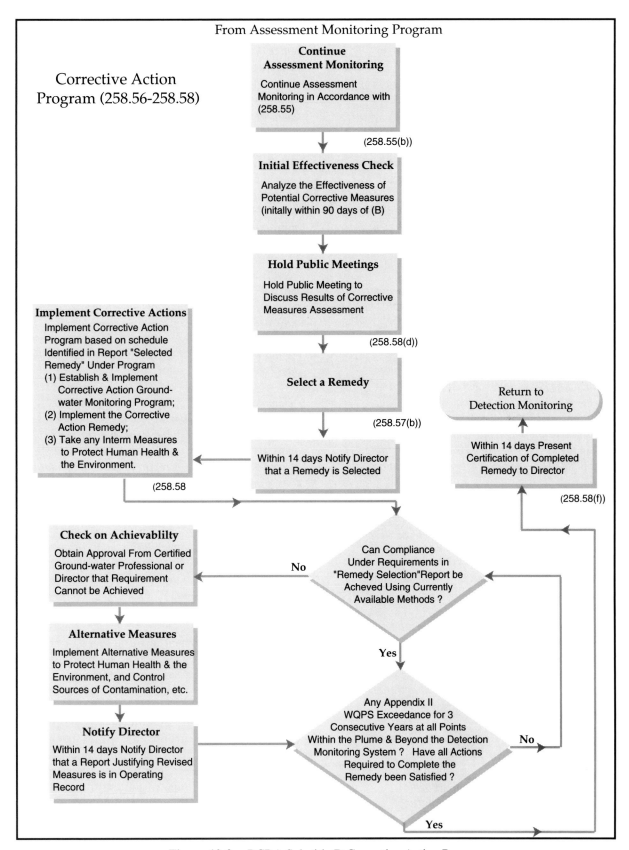

Figure 10-2c RCRA Sub-title D Corrective Action Program

constituents referred to in §258.55(d)(2).

{Note: An alternative monitoring frequency is not an available option in an unapproved state}.

- Establish background concentrations for any constituents detected.

- Establish ground-water protection standards for all constituents detected. The ground-water protection standards shall be set according to the following:

• *The owner or operator must establish a ground-water protection standard for each Appendix II constituent detected in the ground-water. The protection standard shall be:*

- For constituents for which a maximum contaminant level (MCL) has been promulgated under Section 1412 of the Safe Drinking Water Act under 40 CFR 141, the MCL for that constituent,

- For constituents for which MCLs have not been promulgated, the background concentration for the constituent established from the (background) wells,

- For constituents for which the background level is higher than the MCL or health based level (identified under §258.55(i)(1)), the background concentration.

• *The Director of an approved State may establish an alternative ground-water protection standard for constituents for which MCLs have not been established.*

{Note: An alternative ground-water protection standard for constituents without MCLs cannot be established unless in an approved state}.

• *If the concentrations for all Appendix II constituents are shown to be at or below background values, using the statistical methods defined in §258.53(g), for two consecutive sampling events, the owner or operator must notify the State Director and may return to detection monitoring.*

• *If the concentrations of any Appendix II constituents are above background values, but all concentrations are below the ground-water protection standard using the statistical procedures in §258.53(g), the owner or operator must continue assessment monitoring.*

• *If one or more Appendix II constituents are detected at statistically significant levels above the ground-water*

protection standard in any sampling event, the owner or operator, within 14 days of this finding, must place a notice in the operating record identifying the Appendix II constituents that have exceeded the ground-water protection standard and notify the State Director and all appropriate local government officials. The owner or operator must also:

- Characterize the nature and extent of the release by installing additional monitoring wells as necessary,

- Install at least one additional monitoring well at the facility boundary in the direction of contaminant migration and sample the well in accordance with §258.55(d)(2),

- Notify all persons who own land or reside on the land that directly overlies any part of the plume of contamination if contaminants have migrated off-site if indicated by sampling of wells,

- Initiate an assessment of corrective measures within 90 days (see below), or

- Demonstrate that a source other than a MSWLF unit caused the contamination or that a statistically significant increase resulted from an error in sampling, analysis, statistical evaluation, or natural variation in ground-water quality. A report documenting alternate source or error must be certified by a qualified ground-water scientist or approved by the Director of an approved State and be placed in the operating record. If a successful demonstration is made, the owner or operator must continue monitoring in accordance with the assessment monitoring program, and may return to detection monitoring if the Appendix II constituents are at or below background (§258.55(e)).

During the assessment phase, a range of alternative remedies is evaluated by the facility owner or operator; the state responsible then selects the remedy from those evaluated.

While the assessment process is conducted, the facility continues monitoring according to the Phase II requirements. The proposed criteria revisions in §258.56(b), however, allow for ground-water monitoring in addition to the Phase II requirements, as necessary, to characterize the plume of contamination and to demonstrate the effectiveness of the corrective action program. Hence, an assessment monitoring program is addressed within the RCRA Subtitle D regulations for solid waste facilities.

Selection of remedy [§258.57]

• *Based on the results of the corrective measures assessment, the owner or operator must select a remedy that, at a minimum, meets the standards listed in §258.57(b). The owner or operator must notify the State Director, within 14 days of selecting a remedy, that a report describing the selected remedy, and how §258.57(b) is satisfied, has been placed in the operating record.*

• *The owner or operator shall specify as part of the selected remedy, a schedule for initiating and completing remedial activities.*

• *The Director of an approved State may determine that remediation of a release of an Appendix II constituent from a MSWLF unit is not necessary if the owner or operator demonstrates to the satisfaction of the Director of the approved State that:*
- The ground-water is additionally contaminated from a source other than a MSWLF unit and those substances are present in concentrations such that a cleanup of the release from the MSWLF unit would provide no significant reduction in risk to actual or potential receptors, or

- The constituent is present in ground-water that:

-- Is not currently or reasonably expected to be a source of drinking water, and

-- Is not hydraulically connected with waters to which the hazardous constituents are migrating or are likely to migrate in a concentration(s) that would exceed ground-water protection standards.

- Remediation of the release is technically impracticable, or

- Remediation results in unacceptable cross-media impacts.

{Note: Demonstration that remediation of a release is not necessary is not an available option unless in an approved state}.

• *A determination by the Director of an approved State shall not affect the authority of the State to require the owner or operator to undertake source control measures or other measures that may be necessary to eliminate or minimize further releases to the ground-water, or to remediate the ground-water to concentrations that are technically practicable and significantly reduce threats to human health or the environment.*

The U.S. EPA (1989) provides some generic examples of conditions that may require additional monitoring:

• Facilities that have not determined the horizontal and vertical extent of the contaminant plume;

• Locations with heterogeneous or transient ground-water flow regimes;

• Mounding associated with MSWLF units.

Once a contaminant release has been detected and confirmed via the various phases of monitoring, the owner or operator may be requested by state regulatory agencies to implement a more aggressive monitoring program capable of delineating the horizontal and vertical extent of contamination.

Implementation of the Corrective Action Program [§258.58]

• *Based on the schedule established under §258.57(d) for initiation and completion of remedial activities, the owner or operator must:*

- Establish and implement a corrective action ground-water monitoring program that:

-- Meets the requirements of an assessment monitoring program (§258.55),

-- Indicate the effectiveness of the corrective action remedy, and

-- Demonstrate compliance with ground-water protection standard pursuant to §258.58(e).

- Implement the corrective action remedy selected under §258.57, and

- Take any interim measures necessary to ensure the protection of human health and the environment.

• *An owner or operator may determine, based on information developed after implementation of the remedy has begun or other information, that compliance with §258.57(b) is not being achieved. In such cases, the owner or operator must implement other methods or techniques that could practicably achieve compliance with the requirements, unless the owner or operator makes the determination discussed below.*

• *If compliance with §258.57(b) cannot be practically achieved with any currently available methods, the owner or operator must:*

- *Obtain certification of a qualified ground-water scientist or approval by the Director of an approved State that compliance with requirements under §258.57(b) cannot be practically achieved with any currently available methods,*

- *Implement alternate measures to control exposure of humans or the environment to residual contamination,*

- *Implement alternate measures for control of the sources of contamination, or for removal or decontamination of equipment, units, devices, or structures, and*

- *Notify the State Director within 14 days that, prior to implementing the alternate measures, a report justifying the alternate measures, has been placed in the operating record.*

• *All solid wastes that are managed pursuant to a remedy, or an interim measure, shall be managed in a manner:*

- *Protective of human health and the environment, and*

- *Complies with applicable RCRA requirements.*

• *Remedies selected shall be considered complete when:*

- *The owner or operator complies with the ground-water protection standards established under §258.55(h) or (i) at all points within the plume of contamination that lie beyond the ground-water monitoring well system.*

- *Compliance with the ground-water protection standards established under §258.55(h) or (i) has been achieved by demonstrating that concentrations of Appendix II constituents have not exceeded the ground-water protection standard(s) for a period of three consecutive years using the statistical procedures and performance standards in §258.53(g) and (h). The Director of an approved State may specify an alternate length of time during which the owner or operator must demonstrate that concentrations of Appendix II constituents have not exceed the ground-water protection standard(s).*

{Note: An alternate length of time for demonstrating that concentrations have not exceeded ground-water protection standards is not an option in an unapproved state}.

- *All actions required to complete the remedy have been satisfied.*

• *Upon completion of the remedy, the owner or operator must notify the State Director within 14 days that a certification that the remedy has been completed in compliance with the requirements of §258.58(e) has been placed in the operating record. This certification must be signed by the owner or operator and by a qualified ground-water scientist or approved by the Director of an approved State.*

• *Upon completion of the certification, the owner or operator determines that the corrective action remedy has been completed in accordance with the requirements under §258.58 (e), the owner or operator shall be released from the requirements for financial assurance for corrective action under §258.73.*

10.2 CERCLA REGULATIONS

Superfund represents both a legal and procedural process for dealing with past hazardous waste disposal practices that is tied to a series of federal laws the Superfund Amendments and Reauthorization Act, (SARA, no relation to the author!) and the National Contingency Plan (NCP). When evaluating the process of Superfund, many organizational and planning activities are directly related to the phased investigation approach. The first step of the Superfund process is the identification of potentially hazardous sites which may require remedial action and their entry in a data base known as CERCLIS. In early 1990, CERCLIS contained over 35,000 potential Superfund sites. An emergency removal action may be conducted by the EPA at a site at any time after listing on CERCLIS due to environmental conditions that would require rapid response actions or because the situation at the site may potentially worsen unacceptably before a full-scale remedial action later in the Superfund process.

In the pre-remedial process, sites undergo a preliminary assessment (PA) and a site inspection (SI) which usually culminates in a scoring under the hazard ranking system (HRS). Currently, if a site scores over 28.5 on the HRS, it is placed on the National Priority List (NPL) where it becomes eligible for funding of investigative programs (Remedial Investigation/Feasibility Study [RI/FS]) and possible remedial action. Approximately 10% of all sites which are initially identified under

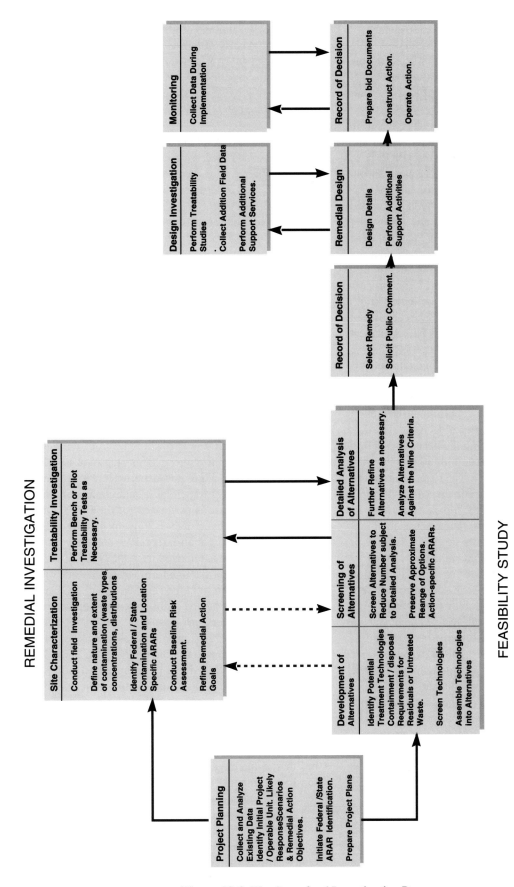

Figure 10-3 The Superfund Investigative Process

CERCLIS are finally listed on the NPL. The Agency for Toxic Substances Disease Registry (ATSDR) also conducts a health assessment once a site is listed on the NPL to determine if an imminent health threat exists or if further community public health studies (e.g., epidemiology and biological monitoring) are necessary.

A listing of a site on the NPL, will cause a site to undergo an RI/FS by either federal superfund contractors or by the potentially responsible parties (PRP's) to determine the nature and extent of contamination, and to evaluate alternatives for remedial action. The RI/FS program is shown in Figure 10-3. This program is divided into a series of tasks and phases. The RI and FS usually overlap in time as shown in Figure 10-3; for example, there can be initial scoping of alternatives while field data are being collected. The RI starts off with the preparation of a series of QA/QC, sampling, and work plans. This process is an evaluation of all data previously collected (e.g., during the SI/PA or by other parties) and an in-depth cost and time proposal for the conduct of the RI/FS. A Phase I investigation, as described in Chapter 2 of this text, can be extended to include a number of these planning documents fundamental to the Superfund program. A preliminary risk assessment, identification of applicable or relevant and appropriate requirements (ARARs), boring and sampling plans, determination of data quality objectives (DQOs), and an initial screening of remedial alternatives often support the work plan. Once the components of the documents are approved by federal and state regulatory agencies, actual investigative work commences at the site.

The majority of this work at the typical Superfund RI project involves the collection of samples for chemical analysis. These results are generally used to determine or estimate the nature and extent of contamination.

Since the passage of the initial Superfund legislation in 1980, engineers and scientists have striven to devise a uniform, yet cost-effective methodology for characterizing sites that are, or are suspected of, releasing contaminants to the environment. The U.S. EPA's initial remedial investigation guidance (EPA, 1985) focused site characterization efforts primarily on determining the area, extent, and magnitude of contamination, i.e., plume delineation. This has commonly necessitated multiple drilling, sampling, and analysis episodes to develop the database required to support a site-specific risk assessment and to evaluate the appropriate remedial actions. In recent years, it has become increasingly apparent that this approach is not an efficient utilization of limited financial and investigative resources.

After the analytical data leave the laboratory, they go through a process of data validation to insure that the data meet the U.S. EPA's QA/QC requirements. The field and analytical data are then used in the RI report to describe the nature and extent of contamination. Another use of analytical results obtained from site samples is in the human health risk assessment or the public health evaluation performed as part of the RI/FS. The stated objective of the risk assessment is to assist the U.S. EPA in remedial alternative decisions which have a public health basis.

Additionally, during this time, the FS progresses through its final evaluation of alternatives, with the result that one alternative is recommended to the U.S. EPA. Two additional risk assessment activities accompany the FS. The first assessment is a determination of preliminary remediation goals (cleanup levels) for contaminants in various media at the site; this determination takes health effects and applicable or relevant ARARs into account. The second assessment is a health based screening of remedial alternatives which accompanies evaluations of long and short-term effectiveness and reduction of toxicity as required by SARA.

Following the completion of the RI/FS, the U.S. EPA issues a Record of Decision (ROD) which states the chosen remedy, justifies its choice and responds to comments received from the public on the RI/FS. The ROD may decide on a no-action alternative. Additionally, a ROD may be issued for a portion or single operable unit at a site.

After the issuance of a ROD, the site proceeds to the remedial design (RD) stage which sets out the details of construction for remediation. This step may be preceded by a conceptual design, and experience shows that most Superfund investigations require additional sampling and analysis over what was performed for the RI/FS. Once the RD is approved, the remedy is implemented as a remedial action (RA). When an effective cleanup has been completed, the site is removed from the NPL. If hazardous materials are left on site in a form where they are still toxic and mobile, the site may be revisited every five years by the EPA to insure that the clean-up remains effective.

The U.S. EPA proposes in the 1989 National Contingency Plan revision to emphasize (53 FR 51423) a "bias for action" and to use the principal of "streamlining" in managing the Superfund program as a whole, and in conducting individual remedial action projects. It is the author's experience that most Superfund actions develop into field data collection and office documentation exercises with little bias for action to place into effect reasonable solutions for dealing with real risks to human health and the environment. This fact is substantiated by the low numbers (if any) Superfund sites with ground-water contamination that have actually been cleaned up in the twelve years since the enactment of Superfund in 1980.

The Superfund process is geared toward establishing guilt of PRPs in a fashion that provides evidence usable in court actions for cost recovery. Field conditions often differ from expected during the prefield planning process. As much of the cost of the RI/FS is wrapped-up in collection and analysis of soil, sediment and water samples preselection of sample locations.

The inflexibility of this type of planning often results in multiple phases of additional rounds of sampling collections. Once the RI/FS process is complete and a Record of Decision (ROD) is issued, additional data is commonly gathered during various design phases. Since the majority of data gathered in the RI/FS is of groundwater quality, the design stage of the RD (after the ROD) often requires additional site aquifer characterization work that could have been easily gathered in the RI. Figure 10-3 illustrates the typical RI/FS project where following the site assessment program as described in Chapters 2 to 9 would greatly ease the RD field assessment programs, and provide for more cost effective site solutions. This more direct program is discussed further after this section.

10.3 STATE REGULATIONS & ASSESSMENT MONITORING

A number of states in the U.S. have instituted comprehensive regulation of solid waste disposal facilities. For example, Massachusetts regulations specific to assessment monitoring have similar assessment components to RCRA and Superfund regulations. The purpose of the Massachusetts Contingency Plan (MCP) is to "insure the protection of health, safety, public welfare and the environment". The MCP provides:

• A list of oils and hazardous materials and a description of the characteristics of hazardous materials which are subject to the MCP;

• Procedures and requirements for notifying the responsible department of a release;

• Procedures and requirements whereby the extent and nature of a release can be addressed;

• Procedures for involvement of potentially responsible parties;

• Procedures for the public involvement in response actions; and

• Cost recovery procedures.

Massachusetts 310 CMR 40.000 provides specific details for each of these points. We will, however, focus on the third point; the procedures and requirements whereby the extent and nature of a release can be addressed.

Section 40.532 of the MCP requires a number of phases of remedial response actions. The sequence of these phases is intended to ensure the comprehensive assessment of the nature, extent, and risk of harm posed by a disposal site. The MCP requires that phases of a remedial response action occur in the following sequence:

• Preliminary Assessment;

• Phase I - Limited Site Investigation;

• Phase II - Comprehensive Site Assessment;

• Phase III - Development of Remedial Response Alternatives and Final Remedial Response Plan; and

• Phase IV - Implementation of the Approved Remedial Response Alternatives.

Each state may develop its own format for assessment of facilities that fail statistical or some specified maximum levels of indicator parameters. It is expected that RCRA Subtitle D will form the basis for such monitoring and assessment activities with some level of variation from state to state.

10.4 ASSESSMENT MONITORING ANALYSIS

The previous subsection described a number of alternative regulatory approaches to assessment of facilities where an exceedance was verified for indicator parameters or health based standards. Each of these approaches have common relationships to a series of goals of site assessments to define the extent of affected ground water. These goals can be defined as follows:

• Definition of physical site characteristics;

• Establishment of background concentrations of parameters;

• Identification of the source of the release;

• Establish the characteristics and extent of the release; and

• Identification of exposure points and risk to human health.

RCRA Subtitle C and D, CERCLA and many state waste management regulations all have these common points within their assessment programs. These waste management regulations then go through various site clean-up procedures that implement remedial actions leading to the ultimate reduction of risk to human health and the environment.

Much of the experience associated with assessment programs (since RCRA solid waste Subtitle D and hazardous waste Subtitle C remediation programs are just underway in 1991) has evolved around the CERCLA Superfund program. Since passage of CERCLA in 1980, the assessment program expenditures in the Superfund program have risen to $1.5 billion per year at over 800 sites (Porter, 1989). However, only approximately fifty sites are reported cleaned up (late 1988 figure) within almost a decade of site remediations. It is believed that few, if any sites with contaminated ground-water aquifers have been confirmed cleaned to health based water quality standards.

The experiences learned from the Superfund program point the direction toward a more effective use of assessment techniques to obtain targeted and timely remediations. The recent (1989) concern of the U.S. Congress on the slow progress of the Superfund clean-up process has brought into focus the inefficiencies of the current RI/FS process. Federal guidance and review documents concentrate on field collection and chemical analysis of large numbers of samples. Remedial Investigations (RI's) conducted pursuant to CERCLA / SARA are primarily designed to delineate the spatial distribution of contaminants with respect to receptor populations and institutional boundaries. This has been traditionally accomplished by assembling a soil and water quality database, the size and scope of which is influenced by non technical factors such as community sensitivity and cost-recovery litigation requirements. Analytical costs can reach to 30% to 40% of the total RI assessment costs and 40% of the remaining costs are devoted to the management and execution of field sampling of the soil and water samples provided to the analytical labs.

Although site-specific hydrogeologic conditions exert the main control on contaminant migration in the subsurface, hydrogeologic data are generally relegated to secondary importance (the remaining estimated 10% to 20% of the RI costs) with respect to the water quality data. The databases resulting from this investigative methodology typically do not adequately support detailed analysis of the predominant contaminant transport mechanisms, and, may be of limited value for evaluating and designing effective remedial actions. The classical geotechnical site characterization process was defined by Ralph B. Peck in six steps back in 1969 (Duplancic and Buckle, 1989):

- Conduct an investigation of sufficient scope to establish the general characteristics of the site;

- Assess the most probable conditions and the deviations from them;

- Develop a design based on the most probable conditions;

- Determine what courses of action should be taken if the conditions deviate from predictions;

- Measure and evaluate actual conditions during construction, and;

- Modify the design, as needed, to suit actual conditions.

While this site characterization process must be modified to provide a basis for reducing facility uncertainties to levels acceptable to regulatory bodies, the general observational approach can from the basis for site assessments.

The observational method to hazardous waste site remediation problems has been described by Wallace (1991) and provides a more direct method to evaluate site for remedial actions. The steps described below seek to identify the most probable model of the site and the reasonable deviations from that model according to the method of Peck (1969).

1. **Gather existing information on general site conditions and set remedial goals and general responses.** The purpose of this step is to decide on a general set of remediation objectives and suitable responses. This step is basically the Phase I investigation as defined by Chapter 2 of this text. This will give direction to the Phase II investigations and analyses that provide the basis for all the assessment and design activities that follow. Existing information developed for the Phase I scope is used to make initial estimates of waste quantities, contaminant concentrations present at site. This assumes that a reasonable level of prior information is available on the facility. The Phase I work should define the scope of work that will be executed in the next step.

2. **Gather information and refine knowledge of general site conditions and nature and extent of contaminants.** This is an information-gathering step to identify the general nature and extent of contamination, receptors, pathways, remedial objectives, treatment discharge standards, aquifer

characteristics, and general site properties within a somewhat extended Phase II scope of work. Phase II investigations are complete when it is possible to differentiate among alternatives, set design criteria, and identify reasonable deviations.

3. **Establish the most probable site conditions and reasonable deviations.** The most probable conditions as defined by the field and office evaluations require significant technical experience in site assessment to know when you have acquired sufficient information. This decision of when to quit is perhaps the most difficult task in using the observational method. The purpose of this third step is to develop a recommendation and a conceptual design leading to the Superfund Record of Decision (ROD). Conceptual designs for handling reasonable deviations from the expected conditions are included in this step. In order to determine the most likely model of the site and the set of reasonable deviations, one must make reasonable assumptions on what potential deviations are likely for the waste disposed and geologic conditions present at site.. This is often a step requiring a significant experience level on such projects.

4. **Design the remedial action based on the most probable conditions and reasonable deviations.** After the remedy selection, typically designs proceed beyond the conceptual level. These consist of remedial designs based on the most probable site conditions, plus designs covering contingencies for the agreed-upon reasonable deviations.

5. **Select quantities to observe during remediation to detect deviations during construction and operation.** The key environmental indicators are selected during this step for observation during remediation for both expected and deviant conditions. These environmental indicators may include chemical parameters, hydraulic head values, geologic strata expected during excavations, etc. The purpose of this step is to identify what will be used to indicate where the actual site conditions do not match the assumed site conditions, and to determine the level of deviation that will trigger a response. Wallace (1991) felt that if a model of the most probable conditions and reasonable deviations cannot be confidently established, or if key indicators cannot be defined, then the observational method may not be

appropriate for the project. In such a situation, alternative choices can be: (1) design the remedy to meet the worst-case scenario (often an expensive choice!), (2) continue to collect and analyze more site data in an attempt to better define the probable site model and deviations (such steps must be careful evaluated to have some opportunity to control costs), or (3) wait until better technology is available (rarely acceptable to State of Federal regulators).

6. **In advance, select a course of action or design modification for each reasonable deviation.** The purpose of this step is to establish contingency plans for what to do if a deviation is detected. This is an extremely important step, to control the potential for errors or unexpected conditions affecting the outcome of the program

7. **Implement the remedial alternative.** Measure the selected parameters during remediation and make the necessary modifications should a deviation occur. Once the remedial action is under way, project staff must work closely with the constructors, looking for deviations. Decisions on changes to the remediation will be made based on detected deviations and the preplanned responses. All deviations and modifications to the design must be fully documented.

The data necessary for an effective characterization of site flow conditions for determination of rate and extent (assessment monitoring) is typically the same data required for design of the ground-water remediation system. In order to develop a cost effective process of linking water quality data with the more traditional geotechnical / hydrogeologic data, a new way of conducting RI/FS investigation is described below using the observational methods tied conceptual geology and hydrology to target more cost effective site assessments and the resulting facility remediations.

The following subsection proposes a phased approach to the site characterizations that link basic observational geotechnical techniques to traditional hydrogeologic site analysis. These methods have often been forgotten in the rush to collect water quality information for Superfund investigations. This methodology would hold-off installation of ground-water monitoring wells until a site is adequately characterized for basic hydrogeologic conditions.

The thesis of this procedure is that significant time and cost savings can be recognized if the focus of assessment is shifted from laboratory analysis of statistically-representative quantities of environmental samples (i.e., "saturation" sampling) to sampling and analysis schemes which are predicated on a thorough understanding of the physical geologic and hydrologic systems (i.e., "smarter" sampling).

Categories of Sites Requiring Remediation

Sites requiring assessment programs can be divided into three major categories. Point sources (PT), multiple point sources (MPS) and large volume/low toxicity (LV/LT) sites. These sites are illustrated in Figure 10-4. PT sites may represent product spills of organics or soil contaminated by heavy metals or residual organics. MPT's represent industrial areas that have had numerous areas of hazardous waste disposal that contribute to a regional

water quality problem. LV/LT sites may be represented by areas containing a municipal landfill that received hazardous wastes or a mine tailings site. Each of these site categories require somewhat different assessment methodologies for the timely implementation of a remedial assessment program.

10.4.1 Traditional RI Methodology

It has become standard practice to begin remedial investigations with generation of massive planning documents for the establishment of monitoring points and the acquisition and analysis of environmental samples (air, soils, sediments, and ground and surface waters). The primary emphasis of this approach is placed on defining the type, magnitude, and spatial and temporal distribution

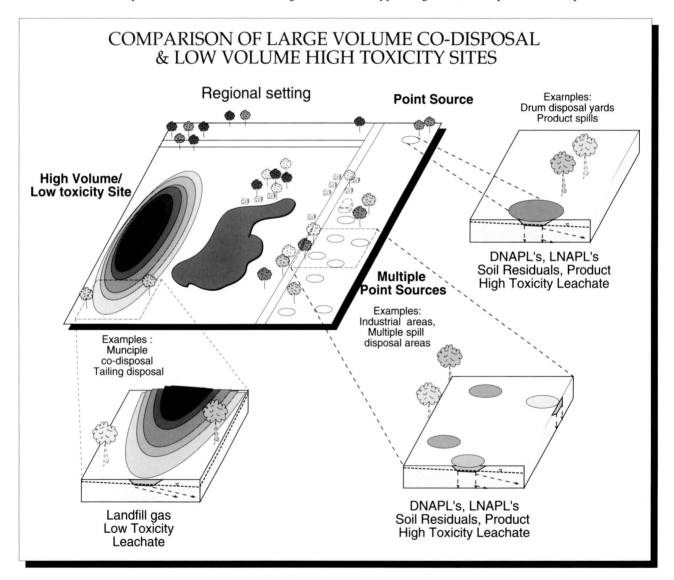

Figure 10-4 Example Sites Requiring Remediation

of contamination (i.e., rate and extent). The parameters which exert the primary control on contaminant transport, such as the media geometry and ambient flux, are estimated during the sampling or monitoring points installation activities.

As a result, the location of these monitoring points is generally predicated on the general knowledge of site conditions derived from topographic gradients, aerial surveys, historical accounts of disposal activities, and a variable amount of site-specific technical data. In those rare cases where moderate to extensive technical databases exist prior to the initiation of the RI, the application of strict, programmatic quality control and documentation requirements often limits the use of this data in the RI.

Frequently, the results of these initial sampling events showed that the contamination has migrated beyond the initial limits of the study area and/or the concentration gradients cannot be sufficiently interpreted within the constraints of the existing database. In either case, the typical response to this situation has been to establish additional sampling points and collect and analyze more environmental samples. Investigations that follow this track rapidly degenerate into "plume chases" wherein the goal gradually shifts away from the definition and quantification of exposure pathways to locating the leading edge of the plume. This approach has been more formally referred to as the plume delineation method (Dowden 1988).

The records of traditional assessment monitoring investigations commonly show three or more major field exercises to define the limits of contamination and the primary exposure pathways. In many of these cases, the site complexities justify a phased or staged analysis. However, in too many of these situations, the continued reliance on repetitive well installation and sampling did not adequately address the site complexities or significantly increase the understanding of the contaminant transport mechanisms. Figure 10-2 column one, shows a flow diagram of the alternative traditional assessment monitoring process known as the "Plume Delineation Method. Column two shows a conceptual analysis method to allow and easily depiction and quantification of site ground-water flow conditions. The accurate depiction and quantification of these flow mechanisms is crucial in estimating potential exposure concentrations beyond the known limits of the plume and in evaluating and designing effective ground-water remedial actions. Without targeted geotechnical and hydrogeologic assessment techniques, the investigations are always required to go back for additional field data for later feasibility and design phases.

10.4.2 Alternative Methodology - Understand the Geology First

An alternative approach to remedial investigations would be to postpone environmental sampling until the site hydrogeologic flow paths and mechanisms are adequately understood. In ground water investigations, the traditional ground-water assessment approach (before Superfund) involves a thorough evaluation of all pre-existing data and aerial photographs, performance of geophysical surveys, drilling of stratigraphic boreholes, installation of

Table 10-1   Conceptual Components of Site Assessments for Remedial Projects

Conceptual Component	Technical Component	Methodology
• Transmitter	- Source evaluation - Characteristics of release	-Historical Records -Air photo historical sequences Leachate/water characteristics
• Media	- Extent of release - Physical Site Investigations	-Assessment phase II Characteristics -Air/surface & ground - Background water backgrounds concentrations
• Receptors	- Exposure points analysis - - Risk assessment	Well census -Exposure assessments -Toxicity assessments

Comparison of Superfund Assessment Techniques

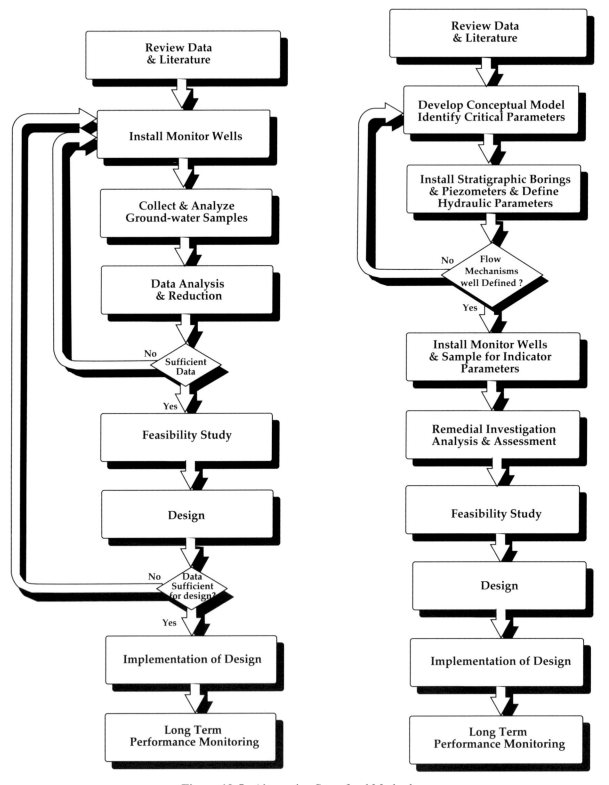

Figure 10-5   Alternative Superfund Methods

piezometers and measurement of heads, performance of in-situ hydraulic conductivity tests, and the evaluation of hydraulic connection between aquifers and confining beds. All of the information derived from these activities can then be utilized to construct stratigraphic and hydrostratigraphic cross sections, vertical and plan view flow nets, and to develop rate and extent of migration predictions.

In complex hydrogeologic settings, numerical modeling may be required to attain the appropriate level of quantitative analysis. These predictions, combined with a relatively small amount of information on the nature and history of the source(s), can be used to develop a rational and efficient environmental sampling and analysis plan. These more linear investigations are shown in the second column of Figure 10-5.

Since the pre-Superfund assessment methods yield more data on the physical controls of the ground water transport system, the relevance and significance of the environmental sampling results can be more readily interpreted. An additional advantage is that the environmental sampling results can be utilized to refine and recalibrate the predictive analyses based on the physical site characterization.

The proposed bias for action and streamlining should also include consideration of the flow diagram (Figure 10-5). This diagram details the activities necessary to fully define the geologic and hydrogeologic conditions before selection of sampling location to confirm real risks to human health and the environment. Once sampling locations were selected, only a VOA scan would be performed. If VOAs were detected at statistically determined levels, full hazardous constituent sampling would be performed only on those wells showing VOAs. Since this proposed bias for action program stresses collection of geologic and hydrogeologic data, the information necessary for design of the remediation would be available at an early point in the remedial process. Decisions within the feasibility study project component on the special situations where it may not be practicable to actively restore ground water could be more easily evaluated and technically defended. The type of geotechnical and geologic information necessary to select the assessment monitoring points would also provide the design information for the pump and treat systems normally required to remediate contaminated aquifers that are affecting human health and the environment. These data include flow nets and target monitoring zones, transmissivity and storage coefficients, cross sections and structural maps of the site. The data gathered during the phased investigation would home in on the geology and hydrogeology first before going into a significant sample collection program.

Figure 10-5 illustrates the process of defining the geology/hydrogeology first and then working out the required sampling program. Few Superfund, or for that matter, state assessment investigations collected too much geotechnical and geologic information, rather, these RI/FS projects (and state analogs to Superfund) are based on extensive soil and ground-water sample collection with limited interpretation of the data and understanding of the site geology and hydrogeology. It is these data collection programs that should be tailored and focused as proposed in the above discussion.

10.5 ASSESSMENT MONITORING TECHNIQUES

The procedures and techniques used in development of an assessment program should be designed to define each of the three components of facility, environment, and potential receptors. This relationship can be summarized in the conceptual model of the television station microwave/air transmitting media/local antenna receivers as illustrated in Figure 10-6 Each of these three components should be thoroughly evaluated to establish the assessment goals outlined in the previous section.

Each of these components can be addressed by following a phased assessment program that follows the detection monitoring phased program with selective additional technical components directed toward determination of extent of affected media and risk to receptors. Specific federal or state assessment programs may have requirement such as QA/QC plans or detailed work plans that must be completed to meet aspects of these regulations. Key technical components of an assessment program that meet the general assessment standards of federal and state programs are presented in the following subsections.

10.5.1 Preliminary Assessment Procedures

The extent of the preliminary assessment procedures are directly dependent on previous analysis performed at the facility in question. If the detection monitoring program was designed on the basis of a detailed hydrogeologic analysis that located the ground-water wells within target flow paths from the facility, additional preliminary assessment procedures should be directed toward source and receptor analysis. Much of the level of effort required in assessment programs is based on the level of complexity of the site. Figure 10-7 (U.S. EPA 1990) illustrates various site complexity issues typical for Superfund projects. The scoping efforts are the most difficult part of the preliminary procedures employed on assessment projects. Table 10-2 (U.S. EPA) provides a first look in development of site complexity. As the site becomes more complex, additional components must be included in the conceptual model and within the scope of work. Simple comparisons of the numbers of checks in a colume can provide an evaluation basis for the estimation of the relative complexity of a particular site.

Most initial state and federal regulations required some form of planning to document the owner/operators assessment program. This document serves a number of purposes:

1. It presents a descriptive procedure for determining the rate of migration, extent, and the constituent composition of the release.

2. It provides a mechanism for obtaining data necessary to the information requirement process of most Federal and State regulations.

3. It provides a mechanism for obtaining data necessary for subsequent corrective actions at facilities.

There are a number of elements that the owner/operator should include in the assessment monitoring plan.

• Narrative discussion of the hydrogeologic conditions at the site; identification of potential contaminant pathways;

• Description of the facility detection monitoring system;

• Description of the approach to be used to make the first determination (false positives rationale);

• Description of the approach to be used to characterize rate and extent of affected media; identification and discussion of investigatory phases;

• Discussion of the number, location, depth and rational of borings that will be used to assess the affected media;

• Information on design and construction of piezometers and wells;

• A description of the sampling and analytical program to be used to obtain and analyze geologic and ground water hydraulic data;

• Description of data collection and analysis procedures the owner/operator plans to employ; and,

• A schedule for the implementation of each phase of the assessment program.

Although planning is important it should not consume inordanant time and cost in its production. Flexibility must be built into the document so that the plan is not defeated by unexpected variations in site conditions. The components of such plans need not require hundreds of pages of text and tables, rather short documents of less than 30 pages of text should be sufficient for the planning of the majority of assessment programs.

10.5.2 Conceptualization Using Regional And Site Geology

In order to define the geologic environment of a specific area, the regional and site–specific geology must

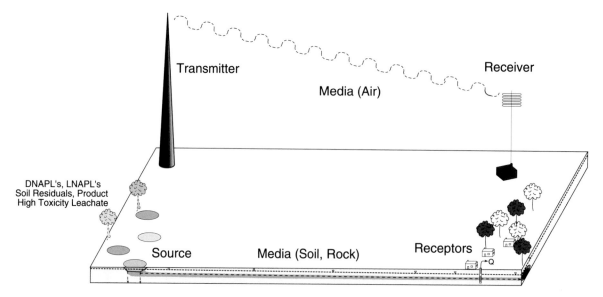

Figure 10-6 Conceptual Basis for Assessments

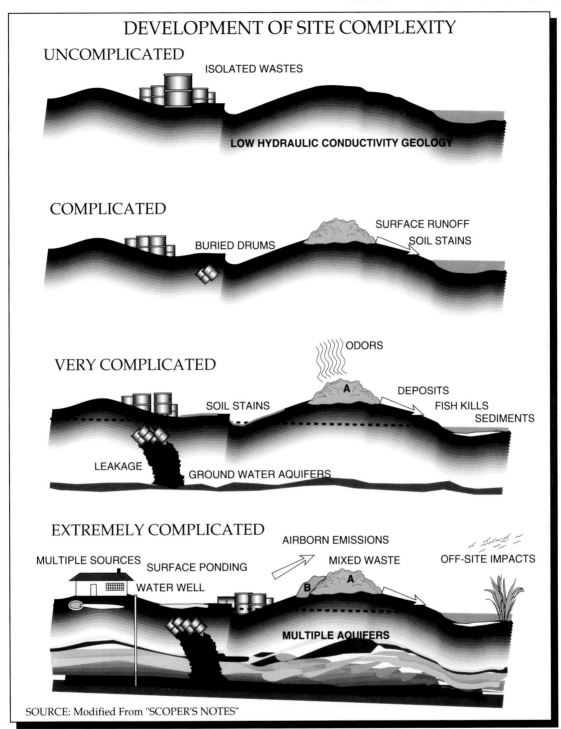

DEVELOPMENT OF SITE COMPLEXITY

UNCOMPLICATED

ISOLATED WASTES

LOW HYDRAULIC CONDUCTIVITY GEOLOGY

COMPLICATED

SURFACE RUNOFF
BURIED DRUMS   SOIL STAINS

VERY COMPLICATED

ODORS
SOIL STAINS   A   DEPOSITS
FISH KILLS
SEDIMENTS
LEAKAGE   GROUND WATER AQUIFERS

EXTREMELY COMPLICATED

AIRBORN EMISSIONS
MULTIPLE SOURCES   SURFACE PONDING   MIXED WASTE   OFF-SITE IMPACTS
WATER WELL   B   A
MULTIPLE AQUIFERS

SOURCE: Modified From "SCOPER'S NOTES"

Figure 10-7 Site Complexity Issues

SOURCE: Scopers Notes U. S. EPA 1989

be thoroughly characterized. These data must establish the properties and features of individual geologic units beneath and near the site, through use of:

- Regional geologic maps and cross–sections that are used to characterize area–wide geologic units and structural features, to depths incorporating probable ground–water

movement;

- Topographic maps used to characterize site–specific topographic relief;

- Stratigraphic maps and cross sections used to characterize detailed site–specific geologic conditions;

Table 10-2   Initial Complexity Issues in Remedial Projects

EXPOSURE RISK FACTORS	RANGE					
significant closest population	(1500m	500m,	100m)			
working drinking water wells	(no		yes)			
offsite sensitive areas	(no		yes)			
adjacent agriculture land use	(no		yes)			

SITE SURFACE FACTORS						
area in acres	(<5,	5-25	>25)			
access (for equipment)	(easy	average	hard)			
topographic variation in feet	(<5	5-20	>20)			
ponds or lagoons	(0	1	>1)			
streams on site	(0	1	>1)			
soil type	(loam	sandy,	rocky)			
rock outcrops	(0	1	>1)			
vegitation on site	(no	sparse,	heavy)			
evident soil erossion	(no	some.	heavy)			
utility easments on site	(no	1	>1)			
safety precautions necessary	(no		yes)			

ESTIMATED MEDIA(S) CONTAMINATED						
soil stains	(no	few	many)			
odors	(no	some	strong)			
wind blown particulate	(no	little	much)			
buildings, structures	(no	few	many)			
water table depth	(<12	13-25	>25)			
offsite complaints ~fishkills)	(no	few	many)			
discolored sediment deposits	(no	few	many)			
multiple aquifers	(no		yes)			

WASTE CONDITION						
drums	(no	few	many)			
container corrosion	(no	little	much)			
storaae tanks	(no	few	many)			
known hiah hazard substances	(no	little	much)			
annual rainfall	(<l0	19-30	>30)			
prerailing wind direction	(NW, W, SW, S, SE)					

EXISTING INFORMATION						
previous site study results	(yes	some	no)			
local official records	(yes	some	no)			
local informal sources	(yes	some	no)			

SUBSURFACE FEATURES						
established geology	(yes	poor	unknown)			
hydrogeology	(Uncomplex		complex)			

SITE COMPLEXITY IS ESTIMATED BY THE NUMBER OF CHECKS APPEARING IN EACH VERTICAL COLUMN. COMPLEX SITES WILL RANGE TO THE RIGHT

Source: Scopers Notes, USEPA 1989

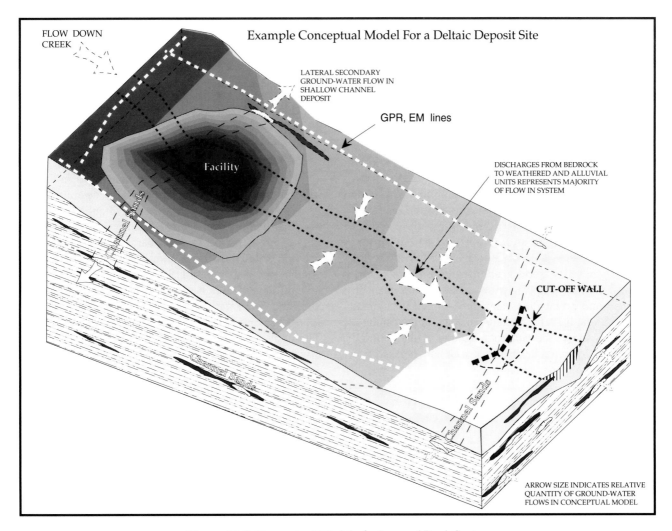

Figure 10-8 Conceptual Model of a Layered Rock System

and,

• Aerial photography, as used to illustrate information such as vegetation, springs, gaining or losing stream conditions, wetland areas, and important elements of geologic structure.

The above data will have been gathered within the field assessment programs. Cross–sections and stratigraphic maps should be prepared so that a clear technical basis is used to derive the conceptual geologic and hydrogeologic models.

The entire process of geologic conceptualization is built- up by using regional data to establish the general geologic setting and then by gradually building an understanding of geologic conditions at the site from the drilling program. Interpretation of data through cross–sections and maps strengthens the conceptual models until a final picture of the geology is established. The final conceptual geologic model is a site–specific representation of the geologic system under and adjacent to the facility or waste disposal area. Figure 8-1 (located in Chapter 8.0) illustrates an overview of the site conceptual model process and the linkage between geology and hydrogeology. Geologic information used to construct conceptual geologic models typically consists of the following:

• The depth, thickness, and area extent of each stratigraphic unit, including weathering horizons;

• Stratigraphic zones and lenses within the near–surface zone of saturation;

• Stratigraphic contacts between significant formations/strata;

• Significant structural features such as discontinuity sets, faults, and folds; and,

• Zones of high hydraulic conductivity or shearing/faulting.

Several cross–sections may be required to depict significant geologic or structural trends and to reflect geologic/structural features in relation to local and regional ground–water flow. The area and vertical extent of the geologic units can be presented in several ways. For complex settings, the most desirable presentation is a series of structure contour maps representing the physical character of the tops and/or bottoms of each unit. Cross–sections and isopach maps can also be used because they are generally good graphic supplements to the structure contour maps. These cross–sections can be combined into a fence diagram (three–dimensional) that can serve as the basis for conceptualization of the geology. Figure 10-8 shows the conceptualization of the geology and hydrogeology of a layered rock with both perched and regional potentiometric surfaces.

Developing a conceptual basis for assessment monitoring as a project starting point cannot be over emphasized!

Conceptualization is a means of achieving a graphic idealization of the actual geologic conditions; the investigator must, therefore, consider the geologic features that would affect ground–water flow and ground–water quality. Minor features that are not important to the overall picture should not be transferred to representations of the conceptual model.

10.5.3 Characteristics Of The Saturated Zone

Each of the significant stratigraphic units in the zone of saturation should be characterized by determination of its hydrogeologic properties such as hydraulic conductivity (vertical and horizontal) and effective porosity. These parameters describe aquifer characteristics that control ground–water movement and, hence, the ability of the aquifer to retain or pass leachate. Both are needed for a general understanding of the hydrogeologic setting at a site, and for completing the conceptual hydrogeologic model for design of ground–water monitoring systems. Typically, the amount of data necessary to complete a conceptual hydrogeologic model will differ for each geologic environment in question.

For example, an aquifer in extensive, homogeneous beach–sand (SP) will require less investigation than a glacial unit consisting of lenticular deposits of outwash interbedded with clayey till (GW/CL). The data gathered, however, will be similar. There are two types of

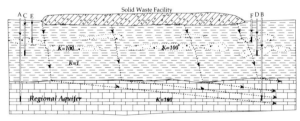

SAND LENSES WITH LITTLE HORIZONTAL DISCHARGE

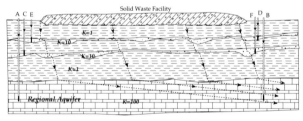

CONTINUOUS SAND LAYERS WITH LITTLE HORIZONTAL DISCHARGE

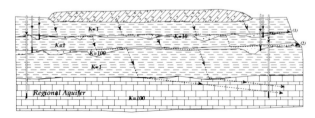

SAND LAYERS WITH SIGNIFICANT HORIZONTAL DISCHARGE

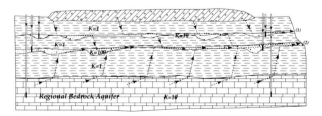

SAND LAYERS DISCHARGING BOTH BEDROCK AND OVERBURDEN

Figure 10-9 Various levels of Ground-Water Flow Discharge

fundamental aquifer characteristics required to define a hydrogeologic conceptual model: characteristics of the aquifer and characteristics of ground–water flow.

10.5.4 Aquifer Characteristics

Aquifer characteristics should be determined for each of the significant stratigraphic units in the zone of saturation including property variations in the geologic unit. Aquifer characteristics, for each hydrogeologic unit, should include:

• Hydraulic conductivity;

• Effective porosity; and,

• Specific yield/storage (as required for the project)

Additional aquifer characteristics such as, transmissivity, and attenuation properties can be derived; and estimated dispersivity values may be necessary for ground–water transport modeling to estimate ultimate (long–term) concentrations at points of exposure. These aquifer characteristics support ground–water flow characteristics used to define the quantity and the direction of ground–water flow as necessary to define the conceptual movement of ground water away from a facility and toward a monitoring point.

Hydraulic Conductivity

Hydraulic conductivity is a measure of the ability of an earth material to transmit water. Several laboratory and field methods can be used to determine the saturated and unsaturated hydraulic conductivity of soils, including tracer tests, auger–hole tests, and pumping tests of wells. Most ground–water systems consist of mixtures of aquifer and aquitard units. Flow path analysis must consider ground–water movements through aquifers and across aquitards. The relative hydraulic conductivity of these units can vary many orders of magnitude; hence, aquifers offer the least resistance to flow. This results in a head loss per unit of distance along a flow line ten to many thousands of times less in aquifers than in aquitards. Therefore, lateral flow in aquitards usually is negligible, and the flow lines concentrate in aquifers and run parallel to aquifer boundaries.

Figure 10-9 illustrates the refraction of ground–water flow lines between sand and clay for various discharge points. If the individual geologic units are relatively isotropic (i.e., horizontal hydraulic conductivity is reasonably equivalent to vertical hydraulic conductivity), flow net construction is relatively straightforward. However, if anisotropic hydraulic conductivity conditions are established, the investigator must use transformation techniques as described by Cedergren (1977) and Freeze and Cherry (1979).

10.5.5 Ground–Water Flow–Directions

The RI field investigations should include a program for precise monitoring of the area and temporal variations in ground–water levels. This program involves the measurement of water levels in piezometers installed for the purpose of investigating the saturated zone. These data are used to define ground–water flow–directions during development of the ground–water monitoring system design. Parameters necessary to measure or ascertain for use in determination of ground–water flow directions include:

• Depth to ground water;

• Potentiometric surface;

• Hydraulic gradients; and,

• Vertical components of flow

Constructing piezometric or potentiometric contour maps from raw data, then combining the data with stratigraphic cross–sections to develop flow nets, can be a tedious effort; however, these derived illustrations are essential for selection of target monitoring zones and design of the assessment monitoring system. The process of flow net construction is fully described in chapter 8.0 of this text. Construction of flow nets offers a direct method for defining the most likely direction of ground–water movement. Flow nets are the basic tool used in site assessment and monitoring well system design to illustrate regional and local flow patterns that may require monitoring.

In using flow nets for ground–water monitoring well design problems, it should be recognized that the solutions are no better than the idealized cross sections drawn. For a given section, however, the flow net can give an accurate solution for flow quantities. To enable proper construction of a flow net for use with a conceptual geologic model, certain hydrologic parameters of the ground–water system must be known, including:

- Distribution of vertical and horizontal head;

- Vertical and horizontal hydraulic conductivity of the saturated zone;

- Thickness of saturated layers; and,

- Boundary conditions.

Head Distribution

Piezometers are used to determine the distribution of head throughout the area of interest. To be valid, head measurements must be time equivalent; that is, all piezometric measurements must be made coincidentally or all measurements must be made for the same ground–water conditions. Piezometers should be spatially distributed and placed at varying depths to determine the existence and magnitude of vertical gradients. If vertical flow components exist, the flow direction cannot be derived simply based on inspection of the potentiometric surface in two dimensions. A three–dimensional representation of the potentiometric surface would be required to interpret the flow direction. Ground water will flow, however, from areas of high hydraulic head to areas of low hydraulic head. Figure 10-9a through 10-9d illustrates the hydraulic heads in a series of piezometers located in recharging and discharging geologic environments where sandy zones show variable levels of interconnection and discharge. In recharging zones, deeper piezometers show lower ground–water levels. This is due to the lowering of equipotentials with depth in recharge areas. In discharging environments, deeper piezometers show higher water levels or heads, as illustrated in Figure 10-9d. Simple homogeneous, isotopic systems, as illustrated, require consideration of depth below the ground–water surface to correctly use the head data. Strong recharge/discharge systems require full understanding of both the vertical and horizontal gradients. Special care should be maintained to not mix piezometer readings from different elevations in generating ground–water contour maps.

Aquifer Thickness and Extent

The thickness of an aquifer or any geologic strata can be determined by evaluation of geologic logs or by geophysical techniques. Geologic logs generated from boreholes show changes in lithology (the characteristics of the geologic material) indicating the relative hydraulic conductivity of materials. Various geophysical techniques, both downhole and surface, can be used to determine the thickness and extent of geologic units. These thicknesses

and extents of geologic materials are used to establish flownets and conceptual models for the facility.

Boundaries and Recharge/Discharge Zones

To aid in the understanding of the ground–water flow–regime and to aid in identification of potential pathways to target for monitoring, the location of any proximate zones of recharge or discharge should be identified within the hydrogeologic conceptual–model. This identification is determined in part by the general site location in the watershed. The investigator probably does not need to quantify such information; rather, a general indication of discharge or recharge characteristics could be used as part of the conceptualization process.

For unconfined aquifers, recharge areas are usually topographic lows. Discharge/recharge areas also indicate relative depth to unconfined ground water. In discharge areas, ground water is found close to or at the land surface; while at recharge areas, there is often a deep unsaturated zone between the unconfined ground–water surface and the land surface. A ground–water contour map can be used to locate these areas.

Recharge and discharge in confined aquifers is typically more complex than for unconfined aquifers. Discharge and recharge may occur where the aquifer is exposed. Some discharge may also occur in the form of upward leakage in areas of upward hydraulic gradient and leaky confinement. Recharge can also occur by downward flow through leaks in the confining layers (aquitards). Topographic information can provide significant insight into ground–water movement; however, topography can also mislead an investigator, when aquifers are confined or perched. Ground water in such aquifers can often move in unexpected directions. The investigator must pay particular attention to potential mounding of ground water and directional complexities that may be covered by low hydraulic conductivity subgrade materials. Caution must always be used in drawing ground-water surface (water table) or poteniometeric level contour maps. Ground-water flow will generally be uniform, you must however, understand the system flow fields. Specific conditions can only be discerned through analysis of lithology and measurement of piezometric heads, which was gathered in the RI field program.

10.5.6 Linkage Of The Conceptual Model And Flownets

The diagram in Figure 10-10 illustrates the selection of the target monitoring zones using facility geologic and hydrologic information gathered during the RI field investigation prior to selection of locations of assessment

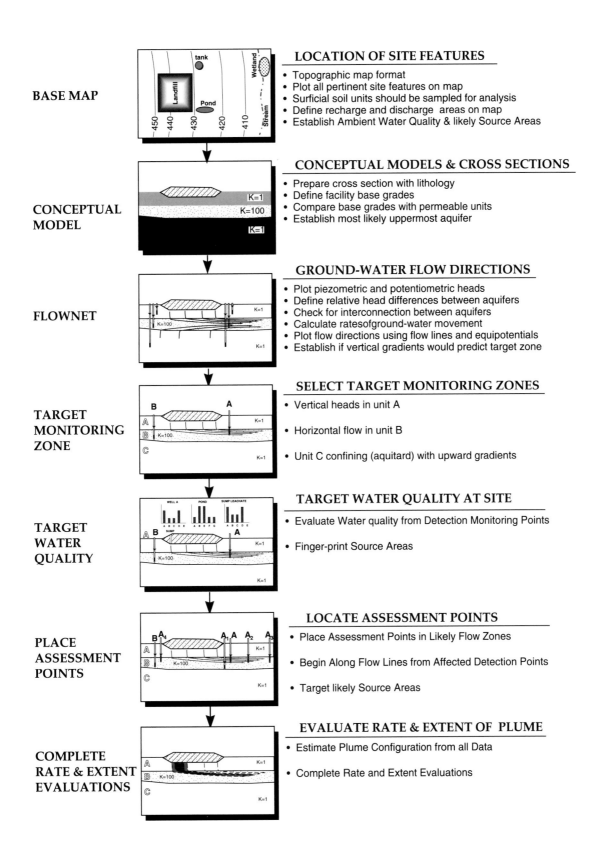

BASE MAP

LOCATION OF SITE FEATURES

- Topographic map format
- Plot all pertinent site features on map
- Surficial soil units should be sampled for analysis
- Define recharge and discharge areas on map
- Establish Ambient Water Quality & likely Source Areas

CONCEPTUAL MODEL

CONCEPTUAL MODELS & CROSS SECTIONS

- Prepare cross section with lithology
- Define facility base grades
- Compare base grades with permeable units
- Establish most likely uppermost aquifer

FLOWNET

GROUND-WATER FLOW DIRECTIONS

- Plot piezometric and potentiometric heads
- Define relative head differences between aquifers
- Check for interconnection between aquifers
- Calculate ratesofground-water movement
- Plot flow directions using flow lines and equipotentials
- Establish if vertical gradients would predict target zone

TARGET MONITORING ZONE

SELECT TARGET MONITORING ZONES

- Vertical heads in unit A
- Horizontal flow in unit B
- Unit C confining (aquitard) with upward gradients

TARGET WATER QUALITY

TARGET WATER QUALITY AT SITE

- Evaluate Water quality from Detection Monitoring Points
- Finger-print Source Areas

PLACE ASSESSMENT POINTS

LOCATE ASSESSMENT POINTS

- Place Assessment Points in Likely Flow Zones
- Begin Along Flow Lines from Affected Detection Points
- Target likely Source Areas

COMPLETE RATE & EXTENT EVALUATIONS

EVALUATE RATE & EXTENT OF PLUME

- Estimate Plume Configuration from all Data
- Complete Rate and Extent Evaluations

Figure 10-10 Summary of Selection of Ground-water Monitoring Locations

monitoring points. The selection process can be described as a series of steps:

- **Step 1:   Locate Site features on a Topographic Map Format**; The location of site feature including previous disposal areas, stained soils, pipelines or underground storage tanks, should be compared to geologic and soils maps to define preferential flow map areas. Any recharge/discharge areas should be defined conceptually. Establish ambient water quality and any likely source areas.

- **Step 2: Cross Section and Conceptual Model Assessment**; Cross sections based on geologic borings should be available for comparison to site features and facility location. Base grades of the actual facility should be plotted on cross sections to establish potential facility and permeable unit contacts or ground-water flow paths. Conceptual models should be constructed to idealize cross sections to establish overall site conditions and to illustrate distribution of permeable /less permeable units.

- **Step 3: Use Flow Net(s) to Define Likely Ground Water Movements:** The flow net will define the water table and piezometric heads between saturated units. The rates of flow along flow paths can be calculated from the flow nets. Vertical gradients can predict target zones by comparison of relative head between units. Inter-connections between aquifers can be predicted by relating hydraulic conductivity to hydraulic heads for the units defined in the conceptual model.

- **Step 4: Select Target monitoring Zones:** The flow zones most likely to represent pathways from the waste disposal area sand show ground-water movement from under or adjacent to the waste disposal areas would represent the target monitoring zone(s) These areas would be selected for assessment monitoring well placement.

- **Step 5: Evaluate Water Quality:** Both the detection monitoring system, background and potential source areas must be "finger printed" for indicator parameters.

- **Step 6: Locate Assessment Wells:** Place assessment monitoring well in likely flow paths or flow zones beginning near the facility and moving along flow lines from the affected detection monitoring points and source areas, (note: DNAPL sites may require outside-inside procedure!).

- **Step 7: Estimate Plume & Complete Program:** Using the entire data set including hydrostratigraphic, geologic and water quality, estimate plume size and depth relationships. The assessment program can now define rate and extent criteria for the facility.

This seven step procedure for defining the target monitoring zones must be flexible to include environmental effects due to seasonal changes in gradient, or other man-induced flow modifications. The target zone might include only a portion of a very thick aquifer, or span several geologic unit as in the case of a thin permeable unconsolidated unit overlying weathered fractured bedrock). These target zones represent the depth-location criteria for placement of monitoring well screens. These principals are further illustrated through Figures 10-11 to 10-15. Figure 10-11 shows the facility layout for a large volume- low toxicity co-disposal facility with potential local source areas. Figure 10-12 illustrates the detection monitoring design with a minimum of two background wells and three "downgradient" wells. Water quality tests showed well number 5 with water quality parameters (shown on Figure 10-13) elevated above background sufficiently to kick the facility into assessment monitoring. Results of water quality testing of the pond, background and the facilities' leachate collection sump are shown on Figure 10-13. The indicator parameter "fingerprint" of well number 5 and the sump agree in parameters and overall chemical concentrations.   Although the concentrations could, (and probably would be in the real world), be significantly less due to dilution; the sump

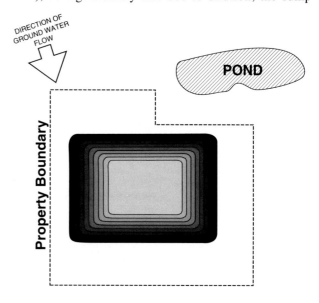

Figure 10-11 Example Site Assessment Project

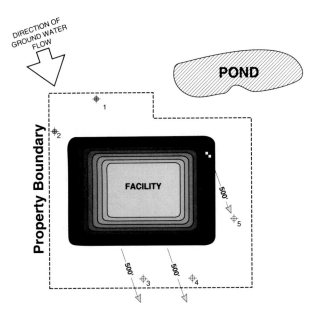

Figure 10-12 Location of Monitoring Points

leachate is the likely source area for the observed parameters in well number 5. Figure 10-14 shows the results of the installation of six assessment wells within the detection monitoring systems' flow path, both up-and down-gradient of the likely source area. These data confirm the rate and extent of the plume and allow the gathering of sufficient data for the remedial program. Figure 10-15 shows as a cross check the results of a regional water quality review to confirm the data gathered in the previous step. Water quality data from various production wells show low levels of various indicator parameters. Although some of these parameters may be undesirable, the facility and pond chemical fingerprints are not evident in the observed water quality from the production wells. This last cross check may not be required in every case, but, it is extremely important to use all data available and not to just focus only on the facility assessment wells. This is not a request to go on a witch-hunt, but rather to include in the evaluation local data points where there is the potential for affecting human health and the environment. Even with the best technology and intentions, unexpected flow path may be present. This is especially true for minimally investigated facilities. Protection of human health and the environment should be the guiding principal in these investigations. If non-facility related parameters with potential negative health effects are observed in potable well, it is the duty of the professional to report these data to the responsible individuals in the soonest possible time.

10.6 SYSTEM DESIGN AND OPERATION

Assessment monitoring programs under either RCRA or Superfund have many similar features. The following discussions first describes the evaluation programs typically experiences with large volume/low toxicity, codisposal facilities. These almost always unlined facilities can show significant leachate plumes of dilute organic and inorganic parameters. These sites are customarily remediated through containment (cover and cut-off areas of discharge from the site) and pump and treat techniques to deal with the resultant ground-water plume. Additional single or multiple point source contaminated sites, as illustrated in Figures 10-4 and 10-7, may require alternative field evaluation techniques, since these low volume/high toxicity sites require more focused assessment procedures. Special considerations relative to DNAPL's and LNAPL's will be reviewed later in this section.

The degree of site characterization necessary to design an effective ground-water extraction system depends on the objectives of the remediation. In cases where well-head treatment is the objective, it may be sufficient simple to characterize contamination to the extent necessary to design the treatment system using volume requirements and water quality parameters.. The site characterization requirements for an aquifer restoration system are likely to be far more extensive. Before designing an aquifer

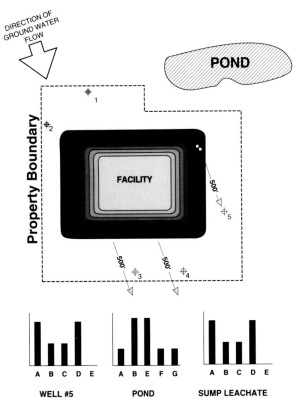

Figure 10-13 Water Quality of Monitoring Points

restoration program, it is generally necessary to characterize:

1) the hydrogeologic properties of the geologic layers involved,
2) the types and distributions of contaminants,
3) the location of sources and their potential for continued contamination of the saturated zone, and
4) the contaminant migration properties of the contamination in the affected aquifers.

In addition, successful operation of the system requires that the actual performance of the implemented design be monitored so that adjustments can be made to optimize performance.

10.6.1 Hydrogeologic Information

Whatever the remedial objective of the ground-water extraction system, its functional purpose is to establish some form of control over ground-water flow in the vicinity of the pumping wells. To design a system of wells that can establish this control requires an adequate understanding of the hydrogeologic characteristics of the site. For the design of an extraction system, the necessary hydrogeologic information includes the stratigraphy, the hydraulic conductivity of aquifer layers, the leakance of semi-confining layers, and the natural distribution of potentiometric head.

Stratigraphy

It is necessary to identify the number and thickness of the aquifers potentially affected by contamination at the site. It is also important to know the lateral extent and continuity of any confining or semi-confining layers as shown in Figure 10-8. Any local gaps in the confining layers that occur within the area of concern can have an important effect on the success of the aquifer remediation. These "open window"-areas are especially important in DNAPL contaminated sites since local holes in clay layers between upper and lower aquifers permit DNAPLs to enter lower uncontaminated aquifer(s).

It is also very important in these efforts to establish the bottom of the hydrostratigraphic zone involved in the remediation, since shallow ground-water control designs can leave deeper unknown contamination uncontrolled. No matter what type of contamination is present at site, one must fully understand the subsurface geology to have any chance to evaluate the potential for aquifer restoration.

Adequate investigation of site stratigraphy requires

installation and appropriate logging of sufficient exploratory borings to characterize the site. The number of borings ultimately required depends on the complexity of the subsurface geologic environment.

Aquifer Hydraulic Properties

In relatively homogeneous aquifers the main reason for investigating the transmissivity is to predict the extent of the capture zone that can be established by wells and the rate of pumping that will be required. While detection monitoring program design would rarely require the expense of a full scale pump test, aquifer remediation requires good field data. Chapters 5 through 7 provide background information setting-up and the overall evaluation of aquifer test data. When the stratigraphy is well defined and good transmissivity estimates can be established through pump tests, depth-averaged hydraulic conductivity can be provided for the design of the

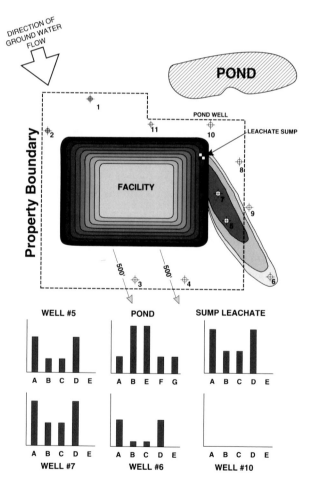

Figure 10-14 Locate Assessment Monitoring Points

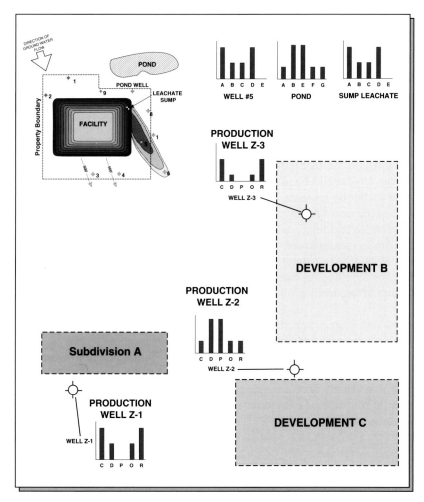

Figure 10-15 Keep a Regional Perspective

extraction system. Depth averaged hydraulic conductivity estimates arising from aquifer test results may be deceptive if the aquifer is vertically heterogeneous. It is the author's opinion that, in general, insufficient effort is expended to determine hydraulic conductivity variations in aquifer systems requiring remediation. Determination of vertical hydraulic conductivity variations within an aquifer unit requires special testing methods. In consolidated open hole bedrock wells this testing can be accomplished by packer testing or stressed flow meter logging. In unconsolidated materials, individual slug tests on wells screened at different depths combined with accurate field logging and grain size/hydraulic conductivity analyses can be used to estimate vertical permeability variations in aquifer properties.

One should use sufficient care in evaluation of aquifers with complex heterogeneity, since conventional aquifer tests can be very difficult to interpret. Long-term aquifer test in which a large number of observation wells are monitored can provide adequate detail for remedial design efforts, dependent on the scope of the test and the ability of the aquifer to transmit pressure head changes throughout the system in question.

A practical system to evaluate complex hydrogeologic systems is the systematic use of phased pilot testing of extraction wells. Since this usually involves incremental design of the extraction system as successive components are installed and tested the design elements must be sufficiently flexible to accommodate changes to the system on the basis of observed hydraulic heads.

The thickness and hydraulic conductivity of semi-confining layers separating individual aquifers in multi-layered hydrogeologic systems can have an important influence on the effectiveness of ground-water extraction systems. Inter-aquifer leakage in response to the pumping of extraction wells can drastically reduce the radius of influence of the wells. Chapter 5.0 reviews pump testing procedures applicable to interaquifer testing so that accommodation can be made to system design elements.

Leakage through semi-confining layers may also

permit contaminants to move between aquifers. Evaluation of storage coefficients is normally of secondary importance in aquifer remediations, since the extraction system operation is usually analyzed as a steady-state phenomenon.

Potentiometric Gradients

Ground-water extraction systems achieve their remedial goals primarily through the manipulation of ground-water flow patterns (hydraulic heads) in the contaminated aquifers. The hydraulic pressure heads or gradients produced by the extraction system must also control or contain the large scale regional gradient that may be moving the contaminants in an undesired direction. These regional gradients may be due to natural recharge and discharge relationships, or they may be the result of nearby production wells that affect local ground-water flow systems.

The design of the extraction system must take both natural and anthropogenic systems into account. Installation and monitoring of piezometers or monitoring wells for water levels must be part of the overall design of the hydraulic control system. Natural gradients can change seasonally, so water levels must be measured enough times during the year to determine gradients for potential seasonal reversals in flow.

10.6.2 Contaminant Source Characteristics

The nature of the ground-water contamination problem is generally established by determination of the number and identity of the contaminant compounds present, their concentration and spatial distribution in the aquifer, and their mobility characteristics. This is the basic goal of the assessment monitoring program.

Identification of Contaminants

This is probably the aspect of the site investigation that is routinely most thoroughly covered in a typical assessment program. It is important in the design of an extraction system because of the different mobility and toxicity of the various contaminants and because of the special handling and treatment problems associated with some compounds.

The vast majority of ground-water remedial assessment projects will be concerned with volatile organics. Whether working with codisposal, point source or multiple point source volatile organics will almost always drive the assessment monitoring risk assessments and ultimate remedial design (unless the site is specific to metallic wastes only). These organic compounds have relatively high mobility as dissolved species in ground water. Do not, however, lose sight that inorganic parameters, many of which are not health based, can provide the "finger print" of the leachate. Establishing the specific water quality "finger print" must represent a main project goal in assessment monitoring. Once establishing the leachate parameter fingerprint, the investigator must stick with using the indicator parameter fingerprint to trace flow path water quality.

In addition to the chemical finger print identity of the compounds, it is also important to determine whether they are present in a non-aqueous phase. If NAPL contamination is involved, it is unlikely that aquifer restoration by a "simple" extraction system will be successful (i.e. lower resident water quality to below health based risk levels) over any reasonable time period (say 5 to 10 years). More than likely hydraulic containment objectives will be the main control technique used for DNAPL contaminated sites.

The presence of LNAPL (floating products) contamination, presents a number of special problems for the investigator; because of interfacial tension effects; NAPLs that are present in an aquifer at relatively low saturation may not flow into a normally constructed monitoring well. Therefore, the presence of NAPLs cannot be ruled out just because they have not been seen as a separate phase in any of the ground-water samples. A reasonably reliable way to detect non aqueous phase contamination is to take soil samples during installation of the piezometer or monitoring wells (e.g. split-spoon or Shelby Tube samples) or as part of a separate soil boring program. Locations for these samples should be based on spill areas established by Phase I investigations. In fractured rock aquifers, diamond rock coring, using air as the fluid to remove cuttings, can be used to collect aquifer samples but this may not conclusively rule out NAPL presence.

Contaminant Distribution and Concentration

Knowledge of the spatial distribution of contaminants in the aquifer is obviously very important for the design of the assessment monitoring system, since the rate and extent values determine the area in which one must establish control of ground-water flow. The general problem an assessment monitoring program must deal with, is essentially to determine the boundaries of the contaminant plume. The "extent"of the rate and extent requirements. This generally requires during the investigation that monitoring wells be installed both inside and outside the contaminated area. This presents the general problem with any traditional (RCRA or Superfund) assessment monitoring design; you must know the location of the

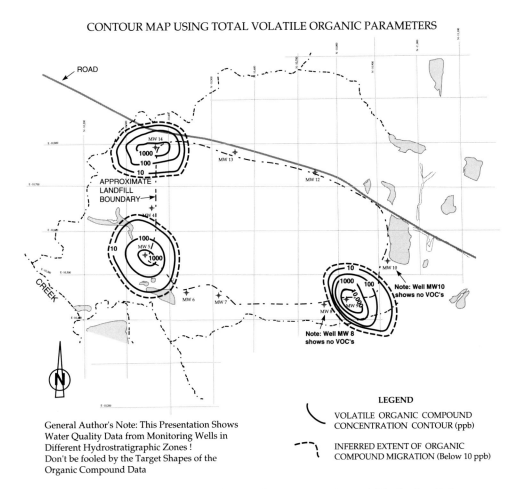

CONTOUR MAP USING TOTAL VOLATILE ORGANIC PARAMETERS

General Author's Note: This Presentation Shows
Water Quality Data from Monitoring Wells in
Different Hydrostratigraphic Zones !
Don't be fooled by the Target Shapes of the
Organic Compound Data

Figure 10-16 Example of Target Isocountours Around Monitoring Point

contaminate plume during the planning stage of these Federal programs. However, you actually don't know the true rate and extent of the plume until the wells are installed and sampled. Often the wells are either installed in the wrong location (out of the plume) or all of the wells are inside the contaminated area, so the plume edge cannot be located.

In the design of plume containment and well-head treatment systems it may not be essential to establish the upgradient extent of the contaminant plume. But, the downgradient and lateral extent of the area in which health-based standards are exceeded must be known for the data to complete "rate and extent" evaluations. In an assessment monitoring program accurate measurement of contaminant concentrations is important for a number of reasons:

• aquifer restoration goals are typically expressed in terms of individual contaminant concentrations, so the boundaries of the contaminated region must be

determined.

• predictions of the contaminant concentrations in the extracted ground-water are required in the design of the treatment system.

• measurement of the progress of aquifer remediation is best handled by periodic comparison of the contaminant concentration distribution using the initial pre-remediation parameter distribution.

Data on concentration distributions defined in assessment monitoring are usually described by drawing contour maps of the contaminant plume. These contaminant contour maps are prepared for defining the initial site conditions and for meeting clean-up goals during the remediation period. Assessment monitoring programs often ignore layered water quality conditions. This is especially true for isocontour presentations developed from single point widely screened monitoring wells. When interpreting concentration contour maps it is important to understand how vertical variations in aquifer contaminant concentration have been accounted for.

Assessment programs result in data presentations that require summarization so that "normal" humans can understand what can develop into a vast data set. Contour maps offer the investigator the tools to evaluate and display reasonable numbers of indicator parameters. Unfortunately isocontour maps can miss-represent the actual indicator parameters, since they are only a two-dimensional presentation of a three-dimensional system. Contour maps are often developed directly from the analytical results without regard to the depth from which the samples were taken, or the length of the screened intervals in the monitoring wells. This presentation can give false impression of the distribution of contaminants in the aquifer. These isocontours presentations also may show "target" contours drawn around monitoring points (see 10-16). This is a direct result of either too little data, incorrect selection of indicator parameters, or having wells located in and out of the containment flow path. If the samples were taken from fully penetrating monitoring wells, the results can be interpreted as a permeability-weighted depth-averaged concentration. Such a parameter concentration measurement is useful in predicting the contaminant concentrations that will be produced by fully penetrating extraction wells. However, concentrations measured in this way will probably be lower than the maximum concentrations in the aquifer and may produce incorrect estimates of total contaminant mass.

Estimates of the total contaminant mass in the aquifer are sometimes used within assessment programs to express the magnitude of the containment problem. Since a goal of aquifer restoration may be to remove 99% of the contaminant mass in the aquifer over some period of time, evaluation of the remediation depends heavily on the accurate estimation of this contaminant mass. The only effective way to do this in an assessment monitoring program is to understand the three-dimensional distribution of contaminant concentration in the aquifer. This may only be obtained by multi-level assessment sampling system if the facility has complex hydrostratigraphic units. Both dissolved and sorbed contaminants must be accounted for in estimates of the total contaminant mass.

Contaminant Mobility Characteristics

One of the most common explanations for the underestimation of the time required for aquifer remediation is that the effects of adsorptive retardation were not accounted for. The retarding effects of contaminant sorption are typically estimated using the concept of multiple pore volumes that must be extracted

before the contamination concentrations are reduced to the necessary regulatory standards.

Quantitative estimates of contaminant retardation developed in assessment programs are usually based on the total organic carbon content of the aquifer materials and tabulated values of partition coefficients for organic compounds. Programs conducting such estimates must include measurement of total soil organic carbon as part of the assessment sampling scope.

Soil samples can also be taken from beneath a contaminant source area to evaluate the partitioning of contaminants between soil and water. Sampling below source areas should, however, be very selective and not further cause additional cross contamination. This data are used to support analysis of the continued leaching of contaminants from the vadose zone into the underlying aquifer. Laboratory testing of site materials for partition coefficients is fairly rare at ground-water contamination sites, but the information obtained from such tests may be quite useful in the calculation of total containment mass (EPA 1990).

Contaminant Sources

Characterization of the sources of contamination is important in the design of any assessment program. While establishing a containment source is relatively easy when viewing a large codisposal landfill with little additional potential source areas, however, one should not lose sight that the main target of the investigation may be entirely, or only partially the source of the observed contamination. Short-sighted assessment programs do a disservice to long term goals of protecting human health and the environment. Continued contaminant discharge from alternative sources can defeat the ground-water cleanup efforts and risk further negative environmental effects. Even with the original source removed, it is important to evaluate the potential for continued leaching of contaminants from the remaining contaminated areas.

10.6.3 PERFORMANCE MONITORING

The performance of aquifer remediation systems must be monitored at regular intervals to ensure that the desired containment is being maintained over the ground-water flow patterns and the resultant movement of contaminants. Monitoring of the hydraulic performance of the system is performed by regular head measurement of water levels in piezometers and monitoring wells throughout the area of remediation. Potentiometric surface maps are then drawn for each of the aquifers of concern to show maintenance of the desired capture zones. In multi-aquifer situations the potentiometric gradients between aquifers should also checked using these maps. Performance monitoring of the aquifer cleanup can be demonstrated in several ways:

- A common, but not very reliable approach, is to monitor the flow rate and contaminant concentrations produced by the extraction wells;

- Integration of the product of flow and concentration over time gives a rough estimate of the mass of contaminants removed. This method does not, however, provide a direct measure of the reduction in contaminant concentrations in the aquifer; and,

- A more accurate method is to take samples simultaneously from enough monitoring wells in the remediation area to permit plume maps to be drawn.

Plume isocontour maps can be used to determine when the aquifer remediation is complete. In general samples taken from the extraction wells may not be reliable for prediction of remedial goals, since these wells generally show considerable dilution of indicator parameters through their hydraulic effects on the aquifer.

In monitoring plume containment systems, it is important that some monitoring wells be located downgradient of the extraction wells. This is necessary so that flow reversal downgradient of the well can be demonstrated to show containment of the contamination. Even though hydraulic head measurements are the primary data sources to demonstrate containment, periodic water quality sampling should be conducted to ensure that contaminants are not escaping by passing under or between the extraction wells. The bottom line to any assessment program is the knowledge of the geologic and hydrogeologic system. The appearance of complete hydraulic capture on the basis of potentiometric heads can be deceptive if the true hydrogeologic nature of the aquifer is not well understood.

10.6.4 Post-Remediation Monitoring

Most assessment monitoring require some form of post remediation monitoring. Performance monitoring of the aquifer restoration sites should continue even after the extraction system has been turned-off after meeting remediation goals. It should be expected to see low contaminant concentrations measured toward the end of the remedial action rebounding after the extraction system pumpage has been terminated. Several reasons for this effect can be established:

- the ground-water flow patterns generated by the

extraction system can cause dilution of the concentrations sampled at monitoring wells;

- residual contaminants stored in low hydraulic conductivity zones of the aquifer sorbed to the aquifer materials, or retained as NAPLs, may cause the concentrations to rise when the extraction is terminated, and,

- the recovery of water levels after system shutdown may re-saturate contaminated soils and thus release to ground water additional parameter mass.

10.7 SOIL SAMPLING AND ANALYSIS DESIGN

Soil sampling and the subsequent decisions on soil remediation represents one of the major actions on the typical "low volume-high toxicity" site investigation. Evaluation of single or multiple spill, hazardous solid waste deposits represents a major component of the remedial investigation as illustrated in Figure 10-4 for all but the least complex site. The goals of collecting samples for VOA's may include source identification, spill delineation, fate and transport modeling, risk assessment, enforcement, remediation, or post-remediation confirmation. The intended purpose of the sampling effort drives the selection of the appropriate sampling approach and the devices to be used in the investigation. Figure 10-11-17 illustrates the design process for a soil sampling investigation.

Soil investigations may include surface area, (two dimensions) or subsurface (three dimensions) environments, definition of a parameter concentration greater or less than a health or regulatory based action limit, "hot spots" where organic or metallic residuals have very high values, or simply the area above and adjacent to a leaking underground storage tank. Statistical approaches used to evaluate the soils data should be established during the development of a sample and analysis design. Statistics typically used in such soil sampling programs may include average analyte concentration and the variance about the mean (statistics that compare whether the observed level is significantly above or below an action level), as well as, temporal and spatial trends. There is an excellent user's guide on quality assurance in soil sampling (Barth et al., 1989) and a document for the development of data quality objectives for remedial response activities provide guidance for setting Data Quality Objectives (DQOs) for soil sampling activities (U.S. EPA, 1987). DQOs can be defined as *"qualitative and quantitative statements of the level of uncertainty a decision maker is willing to accept in making decisions on the basis of environmental data".* Since large errors in one part of an investigation can set the level of error for the rest of the particular sampling effort, the greatest emphasis should be placed on the phase that contributes the largest component of error. As an example, if the error associated with the sample collection or preparation step is large, then the best laboratory quality assurance program will be inadequate for obtaining adequate final data sets (van Ee et al., 1990). Experience shows that in the analysis of soils for VOCs, the greatest sources of error are the sample collection and handling phases. Field sampling personnel should coordinate with laboratory analysts to ensure that samples of a size appropriate to the analytical method are collected.

Over the last five years reconnanance sampling has been routinely evaluated soils in the field through analytical techniques that typically have higher detection limits than laboratory tests. In another words, these tests may only give a rough approximate value or even a yes/no answer for the indicator parameter at some set detection concentration level. However, in this reconnanance mode this less precise methods are OK, since these tests provide sufficient guidance on the location of affected soils. The concept of the DQO for reconnanance sampling and field evaluation provides the necessary error control to use these "field screening" techniques. The benefits of soil field-screening procedures are:

- short turn-around time data to guide more detailed sampling activities;

- target Contract Laboratory Program (CLP) sample collection in critical areas;

- reduced need for a second sampling event of the site; and,

- reduced analytical load on the laboratory.

Although the techniques to target the soil sampling have come into wide-spread use during remedial investigations there are a number of limitations to field-screening procedures:

- some knowledge of the potential VOCs present at the site is needed to accurately identify the indicator parameters;

- methodologies and instruments are in development and procedures for their use are not well documented; and,

- a more stringent level of quality assurance and quality control (QA/QC) must be employed to ensure accurate

and precise measurements.

The potential benefits and limitations associated with soil-screening procedures must be carefully weighed and compared to the data quality objectives necessary for clean-up decisions.

10.7.1 Device Selection Criteria

The soil sampling device and sampling procedures for site evaluation requires the consideration of many project factors including the total number of samples to be collected in the program, the available funds for the sampling effort, soil characteristics as they affect the ability to obtain and analyze the sample, site limitations, ability to sample the target units, whether or not field or laboratory screening procedures are to be used, the size of sample needed, and the required precision and accuracy defined in the DQOs. The total number of samples to be collected during the investigation can greatly affect sampling costs and the time required to complete a site characterization. If many subsurface samples are needed, it may be possible to use techniques such as historic air photo review combined with soil-gas sampling to target areas of contaminated residuals. These data can then be coupled with on-site analysis as an integrated screening technique to reduce the area of interest and thus the number of samples needed. Such a sampling approach has been found to be highly applicable for cases of near-surface contamination of oils and solid metallic waste.

The sampling, sample handling, containerizing, and transport of the soil sample should strive to minimize losses of volatiles for those projects concerned with levels of organic contamination. A common approach for both water and soil sample handling procedures should focus on the avoidance of cross contamination of the sample. Drill cuttings should not be dragged from one hole to another on dirty, (by contaminated soils), drilling tools or rigs. In addition to inter-borehole cross contamination (from one hole to another), the potential for intra-borehole cross-contamination should also be evaluated. These conditions may exist where drill cuttings from shallow waste pits are dragged into lower soil or ground-water sampling zones.

Only a very small volume of organic residuals incorporated into a monitoring well's filter pack can provide years of professional labor and thousands of dollars to explain. Such errors can also cause hundreds of thousands of dollars to be spent on aquifer pump and treat design and system installation where no actual contamination exists, at least, pre-investigation contamination! Soil sampling equipment should be readily decontaminated in the field if it is to be reused on the job site. Decontamination of sampling equipment may require

TABLE 10-3   Primary Soil Characterics for CERCLA Decision-Making Process

Soil Criteria	Measurement Technique / Method (w / Reference)		
	Field	**Laboratory**	**Calculation or Lookup Method**
Bulk density	Neutron probe (ASTM, 1985), Gamma radiation (Blake and Hartage, 1986, Blake, 1965).	Coring or excavation for lab analysis (Blake and Hartage, 1986).	Not applicable.
Soil pH	Measured in field in same manner as in laboratory.	Using a glass electrode in an aqueous slurry (ref. EPRI EN-6637) Analytical Method - Method 9045, SW-846, EPA.	Not applicable.
Texture	Collect composite sample for each soil type. No field methods are available, except through considerable experience of "feeling" the soil for an estimation of % sand, silt, and clay.	ASTM D 522-63 Method for Particle Analysis of Soils. Sieve analysis better at hazardous waste shes because organics can effect hydrometer analysis (Kluate, 1986).	Not applicable.
Depth to ground water	Ground-water monitoring wells or piezometers using EPA approved methods (EPA 1985a).	Not applicable.	Not applicable.
Horizons or stratigraphy	Soil phs dug with backhoe are best. If safety and cost are a concern, soil bores can be collected with either a thin wall sample driver and veilmayer tube (Brown et al. 1990).	Not applicable.	May be possible to obtain information from SCS soil survey for the site.
Hydraulic conductivity (saturated)	Auger-hole and piezometer methods (Amoozeger and Warrick, 1986) and Guelph permeameter (Reynolds & Elrick, 1985; Reynolds & Elrick, 1986).	Constant head and falling head methods (Amoozeger and Warrick, 1986).	Although there are tables available that list the values for the saturated hydraulic conductivity, H should be understood that the values are given for specific soil textures that may not be the same as those on the she.
Water retention (soil water characteristic curves)	Field methods require a considerable amount of time, effort, and equipment. For a good discussion of these methods refer to Bruce and Luxmoore (1986).	Obtained through wetting or drainage of core samples through a series of known pressure heads from low to high or high to low, respectively (Klute, 1986).	Some look-up and estimation methods are available, however, due to high spatial variabiltiy in this characteristic they are not generally recommended unless their use is justified.
Air permeability and water content relationships	None	Several methods have been used, however, all use disturbed soil samples. For field applications the structure of soils are very important, For more information refer to Corey (1986). and van Genuchten (1980).	Estimation methods for air permeability exist that closely resemble the estimation methods for unsaturated hydraulic conductivhy. Example models those developed by Brooks and Corey (1964)
Poroshy (pore volume)		Gas pycnometer (Danielson and Sutherland, 1986). 1986).	Calculated from particle and bulk densities (Danielson and Sutherland,
Climate	Preciphation measured using either Sacramento gauge for accumulated value or weighing gauge or tipping bucket gauge for continuous measurement (Finkelstein et al., 1983; Kite, 1979). Soil temperature measured using thermocouple.	Not applicable.	Data are provided in the Climatic Atlas of the United States or are available from the National Climatic Data Center, Asheville, NC Telephone (704) 259-0682.

Soil characteristics are discussed in general except where specific cases relate to different waste types (i.e., metals, hydrophobic organics or polar organics).

Source: EPA 1991

TABLE 10-4. Ancillary Soil Parameters for the CERCLA Decision-Making Process

Soil Criteria	Field	Laboratory	Calculation or Lookup Method
Organic carbon	Not applicable.	High temperature combustion (either wet or dry) and oxidation techniques (Powell et al.,1989) (Powell, 1990).	Not applicable.
Capacity Exchange Capacity (CEC)	See Rhoades for field methods.	(Rhoades, 1982).	
Erodibility			Estimated using standard equations and graphs (Israelsen et al., 1980) held data for slope, field length, and cover type required as input. Soils data can be obtained from the local Soil Conservation Service (SCS) office.
Water erosion Universal Soil Loss Equation (USLE) or Revised USLE (RUSLE)	Measurement / survey of slope (in ft. rise/ ft run or %), length of field, vegetative cover.	Not applicable.	A Modified Universal Soil Loss Equation (USLE) (Williams, 1975) presented in Mills et al., (1982) and US EPA (1988d) source for equations.
Wind erosion	Air monitoring for mass of containment.Field length along prevailing wind direction.	Not applicable.	The SCS wind loss equation (Israelsen et al., 1980) must be adjusted (reduced) to account for suspended particles of diameter < 1.0mm Cowherd et al., (1985) for a rapid evaluation (~24 hr) of particle emission from a Superfund site.
Vegetative cover	Visual observation and documented using map. USDA can aid in identification of unknown vegetation.	Not applicable.	
Soil structure	Classified into 10 standard kinds - see local SCS office for assistance (Soil Survey Staff, 1990) or Taylor and Ashcroft (1972), p. 310.	Not applicable.	See local soil survey for the site.
Organic carbon partition cooefficient (Koc)	*In* situ-tracer tests (Freeze and Cherry , 1979).	(ASTM E 1195-87, 1988)	Calculated from K , water solubility (Mills et al., 1985; Sims et al., 1986).
Redox couple ratios of waste/soil system	Platium electrode used on lysimeter sample (ASTM, 1987).	Same as field.	Can be calculated from concentrations of redox pairs or 0^2 (Stumm and Morgan, 1981).
Liner soil/water partition coefficient	*In situ* tracer tests (Freeze and Cherry,1979)	Batch experiment ; (Ash et al ., 1973) column tests (van Genuchten and Wierenga, 1986).	Mills et al., 1985.
Soil oxygen content (aeration)	0^2 by membrane electrode 02 diffusion rate by Pt microelectrode (Phene, 1986). 02 by field GC (Smith, 1983).	Same as field.	Calculated from pE (Stumm and Morgan, 1981) or from 0^2 and soil gas diffusion rate.

(Continued)

TABLE 10-4 (Con.). Ancillary Soil Parameters for the CERCLA Decision-Making Process

Soil Criteria	Field	Laboratory	Calculation or Lookup Method
Soil temperature (as it affects volatilization)	Thermotery (Taylor and Jackson, 1986).	Same as field.	Brown and Associates (1980).
Clay mineralogy	Parent material analysis.	X-ray diffraction (Whittig and Allardice, 1986).	
Unsaturated hydraulic conductivity	Unsteady dranage-flux (or instantaneous profile) method and simplified unsteady drainage flux method (Green et al., 1986).The instantaneous profile method was initially developed as a laboratory method (Watson, 1966), however it was adapted to the field (Hillel et al.,1972). Constant-head borehole inflitration (Amoozegar and Wanick,1986).	Not usually done; results very difficult to obtain.	A number of estimation methods exists, each wrth their own set of assumptions and requiremnts. Reviews have been presented by Mualem (1986), and van Gehuchten (in press).
Moisture content	Two types of techniques - indirect and direct. Direct menthods, (i.e., gravimetric sampling), considered the most accurate, with no calibration required. However, methods are destructive to field systems. Methods involve collecting samples, weighing, drying and re-weighing to detemmine field moisture. Indirect methods rely on calibration (Klute, 1986).		
Soil biota	No standard method exists (see model or remedial technology for input or remedial evaluation procedures).	No standard method exists; can use agar plate count using MOSA method 99-3 p. 1462 (Klute, 1986).	

- Soil characteristics are discussed in general except where specific cases relate to different waste types (i.e., metals, hydrophobic organics or polar organics).

Source: EPA 1991

the use of decontamination pads that have impervious liners, wash and rinse troughs, and through careful handling of large equipment. Whenever possible, a liner should be used inside the sampling device to reduce potential cross contamination and carryover to other sampling points. Decontamination procedures take time, require extra equipment, and ultimately increase site characterization costs, however neglect of these important procedures can cause long term confusion of actual field conditions.

Several soil-screening procedures are in use that include headspace analysis of soils using organic vapor analyzers: water (or NaCl-saturated water) extraction of soil, followed by static headspace analysis using an organic vapor analyzer (OVA) or gas chromatograph (GC); colorimetric test kits; methanol extraction followed by headspace analysis or direct injection into a GC; and soil-gas sampling (U.S. EPA, 1988). Field measurements may not provide absolute values but often may be a superior means of obtaining relative values.

Site Characteristics

The facility setting may have important controls on the methods and techniques used in the soil sampling program. For example, the presence of underground utilities, pipes, electrical lines, tanks and leach fields can affect selection of sampling equipment. If the location or absence of these hazards cannot be established, it may be desirable to conduct a non-intrusive survey of the area and select a sampling approach that minimizes hazards. In addition the remoteness of a site and the physical setting may restrict site access, factors as vegetation, steep slopes, rugged or rocky terrain, overhead power lines or other overhead restrictions, and lack of roads can also contribute to access problems. Often one can obtain excellent samples through shallow small scale soil sampling hand tools and backhoe excavation. These methods are often more practical under shallow soil contamination circumstances than a large, hollow-stem auger and drill rig operations. The selection of a sampling device may be influenced by other contaminants of interest such as pesticides, metals, semivolatile organic compounds, radionuclides, and explosives. The presence of ordnance, drums, concrete, voids, pyrophoric materials, and high-hazard radioactive materials may preclude some sampling and may require development of alternate sampling designs, or even reconsideration of project objectives.

Soil Characteristics

The characteristics of the soil material being sampled have a marked effect upon the selection of a sampling device. An investigator must evaluate soil characteristics, the type of VOC's, and the depth at which a sample is to be collected before selection of a proper sampling device. Specific example characteristics that must be considered are discussed in Chapter 4.0

10.7.2 Soil-Gas Measurement

Soil-gas measurements have particular relevance to soil contamination by organic product spills. They can serve a variety of screening purposes in soil sampling and analysis programs, from initial site reconnaissance to remedial monitoring efforts. Soil-gas measurements should be used for screening purposes only, and not for definitive determination of soil-bound VOCs. Field analysis is usually by hand-held detectors, portable GC or GC/MS, infrared detectors, ion mobility spectrometers (IMS), industrial hygiene detector tubes, and, recently, fiber optic sensors.

Soil-gas measurements may several potential applications. Summarized in Table 10-4, these include in situ soil-gas surveying, measurement of headspace concentrations above containerized soil samples, and scanning of soil contained in cores collected from different depths. Currently, no standard protocols exist for soil-gas analysis; many investigators have devised their own techniques, which have varying degrees of efficacy.

Independently, the American Society for Testing and Materials (ASTM) and EPA EMSL-LV are preparing guidance documents for soil-gas measurement.

The required precision and accuracy of site characterization, as defined in the DQOs, affect the selection of a sampling device. After reconnanance field screening efforts targeted sampling can require maximum precision and accuracy. In these cases better results can be obtained from sampling devices that collect an intact core. These devices such as Shelby or Thin Wall Sampling Tubes can provide "undisturbed" samples that can preserve more of the volatile VOCs resident in the non-retentive soil matrices. Solid stem augers and other devices that collect highly disturbed samples and may expose the samples to the atmosphere can be used if lower precision and accuracy can be tolerated. However, hollow-stem augers that can cut a large diameter core can obtain high quality soil samples. As general guidance the closer the expected contaminant level is to the action or detection limit, the more efficient the sampling device should be for obtaining an accurate and representative measurement.

10.7.3 Important Soil Characteristics In Site Evaluations

Tables 10-3 and 10-4 identify methods for collecting and determining data types for soil characteristics either in the field, laboratory, or by calculation methods. Soil characteristics in Table 10-3 are considered the primary indicators that are needed to complete Phase II of the RI/FS process. This is a short, but concise list of soil data types that are needed to make remedial decisions and should be planned for and collected early in the sampling effort. These primary data types should allow for the initial screening of remedial treatment alternatives and preliminary risk assessment evaluations. Many of these characteristics can be obtained relatively inexpensively during periods of early field work when the necessary drilling and sampling equipment are already on site. Investigators should collect all necessary soil characteristics data at the same locations and times soil boring are completed for lithology. Geophysical logging of the well should also be considered as a cost effective method for collecting lithologic information prior to casing the well. Data quality and quantity must also be considered before beginning collection of the appropriate data types.

If the site budget allows, collection of these data types during early periods of field work will improve the database available to make decisions on remedial treatment selection and model-based risk assessments. Advanced planning and knowledge of the need for the soil characteristics should be factored into early site work to reduce overall costs and the time required to reach a decision on the remedial design. A small additional investment to collect these data during initial site assessment tasks is almost always more cost effective than having to conduct additional soil sampling during design project phases.

Further detailed descriptions of the soil characteristics in Tables 10-3 and 10-4 can be found in "Fundamentals of Soil Physics and Applications of Soil Physics" (Hillel, 1980) and in a series of articles by Dragun (1988, 1988a, 1988b). These references provide comprehensive discussions of these characteristics and their influence on ground-water flow in soils, as well as, fate and transport of contaminants.

10.7.4 Soil Characteristics Data Types Required For Modeling

In general, the physical and chemical properties of soils can be expected to widely vary spatially. This variation rarely follows well defined trends and more likely exhibits a random character. However, the random character of many soil properties tends to follow classic statistical distributions (see Chapter 11.0). Soil properties such as bulk density and effective porosity are often normally distributed (Campbell, 1985). Other properties such as saturated hydraulic conductivity, are often found to follow a log-normal distribution. Characterization of a site, when ever possible, should be performed in such a manner as to permit the determination of the statistical characteristics (i.e., mean and variance) and their spatial correlations. Significant advances have been made in understanding and describing the spatial variability of soil properties (Neilsen and Bouma, 1985). Geostatistical methods and techniques (Clark, 1982; Davis, 1986) are available for statistically characterizing soil properties important to contaminant migration. Information gained from a geostatistical analysis of data can be used for three major purposes:

- Determining the heterogeneity and complexity of the site;

- Guiding the data collection and interpretation effort and thus identifying areas where additional sampling may be needed (to reduce uncertainty by estimating error); and,

- Providing data for a stochastic model of fluid flow and contaminant migration.

One of the geostatistical tools useful to help in the interpolation or mapping of a site is referred to as kriging (Davis, 1986). General kriging computer codes are presently available in convenient computer software packages. Application of this type of tool, however, requires an adequate sample size. As a rule of thumb, 50 or more data points are needed to construct the semivariogram required for use in kriging. The benefit of using kriging in site characterization is that it allows one to take point measurements and estimate soil characteristics at any point within the domain of interest, such as grid points, for a computer model. Geostatistical packages are available from the US EPA, Geo-EAS and GEOPACK (Englund and Sparks, 1988, and Yates and Yates, 1990).

10.7.5 Soil Characteristics and Remedial Alternatives

The remediation process consistent between Federal programs (CERCLA and RCRA) and state programs involves the identification, screening and analysis of remedial alternatives at uncontrolled hazardous waste sites (US EPA, 1988c). During screening and analysis process, decisions are made using specific site established soil characteristics for a given remedial alternative. If site-specific values are outside the range required for effective use of a particular alternative, that alternative is less likely to be selected for use in the remedial design. As a general definition "soils" encompass the mass (surface and subsurface) of unconsolidated mantle of weathered rock and loose material lying above solid rock. The soil component can be defined as all mineral and naturally occurring organic material 2 mm or less in particle size; the size normally used to distinguish between soils (e. g., sands, silts, and clays) and gravels. Organic matter is often found as an integral part of the soil.

An additional component that must also be addressed during the sampling effort consists of the non-soil fraction (e.g., sludges, automobile fluff, wood chips, various absorbents and mineral/organic material greater than 2 mm). This component may contain a greater amount of contaminant(s) than the associated soil. At sites in which this occurs reporting contaminant levels only in the soil fraction will ultimately lead to inappropriate and incorrect decisions on project remedial design.

Pollutant behavior in the soil environment is a function of the both pollutant's and soil's physical and chemical properties. Soil sorption (the retention of substances by adsorption or absorption) is related to properties of the pollutants (e.g., solubility's, viscosity, heats of solution, and vapor pressure) and to properties of soils (e.g., clay content, organic content, texture, hydraulic conductivity , pH, particle size, specific surface area, ion exchange capacity, water content, and temperature). The soil components that are most associated with sorption are clay content and organic matter. The soil particle surface

Flowchart for Planning and Implementation of a Soil Sampling and Analysis Project

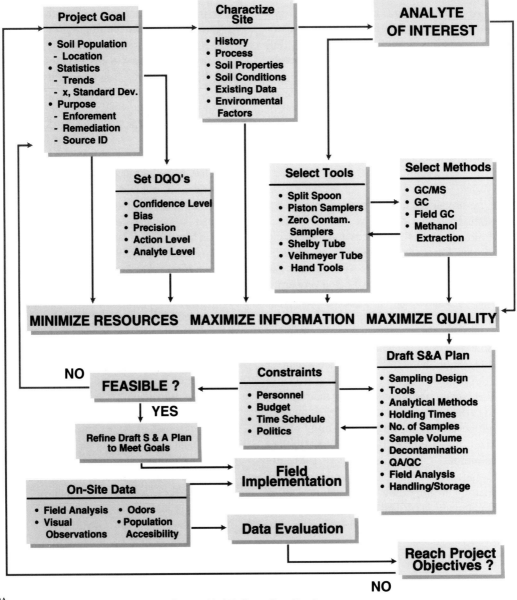

Figure 10-17 Sampling Design

characteristics thought to be most important in adsorption are surface area and cation exchange capacity (U.S. EPA, 1989). The following subsections list specific soil characteristics and their use in remedial alternative selection. Site-and-waste-specific data types that are critical to the effectiveness and ability to implement remedial processes are termed the soil's process-limiting

characteristics. Often, process-limiting characteristics are descriptors of rate-limiting steps in the overall remedial process. In some cases, limitations imposed by process-limiting characteristics can be overcome by adjustment of one or more soil characteristics, such as pH, soil moisture content, temperature and others. In other cases, the level of effort required to overcome these limitations will preclude use of a remedial process. When scoping project tasks related to soil contamination one should include the necessary field and laboratory evaluations to fully establish these characteristics. For waste/site characterization, process-limiting characteristics may be broadly grouped in four categories:

1. Mass transport characteristics;

2. Soil reaction characteristics;

3. Contaminant properties; and,

4. Engineering characteristics

Thorough soil characterization is required to determine site specific values for process-limiting characteristics. Most remedial alternatives will have process-limiting characteristics in more than one category.

Mass Transport Characteristics

Mass transport is the bulk flow, or advection of fluids through soil. Mass transport characteristics are used to calculate potential rates of movement of liquids or gases through soil include:

• Soil texture;

• Unsaturated hydraulic conductivity;

• Dispersivity;

• Moisture content vs. soil moisture tension;

• Bulk density;

• Porosity;

• Saturated hydraulic conductivity;

• Infiltration rate; and,

• Stratigraphy and others.

Mass transport processes are often process-limiting for both in situ and extract-and-treat vadose zone remedial alternatives (see Table 10-4 and 10-5). In-situ alternatives frequently use a gas or liquid mobile phase to move reactants or nutrients through contaminated soil. Alternatively, extract-and-treat processes such as soil vapor extraction (SVE) or soil flushing use a gas or liquid mobile phase to move contaminants to a surface treatment site. For either type of process to be effective, mass transport rates must be large enough to clean up a site within a reasonable time.

Soil Reaction Characteristics

Soil reaction characteristics describe contaminant-soil interactions. Soil reactions include bio-and physicochemical reactions that occur between the contaminants and the site soil. Rates of reactions such as biodegradation, hydrolysis, sorption/desorption, precipitation/dissolution, redox reactions, acid-base reactions, and others are process-limiting characteristics for many remedial alternatives (Table 10-5). Soil reaction characteristics include:

• Kd, specific to the site soils and contaminants;

• Cation exchange capacity (CEC);

• Eh;

• pH;

• Soil biota;

• Soil nutrient content;

• Contaminant abiotic/biological degradation rates;

• Soil mineralogy; and,

• Contaminant properties, described below, and others.

Soil reaction characteristics listed above determine the effectiveness of many remedial alternatives. For example, the ability of a soil to attenuate metals (typically described by Kd), or the rates of biological degradation of a particular organic compound may determine the effectiveness of an alternative that relies on capping and natural attenuation to immobilize and gradually reduce contaminants.

Soil Contaminant Properties

Contaminant properties are critical to contaminant-soil interactions, contaminant mobility, and to the ability of treatment technologies to remove, destroy or immobilize contaminants. Important contaminant properties include:

- Solubility in Water;

- Dielectric constant;

- Diffusion coefficient;

- Molecular weight;

- Vapor pressure density;

- Aqueous solution chemistry, and others.

Soil contaminant properties will determine the effectiveness of many treatment techniques. For example, the aqueous solution chemistry of metal contaminants often dictates the potential effectiveness of stabilization/solidification alternatives.

Soil Engineering Characteristics and Properties

Engineering characteristics and properties of the soil relate both to implementability and effectiveness of the remedial action. Examples of the engineering characteristics and properties of the site include the ability of the treatment method to remove, destroy or immobilize contaminants; the costs and difficulties in installing slurry walls and other containment options at depths greater than 60 feet; the ability of the site to withstand vehicle traffic (trafficability); costs and difficulties in deep excavation of contaminated soil; the ability of soil to be worked for implementation of in situ treatment technologies (known as tilth); and others. Knowledge of site-specific engineering characteristics and properties is therefore required for analysis of effectiveness and implementability of remedial alternatives. These engineering characteristics and properties include, but are not limited to:

- Depth to ground water;

- Thickness of saturated zone;

- Depth and total volume of contaminated soil;

- Trafficability;

- Erodability;

- Tilth; and,

- Soil bearing capacity, shear strength, and "others".

The "others" category can include many varied, very site specific soil characteristics that can greatly affect the performance and ultimate success of the remediation efforts. These site specific criteria must be accounted for within the site assessment so as to develop a targeted remediation specific to the site in question. The only guidance that can be provided in this area is to use experience and technical savvy to sort out these very site specific criteria.

10.7.6 Soil Sampling Objectives

Soil and sediment sampling represents a common tool used in assessment programs to establish "in place" levels of contaminated materials. These soil sampling procedures represent a very different type of problems to investigators as compared to ground-water flow path analysis. While flow path analysis requires consideration of hydraulic head levels in subsurface units, soil and sediments remediation programs represent traditional area sampling where both material sampling theory and geostatistical concepts can be used to locate and optimize sampling points. Even with this basis for soil sampling, targeted sampling programs can greatly improve the cost effectiveness of the program.

Soil sampling designs or strategies should be based on pre-specified sampling objectives. These strategies must, however, be based on some level of site specific data, such as prior disposal patterns, air photo target sites, or known contamination. The sampling design or strategy chosen will largely depend on the objectives of sampling. Without clearly stated program objectives, sampling efforts can lead to unusable or unnecessary data; and result in costly or misdirected remedial actions. Sampling objectives can be reduce to the following actions:

- Detecting evidence of contamination. During initial phases of site investigation the source of contaminated soil may not be known either because the contamination is buried beneath the soil surface or because the contamination has not expressed itself by visible discoloration or other distinctive observational features at the soil surface. This objective only requires that evidence of contamination be determined, not how much is present or its pattern of distribution.

- Estimating the mean, variance, and confidence intervals of soil properties in a volume of soil. One of the primary purposes of soil sampling is to estimate the mean value of a soil mass(s) and associated confidence limits for a pre-specified volume of soil. These data provides input for exposure assessment modeling. Such information may also be necessary to determine whether a level of contamination in site soils exceeds an average "background" soil concentration level or is less than some specified action level.

- Determining the spatial structure of soil properties. site characterization studies require understanding the spatial structure of important soil properties including the spatial distribution of contaminants. Spatial distribution of soil contamination is necessary to optimize removal or remediation of contaminated soils at reasonable costs.

The majority of remedial investigations of hazardous waste sites, (note: even at codisposal sites where a well defined foot-print exists of the disposal area, regulatory practice often requires that confirmation be made that disposal of waste and sediment transport did not occur beyond the foot-print of known disposal areas), require the determination whether or not there is soil or sediment contamination present at the site. Systematic sampling strategy has been reported (Parkhurst, 1984; and Greenberg, 1987) as more efficient than a random sampling strategy in cases where there is:

- no visible evidence of surface contamination;

- contamination is thought to be distributed at random;

- the contamination is believed to be thoroughly dispersed;

- for detecting small localized areas of contaminated soil or "hot spots"; or,

- in finding buried tanks and wastes .

One should, however, evaluate the potential for use of geophysical methods to target the contaminated soils, (especially for tanks and metallic contamination), before conducting an extensive soil sampling program. Disturbed soils can also be evaluated through false color IR air photos to assist in targeting the soils investigation.

By using non-sampling methods to target the potential soil contamination the probability of detecting a soil or sediment hot spot will obviously increase by increasing sampling density, (i.e., the number of sampling locations per unit area) in a surface evaluation program. Targeting the sampling program allows one to use available project funds more effectively. A computer program has been developed for determining the grid spacing (or sampling density) needed to hit an elliptical hot spot of a given size with a specified confidence (Singer, 1972 & 1975). Zirschky and Gilbert (1984) have developed easy to apply

nomographs for determining sample density. Gilbert (1987), defined that the selection of the proper grid spacing depends on the following assumptions:

- The hot spot is circular or elliptical and has no preferred orientation.

- Samples or measurements are taken on a square, rectangular, or triangular grid.

- The distance between sampling locations is much greater than the dimensions of the sampling unit.

- The definition of a hot spot is clear and unambiguous, that is, the types of measurement and the levels of contamination that result in a "detection" are clearly defined.

- There are no measurement miss-classification errors, that is, there are no errors associated with the determination of a detection.

Issues surrounding the use of composite sampling procedures must be addressed within sampling plans for remedial investigations. Since these documents are prepared before the Phase II field investigations, some knowledge of the site must be established during the Phase I program relative to the historic disposal practice. Then some accommodation can be made to soil and sediment sampling methods in the planning of the Phase II field work. In general where the cost of analysis exceeds the cost of sampling, it may be desirable to use composite sampling methods. However, when using composite sampling the criteria for determining whether a hot spot has been hit may have to be decreased by a factor equivalent to the number of sampling units that are composites. For example, if the hot spot criteria for a hit is 100 ppm of some substance in soil, then the criteria would have to be reduced to 20 ppm if five sampling units were composited. If a hit is found in the composite, then the original sampling locations may have to be resampled to determine which one or more of the subsampling locations contributed to the hit. A common practice is to retain a representative uncomposited portion of each of the original sampling units, then if a hit is registered the uncomposited portions of the sampling units could be analyzed individually using the predetermined criteria for a hit. Composite sampling entails a certain level of risk where the criteria set for a hit is at or near the detection limit for the indicator compounds. In these cases the investigator would then have to weigh the risk of not detecting contamination when it is present against the increased coverage offered by taking more samples.

If subsamples are to be composited, the subsamples should have the same dimensions and orientation and be obtained from comparable geologic strata or, if the location of the geologic strata have been masked by soil disturbances, from similar depth intervals below the original land surface. If subsamples do not have the same dimensions and orientation, the average value of a soil property for the composited sample may be biased toward the larger sized sample .

10.7.7 Intended Purpose of Soil Sampling

As a general goal of assessment monitoring programs such as CERCLA or within sections of RCRA, concentrations of hazardous pollutants in soils or sediments, should not exceed levels established as being adequately protective of humans and the environment. From a contaminated soil perspective within assessment programs, one needs to establish the role of soils as sources or sinks for selected pollutants. There are numerous techniques used within RI projects to evaluate contaminated soil as sources of risk to human health and the environment. Specifically, soil sampling efforts can be designed and conducted to:

- determine the extent to which soils act as either sources or sinks for air or water pollutants;

- determine the risk to human health and / or the environment from soil contamination by selected pollutants;

- determine the presence and concentration of specified pollutants in comparison to background levels;

- determine the concentration of pollutants and their spatial and temporal distribution;

- measure the effectiveness of control or removal actions;

- determine the potential risk to flora and fauna from specific soil pollutants;

- identify pollutant sources, transport mechanisms or routes, and potential receptors;

- obtain measurements for validation or use of soil transport and deposition models; and,

- meet the provisions and intent of environmental laws such as the Resource Conservation and Recovery Act (RCRA), the Comprehensive Environmental Response,

Compensation, and Liability Act (CERCLA), the Federal Insecticide, Fungicide, and Rodenticide Act (FIFRA), and the Toxic Substances Control Act (TSCA).

The extreme complexity and variability of soil necessitates a multitude of sampling / monitoring approaches must be incorporated into soil investigations. Both field and laboratory tests are necessary to understand the presence and behavior of soil contaminates. Field tests primarily provide information for soil classification in order to relate near surface environmental conditions. Laboratory tests supply analytical data on the type and quantity of a pollutant present in the soil sampled in the program.

Determination of risk to human health and the environment from contaminated soils involves several investigative steps. The soil/sediment evaluation must determine exposure and dose distributions to the most sensitive populations or receptors via all significant exposure pathways. This may include possible soil-related exposure from other media such as air or water, exposure from the soils themselves either through ingestion, inhalation, or skin absorption, as well as exposure through ingestion of foods contaminated directly or indirectly from the soils. It is important within these investigations to measure or estimate the extent to which the soils act as sources (through contacting air surface and ground-waters) for the pollutant(s) of concern.

If significant quantities of selected pollutants are found to be associated with soils initially and then released slowly over relatively long periods of time, the soils, in essence, act as pollutant sources. If significant quantities of contaminant(s) become permanently attached to soil and remain biologically unavailable, the soils may constitute a sink. Pollutant control needs in these cases may be reduced by the amounts by which the soils reduce the pollutant availability. Underestimating the ability of soils to act as a sink might lead to source control requirements more stringent than necessary, whereas overestimating might lead to less stringent control requirements than necessary.

Underestimating the extent to which soils act as contaminant sources will lead to inappropriate and insufficient controls of other additional sources, whereas overestimating may lead to expensive soil removal to a greater degree than necessary. Remediation through soil removal involves extensive testing of the soils and evaluation of proposed disposal options to evaluate the most cost effective option with the least environmental impact.

Soil sampling to measure the efficacy of control or removal actions must be preceded by the establishment of unacceptable concentrations of pollutants of concern in soil. A critical consideration in this instance will be the depth and surface area extent of the soil sample on the basis of which the soil concentration will be calculated.

10.7.8 Sample Design

Once a site must be sampled, sampling decisions must be made for the soil programs. Initial decisions must be made on the type of sampling design and sampling density. In most cases development of a statistical basis for the sampling is an early goal. The prime objective of a statistically based sampling design are either to provide the necessary site information for a fixed survey cost or to minimize survey cost for a fixed amount of information. Secondary concerns in sampling design are simplicity of resulting data analysis and simplicity of field operations in performing the survey.

A design used in many soil assessment programs is the simple random sample design (see Figure 10-18). Sampling units in these designs are determined by random selection. The random design for site investigation simplifies the statistical analysis; however, it is typically very expensive due to sampling many uncontaminated site areas.

A deviation of simple random sampling is the stratified random design. This design partitions the region to be sampled into subregions (strata) on the basis of suspected differences in level of pollutant, on cost of sampling, on the basis of equal strata areas, or on some combination of the above. Once subregions are established, simple random sample is then taken from each stratum. An example of this design would establish sufficient information to divide the site into strata where the level of pollutant concentrations is either far above action level, near action level, or far below action level. Typically you would expend most of the sampling effort (i.e., high sample density) on the strata that is near action level, so that decisions can be made with a high level of accuracy on the area needing remediation. Stratification ensures that all subregions of the site will be sampled, which may not be the case with a simple random sample of the site.

The stratification design uses scientific or historical knowledge based on a number data sources that the pollutant concentrations are quite different in identifiable segments of the site area. This targeting improves the subsequent estimate of the mean concentration over the entire site. Another criterion that may be useful in stratification for sediment sampling is distance from known point sources.

Simple random sampling and stratified random sampling designs are among a class of designs originally developed for the sampling of units that are discrete objects so that statistical analysis techniques associated with these designs can provide an estimation of population means. The basic designs and statistical procedures associated with surveys of discrete objects are given by Hansen et al., (1953).

Both stratified random sample and simple random sample procedures were developed for sampling of discrete

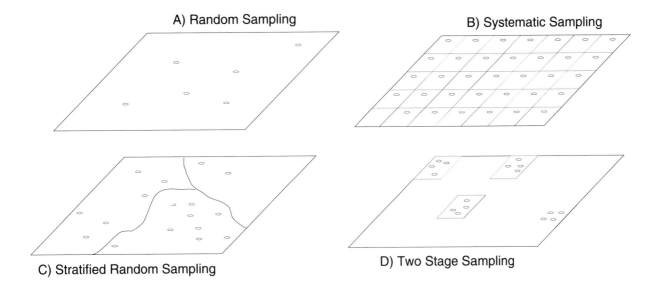

A) Random Sampling   B) Systematic Sampling

C) Stratified Random Sampling   D) Two Stage Sampling

Figure 10-18 Alternative Designs for Soil Sampling Programs

sampling units and may not adequately take into account the spatial continuity and spatial correlation of soil or geologic properties. Samples taken at locations that are close together tend to give redundant information and are therefore wasteful of resources. For this reason, some type of sample selection grid (systematic design) is often used to assure that sample locations will not be close to one another. The grid may be radial, triangular, rectangular, hexagonal, etc.

Systematic grid designs provide many of the advantages of stratification plus the avoidance of redundant samples. They thereby improve statistical precision and power. Investigations of the efficiencies of the grid designs show that the hexagonal grid is the most efficient given certain assumptions about the spatial distribution of the pollutant, but the square or rectangular grid is easier to use in practice. The radial grid may have some advantages in investigating the distribution of a pollutant near a point source.

A grid pattern is best oriented in the direction of flow of the pollutant, which may relate to site topography or a wind rose. Once the sampling density (grid spacing) and the orientation of the grid has been determined, a selection of one sample location will completely determine the locations of all sample locations.

The systematic grid designs are more closely related to the sampling of continuous media such as soil, air, and sediment. Gy (1982) gives an extensive description of techniques for the sampling of the continuous media of particulate materials. The statistical analyses associated with the results of these surveys of continuous media are typically aimed at estimating the spatial distribution of a property of the media such as a pollutant concentration or in finding "hot spots" within the region or site being sampled.

When a grid (systematic)design is used one finds many of the actual sample locations will not be at the grid locations because of the presence of obstructions such as roads, houses, rocks, and trees. If the field crew cannot sample at a specified location, they should have instructions to take a sample at the nearest point in a prespecified direction from the original point, provided that the location is within a specified distance (usually less than half the grid spacing) of the original point.

Many of the statistical techniques used in the analysis of data from surveys of continuous media fall into a category called geostatistics. For sample surveys involving random selection of sampling units, the statistical procedures are usually formed on a probability base provided by the randomization, while in geostatistics, the statistical inferences are based on what is known as a random field model. A good discussion of the nature and differences of these two approaches is given in a paper by Borgman and Quimby (1988).

In general terms, geostatistics is an application of classical statistical theory to geological measurements that takes into account the spatial continuities of geological variables in estimating the distribution of variables. In many ways, geostatistics is for measurements taken in 2-, 3-, and 4-dimensional space (the three spatial dimensions and the time dimension), what time series is for measurements taken in one-dimensional space time. However, a principal use of time series is in forecasting; in geostatistics, the principal emphasis is on interpolation.

10.8 ESTIMATION OF SORPTION PROPERTIES OF SOILS

Site assessment projects often require valuation of chemical fate and movement in both the unsaturated and ground-water zone. In order to establish these data one must understand the chemical contaminant properties, and soil physical and chemical properties unique to the sites under investigation. The RI/FS site characterization and remedial design process often requires knowledge of the the sorptive properties of site soils. The purpose of sorption (adsorption/desorption), estimations and studies is to evaluate the ability of a soil to retain chemicals or contaminants as a parameter for leaching and retardation evaluations, contaminant transport studies (unsaturate and saturated), and site specific modeling efforts at hazardous waste sites. Sorption estimates also can be a major component in evaluating the fate of contaminants in soils leaching to ground water.

The literature discusses many ways of estimating sorption. One must however, not forget that these are only estimation techniques. The basis of these estimation techniques are batch or column studies where graphical isotherms are developed to establish a chemical specific sorption value. Regression equations linking that value to chemical physical parameters can also be developed. This chapter subsection will discuss some of these methods and related limitations.

10.8.1 Sorption Mechanisms

The mechanisms of sorption at the soil-liquid interface is a complex and interrelated process. The complexity and interaction of the various contaminant and soil physical and chemical properties demonstrate that adsorption is a case-specific phenomenon varying with soil composition, soil chemistry, and sorbate chemistry. Estimation of soil adsorption is by the nature of the system under study, a case-by-case issue, where the properties of the various sorbents, and sorbates, representative nature of the soil sample, and the appropriate application of the estimation techniques must be evaluated in the project.

10.8.2 Dominant Sorbent Processes

Sorption of inorganic elemental species and ionic compounds is both a soil mineral and organic material process that requires for accurate estimates batch or column studies. With non-ionic organic compounds, if the organic carbon content (foc) of the soil is greater than 1%, organic material sorption can probably be assumed to be the dominant process; and sorption estimation techniques can be based on partitioning or tabulated Koc values. If the organic carbon content of the soil is less than 1%, some justification should be supplied to demonstrate that the dominant sorbent fraction is organic material before using tabulated Koc values and other estimation techniques based on regressions. If the organic fraction of the soil is not shown to dominate the process, a batch or column study will normally be required to estimate sorption of the soil system.

Inorganic and organic soil constituents are capable of adsorbing organic chemicals. The sorptive effects of inorganic materials for organic chemicals are negligible for soils high in organics (organic material sorption dominates); and sorption estimates such as Kow and Koc are applicable. The assumption that organic matter dominates the adsorption process may not be applicable for all soils due to the low organic content and high clay mineral content found in a large portion of western US soils. When estimating organic chemical sorption for soils with less than 1% foc, it may be necessary to demonstrate that the organic material is the dominant sorbent.

Dragun, (1988) describes a method of comparing the adsorptive effect of inorganic material versus organic carbon using threshold values derived from clay mineral abundance to foc abundance ratios (ratios < 25 imply organic material sorption dominance and Koc estimations would be applicable). This method requires quantification of swelling clay mineral abundance in the soil.

10.8.3 Sorption Determinations From Laboratory Studies

The following section describes laboratory batch, column and desorption studies. Appropriately designed and executed batch and column adsorption determinations are generally considered more reliable than estimations based on Koc or other regressions, because both chemical solute processes and chemical/physical soil processes are incorporated in the determinations. However, batch and column studies are not always practical or expedient within project demands. For this reason the procedures for conducting batch or column tests will be described in the following paragraphs.

Batch Method Adsorption Studies

Generally, laboratory batch adsorption or batch equilibration studies have been used to determine the capacity of a soil to retain a specific chemical. The procedure consists of either agitating a known mass of sorbent with a fixed quantity of solution with the concentration systematically varied, or agitating with a systematically varied soil:solution ratio at a fixed solute concentration. After agitation to equilibrium, the sorbent is separated from the solution by centrifugation or filtration, and the concentration of the solute is determined. The mass adsorbed is assumed to be the difference between initial and final solution concentrations normalized for the solution volume used. This is expressed as mass adsorbed per soil mass tested (ie. concentration of sorbent). The derived adsorption data is plotted against the equilibrated solution concentration, or applied statistically to regression equations such as the Freundlich, or Langmuir equations. These equations describe the distribution of concentrations over the experimental concentration ranges, approximating Kd.

The Draft Technical Resources Document, EPA/530-SW-87, Batch-Type Adsorption Procedures for Estimating Soil Attenuation of Chemicals addresses the specific methodologies and approaches mentioned above. Further, this document addresses experimental conditions and requires the documentation of parameters in the batch design. For both organic and inorganic chemical solutes these parameters include: (1) temperature effects, (2) method of mixing (standardized) effects, (3) pH effects, (4) interference effects (ionic strength and competition), (5) soil:solution ratio effects, (6) sorbent moisture content effects, (7) contact time requirements, and (8) hydrolysis effects. For organics, additional parameters such as dissolved organic carbon, volatility, degradation, and chemical stability may affect sorption, and require an extension of the evaluation effort.

Column Method Adsorption Studies

Since no specific protocol is currently available for column sorption studies, the experimental conditions and parameters required in the Draft Technical Resource Document, EPA/530 SW-87, Batch-Type Adsorption Procedures for Estimating Soil Attenuation of Chemicals shall apply to laboratory column studies. In addition, ASTM Committee D34.02.02, Draft Method for Leaching

Solid Waste in a Column Apparatus, is recommended for column design and procedural considerations. This procedure may not be applicable for volatile organics. Design adaptations for volatile constituents shall be described and documented.

Generally, laboratory column adsorption studies are more complex and time consuming than laboratory batch studies, but can yield more specific information if proper experimental conditions are maintained. There are many methods describing column adsorption studies available in the literature; however, no specific method or protocol is recommended at present. ASTM Committee, D34.02.02, Draft Method for Leaching Solid Waste in a Column Apparatus describes some of the experimental and design considerations.

Typical column studies entail applying a known concentration of test solution to uncontaminated soil which has been compacted within a column apparatus as an approximation of the in situ permeability of the natural soil. An appropriate and constant flow or flux is determined by testing or estimation, and applied to the soil column. The solute concentration is determined for pore volumes ranging from initial to equilibration. A breakthrough curve relating solute concentrations to pore volumes is developed from which adsorptions can be estimated. Additionally, desorption can be evaluated by reverting to an uncontaminated test solution.

Column apparatus design contributes significantly to variations in results. Preferential solute flow along column walls, column size, and improper column packing dominate as causes of variations. Additionally, the same parameters described for batch studies are of equal concern for column studies.

Desorption Batch and Column Studies

Desorption of contaminants from soil can be as important as adsorption. There is no assurance that desorption will be the the direct inverse of adsorption. Hysteresis effects of differential sorption rates and sites are noted in the literature. At present there are no standard protocols available for desorption studies. However, both batch and column methods have been used to generate desorption data employing an appropriate uncontaminated water and contaminated soil..

Determination of Adsorption Coefficience by Estimation

The following section describes, a method for the estimation of adsorption coefficients for organic

compounds and organic elemental or ionic compounds. A discussion of isotherms, the basis of tabulated sorption parameters and regression equations, is included. Descriptions of Kd (soil/water distribution coefficients) and Koc (organic-carbon/water partition coefficient) are given below. The assumptions and limitations of these estimations techniques are also addressed. A brief presentation of Koc estimation methods with appropriate references is included as guidance.

Adsorption Isotherms (Graphical and Regression Representation)

Adsorption is described by two general categories: equilibrium methods and kinetic methods. Equilibrium methods assume that adsorption is attained instantaneously and remains constant. Kinetics methods assume that the relative amount of sorbate versus the solute changes with time. The equilibrium methods are considered valid when:

• The rate of adsorption is much greater than the rate of change of solute concentration due to any other cause, and/or,

• The rate of adsorption is much greater than the rate of ground-water flow (see Travis and Etnier, 1981 for a more detailed review).

The kinetics methods are considered valid when:

• The contaminant is first introduced into the ground-water system,

• The contaminant source is in the form of a spike or a plug; and/or,

• The adsorption rate is considered slower than the ground-water flow rate (see Miller and Weber, 1984).

When a solution contacts a solid, solute transfers from liquid to solid until the concentration of solute in solution is in equilibrium with the concentration of the sorbate. The adsorption isotherm is the relationship between the quantity sorbed per unit mass of sorbent and the concentration of solute. Sorption isotherms therefore graphically describe the relationship (distribution or partition) of a chemical between sorbent and solute under equilibrium conditions over a range of concentrations under constant temperature (isothermal). As discussed, the data are generated from batch or column studies. The most common isotherms used in contamination projects are:

Determination of Isotherm Constants

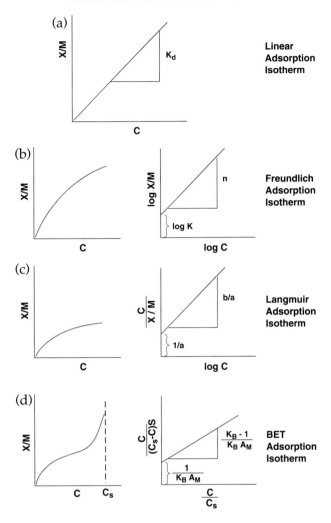

Figure 10-19 Calculation of Isotherm Constants

- Linear Adsorption Isotherm - This equation assumes that there is a linear relationship between the concentrations of solute and sorbate at equilibrium. It is valid for dilute solutions.

- Freundlich Adsorption Isotherm - This empirical equation is based on surface-free energy and monolayer capacity.

- Langmuir Adsorption Isotherm - This equation was originally developed to describe adsorption of a gas to a solid surface, but has been used to describe solid-liquid sorption.

- BET Adsorption Isotherm - This equation describes multilayer adsorption.

Linear equilibrium adsorption assumes the change in sorbate concentration is linearly related to the change in solute concentration. It is described by:

$$S = \frac{X}{M} = K_d C$$ Equation 10-1

where:
S = sorbate concentration,
C = solute concentration,
Kd = distribution coefficient,
X = mass of solute sorbed, and
M = mass of sorbent.

The slope of the above equation is obtained from:

$$\frac{dS}{dC} = K_d$$ Equation 10-2

The distribution coefficient, Kd, is defined as the ratio of the quantity of the sorbate per gram of solid to the amount of the solute at equilibrium. It may be obtained by plotting the results from laboratory batch experiments as shown in Figure 10-19.

The Freundlich, traditional Langmuir, and double-reciprocal Langmuir isotherm equations are empirical equations used to represent the shape of the isotherm describing the distribution of concentrations or adsorption (Kd) for a contaminant/soil system. The Freundlich isotherm regression equation (the oldest and most probably the most widely used equation) represents the dominant isotherm shape, where sorption is concentration dependent:

$$S = \frac{x}{m} = K_f C^{1/n}$$ Equation 10-3

where:
x = the mass of chemical adsorbed (ug),
m = the mass of the adsorbent (g),
C = the concentration of solute (mg/ml),
K, = the Fruendlich constant (ml/g), and
1/n =s a linearity constant

Freundlich equilibrium adsorption describes in general terms the change in sorbate concentration as a nonlinear function of the solute concentration. K and n are experimentally determined constants used within the evaluation. In some equations or derivations describing the formulation, the exponent is written as l/n. If n = 1, the above formula can be reduced to the linear equilibrium adsorption equation; n in practice is typically less than or equal to 1. By taking the logarithm of both sides, this equation is then converted to a linear form

$$\log S = \log(\frac{x}{m}) = \log K_f + \frac{1}{n}\log C \quad \text{Equation 10-4}$$

Figure 10-13 shows a plot of both forms of the equation. Please note that a plot of log S versus log C gives a straight line. The slope of the equation is obtained by taking the derivative to yield:

$$\frac{dS}{dC} = nKC^{n-1} \qquad \text{Equation 10-5}$$

The Freundlich equation is often statistically solved for K_f and the linearity constant (1/n) as a simple linear regression when cast in the logarithmic form (10-14) For cases where the linearity constant is one (1) or assumed to be 1, the adsorption is a linear function and:

$$Kf = \frac{(\frac{x}{m})}{C} \qquad \text{Equation 10-6}$$

where:
  x = the mass of chemical adsorbed (mg),
  m = the mass of the adsorbent (g),
  C = the concentration of solute at equilibrium (mg/ml), and
  Kf = the Fruendlich constant (ml/g)

The above would also represent cases where the isotherm is described by a straight line where sorption sites have a constant affinity for the contaminant (non-concentration dependent). This is often observed for low solute concentration ranges.

Langmuir Equilibrium Adsorption

Langmuir equilibrium adsorption also assumes a nonlinear function between the sorbed concentration and the solute concentration. The Langmuir isotherm is

described by

$$S = \frac{X}{M} = \frac{aC}{1+bC} \qquad \text{Equation 10-7}$$

where:
  a and b = experimentally determined constants.

The equation is sometimes written with a = bK, where b is a constant related to the enthalpy of adsorption and K is the maximum adsorption capacity of the solid (mass solute adsorbed/mass adsorbent for a complete monolayer). The Langmuir isotherm can be rearranged to the following linear form:

$$\frac{C}{S} = \frac{1}{a} + \frac{C}{a/b} \qquad \text{Equation 10-8}$$

Both forms of the equation are shown in Figure 10-15, where the constants are determined from the slope and intercept of the linear plot. The first derivative yields

$$\frac{dS}{dC} = \frac{a}{(1+bC)^2} \qquad \text{Equation 10-9}$$

The Langmuir isotherm equations are also used to represent curves and are useful for empirical curve fitting to data where varying surface site adsorption affinities are evident. The traditional Langmuir is often used for empirical fitting because two constants are present. The Langmuir equations predict maximum adsorption concentrations while the Freundlich equation does not. The traditional Langmuir is as follows:

$$S = \frac{x}{m} = \frac{kbC}{(1+kC)} \qquad \text{Equation 10-10}$$

where:
  x = the mass of chemical adsorbed (ug),
  m = the mass of the adsorbent (g),
  C = the concentration of solute at equilibrium (ug/ml),
  k = a constant maximum adsorption capacity of the solid, and
  b = a constant related to the enthalpy of adsorption

Double-Reciprocal Langmuir

The double-reciprocal Langmuir has been extensively used in site assessment projects for elemental adsorption data. The double-reciprocal Langmuir is an adaptation of the traditional equation for elemental sorption of soils exhibiting two primary adsorbing surface sites. The double-reciprocal Langmuir equations are as follows:

$$\frac{x}{m} = \frac{k_1 b_1 C}{(1 + k_f C)} + \frac{k_2 b_2 C}{(1 + k_2 C)}$$ Equation 10-11

where:

x = the mass of chemical adsorbed (ug),
m = the mass of the adsorbent (g),
C = the concentration of solute at equilibria (μg/ml),
k1 and k2 are constants, and
b1 and b2 are constants (the maximum quantities of the element that can be sorbed by the two surfaces)

The basic assumptions for application of graphic isotherm and regression equations are that the data was derived under equilibrium conditions, under constant temperature, fixation effects are minimal, and the data can be modeled as a regression function. The equations are valid only within the experimental concentration ranges used to determine the sorption.

Brunauer-Emmett-Teller (BET) Adsorption

Brunauer-Emmett-Teller (BET) adsorption describes multi-layer Langmuir adsorption. Multi-layer adsorption occurs in physical or van der Waals bonding of gases or vapors to solids. The BET isotherm, originally used to describe this adsorption, has also been applied to the description of adsorption from soil solution. The adsorption of molecules to the surface of particles forms a new surface layer to which additional molecules adsorb. If it is assumed that the energy of sorption on all successive layers is equal, the BET equation is (Rai et al., 1984):

$$S = \frac{X}{M} = \frac{K_B A_m C}{(Cs - C)[I + (K_{B^{-1}})(\frac{C}{C_s})]}$$ Equation 10-12

where:

K_B = constant related to adsorption energy,
A_m = maximum adsorption density of first layer, and
C_s = concentration of sorbate at saturation.

The above equation reduces to the Langmuir Equation f or KB >> 1 and C/C$_S$ << 1. It may be rearranged to linear form as

Table 10-6 Examples of Regression Equations

No.	Equation	Chemical Classes Represented
1	log K$_{oc}$ = -0.55 log S + 3.64	Wide variety, mostly pesticides (Kenaga and Goring, (1980)
2	log K$_{oc}$ = -0.54 log S + 0.44	Mostly aromatics or polynuclear aromatics; two chlorinated (Karickhof et. al., 1979)
3	log K$_{oc}$ = -0.557 log S + 4.277	Chlorinated hydrocarbons (chiou et al., 1979)
4	log K$_{oc}$ = 0.544 log K$_{ow}$ + 1.377	Wide variety, mostly pesticides (Kenaga and Goring, 1980)
5	log K$_{oc}$ = 0.937 log K$_{ow}$ - 0.006	Aromatics, polynuclear aromatics, triazines and dinitroaniline herbicides (Brown and Flag, 1981)
6	log K$_{oc}$ = 1.00 log K$_{ow}$ - 0.21	Mostly Aromatics or Polynuclear Aromatics; Two Chlorinated (Karickhoff et al., 1979)
7	log K$_{oc}$ = 1.029 log K$_{ow}$ - 0.18	Variety of insecticides, herbicides and fungicides (Rao and Davidson, 1980
8	log K$_{oc}$ = 0.524 log K$_{ow}$ + 0.855	Substituted phenylureas and alkyl-N-phenylcarbamates (Briggs, 1973)

$$\frac{C}{C_S - Cs} = \frac{1}{K_B A_m} + \frac{K_B - 1}{K_B A_m}\frac{C}{C_S}$$ Equation 10-13

A plot of both forms is shown in Figure 10-19, which also shows how the constants are determined.

10.8.4 Kd (soil/water) Distribution Coefficient Description and Koc (organic-carbon/water)

The distribution of chemicals between soil and water (sorption) has often been described by a soil/water distribution coefficient (Kd). Kd is a value relating the observed concentration at the solid-liquid interface expressed for the solid and solution concentrations at equilibrium under specific conditions. Kd is a property for a specific chemical in a specific soil which describes the sorptive capability. Kd is defined as:

$$Kd = \frac{C_{(solid)}}{C_{(solution)}}$$
$\qquad$ Equation 10-14

where:

C (solid) (µg/g) and

C (solution) (mg/ml) are concentrations at equilibrium

As the specific conditions change in the sorption process (such as concentration), the adsorption (Kd) may vary thus generating an isotherm or graphical representation of the sorption distribution. Various regression equations (Freundlich, Langmuir, etc.) have been used to describe this distribution as a single value or to determine concentration of sorbent for a given concentration of sorbate. Kd then describes the sorptive capacity of the soil over specified concentrations ranges and conditions. The distribution coefficient, Kd, is a valid representation of the partitioning between the solution phase and the solid phase only if the partitioning is fast (compared to the flow velocity) and reversible, and the isotherm is linear. Many organic compounds are reported to follow a linear adsorption isotherm. They include: several halogenated aliphatic hydrocarbons (Chiou, et al., 1979), polynuclear aromatic hydrocarbons (Means et al., 1980), di-benzothiophene (Hassett et al., 1980), benzene (Rogers et al., 1980), and halogenated hydrocarbons and some substituted benzene compounds (Wilson et al., 1981). Given these data and the reported frequency of these compounds as ground-water contaminants, organic adsorption from ground waters is most often assumed to follow a linear isotherm and have a constant Kd. Different methods have been used to determine the sorption-desorption behavior of contaminants:

- The most common approach, is laboratory column leach studies where effluent concentrations are measured in order to describe the overall interaction between the liquid phase and the solid matrix;

- A second method is based on field measurement of contaminant concentrations in the soil samples collected at various depths during drilling and in adjacent ground water during subsequent monitoring well sampling; and,

- The third method calculates the distribution coefficient based on the total organic carbon content of the soil.

Koc (organic-carbon/water partition coefficient) for organic compounds is the Kd normalized to the measured foc where sorption is assumed only for the organic material. Koc is defined as:

Table 10-7   Parameters for Regression Equations

No.	Parameter Required	Range of Parameter	Range of K_{oc} Values
1	S (mg/l)	0.0005 - 1,000,000	1 - 1,000,000
2	S (mole fraction)	(0.03 - 410,000)x10^{-9}	80 - 1,000,000
3	S (moles/l)	0.002 - 100,000	30 - 380,000
4	K_{ow}	0.001 - 4,000,000	10 - 1,000,000
5	K_{ow}	100 - 4,000,000	100 - 1,000,000
6	K_{ow}	100 - 4,000,000	100 - 1,000,000
7	K_{ow}	0.3 - 400,000	2 - 250,000
8	K_{ow}	3 - 2,200	10 - 400

$$Koc = \frac{Kd}{foc}$$
$\qquad$ Equation 10-15

Where:

Kd (ml/g) represents adsorption, and

foc = the fraction of organic carbon for a soil

This implies that soil sorption is linear in relation to organic carbon content, which may not be the case due to organic material inhomogeneities. Adsorption is assumed to occur solely from the organic material in the soil.

Koc is considered a property of the chemical, independent of a specific soil. This allows chemical sorption estimations for various soils if the organic material is the dominant sorbent. Koc values developed from laboratory studies have been tabulated in the literature and used to estimate sorption. Tabulated Koc chemical values directly derived from specific soil/chemical studies (batch or column) based on isotherms are considered more reliable than regression equation estimations derived from chemical, structural, or theoretical properties. Tables of Koc values are available in the literature, and have been extensively used to approximate Kd when the foc (organic carbon fraction) is known. Estimations of adsorption (Kd) from tabulated Koc values and Koc approximations from regression equations must be used correctly through a knowledge of method limitations, chemical groups represented and the parameter range limitations inherent in the studies experimental design.

Various methods of estimating and approximating sorption as a Koc from a partitioning processes for organic chemicals (eg. Kow, SAR's, S, pr, etc.) are available in the literature and mentioned in the following sections.

Methods for estimating Kd from an adsorption regression (Freundlich and Langmuir) from experimental data are also discussed. Published values of Kd reported for specific soils can only serve as a gross qualitative comparison to soils because the values are case specific.

Organic matter content in soil has been strongly correlated with adsorption of nonpolar and neutral organic chemicals with low aqueous solubilities and molecular weights less than about 400 atomic weight units. The work of Lambert (1967, 1968) and co-workers (Lambert et al., 1965) demonstrated that the sorption of neutral organic pesticides correlated with the organic carbon content of a given soil. Adsorption of organics has been normalized to the organic carbon content (foc), and the Koc (organic-carbon/water) partition coefficient has been developed to represent sorption. Koc is assumed equal to Kd divided by the fraction of organic carbon in the soil (i.e, Koc = Kd/foc). The distribution coefficient, Kd, is related to organic carbon content fraction according to:

$$Kd = (K_{oc})(f_{oc})$$ Equation 10-16

Koc is the adsorption constant based on organic carbon content and foc is the fraction of total organic carbon content in terms of grams of organic carbon per gram of soil. While Kd for a chemical may vary greatly in different soils, the value of Koc usually does not vary more than a factor of two or three. Using Koc values, Kd can be estimated based on the amount of organic carbon present in the soil. This technique assumes no adsorption by clay mineral complexes; and therefore applies only when organic carbon adsorption dominates. Typically, Koc values have been derived from soil batch or column studies and assumed linearity of the isotherm. The experimental data from those studies has also been used to develop regression equations for Koc values to chemical properties for estimating or predicting Koc for other chemicals.

Koc values have also been estimated by various regression techniques, nominally based on chemical properties (Kow, parachor, aqueous solubility, etc.). Similar to Koc determinations from batch studies, estimation techniques should not be used when soils exceed the experimental limitations (concentration and organic carbon content) used to develop that technique.

Selection of a Regression Estimation Method for Koc

The chemical properties of the contaminant and the data used to derive the regression equations should be fully evaluated in selecting an appropriate regression for estimating Koc. As general guidelines (after Lyman et. al., 1982) the following should be used in the selection process:

- The regression equation used should be derived from chemical compounds similar to the contaminant compound in question.

- The contaminant chemical properties should be evaluated for accuracy and the most accurate used as a basis for selecting a regression equation to estimate Koc. If the chemical properties are judged to be equally accurate, the order for estimating Koc should be octanol-water partition coefficient (Kow) ≥ aqueous solubility (S) > parachor (Pr) >> other methods.

- The foc, Kow, S, pr and other property values for the contaminant/soil system should fall within the range of values used to derive the selected regression equation for estimating a Koc.

Often, experimentally derived values for Koc are unavailable in the literature and partition-based empirical regression estimates of Koc have been used. A large array of regression equations are found in the literature (Lyman et. al., 1982; Bednarz et. al., 1983; Dragun, 1988; .

The reliability of the theoretical method for calculating the adsorption coefficient depends on several factors. Regression equations are experimentally derived from a specific data set that represents particular classes of chemicals and ranges of parameters. The reliability of the Koc value is directly related to the correlation between the data set of the model and the data set from which the regression equation was derived. Therefore, care must be taken when selecting a regression equation. Also, the regression equation method is only valid in porous media with greater than 0.1% organic carbon content (Lyman et al., 1982). Regression equation methods consider organic carbon as the primary adsorbent; however, below an organic carbon content of 0.1%, inorganic surfaces are the dominant adsorbent (Pennington, 1982).

Table 10-5, adapted from Lyman et al. (1982), by Srinvasan and Mercer (1992), includes examples of various regression equations, the source of each equation, and the classes of organic chemicals for which each equation was derived. Table 10-6, also adapted from Lyman et al. (1982), lists the parameters required for each equation, the range of values for each parameter, and the range of Koc values calculated with each equation.

Regression equation selection has been recommended by Lyman et al. (1982) using as a basis data available about the chemical, the chemical classes covered by each regression equation, and the range of Koc and input parameter values covered by each regression equation. In general terms the following points should be considered in the evaluation process:

- One should always place the highest priority to the most accurate data from actual measurements;

- If data are available for all equation input parameters, a regression equation using Kow is preferred to an equation using S;

- If one must make election of a regression equation on the basis of chemical classes, high priority should be given to the equation which was derived from the same chemical class as the chemical being modeled. However, if there is no clear match of chemical classes, Lyman et al. (1982) suggests using equations 1 or 4 from Table 10-6 because they are derived from the widest variety of chemicals;

- Selection of a regression equation on the basis of the range of Koc and input parameter values, the values should be within the range originally covered by the regression equation;

- The use of the parameter values or estimation of Koc values outside the range or the original data set will subject the estimated Koc and, therefore, the estimated retardation coefficient, to greater uncertainty;

- Selection of a regression equation to estimate a Koc should be based on the data used to derive the regressions and the contaminant's chemical properties;

- The value of the chemical property on which the selected regression equation is based should be the most accurate value available;

- If chemical property values are equally accurate then the order of preference would be Kow $\geq$ 5 > pr.

- Regression equations derived from chemicals similar to the compound in question should be used.

- For an appropriate application of Koc estimation, the data used to derive the regression must encompass the case under consideration (foc, Kow, S, pr, etc.).

Some of the regression equations have been experimentally derived from high foc content and low contaminant concentrations; and may not be applicable for Koc estimations at high concentrations typically encountered in contaminated soils.

Estimation of Koc From Kow (n-octanol/water) Partition Coefficient

The n-octanol/water partition coefficient (Kow) has often been used to estimate Koc values for estimating adsorption. The common method for determining the partition coefficient is by mixing the chemical in a two phase octanol/water system and determining the concentrations in one or both phases at equilibrium. Reverse-phase high performance liquid chromatography (HPLC) has also been used to determine partitioning based on retention values and regression with better accuracy for low concentration ranges. Examples for determining Koc from Kow are presented in Lyman (1982) and Dragun (1988), who also list many of the regression equations.

The next step in this process is the determination of Koc. This is usually accomplished using an equation with the general form

$$Log\ Koc = aLog(Sw\ or\ Kow) + b \qquad \text{Equation 10-17}$$

where:
Sw = aqueous solubility of a compound,
Kow = octanol-water partition coefficient (ratio of concentration of solute in octanol phase to concentration of solute in aqueous phase),
a = slope on log-log plot, and
b = minimum value of log Koc (intercept of log-log plot).

Many studies of various chemical groups have been conducted. Karickhoff et al. (1979) related Koc to the octanol-water partition coefficient and to the water solubility by the following relationships:

$$Koc = 0.63 Kow \qquad \text{Equation 10-18}$$

and

$$Koc = -0.54 \log Sw + 0.44 \qquad \text{Equation 10-19}$$

where:
Sw = aqueous solubility of a compound (expressed as a mole fraction).

The water solubilities of the compounds examined ranged from 1 ppb to 1000 ppm. Hassett et al. (1980) found a similar relationship between Koc and Kow for organic energy-related pollutants. Figure 10-20 shows the data relationships from both these studies. Chiou et al.(1979) also investigated the relationship between octanol-water partitioning and aqueous solubilities for a wide variety of chemicals including aliphatic and aromatic hydrocarbons, aromatic acids, organochlorine and organophosphate pesticides, and polychlorinated biphenyls. Their results, shown in Figure 10-20, cover more than eight orders of magnitude in solubility and six orders of magnitude in the octanol-water partition coefficient. The regression equation based on these data is:

$$\log Kow = -0.670 \log Sw + 5.00 \qquad \text{Equation 10-20}$$

where:
Sw = aqueous solubility of a compound, in mmol/l

Brown and Flagg (1981) have extended the work of Karickhoff et al. (1979) by developing an empirical relationship between Kow and Koc for nine chloro-s-triazine and dinitroaniline compounds. They plotted their results, along with those of Karickhoff et al. (1979), as shown in Figure 10-21. The combined data set produces the following correlation:

$$\log Koc = 0.937 \log Ko., -0.006 \qquad \text{Equation 10-21}$$

The correlation between Koc and Kow for the compounds studied by Brown and Flagg (1981) has a larger factor of uncertainty than those studied by Karickhoff et al. (1979).

The octanol-water partition coefficient and the solubility of an organic compound are generally available or may be calculated. Leo et al. (1971) have compiled a list of octanol-water partition coefficients. For organic compounds, Verschueren (1983) provide many common properties. Lyman et al. (1982) provide methods of calculating solubilities and octanol-water partition coefficients.

A wide variety of methods are available in the literature for estimating Kow for organic chemicals based on chemical structures or physical properties. Regression equation correlations with structural activity relations (SARs) using chemical fragments or physical properties such as substitute constants (aqueous solubility, and parachor) are the most dominant. Additionally, specific regressions are available relating Kow to Koc to estimate adsorption.

Estimation of Koc From Solubility (S)

A wide variety of regression equations have been formulated relating aqueous solubility (S) of an organic compound to a Koc value. In these equations, solubility has been expressed in various forms (mol/L ppm, or mole fraction). The most extensive list of organic chemical Koc values determined by solubility regressions and from experimental data is from Kenaga (1980). The reader

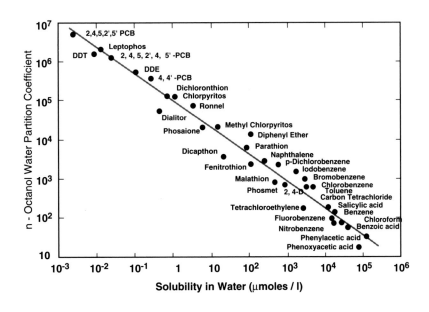

Figure 10-20 Correlation of Aqueous Solubility with Octanol-water Partition Coefficient

should be aware that cited solubility values can have a wide range per compound and should be used with care. Examples are presented for many of the regression equations, listed in Lyman (1982) and Dragun (1988).

Estimation of Koc From SAR's (Structure Activity Relationships)

When Kow and Koc partitioning information and solubility are unavailable, chemical structure activity relationships (SARs) have been used to estimate adsorption. These estimation techniques are based on structural parameters and are not considered as reliable as partitioning information. Substituent constants such as p constants and hydrophobic fragments, and other structural activities such as molecular conductivity indices and parachor properties have been used to estimate Koc. Parachor estimation techniques are the most dominant method.

Inorganic Elemental and Ionic Compounds Adsorption Estimation

Though adsorption to organic material dominates total soil sorption for low solubility organic compounds for soils with greater than 1% organic material, clay mineral (illite, illite/montmorillonite, saponite, vermiculite, chlorite, kaolinite, halloysite, etc) and zeolite mineral adsorption usually dominates for inorganic compounds, cationic species, and ionically charged organic compounds, irrespective of the soil. This adsorption process is pH dependent because cation exchange and protonation are pH dependent. Generally, as pH increases, the adsorption of heavy metal cations such as Pb^{2+}, Cd^{2+}, Zn^{2+}, Cu^{2+}, Hg^{2+}, and Cr^{6+} increases and adsorption decreases for Cr^{6+}, As^{3-}, and Se^{4-}. Ionic organic compound adsorption may also be pH dependent. Ionic organic compound sorption is dependent on whether the compound is an acid or a base and the affinity of the ionic form for organic sites relative to the affinity of non-ionic form for lipophilic sites. Other factors that influence the adsorption of cationic species and organic ionic molecules are moisture content, fixation as insoluble complexes, selective competition, and ionic strength of soil pore water.

Clay particles and clay minerals with high cation exchange capacities (CEC) have been shown to have high adsorption capabilities for cationic species. There is a preferential order of ion replacement progressing from higher valence state to lower. Organic material also demonstrates some ionic exchange capabilities. CEC determinations are usually total soil determinations without segregation of effective sorbents. Bonazountas (1983), documents a method based on soil CEC determinations (uncontaminated) and the mass and valence of contaminant compound for estimating the maximum soil concentration of contaminants. This method does not consider pH, redox, and complexation affects and may only serve as a gross approximation of maximum adsorption. Additionally, CEC determination methodologies have inherent problems in design such as carbonate mineral effects and incomplete ion exchange with some clay minerals.

10.9 GROUND-WATER REMEDIATION

Under the Comprehensive Environmental Response, Compensation, and Liability Act (CERCLA) and the implementing regulation, the National Contingency Plan (NCP), the cleanup goal for contaminated ground water expressed by the U.S. EPA is to return ground water to its beneficial uses within a reasonable time frame where practicable. This goal underlies all remedies designed to address contaminated ground water at Superfund sites. The U.S. EPA (1991) provides the following additional guidelines based on experience and research work from Superfund projects:

1. Adequate site characterization data is essential to understanding and effectively remediating contaminated ground water. A primary finding in the USEPA's studies of actual system performance was that there were rarely sufficient site information available to fully understand how contaminants are moving in the subsurface. An accurate picture of the subsurface environment and

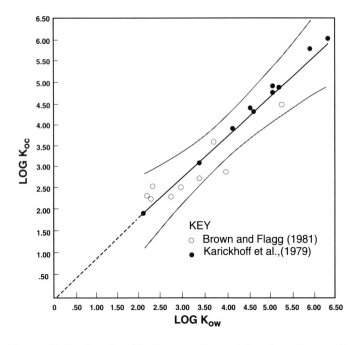

Figure 10-21 Relationship Between K_{oc} and K_{ow} for a Coarse Silt

POTENTIAL EXPOSURE PATHWAYS FOR CO-DISPOSAL FACILITIES

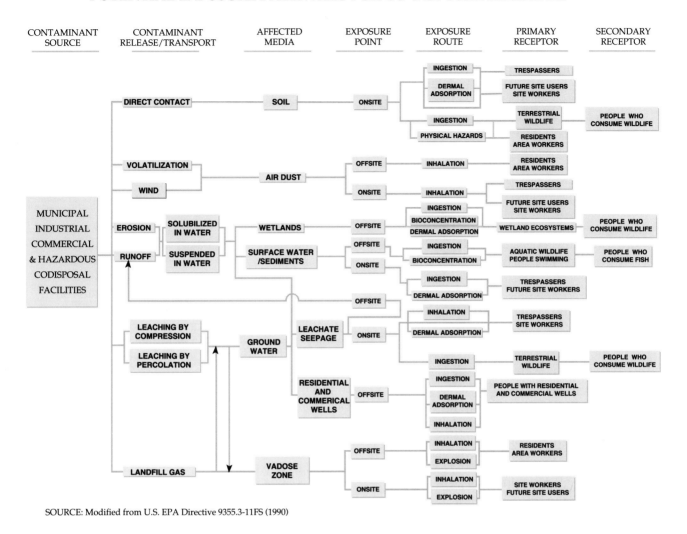

SOURCE: Modified from U.S. EPA Directive 9355.3-11FS (1990)

Figure 10-22 Exposure Pathways for Co-disposal Sites

contaminant form and distribution is ultimately important to properly design aquifer remediations.

2. Its much better to contain aqueous phase plumes early in project activity. The U. S EPA feels it is generally feasible and desirable to implement a containment system; e.g. pumping for gradient control, once the general plume definition phase is complete. It is also the author's opinion that containment represents the only viable remedial alternative in the majority of organic product spills where DNAPLs have penetrated the aquifer. Such containment actions prevent or greatly reduces the contamination from spreading further into uncontaminated ground water, thus limiting the area for which remedial action may be required.

Containment may be necessary until some point in the future where technology has progressed far enough to make active remediation possible.

3. A recent concern of those involved in site assessment projects is the presence of non-aqueous phase liquids (NAPLs). Experience at such sites indicates when contaminants are present as an immiscible phase in the saturated zone, the resultant organic product density and high contamination levels can substantially complicate investigation and aquifer remediations. Where these contaminants are more dense than water site evaluation techniques can become very complicated and costly. One should first evaluate the potential of an immiscible phase being present,

prior to initiating field investigatory work. Figure 10-25 provides a summary of these factors and outlines a process for determining the likelihood that DNAPLs are present in soils or geologic units (USEPA, 1991).

4. As general guidance if NAPLs are encountered in a well or borehole, they should generally be removed directly by pumping the immiscible phase. It is often very difficult to locate dense NAPLs (DNAPLs) due to their ability to move vertically through small discontinuities in the subsurface environment (Cherry, 1991). However, it is more effective to remove the immiscible liquid directly than extract it as it solubilizes into the aqueous phase since much larger volumes of ground water can be potentially contaminated by small volumes of DNAPL's..

5. Extraction systems should be monitored carefully after installation and balanced to improve effectiveness and efficiency. Once a ground-water extraction system is implemented, samples should be collected and analyzed periodically from wells located throughout the contaminated zone to assess the plume response to extraction and determine the need to modify the system during operation.

6. It can be expected that on DNAPL contaminated sites it not be practicable to restore portions of the contaminated ground water with currently available technology. In these cases alternate remedial actions, including containment and shrinking of the contaminated area. It may not be technically practicable to return the ground water contaminated by DNAPL's are to its beneficial uses. However as a general goal it is important to ensure that these areas are managed to prevent expansion and migration into uncontaminated ground water. Hydraulic gradient control of the aquifer to prevent migration of aqueous-phase contamination combined with extraction of the DNAPLs, when technically feasible, provides a currently feasible solution to DNAPL sites.

Additional goals for such sites should include remedial efforts to reduce the size of the contaminated area, to the extent practicable prior to implementing a containment remedy. Based on evidence collected by the U.S. EPA (1991); operating ground-water extraction systems within contaminated areas can be substantially reduced the volume of contaminant resident in groundwater (mass removal) even where portions of the plume may not be fully restored using current technology. This generally involves pumping and treating soluble neutral density phase contamination from the DNAPL zones. Although the long-term costs and liability of maintaining a ground-water containment system, can be high, containment generally represents the only alternative until innovative technologies are developed to clean up those portions of the contained plume.

Approximately 20 percent of the sites on the National Priorities List (NPL) are municipal landfills which may have been used as a co-disposal (both municipal and waste that is now considered hazardous) site prior to passage of the RCRA Subtitle C regulations. These co-disposal sites share many similar attributes and their remediation will involve similar waste management approaches. The National Contingency Plan states the EPA expectation that containment technologies will generally be appropriate for these large volume/low toxicity wastes that poses a relatively low long-term threat or where treatment is impracticable (Sec. 300.430(a)(1)(iii)(B), 55 FR 8846 (March 8,1990)). However, the EPA also expects that treatment should be considered for identifiable areas of highly toxic and/or mobile material that constitute the principal threat(s) posed by the site (Sec.300.430 (a) (1) (iii) (A)). Landfill characteristics for these co-disposal facilities are very similar and the NCP targeted these sites as likely for streamlining the RI/FS process with respect to site characterization, risk assessment, and the development of remedial action alternatives. Assessment tools to assist in scoping activities, are be included in the document Conducting Remedial Investigations Feasibility Studies for CERCLA Municipal Landfill Sites (November 1990, Directive No. 9355.3-11).

10.9.1 Landfill Site Characteristics

In general terms CERCLA municipal landfills are those facilities where a combination of principally municipal, and to a lesser extent hazardous wastes have been co-disposed. The large volume and heterogeneity of waste within these landfills makes treatment of the entire contents generally impracticable. The evaluation of all codisposal site potential threats to human health and the environment resulting may include: (l) leachate generation and ground-water contamination; (2) landfill contents; (4) landfill gases; (4) contamination of surface waters, sediments, and adjacent wetlands; and (5) local site contamination. A conceptual model of the potential pathways of exposure to hazardous substances that may exist at municipal landfill sites is presented in Figure 10-22. Grayed boxes represents the major potential pathways for unlined codisposal sites.

Streamlining Site Characterization

The full characterization of a municipal codisposal landfill can be expedited by focusing field activities on the information needed to (1) sufficiently assess risks posed by the site, and (2) evaluate practicable remedial actions. These goals should drive the investigation process, however, one should not turn the evaluation of risk into a witch hunt to find some potential risk "no matter how hard we have to look". The investigator must concentrate on establishing the ground-water flow system based on the site geologic conditions. This establishes real risk of the site to down gradient receptors. Additionally, site characterization may be streamlined by conducting a limited but mandatory Phase I investigation during scoping of the RI/ FS to assist in identifying necessary Phase II fieldwork. Examples of limited field investigation activities may include evaluating usefulness of an existing monitoring well network or verifying that the landfill construction. Chapter 2.0 describes Phase I programs applicable to RI/FS planning.

Leachate / Ground-water Contamination

Full characterization of a site's geology and hydrogeology are necessary to adequately assess the rate and extent of contamination for risk evaluation and for the design of extraction and treatment systems for leachate and, ground water, as well as, for evaluation of capping options. Ground-water contamination at municipal landfill sites often contains levels of organic matter in the form of complex organic compounds, chlorinated hydrocarbons and metals. Data gathered during the hydrogeologic investigation, however, are similar to those gathered at other types of NPL sites but, density considerations are of lesser importance than typical "spill" type sites.

Leachate generation is of specific concern when characterizing municipal landfill sites. The main factors contributing to leachate quantity include precipitation, as well as recharge from ground water and surface water. Information recommended during characterization program of landfill leachate generally include:

- Surface water drainage patterns;

- Climatological characteristics (e.g., precipitation and evapotranspiration);

- Leachate characteristics (e.g., TCL organics, TAL metals, BOD, COD, pH, TDS, TSS, phosphorus, nitrogen, and oil and grease); and

- Identification of local aquifers and their associated water

levels, flow rates, use potable water and chemistry.

Codisposal sites often appear to have leachate perched within the landfill contents, above the regional ground-water level. However, large codisposal sites that are maintained and constructed according to modern landfill practice actually have very minor perched leachate zones with unsaturated waste both above and below the perched zone. The placement of a limited number of leachate wells in the landfill may provide information regarding the depth, thickness, and types of waste, the moisture content and degree of decomposition of the waste, leachate head levels, the composition of the leachate, and the elevation of the underlying natural soil layer. However, one should not be fooled into believing that a few leachate had values representing a consistent head level across the entire site, since two piezometers located a few 10's of meters apart can have dramatically different leachate levels. This compartmentalization of landfill cells often show highly variable perched leachate levels. Proper precautions should be taken when placing wells into codisposal (or any) landfill since contents and gas generation may create health and safety risks. Additionally, installation of wells through the landfill base may create conduits through which leachate can migrate to lower geologic strata The installation of wells into landfill wastes should be fully considered since they have a propensity to settle between 5 and 15% of the waste depth. Monitoring points often are observed sticking up through the protective casing as the differential settlement of the riser and screens, and the surface protective casing proceeds over time.

Landfill Contents/Hot Spots

The U.S. EPA has expressed guidance on characterization of a municipal landfill's contents as generally not necessary, because, the most practicable technology containment, does not require such specific information on disposal site contents. Certain data, however, are necessary to evaluate containment alternatives. These include:

- Contour maps of pre-landfill ground surface;

- Fill thickness, lateral extent, and age

- Estimate of landfill settlement rate

- Estimate of rate of landfill gas production and landfill gas composition

- Base grade soil characteristics, including hydraulic conductivity, grain size, Atterberg limits, and erosion rates

- Climatic conditions, including frost depth, and the appropriate storm event creating the potential for significant erosion

- Geologic and hydrogeologic characteristics, particularly the hydraulic conductivity of the layer underlying the landfill; the depth to ground water; thickness of waste below the water table; and ground-water flow through the waste, if applicable

- Physical characteristics of any existing cap including thickness, area, slope stability, evidence of freeze-thaw protection, and soil characteristics as well as its ability to reduce surface gas emissions and odors, prevent oxygen intrusion into the refuse, prevent surface water infiltration, provide erosion control, and improve site aesthetics

- Potential future uses of the site (e.g., residential or recreational use)

More extensive characterization activities and development of remedial alternatives (such as thermal treatment or stabilization) may be appropriate for "hot spots" within disposal areas. Hot spots consist of highly toxic and/or highly mobile material and present a potential principal threat to human health or the environment (see NCP Sec. 300.430 (a)(l)(iii)(C)). In the majority of cases CERLA hot spots consist of drummed wastes disposed in pits excavated in municipal waste. The search for these drums often constitutes an area of major disagreements between agency and PRP groups. Unless prior knowledge of the disposal sites of intact drums is available, locating actual drums is extremely time consuming and usually a waste of time. Excavation or treatment of hot spots is generally practicable where the waste type or mixture of wastes is in a discrete, accessible location of a landfill. A hot spot should be large enough that its remediation will significantly reduce the threat posed by the overall site, but small enough that it is reasonable to consider removal and/or treatment options. Where a low to moderate volume of toxic/ mobile waste(e.g., 100,0000 cy or less)poses the principal threat to human health and the environment, these removal and treatment options may be sufficiently cost effective.

Hot spots should be characterized if documentation and/or physical evidence exists to indicate their presence and approximate location. Hot spots may be delineated using geophysical techniques if the drums are sufficiently concentrated to discern from the other metals normally present in landfills. To characterize hot spots, samples

should be collected through soil borings or trenches to determine hot spot waste characteristics, including TAL metals, TCL organics, RCRA waste characteristics (e.g.,TCLP), total Btu content. and bulk weight of the material. Treatability or pilot testing may be required to evaluate treatment alternatives.

Landfill Gas

Decomposition of organic mass in a landfill generates primarily methane and carbon dioxide with minor amounts of trace thiols, and occasionally, hydrogen sulfide. Volatile organic compounds are also present in landfill gases, particularly at co-disposal facilities. The composition, quantity, and generation rates of the gases depend on such factors as refuse quantity and composition, refuse placement methods, age of the disposal unit, landfill depth, refuse moisture content, and amount of oxygen present. Data generated during the site characterization of landfill gases should

- Contour drawings and rate of settlement

- Geologic and hydrogeologic characteristics, including permeability, moisture content, geologic strata, pH, depth to bedrock, and depth to groundwater

- Presence of offsite, subsurface migration

- Surface emissions

Ambient air monitoring has generally been found to be a waste of money in landfill assessment programs once covers have been placed over the fill. Off site air quality has a much larger effect on the results of the program as compared to actual emissions from the facility. However, landfill gas characteristics, including composition, moisture content, quantity, temperature, and methane content are important data for evaluation of potential emission rates.

A baseline risk assessment for codisposal facilities is to evaluate whether a site poses risks to human health and the environment that are significant enough to warrant remedial action. Due to the large volume/low toxicity nature of codisposal facilities options for remedial action are generally limited, to a few alternatives the scope of the baseline risk assessment may be limited to :

- Using the conceptual site model and RI generated data, to perform a qualitative risk assessment that identifies contaminants of concern in the affected media, their concentrations, and their hazardous properties which may pose a risk through the routes of exposure.

- Identifying all pathways that are an obvious threat to human health or the environment (see Figure l0-22) by comparing RI-derived contaminant concentration levels to standards that are potential chemical-specific ARARs for the action. These may include: (1) Non-zero MCLGs and MCLs for ground water and leachate and, (2) State air quality standards for landfill gases. When potential ARARs do not exist for a specific contaminant, risk-based chemical concentrations should be used in the risk assessment calculations.

- Where established standards for one or more observed contaminants in a given medium are clearly exceeded, the basis for taking remedial action is warranted (i.e., quantitative assessments that consider all chemicals, their potential additive effects, or additivity of multiple exposure pathways are not necessary to initiate remedial action).

- In cases where clear exceedance of standards does not occur, a more thorough risk assessment will be necessary prior to initiating any remedial action on the facility.

This streamlined approach recommended by the agency may facilitate early action on the most obvious potential landfill problems—ground water and leachate, landfill gas, and the landfill "hot spots'—while analysis continues on other problems such as potentially affected adjacent wetlands and stream sediments. Remediation of obvious problems easily in the project should be factored into any ongoing risk assessment. Early remedial or protective actions also need to be designed for flexibility so that they will be consistent with subsequent actions. The final remedy selected, should in the final analysis address all observed pathways and contaminants of concern (including environmental risks), for the facility to protect human health and the environment.

The Development Of Alternatives

When evaluating remedial actions for these large volume/low toxicity codisposal facilities only a few realistic alternatives end up on the majority of Records of Decisions (ROD's). Figure 10-23 identifies remedial technologies and process options for achieving various remedial action objectives pertaining to codisposal landfill sites. The following points are recommended by the U.S. EPA (1991) in order to streamline the development of remedial action alternatives:

- The most practicable remedial alternative for landfills is

generally containment. Figure 10-24 is a simplified decision tree for identifying the appropriate type of cap.

- Treatment of soils and wastes may be practicable for hot spots. Consolidation of hot spot materials under a landfill cap is a potential alternative in cases when treatment is not practicable or necessary.

- Extraction and treatment of contaminated ground water and leachate may be required to control offsite migration. One may expect these leachate collection and treatment actions may be required for significant periods of time may be necessary for an indefinite amount of time because of continued contaminant loadings from the landfill.

U.S. EPA guidance recommends the construction of an active landfill gas collection and treatment system adjacent under the following conditions: (l) when adjacent existing or planned structures may be adversely affected through gas migration that may cause either explosion or inhalation hazards, (2) when final use of the site includes allowing public access, or (3) when the landfill produces excessive odors. Most landfills will require at least a passive gas collection (i.e., venting) system to prevent buildup of pressure below the cap and to prevent damage to the vegetative cover.

10.9.2 High Toxicity-Low Volume sites; DNAPL's

The presence of Dense Non-Aqueous Liquids (DNAPLs) in soils and aquifers can control the ultimate success or failure of remediation at a hazardous waste site. Because of the complex nature of the DNAPL transport and fate, however, DNAPL may be undetected by direct methods, leading to poor site assessments and inadequate remedial designs. Sites affected by DNAPL may require a different "paradigm' or conceptual framework, to develop effective characterization and remedial actions. To help site personnel determine if DNAPL-based characterization strategies should be employed at a particular facility, a qualitative method for estimating the potential for DNAPL occurrence was developed. Two sources of existing site information are typically used:

- Historical Site Use Information; and,

- Site Characterization Data.

By using available data, investigative staff can use a flowchart system and a classification matrix for estimating the potential for DNAPL occurrence at a site (USEPA , 1990). Table 10-7 provides example data needs for a

DNAPL contaminated site. If the potential for DNAPL occurrence is low based on site criteria as described in Table 10-8, then conventional site assessment as expressed in previous sections can be used for the evaluation. Remedial actions consistent with near surface soil removal or neutral density leachates may be sufficient for the facility. If the potential for DNAPL is moderate or high, however, a different conceptual approach may be required to account for technical issues associated with DNAPL in the subsurface.

Background

DNAPLs are separate-phase hydrocarbon liquids that are denser than water, for example, chlorinated solvents, wood preservative wastes, coal tar wastes, and pesticides. For many years standard operating practice in a variety of industries resulted in the release of large quantities of DNAPL to the subsurface through tank and pipe leaks or through direct disposal in ponds, pits and surface discharge. Most DNAPLs undergo only limited degradation in the subsurface, and slowly release through

Technologies Frequently Implemented for Remedial Action at CERCLA Municipal Landfills

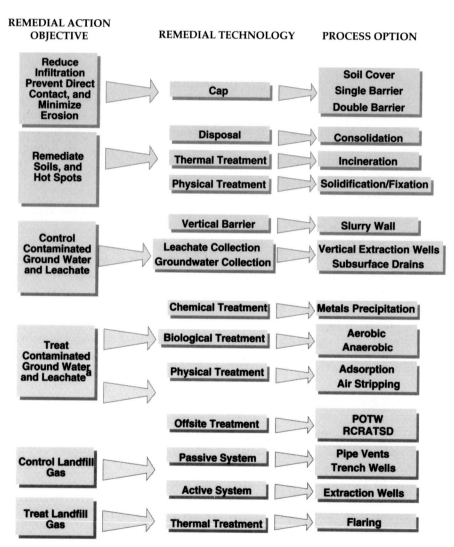

a Other treatment technologies may be appropriate

Source: EPA Directive 9355.3-11FS

Figure 10-23 Remedial Technologies Used at Codisposal Sites

dissolution soluble organic constituents to ground water for long time periods. Dissolution may continue for hundreds of years even with a moderate DNAPL release, before all the DNAPL is dissipated and concentration of soluble organics in ground water return to background levels.

DNAPL exists in two state in the soil/aquifer matrix; free phase DNAPL and residual DNAPL. When release at the surface, free phase DNAPL moves downward through the soil matrix under the force of gravity or laterally along the surface of sloping fine-grained stratigraphic units. As the free phase DNAPL moves, blobs or ganglia are trapped in pores and/or fractures by capillary forces. The amount of the trapped DNAPL, known as residual saturation, is a function of the physical properties of the DNAPL and the

hydrogeologic characteristics of the aquifer medium and ranges up to 50% of total pore volume.

Experience with site contaminated with DNAPL have shown that, once in the subsurface, it is difficult or impossible to recover all the trapped residual DNAPL. The conventional aquifer remediation approach often used with neutral density leachates and floating product spills (LNAPL's), ground-water pump-and-treat, usually removes only a small fraction of residual hydrocarbon. Many DNAPL removal technologies are currently being tested, however, to date there have been no field demonstrations where all of the released DNAPL has been successfully recovered from the subsurface. Even locating DNAPL pools and residuals present in subsurface

LANDFILL COVER SELECTION GUIDE

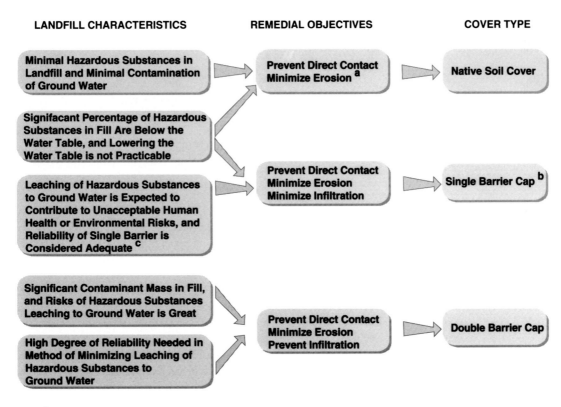

a Primary objective is to prevent direct contact, although the soil cover ccan be designed to reduce infiltration.

b Single barrier caps may include additional layers that provide protection to that barrier.

c Examples include situations where infiltration is not the primary concern and may include sites containing a small volume of contaminant mass, regions with low annual precipitation, or sites where ground water is not being used as a source of drinking water.

Source: U.S. EPA

Figure 10-24 Alternative Cap Designs for Co-disposal Sites

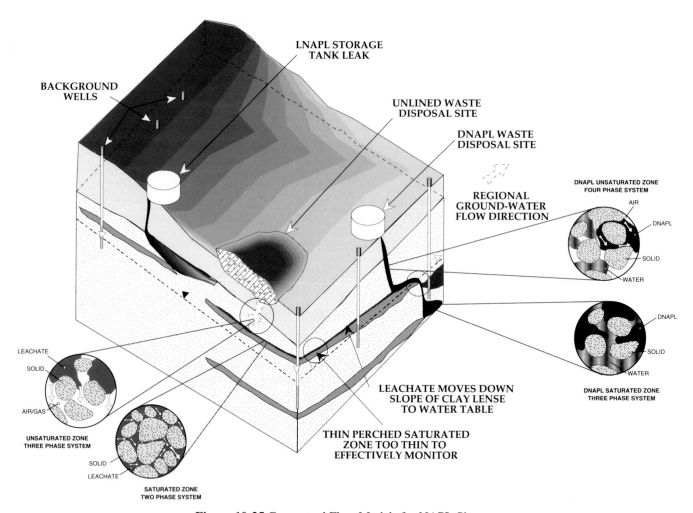

Figure 10-25 Conceptual Flow Models for NAPL Sites

contaminated soils can be very difficult in field investigations. The DNAPL that remains trapped in the soil/aquifer matrix acts as a continuing source of dissolved organics to ground water, preventing the restoration of DNAPL-affected aquifers to drinking water standards for many, perhaps even hundreds of years.

Conceptual Approaches

The major factors controlling DNAPL migration in the subsurface include: 1) the volume of DNAPL released; 2) the area of infiltration; 3) the duration of release; 4) DNAPL properties such as density, viscosity, etc.; and 5) properties of the aquifer media such as pore size, presence of aquitards, and fractures etc. To describe the general transport and fate properties of DNAPL in the subsurface, a series of conceptual models (Waterloo Research Center 1989, USEPA 1991) are presented in the following figures:

Case 1: DNAPL Release to Unsaturated Zone Only:— After release on the surface, DNAPL moves vertically downward under the force of gravity and soil capillarity. Because only a small amount of DNAPL was released, all of the mobile DNAPL is eventually trapped in pores and fractures in the unsaturated zone. Infiltration through the DNAPL zone dissolves some of the soluble organic constituents in the DNAPL, carrying organics to the water table and forming a dissolved organic plume in the aquifer.

Case 2: DNAPL Release to Unsaturated and Saturated Zones:—If enough DNAPL is released at the surface, it can migrate all the way through the unsaturated zone and reach a water-bearing unit. Because the specific gravity of DNAPL is greater than water, it continues downward until the mobile DNAPL is exhausted and is trapped as a

TABLE 10-7   Data needs for DNAPL Sites

Data Pertinent to Evaluating Remedial Actions for Contaminated Ground Water Gaps Identified in the Evaluation of Ground Water Extraction Remedies Phase I and II

Hydrogeologic Information

1. Number of aquifers and degree of hydraulic connection between them;
2. Location and continuity of lower permeability zones;
3. Hydraulic conductivity of each aquifer;
4. Yield of each aquifer;
5. Stratigraphy — continuous coring at select wells;
   For NAPL-contaminated sites, additional information can be important (presented in order of increasing difficulty to obtain):
   • Grain size analysis to obtain rough predictions of NAPL holding and transmitting capacity;
   • Porosity;
   • The relationship between the degree of saturations, the capillary pressure, and the relative permeability for the wetting and nonwetting fluids;
   • Residual saturation for NAPL.
6. Potentiometric gradients.

Contaminant Characteristics

1. Location and definition of the source(s) both horizontally and vertically within the defined site and verification that there are not other major sources outside the defined site;
2. Type of waste and form — likely presence as NAPL;
3. Solubilities of compounds;
4. Miscibility of various contaminants — potential presence as multi-component NAPL;
5. Concentrations of contaminants in soil and ground water — vertical and horizontal extent of contaminant plume;
6. Vertical variations in contaminant concentrations;
7. Sorption of contaminants to soil, evaluated through (in decreasing level of accuracy) column studies, batch analysis of sorbed concentrations, or parition coefficient/ organic carbon content estimate;
8. Presence of NAPLs in ground water samples using interface probe or clear bailer.

Remediation Performance Evaluation

1. Piezometers on downgradient side of extractions wells;
2. Changes in vertical and horizontal extent of contaminant plume as a result of remediation;
3. Hydraulic effects of extraction — water level data;
4. Mass of contaminants removed by individual extractions wells.

Source: U. S EPA, 1989.

TABLE 10-8   Site Data Can Indicate if DNAPL'S are Present

DNAPL in monitoring wells:

Methods for measuring DNAPL thickness or for collecting DNAPL samples include:

1) NAPL/water interface probes that signal a change in the specific conductivity of the borehole fluid,

2) weighted cotton string lowered to the bottom of the well,

3) peristaltic pumps,

4) transparent bottom-loading bailers, and

5) mechanical discrete-depth samplers.

In general, the depth of DNAPL accumulation does not provide quantitative information regarding the amount of DNAPL present.

DNAPL in soil cores or cuttings:

Visual examination of cores or cutting may not be effective for confirming the presence of DNAPL except in cases of gross DNAPL contamination. Methods for enhancing visual inspection of soil samples for DNAPL include:

1) shaking soil samples in a jar with water to separate the DNAPL from the soil and

2) a pint filter test, where soil is placed in a filter funnel, water is added, and the filter is examined for separate phases.

Source: U. S EPA, 1989.

residual hydrocarbon in the porous media. Ground water flowing past the trapped residual DNAPL dissolves soluble components of the DNAPL, forming a dissolved plume downgradient of the DNAPL zone. As with Case 1, water infiltrating down from the source zone also carries dissolved constituents to the aquifer and contributes further to the dissolved plume.

Case 3: DNAPL Pools and Effect of Low-Hydraulic Conductivity Units:—Mobile DNAPL will continue

CONCEPTUAL MODELS FOR DNAPL CONTAMINATION

The major factors controlling DNAPL migration in the subsurface include: 1) the volume of DNAPL released; 2) the area of infiltration; 3) the duration of release; 4) DNAPL properties such as density, viscosity, etc.; and 5) properties of the aquifer media such as pore size, presence of aquitards, and fractures etc. To describe the general transport and fate properties of DNAPL in the subsurface, a series of conceptual models are presented in the following figures:

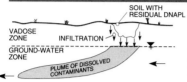

Case 1: DNAPL Release to Unsaturated Zone Only

After release on the surface, DNAPL moves vertically downward under the force of gravity and soil capillarity. Because only a small amount of DNAPL was released, all of the mobile DNAPL is eventually trapped in pores and fractures in the unsaturated zone. Infiltration through the DNAPL zone dissolves some of the soluble organic constituents in the DNAPL, carrying organics to the water table and forming a dissolved organic plume in the aquifer.

Case 2: DNAPL Release to Unsaturated and Saturated Zones

If enough DNAPL is released at the surface, it can migrate all the way through the unsaturated zone and reach a water-bearing unit. Because the specific gravity of DNAPL is greater than water, it continues downward until the mobile DNAPL is exhausted and is trapped as a residual hydrocarbon in the porous media. Groundwater flowing past the trapped residual DNAPL dissolves soluble components of the DNAPL, forming a dissolved plume downgradient of the DNAPL zone. As with Case 1, water infiltrating down from the source zone also carries dissolved constituents to the aquifer and contributes further to the dissolved plume.

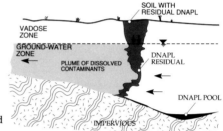

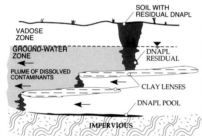

Case 3: DNAPL Pools and Effect of Low-Permeability Units

Mobile DNAPL will continue vertical migration until it is trapped as a residual hydrocarbon (Case 1 and Case 2) or until low-permeability stratigraphic units are encountered which create DNAPL "pools" in the soil/aquifer matrix. In this figure, a perched DNAPL pool fills up and then spills over the lip of the low-permeability stratigraphic unit. The spill-over point (or points) can be come distance away from the original source, greatly complicating the process of tracking the DNAPL migration.

Case 4: Composite Site

In this case study, mobile DNAPL migrates vertically downward through the unsaturated zone and the first saturated zone producing a dissolved constituent plume in the upper aquifer. Although a DNAPL pool is formed on the fractured clay or rock unit, the fractures are large enough to permit vertical migration downward to the deeper aquifer (see Case 5 below). DNAPL pools in a topographic low in the underlying impermeable unit and a second dissolved constituent plume is formed.

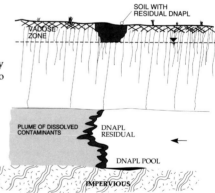

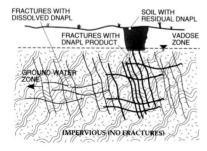

Case 5: Fractured Rock or Fractured Clay System

DNAPL introduced into a fractured rock or fractured clay system follows a complex pathway based on the distribution of fractures in the original matrix. The number, density, size, and direction of the fractures usually cannot be determined due to the extreme heterogeneity of a fractured system and the lack of economical aquifer characterization technologies. Relatively small volumes of DNAPL can penetrate deeply in fractured systems due to the low retention capacity of the fractures and the ability of DNAPL to migrate through very small (-10 micron) fractures. Many clay units, once considered to be relatively impermeable to DNAPL migration, often act as a fractured media with preferential pathways for vertical and horizontal DNAPL migration.

Figure 10-26 Conceptual Models for DNAPL Sites

vertical migration until it is trapped as a residual hydrocarbon (Case 1 and Case 2) or until lower-hydraulic conductivity stratigraphic units are contacted which perch DNAPL "pools" in the soil/aquifer matrix above the layer. In this figure, a perched DNAPL pool fills up and then spills over the lip of the low-hydraulic conductivity stratigraphic unit. The spill-over point (or points) can be some distance away from the original source, thus greatly complicating the investigative process of tracking or containing the DNAPL migration.

Case 4: Composite Site:—In this case study, mobile DNAPL migrates vertically downward through the unsaturated zone and the first saturated zone producing a dissolved constituent plume in the upper aquifer. Although a DNAPL pool is formed on the fractured clay or rock unit, the fractures are large enough to permit vertical migration downward to the deeper aquifer (see

Case 5 below). DNAPL pools in a topographic low in the underlying low hydraulic conductivity unit and a second dissolved constituent plume is formed.

Case 5: Fractured Rock or Fractured Clay System:— DNAPL introduced into a fractured rock or fractured clay system follows a complex pathway based on the distribution of fractures in the original matrix. The number, density, size, and direction of the fractures usually cannot be determined due to the extreme heterogeneity of a fractured system and the lack of economical aquifer characterization technologies. Relatively small volumes of DNAPL can penetrate deeply in fractured systems due to the low retention capacity of the fractures and the ability of DNAPL to migrate through very small (-10 micron) fractures. Clay or glacial till units, once considered to be relatively impermeable to DNAPL migration, may represent

---

TABLE 10-9 DNAPL Organic Parameters

Halogenated Volatiles

Chlorobenzene
1.2 Dichloropropane
1.1 Dichloroethane
1.1 Dichloroethylene
1.2 Dichloroethane
Trans 1.2 Dichloroethylene
Cls 1.2 Dichloroethylene
1.1.1 Trichloroethane
Methylene Chloride
1.1,2 Tricholoroethane
Tricholorethylene
Chloroform
Carbon Tetrachloride
1.1,2.2 Tetrachloroethane
Tetrachloroethylene
Ethylene Dibromide

Halogenated Semi-Volatile

1,4 Dichlorobenzene
1.2 Dichlorobenzene
Aroclor 1242, 1260,
 1254
Chlordane
Dieldrin
2,3,4,6 Tetrachlorophenol
Pentachlorophenol

DNAPL-Related Chemicals

Non-Halogenated
Semi-Volatiles
2 Methyl Napthalene
o-Cresol
p-Cresol
2,4 Dimethylphenol
m-Cresol
Phanol
Napthalene
Benzo(a)Anthra- cene
Flourene
Acenaphthene
Anthracene
Dibanzo(s,h)Anthra
 cena
Flouranthene
Pyrene
Chrysene
2,4 Dinitrophenol

Miscellaneous
Coal Tar
Creosote

preferential pathways for vertical and horizontal DNAPL migration, if these unit contain post-sedimentation stress induced, open fractures.

Site Characterization for DNAPL

Characterization of the subsurface environment at hazardous waste sites containing DNAPL is complex and will likely be expensive. Specific details associated with the volume and timing of the DNAPL release are usually poor or are not available and subsurface heterogeneity is responsible for the complicated and potentially unpredictable migration pathway of subsurface DNAPL transport. The characterization of the subsurface lithology represents the primary goal for such investigations. Random or thoughtless drilling patterns, often employed in such investigations, can significantly spread DNAPL's through the geologic environment. For this reason planning of field drilling and sample collection must be always be alert to the potential for DNAPL contamination, and should provide the flexibility to alter to drilling and sampling design if DNAPL's are observed in the investigation.

Site characterization typically involves a significant investment in ground-water water quality analyses. Although typical chemical analysis of ground water can provide useful information on the distribution of the soluble components of the DNAPL, the presence of other phases of the DNAPL may go unrecognized. The investigation must, therefore, be more detailed to obtain information concerning the phase distribution of the

DNAPL at a site. Site characterization may require analyses on all four phases (aqueous, gaseous, solid, immiscible) to yield the appropriate information (refer to Table 10-9). In brief, data collected in the field program must be compiled, evaluated and used to help identify: where the contaminant is presently located; where it has been; what phases it occurs in; and what potential direction the mobile phases may be moving.

Ground Water

Ground water analyses for organic compounds, in conjunction with ground water flow direction data, has repeatedly been used to: delineate the extent of ground water contamination from DNAPL affected sites; determine the direction of plume migration; and to identify probable DNAPL source area(s). While this approach has been used successfully to characterize the distribution of contaminants in the subsurface, there are limitations. For example, since DNAPL and ground water may flow in different directions, as indicated in Figures 10-26 (case2) and 10-26 (case 3), ground water analyses alone will probably not identify the direction of DNAPL migration.

Industries that can be expected to have high probability of historical DNAPL release are listed in Table 10-11.

Ground water analyses may be useful to identify probable DNAPL source areas if combined with good site historic information, but, estimating the volume of DNAPL in the subsurface is limited using this approach. Soluble phase components of DNAPL are rarely found in excess of

TABLE 10-10 Industrial processes or waste disposal practice with high probability of historical DNAPL release

• Wood preservation (creosote)	• Metal machining
• Old coal gas plants (mid-1800s to mid-1900s)	• Paint removing/stripping
• Electronics manufacturing	• Storage of solvents in underground storage tanks
• Solvent production	
• Pesticide manufacturing	• Storage of drummed solvents in uncontained storage areas
• Airplane maintenance	
• Commercial dry cleaning	
• Instrument manufacturing	• Solvent loading and unloading
• Transformer oil production	
• Transformer reprocessing	• Disposal of mixed chemical wastes in landfills
• Steel industry coking operations (coal tar)	• Treatment of mixed chemical wastes in landfills
• Pipeline compressorstations	
• Metal cleaning/degreasing	• Treatment of mixed chemical wastes in lagoons or ponds
• Tool-and-die operations	

Table 10-11 Important Site / DNAPL Characteristics To Remember During a Site Investigation

• The risk of spreading contaminants increases with the proximity to a potential DNAPL zone. Special precautions should be taken to ensure that drilling does not create pathways for continued vertical migration of free phase DNAPLs. In DNAPL zones, drilling should be suspended when a low-permeability unit or DNAPL is first encountered. Wells should be installed with short screens (<= 5 feet). If required, deeper drilling through known DNAPL zones should be conducted only by using double or triple cased wells to prevent downward migration of DNAPL . As some DNAPL can penetrate fractures as narrow as 10 microns, special care must be taken during all grouting, cementing and well sealing activities conducted in DNAPL zones.

• The subsurface DNAPL distribution may be difficult to delineate accurately at some sites. DNAPL migrates preferentially through selected pathways (fractures, sand layers, etc.) and is affected by small-scale changes in the stratigraphy of an aquifer. Therefore, the ultimate path taken by DNAPL can be very difficult to characterize and predict.

• In most cases, fine-grained aquitards (such as clay or silt units) should be assumed to permit downward migration of DNAPL through fractures unless proven otherwise in the field.

• Drilling in areas known to be DNAPL free should be performed before drilling DNAPL Ones in order to form a reliable conceptual model of site hydrogeology, stratigraphy and potential DNAPL pathways. In areas where it is difficult to form a reliable conceptual model, an "outside-in" strategy may be appropriate: drilling in DNAPL zones is avoided or minimized in favor of delineating the outside plume. Many fractured rock settings may require this approach to avoid opening further pathways for DNAPL migration during site assessment.

• Because the potential risk for exacerbating groundwater contamination problems during drilling through DNAPL zones, the precautions described for Category I should be considered during site assessment. Further work should focus on determining if the site is a "DNAPL site".

• DNAPL is not likely to be a problem during site characterization, and special DNAPL precautions are probably not needed. Floating free-phase organics (LNAPLs), sorption, and other factors can complicate site assessment and remediation activities, however.

Source: U. S. EPA, 1991

10% of the solubility even when organic liquids are known or suspected to be present. The concentration of soluble DNAPL components in the ground water is not only a function of the amount of DNAPL present, but also the chemical and physical characteristics of the DNAPL, the contact area and time between the ground water and DNAPL, and numerous transport and fate parameters (retardation, biodegradation, dispersion, etc.). One technique has been developed using chemical ratios in the ground water as a means of source identification and contaminant fate prediction (Hoag and Marley 1989).

Soil/Aquifer Material

Exploratory Borings

Physical and chemical analyses of soil and aquifer material (drill cuttings, cores) from exploratory borings will provide useful information in the delineation of the horizontal and vertical mass distribution of DNAPL. While simple visual examination for physical presence or absence of contamination might seem like a worthwhile technique, it can be deceiving and does nothing to sort out the various liquid phases and their relationship to each other (Villaume 1985). A quantitative approach is necessary to determine DNAPL distribution. Figure 10-27 illustrates the procedures for the evaluation of DNAPL's in soils.

Drill cuttings or core material brought to the surface from exploratory borings can be screened initially to help delineate the depth at which volatile components from the various phases of the hydrocarbon exists. The organic

METHOD FOR DETERMINING RESIDUAL NAPL BASED ON ORGANIC CHEMICAL CONCENTRATIONS IN SOIL SAMPLES

CALCULATE S_i^e
$Si = X_i S_i$

S_i^e = The actual aqueous dissolved phase concentration in mg/l

X_i = The mole fraction of component i in the DNAPL mixture (obtained from a laboratory analysis of a DNAPL sample)

S_i = The pure-phase solubility of compound i in mg/l (from literature sources)

DETERMINE K_{oc}
The Organic Carbon-water Partition Coefficient

The organic carbon-water partition coefficent is obtained from:
- Literature sources
- From emperical relationships based on K_{oc}, the octanol-water partition coefficient found in the literature (Verschueren, 1983)

DETERMINE f_{oc}
The Fraction of Organic Carbon on Soil

- The fraction of organic carbon on the soil is determined from a laboratory analysis of clean site soil.
- Values for f_{oc} typically range from 0.03 to 0.00017 mg/mg.
- Laboratory values reported in percent should be converted to mg/mg.

DETERMINE OR ESTIMATE
ρb Dry Bulk Density and φw Water Filled Porosity of Soil

- ρb the dry bulk density is obtained or estimated from laboratory soil tests.
- Typical values range from 1.8 to 2.1.
- φw water filled porosity of soil is determined or estimated from laboratory tests.

DETERMINE Kd
The Partition Coefficient (Distribution Coefficient)
$Kd = K_{oc} \cdot F_{oc}$

The partition coefficient between the pore water (ground water0 and the soil solids are determined using the equation:

$$Kd = K_{oc} \cdot F_{oc}$$

CALCULATE C_w
$$C_W = \frac{(C_t \cdot \rho b)}{(Kd \cdot \rho b + \varphi w)}$$

Using Ct, the measured soil chemical concentration in mg/kg, calculate Cw using the equation:

$$C_W = \frac{(C_t \cdot \rho b)}{(Kd \cdot \rho b + \varphi w)}$$

Cw = calculated pore water concentration in mg/l assuming no DNAPL is present

COMPARE C_w and S_i^e

$Cw > S_i^e$ indicates possible presence of DNAPL

$Cw < S_i^e$ indicates possible absence of DNAPL

Figure 10-27 Measuring DNAPL Levels in Soils

vapor analyzer and the HNU are small portable instruments that can detect certain volatile compounds in the air. These methods are used to initially screen subsurface materials for volatile components of DNAPL. Identification of individual compounds and their concentrations may be confirmed by other, more precise, analyses.

Analysis of the soil or aquifer material by more accurate means, such as gas chromatography or high pressure liquid chromatography, will take longer but will provide more specific information on a larger group of organic compounds, i.e. volatile/non-volatile, and on specific compounds. This information is necessary to help fix the horizontal and vertical mass distribution of the contaminant and to help delineate the phase distribution. These analyses do not distinguish between soluble, sorbed or free-phase hydrocarbon, however; a low relative concentration indicates that the contaminant may mainly be present in the gaseous or aqueous phases; and a high relative concentration indicates the presence of sorbed contaminant or free phase liquid either as continuous-phase or residual saturation. A more rigorous set of analyses is required to distinguish between the various phases.

Additional tests to identify the presence of NAPL in soil or aquifer core sample are currently undeveloped and research in this area is warranted. Squeezing and immiscible displacement techniques have been used to obtain the pore water from cores (Patternson 1978). Other methods of phase separation involving vacuum or centrifugation may also be developed for this use. These core analysis techniques have potential to provide valuable field data to characterize NAPL distribution .

Cone Penetrometer

The cone penetrometer (ASTM D3441-86)(69) has been used since the 1940's (see Chapter 5) to supply data on the engineering properties of soils. Cone penetrometer in various forms has found use in site evaluations for remedial actions. The basic procedure is by measurement of the resistance to the cone penetrometer as it is driven vertically into the subsurface. The resistance is interpreted as a measure of pore pressure, and thus provides information on the relative stratigraphic nature of the subsurface.

Features of the cone penetrometer include: a continuous reading of the stratigraphy/ permeability; in-situ measurement; immediate results are available; time requirements are minimal; vertical accuracy of stratigraphic composition is high; ground-water samples can be collected as the cone is advanced into the subsurface.

Extreme care must be used that low hydraulic conductivity units are not punctured by the cone penetrometer to allow DNAPL to drain into lower aquifers. If geologic conditions established in preliminary site work show potential perching units, one should reconsider the use of this method. In all cases holes left from the cone penetration should be quickly grouted to seal this potential cross contamination pathway.

DNAPL

Well Level Measurements

In an effort to delineate the horizontal and vertical extent of the DNAPL at a spill site, it is important to determine the elevation of DNAPL in the subsurface. Monitoring DNAPL elevation over time will indicate the mobility of the DNAPL. There are several methods that can be used to determine the presence of DNAPL in a monitoring well. One method relies on the difference in electrical conductivity between the DNAPL and water. A conductivity or resistivity sensor is lowered into the well and a profile is measured. The interface of the DNAPL is accurately determined when the difference in conductivity is detected between the two fluids. This instrument may also be used to delineate LNAPL. A transparent, bottom-loading bailer can also be used to measure the thickness (and to sample) of DNAPL in a well (Mercer and Cohen 1990). The transparent bailer is raised to the surface and the thickness of the DNAPL is made by visual measurement.

Several laboratory and field studies have been performed which investigate the anomaly between the actual and measured LNAPL levels in ground-waterwells (Hampton and Miller 1988, Hinchee et. al., 1990, Kemblowski and Chiang 1988, 1990). The anomaly between actual and measured NAPL thickness in the subsurface is also applicable to DNAPL, but for different reasons.

The location of the screening interval is the key to understanding both scenarios. First, if the well screen interval is situated entirely in the DNAPL layer, and the hydrostatic head (water) in the well is reduced by pumping or bailing, then to maintain hydrostatic equilibrium, the DNAPL will rise in the well (Mercer and Cohen 1990, Schmidtke and Rovers 1987, Villaume 1985) (refer to Figure 10-28). Secondly, if the well screen extends into the barrier layer, the DNAPL measured thickness will exceed that in the formation by the length of the well below the barrier surface (Mercer and Cohen 1990) (refer to Figure 10-28). Both of these scenarios will result in a greater DNAPL thickness in the well and thus a false indication (overestimate) of the actual DNAPL thickness will result.

One of the main purposes of the monitoring well in a DNAPL investigation is to provide information on the thickness of the DNAPL in the aquifer. Therefore, construction of the well screen should intercept the ground water-DNAPL interface and the lower end of the screen should be placed as close as possible to the impermeable stratigraphic unit.

DNAPL Sampling

Sampling of DNAPL from a well is necessary to perform chemical and physical analyses on the sample. Two of the most common methods used to retrieve a DNAPL sample from a monitoring well are the peristaltic pump and the bailer. A peristaltic pump can be used to collect a sample if the DNAPL is not beyond the effective reach of the pump, which is typically less than 25 ft. The best method to sample DNAPL is to use a double check valve bailer. The key to sample collection is controlled, slow lowering (and raising) of the bailer to the bottom of the well (USEPA 1986c). The dense phase should be collected prior to purging activities.

Soil-Gas Surveys

A soil-gas survey refers to the analysis of the soil air phase as a means to delineate underground contamination from volatile organic chemicals and several techniques have been developed (Marrin and Kerfoot 1987, Thompson and Marrin 1987). This investigative tool is mainly used as a preliminary screening procedure to delineate the area extent of volatile organic compounds in the soil and ground water. This method is quick, less expensive than drilling wells and can provide greater plume resolution (Marrin and Thompson 1987).

Data from a soil-gas survey can be a valuable aid in the development of a more detailed subsurface investigation where ground water monitoring wells and exploratory borings are strategically located for further site characterization. There are limitations to soil-gas surveys (Thompson and Marrin 1987) and data interpretation must be performed carefully (Marrin 1988, Silka 1988). Soil-gas investigations have mainly been conducted to identify the location of the organic contaminants in ground water. At the time of this publication, the scientific literature did not contain information specifically applicable to the delineation of DNAPL from soil-gas Survey data. However, it is surmisable that soil-gas surveys can be used to help delineate DNAPL residual saturation in the unsaturated zone or the location of perched DNAPL reservoirs.

Miscellaneous

The vertical migration of DNAPL in the saturated

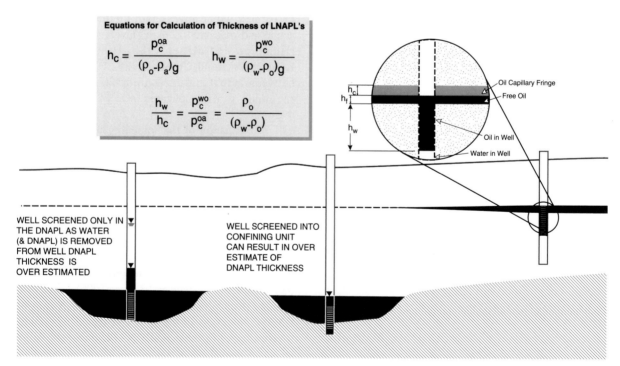

Figure 10-28 Measuring DNAPL Levels in Boreholes

zone will eventually be challenged by a low permeability stratigraphic unit. According to the principles of capillary pressure, the lower permeability unit will exhibit a greater capillary pressure. Displacement of water by DNAPL requires that the hydrostatic force from the mounding DNAPL exceed the capillary force of the low permeability unit. The Hobson formula is used to compute the critical height calculation to overcome the capillary pressure under different pore size conditions (Villaume et. al., 1983).

In an effort to minimize further DNAPL contamination as a result of drilling investigations, precautionary steps should be taken. Penetration of DNAPL reservoirs in the subsurface during drilling activities may offer a conduit for the DNAPL to migrate vertically into previously uncontaminated areas. It is very easy to unknowingly drill through a DNAPL pool and the confining bed it sits on, causing the pool to drain down the hole into a deeper part of the aquifer or into a different aquifer (Mackay and Cherry 1989). Special attention to grouting and sealing details during and after drilling operations will help prevent cross-contamination.

Precautionary efforts should also be considered when a DNAPL reservoir is encountered during drilling operations. The recommended approach is to cease drilling operations and install a well screen over the DNAPL zone and cease further drilling activities in the well. If it is necessary to drill deeper, construction of an adjacent well is recommended. Alternatively, if it is not necessary to screen off that interval, it is recommended to carefully seal off the DNAPL zone prior to drilling deeper. In general term one can follow these guidelines (USEPA 1991) when performing site assessments on DNAPL sites:

• The risk of spreading contaminants increases with the proximity to a potential DNAPL zone. Special precautions should be taken to ensure that drilling does not create pathways for continued vertical migration of free phase DNAPLs. In DNAPL zones, drilling should be suspended when a low-permeability unit or DNAPL is first encountered. Wells should be installed with short screens (<= 5 feet). If required, deeper drilling through known DNAPL zones should be conducted only by using double or triple cased wells to prevent downward migration of DNAPL . As some DNAPL can penetrate fractures as narrow as 10 microns, special care must be taken during all grouting, cementing and well sealing activities conducted in DNAPL zones.

• The subsurface DNAPL distribution may be difficult to delineate accurately at some sites. DNAPL migrates preferentially through selected pathways (fractures, sand layers, etc.) and is affected by small-scale changes in the stratigraphy of an aquifer. Therefore, the ultimate path taken by DNAPL can be very difficult to characterize and predict.

• In most cases, fine-grained aquitards (such as clay or silt units) should be assumed to permit downward migration of DNAPL through fractures unless proven otherwise in the Phase II investigation

Well construction material compatibility with DNAPL should be investigated to minimize downhole material failure. A construction material compatibility review and possible testing will prevent the costly failure of well construction material. The manufacturers of well construction material are likely to have the most extensive compatibility data and information available.

10.10 RISK ASSESSMENT FOR REMEDIAL PROJECTS

Risk assessment is an integral part of the environmental cleanup process. When dealing with a contaminated aquifer or a volume of contaminated soil, risk management decisions are made regarding whether the water and soil genuinely require remediation If the hazard to public health is negligible, a costly program of remediation may he judged unnecessary. If the public health is significantly endangered at a site, risk assessment must again be employed to determine the appropriate Levels or cleanup required to bring risks within acceptable Levels.

Hazardous substance health risks are a product of two factors: 1) the inherent toxicity of the contaminant, and 2) the degree of exposure experienced by the public. Chemicals may cause carcinogenic responses, non carcinogenic responses (such as weight loss, nervous system disorders, sterility, and immune system suppression), both, or have no adverse effect at all. Toxicologic standards are established nationally by the EPA Environmental Criteria and Assessment Office (ECAO). Because these standards have been undergoing a rapid development it is necessary to verify that their application in risk assessments employ the latest values from ECAO. This frequently necessitates on-line database searches of the EPA Integrated Risk Information System (IRIS) to ensure that the toxicologic standards being applied are the most current standards available.

The second component of environmental risk (exposure) must be validated on a site-by-site basis. In general, the more a person is exposed to a contaminant, the greater the risk. The risk assessor must consider all existing and potential pathways which may result in exposure. These pathways include oral intake (drinking water, ingesting soil, eating food which has picked up

INTEGRATION OF RISK ASSESSMENT & RI/FS PROJECTS FOR CO-DISPOSAL SITES

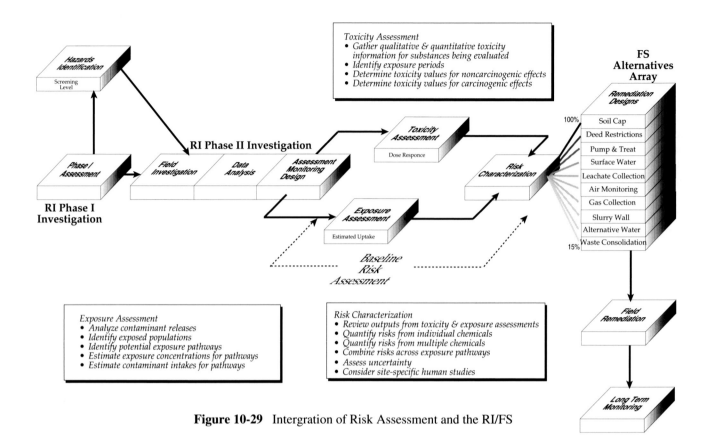

Figure 10-29   Intergration of Risk Assessment and the RI/FS

contaminants), respiratory intake (inhalation of contaminated dust or vapors), and dermal intake (contaminants absorbed through the skin).

To the extent possible, exposure must be quantified, and the population under consideration must he characterized as completely as possible. This requires answers to such questions as: Is there a residential community affected by this contamination? What future activities are going to take place in this area? Does the aquifer serve as a domestic water supply? It is also important to quantify elements of human behavior and physiology as part of the exposure calculations. This involves estimates of body weights, breathing rates, demographics, and occupational patterns.

Other factors which must be considered under exposure involve contaminant concentrations and transport. In order for a site investigation to produce useful results, sampling programs must be designed with risk assessment requirements in mind. This requires evaluation of background concentrations of contaminants, use of

appropriate sample sizes for statistical significance, and delineation of contamination plume boundaries. An understanding of contaminant migration frequently involves use of air dispersion, ground water transport, and geochemical models.

Once the exposure has been characterized and the toxicological properties of the contamination are known, the final phase of risk assessment requires the quantification of site risks (Figure 10-27). This may require describing the risks involved in short-term (acute and subchronic) exposures or the risk from a person's lifetime (chronic) exposure. Risks may also be calculated using approximations of worst-case scenarios (with the highest exposures that are reasonably possible) or using more typical, average scenarios. EPA guidance has recently mandated that risk calculations should be computed using multiple scenarios which express a range of possible situations. If a site is thought to pose a baseline risk that is unacceptably high (strictly a policy decision), then the site must be considered as a candidate for

$$I = C \cdot \frac{CR \cdot EFD}{BW} \cdot \frac{1}{AT}$$

I = intake; the amount of chemical at the exchange boundary (mg/kg body weight-day)

Chemical-Related Variable

C = chemical concentration; the average concentration contacted
over the exposure period (e. g., mg/liter water)

Variables that Describe the Exposed Population

CR = contact rate; the amount of contaminated medium contacted per unit time or event (e g.,
liters/day)
EFD = exposure frequency and duration; describes how long and how often exposure occurs; often
calculated using two terms (EF and ED):
EF = exposure frequency (days/year)
ED = exposure duration (years)
BW = body weight; the average body weight over the exposure period (kg)

Assessment -Determined Variable

AT = averaging time; period over which exposure is averaged (days)

Risk = Intake x Toxicologic Slope Factor

Source: U.S. EPA EPA/540/1-89/002

Figure 10-30   Evaluation of Risk

remediation.

Through the application of risk assessment, the remedial engineer is able to identify the contaminants which are the major contributors to an environmental problem and dismiss the ones which are trivial. Risk assessment is also an indispensable component for an evaluation of the cost/benefit ratios which will result from different remedial alternatives. Finally, risk assessment is now frequently employed to actually establish what the cleanup goals will be. The equations in Figure 10-28 may be rearranged to solve for soil or water concentrations in order to calculate the remediation goals required to achieve acceptable risks.

10.10.1 IDENTIFICATION OF CHEMICALS OF POTENTIAL CONCERN

The purpose of selecting chemicals of potential concern for the risk assessment is to identify those chemicals present at the site that will be the focus of the risk assessment. Prior to selecting chemicals of potential concern, the first step in a risk assessment is to summarize all available data according to the following procedures which are in general accordance with USEPA (1989a, 1990):

• RI data collected and analyzed according to U.S. EPA's Contract Laboratory Program (CLP) procedures are used in the selection of chemicals of concern for this assessment. Most of the RI data must be fully validated for useability within the project.

• In order to select potential chemicals of concern, the RI data is typically divided into groups which describe

TABLE 10-12 Exposure Parameters for Ingestion of Ground Water and Inhalation of Volatiles From Showering by Nearby Residents

Reasonable Average Parameter Case	Average Case	RME Case
Exposure Frequency (days/year) (a)	350	350
Exposure Duration (years) (b)	9	30
Shower Exposure Time (hours/day) (c)	0.28	0.28
Water Ingestion Rate (liters/day) (d)	1.4	2
Body Weight (kg) (f)	70	70
Lifetime (years) (g)	70	70

(a) Residents are assumed to shower with the same frequencies used for ingestion. RME case value is a current default value used by USEPA (1991). Average case value is a standard default value provided by USEPA (1991).

(b) Based on the national averange and upperbound time at one residence (USEPA 1991, 1989a).

(c) Value based on 90th percentile shower time (12 minutes) provided by USEPA (1989a) with 5 minutes included to account for time spent in the shower room after water is turned off.

(d) Values are standard ingestion rates provided by USEPA (1991, 1989a).

(e) Value provided by USEPA (1991) for daily, indoor residential activities.

(f) Standard default value provided by USEPA (1991, 1989a)

(g) Based on USEPA (1991, 1989a) standard assumption for lifetime. This value is used in calculating exposures for potential carcinogens.

environmental conditions relevant to the Baseline RA. For example, a group of background data are used to determine if concentrations of chemicals detected on or downgradient of the site are at naturally occurring levels. Grouping data also helps in determining exposure point concentrations for target populations and in assisting in the identification of areas potentially requiring remediation. These groups are typically described in detail in the Baseline RA by environmental medium.

• Concentration data from multiple samples from the same sample location taken at different times are averaged. Also duplicate samples from the same location and date are averaged prior to averaging among sampling events. If a chemical is detected in one or more sampling rounds at a particular sample location and not in others, the average concentration for the sample location is calculated by averaging the detected concentration(s) with one-half of the contract required quantification limit (CRQL) of the non-detected concentration(s).

• Due to the fact that there are varying chemical-and sample specific detection limits, even within a single medium, samples in which a chemical was not detected are typically compared to the maximum detected concentration for that chemical to determine if the non-detects are included in calculating the mean concentrations. If the detection limit for a non-detected sample is two or more times higher than the maximum detected concentration in that medium, the sample is not included in the calculation of the average for that chemical. This is done to prevent the average from being artificially biased upwards by high detection limits. Contract Laboratory Program (CLP) detection limits are used for non-detects where sample-specific detection limits are not provided (see Chapter 11 for discussions on detection limits).

The summarized chemical data are presented in risk assessments by environmental medium and consist of parameters such as:

- frequency of detection,

- arithmetic mean concentration, and

- range of detected concentrations.

The selection of the chemicals of potential concern for the site is the next step. The purpose of selecting chemicals of potential concern is to eliminate from the risk assessment:

(1) those chemicals that are associated with sampling or laboratory artifacts; and,

(2) those chemicals existing at or below naturally occurring levels at the site.

The following methodology are used in selecting chemicals of potential concern from the summarized data:

- **Evaluate Ambient Water Quality**:—According to USEPA (1989a) guidance, inorganic chemicals present at the site at naturally occurring levels may be eliminated from the quantitative risk assessment. According to this guidance, a statistical method should be used to determine whether chemical concentrations detected at the site are within, or elevated above, background levels. This approach is used to evaluate background concentrations for those sample groupings where a sufficient number of samples (at least three upgradient and three downgradient) is available. The statistical method that is used should be appropriate for the data under evaluation and should account for data below the limit of detection (USEPA 1988a). Justification must be provided in the Baseline RA to support the use of any statistical test.

- **Select a Statistical Basis**:— If a sufficient number of samples is not available for a standard statistical test, alternative statistical methods are considered for the evaluation. For example, using the former approach, if the central tendency (for a parametric test) or rank ordering (for a non-parametric test) of chemical concentrations in potentially site-affected samples significantly exceed those of background samples, then the chemical is considered site-related and included in the Baseline RA. In those cases where relevant background data are not available for a specific chemical, the chemical is conservatively selected as

chemical of potential concern.

- **Compare Site Data to Blank Data**:—Site data are compared to available blank (lab, field, and trip) data as recommended in USEPA (1989a) guidance. According to USEPA (1989a), if the detected concentration in a sample is less than 10 times the blank concentration for common laboratory contaminants (acetone, 2-butanone, methylene chloride, toluene, and the phthalate esters) the chemical is not selected for evaluation in the risk assessment. For those organic or inorganic chemicals that are not considered by USEPA to be common laboratory contaminants (all other compounds), if the detected concentration is less than 5 times (for all other compounds) the maximum detected concentration in the blanks, the chemical is not selected for evaluation in the risk assessment.

10.10.2 TOXICITY ASSESSMENT

The general methodology for the classification of health effects are described within a toxicity assessment the Baseline RA. This assessment is to provide the analytical framework for the characterization of human health impacts. Brief summaries are included in the evaluation of the toxicity of those chemicals U.S. EPA have included quantitatively in this assessment. These summaries may also contain a discussion of the human health and environmental standards, criteria, and guidance developed for evaluation of exposure to these chemicals. In addition, the quantitative dose-response values that are used in the Baseline RA are provided (cancer slope factors and reference doses). For carcinogens, it is important to include cancer slope factors (or unit risks for the inhalation pathway) and the chemicals' U.S. EPA weight-of-evidence classifications for human carcinogenicity. For noncarcinogens, one should provide reference doses (RfDs) (or reference concentrations (RfC's) for the inhalation pathway) and the uncertainty factors used in deriving them. The primary source of these dose-response values are USEPA's Integrated Risk Information System (IRIS) and Health Effects Assessment Summary Tables (HEASTs).

Human Exposure Assessment

In the human exposure assessment portion of the Baseline RA, the most important pathways are evaluated through which individuals may be exposed to chemicals originating from the facility. In this section estimates are necessary for the potential magnitude, frequency, and duration of chemical exposures via the identified exposure pathways. Two overall exposure conditions are typically evaluated:

• Current Site and Surrounding Land Use Condition to evaluate potential risks associated with the site as it exists today, and,

• Future Land Use Condition will evaluate potential risks under future use conditions assuming a particular remedial alternative or if no remedial action is taken (i.e., the no action alternative).

A deterministic reasonable maximum exposure (RME) scenario are evaluated in accordance with USEPA (1991, 1989a) guidance. An average exposure scenario is normally evaluated to indicate the extent of uncertainty in the RME risk estimates. Alternatives to the RME scenario for ground-water pathway, can be based on a number of alternative evaluation techniques (for example using Monte Carlo Simulation), may yield a more accurate estimate of upperbound possible exposure than the deterministic approach typically used in such studies.

Site Characterization

Factors such as population activity patterns in the area, source characteristics, and routes of transport are considered when identifying potentially exposed populations. Based an understanding of the site developed during the RI Phase of work, potential human receptors may also include trespassers on the site, site workers, and nearby workers and residents

Identification of Exposure Pathways

An exposure pathway describes the course a chemical takes from the source to the receptor. An exposure pathway generally consists of four components:

l) a source and mechanism of release,

2) an environmental transport medium,

3) a point of potential exposure with the contaminated medium; and,

4) a route of exposure (e.g., inhalation) at the exposure point.

When all these elements are present, a pathway is considered "complete". Based on the site characterization, potential exposure pathways are identified in the Baseline

RA. Full justification must be provided to support not evaluating a potential pathway quantitatively in the Baseline RA. The typical pathways that will likely to be quantitatively considered in a Baseline RA involve the inhalation of airborne chemicals, ingestion and dermal absorption of chemicals in soil, and ingestion and inhalation of chemicals in ground water. The receiving media, transport of chemicals among media, potential exposure points and exposure routes are discussed in the Baseline RA for each pathway evaluated. The pathways for groundwater exposure are discussed by component below:

• **Source or Mechanism of Release and Environmental Transport Medium**: - Contaminants may enter ground water via leaching from soils and be transported through the aquifer.

• **Point of Potential Contact by Receptor**: - Nearby residents both on-site and off-site may potentially use ground-water wells. Under current conditions, no nearby residents are known to use groundwater for drinking water. However, use of groundwater for drinking water may still possibly occur despite these controls. Therefore, the use of groundwater for ingestion are normally considered a potential point of potential contact under land use conditions. Other uses of ground water by nearby residents (e.g., washing, watering) are also assumed to exist under the current land use conditions and may serve as additional points of contact. Even under a no-action alternative, residents under future land use conditions could potentially be exposed in the same manner as current residents (i.e., through the use of groundwater for drinking water or for other purposes).

• **Route of Exposure**: - Persons could be exposed to chemicals in groundwater via ingestion of drinking water, via inhalation of chemicals that have volatilized from ground water during use (e.g., while showering, cooking, watering the lawn); and via dermal contact with ground water during in-home use (e.g., while bathing, washing dishes).

• **Completeness of Pathways** :- Based on the above components, human receptors could be impacted by exposure to chemicals in groundwater, therefore, the magnitude of possible exposures to groundwater and potential effects from such exposures are typically evaluated based on data collected during the RI.

When applicable, three exposure pathways are evaluated for each of the well groupings described above. Exposures via ingestion of drinking water and inhalation of volatiles are evaluated quantitatively. Exposures via dermal absorption must be evaluated qualitatively, as dermal exposure through indoor water use is expected to be much lower than the more direct exposures of ingestion and inhalation. This is due to the fact that chemicals are more likely to remain in the water or air phase, and because dermal exposures to tap water are of short duration.

Quantification of Exposure Point Concentrations

This section of the baseline risk assessment quantifies exposure point concentrations for the selected pathways. Exposure point concentrations for each pathway are determined using the RI data following USEPA (1989a) guidance. In many cases where no sampling data are available, one must rely on fate and transport models to evaluate the exposure point concentrations. Indoor air volatile organic concentrations associated with showering can be estimated using a model developed by Foster and Chrostowski (1987).

Two sets of concentrations are typically estimated for each pathway, one in accordance with USEPA (1991, 1989a) guidance for evaluation of USEPA's reasonable maximum exposure (RME) case, another for evaluation of a conservative average case. In addition, for the ground-water pathway an alternative RME case can be evaluated. The RME case concentrations are based on the 95 percent upper confidence limit on the population mean concentrations (or the maximum concentrations, whichever is smaller), and conservative average case concentrations are typically based on population mean values. The 90th percentile concentrations is used for the ground-water alternative RME case.

Estimation of Chemical Intakes

The procedure for the calculation of intakes of chemicals by potentially exposed populations is provided below. To determine these intakes, assumptions concerning chemical concentrations, exposed populations, and exposure conditions, such as frequency and duration of exposure, are used together with intake parameters developed within the section. Tables containing exposure parameters for each pathway are presented and discussed in detail in the Baseline RA. As discussed earlier, one exposure scenario are evaluated for each pathway. This scenario is the reasonable maximum exposure (RME) case identified by USEPA (199la, 1989a) for use in Baseline RAs, the second, a conservative average case scenario. As an example drinking water exposures can be calculated using the following equation:

Equation 10-22

where
  CDI = chronic daily intake (mg/kg-day),
  Cw = chemical concentration in groundwater (mg/l),
  IR = water ingestion rate (l/day),
  EF = exposure frequency (days/year),
  ED = exposure duration (years),
  BW = body weight (kg),
  Days = conversion factor (365 days/year), and
  AT = averaging time (lifetime for carcinogens, exposure duration for noncarcinogens).

Exposures associated with inhalation of volatile organic chemicals released while showering can be calculated using the following equation:

Equation 10-23

where:
  IEC = inhalation exposure concentration (mg/m^3),
  Ca = chemical concentration in shower room air (mg/m^3),
  ET = shower exposure time (hours/day),
  EF = shower exposure frequency (days/year),
  ED = exposure duration (years), and
  AT = averaging time (lifetime for carcinogens, exposure duration for noncarcinogens).

Table 10-12 presents the exposure parameters which have been developed for ground-water ingestion and inhalation of volatiles while showering. The upperbound RME values are based on specific parameter values identified by USEPA (1991, 1989a) for use in Baseline RA's. The average case parameter values are based on data provided in the scientific literature. The selected parameter values reflect both the current and future case; therefore, intakes (and the potential for risks) for these two land use conditions are identical.

Where regulatory guidance has specified RME values, such as the USEPA guidance documents (e.g., USEPA l991a) should be used in the evalution. In the absence of such guidance, the most relevant and recent scientific literature are used to develop parameter values.

10.10.3 RISK CHARACTERIZATION

In the risk characterization section of a baseline risk assessment, the chemical intake estimates derived in the exposure assessment are integrated with the dose-response health criteria values. The results of the risk characterization provides estimates of the "upper-bound" individual cancer risk estimates for potential carcinogens and a hazard index for noncarcinogens. The individual lifetime excess cancer risk for a chemical exhibiting carcinogenic effects are calculated by multiplying the exposure-weighted medium concentration by the unit risk factor or by multiplying estimated chronic daily chemical intakes averaged over a 70-year lifetime by the upper-bound cancer slope factor. To estimate the combined impact to an individual from exposure to all carcinogens, cancer risk estimates are summed within each exposure pathway, and where applicable, will also be summed across the exposure pathways. The additivity of upper-bound risks is assumed based on guidance from USEPA (1986a, c, 1989a).

For noncarcinogens, potential risks are typically calculated by means of a hazard index technique, as recommended by USEPA (1989a, 1986c). For each noncarcinogenic chemical,one calculates the ratio of the chronic daily intake to the reference dose or the ratio of the air concentration to the reference concentration. The effects of all the noncarcinogens taken together are computed by summing the individual ratios. This sum, known as the hazard index, indicates the potential for adverse effects for the total mixture of noncarcinogens. A hazard index greater than a threshold level of one will normally trigger a more detailed evaluation in which hazard indices for chemicals affecting similar target organs are calculated. If a target organ specific hazard index exceeds one, there may be a concern for potential health effects (USEPA 1989).

10.10.4 ECOLOGICAL RISK ASSESSMENT

The ecological risk assessments when necessary can evaluate potential risks to aquatic life and terrestrial animals and plants (USEPA 1989c). This type of assessment may include four major components:

1) identification of potential receptors,

2) exposure assessment,

3) toxicity assessment, and

4) risk characterization.

Potential receptors are identified based on contacts with state and federal biologists, information provided in the RI, and obtained during a site visit. Endangered, threatened, and special status species are identified. Wetland areas should also be identified. Potential exposure pathways for selected representative species are evaluated. The toxicity assessment will identify toxicity reference values and surface water and sediment quality criteria for the chemicals of concern. Toxicity information from the scientific literature are used where no standards or guidance values exist. These values are used, along with estimated exposures, to characterize potential risks associated with the site.

10.10.5 Risk Assessments as the Basis For Remeadiation

On both a regional and state basis, federal and state regulatory agencies take different approaches to risk assessment. This produces a program with many inconsistencies and overlapping guidance documents. Often, one EPA region may be driven more by policy considerations than by technical and scientific concerns, and internal guidance documents are not followed. Each variation builds on the next to make the overall program less efficient and lacking in a consistent, unified methodology.

The application of EPA cleanup policy has been inconsistent over the years. Recently, Oak Ridge National Laboratory evaluated cleanup levels for remedial action at 42 Superfund sites, based on a review of RODs.'s The review revealed that feasibility and cost appear to be"the two major factors in determining how closely a specific cleanup approaches current legal mandates." However, PRP participation, state acceptance and the influence of public opinion are additional factors in selecting site cleanup standards.

Because federal and state ARARs do not exist in most cases for soils and sediments, risk assessments are the basis for decisions on these media. According to the Oak Ridge study, although the numerical goals for soil cleanup vary, a common end point is the 10^{-5} or 10^{-6} excess cancer risk. For ground water and surface water, the remedy considers vulnerability, value and water use; a wide variety of criteria and standards are addressed, sometimes all at one site. According to the Oak Ridge study, the target risk range is from 10^{-4} to 10^{-6} excess cancer risk. It concludes that "clean becomes whatever can be done at a reasonable cost with the technology available and that are accepted by the public. "

There are several reasons why Superfund cleanup

decisions appear to be inconsistent. In some EPA regions, regional staff are unfamiliar with the risk assessment process. Because of this risk assessment is not uniformly applied to the FS when evaluating proper cleanup measures. Often only engineering alternatives are considered, and the step from risk to abatement is skipped because of a lack of complete understanding of the problem. If you don't address risks in the FS, then there is no clear and coherent approach to selection of remediation alternatives. Without risk assessment the remedial actions proposed are overly conservative and will overestimate the amount of work to be done to remediate the site. Other reasons for inconsistent clean up goals include the following:

- EPA often uses extremely conservative worst case exposure assumptions to justify the remedy selected, but the assumptions do not reflect what actually is occurring in the exposed population;

- There is no acceptable methodology to determine soil exposure and cleanup levels; and

- There is minimal peer review of enforcement risk assessments. Only 10 to 20 percent of risk assessments are reviewed by members of the technical support group within Superfund, or by regional health and environmental assessment offices; the balance of enforcement risk assessments generally are reviewed by remedial project managers and EPA oversight contractors.

Risk assessors and regulators also must come to grips with the uncertainties surrounding the selection of exposure scenarios. Hypothetical worst-case exposure scenarios for example, the existence of an on-site drinking-water well that withdraws water from the most contaminated portion of the plume drive perceived risks at sites that ultimately require remediation efforts. However, in the proposed NCP and the new Superfund Public Health Evaluation Manual, EPA will provide new guidance on how to develop a more realistic exposure scenario at Superfund sites.

Risk assessment must be put in perspective as a critical bridge between the RI and the FS, enabling remedial action to be implemented. Under CERCLA's five-year re-opener clause, re-evaluating site conditions every five years not only permits assessments of effectiveness but also enables new scientific information to be used in judging remedial actions in progress. The importance of PRP participation in the risk assessment process cannot be overemphasized. PRPs conducting risk assessments or actively involved in negotiations can

constrain the risk assessment process to focus on "real world" rather than perceived levels of exposure and risk.

10.11 BIAS FOR ACTION ON SUPERFUND AND RCRA REMEDIAL PROJECTS

The recent concern of Congress on the slow progress of Superfund clean-up process has brought into focus the inefficiencies of the current RI/FS process. Federal guidance and review documents concentrate on field collection and chemical analysis of large numbers of samples. Remedial Investigations (RIs) conducted pursuant to CERCLA/SARA are primarily designed to delineate the spatial distribution of contaminants with respect to receptor populations and institutional boundaries. This has been traditionally accomplished by assembling a soil and water quality data base, the size and scope of which is influenced by nontechnical factors such as community sensitivity and cost-recovery litigation requirements.

Although site-specific hydrogeologic conditions exert the main control on contaminant migration in the subsurface, hydrogeologic data are generally relegated to secondary importance with respect to the water quality data. The data bases resulting from this investigative methodology typically do not adequately support detailed analysis of the predominant contaminant transport mechanisms, and, as such, are of limited value for evaluating and designing effective remedial actions. As such the main focus of Superfund is collection of the wrong data for effective aquifer clean-up.

The data necessary for an effective characterization of site flow conditions for determination of rate and extent(assessment monitoring) is typically the same data required for design of the ground-water remediation system. In order to develop a cost effective process of linking water quality data with the more traditional geotechnical/hydrogeologic data, a new way of conducting RI/FS investigation is proposed using conceptual geology and hydrology to target monitoring program under Superfund.

Since the passage of the initial Superfund legislation in 1980, engineers and scientists have striven to devise a uniform, yet cost-effective methodology for characterizing sites that are, or are suspected of, releasing contaminants to the environment. The U.S. EPA's initial Remedial Investigation guidance (EPA, 1985) focused site characterization efforts primarily on determining the area, extent, and magnitude of contamination, i.e., plume delineation. This has commonly necessitated multiple drilling, sampling, and analysis episodes to develop the data base required to support a site specific risk assessment and to evaluate the appropriate remedial actions. In recent years, it has become increasingly apparent that this approach is not an efficient utilization of limited financial and investigative resources.

The U.S. EPA proposes in the 1989 National Contingency Plan revision to emphasize (53 FR 51423) a "bias for action" and to use the principal of "streamlining" in managing the Superfund program as a whole, and in conducting individual remedial action projects. It is our experience that most Superfund actions develop into field data collection and office documentation exercises with little bias for action to place into effect reasonable solutions for dealing with real risks to human health and the environment. This fact is substantiated by the low numbers (if any) Superfund sites with ground water contamination that have actually been cleaned up in almost ten years since the enactment of Superfund in 1980. Overextended RI studies running five to ten years are now becoming commonplace in the Superfund program.

The need for more time- and cost-effective investigative methodologies is implicitly reflected in the congressionally mandated clean-up schedules that are embodied in the Superfund Amendment and Reauthorization Act of 1986 (SARA). The U.S. EPA's interpretation of this schedule (Porter, 1988) suggests that future Remedial investigations (RI). and Feasibility Studies (FS) must be completed in 12 to18 months. This is in contrast to the existing situation where RIs alone range in duration from 18 months to 6 years. Clearly there is a need to adopt investigative methodologies which will minimize the need for multiple and repetitive data gathering exercises without compromising the overall objectives of the program. Promotion of a bias for action is an important goal in Superfund investigations. In order to meet the goals for streamlining the process we would recommend as a first step to limit RI/FS projects to eighteen months maximum. This could be completed using the following concepts:

- Use Predefined Generic Scopes of Work;

- Use Predefined Work, Health & Safety, and QA plans

- Base the RI on gathering site data directed toward design of known site remedial solutions;

- Use VOC indicator parameters as the basis for assessment of codisposal sites;

- Make predetermination of site remedial solutions based on experience of other large volume codisposal sites ; and

- Use risk assessment techniques as the basis for selection of remediation technologies

Predefined Scopes of Work

A generic Scope of Work than has been accepted by U.S. EPA Region V and Region VI as the basis for RI/FS investigations on five codisposal sites (Appendix B3). This scope of work was then used to generate detailed work, Health & Safety and QA plans for the sites. These scopes of work provide a more efficient alternative by characterizing and quantifying the hydrogeologic controls of the hydrogeologic system prior to initiating water quality assessment activities. This results in more rational placement of, and consequently fewer, monitoring points and minimizes the inefficiencies associated with multiple field investigations. This approach utilizes water quality data as an independent means of confirming and refining the understanding of the ground-water system. The data base resulting from this proposed approach can also be utilized to support more accurate and realistic estimates of exposure concentrations. If the down gradient flow paths are adequately characterized one can complete a targeted, more representative sampling of the environment

10.11.1 Predefined Work, Health & Safety, and QA plans

It has become standard practice to begin remedial investigations with generation of massive planning documents for the establishment of monitoring points and the acquisition and analysis of environmental samples (air, soils, sediments, and ground-and surface-waters). The primary emphasis of this approach is placed on defining the type, magnitude, and spatial and temporal distribution of contamination(ie. rate & extent). The parameters which exert the primary control on contaminant transport, such as the media geometry and ambient flux, are estimated during the sampling or monitoring points installation activities.

As a result, the location of these monitoring points is generally predicated on the general knowledge of site conditions derived from topographic gradients, aerial surveys, historical accounts of disposal activities, and a variable amount of site-specific technical data. In those rare cases where moderate to extensive technical data bases exist prior to the initiation of the RI, the application of strict, programmatic quality control and documentation requirements often limits the use of this data in the RI.

Frequently, the results of these initial sampling events showed that the contamination has migrated beyond the initial limits of the study area and/or the concentration gradients cannot be sufficiently interpreted within the constraints of the existing data base. In either case, the typical response to this situation has been to establish additional sampling points and collect and analyze more environmental samples. Investigations that follow this track rapidly degenerate into "plume chases" wherein the goal gradually shifts away from the definition and quantification of exposure pathways to locating the leading edge of the plume. This approach has been more formally referred to as the Plume Delineation method (Dowden 1988).

The records of Superfund investigations commonly show three or more major field exercises to define the limits of contamination and the primary exposure pathways. In many of these cases, the site complexities justify a phased or staged analysis. However, in too many of these situations the continued reliance on repetitive well installation and sampling did not adequately address the site complexities or significantly increase the understanding of the contaminant transport mechanisms. Figure 10-5 shows a flow diagram of the iterative traditional Superfund process. The accurate depiction and

quantification of these flow mechanisms is crucial in estimating potential exposure concentrations beyond the known limits of the plume and in evaluating and designing effective ground-water remedial actions. Without targeted geotechnical and hydrogeologic assessment techniques the investigations are always required to go back for additional field data for later feasibility and design phases.

This repetitive planning/sampling/planning cycles, each with individual Agency, consultant/contractor, PRP reviews are the primary reason behind the long RI/FS process. We recommend that an alternative approach be used to prepare the required planning documents that adequately describe to work to be performed during the RI/FS. The planning documents can be made relatively uncomplex if the goal of the RI/FS investigation changes from a sample collection program to an investigation that is directed toward understanding the geology & hydrogeology of the site. A generic work plan can addresses this goal and would be applicable to a vast proportion of codisposal site currently on Superfund.

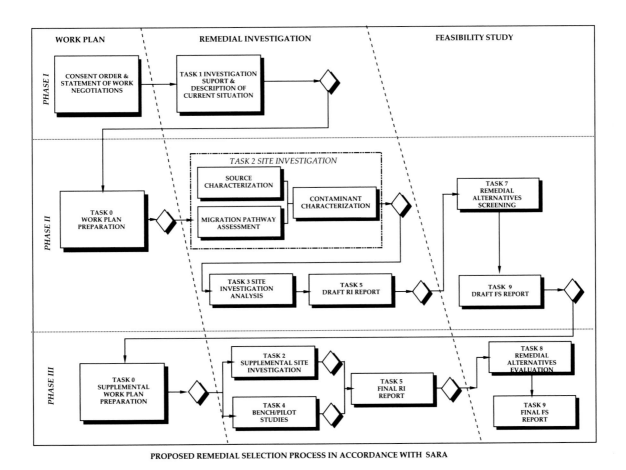

PROPOSED REMEDIAL SELECTION PROCESS IN ACCORDANCE WITH SARA

Figure 10-31 Streamlined RI/FS Investigation

10.11.2 Base RI on Goals to Design the Remeadiation

An alternative approach to remedial investigations would be to postpone environmental sampling until the site hydrogeologic flow paths and mechanisms are adequately understood. In ground water investigations, the traditional ground-water assessment approach (before superfund) involves a thorough evaluation of all pre-existing data and aerial photographs, performance of geophysical surveys, drilling of stratigraphic boreholes, installation of piezometers and measurement of heads, performance of in-situ hydraulic conductivity tests, and the evaluation of hydraulic connection between aquifers and confining beds. All of the information derived from these activities can then be utilized to construct stratigraphic and hydrostratigraphic cross-sections, vertical and plan-view flow nets, and to develop rate and extent of migration predictions.

In complex hydrogeologic settings, numerical modeling may be required to attain the appropriate level of quantitative analysis. These predictions, combined with a small amount of information on the nature and history of the source(s), can be used to develop a rational and efficient environmental sampling and analysis plan. These more linear investigations are shown in the second colume of Figure 10-5.

Since the pre-superfund ground-water assessment methods yield more data on the physical controls of the ground-water transport system, the relevance and significance of the environmental sampling results can be more readily interpreted. An additional advantage is that the environmental sampling results can be utilized to refine and recalibrate the predictive analyses based on the physical site characterization.

The proposed bias for action and streamlining should also include consideration of the flow diagram (Figure 10-28). This diagram details the activities necessary to fully define the geologic and hydrogeologic conditions before selection of sampling location to confirm real risks to human health and the environment. Once sampling locations were selected, only a VOA scan would be performed. If VOAs were detected at statistically determined levels full hazardous constituent sampling would be performed only on these wells showing VOAs. This simple screening procedure would save the EPA tens of millions of dollars annually in unnecessary analytical work with no loss in program effectiveness.

Since this proposed bias for action program stresses collection of geologic and hydrogeologic data, the information necessary for design of the remediation would be available at an early point in the RI/FS. Decisions within the FS on the special situations where it may not be practicable to actively restore ground water could be more easily evaluated and technically defended. The type of geotechnical and geologic information necessary to select the assessment monitoring points would also provide the design information for the pump and treat systems normally required to remediate contaminated aquifers that are affecting human health and the environment. These data include flow nets and target monitoring zones, transmissivity and storage coefficients, cross sections and structural maps of the site. The data gathered during the phased investigation would home-in on the geology and hydrogeology first before going into a significant sample collection program. Figure 10-31 illustrates the process of defining the geology / hydrogeology first then working out the required sampling programs for the remedial investigation.

10.11.3 Use VOC's as the Primary Indicators for Assessment Monitoring

A major step in streamlining process of Superfund should be to switch from the current main activity of sample collection of soils and ground water for extensive organic analytical work, to collection of sufficient geologic, geotechnical and hydrogeologic information to determine the flow paths from the disposal areas to downgradient users or discharge points. A sufficient base of technical literature (Plumb 1988) has been developed to defend to concept that federal funds currently being spent on all inclusive Hazardous Substances Lists (HSL) would be much better directed toward an understanding of the geologic and hydrogeologic conditions that permit movement of mobile parameters away from the waste disposal areas. Once full understanding is established of the transmitting media and the hydraulic conditions affecting the contaminated ground water movement, a limited chemical sampling phase using VOC's as indicator parameters could quickly establish true risk of the site to human health and the environment.

The following discussion proposes a phased approach to these site characterization that links basic geotechnical techniques to traditional hydrogeologic site analysis. These methods have often been forgotten in the rush to collect water quality information for Superfund investigations. This methodology would hold off installation of verifying ground-water monitoring well systems until a site was adequately characterized for basic hydrogeologic conditions.

The thesis of this proposal is that significant time and cost savings can be recognized on codisposal site assessments if the focus of RI is shifted from laboratory analysis of statistically-representative quantities of environmental samples (i.e., "saturation" sampling) to sampling and analysis schemes which are predicated on a thorough understanding of the physical geologic and hydrologic systems (i.e "smarter" sampling).

10.11.4 Predetermination of Site Remedial Solutions

The universe of remeadiations for codisposal is relatively limited both in alternatives and in effectiveness of the solutions. A review was conducted on ROD's completed after 1986 on codisposal sites. The bulleted items below show the range of remedial alternatives included in the 44 ROD's completed during these years. A codisposal site remedial profile consisting of a 90% solution would include the following:

- Soil Cap 95%
- Access/deed restrictions 55%
- Ground water pump & Treat 50%
- Surface water diversion & collection 35%
- Leachate collection 28%
- Air Monitoring 25%
- Gas collection 25%
- Slurry Wall 20%
- Alternate water supply 18%
- On-site waste consolidation 15%

All the rest of the many remedial alternatives and technologies were less than 7% of the total for the codisposal sites ROD's. This analysis provides a rational direction for the RI/FS and design of the remediation for these large volume low toxicity sites. The use of these alternatives would probable be applied on any large volume site, no matter if the site consisted of radioactive mine tailings or municipal solid waste theat took liquid waste.

10.12 CONCLUSIONS

The application of an effective assessment monitoring program must be based on targeted programs that have a heavy reliance on site geologic and hydrogeologic data. The assessment monitoring system should be only installed after full knowledge is developed on the targeted monitoring zones.

REFERENCES

Anderson, M.R., R.L. Johnson and J.F. Pankow, 1987. The Dissolution of Residual Dense Non-Aqueous Phase Liquid (DNAPL) from a Saturated Porous Medium, Proceedings: Petroleum Hydrocarbons and Organic Chemicals in Groundwater, National Water Well Association, Houston, Texas, Nov. 17-19,

Banerjee, S., Solubility of Organic Mixtures in Water. Environmental Science & Technology, Vol. 18 (8), p 587-591.

Bednarz, R. L, Fink, L E., and Roth, W. L, 1983. Chemical Environmental Fate Data Acquisition, Project C-1 Exposure Evaluation. Toxic Chemical Evaluation Section, Michigan Department of Natural Resources, Environmental Protection Bureau.

Brady, N.C., 1974. The Nature and Properties of Soils, 8th ed., MacMillan Publishing Co., Inc., New York.

Briggs, G.G., 1973. A simple relationship between soil adsorption of organic chemicals and their octanol/water partition coefficients, Proceedings of the 7th British Insecticide and Fungicide Conference, vol. 1, The Boots Company Ltd., Nottingham, Great Britain.

Brown, D.S., and E.W. Flagg, 1981. Empirical prediction of organic pollutant sorption in natural sediments, Journal of Environmental Qualit.v, 10(3), pp. 382-386.

Buckman H., and Brady, N., 1969. The Nature and Properties of Soils, 7th Ed. Macmillan.

Chatzis, I., M.S. Kuntamukkula, and N.R. Morrow, "Blob-size Distribution as a Function of Capillary Number in Sandstones," Paper 13213, presented at the SPE Annual Tech. Conference and Exhibition, Houston, Texas, 1984.

Cherry, J.A., R.W. Gillham, and J.F. Barker, 1984. Contaminants in groundwater: Chemical processes, Studies in Geo physics: Ground water Contamination, National Academy Press, Washington, DC, pp. 46-64.

Cherry, J.A., written communication to EPA DNAPL Workshop, R.S. Kerr Environmental Research Laboratory, Dallas, Texas, April, 1991.

Chiou, C.T., L.J. Peters, and V.H. Freed, 1979. A physical concept of soil-water equilibrium for nonionic organic compounds, Science, 206, pp. 83 1 -832.

Chiou, C. T., 1981. Partition Coefficient and Water Solubility in Environmental Chemistry. Hazard Assessment of Chemicals: Current Development, Vol. 1.

Connor, J.A., C.J. Newell, D.K. Wilson, 1989. Assessment, Field Testing, Conceptual Design for Managing Dense Nonaqueous Phase Liquids

(DNAPL) at a Superfund Site, Proceedings of Petroleum Hydrocarbons and Organic Chemicals in Groundwater, presented at Prevention, Detection and Restoration Conference and Exposition, Houston, Texas, Vol. 1, p 519-533,

Domenico, P.A. and F.W. Schwartz, 1990. Physical and Chemical Hydrogeology, Wiley, New York,

Dragun, J., 1988. Soil Chemistry of Hazardous Materials. Hazardous Materials Research Institute, Maryland.

Ellgehausen, D'Hondt and Fuerer, 1981. Reversed-Phase Chromatography as a General Method for Determining octanol/water Partition Coefficients. Pestc. Sci.

Feenstra, S., and J.A. Cherry, 1988. Subsurface Contamination by Dense Non-Aqueous Phase Liquids (DNAPL) Chemicals, International Groundwater Symposium, International Association of Hydrogeologist, Halifax, N.S. May 1-4.

Feenstra, S., D.M. MacKay, and J.A. Cherry, 1991. A Method for Assessing Residual NAPL Based on Organic Chemical Concentrations in Soil Samples, Groundwater Monitoring Review, Vol 11, No. 2.

Freeze, R. & Cherry, J., 1989. What has gone wrong, Groundwater 27(4):458-464.

Freeze, R.A., and J.A. Cherry, 1979. Ground water, Prentice Hall, Inc., New Jersey, pp. 402-413.

Fujita, 1., and Fujita, H., 1964. A New Substituent Constant, Derived from Partition Coefficients. Jour. Amer. Chem. Soc.

Gilbert, R. O. 1987. Statistical methods for environmental pollution monitoring. Van Nostrand Reinhold, New York.

Gillham, R.W., and J.A. Cherry, 1982. Contaminant transport by groundwater in nonindurated deposits, Recent Trends in Hydrogeology, T.N. Narasimhan, ed., Geological Society of America Special Publication 189, Boulder, Colorado, pp. 3 1-62.

Goldberg, 1982. A method of Calculating the Pesticide Partition Between Sediment and Water for Environmental Systems. Sci. Total Environ.

Greenberg, M. R. 1987. Sampling strategies for finding contaminated land. Applied Geography 7:197-202.

Guenther, W. C. 1981. Sample size formulas for normal theory T tests. Amer. Stat. 35(4):243244.

Hall, C., 1988. Practical limits to pump and treat technology for aquifer remediation, Presentation at Groundwater Quality Protection Preconference Workshop Proceedings, Water Pollution Control Federation, Dallas, Te~as, pps. 7~12.

Hassett, J.J., J.C. Means, W.L. Banwart, S.G. Wood, S. Ai, and A. Khan, 1980. Sorption of dibenzothiophene by soils and sediments, Journal of Fnvironmental Quality, 9, pp. 184-186.

Hinchee, R.E. and H.J. Reisinger, 1987. A Practical Application of Multiphase Transport Theory to Groundwater Contamination Problems, Ground Water Monitoring Review, p 84-92, Winter.

Hounslow, A.W., 1983. Adsorption and movement of organic pollutants, Proceed ings of the Third National Symposium of Aquifer Restoration and Ground Water Monitoring, NWWA, Worthington, Ohio, pp. 334-346.

Huling, S.G. and Weaver, J.W., 1991. Dense Nonaqueous Phase Liquids, EPA Groundwater Issue Paper, EPA/540/4-91-002, March.

Hunt, J.R., N. Sitar, K.D. Udell, 1991. Nonaqueous Phase Liquid Transport and Cleanup, Water Resources Research, Vol. 24(8), p 1247-1258.

Karickhoff, S. W., 1984. Organic Pollutant Sorption in Aquatic Systems. Journal of Hydraulic Engineering: Vol. 110 (6).

Karickhoff, S.W., D.S. Brown, and T.A. Scott, 1979. Sorption of hydrophobic pollutants on natural sediments, Water Research, 13, pp. 241-248.

Karickhoff, S.W., D.S. Brown, and T.A. Scott, 1979. Water Resources, 13:241-248.

Keller, C.K. G. Van der Kamp and J.A. Cherry, 1988. Hydrogeology of Two Saskatchewan Tills, Journal of Hydrology, p 97-121.

Kenaga, E., 1980. Predicted Bioconcentration Factors and Soil Sorption Coefficients of Pesticides and Other Chemicals. Ecotoxicology and Environmental Safety.

Kenaga, E.E., and C.A.I. Goring, 1980. Relationship between water solubility, soil sorption, octanolwater partitioning, and concentration of chemicals in biota, Aquatic Toxicology, ASTM STP 707, J.G. Eaton, P.R. Parrish, and A.C. Hendricks, eds., American Society for Testing and Materials, pp. 78-115.

Koch, A., 1983. Molecular Conductivity Index for assessing Ecotoxicological Behavior of Organic Compounds. Toxicol. Environ. Chem.

Kueper, B.H. and Frind, E.O., 1988. An Overview of Immiscible Fingering in Porous Media," Journal of Contaminant Hydrology, Vol. 2, p 95-110

Lambert, S.M., P.E. Porter, and H. Schieferstein, 1965. Movement and sorption of chemicals applied to soil, Weeds, 13, pp. 185-190.

Lambert, S.M., P.E. Porter, and H. Schieferstein, 1967. Functional relationship between sorption in soil and chemical structure, Journal of Agricultural Food Chemistry, 15(4), pp. 572-576.

Lambert, S.M., P.E. Porter, and H. Schieferstein, 1968. Omega, a useful index of soil sorption equilibrium, Journal of Agricultural Food Chemistry, 16(2), pp. 340-343.

Leo, A., C. Hansch, and D. Elkins, 1971. Partition

coefficients and their uses, Chemical Review, 71, pp. 525-621.

Lyman, W. J., Reehl, W. F., and Rosenblatt, D. H., 1982. Handbook of Chemical Property Estimation Methods, Environmental Behavior of Organic Compounds. McGrawHill, Co.

Lyman, W.J., W.F. Riehl, D.H. Rosenblatt, eds., 1982. Hand book of Chemical Property Estimation Methods: Environment Behavior of Organic Compounds, McGraw-Hill Book Company.

Mackay, D., 1989. Characterization of the distribution and behavior of contaminants in the subsurface, Proceedings of the National Research Council Water Science and Technology Board Colloquium, Washington, D.C.

Mackay, D.M. and J.A. Cherry, 1989. Groundwater Contamination: Pump and Treat Remediation, Environmental Science & Technology, Vol 23, No. 6, p 630-636.

Mackay, D.M., P.V. Roberts, and J.A. Cherry, 1985. Transport of Organic Contaminants in Ground Water," Environmental Science & Technology, Vol. 19, No. 5, p 384-392.

Mason, B. J. 1983. Preparation of soil sampling protocol: techniques and strategies. Environmental Monitoring Systems Laboratory, Las Vegas, Nevada. EPA-600/4-83-20.

McCarty, P. L, Reinhard, M., Rittman, B., 1981. Trace Organics in Groundwater: Processes Affecting Their Movement and Fate in the Subsurface. Environmental Science & Technology: Vol 15.

Marrin, D. L. and G. M. Thompson. 1987. Gaseous Behavior of TCE Overlying a Contaminated Aquifer, Ground Water, Vol. 25, No. 1, pp. 21-27

Marrin, D. L., and H. Kerfoot. 1988. Soil-gas surveying Techniques, Environmental Sciences & Technology, Vol. 22, No. 7, pp. 7 40-745.

Marrin, D. L. 1988. Soil-gas Sampling and Misinterpretation, Ground Water Monitoring Review, pp51-54.

Means, J.C., S.G. Wood, J.J. Hassett, and W.L. Banwart, 1980. Sorption of polynuelear aromatie hydrocarbons by sediments and soils, Environmental Science and Technology, 14(12), pp. 15241 528.

Mercer, J.W. and R.M. Cohen, 1990. A Review of Immiscible Fluids in the Subsurface, Properties, Models, Characterization and Remediation," Journal of Contaminant Hydrology, Vol 6, p 107-163.

Miller, C.T., and W.J. Weber, Jr., 1984. Modeling organie eontaminants partitioning in groundwater systems, Ground Water, 22(5), pp. 584-592.

Mingelgrin, U. and Gerstl, Q, 1983. Reevaluation of Partitioning as a Mechanism of Nonionic Chemicals Adsorption in Soils. Journal of Environmental Quality: Vol 12, Jan.

Moriguchi, 1975. Quantitative Structure-Activity Studies. 1. Parameters Relating to Hydrophobicity. Chem. Pharm. Bull. 23.

Ng, K. M. et. al., 1978. Visualization of Blob Mechanics in Flow Through Porous Media, Chemical Eng. Sciences, Vol. 33, pp 1009-1017.

Olsen, R.L. and A. Davis, 1990. Predicting the Fate and Transport of Organic Compounds in Groundwater," HMC, May/June, p 40-64.

Parkhurst, D. F. 1984. Optimum sampling geometry for hazardous waste sites. Environ. Sci. Technol. 18:521-523.

Pennington, D., 1982. Retardation Factors In Aquifer Decontamination of Organics, proceedings of the second annual Symposium and Exposition on Aquifer Restoration and Ground Water Monitoring, NWWA, pp. 1-5.

Rai, D., J.M. Zaehara, A.P. Sehwab, R.L. Schmidt, D.C. Girvin, and J.E. Rogers, 1984. Chemical Attenuation Rates Coeffieients, And Constants In Leachate Migration, volume 1: a critical review, Electric Power Research Institute Report EPRI EA-3356.

Rao, P.S.C., and J.M. Davidson, 1980. Estimation Of Pesticide Retention And Transformation Parameters Required In Nonpoint Source Pollution Models, M R. Overeash and J.M. Davidson, eds., Environmental Impact of Nonpoint Source Pollutants, Ann Arbor Seience Publishing Inc., Ann Arbor, Michigan, pp. 23-67.

Reinbold, K.A., J.J. Hassett, J.C. Means, and W.L. Banwart, 1979. Adsorption of Energy-Related Organic Pollutants: A Literature Review. Environmental Research Laboratory, U.S. Environmental Protection Agency, Athens, Georgia, EPA600/3-79-086, 178 pp.

Reynolds, W.D., 1978. Column studies of strontium and cesium transport through a granular geologic porous medium, M.Sc. thesis, University of Waterloo, Ontario, Canada, 149 pp.

Rogers, R.D., J.C. McFarlane, and A.J. Cross, 1980. Adsorption and desorption of benzene in two soils and montmorillonite clay, Environmental Science andTechnology, 14, pp. 457-460.

Schwille, F. 1988. Dense Chlorinated Solvents in Porous and Fractured Media: Model Experiments, (English Translation), Lewis Publishers, Ann Arbor, Michigan.

Silka, L. 1988. Simulation of Vapor Transport Through the Unsaturated Zone- Interpretation of Soil-gas Surveys, Ground-water Monitoring Review, pp. 115-123.

Singer, D. A. 1972. ELIPGRID: A Fortran IV program for calculating the probability of success in locating elliptical targets with square, rectangular and hexagonal grids. Geocom Programs 4-1-16.

Singer, D. A. 1975. Relative Efficiencies Of Square And Triangular Grids In The Search For Elliptically Shaped Resource Targets. J. Research U. S. Geological Survey 3:163-167.

Sitar, N. J.R. Hunt, and J.T. Geller, 1990. Practical Aspects of Multiphase Equilbria in Evaluating the Degree of Contamination," published in the proceedings of the International Association of Hydrogeologists Conference on Subsurface Contamination by Immiscible Fluids, April 18-20, Calgary, Alb.

Starks, T. H. 1986. Determination of support in soil sampling. Mathematical Geology 18:529-37.

State of California, Department of Health Services, Toxic Substance Control Division, 1986. The California Site Mitigation Decision Tree Manual. State of California.

Sundstrom, D.W., and H.E. Klei, 1979. Wastewater Treatment, Prentice-Hall, Inc., Englewood Cliffs, N.J., 444 pp.

Travis, C. & Doty, C., 1990. Can contaminated aquifers at Superfund sites be remediated?, U Environ. Sci. Technol. 24(10):1464.

Travis, C.C., and E.L. Etnier, 1981. A survey of sorption relationships for reactive solutes in soil, Journal of Environmental Quality, 10(1), pp. 8

U. S. Environmental Protection Agency, 1980. Adsorption Movement, and Biological Degradation of Large Concentrations of Selected Pesticides in Soils. Solid and Hazardous Waste Research Division Municipal Environmental Research Laboratory: EPA600/2-80-1 24

U. S. Environmental Protection Agency, 1986. Test Methods for Evaluating Solid Waste, Physical/Chemical Methods, 3rd Ed. Office of Solid Waste and Emergency Response. EPA, SW-846.

U.S. Environmental Protection Agency. 1986. Test Methods for Evaluating Solid Waste. Third edition. Office of Solid Waste and Emergency Response. SW-846. November 1986.

U. S. Environmental Protection Agency, 1987. Batch-type Adsorption Procedures for Estimating Soil Attenuation of Chemicals, a Draft Technical Resource Document. Office of Solid Waste and Emergency Response: EPA/530-SW-87-006.

U.S. Environmental Protection Agency, 1988. Contaminant Transport in Fractured Media: Models for Decision Makers, EPA/600/SF-88/002, Oct.

U.S. Environmental Protection Agency. 1989. Methods for attainment of cleanup standards. Volume 1: Soils and solid media. Office of Policy, Planning, and Evaluation. EPA 230/02-89042. February.

U.S. Environmental Protection Agency. 1989. Soil Sampling Quality Assurance User's Guide. Second edition. Environmental Monitoring Systems Laboratory. EPA1600/8-89/046. March.

U.S. Environmental Protection Agency, 1989. Evaluation of Groundwater Extraction Remedies, Vol. 1 Summary Report, EPA/540/2-89/054.

U. S. Environmental Protection Agency., 1991, Directive on "Considerations in Ground-Water Remediation at Superfund Sites—Update" form Don R. Clay, Office of Solid Waste and Emergency Response.

U.S. Environmental Protection Agency, 1991, Estimating Potential for Occurrence of DNAPL at Superfund Sites, OSWER, 9p.

Verschueren, K. 1983. Handbook of Environmental Data on Organic Chemicals, Van Nostrand Reihhold, New York.

Verschueren, K., 1983. Handbook of Environmental Data on Organic Chemicals, Van Nostrand Reinhold Co., New York.

Villaume, J.F. 1985. Investigations at Sites Contaminated with Dense Nonaqueous Phase Liquids (DNAPLs), Groundwater Monitoring Review, Vol 5, No. 2, p 60-74.

Waterloo Centre for Groundwater Research, 1989. Univeristy of Waterloo Shourt Course, "Dense Immiscible Phase liquid Contaminants in Porous and Fractured Media," Kitchener, Ontario, Canada, Nov. 6-9,

Weber, W.J., Jr., 1972. Physicochemical Processes for Water Qualily~ Wiley-lnterscience, l~ew York, 640 pp.

Weed, S.B., and J.W. Weber, 1974. Pesticide-organic matter interactions, [In] Guenzi, W.D. (ed), Pesticides in Soil and U'ater, Soil Science Society of America, Inc., Madison, Wisconsin.

Wilson, J.L. and S.H. Conrad, 1984. Is Physical Displacement of residual Hydrocarbons a Realistic Possibility in Aquifer Restoration?, Proceedings of the NWWA/API Conference on Petroleum Hydrocarbons and Organic Chemicals in Ground-water Prevention, Detection and Restoration, Houston, Texas, p 274-298, Nov. 5-7.

Wilson, J.T., C.G. Enfield, W.J. Dunlap, R.L. Cosby, D.A. Foster, and L.B. Baskin, 1981. Transport and fate of selected organic pollutants in a sandy soil, Journal of Environmental Quality, 10, pp. 501-506.

Yfantis, E. A., G. T. Flatman, and J. V. Behar. 1987. Efficiency of kriging estimation for square, triangular, and hexagonal grids. Mathematical

Geology 19:183-205.

Zirschky, J., and R. O. Gilbert. 1984. Detecting hot spots at hazardous waste sites. Chem. Engin. 91:97-100.

CHAPTER 11

ORGANIZATION AND ANALYSIS OF WATER QUALITY DATA

Water quality analyses and interpretative data summaries are important to Phase II site characterization efforts, but these data are even of greater importance under detection and assessment activities associated with facility compliance. Detection monitoring efforts are performed to verify attainment of performance objectives; and assessment monitoring is made of efforts to identify facility non–compliance, in terms of nature, location, and extent of contamination.

Hydrogeologists and others who make use of water analyses must incorporate individual values, or large numbers of analyses (data sets) into their interpretations. On the basis of on these interpretations, final decisions are made regarding detection and assessment monitoring programs. In the last 15 years, few aspects of hydrogeology have expanded more rapidly than interpretation of water quality assessment at and around industrial plants and waste management facilities. The expansion of water quality programs was based on two factors (Davis, 1988):

• Improvements in analytical methods have greatly increased our ability to accurately and precisely analyze a vast number of trace elements and organic compounds in water. An automation of analytical processes now allows statistically significant studies of constituents that formerly were beyond the analytical detection capabilities of all but the most sophisticated instrumentation.

• The expansion of water chemistry technology has occurred in response to public and professional concern about health, particularly as related to analyses of radionuclides and trace–level organic hydrocarbon compounds.

As a result, many comprehensive programs for monitoring water quality at waste management facilities have resulted in analyses of thousands of individual

parameters. Interpretation of such massive quantities of data must include attempts to determine correlations among the parameters and demonstration of correlations that exist between water quality parameters and the hydrogeology of the site. Comparison of water quality in upgradient (background) and downgradient wells may also be necessary as part of detection monitoring programs. In the Superfund program, data are being collected by EPA regional offices, states, other federal agencies, PRPs, and contractors. The data are used to support the following functions:

• Waste site characterization

• Risk assessment;

• Evaluation of clean up alternatives;

• Monitoring of remedial actions; and,

• Monitoring post-cleanup conditions

In general terms report of water quality should contain an organized evaluation of the data, including graphics, as necessary, to illustrate important environmental relationships. The recommended procedure for assessment of water quality data, in baseline and detection monitoring, is illustrated on Figure 11–1.

The interpretative techniques and correlation procedures, described herein, do not require extensive application of chemical principles. The procedures range from simple comparisons and inspection of analytical data to more extensive statistical analyses. Typically a first step in evaluating ground-water quality is to review existing hydrogeologic information and try to define subsurface stratigraphy and ground-water flow. Most regulations require comparisons of data between upgradient to downgradient conditions. This usually only is useful in homogeneous aquifers that have very rapid flow (i.e.,

11-1

GENERAL WATER QUALITY ASSESSMENT PROCEDURE

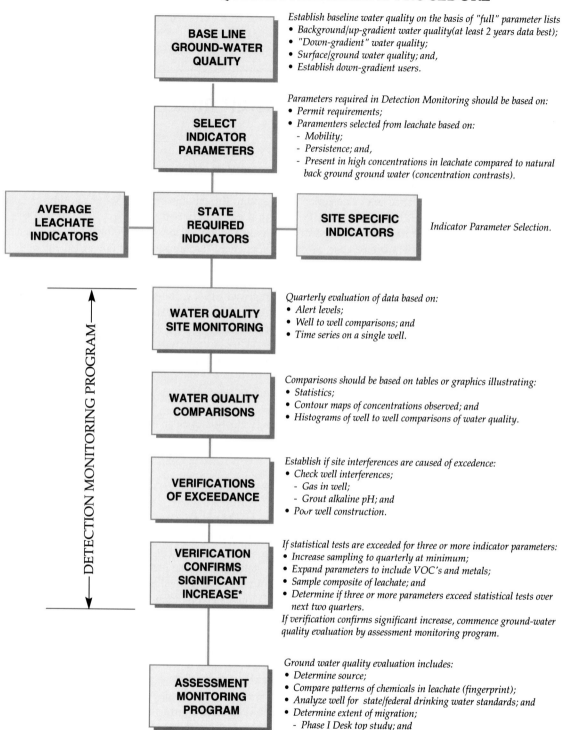

BASE LINE GROUND-WATER QUALITY

Establish baseline water quality on the basis of "full" parameter lists
• Background/up-gradient water quality(at least 2 years data best);
• "Down-gradient" water quality;
• Surface/ground water quality; and,
• Establish down-gradient users.

SELECT INDICATOR PARAMETERS

Parameters required in Detection Monitoring should be based on:
• Permit requirements;
• Paramenters selected from leachate based on:
  - Mobility;
  - Persistence; and,
  - Present in high concentrations in leachate compared to natural back ground ground water (concentration contrasts).

AVERAGE LEACHATE INDICATORS   **STATE REQUIRED INDICATORS**   **SITE SPECIFIC INDICATORS**   Indicator Parameter Selection.

WATER QUALITY SITE MONITORING

Quarterly evaluation of data based on:
• Alert levels;
• Well to well comparisons; and
• Time series on a single well.

WATER QUALITY COMPARISONS

Comparisons should be based on tables or graphics illustrating:
• Statistics;
• Contour maps of concentrations observed; and
• Histograms of well to well comparisons of water quality.

VERIFICATIONS OF EXCEEDANCE

Establish if site interferences are caused of excedence:
• Check well interferences;
  - Gas in well;
  - Grout alkaline pH; and
• Poor well construction.

VERIFICATION CONFIRMS SIGNIFICANT INCREASE*

If statistical tests are exceeded for three or more indicator parameters:
• Increase sampling to quarterly at minimum;
• Expand parameters to include VOC's and metals;
• Sample composite of leachate; and
• Determine if three or more parameters exceed statistical tests over next two quarters.
If verification confirms significant increase, commence ground-water quality evaluation by assessment monitoring program.

ASSESSMENT MONITORING PROGRAM

Ground water quality evaluation includes:
• Determine source;
• Compare patterns of chemicals in leachate (fingerprint);
• Analyze well for state/federal drinking water standards; and
• Determine extent of migration;
  - Phase I Desk top study; and
  - Phase II Field Investigation & Phase III Assessment Monitoring.

DETECTION MONITORING PROGRAM

* Subtitle D regulations have different verification procedures.

Figure 11-1 General Monitoring Procedure

lOO's feet/year). As will be fully explained in the following sections more than one upgradient well is necessary to account for subsurface spatial affects present on most sites. When facilities are located over low hydraulic conductivity soils and rock that are heterogeneous in composition additional spatial variability considerations must be addressed in the evaluation of water quality. Upgradient to downgradient comparisons for natural constituents may not be possible for those sites where downward gradients predominate. These situations require sufficient background sampling points to establish the ambient spatial and seasonal variability. Landfills along hillsides often have recharge and discharge conditions that create different chemical evolution pathways and natural differences in upgradient to downgradient ground-water quality (Freeze and Cherry, 1979). In some cases, wells can be located "sidegradient"

(along the downgradient directions of ground-water flow) at these sites if enough land is available to eliminate concerns about landfill impacts. The Federal regulations recognize that if a site is located on a ridge, for example, where there is no upgradient wells available, then wells can be compared to themselves. This comparison is called a trend analysis or intra-well comparisons.

Natural ground-water quality is known to vary both spatially — between wells — and temporally — at a single well. Anthropogenic (or man made) effects also contribute to the variability observed in water quality data. To evaluate the potential releases from a facility to ground water the sources of natural variability must be fully understood and the additional interrelationships of human activities to ground-water quality can greatly affect understanding of site conditions. Sources of variability and error in ground-water data are listed in Figure 11-2.

SOURCES OF VARIABILITY AND HUMAN ERROR IN GROUND-WATER QUALITY DATA

SPATIAL	TEMPORAL	WELL CONSTRUCTION	SAMPLING
GEOLOGIC PROPERTIES • Lithologic composition sorting and grain size • Structure of lithologic units • Bedding planes • Fractures (joints and faults) • Soil development • Properties of vadose zone **HYDRAULIC CONDITIONS** • Location of recharge/ discharge zones • Proximity of water • Presence of aquitards • Pumping **OTHER** • Other chemical sources • Non-point source inputs	**TRENDS** **SEASONAL** • Recharge • Irrigation • Fertilization • Pesticide/herbicide application • Frozen ground **PERIODIC** • Short term precipitation • Pumping • River flooding	**DRILLING PROCESS** • Drilling fluids • Type of borehole • Inter-aquifer transport of materials **WELL DESIGN** • Casing and screen material • Diameter • Screen length, depth, slot size • Filter pack material • Annular seal **WELL DEVELOPMENT**	**COLLECTION** • Purging methods • Purging rate/duration • Sampling apparatus • Cross-contamination between wells • Field versus laboratory measurements • Sample preparation filtering/container/ preservatives/storage time • Operator error • Incomplete well development **ANALYTIC ERROR** • Analytic methods • Operator experience • Instrument calibration • Interference from other constituents • Holding times • Clerical/transcription errors

SOURCE: Modified from Doctor et al, 1985

Figure 11-2   Sources of Variability in Ground Water

Natural spatial variability of ground-water quality is often due to variations in lithology within both aquifers and confining units (Sen, 1982). Soil and rock heterogeneity may cause the chemical composition of ground water to vary even at short distances. As described in previous chapters of this manual, spatial variation in water quality data may be additionally affected by variations in well installation and development methods, as well as, the sampling techniques used in the program (Doctor et al, 1985a).

Temporal or seasonal effects are usually associated with annual cycles in precipitation recharge events, to shallow, unconfined aquifers and additionally to areas where surface water and aquifer interactions are significant (Harris et al, 1987). Also seasonal pumping for irrigation and high summer recharge from non-point pollution sources may be causes for seasonal fluctuations in background water quality (Doctor et al, 1985a). A literature review on seasonality in ground-water data is presented by Montgomery et al (1987).

The relative importance of these sources of variability is clearly site specific. Doctor (1985a) described natural temporal and spatial variability was greater in magnitude than sampling and analytic error, unless gross sample contamination or mishandling of the samples occurs (Doctor et al, 1985a). Goals and procedures used in developing a monitoring program (i.e., baseline or detection), tasks, and descriptions of tasks are illustrated in Figure 11–1.

11.1 BASELINE WATER QUALITY

Characterizing the existing, or baseline, quality of ground water is an important task for a number of reasons. First, existing drinking water quality standards normally defines the baseline ground–water conditions, against which risks to human health and the environment is evaluated. Second, existing ground–water quality, in part, determines current uses and affects potential future uses of the water. In addition, determining ground-water uses is an important initial step in identifying potential exposure pathways downgradient from the site.

In evaluating the background water quality for an area, the investigation must consider possible background concentrations of the selected indicator chemicals, and the background concentrations of other potential constituents of leachate. Existing chemical parameters associated with indicator chemicals (i.e., chloride or iron) or other RCRA hazardous constituents may be due to natural geologic conditions in the area; prior releases from the old, unlined landfills; or prior or current releases from other upgradient sources. Evaluation of water quality parameters in ground water is necessary to establish an existing baseline of ground–water quality to which the incremental effects of a potential release can be added.

Measuring ambient concentrations of every RCRA–listed hazardous constituent is not feasible during most baseline studies. To adequately assess background ground–water quality, the investigation should attempt to identify other potential sources in the area (e.g., CERCLA sites, RCRA facilities, municipal landfills, agricultural areas, or NPDES discharges to surface waters); and to identify which constituents are most likely to originate from each source. Some of the background chemicals may also be site–specific indicator parameters, particularly if the facility has experienced a prior release. When determining which chemicals to include on a list of background parameters, the investigator should include all indicator chemicals described as baseline, water–quality parameters in the next section.

Where sufficient data from historical monitoring are unavailable, the investigator may install a ground–water monitoring system, or expand an existing system, in order to adequately assess the background quality of ground water. The design of a monitoring program should be based on guidance in the previous chapter and a minimum, background water quality should be, as based upon at least two separate sampling rounds of existing or newly–installed monitoring wells.

For facilities that have experienced a prior release, the investigation should also establish the results of any sampling, monitoring, or hydrogeological investigations conducted in connection with the release (if available), and should provide references to any reports prepared in connection with that release.

11.2 SELECTION OF INDICATOR PARAMETERS

The United Nations Statistical Office defines "environmental statistics" as "multidisciplinary in nature encompass(ing) the natural sciences, sociology, demography and economics. In particular environmental statistics: (a) cover natural phenomena and human activities that affect the environment and in turn affect human living conditions; (b) refer to the media of the natural environment, i.e., air, water, land/soil, and to the man made environment which includes housing, working conditions and other aspects of human settlements".

Environmental indicators are environmental statistics, or aggregations of environmental statistics, that are used in some specific decision making context to demonstrate environmentally significant trends or relationships. An environmental indicator can be a representative indicator which is selected by some procedure such as expert opinion or multivariate statistical methods to reflect the behavior of a larger number of variables, or it can be a composite indicator which aggregates a number of variables into a single quantity (i.e., an index).

The concept of the "indicator parameter" forms the basis for water quality sampling programs. Since an investigator cannot include all chemical parameters that may be present in a natural or contaminated ground–water system, a selection process must be used to bring the spectrum of chemical parameters down to a workable number. These indicator parameters are selected to provide a representative value that can be used to establish performance of a facility (detection) or quantify rate and extent of contamination (assessment).

Each chemical analysis, with its columns of parameter concentrations reported to two or three significant figures, has an authoritative appearance which can be misleading. Indicator parameters in general terms must represent the movement of ground water or change in water quality in a clear–cut and understandable descriptive presentation.

11.2.1 Detection Monitoring Indicator Parameters

Detection monitoring programs require that individual chemical parameters be selected to represent the natural quality of the water, as well as, the chemical parameters that may be changed or adversely affected through facility operation. These parameters, called "indicators," are selected with consideration of a number of criteria:

- Required by permit, State or Federal regulation, or regulatory guidance;

- Are mobile (i.e., likely to reach ground water first and be relatively unretarded with respect to ground–water flow), stable, and persistent;

- Do not exhibit significant natural variability in ground water at the site;

- Are correlative with constituents of the wastes that are known to have been disposed at the site;

- Are easy to detect and are not subject to significant interferences due to sampling and analysis;

- Are not redundant (i.e., one parameter may sufficiently represent a wider class of potential contaminants);

- Do not create difficulties during interpretation of analyses (e.g., false positives or false negatives, caused by common constituents from the laboratory and field).

Selection of indicator parameters should consider natural levels of constituents in the detection process. Because chemical indicators include naturally occurring chemicals, Table 11–1 provides an example indicator parameter list with ranges of values occurring in natural aquifers, as well as, the persistent and mobile parameters typically present in leachates from sanitary landfills:

TABLE 11-1 Example Indicator Parameters for Sanitary Landfills

Indicators of Leachate	Ranges in Natural Aquifers
TOC (filtered)	1 – 10 ppm
pH	6.5 – 8.5 units
Specific conductance	100 – 1,000 mm/cu
Manganese (Mn)	0 – 0.1 ppm
Iron (Fe)	0.01 – 10 ppm
Ammonium (NH_4 as N)	0 – 2 ppm
Chloride (Cl)	2 – 200 ppm
Sodium (Na)	1 – 100 ppm
Volatile organics*	<40 ppb

*(USEPA Method 624)

The above indicators represent a restricted selection of parameters measurable in an aquifer, and limit the ability of an investigator to assess baseline water quality. However, they are the most likely parameters to undergo change when ground water is effected by chemical releases from a solid waste facility.

11.2.2 "Complete" Detection Parameter List for Sanitary Landfills

Although individual definitions vary, a "complete" analysis of ground water includes those natural constituents which occur commonly in concentrations of 1.0 ppm or more in ground water. Depending on the hydrogeologic setting, a "complete" analysis is shown in Table 11–2.

TABLE 11-2 A "Complete" Water Quality Parameter List

Ammonia (as N)	Total Organic Carbon (TOC)
Bicarbonate (HCO3)	pH
Calcium	Arsenic (As)
Chloride	Barium (Ba)
Fluorides (F⁻)	Cadmium (Cd)
Iron (Fe)	Chromium (Cr⁺⁺⁺)
Magnesium (Mg)	Cyanide (Cn)
Manganese, (Mn²⁺)	Lead (Pb)
Nitrate (as N)	Mercury (Hg)
Potassium (K)	Selenium (Se)
Sodium (Na⁺)	Silver (Ag)
Sulfate (SO₄)	Nitrogen, dissolved (N²)
Silicon (H₂SiO₄)	Oxygen, dissolved (O²)
Chemical Oxygen Demand (COD)	
Total Dissolved Solids (TDS)	
The volatile organic compounds (VOCs) established in Method 624	

In general, the investigator should examine closely the water quality results if these indicators are above the natural ranges of ground waters given above. The concentration of total volatile organics (40 ppb) was established from tolerance intervals on numerous upgradient wells at 17 facilities (Hurd, 1986) and include cross-contamination interferences from the collection and analysis process.

11.3 ANALYTIC SERVICES LABORATORIES

The importance of laboratory selection for evaluation of water quality samples cannot be over stressed. Significant legal and technical decisions which will determine the success of the environmental monitoring program will depend on the quality of the lab's work. The choice of a laboratory may ultimately make the difference between a successful project and one that falls into a pattern of persistent failure, frustration, later recrimination and resampling.

The general requirement of a laboratory program is to determine the types and concentrations present of both inorganic and organic indicator parameters. Depending on the project requirements, specific laboratory testing methodologies have been approved within the project scope or are specifically required. For example, under Subtitle C of RCRA, analytical methods contained in Test Method's for Evaluating Solid Waste, Physical Chemical Methods (SW-846) are specified.

Under the federal CERCLA/SARA program, Contract Laboratory Program (CLP) was established by the EPA in

fiscal year 1980 as one of the first organizational activity. The CLP program provides standard analytical services and is designed to obtain consistent and accurate results of demonstrated quality through use of extensive QA/QC procedures.

The selection of an analytical laboratory services depends primarily on the client needs and the intended end-use of the analytical data. While laboratories performing analytical services must use well defined analytical methods as contained in "standard methods documents" and to employ method-specified quality control procedures, the choice of which laboratory may be based on other factors, as are described in the following sections. Laboratory analyses are critical in determining project direction, therefore the reliability of the analytical data is essential. The use of quality control/quality assurance must be an integral part of laboratory operations and an important element in each phase of the technical review of data and reports.

11.3.1 Steps In A Lab Evaluation

The first step in the laboratory selection process is for the client or the consulting engineer to organize a detailed document defining the analytical and quality control (QC) requirements of the program within the project scope of work. A typical laboratory would be assigned the responsibility to:

• Evaluate the scope of the project;
• Confirm its capacity to comply to the program;
• Resolve identified discrepancies in the scope of work requirements;
• Propose viable analytical alternatives consistent with the Data Quality Objectives, (DQO's) of the program; and
• Confirm project commitment to within the specified turn-around times.

Assessment monitoring programs often require that a Quality Assurance Project Plan (QAPP) be approved by the responsible regional EPA office, state regulatory, or other regulatory agency. The QAPP documentation describes:

• The full scope of the project field and laboratory activities;
• The analytic methods to be used with their QC requirements, and;
• Projects reporting and documentation standards.

An experienced laboratory will normally perform a complete and independent assessment of the QAPP, and document the laboratories complete understanding of project responsibilities.

Very large or complex projects, may require data collection activity over a broad spectrum of soil and water analyses that may require multiple laboratories. These very large projects can be handled in several ways, 1) contract with additional laboratories as needed to encompass the full scope of the project, or 2) contracting with a primary or lead laboratory. The lead laboratory, then has the direct responsibility to obtain subcontracting laboratory services. This is not a job for amateurs, since, as additional laboratories are added to the project, complexities mount rapidly that require significant experienced project management efforts.

11.3.2 The Sops And Qapp

The majority of analytical laboratories have standard procedures for how the laboratory conducts its analytical quality and reporting programs just as consulting firms have standard operating procedures (SOP's) for field testing procedures. Sample and data pathways (Figure 11-2) should form part of the documents provided for review from the laboratory. Simple listing of analytical procedures tell only part of the necessary documentation, sample preparation and instrumentation procedures should refer to approved methods (as designated in the QAPP, or work plan). Procedures for sample handling and storage, sample

TABLE 11-3 Laboratory Quality Assurance Program Plan Guidelines

- Title Page
- Table of Contents
- Laboratory and Quality Assurance Qrganization
- Facilities and Equipmen
- Personnel Training and Qualifications
- Laboratory Safety and Security
- Sample Handling and Chain-of-Custody
- Analytical Procedures
- Holding Times and Preservatives
- Equipment Calabration and Maintance
- Detection Limits
- Quality Control Objectives for Accuracy, Precision and Completeness
- Analysis of Quality Control Samples and Documentation
- Data Reduction and Evaluation
- Internal Laboratory Audits and Approvals from Other Agencies
- Quality Assurance Reports to Management
- Document Control

tracking, bottle and glassware decontamination, document control, etc. are described in the non-analytical (SOP's).

As with any quality assurance program documents, the laboratory SOP's should employ formal document control procedures, so that revision numbers and dates are presented on each page. All SOPs should include the staff position performing the task, the specific analytical and quality procedures involved, and the individual responsible for resolving difficulties before taking corrective action when out-of-control events occur. Formal approval by the designated QA Manager and/or laboratory manager should appear on the SOP Permanent training documentation include each staff member's review and understanding of the SOP's. All copies of earlier revisions of SOPs should also be retained within the laboratory documentation system.

The QAPP is the document that brings together the laboratory QA/QC plans and SOP's and specific project requirements. The QAPP should include, at a minimum, the information presented in Table 11-3. Laboratory quality systems must pay particular attention to data quality assessment and corrective action procedures. The document through reference to the laboratory SOP's and QA/QC program specifically address the laboratory's mechanisms for a program of QC samples analyzed at the appropriate or predetermined frequencies. The QC sampling requirements within the quality assurance program are usually client, method or contract-dependent. The QA Plan should specify the mechanisms by which the laboratory identifies these requirements.

Control and reporting of analytical results are an important part of an environmental laboratories' responsibilities. Laboratory data quality assessment procedures should include:

- A general description of all data review levels;
- The responsibilities at each level;
- Examples of the documentation accompanying the assessment;
- The analytical data quality criteria used by the reviewers; and,
- The final accountability or "sign-off" on the data report.

The control and reporting section should also address the use of data qualifiers (tags), and whether or not it is the laboratory's policy to adjust results based on discovery data or observed blank sample contamination. Since very low levels of organic parameters can cause significant data evaluation problems the "adjustment" of data by the labs must be well known by project staff responsible for data interpretation. As general guidance data tags are generally preferred over data adjustment.

11.3.3 Custody And Chain-Of-Laboratory Security

Environmental laboratories should be restricted to authorized personnel only. Security should extend to sample and data storage areas even for the smallest laboratories. The work plan applicable to the project should contain specific chain-of-custody requirements. The basic components for maintaining sample chain-of-custody are:

• Samples must be delivered into the possession of an authorized laboratory staff member, by the sample handling or transporting organization (such as for example Fed. Ex or specific sampling teams), or;

• Samples must be within the authorized staff member's line-of-sight; or; and,

• Samples must be locked in a secured storage area with restricted access.

Samples should be kept in locked storage with restricted access when not being processed (refrigerated, as required). The chain-of-custody form is used to document the transfer of these any sample fractions (such as splits, extracts or digestates) as part of the permanent sample processing record.

11.3.4 Facility And Equipment

A quality assurance program typically contains documentation on equipment maintenance and calibrations, analytical laboratories must maintain such documentation as part of its QA/QC program. Standards used in the analytical process must also be traceable to a certified source such as, the EPA, the National Institute of Standards and Technology (NIST) or commercial sources.

A very important part of the success of an environmental sampling program for state or federal regulatory programs is the turn-around time of the sample. This turn-around is defined as the time from field sample collection to receiving QA/QC'ed confirmed analytical results usable for evaluating the performance of the facility. Turn-around times provided by laboratories are typically based on the current sample load and capacity, average turnaround times for data delivery, and history in meeting sample holding times. Holding time is the maximum allowable time between sample collection and analytical testing. Each chemical parameter has some holding time attached to the sample such as 24 HRS, two weeks or 30 days. For most environmental monitoring projects, data for analytical samples not meeting the required holding times will cause the results to be rejected or at best qualified. Exceeding

holding times has caused many environmental programs to get into very serious trouble with both permits requirements and stipulated penalties for the project deliverables.

Analytical laboratories are often plagued by persistent low levels of organic parameters such as methylene chloride, or acetone. These parameters are common laboratory chemicals used in various organic extraction processes. As such these organics often show up in analytical results as low background levels. Some laboratories commonly subtract these values from results; other laboratories report the values and let the investigator explain the results to regulatory agencies; others tag the data as background for the lab. Whatever method used by the laboratory, expect to see such low levels of common laboratory organic chemicals in analytical results. The laboratory should report in QA/QC plans how they deal with such data.

The laboratory may purchase reagent-grade water or produce its own, using a water purification system. A logbook should also be maintained to document checks for water purely, whatever the source. The product water should also be the source for QC method blanks (i.e., samples) in order to verify the absence of organic and inorganic constituents.

11.3.5 Data Accuracy And Availability

Reliability of laboratory generated environmental data depends on a whole series of program procedures that include proficiency test samples, mechanisms for handling data inquiries, QA reporting to management, organized ways of handling corrective action, long term data, storage and access. Initially, analytical results must be reviewed in reelationship to the other analytes reported for the project. The purpose of the type of review is to attempt to identify trends, anomalies or interferences that can miss lead or bias the overall use of the data. The technical review process begins with an initial review of the testing program and the overall project requirements. Once samples are analyzed according to project plans, and analytical results generated, the laboratory should conduct an initial math check, QC review and a laboratory supervisor's technical release of the data. Reviewers consider the relative accuracy and precision of each analyte when interpreting the analytical data. Several alternative methods are avaliable for entering results into a data base. Procedures such as double key entry and internal computer error-checking routines are employed to compare both data entries and generate an exceptions report. Data must be reviewed by qualified staff before changing any analytical or field generated results. These procedures along with those described below are used to establish the reliability of the results before moving to evaluation of the actual project data sets.

Data Inquires

The mechanisms in place for handling data inquiries are often vital to the success of a project. No matter the length or the extent of the program, data inquiries will happen on a reoccurring basis. In general, procedures used in the laboratory should describe how the data are requested from storage, the individual(s) responsible for resolving the inquiry, and the standard response time.

Expect to see questionable data coming from even the best analytical laboratories. The laboratory should have an SOP in place for responding to client inquiries, both technical and administrative (invoicing, sample shipping logistics, requests for additional copies, etc.).

QA Reports to Management and Corrective Action

When an out-of-control incident is observed on water quality samples from a environmental monitoring program, it is essential that the event be documented and a form of corrective action be taken. Out-of-control events may be:

- Isolated to individual QC sample recoveries or calibration criteria failures; or,
- They may be systematic—having widespread effect on the analytical data generation system.

The sample once triggering an out-of-control action, may for example require reextraction, or may require qualification with a notice to the data end-user that identifies the criteria that were not met and the effect on data acceptability. When sufficient sample volume is not available to reprocess a sample a resampling may be required for an extreme out-of-control event.

Laboratory records should be archived so that individual reports or project files can be easily retrieved. As with any QA program access to data must be restricted to specified individuals. If data are also stored on magnetic tape, the tapes should be similarly protected with back-up copies stored at a second location. As part of the quality assurance program resumes and qualifications of key technical staff must be maintained along with training records for the staff.

11.4 MDLS, PQLS, IDLS, EMLRLS

Site assessment projects generate much analytical data that may be reported by the laboratory in numerous ways.

These reported values many times reference some form of detection limits including: Method Detection Limits (MDL's), Instrument Detection Limits (IDL's), Practical Quantitation Limits (PQL's) or Reporting Limits (RL's). Each of these limits evolves around a detection limit. The above detection limits are only a way of statistically expressing how low a particular measuring system can measure. There are a number of ways to evaluate the limit of detection of a particular measuring device. One, for example, could take an object whose weight is known accurately such as a 10 pound weight. The 10 pound object is weighed a series of times using a typical spring loaded scale. The results of this process will vary depending on the temperature in the room, how you place the object on the scale, how accurately you can read the results, who reads the results and on the quality of the scale (Jarke, 1989). This is called "variability" of the measuring device.

If for example your results were: 10.2, 10.4, 10.7, 9.1, 9.8, 9.3, 10.0,

the average value is 10.07 pounds and the standard deviation is 0.4461. In such exercises it is a good practice to carry more figures than are really significant until you make your final calculation, then report only those figures that are significant.

The USEPA's definition of "method detection limit" (MDL), (40CFR Part 136 Appendix B), describes the detection limit for this scale as 1.1 pounds. This results in a number so that any value less than these 1.1 pounds cannot be determined to be different from zero. Even if the scale shows a value, the significance of this value remains questionable. To obtain a lower MDL result than the 1.1 pounds one must go to a scale with a much lower detection limit to get to an accurate or reliable value.

The example of the simple weight scale is similar in many respects to any measuring device since every measuring device has a detection limit and every device's detection limit is different depending on who, what, how, when and where it is used. Since all these components can vary detection limits are not constants especially for analytical instruments.

Every instrumental measuring device used an analytical laboratory has an inherent minimum limit of detection, as described above. This limit of detection is usually referred to as the specific "instrument detection limit (IDL)". For simple devices the IDL is based on the smallest unit of measurement the device is capable of reporting. For example, if a ruler has markings of sixteenth of an inch, the IDL (if based on one-half of the smallest unit of measure) would be one thirty-second of an inch. While the overall concept of IDL and MDL are quite similar, IDLs for instruments are generally far below the experimentally determined MDLs. The analytical instrument can be optimized for a specific parameter, for the desired need with fewer and more easily controlled sources of variability within the IDL procedure. MCL determinations include many more sources of variability and therefore have higher experimentally determined MDL's.

In 1980, USEPA began to administer the Resource Conservation and Recovery Act (RCRA). One of the requirements of this new law was that landfills begin to monitor ground water. The agency establish "method detection limits (MDL)"in 40CFR Part 136 Appendix B to assure that analytical laboratories were conducting the testing at an acceptable level. This regulation requires that each analytical laboratory must establish MDLs on a routine basis for every analyte, for every analyst and for every instrument. The goal of the regulation was to demonstrate that the analytical laboratories could obtain results as good as, or better than those published with many of the USEPA methods. The U.S. EPA MDL studies are always performed in highly purified water, with only a single known analyte added. The resultant MDLs, therefore, reflect the best performance a laboratory is capable of under the best conditions. Site assessment projects produce environmental samples that do not contain a single known analyte in highly purified water. Rather samples are delivered to the laboratory containing many types of organic and inorganic parameters, sometimes residing in a significantly concentrated liquid. This produces a matrix effect that can significantly raise MDL's many times over U.S. EPA reported values. As additional sources of variability presented by real samples, can include sampling, and site location variability, and the interferences that can be caused by other compounds in the sample other than the target compound. As one can imagine the effective MDL for these field samples can be many times larger than those used in establishing laboratory performance.

Although the 40 CFR Part 186 Appendix B requirements to establish MDL's are clearly explained, there is little standard in how the regulation is applied at analytical laboratories. A full spectrum of applications of MCL's is observed applied in analytical laboratory work (Jarke, 1989):

• Laboratories perform MDL studies that meet or exceed the published values, but use the published values in their reports.

• Laboratories do not perform MDL studies and assume that if they are using a USEPA approved method, then the published MDLs can be used in reporting without performing the MDL study;

• Laboratories perform MDL studies and use these as the reporting limits in their reports; and,

• Laboratories either do or do not perform MDL studies, but use reporting limits that are significantly different from the USEPA published MDLs, such as, Practical Quantitation Limits (PQLs), or Reporting Limits (RLs).

Site assessment water quality evaluations should be based on using analytical laboratories that have performed MDL studies to verify that they can perform a method, and provide QA/QC data how well they are performing that method.

The definition of MDL includes the phrase the minimum concentration of a substance that can be measured and reported with 99% confidence that the analyte concentration is greater than zero. Using the definition above, if an analytical laboratory reports all results above experimentally determined MDLs, 1% of reported values are false positives. False positives are statistically valid reported values. They appear to be real values, but in reality are not. Therefore, many laboratories that perform environment programs have recognized the need to set meaningful reporting limits. The Contract Lab Program (CLP), organized by USEPA to control site remediation analytical programs, have also recognized the false MDL rates for analytical data. The methods published for the CLP program use the concept of the PQL (Practical Quantification Limit).

PQL is considered by the EPA as the concentration that can be reliably determined within specified limits during routine laboratory operation and is defined as either 5-10 times the MDL or 5-10 times the standard deviation used in calculating the MDL. This definition of a reporting limit still raises technical questions but can be determined experimentally using statistical procedures proposed by Gibbons (1988) and presented by Grams (1989).

Additional terms have been proposed to address the ability of analytical to evaluate low levels of chemical parameters. The Environmental Committee of the American Chemical Society (ACS), published a report in 1983 addressing the issue of reporting limits and detection limits. Figure 11-3 shows graphically this idea. The Committee used instead of MDL, 'limit of detection (LOD)'. A new value, limit of quantitation (LOQ), was defined as 10 or ten times the standard deviation used in the MDL calculation. This value is equal to approximately 3 times the MDL defined in 40CFR 136 and is equal to the PQL. The ACS Committee reasoned that data above the LOQ could be reported quantitatively. The region between the LOD and LOQ contained results of ~108s uncertain quantitation.

In summary, MDLs should be used in establishing the capability of a laboratory to perform a particular test method in accordance to regulations applicable to the project. The reporting limits (RL) should be established by first determining the intended uses of the data. Reporting any value above the MDL means that some analytical values will still be false positives because they fall in the region of less certain quantitation. Each of these detection limit definitions can be summarized using the weight scale example (Jarke 1989):

• The IDL is the same as the pound scale markings;

• The MDL is determined to be 1.4 pounds based on one person (observer) using a single scale;

• The PQL would represent statistically what multiple scales being used by multiple people (observers) could achieve; and similarly,

• The RL would be a constant value that is above the statistical variation of all people using all similar type scales.

Each type of limit is based on the population observing the operation, from the smallest, the IDL, where no one is observing, to the single observer (MDL), finally, to the whole population of observers (PQL and RL).

11.4.1 Sample Dilution

In site assessment projects it is often necessary to dilute samples to either eliminate instrument or analyte interferences, or to bring down large concentrations to within instrument scale. This reduces the occurrences of

"blown columns" during gas chromatographic analysis. Diluting a sample affects fundamentally the MDL first. That is, if the MDL times the dilution factor is still equal to or less than the reporting limit, then the reporting limit remains unchanged. If, however, the effect of diluting the sample results in an MDL above the reporting limit, then a new reporting limit must be established. This may seem in conflict with the previous discussion. However, if a laboratory is using MDLs as their reporting limits, then as the sample is diluted, the MDL and reporting limits both change because they are equal. If, however, a laboratory is using the concept of a reporting limit that is larger than the MDL, then the dilution factor should only effect the MDL until it reaches the value of the reporting limit and then any further dilution should effect the two simultaneously. The client should only be aware of dilution when they effect the reporting limit as indicated in the later case.

11.4.2 Low Level Organic Chemical Results

Evaluation of low levels organics in ground water presents one of the more common problems in environmental monitoring programs. The difficulties associated with interpreting low level analytical results for organic chemicals can be divided into three broad categories:

• Deficiencies in sampling and analytical methods;

• Background levels for compounds that are commonly present in homes, industrial facilities, transportation facilities and analytical laboratories; and,

• Varying significance, as well as, incomplete data on the significance of organic compounds to public health and the environment.

All sampling and analytical methods commonly used for environmental monitoring are subject to variability and error. Replicate samples taken in the field from a single well, or samples split in the laboratory will not produce identical analytical results due to:

• imperfect sampling procedures;

• inability to maintain perfectly constant conditions around a sample point; and

• absence of perfect homogeneity in the sample material.

Replicate analysis on the same sample by the same method and even by the same analyst will not necessarily

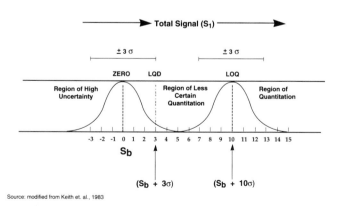

Source: modified from Keith et. al., 1983

Figure 11-3 Relationship of LOD and LOQ

produce identical analytical results. At concentrations near the analytical limits of detection (e.g., typically 1.0 - 10 mg/l for Chromatography/Mass Spectrometry (GC/MS) and lower for Gas Chromatrography (GC) with other detectors variations are likely to be most pronounced.

It may be practically impossible to produce two samples that are identical. For ground-water samples, conditions in a well will vary slightly between consecutive sampling events, or even during a single sampling process. When a sample is split after sampling, the two splits may not be exposed to the atmosphere in exactly the same way, for exactly the same lengths of time. Furthermore, the slightest amount of suspended solids or turbidity will most likely result in two samples that are also not identical. Soil samples can show an extreme lack of sample homogeneity even from samples taken a foot away from an X-Y-Z coordinates.

The key to evaluation of sampling and analytical data, therefore, is to be cognizant of the types and extent of variability inherent in sampling and analytical methods and to take into account all available QA/QC data when interpreting results.

11.4.3 Background Water Quality Evaluation

Background refers to chemical parameters introduced into a sample from natural and man related sources other than those which are the subject of the monitoring program. The problem of background changes in water quality is similar to that of analytical method variability in that it seldom is practical to eliminate it completely. There are many opportunities for a water sample to be exposed to detectable levels of both organic and inorganics at the low detection limits currently available for chemical analysis. As with the problem of method variability, the solution to background sample contamination is to first define to the extent practical the natural variability of the system, then combine these data with documentation of background levels to make reliable interpretations of analytical results.

To give a few examples, some of the most common compounds found as background levels in environmental samples are volatile organics and phthalates. Sources of these compounds include homes, transportation facilities and analytical laboratories. Some specific examples are included in Table 11-4 below.

Table 11-4 Examples of Laboratory and Cross-contamination Compounds

Compound	Typical Sources
1. chloroform	chlorination of drinking water
2. phthalates	plasticizers used in numerous household and industrial products including pipes, shower curtains, car seats and many bottles and containers, etc.
3. methylene chloride	common in paint strippers, house hold solvents, septic system cleaners and spray propellants; used extensively in laboratory procedures
4. other solvents (e.g., trichloroethylene, septic system tetrachloroethylene, cleaners, and to a limited extent toluene, dichloroethane	household cleaners, paints and paint strippers, in laboratory procedures
5. trichlorofluoromethane	common refrigerant (freon) found in freezers, refrigerators and air conditioners

Many laboratories will not even report some of these compounds (e.g., methylene chloride) below certain levels (usually the 15 to 30 ppb range) because of assumed laboratory background levels.

11.5 MONITORING SITE WATER QUALITY

Ground–water data collected during site characterization and detection monitoring is typically restructured, or simplified, and must be presented in a manner that facilitates verification and interpretation. All analytical data (physical and chemical) are reported through transmittal sheets of laboratory analysis. The data are then compiled into tables and graphic formats that facilitate understanding and correlation of the information. At the very beginning of assessment activities, the investigator should establish common data requirements listing and standard reporting format(s).

A list of all data should be provided for each sampling event and updated as new data become available. The data should include the following: well identification number or alphanumeric designation, date of analysis, name of laboratory, units of measurement, limits of detection, and chemical concentrations. The data are then categorized and organized into the established format to allow quick reference to specific values. Compilation and evaluation of laboratory data into summary reports must be performed without transcription errors. All of this task is made most achievable by use of standard formatting procedures.

11.5.1 Reporting

Laboratory results for a given analyte generally are presented as a quantified value, or as ND (not detected). All chemical data should be presented according to this protocol. Results are reported either as a quantified concentration or as less than ("<") the method detection limit or threshold value (thus, ND results are shown as "<" on the summary report).

To the extent feasible, all laboratory results should be reported in a manner similar to that described above.

11.5.2 Significant Digits

The number of significant digits reported by the laboratories reflects the precision of the analytical method used. Rounding of values is generally inappropriate because it decreases the number of significant digits and alters the apparent precision of the measurements. Therefore, the investigator retains the number of significant digits in the transcription, evaluation, and compilation of data into secondary reports. Variation in the number of significant digits reported for a given analyte may be unavoidable if there is an order of magnitude change in the concentration of a chemical species from one round of sampling to the next, or if the precision of the analytical methodology differs from one round to the next.

11.5.3 Outliers

Unusually high, low, or otherwise unexpected values (i.e., outliers) can be attributed to a number of conditions, including:

• Sampling errors or field contamination.

• Analytical errors or laboratory contamination.

• Recording or transcription errors.

• Faulty sample preparation or preservation, or shelf–life exceedance.

• Extreme, but accurately detected, environmental conditions (e.g., spills, migration from facility, etc.).

Gross outliers may be identified by informal visual scanning of the data. This exercise is facilitated by printouts of high and low values. Formal statistical tests also are available for identification of outliers (Dunn and Clark, 1974). When feasible, outliers are corrected (e.g., in the case of transcription errors) and documentation and validation of the reasons for outliers are performed (e.g., review of field blank, trip blank, QA duplicate–sample results, and laboratory QA/QC data). Results of the field and laboratory QA/QC results, as well as field and laboratory logs of procedures and environmental conditions, are invaluable in assessing the validity of reported but suspect concentrations. Outliers that reasonably can be shown not to reflect true or accurate environmental conditions are eliminated from statistical analyses, but are permanently flagged and continue to be reported within summaries of data.

11.5.4 Units Of Measure

Units of measure are recorded for each parameter in the laboratory reports. Special care is taken not to confuse "μg/L" measurements with "mg/L" when compiling, transcribing, or reporting the data.

11.6 COMPARISONS OF WATER QUALITY

The type of interpretation most commonly required of hydrogeologists is preparation of a report summarizing the water quality in an aquifer, a drainage basin, or some other unit that is under study. The author of such a report is confronted with large amounts of data from a few sources, and this information must be interpretable. The finished report must convey water quality information in ways in which it will be understandable by staff of the regulatory agency and technical management.

As an aid to interpreting chemical analyses, several approaches will be discussed that can serve to identify chemical relationships and to predict chemical changes in space and in time. Different types of visual aids, which often are useful in reports, will be described. The basic methods used during interpretation are inspection and simple mathematical or statistical treatment to identify relationships among chemical analyses; procedures for extrapolation of data in space and time; and preparation of graphs, maps, and diagrams to illustrate the relationships.

11.6.1 Inspection And Comparison

A simple inspection of a group of chemical analyses generally will allow distinction of obviously interrelated parameter subgroups. For example, it is easy to group waters that have dissolved solids or chloride concentrations falling within certain ranges. The consideration of dissolved solids, however, should include consideration of the kinds of ions present as well.

Simple visual review of tabulated water quality data is probably the most frequently used technique by regulatory agencies, to decide if a particular facility is contributing to ground–water contamination. Such analyses commonly exclude consideration of geologic and hydrogeologic conditions at the site. However, placement of water–quality data on maps and cross–sections provides a powerful tool for integration of all chemical and hydrogeologic conditions. These data can be arrayed on maps and cross–sections in a number of ways to enhance interpretation of flow paths and ground–water movement.

Figure 11–4 shows a typical, tabular array of water–quality data. Because such a format requires significant efforts to assimilate, it is recommended that alternative formats be employed to display data whenever appropriate for detailed understanding of water quality information. Water quality display formats in increasing complexity can be divided into the following categories:

- Tabular presentation
- Contour maps
- Time series displays
- Histograms
- Box Plots
- Stiff diagrams
- Scholler diagrams
- Trilinear diagrams
- Correlation coefficients
- Probability plots

Each of these formats can have useful application for understanding variations in water quality and categorization of ground water. Tabular presentations are a necessary evil, the associated tedium of which can be eased by use of summaries and averages. Particular care should be used in proof–reading sets of compiled or merged data since massive arrays of data almost always contain errors of transcription. Computer–based spread–sheets can decrease time for data reduction, however, any transcription of data must be carefully checked, and rechecked for accuracy.

11.6.2 Contour Maps

Presentation of water quality base maps has been traditionally handled through contouring of data. The technique of mapping of ground–water quality by drawing lines (isocontours) of equal concentration (isograms) of dissolved solids, or of single ions, has been used in the scientific literature for more than 70 years (Hem, 1970). The applicability of constructing isogram maps depends on several factors:

- Homogeneity of water composition with depth; and

- Parameter concentration increment between measuring points.

Restriction of sampling point density (i.e., insufficient data points), in either vertical or horizontal directions, will limit the usefulness of this technique; however, the detection or assessment monitoring system at a typical facility developed using procedures contained in this manual should provide sufficient data points for constriction of isocontour maps. Contour maps can contain either closed isopleths, as shown in Figure 11–5, or open gradient–lines, as shown in Figure 11–6. Both of these contour maps show isocontours of chlorides. Because chlorides are typically not affected by precipitation, or by other reactions that would lower concentrations (decreasing only by dispersion and/or dilution), this parameter serves as one of the best inorganics to use in contour formats.

Additional parameters such as conductivity, temperature, COD, or any dissolved parameter with sufficient data density can also be displayed on contour maps. On occasion, lumped organic parameters, such as total chlorinated hydrocarbons, can also be contoured. Figure 11–7 shows such a presentation. Organic parameters in ground water are difficult to contour effectively due to the typically wide ranges observed in water quality tests. However, water quality data from highly concentrated sources, such as product spills or very large volume, low concentration organic sources (such as an unlined codisposal facility), may be amenable to such presentations. Questionable data should always be represented by dashed lines on the illustration.

11.6.3 Time–Series Formats

In water quality evaluations there is always a continuing interest in observing parameter concentration change over time. To record such data, the standard approach is to make a series of observations at fixed intervals of time; this describes the time series format. Such time series formats have the objective to obtain an understanding of past events by determining the structure of the data, or to predict the future by extrapolating from past data. Those responsible for managing data collection systems can appreciate the difficulties of collection of environmental data at regular time intervals.

EXAMPLE OF TABULAR ARRAY OF WATER QUALITY DATA

WELLS	1C	2C	3B	3C	4C	5C	6C
CALCIUM	1500	2300	370	540	130	75	2300
MAGNESIUM	4300	2200	260	430	53	29	1200
SODIUM	43200	14100	1900	2600	196	150	5100
POTASSIUM	520	53	9.4	12	3.4	73	14
IRON	6.1	3.6	290	7.3	11	20	14
MANGANESE	35	13	2.9	0.35	0.47	0.34	18
NICKEL	1.4	0.45	0.26	<0.05	<0.05	<0.05	0.18
CHLORIDE	72000	30000	3600	5300	390	130	15000
SULFATE	6800	3180	530	760	110	94	1700
TOTAL PHOSPHOROUS (AS PO4)	0.08	0.63	8.1	0.41	0.43	0.74	<0.04
NITRATE	<0.1	<0.1	<0.1	<0.1	<0.1	<0.1	<0.1
NITRITE	0.07	0.03	<0.01	<0.01	0.01	0.01	0.01
FLUORIDE	<0.01	<0.01	0.32	0.28	0.43	0.35	0.11
SILICA (as SiO2)	16	24	22	25	27	20	24
CALCIUM HARDNESS	1900	2600	1100	1600	320	15	7100
MAGNESIUM HARDNESS	30100	20400	1600	2100	350	140	5900
TOTAL HARDNESS	32000	23000	2700	3700	670	150	13000
CARBONATE ALKALINITY (as CaCO3)	1	<0.1	<0.1	<0.1	<0.1	250	<0.1
BICARBONATE ALKALINITY (as CaCO3)	550	480	400	350	410	240	380
HYDROXIDE ALKALINITY (as CaCO3)	<1.0	<0.1	<0.1	<0.1	<0.1	<0.1	<0.1
TOTAL ALKALINITY (as CaCO3)	550	480	400	350	410	490	380
TOTAL DISSOLVED SOLIDS	130000	57000	7100	11000	1300	630	30000
pH, FIELD	7.27	7.34	7.69	7.63	8.07	9.02	7.35
pH, LAB	6.6	6.9	7.1	7.6	8.0	10.2	7.2
SPECIFIC CONDUCTANCE, umhos/cm, FIELD	off scale	off scale	15200	17300	1900	800	off scale
SPECIFIC CONDUCTANCE, umhos/cm, LAB	137000	72500	12000	17400	2070	1100	41100
TEMPERATURE, C, FIELD	17	14	15	16	17	19	16
TURBIDITY, NTU	570	72	1400	100	120	350	140
COLOR, APHA units	4	1	1	2	1	2	2

NOTE: ALL ANALYSES IN MG/L UNLESS OTHERWISE NOTED

Figure 11-4 Typical Water Quality Tabular Data Set

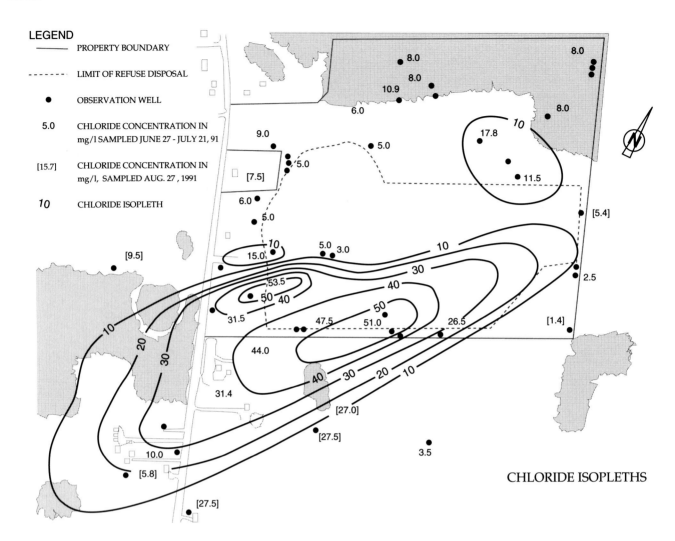

LEGEND

————————	PROPERTY BOUNDARY
- - - - - - - -	LIMIT OF REFUSE DISPOSAL
●	OBSERVATION WELL
5.0	CHLORIDE CONCENTRATION IN mg/l SAMPLED JUNE 27 - JULY 21, 91
[15.7]	CHLORIDE CONCENTRATION IN mg/l, SAMPLED AUG. 27 , 1991
10	CHLORIDE ISOPLETH

CHLORIDE ISOPLETHS

Figure 11-5 Isocontours - Closed Isopleths

Source: Modified from CRA

The variable (the data point) may be directly related to a defined time interval, such as, the high and low temperature for the day. Environmental data may also be continuously changing as would apply for measurements of hydraulic head in a piezometer. These observations are actually samples of instantaneous values, but are expressed as averages over the measured time interval. Readings taken once per day of a rapidly changing variable only establish a single point on a curve that can vary significantly until the next measurement. Fortunately water quality variables obtained from ground water do not vary significantly on a short term basis due to the typically slow movement of ground water in granular aquifers. Fractured or Karst bedrock may, however, show much faster "reaction times" both in hydraulic head level changes and variations in water quality. Most detection ground-water monitoring programs sample on a quarterly basis. While some cases can be made for somewhat shorter (or longer) sampling periods based on ground-water flow rates, these four-times per year sampling programs represent a "standard" period for time series analysis.

The first step in evaluating time series is to determine if any "structure" exists in the data. Structure can be defined as the data behavior at a particular point in time will be at least partially predicted by its value at other times. These structure elements in the data can be evaluated by:

• Defining a trend in the data, (i.e., does the data increase or decrease with time), using straight lines, higher order polynomials or exponential curves; or,

TOTAL DISSOLVED SOLIDS

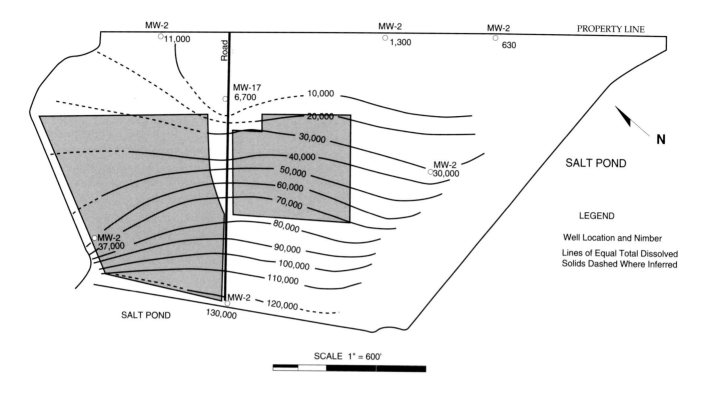

Figure 11-6 Isocontours - Open Isopleths

Source: Modified from Hydro-Search

- Testing for isolated events or unexpected departures from the normal behavior of the data set. This has specific application for detection monitoring programs where departures from long term trends can kick environmental programs into assessment actions.

Water quality at a single collection point such as a well or spring should be expected to change with time. Even with the generally slow movement of ground water, long term detection or assessment monitoring programs can show gradual changes in water quality. These changes can be best illustrated by time–series presentations. Time–series diagrams can be used to compare individual parameters with time (i.e., compare water quality in a well against itself), or can illustrate changes in multiple parameters with time, or to illustrate changes with time for a common parameter in multiple wells. Figure 11-8 shows a comparison of total dissolved solids in a number of wells. All six wells are compared to each other at any displayed point in time. The figure also shows changes in TDS with time for each of the wells. Time–series presentations can be generally ineffective if too much data is presented on one plot. Figure 11-9 shows a time series plot for chloride

in 8 wells; although only a single parameter is displayed, the variable y scales used in the presentation make interpretation of trends difficult. Time series presentations are most effective when single parameters are compared, as shown in Figure 11-10. This illustration includes water level elevations with chloride concentrations. Whether or not the water level elevation is related to the chloride concentration is a separate question, however, the data is displayed in an easily understood format.

A similar data set is presented in box and whisker plots (or "box plots") on Figure 11-11. Box plots are useful statistical tools for evaluating changes in water quality. Complicated site evaluations may require a series of box plots. For example all wells screened in a hydrostratigraphic unit may be combined on a single plot, or data from a number of well "nests" may be shown on one plot to illustrate vertical trends in water quality. The box plot can be considered as an economical graphical method of presenting the constituent summary statistics. The boxes are constructed using the median (middle value of the data) and the interquartile range (the range of the middle fifty percent of the data) These plots separate the results of each well and can clearly show the difference in

CONTOUR MAP USING TOTAL VOLATILE ORGANIC PARAMETERS

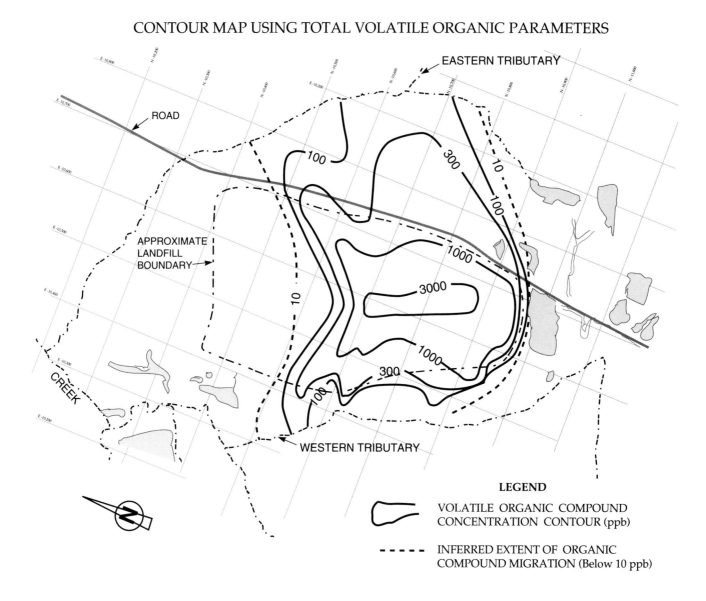

Figure 11-7 Isocontours - Total Organic Parameters

Source: Modified from Golder Associates Inc.

the data distributions. These plots are generated by ranking the data and may be constructed in a number of different ways (McGill et al, 1978). Some box plots constructed by various software programs use the median and the F-spread. The F-spread or fourth spread is a function of the data distribution and measures the variability in the water quality results, similar to the standard deviation. Hoaglin et. al., (1983) provides a full discussion of these order statistics. The median and

interquartile range (IQR) are analogous to the more commonly used mean and standard deviation of a set of data. The mean and median are measures of "central tendency" or "location", whereas the standard deviation and IQR are measures of "variability."

Typically a first step in evaluating ground water quality for box plot presentations is to review existing hydrogeologic information and to try to define subsurface stratigraphy and ground-water flow. The next logical step,

TIME SERIES PRESENTATION FOR
TOTAL DISSOLVED SOLIDS

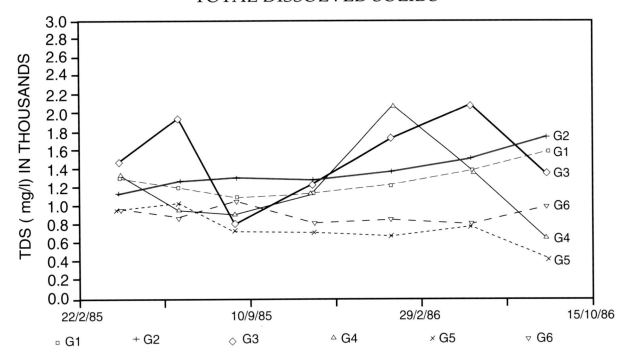

Figure 11-8 Time Series Comparisons for Six Wells

is to graph the chemical data as concentration versus time series plots.

Figure 11-12 shows a chart where the mean values (solid circle), plus or minus one standard deviation error bars (vertical line) is plotted for each well next to each box plot. The plots show the mean for the data is consistently greater than the median. Two standard deviations for the data is also larger than the IQR. High values otherwise described as "outliers" inflate the estimate of the mean and standard deviation in these statistical plots. The median and IQR, are based on ranks, and are not particularly sensitive to outlying values. Similar to Figure 11-11, the high variability in the impacted data is shown by the wide error bars.

The box plots are considered more powerful in illustrating impacted water quality than simple error bar plots because they contain more information about the actual data distribution. The error bar plots, however, can be applied to parametric statistics evaluations.

11.6.4 Histograms

The histogram is a two-dimensional graph, where one axis represents the data where and the other is the number of samples that have that value. The Y axis of the plot is frequency expressed in terms of the percentage of total samples, rather than as an absolute count. The process of creating a histogram is primarily a counting process. A number of classes or groupings are defined in terms f subranges of the numeric value. These may be set to cover the complete range of the project data, or a restrictive range derived from the mean and standard deviation, or from knowledge of data ranges from previous project data evaluations. With many computer based spreadsheet programs offering automated histogram production these project data can be quickly plotted in a histographic format to evaluate the appearance of the figure.

Even with automated histogram production the basic usefulness of the display can be enhanced by changing the parameters that influence the appearance of the histogram. These parameters are (Green 1985):

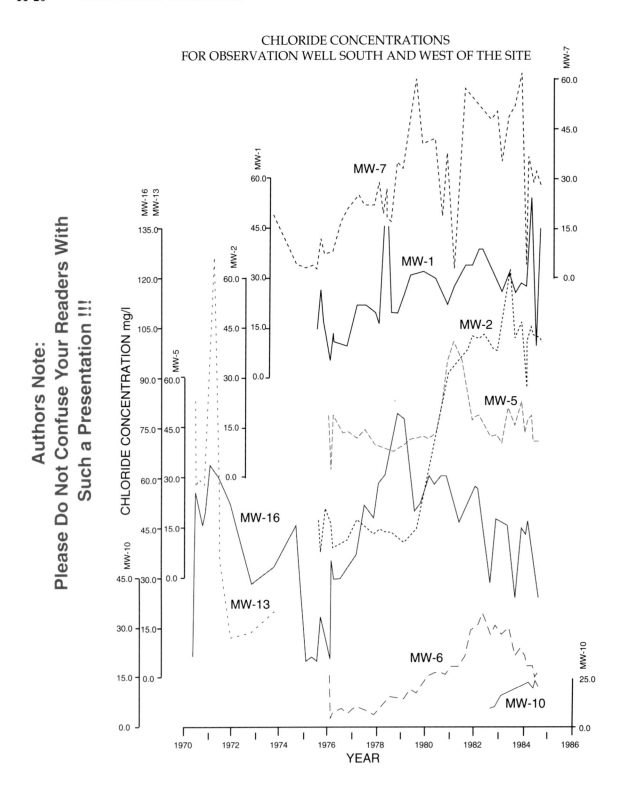

Figure 11-9   Time Series Comparisons With Sliding Scales

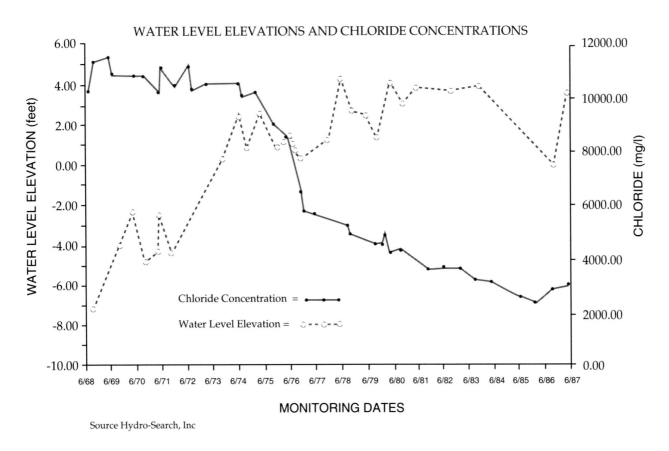

Source Hydro-Search, Inc

Figure 11-10 Time Series Comparisons for Chloride and Water Levels

- The "range" which include the minimum and maximum included values;
- The number of classes used in the counting;
- The size of a class such as the range of numeric values treated as a unit in the value counting; and,
- Transformations of numeric values that include scaling, logarithmic, and exponential.

As general guidance, at least one histogram should be produced that covers the complete range of data values to evaluate samples outside the main distribution of data sets. It is recommended that all extreme values be investigated as errors or true anomalies. A common problem with histograms that have broad ranges; is that the resultant figure will have poor resolution. The majority of the results in these displays are combined into one or two classes, obscuring the details of the distribution. If you exclude the outliers you can obtain better resolution of the main data sets.

Plotting of chemical data, as a series of comparative histograms (or bar graphs), has been a traditional methodology for representation of variability in water quality. Most of the traditional methods are designed to represent the total concentration of solutes and the proportions assigned to each ionic species (for one analysis or group of analyses). The units, in which concentrations are expressed, in these traditional diagrams are milliequivalents per liter (meq). Hem (1970) provides descriptions of bar graphs, radiating vector–plots, circular diagrams, and stiff diagrams. These methods will not be discussed herein. Water–quality data collected during detection or assessment monitoring programs traditionally have not been portrayed in a format of "whole" analysis, i.e., with anions and cations given in units of milliequivalents per liter. Rather, results of water–quality analyses are presented in milligrams or micrograms per liter and presented in formats including only a few parameters. These data, especially volatile organics and hazardous metals, have been displayed as histogram "fingerprints" illustrating variations in water quality. Figure 11-13 shows a series of histograms of hazardous metals, obtained from analyses of water in individual wells. Similar histograms have also been used to track plumes of volatile organic, and to compare relative

TIME SERIES - CONCENTRATION PLOTS

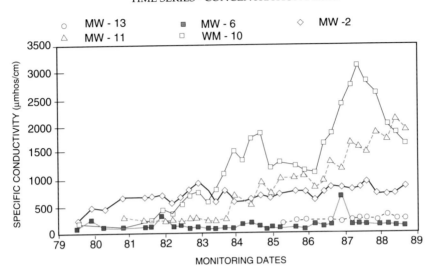

BOX AND WHISKER PLOTS

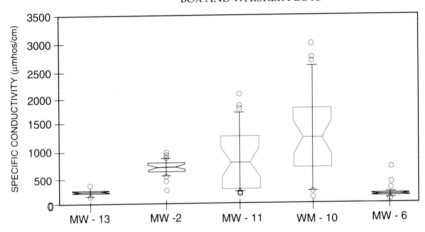

Figure 11-11   Time Series Comparisons With Box Plots

Source: Modified from Fisher and Potter, 1989

proportions of organic species in water from individual wells.

Tabular summaries of constituents are another form of comparative histograms. Figure 11-14 shows a summary table used to compare organic parameters observed in leachate with organic parameters observed in off–site wells. Many of the constituents in the "fingerprint" of the landfill leachate are different from those in the off–site monitoring well, and thus tend to indicate a non-relationship. Care must be taken to use indicator

parameters that will not change with time and therefore provide a misinterpretation of the water quality fingerprint.

Additional graphical displays of histograms are shown on Figure 11-14 (a), (b), (c) and (d). The data shown on Figure 11-14 a, b illustrate over a thousand observation of specific conductivity for two wells. These histograms can be compared to a Log-Normal distribution (Figure 11-15c), and Normal distribution (Figure 11-15d). The histogram construction format for large numbers of observations can be used to investigate the probability distribution of the data. In general terms the histogram plots values where the higher the bar, the greater the probability that additional measurements will fall in this range, therefore, the more sample values incorporated into the histogram, the closer the graph is to the "true" population distribution. Many statistical tests used in evaluation of water quality data require knowledge if the data comes from a normally distributed population. The plotted data distribution illustrated on the histogram can be compared to a normally distributed data set. This provides a qualitative evaluation of the assumption of a normally distributed population is truly represented in the displayed environmental data.

The example project data sets Figure 11-16 shows neither of the wells have normally distributed data; both sets of data are "skewed" to the right. Since the data are not symmetric about the mean, the distribution is the considered to be positively skewed. The lognormal distribution is also skewed right as shown. Natural log scale transformations of positively skewed environmental data, can make the data appear more normally distributed. Although histograms represent a good visual tool for evaluation of the probability of the environmental data, Benjamin and Cornell, (1970) point out that normal

probability plots give a better representation of the data and are easy to construct

Normal probability plots provide an excellent technique to graphically compare environmental data to the normal distribution. Figure 11-16 shows a normal probability plot for the same data as in Figure 11-15 (a). These are constructed by first ordering the raw data from smallest to largest. Let x [1] < x [2] <....< x [n] denote the ordered data. The X [i] are called the order statistics of the data. The x [i] are then plotted on normal probability paper versus the corresponding plotting position $(1/n+1) * 100$. If the data are from a normal distribution, the plotted points should lie approximately on a straight line(Fisher and Potter, 1989).

Figure 11-16 (a) illustrates that the data does not plot as a straight line, hence, the assumption of normality is in question. Transforming the data to log scale and then reploting as shown on Figure 11-16 (b) does not provide for a straighter line, and therefore we cannot conclude that the lognormal distribution is more appropriate for demonstrating normality. In summary, environmental data from a waste disposal facility can be visually presented in a number of ways that assist inspection of the data sets:

ONE STANDARD DEVIATION ERROR BAR PLOT
COMPARED TO BOX AND WHISKER PLOTS

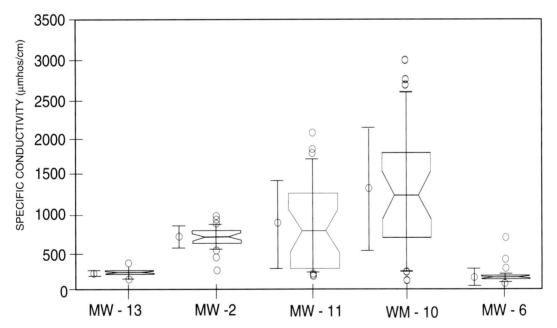

Figure 11-12 Error Bar Plot Compared to Box and Whisker Plots

Source: Modified from Fisher and Potter, 1989

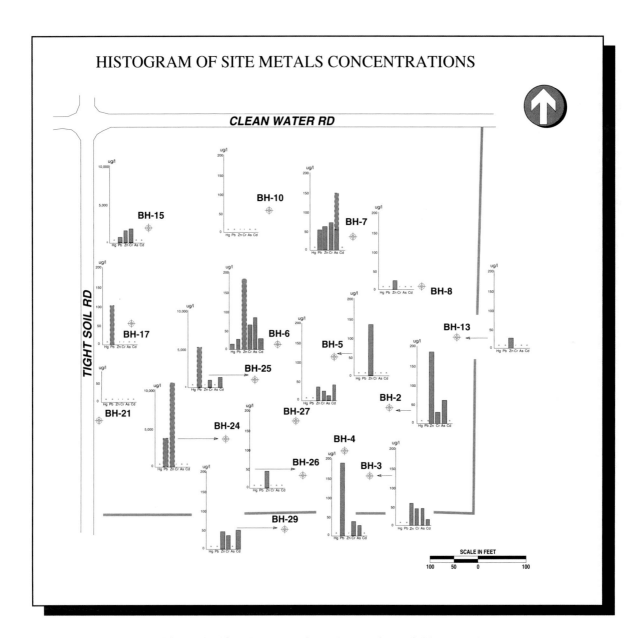

Figure 11-13   Histogram of Metals Data for Individual Boreholes

Source: Modified from Golder and Associates

(1) time versus concentration plots;
(2) box and whisker plots;
(3) one standard deviation error bar charts;
(4) histograms; and,
(5) normal probability plots.

The first two graphical tools can clearly illustrate qualitatively the relative water quality between wells (known as Inter-well comparisons). The error bar charts may be valuable when working with parametric statistics Histograms may be used to view the probability distribution of the data. When evaluating the assumption of normality, normal probability plots are commonly prepared to observe deviations from normality. The example data illustrate several important points.

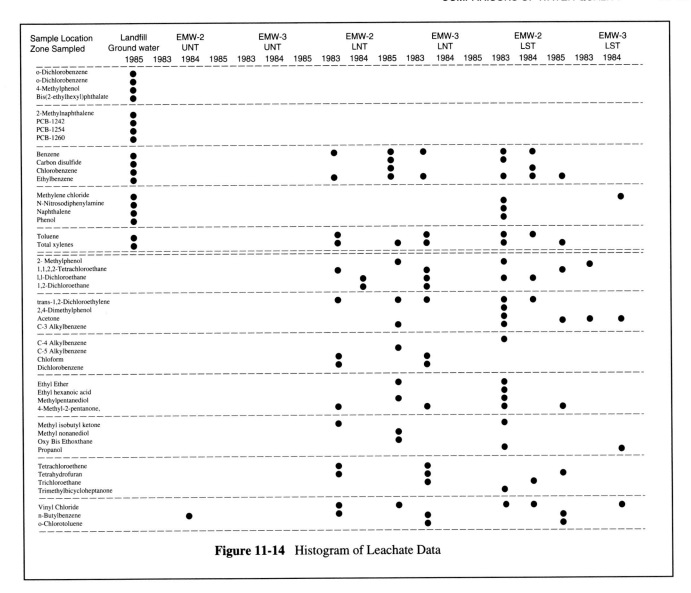

Figure 11-14 Histogram of Leachate Data

- Data outliers tend to inflate the mean and standard deviation of the data;

- The median and interquartile range are good estimates of the central tendency and variation of data sets, particularly when outliers are present.

- Large data variability (IQR) is usually associated with high medians, (i.e., impacted wells). Natural temporal variability is much lower than the variability observed when contamination is present.

- In those occasions where histograms and normal probability plots show that ground-water data may not be normally distributed, the median and IQR may be better estimates of the central tendency and variability of the data.

11.6.5 Trilinear Diagrams

If one considers only the major, dissolved ionic–constituents, in milliequivalents per liter, and lumps potassium and sodium together and fluoride and nitrate with chloride, the composition of most natural water can be illustrated in terms of three cationic and three anionic species. If the values are expressed as percentages of the total milliequivalents per liter of cations and anions, the composition of the water can be represented conveniently by a trilinear plotting technique.

The simplest trilinear plots utilize two triangles — one for anions and one for cations. Each vertex represents 100 percent of a particular ion or group of ions. The composition of cations is indicated by a point plotted in the cation triangle, and the composition of anions by a point plotted in the anion triangle. The coordinates at each point

HISTOGRAMS OF SPECIFIC CONDUCTIVITY (mmhos/cm)

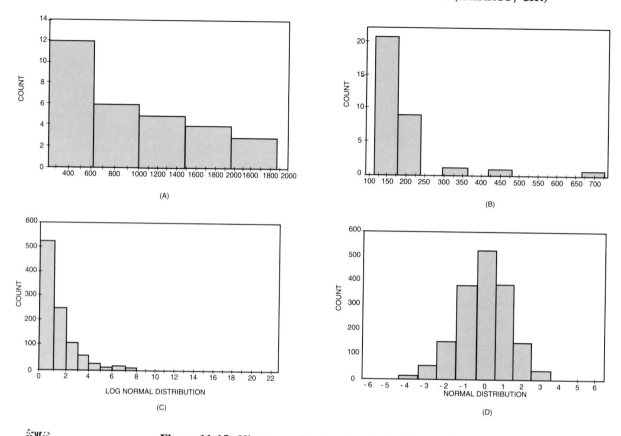

Figure 11-15 Histograms Used to Check Data Normality

Source: Modified from Fisher and Potter, 1989

add to 100 percent. Most trilinear diagrams are in the form of two triangles bracketing a diamond shaped plotting field, as first described by Piper (1944).

The trilinear diagram constitutes a useful tool for interpretation of water–analysis. Most of the graphical procedures described here are of value in pointing out features of analyses and arrays of data that need closer study. The graphs themselves do not constitute an adequate means of making such studies, however, unless they can demonstrate that certain relationships exist among individual samples. The trilinear diagrams sometimes can be used for this purpose.

Figure 11-17 is a trilinear diagram derived from analyses of water from San Francisco Bay and the Newark Aquifer. In any illustration of water quality data, a diagram should aid interpretation by providing a visual clarification of trends, or a comparison of differences in water quality.

Trilinear diagrams have become so popular that computer programs have been written to automatically calculate and display the data.

11.7 STATISTICAL TREATMENT OF WATER QUALITY DATA

Various procedures, such as averaging, determining frequency distributions, and making simple or multiple correlations, are widely used in interpretation of water analyses. More sophisticated applications of statistical methods, particularly procedures that utilize digital computers, are being applied more and more frequently. Some potential applications of these statistical techniques will be covered in the following sections. It is essential

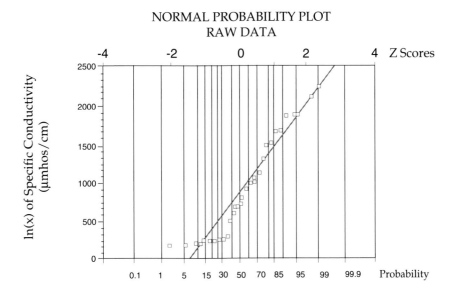

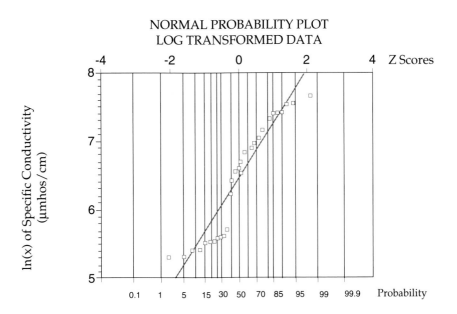

Figure 11-16 Probability Plot for Raw and Log Transformed Data

that proper consideration be given to chemical principles during application of statistical tests of data sets. Most data are evaluated using simple averaging and standard deviations, as shown on Figure 11-18, however, more complex statistical treatments of analytical data are necessary when state and federal performance standards are required for a waste disposal facility.

Federal regulations have been established for statistical determination of compliance for RCRA facilities. Both existing and new hazardous waste facilities are covered by

Subtitle C of the Resource Conservation and Recovery Act (RCRA) and regulated by 40 CFR Parts 264 and 265. When first issued, Part 264 Subpart F required that Cochran's Approximation to the Behrens Fisher Student's t-test (CABF) or an alternative statistical procedure approved by EPA be used to determine whether there is a statistically significant exceedance. This Part 264 Subpart F regulation, and in particular the CABF procedure, generated significant technical criticism over use of these statistical procedures for use with ground-water quality

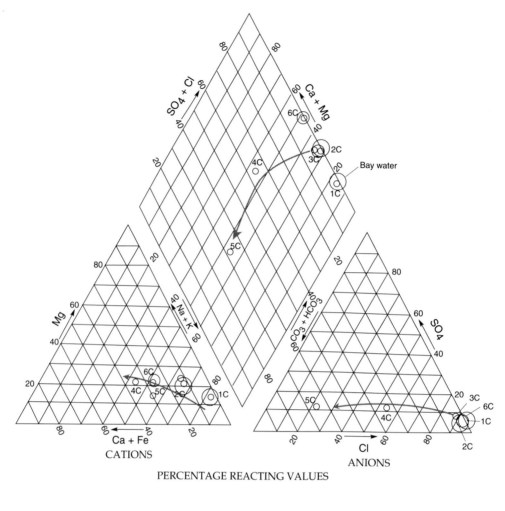

PERCENTAGE REACTING VALUES

TOTAL DISSOLVED SOLIDS

○   0 - 10,000

○   10,000 - 20,000

◯   20,000 - 50,000

◯   > 50,000

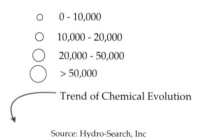   Trend of Chemical Evolution

Source: Hydro-Search, Inc

TRILINEAR DIAGRAM DEPICTING WATER QUALITY AND TRENDS

Figure 11-17   Trilinear Data Sets Used for Comparisons of Water Quality

data and EPA proposed a new regulation in response to these concerns (EPA, August 24,1987). The proposed regulation was revised based on comments EPA received and was then made final (EPA, October 11, 1988).

The final regulation (Oct. 1988) describes five performance standards that a statistical procedure must meet. The Federal regulations do recommend four types of

statistical procedures to evaluate performance of RCRA facilities for releases to ground water. In addition EPA has issued (Oct. 1991) amendments to Subtitle D of RCRA to include criteria for municipal solid waste landfills (MSWLF's). 'The number of samples collected to establish ground-water quality data must be consistent with the appropriate statistical procedures (discussed below).

The sampling procedures are defined in the applicable sections for detection monitoring (§258.54(b)), assessment monitoring (§258.55(b)), and corrective action (§258.56(b)).

• *Owner or operator must specify in the operating record one of the following statistical methods to be used in evaluating ground-water monitoring data for each hazardous constituent. The statistical test chosen shall be conducted separately for each hazardous constituent in each well.*

- A parametric analysis of variance followed by multiple comparisons procedures to identify statistically significant evidence of contamination.
- An analysis of variance based on ranks followed by multiple comparisons procedures to identify statistically significant evidence of contamination.
- A tolerance or prediction interval procedure in which an interval for each constituent is established from the distribution of the background data, and the level of each constituent in each compliance well is compared to the upper tolerance or prediction limit.

Groundwater Level Data

WELL	Sampling Date 04/08/85	6/18/85	8/22/85	9/17/85	11/20/85	2/19/86	5/23/86	8/19/86	N	Mean Water Elevation	S.D.
GEI-1	401.43	400.85	399.26	398.64	399.18	400.43	399.10	397.89	8	399.60	1.12
GEI-1D	---	---	---	---	---	400.63	399.00	398.00	3	399.21	1.08
GEI-1S	---	---	---	---	---	400.34	399.09	397.75	3	399.06	1.06
GEI-2	401.42	400.63	398.63	398.07	398.80	400.12	398.87	397.62	8	399.27	1.23
GEI-3	402.04	401.14	398.80	398.16	400.07	400.65	399.44	398.03	8	399.79	1.35
GEI-4	402.48	401.57	399.34	398.57	400.23	401.06	399.77	398.56	8	400.20	1.33
GEI-5	402.36	401.53	399.57	398.82	399.86	401.01	399.63	398.66	8	400.18	1.23
GEI-6	402.37	401.61	399.61	398.87	399.90	400.96	399.58	398.66	8	400.20	1.24
GEI-7	402.14	401.42	399.81	398.94	399.95	401.05	399.38	398.71	8	400.18	1.15
GEI-8	402.27	401.56	399.90	398.96	400.12	400.98	399.49	398.65	8	400.24	1.19
GEI-9	403.00	402.25	399.81	398.37	400.21	---	---	---	5	400.73	1.68
GEI-10	403.04	402.27	400.19	394.40	396.13	---	---	399.19	6	399.20	3.10
GEI-11	403.04	402.23	400.02	396.10	398.06	---	---	---	5	399.89	2.57
GEI-12	403.08	402.16	399.82	397.72	400.11	401.55	398.80	399.00	8	400.28	1.72
GEI-13	---	---	---	---	---	400.38	399.01	397.63	3	399.01	1.12
GEI-14	---	---	---	---	---	400.29	399.08	399.04	3	399.47	0.58
GEI-15	---	---	---	---	---	400.64	399.10	397.93	3	399.22	1.11
GEI-16	---	---	---	---	---	400.53	399.20	397.95	3	399.23	1.05
GEI-17D	---	---	---	---	---	400.30	398.89	397.39	3	398.86	1.19
GEI-17S	---	---	---	---	---	400.22	399.01	397.59	3	398.94	1.07

Groundwater Data
Total Dissolved Solids (TDS) mg/l

WELL	Sampling Date 04/08/85	6/18/85	8/22/85	11/20/85	2/19/86	5/23/86	8/19/86	N	Mean Concent.	S.D.
GEI-1	1280	1210	1120	1170	1290	1450	1670	7	1318	189
GEI-1D	---	---	---	---	530	600	539	4	417	242
GEI-1S	---	---	---	---	980	1210	1780	3	1323	336
GEI-2	1130	1270	1340	1300	1430	1580	1810	7	1455	189
GEI-3	1480	1960	820	1290	1780	2160	1420	7	1572	449
GEI-4	1320	960	930	1160	2150	1460	726	7	1231	469
GEI-5	950	1040	740	760	720	850	508	7	770	159
GEI-6	960	890	1060	850	900	830	1060	7	932	94
GEI-7	970	990	960	710	670	860	683	7	812	131
GEI-8	660	680	1120	690	660	690	691	7	770	175
GEI-9	470	490	500	---	---	---	---	3	495	5
GEI-10	390	400	420	420	---	---	384	5	406	15
GEI-11	560	530	550	570	---	---	---	4	550	16
GEI-12	650	460	480	480	430	490	441	7	464	22
GEI-13	---	---	---	---	1040	1010	951	3	1000	37
GEI-14	---	---	---	---	2180	2540	2500	3	2407	161
GEI-15	---	---	---	---	1320	---	1390	2	1355	35
GEI-16	---	---	---	---	1400	1330	1350	3	1360	29
GEI-17D	---	---	---	---	1440	1360	1480	3	1427	50
GEI-17S	---	---	---	---	2370	1680	1680	3	1910	325

Chloride Concentrations (mg/l)

WELL	Sampling Date 04/08/85	6/18/85	8/22/85	11/20/85	2/19/86	5/23/86	8/19/86	N	Mean Concent.	S.D.
GEI-1	250	250	200	210	360	310	390	7	286.67	72.26
GEI-1D	---	---	---	---	19	33	25	4	19.25	12.17
GEI-1S	---	---	---	---	140	180	440	3	253.33	133.00
GEI-2	240	260	290	320	330	340	420	7	326.67	49.55
GEI-3	200	280	97	220	330	410	230	7	261.17	97.35
GEI-4	180	91	86	180	540	300	91	7	214.67	164.05
GEI-5	130	160	86	110	110	140	37	7	107.17	39.28
GEI-6	97	110	130	100	130	110	250	7	138.33	51.13
GEI-7	120	96	94	52	56	77	40	7	69.17	21.28
GEI-8	6	9	9	8	8	8	7	7	8.17	0.69
GEI-9	5	7	10	5	---	---	---	4	7.33	2.05
GEI-10	3	4	4	3	---	---	4	5	3.75	0.43
GEI-11	6	8	8	8	---	---	---	4	8.00	0.00
GEI-12	1	12	9	11	10	2	21	7	10.83	5.58
GEI-13	---	---	---	---	190	140	150	3	160.00	21.60
GEI-14	---	---	---	---	550	510	600	3	553.33	36.82
GEI-15	---	---	---	---	160	170	190	3	173.33	12.47
GEI-16	---	---	---	---	140	110	110	3	120.00	14.14
GEI-17D	---	---	---	---	290	260	260	3	270.00	14.14
GEI-17S	---	---	---	---	680	420	360	3	486.67	138.88

Figure 11-18 Simple Average Table Data Set

- A control chart approach that gives control limits for each constituent.
- Another statistical test method that meets the performance standards discussed immediately below.

• *Any statistical method chosen shall comply with the following performance standards, as appropriate:*

- The statistical method shall be appropriate for the distribution of chemical parameters or hazardous constituents.
- If an individual well comparison procedure is used to compare an individual compliance well constituent concentration with background constituent concentrations or a ground-water protection standard, the test shall be done at a Type I error level no less than 0.01 for each testing period. If a multiple comparisons procedure is used, the Type I experiment wise error rate for each testing period shall be no less than 0.05; however the Type I error of no less than 0.01 for individual well comparisons must be maintained.
- If a control chart approach is used to evaluate ground-water monitoring data, the specific type of control chart and its associated parameter values shall be protective of human health and the environment.

- If a tolerance interval or a prediction interval is used to evaluate ground-water monitoring data, the levels of confidence and, for tolerance intervals, the percentage of the population that the interval must contain, shall be protective of human health and the environment.

- The statistical method shall account for data below the limit of detection with one or more statistical procedures that are protective of human health and the environment. Any practical quantitation limit (PQL) that is used in the statistical method shall be the lowest concentration level that can be reliably achieved within specified limits of precisions and accuracy during routine laboratory operating conditions that are available to the facility.

- If necessary, the statistical method shall include procedures to control or correct for seasonal and spatial variability as well as temporal correlation in the data.

• *The owner or operator must determine whether or not there is a statistically significant increase over background values for each parameter or constituent required in the particular ground-water monitoring program that applies to the MSWLF unit.*

- In determining whether a statistically significant increase has occurred, the owner or operator must

compare the ground-water quality of each parameter or constituent at each monitoring well to the background value of that constituent.

- Within a reasonable period of time after completing sampling and analysis, the owner or operator must determine whether there has been a statistically significant increase over background at each monitoring well.'

The statistical test requirements are the same as the RCRA Subtitle C final regulation since the solid waste rules recommends the same four types of procedures.

The performance standards in these Federal rules allow flexibility in designing statistical procedures to site specific considerations. Selection of an appropriate statistical test must be made based on the quality of the data available, the hydrogeology of the site and the theoretical properties of the test. As expressed in previous sections, ground-water quality data can be expected to vary temporally and spatially due to natural effects, and the results are also affected by sampling and analytic errors. Due to natural variability observed in ground water, the determination of a significant change in water quality is linked to statistical probability theory.

In order to define if there has been a significant change in water quality, comparison must be made between supposedly "clean" background data and possibly impacted data. Both of these ground-water classes are subject to temporal and spatial variability as well as sampling and analytic error. Hence, the problem becomes one of evaluation of variable water quality in time and space with potential statistical inferences. A statistical "hypothesis" is used to compare water quality:

• *Null Hypothesis:* Ho: No Contamination exists therefore the facility is in compliance

• *Alternative Hypothesis:* H1: Contamination exists; facility is in violation

A statistical test is made on the null hypothesis and a conclusion is reached that either the facility is or is not in violation. The null hypothesis starts out with the assumption that there is no real difference between the quality of up and down gradient ground water. The assumption is that they are all from the same population. Thus, the difference between the means of the two samples would be just one possible difference from the theoretical distribution where the mean difference is zero. The assumption is called the NULL HYPOTHESIS since it attempts to nullify the difference between the two sample means by suggesting or forming a hypothesis of no statistical difference. If the statistical difference between the two sample means turns out to be too big to be explained by the kind of variation that would often occur by chance between random samples, then one must reject it (the Null Hypothesis), since it will not explain our observations. The typical alternative hypothesis would be that the two water quality population means are not equal. In this context, a "violation" implies that water quality is significantly different from background. Figure 11-19 illustrates the two types of errors associated with hypothesis testing.

Significant technical discussions surround whether a site has observed a "false positive" indicating contamination. Type I error (false positive) occurs when a site (or well) is actually in compliance but the statistical test is triggered that decides it to be in violation. The probability of a Type I error is defined as the controllable "significance level" of the test. Usually, this (α) is set at 0.05, giving a 1/20 chance that a "false positive" conclusion of contamination will occur.

Type II error (false negative) occurs when contamination exists but is not detected. The probability of a false negative conclusion, (β), is more difficult to control, is often difficult to calculate, and is dependent on many factors that may include sample size, the overall magnitude of "change" in parameter concentration, and choice of statistic tested in the decision process.

Statistical hypothesis testing can be divided into two general categories: *parametric*, those which rely on the estimation of parameters of a probability distribution (usually the mean and standard deviation of the normal distribution) and *nonparametric*, those which do not fit a

normal distribution. Nonparametric methods usually rely on test statistics developed from the ordered ranks of the data. The simplest nonparametric evaluation is the median, or middle value of a data set. Both parametric and nonparametric statistical tests are reviewed in the context of ground-water monitoring events in later subsections of this chapter.

In general terms the type of statistical test to use of a facility regulated under Federal laws should be consistent with USEPA (1988) 40CFR Part 264: Statistical methods for evaluating ground-water monitoring from hazardous waste facilities; final rule. *Federal Register*, **53**, **196**, 39720-39731. Both RCRA solid waste (Subtitle D) and hazardous waste (Subtitle C) are keyed into this code.

11.7.1 Data Independence

Independence of data collected in environmental programs must be evaluated by determining if the data show serial correlation. Serial correlation of ground-water sampling data is most likely to occur from very slow ground-water flow. Even with reasonably permeable aquifers (say with hydraulic conductivity > 10^{-3} cm/sec) low gradients can slow down ground-water velocity to less than 20 feet per year. When ground-water quality measurements are collected too frequently to be independent of each other, one can observe serial correlation in the data. Independence can often be achieved by increasing the time between observations. Several tests have been reported in the literature to evaluate the presence of serial correlation in ground-water quality data. Montgomery et al (1987) chose the Lag 1 autocorrelation function (ACF). Goodman and Potter (1987) also used this method as well as the nonparametric autorun test (AR). The application of the ACF test to ground-water quality data is described in detail by Harris et al, 1987. The AR test is applied to hydrologic data by Sen, 1979. Most advanced statistics texts and computer packages include these tests.

Serial correlation may exist in ground-water quality data even though the sampling was at intervals of three months. The reality of most sampling programs dictate the sampling period required by regulatory standards or permit requirements. It is probably sufficient that one is aware of the potential difficulties associated with serial correlation of the data so that independence of the observations can be checked to help in the selection of the statistical test used in the evaluation of the data.

STATISTICAL DECISION

	IN COMPLIANCE	IN VIOLATION
IN COMPLIANCE	Good Decision $1-\alpha$	False Positive Decision α Type I error
IN VIOLATION	False Negative Decision β Type II error	Good Decision $1-\beta$ Power of Test

TRUE SITUATION

Figure 11-19 Statistical Error in Hypothesis Testing

11.7.2 Data Normality

The normal distribution is perhaps the single most important and widely used probability modal in applied statistics. This is because many real systems fluctuate "normally" about a central mean; i.e., measurement error of a random variable is symmetric about a "true" mean and has a greater probability of being small (close to the mean) than large "in the tail of the distribution".

EPA's statistical RCRA regulations (40 CFR 264 Subpart F) which do not require tests for normality or other distributional assumptions unless:

1. A data transformation is made, or

2. Nonparametric statistical tests are applied.

Data transformations are commonly used to "normalize" skewed data for parametric tests. Many environmental systems are modeled using the lognormal distribution because, (1) it has a lower bound of zero, and (2) is positively skewed, allowing high values to be included (Benjamin and Cornell, 1970).

The hypothesis of normality can be evaluated through any number of statistical "goodness-of-fit" tests. These tests are used to mathematically compare the shape of the normal distribution to the data set. Care should be taken to only apply these tests to independent, stationary data sets.

In the ground-water quality literature, Montgomery et al (1987) tested the normality of ground-water quality data using graphical methods, the chi square test and the skewness test. Harris et al (1987) recommends the skewness test for general use with ground-water quality data.

11.7.3 Evaluation Of Ground Water Contamination

A main objective of a ground–water detection monitoring program is to determine if the facility is effecting ground water. Owner/operators are required in Federal rules to place detection monitoring wells in both upgradient (background) and downgradient locations around the facility, and to monitor those wells at regular intervals, typically quarterly or twice per year, for a series of "indicator parameters." Subtitle D defines in [§258.53]—Ground-water sampling and analysis requirements, that "The ground-water monitoring program

must include consistent sampling and analysis procedures that are designed to ensure monitoring results that provide an accurate representation of ground-water quality at the background and downgradient wells. The owner or operator must notify the State Director that the sampling and analysis program documentation have been placed in the operating record and the program must include procedures and techniques specified in §258.53(a)(1) through (5).

• The ground-water monitoring program must include sampling and analytical methods that are appropriate for ground-water sampling and that accurately measure hazardous constituents and other monitoring parameters in ground-water samples. Ground-water samples shall not be field filtered prior to laboratory analysis.

• Ground-water elevations must be measured in each well immediately prior to purging each time ground-water is sampled. Ground-water elevations must be measured within a period of time short enough to avoid temporal variations in ground-water flow that could preclude accurate determination of ground-water flow rate and direction.

• The owner or operator must establish background ground-water quality in a hydraulically upgradient or background well(s) for each of the monitoring parameters or constituents required in the particular ground-water monitoring program that applies to the MSWLF unit.

• The number of samples collected to establish ground-water quality data must be consistent with the appropriate statistical procedures (discussed below). The sampling procedures are defined in the applicable sections for detection monitoring (§258.54(b)), assessment monitoring (§258.55(b)), and corrective action (§258.56(b))."

The logic of this sampling strategy is that upgradient water quality represents the background conditions for that particular region, and downgradient water quality represents background water quality, plus any influence produced by the facility. Section 258.40—design states "The relevant point of compliance specified by the Director of an approved State shall be no more than 150 meters from the waste management unit boundary and shall be located on land owned by the owner of the

MSWLF unit.". This sets the stage for defining a boundary zone in which to locate the monitoring wells.

The detection monitoring program as described in subtitle D for solid waste in [§258.54] then defines the following:

• *Detection monitoring is required at MSWLF units at all ground-water monitoring wells specified in §258.51(a)(1) and (2) [background and downgradient]. At a minimum, the constituents from Appendix I must be included in the program.*

  - The Director of an approved State may delete any of the Appendix I constituents if it can be shown that the removed constituents are not reasonably expected to be in or derived from the waste contained in the unit.

  - A Director of an approved State may establish an alternative list of inorganic indicator parameters for a MSWLF unit, in lieu of some or all of the heavy metals in Appendix I.

{Note: Deletion of Appendix I constituents or establishing alternative inorganic parameters is not possible unless the state is approved by the USEPA}.

• *The monitoring frequency shall be at least semi-annual during the active life of the facility (including closure) and the post-closure period. A minimum of four independent samples from each well (background and downgradient) must be collected and analyzed during the first semi-annual sampling event. At least one sample from each well (background and downgradient) must be collected and analyzed during subsequent semi-annual events. The Director of an approved State may specify an alternative frequency during the active life (including closure) and the post-closure care period. The alternative frequency shall be no less than annual.*

{Note: An alternative monitoring frequency is not an option unless the state is approved by the USEPA}.

• *If the owner determines that there is a statistically significant increase over background for one or more of the constituents in Appendix I or an approved alternative list, the owner or operator must:*

  - Within 14 days of the finding, place a notice in the operating record indicating which constituents have shown statistically significant changes from background levels, and notify the State Director that the notice was placed in the operating record.

  - Establish an assessment monitoring program within 90 days unless the owner or operator can demonstrate that a source other than a MSWLF unit caused the contamination or that a statistically significant increase resulted from an error in sampling, analysis, statistical evaluation, or natural variation in ground-water quality. A report documenting alternate source or error must be certified by a qualified ground-water scientist or the Director of an approved State and placed in the operating record.

As described in Chapter 9.0 the issue of the regulatory concept of the simple upgradient and downgradient model is rarely observed in real world monitoring programs, however, the use of background wells as representative of upgradient water can be used in statistical comparisons. In many cases, particularly when adequate background data are available prior to the installation of the facility, intra–well comparisons may be the most successful technique to use (i.e., each well compared to its own history). The major advantage of this approach is that it eliminates the spatial component of variability from the comparison. One is left with evaluating the local effects on the well installation such as, construction, maintenance and near-by vertical interferences to water quality, (such as wells located near roads that are salted during winter, local spills, etc.). The statistical methodology is illustrated on Figure 11-20, which graphically portrays the variable bases for statistical comparisons between wells, and for intra–well statistical comparisons.

Detection monitoring programs at waste disposal facilities requires not only that a release to the environment has occurred, but also that the release observed is directly due to discharges from the facility. Water quality standards are commonly used as basis for judging if a release to the environment. Yet, even a water quality standard exceedance must be compared to background water quality to conclude that the facility is responsible for the ground-water impact. In reviewing facility data one can expect a number of water quality standards (especially non-organic parameters) to be exceeded in natural ground water. Thus, comparison of downgradient water quality to "known" background water quality is an important part of

DETECTION, ASSESSMENT MONITORING AND CORRECTIVE ACTION FOR SOLID WASTE FACILITIES UNDER RCRA SUBTITLE D

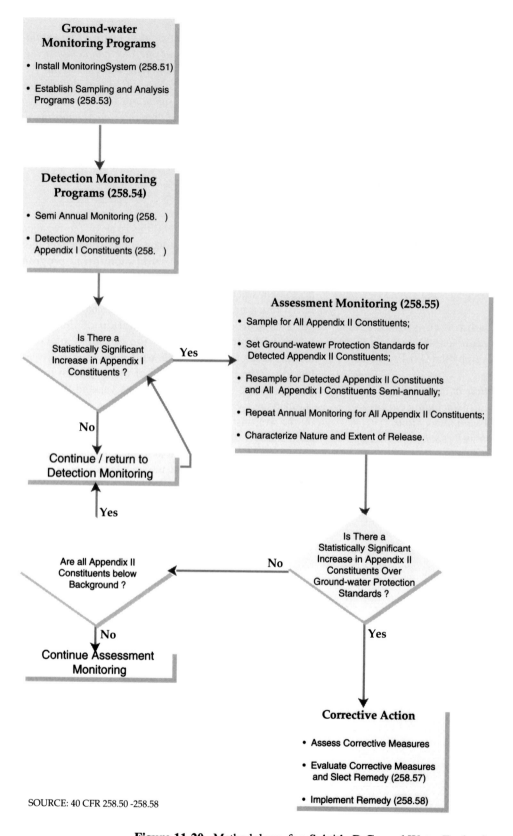

SOURCE: 40 CFR 258.50 -258.58

Figure 11-20 Methodology for Subtitle D Ground Water Evaluations

any detection monitoring program. It is imperative that detection monitoring programs not rely on only one background well as the basis for comparisons to downgradient wells. Spatial variability within the geologic environment can significantly affect the statistical comparisons necessary for detection monitoring programs. Evaluation and knowledge of ground-water flow and geology sufficient to design detection monitoring systems, together with time series graphs of ground-water quality, may clearly show a release. Statistical tests applicable to water quality evaluations currently recommended by EPA with other methods proposed in the water quality literature can be used as evaluation tools for the following example facilities:

- existing municipal solid waste landfills (MSWLF's), new facilities and existing facilities with historically clean water quality;

- RCRA Subtitle C hazardous waste disposal sites, industrial waste disposal sites, land disposal sites for waste water; and,

- Superfund type evaluations for assessment and aquifer remediation projects for federal and state clean-up programs.

11.7.4 Types of Statistical Tests

There are four general categories of statistical methods used for facility compliance comparisons with ground-water quality RCRA regulations:

- tests of central tendency (location)
- tests of trend,
- prediction, tolerance, and confidence intervals, and,
- control charts.

Statistical tests of central tendency are used to compare the mean or median of two or more sets of data and establish if they are significantly different. Tests of trend evaluate significant increase or decrease in water quality over time. Prediction and tolerance intervals are statistical methods that set limits for "acceptable" background water quality based on historic data sets. These interval tests also can be used to define the number of background measurements required to fully establish background water quality. Confidence intervals also set limits for "average" background water quality. Control charts are widely used graphical methods for industrial engineering quality control and are similar to the prediction, tolerance and confidence intervals. .

To evaluate water quality data a series of questions must be formulated to select the appropriate statistical tests to evaluate the data. These statistical tests are designed to evaluate whether or not a significant difference exists between the historical mean/median of background water quality and the mean/median of each downgradient well. The ability of these tests to detect ground-water contamination quickly, (i.e., when applied quarterly with detection within one or two quarters after a release occurs), depend on the choice of the statistical test and other data set factors including:

- the length of the unaffected water quality record;
- the variability in the data sets, and,
- the magnitude of the increase in concentration due to the release.

Regulatory requirements for "immediate" detection of releases to the environment can be extremely difficult to demonstrate even with the monitoring well screen located directly within the ground-water flow path from the facility. Also of interest in a detection monitoring program is to establish if specific water quality standards have been exceeded. If a sample value exceeds a regulatory standard such as Maximum Contaminant Level (MCL) under the Clean Water Act, (accounting for sampling and analytic variability), then a determination of a violation may be made by State or Federal regulators. In this situation a "violation" means only that a mandated concentration level has been exceeded, not that certain actions must be taken. The problem of regulatory violation of a standard is acute when State or Federal standards are at, or approach, the level of detection of the contaminant, as is the case with some volatile organic compounds. Possible decision approaches may include:

- A regulatory mandated "hard" limit where no data should exceed the water quality standard with consideration given to sampling and laboratory error; or,

- The more flexible historic mean concentration at a well where the water quality standard should not exceed this regulatory limit; or,

- The moving window approach where the last-year's mean concentration should not exceed the limit; or,

- The statistical limits where 95 percent of the population must be below the standard.

The above provide a number alternative decision paths for water quality evaluation or for making decisions on analytical sample values close to a water quality standard.

Comparison of downgradient water quality to the standard is conceptually straight forward. The indicator parameter is plotted on a simple item series graph and compared to the concentration called out in the water quality standard. Background concentrations should be plotted to evaluate if the background water quality levels exceed the particular standard. If background parameter levels do exceed the relevant standard, the downgradient well parameter concentrations must be evaluated statistically against the background rather than comparing the well data to the water quality standard. In those cases where background does not exceed the standard, downgradient concentration cable compared to t⸍ standard. Parameter trends, (especially for i⸍ indicator variables) can serve as a useful manag for implementing corrective actions befor⸍ drinking water standards are exceeded. Once wat⸍ standards are clearly exceeded and verified throu⸍ resampling, a release to the environment is confirmed. In the second and third approaches, confidence limits on the mean (where the standard must be below the lower confidence level) can provide the evaluation tool. The last approach can effectively use tolerance interval tests for evaluation of exceedances.

After evaluation of the issues or questions to be answered statistically, the next step is to choose a specific test that answers the question. The test must not only have an appropriate experimental design (i.e., answer the right question) but the implicit assumptions of the test must not be grossly violated. As previously discussed, ground-water quality data may grossly violate the assumption of normality, even after appropriate data transformation.

A detection monitoring data evaluation must be based on the variable regulatory issues that can change from State to State and from site to site. Some facilities may have specific permit requirements for statistical tests or specified parameter lists that can be significantly different from site to site within a single state. In the following sections potential releases to ground water can be evaluated three general types of statistical methods:

(1) tests of central tendency (location);
(2) tests of trend; and,
(3) prediction, tolerance, and confidence intervals.

In each subsection emphasis is placed on the situations where the type of test is appropriate. The types of water quality questions these tests can answer are discussed.

Tests of Central Tendency (Location)

The statistical mean and median of water quality data sets are the most common estimates of central tendency. Tests that compare the mean or median of two or more sets of data are termed "tests of central tendency or tests of location."

The U.S. EPA had previously required the use of the Cochran's Approximation to the Student's t-test be applied between pooled background water quality data and each downgradient compliance well. Significant criticism of this procedure (see EPA, Oct. 11, 1988; Miller and ⸍⸍⸍ and McBean and Rouas, 1984) ⸍ge to a parametric one-way ⸍A) or the nonparametric ⸍Vallis test (EPA 1988). ⸍est also suffers from a high ⸍⸍nen many multiple comparisons must ⸍⸍ ⸍⸍⸍⸍ ⸍or sites with for example more than 5 or 6 compliance wells (see Fisher and Potter, 1989)

Tests of Trend

Tests of trend are commonly used in detection monitoring programs to evaluate whether water quality parameter values are increasing or decreasing with time. Trend analysis is also useful for evaluating changes in background water quality. Trends in data could be observed as a gradual increase (usually modeled as a linear function) or a step function or even cyclical on a seasonal basis.

Trend evaluations have traditionally been performed by inspection of graphed time concentration plots. Time series plots also can be used in conjunction with box plots to evaluate trends and seasonal fluctuations. A number of statistical methods can be applied to data sets to evaluate for trends and seasonality. Example procedures such as the Mann-Kendall test for trend evaluates the relative magnitudes of the concentration data with time (Goodman 1987). The length of time necessary to obtain adequate trends; for long term trends two years of data are recommended (Doctor et al 1986), for seasonal trends much longer period data set may be necessary; Goodman (1987) using a modified Mann-Kendall test found that at least 10 years of quarterly data were required for obtaining adequate power to detect seasonal trends. Although few facilities have such a long period of data, the long (post) closure requirements in State and Federal regulations of 10 to 30 years will make such evaluations for seasonal trends possible.

Statistical trend tests alone cannot be used to determine compliance with ground-water quality regulations. These tests can only answer the question "Does a positive or negative trend exist?" The presence of a minor trend should not be construed to mean there has been a release from the facility. Therefore if a test of trend is used to support the hypothesis of a release, the results must be linked to exceedance of water quality standards and to likelihood of the release based on review of potential cross-contamination and interferences.

Tests of trend have been commonly used in evaluating the expected effectiveness of remedial action. However, tests of trend should not be used to predict when a target concentration will be reached since aquifer restoration is usually not a linear but rather an asymptotic process.

A common use of trend tests is to evaluate if background water quality is significantly (gradually) changing in time. Hence, the background water quality represents a moving window that will be compared to down-gradient water quality. In this case, the background trend should be removed prior to further analysis (Harris et al, 1987). An apparent trend at a downgradient well cannot be confirmed as evidence of contamination, unless it can be shown that the same trend does not exist in background or upgradient wells.

The nonparametric analogs to the linear regression f-test are Kendall's Tau statistic and Spearman's (Rho) rank correlation coefficient. Usually Kendall's Tau is chosen for water quality data because the test statistic approaches normality at smaller sample sizes than Spearman's Rho (Montgomery et al, 1987).

Linear regression is considered a powerful technique of trend, but analysts tend to delete outlying values without physical justification to get a "good fit." to the data. Also, some users will wrongly try to make predictions of when concentration will return to normal or when a standard will be exceeded. Reviewers should make sure that deletion of data is 'physically' justified. Also any predictions made with the regression line should be interpreted as no more than a best guess.

Fisher and Potter (1989) reviewed statistical tests for applicability for use in detecting facility ground-water contamination events. They found that tests of central tendency both parametric and nonparametric has severe limitations. At least for the cases reviewed natural spatial variability did not permit ANOVA results to discern between natural variations in mean and those due to potential contamination. They also observed that ground water quality data often violated the parametric assumptions of normality for both raw and log transformed data sets. Even nonparametric tests of central tendency (such as Kruskal-Wallis) are not recommended for detecting contamination but rather should be used for

evaluating spatial variability (Fisher and Potter 1989). Statistical tests based on trend can be used in conjunction with other data evaluation techniques to support the conclusion of observed contamination. Prediction interval tests were recommended by Fisher and Potter (1989) as the most theoretically sound approach to setting background levels and in the author's opinion interval statistical tests represent the most applicable methods for evaluating detection monitoring programs. As such, the remaining discussion of statistics will concentrate on interval tests.

Confidence, Tolerance and Prediction Intervals

Statistical intervals tests are used to set limits using background water quality or for intra-well comparisons. Measurements may be compared to the upper bound of the interval to determine if there has been a release. Both the upper and lower bounds are considered for parameters, such as, pH that may increase or decrease depending on the type of contamination.

Confidence limits on the mean define an interval within which the true mean of the population will fall (90, 95, 99 percent) of the time.

Tolerance limits define a range within which some proportion of the population will fall (90, 95, 99 percent) of the time. Usually this proportion is also 90, 95 or 99 percent.

Prediction limits define an interval within which it can be stated that the next k measurements will fall (90,95, 99 percent) of the time. Hahn (1970) explains the difference between these limits.

"A typical astronaut, who has been assigned to a specific number of flights, is generally not very interested in what will happen on the average in the population of all space flights, of which his happens to be a random sample (confidence interval on the mean), or even what will happen in at least 99 percent of such flights (tolerance interval). His main concern is the worst that will happen in the (next) one, three or five flights in which he will personally be involved (prediction interval)".

All three (parametric) intervals are symmetric and are calculated based on the models x $\pm$ ks (two-sided) or x + ks (one-sided) where k is a constant obtained from tabulated values. The environmental use of these statistics depends on the validity of the assumptions of normality, stationarity and independence. Table 11-5 lists questions asked from a regulatory perspective and the appropriate method(s) in each case. A common mistake observed is the use confidence limits when tolerance intervals or prediction limits will answer the question of concern.

TABLE 11-5 Application of intervals to regulatory questions

QUESTIONS	METHOD
1. What is a reasonable upper limit for background water QUALITY?	Tolerance
2. Are downgradient concentrations outside the allowable range of background water quality?	Tolerance Prediction
3. Do new measurements at downgradient wells come from the background population?	Prediction
4. Has a standard been exceeded based on average water quality over a time period?	Confidence
5. Has a standard been exceeded more than a specified percent of the time?	Tolerance
6. Within what range can we state the mean/median of background water quality falls?	Confidence

In theory question 6 of Table 11-5 is not applicable for determining compliance with ground-water quality regulations. This is because we are not interested in comparisons to average water quality, but rather on comparison of compliance well data to the population of background data.

Assumptions:—The following statistical methods are suitable for data that are normally distributed, rare event data that have a Poisson distribution, or data for which the distribution is unknown and/or atypical (i.e., non–parametric). In addition, normally distributed data that have some proportion (less than 90%) below a detection limit (i.e., nondetects) can also be accommodated. We will refer to such data as "censored" and the corresponding distribution as a censored– normal –distribution. Given these four choices (i.e., normal, censored–normal, Poisson, or non–parametric), the only case that is not covered is when nothing is detected in the background, water–quality samples (see Figure 11-21). For this case, we will present a method by which practical quantification–limits can be obtained based on an analyte present in a laboratory calibration–study. These limits may, in turn, be used as criteria for decisions in detection monitoring (i.e., limits), or substituted for nondetects in other analyses.

In addition to selecting the proper sampling distribution, the most critical assumption underlying all of the statistical methods to be presented is the independence of the data.

These models strictly assume that observations are the result of a random–sampling and that each observation is an independent random–sample from the parent population. In the context of ground–water monitoring, this assumption rules out the use of replicate samples, daily samples, and perhaps even monthly samples, given how slowly ground–water moves. As such, the author strongly recommends adoption of a quarterly sampling program as the basis for the detection monitoring program.

Statistical Overview:—In detection monitoring programs, the investigator obtains a sample from a monitoring well and must decide whether the facility has had an impact on concentrations of a series of indicator parameters. It is critically important to realize that each new measurement is not a mean value, but rather, a single, new observation in a supposedly dynamic (flowing) ground–water system. As such, statistical methods for the comparison of mean values (e.g., student's t–test) do not apply. From a statistical perspective, the problem is, therefore, to estimate the probability that each new datum was drawn from the population of pristine, background water–quality, for which we only have estimates of mean and variance, as obtained from a limited number of upgradient measurements. If the investigator knew that a particular indicator parameter was normally distributed and somehow had the privileged information of knowing the population mean (μ) and variance (σ^2), then he/she could construct the interval $\mu \pm 1.96\sigma$, which would contain

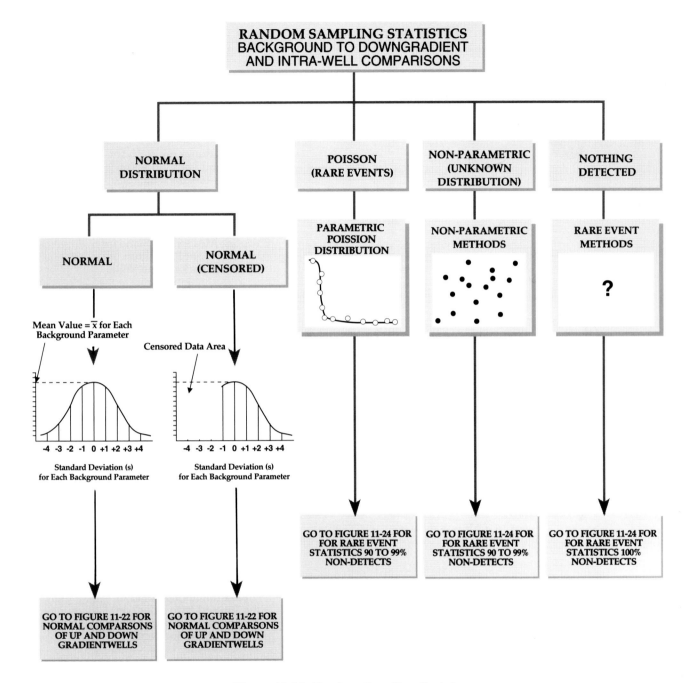

Figure 11-21 Random Sampling Statistics

95% of all individual measurements (not means), which were drawn from that population, and the job would be finished. However, in practice, the investigator never knows the values of μ and σ but only has the sample–based estimates ($\bar{x}$ and s), obtained from n independent, upgradient measurements. As such, uncertainty is two–fold. First, a range of possible values exists when sampling known parameters with a normal distribution,

and second, there is a range of possible means ($\bar{x}$) and standard deviations (s) that could be obtained from drawing a sample, of size n, from a normally–distributed population with mean (μ) and variance (σ^2). This latter source of uncertainty will require a multiplier that is larger than 1.96 if one requires reasonable confidence that 95% of the population is contained within the interval. As the number of background water quality measurements

STATISTICAL PROCEDURE FOR EVALUATION OF BACKGROUND AND DOWNGRADIENT
WATER QUALITY FOR PARAMETERS THAT HAVE NORMALLY DETECTABLE VALUES

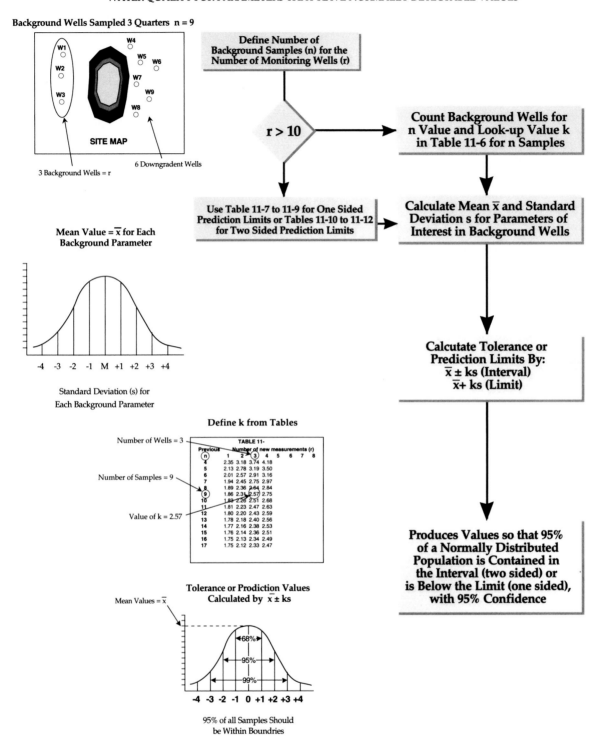

Figure 11-22 Statistics for Parameters That Have Detectable Values

approaches infinity, however, the multiplier once again approaches 1.96. When the sample size is small, say n = 8, the required multiplier for 95% confidence that 95% of the population is contained is 3.732 (i.e., $\bar{x} \pm 3.732$ s). For a sample size of n = 30, 95% confidence is achieved using a multiplier of 2.549; and n = 100, the multiplier is 2.233.

These intervals are known as two–sided tolerance intervals in the statistical literature, and are largely due to the work of Wald and Wolfowitz (1946). When one is only concerned with values that are large, one–sided tolerance limits can be constructed as $\bar{x} + ks$, where the multiplier k is somewhat smaller than the previous two–sided tolerance–limit factors. Figure 11-22 shows the statistical procedure for evaluation of background and downgradient water quality for parameters that normally have detectable values. For example, for n = 8, the one–sided tolerance limit is obtained as $\bar{x} + 3.188$ s; for n = 30, k = 2.220; and for n = 100, k = 1.927. Table 11–6 presents values of the multiplier k for n = 4 to 100 that are required to have 95% confidence that 95% of a normally distributed population are contained in the interval (i.e., two–sided) or is below the limit (i.e., one–sided).

Although tolerance intervals are generally quite useful in quality control problems, which are similar to ground–water detection monitoring, even more precise probability statements are possible. For example, in the context of ground–water monitoring, one is generally less interested in what can happen in 95% of all possible samples, and more interested in what can happen on the next round of sampling, for which measurements are to be obtained from the monitoring wells at the facility. Because one knows the number of future comparisons (i.e., monitoring wells), we can construct an interval (two–sided) or limit (one–sided) that will contain the next r measurements with 95% confidence. If r, in this case the number of monitoring wells, is reasonably small, it will provide a more conservative test than the corresponding 95% confidence and 95% coverage tolerance–interval. For example, for a facility with n = 8 background measurements and r = 3 monitoring wells, the multiplier for a one–sided 95% confidence and 95% coverage tolerance–limit is k = 3.188; whereas the corresponding factor for a 95% prediction limit is only k = 2.80. However, if the facility had r = 10 monitoring wells, the tolerance–limit factor is, of course, unchanged; but the prediction limit factor is now k = 3.71, which is considerably larger than the corresponding tolerance–limit factor. In general, for 95% confidence and 95% coverage, tolerance–intervals will be more conservative for facilities with r > 10 monitoring wells, and prediction limits will be more conservative for facilities with r ≤ 10 monitoring wells. Given the large number of detection monitoring wells at most modern waste disposal facilities, tolerance

Table 11-6 Factors (k) for Constructing Two-sided and One -sided Normal Tolerance Limits ($\bar{x} \pm ks$ and $\bar{x} + ks$) 5% Confidence that 95% of the Distribution is Covered

n	Two-sided	One-Sided
4	6.370	5.144
5	5.079	4.210
6	4.414	3.711
7	4.007	3.401
8	3.732	3.188
9	3.532	3.032
10	3.379	2.911
11	3.259	2.815
12	3.169	2.736
13	3.081	2.670
14	3.012	2.614
15	2.954	2.566
16	2.903	2.523
17	2.858	2.486
18	2.819	2.453
19	2.784	2.423
20	2.752	2.396
21	2.723	2.371
22	2.697	2.350
23	2.673	2.329
24	2.651	2.309
25	2.631	2.292
30	2.549	2.220
35	2.490	2.166
40	2.445	2.126
50	2.379	2.065
60	2.333	2.022
80	2.272	1.965
100	2.233	1.927

intervals may well be the method of choice. Tables 11–7 through 11–9 contain factors for computing one–sided 95% prediction limits based on background samples of n = 4 to 100 and number of monitoring wells of r = 1 to 100. Corresponding two–sided 95% prediction interval factors are provided in Tables 11-10 through 11-12 As in the case of tolerance intervals and limits, these factors are applied as $x \pm ks$ (interval) and $x + ks$ (limit). A detailed description of prediction limits in the context of ground–water monitoring problems is provided by Gibbons (1987a).

Selecting the Number of Background Samples:—A common question asked of statisticians is: "How many background samples do I need?" This question became even more common when, in previous regulations (40 CFR

Table 11-7. One-Sided 95% Poisson Prediction Limits For r Additional Samples Given Background Sample of Size n.

Previous n	\multicolumn{15}{c}{Number of new measurements (r)}														
	1	2	3	4	5	6	7	8	9	10	11	12	13	14	15
4	2.63	3.56	4.18	4.67	5.08	5.43	5.74	6.03	6.29	6.53	6.76	6.97	7.17	7.36	7.54
5	2.34	3.04	3.49	3.83	4.10	4.34	4.54	4.73	4.89	5.04	5.18	5.31	5.44	5.55	5.66
6	2.18	2.78	3.14	3.42	3.63	3.82	3.97	4.11	4.24	4.35	4.46	4.56	4.65	4.73	4.81
7	2.08	2.62	2.94	3.17	3.38	3.51	3.65	3.76	3.87	3.96	4.05	4.13	4.20	4.27	4.34
8	2.01	2.51	2.80	3.01	3.18	3.32	3.43	3.54	3.63	3.71	3.79	3.86	3.92	3.98	4.04
9	1.96	2.43	2.70	2.90	3.05	3.18	3.29	3.38	3.48	3.54	3.60	3.67	3.72	3.78	3.83
10	1.92	2.37	2.63	2.82	2.96	3.08	3.18	3.26	3.34	3.41	3.47	3.53	3.58	3.63	3.68
11	1.89	2.33	2.58	2.75	2.89	3.00	3.09	3.17	3.25	3.31	3.37	3.42	3.47	3.52	3.56
12	1.87	2.29	2.53	2.70	2.83	2.93	3.02	3.10	3.17	3.23	3.29	3.34	3.39	3.43	3.47
13	1.85	2.26	2.49	2.66	2.78	2.88	2.97	3.04	3.11	3.17	3.22	3.27	3.32	3.36	3.40
14	1.83	2.24	2.46	2.62	2.74	2.84	2.93	3.00	3.06	3.12	3.17	3.22	3.26	3.30	3.34
15	1.82	2.21	2.44	2.59	2.71	2.81	2.89	2.96	3.02	3.07	3.12	3.17	3.21	3.25	3.28
16	1.81	2.20	2.41	2.57	2.68	2.78	2.85	2.92	2.98	3.04	3.09	3.13	3.17	3.21	3.24
17	1.80	2.18	2.40	2.54	2.66	2.75	2.83	2.89	2.95	3.01	3.05	3.09	3.13	3.17	3.21
18	1.79	2.17	2.38	2.53	2.64	2.73	2.80	2.87	2.93	2.98	3.02	3.07	3.10	3.14	3.17
19	1.78	2.16	2.36	2.51	2.62	2.71	2.78	2.85	2.90	2.95	3.00	3.04	3.08	3.11	3.14
20	1.77	2.14	2.35	2.49	2.60	2.69	2.76	2.83	2.88	2.93	2.98	3.02	3.05	3.09	3.12
21	1.77	2.13	2.34	2.48	2.59	2.67	2.75	2.81	2.86	2.91	2.96	3.00	3.03	3.07	3.10
22	1.76	2.13	2.33	2.47	2.57	2.66	2.73	2.79	2.85	2.89	2.94	2.98	3.01	3.05	3.08
23	1.75	2.12	2.32	2.46	2.56	2.65	2.72	2.78	2.83	2.88	2.92	2.96	3.00	3.03	3.06
24	1.75	2.11	2.31	2.45	2.55	2.63	2.70	2.76	2.82	2.87	2.91	2.95	2.98	3.01	3.04
25	1.74	2.10	2.30	2.44	2.54	2.62	2.69	2.75	2.81	2.85	2.89	2.93	2.97	3.00	3.03
26	1.74	2.10	2.29	2.43	2.53	2.61	2.68	2.74	2.79	2.84	2.88	2.92	2.95	2.99	3.01
27	1.74	2.09	2.29	2.42	2.52	2.61	2.67	2.73	2.78	2.83	2.87	2.91	2.94	2.97	3.00
28	1.73	2.09	2.28	2.42	2.52	2.60	2.66	2.72	2.77	2.82	2.86	2.90	2.93	2.96	2.99
29	1.73	2.08	2.28	2.41	2.51	2.59	2.66	2.71	2.77	2.81	2.85	2.89	2.92	2.95	2.98
30	1.73	2.08	2.27	2.40	2.50	2.58	2.65	2.71	2.76	2.80	2.84	2.88	2.91	2.94	2.97
31	1.72	2.07	2.27	2.40	2.50	2.58	2.64	2.70	2.75	2.79	2.83	2.87	2.90	2.93	2.96
32	1.72	2.07	2.26	2.39	2.49	2.57	2.64	2.69	2.74	2.79	2.83	2.86	2.89	2.92	2.95
33	1.72	2.07	2.26	2.39	2.49	2.56	2.63	2.69	2.74	2.78	2.82	2.85	2.89	2.92	2.94
34	1.72	2.06	2.25	2.38	2.48	2.56	2.62	2.68	2.73	2.77	2.81	2.85	2.88	2.91	2.94
35	1.71	2.06	2.25	2.38	2.48	2.55	2.62	2.67	2.72	2.77	2.81	2.84	2.87	2.90	2.93
36	1.71	2.06	2.25	2.37	2.47	2.55	2.61	2.67	2.72	2.76	2.80	2.83	2.87	2.90	2.92
37	1.71	2.06	2.24	2.37	2.47	2.54	2.61	2.66	2.71	2.76	2.79	2.83	2.86	2.89	2.92
38	1.71	2.05	2.24	2.37	2.46	2.54	2.60	2.66	2.71	2.75	2.79	2.82	2.86	2.B8	2.91
39	1.71	2.05	2.24	2.36	2.46	2.54	2.60	2.66	2.70	2.75	2.78	2.82	2.85	2.88	2.91
40	1.71	2.05	2.23	2.36	2.46	2.53	2.60	2.65	2.70	2.74	2.78	2.81	2.85	2.87	2.90
41	1.70	2.05	2.23	2.36	2.45	2.53	2.59	2.65	2.69	2.74	2.77	2.81	2.84	2.87	2.90
42	1.70	2.04	2.23	2.35	2.45	2.53	2.59	2.64	2.69	2.73	2.77	2.80	2.84	2.87	2.89
43	1.70	2.04	2.23	2.35	2.45	2.52	2.59	2.64	2.69	2.73	2.77	2.80	2.83	2.86	2.89
44	1.70	2.04	2.22	2.35	2.44	2.52	2.58	2.64	2.68	2.73	2.76	2.80	2.83	2.86	2.88
45	1.70	2.04	2.22	2.35	2.44	2.52	2.58	2.63	2.68	2.72	2.76	2.79	2.82	2.85	2.88
46	1.70	2.04	2.22	2.34	2.44	2.51	2.58	2.63	2.68	2.72	2.76	2.79	2.82	2.85	2.88
47	1.70	2.03	2.22	2.34	2.44	2.51	2.57	2.63	2.67	2.72	2.75	2.79	2.82	2.85	2.87
48	1.70	2.03	2.22	2.34	2.43	2.51	2.57	2.62	2.67	2.71	2.75	2.78	2.81	2.84	2.87
49	1.69	2.03	2.21	2.34	2.43	2.51	2.57	2.62	2.67	2.71	2.75	2.78	2.81	2.84	2.86
50	1.69	2.03	2.21	2.34	2.43	2.50	2.57	2.62	2.67	2.71	2.74	2.78	2.81	2.84	2.86
60	1.68	2.02	2.20	2.32	2.41	2.48	2.55	2.60	2.64	2.68	2.72	2.75	2.78	2.81	2.84
70	1.68	2.01	2.19	2.31	2.40	2.47	2.53	2.58	2.63	2.67	2.70	2.74	2.77	2.79	2.82
80	1.67	2.00	2.18	2.30	2.39	2.46	2.52	2.57	2.62	2.66	2.69	2.72	2.75	2.78	2.80
90	1.67	2.00	2.17	2.29	2.38	2.45	2.51	2.56	2.61	2.65	2.68	2.71	2.74	2.77	2.79
100	1.67	1.99	2.17	2.29	2.38	2.45	2.51	2.56	2.60	2.64	2.67	2.71	2.73	2.76	2.79

Tables 11-7 to 11-12 : Source Gibbons 1990 and Factor Equation $\text{Factor} = t(n-1, 1-\alpha/r)\sqrt{1+1/n}$

Table 11-8.   Values of t For Obtaining One-Sided 95% Poisson Prediction Limits For r Additional Samples Given A Background Sample Of Size n.

Previous n	Number of new measurements (r)														
	16	17	18	19	20	21	22	23	24	25	26	27	28	29	30
4	6.89	7.04	7.18	7.32	7.45	7.58	7.70	7.82	7.94	8.05	8.16	8.27	8.37	8.47	8.57
5	5.26	5.35	5.44	5.52	5.60	5.67	5.75	5.82	5.88	5.95	6.01	6.08	6.14	6.20	6.25
6	4.53	4.59	4.65	4.72	4.77	4.83	4.88	4.93	4.98	5.03	5.08	5.12	5.16	5.21	5.25
7	4.11	4.17	4.22	4.27	4.32	4.36	4.40	4.45	4.49	4.52	4.56	4.60	4.63	4.66	4.70
8	3.85	3.90	3.95	3.99	4.03	4.07	4.10	4.14	4.17	4.21	4.24	4.27	4.30	4.33	4.35
9	3.68	3.72	3.76	3.80	3.83	3.87	3.90	3.93	3.96	3.99	4.02	4.05	4.07	4.10	4.12
10	3.55	3.59	3.62	3.66	3.69	3.72	3.75	3.78	3.81	3.83	3.86	3.88	3.91	3.93	3.95
11	3.45	3.48	3.52	3.55	3.58	3.61	3.64	3.67	3.69	3.72	3.74	3.76	3.78	3.81	3.83
12	3.37	3.40	3.44	3.47	3.50	3.52	3.55	3.58	3.60	3.62	3.65	3.67	3.69	3.71	3.73
13	3.31	3.34	3.37	3.40	3.43	3.45	3.48	3.50	3.53	3.55	3.57	3.59	3.61	3.63	3.65
14	3.26	3.29	3.32	3.35	3.37	3.40	3.42	3.45	3.47	3.49	3.51	3.53	3.55	3.57	3.58
15	3.21	3.24	3.27	3.30	3.33	3.35	3.37	3.40	3.42	3.44	3.46	3.48	3.49	3.51	3.53
16	3.18	3.21	3.23	3.26	3.29	3.31	3.33	3.35	3.37	3.39	3.41	3.43	3.45	3.47	3.48
17	3.15	3.17	3.20	3.23	3.25	3.27	3.30	3.32	3.34	3.36	3.38	3.39	3.41	3.43	3.44
18	3.12	3.15	3.17	3.20	3.22	3.24	3.27	3.29	3.31	3.33	3.34	3.36	3.38	3.39	3.41
19	3.09	3.12	3.15	3.17	3.20	3.22	3.24	3.26	3.28	3.30	3.32	3.33	3.35	3.36	3.38
20	3.07	3.10	3.13	3.15	3.17	3.20	3.22	3.24	3.25	3.27	3.29	3.31	3.32	3.34	3.35
21	3.05	3.08	3.11	3.13	3.15	3.17	3.19	3.21	3.23	3.25	3.27	3.28	3.30	3.32	3.33
22	3.04	3.06	3.09	3.11	3.13	3.16	3.18	3.20	3.21	3.23	3.25	3.26	3.28	3.29	3.31
23	3.02	3.05	3.07	3.10	3.12	3.14	3.16	3.18	3.20	3.21	3.23	3.25	3.26	3.28	3.29
24	3.01	3.03	3.06	3.08	3.10	3.12	3.14	3.16	3.18	3.20	3.21	3.23	3.24	3.26	3.27
25	3.00	3.02	3.05	3.07	3.09	3.11	3.13	3.15	3.17	3.18	3.20	3.21	3.23	3.24	3.26
26	2.99	3.01	3.03	3.06	3.08	3.10	3.12	3.14	3.15	3.17	3.19	3.20	3.22	3.23	3.24
27	2.98	3.00	3.02	3.05	3.07	3.09	3.11	3.12	3.14	3.16	3.17	3.19	3.20	3.22	3.23
28	2.96	2.99	3.01	3.04	3.06	3.08	3.09	3.11	3.13	3.15	3.16	3.18	3.19	3.21	3.22
29	2.96	2.98	3.00	3.03	3.05	3.07	3.09	3.10	3.12	3.14	3.15	3.17	3.18	3.19	3.21
30	2.95	2.97	3.00	3.02	3.04	3.06	3.08	3.09	3.11	3.13	3.14	3.16	3.17	3.18	3.20
31	2.94	2.96	2.99	3.01	3.03	3.05	3.07	3.08	3.10	3.12	3.13	3.15	3.16	3.18	3.19
32	2.93	2.96	2.98	3.00	3.02	3.04	3.06	3.08	3.09	3.11	3.12	3.14	3.15	3.17	3.18
33	2.93	2.95	2.97	2.99	3.01	3.03	3.05	3.07	3.09	3.10	3.12	3.13	3.14	3.16	3.17
34	2.92	2.94	2.97	2.99	3.01	3.03	3.04	3.06	3.08	3.09	3.11	3.12	3.14	3.15	3.16
35	2.91	2.94	2.96	2.98	3.00	3.02	3.04	3.06	3.07	3.09	3.10	3.12	3.13	3.14	3.16
36	2.91	2.93	2.95	2.98	3.00	3.01	3.03	3.05	3.07	3.08	3.10	3.11	3.12	3.14	3.15
37	2.90	2.93	2.95	2.97	2.99	3.01	3.03	3.04	3.06	3.07	3.09	3.10	3.12	3.13	3.14
38	2.90	2.92	2.94	2.97	2.98	3.00	3.02	3.04	3.05	3.07	3.08	3.10	3.11	3.12	3.14
39	2.89	2.92	2.94	2.96	2.98	3.00	3.02	3.03	3.05	3.06	3.08	3.09	3.11	3.12	3.13
40	2.89	2.91	2.94	2.96	2.98	2.99	3.01	3.03	3.04	3.06	3.07	3.09	3.10	3.11	3.13
41	2.89	2.91	2.93	2.95	2.97	2.99	3.01	3.02	3.04	3.05	3.07	3.08	3.10	3.11	3.12
42	2.88	2.91	2.93	2.95	2.97	2.98	3.00	3.02	3.03	3.05	3.06	3.08	3.09	3.10	3.12
43	2.88	2.90	2.92	2.94	2.96	2.98	3.00	3.01	3.03	3.05	3.06	3.07	3.09	3.10	3.11
44	2.88	2.90	2.92	2.94	2.96	2.98	2.99	3.01	3.03	3.04	3.06	3.07	3.08	3.10	3.11
45	2.87	2.89	2.92	2.94	2.96	2.97	2.99	3.01	3.02	3.04	3.05	3.07	3.08	3.09	3.10
46	2.87	2.89	2.91	2.93	2.95	2.97	2.99	3.00	3.02	3.03	3.05	3.06	3.07	3.09	3.10
47	2.87	2.89	2.91	2.93	2.95	2.97	2.98	3.00	3.02	3.03	3.04	3.06	3.07	3.08	3.10
48	2.86	2.89	2.91	2.93	2.95	2.96	2.98	3.00	3.01	3.03	3.04	3.05	3.07	3.08	3.09
49	2.86	2.88	2.90	2.92	2.94	2.96	2.98	2.99	3.01	3.02	3.04	3.05	3.06	3.08	3.09
50	2.86	2.88	2.90	2.92	2.94	2.96	2.97	2.99	3.01	3.02	3.03	3.05	3.06	3.07	3.09
60	2.84	2.86	2.88	2.90	2.92	2.93	2.95	2.97	2.98	3.00	3.01	3.02	3.04	3.05	3.06
70	2.82	2.84	2.86	2.88	2.90	2.92	2.93	2.95	2.96	2.98	2.99	3.00	3.02	3.03	3.04
80	2.81	2.83	2.85	2.87	2.89	2.90	2.92	2.94	2.95	2.97	2.98	2.99	3.00	3.02	3.03
90	2.80	2.82	2.84	2.86	2.88	2.90	2.91	2.93	2.94	2.96	2.97	2.98	2.99	3.01	3.02
100	2.79	2.82	2.84	2.85	2.87	2.89	2.90	2.92	2.93	2.95	2.96	2.97	2.99	3.00	3.01

Table 11-9 Values of t For Obtaining One-Sided 95% Poisson Prediction Limits For r Additional Samples Given A Background Sample of Size n.

Previous n	30	35	40	45	50	55	60	65	70	75	80	85	90	95	100
						Number 0f new measurements (r)									
4	8.57	9.04	9.46	9.85	10.21	10.55	10.87	11.17	11.45	11.72	11.98	12.23	12.47	12.70	12.92
5	6.25	6.52	6.76	6.97	7.17	7.36	7.53	7.69	7.84	7.98	8.12	8.25	8.38	8.50	8.61
6	5.25	5.44	5.60	5.76	5.89	6.0 2	6.14	6.25	6.35	6.45	6.54	6.63	6.71	6.79	6.87
7	4.70	4.85	4.98	5.10	5.21	5.31	5.40	5.48	5.56	5.64	5.71	5.78	5.84	5.90	5.96
8	4.35	4.48	4.59	4.69	4.78	4.87	4.94	5.02	5.08	5.14	5.20	5.26	5.31	5.36	5.41
9	4.12	4.23	4.33	4.42	4.50	4.57	4.64	4.70	4.76	4.81	4.86	4.91	4.96	5.00	5.04
10	3.95	4.06	4.15	4.22	4.30	4.36	4.42	4.48	4.53	4.58	4.62	4.67	4.71	4.74	4.78
11	3.83	3.92	4.00	4.08	4.14	4.20	4.26	4.31	4.36	4.40	4.44	4.48	4.52	4.55	4.59
12	3.73	3.82	3.89	3.96	4.02	4.08	4.13	4.18	4.22	4.26	4.30	4.34	4.37	4.41	4.44
13	3.65	3.73	3.81	3.87	3.93	3.98	4.03	4.08	4.12	4.15	4.19	4.23	4.26	4.29	4.32
14	3.58	3.66	3.73	3.80	3.85	3.90	3.95	3.99	4.03	4.07	4.10	4.13	4.16	4.19	4.22
15	3.53	3.61	3.67	3.73	3.79	3.84	3.88	3.92	3.96	3.99	4.03	4.06	4.09	4.11	4.14
16	3.48	3.56	3.62	3.68	3.73	3.78	3.82	3.86	3.90	3.93	3.96	3.99	4.02	4.05	4.07
17	3.44	3.52	3.58	3.64	3.69	3.73	3.77	3.81	3.84	3.88	3.91	3.94	3.96	3.99	4.01
18	3.41	3.48	3.54	3.60	3.65	3.69	3.73	3.77	3.80	3.83	3.86	3.89	3.92	3.94	3.96
19	3.38	3.45	3.51	3.56	3.61	3.65	3.69	3.73	3.76	3.79	3.82	3.85	3.87	3.90	3.92
20	3.35	3.42	3.48	3.53	3.58	3.62	3.66	3.69	3.73	3.76	3.79	3.81	3.84	3.86	3.88
21	3.33	3.40	3.46	3.51	3.55	3.59	3.63	3.66	3.70	3.73	3.75	3.78	3.80	3.83	3.85
22	3.31	3.38	3.43	3.48	3.53	3.57	3.60	3.64	3.67	3.70	3.73	3.75	3.77	3.80	3.82
23	3.29	3.35	3.41	3.46	3.50	3.54	3.58	3.61	3.64	3.67	3.70	3.72	3.75	3.77	3.79
24	3.27	3.34	3.39	3.44	3.48	3.52	3.56	3.59	3.62	3.65	3.68	3.70	3.72	3.75	3.77
25	3.26	3.32	3.38	3.42	3.47	3.50	3.54	3.57	3.60	3.63	3.66	3.68	3.70	3.72	3.75
26	3.24	3.31	3.36	3.41	3.45	3.49	3.52	3.55	3.58	3.61	3.64	3.66	3.68	3.70	3.72
27	3.23	3.29	3.35	3.39	3.43	3.47	3.51	3.54	3.57	3.59	3.62	3.64	3.67	3.69	3.71
28	3.22	3.28	3.33	3.38	3.4 2	3.46	3.49	3.52	3.55	3.58	3.60	3.63	3.65	3.67	3.69
29	3.21	3.27	3.32	3.37	3.41	3.44	3.48	3.51	3.54	3.56	3.59	3.61	3.63	3.65	3.67
30	3.20	3.26	3.31	3.36	3.40	3.43	3.47	3.50	3.52	3.55	3.58	3.60	3.62	3.64	3.66
31	3.19	3.25	3.30	3.34	3.38	3.42	3.45	3.48	3.51	3.54	3.56	3.58	3.61	3.63	3.65
32	3.18	3.24	3.29	3.33	3.37	3.41	3.44	3.47	3.50	3.53	3.55	3.57	3.59	3.61	3.63
33	3.17	3.23	3.28	3.33	3.37	3.40	3.43	3.46	3.49	3.52	3.54	3.56	3.58	3.60	3.62
34	3.16	3.22	3.27	3.32	3.36	3.39	3.42	3.45	3.48	3.51	3.53	3.55	3.57	3.59	3.61
35	3.16	3.21	3.26	3.31	3.35	3.38	3.42	3.44	3.47	3.50	3.52	3.54	3.56	3.58	3.60
36	3.15	3.21	3.26	3.30	3.34	3.37	3.41	3.44	3.46	3.49	3.51	3.53	3.55	3.57	3.59
37	3.14	3.20	3.25	3.29	3.33	3.37	3.40	3.43	3.45	3.48	3.50	3.52	3.54	3.56	3.58
38	3.14	3.19	3.24	3.29	3.32	3.36	3.39	3.42	3.45	3.47	3.49	3.52	3.54	3.56	3.57
39	3.13	3.19	3.24	3.28	3.32	3.35	3.38	3.41	3.44	3.46	3.49	3.51	3.53	3.55	3.57
40	3.13	3.1 8	3.23	3.27	3.31	3.35	3.38	3.41	3.43	3.46	3.48	3.50	3.52	3.54	3.56
41	3.12	3.18	3.23	3.27	3.31	3.34	3.37	3.40	3.43	3.45	3.47	3.49	3.51	3.53	3.55
42	3.12	3.1 7	3.22	3.26	3.30	3.33	3.37	3.39	3.42	3.44	3.47	3.49	3.51	3.53	3.54
43	3.11	3.17	3.22	3.26	3.30	3.33	3.36	3.39	3.41	3.44	3.46	3.48	3.50	3.52	3.54
44	3.11	3.16	3.21	3.25	3.29	3.32	3.35	3.38	3.41	3.43	3.45	3.48	3.49	3.51	3.53
45	3.10	3.16	3.21	3.25	3.29	3.32	3.35	3.38	3.40	3.43	3.45	3.47	3.49	3.51	3.53
46	3.10	3.15	3.20	3.24	3.28	3.31	3.34	3.37	3.40	3.42	3.44	3.46	3.48	3.50	3.52
47	3.10	3.15	3.20	3.24	3.28	3.31	3.34	3.37	3.39	3.42	3.44	3.46	3.48	3.50	3.51
48	3.09	3.15	3.19	3.24	3.27	3.31	3.34	3.36	3.39	3.41	3.43	3.45	3.47	3.49	3.51
49	3.09	3.14	3.19	3.23	3.27	3.30	3.33	3.36	3.38	3.41	3.43	3.45	3.47	3.49	3.50
50	3.09	3.14	3.19	3.23	3.26	3.30	3.33	3.35	3.38	3.40	3.43	3.45	3.46	3.48	3.50
60	3.06	3.11	3.16	3.20	3.23	3.27	3.30	3.32	3.35	3.37	3.39	3.41	3.43	3.45	3.46
70	3.04	3.09	3.14	3.18	3.21	3.24	3.27	3.30	3.32	3.34	3.37	3.39	3.40	3.42	3.44
80	3.03	3.08	3.12	3.16	3.2 0	3.23	3.26	3.28	3.31	3.33	3.35	3.37	3.38	3.40	3.42
90	3.02	3.07	3.11	3.15	3.18	3.21	3.24	3.27	3.29	3.31	3.33	3.35	3.37	3.39	3.40
100	3.01	3.06	3.10	3.14	3.17	3.20	3.23	3.26	3.28	3.30	3.32	3.34	3 36	3.38	3.39

Table 11-10 Factors For Obtaining Two-Sided 95% Prediction Limits For r Additional Samples Given a Background Sample of Size n.

Previous n	Number of new mesurements (r)														
	1	2	3	4	5	6	7	8	9	10	11	12	13	14	15
4	3.56	4.67	5.43	6.03	6.53	6.97	7.36	7.71	8.03	8.33	8.61	8.88	9.13	9.36	9.59
5	3.04	3.83	4.34	4.73	5.04	5.31	5.55	5.76	5.95	6.13	6.29	6.45	6.59	6.72	6.85
6	2.78	3.42	3.82	4.11	4.35	4.56	4.73	4.89	5.03	5.15	5.27	5.38	5.48	5.58	5.67
7	2.62	3.17	3.51	3.76	3.96	4.13	4.27	4.40	4.51	4.61	4.71	4.80	4.88	4.95	5.02
8	2.51	3.01	3.32	3.54	3.71	3.86	3.98	4.09	4.19	4.27	4.35	4.43	4.50	4.56	4.62
9	2.43	2.90	3.18	3.38	3.54	3.67	3.78	3.87	3.96	4.04	4.11	4.18	4.24	4.29	4.34
10	2.37	2.82	3.08	3.26	3.41	3.53	3.63	3.72	3.80	3.87	3.93	3.99	4.05	4.10	4.15
11	2.33	2.75	3.00	3.17	3.31	3.42	3.52	3.60	3.67	3.74	3.80	3.85	3.91	3.95	4.00
12	2.29	2.70	2.93	3.10	3.23	3.34	3.43	3.51	3.58	3.64	3.70	3.75	3.79	3.84	3.88
13	2.26	2.66	2.88	3.04	3.17	3.27	3.36	3.43	3.50	3.56	3.61	3.66	3.71	3.75	3.79
14	2.24	2.62	2.84	3.00	3.12	3.22	3.30	3.37	3.43	3.49	3.54	3.59	3.63	3.67	3.71
15	2.21	2.59	2.81	2.96	3.07	3.17	3.25	3.32	3.38	3.43	3.48	3.53	3.57	3.61	3.64
16	2.20	2.57	2.78	2.92	3.04	3.13	3.21	3.27	3.33	3.39	3.43	3.48	3.52	3.56	3.59
17	2.18	2.54	2.75	2.89	3.01	3.09	3.17	3.24	3.29	3.35	3.39	3.44	3.47	3.51	3.54
18	2.17	2.53	2.73	2.87	2.98	3.07	3.14	3.20	3.26	3.31	3.36	3.40	3.44	3.47	3.50
19	2.16	2.51	2.71	2.85	2.95	3.04	3.11	3.18	3.23	3.28	3.32	3.36	3.40	3.44	3.47
20	2.14	2.49	2.69	2.83	2.93	3.02	3.09	3.15	3.20	3.25	3.30	3.33	3.37	3.40	3.44
21	2.13	2.48	2.67	2.81	2.91	3.00	3.07	3.13	3.18	3.23	3.27	3.31	3.34	3.38	3.41
22	2.13	2.47	2.66	2.79	2.89	2.98	3.05	3.11	3.16	3.21	3.25	3.29	3.32	3.35	3.38
23	2.12	2.46	2.65	2.78	2.88	2.96	3.03	3.09	3.14	3.19	3.23	3.26	3.30	3.33	3.36
24	2.11	2.45	2.63	2.76	2.87	2.95	3.01	3.07	3.12	3.17	3.21	3.25	3.28	3.31	3.34
25	2.10	2.44	2.62	2.75	2.85	2.93	3.00	3.06	3.11	3.15	3.19	3.23	3.26	3.29	3.32
26	2.10	2.43	2.61	2.74	2.84	2.92	2.99	3.04	3.09	3.14	3.18	3.21	3.25	3.28	3.31
27	2.09	2.42	2.61	2.73	2.83	2.91	2.97	3.03	3.08	3.12	3.16	3.20	3.23	3.26	3.29
28	2.09	2.42	2.60	2.72	2.82	2.90	2.96	3.02	3.07	3.11	3.15	3.19	3.22	3.25	3.28
29	2.08	2.41	2.59	2.71	2.81	2.89	2.95	3.01	3.05	3.10	3.14	3.17	3.21	3.24	3.26
30	2.08	2.40	2.58	2.71	2.80	2.88	2.94	3.00	3.04	3.09	3.13	3.16	3.19	3.22	3.25
31	2.07	2.40	2.58	2.70	2.79	2.87	2.93	2.99	3.04	3.08	3.12	3.15	3.18	3.21	3.24
32	2.07	2.39	2.57	2.69	2.79	2.86	2.92	2.98	3.03	3.07	3.11	3.14	3.17	3.20	3.23
33	2.07	2.39	2.56	2.69	2.78	2.85	2.92	2.97	3.02	3.06	3.10	3.13	3.16	3.19	3.22
34	2.06	2.38	2.56	2.68	2.77	2.85	2.91	2.96	3.01	3.05	3.09	3.12	3.15	3.18	3.21
35	2.06	2.38	2.55	2.67	2.77	2.84	2.90	2.96	3.00	3.04	3.08	3.12	3.15	3.17	3.20
36	2.06	2.37	2.55	2.67	2.76	2.83	2.90	2.95	3.00	3.04	3.07	3.11	3.14	3.17	3.19
37	2.06	2.37	2.54	2.66	2.76	2.83	2.89	2.94	2.99	3.03	3.07	3.10	3.13	3.16	3.19
38	2.05	2.37	2.54	2.66	2.75	2.82	2.88	2.94	2.98	3.02	3.06	3.09	3.12	3.15	3.18
39	2.05	2.36	2.54	2.66	2.75	2.82	2.88	2.93	2.98	3.02	3.05	3.09	3.12	3.15	3.17
40	2.05	2.36	2.53	2.65	2.74	2.81	2.87	2.93	2.97	3.01	3.05	3.08	3.11	3 14	3.17
41	2.05	2.36	2.53	2.65	2.74	2.81	2.87	2.92	2.97	3.01	3.04	3.08	3.11	3.13	3.16
42	2.04	2.35	2.53	2.64	2.73	2.80	2.87	2.92	2.96	3.00	3.04	3.07	3.10	3.13	3.15
43	2.04	2.35	2.52	2.64	2.73	2.80	2.86	2.91	2.96	3.00	3.03	3.07	3.10	3.12	3.15
44	2.04	2.35	2.52	2.64	2.73	2.80	2.86	2.91	2.95	2.99	3.03	3.06	3.09	3.12	3.14
45	2.04	2.35	2.52	2.63	2.72	2.79	2.85	2.90	2.95	2.99	3.02	3.06	3.09	3.11	3.14
46	2.04	2.34	2.51	2.63	2.72	2.79	2.85	2.90	2.94	2.98	3.02	3.05	3.08	3.11	3.13
47	2.03	2.34	2.51	2.63	2.72	2.79	2.85	2.90	2.94	2.98	3.02	3.05	3.08	3.10	3.13
48	2.03	2.34	2.51	2.62	2.71	2.78	2.84	2.89	2.94	2.98	3.01	3.04	3.07	3.10	3.12
49	2.03	2.34	2.51	2.62	2.71	2.78	2.84	2.89	2.93	2.97	3.01	3.04	3.07	3.10	3.12
50	2.03	2.34	2.50	2.62	2.71	2.78	2.84	2.89	2.93	2.97	3.00	3.04	3.06	3.09	3.12
60	2.02	2.32	2.48	2.60	2.68	2.75	2.81	2.86	2.90	2.94	2.97	3.01	3.03	3.06	3.08
70	2.01	2.31	2.47	2.58	2.67	2.74	2.79	2.84	2.88	2.92	2.95	2.98	3.01	3.04	3.06
80	2.00	2.30	2.46	2.57	2.66	2.72	2.78	2.83	2.87	2.91	2.94	2.97	3.00	3.02	3.05
90	2.00	2.29	2.45	2.56	2.65	2.71	2.77	2.82	2.86	2.89	2.93	2.96	2.98	3.01	3.03
100	1.99	2.29	2.45	2.56	2.64	2.71	2.76	2.81	2.85	2.89	2.92	2.95	2.98	3.00	3.02

Table 11-11 Factors For Obtaining Two-Sided 95% Prediction Limits For r Additional Samples Given a Background Sample of Size n.

Prevlous n	| Number ot new measurements (r)														
	16	17	18	19	20	21	22	23	24	25	26	27	28	29	30
4	9.80	10.01	10.21	10.40	10.58	10.76	10.93	11.10	11.26	11.42	11.57	11.72	11.87	12.01	12.15
5	6.97	7.09	7.20	7.30	7.40	7.50	7.59	7.68	7.77	7.86	7.94	8.02	8.10	8.17	8.25
6	5.75	5.83	5.91	5.98	6.05	6.1 2	6.19	6.25	6.31	6.37	6.42	6.48	6.53	6.58	6.63
7	5.09	5.15	5.21	5.27	5.32	5.38	5.43	5.48	5.52	5.57	5.61	5.65	5.69	5.73	5.77
8	4.67	4.73	4.78	4.83	4.87	4.92	4.96	5.00	5.04	5.07	5.11	5.15	5.18	5.21	5.24
9	4.39	4.44	4.49	4.53	4.57	4.61	4.64	4.68	4.71	4.74	4.78	4.81	4.83	4.86	4.89
10	4.19	4.23	4.27	4.31	4.35	4.38	4.42	4.45	4.48	4.51	4.53	4.56	4.59	4.61	4.64
11	4.04	4.08	4.11	4.15	4.18	4.21	4.24	4.27	4.30	4.33	4.35	4.38	4.40	4.43	4.45
12	3.92	3.96	3.99	4.02	4.05	4.08	4.11	4.14	4.16	4.19	4.21	4.24	4.26	4.28	4.30
13	3.82	3.86	3.89	3.92	3.95	3.98	4.00	4.03	4.05	4.08	4.10	4.12	4.14	4.16	4.18
14	3.74	3.78	3.81	3.84	3.86	3.89	3.92	3.94	3.96	3.99	4.01	4.03	4.05	4.07	4.09
15	3.68	3.71	3.74	3.77	3.79	3.82	3.84	3.87	3.89	3.91	3.93	3.95	3.97	3.99	4.01
16	3.62	3.65	3.68	3.71	3.74	3.76	3.78	3.81	3.83	3.85	3.87	3.89	3.90	3.92	3.94
17	3.58	3.60	3.63	3.66	3.68	3.71	3.73	3.75	3.77	3.79	3.81	3.83	3.85	3.86	3.88
18	3.53	3.56	3.59	3.61	3.64	3.66	3.68	3.71	3.73	3.75	3.76	3.78	3.80	3.82	3.83
19	3.50	3.53	3.55	3.58	3.60	3.62	3.65	3.67	3.68	3.70	3.72	3.74	3.76	3.77	3.79
20	3.47	3.49	3.52	3.54	3.57	3.59	3.61	3.63	3.65	3.67	3.68	3.70	3.72	3.73	3.75
21	3.44	3.46	3.49	3.51	3.54	3.56	3.58	3.60	3.62	3.63	3.65	3.67	3.69	3.70	3.72
22	3.41	3.44	3.46	3.49	3.51	3.53	3.55	3.57	3.59	3.61	3.62	3.64	3.65	3.67	3.68
23	3.39	3.41	3.44	3.46	3.49	3.51	3.53	3.54	3.56	3.58	3.60	3.61	3.63	3.64	3.66
24	3.37	3.39	3.42	3.44	3.46	3.48	3.50	3.52	3.54	3.56	3.57	3.59	3.60	3.62	3.63
25	3.35	3.38	3.40	3.42	3.44	3.46	3.48	3.50	3.52	3.53	3.55	3.57	3.58	3.60	3.61
26	3.33	3.36	3.38	3.40	3.42	3.44	3.46	3.48	3.50	3.52	3.53	3.55	3.56	3.58	3.59
27	3.32	3.34	3.36	3.39	3.41	3.43	3.45	3.46	3.48	3.50	3.51	3.53	3.54	3.56	3.57
28	3.30	3.33	3.35	3.37	3.39	3.41	3.43	3.4 5	3.46	3.48	3.50	3.51	3.53	3.54	3.55
29	3.29	3.31	3.34	3.36	3.38	3.40	3.42	3.43	3.45	3.47	3.48	3.50	3.51	3.52	3.54
30	3.28	3.30	3.32	3.34	3.36	3.38	3.40	3.4 2	3.44	3.45	3.47	3.48	3.50	3.51	3.52
31	3.27	3.29	3.31	3.33	3.35	3.37	3.39	3.41	3.42	3.44	3.45	3.47	3.48	3.50	3.51
32	3.25	3.28	3.30	3.32	3.34	3.36	3.38	3.39	3.41	3.43	3.44	3.46	3.47	3.48	3.50
33	3.24	3.27	3.29	3.31	3.33	3.35	3.37	3.38	3.40	3.42	3.43	3.44	3.46	3.47	3.48
34	3.23	3.26	3.28	3.30	3.32	3.34	3.36	3.37	3.39	3.41	3.42	3.43	3.4 5	3.46	3.47
35	3.23	3.25	3.27	3.29	3.31	3.33	3.35	3.36	3.38	3.39	3.41	3.42	3.44	3.45	3.46
36	3.22	3.24	3.26	3.28	3.30	3.32	3.34	3.35	3.37	3.39	3.40	3.41	3.43	3.44	3.45
37	3.21	3.23	3.25	3.27	3.29	3.31	3.33	3.35	3.36	3.38	3.39	3.40	3.42	3.43	3.44
38	3.20	3.23	3.25	3.27	3.29	3.30	3.32	3.34	3.35	3.37	3.38	3.40	3.41	3.42	3.44
39	3.20	3.22	3.24	3.26	3.28	3.30	3.31	3.33	3.35	3.36	3.37	3.39	3.40	3.41	3.43
40	3.19	3.21	3.23	3.25	3.27	3.29	3.31	3.32	3.34	3.35	3.37	3.38	3.39	3.41	3.42
41	3.18	3.21	3.23	3.25	3.2 7	3.28	3.30	3.32	3.33	3.35	3.36	3.37	3.39	3.40	3.41
42	3.18	3.20	3.22	3.24	3.26	3.28	3.29	3.31	3.33	3.34	3.35	3.37	3.38	3.39	3.41
43	3.17	3.19	3.21	3.23	3.25	3.27	3.29	3.30	3.32	3.33	3.35	3.36	3.37	3.39	3.40
44	3.17	3.19	3.21	3.23	3.25	3.27	3.28	3.30	3.31	3.33	3.34	3.36	3.37	3.38	3.39
45	3.16	3.18	3.20	3.22	3.24	3.26	3.28	3.29	3.31	3.32	3.34	3.35	3.36	3.37	3.39
46	3.16	3.18	3.20	3.22	3.24	3.25	3.27	3.29	3.30	3.32	3.33	3.34	3.36	3.37	3.38
47	3.15	3.17	3.19	3.21	3.23	3.25	3.27	3.28	3.30	3.31	3.33	3.34	3.35	3.36	3.38
48	3.15	3.17	3.19	3.21	3.23	3.24	3.26	3.28	3.29	3.31	3.32	3.33	3.35	3.36	3.37
49	3.14	3.17	3.19	3.20	3.22	3.24	3.26	3.27	3.29	3.30	3.32	3.33	3.34	3.35	3.37
50	3.14	3.16	3.18	3.20	3.22	3.24	3.25	3.27	3.28	3.30	3.31	3.32	3.34	3.35	3.36
60	3.11	3.13	3.15	3.17	3.18	3.20	3.22	3.23	3.25	3.26	3.27	3.29	3.30	3.31	3.32
70	3.08	3.11	3.12	3.14	3.1 6	3.18	3.19	3.21	3.22	3.24	3.2 5	3.26	3.27	3.28	3.30
80	3.07	3.09	3.11	3.13	3.1 4	3.16	3.17	3.1 9	3.20	3.22	3.2 3	3.24	3.25	3.26	3.28
90	3.05	3.08	3.09	3.11	3.1 3	3.15	3.16	3.1 7	3.19	3.20	3.21	3.23	3.24	3.25	3.26
100	3.04	3.06	3.08	3.10	3.12	3.13	3.15	3.16	3.18	3.19	3.20	3.22	3.23	3.24	3.25

Table 11-12   Factors For Obtaining Two-Sided 95% Prediction Limits For r Additional Sample Given a Background Sample of Size n.

Preview						Number of new measurements (r)								
n	30	35	40	45	50	55	60	65	70	75	80	85	90	95	100
4	12.15	12.80	13.40	13.94	14.45	14.92	15.37	15.79	16.19	16.57	16.93	17.28	17.62	17.94	18.25
5	8.25	8.59	8.90	9.18	9.43	9.67	9.89	10.10	10.29	10.48	10.66	10.83	10.99	11.14	11.29
6	6.63	6.86	7.06	7.25	7.42	7.57	7.72	7.85	7.98	8.10	8.21	8.32	8.42	8.52	8.61
7	5.77	5.95	6.10	6.24	6.37	6.49	6.59	6.69	6.79	6.88	6.96	7.04	7.12	7.19	7.26
8	5.24	5.39	5.52	5.63	5.74	5.83	5.92	6.00	6.08	6.15	6.21	6.28	6.34	6.40	6.45
9	4.89	5.02	5.13	5.23	5.31	5.40	5.47	5.54	5.60	5.66	5.72	5.78	5.83	5.87	5.92
10	4.64	4.75	4.85	4.93	5.01	5.09	5.15	5.21	5.27	5.32	5.37	5.42	5.47	5.51	5.55
11	4.45	4.55	4.64	4.72	4.79	4.86	4.92	4.97	5.02	5.07	5.11	5.16	5.20	5.24	5.27
12	4.30	4.40	4.48	4.55	4.62	4.68	4.73	4.78	4.83	4.88	4.92	4.96	4.99	5.03	5.06
13	4.18	4.27	4.35	4.42	4.48	4.54	4.59	4.64	4.68	4.72	4.76	4.80	4.83	4.86	4.89
14	4.09	4.1 7	4.25	4.31	4.37	4.42	4.47	4.52	4.56	4.60	4.63	4.67	4.70	4.73	4.76
15	4.01	4.09	4.16	4.22	4.28	4.33	4.37	4.41	4.45	4.49	4.53	4.56	4.59	4.62	4.65
16	3.94	4.02	4.08	4.14	4.20	4.25	4.29	4.33	4.37	4.40	4.44	4.47	4.50	4.53	4.55
17	3.88	3.96	4.02	4.08	4.1 3	4.18	4.22	4.26	4.30	4.33	4.36	4.39	4.42	4.45	4.47
18	3.83	3.90	3.97	4.02	4.0 7	4.12	4.16	4.20	4.23	4.27	4.30	4.33	4.35	4.38	4.40
19	3.79	3.86	3.92	3.97	4.02	4.07	4.11	4.14	4.18	4.21	4.24	4.27	4.29	4.32	4.34
20	3.75	3.82	3.88	3.93	3.98	4.02	4.06	4.10	4.13	4.16	4.19	4.22	4.24	4.27	4.29
21	3.72	3.78	3.84	3.89	3.94	3.98	4.02	4.05	4.09	4.12	4.15	4.17	4.20	4.22	4.24
22	3.68	3.75	3.81	3.86	3.90	3.95	3.98	4.02	4.05	4.08	4.11	4.13	4.16	4.18	4.20
23	3.66	3.72	3.78	3.83	3.87	3.91	3.95	3.98	4.01	4.04	4.07	4.10	4.12	4.14	4.16
24	3.63	3.70	3.75	3.80	3.84	3.88	3.92	3.95	3.98	4.01	4.04	4.06	4.09	4.11	4.13
25	3.61	3.67	3.73	3.78	3.82	3.86	3.89	3.93	3.96	3.98	4.01	4.03	4.06	4.08	4.10
26	3.59	3.65	3.71	3.75	3.80	3.83	3.87	3.90	3.93	3.96	3.98	4.01	4.03	4.05	4.07
27	3.57	3.63	3.69	3.73	3.77	3.81	3.85	3.88	3.91	3.93	3.96	3.98	4.01	4.03	4.05
28	3.55	3.61	3.67	3.71	3.75	3.79	3.83	3.86	3.89	3.91	3.94	3.96	3.98	4.00	4.02
29	3.54	3.60	3.65	3.70	3.74	3.77	3.81	3.84	3.87	3.89	3.92	3.94	3.96	3.98	4.00
30	3.52	3.58	3.63	3.68	3.72	3.76	3.79	3.82	3.85	3.87	3.90	3.92	3.94	3.96	3.98
31	3.51	3.57	3.62	3.66	3.70	3.74	3.77	3.80	3.83	3.86	3.88	3.90	3.92	3.94	3.96
32	3.50	3.55	3.61	3.65	3.69	3.73	3.76	3.79	3.81	3.84	3.86	3.89	3.91	3.93	3.95
33	3.48	3.54	3.59	3.64	3.68	3.71	3.74	3.77	3.80	3.83	3.85	3.87	3.89	3.91	3.93
34	3.47	3.53	3.58	3.62	3.66	3.70	3.73	3.76	3.79	3.81	3.84	3.86	3.88	3.90	3.92
35	3.46	3.52	3.57	3.61	3.65	3.69	3.72	3.75	3.77	3.80	3.82	3.84	3.86	3.88	3.90
36	3.45	3.51	3.56	3.60	3.64	3.67	3.71	3.73	3.76	3.79	3.81	3.83	3.85	3.87	3.89
37	3.44	3.50	3.55	3.59	3.63	3.66	3.70	3.72	3.75	3.78	3.80	3.82	3.84	3.86	3.88
38	3.44	3.49	3.54	3.58	3.62	3.65	3.69	3.71	3.74	3.76	3.79	3.81	3.83	3.85	3.86
39	3.43	3.48	3.53	3.57	3.61	3.64	3.68	3.70	3.73	3.75	3.78	3.80	3.82	3.84	3.85
40	3.42	3.47	3.52	3.56	3.60	3.64	3.67	3.69	3.72	3.74	3.77	3.79	3.81	3.83	3.84
41	3.41	3.47	3.51	3.56	3.59	3.63	3.66	3.69	3.71	3.74	3.76	3.78	3.80	3.82	3.83
42	3.41	3.46	3.51	3.55	3.59	3.62	3.65	3.68	3.70	3.73	3.75	3.77	3.79	3.81	3.82
43	3.40	3.45	3.50	3.54	3.58	3.61	3.64	3.67	3.69	3.72	3.74	3.76	3.78	3.80	3.82
44	3.39	3.45	3.49	3.53	3.57	3.60	3.63	3.66	3.69	3.71	3.73	3.75	3.77	3.79	3.81
45	3.39	3.44	3.49	3.53	3.56	3.60	3.63	3.65	3.68	3.70	3.72	3.74	3.76	3.78	3.80
46	3.38	3.43	3.48	3.52	3.56	3.59	3.62	3.65	3.67	3.70	3.72	3.74	3.76	3.77	3.79
47	3.38	3.43	3.48	3.52	3.55	3.58	3.61	3.64	3.67	3.69	3.71	3.73	3.75	3.77	3.78
48	3.37	3.42	3.47	3.51	3.55	3.58	3.61	3.63	3.66	3.68	3.70	3.72	3.74	3.76	3.78
49	3.37	3.42	3.46	3.50	3.54	3.57	3.60	3.63	3.65	3.68	3.70	3.72	3.74	3.75	3.77
50	3.36	3.41	3.46	3.50	3.53	3.57	3.60	3.62	3.65	3.67	3.69	3.71	3.73	3.75	3.76
60	3.32	3.37	3.42	3.46	3.49	3.52	3.55	3.58	3.60	3.62	3.64	3.66	3.68	3.70	3.71
70	3.30	3.35	3.39	3.43	3.46	3.49	3.52	3.54	3.57	3.59	3.61	3.63	3.65	3.66	3.68
80	3.28	3.33	3.37	3.41	3.44	3.47	3.50	3.52	3.54	3.57	3.59	3.60	3.62	3.64	3.65
90	3.26	3.31	3.35	3.39	3.42	3.45	3.48	3.50	3.53	3.55	3.57	3.58	3.60	3.62	3.63
100	3.25	3.30	3.34	3.38	3.41	3.44	3.46	3.49	3.51	3.53	3.55	3.57	3.59	3.60	3.62

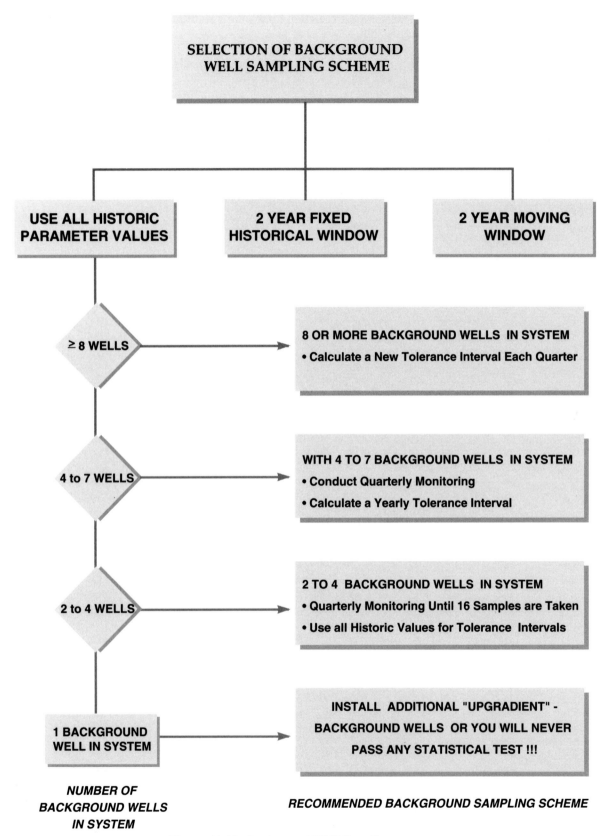

Figure 11-23   Background Well Sampling

Part 264), owner/operators were required to demonstrate that alternate statistical procedures balanced false positive and false negative results. Fortunately, this requirement is no longer a part of the new RCRA (1989) statistical regulation. The reason difficulty arose is that the false negative rate, or one–minus statistical power required to define number of samples, is dependent on three things: the false positive rate (selected a–priori by the regulation; that is, 5% for the facility as a whole), the number of background measurements, and the effect size. The effect size describes the smallest difference (typically described in standard deviation units) that is environmentally meaningful.

Since the USEPA has not provided such a minimum effect size, it is completely impossible to derive a sample size that will properly balance false positive and false negative rates of tolerance or prediction limits. As such, it is suggested that the number of samples be selected on the basis of the size of the multiplier. For example, a 95% confidence 95% coverage one–sided tolerance interval multiplier goes from 7.656 to 5.144 for a change in background sample size of 3 to 4; but only 2.566 to 2.523 for a change in background sample size of n = 15 to n = 16. Even doubling the background sample size to n = 30 only decreases the multiplier by approximately 1/3 of a standard deviation unit (i.e., k = 2.220). As such, we recommend a background sample size in the range of 16 to 32. If more observations are available, they should, of course, also be included because they will provide even more precise estimates of μ and $\sigma2$.

The previous line of reasoning also suggests how the background sample should be selected. The three choices are: (1) fixed sample of n time–series historical measurements (e.g., 2 years fixed window), (2) the n most recent measurements (e.g., a 2 year moving window), (3) all available historical measurements. When using tolerance or prediction limits or intervals, we recommend the third option, as shown on Figure 11-23.

Finally, how shall the n background measurements be selected? Should we obtain n measurements from a single background well, a single measurement from n different background wells, or something in between? In the case of upgradient versus downgradient comparisons, there are two possible strategies. For a facility with eight or more upgradient or background wells, the eight measurements could be used to construct a new tolerance limit specific to each quarterly monitoring event (Figure 11-23). With eight background measurements, the one–sided 95% confidence and 95% coverage tolerance limit multiplier would be 3.188. Although the multiplier is somewhat larger than the value of 2.523 obtained for n = 16 background samples, this strategy eliminates the temporal component of variability, and will, therefore, yield a

smaller standard deviation than if historical measurements are pooled. The result is an effective detection monitoring program that is supported by the new USEPA statistical rule.

For facilities with less than 8 background water quality wells, we suggest that at least 4 upgradient or background wells be monitored quarterly so that after one year of monitoring, 16 background measurements will be available. Furthermore, four upgradient wells, if widely spaced, will provide a reasonable characterization of the spatial component of variability at the facility. In general two years of quarterly data collection are recommended for the background to provide a reasonable idea of the seasonal effects on ground-water quality including ground-water level fluctuations and ambient anthropogenic changes. As the ground-water surface fluctuates, different soil and rock types may be exposed compared to other times and possibly influence the hydrogeochemistry.

The eight samples also will provide enough degrees of freedom to yield a prediction limit formula of x + 3.16s, where x is the mean of 8 background samples and s is the standard deviation. This closely resembles the QA/QC practice of commercial labs that use 3 standard deviations to determine if data are within acceptable limits. The 3.16 factor is derived from statistical tables that compare the number of background samples (8) to the number of new observations (K=l) at the one-sided 99 percent prediction limit. Values for pH will be determined using the two-sided 99 percent prediction limit tables (Gibbons, 1987).

As a last word on mimium RCRA (Subtitle C) standards of one upgradient and three downgradient wells, if only a single upgradient well were installed, as required in the previous RCRA regulations, differences between upgradient and downgradient water quality would be completely confounded with spatial variability; that is, is the difference between upgradient and downgradient measurements due to the influence of the facility or simply the difference that one would find by drilling any two holes in the ground? There is, of course, no answer to this question if there is only a single background well.

Non–detects:—It is very common in ground–water detection monitoring to obtain samples that cannot be properly quantified because of low–level indicator concentrations that are less than the limit of detection of the analytical instrument. This condition can make the direct application of the previously described statistical prediction and tolerance limits and intervals problematic because the usual sample statistics x and s are no longer valid estimates of μ and s. Statistically, these distributions are termed "censored." In this section, we consider three different levels of "censoring": (1) up to 90%

non–detects, (2) 91% to 99% non–detects, and (3) nothing detected as shown in Figure 11-24.

Case 1 – Up to 90% Non–detects

When at least 10% of the ground–water samples has a measurable (detectable) value of a particular indicator parameter, the mean and variance of the distribution can be approximated using a method due to Aitchison (1955).

The adjusted mean value is given by:

$$\bar{x} = \left(1 - \frac{n_0}{n}\right)\bar{x}' \qquad \text{Equation 11-1}$$

Where x' is the average of the n, detected values, no is the number of non–detects and n = n1 + no is the total number of samples. The adjusted standard deviation is:

$$s = \sqrt{\left(1 - \frac{n_0}{n}\right)s'^2 + \frac{n_0}{n}\left(1 - \frac{n_0 - 1}{n - 1}\right)\bar{x}'^2} \qquad \text{Equation 11-2}$$

Where s' is the standard deviation of the n1 detected measurements. The normal tolerance and prediction limits can then be computed as previously described, using the total sample size n to obtain the appropriate tabled multiplier.

Case 2: Compounds Detected in 1% to 10% of All Background Samples (VOCs)

When the detection frequency is less than 10%, the previous method of obtaining adjusted mean and variance estimates no longer applies. With limited data, it is difficult to know just what to do. What further complicates this problem is that one of the most important classes of detection monitoring compounds are the Volatile Organic Priority Pollutants (VOCs) which typically have detection frequencies in this range.

To date, the only applicable statistical approach to setting site specific limits for these compounds are described by Gibbons (1987b). This procedure is based on tolerance and prediction limits for the Poisson distribution; a distribution that has been widely used for the analysis of

rare events such as suicide, mutation rates, or atomic particle emission. These limits can be applied either to detection frequencies (i.e., number of detected compounds per scan) or to the actual concentrations when recorded in parts per billion (ppb). In the latter case, it is assumed that a measurement of 20 ppb of benzene represents a count of 20 molecules of benzene for each billion molecules of water examined. To the extent that this is an accurate description of the true physical measurement process, Poisson prediction and tolerance limits provide a reasonable approximation that appears to be sufficiently accurate for most practical purposes. This sentiment is echoed in the 1988 USEPA statistical RCRA regulation:

Tolerance intervals and prediction intervals have not been widely used by the USEPA to evaluate ground–water monitoring data. However, these statistical methods to evaluate ground–water monitoring data, especially in evaluating certain classes of chemical compounds (e.g., volatile organic compounds) may be the best way to evaluate data that is below the limit of analytical detection.

n the case of VOCs, the 95% Poisson prediction limit is computed as follows (Figure 11-24):

1.   For each USEPA Method 624 volatile organic priority pollutant scan, sum the detected concentrations of the 27 compounds listed in Table 11-13, substituting the published method detection limit (MDL) (see Table 11–13 also) for those compounds that were not detected. For example, if none of the compounds were detected, the sum for that scan is 154 ppb.

2. Compute the 95% Poisson prediction limit as:

$$\frac{y}{n} + \frac{t^2}{2n} + \frac{t}{n}\sqrt{y(1+n) + \frac{t^2}{4}} \qquad \text{Equation 11-3}$$

Where y is the total ppb for all n background scans (i.e., the sum of n individual scan totals), n is the number of background scans, t is the (1 – 0.05/r) 100% point of student's t distribution on n – 1 degrees of freedom (see Tables 11-14 through 11-16), and r is the number of monitoring wells.

Scan totals for each monitoring well (computed as in Step 1) are then compared to the limit value computed in Step 2.

STATISTICAL METHODS FOR PARAMETERS WITH 90 TO 100 PERCENT NON-DETECTS

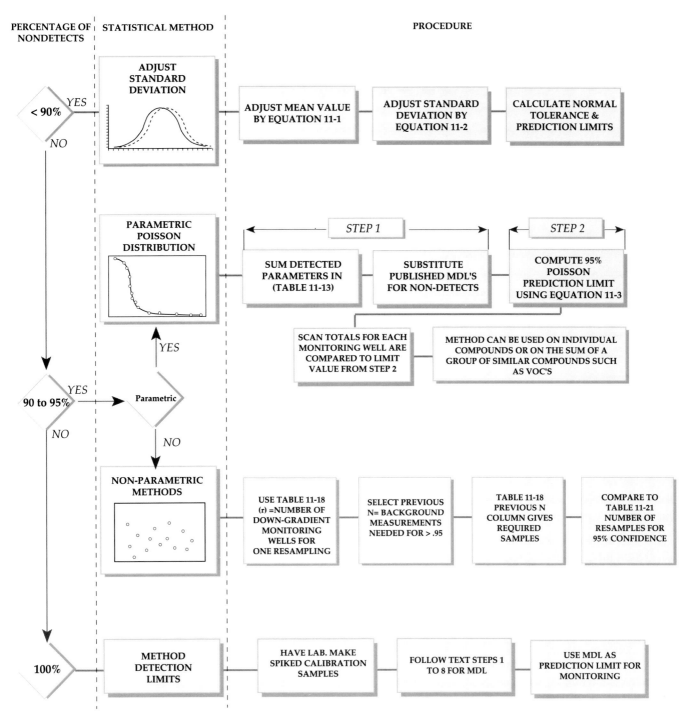

Figure 11-24   Methods for 90-100 Percent Non-detects

Table 11-13 Method 624 Volatile Organic Compounds and Published Method Detection Limits.

Compound	Reported MDL
Benzene	4.4
Bromodichloromethane	2.2
Bromoform	4.7
Bromomethane	10.0
Carbon Tetrachloride	2.8
Chlorobenzene	6.0
Chloroethane	10.0
Chloroform	1.6
Chloromethane	10.0
Dibromochloromethane	3.1
1,1 -Dichloroethane	4.7
1,2-Dichloroethane	2.8
1,1-Dichloroethene	2.8
trans-1 ,2-Dichloroethene	10.0
1,2-Dichloropropane	6.0
cis-1,3-Dichloropropene	5.0
trans-1 ,3-Dichloropropene	10.0
Ethyl Benzene	7.2
Methylene Chloride	2.8
1,1,2,2-Tetrachloroethane	6.9
Tetrachloroethene	4.1
Toluene	6.0
1,1 ,1-Trichloroethane	3.8
1,1,2-Trichloroethane	5.0
Trichloroethene	1.9
Trichlorofluoromethane	10.0
Vinyl Chloride	10.0

all values reported In µg/L.

For other compounds that have detection frequencies in the range of 1% to 10%, the same strategy may be applied, either individually or on the sum of a group of similar compounds, as in the case of the VOCs.

The Non–parametric Approach

The previous discussion has been based on the assumption that the distribution of the parameter(s) of interest is known and has a parametric form (i.e., normal, censored normal, or Poisson). In some cases, however, this assumption is unreasonable and a "distribution free" statistical method may be required. In the context of ground–water monitoring, Gibbons (1988a) has adapted the non–parametric prediction limit originally described by Chou and Owen (1986). In contrast to the parametric approach in which we estimate a limit value from the mean and standard deviation of a sample of n previous measurements, the non–parametric approach identifies the

required number of samples (n) such that the maximum value of those samples is the 95% prediction limit. Gibbons (1988a) further generalizes the procedure to include the effects of resampling (i.e., taking a verification sample following a statistically significant ground–water monitoring result), and defines a full set of detection monitoring statistics using a non-parametric approach (Gibbons 1990).

Over the last several years there has been considerable controversy over the appropriate statistical methodology for evaluating the impact of waste disposal facilities on ground-water quality. The regulatory methodology uses comparisons between a set of historical upgradient background measurements and an individual new monitoring measurement, separately in each of a series of downgradient wells (see Figure 9-1 for a conceptual model of the regulatory standard for the up-gradient/down-gradient system). A single comparison between an up-gradient and down-gradient well that results in statistical significance at the 5% level is taken as evidence that the facility is impacting ground water, regardless of the number of comparisons made. Gibbons (1990) reported several additional complications of the simple statistical comparisons of water quality data sets:

- several commonly used variables (i.e., indicator parameters) have markedly non-normal distributions, that are often resistant to simple data transformation.

- these distributions are often censored, in the sense that a proportion of the observations is measured below an established method detection limit.

The commonly used indicators of contamination for both RCRA Subtitle D (solid waste), and Subtitle C (Hazardous waste) are volatile organic priority pollutant compounds. For these indicator parameters the proportion of nondetected values is often in excess of 95%, ruling out the application of traditional statistical methods. Statistical procedures that use prediction and tolerance intervals are described below for both detection and compliance monitoring of hazardous waste disposal facilities by the new USEPA statistical regulation (USEPA, 1988). These statistical procedures are recommended for use where these rare event data sets are important to facility compliance. An important part in use of these methods is the ability to resample those wells that exceed the prediction limit on the initial sample. A well that produces a measurement below the prediction limit on the initial sample can either exceed or not exceed the limit on a potential resample and still fulfill the requirement that at least one of m measurements did not exceed the limit. The probability estimates that are computed here do not depend on the order of the m results

Table 11-14 One-Sided 95% Poisson Prediction Limits for r Additional Samples Given Background Sample of Size n.

Previous n	\multicolumn{15}{c}{Number of new measurements (r)}														
	1	2	3	4	5	6	7	8	9	10	11	12	13	14	15
4	2.35	3.18	3.74	4.18	4.54	4.86	5.14	5.39	5.62	5.84	6.04	6.23	6.41	6.58	6.74
5	2.13	2.78	3.19	3.50	3.75	3.96	4.15	4.31	4.47	4.60	4.73	4.85	4.96	5.07	5.17
6	2.01	2.57	2.91	3.16	3.36	3.53	3.68	3.81	3.93	4.03	4.13	4.22	4.30	4.38	4.45
7	1.94	2.45	2.75	2.97	3.14	3.29	3.41	3.52	3.62	3.71	3.79	3.86	3.93	4.00	4.06
8	1.89	2.36	2.64	2.84	3.00	3.13	3.24	3.33	3.42	3.50	3.57	3.64	3.70	3.75	3.81
9	1.86	2.31	2.57	2.75	2.90	3.02	3.12	3.21	3.28	3.35	3.42	3.48	3.53	3.58	3.63
10	1.83	2.26	2.51	2.68	2.82	2.93	3.03	3.11	3.18	3.25	3.31	3.36	3.41	3.46	3.50
11	1.81	2.23	2.47	2.63	2.76	2.87	2.96	3.04	3.11	3.17	3.22	3.28	3.32	3.37	3.41
12	1.80	2.20	2.43	2.59	2.72	2.82	2.91	2.98	3.05	3.11	3.16	3.21	3.25	3.29	3.33
13	1.78	2.18	2.40	2.56	2.68	2.78	2.86	2.93	3.00	3.05	3.11	3.15	3.20	3.23	3.27
14	1.77	2.16	2.38	2.53	2.65	2.75	2.83	2.90	2.96	3.01	3.06	3.11	3.15	3.19	3.22
15	1.76	2.14	2.36	2.51	2.62	2.72	2.80	2.86	2.92	2.98	3.02	3.07	3.11	3.15	3.18
16	1.75	2.13	2.34	2.49	2.60	2.69	2.77	2.84	2.89	2.95	2.99	3.04	3.07	3.11	3.15
17	1.75	2.12	2.33	2.47	2.58	2.67	2.75	2.81	2.87	2.92	2.97	3.01	3.05	3.08	3.11
18	1.74	2.11	2.31	2.46	2.57	2.65	2.73	2.79	2.85	2.90	2.94	2.98	3.02	3.06	3.09
19	1.73	2.10	2.30	2.44	2.55	2.64	2.71	2.77	2.83	2.88	2.92	2.96	3.00	3.03	3.07
20	1.73	2.09	2.29	2.43	2.54	2.62	2.70	2.76	2.81	2.86	2.90	2.94	2.98	3.01	3.04
21	1.72	2.09	2.28	2.42	2.53	2.61	2.68	2.74	2.80	2.84	2.89	2.93	2 96	3.00	3.03
22	1.72	2.08	2.28	2.41	2.52	2.60	2.67	2.73	2.78	2.83	2.87	2.91	2.95	2.98	3.01
23	1.72	2.07	2.27	2.40	2.51	2.59	2.66	2.72	2.77	2.82	2.86	2.90	2.93	2.96	2.99
24	1.71	2.07	2.26	2.40	2.50	2.58	2.65	2.71	2.76	2.81	2.85	2.89	2.92	2.95	2.98
25	1.71	2.06	2.26	2.39	2.49	2.57	2.64	2.70	2.75	2.80	2.84	2.88	2.91	2.94	2.97
26	1.71	2.06	2.25	2.38	2.48	2.57	2.63	2.69	2.74	2.79	2.83	2.86	2.90	2.93	2.96
27	1.71	2.06	2.25	2.38	2.48	2.56	2.63	2.68	2.73	2.78	2.82	2.85	2.89	2.92	2.95
28	1.70	2.05	2.24	2.37	2.47	2.55	2.62	2.68	2.73	2.77	2.81	2.85	2.88	2.91	2.94
29	1.70	2.05	2.24	2.37	2.47	2.55	2.61	2.67	2.72	2.76	2.80	2.84	2.87	2.90	2.93
30	1.70	2.05	2.23	2.36	2.46	2.54	2.61	2.66	2.71	2.76	2.80	2.83	2.86	2.89	2.92
31	1.70	2.04	2.23	2.36	2.46	2.54	2.60	2.66	2.71	2.75	2.79	2.82	2.86	2.89	2.91
32	1.70	2.04	2.23	2.36	2.45	2.53	2.60	2.65	2.70	2.74	2.78	2.82	2.85	2.88	2.91
33	1.69	2.04	2.22	.235	2.45	2.53	2.59	2.65	2.70	2.74	2.78	2.81	2.84	2.87	2.90
34	1.69	2.03	2.22	2.35	2.44	2.52	2.59	2.64	2.69	2.73	2.77	2.81	2.84	2.87	2.90
35	1.69	2.03	2.22	2.34	2.44	2.52	2.58	2.64	2.69	2.73	2.77	2.80	2.83	2.86	2.89
36	1.69	2.03	2.22	2.34	2.44	2.51	2.58	2.63	2.68	2.72	2.76	2.80	2.83	2.86	2.88
37	1.69	2.03	2.21	2.34	2.43	2.51	2.57	2.63	2.68	2.72	2.76	2.79	2.82	2.85	2.88
38	1.69	2.03	2.21	2.34	2.43	2.51	2.57	2.63	2.67	2.72	2.75	2.79	2.82	2.85	2.87
39	1.69	2.02	2.21	2.33	2.43	2.50	2.57	2.62	2.67	2.71	2.75	2.78	2.81	2.84	2.87
40	1.68	2.02	2.21	2.33	2.43	2.50	2.56	2.26	2.67	2.71	2.75	2.78	2.81	2.84	2.87
41	1.68	2.02	2.20	2.33	2.42	2.50	2.56	2.62	2.66	2.70	2.74	2.78	2.81	2.84	2.86
42	1.68	2.02	2.20	2.33	2.42	2.50	2.56	2.61	2.66	2.70	2.74	2.77	2.80	2.83	2.86
43	1.68	2.02	2.20	2.32	2.42	2.49	2.56	2.61	2.66	2.70	2.74	2.77	2.80	2.83	2.85
44	1.68	2.02	2.20	2.32	2.42	2.49	2.55	2.61	2.65	2.69	2.73	2.77	2.80	2.82	2.85
45	1.68	2.02	2.20	2.32	2.41	2.49	2.55	2.60	2.65	2.69	2.73	2.76	2.79	2.82	2.85
46	1.68	2.01	2.20	2.32	2.41	2.49	2.55	2.60	2.65	2.69	2.73	2.76	2.79	2.82	2.84
47	1.68	2.01	2.19	2.32	2.41	2.48	2.55	2.60	2.65	2.69	2.72	2.76	2.79	2.82	2.84
48	1.68	2.01	2.19	2.32	2.41	2.48	2.54	2.60	2.64	2.68	2.72	2.75	2.78	2.81	2.84
49	1.68	2.01	2.19	2.31	2.41	2.48	2.54	2.60	2.64	2.68	2.72	2.75	2.78	2.81	2.84
50	1.68	2.01	2.19	2.31	2.40	2.48	2.54	2.59	2.64	2.68	2.72	2.75	2.78	2.81	2.83
60	1.67	2.00	2.18	2.30	2.39	2.46	2.52	2.58	2.62	2.66	2.70	2.73	2.76	2.79	2.81
70	1.67	1.99	2.17	2.29	2.38	2.45	2.51	2.56	2.61	2.65	2.68	2.72	2.75	2.77	2.80
80	1.66	1.99	2.17	2.28	2.37	2.45	2.51	2.56	2.60	2.64	2.67	2.71	2.74	2.76	2.79
90	1.66	1.99	2.16	2.28	2.37	2.44	2.50	2.55	2.59	2.63	2.67	2.70	2.73	2.75	2.78
100	1.66	1.98	2.16	2.28	2.36	2.44	2.49	2.54	2.59	2.63	2.66	2.69	2.72	2.75	2.77

Tables 11-14 to 11-16 : Source Gibbons 1990 and Factor Equation   Factor $= t(n-1, 1-\alpha/r)$

Table 11-15 Factors For Obtaining Two-Sided 95% Prediction Limits For r Additional Samples Given a Background Sample of Size n.

Previous n	Number ot new measurements (r)														
	16	17	18	19	20	21	22	23	24	25	26	27	28	29	30
4	9.80	10.01	10.21	10.40	10.58	10.76	10.93	11.10	11.26	11.42	11.57	11.72	11.87	12.01	12.15
5	6.97	7.09	7.20	7.30	7.40	7.50	7.59	7.68	7.77	7.86	7.94	8.02	8.10	8.17	8.25
6	5.75	5.83	5.91	5.98	6.05	6.1 2	6.19	6.25	6.31	6.37	6.42	6.48	6.53	6.58	6.63
7	5.09	5.15	5.21	5.27	5.32	5.38	5.43	5.48	5.52	5.57	5.61	5.65	5.69	5.73	5.77
8	4.67	4.73	4.78	4.83	4.87	4.92	4.96	5.00	5.04	5.07	5.11	5.15	5.18	5.21	5.24
9	4.39	4.44	4.49	4.53	4.57	4.61	4.64	4.68	4.71	4.74	4.78	4.81	4.83	4.86	4.89
10	4.19	4.23	4.27	4.31	4.35	4.38	4.42	4.45	4.48	4.51	4.53	4.56	4.59	4.61	4.64
11	4.04	4.08	4.11	4.15	4.18	4.21	4.24	4.27	4.30	4.33	4.35	4.38	4.40	4.43	4.45
12	3.92	3.96	3.99	4.02	4.05	4.08	4.11	4.14	4.16	4.19	4.21	4.24	4.26	4.28	4.30
13	3.82	3.86	3.89	3.92	3.95	3.98	4.00	4.03	4.05	4.08	4.10	4.12	4.14	4.16	4.18
14	3.74	3.78	3.81	3.84	3.86	3.89	3.92	3.94	3.96	3.99	4.01	4.03	4.05	4.07	4.09
15	3.68	3.71	3.74	3.77	3.79	3.82	3.84	3.87	3.89	3.91	3.93	3.95	3.97	3.99	4.01
16	3.62	3.65	3.68	3.71	3.74	3.76	3.78	3.81	3.83	3.85	3.87	3.89	3.90	3.92	3.94
17	3.58	3.60	3.63	3.66	3.68	3.71	3.73	3.75	3.77	3.79	3.81	3.83	3.85	3.86	3.88
18	3.53	3.56	3.59	3.61	3.64	3.66	3.68	3.71	3.73	3.75	3.76	3.78	3.80	3.82	3.83
19	3.50	3.53	3.55	3.58	3.60	3.62	3.65	3.67	3.68	3.70	3.72	3.74	3.76	3.77	3.79
20	3.47	3.49	3.52	3.54	3.57	3.59	3.61	3.63	3.65	3.67	3.68	3.70	3.72	3.73	3.75
21	3.44	3.46	3.49	3.51	3.54	3.56	3.58	3.60	3.62	3.63	3.65	3.67	3.69	3.70	3.72
22	3.41	3.44	3.46	3.49	3.51	3.53	3.55	3.57	3.59	3.61	3.62	3.64	3.65	3.67	3.68
23	3.39	3.41	3.44	3.46	3.49	3.51	3.53	3.54	3.56	3.58	3.60	3.61	3.63	3.64	3.66
24	3.37	3.39	3.42	3.44	3.46	3.48	3.50	3.52	3.54	3.56	3.57	3.59	3.60	3.62	3.63
25	3.35	3.38	3.40	3.42	3.44	3.46	3.48	3.50	3.52	3.53	3.55	3.57	3.58	3.60	3.61
26	3.33	3.36	3.38	3.40	3.42	3.44	3.46	3.48	3.50	3.52	3.53	3.55	3.56	3.58	3.59
27	3.32	3.34	3.36	3.39	3.41	3.43	3.45	3.46	3.48	3.50	3.51	3.53	3.54	3.56	3.57
28	3.30	3.33	3.35	3.37	3.39	3.41	3.43	3.4 5	3.46	3.48	3.50	3.51	3.53	3.54	3.55
29	3.29	3.31	3.34	3.36	3.38	3.40	3.42	3.43	3.45	3.47	3.48	3.50	3.51	3.52	3.54
30	3.28	3.30	3.32	3.34	3.36	3.38	3.40	3.4 2	3.44	3.45	3.47	3.48	3.50	3.51	3.52
31	3.27	3.29	3.31	3.33	3.35	3.37	3.39	3.41	3.42	3.44	3.45	3.47	3.48	3.50	3.51
32	3.25	3.28	3.30	3.32	3.34	3.36	3.38	3.39	3.41	3.43	3.44	3.46	3.47	3.48	3.50
33	3.24	3.27	3.29	3.31	3.33	3.35	3.37	3.38	3.40	3.42	3.43	3.44	3.46	3.47	3.48
34	3.23	3.26	3.28	3.30	3.32	3.34	3.36	3.37	3.39	3.41	3.42	3.43	3.4 5	3.46	3.47
35	3.23	3.25	3.27	3.29	3.31	3.33	3.35	3.36	3.38	3.39	3.41	3.42	3.44	3.45	3.46
36	3.22	3.24	3.26	3.28	3.30	3.32	3.34	3.35	3.37	3.39	3.40	3.41	3.43	3.44	3.45
37	3.21	3.23	3.25	3.27	3.29	3.31	3.33	3.35	3.36	3.38	3.39	3.40	3.42	3.43	3.44
38	3.20	3.23	3.25	3.27	3.29	3.30	3.32	3.34	3.35	3.37	3.38	3.40	3.41	3.42	3.44
39	3.20	3.22	3.24	3.26	3.28	3.30	3.31	3.33	3.35	3.36	3.37	3.39	3.40	3.41	3.43
40	3.19	3.21	3.23	3.25	3.27	3.29	3.31	3.32	3.34	3.35	3.37	3.38	3.39	3.41	3.42
41	3.18	3.21	3.23	3.25	3.2 7	3.28	3.30	3.32	3.33	3.35	3.36	3.37	3.39	3.40	3.41
42	3.18	3.20	3.22	3.24	3.26	3.28	3.29	3.31	3.33	3.34	3.35	3.37	3.38	3.39	3.41
43	3.17	3.19	3.21	3.23	3.25	3.27	3.29	3.30	3.32	3.33	3.35	3.36	3.37	3.39	3.40
44	3.17	3.19	3.21	3.23	3.25	3.27	3.28	3.30	3.31	3.33	3.34	3.36	3.37	3.38	3.39
45	3.16	3.18	3.20	3.22	3.24	3.26	3.28	3.29	3.31	3.32	3.34	3.35	3.36	3.37	3.39
46	3.16	3.18	3.20	3.22	3.24	3.25	3.27	3.29	3.30	3.32	3.33	3.34	3.36	3.37	3.38
47	3.15	3.17	3.19	3.21	3.23	3.25	3.27	3.28	3.30	3.31	3.33	3.34	3.35	3.36	3.38
48	3.15	3.17	3.19	3.21	3.23	3.24	3.26	3.28	3.29	3.31	3.32	3.33	3.35	3.36	3.37
49	3.14	3.17	3.19	3.20	3.22	3.24	3.26	3.27	3.29	3.30	3.32	3.33	3.34	3.35	3.37
50	3.14	3.16	3.18	3.20	3.22	3.24	3.25	3.27	3.28	3.30	3.31	3.32	3.34	3.35	3.36
60	3.11	3.13	3.15	3.17	3.18	3.20	3.22	3.23	3.25	3.26	3.27	3.29	3.30	3.31	3.32
70	3.08	3.11	3.12	3.14	3.1 6	3.18	3.19	3.21	3.22	3.24	3.2 5	3.26	3.27	3.28	3.30
80	3.07	3.09	3.11	3.13	3.1 4	3.16	3.17	3.1 9	3.20	3.22	3.2 3	3.24	3.25	3.26	3.28
90	3.05	3.08	3.09	3.11	3.1 3	3.15	3.16	3.1 7	3.19	3.20	3.21	3.23	3.24	3.25	3.26
100	3.04	3.06	3.08	3.10	3.12	3.13	3.15	3.16	3.18	3.19	3.20	3.22	3.23	3.24	3.25

Table 11-16 Values of t For Obtaining One-Sided 95% Poisson Prediction Limits For r Additional Samples Given A Background Sample of Size n.

Previous n	Number of new measurements (r)														
	30	35	40	45	50	55	60	65	70	75	80	85	90	95	100
4	8.57	9.04	9.46	9.85	10.21	10.55	10.87	11.17	11.45	11.72	11.98	12.23	12.47	12.70	12.92
5	6.25	6.52	6.76	6.97	7.17	7.36	7.53	7.69	7.84	7.98	8.12	8.25	8.38	8.50	8.61
6	5.25	5.44	5.60	5.76	5.89	6.0 2	6.14	6.25	6.35	6.45	6.54	6.63	6.71	6.79	6.87
7	4.70	4.85	4.98	5.10	5.21	5.31	5.40	5.48	5.56	5.64	5.71	5.78	5.84	5.90	5.96
8	4.35	4.48	4.59	4.69	4.78	4.87	4.94	5.02	5.08	5.14	5.20	5.26	5.31	5.36	5.41
9	4.12	4.23	4.33	4.42	4.50	4.57	4.64	4.70	4.76	4.81	4.86	4.91	4.96	5.00	5.04
10	3.95	4.06	4.15	4.22	4.30	4.36	4.42	4.48	4.53	4.58	4.62	4.67	4.71	4.74	4.78
11	3.83	3.92	4.00	4.08	4.14	4.20	4.26	4.31	4.36	4.40	4.44	4.48	4.52	4.55	4.59
12	3.73	3.82	3.89	3.96	4.02	4.08	4.13	4.18	4.22	4.26	4.30	4.34	4.37	4.41	4.44
13	3.65	3.73	3.81	3.87	3.93	3.98	4.03	4.08	4.12	4.15	4.19	4.23	4.26	4.29	4.32
14	3.58	3.66	3.73	3.80	3.85	3.90	3.95	3.99	4.03	4.07	4.10	4.13	4.16	4.19	4.22
15	3.53	3.61	3.67	3.73	3.79	3.84	3.88	3.92	3.96	3.99	4.03	4.06	4.09	4.11	4.14
16	3.48	3.56	3.62	3.68	3.73	3.78	3.82	3.86	3.90	3.93	3.96	3.99	4.02	4.05	4.07
17	3.44	3.52	3.58	3.64	3.69	3.73	3.77	3.81	3.84	3.88	3.91	3.94	3.96	3.99	4.01
18	3.41	3.48	3.54	3.60	3.65	3.69	3.73	3.77	3.80	3.83	3.86	3.89	3.92	3.94	3.96
19	3.38	3.45	3.51	3.56	3.61	3.65	3.69	3.73	3.76	3.79	3.82	3.85	3.87	3.90	3.92
20	3.35	3.42	3.48	3.53	3.58	3.62	3.66	3.69	3.73	3.76	3.79	3.81	3.84	3.86	3.88
21	3.33	3.40	3.46	3.51	3.55	3.59	3.63	3.66	3.70	3.73	3.75	3.78	3.80	3.83	3.85
22	3.31	3.38	3.43	3.48	3.53	3.57	3.60	3.64	3.67	3.70	3.73	3.75	3.77	3.80	3.82
23	3.29	3.35	3.41	3.46	3.50	3.54	3.58	3.61	3.64	3.67	3.70	3.72	3.75	3.77	3.79
24	3.27	3.34	3.39	3.44	3.48	3.52	3.56	3.59	3.62	3.65	3.68	3.70	3.72	3.75	3.77
25	3.26	3.32	3.38	3.42	3.47	3.50	3.54	3.57	3.60	3.63	3.66	3.68	3.70	3.72	3.75
26	3.24	3.31	3.36	3.41	3.45	3.49	3.52	3.55	3.58	3.61	3.64	3.66	3.68	3.70	3.72
27	3.23	3.29	3.35	3.39	3.43	3.47	3.51	3.54	3.57	3.59	3.62	3.64	3.67	3.69	3.71
28	3.22	3.28	3.33	3.38	3.4 2	3.46	3.49	3.52	3.55	3.58	3.60	3.63	3.65	3.67	3.69
29	3.21	3.27	3.32	3.37	3.41	3.44	3.48	3.51	3.54	3.56	3.59	3.61	3.63	3.65	3.67
30	3.20	3.26	3.31	3.36	3.40	3.43	3.47	3.50	3.52	3.55	3.58	3.60	3.62	3.64	3.66
31	3.19	3.25	3.30	3.34	3.38	3.42	3.45	3.48	3.51	3.54	3.56	3.58	3.61	3.63	3.65
32	3.18	3.24	3.29	3.33	3.37	3.41	3.44	3.47	3.50	3.53	3.55	3.57	3.59	3.61	3.63
33	3.17	3.23	3.28	3.33	3.37	3.40	3.43	3.46	3.49	3.52	3.54	3.56	3.58	3.60	3.62
34	3.16	3.22	3.27	3.32	3.36	3.39	3.42	3.45	3.48	3.51	3.53	3.55	3.57	3.59	3.61
35	3.16	3.21	3.26	3.31	3.35	3.38	3.42	3.44	3.47	3.50	3.52	3.54	3.56	3.58	3.60
36	3.15	3.21	3.26	3.30	3.34	3.37	3.41	3.44	3.46	3.49	3.51	3.53	3.55	3.57	3.59
37	3.14	3.20	3.25	3.29	3.33	3.37	3.40	3.43	3.45	3.48	3.50	3.52	3.54	3.56	3.58
38	3.14	3.19	3.24	3.29	3.32	3.36	3.39	3.42	3.45	3.47	3.49	3.52	3.54	3.56	3.57
39	3.13	3.19	3.24	3.28	3.32	3.35	3.38	3.41	3.44	3.46	3.49	3.51	3.53	3.55	3.57
40	3.13	3.1 8	3.23	3.27	3.31	3.35	3.38	3.41	3.43	3.46	3.48	3.50	3.52	3.54	3.56
41	3.12	3.18	3.23	3.27	3.31	3.34	3.37	3.40	3.43	3.45	3.47	3.49	3.51	3.53	3.55
42	3.12	3.1 7	3.22	3.26	3.30	3.33	3.37	3.39	3.42	3.44	3.47	3.49	3.51	3.53	3.54
43	3.11	3.17	3.22	3.26	3.30	3.33	3.36	3.39	3.41	3.44	3.46	3.48	3.50	3.52	3.54
44	3.11	3.16	3.21	3.25	3.29	3.32	3.35	3.38	3.41	3.43	3.45	3.48	3.49	3.51	3.53
45	3.10	3.16	3.21	3.25	3.29	3.32	3.35	3.38	3.40	3.43	3.45	3.47	3.49	3.51	3.53
46	3.10	3.15	3.20	3.24	3.28	3.31	3.34	3.37	3.40	3.42	3.44	3.46	3.48	3.50	3.52
47	3.10	3.15	3.20	3.24	3.28	3.31	3.34	3.37	3.39	3.42	3.44	3.46	3.48	3.50	3.51
48	3.09	3.15	3.19	3.24	3.27	3.31	3.34	3.36	3.39	3.41	3.43	3.45	3.47	3.49	3.51
49	3.09	3.14	3.19	3.23	3.27	3.30	3.33	3.36	3.38	3.41	3.43	3.45	3.47	3.49	3.50
50	3.09	3.14	3.19	3.23	3.26	3.30	3.33	3.35	3.38	3.40	3.43	3.45	3.46	3.48	3.50
60	3.06	3.11	3.16	3.20	3.23	3.27	3.30	3.32	3.35	3.37	3.39	3.41	3.43	3.45	3.46
70	3.04	3.09	3.14	3.18	3.21	3.24	3.27	3.30	3.32	3.34	3.37	3.39	3.40	3.42	3.44
80	3.03	3.08	3.12	3.16	3.2 0	3.23	3.26	3.28	3.31	3.33	3.35	3.37	3.38	3.40	3.42
90	3.02	3.07	3.11	3.15	3.18	3.21	3.24	3.27	3.29	3.31	3.33	3.35	3.37	3.39	3.40
100	3.01	3.06	3.10	3.14	3.17	3.20	3.23	3.26	3.28	3.30	3.32	3.34	3 36	3.38	3.39

they only require that the m samples are independent and are drawn from a common but unspecified distribution. Gibbons (1990) provides the statistical basis of the tables for defining the probability that one of a particular number of samples will be below the maximum of n background measurements at each of k monitoring wells.

These approximated probabilities have been computed by Gibbons (1990) for selected values of n ranging from 4 to 100, selected values of k ranging from 1 to 100, and m ranging from 1 to 3 (i.e., 0 through 2 resamplings). The results are displayed in Tables 11-17 to 11-19 (0, 1, and 2, respectively). The accuracy of the approximation is presented in Table 11-20 (note, the approximation was not required for m = 1, see Table 11-18). Table 11-20 shows that for n > 10, the approximation is virtually identical to the exact values. These results are consistent for k = 5 and 10 (i.e., 5 or 10 monitoring wells) and m = 2, 3, and 4 (i.e., 1, 2, or 3 resamplings). Even at n = 5, when the exact probability was greater than .9, the approximate value differed by no more than 1.5%.

Table 11-21 provides the number of resamplings required to achieve at least a 95% confidence level for combinations of n = 4 to 100 and k = 1 to 100. Tables 11-17 and 11-21 shows that without the action of resampling the failing well, the necessary background sample sizes are generally too large to be of much practical value; however, a single resampling decreases the required number of background samples to within a reasonable range for most waste disposal facilities.

Single background wells do not provide a sufficient statistical basis for detection monitoring programs at waste disposal sites. Typically with small numbers of background measurements and large numbers of downgradient monitoring wells, (unfortunately a common monitoring problem), the required number of resamplings will be quite large, and will lead to a detection monitoring program with a high false negative rate. Table 11-18 illustrates the effect of various combinations of numbers of background samples and numbers of monitoring wells on the integrity of detection monitoring decisions. Given the requirements of RCRA Subtitle D ground-water monitoring with set time limits for the review of water quality data, it will be difficult having sufficient time for more than a single resample (i.e., m = 2, see Table 11-18).

General Procedure:--In this discussion, a nonparametric approach to constructing prediction limits is described. Specifically, interest is in the probability that at least one of m future measurements in each of k monitoring wells will not exceed the maximum of n previous samples. For example, consider a hypothetical (but not untypical) facility with two upgradient wells and five downgradient wells for

which quarterly monitoring produces relatively independent ground-water measurements. In terms of the upgradient wells, two years of quarterly monitoring have taken place yielding 16 background measurements. This evaluation procedure requires the assumptions that:

• The distribution of the indicator parameter is continuous;

• The distribution of water quality is the same in background and monitoring locations, and;

• The measurements are independent,

The probability that the five new monitoring values (i.e., one at each of the five downgradient wells) will be less than the maximum of the 16 background measurements is .762, (obtained from Table 11-17 where no resampling is performed and n=16 and k=5), This result has a high false positive rate (i.e., 1 - .762 = .238 or 23.8% of the time a false positive will be obtained). However, assume that in this example in the 23.8% of the cases in which a false positive result is obtained, the owner/operator is permitted to resample the well and if the new measurement is below the maximum of the 16 background values, the facility could return to normal detection monitoring following the procedure shown on Figure 11-25. With a single resampling, we obtain the probability that at least one out of two measurements at each of the five monitoring wells will be less than the maximum of the 16 background measurements. This probability is given in Table 11-18 by looking up k=5 and n=16 as .968 or a false positive rate of only 3.2%. This rate provides both an acceptable false positive rate and is fully protective of the environment.

Case X:—A new landfill has one year to obtain four quarterly measurements in a single well before operation. State law requires a single resampling. How many monitoring wells should the owner/operator install in order to obtain 95% confidence that at least one of the two measurements at each monitoring well will not exceed the largest of the four background measurements? Inspection of Table 11-18 reveals that it is hopeless; even with a single monitoring well, we can at most have .933 confidence. Inspection of Table 11-21 reveals that even with as few as three monitoring wells, three resamplings would be required to achieve a 95% confidence level. Furthermore, from a hydrogeological perspective, characterizing background water quality with a single background well is not a very reasonable choice. From a statistical perspective, a single background well produces a confound between contamination and spatial variability that cannot be resolved without additional background wells. Finally, the selection

Table 11-17   Probability That a Single Sample Will Be Below the Maximum of n Background Measurements at Each of k Monitoring Wells

Previous	Number or Monitoring Wells (k)														
n	1	2	3	4	5	6	7	8	9	10	11	12	13	14	15
4	.800	.667	.571	.500	.444	.400	.364	.333	.308	.286	.267	.250	.235	.222	.211
5	.833	.714	.625	.556	.500	.455	.417	.385	.357	.333	.312	.294	.278	.263	.250
6	.857	.750	.667	.600	.545	.500	.462	.429	.400	.375	.353	.333	.316	.300	.286
7	.875	.778	.700	.636	.583	.538	.500	.467	.437	.412	.389	.368	.350	.333	.318
8	.889	.800	.727	.667	.615	.571	.533	.500	.471	.444	.421	.400	.381	.364	.348
9	.900	.818	.750	.692	.643	.600	.562	.529	.500	.474	.450	.429	.409	.391	.375
10	.909	.833	.769	.714	.667	.625	.588	.556	.526	.500	.476	.455	.435	.417	.400
11	.917	.846	.786	.733	.687	.647	.611	.579	.550	.524	.500	.478	.458	.440	.423
12	.923	.857	.800	.750	.706	.667	.632	.600	.571	.545	.522	.500	.480	.462	.444
13	.929	.867	.812	.765	.722	.684	.650	.619	.591	.565	.542	.520	.500	.481	.464
14	.933	.875	.824	.778	.737	.700	.667	.636	.609	.583	.560	.538	.519	.500	.483
15	.937	.882	.833	.789	.750	.714	.682	.652	.625	.600	.577	.556	.536	.517	.500
16	.941	.889	.842	.800	.762	.727	.696	.667	.640	.615	.593	.571	.552	.533	.516
17	.944	.895	.850	.810	.773	.739	.708	.680	.654	.630	.607	.586	.567	.548	.531
18	.947	.900	.857	.818	.783	.750	.720	.692	.667	.643	.621	.600	.581	.562	.545
19	.950	.905	.864	.826	.792	.760	.731	.704	.679	.655	.633	.613	.594	.576	.559
20	.952	.909	.870	.833	.800	.769	.741	.714	.690	.667	.645	.625	.606	.588	.571
25	.962	.926	.893	.862	.833	.806	.781	.758	.735	.714	.694	.676	.658	.641	.625
30	.968	.937	.909	.882	.857	.833	.811	.789	.769	.750	.732	.714	.698	.682	.667
35	.972	.946	.921	.897	.875	.854	.833	.814	.795	.778	.761	.745	.729	.714	.700
40	.976	.952	.930	.909	.889	.870	.851	.833	.816	.800	.784	.769	.755	.741	.727
45	.978	.957	.937	.918	.900	.882	.865	.849	.833	.818	.804	.789	.776	.763	.750
50	.980	.962	.943	.926	.909	.893	.877	.862	.847	.833	.820	.806	.794	.781	.769
60	.984	.968	.952	.937	.923	.909	.896	.882	.870	.857	.845	.833	.822	.811	.800
70	.986	.972	.959	.946	.933	.921	.909	.897	.886	.875	.864	.854	.843	.833	.824
80	.988	.976	.964	.952	.941	.930	.920	.909	.899	.889	.879	.870	.860	.851	.842
90	.989	.978	.968	.957	.947	.937	.928	.918	.909	.900	.891	.882	.874	.865	.857
100	.990	.980	.971	.962	.952	.943	.935	.926	.917	.909	.901	.893	.885	.877	.870

Previous	Number of Monitoring Wells (k)														
n	20	25	30	35	40	45	50	55	60	65	70	75	80	90	100
4	.167	.138	.118	.103	.091	.082	.074	.068	.062	.058	.054	.051	.048	.043	.038
5	.200	.167	.143	.125	.111	.100	.091	.083	.077	.071	.067	.062	.059	.053	.048
6	.231	.194	.167	.146	.130	.118	.107	.098	.091	.085	.079	.074	.070	.062	.057
7	.259	.219	.189	.167	.149	.135	.123	.113	.104	.097	.091	.085	.080	.072	.065
8	.286	.242	.211	.186	.167	.151	.138	.127	.118	.110	.103	.096	.091	.082	.074
9	.310	.265	.231	.205	.184	.167	.153	.141	.130	.122	.114	.107	.101	.091	.083
10	.333	.286	.250	.222	.200	.182	.167	.154	.143	.133	.125	.118	.111	.100	.091
11	.355	.306	.268	.239	.216	.196	.180	.167	.155	.145	.136	.128	.121	.109	.099
12	.375	.324	.286	.255	.231	.211	.194	.179	.167	.156	.146	.138	.130	.118	.107
13	.394	.342	.302	.271	.245	.224	.206	.191	.178	.167	.157	.148	.140	.126	.115
14	.412	.359	.318	.286	.259	.237	.219	.203	.189	.177	.167	.157	.149	.135	.123
15	.429	.375	.333	.300	.273	.250	.231	.214	.200	.187	.176	.167	.158	.143	.130
16	.444	.390	.348	.314	.286	.262	.242	.225	.211	.198	.186	.176	.167	.151	.138
17	.459	.405	.362	.327	.298	.274	.254	.236	.221	.207	.195	.185	.175	.159	.145
18	.474	.419	.375	.340	.310	.286	.265	.247	.231	.217	.205	.194	.184	.167	.153
19	.487	.432	.388	.352	.322	.297	.275	.257	.241	.226	.213	.202	.192	.174	.160
20	.500	.444	.400	.364	.333	.308	.286	.267	.250	.235	.222	.211	.200	.182	.167
25	.556	.500	.455	.417	.385	.357	.333	.312	.294	.278	.263	.250	.238	.217	.200
30	.600	.545	.500	.462	.429	.400	.375	.353	.333	.316	.300	.286	.273	.250	.231
35	.636	.583	.538	.500	.467	.437	.412	.389	.368	.350	.333	.318	.304	.280	.259
40	.667	.615	.571	.533	.500	.471	.444	.421	.400	.381	.364	.348	.333	.308	.286
45	.692	.643	.600	.562	.529	.500	.474	.450	.429	.409	.391	.375	.360	.333	.310
50	.714	.667	.625	.588	.556	.526	.500	.476	.455	.435	.417	.400	.385	.357	.333
60	.750	.706	.667	.632	.600	.571	.545	.522	.500	.480	.462	.444	.429	.400	.375
70	.778	.737	.700	.667	.636	.609	.583	.560	.538	.519	.500	.483	.467	.437	.412
80	.800	.762	.727	.696	.667	.640	.615	.593	.571	.552	.533	.516	.500	.471	.444
90	.818	.783	.750	.720	.692	.667	.643	.621	.600	.581	.562	.545	.529	.500	.474
100	.833	.800	.769	.741	.714	.690	.667	.645	.625	.606	.588	.571	.556	.526	.500

Table 11-18   Probability That At Least One Out of Two Samples Will Be Below the Maximum of n Background Measurements At Each of k Monitoring Wells

Previous n	\multicolumn{15}{c}{Number of Monitoring Wells (k)}														
	1	2	3	4	5	6	7	8	9	10	11	12	13	14	15
4	.933	.871	.813	.759	.708	.661	.617	.576	.537	.502	.468	.437	.408	.381	.355
5	.952	.907	.864	.823	.784	.746	.711	.677	.645	.614	.585	.557	.530	.505	.481
6	.964	.930	.897	.865	.834	.804	.775	.748	.721	.695	.670	.646	.623	.601	.580
7	.972	.945	.919	.893	.869	.844	.821	.798	.776	.754	.734	.713	.693	.674	.655
8	.978	.956	.935	.914	.894	.874	.854	.835	.817	.799	.781	.764	.747	.730	.714
9	.982	.964	.946	.929	.912	.896	.879	.863	.848	.832	.817	.802	.788	.773	.759
10	.985	.970	.955	.941	.927	.912	.899	.885	.872	.858	.845	.833	.820	.808	.795
11	.987	.975	.961	.950	.938	.926	.914	.902	.890	.879	.868	.8s7	.846	.835	.824
12	.989	.978	.967	.957	.946	.936	.926	.915	.905	.895	.886	.876	.866	.857	.847
13	.990	.981	.972	.962	.953	.944	.935	.926	.917	.909	.900	.892	.883	.875	.866
14	.992	.983	.975	.967	.959	.951	.943	935	.927	.920	.912	.904	.897	.889	.882
15	.993	.985	.978	.971	.964	.957	.950	.943	.936	.929	.922	.915	.909	.902	.895
16	.993	.987	.981	.974	.968	.961	.955	.949	.943	.937	.930	.924	.918	.912	.906
17	.994	.988	.983	.977	.971	.965	.960	.954	.949	.943	.938	.932	.927	.921	.916
18	.995	.990	.984	.979	.974	.969	.964	.959	.954	.949	.944	.939	.934	.929	.924
19	.995	.990	.986	.981	.976	.972	.967	.963	.958	.953	.949	.944	.940	.935	.931
20	.996	.991	.987	.983	.979	.974	.970	.966	.962	.958	.953	.949	.945	.941	.937
25	.997	.994	.991	.989	.986	.983	.980	.977	.975	.972	.969	.966	.964	.961	.958
30	.998	.996	.994	.992	.990	.988	.986	.984	.982	.980	.978	.976	.974	.972	.970
35	.998	.997	.996	.994	.993	.991	.990	.988	.987	.985	.984	.982	.981	.979	.978
40	.999	.998	.997	.995	.994	.993	.992	.991	.990	.988	.987	.986	.985	.984	.983
45	.999	.998	.997	.996	.995	.994	.994	.993	.992	.991	.990	.989	.988	.987	.986
50	.999	.998	.998	.997	.996	.995	.995	.994	.993	.992	.992	.991	.990	.989	.989
60	.999	.999	.998	.998	.997	.997	.996	.996	.995	.995	.994	.994	.993	.993	.992
70	1.00	.999	.999	.998	.998	.998	.997	.997	.996	.996	.996	.995	.995	.995	.994
80	1.00	.999	.999	.999	.998	.998	.998	.998	.997	.997	.997	.997	.996	.996	.995
90	1.00	1.00	.999	.999	.999	.999	.998	.998	.998	.998	.997	.997	.997	.997	.996
100	1.00	1.00	.999	.999	.999	.999	.999	.998	.998	.998	.998	.998	.998	.997	.997

Previous n	\multicolumn{15}{c}{Number of Monitoring Wells (k)}														
	20	25	30	35	40	45	50	55	60	65	70	75	80	90	100
4	.252	.178	.126	.089	.063	.045	.032	.022	.016	.011	.008	.006	.004	.002	.001
5	.377	.295	.231	.181	.142	.111	.087	.068	.054	.042	.033	.026	.020	.012	.008
6	.483	.403	.336	.280	.233	.195	.162	.135	.113	.094	.078	.065	.055	.038	.026
7	.569	.494	.430	.373	.324	.281	.244	.212	.184	.160	.139	.121	.105	.079	.060
8	.638	.570	.510	.455	.407	.364	.325	.291	.260	.232	.207	.185	.166	.132	.106
9	.693	.632	.577	.526	.480	.438	.400	.365	.333	.303	.277	.253	.230	.192	.160
10	.737	.683	.633	.586	.543	.503	.466	.432	.400	.371	.343	.318	.295	.253	.217
11	.773	.724	.679	.637	.597	.560	.525	.492	.461	.432	.405	.380	.356	.313	.275
12	.802	.759	.718	.679	.643	.608	.576	.545	.515	.488	.461	.437	.413	.370	.331
13	.826	.787	.750	.715	.682	.650	.620	.591	.563	.537	.512	.488	.465	.423	.389
14	.846	.811	.778	.746	.716	.686	.658	.631	.605	.580	.5s7	.534	.512	.471	.433
15	.863	.832	.801	.772	.744	.717	.691	.666	.642	.619	.597	.575	.554	.515	.478
16	.877	.849	.821	.795	.769	.744	.720	.697	.675	.653	.632	.612	.592	.554	.519
17	.889	.864	.839	.814	.791	.768	.746	.724	.703	.683	.663	.644	.625	.590	.556
18	.900	.876	.854	.831	.810	.789	.768	.748	.729	.710	.691	.673	.656	.622	.590
19	.909	.888	.867	.846	.826	.807	.788	.769	.751	.733	.716	.699	.683	.651	.620
20	.917	.897	.878	.859	.841	.823	.805	.788	.771	.754	.738	.722	.707	.677	.648
25	.945	.931	.918	.905	.892	.880	.867	.855	.843	.831	.819	.807	.796	.774	.752
30	.960	.951	.941	.932	.922	.913	.904	.895	.886	.877	.868	.860	.851	.834	.817
35	.970	.963	.956	.949	.942	.935	.928	.921	.914	.907	.900	.893	.887	.874	.860
40	.977	.971	.966	.960	.955	.949	.944	.938	.933	.927	.922	.917	.911	.901	.890
45	.982	.977	.973	.968	.964	.959	.955	.950	.946	.942	.937	.933	.929	.920	.912
50	.985	.981	.978	.974	.970	.967	.963	.959	.956	.952	.949	.945	.941	.934	.927
60	.989	.987	.984	.982	.979	.976	.974	.971	.969	.966	.964	.961	.959	.954	.948
70	.992	.990	.988	.986	.984	.983	.981	.979	.977	.975	.973	.971	.969	.965	.962
80	.994	.992	.991	.990	.988	.987	.985	.984	.982	.981	.979	.978	.976	.973	.970
90	.995	.994	.993	.992	.990	.989	.988	.987	.986	.985	.983	.982	.981	.979	.976
100	.996	.995	.994	.993	.992	.991	.990	.989	.988	.987	.987	.986	.985	.983	.981

Table 11-19 Probability That At Least One Out of Three Samples Will Be Below the Maximum of n Background Measurements At Each of k Monitoring Wells

Previous n	1	2	3	4	5	6	7	8	9	10	11	12	13	14	15
4	.971	.944	.917	.891	.865	.840	.816	.793	.770	.748	.727	.706	.686	.666	.647
5	.982	.965	.947	.930	.914	.898	.882	.866	.850	.835	.827	.806	.791	.777	.763
6	.988	.976	.965	.953	.942	.931	.920	.909	.898	.887	.877	.866	.856	.846	.836
7	.992	.983	.975	.967	.959	.951	.943	.935	.927	.920	.912	.904	.897	.889	.882
8	.994	.988	.982	.976	.970	.964	.958	.953	.947	.941	.935	.930	.924	.918	.913
9	.995	.991	.986	.982	.977	.973	.969	.964	.960	.955	.951	.947	.942	.938	.934
10	.997	.993	.990	.986	.983	.979	.976	.972	.969	.966	.962	.959	.955	.952	.949
11	.997	.995	.992	.989	.986	.984	.981	.978	.976	.973	.970	.968	.965	.962	.960
12	.998	.996	.993	.991	.989	.987	.985	.983	.980	.978	.976	.974	.972	.970	.968
13	.998	.996	.995	.993	.991	.989	.988	.986	.984	.982	.981	.979	.977	.975	.974
14	.999	.997	.996	.994	.993	.991	.990	.988	.987	.985	.984	.982	.981	.980	.978
15	.999	.998	.996	.995	.994	.993	.991	.990	.989	.988	.987	.985	.984	.983	.982
16	.999	.998	.997	.996	.995	.994	.993	.992	.991	.990	.989	.988	.987	.986	.985
17	.999	.998	.997	.996	.996	.995	.994	.993	.992	.991	.990	.990	.989	.988	.987
18	.999	.998	.998	.997	.996	.995	.995	.994	.993	.993	.992	.991	.990	.990	.989
19	.999	.999	.998	.997	.997	.996	.995	.995	.994	.994	.993	.992	.992	.991	.990
20	.999	.999	.998	.998	.997	.997	.996	.995	.995	.994	.994	.993	.993	.992	.992
25	1.00	.999	.999	.999	.998	.998	.998	.998	.997	.997	.997	.996	.996	.996	.995
30	1.00	1.00	.999	.999	.999	.999	.999	.999	.998	.998	.998	.998	.998	.997	.997
35	1.00	1.00	1.00	1.00	.999	.999	.999	.999	.999	.999	.999	.999	.998	.998	.998
40	1.00	1.00	1.00	1.00	1.00	1.00	.999	.999	.999	.999	.999	.999	.999	.999	.999
45	1.00	1.00	1.00	1.00	1.00	1.00	1.00	1.00	.999	.999	.999	.999	.999	.999	.999
50	1.00	1.00	1.00	1.00	1.00	1.00	1.00	1.00	1.00	1.00	1.00	.999	.999	.999	.999
60	1.00	1.00	1.00	1.00	1.00	1.00	1.00	1.00	1.00	1.00	1.00	1.00	1.00	1.00	1.00
70	1.00	1.00	1.00	1.00	1.00	1.00	1.00	1.00	1.00	1.00	1.00	1.00	1.00	1.00	1.00
80	1.00	1.00	1.00	1.00	1.00	1.00	1.00	1.00	1.00	1.00	1.00	1.00	1.00	1.00	1.00
90	1.00	1.00	1.00	1.00	1.00	1.00	1.00	1.00	1.00	1.00	1.00	1.00	1.00	1.00	1.00
100	1.00	1.00	1.00	1.00	1.00	1.00	1.00	1.00	1.00	1.00	1.00	1.00	1.00	1.00	1.00

Previous n — Number of Monitoring Wells (k)

Previous n	20	25	30	35	40	45	50	55	60	65	70	75	80	90	100
4	.560	.484	.419	.363	.314	.271	.235	.203	.176	.152	.131	.114	.098	.074	.055
5	.697	.637	.582	.532	.486	.444	.406	.371	.339	.310	.283	.259	.237	.198	.165
6	.787	.741	.698	.658	.619	.583	.549	.518	.487	.459	.432	.407	.384	.340	.302
7	.846	.811	.778	.746	.716	.686	.658	.631	.605	.580	.557	.534	.512	.471	.433
8	.886	.859	.833	.808	.784	.761	.738	.716	.694	.674	.653	.634	.615	.579	.544
9	.913	.892	.872	.853	.833	.815	.796	.778	.761	.744	.727	.711	.695	.664	.634
10	.932	.916	.900	.885	.869	.854	.839	.825	.810	.796	.783	.769	.756	.730	.705
11	.946	.934	.921	.908	.896	.884	.871	.860	.848	.836	.825	.814	.802	.781	.759
12	.957	.946	.936	.926	.916	.906	.896	.886	.876	.867	.857	.848	.839	.820	.803
13	.965	.956	.948	.939	.931	.923	.915	.906	.898	.890	.882	.875	.867	.851	.836
14	.971	.964	.957	.950	.943	.936	.929	.922	.915	.909	.902	.895	.889	.876	.863
15	.976	.970	.964	.958	.952	.946	.941	.935	.929	.923	.918	.912	.907	.896	.885
16	.980	.975	.969	.965	.960	.955	.9so	.945	.940	.935	.930	.925	.921	.911	.902
17	.983	.978	.974	.970	.966	.961	.957	.953	.949	.945	.940	.936	.932	.924	.916
18	.985	.981	.978	.974	.970	.967	.963	.959	.956	.952	.949	.945	.942	.935	.928
19	.987	.984	.981	.978	.974	.971	.968	.965	.962	.959	.956	.952	.949	.943	.937
20	.989	.986	.983	.980	.978	.975	.972	.969	.967	.964	.961	.959	.956	.950	.945
25	.994	.992	.991	.989	.988	.986	.985	.983	.982	.980	.979	.977	.976	.973	.970
30	.996	.995	.995	.994	.993	.992	.991	.990	.989	.988	.987	.986	.985	.984	.982
35	.998	.997	.996	.996	.995	.995	.994	.994	.993	.992	.992	.991	.991	.989	.988
40	.998	.998	.998	.997	.997	.996	.996	.996	.995	.995	.994	.994	.994	.993	.992
45	.999	.999	.998	.998	.998	.997	.997	.997	.997	.996	.996	.996	.995	.995	.994
50	.999	.999	.999	.999	.998	.998	.998	.998	.997	.997	.997	.997	.997	.996	.996
60	.999	.999	.999	.999	.999	.999	.999	.999	.998	.998	.998	.998	.998	.998	.997
70	1.00	1.00	1.00	.999	.999	.999	.999	.999	.999	.999	.999	.999	.999	.999	.998
80	1.00	1.00	1.00	1.00	1.00	1.00	.999	.999	.999	.999	.999	.999	.999	.999	.999
90	1.00	1.00	1.00	1.00	1.00	1.00	1.00	1.00	1.00	.999	.999	.999	.999	.999	.999
100	1.00	1.00	1.00	1.00	1.00	1.00	1.00	1.00	1.00	1.00	1.00	1.00	1.00	.999	.999

Table 11-20 Comparison of Exact and Approximate Results

| | 1 Resample | | | | 2 Resample | | 3 Resample | |
| | K=5 | | K=10 | | K=5 | | K=5 | |
N	exact	approx	exact	approx	exact	approx	exact	approx
5	.823	.784	.726	.614	.928	.914	.967	.961
10	.933	.927	.882	.872	.984	.983	.995	.995
20	.979	.979	.961	.658	.997	.997	1.00	1.00
50	.996	.996	.993	.992	1.00	1.00	1.00	1.00
100	.999	.999	.998	.998	1.00	1.00	1.00	1.00

of monitoring wells should be based on a detailed hydrogeological understanding of the site and not simply statistical optimization. Conversely, placing large numbers of monitoring wells on a site may be of some hydrogeological interest, but it will make it almost impossible to develop a statistically rigorous detection monitoring program.

Case XX:—Consider a facility with five upgradient wells monitored quarterly for one year (n = 20), and 12 down- gradient monitoring wells (k = 12). In the event of a statistical failure, the owner/operator is permitted to resample the well in question, and if the resample does not exceed the statistical limit, he may continue with normal detection monitoring in the following quarter. Benzene is of particular interest at this facility due to high levels of benzene in the leachate. Of the 20 background measurements, there were 17 nondetects and three detected values of 10, 12, and 16 ppb, respectively. The upper prediction limit for benzene is, therefore, 16 ppb. Inspection of Table 11-18 reveals that with n = 20, k = 12, and m = 2, we can have 94.9% confidence that either the initial measurement or the resample will not exceed 16 ppb in each of the 12 down-gradient monitoring wells. For example, if 11 of the 12 downgradient wells yielded nondetects for benzene and one well yielded a value of 20 ppb, then we would only resample that single well. If the resample exceeded 16 ppb, a statistical failure would be recorded; otherwise, the owner/operator would continue with normal detection monitoring on the next quarterly sampling event. It is important to realize that we do not have to resample the 11 wells that did not exceed the limit on the first sample. The prediction limit is for at least one of two samples; therefore, the requirement is met if the first sample does not exceed the limit.

Resampling as a Integral Part of Monitoring

Nonparametric prediction limits provide a straightforward approach to detection monitoring for a wide variety of potential indicator parameters at waste disposal facilities. With the use of resampling as a fundamental part of the probability statement, remarkably small background sample sizes are required to provide simultaneous 95% upper predictor limits for relatively large numbers of monitoring wells. Without taking resampling into consideration, however, nonparametric prediction limits require background sample sizes that are generally too large to be of practical value. This point also has been raised in the new statistical rule (USEPA, 1988).

Any ground water sampling program is very sensitive to the independence of the samples. Independence of data must be maintained to statistically evaluate the water quality data. Resampling also requires that the repeated samples must be independent. This rules out possible strategies in which a single sample is split into aliquots and simply reanalyzed. In areas in which ground water moves very slowly, it may be difficult to get repeat samples in a reasonable time-frame to both insure independence and protect the environment. These resampling procedures are sensitive to both the method of sampling and the purging procedures used in the sample collection. It is clear that aquifer units represent the optimum zone to obtain ground-water samples for detection monitoring programs. They can provide a more rapid flow pathway for evaluation of discharges from facilities and offers faster recharge of target monitoring zones for resampling efforts. The potential effect of correlated measurements on this method does not, however, negatively effect environmental protection. Since one would only resample elevated initial measurements, if the resample and the initial sample are correlated, the resample would be expected to have an increased probability of also being elevated and have an increased likelihood of exceeding the limit. The result of this serial correlation is an increased false positive rate and a statistical result that is more protective of the

Table 11-21. Number of Resamples Required To Have At Least 95% Confidence That At Least One Measurement Will Be Below the Maximum of n Background Measurements At Each of k Monitoring Wells

Previous n	Number of Monitoring Wells (k)														
	1	2	3	4	5	6	7	8	9	10	11	12	13	14	15
4	2	3	3	4	4	4	5	5	5	5	>5	>5	>5	>5	>5
5	1	2	3	3	3	3	4	4	4	4	4	4	5	5	5
6	1	2	2	2	3	3	3	3	3	3	4	4	4	4	4
7	1	2	2	2	2	2	3	3	3	3	3	3	3	3	3
8	1	1	2	2	2	2	2	2	3	3	3	3	3	3	3
9	1	1	2	2	2	2	2	2	2	2	2	3	3	3	3
10	1	1	1	2	2	2	2	2	2	2	2	2	2	2	3
11	1	1	1	2	2	2	2	2	2	2	2	2	2	2	2
12	1	1	1	1	2	2	2	2	2	2	2	2	2	2	2
13	1	1	1	1	1	2	2	2	2	2	2	2	2	2	2
14	1	1	1	1	1	1	2	2	2	2	2	2	2	2	2
15	1	1	1	1	1	1	2	2	2	2	2	2	2	2	2
16	1	1	1	1	1	1	1	2	2	2	2	2	2	2	2
17	1	1	1	1	1	1	1	1	2	2	2	2	2	2	2
18	1	1	1	1	1	1	1	1	1	2	2	2	2	2	2
19	0	1	1	1	1	1	1	1	1	1	2	2	2	2	2
20	0	1	1	1	1	1	1	1	1	1	1	2	2	2	2
25	0	1	1	1	1	1	1	1	1	1	1	1	1	1	1
30	0	1	1	1	1	1	1	1	1	1	1	1	1	1	1
35	0	1	1	1	1	1	1	1	1	1	1	1	1	1	1
40	0	0	1	1	1	1	1	1	1	1	1	1	1	1	1
45	0	0	1	1	1	1	1	1	1	1	1	1	1	1	1
50	0	0	1	1	1	1	1	1	1	1	1	1	1	1	1
60	0	0	0	1	1	1	1	1	1	1	1	1	1	1	1
70	0	0	0	1	1	1	1	1	1	1	1	1	1	1	1
80	0	0	0	0	1	1	1	1	1	1	1	1	1	1	1
90	0	0	0	0	1	1	1	1	1	1	1	1	1	1	1
100	0	0	0	0	0	1	1	1	1	1	1	1	1	1	1

Previous n	Number of Monitoring Wells (k)														
	20	25	30	35	40	45	50	55	60	65	70	75	80	90	100
4	>5	>5	>5	>5	>5	>5	>5	>5	>5	>5	>5	>5	>5	>5	>5
5	5	>5	>5	>5	>5	>5	>5	>5	>5	>5	>5	>5	>5	>5	>5
6	4	5	5	5	5	5	>5	>5	>5	>5	>5	>5	>5	>5	>5
7	4	4	4	4	4	5	5	5	5	5	5	5	5	>5	>5
8	3	3	4	4	4	4	4	4	4	4	5	5	5	5	5
9	3	3	3	3	4	4	4	4	4	4	4	4	4	4	4
10	3	3	3	3	3	3	3	4	4	4	4	4	4	4	4
11	3	3	3	3	3	3	3	3	3	3	4	4	4	4	4
12	2	3	3	3	3	3	3	3	3	3	3	3	3	3	4
13	2	2	3	3	3	3	3	3	3	3	3	3	3	3	3
14	2	2	2	3	3	3	3	3	3	3	3	3	3	3	3
15	2	2	2	2	2	3	3	3	3	3	3	3	3	3	3
16	2	2	2	2	2	2	3	3	3	3	3	3.	3	3	3
17	2	2	2	2	2	2	2	2	3	3	3	3	3	3	3
18	2	2	2	2	2	2	2	2	2	2	3	3	3	3	3
19	2	2	2	2	2	2	2	2	2	2	2	2	3	3	3
20	2	2	2	2	2	2	2	2	2	2	2	2	2	2	3
25	2	2	2	2	2	2	2	2	2	2	2	2	2	2	2
30	1	1	2	2	2	2	2	2	2	2	2	2	2	2	2
35	1	1	1	2	2	2	2	2	2	2	2	2	2	2	2
40	1	1	1	1	1	2	2	2	2	2	2	2	2	2	2
45	1	1	1	1	1	1	1	1	2	2	2	2	2	2	2
50	1	1	1	1	1	1	1	1	1	2	2	2	2	2	2
60	1	1	1	1	1	1	1	1	1	1	1	1	1	1	2
70	1	1	1	1	1	1	1	1	1	1	1	1	1	1	1
80	1	1	1	1	1	1	1	1	1	1	1	1	1	1	1
90	1	1	1	1	1	1	1	1	1	1	1	1	1	1	1
100	1	1	1	1	1	1	1	1	1	1	1	1	1	1	1

THE TWO STEP STATISTICAL APPROACH

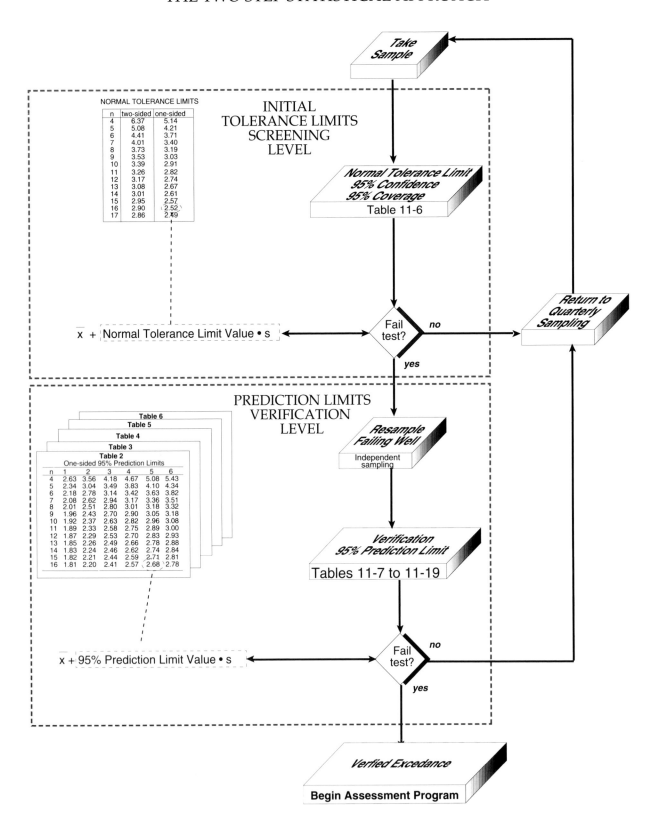

Figure 11-25 Evaluation Procedure where Resampling is Allowed

environment. In actual field practice, however, resamples are generally not taken within days of the initial sample. With the amount of time required for transport, analytical work, and sampling, the resample is typically obtained within a few weeks or even a month of the initial sample, and as such, serial dependence may be less of a problem.

In the examples considered here, we have only dealt with the case of a single variable (i.e., indicator parameter). This is rarely, if ever, the case in practice. When multiple variables are evaluated, the number of future measurements (k) is equal to the number of monitoring wells multiplied by the number of variables, assuming that the variables are independent. To the extent that the variables provide redundant information (i.e., correlated), the value of k decreases to an intermediate value between the number of monitoring wells and the number of monitoring wells multiplied by the number of variables. There is no direct way of estimating this intermediate value of k. For this reason, it is critically important to limit both the numbers of monitoring wells and the number of indicator parameters. The following examples illustrate the use of these Tables.

Examples - Case XXX:—A facility with 10 monitoring wells is only permitted a single resampling by state law. How many background measurements must be obtained such that the maximum will have 95% probability of exceeding at least one of the two measurements at each of the 20 monitoring wells? This question can be answered by reviewing Table 11-18 under the column k = 10. A background sample size of n = 18 yields λ = .949, and n = 19 yields λ = .953. Alternatively, inspection of Table 11-21 reveals that for k = 10, n = 19 is the smallest number of background measurements that achieves 95% confidence with a single resampling.

Case XXXX:—The 95%, non–parametric prediction –limit can be easily obtained with the aid of Table 11-21. For example, assume that we have a facility with r = 15 downgradient monitoring wells. Furthermore, let us also assume that if we fail a detection monitoring test, we are permitted to resample the well before any further action is taken; and if the repeat sample does not fail the test, we return to normal detection monitoring. How many background samples (n) are we required to obtain, such that the maximum observed measurement of those n samples will contain the next r = 15 monitoring measurements given the possibility of a single resample of any well that fails the initial test? Inspection of Table 11–21 reveals that a background sample of n = 20 measurements provides 93.7% confidence, and a background sample of n = 25 provides 95.8% confidence. The answer is, therefore, n = 23 (based on linear interpolation) background samples must be taken in order to insure that the maximum of those 23 samples will not be exceeded by the next 10 monitoring measurements (i.e., one at each downgradient well), given that we can resample any well that fails the initial test.

Alternate experimental designs for ground-water detection monitoring, and the nonparametric prediction limit procedure described above can also be applied to data collected from these designs. A common occurance in monitoring facilities in variable geologic conditions is finding that the levels of an indicator parameter in a particular downgradient well are considerably beyond the range of the background levels, however, the priority pollutant compound fingerprint of the leachate does not appear at the well. In greenfield siting projects where no previous disposal operations have occurred at the facility and other potential sources are remote, such water quality results must be ascribed to spatial variability. This type of purely spatial variability implies that the water quality in the relatively small number of background or upgradient wells is not from the same water mass as the water quality in one or more of the downgradient locations. As previously discussed, this condition violates both the regulatory conceptual model of a detection monitoring system and a critical assumption of the statistical model, namely, that in the absence of contamination, the ground-water quality measurements in background and monitoring wells are sampled from the same population. Under these conditions two alternatives can be used:

- a historical background for the aberrant downgradient well or wells may be established by selecting the maximum of the first n measurements for that well. New monitoring values for the well may then be compared to this maximum with resampling. In this case, k = 1, and m and n can be selected to satisfy the condition >-.95 using Table 11-19, or;

- when upgradient and downgradient wells are installed in pairs, their difference can, in some cases, be used as the unit of measurement and the maximum of n such differences may be used as the basis of comparison for a future difference (Loftis et al., 1987).

One must verify that the downgradient wells used in evaluating these differences are not already influenced by the facility; otherwise, the environment will not be fully protected and existing contamination may go undetected. For this reason, Gibbons (1990) believes the above second experimental approach may better suited to facilities in which data are available prior to operation.

Background sampling questions often arise in ground-water monitoring programs. Should the background measurements (n) consist of:

• a fixed one or two year set of measurements obtained at the time of permitting,

• a one or two year moving window that incorporates possible systematic changes in background water quality, or,

• all available background measurements.

The latter approach can provide the optimum choice among the above three alternatives, because the environment is protected from short-run fluctuations that may dominate a moving window, and the increasing size of the background sample provides the owner/operator with increased confidence that the data represents a large base of data. However, gradual changes in background water quality estimates may occur over time in the procedure of adding additional observations to the background sample. If these changes in the data reflect natural variability in background conditions, this using all background measurements can be an advantage; however, if these changes represent possible impact from the facility due to changes in gradient, this can also be a poor choice.

In summary, the major advantage of the non-parametric approach is that it only assumes that the samples are independent and measured on a continuous scale, but no particular distribution is specified. Furthermore, as long as at least one ground–water sample has a measurable value, the limit is defined. When nothing is detected in the n background samples, an alternative approach must be taken as given in Case 3 by using MCL's

Case 3: What to do When Nothing is Detected

Statistical methods are of little use without measurable data. Nevertheless, it is surprisingly common to observe a background collection of 10 or so measurements for which nothing was detected. What do we do? Is the tolerance or prediction limit zero? Is it the method detection limit? The answer to this question can only be found by examining the specifics of the analytic measurement process itself (see Gibbons 1988b). Interestingly, the analyst's decision as to whether or not a particular substance is present in a particular sample is based on the application of statistical decision rules. Even more remarkable, these decision rules are based on tolerance limits (see Currie 1968) for analyte absent or single concentration detection limit studies and prediction limits for calibration designs (i.e., a series of different spiking

concentrations in the range of the method detection limit (MDL) (see Hubaux and Vos, 1970; and Clayton et al., 1987).

These method detection limits are defined as the point at which the false positive and false negative rates are both less than 5%, for a test of the null–hypothesis that the concentration of the analyte in the sample is zero. To compute the method detection limit from "spiked" calibration samples, use the method of Clayton et al. (1987) as follows:

1. Select four concentrations in the range of the hypothesized MDL. For example, for Benzene we might select concentrations of 4, 8, 12, and 16 mg /L.

2. Prepare 16 samples; that is 4 replicates at each of the 4 spiking concentrations.

3. Several compounds may be examined simultaneously by including them in the same samples; however, the order of their concentrations should be randomized so that one sample does not contain all of the lowest concentrations and another sample all of the highest concentrations.

4. Introduce these 16 samples in the usual daily workload of two or more analysts (e.g., two analysts would receive eight samples each). It is essential that the analysts be completely blind to which compounds are present in the samples and their respective spiking concentrations, and that they simply be instructed to perform the standard analytic method in question (e.g., Method 624 VOC Scan).

5. The results of the analysis should be recorded as the square root of the ratio of the compound peak area to the internal standard; that is,

$$\text{response signal} = \sqrt{\frac{\text{peak area count}}{\text{internal standard area count}}} = y_i \qquad \text{Equation 11-4a}$$

Transform the spiking concentration

$$x_i = \sqrt{x^*_i - 0.1} - \sqrt{0.1} \qquad \text{Equation 11-4b}$$

Where x^* is the original spiking concentration

6. For each compound, compute the slope of the regression line of the instrument response signal (y) against the targeted concentration (x) as:

$$b = \frac{\sum_{i=1}^{16}(x_i - \bar{x})(y_i - y)}{\sum_{i=1}^{16}(x_i - \bar{x})^2}$$

Equation 11-5

Where x is the average of the four target concentrations, and y is the average of the 16 instrument response signals as defined above.

7. For each compound, compute the variance of deviations from the regression line as:

$$s_{y.x}^2 = \sum_{i=1}^{16}\left(y_i - \hat{y}_i\right)^2 / (16-2)$$

Equation 11-6

Where $\hat{y}_i = y_i + b (x_i - x)$ is the predicted instrument response for target concentration xi.

8. The method detection limit for n = 16 samples is then computed as:

$$MDL^* = (3.46 s_{y.x} / b)\sqrt{1 + 1/16 + \bar{x}^2 / \sum_{i=1}^{16}(x_i - \bar{x})^2}$$

Equ. 11-7a

Where 3.46 is the a = b = .95 percentage point of the noncentral t distribution on 16 − 2 = 14 degrees of freedom. To express MDL in the n original metric (example mg/l), compute

$$MDL = (MDL^*)^2 + 0.632455\ MDL^*$$

Equation 11-7b

In the absence of any detected values, this estimated MDL can be used as the corresponding prediction limit for detection monitoring.

We note that there are existing published MDLs for many compounds, including the Method 624 VOCs. More recently, Practical Quantitation Limits (PQLs) have also published in the Method SW–846 regulation. These national values were established under idealized conditions in which both presence and spiking concentration were known to the analyst and very questionable statistical computations were performed (see Gibbons 1988b). As such, it is quite reasonable that such levels will not be reached in routine laboratory practice and that the procedure described here will provide more realistic estimates that are consistent with attainable standards in the routine application of these methods. This view is reiterated in the new 40 CFR 264 statistical regulation:

The Appendix IX rule (52 FR 25942, July 9, 1987) listed practical quantification limits (PQLs) that were established from "Test Methods for Evaluating Solid Waste" (SW–846). SW–846 is the general RCRA analytical methods manual, currently in its third edition. The PQLs listed were USEPA's best estimate of the practical sensitivity of the applicable method for RCRA ground–water monitoring purposes. However, some of the PQLs may be unattainable because they are based on general estimates for the specific substance. Furthermore, due to site specific factors, these limits may not be reached. For these reasons the Agency feels that the PQLs listed in Appendix IX are not appropriate for establishing a national baseline value for each constituent for determining whether a release to ground water has occurred. Instead, the PQLs are viewed as target levels that chemical laboratories should try to achieve in their analyses of ground water. In the event that a laboratory cannot achieve the suggested PQL, the owner or operator may submit a justification stating the reasons why these values cannot be achieved (e.g., specific instrument limitations). After reviewing this justification, the Regional Administrator may choose to establish facility specific PQLs based on the technical limitations of the contracting laboratory. Thus, USEPA may allow under 264.97(h) owners or operators to propose facility specific PQLs. These PQLs may be used with the statistical methods listed in 264.97.

Summary

The statistical methods described here provide a series of general tools by which detection monitoring programs can be designed using indicator parameters that vary from 100% detection to no detection. The methods are completely site–specific with the exception of the case of no detection for which they are specific to the monitoring laboratory responsible for the routine analysis of the ground–water samples. The statistical procedures are parametric and non–parametric forms of prediction and tolerance intervals and limits, and as such, are consistent with the new RCRA 40 CFR Part 264 statistical regulation. The facility–wide false positive rate is restricted to 5%; therefore, quarterly monitoring should statistically result in one false positive decision every five years. Resampling of the well or wells in question should produce even fewer false positive results. Using the suggested sample sizes should produce false negative rates of less than 5% for monitoring requirements in excess of 2 to 3 standard deviation units above the background mean. False positive and false negative results are, therefore, balanced for even modest deviations from background water quality levels.

11.8 VERIFICATION OF EXCEEDANCE

When water quality data from a detection monitoring program shows a statistically significant increase, a series of steps should be performed to determine if the increase is due to a release from the facility or from non–facility based interferences. Three major components of potential exceedance can be defined as:

- Site interference
  - Landfill gas present in well
  - Grout alkaline pH interferences
  - Poor well construction or maintenance
  - Background increase due to upgradient discharge
  - Natural aquifer interferences

- Laboratory interference
  - Transcription errors
  - Laboratory contamination
  - Method or chemical interferences

- A facility release exceedance

Each of the above potential pathways can cause a site to go into an exceedance verification analysis. Some of the interferences are relatively simple to define such as transcription errors and laboratory chemical interferences (e.g., persistent low levels of methylene chloride).

Other interferences are difficult to define and quantify, such as landfill gas cross–contamination or poor well construction. Each of the site interferences will be briefly discussed in the following subsections.

11.8.1 Site Evaluation Procedure if Statistical Tests are Triggered

If three or more parameters tested fail the statistical test, then the following steps will be followed to confirm the statistically increase:

- Review Procedures Sampling and laboratory procedures may be reviewed to determine if there was a systematic error. Field notes, sample team interviews, chain-of-custody, trip and field blanks, and lab quality control data will be reviewed for these types of errors. Laboratory performance verification is directed toward determining the quality of individual analyses. The interpretative technique is designed to review all

parameters together to form an overall picture of the data quality. This review process is sometimes an acquired skill, other times individuals may never be able to develop the cognitive skills to evaluate large data bases by visual inspection of the information. There are however, a number of computer generated reports described in the next section that can assist in data evaluation. The historical comparison of analytical and field results represents one for the most important potential data sources for the review process. Table 11-17 shows an example of this data format. One should not rely on memory of past results as compared to current data to form conceptual ideas of the changes in water quality. This technique should not be used without access to easily reviewed historical data arrays. Historical comparison reports can provide data at a sampling location for full historic record or for a defined period of time. These reports are tabular arrays of the trend of the data, much in the same sense as a graphical trend analysis of a particular variable. The historical comparison report however, allows the reviewer to compare the entire data set as a whole. Anomalous data and specific indicator parameters can be targeted for further evaluation through additional sampling point comparisons. Data analysis may include other data relationships within the water quality review process :

- Anion-cation balances and relationships to electric conductivity(EC);

- Comparison of theoretical and measured electric conductivity/total dissolved solids (EC/TDS);

- Demand parameter relationships (biochemical oxygen demand [BOD]/chemical oxygen demand[COD]/total organic carbon[TOC] ratios);

- Evaluate trace elements data in terms of potential inter-elemental interferences;

- "Logical" volatile organic compounds (VOC) degradation patterns(landfill age vs. solvent breakdown product appearances;

- Confirm the presence or absence of common laboratory contaminants such as : solvents, phthalates, methylene chloride;

- Interpretation of data relative to detection limits and dilutions;

- Close scrutiny of rare event analytes;

- Relationship of detected analytes to potential sources contamination (e.g., elevated lead and cadmium near highways or due to the acid digestion of a turbid ground-water sample).

- Natural Inter–Aquifer Interferences: Water quality obtained from aquifers can typically exceed standards for drinking–water quality for many inorganic and some organic indicators. Interferences with detection monitoring programs have been documented for chlorides (shale bedrock discharge). Arsenic has been reported in bedrock wells from Massachusetts; and even organic indicator parameters such as phenols and TOX occur naturally in swamps and near the sea shore. The best procedure to follow in assessing these potentially natural indicator parameters is to have adequate, upgradient or background monitoring wells. Sufficient numbers of background wells should assist in establishing ambient water quality and provide a better statistical basis for water quality comparisons.

Expression of the relationships among ions, or of one constituent to the total concentration (such as use of mathematical ratios), is often helpful in identifying similarities and differences among different samples of ground water. These differences can define interferences by indicator parameters that may be naturally present in the aquifer. For most comparisons of this type, concentrations, expressed in terms of milliequivalents per liter or moles per liter, are the most useful.

Ratios can be useful to establish chemical similarities among waters; for example, by grouping analyses representing a single geologic terrain, upgradient and downgradient waters in a single aquifer, or a water–bearing zone. Fixed rules for selection of the most significant parameters to compare cannot be given, but the investigator should consider the sources of ions and the chemical behavior of the parameter.

Distinguishing between two sources of contamination usually involves comparison of different ionic ratios, using a variety of graphical methods (i.e., Stiff, Piper, or Scholler diagrams).

- Compare Gas Data: The presence of gas from landfills in monitoring wells is probably one of the most significant contamination problems observed at solid waste sites. The typical gas contaminated well shows persistent, low concentrations of vinyl chloride. The well may be sampled through bailing or by bladder pumps. Typically, bailed wells contain higher concentrations of contamination by gas because the bailer passes through the gas as it is removed from the well. When the screened interval in a monitoring well is above the potentiometric surface, landfill gas can enter the well.

Figure 11-26 Sampling used for Ground-Water Comparisons

These gas cross–contamination problems, however, do not relate to true ground–water contamination. In a number of cases, hand–bailed wells showing vinyl chloride hits produced non–detect VOAs once a bladder pump was installed. To define if gas is causing contamination of samples, the following procedure should be considered:

• If persistent low levels of VOCs are observed from a well, answer the following questions:

  – Is the well screened partially in the unsaturated zone?
  – Is a bailer used to obtain a sample?
  – Can landfill gas be detected in the well with a methane monitor?
  – Is vinyl chloride observed in the results of analyses?
  – Is the well casing cracked or broken in the unsaturated zone (observe with down hole TV camera)?

Landfill gas and condensate may be compared to ground-water quality data. The patterns of "fingerprint" VOCs in gas and condensates can be compared to ground-water quality data patterns of VOCs. Also, VOCs detected in well headspace will be evaluated using Henry's Law to determine whether VOCs have migrated from the vapor phase to the liquid phase or vice versa. The presence of

landfill gas cross–contamination would be suspected if any of the above were observed at a VOC detect well.

• Grout Alkaline Ph Interferences: High pH readings are unusual in ground–water. Typically, pH above 8.5 would be considered uncommon in natural aquifers (Hem, 1967). However, monitoring wells have been observed producing water in which pH approaches 14. For wells contaminated with cement grout, pH values of 11 to 12 are typical. Grout contaminated wells are a result of poor well construction, such as not separating the well screen intake area sufficiently from the annular seals of cement/bentonite grout. Some examples of likely environments for grout contamination are:

- Wells located in low hydraulic conductivity units with strong vertical gradients (i.e., grout or alkaline water moves vertically downward).

- Grout injected into the screened area of the well.

- Bentonite seals are either too thin or ineffective.

- Fractured rock provides channels for alkaline water to move around bentonite seals.

Figure 11-27 Frozen Ground May Cause Increased Gas Migration

All of the above causes of well contamination can be remedied by proper construction and development of the well using ASTM standard well–designs. For those wells already showing high pH, the investigation should redevelop the well until a reasonable pH is obtained. Normal purging of wells during sampling (in low hydraulic conductivity environments) can require many years to reduce pH to background levels. Such wells may have to be replaced with properly constructed wells, or continuously pumped at low discharge until acceptable pH readings are obtained. Low pumping rates, however, may not be effective because high pH readings may return after purging ceases.

- Review Screened Intervals: The spatial and seasonal differences may be evaluated by reviewing the screened intervals of the well construction logs, the water levels (i.e., falling or rising water levels that may result in contact with different soil and rock type), and the geochemistry of the soil and/or rock screened and gravel packed.

- Poor Well Construction And Poor Maintenance: Interferences due to inter–aquifer connections or surface water entering the screened zone of the well along the annular space can cause excessive concentrations of indicator parameters. Proper well construction based on monitoring flow paths can target the uppermost aquifer for monitoring. Damaged wells can often be identified through a program of maintenance and inspection. Each monitoring well should be visually inspected before sampling to document the condition of the installation.

- Background Increase Due To Discharge: Resolving an exceedance of indicator parameters may require inspection of upgradient areas for visible signs of spills, or use of chemicals that can cause high concentrations of the indicator parameters. For example, road salt, on upgradient recharge areas, can cause a dramatic increase in chlorides in permeable, surficial aquifers. Because chloride is mobile and is present in leachate from landfills, road salt can give false–positive results in downgradient wells. This interference is especially troublesome with poorly defined monitoring systems in which upgradient and downgradient relationships have not been properly identified.

- Historical Review: A historical review of land use patterns may be conducted. The potential for industrial and municipal sludges deposited in the floodplain of the river will be evaluated in addition to the prior uses of the landfill property.

11.8.2 Verification Of Release

If concentrations of indicator parameters exceed a pre–determined threshold, steps must be taken to confirm the significant increase. These steps can be reduced to:

- Resample Well If the previous evaluation steps above do not satisfactorily explain a statistical exceedance, then the well should be resampled for the specific parameters.

- Evaluate Sources Evaluation may be made of the data (if it is not a systematic error) by comparing the additional parameters analyzed along with the required parameters to the various potential sources of ground-water quality impacts. The potential sources include leachates, landfill gas, gas condensates, naturally occurring changes (i.e., drought, flooding, spatial or seasonal variations), offsite sources, or prior land use impacts. The primary task is to determine if leachate has affected ground-water quality. To determine the potential source of observed contaminants several tasks should be performed:

- Sample leachate and verify the "fingerprint" of the leachate.
- Compare fingerprint of the leachate with observed parameters.

Sample the leachate and affected ground water well for the comparison of major cations and anions. Leachate will be sampled in the risers nearest the ground-water monitoring well(s) that failed the statistical test. Ground-water data may be compared to leachate chemistry To determine if leachate has affected ground-water chemistry. Patterns in leachate chemistry will be compared to patterns in ground-water chemistry. For example, bedrock ground water contains a much higher ratio of sodium to chloride than does leachate (sodium is two-thirds the concentration as chloride indicating a sodium chloride origin of natural salt). If leachate was released, the ratio would shift and decrease to that resembling leachate. Also, multivariate statistical tests can be used for this evaluation. A geochemical analysis may be done using trilinear diagrams. An example of a trilinear plot is presented (Figure 11-17) of leachate and ground-water data from another facility to determine if mixing of the two types have occurred. Leachate data plotted on the trilinear plot in a similarly pattern to that observed in Baedecker and Back, 1979. Leachate cations shifted to sodium and potassium while ground water primarily was of calcium-magnesium type cations (as seen in Baedecker, 1979; Johnson and Cartwright, 1980). Leachate also is generally dominated by alkalinity, chloride, sodium, potassium, ammonia nitrogen and iron. Most ground water particularly is depleted in potassium and ammonia nitrogen. The leachate evaluation may require several additional actions in the facility monitoring program:

- Expand parameters to include additional VOCs and metals.

- Determine if three or more parameters exceed statistical thresholds during the next two quarters of sampling.

If VOCs exceed two times the PQL, then the affected monitoring well should be sampled for landfill gas in the headspace above the ground-water level in the well. VOCs found in the vapor phase can be compared to VOCs found in the ground-water sample and landfill gas sampled from nearby gas collection piezometers. Landfill gas condensates also should be sampled for VOCs.

If confirmation of the increase occurs, then the investigation should initiate an assessment monitoring program. Assessment monitoring may include drilling of additional wells, expansion of analytical parameters, and additional sampling of the detection monitoring wells. Assessment monitoring is specifically addressed in this manual in Chapter 10.0 where the techniques evaluate the rate and extent of leachate migration. If no Phase I and Phase II investigations were previously performed before beginning an assessment program, the investigator should use the phased sequence of defining target pathways. The assessment monitoring system should be fully based on knowledge of the site geology and hydrogeology.

11.9   EDITING AND MANAGING DATA

Analytical data from environmental programs generates massive amounts of tabular data. Evaluations of these large data sets require special techniques to manage, edit, and analyze data. Previous sections of this chapter discussed the graphical techniques appropriate for the visual display of water quality data. There are, however, basic issues of data error control and data base management that go beyond the straight-forward display of analytical results. These techniques are required to deal with the large volume of analytical result.

All scientific data is gathered in ways that result in the data that is not absolutely correct. Every datum has some inherent uncertainty. There are three different types of errors associated with any collected data:

- systematic errors of determination,

- random errors of determination, and,

- blunders.

Systematic errors consistently bias data away from the "correct" value in one direction. Data analyzed in microgram per liter and reported in milligrams is one example; incomplete digestion of metals samples is another. Random errors are those that are as likely to bias a datum in one direction or the other. Reading the needle on a pH meter is as likely to cause a pH 0.1 unit higher as 0.1 unit lower, and rainfall estimates are as likely to be 1 inch/month too high as 1 inch/month too low. Blunders are gross errors — reading the wrong scale, cross-contamination of a sample using the wrong preservative on a water sample, or sampling the wrong well. These blunders may also include components of both systematic or random errors in addition to the simple blunder errors.

Statistically, random errors are somewhat easier to determine than systematic errors. Blunders are often easiest to determine, especially when they violate a "boundary condition" such as a pH of 356.7 or a 17,547 foot deep "surficial aquifer" well.

There are a number of methods of finding blunders in reported data. The first of these is a set of physical bounds checks done as the data is first received. If the data is received in a digital format, the data base can be automatically filtered for a number of key bounds checks. These checks would include pH not less than 1.0 and not greater than 14.0, temperatures above 1000F or less than 00F, total depth of well not less than cased depth, and potentially many others. Additional comparative data manipulation techniques can be used to further evaluate large data sets. Parameters can be ranked and the uppermost and lowermost 10% then compared to file values as a check against data entry error. Any errors determined through these checks are corrected, if possible, before further checks of data are done.

Random errors, other than blunders, can be evaluated by outlier determination and these points marked for further examination. Since environmental data often does not fit a normal distribution, one cannot use parametric outlier tests that assume normality. Instead, nonparametric outlier checks may be usable for these evaluations. These tests include parametric checks that have been made distribution independent by the substitution of medians for means and quarter-spreads for standard deviations.

Commercially written database management systems are normally used as a "front end" for maintenance of environmental water quality data. These software programs allow one to manipulate incoming data, do edit check, and reformat the data into a form that allow other commercial software products to further graph and hardcopy display the required data for report generation. Extreme care must be maintained in protecting these large data sets. Multiple copies, and various levels of working and protected copies should be maintained for use by project staff as part of QA/QC procedures.

Reporting Water Quality Data to Agencies

All State and Federal regulations require some form of reporting to confirm that the monitoring system is working as required by the codes. Some regulations require the reporting of tabular sets of data on forms or through a formatted electronic media. In general terms all data should be fully reviewed before transmittal to regulatory organizations. A simple set of guidelines can ease potential errors and embarrassment when submitting water quality data on your facility:

- Read your permit or waste discharge requirements and then follow them;

- Format data as required in a manner that communicates the data most effectively (so everyone reaches the same conclusions);

- If the State requires reporting of exceedances, format the response in a neutral manner;
  - Talk about the specific exceedance issues;
  - Relate progress made on defining causes of the exceedance(s);
  - Propose schedules for establishment of the cause of the exceedance or schedules for the remedial actions required;
  - Provide a summary statement on the level of concern

- Maintain consistency and continuity between quarterly reports:
  - Parameters exceedance changes from quarter to quarter,
  - New personnel should review past data,
  - Always cross-check reports from quarter to quarter.

- Explain why and what you will do with the data

- Maintain technical standards and textural reporting consistency between sites; you should always maintain consistent "standard" format for reporting water quality data.

REFERENCES

40 CFR Part 264 and FR Vol. 53 No. 196 pp. 3972039731, October 11, 1988

Benjamin J.R. and C. A. Cornell. 1970. Probability. Statistics. and Decision for Civil Engineers. New York: McGraw-Hill Book Company

Chou, Y.M.; and D.B. Owen. 1986. One–sided distribution–free simultaneous prediction limits for future samples. Journal of Quality Technology, 18, pp. 96–98.

Clayton, C.A.; J.W. Hines; and P.D. Elkins. 1987. Detection limits with specified assurance probabilities. Analytical Chemistry, 59, pp. 2506–2514.

Currie, L.A. 1968. Limits for qualitative decision and quantitative determination. Analytical Chemistry, 40, pp. 586–593.

Davis, C. B. and R. J. McNicols 1988. Statistical Issues and Problems in Ground-water Detection Monitoring at Hazardous Waste Facilities, Ground Water Monitoring Review v. 7, pp72-76.

Doctor et al 1986. Draft - Statistical Comparisons of Ground-Water Monitoring Data. Ground-Water Plans and Statistical Procedures to Detect Leaking at Hazardous Waste Facilities, PNL-5754, Pacific Northwest Laboratories, Richland Washington 55112.

Doctor, P.G., R.O. Gilbert, R.A. Saar, and G. Duffield. 1985a (February 14). An Analysis of Sources of Variation in Ground-Water Monitoring Data of HazardousWaste Sites. Milestone 1. Revised Draft. EPA Contract No. 68-01-6871. Battelle, Pacific Northwest Laboratories. Richland, WA.

Fisher and Potter, 1989. Methods for Determining Compliance with Groundwater Quality Regulations at Waste Disposal Facilities, Wisconsin Dept. of Natural Resources, Jan. 120pp.

Freeze, R.A., and J.A. Cherry. 1979. Ground water. Prentice–Hall, Inc., NJ, 604 pp.

Gibbons, R. D. 1990. A General Statistical Procedure for Ground-Water Detection Monitoring at Waste Disposal Facilities. Ground Water, 28, pp. 235–243.

Gibbons, R.D. 1987a. Statistical prediction intervals for the evaluation of ground water quality. Ground Water, 25, pp. 455–465.

Gibbons, R.D. 1987b. Statistical models for the analysis of volatile organic compounds in waste disposal facilities. Ground Water, 25, pp. 572–580.

Gibbons, R.D. 1988a. A general statistical procedure for ground water detection monitoring at waste disposal facilities. Ground Water, submitted for publication.

Gibbons, R.D. and Baker, J.A., 1991, "The Properties of Various Statistical Prediction Intervals for Groundwater Detection Monitoring", Environmental

Science and Health A26(4), pp. 535-553.

Gibbons, R.D.; F.H. Jarke; and K.P. Stoub. 1988b. Method detection limits. Ground Water, submitted for publication.

Goodman, Iris and Kenneth Potter. 1987. Graphical and Statistical Methods to Assess the Effects of Landfills on Groundwater Quality. Report to Wisconsin Department of Natural Resources, Bureau of Solid and Hazardous Waste.Harris et al, 1987

Goodman, Iris. 1987. Graphical and Statistical Methods to Assess the Effect of Landfills on Groundwater Quality. M.S. Thesis. University of Wisconsin Madison.

Green W. R. 1985. Computer-Aided Data Analysis, A Practical GuideJohn Wiley and Sons, 268 p.

Hahn, Gerald J. 1970a. Statistical InteNals for a Normal Population, Part 1. Examples & Applications. Journal of Quality Technology. Vol. 2, No. 3 (July): 1 15-125.

Hamilton, L.F., and Simpson, S.G. Calculations of Analytical Chemistry McGraw-Hill Book Company, 1960.

Harris, Jane, Jim C. Loftis and Robert H. Montgomery. 1987. Statistical Methods for Characterizing Ground-Water Quality. Ground Water. Vol. 25, No. 2 (March-April): 185-193EPA, Oct. 11, 1988;

Hem, J. D. 1970, Study and Interpretation of the Chemical Characteristics of Natural Water, U.S. Geological Survey Water-Supply Paper 1473, 363 p.

Hoaglin, D.C., Mosteller, F., and Tukey, J.W. Understanding Robust and Exploratory Data Analysis, John Wiley & Song, Inc. 1983.

Hoaglin, D.C., Mosteller, F., and Tukey, J.W. Understanding Robust and Exploratory Data Analysis, John Wiley & Song, Inc. 1983.

Hubaux, A.; and G. Vos. 1970. Decision and detection limits for linear calibration curves. Analytical Chemistry, 42, pp. 849–855.

Jarke, 1990, Is it Possible to Understand MDLs, PQLs, IDLs, EMLRLs, etc. Lab Notes, 2pp.

Loftis, Jim C., Jane Harris and Robert H. Montgomery. 1987. Detecting Changes in Ground Water Qaulity at Regulated Facilities. Monitoring Review. (Winter1987): 72-76.

McBean, Edward and Frank A. Rovers. 1984. Alternatives for Handling Detection Limit Data in Impact Assessments. Ground Water Monitoring Review. Vol. 4, No. 2 (Spring): 42-44.

McGill, Robert, John W. Tukey and Wayne A. Larsen. 1978. Variations of Box Plots. The American Statistician. Vol. 32, No. 1 (Febnuary): 12-16.

Meierer, R. E. and R. J Whitehead. 1989. Making an On-Site Evaluation of an Analytical Services Laboratory. Enviromental Claimes Journal, Vol. 1, No. 4, p 503-

515.

Miller, M.D. and F.C. Kohout. (Undated.) RCRA Ground Water Monitoring Statistical Comparisons: A Better Version of Student's T-Test. Mobil Research and Development Corporation. Paulsboro, New Jersey.

Montgomery, Robert H. and Jim C. Loftis. 1987. Applicability of the T-Test for Detecting Trends in Water Quality Variables. Water Resources Bulletin. Vol. 23, No. 4: 653-662.

Montgomery, Robert H., Jim C. Loftis and Jane Harris. 1987. Statistical Characteristics of Ground-Water Quality Variables. Ground Water. Vol. 25, No. 2 (March-April): 176-184.

Piper A. M. 1944. A Graphic Procedure in the Geochemical Interpretation of Water Analysis. Trans Amer. Geophys. Union, 25, pp 914-923.

Rosner, B. Technometrics, 17, 221-227, 1975.

Sen, Z. 1979. Application of the Autorun Test to Hydrologic Data. Journal of llydrology. Vol. 42: 1-7.

Sen, Z. 1982. Discussion of Statistical Considerations and Sampling Techniques for Ground-Water Quality Monitoring by J.D. Nelson and R.C. Ward. Vol. 20: 494-495.

Silver, Carl A. 1986a. Statistical Approaches to Groundwater Monitoring. Open File Report #7. University of Alabama, Environmental Institute for Waste Management Studies.

Test Method's for Evaluating Solid Waste, Physical Chemical Methods (SW-846)

U.S. Environmental Protection Agency. 1988. 40 CFR Part 264: Statistical methods for evaluating ground water monitoring from hazardous waste facilities; final rule. Federal Register, 53, 196, pp. 39720–39731.

Wald, A.; and J. Wolfowitz. 1946. Tolerance limits for a normal distribution. Ann. Math. Statistics, 17, pp. 208–215.

CHAPTER 12

REPORTING

This chapter reviews both proposal preparation guidelines and the overall reporting of technical data in reports and long term documentation. These guidelines were also prepared to assist engineering and science staff with the assessment of proposals written for site assessments. The guidelines provide formats recommended so that the scope, schedule, and costs of technical services are full documented so that during later execution of the project both the client and the consultant know fully well the expected product of the investigation and the project deliverables.

Requests for Proposals (RFP's) are typically the first step in a project. Although the client may have spent tens, hundreds, or even thousands of hours preparing the RFP, the consultant's first contact with a specific project is likely to be through the RFP transmitted from the client. Addressing the specific points of the RFP is where the proposal comes in to the office. Proposals represent legal agreements between client and consultants; hence, the various components of a proposal are important for documenting the details as in a formal contract. As with any contract the terms and limitations are important for protecting both the client and the consultant. The proposal typically consists of four major parts:

1. Introduction. Here the consultant briefly sets down everything needed to inform the reader about the problem being presented. The proposal should be identified. The consultant should explain how the idea came to him. The subject matter of the proposed project should be clearly identified. There should also be comments on the importance of the problem. The consultant's qualifications for the

work should be presented. If this can be done briefly, there should be a preview of the proposal.

2. Body. Here we would expect that the consultant develop the substance of the proposal in all necessary detail. This should be the longest section of the proposal. This section would typically be called the "Scope of Work". For topical contents, see the next section.

3. Conclusion. Here the text should consolidate what the consultant has accomplished in the body of the proposal. He should do whatever might help to encourage acceptance and precipitate action. Again, the consultant may stress the importance of the problem. The consultant may also suggest an interview or assert willingness to modify portions of the proposal if required by unforeseen events.

4. Attachments. Here the consultant can insert display matter and back-up matter that would interrupt the main presentation in the body of the proposal. Possibilities for inclusion here are testimonial letters from previous clients, flow charts of the intended work program, and descriptions of past projects.

This guideline focuses on the components of a typical hydrogeological/ geotechnical proposal for which a request for proposal (RFP) has been prepared and for which the estimate of the consultant's fee is based on a time and materials contractual arrangement.

12.1 MEET WITH THE CONSULTANT

A successful proposal effort should reflect the needs of the client and demonstrate that our expectations are clearly understood. Therefore, a clear understanding of the clients' s thinking and expectations is essential to your project. The basic expectations are set forth in the RFP, but frequently the RFP is only the beginning of the proposal effort.

After distribution of the RFP, a client would expect that the consultant will contact the responsible engineer. This discussion should clarify the client's expectations so that the proposal reflects that understanding.

12.2 PROPOSAL CONTENT

A typical proposal may conform to the following annotated outline or can vary to a greater of lesser degree. A client may have many consulting companies currently under master agreement contracts, although variations in proposal content should be expected; however, major deviations from an acceptable standard format should be avoided.

12.2.1 Title Page

The title page should show the title, local division and address, proposal number, and the date of issue (see Figure 12-1).

12.2.2 Transmittal Cover Letter

The cover letter of a proposal should be relatively direct; in most instances, it should not exceed one to two pages. The cover letter should transmit the proposal, citing its title and number of copies submitted. Reference to the request for proposal and any subsequent contact which may have had a bearing on understanding of the client's requirements should also be cited. The cover letter should also highlight those aspects of the proposal that may affect client's attitude toward the proposal. The consultant, for example, may wish to direct client's attention to the strengths of the proposal, or they may wish to explain various alternatives or modifications the consultant thinks the client should consider in selecting a consultant that will serve the needs the best. It should be signed by an authorized representative of the consultant who has legal authority to sign. For a sample cover letter, see Figure 12-2.

12.2.3 Table of Contents

A simple table of contents should be prepared, containing the major outline headings listed below and adding subheadings as necessary. A list of figures and tables may also be included. See Figure 12-3 for a sample table of contents.

12.2.4 Introduction

This section should contain a basic discussion of the proposed work, answering the question, "What is the purpose of the project?" It should include the request for proposal date and repeat the requirements set forth in the RFP. It should indicate general understanding of the problem
and the consultant should briefly describe any possible difficulties. This section should also contain a general description of the objectives of the proposed project.

12.2.5 Scope of Work

This section should set forth the limits of the technical services by explaining in general what services are to be provided. The scope statement sets boundaries, and states what is to be done within these boundaries. In other words, the scope statement establishes the depth, breadth, and means of the consultants approach. Also, for everyone's protection, the consultant may include in it negative observations covering what he will not do. For example, the scope of work section of a hydrogeological proposal might discuss:

- Geophysical Survey
- Field Drilling Program
- Laboratory Testing
- Installation of Monitoring Wells
- Permeability Tests
- Evaluation
- Report

Individual scopes of work must be specific to the site in question; guidance for components of work are referred to the Site Assessment Manual.

12.2.6 Approach

This section should state how the work will be done by the consultant by providing a detailed description (methodology) of the proposed work. This section is often

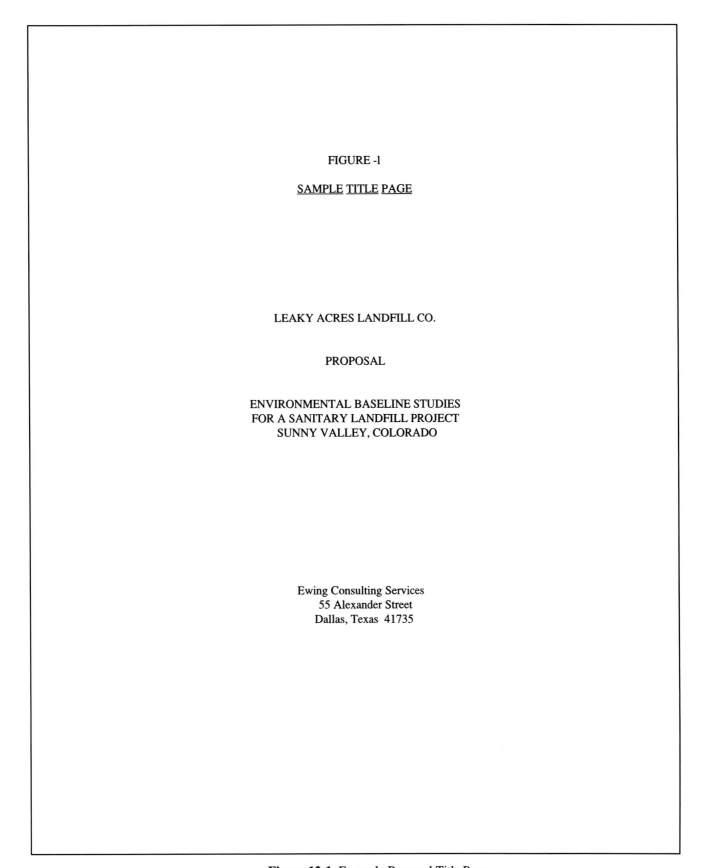

FIGURE -1

<u>SAMPLE</u> <u>TITLE</u> <u>PAGE</u>

LEAKY ACRES LANDFILL CO.

PROPOSAL

ENVIRONMENTAL BASELINE STUDIES
FOR A SANITARY LANDFILL PROJECT
SUNNY VALLEY, COLORADO

Ewing Consulting Services
55 Alexander Street
Dallas, Texas 41735

Figure 12-1 Example Proposal Title Page

SAMPLE COVER LETTER

August 14, 1989
Our Ref: WAJ : CK
Your Ref: TEB:JA:1400-3

Confidential

Leaky Acres Landfill Co.
34 Jones Street, 5th Floor
Portland, Maine 54321

Attention- Mr. I. M. Smart

Gentlemen:
PROPOSAL
ENVIRONMENTAL BASELINE STUDIES
 FOR A SANITARY LANDFILL PROJECT
SUNNY VALLEY, COLORADO

In response to your invitation dated July 29, 1991, Ewing Consulting Services is pleased to submit three (3) copies of this proposal to conduct an environmental and hydrogeologic baseline study for the Sanitary Landfill Project in the Sunny Valley, Colorado.

The proposal has been prepared in accordance with the Brief submitted to us and our subsequent discussion on August 2, 1991, regarding your requirements for this project.

I wish to bring to your attention our recommendation that the vegetative and avian studies have been rescheduled one month earlier than was requested. We feel that a more representative estimate of plant cover can be obtained in May than in June, and our experience in conducting avian surveys in Colorado and elsewhere indicates that they can be conducted in May without any loss in accuracy or Government acceptability of the results. We anticipate these changes in the sampling program will benefit the project by permitting the environmental report to be completed ahead of its present schedule.

It has been our pleasure to provide Leaky Acres Landfill Co. with this proposal. If we can be of further service to you, please feel free to contact me.

Yours faithfully,

EWING CONSULTING SERVICES

J. R. Ewing
Principal-in-Charge

Figure 12-2 Example Proposal Transmittal Letter

SAMPLE TABLE OF CONTENTS

Figure 12-3 Example Proposal Table of Contents Page

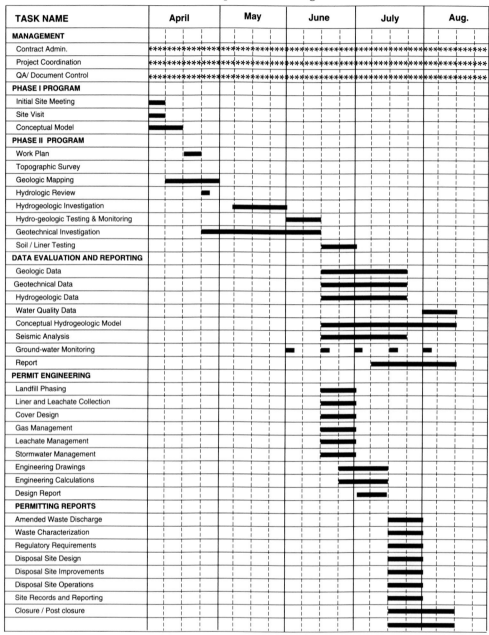

Landfill Phase II Expansion Investigation Schedule

Figure 12-4 Example of a Simple Proposal Schedule

subdivided according to phases/tasks and/or disciplines to be considered.

All but the simplest projects require a breakdown into tasks. One task may consist of initial exploration and planning, another of a search of the literature, another of correspondence and interviews, and so on. To try to do everything at one time produces confusion and dispersion of effort. Furthermore, one or several tasks may have to be completed before others can be started. Breaking a project into its component tasks also gives some assurance that the total job will be done in an orderly manner and completed on schedule.

The guidelines for RFPs provide significant direction as to the scope of work; however, the methodology often requires definition by the consultant as to test methods.

12.2.7 Schedule

Each proposal should include a program plan or schedule chart (Figure 12-3 example is shown). With few exceptions, work proposals and agreements specify the calendar period within which projects are to be completed. The period may extend for days, months, or years. Further, the schedule may stipulate that portions of the work are to be done in a stated order and are to be completed by a given date. For completed projects, time and effort must be carefully allocated. It is common practice to prepare a time-based flow chart on a wall. Even limited projects require planning for efficient performance to guarantee that work will be completed by target or deadline dates. The amount of detail to be shown should govern the type of presentation. The standard presentation is a bar chart with markers to indicate significant events. With the increased use of PERT, some companies are using PERT charts in place of bar charts. Either method is acceptable for the typical site assessment project.

Normally, the text will only supplement the schedule. Therefore, the amount of text in this section will be limited, even in a large program. In any program longer than three months, check-points and review points should be noted. This will indicate to the owner that tight operational and fiscal control of the program is planned. If the program is extended, i.e., two to five years, large expenditures should be phased-in with the the clients funding procedures.

The project schedule answers the question, "When will it be done?" Adequate schedule is often the deciding factor in awarding a contract. The project schedule is most effectively conveyed in the proposal by means of a figure such as that shown on Figure 12-4. The schedule should be simple and easily understandable, conveying the message that the project will be completed in the time specified. The text portion of the proposal should refer to the figure and highlight the importance of the schedule to the overall project.

12.2.8 Project Organization and Schedule

The project organization, showing who will do the job, should be presented to client in terms of a chart showing how the Principal Investigators have been organized to help the owner's Engineer accomplish his objectives. The Project Director, Principal Investigator, and Project Manager should be introduced and their experience discussed. All staff should be detailed within this section (see Figure 12-5). Qualifications and the vita of key personnel should be appended to the proposal. However, if the consultant cannot guarantee the availability of such persons for the project, he should not use wording that can be construed as a firm promise. Otherwise, the contract may be broken on the grounds of misrepresentation.

12.2.9 Previous Experience

Lack of experience in someone offering a proposal makes a client pause, whereas successful accomplishment in the past promises success in the future. Whenever possible, therefore, the consultant should cite earlier successes with similar problems when preparing the new proposal. It is not essential that the past and present problems be identical—only that they overlap and have major points in common. He should include dates, contract numbers, names, and addresses so that the client can verify the statements.

12.2.10 Products of the Project

Stating what the consultant is to produce throughout a project and at its termination commits the writer of the proposal to definite performance and productiveness. In terms of the typical field investigation, details on the numbers, locations, expected depths, and reasons for the exploration provides a clear vision of the expected program. The technical reasons for every borehole should be described in the proposal. While these data may (or even should) change due to unknown field conditions, the reasons for gathering data is an important part of technical proposals. Sometimes the consultant may find it burdensome and costly to satisfy all points of the client's RFP. However, if he succeeds in dodging this issue, and does not specify exactly what will be produced and supplied, the consultant may be placing himself in a still worse situation. The owner may "construe" the proposal as promising delivery of a working model, or prototype equipment, or as promising follow-up maintenance and consulting services free of extra cost. Therefore, a clear and specific statement of products to be delivered, while it commits the consultant, also sets bounds on what are to be delivered. Often it is beneficial to clearly describe in the proposal the deliverables of the project, on a task by task basis so there is not misunderstanding of exactly what will be the products of the work.

12.2.11 Costs

The consultant should detail the expected costs associated with the execution of the project using actual labor charges of the people that will be used on the project. Multiple tables should be constructed defining individual costs for labor, reimbursables, and support costs as are shown on Figures 12-6 and 12-7. The spreadsheet formats allow for rapid costing of the project tasks. Since these spreadsheet programs in the PC and Macintosh system are available in every engineering office, the use of such programs are mandatory for the project costing efforts.

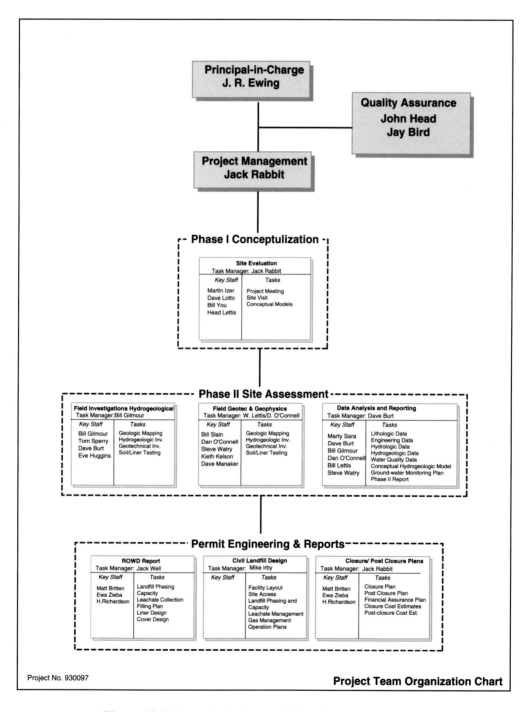

Figure 12-5 Example of a Proposal Organization Chart

12.3 DOCUMENTATION

12.3.1 INTRODUCTION

Much of this Site Assessment Manual revolves around the documentation typically applied to site assessments. Proposals for technical work, quality assurance, and both progress and final reports make up much of the documentation of a site assessment on a waste disposal facility.

The proposal effort is important for both the consultant and the facility engineer to understand the various technical, schedule, and financial elements of the proposed scope of work. The proposal is a legal document committing the consultant to perform the agreed work elements within the schedule given for the cost estimate. The owner is committed to pay the consultant if the agreed price for all elements are performed satisfactorily. This straight forward concept, however, often becomes quite complex in execution. The Subsections 12.1 and 12.2 information provides detailed information on the form and

content of proposals to assist in the review of such documents.

The requirement to deal with both real and perceived risk from both sanitary and hazardous waste facilities makes the technical quality of studies performed by consultants of the utmost importance to the long-term success of the owner.

Clearly, the only time-tested solution to maintenance of technical quantity to reduce environmental risk is to apply a quality assurance program to the site based work performed by our consultants. Quality assurance has been used since the 1950's in aerospace, and later was used for nuclear power plant siting and high level radioactive waste repository site investigations. Many owners and operators currently uses modified quality programs in analytical laboratory programs.

Quality assurance provides a planned and systematic approach of developing and applying actions necessary to assure the quality of output from work activities. It helps to avoid omissions and oversight that could adversely affect quality and, hence, increase long-term risk to the owner.

TABLE 1A - MANPOWER LOADING SELEKA FARMS PROPOSAL
ESTIMATED LABOR COST BY TASK

PERSONNEL	PROJECT ASSIGNMENT	COST/HR	1	2	3	4	5	6	7	8	9	10	TOTAL COSTS DOLLARS
J. R. Ewing	Principal Hydrogeologist	$150	xxxx	xxxx	xxxx	xxxx	xxxx	xxxx	xxxx	xxxx	xxxx	xxxx	xxxxxx
R. Ewing	Senior Hydrogeologist	$100	xxxx	xxxx	xxxx	xxxx	xxxx	xxxx	xxxx	xxxx	xxxx	xxxx	xxxxxx
P. Rabbit	External Consultant	$130	xxxx	xxxx	xxxx	xxxx	xxxx	xxxx	xxxx	xxxx	xxxx	xxxx	xxxxxx
M. Mouse	Project Hydrogeologist	$75	xxxx	xxxx	xxxx	xxxx	xxxx	xxxx	xxxx	xxxx	xxxx	xxxx	xxxxxx
D. Duck	Staff Geologist	$65	xxxx	xxxx	xxxx	xxxx	xxxx	xxxx	xxxx	xxxx	xxxx	xxxx	xxxxxx

xxxxxxx

TABLE 1B - MANPOWER LOADING SELEKA FARMS PROPOSAL
ESTIMATED LABOR HOURS BY TASK

PERSONNEL	PROJECT ASSIGNMENT	COST/HR	1	2	3	4	5	6	7	8	9	10	TOTAL HOURS
J. R. Ewing	Principal Hydrogeologist	$150	xx	xx	xx	x	xx	xx	xx	xx	xx	xx	xxx
R. Ewing	Senior Hydrogeologist	$100	xx	xx	xx	xx	xx	xx	xx	xx	x	xx	xxx
P. Rabbit	External Consultant	$130	xx	xx	xx	xx	x	xx	xx	xx	xx	xx	xxx
M. Mouse	Project Hydrogeologist	$75	x	xx	xxx	xxx	xx	xxx	xxx	xxx	xx	xx	xxxx
D. Duck	Staff Geologist	$65	x	xxx	xx	xxx	xx	xx	xxx	xxx	xxx	xxx	xxxx

xxxxx

Figure 12-6 Example of a Proposal Manpower Spreadsheet

Quality assurance for geologic investigations could be defined as follows: the planned and systematic actions necessary to provide adequate confidence that data are valid, have integrity, and are preserved and retrievable. The goal of quality assurance is to help achieve success. Its purpose is not to replace established work practices, but to supplement them as necessary to produce an effective control program.

The quality assurance standard, which the author believes would fit the typical owner's program, is ANSI/ASME NQA-l. Based on NQA-l, nine principles of quality assurance have been identified for geologic / hydrogeologic investigations. The principles include practices that are easily associated with geologic work activities:

Planning and organization of activities; Preparation and control of procedures; Training and qualification of personnel; Control and handling of samples; Acquisition and protection of data; Peer review; Identification and correction of deficiencies; Use and control of records; and Control of purchased items and services.

Most investigations do not require a fully formalized QA program be used on non-RI/FS projects; however, we plan to require the addition of QA elements in future revisions of this manual.

Reports on the completed scope of work are a major subsection element. Details on format, scope, and content are included in the Subsection 3 area. In general, the Author believes flexibility is important in the documentation of many site assessments; however, all field based drilling, logging, and well / piezometer installation must be fully documented to stand the test of time. Sample logging and field forms in addition to well / piezometer completion diagrams are provided in Chapter 7.0 of this text.

Requirements for progress reports are also important as the necessity for multi-task projects become more typical. Documentation often walks a fine line between not

TABLE 1C - REIMBURSABLES - SELEKA FARMS PROPOSAL
ESTIMATED COST BY TASK

PERSONNEL	UNITS	1	2	3	4	5	6	7	8	9	10	TOTAL COSTS DOLLARS
Airfares	Trip	x	-	-	x	-	x	-	x	-	x	xx
Local Transportation	Km	xxxx	xxxx	xxxx	xxxx	xxxx	xxxx	xxxx	xxxx	xxxx	xxxx	xxxxxx
Vehicle Rental	Week	x	x	x	x	x	x	x	x	x	x	xx
Subsistance	Week	x	x	x	x	x	x	x	x	x	x	xx
Equipment	Week	-	x	x	-	x	x	x	x	-	x	xxx
Freight	L.S.	x	x	x	x	x	x	x	x	x	x	xx
Local Labor	Week	x	x	x	x	x	x	x	x	x	x	xx
Field Supplies	Week	-	x	x	-	x	x	x	x	-	x	xx
Computing	L.S.	x	x	x	x	x	x	x	x	x	x	xxx
Phone/Telex	Week	x	x	x	x	x	x	x	x	x	x	xx
Courier	No.	-	x	x	-	x	x	x	x	-	x	xx
Equipment Insurance	L.S.	x	x	x	x	x	x	x	x	x	x	xx
Geophysical Equipment	Week	x	x	x	x	x	x	x	x	x	x	xxx
Geophysical Borehole Logger	Week	-	x	x	-	x	x	x	x	-	x	xx
Drafting	Hrs	x	x	x	x	x	x	x	x	x	x	xx
											TOTAL	**xxxxx**

Figure 12-7 Example of a Proposal Rebillables Spreadsheet

TABLE 12-1 Quality Assurance Principles for Geologic Investigations

A. PLANNING AND ORGANIZATION OF ACTIVITIES
B. PREPARATION AND CONTROL OF PROCEDURES
C. TRAINING AND QUALIFICATION OF PERSONNEL
D. CONTROL AND HANDLING OF SAMPLES
E. ACQUISITION AND PROTECTION OF DATA
F. PEER REVIEW
G. IDENTIFICATION AND CORRECTION OF DEFICIENCIES
H. USE AND CONTROL OF RECORDS
I. CONTROL OF PURCHASED ITEMS AND SERVICES

enough and too much. In practice, field activities rarely are documented to the extent that it becomes a burden.

12.4 QUALITY ASSURANCE FOR SITE ASSESSMENT INVESTIGATIONS

12.4.1 Introduction

Investigations associated with site characterization include a variety of work activities from which data are obtained and analyzed, leading to site selection and permitting of landfills. The quality of data and the validity of the analyses will have a significant influence on the site selection and permitting processes. Obtaining acceptable quality will depend, of course, on how well those work activities are performed. The application of appropriate principles of quality assurance will assist in achieving adequate performance, particularly if those principles are effectively integrated into the work activities.

Nine principles of quality assurance for hydrogeologic investigations have been selected (see Table 12-1). Application requires a knowledge of the activity's objective(s), work tasks, constraints, potential sources of loss and failure, and the consequences of loss and failure. Also required is an understanding of the principles and how they are used.

This process has four steps. The first is to define the technical and programmatic objectives and requirements of the work activity. The second is to identify within the work activity sources of risk and then to evaluate those risks in terms of probability and consequences. The third step is to select the principles of quality assurance that will provide the control needed for successful completion of the activity. These steps should be carried out early so that

quality assurance can be more effectively integrated into the work. The fourth step is implementing the selected principles. Hydrogeological studies involve the following types of activities:

• Gathering and evaluating historical data and information;

• Making regional and site geologic, hydrogeologic, and geophysical surveys;

• Collecting samples through in-situ geologic sampling;

• Gathering and evaluating test and measurement data through laboratory tests and field analyses;

• Performing field experiments and demonstrations; and

• Processing and analyzing geologic data.

Much of the data and information obtained from these activities ultimately becomes the technical bases upon which site selections will be made. The quality and uniformity of that data are extremely important.

All scientific disciplines have established practices that provide control over various sources of loss and failure that can occur when carrying out experiments and studies. Such practices help to assure the scientists that scientific activities are properly planned and executed so that useful (quality) data are obtained. Comparison of practices used by scientists with quality assurance practices often shows that both types are clearly related. Thus, quality assurance is used by scientists to some degree whether or not they are consciously doing so.

Quality assurance provides a planned and systematic approach for developing and applying actions necessary to assure the quality of output from work activities. It helps to avoid omissions and oversight that could adversely affect quality. A generally accepted definition of quality assurance is as follows: The planned and systematic actions necessary to provide adequate confidence that a material, component, system, or facility will perform satisfactorily in service. Based on this definition, quality assurance for geologic investigations could be defined as follows: The planned and systematic actions necessary to provide adequate confidence that data are valid, have integrity, and are preserved and retrievable.

The goal of quality assurance is to help achieve success. Its purpose is not to replace established work practices, but to supplement them as necessary to produce an effective control program. The sources for identifying the planned and systematic actions needed for our program are quality assurance standards. They provide the requirements established or criteria for effectively carrying out such functions associated with work activities as:

1. Organizing roles and responsibilities of people;

2. Preparing and qualifying procedures;

3. Training and qualifying personnel;

4. Obtaining and preserving information (data);

5. Assessing and improving performance; and

6. Purchasing items and services.

Standards have been developed for nuclear energy and aerospace programs and, although names and arrangements of the standards vary, they all use the same basic logic for controlling work activities.

The quality assurance standard which would fit the typical hydrogeologic program is ANSI/ASME NQA-1. Based on NQA-1, nine principles of quality assurance were identified for geologic/hydrogeologic investigations. The principles are outlined below and they include practices that are easily associated with geologic work activities. Table 12-1 is a list of the principles. Quality Assurance programs developed for hydrogeologic investigations and based on these nine principles should provide a program that is in compliance with NQA-1. The following section gives more detail on each of the nine components given in Table 12-1.

Implementation of an appropriate quality program will provide not only a more reliable gathering of site data and, hence, reduced corporate risk, but also can be an effective marketing and permitting tool. Since these sites can be involved in litigation many years into the future, a quality program with retrievable documentation is the most cost-effective solution for field based studies.

12.4.2 Planning and Organization of Activities

Planning and organizing work activities is identified as a principle because it is essential that there be a conscious and deliberate planning and organizing effort for each important work activity. Although this principle emphasizes a function that usually precedes the actual conduct of work, there is need to maintain a continuing planning effort until completion of a work activity. This need, unfortunately, can be too easily de-emphasized or overlooked, particularly during the work performance stages where cost and schedule priorities may dominate.

The primary purpose of planning and organizing is the clear and specific identification of work requirements and objectives. Planning and organizing should clearly identify objectives, participants, organizational interfaces, responsibilities, restraints or limitations, and the expected result of the work.

12.4.3 Preparation and Control of Procedures

Work activities are carried out in a planned, systematic, and controlled manner so that the products or end results will conform to expected outcomes. The process used to produce such an outcome often involves discrete actions taken in a specific order. Any change in an action or in the order without a valid reason most likely will result in an unsatisfactory outcome. To control the processes and avoid errors leading to unsatisfactory results, procedures are written that provide guides for those doing the work. The significance of having a procedure for a specific activity depends on several factors such as the importance of the results to the overall success of the activity or larger project/program, the degree to which an error has an adverse effect on the end result, the state of the training and knowledge of those doing the work, the need to document the process used, and the need to substantiate the technical basis of the process used. To be effective and to help provide credibility to the activity being performed, procedures should be well-written, complete, and correct.

12.4.4 Training and Qualification of Personnel

An important factor affecting all work activities is the training and qualification of those doing the work. There are very few, if any, organizations engaged in work activities that do not provide some type of training. Training can vary from direct, on-the-job training by a more experienced worker to a formal program involving both classroom and on-the-job training. The extent of training required depends on the complexity of the work, education and previous experience requirements, the economics involved, and the overall importance of the work to meeting the goals of the organization. Closely associated with training is the concept of qualifying people for a job before beginning work.

Qualification includes not only specific training, but also the review and verification of applicable education and experience. Using adequately trained and qualified people should be a requirement for all geologic investigations.

12.4.5 Control and Handling of Samples

A source of data important to site hydrogeologic characterization is the analysis and testing of geologic materials associated with a landfill site. The reliability of the physical and chemical characteristics determined for each type of material depends on the validity of the processes used to analyze those materials and to make experimental tests. As important, or even more important, is the integrity of the samples used. Invalid results will be obtained from samples that do not truly represent the materials from which they were taken. Loss of sample integrity can occur from inadequate sampling procedures and from improper control and handling practices once the samples have been taken.

12.4.6 Acquisition and Protection of Data

The practices used to assure proper acquisition and protection of data are of importance to essentially all facets of geologic investigations. Data are interpreted to include any site-related measurements or recordable observations acquired as a part of geophysical, geochemical, or hydrological studies, plus the results of any associated laboratory analyses. In the broadest sense, data include any information generated for use in the technical assessment of site-related evaluations or experiments. The success of site explorations, experiments, or any of the R&D oriented site studies required to determine site landfill suitability depends on obtaining data that are applicable, sufficiently accurate, readily identifiable, retrievable, and suitably preserved.

12.4.7 Peer Review

A way to validate practices is through independent peer review. Peer reviews can help validate technical adequacy and, perhaps more importantly, can validate the application of established practices. For those situations when practices are new or beyond the state-of-the-art, independent peer review is essential.

12.4.8 Identification and Correction of Deficiencies

Deficiencies, as discussed here, include failures, defects, errors, deviations from specified requirements, and other conditions considered adverse to quality. Uniform and well-defined practices are required to assure proper control and disposition of deficiencies. These practices should emphasize the timely or prompt identification and correction of all deficiencies or conditions that might adversely affect the quality of continuing geologic investigations.

12.4.9 Use and Control of Records

The use and control of records are key in providing documentary evidence of technical adequacy and quality for all hydrogeologic investigations.

Records provide the direct evidence and support for the necessary technical interpretations, judgments, and decisions for site selection. Records preparation and use must be an integral part of ongoing work activities. These records must directly support current or ongoing technical studies and activities and must provide the historical evidence needed for later reviews or analyses, particularly those which might be anticipated as a part of the landfill permitting.

12.4.10 Control of Purchased Items and Services

When items and services are required that are not available on-site, purchase of those items and services is usually necessary, which requires that procedures be established and followed to obtain the necessary quality. This will require that procurement actions follow uniform written procurement instructions which provide a systematic approach to the procurement process. It is essential that technical and quality requirements be properly evaluated and control practices be established according to quality needs. Evaluation should consider various factors, including cost and schedule effects, failure consequences, methods of acceptance, programmatic importance, and applicable codes or standards.

12.5 REPORTS

Report writing is a critical portion of any profession, and this is particularly true in hydrogeology where word communication often is the only way a reader or client can "see" the problem.

1. Outline: First, make a thorough outline of your report—to third and fourth order headings. Rework the outline and make certain that a logical sequence is followed and that one entry tends to lead into another.

2. In writing a hydrogeological report, you should have three books before you:

   a. E.B. White's "A Style Manual";

   b. Suggestions to Authors published by U.S. Geological Survey; and

   c. A recent issue of a Geological Society of America Bulletin, Ground Water, or the Journal of Sedimentary Petrology. Almost all questions in form that arise can be answered by examining current articles, and the Geological Society of America Bulletins and Journal of Sedimentary Petrology are doubtless the best.

Most clients expect both progress and final reports will be submitted during the course of an investigation.

12.5.1 Progress Reports

Time and distance normally separate consultants from those who will make use of their findings. An accounting of progress, whether submitted as a bound report or as a letter, helps to keep the owner in touch with the work being done.

The main and obvious function of any progress report is to give the company an accounting of the work that has been done. It explains how the consultant has spent billable hours and owner/operator's money and what has been accomplished as a result of the investment. Though this purpose is dominant, do not lose sight of four other purposes that are discharged by a progress letter or report:

1. It enables the owner/operator to check on progress, direction of development, emphasis of the investigation, and general conduct of the research. Thus, owner/operator can alter the course of the work before too much time and money have been invested.

2. It enables the consultant to estimate work done and work remaining with respect to the total time and effort available.

3. It compels consultants to shape their material and focus their attention.

4. It provides a sample report that helps both the owner and the consultants to decide upon the tone, content, and plan of the final report.

Work is done and progress is made with the passage of time. This fact gives us a useful clue to the basic nature of

---

Example Progress Report Format

First Progress Report

(l) Introduction
(2) Project Description
(3) Work done in the period
     just closing
(4) Work planned for the next
     work period
(5) Work planned for periods
     thereafter
(6) Overall appraisal of
     progress to date
(7) Overall appraisal of progress
     to date

Second Progress Report

(l) Introduction
(2) Project Description
(3) Summary of work done in
     the preceding period(s)
(4) Work done in the period just
     closing
(5) Work planned for the next
     work period
(6) Work planned for periods
     thereafter

Figure 12-8 Example of a Progress Report Format

every progress report, whether or not the particular format makes time the dominant element. However, the progress-with-time philosophy is clearly evident in the following scheme of reporting progress.

The opening parts, Introduction and Project Description, seldom vary from one progress report to another in a series dealing with the same project. In fact, these two parts of a succeeding progress report are often

---

PROJECT ACTIVITY AND FINANCIAL STATUS REPORT
CANYON LANDFILL PROJECT

Phase IIA - Site Feasibility Studies - Addendum 102. 102A (2-WE, 2A-WE)
Project Activity And Financial Status- Report (of 8/29/89)

Activity Description	Percent Complete	Status Budget	Financial Status Expenditure	Over/Under Amount
Task 1: Geotechnical Investigation				
IA: Field Investigation	100	33,381	36,291	-2970
IB: Laboratory Analyses	75	20,553	18.084	2474
IC: Engineering Analyse~	50	9,080	7,979	1101
	Total	62,959	62.354	605
Task 2: Geologic Investigation				
2A: Data Collection And Review	100	2,551	3,367	816
2E: Photogeologic Interpretation	100	1,152	2,220	1068
2C: Geologic Mapping	100	8,664	7,946	718
2D: Backhoe Pit Excavation and Inspection	100	35,837	33,935	1902
2E: Seismic Analysis	100	3,180	3,326	146
2F: Data Analysis And Report Preparation	95	18,272	29,593	11,321
2G: Red Mountain Fault Trenching	100	30,158	30,676	-518
	Total	99,814	111,063	11,249
Task 3: Hydrogeologic Investigation				
3A: Alluvial Aquifer Wells	100	28,643	28,331	312
3B: Pico Formation Wells	30	83,164	7.878	75,286
3C: G.W. And S.W. Sampling	80	7,810	2.221	5.589
3D: Percolation Tests	100	5,146	4.376	770
3E: Aquifer Tests	50	8,770	6,188	2,582
3F: Laboratory Analysis	50	12,005	1.707	10.298
3G: Data Analysis And Report Preparation	30	25,611	9,980	15.631
	Total	171,149	60,681	110.468
Task 4: Other Investigations				
4A: CADD Topography	100	2,384	4.992-	2.608
4B: Winter Biological Survey	100	1.358	1.644	286
	Total	3,742	6,636	2,894
	Total	337,664	240,734	96,930

Figure 12-9 Example of a Financial Status Spreadsheet

created by correcting (updating) these parts of the preceding progress report.

If changes have been made in the contractual agreement, the Project Description (also called Work Statement or Contractual Requirements) has to be updated to accord with the most recent agreement. This description spells out what the researcher is required to do and produce. That is, it represents the total job he is prepared to undertake.

The "work done" sections are of chief interest in most progress reports, for these sections describe, for better or worse, what has been accomplished during the work period (or periods) just closing. Progress, schedule, and financial status documents are important components to progress reports. Figures 12-8 and 12-9 are easy to prepare and provide documentation that the consultant is "in control" of the costs of the project.

For our purposes here, we can regard every progress report as having three main ingredients in the body: (l) the consumption or passage of time and the order of events throughout that time; (2) the allocation of effort from the task point of view; and (3) the activity or problem point of view.

12.5.2 Requirements for Independent Technical Reviews

For written reports or recommendations produced as final contractual deliverables, a detailed technical review will be performed by a qualified reviewer who is independent from the preparation of the document. At a minimum, independent technical reviews shall be documented by a standard review memo. Guidelines for using the memo are as follows:

• The review draft is complete when presented for review. Incorporation of informal comments sought when possible during the development of the review draft, but no informal comments are accepted after the draft is submitted for review. All calculations are documented and checked prior to final review; where standard engineering practices are involved, checks are performed to the full extent required. Where appropriate, calculation sheets include a brief description of the scope of the checks made and the methods used. Calculation sheets or drawings are signed and dated by the preparer and checker, and are attached to the review form when presented to the reviewer.

• The reviewer ensures that all calculations have been adequately checked to the extent appropriate for the

standard practices that may be involved, and as appropriate for the type and purpose of the calculations. The reviewer signs and date the calculation sheets or drawings if acceptable.

• Review comments may be mandatory or non-mandatory. All mandatory comments shall be numbered in the reviewed text; the review memo should be marked appropriately whenever mandatory comments have been made. Unnumbered comments shall be considered non-mandatory and may be incorporated at the discretion of the author. All mandatory comments must be documented and resolved.

• The review form is filed along with the as-reviewed draft, completed calculation check-sheets, and the final approved text.

Independent technical review requirements do not apply categorically to interim recommendations or other types of technical communications that are based on work in process, provided that any associated uncertainties are addressed by appropriate disclaimers. However, independent technical reviews of these and other types of documents may be requested at the Project Manager's discretion at any time.

12.5.3 Control of Documents

Records preparation and control must be an integral part of ongoing work activities. The following items provide a series of steps for the use of records associated with site investigations. These points are important for the management of projects and the control of the completed data and reports. One should establish documentation and management actions to achieve the following:

1. Procedures for recording pre-engagement activities and communications.

    a. Require that working files adequately record client-furnished information and other data on which the project proposal was based.

    b. Identify guidelines and responsibility for project acceptance and for proposal review, as well as, procedures for recording such approvals.

2. Provide procedures for recording the performance and review of project activities.

a. Require that project files provide a reasonably complete chronological record of project activity, including planning documents, telephone calls, letters conferences, calculations, field and laboratory data, and progress reports.

b. Require that all information originating within the firm be identified as to job, date and name of persons originating each work item.

c. Develop guidelines for review process.

3. Provide procedures for retiring, storing and/or purging project files.

a. Assign responsibility for closing an active file.

b. Identify appropriate form and content for closed files.

c. Provide for duplicate storage or other method of safekeeping of essential documents.

d. Develop guidelines for terminating or retiring files.

12.5.4 Final Report Content - Subsurface Investigation

Introduction

Detailed subsurface investigations must be of sufficient intensity to determine the conditions that may influence the design and the construction of the monitoring system. The extent of geologic investigation required for a particular site depends on: (1) complexity of the site conditions, (2) size of the landfill construction, and (3) potential damage if there is functional failure in the liner.

Detailed exploration and location of the monitoring system typically consists of three components: (1) determining and interpreting subsurface conditions, (2) taking samples for soil and rock tests, and (3) installing the monitoring wells.

During the first work component, test holes must be put down and logged in the foundation and borrow areas. These test holes must be deep enough to penetrate all pertinent materials. The number and spacing of test holes must be adequate for correlation in both longitudinal and transverse directions for complete interpretation of any condition that may influence the local permeabilities. Geologic structural features, such as faults, folds, and joints, should be documented and information must be obtained on soils to classify them and to determine their location, thickness, and extent. Test holes can be put down by drilling or by excavating pits or trenches for areas of shallow soils.

In the second work component of the detailed site investigation, the data gathered in the first work component are analyzed on the site and behavior characteristics and engineering significance of the materials and conditions are evaluated. From this analysis and evaluation, the geologist and the engineer determine what materials are to be sampled and what laboratory analyses are needed. This determines the kind, number, and size of samples needed. The necessary samples are obtained by using appropriate sampling procedures. Any additional or special in-place field tests should be made.

In practice, these work components are not usually consecutive. Where the geologist is familiar with the area and knows generally the materials and conditions that will be found, sampling and field testing proceed concurrently with the drilling program.

The third component, location of the monitoring wells, would proceed from the knowledge gained from the first two work components. The Federal RCRA requirement of three downgradient and one upgradient monitoring wells are the minimum acceptable for sanitary landfills. In practice, the owner/operator often has numerous piezometers and monitoring wells in excess of the minimums. Actual numbers of monitoring wells can be expected to range from the required three to in excess of a dozen(s) of wells.

Location of Monitoring Wells for Detection Monitoring

The actual location and depth of wells for detection monitoring must be based on two major factors:

1. The direction of groundwater flow; and
2. The first or "uppermost" aquifer depth.

The wells to be installed under this phase two detection monitoring work must be designed to intercept potential landfill leachate down- gradient of the waste area in the first continuous saturated and permeable zone adjacent to the waste disposal area. All the technical work to be performed in this phase two effort must be directed toward this goal. The final performance assessment of the consultant and peer review will be directed toward determining if the consultant has met this goal.

Field Investigation Report Content

The field based investigation is aimed at several goals:

- Documenting field conditions found at the site; and
- Documenting the location and acceptability of the monitoring wells.

These goals may differ for investigations directed toward expansions or permitting new sites and those required within RI/FS projects. However, reporting of technical observations in final reports have many similar components as given in the following sections.

Reports on groundwater also differ greatly in their subject matter and objectives, and they have a highly diversified group of readers. Although no general outline can be given that will be suitable for all, a report, which gives a systematic description of the ground water conditions in a specified area, is the most typical and should comprise the following parts:

Introduction:--Introductions to site assessment reports seldom have subheadings. However, analysis of these reports reveals that the contents of the introduction could be sorted and placed under these headings:

- Reasons for the investigation;
- Some historical background;
- Purpose and objectives of the investigation;
- Scope and limitations of the work; and
- Plan of development (if unconventional).

In other words, the introduction provides the reader with a backdrop against which the new data is to be seen and appraised.

Setting:--"Setting" should include regional and local description of geographic matters, such as climate, rainfall, flooding, runoff and drainage, attitude, land use and agricultural soils, surface geology, topography, and other factors.

A description of the general characteristics of the site is also an important step. Because surface attributes of the facility will directly and indirectly affect the subsurface environment, these attributes need to be identified. Information is needed on the climactic factors of precipitation, temperature, and evapotranspiration. Locational factors for which information is needed include topography, accessibility, site size, proximity to surface water, and proximity to population centers. Additional comments on each of these factors are as follows:

1. Precipitation: Precipitation, in most cases, will dictate both the amount and rate at which leachate from a site moves into the groundwater. This is especially important for solid waste disposal areas. Precipitation also affects the recharge rate of an aquifer.

2. Temperature: Surface temperature will become an important factor in determining the feasibility of certain surface treatment strategies.

3. Evapotranspiration: The amount of water lost to the atmosphere through transpiration and evaporation can be important when considering the ultimate use of a site. For example, surface capping of a landfill relies in part on evapotranspiration from the vegetative cover to reduce the amount of water infiltrating to the solid waste.

4. Topography: The general topography of the site will affect the infiltration rate and the feasible solutions. Areas of steep gradients will have little infiltration and will be subject to high erosion rates.

5. Accessibility: Related to topography is the accessibility of the site. Areas of rugged terrain or limited access will present not only construction problems but could also hamper any subsequent operation and maintenance activities.

6. Site Size: The size of the site refers to the actual surface areal extent. This factor will affect the fill design solutions in that many of the technologies are size specific.

7. Proximity to Groundwater Users: All potential users of groundwater adjacent to the site should be identified in the site assessment study.

Facility Description:--"Facility Description" should locate the site on a regional map (first figure), describe the site features, land use, land surface elevations, local drainage patterns, and slopes.

Existing Monitoring Well Locations and Procedures:--Identification of existing monitoring well locations and the parameters monitored can save both study time and costs. In addition to providing immediate data, existing monitoring wells can, in some cases, become permanent parts of a monitoring network or converted to piezometers.

Regional Hydrogeology and Ground-water Quality:--"Regional Hydrogeology and Groundwater Quality" should include a discussion of the major aquifers, their usage, stratigraphy, and groundwater conditions—gradient directions and magnitudes, and elevation of the ground-water surface. An indication of regional ground-water quality and related issues should be included.

This section should include a systematic description of each of the successive geologic formations in the area, including its water-bearing properties and the quality of its water. These should also be either detailed descriptions of subareas, such as counties, townships, or geomorphic units; or a detailed discussion of the hydrology of the area as a whole, with emphasis on recharge, discharge, and potential availability of groundwater.

Results: -- Site Hydrogeology and Ground-water Quality:--"Site Hydrogeology and Ground-water Quality" should provide locations of borings and piezometers (second figure), and include discussions of the site stratigraphy, subsurface profiles (additional figures), and shallowest water-bearing soils. Shallow groundwater conditions—gradient directions and magnitude, elevation of water table or piezometric surface—should be covered in detail. This section should also include a description of local groundwater quality, with significant anomalies and frequencies, and seasonal fluctuations noted. Some areas of needed information and reasons for their interest are as follows:

1. *Geologic Setting and Generalized Soil Profiles:* Determination of the types of soils is important for determining the capacity of the leachate to move through the subsurface. Certain soils will possess higher tendencies to attenuate the leachate(s) (through adsorption, precipitation, filtration, etc.) than others. Likewise, certain soils will be amenable to certain leachate collection strategies while others are not.

2. *Soil Physical/Chemical Characteristics*: Once a general soil type has been identified, it is then necessary to characterize this type both physically and chemically. Physical characterization of the soil type will provide information on the ability of the soil to filter landfill leachate. The physical characterization will also give an idea as to the "workability" of the soil for different containment designs. Chemical characterization will provide information on the ability of the soil to chemically remove a given chemical through adsorption, precipitation, etc.

3. *Depth to Ground Water and Bedrock*: The depth to ground water in conjunction with the soil physical/chemical characterization will give insight as to how long generated leachate will take to actually reach the aquifer, if in fact it will. If sufficient depth to the ground water exists in a highly attenuating soil, minimal ground water impacts can be expected. The depth to bedrock is needed to assess the feasibility of some leachate containment strategies.

4. *Ground-water Flow Patterns and Volumes*: The flow patterns and volume of groundwater threatened will play a vital role in determining the feasible solutions to the problem.

5. *Recharge Areas and Rates*: Identifying recharge areas and rates will play an important role in aquifer protection plans. Information on recharge areas and rates for the case of an existing site will be important for a number of reasons. First, it should be determined whether or not the source(s) is in a recharge area. Solutions to problems located in recharge areas will most likely be more elaborate that those not located in recharge areas, and should include leachate removal, if possible. Second, recharge rates will give insight into the rate of leachate movement and leachate dilution.

6. *Aquifer Characteristics*: Identification of aquifer characteristics will be essential for any analysis of groundwater flow and chemical transport. This information becomes extremely important if ground-water modeling studies are to be initiated.

7. *Background Water Quality Data*: Background water quality data is important in determining the severity of the problem and the appropriate remedial actions.

Interpretation or Discussion:--In the Interpretation or Discussion section, the consultant reviews and analyzes the data presented in the preceding section. The discussion is one of the most important sections of physical research reports—and, unfortunately, often one of the most poorly handled. The consultant should, therefore, take special pains with this section. He must try to step back from the work and findings and reflect upon them before putting the whole story together.

In site assessment reports, the discussion sometimes does little more than translate numbers into words. Justification for this practice may lie in the fact that all report users are not equally capable of digesting data presented in tabular or graphic form. At other times, and more commonly, the discussion must do far more than provide a prose recapitulation. Are there blanks in the data? Were some of the data "projected" from actual test data? Were the data subjected to mathematical processing? Do the data contain puzzling irregularities? Were all the data taken with the same equipment? Do the data, when considered in their entirety, lead to a significant conclusion? Were there recognized inaccuracies in measuring and recording?

Points such as these must be dealt with in the discussion. Analysis and interpretation are frequently needed to qualify and illuminate "raw" data.

"Assessment," in any case, should combine the hydraulic analysis of the site hydrogeology and the analysis of the site's soils and geology. A reasonable picture of the potential for contaminant migration from the waste facility from seepage and horizontal or vertical movement through and off the site should be produced. Imminent and potential ground-water quality risks to the public must be identified.

An important part of the interpretation is if the observed geologic/hydrogeologic features match the conceptual model generated during Phase I studies. If not, why not; and how should the conceptual model be changed to fit the observed geology, hydraulic heads, aquifer characteristics, and water quality. This section is where the consultant can provide real insight into the monitoring well system design.

Summary, Conclusions, and Recommendations:-- In the Summary section, you extract and extrapolate from the preceding sections of the report what you consider the outcome and final worth of the investigation. You may comment on the reliability and validity of the results, and you may suggest new approaches to the problem for later use.

"Recommendations" should refine the quality of the hydraulic and water-quality analysis and evaluate the synthesis of hydrogeologic and geologic analysis, and work towards statistical significance, if needed. Recommendations should be presented in a natural and supportive way so that the client is logically drawn toward them. Typically, the consultant must answer requirements of the state issuing the permit. These requirements often are as follows:

1. Determine the availability, quality, and quantity of on-site soil for cover material;

2. Evaluate the influence that geologic factors, such as ledge, would have on the ease of excavation and

potential for groundwater and surface water pollution;

3. Determine the maximum high ground-water surface elevation and ground-water flow patterns;

4. Determine background water quality of ground water at the site and of surface water at the most likely discharge area; and

5. Evaluate the importance of the groundwater resource which might be affected by the operation of a landfill facility.

12.6 DOCUMENTATION

A. All materials in the report should be relevant to the purpose of the report.

B. All statements should be documented by references or by accurate field observations.

C. Areal photos (originals or suitable copies) should be included to document any discussion on landslides and faults.

D. The method(s) of field analysis should be discussed in a lucid manner.

E. All mapping should be done on a base with satisfactory horizontal and vertical control—in general, a detailed topographic map. The nature and source of the base map should be specifically indicated. For operating landfills, the base map should be the same as that to be used for the tentative map or grading plan.

F. Mapping by the geologist should reflect careful attention to the lithogy, structural elements, and three-dimensional distribution of the earth materials exposed or inferred within the area. In most hillside areas, these materials will include both bedrock and surficial deposits. A clear distinction should be made between observed and inferred features and relationships.

G. A detailed large-scale map normally will be required for a report on a small expansion, as well as for a report on a smaller area in which the geologic relationships are not simple.

H. Where three-dimensional relationships are significant but cannot be described satisfactorily in words alone, the report should be accompanied by one or more appropriately positioned structure sections.

The locations of test holes and other specific sources of subsurface information should be indicated in the text of the report or, better, on the map and any sections that are submitted with the report.

12.6.1 Captions, Titles, and Headings

We use the term captions to refer to the various headings and titles that may be used to display the report's coordination and subordination to the reader. The caption itself is a phrase which describes what is discussed in the paragraph or paragraphs that follow it. Coordination and subordination are shown through captions by a consistent use of various typefaces and positions for different level captions.

There are several systems used for headings in a geologic report. The main or first heading is centered and is in capital letters; the second heading is centered, first letter capitalized and remainder are in lower case; third heading is the same as the second heading only it is underlined; fourth order headings are indented three spaces from the left margin, followed by a colon and the text follows the colon after a double space. This heading is also underlined. Example:

INTRODUCTION

Climate

Rainfall

Distribution: Rainfall is principally in the winter months, but on rare occasions...

Quite often, in order to facilitate reference back and forth in a report or book, a numbering system is combined with the captions. The three systems in most common use are the traditional outline system, the century-decade-unit system (often called the Navy System), and the multiple-decimal system. The schematics traditional and multiple-decimal systems that follow illustrate the systems acceptable to most clients, are shown on Figure 12-10.

12.6.2 Figures and Plates

There are several views on what the difference is between a figure and a plate. The authors' view is that all items in the text are considered to be Figures and those placed in a jacket at the end of the report, Plates.

Every figure or plate must be accompanied by a complete title and description beneath the figure. A somewhat abbreviated or shortened title is used in the list of illustrations. Don't simply copy the caption beneath each of your plates and figures into the Illustrations compilation. Shorten them. All figures should read right-handed (i.e., the bottom of the figure is toward the reader or on the right-hand side of the page). North should almost always be at the top of the figure.

A good report makes constant reference to the figures and plates in the body of the text and an illustration should never be included that is not referred to in the written text. Where the illustrations are included in the body of the text, they should appear on the page following the page where they are first mentioned. If two (or four or five) are mentioned on a single page, they should appear in proper sequence following that page.

Citing Figures and Plates in Your Text: If reference to a figure or plate is made parenthetically (Fig. 6), then the abbreviations Fig., Pl., or Figs. 6 and 7 or Pls. 8 and 9 are used. If the writer refers to a Figure or Plate and it is expressed as part of the sentence, then the word is spelled out. For example: "On Plate 7 the dendritic pattern is evident and on Figure 3 the location is shown." In all instances, capital "F" or ,,p ,,

Arrangement of this page should be much like the Contents page. A brief sample is outlined below, and NOTE that only first letter of first word and of proper nouns are capitalized:

"Figures" should include a vicinity map, site plan with boring and monitoring well locations, two subsurface profiles (N-S and E-W), and a local geohydrologic column at a minimum. The site plan should show the location of the subsurface profiles, and the direction and magnitude of the gradient. If ground-water anomalies are clearly present, additional figures may highlight this through anomaly maps which overextend the size of anomalies because of sparse and distant control points.

12.6.3 References Cited

This section may also be entitled "List of References" or "References." The term "Bibliography" is customarily used when references other than those actually cited in the

Traditional outline system:

TITLE CAPTION

   I. FIRST-LEVEL CAPTION
      A. Second-Level Caption
         1. Third-level caption
         2. Third-level caption
      B. Second-Level Caption
  II. FIRST-LEVEL CAPTION
      A. Second-Level Caption
         1. Third-level caption
         2. Third-level caption
      B. Second-Level Caption

Multiple-decimal system:

TITLE CAPTION

  1. FIRST-LEVEL CAPTION
      1.1 Second-Level Caption
         1.1.1 Third-level caption
         1.1.2 Third-level caption
      1.2 Second-Level Caption
  2. FIRST-LEVEL CAPTION
      2.1 Second-Level Caption
         2.1.1 Third-level caption
         2.1.2 Third-level caption
      2.2 Second-Level Caption

When all plates are listed, then start, same page, with figures, then tables.

Figure 12-10 Report Table of Contents Alternatives

report are listed for the reader. Generally, only those references cited in the text are included in this section. The purpose of the bibliography references is to show the reader the source of the reference and to enable him to locate supplementary material. Also, citing a reference in the text enables the reader to recognize that the work is other than your own.

"References" should include all cited literature, the consultant's and client's files, conversations with agency personnel, and client reports. Confidential reports completed for other clients with nearby facilities should not be explicitly cited.

Several different systems of recording bibliographic references are in use, but the most commonly accepted one is the form of the Geological Society of America. The form followed by the G.S.A. is:

1. Name of author, last name first and then initials.

2. Year of publication, set off with a comma.

3. Title of paper, in full. First letter of first word and only geographical and formation names have first letter capitalized. At end of title, a colon is used.

4. Series of publication, or publisher, with standard abbreviations. (Standard abbreviations are listed in "Suggestions to Authors.")

5. Volume and page reference.

The items are listed alphabetically by authors. Where more than one paper by the same author is cited, each is listed chronologically under his name. On the second listing, the name is not rewritten but is indicated by a solid line. Below are some typical examples of a reference listing:

Daly, R.A., 1933, Igneous rocks and the depths of the earth: McGraw Hill, New York, 598 pp.

Donnelly, M., 1934, Geology and mineral resources of the Julian district, San Diego Co., California: Calif. Jour. Mines Geol., v. 30, p. 331-370.

King, W.B.R., 1951, The influence of geology on military operations in northwest Europe: Adv. Sci., v. 8, n. 30, p. 131-137.

McIntyre, D.B., 1951, Lineation in Highland schists: Geol. Mag., v . 88, p . 150- 15 1 .

1954, The Moine Thrust, its discovery, age, and tectonic significance : Geol . Associ . Proc ., v. 65 , p. 203-233 .

Scheidegger, A.E., 1953, Examination of the physics of theories of orogenesis: Geol. Soc. America Bull., v. 65, p. 127-150.

Alternatively the title name may have the first letter of each word capitalized (called the "Title Case"). References are used in the text by enclosing the author's name, the date of publication, and, where necessary, the page numbers in parentheses. NEVER USE THE FOOTNOTE SYSTEM OF REFERENCING. Below are several examples of how references may appear in a text.

1. The rocks fit into the eugeosynclinal category, as defined by Stille (1941) and Kay (1951).

2. Flint (1953) referred to casehardening of friable limestone to explain the origin of certain limestone ridges in Okinawa.

3 . The Military Geology unit of the U. S . Geological Survey (Hunt, 1950) made a substantial contribution during World War II.

4. With regard to this important point, Allen C. Tester (Personal Communication) states: "the geological significance, etc..."

5 . ... such crystallization is shown by the orbicular and nodular structure (Miller, 1938, p . 1224 ; Lawson, 1904) .

6. ... so that they were not covered by a thick volcanic cover (Lindgren , 19 1 1, p . 37-39)

7. The name applied by Miller (1937) will be used.

"Appendices" should include any available field exploration and soil laboratory testing, groundwater monitoring network, groundwater sampling and analysis plans, and analyses of groundwater samples. Specific tables, such as logs of borings, summary of groundwater level measurements, well completion data, soil laboratory tests, field water-quality analyses, laboratory water-quality analyses, and chain of custody records, are appropriate. Personnel involved and procedures utilized in the various phases of the project should be identified. Examples include notations such as "water levels were measured with a steel tape by...;" "field measurements were made with a

COMMON MISTAKES

RESEARCH INTO PROFESSIONAL LIABILITY CLAIMS HAS SHOWN THE FOLLOWING TO BE
COMMON MISTAKES WHICH CAUSE A MAJORITY OF CLAIMS.

A. WORK BEGINS WITHOUT A WRITTEN CONTRACT

- This could lead to exaggerated claims of what you agreed to do.

- Without a written contract, it may be too late to negotiate contractual terms which minimize your risk.

- If you must proceed on verbal authorization, it is extremely important to confirm the scope of work in writing, and other contract terms where possible.

B. COST CUTTING CLIENTS/FINANCIAL RESPONSIBILITY

- Experience shows there is a high correlation between the financial irresponsibility of clients or subcontractors, and the frequency of professional liability claims.

- When cost cutting takes place, be sure you DOCUMENT what the consequences may be to the cost cutting.

- When contracting with a subsidiary, make the parent a party to the contract or require the parent company to guarantee the subsidiary's obligation.

C. LACK OF RECOMMENDATIONS / POSSIBLE CONSEQUENCES

- Experience in the defense of engineers in negligence suits demonstrates that well drafted, well kept, complete records are the strongest defense weapons.

- Put in writing any recommendation you make, particularly a matter of serious consequence and obtain, if possible, a written response.

- Confirm in writing decisions reached in the course of conversation and conferences.

- Remember, some day, your writing may be seen by judge and jury. The technical language used should leave NO room for misinterpretation by any member of the same profession called upon to explain what you meant to a layman.

- Write nothing which can impose responsibility on you for the work of another.

Source: Dames and Moore Professional Liability Seminar, 1985

D. FACT vs. OPINION

- If a conclusion is stated as a fact, and it turns out to be wrong (no matter how carefully you arrived at it), you may be liable merely because you stated it as a fact.

- Avoid stating facts which are actually your professional opinion.

E. PREDICTABLY TROUBLESOME PROJECTS (EXAMPLE - RESIDENTIAL)

- The potential of lawsuits is greater because there are many more potential plaintiffs.

- The income on these projects is usually small compared to the potential liability.

- The general contractors for residential projects are more often cost cutters than commercial contractors.

F. FAILURE TO USE MODEL CONTRACTS

- While it is not always possible to use the model contract in full, its provisions should always be referred to. Your company's Schedule of Charges and General Conditions should always be incorporated.

- A client may request the limitation of liability clause be removed because the "risk is not there." Try to explain that if the risk is not there, then there is no harm in leaving the clause in.

G. GOOD SAMARITAN OBSERVATIONS

- Should you volunteer an observation, which is outside your scope (and for which you are not being paid), you become liable if that observation is used and it turns out to be wrong.

H. "I'M INSURED/IT WON'T HAPPEN TO ME"

- The purpose of insurance is to provide coverage for CATASTROPHIC losses.

- Most consultant insurance programs have several layers, of which the first $100,000 of any one loss is paid by the consultant. The remaining may be paid by insurance, but the insurance companies may either drop your firm or make the insurance much more expensive.

Figure 12-11 Common Mistakes of Professional Liability Claims

conductivity meter, pH meter, and thermometer by...;" etc. Data should be reported to significant figures. Basic data collected during the investigation must be provided so that client can perform an independent assessment of the information generated by the investigation.

The draft and final reports should be thoroughly reviewed for technical accuracy and editorial consistency. Reports should be checked for numerical, citation, and typographical errors; as well as tone, consistency, emphasis, and logic. A poorly-written report will erode credibility, diminish future opportunities, and cause later expense and embarrassment.

12.6.4 Report Presentation

The client should be given sufficient copies of the draft final report to review. The owner/operator may correct any facility-related or procedural errors that the consultant has made; hence, this draft must be as clear and accurate as possible.

It is generally more vivid, and consequently more effective, for the consultant to review the report in the physical presence of the client.

12.7 PROFESSIONAL LIABILITY CONTROL

As the closing to this manual the presentation of what could be termed "loss limitation" All professionals in the geosciences want to reduce needless exposure to professional liability. Lawsuits and other form of litigation rairly benefit the professional involved in the legal action. Control of professional liability must be a daily inbred set of actions that begins at the proposal stage and works itself throughout the project and final report One can assume responsibility for items that are beyond the Scope of Work if observations or recommendation are made either in the field or in the office through report text, even when you are only doing a favor for the client in providing such information. A series of common mistakes that have caused professional liability claims are provided on Figure 12-11. These common mistakes cause the majority of the professional liability claims, and should be committed to memory by every professional geoscientist practicing in the environmental market.

Since loss prevention begins at the proposal stage the following three subsectional provide check-off lists for reviewing both proposals and the reports. These lists (Dames & Moore 1985) provide a firm handle on the potential common mistakes made in professional liability claims

12.7.1 Proposal and contract Check off-list

1. Has the type of study (e.g., preliminary, feasibility, investigation) been clearly defined?

2. Have the project requirements and conditions been clearly defined?

3. Were the purposes of the study clearly defined?

4. Are the scopes of investigation and testing appropriate for the types and nature of the project and conclusions to be reached?

5. Is the scope of Services clearly defined?

6. Are the items of work clearly and specifically spelled out?

7. Has a clause regarding warranty been included?

8. Are clauses delineating responsibility of the Engineer and/or Engineering Geologist included?

9. Have limitations of the study been discussed?

10. Have "taboo" words such as: complete, thorough, supervise, certify, assure, control, etc. been avoided?

11. Have limitations of cost estimates been discussed?

12. Was the proposal well worded to avoid misunderstanding?

13. Was limitation of liability addressed?

14. What is your overall evaluation of the proposal?

12.7.2 Part A—Loss Prevention Wording and Clauses

1. Has a clause regarding warranty been included in the Report (i.e., "Report prepared in accordance with generally accepted soil and foundation engineering practice. No other warranty expressed, . . ." etc.)?

2. Are clauses indicating the accuracy of location and elevation of borings, test pits, samples, etc., included in report?

3. Are clauses regarding groundwater variation, conditions at and between borings or samples included in the report?

4. Are clauses regarding reinterpretation included in the report such as limitation of data for particular site and time, review of plans and specifications, and limitation of data to use of current client?

5. Are clauses in the report regarding changed conditions during construction?

6. Are clauses in the report recommending use of the soils engineer during construction?

7. Are appropriate notes on the boring logs and drawings regarding ground water, interpolation between borings, nature and exactness of descriptions, etc?

8. Have "taboo" words been eliminated from the text; for example, words such as supervise, certify, assure, control, required, etc.?

9. Does the report volunteer any recommendations that are not a part of the scope of work?

10. Has the report indicated the inexactness of our science by eliminating words such as complete, final, obvious, best, etc., and by using rounded numbers for settlement, shear strength, permeability, etc.?

11. Does the report limit the scope of work to satisfy intended purposes?

12. Are recommendations consistent with the scope of work?

13. Has the report indicated that it was prepared only for intended purpose, such as design, feasibility, etc.?

14. Has the nature and extent of the project been described adequately?

15. What is your overall evaluation of the loss prevention part of this report?

12.7.3 Part B - Technical Matters

1. Was the type of project, its nature and the scope of work discussed?

2. Were the limits of the site and the existing site conditions discussed?

3. Was the general geology and soil stratigraphy discussed?

4. If applicable, were ground water/seepage and seismicity addressed?

5. Did the Engineer and/or Engineering Geologist limit himself to his own area of expertise?

6. Were drilling, sampling and testing done in accordance with the standard procedures and/or adequately described?

7. Were the sampling and testing sufficient to draw conclusions when integrated with general site conditions and prior site experience?

8. Was all test data adequately considered or used in the analysis?

9. Did the recommendations and conclusions reflect the field and laboratory data obtained?

10. Where applicable, was there a discussion of the feasible solutions in terms of financial or practical consequences?

11. If applicable, were alternates discussed?

12. If applicable, was the seismicity addressed?

13. Were recommendations, conclusions, data and information presented for all significant features?

14. Were the methods of analysis appropriate, well documented, and adequate?

15. Was the report technically up-to-date?

16. What is your overall evaluation of the technical part of this report?

12.7.4 Part C—Report Writing and Presentation

1. Were statements clear and concise?

2. Was the report self-contained with complete references to auxiliary work?

3. Were the choice of words and explanation of terms consistent with the client, and user of the report (layman, architect, engineer, etc.)?

4. Were the exhibits, data, and written material integrated to form a well-organized presentation?

5. Were the standard clauses wordings well-integrated at the appropriate places in the report?

6. Were the choice of words, grammar, sentence structure, and length appropriate?

7. Were the drawings clear with sufficient notes for a quick grasp of the information portrayed?

8. Were the data sheets or tabular data clear and understandable?

9. Did the report exhibit competence and completeness and create an atmosphere of confidence, such that good client relations would exist under possible future unforeseen difficulties on the project?

10. What is your overall evaluation of the composition part of this report?

Part D - Overall Evaluation

Your overall evaluation may differ from that indicated by the previous three parts. Provide an overall rating of the report.

12.8 Conclusions

The use of the correct wording in reports and proposals are developed over a long professional career of many individual letters, proposals and reports. The guidance provided above can provide a shorter and less error filled route for the professional.

APPENDIX A

GLOSSARY OF TECHNICAL TERMS

Absorption: The process by which one substance is taken into and included within another substance, as the absorption of water by soil or nutrients by plants. (2)

Acidity, total: The total acidity in a soil or clay. usually it is estimated by a buffered salt determination of (cation exchange minus exchangeable bases) = total acidity. (l)

Adsorption: The increased concentration of molecules or ions at a surface, including exchangeable cations and anions on soil particles. (2)

Aggregation: The act of soil particles cohering so as to behave mechanically as a unit. (2)

Air-dry: (a) The state of dryness (of a soil) at equilibrium with the moisture content in the surrounding atmosphere. The actual moisture content will depend upon the relative humidity and the temperature of the surrounding atmosphere. (b) To allow to reach equilibrium in moisture content with the surrounding atmosphere. (l)

Alkaline soil: Any soil having a pH > 7.0. (1)

Anisotropic mass: A mass having different properties in different directions at any given point. (5)

Annular Space; Annulus: - The space between two concentric tubes or casings, or between the casing and the borehole wall. This would include the space(s) between multiple strings of tubing/casings in a borehole installed either concentrically or multi-cased adjacent to each other. (9)

Aquiclude: A body of relatively impermeable soil or rock that is capable of absorbing water slowly but does not transmit it rapidly enough to supply a well or spring.

Aquifer: - A geologic formation, group of formations, or part of a formation that is saturated, and is capable of providing a significant quantity of water. (9)

Aquitard: - A lithologic unit that impedes ground water movement and does not yield water freely to wells or springs but that may transmit appreciable water to or from adjacent aquifers. Where sufficiently thick, may act as a ground water storage zone. Synonymous with confining unit. (9)

Area: Property contained within the boundaries.

Assessment (Investigation): The study of a particular area or region for defining the appropriateness of the area for waste disposal.

Assessment Monitoring: - An investigative monitoring program that is initiated after the presence of a contaminant in ground water has been detected. The objective of this program is to determine the concentration of constituents that have contaminated the ground water and to quantify the rate and extent of migration of these constituents. (9)

ASTM Cement Types: - Portland cements meeting the requirements of ASTM C 150 (Standard Specifications for Portland Cement). Cement types have slightly different formulations that

A-1

result in various characteristics which address different construction conditions and different physical and chemical environments. They are as follows:

ASTM Type I (Portland): - A general-purpose construction cement with no special properties.

ASTM Type II (Portland): - A construction cement that is moderately resistant to sulfates and generates a lower heat of hydration at a slower rate than ASTM Type I.

ASTM Type III (Portland; high early strength): - A construction cement that produces a high early strength. This cement reduces the curing time required when used in cold environments, and produces a higher heat of hydration than ASTM Type I.

ASTM Type IV (Portland): - A construction cement that produces a low heat of hydration (lower than ASTM Types I and II) and develops strength at a slower rate.

ASTM Type V (Portland): - A construction cement that is a high sulfate resistant formulation. Used when there is severe sulfate action from soils and ground water. (9)

Available water: The portion o£ water in a soil that can be readily absorbed by plant roots. Considered by most workers to be that water held in the soil against a pressure of up to approximately 15 bars. (1)

Average: Arithmetic mean.

Bailer: - A hollow tubular receptacle used to facilitate withdrawal of fluid from a well or borehole. (9)

Ballast: - Materials used to provide stability to a buoyant object (such as casing within a borehole filled with water). (9)

Bar: A unit of pressure equal to one million dynes per square centimeter. (1)

Base-saturation percentage: The extent to which the adsorption complex of a soil is saturated with exchangeable cations other than hydrogen. It is expressed as a percentage of the total cation-exchange capacity. (1)

Baseline: A surveyed condition which serves as a reference point to which later surveys are coordinated or correlated.

Bearing capacity: Ability of a material to support a load normal to the surface. (6)

Bedrock: The more or less continuous body of rock which underlies the overburden soils. (7)

Bentonite Clay: - An altered deposit of volcanic ash usually consisting of sodium montmorillonite clay. (9)

Blow-In: The inflow of ground water and unconsolidated material into a borehole or casing caused by differential hydraulic heads; that is, caused by the presence of a greater hydraulic head outside of a borehole/casing than inside. (9)

Borehole Log: The record of geologic units penetrated, drilling progress, depth, water level, sample recovery, volumes and types of materials used, and other significant facts regarding the drilling of an exploratory borehole or well. (9)

Borehole: A circular open or uncased subsurface hole created by drilling. (9)

Bridge: An obstruction within the annulus which may prevent circulation or proper emplacement of annular materials. (9)

Bulk density, soil: The mass of dry soil per unit bulk volume. The bulk volume is determined before drying to constant weight at 1_05 degrees Centigrade. (1)

Bulk specific gravity: The ratio of the bulk density of a soil to the mass of unit volume of water. (1)

Bulk volume: The volume, including the solids and the pores, of an arbitrary soil mass. (1)

California bearing ratio: The ratio of: (1) The force per unit area required to penetrate a soil mass with a 3 square inch (8 cm) circular piston (approximately 1.95 inch (51 mm) diameter) at the rate of 0.05 inches (1.3 mm)/minute, to (2) That required for corresponding penetration of a standard material. The ratio is usually determined at 0.1 inch (12.7 mm). Corps of Engineers' procedures require determination of the ratio at 0.1 inch and 0.2 inch (5.1 mm). Where the ratio of 0.2 inch is consistently higher that at

0.1 inch, the ratio at 0.2 inch is used. (5)

Capable fault: A fault defined by the Nuclear Regulatory Commission as one that is "capable" of "near future" movement; in general, a fault on which there has been movement within the last 35,000 years. The definition was developed for use in the siting of nuclear power plants.

Capillary attraction: A liquid's movement over or retention by a solid surface due to the interaction of adhesive and cohesive forces. (l)

Capillary conductivity: (Obsolete) See soil water - B: hydraulic conductivity.

Capillary fringe: A zone just above the water table (zero gauge pressure) that remains almost saturated. (The extent and the degree of definition of the capillary fringe depends upon the size-distribution of pores). (l)

Capillary migration (capillary flow): The movement of water by capillary action. (5)

Capillary potential: The amount of work that must be done per unit of pure water in order to transport reversibly and isothermally an infinitesimal quantity of water, identical in composition to the soil water, from a pool at the elevation and the external gas pressure of the point under consideration, to the soil water.

Casing, Protective: A section of larger diameter pipe that is emplaced over the upper end of a smaller diameter monitoring well riser or casing to provide structural protection to the well and restrict unauthorized access into the well. (9)

Casing, Surface: Pipe used to stabilize a borehole near the surface during the drilling of a borehole that may be left in place or removed once drilling is completed. (9)

Casing: Pipe, finished in sections with either threaded connections or bevelled edges to be field welded, which is installed temporarily or permanently to counteract caving, to advance the borehole, and/or to isolate the zone being monitored. (9)

Cation-exchange: The interchange between a cation and solution and another cation on the surface of any surface-active material such as clay colloid or organic colloid. (l)cation-exchange capacity (CEC): The sum total of exchangeable cations that a soil can absorb. Expressed in milliequivalents per 100 grams or per gram of soil (or of other exchangers such as clay). (l)

Caving; Sloughing: The inflow of unconsolidated material into a borehole which occurs when the borehole walls lose their cohesive strength. (9)

Cement; Portland Cement: Commonly known as Portland cement. A mixture that consists of calcareous, argillaceous, or other silica, alumina-, and iron-oxide-bearing materials that is manufactured and formulated to produce various types which are defined in ASTM C 150. Portland cement is also considered a hydraulic cement because it must be mixed with water to form a cement-water paste that has the ability to harden and develop strength even if cured under water (see ASTM Cement Types). (9)

Centralizer: A device that assists in the centering of a casing or riser within a borehole or another casing. (9)

Channels: Voids that are significantly larger than packing voids. They are generally cylindrical shaped and smooth walled, have regular conformation, and have relatively uniform cross-sectional size and shape. (4)

Circulation: Applies to the fluid rotary drilling method; drilling fluid movement from the mud pit, through the pump, hose and swivel, drill pipe, annular space in the hole and returning to the mud pit. (9)

Clay films: Coating of clay on the surfaces of soil peds and mineral grains and in soil pores. (Also called clay skins, clay flows, illuviation cutans, argillans or tonhautchen.)

Clay mineral: Naturally occurring inorganic crystalline material found in soils and other earthy deposits, the particles being clay sized; that is, > 0.002 mm in diameter. (1)

Clay: (a) A soil separate consisting of particles > 0.002 mm in equivalent diameter. (b) A textural class. (l)

Clod: A compact, coherent mass of soil ranging in size from 5 to 100 mm to as much as 20 to 25 cm and is produced artificially usually by the activity of man by plowing, digging, etc., especially when these operations are performed on soils that are either too wet or too dry for normal tillage operations.

Coarse fragments: Rock or mineral particles > 2.0 mm in diameter. (l)

Coarse texture: The texture exhibited by sand, loamy sands, and sandy loams except very fine sandy loams. (l)

Cohesionless soil: A soil that when unconfined has little or no strength when air-dried and that has little or no cohesion when submerged. (S)

Cohesive soil: A soil that when unconfined has considerable strength when air-dried and that has significant cohesion when submerged. (5)

Colloidal particles: Particles that are so small that the surface activity has an appreciable influence on the properties of the particle. (l)

Compaction curve (Proctor curve) (moisture-density curve): The curve showing the relationship between the dry unit weight (density) and the water content of a soil for a given compactive effort. (5)

Compaction test (moisture-density test): A laboratory compacting procedure whereby a soil at a known water content is placed in a specified manner into a mold of given dimensions, subjected to a compactive effort of controlled magnitude, and the resulting unit weight is determined. The procedure is repeated for various water contents sufficient to establish a relation between water content and unit weight. (5)

Compaction: The densification of a soil by means of mechanical manipulation. (5)

Compressibility: Property of a soil or rock pertaining to its susceptibility to decrease in volume when subjected to load. (5)

Compression curve: See pressure-void ratio curve.

Compressive strength (unconfined or uniaxial compressive strength): The load per unit area at which an unconfined cylindrical specimen of soil or rock will fail in a simple compression test. Commonly the failure load is the maximum that the specimen can withstand in the test. (5)

Conceptual model: A written or illustrated visualization of geologic/hydro-geologic/environmental conditions of a particular area.

Conductance (Specific): A measure of the ability of the water to conduct an electric current at 770 F (250C). It is related to the total concentration of ionizable solids in the water. It is inversely proportional to electrical resistance. (9)

Conductivity, hydraulic: See soil water.

Confining Unit: A term that is synonymous with "aquiclude," "aquitard," and "aquifuge"; defined as a body of relatively low permeable material stratigraphically adjacent to one or more aquifers. (9)

Consistency: (a) The resistance of a material to deformation or rupture. (b) The degree of cohesion or adhesion of the soil mass. (l)

Consolidation test: A test in which the specimen is laterally confined in a ring and is compressed between porous plates. (5)

Consolidation-time curve (time curve) (consolidation curve) (theoretical time curve): A curve that shows the relation between: (1) The degree of consolidation, and (2) The elapsed time after the application of a given increment of load. (5)

Consolidation: The gradual reduction in volume of a soil mass resulting from an increase in compressive stress. (a) initial consolidation (initial compression): A comparatively sudden reduction in volume of a soil mass under an applied load due principally to expulsion and compression of gas in the soil voids preceding primary consolidation. (b) primary consolidation (primary compression) (primary time effect): The reduction in volume of a soil mass caused by the application of a sustained load to the mass and due principally to a squeezing out of water from the void spaces of the mass and accompanied by a transfer of the load from the soil water to the soil solids. (c) secondary consolidation (secondary compression) (secondary time effect): The reduction in volume of a soil mass caused by the application of a sustained load to the mass and due principally to the adjustment of the internal structure of the soil mass after most of the load has been transferred from the soil water to the soil solids. (5)

Contaminant: An undesirable substance not normally present or an unusually high concentration of a naturally occurring substance in water or soil. (9)

Creep: Slow mass movement of soil and soil material down relatively steep slopes primarily under the influence of gravity, but facilitated by saturation with water and by alternate freezing and thawing. (1)

Crust: A surface layer on soils, ranging in thickness from a few millimeters to perhaps as much as an inch, that is much more compact, hard, and brittle, when dry, than the material immediately beneath it. (1)

Cutan: A modification of the texture, structure, or fabric at natural surfaces in soil materials due to concentration of particular soil constituents or 'in-situ' modification of the plasma; cutans can be composed of any of the component substances of the soil material. (4)

d-10: The diameter of a soil particle (usually in millimeters) at which 10% by weight of the particles of a particular sample are finer. (9) Synonymous with the effective size or effective grain size. (9)

d-60: The diameter of a soil particle (usually in millimeters) at which 60% by weight of the particles of a particular sample are finer. (9)

Darcy's law: (a) A law describing the rate of flow of water through porous media. (Named for Henry Darcy of Paris who formulated it in 1856 from extensive work on the flow of water through sand filter beds.)

Deflocculate: (a) To separate the individual components of compound particles by chemical and/or physical means. (b) To cause the particles of the disperse phase of a colloidal system to become suspended in the dispersion medium. (1)

Deformation: A change in the shape or size of a solid body. (7)

Degradation: The breakdown of substances by biological action. (2)

Degree of consolidation (percent consolidation): The ratio, expressed as a percentage of: (1) The amount of consolidation at a given time within a soil mass, to (2) The total amount of consolidation obtainable under a given stress condition. (5)

Degree of saturation: The extent or degree to which the voids in rock contain fluid (water, gas, or oil). Usually expressed in percent related to total void or pore space. (7)

Deposit: Material left in a new position by a natural transporting agent such as water, wind, ice, or gravity, or by the activity of man. (1)

Depression curve: Record of profile of water table as a result of pumping. (6)

Detection Monitoring: A program of monitoring for the express purpose of determining whether or not there has been a contaminant release to ground water. (9)

Detection monitoring: A program of monitoring for the express purpose of determining whether or not there has been a contaminant release to ground water.

Differential water capacity: The absolute value of the rate of change of water content with soil water pressure. The water capacity at a given water content will depend on the particular desorption or adsorption curve employed. Distinction should be made between volumetric and specific water capacity.

Direct methods: Methods (e.g., boreholes and monitoring wells) which entail the excavation or drilling, collection, observation, and analysis of geologic materials and water samples.

Discontinuity: (a) Boundary between major layers of the Earth which have different seismic velocities. (b) Interruption of the homogeneity of a rock mass (e.g. joints, faults, etc.). (5)

Disperse: (a) To break up compound particles, such as aggregates, into the individual component particles. (b) To distribute or suspend fine particles, such as clay, in or throughout a dispersion medium, such as water . (l)

Disposal: The discharge, deposit, injection, dumping, spilling, leaking, or placing of any solid waste or hazardous waste into or on any land or water so that such solid waste or hazardous waste or any constituent thereof may enter the environment or be emitted into the air or discharged into any waters, including ground waters.

Dissolution: The process where soluble organic components from DNAPL dissolves in groundwater or infiltration and forms a groundwater contaminant plume. the duration of remediation measures (either clean-up or containment) is determined by the 1) the rater of the dissolution process that can be achieved in the field, and 2) the mass of soluble components in the residual DNAPL trapped in the aquifer. (10)

Disturbed samples: Soil samples obtained in a manner which destroys the original orientation and some of the physical properties of the naturally disposed material. (6)

DNAPL Entry Location: The area where DNAPL has entered the subsurface. (10)

DNAPL Site: A site where DNAPL has been released and is now present in the subsurface as an immiscible phase. (10)

DNAPL Zone: The portion of a site affected by free-phase or residual DNAPL in the subsurface either the vadose zone or saturated zone). The DNAPL zone has organics in the vapor phase (unsaturated zone), dissolved phase (both unsaturated and saturated zone), and DNAPL phase (both unsaturated and saturated zone). (10)

DNAPL: A Dense Non-aqueous Phase Liquid. Also known as free product or a sinking plume (sinker). (10)

Downhole geophysics: Techniques that use a sensing device that is lowered into a borehole for the purpose of characterizing geologic formations and their associated fluids. The results can be interpreted to determine lithology, resistivity, bulk density, porosity, permeability, and moisture content and to define the source, movement, and physical/chemical characteristics of ground water.

Draw down curve: The trace of the top surface of the water table in an aquifer or of the free water surface, when a new or changed means of extraction of water takes place. (6)

Drill Cuttings: Fragments or particles of soil or rock, with or without free water, created by the drilling process. (9)

Drilling Fluid: A fluid (liquid or gas) that may be used in drilling operations to remove cuttings from the borehole, to clean and cool the drill bit, and to maintain the integrity of the borehole during drilling. (9)

Dry-weight percentage: The ratio of the weight of any constituent (of a soil) to the oven-dry weight of the soil. See oven-dry soil. (l)

Ductility: The condition in which material can sustain permanent deformation without losing its ability to resist load. (7)

Duripan: A mineral soil horizon that is cemented by silica, usually opal or micro-crystalline forms of silica, to the point that air-dry fragments will not slake in water or HCl. A duripan may also have accessory cement such as iron oxide or calcium carbonate. (l)

Effective Solubility: The actual aqueous solubility of an organic constituent in groundwater that is in chemical equilibrium with a mixed DNAPL (a DNAPL containing several organic constituents). The effective solubility of a particular organic chemical can be estimated by multiplying its mole fraction in the DNAPL mixture by its pure phase solubility. (10)

Elastic limit: Point on stress-strain curve at which transition from elastic to inelastic behavior takes place. (7)

Eolian: Pertaining to material transported and deposited by the wind. Includes earth materials ranging from dune sands to silt loess deposits. (3)

Ephemeral: A stream or portion of a stream which flows only in direct response to precipitation. It receives little or no water from springs and no long–continued supply from melting snow or other sources. Its channel is at all times above the water table. The term may be arbitrarily restricted to streams which do not flow continuously during periods of 1 month.

Equivalent diameter: In sedimentation analysis, the diameter assigned to a non-spherical particle, it being numerically equal to the diameter of a spherical particle of the same density and velocity of fall. (1)

Erode: To wear away or remove the land surface by wind, water, or other agents. (1)

Evapotranspiration: The combined loss of water from a given area, and during a specified period of time, by evaporation from the soil surface and by transpiration from plants. (1)

Fabric (soils): The physical constitution of a soil material as expressed by the spatial arrangement of the solid particles and associated voids. Fabric is the element of structure which deals with arrangement. (4)

Failure (in rocks): Exceeding the maximum strength of the rock or exceeding the stress or strain requirement of a specific design. (7)

Fatal flaws: A "fatal flaw" is herein defined as any site or near–site condition that could render the proposed facility unlicensable which could significantly impact the cost–profit ratio of operation, or which might later lead to non–compliance with performance objectives of the operating permit.

Fault: A fracture or fracture zone along which there has been displacement of the two sides relative to one another parallel to the fracture (this displacement may be a few centimeters or many kilometers). (7)

Field capacity (field moisture capacity): (Obsolete in technical work.) The percentage of water remaining in a soil 2 or 3 days after having been saturated and after free drainage has practically ceased. (The percentage may be expressed on the basis of weight or volume.) See moisture tension. (1)

Fill: Man-made deposits of natural soils or rock products and waste materials. (5)

Film water: A layer of water surrounding soil particles and varying in thickness from 1 or 2 to perhaps 100 or more molecular layers. Usually considered as that water remaining after drainage has occurred because it is not distinguishable in saturated soils. (1)

Fine texture: Consisting of or containing large quantities of the fine fractions, particularly of silt and clay. (Includes all clay loams and clays; that is, clay loams, sandy clay loam, silty clay loam, sandy clay, silty clay, and clay textural classes. Sometimes subdivided into clayey texture and moderately fine texture.) See soil texture. (1)

Firm: A term describing the consistency of a moist soil that offers distinctly noticeable resistance to crushing but can be crushed with moderate pressure between the thumb and forefinger. See consistency. (1)

Fissure flow: Flow of water through joints and larger voids. (5)

Flow curve: The locus of points obtained from a standard liquid limit test and plotted on a graph representing water content as ordinate on an arithmetic scale and the number of blows as abscissa on a logarithmic scale. (5)

Flow line: The path that a particle of water follows in its course of seepage under laminar flow conditions. (5)

Flow net: A graphical representation of flow lines and equipotential (piezometric) lines used in the study of seepage phenomena. (5)

Flow Path: Represents the area between two flow lines along which ground water can flow. (9)

Flow tubes: Area between two adjacent flow lines (Freeze and Cherry, 1979).

Flush Joint or Flush Coupled: Casing or riser with ends threaded such that a consistent inside and outside diameter is maintained across the threaded joints or couplings. (9)

Fracture: A break in a rock formation due to structural stresses. Faults, shears, joints, and planes of fracture cleaveage are all types of fractures.

Fracture: A break in the mechanical continuity of a body of rock caused by stress exceeding the strength of the rock. Includes joints and faults. (5)

Fragipan: A natural subsurface horizon with high bulk density relative to the solum above, seemingly cemented when dry, but when moist showing a moderate to weak brittleness. The layer is low in organic matter, mottled, slowly or very slowly permeable to water, and usually shows occasional or frequent bleached cracks forming polygons. It may be found in profiles of either cultivated or virgin soils but not in calcareous material. (l)

Free water (gravitational water) (ground water) (phreatic water): Water that is free to move through a soil or rock mass under the influence of gravity. (5)

Free-Phase DNAPL: Immiscible liquid exiting in the subsurface with a positive pressure such that it can flow into a well. If not trapped in a pool, free phase DNAPL will flow vertically through an aquifer or laterally down sloping fine-grained stratigraphic units. Also called mobile DNAPL or continuous phase DNAPL. (10)

Friable: A consistency term pertaining to the ease of crumbling of soils. See consistency. (l)

Gas pressure potential: This potential component is to be considered only when external gas pressure differs from atmospheric pressure as, e.g., in a pressure membrane apparatus. A specific term and definition is not given.

Geohydrology: Science of the occurrence, distribution, and movement of water below the surface of the Earth. (6)

Geomorphology: The description of the present exposed surfaces of the crust of the Earth, and seeks to interpret these surfaces in terms of natural processes (chiefly erosion) which lead or have led to their formation. (6)

Geophysical borehole logging: See Downhole Geophysics.

Geophysics: The study of all the gross physical properties of the Earth and its parts, particularly associated with the detection of the nature and shape of unseen subsurface rock bodies by measurement of such properties and property contrasts. Small scale applied geophysics is now a major aid in geological reconnaissance. (6)

Geotechnical: Pertaining to Geotechnics, which is the application of scientific methods to problems in engineering geology. (6)

Glacial drift: Rock debris that has been transported by glaciers and deposited, whether directly from the ice or from the meltwater. The debris may or may not be heterogeneous. (l)

Glacial geology: The study of the direct effects of the formation and flow under gravity of large ice masses on the Earth's surface. Glaciology is concerned with the physics of ice masses. (6)

Glacial outwash: Stratified sand and gravel produced by glaciers and carried, sorted, and deposited by water that originated mainly from the melting of glacial ice. Outwash deposits may occur in the form of valley fills (valley trains and/or outwash terraces) or as widespread outwash plains. (3)

Glacial till: Unsorted and unstratified glacial drift, generally unconsolidated, deposited directly by a glacier without subsequent reworking by water from the glacier, and consisting of a heterogeneous mixture of clay, silt, sand, gravel, and boulders varying widely in size and shape. (3)

Glaciofluvial deposits: Material moved by glacier and

subsequently sorted and deposited by streams flowing from the melting ice. (3)

Glaciolacustrine deposits: Material ranging from fine clay to sand derived from glaciers and deposited in glacial lakes by water originating mainly from the melting of glacial ice. Many are bedded or laminated with varves. (3)

Gleying: Formation of gray or green material in soil when stagnation of water results in exclusion of air and reduction of iron. (6)

Gradation (grain-size distribution) (texture): The proportions by mass of a soil or fragmented rock distributed in specified particle-size ranges. (5)

Grading: A 'well-graded' sediment containing some particles of all sizes in the range concerned. Distinguish from 'well sorted', which describes a sediment with grains of one size. (6)

Grain-size analysis (mechanical analysis) (particle-size analysis): The process of determining grain-size distribution. (5)

Granule: A natural soil aggregate or ped which is relatively nonporous. See soil structure and soil structure types. (l)

Gravel Pack: Common nomenclature for the preferred terminology, primary filter of a well (see primary filter pack). (9)

Gravel: Round or semirounded particles of rock that will pass a 3-in. (76.2 mm) sieve and be retained on a No. 4 (4.75 mm) U.S. standard sieve. (5)

Gravitational potential: See soil water.

Gravitational potential: The amount of work that must be done per unit quantity of pure water in order to transport reversibly and isothermally an infinitesimal quantity of water, identical in composition to the soil water, from a pool at a specified elevation and at atmospheric pressure to a similar pool at the elevation of the point under consideration.

Greenfield development: A new disposal facility on an area previously not developed for this purpose.

Ground water level: The level below which the rock and subsoil, to unknown depths, are saturated. (7)

Ground water regime (ground water): Water below the land surface in a zone of saturation.

Ground water: The portion of the total precipitation which at any particular time is either passing through or standing in the soil and the underlying strata and is free to move under the influence of gravity. (l)

Grout Shoe: A "plug" fabricated of relatively inert materials that is positioned within the lowermost section of a permanent casing and fitted with a passageway, often with a flow check device, through which grout is injected under pressure to fill the annular space. After the grout has set, the grout shoe is usually drilled out. (9)

Grout: A low permeability material placed in the annulus between the well casing or riser pipe and the borehole wall (i.e., in a single cased monitoring well), or between the riser and casing (i.e., in a multi-cased monitoring well), to maintain the alignment of the casing and riser and to prevent movement of ground water or surface water within the annular space. (9)

Hardpan: A hardened soil layer, in the lower A or in the B horizon, caused by cementation of soil particles with organic matter or with materials such as silica, sesquioxides, or calcium carbonate. The hardness does not change appreciably with changes in moisture content and pieces of the hard layer do not slake in water. (l)

Head (Static): The height above a standard datum of the surface of a column of water (or other liquid) that can be supported by the static pressure at a given point. The static head is the sum of the elevation head and the pressure head. (9)

Head (Total): The sum of three components at a point: (1) elevation head, he, which is equal to the elevation of the point above a datum; (2) pressure head, hp, which is the height of a column of static water that can be supported by the static pressure at the point; and (3) velocity head, hv, which is the height the kinetic energy of the liquid is capable of lifting the liquid. (9)

Head: The energy, either kinetic or potential, possessed by each unit weight of a liquid, expressed as the vertical height through which a unit weight would have to fall to release the average energy possessed. It is used in various compound terms such as pressure head, velocity head, and loss of head. (2)

Heave: Upward movement of soil caused by expansion or displacement resulting from phenomena such as: moisture absorption, removal of overburden, driving of piles, frost action, and loading of an adjacent area. (5)

Heterogeneity: Having different properties at different points. (7)

Holocene: An epoch of the Quarternary period, from the end of the Pleistocene, approximately 10,000 years ago, to the present time; also, the corresponding series of rocks and deposits. When the Quarternary is designated as an era, the Holocene is considered to be a period.

Homogeneous mass: A mass that exhibits essentially the same physical properties at every point throughout the mass. (5)

Horizon: See soil horizon.

Hydration: The physical binding of water molecules to ions, molecules, particles, or other matter. (2)

Hydraulic Conductivity: The volume of water at the existing kinematic viscosity and density that will move in unit time under unit gradient through a unit area measured at right angles to the direction of flow. The proportionality factor in Darcy's law as applied to the viscous flow of water in soil, i.e., the flux of water per unit gradient of hydraulic potential. For the purpose of solving the partial differential equation of the non-steady-state flow in unsaturated soil it is often convenient to introduce a variable termed the soil water diffusivity. (9)

Hydraulic gradient: The loss of hydraulic head per unit distance of flow. (5)

Hydraulic head: The elevation with respect to a specified reference level at which water stands in a piezometer connected to the point in question in the soil. Its definition can be extended to soil above the water table if the piezometer is replaced by a tensiometer. The hydraulic head in systems under atmospheric pressure may be identified with a potential expressed in terms of the height of a water column. More specifically it can be identified with the sum of gravitational and capillary potentials, and may be termed the hydraulic potential.

Hydrogeology: The study of the natural (and artificial) distribution of water in rocks, and its relationship to those rocks. Inasmuch as the atmosphere is a continuation of the hydrosphere, and is in physical and chemical balance with it, there is a close connection with meteorology. (6)

Hydrologic Unit: Geologic strata that can be distinguished on the basis of capacity to yield and transmit fluids. Aquifers and confining units are types of hydrologic units. Boundaries of a hydrologic unit may not necessarily correspond either laterally or vertically to lithostratigraphic formations. (9)

Hydrostatic pressure: A state of stress in which all the principal stresses are equal (and there is no shear stress). (7)

Hygroscopic water: Water adsorbed by a dry soil from an atmosphere of high relative humidity, water remaining in the soil after "air-drying" or water held by the soil when it is in equilibrium with an atmosphere of a specified relative humidity at a specified temperature, usually 98% of relative humidity at 25 degrees Centigrade. (l)

Igneous rock: Rock formed from the cooling and solidification of magma, and that has not been changed appreciably since its formation. (l)

Immobilization: The conversion of an element from the inorganic to the organic form in microbial tissues or in plant tissues. (1)

Impervious: Resistant to penetration by fluids or by roots. (l)

Indirect methods: Methods which include the measurement or remote sensing of various physical and/or chemical properties of the earth (e.g., electromagnetic conductivity, electrical resistivity, specific conductance, geophysical logging, aerial photography). (l)

Infiltration rate: A soil characteristic determining or describing the maximum rate at which water can enter the soil under specified conditions, including the presence of an excess of water. (l)

Infiltration rate: The rate at which a soil under specified conditions can absorb falling rain or melting snow; expressed in depth of water per unit time (cm/sec; in/hr).

Infiltration: The downward entry of water into the soil. (l)

Integral sampling: A technique of core drilling which provides knowledge of the original orientation of the samples recovered. (6)

Intergrade: A soil that possesses moderately well-developed distinguishing characteristics of two or more genetically related soil Great Groups. (l)

Intermittent: (1) Stream which flows but part of the time, as, after a rainstorm, during wet weather, or during but part of the year. (2) One which flows only at certain times when it receives water from springs (spring fed) or from some surface source (surface fed) such as melting snow in mountainous areas.

Internal friction (shear resistance): The portion of the shearing strength of a soil or rock that is usually considered to be due to the interlocking of the soil or rock grains and the resistance to sliding between the grains. (5)

Ion exchange: A chemical process involving reversible interchange of ions between a liquid and a solid but no radical change in structure of the solid. (2)

Isochrome: A curve showing the distribution of the excess hydrostatic pressure at a given time during a process of consolidation. (5)

Isocontours: A line drawn on a map to indicate equal concentrations of a solute in ground water.

Isograms: A general term proposed by Galton (1889, p. 651) for any line on a map or chart connecting points having an equal numerical value of some physical quantity (such as temperature, pressure, or rainfall); an isopleth.

Isomorphous substitution: The replacement of one atom by another of similar size in a crystal lattice without disrupting or changing the crystal structure of the mineral. (l)

Isopleth: A general term for any map showing the areal distribution of some variable quantity in terms of lines of equal or constant value; e.g., an isopach map.

Isotropic: Having the same properties in all directions. (6)

Jetting: When applied as a drilling method, water is forced down through the drill rods or casing and out through the end aperture. The jetting water then transports the generated cuttings to the ground surface in the annulus of the drill rods or casing and the borehole. The term jetting may also refer to a development technique (see well screen jetting). (9)

Joint: A break of geological origin in the continuity of a body of rock occurring either singly, or more frequently in a set or system, but not attended by a visible movement parallel to the surface of discontinuity. (7)

Kame: A moundlike hill of ice-contact glacial drift, composed chiefly of stratified sand and gravel. (3)

Karst: A type of topography that is characterized by closed depressions or sink holes, and is dependent upon underground solution and the diversion of surface waters to underground routes. It is formed over limestone, dolomite, gypsum and other soluble rocks as a result of differential solution of these materials and associated processes of subsurface drainage, cave formation, subsidence, and collapse. (3)

Laminar flow (streamline flow) (viscous flow): Flow in which the head loss proportional to the first power of the velocity. (S)

Landform: Any physical, recognizable form or feature of the Earth's surface, having a characteristic shape, and produced by natural causes. (3)

Landscape: All the natural features, such as fields, hills, forests, and water that distinguish one part of the Earth's surface from another part; usually that portion of land or territory which the eye can comprehend in a single view, including all of its natural characteristics. The distinctassociation of landforms, especially as modified by geologic forces, that can be seen in a single view. (3)

Leach: To cause water or other liquid to percolate through soil. (2)

Line of seepage (seepage line) (Phreatic line): The upper free water surface of the zone of seepage. (5)

Liquefaction: Act or process of liquefying or of rendering or becoming liquid; reduction to a liquid state. (2)

Liquid limit: The minimum percentage (by weight) of moisture at which a small sample of soil will barely flow under a standard treatment. Synonymous with "upper plastic limit". See plastic limit. ~1)

Liquidity index (water-plasticity ratio) (relative water content): The ratio, expressed as a percentage, of (l) The natural water content of a soil minus its plastic limit, to (2) its plasticity index. (5)

Lithologic: Pertaining to the physical character of a rock. (3)

Loading: The time rate at which material is applied to a treatment device involving length, area, volume or other design factor. (l)

Loess: Material transported and deposited by wind and consisting of predominantly silt-sized particles. (l)

Loss of Circulation: The loss of drilling fluid into strata to the extent that circulation does not return to the surface. (9)

Lysimeter: (a) A device for measuring percolation and leaching losses from a column of soil under controlled conditions. (b) A device for measuring gains (precipitation and condensation) and losses (evapotranspiration) by a column of soil. (l)

Manometer: An instrument for measuring pressure. It usually consists of a U-shaped tube containing a liquid, the surface of which in one end of the tube moves proportionally with changes in pressure on the liquid in the other end. Also, a tube type of differential pressure gauge. (2)

Matric potential: See potential, soil water.

Mechanical analysis: (Obsolete) See particle-size analysis and particle-size distribution.

Mesh: One of the openings or spaces in a screen. The value of the mesh is usually given as the number of openings per linear inch. This gives no recognition to the diameter of the wire and thus mesh number does not always have a definite relation to the size of the hole. (2)

Metamorphic rock: Rock derived from pre-existing rocks but that differ from them in physical, chemical, and mineralogical properties as a result of natural geological processes, principally heat and pressure, originating in the Earth. The pre-existing rocks may have been igneous, sedimentary, or another form of metamorphic rock. (l)

Modulus of elasticity (modulus of deformation): The ratio of stress to strain for a mineral under given loading condition; numerically equal to the slope of the tangent or the secant of a stress-strain curve. (S)

Moisture content (water content): The ratio, expressed as a percentage, of: (a) The weight of water in a given soil mass, to (b) The weight of solid particles. (S)

Moisture-retention curve: A graph showing the soil moisture percentage (by weight or by volume) versus applied tension (or pressure). Points on the graph are usually obtained by increasing (or decreasing) the applied tension or pressure over a specified range. (l)

Moraine: An accumulation of drift, with an initial

topographic expression of its own, built chiefly by the direct action of glacial ice. Examples are end, ground, lateral, recessional, and terminal moraines. (3)

Morphology: See soil morphology.

Mud Pit: Usually a shallow, rectangular, open, portable container with baffles into which drilling fluid and cuttings are discharged from a borehole and that serves as a reservoir and settling tank during recirculation of the drilling fluids. Under some circumstances, an excavated pit with a lining material may be used. (9)

Multi-Cased Well: A well constructed by using successively smaller diameter casings with depth. (9)

Multiport systems: A single hole device in which points are installed that are capable of sampling or measuring at multiple levels within a formation or series of formations.

N-value: The number of blows required to drive the sampler of the Standard Penetration test its last 12 inches (300 mm). (~)

Neat Cement: A mixture of Portland cement (ASTM 150) and water. (9)

Negative pressure: A pressure less than the local atmospheric pressure at a given point. (3)

Normally consolidated soil deposit: A soil deposit that has never been subjected to an effective pressure greater than the existing overburden pressure. (5)

Observation Well: Typically, a small diameter well used to measure changes in hydraulic heads, usually in response to a nearby pumping well. (9)

Oil Air Filter: A filter or series of filters placed in the air flow line from an air compressor to reduce the oil content of the air. (9)

Oil Trap: A device used to remove oil from the compressed air discharged from an air compressor. (9)

Open gradient–lines: Data not sufficiently extensive

to complete contour lines through an area.

Optimum moisture content (optimum water content): The water content at which a soil can be compacted to a maximum dry unit weight by a given compactive effort. (S)

Osmotic potential: The amount of work that must be done per unit quantity of pure water in order to transport reversibly and isothermally an infinitesimal quantity of water from a pool of pure water, at a specified elevation and at atmospheric pressure, to a pool of water identical in composition to the soil water (at the point under consideration), but in all other respects being identical to the reference pool.

Osmotic pressure: The pressure to which a pool of water, identical in composition to the soil water, must be subjected in order to be in equilibrium, through a semipermeable membrane, with a pool of pure water (semipermeable means permeable only to water) . May be identified with the osmotic potential defined above.

Outwash plain: An extensive lowland area forming the surface of a body of coarse textured, glaciofluvial material. An outwash plain is commonly smooth; where pitted, due to melt-out of incorporated ice masses, it is generally low in relief. (3)

Oven-dry soil: Soil which has been dried at 105 degrees Centigrade until it reaches constant weight. (l)

Overburden load: The load on a horizontal surface underground due to the column of material located vertically above it. (7)

Overburden: The loose soil, sand, silt, or clay that overlies bedrock. (7)

Overconsolidated soil deposit: A soil deposit that has been subjected to an effective pressure greater than the present overburden pressure. (5)

Oxidation-reduction potential: The potential required to transfer electrons from the oxidant to the reductant and used as 2 qualitative measure of the state of oxidation in wastewater treatment systems. (2)

Packer: A transient or dedicated device placed in a well that isolates or seals a portion of the well, well annulus, or borehole at a specific level. (9)

Parent material: The unconsolidated and more or less chemically weathered mineral or organic matter from which the solum of soil is developed by pedogenic processes. (l)

Particle density: The mass per unit volume of the soil particles. In technical work, usually expressed as grams per cubic centimeter. See bulkeny, soil. (l)

Particle size: The effective diameter of a particle measured by sedimentation, sieving, or micrometric methods. (l)

Particle-size analysis: Determination of the various amounts of the different separates in a soil sample, will usually be sedimentation, sieving, micrometry, or combinations of these methods. (l)

Particle-size distribution: The amounts of the various soil separates in a soil sample, usually expressed as weight percentages. (l)

Ped: An individual natural soil aggregate consisting of a cluster of primary particles, and separated from adjoining peds by surfaces of weakness which are recognizable as natural voids or by the occurrence of cutans. (4)

Pedologic: One of the disciplines of soil science, the study of soil morphology, genesis, and classification. It is sometimes used as a synonym of soil science.

Pedology: (a) The description of those parts of the present Earth surface which have become weathered or otherwise modified 'in-situ' by solar energy and by the effects of organisms to form a soil which is of primary importance to man in agriculture. (6) (b) The science of soils, that is, the study of the origin, classification, description and use of natural soil bodies. (4)

Pedon: A three-dimensional body of soil with lateral dimensions large enough to permit the study of horizon shapes and relations. Its area ranges from 1 to 10 square meters. Where horizons are intermittent or cyclic, and recur at linear intervals of 2 to 7 m, the pedon includes one-half of the cycle. Where the cycle is less than 2 m, or all horizons are continuous and of uniform thickness, the pedon has an area of approximately 1 square meter. If the horizons are cyclic, but, recur at intervals greater than 7 m, the pedon reverts to the 1 square meter size, and more than one soil will usually be represented in each cycle. (l)

Penetrability: The ease with which a probe can be pushed into the soil. (May be expressed in units of distance, speed, force, or work depending on the type of penetrometer used. (1)

Penetration resistance (standard penetration resistance) (Proctor penetration resistance): (a) A number of blows of a hammer of specified weight falling a given distance required to produce a given penetration into soil of a pile, casing, or sampling tube. (b) Unit load required to maintain constant rate of penetration into soil of a probe or instrument. (c) Unit load required to produce a specified penetration into soil at a specified rate of a probe or instrument. For a Proctor needle, the specified penetration is 2.5 in. (63.5 mm) and the rate is 0.5 in. (12.7mm)/sec. (5)

Penetration resistance curve (Proctor penetration curve): The curve showing the relationship between: (a) The penetration resistance, and (b) The water content. (5)

Percent compaction: The ratio, expressed as a percentage, of: (a) Dry unit weight of a soil, to (b) Maximum unit weight obtained in a laboratory Compaction test. (5)

Percent saturation (degree of saturation): The ratio, expressed as a percentage, of: (a) The volume of water in a given soil or rock mass, to (b) The total volume of intergranular space (voids). (5)

Perched water table: A water table usually of limited area maintained above the normal free water elevation by the pressure of an intervening relatively impervious confining stratum. (5)

Percolation: The flow or trickling of a liquid downward through a contact or filtering medium. The liquid may or may not fill the pores of the medium. (2)

Perennial: Streams that flow throughout the year and from source to mouth.

Permanent strain: The strain remaining in a solid with respect to its initial condition after the application and removal of stress greater than the yield stress (commonly also called "residual" strain). (7)

Permeability, soil: (a) The ease with which gases, liquids, or plant roots penetrate or pass through a bulk mass of soil or a layer of soil. Since different soil horizons vary in permeability, the particular horizon under question should be designated. (b) The property of a porous medium itself that relates to the ease with which gases, liquids, or other substances can pass through it. Previously, frequently considered the "k" in Darcy's law. See Darcy's law and soil water. (l)

pH, soil: The negative logarithm of the hydrogen-ion activity of a soil. The degree of acidity (or alkalinity) of a soil as determined by means of a glass, quinhydrone, or other suitable electrode or indicator at a specified moisture content or soil-water ratio, and expressed in terms of the pH scale. (l)

Physical properties (of soil): Those characteristics, processes, or reactions of a soil which are caused by physical forces and which can be described by, or expressed in, physical terms or equations. Sometimes confused with and difficult to separate from chemical properties; hence, the terms "physical-chemical" or "physiochemical". Examples of physical properties are bulk density, water-holding capacity, hydraulic conductivity, porosity, pore-size distribution, etc. (l)

Piezometers: Generally a small diameter, non–pumping well used to measure the elevation of the water table or potentiometric surface.

Piezometric surface: (a) The surface at which water will stand in a series of piezometers. (5) (b) An imaginary surface that everywhere coincides with the static level of the water in the aquifer. (7)

Piping: An underground flow of water with a sufficient pressure gradient to cause scour along

a preferred path. (6)

Piston sampler: A tube with an internal piston used for obtaining relatively undisturbed samples from cohesive soils. (6)

Plane of weakness: Surface or narrow zone with a (shear or tensile) strength lower than that of the surrounding material. (7)

Plane stress (strain): A state of stress (strain) in a solid body in which all stress (strain) components normal to a certain plane are zero. (7)

Plastic equilibrium: State of stress within a soil or rock mass or a portion thereof, which has been deformed to such an extent that its ultimate shearing resistance is mobilized. (5)

Plastic flow (plastic deformation): The deformation of a plastic material beyond the point of recovery, accompanied by continuing deformation with no further increase in stress. (S)

Plastic limit: (a) The water content corresponding to an arbitrary limit between the plastic and the semisolid states of consistency of a soil. (b) Water content at which a soil will just begin to crumble when rolled into a thread approximately 1/8 in. (3.2 mm) in diameter. (5)

Plasticity range: The range of moisture weight percentage within which a small sample of soil exhibits plastic properties. (l)

Plasticity: The property of a soil or rock which allows it to be deformed beyond the point of recovery without cracking or appreciable volume change. (5)

Plume: The zone of contamination containing organics in the dissolved phase. The plume usually will originate from the DNAPL zone and extend downgradient for some distance depending on site hydrogeologic and chemical conditions. To avoid confusion, the term "DNAPL plume" should not be used to describe a DNAPL pool; "plume should be used only to refer to dissolved-phase organics. (10)

Pool And Lens: A zone of free-phase DNAPL at the bottom of an aquifer. A lens is a pool that rests on a fine-grained stratigraphic unit of limited areal extent. DNAPL can be recovered from a pool or a lens if a well is placed in the right location.(10)

Pore-size distribution: The volume of the various sizes of pores in a soil. Expressed as percentages of the bulk volume (soil plus pore space). (l)

Porosity: The ratio, usually expressed as a percentage, of: (a) The volume of voids of a given soil or rock mass, to (b) The total volume of the soil or rock mass. (5)

Potential, soil water: See soil water.

Potentiometric Surface: An imaginary surface representing the static head of ground water. The water table is a particular potentiometric surface. NOTE: Where the head varies with depth in the aquifer, a potentiometric surface is meaningful only if it describes the static head along a particular specified surface or stratum in that aquifer. More than one potentiometric surface is required to describe the distribution of head in this case. (9)

Preconsolidation pressure (prestress): The greatest effective pressure to which a soil has been subjected. (5)

Pressure surface: The level of the water surface in an (imaginary) vertical well connecting with an aquifer. (5)

Pressure-void ratio curve (compression curve): A curve representing the relationship between effective pressure and void ratio of a soil as obtained from a consolidation test. The curve has a characteristic shape when plotted on semilog paper with pressure on the log scale. The various parts of the curve and extensions to the parts have been designated as recompression, compression, virgin compression, expansion, rebound, and other descriptive names by various authorities. (5)

Primary Filter Pack: A clean silica sand or sand and gravel mixture of selected grain size and gradation that is installed in the annular space between the borehole wall and the well screen, extending an appropriate distance above the screen, for the purpose of retaining and stabilizing the particles from the adjacent strata. The term is used in place of "gravel pack. (9)"

Primary state of stress: The stress in a geological formation before it is disturbed by manmade works. (7)

Principal stress (strain): The stress (strain) normal to one of three mutually perpendicular planes on which the shear stresses (strains) at a point in a body are zero. (7~

Profile, soil: A vertical section of the soil through all its horizons and extending into the parent material. (l)

PTFE Tape: Joint sealing tape composed of polytetrafluoroethylene. (9)

Puddled soil: A soil in which structure has been mechanically destroyed, which allows the soil to run together when saturated with water. A soil that has been puddled occurs in a massive nonstructural state. (2)

Pyroclastic: Pertaining to fragmental materials produced by usually explosive, aerial ejection of clastic particles from a volcanic vent. (3)

Reaction, soil: The degree of acidity or alkalinity of a soil, usually expressed as a pH value. Descriptive terms commonly associated with certain ranges in pH are: extremely acid, less than 4.5; very strongly acid, 4.5 to 6.0; slightly acid, 6.1 to 6.5; neutral, 6.6 to 7.3; slightly alkaline, 7.4 to 7.8; moderately alkaline, 7.9 to 8.4; strongly alkaline, 8.5 to 9.0; and very strongly alkaline, greater than 9.1. (1)

Recharge: Natural or artificial replenishment of an aquifer.

Regolith: All unconsolidated earth materials above the solid bedrock. (3)

Relative consistency: Ratio of: (a) The liquid minus the natural water content, to (b) The plasticity index. (5)

Relative density: The ratio of: (a) The difference between the void ratio of a cohesionless soil in the loosest state and any given void ratio, to (b) The difference between the void ratios in the loosest and the densest states. (5)

Remolded soil: Soil that has had its natural structure modified by manipulation. (5)

Residual Saturation: The fraction of available pore space containing residual DNAPL s, or the saturation level where free-phase DNAPL becomes residual DNAPL. In the vadose zone, residual saturation range u to 20% of total pore volume while in the saturated zone residual saturations range up to 50% of total pore volume. (10)

Residual shrinkage: The decrease in the bulk volume of soil in addition to that caused by the loss of water. (l)

Residual stress: Stress remaining in a solid under zero external stress after some process that causes the dimensions of the various parts of the solid to be incompatible under zero stress; for example, (a) Deformation under the action of external stress when some parts of the body suffer permanent strain, or (b) Heating or cooling of a body in which the thermal expansion coefficient is not uniform throughout the body. (5)

Residual: Immiscible phase liquid held in the pore spaces or fractures by capillary forces (negative pressure on DNAPL), Residual will remain trapped within the pore of the porous media unless the viscous forces (caused by the dynamic force of water against the DNAPL) are greater than the capillary forces holding the DNAPL in the pore. At most sites the hydraulic gradient required to mobilize all of the residual trapped in an aquifer is usually much greater than can be produced by wells or trenchers. (10)

Retentivity profile, soil: A graph showing the retaining capacity of a soil as a function of depth. The retaining capacity may be for water, for water at any given tension, for cations, or for any other substances held by soils. (l)

Riser: The pipe extending from the well screen to or above the ground surface. (9)

Rock: Natural solid mineral matter occurring in large masses or fragments. (5)

S-matrix: The material (plasma and/or skeleton grains and associated voids) within the simplest (primary) peds, or composing apedal soil materials that does not occur as pedalogical features other than plasma separations; it may be absent in some soil materials, for example those that consist entirely of pedological features. (4)

Sand: (a) A soil particle between 0.05 and 2.0 mm in diameter. (b) Any one of five soil separates, namely: very coarse sand, coarse sand, medium sand, fine sand, and very fine sand. See soil separates. (c) A soil textural class. See soil texture. (1)

Saturated: The subsurface zone in which all openings are full of water.

Saturation: A condition reached by a material, whether it be in solid, gaseous, or liquid state, that holds another material within itself in a given state in an amount such that no more of such material can be held within it in the same state. The material is then said to be saturated on in a condition of saturation. (2)

Scope of work: A written description of site assessment work to be performed within an investigation.

Secondary Filter Pack: A clean, uniformly graded sand that is placed in the annulus between the primary filter pack and the over-lying seal, or between the seal and overlying grout backfill or both, to prevent movement of seal, or grout, or both into the primary filter pack. (9)

Secondary porosity: The porosity developed in a rock after its deposition or emplacement, through such processes as solution or fracturing.

Sediment Sump: A blank extension beneath the well screen used to collect fine-grained material from the filter pack and adjacent strata. The term is synonymous with rat trap or tail pipe. (9)

Sedimentation: The process of subsidence and deposition of suspended matter carried by water, wastewater, or other liquids, by gravity. It is usually accomplished by reducing the velocity of the liquid below the point at which it can transport the suspended material. (2)

Seepage (percolation): The slow movement of gravitational water through the soil or rock. (5)

Seepage force: The force transmitted to the soil or rock grains by seepage. (5)

Sensitivity: Ratio of disturbed to undisturbed shear strength of a soil. (6)

Shear failure (failure by rupture): Failure in which movement cause by shearing stresses in a soil or rock mass is of sufficient magnitude to destroy or seriously endanger a structure. (a) General shear failure: Failure in which the ultimate strength of the soil or rock is mobilized along the entire potential surface of sliding before the structure supported by the soil or rock is impaired by excessive movement. (b) Local shear failure: Failure in which the ultimate shearing strength of the soil or rock is mobilized only locally along the potential surface of sliding at the time the structure supported by the soil or rock is impaired by excessive movement. (5)

Shear force: A force directed parallel to the surface element across which it acts. (7)

Shear plane: A plane along which failure of material occurs by shearing. (7)

Shear strain: The change in shape, expressed by the relative change of the right angles at the corner of what was in the undeformed state an infinitesimally small rectangle or cube. (7)

Shear Strength: A measure of the shear or gel properties of a drilling fluid or grout; also, the maximum resistance of a soil or rock to shearing stresses. (9)

Shear stress: Stress directed parallel to the surface element across which it acts. (7)

Shrinkage limit: The maximum water content at which a reduction in water content will not cause a decrease in volume of the soil mass. (5)

Silt: (a) A soil separate consisting of particles between 0.005 and 0.002 mm in equivalent diameter. See

soil separates. (b) A soil texture class. See soil texture. (l)

Single-Cased Well: A monitoring well constructed with a riser but without an exterior casing. (9)

Site assessment: A formal means of exploring and characterizing a proposed waste management facility or location so that all physical factors are identified and so quantified as to serve as the basis of an environmentally sound design and operational plan.

Skeleton grains: The individual grains larger than colloidal size (> 0.002 mm) of a soil material; they consist of mineral grains originally present in the parent material and resistant siliceous and organic bodies. (4)

Slickensides: Polished and grooved surfaces produced by one mass sliding past another. (l)

Soil air: The soil atmosphere; the gaseous phase of the soil, being that volume not occupied by solid or liquid. (l)

Soil auger: A tool for boring into the soil and withdrawing a small sample for field or laboratory observation. Soil augers may be classified into several types as follows: (a) Those with worm-type bits, unenclosed; (b) Those with worm-type bits enclosed in a hollow cylinder; and (c) Those with a hollow cylinder with a cutting edge at the lower end. (l)

Soil fabric: The physical constitution of a soil material as expressed by the spatial arrangement of the solid particles and associated voids. (4)

Soil horizon: A layer of soil or soil material approximately parallel to the land surface and differing from adjacent genetically related layers in physical, chemical, and biological properties or characteristics such as color, structure, texture, consistency, kinds and numbers of organisms present, degree of acidity or alkalinity, etc. (l)

Soil mechanics: The science dealing with all phenomena which affect the action of soil in a capacity in any way associated with engineering. (8)

Soil mineral: (a) Any mineral that occurs as a part of or in the soil. (b) A natural inorganic compound with definite physical, chemical, and crystalline properties (within the limits of isomorphism), that occurs in the soil. (l)

Soil moisture: Water contained in the soil. (l)

Soil morphology: (a) The physical constitution, particularly the structural properties, of the soil profile as exhibited by the kinds, thickness, and arrangement of the horizons in the profile, and by the texture, structure, consistency, and the porosity of each horizon. (b) The structural characteristics of the soil or any of its parts. (l)

Soil physics: The organized body of knowledge concerned with the physical characteristics of soil and with the methods employed in their determinations. (5)

Soil piping or tunneling: Accelerated erosion which results in subterranean voids and tunnels. (l)

Soil science: That science dealing with soils as a natural resource on the surface of the Earth including soil formation, classification, and mapping, and physical, chemical, biological, and fertility properties of soil per se; and these properties in relation to their management. (l)

Soil separates: Mineral particles, < 2.0 mm in equivalent diameter, ranging between specified size limits. The names and size limits of separates recognized in the U.S.D.A. system are: very coarse sand, 2.0 to 1.0 mm; coarse sand, 1.0 to 0.5 mm; medium sand, 0.5 to 0.25 mm; fine sand, 0.25 to 0.10 mm; very fine sand, 0.10 to 0.05 mm; silt, 0.05 to 0.002 mm; and clay, < 0.002 mm. The U.S.C.S. particle and size range are as follows: coarse sand, 2.0 to 4.76 mm; medium sand, 0.42 to 2.0 mm; fine sand, 0.074 to 0.42 mm; fines (silt and clay), < 0.074 mm. (Note: U.S.C.S. silt and clay designations are determined by response of the soil to manipulation at various water contents rather than by measurement of size.)

Soil series: The basic unit of U.S.D.A. soil classification being a subdivision of a family and consisting of soils which are essentially alike in all major profile characteristics except the texture of the A horizon. (l)

Soil solution: The aqueous liquid phase of the soil and its solutes. (l)

Soil structure: The combination or arrangement of primary soil particles into secondary particles, units, or peds. These secondary units may be, but usually are not, arranged in the profile in such a manner as to give a distinctive, characteristic pattern. The secondary units are characterized and classified on the basis of size, shape, and degree of distinctness into classes, types, and grades, respectively. (l)

Soil suction: A measure of the force of water retention in unsaturated soil. Soil suction is equal to a force per unit area that must be exceeded by an externally applied suction to initiate water flow from the soil. Soil suction is expressed in standard pressure terms. (2)

Soil texture: The relative proportion of the various soil separates in a soil as described by the classes of soil texture. (l)

Soil water diffusivity: The hydraulic conductivity divided by the differential water capacity (care being taken to be consistent with units), or the flux of water per unit gradient of moisture content in the absence of other force fields.

Soil water pressure: The pressure (positive or negative), relative to the external gas pressure on the soil water, to which a solution identical in composition to the soil water must be subjected in order to be in equilibrium through a porous permeable wall with the soil water. May be identified with the capillary potential defined above.

Soil water: A general term emphasizing the physical rather than the chemical properties and behavior of the soil solution.

Soil-moisture tension: See moisture tension (or pressure).

Soil: (a) The unconsolidated mineral material on the immediate surface of the Earth that serves as a natural medium for the growth of land plants. (b) The unconsolidated mineral matter on the surface of the Earth that has been subjected to and influenced by genetic and environmental factors of: parent material, climate (including moisture and temperature effects), macro- and microorganisms, and topography, all acting over a period of time and producing a product, soil, that differs from the material from which it is derived in many physical, chemical, biological, and morphological proper ties and characteristics. (l)

Solid waste disposal facilities: A facility or part of a facility at which solid waste is intentionally placed into or on any land or water, and at which waste will remain after closure.

Specific gravity (of solids): Ratio of: (a) The weight in air of a given volume of solids at a stated temperature, to (b) The weight in air of an equal volume of distilled water at a stated temperature. (l)

Specific retention: Ratio of volume of suspended water to volume of associated voids. (6)

Specific surface: The surface area per unit of volume of soil particles. (5)

Specific yield: Ratio of voids not occupied by suspended water to the total volume of the associated area. (6)

Stability: The condition of a structure or a mass of material when it is able to support the applied stress for a long time without suffering any significant deformation or movement that is not reversed by the release of stress. (7)

Standard Penetration Test (SPT): The most commonly used in situ test to measure in relative terms the resistance of soil to deformation by shearing. (6)

Static Water Level: The elevation of the top of a column of water in a monitoring well or piezometer that is not influenced by pumping or conditions related to well installation, hydrologic testing, or nearby pumpage. (9)

Stereograms: A graphic diagram on a plane surface, giving a three–dimensional representation, such as projecting a set of angular relations; e.g., a block diagram of geologic structure, or a stereographic projection of a crystal.

Storativity: The volume of water an aquifer releases from or takes into storage per unit surface area of the aquifer per unit change in head. It is equal to the product of specific storage and aquifer thickness. In an unconfined aquifer, the storativity is equivalent to the specific yield. Also called storage coefficient.

Strain (linear or normal): The change in length per unit of length in a given direction. (5)

Stratified: Arranged in strata, or layers. The term refers to geologic material. Layers in soils that result from the processes of soil formation are called horizons; those inherited from the parent material are called strata. (3)

Strength: Maximum stress which a material can resist without failing for any given type of loading. (7)

Stress: The force per unit area acting within the soil mass.

Structure: One of the larger features of a rock mass, like bedding, foliation, jointing, cleavage, or brecciation; also the sum total of such features as contrasted with texture. Also, in a broader sense, it refers to the structural features of an area such as anticlines or synclines. (7) See also soil structure.

Subsidence: The downward displacement of the overburden (rock or soil, or both) lying above an underground excavation or adjoining a surface excavation. Also the sinking of a part of the earth's crust. (7)

Subsoil: In general concept, that part of the soil below the depth of plowing. (2)

Summation curve, particle size: A curve showing the accumulative percentage by weight of particles within increasing (or decreasing) size limits as a function of diameter; the percent by weight of each size fraction is plotted accumulatively on the ordinate as a function of the total range of diameter represented in the sample plotted on the abscissa. (l)

Surface sealing: The orientation and packing of dispersed soil particles in the immediate surface layer of the soil, rendering it relatively impermeable to water. (l)

Swelling pressure: Pressure exerted by confined swelling clays when moisture content is increased. (6)

Tamper: A heavy cylindrical metal section of tubing that is operated on a wire rope or cable. It slips over the riser and fits inside the casing or borehole annulus. It is generally used to tamp annular sealants or filter pack materials into place and prevent bridging. (9)

Target Monitoring Zone: The ground water flow path from a particular area or facility into which monitoring wells will be screened. The target monitoring zone should be a stratum (strata) in which there is a reasonable expectation that a vertically placed well will intercept migrating contaminants. (9)

Target: In detection monitoring programs, the ground water flow path from a particular area or facility into which monitoring wells will be screened. The target monitoring zone should be a stratum (strata) in which there is a reasonable expectation that a vertically placed well will intercept migrating contaminants.

Tensile strength (unconfined or uniaxial tensile strength): The load per unit area at which an unconfined cylindrical specimen will fail in a simple tension (pull) test. (5)

Tensiometer: A device for measuring the negative pressure (or tension) of water in soil in situ; a porous, permeable ceramic cup connected through a tube to a manometer or vacuum gauge.

Tension, soil water: The expression, in positive terms, of the negative hydraulic pressure of soil water. (2)

Test Pit: A shallow excavation made to characterize the subsurface. (9)

Time–series: A series of statistical data collected at regular intervals of time; a frequency distribution in which the independent variable is time.

Total potential (of soil water): The amount of work that must be done per unit quantity of pure water in order to transport reversibly and isothermally an infinitesimal quantity of water from a pool of pure water, at a specified elevation and at atmospheric pressure, to the soil water (at the point under consideration). The total potential (of soil water) consists of the following:

Total pressure: The pressure (positive or negative), relative to the external gas pressure on the soil water, to which a pool of pure water must be subjected in order to be in equilibrium through a semipermeable membrane with the soil water. Total pressure is thus equal to the sum of soil water pressure and osmotic pressure. Total pressure may also be derived from the measurement of the partial pressure of the water vapor in equilibrium with the soil water. May be identified with the total potential defined above when gravitational and external gas pressure potentials can be neglected.

Transects: In ecology, a sample area (usually elongate or linear) chosen as the basis for studying a particular assemblage of organisms.

Transmissivity: Rate of transmission of water through unit width of an aquifer under unit hydraulic gradient. (6)

Transmissivity: The rate at which water of the prevailing kinematic viscosity is transmitted through a unit width of an aquifer under a unit hydraulic gradient. It equals the hydraulic conductivity multiplied by the aquifer thickness.

Transmissivity: The rate at which water of the prevailing kinematic viscosity is transmitted through a unit width of the aquifer under a unit hydraulic gradient. NOTE: It is equal to an integration of the hydraulic conductivities across the saturated part of the aquifer perpendicular to the flow paths. (9)

Transpiration: Water loss from leaves and other plant organs to the atmosphere. (6)

Tremie Pipe: A pipe or tube that is used to transport filter pack materials and annular sealant materials from the ground surface into the borehole annulus or between casings and casings or riser pipe of a monitoring well. (9)

Triaxial compression: Compression caused by the application of normal stresses in three perpendicular directions. (7)

Triaxial shear test (triaxial compression test): A test in which a cylindrical specimen of soil or rock encased in an impervious membrane is subjected to a confining pressure and then loaded axially to failure. (5)

Triaxial state of stress: State of stress in which none of the three principal stresses is zero. (S)

Tuff: Volcanic ash usually more or less stratified and in various states of consolidation. (l)

Ultimate bearing capacity: The average load per unit of area required to produce failure by rupture of a supporting soil or rock mass. (5)

Unconsolidated-undrained test (quick test): A soil test in which the water content of the test specimen remains practically unchanged during the application of the confining pressure and the additional axial (or shearing) force. (5)

Undisturbed sample: A soil sample that has been obtained by methods in which every precaution has been taken to minimize disturbance to the sample. (5)

Uniaxial (unconfined) compression: Compression caused by the application of normal stress in a single direction. (7)

Uniformity Coefficient: The size ratio of the 60% finer (d-60) grain size to the 10% (d-10) finer grain size of a sample of granular material (refer to ASTM Standard Test Method D 2487). (9)

Uniformly Graded: A quantitative definition of the particle size distribution of a soil which consists of the majority of particles being of the same approximate diameter. A granular material is considered uniformly graded when the uniformity coefficient is less than about 5 (refer to ASTM Standard Test Method D 2487). Analogous with the geologic term well sorted. (9)

Unsaturated flow: The movement of water in a soil which is not filled to capacity with water. (l)

Uplift: The hydrostatic force of water exerted on or underneath a structure, tending to cause a displacement of the structure. (7)

Uppermost aquifer: The geologic formation nearest the natural ground surface that is an aquifer, as well as lower aquifers that are hydraulically interconnected with this aquifer within the facility's property boundary.

Vane shear test: An in-place shear test in which a rod with thin radial vanes at the end is forced into the soil and the resistance to rotation of the rod is determined. (5)

Vapor pressure: (a) The pressure exerted by a vapor in a confined space. It is a function of the temperature. (b) The partial pressure of water vapor in the atmosphere. (c) Partial pressure of any liquid. (2)

Vented Cap: A cap with a small hole that is installed on top of the riser. (9)

Viscosity: The cohesive force existing between particles of a fluid which causes the fluid to offer resistance to a relative sliding motion between particles. (2)

Void ratio: The ratio of: (a) The volume of void space, to (b) The volume of solid particles in a given soil mass. (S)

Voids: Entities which are interconnected with each other either through voids of dissimilar size and shape, through narrow necks, or through intersection with voids of similar size and shape. (4)

Volumetric shrinkage (volumetric change): The decrease in volume, expressed as a percentage of the soil mass when dried, of a soil mass when the water content is reduced from a given percentage to the shrinkage limit. (5)

Washout Nozzle: A tubular extension with a check valve utilized at the end of a string of casing through which water can be injected to displace drilling fluids and cuttings from the annular space of a borehole. (9)

Water content: The amount of water lost from the soil upon drying to constant weight at 105 degrees Centigrade; expressed either as the weight of water per unit weight of dry soil or as the volume of water per unit bulk volume of soil. The relationships between water content and soil water pressure can be referred to as the soil moisture characteristic curve. Depending upon whether the curve is determined with decreasing or increasing water content one may designate it as a desorption or adsorption curve, respectively.

Water in soil is subject to several force fields originating from: the presence of the soil solid phase; the dissolved salts; the action of external gas pressure; and, the gravitational field. These effects may be quantitatively expressed by assigning an individual component potential to each. The sum of these potentials is designated the total potential of soil water and may be identified with the partial specific Gibb's free energy of the soil water relative to free pure water at the same temperature. It should be noted that soil water is understood to be the equilibrium solution in the soil; pure water refers to the chemically pure compound HOH. (l)

Water table: The ground water surface in an unconfined aquifer at which the pressure is equal to that of the atmosphere; the surface between the zone of saturation and the zone of aeration. (9). The upper surface of ground water or that level below which the soil is saturated with water; locus of points in soil water at which the hydraulic pressure is equal to atmospheric pressure. (l)

Water-Cement Ratio: The amount of mixing water in gallons used per sack of cement. (9)

Water-holding capacity: The smallest value to which the water content of a soil or rock can be reduced by gravity drainage. (5)

Water-retention curve: See moisture-retention curve.

Weathering: All physical and chemical changes produced in rocks, at or near the earth's surface, by atmospheric agents. (l)

Weep Hole: A small diameter hole (usually 1/4 in.) drilled into the protective casing above the

ground surface that serves as a drain hole for water that may enter the protective casing annulus. (9)

Well Completion Diagram: A record that illustrates the details of a well installation. (9)

Well Screen Jetting (Hydraulic Jetting): When jetting is used for development, a jetting tool with nozzles and a high pressure pump is used to force water outwardly through the screen, the filter pack, and sometimes into the adjacent geologic unit. (9)

Well Screen: A filtering device used to retain the primary or natural filter pack; usually a cylindrical pipe with openings of a uniform width, orientation, and spacing. (9)

Yield stress: The stress beyond which the induced deformation is not fully annulled after complete destressing. (7)

Zero air voids curve (saturation curve): The curve showing the zero air voids unit weight as a function of water content. (1)

Zone of aeration: that part of the ground in which the voids are not continuously saturated. (l)

Zone of Saturation: A hydrologic zone in which all the interstices between particles of geologic material or all of the joints, fractures, or solution channels in a consolidated rock unit are filled with water under pressure greater than that of the atmosphere. (9)

SOURCES

(l) Soil Science Society of America. 1975. Glossary of Soil Science Terms. Madison, Wisconsin. 35 pp.

(2) Small Scale Waste Management Project (SSWMP). 1978. Management of Small Waste Flows. EPA 600/2-78-173, U.S. Environmental Protection Agency, Cincinnati, Ohio. 764 pp.

(3) Soil Conservation Service. 1977. Glossary of Selected Geologic and Geomorphic Terms.

Western Technical Service Center, Portland, Oregon. 24 pp.

(4) Brewer, R. 1976. Fabric and Mineral Analysis of Soils. R.E. Krieger Publishing Co., Huntington, New York. 482 pp.

(5) ASTM Committee D-18. 1979. Tentative Definitions of Terms and Symbols Relating to Soil Mechanics, ASTM D 653-42T. Annual Book of ASTM Standards, Part 19, Amer. Soc. for Testing and Materials, Philadelphia, Pennsylvania.

(6) Institution of Civil Engineers. 1976. Manual of Applied Geology for Engineers. Institution of Civil Engineers, London. 378 pp.

(7) International Society for Rock Mechanics. 1972. Final Document on Terminology, English Version. Comm. on Terminology, Symbols and Graphic Representation. 19 pp.

(8) Taylor, Donald W. 1948. Soil Mechanics. John Wiley and Sons, Inc. 700 pp.

(9) ASTM Committee D-18. 1990. D5092 Annual Book of ASTM Standards, Part 19, Amer. Soc. for Testing and Materials, Philadelphia, Pennsylvania.

(10) Nonaqueous Phase Liquids, 1991. EPA Groundwater Issue paper EPA/540/4-91-002.

APPENDIX B-1

KEY ELEMENTS FOR THE DEVELOPMENT OF
PHASE I HYDROGEOLOGICAL INVESTIGATION
REQUEST FOR PROPOSAL

PURPOSE:

The purpose of this document is to highlight the key elements to be included in a Phase I Hydrogeologic Investigation Request for Proposal (RFP). The goal of the Phase I investigation would be to define the field work necessary for completion of a ground water monitoring system.

SCOPE:

The following list itemizes the recommended contents of an RFP for a standard Phase I Hydrogeologic Investigation leading toward development of a ground water monitoring system. The more extensive Phase I study for remedial investigations of greenfield siting is not addressed in the scope of this document.

INTRODUCTION:

An actual RFP must reflect site specific details. For instance, a greenfield study would include a search for all items on the list. Where a facility already exists, the list must be modified to exclude existing information from the scope of work. In this case, the RFP would include (1) a summary of existing and reviewed data in the introduction, and (2) a task to verify any existing data that cannot be substantiated by the current documentation.

Where previous geotechnical and/or hydrogeological investigations have been conducted, it is recommended that these studies be reviewed prior to developing the RFP. These documents can direct the scope of further investigations and enhance or supplement available site information.

KEY ELEMENTS:

The following list presents the key components of a Phase I RFP. In the actual preparation of an RFP, it may not be necessary to include the rationale for an objective or task. However, the consultant and the owner/operator project manager should state in a clear and concise manner the reasons for each component of the RFP.

- OPENING STATEMENT

Invite a firm to develop a proposal to do the work specified in the scope of work to follow. The opening statement should provide details on the following items:

a. Introductory paragraph
b. Summary of the RFP tasks
c. Deadline dates and submittal locations
d. Staff responsible for the work

- INTRODUCTION

a. Describe site location and physical features
b. Summarize known geology/hydrogeology
c. Summarize existing wells, borings, piezometers, test pits, etc.

Rationale: To provide the potential firm with a general idea of the physical setting and known site characteristics significant to understanding the scope of work.

- SCOPE OF WORK

a. Project Objectives

Provide project objectives prior to describing tasks. The following statements are the primary objectives of a Phase I study:

1. Develop an understanding of the regional and site specific geology and hydrogeology as understood by the existing literature.

2. Determine if and how the potential for leachate migration from the site might impact ground water.

3. Develop a preliminary conceptual model of the hydrogeological system.

4. Determine the scope of work for a Phase II study.

Rationale: Project objectives provide direction for the individual tasks itemized in the scope of work.

b. Project Tasks

List project tasks in logical order. The following discussions present the primary Phase I tasks:

Task 1: Literature Review — Conduct a detailed literature review of regional and site specific (if any) geology, hydrogeology, and surface water hydrology. Key data to recover, if available include:

Land Use — Topographic maps, air photos

Climatology — Ten year average annual temperature and precipitation, and seasonal variations; landfill water balance

Surface Water Hydrology — Drainage patterns, runoff volumes, delineation of the 100 year flood plain

Geology — Soils maps, geologic maps, stratigraphic column

Hydrogeology —– Area boring/coring logs, cross sections, or regional stratigraphic columns illustrating hydrostratigraphic units, isopach of uppermost aquifer and significant confining units, regional averages and previously defined site specific values for hydraulic conductivity, transmissivity, storage coefficient or specific yield, thickness of uppermost aquifer and confining unit(s), gradient, flow direction, recharge source, general water quality (i.e., background), surface water/ground water interrelationships

Rationale: The literature review for the above listed data will (1) identify potential hydrogeologic complexities and contingencies that may require further analysis in later phased investigations, and (2) provide the basis for the development of the field work in a Phase II investigation.

Task 2: Conduct a site reconnaissance to identify significant surface characteristics, structural features, surface drainage, erosional conditions, and vegetation.

Rationale: This information is basic to an evaluation of the site specific geology, hydrogeology, and surface water hydrology.

Task 3: Conduct a well inventory of all wells within a one–half mile radius of the site. The inventory should include the following data if it exists: well locations, owner's name, total depth, aquifer tapped, depth of water, pumping rates, and water quality. Conduct an integrity check of existing site piezometers and monitoring wells. This check should include water level measurements.

Rationale: This data is used to assess the potential impact of leachate migration and to determine adequacy of existing monitoring wells and/or piezometers.

Task 4: Assimilate data recovered in Tasks 1–3 to develop the following items (if not already available): (1) site or regional topographic map; (2) description of the depositional history of the area; (3) delineation of uppermost aquifer and significant confining units in cross sections or fence diagrams (as necessary); (4) estimates of flow paths, gradients, and potential directions; and (5) description of water quality and water levels.

Rationale: These tools are used to define a conceptual model of the hydrogeologic regime.

Task 5: Develop a preliminary conceptual model of the hydrogeologic regime.

Rationale: A conceptual model is a picture or description of the physical system that the geohydrologist forms in his/her mind. Developing such a picture is essential to fully understanding potential ground water flow paths and, hence, determine appropriate locations for exploratory boreholes, piezometers, and monitoring wells. The key to developing a conceptual model is to "incorporate the essential features of the physical system under study. With this constraint, the conceptualization is tailored to an appropriate level of detail or sophistication for the problem under study"* (i.e., detection versus assessment or RI).

Task 6: Prepare a clear, comprehensive report. The Phase I report is to include the purpose of the investigation (objectives), the scope (tasks), summary of results, conclusions, recommendations, and the data in appendices. This report should not exceed 30 pages in length.

Task 7: Prepare the scope of work for the Phase II investigation** using the preliminary conceptual model and data developed in this Phase I study as an appendix to the Phase I report.

Rationale: The technical comprehensiveness and adequacy of the existing information recovered in this Phase I determines the need for or the extent of a Phase II study.

* Site Assessment Manual.

**In cases where the hydrogeology is relatively non–complex, a monitoring system can be designed solely on the basis of the Phase I investigative results.

APPENDIX B-2

EXAMPLE

MODEL STATEMENT OF WORK

PHASES I AND II

MODEL STATEMENT OF WORK

PHASES I AND II

HYDROGEOLOGIC INVESTIGATION

1.0 INTRODUCTION

_____ is interested in receiving proposals for conducting preliminary engineering for the expansion of a solid waste landfill. Work will consist of two projects: (1) completion of a hydrogeologic investigation, and (2) conceptual design of the sanitary landfill. This Request for Proposal (RFP) details the Scope of Work required to be addressed in the consultant's proposal for the hydrogeologic investigation.

1.1 PROJECT DESCRIPTION

[Engineer fill in Site Description]

GENERAL PHYSICAL FEATURES

[General Physical Features.]

GENERAL GEOLOGY

[General Geology, if known.]

HYDROGEOLOGY

[General Hydrogeology, if known.]

The policy of the client is to gather sufficient information to assess site hydrogeologic conditions, identify potential leachate pathways, and select appropriate placement of monitoring wells capable of defining the landfill's potential impact on the uppermost aquifer. This identification of ground water pathways and the successful placement of monitoring wells in these pathways represents the main goal of a site hydrogeologic analysis.

A hydrogeologic analysis is composed of a series of phases. This RFP for hydrogeologic investigations is similar to a performance specification were deliverables are well known and scheduled throughout the various phases of the study. The Phase I activities are primarily desk–based; summarizing literature, reports, and raw data into a literature whole with a maximum length of 30–40 pages of main text. Supporting documentation should be placed in appendices.

2.0 PHASE I – DESK STUDY

2.1 LITERATURE REVIEW – TASK 1

Reviewing published literature and the client's data is always required in Phase I project execution. This review may put limits on the extent, depth, and location of required borings and monitoring wells; define the type and number of soil and water quality analyses; and identify potential hydrogeologic complexities and contingencies that may require further analysis in later phased investigations. The latter may include the presence of faults, multiple and pre–existing land uses, potential for reversing ground water gradients, likelihood of variable background water quality, and uncertain stratigraphy.

The client would expect that the literature will be sufficiently reviewed so that important information about the site will not go unnoticed.

An important source of site information, in addition to appropriate state and federal agencies, is the client files of boring logs, engineering reports, and drawings. It is expected that the consultant will avail himself of pertinent site data from our District Engineering files.

2.2 AIR PHOTO COVERAGE

Photographs are available either as contact prints or enlargements at scales ranging from 1:20,000 to 1:4,000. Where the photographs have been taken with sufficient overlap, they may be used with a stereoscope to obtain a three–dimensional view of the terrain. We expect that air photos be obtained in stereo coverage for at least one date, and preferably for numbers of dates, for a site that has numerous air photo flight dates. Two copies of the stereoscopic coverage photos should be submitted to the client for distribution to District and Corporate staff.

2.3 SURFACE WATER

The surface water hydrology baseline study will describe drainage systems, flow characteristics, water quality of streams and water bodies, and aid in determining ground water/surface water relationships at the site. This information will document the baseline conditions and form the basis of assessing environmental impacts.

2.3.1 100 Year Flood Zones

The client's policy is to avoid disposing of solid waste in areas subjected to flooding. State regulations require documentation that the proposed site will be above the 100 year flood zone. These data may be obtained by review of the most recent U.S. Geological Survey, Army Corps of Engineers, or the Federal Insurance Administration 100 Year Frequency Floodplain Map for the area (if available).

The consultant should prepare a site plan utilizing the client's most recent aerial topography to delineate areas subject to flooding during 100 year flood events.

2.3.2 Wetlands, Stream Flow, and Runoff

One of the client's most important environmental policies is the prevention of surface water pollution. This policy is directed at protection of wetlands and control of runoff from waste disposal areas into surface water bodies. Consultants, therefore, must include on base maps locations of wetlands and all surface water found in the area of the proposed site within a one–half mile radius of the site's boundary.

As part of the available data assessment, stream and river flows — where monitored by state and federal agencies — should be documented in this report.

The final goal of this section is to define, under state regulations, if the landfill is:

   a.   Located on a flood plain; and
   b.   Located within 300 feet of any classified body of water.

2.4 CLIMATIC DATA

In ground water investigations, records of precipitation, temperatures, wind movement, evaporation, and humidity may be essential or useful supplemental data. Climatic data are used primarily for estimating the seasonal variations and amounts of precipitation which may be available for ground water recharge.

The consultant should present precipitation (10 year average) and other pertinent climatological data, as required, to complete surface runoff calculations and landfill water budgets.

2.5 GROUND WATER

The client's greatest environmental consideration in landfill siting is the regional and site specific hydrogeology. The consultant must fully document that the landfill is sited in a manner which protects ground water quality.

The primary purposes of the ground water study (existing data review) in Phase I is the following:

• Define and quantify, to the extent practical, the overall ground water flow systems (occurrence, recharge, discharge, direction, and rate of movement) at the site areas and the vicinity;

• Define the present ground water quality in the areas of concern;

• Determine the maximum high ground water table elevation; and

• Evaluate the importance of the ground water resource which might be affected by the operation of a landfill facility.

2.5.1 Geology/Hydrogeology

This Phase I task requires that all available data is reviewed and assimilated into a report that describes the known regional geology and hydrogeology for the site area. Sections on regional geology and hydrogeology should contain all pertinent information on federal, state, and university studies of the geology and hydrogeology for the site area.

Maps should support the text with appropriate regional cross sections. This Phase I review should "set the stage" for Phase II field tasks. The location and logs of existing test pits, borings, or local water well logs should be included for supplemental information. Particular attention should be paid to primary sand and gravel recharge areas of significant ground water aquifers.

2.5.2 Well Inventory

The locations of all wells (differentiated between public water, domestic, industrial, and other) and springs should be shown on a base map with a scale of 1 inch = 500 feet within 2,000 feet of the property boundaries.

Supporting information on the proximity and withdrawal rates of users and the availability of alternative drinking water supplies must be included in the review documentation. The well inventory should include such available data as: location, owner, surface elevation, aquifer tapped, water quality, well depth, casing size and depth, depth to water, estimated rate of pumpage or use, and date of inventory.

2.6 SOILS AND VEGETATIVE COVER

Soil maps and reports are not usually as readily available as topographic and geologic maps, and vary more widely with respect to the quantity and quality of the information they contain. Soil maps and reports supply information on soil characteristics and surface gradients which influence runoff and infiltration.

The consultant should contact appropriate data sources and present a plan documenting site soil characteristics based upon readily available data.

2.7 BASIC DATA CHECKLIST

The consultant should compile and present normally available data in published and unpublished reports and records as an aid in planning the data to be obtained by field investigations and tests. The following data list is suggested given the extent of existing readily available data.

The client requests that the consultant present the following noted data in the Phase I report. If additional data is obtained, the consultant should include it as appropriate.

PHASE I BASIC DATA CHECKLIST

A. Maps, Cross Sections, and Fence Diagrams

1. Planimetric
2. Topographic
3. Geologic

    (a) Structure
    (b) Stratigraphy
    (c) Lithology

4. Hydrologic
    (a) Location of wells, observation holes, and springs
    (b) Ground water table and piezometric contours
    (c) Depth to water
    (d) Quality of water
    (e) Recharge, discharge, and contributing areas

5. Vegetative cover
6. Soils
7. Aerial photographs

B. Data on Wells, Observation Holes, and Springs

1. Location, depth, diameter, types of well, and logs
2. Static and pumping water level, hydrographs, yield, specific capacity, quality of water
3. Present and projected ground water development and use
4. Corrosion, incrustation, well interference, and similar operation and maintenance problems
5. Location, type, geologic setting, and hydrographs of springs
6. Observation well networks
7. Water sampling sites

C. Aquifer Data

1. Type, such as unconfined, artesian, or perched
2. Thickness, depths, and formational designation
3. Boundaries
4. Transmissivity, storativity, and permeability
5. Specific retention
6. Discharge and recharge
7. Ground and surface water relationships
8. Aquifer models

D. Climatic Data

1. Precipitation
2. Temperature
3. Evapotranspiration
4. Wind velocities, directions, and intensities

E. Surface Water

1. Use
2. Quality
3. Runoff distribution, reservoiric capacities, inflow and outflow data
4. Return flows, section gain or loss
5. Recording stations

2.8 ENVIRONMENTAL REVIEW

The Federal Regulations contain a criterion which required that solid waste facilities or management practices not harm or threaten any plants, fish, or wildlife in danger of extinction. In general, states have developed location standards which would prevent the siting of landfills in areas which would cause or contribute to the endangerment of threatened species. _____ thoroughly investigates proposed sites for potential endangered species or practices which may result in the destruction or adverse modification of the critical habitat of endangered or threatened species identified in 50 CFR Part 17. The consultant must assess if the proposed facility will impact endangered species. This review of available data must be performed by a Certified Ecologist. This environmental review section may be subcontracted to an experienced ecologist. The Phase I botanical work should include a review of air photos to survey botanical zones.

Many states also require mapping of archaeological and historic sites as part of the overall permit application. _____ also promotes the protection of archaeological and historic sites. We, therefore, require the mapping of archaeological and historic sites within 2,000 feet of the perimeter of the facility.

The Phase I environmental review is based on document ranges of endangered or threatened species supplemented by a field visit. Archaeological and historic sites are also reviewed by literature surveys, review of air photos, and a site visit. Detailed site data will be generated within the Phase II field investigation scope of work.

2.9 SITE VISIT – TASK 2

The consultant should plan to visit the site to gain a thorough understanding of current site conditions.

Terminology, units, abbreviations, balance, and emphasis in the final report. The draft and final report should be read thoroughly by the Project Manager and Senior Hydrogeologist assigned to the project.

The consultant should have, upon the conclusion of the Phase I investigation, an understanding of the site's basic geologic and hydrogeologic environment. This information must be used to formulate a specific Phase II investigation to fully assess the proposed site's hydrogeological conditions.

3.0 PHASE II – SUBSURFACE HYDROGEOLOGIC INVESTIGATION

3.1 INTRODUCTION

Subsurface hydrogeologic investigations must be of sufficient intensity to determine the conditions that may influence the design and the construction of both the landfill and the ground water monitoring system. The extent of geologic investigation required for a particular site depends on: (1) complexity of the site conditions, (2) size of the landfill construction, and (3) potential damage if there is functional failure in the liner.

Detailed geotechnical/hydrogeologic exploration and location of the monitoring system would typically consist of four tasks: (1) determining and interpreting subsurface conditions, (2) taking samples for soil and rock tests, (3) installing the monitoring wells, and (4) preparation of a final report detailing all findings.

3.2 GEOLOGICAL MAPPING – TASK 2–1

Conduct a geologic mapping program on the surficial material present at the site. If sufficiently detailed maps are available for the assessment, this task may not be required under the Phase II scope. The primary purpose of the mapping is to aid in monitoring well locations. The maps are expected to be extended on the basis of the later drilling program.

3.3 GEOPHYSICAL SURVEY – TASK 2–2

Geophysics can often be employed during the hydrogeological investigation to solve specific problems or merely to provide information cheaply and rapidly.

On the basis of geophysical surveys, depth to bedrock, water tables, and anomalous geologic conditions can be assessed for later drilling. Seismic velocities can also be used for later ripability analyses.

The client expects that geophysical seismic refraction methods will assist in selection of locations for geotechnical borings. The seismic survey should tie into existing borings for control of depth and for correlation of lithologies with velocities. The estimated seismic survey lengths should be costed on the basis of [ ] feet. All geophysical work should be performed and interpreted by adequately trained geophysicists.

3.4 FIELD DRILLING PROGRAM – TASK 2–3

During Phase II, test holes must be put down and logged in the landfill's foundation and potential borrow areas. Selection of drilling locations must be based on geophysical interpretation and results of Phase I studies. These test holes must be deep enough to penetrate all pertinent materials. The number and spacing of test holes must be adequate for correlation in both longitudinal and transverse directions for complete interpretation of any condition that may influence the local permeabilities. Geologic structural features, such as faults, folds, and joints, must be obtained during drilling, and sufficient information is required on soils to classify them and to determine their location, thickness, and extent. Test holes can be put down by drilling, or by excavating pits or trenches in areas of shallow soils.

The data gathered during this drilling program are analyzed on the site, and behavior characteristics and engineering significance of the materials and conditions are evaluated. From this analysis and evaluation, the geologist and the engineer determine what materials are to be sampled and what laboratory analyses are needed. The necessary samples are obtained by using appropriate sampling procedures. Any additional or special in–place field tests should be completed at this time. Because we believe weathered and fractured bedrock may represent a major hydraulic component of ground water flow, holes should be advanced at least into unweathered bedrock. These holes should be diamond cored with at least an NX–sized core barrel. For estimating costs, provide footage costs for coring and estimate feet of coring per hole. Unconsolidated materials should be sampled every 5 feet and at changes in materials to obtain a clear picture of the lithology of the overburden.

3.4.1 Logging Test Holes

Logging is the recording of data concerning the materials and conditions in individual test holes. It is imperative that logging be accurate to provide a true picture of subsurface conditions. It is equally imperative that recorded data be concise and complete and be presented in descriptive terms that are readily understood and evaluated in the field, laboratory, and design office.

The basic element of logging is a geologic description of the material between specified depths or evaluations. This description includes such items as name, texture, structure, color, mineral content, moisture content, relative permeability, age, and origin. To this must be added any information that indicates the engineering properties of the material. Examples are gradation, plasticity, and the unified soil classification symbol determined by field identification. In addition, the results of any field test, such as the standard penetration test, must be recorded along with the specific vertical interval that was tested.

After a hole is logged, it should be plotted graphically to scale and properly located both vertically and horizontally on the applicable cross section or profile. Correlation and interpretation of these graphic logs indicate the need for any additional test holes and their location, and permit the plotting of stratigraphy and structure and the development of complete geologic profiles. Analysis of the geologic profile frequently gives more information on the origin of the deposits. It should be noted that driller's logs will not be acceptable to the client.

The client expects that a professional geologist or soils engineer will be present at the drill rig at all times that drilling is underway for logging and supervising drilling operations. All logging should be completed on Field Assessment Forms. Copies of these forms can be obtained at Regional and District offices.

3.4.2 Laboratory Tests – Task 2–4

Laboratory tests of soils collected during the field drilling task defines four basic properties of soils and their suitability for use on landfill sites. These soil characteristics can be summarized:

1. Strength
2. Compressability
3. Permeability
4. Chemistry

The following laboratory tests are designed to provide information on the quantity and physical characteristics of both borrow and landfill cut areas. Guidelines for the numbers of tests are given below; however, good engineering judgment must apply to adequately describe the quality and quantity of the individual soils located on the property.

The investigations will generally consist of (1) either drilling and sampling soil borings or excavating shallow test pits, (2) obtaining bulk samples of representative soils, and (3) evaluating these samples for use as borrow material.

Soil samples will be evaluated by performing the following laboratory tests:

A.   Particle–size analysis of soils (ASTM D422–63).

B.   Liquid and plastic limits (ASTM D4318–83).

C.   Classification of soils for engineering purposes (ASTM D248769).

D.   Moisture/density relationship utilizing five compaction points (standard proctor) to determine the maximum dry density and ultimate moisture content (ASTM D698–78).

E.   Organic content by the burn–off method (ASTM D6341–79).

F.   Permeability at 90a of standard proctor density within 12 percentage points of optimum moisture content. A falling head test method using back pressure should be done in general accordance with Corps of Engineers' Manual EM 1110–2–1906, "Laboratory Soils Testing," Appendix VII.7.

G.   Strength test: "Test for unconsolidated, undrained strength of cohesive soils in triaxial compression (ASTM D2850–70), or in general in accordance with Appendix X of the Corps of Engineers' Manual EM 1110–2–1906 for non–cohesive soils.

H.   Consolidation tests: Test for one–dimensional consolidation properties of soils (ASTM D2435–70).

I.   Cation Exchange Capacity Tests. The USEPA tests methods 9080 and 9081 were adopted from standard methods for soil analysis used by the American Society of Agronomy (Chapman, H.D., Cation Exchange Capacity in C.A. Black [ed.] Methods of Soil Analysis, American Society of Agronomy, Madison, WI, Part 2 [1965]).

J.   Soil pH. By electrometric method reference in (I) above.

The particle–size, liquid and plastic limit tests should be costed at three per boring. Care must be taken to select particle–size test samples for those zones being selected for piezometer monitoring. The classification of soils for engineering purposes must be made initially during drilling and reviewed during laboratory testing to confirm the classification. Moisture density tests should be made at the rate of one per each material type found during drilling, with more tests performed on suitable liner material. Estimate ten tests total for moisture density. Both organic content and permeability tests should be run for each material type, unless sound engineering judgment suggests that additional tests are required. For purposes of this proposal, estimate ten tests for organic content and laboratory permeability. Consolidation tests should also be made for each material type. Cation exchange tests, which determine the potential capacity of the soil to exchange cations with leachate, should be run on every different material found at the proposed site. Soil pH tests should be run in the laboratory program for each material found at the site.

3.4.3 Installation of Monitoring Wells and Piezometers – Task 2–5

The estimated borings required in this program provide an opportunity to assess the hydrogeologic environment to:

1.   Define in–place permeability of each geologic unit;
2.   Assess the directions of ground water flow;
3.   Calculate the rates or velocity of flow and estimated flux;
4.   Determine the maximum high ground water table elevation and aquifer types;
5.   Define ground water recharge and discharge areas; and
6.   Determine background water quality.

In order to convert the soil borings into reliable monitoring wells and piezometers, the consultant must construct and document these installations according to_____ specifications. Deviations from these specifications without written approval from the client will require the consultant to reinstall the installation at their expense. Professional field control of installation of monitoring wells and piezometers cannot be overemphasized.

The client's general plan for installation of wells and piezometers would require that soil borings near the edge of the landfill, that would not be covered by landfilling operations, would be 2 inch PVC. Installation must be according to the _____ specification in Appendix A.

These monitoring wells should be completed in the uppermost aquifer. The uppermost aquifer would be the first continuous permeable zone under the base of the proposed landfill which could supply economic quantities of water to wells or springs off–site. Since the site has several potential upgradient areas, one or two monitoring wells should be located in the upgradient directions and two to three monitoring wells sited in the downgradient directions. The actual location and depth of wells for this ground water monitoring must be based on two major factors:

1.   The direction of ground water flow; and
2.   The first or "uppermost" aquifer depth.

The wells to be installed under this monitoring phase must be designed to intercept potential landfill leachate downgradient of the waste area in the first continuous saturated and permeable zone adjacent to the waste disposal area. All the technical work to be performed in this Phase II effort must be directed toward this goal. The final performance assessment of the consultant and peer review will be directed toward determining if the consultant has met this goal.

The second type of ground water assessment installation is 3/4 inch PVC piezometers. The borings not selected for 2 inch monitoring well installations should be completed as small diameter piezometers where only water levels will be taken. The construction of these 3/4 inch PVC piezometers should be according to the THE CLIENT specification for monitoring well installation with a number of acceptable modifications:

1.   The piezometers should be of 3/4 inch PVC with machine cut slots to obtain the open screened area. The screened length should be based on drilling observations of permeable zones and should adequately monitor pressure heads within individual zones.

2.   The location or depth of the piezometers can be varied from the uppermost aquifer zone. Good engineering judgment should be used to obtain the required hydraulic data to reach the six goals previously listed in this section. The consultant should select adequate numbers of piezometers to assess permeability, flow directions, ground water velocity, water levels, and vertical and horizontal gradients for the geologic materials located on site. This may include both consolidated and unconsolidated materials for primary and secondary porosity.

3.   Gravel or filter sand packing of the screens, construction details, bentonite seals, grouting, protective casings, and surveying should be equivalent to the the client's specification for monitoring wells.

4.   Small diameter piezometers can be constructed in central areas where landfilling may destroy the piezometer at some point during the life of the facility.

3.4.4 Performance Tests – Task 2–6

The client requires that all piezometers and monitoring wells be tested for permeability through appropriate hydrogeologic methods. The field permeability testing of geologic materials through variable head or constant head techniques represents a relatively quick method that will establish defensible field permeability for each of the geologic materials present on site. THE CLIENT expects the consultant to be fully conversant with these techniques and adequately document the performance of the test. These tests can be performed quickly; however, early time data must be taken to establish reliable values for permeability for each monitoring well and piezometer.

3.5 FIELD ECOLOGY SURVEY – TASK 2–7 (if required)

The purpose of conducting a summer ecological survey is to take advantage of the season to obtain information that will be used to address the issue of threatened and endangered plant and animal species.

The summer biological field survey should consist of three field activities:

- Botanical survey to refine previous air–photo–based map surveys;
- Early morning bird surveys; and
- Small mammal trapping.

A vegetation map should be produced based on an examination of aerial photographs of the site (obtained during Phase I). All areas in which vegetation types could not be readily identified from aerial photographs will be checked in the field. In addition, all mapped boundaries between adjacent vegetation types will be reviewed in the field for accuracy. Known or potential habitat for sensitive plant species should be identified during the survey.

Bird surveys should be conducted for 2–4 hours on two mornings. In order to observe a high proportion of the species occurring at the site, these surveys will be conducted primarily in the vicinity of springs, drainages (including ephemeral and intermittent streams), rocky hillsides, and areas of woodland. All birds seen and heard should be identified to species and, whenever possible, to sex.

A minimum of 25 Sherman live traps should be set on two nights to determine the occurrence of small mammals. In order to detect a high proportion of species present at the site, traps should be placed primarily in the vicinity of springs, drainages, rocky hillsides, and areas of woodland.

All small mammals should be identified as to species; sex; and, whenever possible, age. In addition, all wildlife species observed during these three field activities must be recorded.

As with the previous Phase I activity, only Certified Ecologists are acceptable for performing this work.

3.6 HISTORY AND ARCHAEOLOGY – TASK 2–8 (if required)

A professional archaeologist should be retained to document the cultural importance of the area. A regional overview of the area has been developed in Phase I based primarily on literature review. In addition, a Phase II reconnaissance survey should be conducted to confirm the presence or absence of significant historical or archaeological sites on the site.

The reconnaissance survey includes the deployment of multiple transects which involve systematic examination of samples of subsurface conditions at 15 meter intervals. Standard survey procedures also include visual examination of all disturbed earth areas such as road cuts, farmers' fields, spoil from rodent burrows, etc. These procedures are necessary in heavily forested areas with heavy ground cover conditions.

These procedures should result in the field check of known or suspected locations of archaeological and historical remains. They may also result in the discovery of previously unknown remains. The precise location of these new sites, together with maps which indicate areas which failed to produce evidence of remains, will be prepared. All recovered archaeological and historical debris should be washed, marked, and cataloged, and will be deposited in the appropriate museum of anthropology.

3.7 PHASE II REPORT – TASK 2–9

"Introduction" should include the background, contract and proposal citation, and objectives of the assessment. Abbreviations for the client and facility to be used throughout the report should be included in this section.

"Setting" should include regional and local description of geographic matters, such as climate, rainfall, flooding, runoff and drainage, attitude, land use and agricultural soils, surface geology, topography, and other pertinent factors.

"Facility Description" should locate the site on a regional map (first figure), describe the site features, land use, land surface elevations, local drainage patterns, and slopes.

"Regional Hydrogeology and Groundwater Quality" should include a discussion of the major aquifers, their usage, stratigraphy, ground water conditions — gradient directions and magnitudes, elevation of the ground water surface. An indication of regional ground water quality and related issues should be included.

"Site Hydrogeology and Ground Water Quality" should provide locations of borings and piezometers (second figure), and include discussions of the site stratigraphy, subsurface profiles (additional figures) and shallowest water–bearing soils. Shallow ground water conditions — gradient directions and magnitude, elevation of water table or piezometric surface — should be covered in detail.

This section should also include a description of local ground water quality, with significant anomalies and frequencies, and seasonal fluctuations noted.

"Assessment" should combine the hydraulic analysis of the site hydrogeology, and the analysis of the site's soils and geology. A reasonable picture of the potential for contaminant migration from the waste facility from seepage and horizontal or vertical movement through and off the site should be produced. Imminent and potential ground water quality risks to the public must be identified.

"Recommendations" should refine the quality of the hydraulic and water quality analysis and evaluate the synthesis of hydrogeologic and geologic analysis, and work towards statistical significance if needed. Recommendations should be presented in a natural and supportive way so that the client is logically drawn toward them.

The consultant must answer the following requirements:

a. Determine the availability, quality, and quantity of on–site soil for cover material;

b. Evaluate the influence that geologic factors would have on the ease of excavation and potential for ground water and surface water pollution;

c. Determine the maximum high ground water table elevation, and ground water flow patterns;

d. Determine background water quality of ground water at the site and of surface water at the most likely discharge area; and

e. Evaluate the importance of the ground water resource which might be affected by the operation of a landfill facility.

"Figures" should include a vicinity map, site plan with boring and monitoring well locations, two subsurface profiles (N–S and E–W), and a local geohydrologic column at a minimum. The site plan should show the location of the gradient. If ground water anomalies are clearly present, additional figures may highlight this through anomaly maps which over–extend the size of anomalies because of sparse and distant control points.

"References" should include all cited literature, the consultant's and client's files, conversations with agency personnel, and client reports. Confidential reports completed for other clients with nearby facilities should not be explicitly cited.

"Appendices" should include any available field exploration and soil laboratory testing, ground water monitoring network, ground water sampling and analysis plans, and analyses of ground water samples. Specific tables, such as logs of borings, summary of ground water level measurements, well completion data, soil laboratory tests, field water quality analyses, laboratory water quality analyses, and chain of custody records, are appropriate. Personnel involved and procedures utilized in the various phases of the project should be identified. Examples include notations such as "water levels were measured with a steel tape by...;" "elevations were determined with a topographic survey by...;" "field measurements were made with a conductivity meter, pH meter, and thermometer by...;" etc. Data should be reported to significant figures. Basic data collected during the investigation must be provided so that the client can perform an independent assessment of the information generated by the investigation.

The draft and final reports should be thoroughly reviewed for technical accuracy and editorial consistency. Reports should be checked for numerical, citation, and typographical errors as well as tone, consistency, emphasis, and logic. A poorly written report will erode credibility, diminish future opportunities, and cause later expense and embarrassment.

3.8 REPORT PRESENTATION

The client should be given five copies of the draft final report to review. The client may correct any facility–related or procedural errors that the consultant has made; hence, this draft must be as clear and accurate as possible.

It is generally more vivid, and consequently more effective, for the consultant to review the report in the physical presence of the client.

APPENDIX B-3

MODEL STATEMENT OF WORK FOR CONDUCTING A

REMEDIAL INVESTIGATION AND FEASIBILITY STUDY AT THE

EXAMPLE LANDFILL SITE,

[COUNTY], [STATE]

This document presents the Statement of Work (SOW) to conduct a Remedial Investigation and Feasibility Study (RI/FS) at the Example Sanitary Landfill NPL site located in [County], [State]. The purpose of this SOW is to provide the direction and intent of the RI/FS. Within 60 days of the effective date of the Consent Order, a RI/FS Work Plan will be submitted which will provide detailed guidance on the execution of the RI/FS.

The purpose of the RI is to determine the nature and extent of contamination at the Example Landfill site. The purpose of the FS is to develop and evaluate appropriate remedial action alternatives based on the RI data and report. All personnel, materials, and services required to perform the RI/FS will be provided by the Potentially Responsible Parties (PRP).

The tasks described herein are grouped into the following three categories:

- Plans and Management;
- Remedial Investigation (RI); and
- Feasibility Study (FS).

The Work Plan that will be developed pursuant to this SOW will present a phased, iterative approach that recognizes the interdependency of the RI and FS. The overall organization and interactive nature of this approach are illustrated in Figure 1. Please note that the activity sequence depicted in Figure 1 is not consistent with the topical sequence of presentation in this SOW.

B3-1

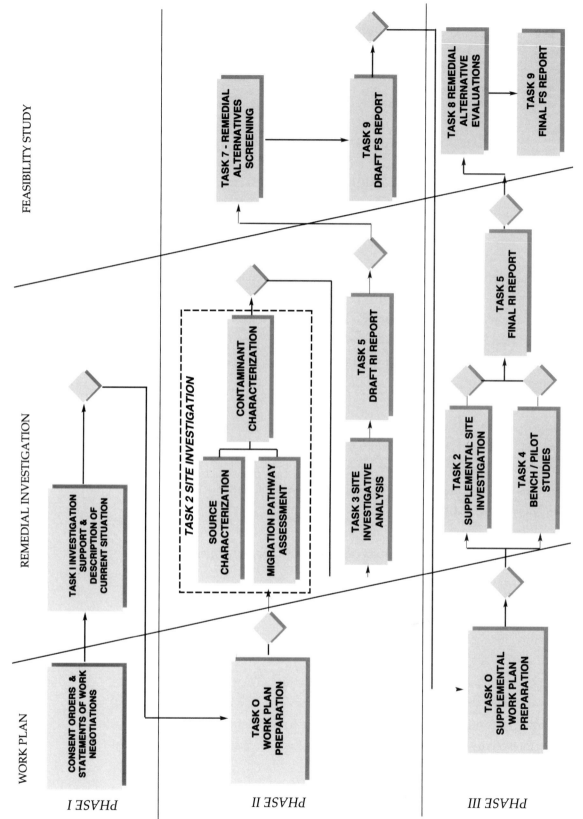

PROPOSED REMEDIAL SELECTION PROGRESS IN ACCORDANCE WITH SARA

The primary intent of the phased approach is to minimize the need for conducting post–FS or supplemental RI/FS activities by thorough characterization of the migration pathways and early identification of the site specific data requirements associated with the applicable remedial technology.

Brief discussions of the major RI/FS tasks are presented, by major topical categories, in the following sections.

PLANS AND MANAGEMENT

TASK 0 – WORK PLAN PREPARATION

A RI/FS Work Plan will be prepared for the Example Landfill site that details the technical approach, personnel requirements, and schedule for each task described in this SOW. The schedule will show the implementation of tasks and submission of deliverables in weeks subsequent to regulatory (e.g., USEPA and [State Agency]) approval and acceptance of prior deliverables. Incorporated in this Work Plan will be the following specific plans: Sampling Plan

A Sampling Plan that addresses all data acquisition activities will be prepared. The plan will contain a statement of sampling objectives, specification of equipment, required analyses, sample types, and sample locations and frequency. The plans will address specific hydrologic, hydrogeologic, and air transport characterization methods including, but not limited to, geologic mapping, geophysics, field screening, drilling and well installation, flow determination, and sampling. The application of these methods will be described for each major subtask within the site investigation (e.g., waste characterization, migration pathway assessment, and contaminant characterization). The plan will also identify the data requirements of specific remedial technologies which may be necessary to evaluate remedial alternatives in the FS.

Quality Assurance Project Plan

A Quality Assurance Project Plan (QAPP), prepared in accordance with current USEPA guidance, will be submitted with the Sampling Plan. The QAPP will specify the analytical methodologies and protocols to be used at the various stages of the site investigation. Specific methodologies will be defined for field screening of samples, waste and contaminant characterizations, and bench and pilot treatability testing. Currently, it is anticipated that the USEPA Contract Laboratory Program protocols will be followed for all waste and contaminant characterization analyses; however, unforeseeable circumstances may necessitate the use of special analytical methods.

Health and Safety Plan

A Health and Safety Plan will be prepared to address hazards that the investigation activities may present to the investigation team and to the surrounding community. The plan will conform to all applicable regulatory requirements and guidance, including the USEPA Standard Operating Safety Guides, and will detail personnel responsibilities, protective equipment, procedures and protocols, decontamination, training, and medical survellance. The plan will identify problems or hazards that may be encountered and their solutions. Procedures for protecting third parties, such as visitors or the surrounding community, will also be provided.

Endangerment Assessment Plan

An Endangerment Assessment Plan will be developed for quantifying the risks posed by the Example Landfill and for analyzing the public health impacts of the remedial alternatives. The methodology presented in this plan will conform to the Superfund Public Health Evaluation Manual (ICF, 1986).

Data Management Plan

A Data Management Plan will be developed to document and track investigation data and results. The plan will identify and establish laboratory and data documentation materials and procedures, project file requirements, and project–related progress reporting procedures and documents.

ATSDR Health Assessment

The Work Plan for the site should also provide for collection of adequate information to support an ATSDR Health Assessment which is required by SARA. Since the health assessment will be prepared by ATSDR, all draft Work Plans and support documents will be submitted for ATSDR review and comment to ensure that their needs and requirements are being met. In the event that the health assessment has already been completed by ATSDR, the RI report will include and address the findings of that report.

As shown on Figure 1, the preparation of the project plan will be preceded by an evaluation of the existing information and initiation of investigative support activities (Task 1).

The Work Plan will be submitted in accordance with the schedule defined in Section VIII (Work to be Performed) of the Consent Order. Specifically, the RI/FS Work Plan will be developed and implemented in conformance with all provisions of the Consent Order and this SOW, and the standards set forth in the following statutes, regulations, and guidance:

- Section 121 of SARA;

- USEPA "Guidance on Remedial Investigations under CERCLA," dated May 1985, as amended;

- USEPA "Guidance on Feasibility Studies under CERCLA," dated May 1985, as amended;

- National Contingency Plan, dated November 1985, as amended; and

- Any additional guidance documents provided by the USEPA.

REMEDIAL INVESTIGATION

Objectives and Scope

The objectives of the RI are to:

- Characterize the source(s) of potential contamination;

- Characterize the hydrogeologic setting to determine most likely contaminant migration pathways and physical features that could effect potential remedial actions;

- Determine the migration rates, extent, and characteristics of any contamination that may be present at the site; and

• Gather data and information to the extent necessary and sufficient to quantify risk to public health and the environment and to support the development and evaluation of viable remedial alternatives in the FS.

The remedial investigation consists of six tasks:

Task 1: Description of Current Situation and Investigative Support

Task 2: Site Investigation

Task 3: Site Investigation Analysis

Task 4: Bench/Pilot Testing Studies

Task 5: Reports

Task 6: Community Relations Support

A description of each of these tasks is presented in the following section.

TASK 1 – INVESTIGATIVE SUPPORT AND DESCRIPTION OF CURRENT SITUATION

Site Mapping

An accurate topographic map of appropriate working scale will be prepared. The base map will have a scale of 1 inch to 100 feet (1″ – 100′) and 2–foot contour intervals. A base map of the site will be prepared from this topographic map. The base map will illustrate the locations of wetlands, floodplains, water, features, drainage patterns, tanks, buildings, utilities, paved areas, easements, right–of–ways, and other pertinent features. Larger scale maps will be produced from the base mapping, as necessary.

Surveying will be required to establish horizontal and vertical controls for sites of the work relative to the National Geodetic Vertical Datum of 1929. In addition to the topographic map, a grid plan will be prepared using the base map and grid overlay at a nominal scale of the map. This grid plan will show the location of existing monitoring wells, sampling locations, and water supply wells.

A legal description of the property will be reviewed and field verified. The intent is not to perform a boundary survey, but to locate the boundaries so that the future activities do not carry over onto adjacent property without proper permission.

Meets and Bound

A legal description of the site will be assembled from existing county and township records and results of the site survey.

Access Arrangements

The necessary arrangements will be made to facilitate access to the site and surrounding parcels. These arrangements will include negotiating access agreements with the appropriate landowners and the demarcation clearance of all buried utilities and construction of access roads.

Preparation of Support Facilities

Arrangements will be made to construct the appropriate support facilities and/or procure the equipment necessary to perform a hazardous site investigation. This includes preparation of decontamination facilities, utility hook-ups, and site access control stations.

Description of Current Situation

The background information pertinent to the site and the environmental concerns will be described and the purpose of the RI will be further detailed. The data gathered during previous investigations will be reviewed and evaluated. Regional information will be obtained from available U.S. and State Geological Survey reports. The existing site information that will be reviewed will include but will not necessarily be limited to:

- State Geological Survey files
- County Soil Conservation Service reports
- Various consultant files
- Aerial photographs
- Historical water quality data
- Disposal records (if available)

In addition to this literature search, on-site activities may be used to confirm and/or update certain information. For example, existing monitoring wells may be inspected to determine if they are functional. Also, the location and status of selected water supply wells may be field verified.

Information and data that are gathered during these initial steps will be used to generate a Preliminary Site Evaluation Report that will address the following:

Site Background

A summary of pertinent boundary conditions, general site physiography, hydrology, and geology. A complete site history as it pertains to waste disposal activities and ownership transfer will also be prepared.

Nature and Extent of the Problem

A summary of actual or potential on–site and off–site health and environmental affects will be prepared. Threats or potential threats to public health will be emphasized.

History of Response Actions

A history of response actions as conducted by local, state, or private parties will be prepared.

Definition of Boundary Conditions

Site boundary conditions will be established to limit the areas of investigation. The boundaries will be set so that the on–site activities will cover the contaminated media in sufficient detail to support the FS. Boundaries for site access control and site security will also be identified. The boundaries of the study area may or may not correspond to the property boundaries.

Identify Potential Receptors

Private and public water supply wells within a 2–mile radius of the site will be identified. If possible, well construction details for these wells and other private water supply wells, which may have been previously sampled, will be obtained. A table summarizing the known construction details will be prepared and submitted with the original drilling logs, as available.

<u>Develop Site Conceptual Model</u>

A description of the physical site conditions including geology, meteorology, hydrology, and hydrogeology will be developed from the existing site data. This description will constitute the site conceptual model. All subsequent site investigation activities will be focused on refining and validating this model. The conceptual model will focus on the groundwater flow system and will be based on the depositional history, inferred recharge and discharge mechanisms, estimated topographic and hydraulic gradients, and existing and past land use patterns.

As shown on Figure 1, the Investigative Support and Description of Current Situation (Task 1) will be conducted prior to, or concurrent with, the Work Plan Preparation (Task 0). The Preliminary Site Evaluation Report will be submitted as supporting documentation with the Work Plan.

TASK 2 – SITE INVESTIGATIONS

Investigations necessary to characterize the site and its actual or potential hazard to public health and the environment will be conducted. The investigations will result in data of adequate technical content to support the development and evaluation of remedial alternatives during the FS. Investigation activities will focus on problem definition and data to support the screening of remedial technologies, alternative development and screening, and detailed evaluation of alternatives.

The site investigation activities will follow the plans set forth in Task 0. All sample analyses will be conducted at laboratories following USEPA protocols or their equivalents. Strict chain–of–custody procedures will be followed, and all samples will be located on the site map (and grid system) established under Tasks 0 and 1. A description of the types of investigations that will be conducted is presented below.

<u>Source Characterization</u>

An investigation will be carried out to characterize the physical and chemical aspects of the waste materials and the materials in which they are contained. The investigation of these source areas will involve obtaining data related to:

- Waste characteristics (type, quantity, chemical and physical properties, and concentrations); and

- Facility characteristics (type and integrity of containment, leachate collection systems, and drainage control).

It is anticipated that this information will be obtained from a combination of existing site information, field inspections, and site sampling activities. Field investigations may be needed to determine the integrity of the landfill covers.

The source characterization will culminate in the preparation and submittal of a technical memorandum. This memorandum will summarize the findings of the source characterization and will recommend parameters, or classes of parameters, which will be the focus of subsequent contaminant characterization studies.

Migration Pathway Assessment

The migration pathways at the Example Landfill site will be physically characterized through the following types of investigations:

Hydrogeologic: A hydrogeologic study will be performed to further evaluate the subsurface geology and characteristics of the water bearing formations. This study will define the site hydrostratigraphy, controlling geologic features, zones of preferential groundwater transmission, and the distribution of hydraulic heads within the water bearing formations. The results of this study will be combined with the existing site data described in the preliminary site evaluation report and the results of the source characterization to define the groundwater flow patterns and to predict the vertical and lateral extent of contaminant migration. These predictions will form the rationale for locating and designing monitor wells and the subsequent contaminant characterization.

Hydrologic: Drainage patterns and runoff characteristics will be evaluated for the potential of erosional transport. Staff gauges may also be used to evaluate the hydraulic connection between surface water bodies and the groundwater flow system and to determine the potential for sediment transport.

Soils and Sediments: The physical characteristics of the site soils and aquatic sediments will be evaluated. Some elements of this investigation may overlap with the Hydrogeologic and the Hydrologic investigations.

Air: The potential for airborne particle transport will be evaluated to determine if an atmospheric testing program (over and above that required for assuring the personal protection of the site workers and surrounding community) should be initiated at later project stages.

It is anticipated that this information will be derived from a combination of existing data and information, and data resulting from the field investigations.

Again, the Migration Pathway Assessment will culminate in the preparation and submittal of a technical memorandum describing the findings of the assessment. This memorandum will also contain specific recommendations for the location and design of monitoring stations (i.e., wells, air quality samplers, surface water samplers, etc.).

Contaminant Characterization

Data generated from the Pathway Assessment and Source Characterization will be used to design an environmental sampling and analysis program. The objective of this program is to evaluate the extent and magnitude of contaminant migration along the pathways of concern at the Example Landfill site.

Monitoring points will be installed in each media previously identified as a migration pathway. This monitoring network may incorporate several of the piezometers and/or gauge installed during the Pathway Assessment.

The analytical parameters list used in this subtask will be based on the data collected during the source characterization. The selection of parameters or classes of parameters (i.e., volatile organics, metals, PCBs/pesticides) will be based upon their source concentration and their persistence and mobility within the most likely pathway of migration. Provisions will be made for conducting full Hazardous Substance List (HSL) analyses at those monitoring stations where there is a reasonable anticipation of detecting a complex contaminant profile. All samples will be collected, handled, and analyzed in accordance with the protocols and procedures described in the site Sampling and Quality Assurance Project Plans.

As shown on Figure 1, provisions will be made for conducting additional site investigation activities after the completion of the Remedial Alternatives Screening (Task 7). These supplemental investigations are intended to further characterize the sources, pathways, and/or contaminants and to satisfy the specific data requirements of the applicable remedial actions. The plans for these investigations and the Bench/Pilot Studies (Task 4) will be prepared and submitted for agency comment and approval after the completion of Task 7.

TASK 3 – SITE INVESTIGATION ANALYSIS

An analysis of all data collected during this investigation will be made to assure that the quality (e.g., QA/QC procedures have been followed) and quantity of data adequately support the Endangerment Assessment and FS. A summary of the analysis will be submitted to USEPA and [State] as the preliminary data transmittal.

The results of the site investigations will be organized and presented in a report. The data from the investigation will be used to develop a summary of the type and extent of on–site contamination.

Based upon the specific chemicals and ambient levels at the site, the number and location of the surrounding population, and migration pathways, an endangerment assessment will be conducted by the responsible parties to evaluate the actual or potential threat to human health, welfare, or the environment. Actual or potential risks will be quantified whenever possible. A general approach for the endangerment assessment is offered as follows:

• Select target chemicals for evaluation, based on the likelihood of their being major contributors to the risks associated with the site.

• Conduct exposure assessments which include identifying acute and chronic hazards of concerns and also identifying populations at risk.

• Conduct an evaluation of existing toxicity information to determine the potential impacts of the site contaminants as they regard acute and chronic effects; specific effects such as carcinogenicity, reproductive dysfunction, teratogenicity, neurotoxicity, and other metabolic alterations; and environmental effects of aquatic and terrestrial toxicities.

• Assess impact by identifying acceptable exposure guidelines or standards, comparing estimated doses with these guidelines or standards, and for target chemicals at the site that have received carcinogen assessments by USEPA, utilize USEPA's evaluations to drive estimates of increased cancer risks.

This assessment will be conducted in accordance with the procedures described in the

Superfund Public Health Evaluation Manual (ICF, 1986).

TASK 4 – BENCH/PILOT TESTING STUDIES

If necessary, bench and piloting scale testing studies will be performed to determine the applicability of selected remedial technologies to site specific conditions. These may include treatability and cover studies, aquifer testing, and/or material compatibility testing. As shown on Figure 1, these studies will be conducted in the later stages of the RI after the initial screening of remedial technologies and actions.

TASK 5 – REPORTS

Progress Reports

Monthly progress reports will be prepared to describe the technical progress of the RI. These reports should be submitted to the USEPA and [State] by the tenth business day of each month, following the commencement of the work detailed in the RI/FS Work Plan. The monthly progress reports should include the following information:

- All sampling and testing results and all other raw data produced during the month pursuant to the implementation of the Consent Order;

- A description of activities completed during the past month pursuant to the Consent Order, as well as such actions and plans which are scheduled for the next month pursuant to the Consent Order;

- A description of difficulties encountered during the reporting period and the actions taken to rectify the problems.

- Target and actual completion dates for each element of activity, including the project completion, and an explanation of any deviation from the schedules provided in the RI/FS Work Plan; and

- Changes in key personnel.

Technical Memorandums

The results of specific remedial investigation activities such as the migration assessment, source characterization, endangerment assessment, etc., will be submitted to the USEPA and [State] throughout the RI process. These memorandums will only be submitted in draft form. All responses to USEPA and [State] comments regarding issues pertaining to the memorandums will be addressed in letters from the Respondent Project Coordinator to the USEPA Remedial Project Manager and will be summarized in the draft RI report. The specific technical memorandums and their associated schedule of submittal will be identified in the project Work Plan (Task 0).

Remedial Investigation Report

A final report covering the remedial investigations will be prepared. The report will characterize the site and summarize the data collected and conclusions drawn from investigative Tasks 1 through 3. The report will be submitted in draft form for review and comment. Upon receipt of comments, a draft final report will be prepared and submitted. The RI report will not be considered final until a letter of approval is issued by the USEPA Remedial Project Manager.

TASK 6 – COMMUNITY RELATIONS SUPPORT

A community relations program will be implemented conjunctively by the USEPA and the [State]. The Responsible Parties will cooperate with the USEPA and the [State] in providing RI/FS information to the public. The Responsible Parties will, at the request of the USEPA or [State], participate in the preparation of all appropriate information disseminated to the public, and in public meetings which may be held or sponsored by the USEPA or the [State] to explain activities at, or concerning, the site, including the findings of the RI/FS.

Community relations support will be consistent with Superfund community relations policy, as stated in the "Guidance for Implementing the Superfund Program" and Community Relations in Superfund – A Handbook.

FEASIBILITY STUDY

Scope

The purpose of the FS for the Example Landfill NPL site is to develop alternative remedial actions, based upon the results of the RI, which will mitigate impacts to public health and welfare and the environment.

The FS will conform to Section 121 of SARA; the NCP, as amended; and the FS Guidance, as amended; and is comprised of the following four tasks:

> Task 7: Remedial Alternatives Screening
>
> Task 8: Remedial Alternatives Evaluation
>
> Task 9: Feasibility Study Report
>
> Task 10: Additional Requirements

The intent and purpose of each of these tasks is outlined in the following sections. The technical approach and schedule for each of these tasks will be detailed in the RI/FS Work Plan (Task 0).

TASK 7 – REMEDIAL ALTERNATIVES SCREENING

This task constitutes the first stage of the FS and is comprised of six inter–related subtasks. The objective of this task is to develop and evaluate remedial alternatives for additional screening and evaluation. The results of the Public Health Evaluation will be considered throughout this evaluation process.

Subtask 7A – Preliminary Remedial Technologies

A master list of potentially feasible technologies will be developed. These technologies will include both on–site and off–site remedies, depending on the magnitude of the problem documented during the RI. In addition to site conditions, the master list will also be screened based on waste characteristics, technical requirements to eliminate or modify those technologies that may prove extremely difficult to implement, will require unreasonable time periods, will rely on insufficient developed technology. Emerging technologies that are being evaluated through the USEPA's SITE program will also be evaluated, if that information is available. The results of this task will be summarized in a technical memorandum which will be submitted to the USEPA and the [State].

Subtask 7B – Development of Alternatives

Based on the results of the RI and consideration of preliminary remedial technologies, a limited number of alternatives, which are based on objectives established for the response, will be developed.

1. Establishment of Remedial Response Objectives

Site–specific objectives for the response will be established. These objectives will be based on public health and environmental concerns for the Example Landfill site, the description of the current situation, information gathered during the RI, Section 300.68 of the National Contingency Plan (NCP), the USEPA's interim guidance, and the requirements of any other applicable USEPA, Federal, and [State] environmental standards, guidance and advisories as defined under Section 121 of SARA. Preliminary cleanup objectives will be developed under formal consultation with the USEPA and the [State].

2. Alternatives Remedial Actions

Combinations of identified technologies will be assembled into alternative remedial actions. To the extent it is both feasible and appropriate, alternatives and other appropriate considerations will be developed into a comprehensive, site–specific approach. Alternatives that will be developed will include the following:

a. Treatment alternatives for source control that eliminates the need for long–term management (including monitoring).

b. Alternatives involving treatment as a principal element to reduce the toxicity, mobility, or volume of waste.

3. Additional Alternatives

At least two additional alternatives will be developed, including the following:

a. An alternative that involves containment of waste with little or no treatment but provides protection of human health and the environment primarily by preventing the exposure or reducing the mobility of the waste.

b. A no action alternative.

For groundwater response actions, a limited number of remedial alternatives will be developed within a performance range that is defined in terms of a remediation level.

The targeted remediation level will be within the risk range of 10E–04 to 10E–07 for maximum lifetime risk and includes different rates of restoration. If feasible, one alternative that would restore groundwater quality to a 10E–06 risk for maximum lifetime risk level within five years will be configured.

The remedial action alternatives developed for the Example Landfill site may involve both source control and groundwater response actions. In these instances, the two elements may be formulated together so that the comprehensive remedial action is effective and the elements are complimentary. However, because each element has different requirements, they will be detailed separately in the development and analysis of alternatives.

Subtask 7C – Initial Screening of Alternatives

1. Initial Screening

The alternatives developed under Subtask 7B will be subjected to an initial screening to narrow the list of potential remedial actions for detailed analysis. The rationale for eliminating any alternative will be included. The considerations that will be used in the Initial Screening include:

a. Effectiveness: Alternatives will be evaluated to determine whether they adequately protect human health and the environment; attain Federal and [State] ARARs or other applicable criteria, advisories, or guidance; significantly and permanently reduce the toxicity, mobility, or volume of the hazardous constituents; are technically reliable; or are effective in other respects. The consideration of reliability will include the potential for failure and the need to replace the remedy.

b. Implementability: Alternatives will be evaluated as to the technical feasibility and availability of the technologies that each alternative would employ; the technical and institutional ability to monitor, maintain, and replace technologies over time; and the administrative feasibility of implementing the alternative.

c. Cost: The cost of construction and any long–term costs to operate and maintain the alternative will be evaluated. This evaluation will be based on conceptual costing information and not a detailed cost analysis. At this stage of the FS, cost will be used as a factor when comparing alternatives that provide similar results, but will not be a consideration when comparing treatment and non–treatment alternatives. However, cost will be a factor in the final remedial selection process as described in Subtask 8B, Section 1, paragraphs (c) and (d).

2. Preservation of Alternatives

The initial screening of alternatives incorporating treatment will be conducted with the intent of:

a. Preserving the most promising alternatives as determined by their likely effectiveness and implementability.

b. Preserving for further analysis a range of alternatives as described previously in Subtask 7B(2).

Innovative alternative technologies will be carried through the screening if there is a reasonable belief that they offer either potential for better treatment performance or implementability; fewer or lesser adverse impacts than other available approaches; or lower costs for similar performance than the demonstrated technologies.

The containment and no–action alternatives will be carried through the screening process to the detailed analysis.

Subtask 7D – Alternatives Array Document

To obtain ARARs from the [State], a detailed description of alternatives (including the extent of remediation, contaminant levels to be addressed, and method of treatment) will be prepared. This document will also include a brief site history and background, a site characterization that indicates the contaminants of concern, migration pathways, receptors, and other pertinent site information. A copy of this Alternative Array Document will be submitted to the USEPA and the [State], along with the request for a notification of the standards. No further work on the RI or FS will be performed until the ARARs are obtained from the [State], along with comments on the Alternatives Array Document.

Subtask 7E – Community Relations Program

A program for community relations support will be developed. The program will be consistent with the Community Relations Plan developed under Task 2 and with the conditions set forth in the Administrative Order.

Subtask 7F – Data Requirements

Data requirements that are specific to the relevant and applicable technologies will be identified. These requirements will be focused on providing data that is needed for detailed evaluation and development of a preferred alternative.

TASK 8 – REMEDIAL ALTERNATIVES EVALUATION (USEPA TASK 9

Subtask 8A – Detailed Analysis of Alternatives

1. Evaluation of Alternatives

The action–specific Federal and [State] ARARs and other criteria, advisories, and guidance to be used in the analysis and selection of a remedy will be identified and described. Alternatives will be analyzed in sufficient detail so that the remedies can be selected from a set of defined and discrete hazardous waste management approaches.

The information needed to develop and evaluate each alternative will be developed. The alternatives will be evaluated under the general factors of effectiveness, implementability, and cost using the more specific component measures such as protectiveness, compliance with ARARs, reliability, and technical feasibility. The detailed analysis of each alternative will include both short–term and long–term considerations for effectiveness, implementability, and cost.

2. Comparison of Alternatives

Under this subtask, the alternatives will be compared to each other using the full array of evaluation factors appropriate for the Example Landfill site. Component measures of effectiveness will include the degree to which the alternative is protective of human health and the environment. Where ARAR health based standards are established and applicable, they will be used to establish the minimum level of protection at the site. Where such levels do not exist, risk assessments will be used to establish site appropriate levels. The reliability of the remedy, including the potential need for the cost of replacement, will be used as another important element in measuring effectiveness. Site specific measures may also include other health risks borne by the affected population, population sensitivities, and the impacts on environmental receptors. If a groundwater response action is appropriate for the site, the potential for the spread of the contaminant plume and the technical limits of aquifer restoration will be used as measures of effectiveness. Another important measure of effectiveness is the degree to which the mobility, toxicity, or volume of the hazardous substance, pollutant, or contaminant is reduced.

Component measures of implementability that will be considered include the technical feasibility of the alternative; the administrative feasibility of implementing this alternative; and the availability of any needed equipment, specialists, or off–site capacity. Specific measures for groundwater remedial actions will include the feasibility of providing an alternate water supply to meet current groundwater needs, the potential need for groundwater, and the effectiveness and reliability of institutional controls.

Component measures of cost that will be used in this comparison will include short–term capital and operational costs and any long–term operation and maintenance costs. Present worth analyses will be used to compare the alternatives.

Subtask 8B – Preferred Remedy

The preferred remedy will be described within a chapter of the FS report. The criteria for remedy selection will be:

1. The appropriate remedy will be recommended from among those alternatives that meet the following findings:

   a. The alternative should be protective of human health and the environment. The alternative will meet the ARARs or health based levels that are established through risk assessments when ARARs do not exist or are waived.

   b. Except under circumstances listed in the NCP, the alternative should attain all ARARs that have been identified for the site.

   c. The alternative should be cost effective, accomplishing a level of protection that cannot be achieved by less costly methods.

   d. The alternative will utilize treatment technologies and permanent solutions to the maximum extent practicable as determined by technological feasibility, availability, and cost effectiveness.

2. The preferred remedy will reflect the following preferences:

   a. Remedies that involve treatment that significantly reduces the toxicity, mobility, or volume of hazardous   constituents as a principal element.

   b. Remedies that minimize the requirement for long–term management of residuals.

3. An alternative that is preferred, but does not meet the Federal or [State] public health or environmental ARARs, will be selected only when:

   a. The alternative is an interim remedy and will become part of a more comprehensive final remedy that will meet the Federal and [State] ARARs.

   b. Compliance with the ARAR will result in a greater risk to human health and the environment than the alternative options.

   c. Compliance with the requirements is technically impractical.

   d. The alternative will attain a standard of performance that is equivalent to that required under the otherwise applicable standard, requirement, or limitation through the use of another method or approach.

   e. The [State] has not consistently applied, or demonstrated the intention to consistently apply, the requirement at other similar facilities across the state.

4. The evaluation of alternatives to select the appropriate remedy will, in addition to meeting the required findings in Section 300.68(h)(1) of the NCP and reflecting the preferences in Section 300.68(h)(2) of the NCP, also consider and weigh the full range of factors in Section 300.68(e)(2) of the NCP. The selected alternative will represent the best balance across all evaluation criteria.

TASK 9 – FINAL FS REPORT

The FS will be documented in a draft report which will be submitted for review and comment. Upon receipt of comments, a draft final FS report will be prepared and submitted. The FS report will not be considered final until a letter of approval is issued by the USEPA Remedial Project Manager. Deliverables and technical memorandums prepared previously will be summarized and referenced in order to limit the size of the report. However, the report will completely document the FS and the process by which the recommended remedial alternative was selected.

APPENDIX C-1

CERCLA

CHEMICAL PARAMETER LISTS

CERCLA

In 1980 Congress passed the Comprehensive Environmental Response, Compensation and Liability Act (CERCLA) commonly known as Superfund. This law provided broad Federal authority to respond directly to releases or threatened releases of hazardous substances that may endanger public health or welfare of the environment. The law provides the EPA with the authority to respond directly or to compel potentially responsible parties (PRP's) to respond through litigation and negotiations. Under the 1980 law, a trust fund was established, financed through a special tax on the chemical and petroleum industries. This trust fund, known as the Superfund, was used to finance site remediation when no viable PRP's are found or when PRP's fail to take necessary response actions.

In 1986 CERCLA was increased and amended. The major goals of the amendments, known as The Superfund Amendments Second Reauthorization Act (SARA) were a faster pace of clean-up, more public participation, and more rigid and clearly defined clean-up standards, with an emphasis on achieving remedies that permanently and significantly reduce the mobility, toxicity, or volume of wastes.

There are two types of reporting requirements under Superfund: spill reporting and facility notification requirements. The first reporting requirement under CERCLA 103(a) relates to actual releases (including spills) of hazardous substances; the second, under CERCLA 103(c), relates to facilities where hazardous wastes have been disposed of and where such releases might potentially occur.

Separate reporting requirements were created under Title III of SARA, known as the Emergency Planning and Community Right-to-Know Act of 1986, requiring you to provide (1) immediate notice for accidental releases of hazardous substances and extremely hazardous substances; (2) information to local emergency planning committees for the development of emergency plans; and (3) Material Safety Data Sheets, emergency and hazardous chemical inventory forms, and toxic chemical release forms.

A. Contract Laboratory Program

The EPA's Contract Laboratory Program (CLP) was established in 1980 to support EPA enforcement activities under CERCLA. The analytical results generated under the CLP are used to help determine the severity of site contamination and whether a site should be placed on the National Priority List. The National Priority List designates the nation's worst-known sites contaminated with hazardous substances. Only sites included on this list are eligible for long-term remedial action under the Superfund project. CLP protocol is generally used for all PRP funded Remedial Investigation/Feasibility Studies (RI/FS).

ORGANIC CLP - TARGET COMPOUND LIST

CAS Number	Volatile Organics Compound	Contract Required Quantitation Limits* Water µg/ 1	Low Soil/ Sediment (b) µg/kg
74-87-3	Chloromethane	10	10
74-83-9	Bromomethane	10	10
75-01-4	Vinyl Chloride	10	10
75-00-3	Chloroethane	10	10
75-09-2	Methylene Chloride		
67-64- 1	Acetone	10	10
75-15-0	Carbon Disulfide	5	5
75-35-4	1, l-Dichloroethene	5	5
75-34-3	1, l-Dichloroethane	5	5
540-59-0	1,2-Dichloroethene (total)	5	5
67-66-3	Chloroform	5	5
107-06-2	1,2-Dichloroethane	5	5
78-93-3	2-Butanone	10	10
71-55-6	1,1,1 -Trichloroethane	5	5
56-23-5	Carbon Tetrachloride	5	5
108-05-4	Vinyl Acetate	10	10
75-27-4	Bromodichloromethane	5	5
79-34-5	1, 1,2,2-Tetrachloroethane	5	5
78-87-5	1,2-Dichloropropane	5	5
10061-02-6	trans- 1 ,3-Dichloropropene	5	5
79-01-6	Trichloroethene	5	5
124-48-1	Dibromochloromethane	5	5
79-00-5	1, 1,2-Trichloroethane	5	5
71-43-2	Benzene	5	5
10061-01-5	cis-1,3-Dichloropropene	5	5
75-25-2	Bromoform	5	5
108-10-1	4-Methyl-2-Pentanone	10	10
591-78-6	2-Hexanone	10	10
127- 18-4	Tetrachloroethene	5	5
108-88-3	Toluene	5	5
108-90-7	Chlorobenzene	5	5
100-41-4	Ethylbenzene	5	5
100-42-5	Styrene	5	5
1330-20-7	Xylene (total)	5	5

(a) Medium Water Contract Required Quantitation Limits (CRQL) for Volatile TCL Compounds are 125 times the individual Low Soil/Sediment CRQL.

* Specific quantitation limits are highly matrix dependent. The quantitation limits listed herein are provided for guidance and may not always be achievable.

Quantitation limits listed for soil/sediment are based on wet weight. The quantitation limits calculated by the laboratory for soil/sediment, calculated on dry weight basis as required by the contract, will be higher.

Semi-Volatile Compounds:

CAS Number	Compound	Contract Required Quantitation Limits*	
		Water µg/ 1	Low Soil/ Sediment (b) µg/kg
108-95-2	Phenol	10	330
111-44-4	bis(2-Chloroethyl)ether	10	330
95-57-8	2-Chlorophenol	10	330
541-73-1	1,3-Dichlorobenzene	10	330
106-46-7	1,4-Dichlorobenzene	10	330
100-51-6	Benzyl alcohol	10	330
95-50-1	1,2-Dichlorobenzene	10	330
95-48-7	2-Methylphenol	10	330
108-60-1	bis(2-chloroisopropyl)ether	10	330
106-44-5	4-Methylphenol	10	330
621-64-7	N-Nitroso-di-n-propylamine	10	330
67-72-1	Hexachloroethane	10	330
98-95-3	Nitrobenzene	10	330
78-59-1	Isophorone	10	330
88-75-5	2-Nitrophenol	10	330
105-67-9	2,4-Dimethylphenol	10	330
65-85-0	Benzoic acid	50	1600
111-91-1	bis(2-Chloroethoxy)methane	10	330
120-83-2	2,4-Dichlorophenol	10	330
120-82-1	1,2,4-Trichlorobenzene	10	330
91-20-3	Naphthalene	10	330
106-47-8	4-Chloroaniline	10	330
87-68-3	Hexachlorobutadiene	10	330
59-50-7	4-Chloro-3-methylphenol	10	330
91-57-6	2-Methylnaphthalene	10	330
77-47-4	Hexachlorocyclopentadiene	10	330
88-06-2	2,4,6-Trichlorophenol	10	330
95-95-4	2,4,5-Trichlorophenol	50	1600
91-58-7	2-Chloronaphthalene	10	330
88-74-4	2-Nitroaniline	50	1600
131 ~ 3	Dimethylphthalate	10	330
208-96-8	Acenaphthylene	10	330
606-20-2	2,6-Dinitrotoluene	10	330
99-09-2	3-Nitroaniline	50	1600
83-32-9	Acenaphthene	10	330
51-28-5	2,4-Dinitrophenol	50	1600
100-02-7	4-Nitrophenol	50	1600
132-64-9	Dibenzofuran	10	330
121-14-2	2,4-Dinitrotoluene	10	330
84-66-2	Diethylphthalate	10	330
7005-72-3	4-Chlorophenyl-phenylether	10	330
86-73-7	Fluorene	10	330
100-01-6	4-Nitroaniline	50	1600
534-52-1	4,6-Dinitro-2-methylphenol	50	1600

86-30-6	N-Nitrosodiphenylamine (1)	10	330
101-55-3	4-Bromophenyl-phenylether	10	330
118-74-1	Hexachlorobenzene	10	330
87-86-5	Pentachlorophenol	50	1600
85-01-8	Phenanthrene	10	330
120-12-7	Anthracene	10	330
84-74-2	Di-n-butylphthalate	10	330
206-44-0	Fluoranthene	10	330
129-00-0	Pyrene	10	330
85-68-7	Butylbenzylphthalate	10	330
91-94-1	3,3'-Dichlorobenzidine	20	660
56-55-3	Benzo(a)anthracene	10	330
218-01-9	Chrysene	10	330
117-81-7	bis(2-Ethylhexyl)phthalate	10	330
117-84-0	Di-n-octylphthalate	10	330
205-99-2	Benzo(b)fluoranthene	10	330
207-08-9	Benzo(k)fluoranthene	10	330
50-32-8	Benzo(a)pyrene	10	330
193-39-5	Indeno(1,2,3-cd)pyrene	10	330
53-70-3	Dibenz(a,h)anthracene	10	330
191-24-2	Benzo(g,h,i)perylene	10	330

(l) Cannot be separated from Diphenylamine

(b) Medium Soil/Sediment Contract Required Quantitation Limits (CRQL) for Semi-Volatile TCL Compounds are 60 times the individual Low Soil/Sediment CRQL.

* Specific quantitation limits are highly matrix dependent. The quantitation limits listed herein are provided for guidance and may not always be achievable.

Quantitation limits listed for soil/sediment are based on wet weight. The quantitation limits calculated by the laboratory for soil/sediment, calculated on dry weight basis as required by the contract, will be higher.

The 1985 Contract List also included the following compound:

92-87-5	Benzidine	80ug/1	2600ug/kg

The use of gel permeation clean-up (GPC) is strongly recommended for the preparation of soil/sediments under the TCL/Routine Analytical Services (RAS) protocols. It should be noted that the use of GPC increases the CRQL by a factor of two (2).

Combination of Base Neutral and Acid extracts for water samples increase the CRQL's by a factor of two (2).

Pesticides/PCB's

Semi-Volatile Compounds:

CAS Number	Compound	Contract Required Quantitation Limits*	
		Water µg/ 1	Low Soil/ Sediment (b) µg/kg
319-84-6	alpha-BHC	0.05	8.0
319-85-7	beta-BHC	0.05	8.0
319-86-8	delta-BHC	0.05	8.0
58-89-9	gamma-BHC (Lindane)	0.05	8.0
76-44-8	Heptachlor	0.05	8.0
309-00-2	Aldrin	0.05	8.0
1024-57-3	Heptachlor epoxide	0.05	8.0
959-98-8	Endosulfan I	0.05	8.0
60-57-1	Dieldrin	0.10	16.0
72-55-9	4,4'-DDE	0.10	16.0
72-20-8	Endrin	0.10	16.0
33213-65-9	Endosulfan II	0.10	16.0
72-54-8	4,4'-DDD	0.10	16.0
1031-07-8	Endosulfan sulfate	0.10	16.0
50-29-3	4,4'-DDT	0.10	16.0
72-43-5	Methoxychlor	0.5	80.0
53494-70-5	Endrin ketone	0.10	16.0
5103-71-9	alpha-chlordane	0.5	80.0
5103-74-2	gamma-chlordane	0.5	80.0
8001-35-2	Toxaphene	1.0	160.0
12674 ~ 2	Aroclor-1016	0.5	80.0
11104-28-2	Aroclor-1221	0.5	80.0
11141-16-5	Aroclor-1232	0.5	80.0
53469-21-9	Aroclor-1242	0.5	80.0
12672-29-6	Aroclor-1248	0.5	80.0
11097-69-1	Aroclor-1254	1.0	160.0
11096-82-5	Aroclor-1260	1.0	160.0

(c) Medium Soil/Sediment Contract Required Quantitation Limits (CRQL) for Pesticide/PCB TCL compounds are 15 times the individual Low Soil/Sediment CRQL.

Specific quantitation limits are highly matrix dependent. The quantitation limits listed herein are provided for guidance and may not always be achievable.

Quantitation limits listed for soil/sediment are based on wet weight. The quantitation limits calculated by the laboratory for soil/sediment, calculated on dry weight basis as required by the contract, will be higher.

Inorganic CLP-Target Analyte List:

Metals

CAS Number	Compound	Water μg/ l	Contract Required Quantitation Limits* Low Soil/ Sediment (b) μg/kg
7429-90-5	Aluminum	200	200
7440-36-0	Antimony	60	60
7440-38-2	Arsenic	10	10
7440-39-3	Barium	200	200
7440-41-7	Beryllium	5	5
7440-43-9	Cadmium	5	5
7440-70-2	Calcium	5000	5000
7440-47-3	Chromium	10	10
7440-48-4	Cobalt	50	50
7440-50-8	Copper	25	25
7439-89-6	Iron	100	100
7439-92-1	Lead	5	5
7439-95-4	Magnesium	5000	5000
7439-96-5	Manganese	15	15
7439-97-6	Mercury	0.2	0
7440-02-0	Nickel	40	40
7440-09-7	Potassium	5000	5000
7782-49-2	Selenium	5	5
7440-22-4	Silver	10	10
7440-23-5	Sodium	5000	5000
7440-28-0	Thallium	10	10
7440-62-2	Vanadium	50	50
7440-66-6	Zinc	20	20

Conventionals

57-12-5	Cyanide	25	500

B. Hazardous Substances Priority List

A requirement of SARA is that EPA and the Agency for Toxic Substances and Disease Registry (ATSDR) prepare a list of hazardous substances which are most often found at facilities on the CERCLA National Priorities List and which the agencies determine are posing the most significant potential threat to human health. The first priority list of 100 substances was published on April 17, 1987. The second list was published October 20, 1988.

List of First 100 Hazardous Substances

CAS Number	Substance Name

Priority Group 1

CAS Number	Substance Name
50-32-8	Benzo(a)pyrene
53-70-3	Dibenzo(a,h)anthracene
56-55-3	Benzo(a)anthracene
57- 12-5	Cyanide
60-57-1, 309-00-2	Dieldrin/aldrin
67-66-3	Chloroform
71-43-2	Benzene
75-01-4	Vinyl chloride
75-09-2	Methylene chloride
76-44-8, 1024-57-3	Heptachlor/heptachlor epoxide
79-01-6	Trichloroethylene
86-30-6	N-Nitrosodiphenylamine
106-46-7	1,4-Dichlorobenzene
117-81-7	Bis(2-ethylhexyl)phthalate
127-18-4	Tetrachloroethylene
205-99-2	Benzo(b)fluoranthene
218-01-9	Chrysene
1746-0 1-6	2,3,7,8-Tetrachlorodibenzo-p-dioxin
7439-92- 1	Lead
7440-02-0	Nickel
7440-38-2	Arsenic
7440-4 1-7	Beryllium
744°-43-9	Cadmium
7440-47-3	Chromium
11096-82-5, 11097-69-1, 1232,	PCBs-Aroclor 1260, 1254, 1248, 1242,
12672-29-6, 53469-21-9, 11141-16-5, 11104,28-2, 12674-11- 2	1221, 1016

Priority Group 2

CAS Number	Substance Name
56-23-5	Carbon tetrachloride
57-74-9	Chlordane
62-75-9	N-Nitrosodimethylamine
72-55-9, 50-29-3, 72-54-8	4,4'-DDE, DDT, DDD
75-00-3	Chloroethane
75-27-4	Bromodichloromethane
75-35-4	1, l-Dichloroethane
78-59- 1	Isophorone
78-87-5	1,2-Dichloropropene
79-00-5	1, 1,2-Trichloroethane
79-43-5	1, 1,2,2-Tetrachloroethane
87-86-5	Pentachlorophenol

91-94- 1	3,3'-Dichlorobenzidine
92-87-5	Benzidine
107-06-2	1,2-Dichloroethane
108-88-3	Toluene
108-95-2	Phenol
111-44-4	Bis(2-chloroethyl) ether
121 - 14-2	2,4-Dinitrotoluene
319-64-6, 58-89-9,	BHC-alpha, gamma, beta, delta
319-65-7, 319-86-8, 542-86-1	Bis(chloromethyl) ether
621-64-7	N-Nitrosodi-n-propylami ne
7439-97-6	Mercury
7440-66-6	Zinc
7782-49-2	Selenium

Priority Group 3

71-55-6	1,1,1 -Trichloroethane
74-87-3	Chloromethane
75-21-8	Oxirane
75-25-2	Bromoform
75-34-3	1, l-Dichloroethane
84-74-2	Di-n-butyl phthalate
88-06-2	2,4,6-Trichlorophenol
9 1-20-3	Naphthalene
100-41-4	Ethylbenzene
1 07-02-8	Acrolein
107-13-1	Acrylonitrile
108-90-7	Chlorobenzene
1 18-74-1	Hexachlorobenzene
122-66-7	1,2-Dinitrotoluene
124-48-1	Chlorodibromomethane
156-60-5	1,2-Trans-dichloroethene
193-39-5	Indeno(1,2,3-cd)pyrene
606-20-2	2,6,Dinitrotoluene
1330-2-07	Total xylenes
7621-93-4,72-20-8	Endrin aldehyde/endrin
7440-22-4	Silver
7440-50-8	Copper
7664-41-7	Ammonia
8001-35-2	Toxaphene

Priority Group 4

5 1-28-5	2,4-Dinitrophenol
59-50-7	p-Chloro-m-cresol
62-53-3	Aniline
65-85-0	Benzoic acid
67-72-1	1 Hexachloroethane
74-83-9	Bromethane
75-15-0	Carbon disulfide

75-69-4	Fluorotrichloromethane
75-71-8	Dichlorodifluoromethane
78-93-3	2-Butanone
84-66-2	Diethyl phthalate
85-01-8	Phenanthrene
87-68-3	Hexachlorobutadiene
95-48-7	2-Methylphenol
95-50-1	1,2-Dichlorobenzene
105-67-9	2,4-Dimethylphenol
108-10-1	4-Methyl-2-pentanone
120-82-1	1,2,4-Trichlorobenzene
120-83-2	2,4-Dichlorophenol
123-91-1	1,4-Dioxane
131 ~ 3	Dimethyl phthalate
206-44-0	Fluoranthene
534-52-1	4,6-Dinitro-2-methylphenol
541-73-1	1,3-Dichlorobenzene
7440-28-0	Thallium

LIST OF SECOND 100 HAZARDOUS SUBSTANCES

CAS No.   **Substance Name**

Priority Group 1

5 1-75-2	Mechlorethamine
77-47-4	Hexachlorocyclopentadiene
100-42-5	Styrene
108-05-4	Vinyl acetate
115-29-7	Endosulfan (alpha, beta, sulfate)
1 18-96-7	2,4,6-Trinitrotoluene
120-12-7	Anthracene
129-00-0	Pyrene
302-01-2	Hydrazine
591-78-6	2-Hexanone
1332-21-4	Asbestos
15 17-48-3	Plutonium
7439-96-5	Manganese
7440-14-4	Radium and compounds
7440-29-1	Thorium and compounds
7440-3 1-5	Tin
7440-36-0	Antimony
7440-39-3	Barium and compounds
7440-42-6	Boron and compounds
7440-48-4	Cobalt and compounds
7440-61-1	Uranium and compounds
8001-58-9	Cresote
10043-92-2	Radon and compounds
10061-02-6	trans- 1 ,3-Dichloropropene
16984-48-8	Fluorides/fluorine/hydrogen fluoride

Priority Group 2

	Chlorodibenzodioxins
50-00-0	Formaldehyde
56-38-2	Parathion (DNTP)
67-64- 1	Acetone
75-44-5	Phosgene
83-32-9	Acenaphthene
86-73-7	Fluorene
91-57-6	2-Methylnaphthalene
96-18-4	1,2,3-Trichloropropane
100-02-7	4-Nitrophenol
1 06-47-8	4-Chloroaniline
106-93-4	1,2-Dibromoethane
110-75-8	2-Chloroethyl vinyl ether
117-84-0	Di-n-octyl phthalate
132-64-9	Dibenzofuran
156-59-2	cis-1,2-Dichloroethylene
191-24-2	Benzo(g,h,i)perylene
207-08-9	Benzo(k) fluoranthene
208-96-6	Acenaphthylene
298-04-4	Disulfoton
505-60-2	Mustard gas
1912-24-9	Atrazine
7440-62-2	Vanadium
7446-09-5	Sulfur Dioxide
14797-55-8	Nitrates/nitr

Priority Group 3

-	Polybrominated-biphenyls
-	Chlorodibenzofurans
63-25-2	Sevin (Carbaryl)
68-12-2	Dimethyl formamide (DMF)
72-43-5	Methoxychlor
74-97-5	Bromochloromethane
75-45-6	Chlorodifluoromethane
85-68-7	Butylbenzyl phthalate
88-74-4	o-Nitroaniline
88-75-5	2-Nitrophenol
93-72-1	2,4,5-TP acid (Silvex)
93-76-5	2,4,5-T
94-75-7	2,4-D, salts and esters
95-57-8	2-Chlorophenol
95-95-4	2,4,5-Trichlorophenol
96-12-8	Dibromochloropropene
100-51-6	Benzyl alcohol
101-14-4	4,4'-Methylene-bis-(2-chloraniline)
1 09-66-0	n-Pentane
11 0-54-3	Hexane
121-82-4	RDX (Cyclonite)
540-59-0	1,2-Dichloroethylene

2385-85-5	Mirex
7783-06-4	Hydrogen sulfide
26471-62-5	Toluene diisocyanite

Priority Group 4

67-56-1	Methanol
76-13-1	1, 1 ,2-Trichloro- 1 ,2,2-trifluoroethane
80-62-6	Methyl methacrylate
99-09-2	m-Nitroaniline
99-35-4	1,3,5-Trinitrobenzene
101-55-3	l-Bromo-4-phenyoxy benzene
106-44-5	4-Methylphenol
107-21-1	Ethylene glycol
108-94-1	Cyclohexanone
109-99-9	Tetrahydrofuran
11 1-65-9	Octane
111-91- 1	bis(2-Chloroethoxy)methane
121-75-5	Malathion
140-57-8	Aramite
142-82-5	Heptane
479-45-8	Trinitrophenylmethylnitramine
608-93-5	Pentachlorobenzene
13 19-77-3	Cresols
7005-72-3	4-Chlorophenyl phenyl ether
7439-98-7	Molybdenum
7440-24-6	Strontium
7664-93-9	Sulfuric acid
10028- 17-8	Tritium
10061-01-5	cis-1,3-Dichloropropene
25 154-55-6	Nitrophenol

APPENDIX C-2

RCRA SUBTITLE-D

CHEMICAL PARAMETER LISTS

RCRA SUBTITLE D (SOLID WASTE -Proposed Rule

On August 30, 1988, EPA proposed a series of regulations for municipal solid waste landfills under Subtitle D of RCRA. Schedule E, Part 258, of the proposed ruling covers ground-water monitoring and corrective action. EPA is establishing minimum requirements in these areas and delegating further authority to the individual states. There is a maximum five year phase-in period associated with this ruling. On Wednesday October 9, 1991 the EPA issued the 40 CFR Parts 257 and 258 Solid Waste Disposal Facility Criteria; Final Rule The parameter lists changed for both the Appendix I and Appendix II lists

EPA is delegating authority to the State and is expecting the States to establish specific criteria on:

o number and location of monitoring wells
o sampling frequency
o statistical determination of contamination
o individual compound ground-water trigger levels

EPA's minimum requirements for ground-water monitoring at Subtitle D facilities consist of a two-phase program. Each facility must monitor, at a minimum semi-annually, the Phase I list of parameters. The Phase I list of parameters As proposed on August 30, 1988, consisted of 15 groundwater quality indicators (conventionals), 9 metals and 46 volatile organics.   On October 9, 1991 the EPA released a revised list of parameters that consisted of only metels and volatile organics in the Phase I detection monitoring program. The final Appendix II list issued on October 9, 1991 contains 214 parameters.

A statistical change from background on two or more Phase I ground-water quality parameters or one or more Phase I metal or volatile organic parameters may force the owner/operator into Phase II sampling. The Phase II (Appendix II) list of parameters now contains 214 organic and inorganic parameters, many of them not considered to be prevalent at solid waste facilities.

Proposed Sub-title D Phase I Parameters From August 30 1988

(l) Ammonia (as N)
(2) Bicarbonate (HC02)
(3) Calcium
(4) Chloride
(5) Iron
(6) Magnesium
(7) Manganese, dissolved
(8) Nitrate (as N)
(9) Potassium
(10) Sodium
(11) Sulfate (S04)
(12) Chemical Oxygen Demand (COD)
(13) Total Dissolved Solids (TDS)

(14) Total Organic Carbon (TOC)
(15) pH
(16) Arsenic
(17) Barium
(18) Cadmium
(19) Chromium
(20) Cyanide
(21) Lead
(22) Mercury
(23) Selenium
(24) Silver
(25) The volatile organic compounds (VOCs) listed in Appendix I of this part.

Proposed August 1988 Volatile Organic Constituents for Ground-water Monitoring
Phase I - SW-846 Method 8240

Acetone
Acrolein
Acrylonitrile
Benzene
Bromochloromethane
Bromodichloromethane
cis-1, 3 Dichloropropene
Trans 1, 3 Dichloropropene
1,4 Difluorobenzene
Ethanol Ethylbenzene
Ethyl methacrylate
4 Bromofluorobenzene
Bromoform
Bromomethane
2 Butanone (Methyl ethyl ketone)
Carbon disulfide
Carbon Tetrachloride
Chlorobenzene
Chlorodibromomethane
Chloroethane
2 Chloroethyl vinyl ether
Chloroform

Chloromethane
Dibromomethane
1,4 Dichloro - 2-butane
Dichlorodifluoromethane
1,1 Dichloroethane
1,2 Dichloroethane
2 Hexanone
Iodomethane
Methylene chloride
4 Methyl - 2-pentanone
1,1 Dichloroethane
trans 1,2 Dichloroethane
Styrene
1,1,2,2 Tetrachloroethane
Toluene
1,1, 1 Trichloroethane
1,1,2 Trichloroethane
Trichloroethene
Trichlorofluoromethane
1,2,3 Trichloropropane
Vinyl acetate
Vinyl Chloride
Xylene

Final Sub-title D Phase I Parameters From October 9, 1991

(1) Antimony
(2) Arsenic
(3) Barium
(4) Beryllium
(5) Cadmium
(6) Chromium
(7) Cobalt
(8) Copper
(9) Lead
(10) Nickel
(11) Selenium
(12) Silver
(13) Thallium
(14) Vanadium
(15) Zinc

The volatile organic compounds (VOCs) listed in Appendix I of this part.

Final Volatile Organic Constituents for Ground-water Monitoring, October 9, 1991
Phase (Appendix) I - SW-846 Method 8240

COMMON NAME	CAS RN
(16) Acetone	67-64-1
(17) Acrylonitrile	107-13-1
(18) Benzene	71-43-2
(19) Bromochloromethane	74-97-5
(20) Bromodichloromethane	75-27-4
(21) Bromoform, Tribromomethane	75-25-2
(22) Carbon disulfide	75-15-0
(23) Carbon tetrachloride	56-23-5
(24) Chlorobenzene	108-90-7
(25) Chloroethane , ethyl chloride	75-00-3
(26) Chloroform, Trichloromethane	67-66-3
(27) Dibromochloromethane; Chlorodibromomethane	124-48-1
(28) 1, 2 Dibromo-3-chloropropane DBCP	96-12-8
(29) 1, 2-Dibromomethane; Ethylene dibromide, EDB	106-93-4
(30) o-Dichlorobenzene; 1, 2-Dichlorobenzene	95-50-1
(31) *p*-Dichlorobenzene; 1, 4-Dichlorobenzene	106-46-7
(32) trans-1, 4-Dichloro-2-butene	110-57-6
(33) 1,1 Dichloroethane; Ethylidene chloride	75-34-3
(34) 1,2 Dichloroethane ; Ethylene dichloride	107-06-2
(35) 1,1 Dichloroethylene; 1,1 Dichloroethene	75-35-4
(36) cis-1,2 Dichloroethylene; cis-1,1 Dichloroethene	156-59-2
(37) trans-1, 2 Dichloroethylene; trans-1, 2 Dichloroethene	156-60-5
(38) 1, 2-Dichloropropane	78-87-5
(39) cis-1,3 Dichloropropene	10061-01-5

(40) Trans-1, 3-Dichloropropene   10061-02-6
(41) Ethylbenzene   100-41-4
(42) 2 Butanone (Methyl butyl ketone)   591-78-6
(43) Methyl bromide; Bromomethane   74-83-9
(44) Methyl chloride; Chloromethane   74-87-3
(45) Methylene bromide; Dibromomethane   74-95-3
(46) Methylene chloride; Dichloromethane 75-09-2
(47) Methyl ethyl ketone; MEK, 2-Butanone 78-93-3
(48) Methyl iodide; Iodomethane   74-88-4
(49) 4 Methyl- 2-pentanone ; Methyl isobutyl ketone 108-10-1
(50) Styrene   100-42-5
(51) 1,1,1,2-Tetrachloroethane   630-20-6
(52) 1,1,2,2-Tetrachloroethane   79-34-5
(53) Tetrchloroethylene   127-18-4
(54) Toluene   108-88-3
(55) 1,1,1 Trichloroethane   71-55-6
(56) 1,1,2 Trichloroethane   79-00-5
(57) Trichloroethylene; Trichloroethene 79-01-6
(58) Trichlorofluoromethane; CFC-11   75-69-4
(59) 1, 2, 3, Trichlorpropane   96-18-4
(60) Vinyl acetate   108-05-4
(61) Vinyl Chloride   75-01-4
(62) Xylene   1330-20-7

Subtitle D - Final October 9, 1991 Phase II Parameters

SYSTEMATIC NAME	CAS RN	COMMON NAME
Acenaphthylene	206-96-B	Acenaphthalene
Acenaphthylene, 1,2-dihydro-	83-32-9	Acenaphthene
Acetamide, N-(4-ethoxphenyl)-H	62-44-2	Phenacetin
Acetamide, N-9H-fluoren-2-yl	53-96-3	2-Acetylaminofluorene
Acetic acid ethenyl ester	106-05-4	Vinyl acetate
Acetic acid (2,4,5-trichloro-phenoxy)-	93-76-5	2,4,5-T
Aceticacid(2,4-dichloro-phenoxy)-	94-75-7	2,4-Dichlorophenoxy-aceticaci
Acetronitrite	75-05-8	Acetonitrile
Anthracene	120-12-7	Anthracene
Antimony	7440-36-0	Antimony (total)
Aroclor 1016	12674 ~ 2	Aroclor 1016
Aroclor 1221	11104-26-2	Aroclor 1221
Aroclor 1232	11141-16-5	Aroclor 1232
Aroclor 1242	53468-21-9	Aroclor 1242
Aroclor 1248	12672-29-6	Aroclor 1248
Aroclor 1254	11097-69-1	Aroclor 1254
Aroclor 1260	11096-82-5	Aroclor 1260
Arsenic	7440-38-2	Arsenic (total)
Barium	7440-39-3	Barium (total)

Benz[a]anthracene,7,12,-dimethyl	57-97-6	7,12-Dimethylbenz[a]antracene
Benz~]aceanthrylene,1,2-dihydo-3-methyl	56-49-5	3-Methylcholanthrene
Benz[e]acephenanthrylene	205-99-2	Benzo[b]fluoranthene
Benzamide,3,5-dichloror-N- (1, 1 -dimethyl-2-propynl)-	23950-58-5	Benzo[b]fluoranthene
Benzamide,3,5-dichloro-N- (1, l-dimethyl-2-propynyl)-	23950-58-5	Pronamide
Benz[a]anthracene	56-55-3	Benx[a]anthracene
Benzenamine, 2-methyl-5-nitro	99-55-8	5-Nitro-o-toluidine
Benzenamine, 2-nitro	88-74-4	2-Nitroaniline
Benzenamine, 3-nitro	99-09-2	3-Nitroaniline
Benzenamine, 4-chloro	106-47-8	p-Chloroaniline
Benzenamine, 4-nitro-	100-01-6	p-nitroaniline
Benzenamine, N-nitroso-N-phenyl	86-30-8	N-Nitrosodi phenylmamine
Benzenamine, N-phenyl-	122-39-4	Diphenylamine
Benzenamine, N,N-dimethy-4-(phenylazo)-		60-11- 7p-Dimethylamino-azobenzene
Benzene	71-43-2	Benzene
Benzene, l-bromo-4-phenoxy-	101-55-3	4-Bromophenyl phenyl ether
Benzene, l-chloro-4-phenoxy-	7005-72-3	4-Chlorophenyl phenyl ether
Benzene, l-methyl-2,4-dinitro	121-14-2	2,4-Dinitrotoluene
Benzene, 1,1'-(2,2,2-trichloro- ethylidene)bis[4-chloro-	50-29-3	DDT
Benzene, 1,1'-(2,2,2-trichloro- ethylidene)bis[4-methoxy	72-43-5	Methoxychlor
Benzene, 1,1'-(2,2-dichloro- ethylidene)bis[4-chloro-	72-54-8	DDD
Benzene 1,1'-(2,2-dichloro- ethenylidene)bis[4-chloro	72-55-9	DDE
Benzene 1,2-dichloro-	95-50-1	o-Dichlorobenzene
Benzene 1,2,4-trichloro	120-82-1	1,2,4-trichlorobenzene
Benzene 1,2,4,5-tetrachloro	95-94-3	1,2,4,5-Tetrachloro-benzene
Benzene 1,3-Dichloro-	541-73-1	M-Dichlorobenzene
Benzene 1,4-dichloro-	106-46-7	p-Dichlorobenzene
Benzene 1,3, 5,-dinitro-	99-35-4	sym-Trinitrobenzene
Benzene 1,4-dinitro-	100-25-4	meta-Dinitrobenzene
Benzene, 2-methyl-1,3-dinitro-	606-20-2	2,6-Dinitrotoluene
Benzene, chloro	108-90-7	Chlorobenzene
Benzene, dimethyl-	1330-20-7	Xylene (total)
Benzene, ethenyl-	100-42-5	Styrene
Benzene, ethyl-	100-41-4	Ethyl benzene
Benzene, hexachloro-	118-74-1	Hexachlorobenzene
Benzene, methyl	106-88-3	Toluene
Benzene, nitro-	96-95-3	Nitrobenzene
Benzene, pentachloro-	606-93-5	Pentachlorobenzene
Benzene, pentachloronitro-	82-68-8	Pentachloronitro benzene
Benzeneacetic acid, 4-chloro- a-(4-chlorophenyl)-a-hydroxy-,ethyl ester	510-15-6	Chlorobenzilate
1, 4-Benzenediamine	106-50-3	p-Phenylenediamine
Benzenediamine, 2-methyl	95-530-4	o-Toluidine
1,2-Benzenedicarboxylic acid, bis(2-ethylhexyl) ester	117-81-7	Bis(2-ethylhexyl) phthalate
1,2-Benzenedicarboxylic acid, butyl	85-68-7	Butyl benzyl phthalate

phenylmethyl ester

1,2-Benzenedicarboxylic acid, dibutyl ester	84-74-2	Di-n-butyl phthalate
1,2-Benzenedicarboxylic acid, diethyl ester	84-66-2	Diethyl phthalate
1,2-Benzenedicarboxylic acid, dimethyl ester	131-1-3	Dimethyl phthalate
1,2-Benzenedicarhoxylic acid, dioctyl ester	117-84-0	Di-n-octyl phthalate
Benzenemethanol	100-51-8	Benzyl alcohol
1,3-Benodioxole, 5 ~ propenyl)-	120-58-1	Isosafrole
1,3-Benzodioxole, 5-(2-propenyl)-	94-59-7	Safrole
Benzo[k]fluoranthene	207-08-9	Benzo[k]fllloranthene
Benzo[ghi]perylene	191-24-2	Benzo(ghi)perylene
Benzo[a]pyrene	50-32-8	Benzo[a]pyrene
Beryllium	7440-41-7	Beryllium (total)
1, 1 '-Biphen[yi]4,4'-diamine, 3,3'-dichloro-	91-94-1	3,3'-Dichlorobenzidine
l,l'-Biphen[yi]-4,4'-diamine, 3,3'-dimethyl	119-93-7	3,3'-Dimethylbenzidine
l,l'-Biphenyl[-4-amine	92-67-1	4-Aminobiphenyl
1,3-Butadiene, 1,1,2,3,4,4-hexachloro-	87-68-3	Hexachorobutadiene
1,3-Butadiene, 2-chloro-	126-99-8	2-Chlro-1,3-butylamine
l-Butanamine, N-butyl-N-nitroso-	924-16-3	N-Nitrosodi-n-butylamine
2-Butanone	78-93-3	Methyl ethyl ketone
2-Butene, 1,4-dichloro-,(E)-	110-57-6	trans-1,4—Dichloro-2-butene
Cadmium	7440-43-9	Cadmium (total)
Carbamothioic acid	2303-16-4	Diallate
Carbon disulfide	75-15-0	Carbon disulfide
Chromium	7740-47-3	Chromium (total)
Chrysene	218-01-9	Chrysene
Cobalt	7440-48-4	Cobalt (total)
Copper	7440-50-8	Copper (total)
Cyanide	57-12-5	Cyanide
Cyclohexane, 1,2,3,4,5,6- hexachloro-(la,2a,3B,4a,5B,6B)-	319-84-6	alpha-BHC
Cyclohexane, 1,2,3,4,5,6- hexachloro-,(la,2B,3a,4B,5a,6B)-	319-85-7	beta-BHC
Cyclohexane, 1,2,3,4,5,6- hexachloro-,(la,2a,3a,4B,5a,6B)-	319-86-8	delta-BHC
Cyclohexane, 1,2,3,4,5,6- hexachloro-(la,2a,3B,4a,5a,6B)-	58-89-9	gamma-BHC
2-Cyclohexen-l-one,3,5,5-trimethyl	78-59-1	Isophorone
1,3-Cyclopentadiene, 1,2,3,4,5,5-hexachloro-	77-47-4	Hexachlorocyclopent-adlene
Dibenz[a,h]anthracene	53-70-3	Dibenz[a]anthracene
2,7:3,6-Dimethanonaphth[2,3-b]oxirene, 3,4,5,6,9,9-hexachloro- la,2,2a,3,6,6a,7,7a-octahydro, laa,2B,2aa,3B,6B,6aa,7B,7aa)-	60-57-1	Dieldrin
2,7:3,6-Dimethanonaphth[2,3-b]oxirene, 3,4,5,6,9,9-hexachloro- la,2,2a,3,6,6a,7,7a-octahydro, laa,2B,2aB,3a,6a,6aB,7B,7aa)-	72-20-8	Endrin
1,4:5,8-Dimethanonaphthalene, 1,2,3,4, 10, lO-hexachloro- 1,4,4a,5,8,8a-hexahydro- 1 aa,4a,4aB,5a,8a,8aB)-	309-00-2	Aldrin
1,4:5,8-Dimethanonaphthalene,	465-73-6	Isodrine

1,2,3,4,10,10-hexachloro- 1,4,4a,5,8,8a-hexahydro-, laa,4a,4aB,5B,8B,8aB)-		
Ethanamine, N-ethyl-N-nitroso	55-18-5	N-Nitrosodiethylamine
Ethanamine, N-methyl-N-nitroso-	10595-95-6	N-Nitrosomethyl ethylame
Ethane, l,l-dichloro-	75-34-3	l,l-Dichloroethane
Ethane, l,l'-[methylenebis(oxy)]bis[2-chloro	111-91-1	Bis(2-chloroethoxy)methane
Ethane, 1,1'-oxybis[2-chloro ~	44-4	Bis(2-chloroethyl)ether
Ethane, l,l'-trichloro-	71-55-6	l,l,l-Trichloroethane
Ethane, 1,1,1,2-tetrachloro-	630-20-6	1,1,1,2-Tetrachloroethane
Ethane, 1,1,2-trichloro-	79-00-5	1,1,2-Trichloroethane
Ethane, 1,1,2,2-tetrachloro-	79-34-5	1,1,2,2-Tetrachloroethane
Ethane, 1,2-dibromo-	106-93-4	1,2-Dibromoethane
Ethane, 1,2-dichloro-	107-06-2	1,2-Dichloroethane
Ethane, chloro-	75-00-3	Chloroethane
Ethane, hexachloro-	67-72-1	Hexachloroethane
1,2-Ethanediamine, N,N-dimethyl- N'-'2-pyridinyl-N'-(2-thienylmethyl)-	91-80-5	Methapyrilene
Ethanone, l-phenyl-	98-86-2	Acetophenone
Ethene, l,l-dichloro-	75-35-4	l,l-Dichloroethylene
Ethene, 1,2-dichloro-(E)-	156-60-5	trans-1,2-Dichloro Ethene
Ethene, 1,2-dichloro-(Z)-	156-59-2	trans-1,2-Dichloro Ethene
Ethene, chloro-	75-01-4	Vinyl Chloride
Ethene, tetrachloro-	127-18-4	Tetrachloroethene
Ethene, trichloro-	79-01-6	Trichloroethene
Fluoranthene	206-44-0	Fluoranthene
9H-Fluorene	86-73-7	Fluorene
2-Hexanone	501-78-6	2-Hexanone
Indeno[1,2,3-cd]pyrene	193-39-5	Indeno(1,2,3-cd)pyrene
Lead	7439-92-1	Lead (total)
Mercury	7439-97-6	Mercury (total)
Methanamine, N-methyl-N-nitroso	62-75-9	N-Nitrosodimethylamine
Methane, bromo-	74-83-9	Bromomethane
Methane, bromodichloro-	75-27-4	Bromodichloromethane
Methane, chloro-	74-87-3	Chloromethane
Methane, dibromo-	74-95-3	Dibromomethane
Methane, dibromochloro-	124-48-1	Chlorodibromomethane
Methane, dichloro-	74-09-2	Dichloromethane
Methane, dichlorodifluoro-	75-71-8	Dichlorodifluoro methane
Methane, iodo-	74-88-4	Iodomethane
Methane, tetrachloro-	56-23-5	Carbon tetrachloride
Methane, tribromo-	75-25-2	Tribromomethane
Methane, trichloro-	67-66-3	Chloroform
Methane, trichlorofluoro-	75-69-4	Trichloromonofluoro methan
Methanesulfonic acid, methyl ester	66-27-3	Methyl methane sulfonate
Ethyl Methanesulfonate, ethyl ester	62-50-0	Methanesulfonic acid, Ethyl ester
4,7-Methano-lH-indene- 1,2,4,5,6,7,8,8-octachloro- 2,3,3a,4,7,7a-hexahydro	57-74-9	Chlordane
4,7-Methano-lH-indene-	76-44-8	Heptachlor

1,4,5,6,7,8,8-heptachloro-3a,4,7,7a-tetrahydro-2,5-Methano-2H-indeno[1,2-b] oxirene, 2,3,4,5,6,7,7-heptachloro-la,lb,5,5a,6,6a-hexahydro-,(laa, lbB,2a,5a,5aB,6B,6aa)	1024-57-3	Heptachlor epoxide
6,9-Methano-2,4,3-benzo-dioxathiepin, 6,7,8,9,10,10-hexachloro- 1,5,5a,6,9,9a-hexahyrdo-,3-oxide(3a,5aB,6a,9a,9aB)	959-96-8	Endosulfan I
6,9-Methano-2,4,3-benzo-dioxathiepin, 6,7,8,9,10,10-hexachloro, 1,5,5a,6,9,9a-hexahydro-, 3-oxide, (3a,5aa,6B,9B,9aa)	33213-65-9	Endosulfan II
1,3,4-Methano-2H-cyclobutal[cd]pentalen-2-one, 1, la,3,3a4,5,5,5a,5b"6-decachloro-octahydro-	143-50-0	Kepone
1,2,4-Methanocyclopental[cd] pentalene-5-carboxaldehyde, 2,2a,3,3,4,7-hexachloro-decahydro-,(la,2B,2aB,4B,4aB,SB,6aB,6aB,7R)	7421-93-4	Endrin aldehyde
6,9-Methano-2,4,3-benzo-dioxathiepin, 6,7,8,9,10,10-hexachloro- 1,5,5a,6,9,9a-hexahyrdo-3-3-dioxide	1031-07-8	Endosulfan sulfate
l-Naphthalenamine	134-32-7	l-Naphthylamine
2-Naphthalenamine	91-59-8	2-Naphthylamine
Naphthalene	91-20-3	Naphthalene
Naphthalene, 2-chloro-	91-58-7	2-Chloronaphthalene
Naphthalene, 2-methyl-	91-57-6	2-Methylnaphthalene
1,4-Naphthalenedione	130-15-4	1,4-Naphthoquinone
Nickel	7440-02-0	Nickel (total)
2-Pentanone, 4-methyl-	108-10-1	4-Methyl-2-pentanone
Phenanthrene	85-01-8	Phenanthrene
Phenol	108-95-2	Phenol
Phenol,2 ~ methylpropyl)-4,6-dinitro	88-85-7	2-sec-Butyl-4,6-dinitro-pheno
Phenol, 2-chloro-	95-57-8	2-Chlorophenol
Phenol, 2-methyl-	95-48-7	ortho-Cresol
Phenol, 3-methyl-	108-39-4	*m*-Cresol
Phenol, 2-methyl-4,6-dinitro-	534-52-1	4,6-Dinitro-o-cresol
Phenol, 2-nitro-	88-75-5	2-Nitrophenol
Phenol, 2,3,4,6-tetrachloro-	58-90-2	2,3,4,6-Tetrachloro phenol
Phenol,2,4-dichloro-	120-83-2	2,4-Dichlorophenol
Phenol, 2,4-dimethyl-	105-67-9	2,4-Dimethylphenol
Phenol, 2,4-dinitro-	51-28-5	2,4-Dinitrophenol
Phenol, 2,4,5-trichloro-	95-95-4	2,4,5-Trichlorophenol
Phenol, 2,4,6-trichloro-	88-06-2	2,4,6-Trichlorophenol
Phenol, 2,6-dichloro-	87-65-0	2,6-Dichlorophenol
Phenol, 4-chloro-3-methyl	59-50-7	p-Chloro-m-cresol
Phenol, 4-methyl	106-44-5	para-Cresol
Phenol, 4-nitro-	100-02-7	4-Nitrophenol
Phenol, pentachloro-	87-86-5	Pentachloropheno
Phosphorodithioic acid, 0,0-diethyl S-[(ethylthio)	298-02-2	Phorate

methyl] ester

Phosphorodithioic acid, 0,0-diethyl	298-04-4	Disulfoton
Phosphorothioic acid, 0-[4-[(dimethylamino) sulfonyl)]phenyl] 0,0-di-methyl ester	52-85-7	S-[2-(ethylthio)ethyl] ester Famphur
Phosphorothioic acid, 0,0-triethylester	126-68-1	0, 0, 0-Triethyl phosphorothioate
Phosphorothioic acid, 0,0-diethyl 0-(4-nitrophenyl) ester	56-38-2	Parathion
Phosphorothioic acid, 0,0-diethyl 0-pyrazinyl ester	297-97-2	O,O-Diethyl 0,2-pyrazinyl phosphorothioate
Phosphorodithioic acid, 0,0-dimethyl S-[2-methylamino)-2-oxoethyl] ester	60-51-5	Dimethoate
Phosphorothioic acid, 0,0-dimethyl 0-(4-nitrophenyl) ester	296-00-0	Methyl parathion
Piperidine, l-nitroso	100-75-4	N-Nitrosopiperidine
l-Propanamine, N-nitroso-N-propyl-	621-64-7	Di-n-propylnitrosamine
Propane, 1,2-dibromo-3-chloro	96-12-8	1,2-Dibromo-3-chloro-propan,
Propane, 1,2-dichloro-	78-87-5	1,2-Dichloropropane
Propane, 1,3-dichloro-	142-28-9	1,3-Dichloropropane
Propane, 2,2-dichloro-	594-20-7	2,2-Dichloropropane
Propane, 1,1-dichloro-	563-58-6	1,1-Dichloropropane
Propane, 12,3-trichloro-	96-18-4	1,2,3-Trichloropropane
Propane, 2,2'-oxybis[1-chloro-	106-60-1	Bis(2-chloroisopropyl)ether
Propanenitrile	107-12-0	Ethyl cyanide
Propanoic acid, 2-(2,4,5-trichlorophenoxyl)-	93-72-1	Silvex
l-Propanol, 2-methyl-	78-83-1	Isolcutyl alcohol
2-Propanone	67-64-1	Acetone
2-Propenal	107-02-8	Acrolein
l-Propene, 1,1,2,3,3,3-hexachloro-	1888-71-7	Hexachloropropene
l-Propene, 1,3-dichloro-,(E)-	10061-02-6	trans-1,3-Dichloropropene
l-Propene, 1,3-dichloro-,(Z)-	10061-01-5	cis-1,3-Dichloropropene
l-Propene, 1,3-chloro-	107-05-1	3-Chloropropene
2-Propenenitrile, 2-methyl-	126-98-7	Methacrylonitrile
2-Propenenitrile	107-13-1	Acrylonitrile
2-Propenoic acid, 2-methyl-,ethyl ester	97-63-2	Ethyl Methacrylate
2-Propenoic acid, 2-methyl-,methyl ester	80-62-6	Methyl methacrylate
Pyrene	129-00-0	Pyrene
Pyridine, l-nitroso	930-55-2	N-Nitrosopyrrolidine
Selenium	7782-49-2	Selenium (total)
Silver	7440-22-4	Silver (total)
Sulfide	18496-25-8	Sulfide
Sulfurous acid, 2-chloroethyl 2-[4-(1, l-dimethylethyl) phenoxy)-l-methy-lethyl ester	140-57-8	Aramite
Thallium	7440-28-0	Thallium (total)
Tin	7440-31-2	Tin (total)
Toxaphene	8001-35-2	Toxaphene
Vanadium	7440-62-2	Vanadium (total)
Zinc	7440-66-6	Zinc (total)

APPENDIX C-3

RCRA APPENDIX IX

CHEMICAL PARAMETER LISTS

RCRA Groundwater Assessment - Appendix IX

The Appendix VIII groups of chemicals was intended to be a comprehensive list of chemicals that could exist in hazardous waste, were considered to be a health hazard, and should therefore be regulated. Chemicals were listed on Appendix VIII as they would exist in a pure state, as opposed to the forms they would be expected to take after being dispersed in the environment. No attempt was made to examine factors such as amount of production or environmental fate.

Therefore, Appendix VIII contains both prevalent, mobile, and toxic chemicals that present major risks in groundwater at hazardous waste sites, as well as chemicals which do not present such risks because of factors such as low prevalence or instability in water.

Furthermore, the Appendix VIII list of chemicals has a variety of analytical problems associated with it included ambiguous listings, compound categories, constituents unstable in water, unavailable reference standards, lack of standardized test methods and technical problems.

In response to these problems EPA promulgated on July 9, 1987 a final ruling replacing the use of the Appendix VIII list of chemicals with a new list, Appendix IX. In summary, the Appendix IX list contains these chemicals from the Appendix VIII list that are amenable to SW846 analytical techniques plus seventeen additional compounds routinely analyzed under the Superfund program. Appendix IX contains 233 compounds.

The Appendix IX list also contains Practical Quantitation Limits (PQL) for each constituent using the suggested method. PQLs are the lowest concentrations of the analyte in groundwater which can be accurately determined using the indicated method under routine laboratory operating conditions. In many cases the PQL is based on a general estimate for the method not on experimentation. The PQLs are <u>NOT</u> part of the regulation.

APPENDIX IX

COMPOUND NAME	CAS No.	Suggested Methods	PQL (µg/L)
Acenaphthene	83-32-9	8100	200
		8270	10
Acenaphthylene	208-96-8	8100	200
		8270	10
Acetone	67-64-1	8240	100
Acetophenone	98-86-2	8270	10
Acetonitrile;Methyl cyanide	75-05-8	8015	100
2-Acetylaminofluorene;2 AAF	53-96-3	8270	10
Acrolein	107-02-8	8030	5
		8240	5
Acrylonitrile	107-13-1	8030	5
		8240	5
Aldrin	309-00-2	8080	0.05
		8270	10
Allyl chloride	107-05-1	8010	5
		8240	100
4-Aminobiphenyl	92-67-1	8270	10
Aniline	62-53-3	8270	10
Anthracene	120-12-7	8100	200
		8270	10
Antimony	(Total)	6010	300
		7040	2000
		7041	30
Aramite	140-57-8	8270	10
Arsenic	(Total)	6010	500
		7060	10
		7061	20
Barium	(Total)	6010	20
		7080	1000
Benzene	71-43-2	8020	2
		8240	5
Benzo[a]anthracene;Benzanthracene	56-55-3	8100	200
		8270	10
Benzo[b]fluoranthene	205-99-2	8100	200
		8270	10
Benzo[k]fluoranthene	207-08-9	8100	200
		8270	10
Benzo[gh]perylene	191-24-2	8100	200
		8270	10
Benzo[a]pyrene	50-32-8	8100	200
		8270	10
Benzyl alcohol	100-51-6	8270	20

APPENDIX IX Cont.

COMPOUND NAME	CAS No.	Suggested Methods	PQL (µg/L)
Beryllium	(Total)	6010	3
		7090	50
		7091	2
alpha BHC	319-84-6	8080	0.05
		8250	10
beta BHC	319-85-7	8080	0.05
		8250	40
delta BHC	319-86-8	8080	0.1
		8250	30
gamma BHC;Lindane	58-89-9	8080	0.05
		8250	10
Bis(2-chloroethoxy)methane	111-91-1	8270	10
Bis(2-chloroethyl)ether	111-44-4	8270	10
Bis(2-chloro- l-methyl)ether,2,2'-Di-chlorodiisopropyl ether	108-60-1	8010	100
		8270	10
Bis(2-ethylhexyl)phthalate	117-81-7	8060	20
		8270	10
Bromodichloromethane	75-27-4	8010	1
		8240	5
Bromoform;Tribromomethane	75-25-2	8010	2
		8240	5
4-Bromophenyl phenyl ether	101-55-3	8270	10
Butyl benzyl phthalate; Benzyl butyl phthalate	85-68-7	8060	5
		8270	10
Cadmium	(Total)	6010	40
		7130	50
		7131	1
Carbon disulfide	75-15-0	8240	5
Carbon tetrachloride	56-23-5	8010	1
		8240	5
Chlordane	57-74-9	8080	0.1
		8250	10
p-Chloroaniline	106-47-8	8270	20
Chlorobenzene	108-90-7	8010	2
		8020	2
		8240	5
Chlorobenzilate	510-15-6	8270	10
p-Chloro-m-cresol	59-50-7	8040	5
		8270	20
Chloroethane;Ethyl chloride	75-00-3	8010	5
		8240	1 0
Chloroform	67-66-3	8010	0.5
		8240	5
2-Chloronaphthalene	91-58-7	8120	10
		8270	10

APPENDIX IX Cont.

COMPOUND NAME	CAS No.	Suggested Methods	PQL (µg/L)
2-Chlorophenol	95-57-8	8040	5
		8270	10
4-Chlorophenyl phenyl ether	7005-72-3	8270	10
Chloroprene	126-99-8	8010	50
		8240	5
Chromium	(Total)	6010	70
		7190	500
		7191	10
Chrysene	218-01-9	8100	200
		8270	10
Cobalt	(Total)	6010	70
		7200	500
		7201	10
Copper	(Total)	6010	60
		7210	200
m-Cresol	108-39-4	8270	10
o-Cresol	95-48-7	8270	10
p-Cresol	106-44-5	8270	10
Cyanide	57-12-5	9010	40
2,4-D,2,4,-Dichlorophenoxyacetic acid	94-75-7	8150	10
4,4'-DDD	72-54-8	8080	0.1
		8270	10
4,4'-DDE	72-55-9	8080	0.05
		8270	10
4,4'-DDT	50-29-3	8080	0.1
		8270	10
Diallate	2303-16-4	8270	10
Dibenz[a,h]anthracene	53-70-3	8100	200
		8270	10
Dibenzofuran	132-64-9	8270	10
Dibromochloromethane;Chlorodibromomethane	124-48-1	8010	5
		8240	1
1,2-Dibromo-d-chloropropane,DBCP	96-12-8	8010	100
		8240	5
		8270	10
1,2-Dibromoethane,Ethylene dibromide	106-93-4	8010	10
		8240	5
Di-n-butyl phthalate	84-74-2	8060	5
		8270	10
o-Dichlorobenzene	95-50-1	8010	2
		8020	5
		8120	10
		8270	10
m-Dichlorobenzene	541-73-1	8010	5
		8020	5
		8120	10
		8270	10

APPENDIX IX Cont.

COMPOUND NAME	CAS No.	Suggested Methods	PQL (μg/L)
p-Dichlorobenzene	106-46-7	8010	2
		8020	5
		8120	15
		8270	10
3,3' Dichlorobenzidine	91-94-1	8270	20
trans-1,4-Dichloro-2-butene	110-57-6	8240	5
Dichlorodifluoromethane	75-71-8	8010	10
		8240	5
l,l-Dichloroethane	75-34-3	8010	1
		8240	5
1,2-Dichloroethane;Ethylene dichloride	107-06-2	8010	0.5
		8240	5
l,l-Dichloroethylene;Vinylidene chloride	75-35-4	8010	1
		8240	5
trans-1,2-Dichloroethylene	156-60-5	8010	1
		8240	5
2,4-Dichlorophenol	120-83-2	8040	5
		8270	10
2,6-Dichlorophenol	87-65-0	8270	10
1,2-Dichloropropane	78-87-5	8010	0.5
		8240	5
cis-1,3-Dichloropropene	10061-01-5	8010	20
		8240	5
trans-1,3-Dichloropropene	10061-02-6	8010	5
		8240	5
Dieldrin	60-57-1	8080	0.05
		8270	10
Diethyl phthalate	84-66-2	8060	5
		8270	10
O,O-Diethyl 0-2-pyrazinyl phosphorothioate-Thionazin	297-97-2	8270	10
Dimethoate	60-51-5	8270	10
p-(Dimethylamino)azobenzene	60-11-7	8270	10
7/12-Dimethylbenz[a]anthracene	57-97-6	8270	10
3,3'-Dimethylbenzidine	119-93-7	8270	10
alpha,alpha-Dimethylphenethylamine(2)	122-09-8	8270	10
2,4-Dimethylphenol	105-67-9	8040	5
		8270	10
Dimethyl phthalate	131-11-3	8060	5
		8270	10
m-Dinitrobenzene	99-65-0	8270	10
4,6-Dinitro-o-cresol	534-52-1	8040	150
		8270	50
2,4-Dinitrophenol	51-28-5	8040	150
		8270	50

APPENDIX IX Cont.

COMPOUND NAME	CAS No.	Suggested Methods	PQL (µg/L)
2,6-Dinitroluene	606-20-2	8090	0.1
2,4-Dinitrotoluene	121-14-2	8090	0.2
		8270	10
Dinoseb; DNBP, 2-sec-Butyl-4,6-dinitro phenol	88-85-7	8150	1
		8270	10
Di-n-octyl phthalate	117-84-0	8060	30
		8270	10
1,4-Dioxane	123-91-1	8015	150
Diphenylamine	122-39-4	8270	10
Disulfoton	298-04-4	8140	2
		8270	10
Endosulfan I	959-98-9	8080	0.1
		8250	10
Endosulfan II	33213-65-9	8080	0.05
Endosulfan sulfate	1031-07-8	8080	0.5
		8270	10
Endrin	72-20-8	8080	0.1
		8250	10
Endrin aldehyde	7421-93-4	8080	0.2
		8270	10
Ethylbenzene	100-41-4	8020	2
		8240	5
Ethyl methacrylate	97-63-2	8015	10
		8240	5
		8270	10
Ethyl methanesulfonate	62-50-0	8270	10
Famphur	52-85-7	8270	10
Fluoranthene	206-44-0	8100	200
		8270	10
Fluorene	86-73-7	8100	200
		8270	10
Heptachlor	76-44-8	8080	0.05
		8270	10
Heptachlor epoxide	1024-57-3	8080	1
		8270	10
Hexachlorobenzene	118-74-1	8120	0
		8270	10
Hexachlorobutadiene	87-68-3	8120	5
		8270	10
Hexachlorocyclopentadiene	77-47-4	8120	5
		8270	10
Hexachloroethane	67-72-1	8120	0
		8270	10
Hexachlorophene	70-30-4	8270	10
Hexachloropropene	1888-71-7	8270	10
2-Hexanone	591-78-6	8240	50

APPENDIX IX Cont.

COMPOUND NAME	CAS No.	Suggested Methods	PQL (µg/L)
Indeno(1,2,3-cd)pyrene	193-39-5	8100	200
		8270	10
Isobutyl alcohol	78-83-1	8015	50
Isodrin	465-73-6	8270	10
Isophorone	78-59-1	8090	60
		8270	10
Isosafrole	120-58-1	8270	10
Kepone	143-50-0	8270	10
Lead	(Total)	6010	40
		7420	1000
		7421	10
Mercury	(Total)	7470	2
Methacrylonitrile	126-98-7	8015	5
		8240	5
Methapyrilene (1)	91-80-5	8270	10
Methoxychlor	72-43-5	8080	2
		8270	10
Methyl bromide;Bromomethane	74-83-9	8101	20
		8240	10
Methyl chloride; Chloromethane	74-87-3	8010	1
		8240	10
3-Methylcholanthrene	56-49-5	8270	10
Methylene bromide; Dibromomethane	74-95-3	8010	5
		8240	5
Methylene chloride; Dichloromethane	75-09-2	8010	5
		8240	5
Methyl ethyl ketone; MEK	78-93-3	8015	10
		8240	100
Methyl iodide;iodomethane	74-88-4	8010	40
		8240	5
Methyl methacrylate	80-62-6	8015	2
		8240	5
Methyl methanesulfonate	66-27-3	8270	10
2-Methylnaphthalene	91-57-6	8270	10
Methyl parathion; Parathion methyl	298-00-0	8140	0.5
		8270	10
4-Methyl-2-pentanone, Methyl isobutyl ketone	108-10-1	8015	5
		8240	50
Naphthalene	91-20-3	8100	200
		8270	10
1 ,4-Naphthoquinone	130-15-4	8270	10
l-Naphthylamine	134-32-7	8270	10
2-Naphthylamine	91-59-8	8270	10
Nickel	(Total)	6010	50
		7520	400
o-Nitroaniline	88-74-4	8270	50

APPENDIX IX Cont.

COMPOUND NAME	CAS No.	Suggested Methods	PQL (µg/L)
m-Nitroaniline	99-09-2	8270	50
p-Nitroaniline	100-01-6		50
Nitrobenzene	98-95-3	8090	40
		8270	10
o-Nitrophenol	88-75-5	8040	5
		8270	10
p-Nitrophenol	100-02-7	8040	10
		8270	50
4-Nitroquinoline l-oxide (1)	56-57-5	8270	10
N-Nitrosodi-n-butylamine	924-16-3	8270	10
N-Nitrosodiethylamine	55-18-5	8270	10
N-Nitrosodimethylamine	62-75-9	8270	10
N-Nitrosodiphenylamine	86-30-6	8270	10
N-Nitrosodipropylamine;Di-n-propyl-nitrosamine	621-64-7	8270	10
N-Nitrosomethylethylamine	10595-95-6	8270	10
N-Nitrosomorpholine	59-89-2	8270	10
N-Nitrosopiperidine	100-75-4	8270	10
N-Nitrosopyrrolidine (1)	930-55-2	8270	10
5-Nitro-o-toluidine	99-55-8	8270	10
Parathion	56-38-2	8270	10
Polychlorinated biphenyls; PCBs (3)		8080	50
		8250	100
Polychlorinated dibenzo-p-dioxins (4) PCDDs		8280	0.01
Polychlorinated dibenzofurans; PCDFs (5)		8280	0.01
Pentachlorobenzene	608-93-5	8270	10
Pentachloroethane (2)	76-01-7	8240	5
		8270	10
Pentachloronitrobenzene	82-68-8	8270	10
Pentachlorophenol	87-86-5	8040	5
		8270	50
Phenacetin	62-44-2	8270	10
Phenanthrone	85-01-8	8100	200
		8270	10
Phenol	108-95-2	8040	1
		8270	10
p-Phenylenediamine	106-50-3	8270	10
Phorate	298-02-2	8140	2
		8270	10
2-Picoline	109-06-8	8240	5
		8270	10
Pronamide	23950-58-5	8270	10
Propionitrile, Ethyl cyanide	107-12-0	8015	60
		8240	5
Pyrene	129-00-0	8100	200
		8270	10
Pyridine (6)	110-86-1	8240	5
		8270	10

APPENDIX IX Cont.

COMPOUND NAME	CAS No.	Suggested Methods	PQL (µg/L)
Safrole	94-59-7	8270	10
Selenium	(Total)	6010	750
		7740	20
		774 1	20
Silver	(Total)	6010	70
		7760	100
Silvex,2,4,5-TP	93-72-1	8150	2
Styrene	100-42-5	8020	1
		8240	5
Sulfide	18496-25-8	9030	10,000
2,4-T;2,4,5-Trichlorophenoxyacetic acid	93-76-5	8150	2
2,3,7,8-TCDD;2,3,7,8-Tetrachloro-dibenzo-p-dioxin	1746-01-6	8280	0.005
1,2,4,5-Tetrachlorobenzene	95-94-3	8270	10
1,1,1,2-Tetrachloroethane	630-20-6	8010	5
		8240	5
1,1,2,2-Tetrachloroethane	79-34-5	8010	0,5
		8240	5
Tetrachloroethylene,Perchloroethylene Tetrachloroethene	127-18-4	8010	0.5
		8240	5
2,3,4,6-Tetrachlorophenol	58-90-2	8270	10
Tetraethyl dithiopyrophosphate; Sulfotepp (2)	3689-24-5	8270	10
Thallium	(Total)	6010	400
		7840	1000
		7841	10
Tin	(Total)	7870	8000
Toluene	108-88-3	8020	2
		8240	5
o-Toluidine	95-53-4	8270	10
Toxaphene	8001-35-2	8080	2
		8250	10
1,2,4-Trichlorobenzene	120-82-1	8270	10
1,1,1-Trichloroethane;Methylchloroform	71-55-6	8240	5
1,1,2-Trichloroethane	79-00-5	8010	0
		8240	5
Trichloroethylene;Trichloroethene	79-01-6	8010	1
		8240	5
Trichlorofluoromethane	75-69-4	8010	10
		8240	5
2,4,5-Trichlorophenol	95-95-4	8270	10
2,4,6-Trichlorophenol	88-06-2	8040	5
		8270	10
1,2,3-Trichloropropane	96-18-4	8010	10
		8240	5
O,O,O-Triethyl phosphorothioate (2)	126-68-1	8270	10
sym-Trinitrobenzene (1)	99-35-4	8270	10

APPENDIX IX Cont.

COMPOUND NAME	CAS No.	Suggested Methods	PQL (μg/L)
Vanadium	(Total)	6010	80
		7910	2000
		791 1	40
Vinyl acetate	108-05-4	8240	5
Vinyl chloride	75-01-4	8010	2
		8240	10
Xylene (total)	1330-20-7	8020	5
		8240	5
Zinc	(Total)	6010	20
		7950	50

APPENDIX C-4

SAFE DRINKING WATER ACT

CHEMICAL PARAMETER LISTS

SAFE DRINKING WATER ACT

The Safe Drinking Water Act (SDWA) of 1974 is the basis for the comprehensive regulation of drinking water. Two major regulatory programs are contained in the Act, one related to public water supplies, and the other to underground well injections. The SDWA regulates both primary and secondary drinking water contaminants.

In 1986 Congress passed major amendments in Section 1412 to the SDWA as a result of the growing public concern over contamination of public drinking water supplies and a lack of adequate federal standards. The amendments required the EPA to set enforceable standards (called Maximum Contaminant Levels or MCLs) for contaminants in drinking water based upon the level of removal that can be achieved using the best available technology to treat contaminated water. Maximum Contaminant level Goals (MCLGs) and National Primary Drinking Water Regulations (NPDWRs) for a list of 83 contaminants (53 FR 1892, January 22, 1988).

A. **Primary Drinking Water Standards**

The SDWA requires EPA to establish National Primary Drinking Water Regulations (NPDWR's) that apply to public drinking water systems and that specify contaminants which may have an adverse effect on the health of persons. The NPDWR's are to include Maximum Contaminant Levels (MCL's).

The initial list of NPDWR's and their associated MCL's are as follows:

Chemicals	MCL (mg/l)
Arsenic	0.05
Barium	1.0
Cadmium	0.010
Chromium	0.05
Lead	0.05
Mercury	0.002
Nitrate (as N)	10.0
Selenium	0.01
Silver	0.05
Endrin	0.0002
Lindane	0.004
Methoxychlor	0.1
Toxaphene	0.005
2,4-D	0.1
2,4,5-TP	0.01
Total Trihalomethanes	0.10
Turbidity	1 NTU/day
Coliform	1/100 ml
Gross Alpha/Beta	15 pCi/l
Radium 226/228	5 pCi/l

The new standard-setting process resulting from the 1986 SDWA amendments requires EPA to list those contaminants that, based upon public health concerns, need to be regulated, and within 24 months of listing, will have to propose NPDWR and MCL's, which must be finalized within another 12 months.

On July 8, 1987, EPA promulgated NPDWR's for certain volatile synthetic organic chemicals. The NPDWR's include MCL's, monitoring, reporting and public notification requirements for eight VOC's. Additionally, EPA published the maximum contaminant level goal (MCLG) for p-dichlorobenzene. As per the 1986 amendments, these NPDWR's are based upon the best available technology (BAT). Those compounds and MCL's are:

Chemical	MCL (ug/l)
Benzene	5.0
Carbon Tetrachloride	5.0
1,2-Dichloroethane	5.0
Trichloroethylene	5.0
p-Dichlorobenzene	75.0
1, l-Dichloroethylene	7.0
1,1, l-Trichloroethane	200.0
Vinyl Chloride	2.0
p-Dichlorobenzene	75.0 (MCLG)

Monitoring requirements for the above list of chemicals varies from quarterly to once per five years based on whether VOC's have been detected in the initial sampling and the vulnerability of the system. Initial monitoring was to be accomplished either January 1, 1988, January 1, 1989 or January 1, 1991 depending on the system size (population served). Analytical methods to be followed are: 502.1, 502.2, 503.1, 504, 524.1, or 524.2.

On July 8, 1987 when EPA finalized MCL's on the eight VOC's described above, they also requested each source to monitor, one time only, for 51 unregulated contaminants. The purpose of this monitoring effort was to gather information on detectibility of VOC's for potential regulation in the future.

Those 51 contaminants are:

- Monitoring required for all systems (34 contaminants)

Bromobenzene	1, l-Dichloroethane
Bromodichloromethane	1 ,2-Dichloropropane
Chloroform	1, 1,2,2-Tetrachloroethane
Chlorodibromomethane	Ethylbenzene
Bromoform	1,3-Dichloropropane
trans- 1 ,2-Dichloroethylene	Styrene
Chlorobenzene	Chloromethane
m-Dichlorobenzene	Bromomethane
Dichloromethane	1,2,3-Trichloropropane
cis- 1 ,2-Dichloroethylene	1, 1, 1,2-Tetrachloroethane
o-Dichlorobenzene	Chloroethane
Dibromomethane	1, 1,2-Trichloroethane
1, l-Dichloropropene	2,2-Dichloropropane
Tetrachloroethylene	o-Chlorotoluene
Toluene	p-Chlorotoluene
p-Xylene	Bromobenzene
o-Xylene	1,3-Dichloropropene
m-Xylene	

- Monitoring required for vulnerable systems (2 contaminants)

  Ethylene Dibromide (EDB)
  1,2-Dibromo-3-chloropropane (DBCP)

- Monitoring required at State discretion (15 contaminants)

  1,2,4-Trimethylbenzene
  1,2,4-Trichlorobenzene
  1,2,3-Trichlorobenzene
  n-Propylbenzene
  n-Butylbenzene
  Naphthalene
  Hexachlorobutadiene
  1,3,5-Trimethylbenzene
  p-Isopropyltoluene
  lsopropylbenzene
  Tert-butylbenzene
  Sec-butylbenzene
  Fluorotrichloromethane
  Dichlorodifluoromethane
  Bromochloromethane

The 1986 Amendments specify a list of 83 specific drinking water contaminants for which EPA must publish MCLG's and promulgate NPDWR's on a specified schedule. EPA can make up to seven substitutions to this list prior to its publication.

The list of 83 contaminants, including its seven substitutions was published on January 22 1988. That list incorporates the 8 VOC's regulated on July 8, 1987 and the initial list of 22 NPDWR's that existed prior to the 1986 Amendments. The list of 83 is as follows:

Volatile Organic Chemicals

Trichloroethylene	Benzene
Tetrachloroethylene	Chlorobenzene
Carbon tetrachloride	Dichlorobenzene
1,1, l-Trichloroethane	Trichlorobenzene
1,2-Dichloroethane	1, l-Dichloroethylene
Vinyl Chloride	trans-1,2-Dichloroethylene
Methylene Chloride	Cis-1,2-Dichloroethylene

Microbiology and Turbidity

Total Coliforms	Viruses
Turbidity	Standard Plate Count
Giardia Lamblia	Legionella

Inorganics

Arsenic	Molybdenum
Barium	Asbestos
Cadmium	Sulfate
Chromium	Copper
Lead	Vanadium
Mercury	Sodium
Nitrate	Nickel
Selenium	Zinc
Silver	Thallium
Fluoride	Beryllium
Aluminum	Cyanide
Antimony	

Organics

Endrin	1, 1,2-Trichloroethane
Lindane	Vydate
Methoxychlor	Simazine
Toxaphene	PAH's
2,4-D	PCB's
2,4,5-TP	Atrazine
Aldicarb	Phthalates
Chlordane	Acrylamide
Dalapon	Dibromochloropropane (DBCP)
Diquat	1,2-Dichloropropane
Endothall	Pentachlorophenol
Glyphosate	Pichloram
Carbofuran	Dinoseb
Alachlor	Ethylene Dibromide (EDB)
Epichlorohydrin	Dibromethane
Toluene	Xylene
Adipates	Hexachlorocyclopentadiene
2,3,7,8-TCDD (Dioxin)	

Radionuclides

Radium 226 and 228	Gross alpha particle
Beta particle and photon	activity
radioactivity	Uranium
Radon	

The 1986 Amendments also require EPA to publish a list of additional contaminants that are known or anticipated to occur in drinking water and which may require regulation under the Act. EPA was required to publish the first of these lists, the "Drinking Water Priority List" or DWPL on January 1, 1988 and subsequent lists every three years thereafter.

EPA established strict criteria for contaminants to be listed on the DWPL. The DWPL published in January 1988 contained contaminants from six groups. These groups are: substitutes from the SDWA list of 83 contaminants; disinfectants and their by-products; SARA priority list, pesticides registered under FIFRA; monitored but currently unregulated contaminants; substances reported frequently and/or at high concentration in reported surveys or from other sources.

The 53 contaminants on the final DWPL are as follows:

1, 1, 1,2-Tetrachloroethane
1, 1,2,2-Tetrachloroethane
1, 1 -Dichloroethene
1, l-Dichloropropene
1,2,3-Trichloropropene
1,3-Dichloropropane
1,3-Dichloropropene
2,2-Dichloropropane
2,4,5-T
2,4-Dinitrotoluene
Aluminum
Ammonia
Boron
Bromobenzene
Bromochloroacetonitrile
Bromodichloromethane
Bromoform
Bromomethane
Chloramine
Chlorate
Chlorine
Chlorine Dioxide
Chlorite
Chloroethene
Chloroform
Chloromethane
Chloropicrin
Cryptosporidium

Cyanazine
Cyanogen
Chloride
Dibromoacetonitrile
Dibromochloromethane
Dibromethane
Dicamba
Dichloroacetonitite
ETU
Hypochlorite
Isophorone
Methyl Tert-butyl ether
Metolachlor
Metribuzin
Molybdenum
Ozone byproducts
Silver
Sodium
Strontium
Trichloroacetonitrile
Trifluralin
Vanadium
Zinc
o-Chlorotoluene
p-Chlorotoluene
Halogenated acids, alcohols,
aldehydes, ketones and other
nitriles

On May 22, 1989 EPA proposed several additional contaminant lists to be monitored under the SDWA, in accordance with the 1986 Amendments.

EPA is proposing maximum contaminant level goals (MCLG's) and maximum contaminant levels (MCL's) for 30 synthetic organic chemicals and 8 inorganic chemicals under the National Primary Drinking Water Regulations. MCLG's are nonenforceable health goals. MCL's are enforceable standards which EPA considers to be feasible with regard to technology, treatment techniques, field and laboratory conditions and cost. Monitoring frequency requirements vary based on water system size (community served) and contaminant.

The May 22nd list of 38 contaminants is as follows:

CONTAMINANT	PROPOSED MCLG (mg/l)	CURRENT MCL (mg/l)	PROPOSED MCL (mg/l)
INORGANICS			
Asbestos	7MFL[1]	—	7MFL[1]
Barium	5.	1	5
Cadmium	0.005	0.01	0.005
Chromium	0.1	0.05	0.1
Mercury	0.002	0.002	0.002
Nitrate2	10	10	10
Nitrite2	1	—	1
Selenium	0.05	0.01	0.05
Silver	—	0 05	--[3]
VOLATILE ORGANICS			
cis-1,2-Dichloroethylene	0.07	—	0.07
1,2,-Dichloropropane	0	—	0.005
Ethylbenzene	0.07	—	0.7
Monochlorobenzene	0.1	—	0.1
o-dichlorobenzene	0.6	—	0.6
Styrene	0/0.1	—	0.005/0.1
Tetrachloroethylene	0	—	0.005
Toluene	1	—	1
trans- 1 ,2-dichloroethylene	0.1	—	0.1
Xylenes	10	—	10

PESTICIDES/PCBS

Alachlor	0	—	0.002
Aldicarb	0.01	—	0.01
Aldicarb sulfoxide	0.01	—	0.01
Aldicarb sulfone	0.04	—	0.04
Atrazine	0.003	—	0.003
Carbofuran	0.04	—	0.04
Chlordane	0	—	0.002
Dibromochlororopane	0	—	0.0002
2,4-D	0.07	0.1	0.07
Ethylene dibromide	0	—	0.00005
Heptachlor	0	—	0.0004
Heptachlor epoxide	0	—	0.0002
Lindane	0.0002	0.004	0.0002
Methoxychlor	0.4	0.1	0.4
PCBs	0		0.0005
Pentachlorophenol	0.2	—	0.2
Toxaphene	0.	0.005	0.005
2,4,5-TP (Silvex)	0.05	0.01	0.05

WATER TREATMENT CHEMICALS

Acrylamide	0	—	TT[4]
Epichlorohydrin	0	—	TT[4]

1 MFL = Million Fibers per Liter longer than 10 um
2 The MCLG and MCL for total nitrate and nitrite is 10 mg/l (as N)
3 Deleted as primary regulation; proposed as secondary
4 TT = treatment technique requirement

The May 22, 1989 proposal also outlines suggested monitoring requirements for approximately 100 "unregulated" organic chemicals and 6 inorganic contaminants. These unregulated contaminants are divided into two groups. The first group, containing 29 contaminants applies only to those water systems that are vulnerable to contamination of the listed constituents. The State must conduct a vulnerability assessment for each contaminant. The second group of contaminants are to be monitored at the State's discretion.

UNREGULATED CONTAMINANTS MONITORING FOR PRIORITY #1

CONTAMINANTS VULNERABLE SYSTEMS

Contaminant	Method
SOCs:	
Hexachlorobenzene	505,508
Dalapon	515 .1
Dinoseb	515. 1
Picloram	515. 1
Oxamyl (vydate)	531.1
Simazine	505,507
Glyphosate	547
Hexachlorocyopentadiene	505,525
PAHs	550,550.1,525
Phthalates	506,525
2,3,7,8-TCDD (Dioxin)	513
Aldrin	505,508
Dieldrin	505,508
2,4-D	515.1
Dicamba	515. 1
2,4,5-T	515.1
Carbaryl	531. 1
3-Hydroxycarbofuran	531. 1
Methomyl	531. 1
Butachlor	505,507
Metolachlor	505,507
Propachlor	505,507
Metribuzin	507

IOCs:	
Antimony	Graphite Furnace Atomic Absorption; Inductively Coupled Plasma.
Beryllium Spectrometry	Atomic Absorption; Inductively Coupled Mass
	Plasma; Spectrophotometric.
Nickel Spectrometry	Atomic Absorption; Inductively Coupled Mass
	Plasma; Graphite Furnace Atomic Absorption.
Sulfate	Colorimetric.
Thallium	Graphite Furnace Atomic Absorption; Inductively

Coupled   Mass Spectrometry Plasma.
Cyanide   Spectrophotometric.

(Monitoring required for all contaminants for which systems are determined by the State to be vulnerable.)

UNREGULATED CONTAMINANTS MONITORING FOR PRIORITY #2
CONTAMINANTS-STATE DISCRETION

Contaminants Analyzed Using Method 507:

Ametryn	EPN	MGK 326
Aspon	EPTC	Molinate
Atraton	Ethion	Napropamide
Azinphos methyl	Ethoprop	Norflurazon
Bolstar	Ethyl parathion	Pebulate
Bromacil	Famphur	Phorate
Butylate	Fenamiphos	Phosmet
Carboxin	Fenarimol	Phometon
Chlorpropham	Fenitrothion	Prometryn
Coumophos	Fensulfothion	Pronamide
Cycloate	Fenthion	Propazine
Demeton-O	Fluridone	Simetryn
Demeton-S	Fonofos	Stirofos
Diazinon	Hexazinone	Tebuthiuron
Dichlofenthion	Malathion	Terbacil
Dichlorvos	Merphos	Terbufos
Diphenamid	Methyl paraoxon	Terbutryn
Disulfoton	Methyl parathion	Triademefon
Disulfoton sulfone	Mevinphos	Tricyclazole
Disulfoton sulfoxide	MGK 264	Vernolate

Contaminants Analyzed Using Method 508:

Chlomeb	Endosulfan II
Chlorobenzilate	Endosulfan sulfate
Chloropropylate	Endrin aldehyde
Chlorothalonil	Etridiazole
Chlorpyrifos	HCH-alpha
DCPA	HCH-beta
4,4'-DDD	HCH-delta
4,4'-DDE	HCH-gamma
4,4'-DDT	cis-Permethrin
Dichloran	trans-Permethrin
Endosulfan I	Trifluralin

Contaminants Analyzed Using Other Methods:
Diquat - Method 549 Endothall - Method 548.

(Monitoring for these contaminants is at the discretion of the State.)

B. Secondary Drinking Water Regulations

The National Secondary Drinking Water Regulations of the SDWA control contaminants in drinking water that primarily affect the aesthetic qualities relating to the public acceptance of drinking water. Goals have been established for these secondary contaminants by EPA. The States may establish higher or lower levels based upon local conditions.

The Secondary drinking water list and maximum contaminant levels for public water systems are as follows:

Contaminant	Level
Chloride	250 mg/l
Color	15 color units
Copper	1 mg/l
Corrosivity	Noncorrosive
Fluoride	2.0 mg/l
Foaming agents	0.5 mg/l
Iron	0.3 mg/l
Manganese	0.05 mg/l
Odor	3 threshold odor number
pH	6.5-8.5
Sulfate	250 mg/l
Total dissolved solids (TDS)	500 mg/l
Zinc	5 mg/l

On May 22, 1989, EPA proposed that nine (9) additional contaminants be monitored as secondary maximum contaminants. Seven (7) are based on taste or odor detection levels, and two (2) on cosmetic or aesthetic effects.

The proposed contaminants and MCL's are:

Contaminants	SMCLs
Aluminum	0.05 mg/l
o-Dichlorobenzene	0.01 mg/l
p-dichlorobenzene	0.005 mg/l
Ethylbenzene	0.003 mg/l
Pentachlorophenol	0.003 mg/l
Silver	0.009 mg/l
Styrene	0.001 mg/l
Toluene	0.004 mg/l
Xylene	0.002 mg/l

Appendix D-1

EXAMPLE

SPECIFICATIONS FOR CONSTRUCTING

A FIRE PROTECTION WELL

GENERAL CONDITIONS

The term "Owner" shall mean either the Owner or the Owner's Representative.

Section 1: Scope of Work

The work to be performed here under includes the furnishing of all labor, material, transportation, tools, supplies, plant, equipment and appurtenances, unless hereinafter specifically excepted, necessary for the complete and satisfactory construction, disinfecting and testing of one (1) reverse rotary drilled gravel envelope well as herein specified.

Section 2: Contractor's Qualifications

The Contractor shall have been engaged in the business of constructing reverse rotary-drilled gravel-envelope wells of diameter, depth, and capacity similar to the proposed well for a period of at least three (3) years. The Contractor shall submit a list of three or more well owners for whom the Contractor has drilled similar wells. The list shall include the owner's name and address, the casing diameter and depth, the well's maximum capacity and the well's specific capacity. The Contractor shall also submit a list of his last 3 major jobs.

Section 3: Competent Workmen

The Contractor shall employ only competent workmen for the execution of his work and all such work shall be performed under the direct supervision of an experienced well driller satisfactory to the Owner. The Contractor shall list the position and experience of all drilling personnel to work on this contract.

Section 4: Permits, Certificates, Laws and Ordinances

The Contractor shall, at his own expense, procure all permits, certificates and licenses required of him by law for the execution of his work. He shall comply with all federal, state or local laws, ordinances or rules and regulations relating to the performance of the work. The Owner has obtained the proper State Appropriation Well Permits.

Section 5: Location

The well to be hereunder is to be located in the NW 1/4 of the SE 1/4 of Section 30, T27N, R22W, in _____ County, _____ This locality is about _____ miles north of the town of _____ near State Highway ___.

Section 6: Local Conditions

Information regarding subsurface conditions is given on the attached log and is intended to assist the Contractor in preparing his bid. However, the Owner does not guarantee its accuracy, nor that it is necessarily indicative of conditions to be encountered in drilling the well to be constructed hereunder, and the Contractor shall satisfy himself regarding all local conditions affecting his work by personal investigation and neither the information contained in this section nor that derived from maps or plans, or from the Owner or his agents or employees shall act to relieve the Contractor from any responsibility hereunder or from fulfilling any and all of the terms and requirements of his contract.

Section 7: Boundaries of Work

The Owner shall provide land or rights-of-way for the work specified in this contract and make suitable provisions for ingress and egress, and the Contractor shall not enter on or occupy with men, tools, equipment of material, nor shall the Contractor discharge water either directly or indirectly to or on any ground outside the property of the Owner without the written consent of the Owner of such ground. Other contractors and employees or agents of the Owner may for all necessary purposes enter upon the work and premises used by the Contractor, and the Contractor shall conduct his work so as not to impede unnecessarily any work being done by others on or adjacent to the site.

Section 8: Protection of the Site

Excepting as otherwise provided herein, the Contractor shall protect all structures, walks, pipelines, trees, shrubbery, lawns/ etc., during the progress of his work; shall remove from the site all cuttings, drillings, debris and unused materials; and shall, upon completion of the work}~, restore the site as nearly as possible to its original condition, including the replacement, at the contractor's expense, of any facility or landscaping which has been damaged beyond restoration to its original condition or destroyed. Water pumped from the well shall be conducted to a place where it will be possible to dispose of the water without damage to property or the creation of a nuisance. It shall be assumed that the Contractor has inspected the site.

Section 9: Capping the Well

1. At all times during the progress of the work, the Contractor shall use all reasonable measures to prevent either tampering with the well or the entrance of foreign matter into it. The Contractor shall be responsible for any objectionable material that may fall into the well and its consequences until the completion and acceptance of the work by the Owner.

2. After testing the well, the Contractor shall furnish and install a cap on the well. The cap shall consist of 1/4-inch steel plate cut to the OD of the casing and tack-welded to cover the top of the well. A 1 inch-diameter threaded nipple and pipe cap shall be attached by threads or welding to a matching hole drilled in the center of the cap.

Section 10: Standby Times

Measurements of standby time will be made only for inactive periods resulting from the Contractor being notified by written communication to cease operations during normal working hours. Idle time required for maintenance of equipment or caused by failure of equipment shall not be measured for standby time. Standby time will be paid for a maximum of 8 hours per day, regardless of Contractor's actual work schedule. Standby time will not be paid for time on Saturdays, Sundays, or national holidays on which work is not customarily performed unless the Contractor had previously agreed to work on such days. Payment for standby time will be made at the unit price per hour quoted in Bid Item 9.

Section 11: Abandoned Hole

If the well fails to conform to specifications and because of Contractor's fault he is unable to correct the condition at his own expense or negotiate a mutually acceptable cost reduction for specification deviations, it shall be considered an abandoned hole, and Contractor shall immediately start a new well at a nearby location designated by the owner. The abandoned hole shall be treated as follows:

1. Contractor may salvage as much casing and screen from the initial well as possible and use in a new well if it is not damaged.

2. Salvage material shall remain the property of Contractor.

3. The initial hole shall be filled with clayey material to within 15 feet of the land surface and then a 10-foot-long concrete plug should be formed in the hole.

4. Casing remaining in the hole shall be cut off at least 5 feet below ground surface. The remaining 5 feet of hole shall be filled with native top soil.

5. No payment will be made for work done on an abandoned hole or for salvaging materials and sealing the hole.

If the well fails to conform to the Owner's needs, and the Contractor has faithfully fulfilled his obligations under this contract, the Owner may direct the Contractor to abandon the hole. In this event, negotiations between the two parties for salvage and sealing the hole will determine payment.

Section 12: Sub-Contractors

None of this work may be sublet without the written consent of the Owner. If any part is sublet the subcontractor shall be considered as an employee of the Contractor.

Section 13: Records

The Contractor shall keep records providing the following information:

1. A log of the formation drilled from surface to total depth showing each change in formation.

2. The final well log shall show: diameter, wall thickness, depths, and quantities of casings and screens installed; details of reducing sections; type, aperture size, and pattern of perforations; borehole diameters; cemented sections; gradation of gravel envelope; quantity of gravel initially installed, quantity of gravel added during development operations; quantity of material removed during development operations; and all other pertinent details.

3. A record of drilling fluid properties at 4-hour intervals. The record shall show weight, funnel viscosity, 30-minute water loss, cake thickness, and sand content.

4. Development and test records shall be maintained on an hourly basis, showing production rate, static water level, pumping level, drawdown, production of sand, and all other pertinent information concerning method of development.

These records shall be provided to the Owner upon demand at any time during or at the completion of the well drilling. Two copies of the Water Well Driller's Report submitted to the State of _____ upon completion of the well shall be provided to the Owner as well within 1 week of completion.

Section 14: Arbitration

Any dispute regarding performance of work, quality of workmanship or materials and interpretation of the intent and content of these specifications shall immediately be brought to the attention of all parties concerned. If disputes cannot be resolved through arbitration between Contractor and Owner, a mutually acceptable third party shall be appointed to resolve said differences.

REFERENCE LOGS

Owner: Inner States Power
Location: NE 1/4 of SW 1/4, Sec. 30, T27N, R22W,
                _____ County, _____, 1200 feet west
                of well proposed herein
Contractor: Keyes Well Drilling Company
Date

Completed: September, 1960

Well Log

Depth (ft)	Material
0 55	Drift, undifferentiated.
55- 75	Sand & Gravel, yellow-brown, silty.
75-215	Sand, Brown to tan, clean to silty, very fine to medium.
215-335	Sand, tan, fine with some gravel.
335-380	Sand, tan, fine to coarse, poorly sorted
380-390	Gravel, tan, very fine.
390-435	Sand, tan to brown, clean to silty.
435-450	Sand to gravel, tan, clean.
450-480	Jordan Sandstone, medium ss, well sorted, well rounded & frosted grains, quartzose, very clean.

Owner: Inner States Power
Location: NW 1/4 of SE 1/4, Sec. 30, T27N, R22W,
                about 50 feet west of well proposed
                herein,
                _____ County, _____
Contractor: Layne Company
Date

Completed: December, 1974

Depth (ft)	Material
0- 8	Red sandy clay, gravel.
8-248	Red sand and small gravel cemented seams
248-358	Gray firm sand.
358-358ft. 6 in.	Cemented sand
358ft. 6in.-373	Soft gray-black clay
373-382	Gray sandy clay
382-398	Gray clay, rock seams
398-403	Lime rock, harder

<u>SPECIFICATIONS</u>

1. <u>DRILLING</u>: A 24-inch hole shall be drilled to an estimated depth of 360 feet more or less, as determined by the Owner. A pilot bore or test hole, if used, shall be of such diameter as selected by the Contractor, but shall be of such size and depth as to adequately determine the character of the underground formation.

   Samples of the subsurface formations shall be obtained every 10 feet and preserved in sample bags for inspection by the Owner.

   The drilling shall be done with a first-class, high-speed, reverse rotary drilling unit, subject to the approval of the Owner. No unnecessary delays nor work stoppages due to negligence or willful miss-operation of the Contractor will be tolerated and the Contractor shall be held responsible and payment withheld for damage done to the well due to such negligence or miss-operation. The equipment shall be of the proper type and shall be in good condition so that the work performed can be done without any interruption arising from defective or improper equipment; however, the Contractor shall not be held liable for stoppage or delays due to mechanical failures beyond his control.

   Water for drilling and developing purposes shall be agreed upon between the Contractor and the Owner prior to drilling. It is expected that a nearly-continuous 25 gpm water supply would be available from a well approximately 50 feet from the well proposed herein.

   If at the completion of developing and test pumping, the well yield is insufficient for its intended use, the Owner may require the well to be deepened to about 480 feet and through the Jordan Sandstone using a cable tool churn drill) drilling rig. A cable tool rig is the preferred drilling method below about 360 feet to avoid invading the glacial outwash aquifer with clay particles. Approximately 60 feet of 12-inch I.D. casing shall be installed in this hole with at least a 3-foot overlap into the 16-inch O.D. casing. The annulus between the 16-inch O.D. casing and the 12 inch I.D. casing shall be plugged with a swaged lead seal.

   The 10-inch in diameter open-hole portion of the deepened well tapping the Jordan Sandstone shall be developed by the appropriate means. If the open hole is enlarged through blasting with explosives, care must be taken to avoid damaging the well screens in the glacial outwash aquifer. Only small quantities of explosives may be used at any one time and only with the approval of the Owner.

   A graphical description of the casing and well design is shown on Plate 1 attached to the Specifications.

2. <u>CASING</u>: The casing shall be only new Standard Pipe (Schedule 40) Class "A", electric welded, single-wall casing, complying with the requirements of "A.S.T.M. Standard Specifications A 283-Grade B, with any subsequent amendments thereto or A.S.T.M. A-242-55" (Grade 2 Kaisaloy or equivalent).

   The surface water shut-off casing, if any, shall be at the Contractor's discretion.

   The well casing shall consist of 16.000 inch outside diameter pipe casing. The casing shall have an inside diameter of 15.250 inches, a wall thickness of 0.375 inches and shall weight 62.58 pounds per lineal foot. The string of casing shall be about 300 feet in total length. All well screens put into the casing string shall be Stainless Steel 304 Johnson or its equivalent or better. The slot size of the screens will be determined by the Owner. Intermediate joints in screen sections shall be made by welding of a material and type recommended by the well screen manufacturer. The well screens shall be about 60 feet in total length.

   The bottom of the 16-inch O.D. tail pipe below the lowest well screen shall be plugged with two (2) feet of

Portland cement. Cement will be used for the bottom of the 16-inch O.D. casing so that the well can be drilled deeper if necessary.

If any auxiliary casing is used in the bottom of the well it shall be 12-inch I.D. pipe with an outside diameter of 12.750 inches, a wall thickness of at least 0.375 inches, and a weight of at least 49.56 pounds per lineal foot. If used, an estimated 60 feet of 12-inch I.D. casing would be installed from about 357 to about 417 feet.

3. GRAVEL: The gravel for the gravel envelope shall be clean, rounded, water-washed quartz or granitic gravel, free from silt, clay and other deleterious material.

Crushed or granular rock will not be permitted. Gravels shall be as near uniform sizes as possible and shall consist of a uniform size to be determined by the Owner. Sufficient gravel shall be furnished for initial graveling of the well and such additional' gravel as the well may take during development activities. The gravel shall be installed after l2 hours advance notification to the Owner and shall be approved by the Owner prior to installation.

4. MUD: Drilling fluids are expected to consist of water. However, if a higher viscosity fluid is needed, only a time-degradable drilling mud such as Revert manufactured by Johnson) plus any additives normally used with Revert shall be permitted in the well.

5. LOG AND RECORD: After the hole has been drilled .to the desired depth, the Owner, at his option, may request an electric log of the entire well from the ground surface to the bottom. The contractor shall be paid for his idle time, if during normal working hours, in accordance to Bid Item 9 "Standby time."

6. PLUMBNESS: The completed well shall be straight and plumb. Plumbness of the hole is of primary importance. To assure plumbness, drill collars of sufficient size and length are required. The alignment of the completed well shall not deviate more than 6 inches per 100 feet of depth.

The Contractor may use any appropriate means, upon the approval of the Owner, to correct any misalignment. Failure or inability of a Contractor to properly plumb the well shall be sufficient cause of its rejection.

7. PLACING OF WELL CASING: After the hole has been drilled no less than 24 inches in diameter to a depth of 360 feet, the well casing shall then be installed. The bottom of the casing shall be held slightly above the bottom of the reamed hole by suspending the casing from the ground surface until the gravel has been placed.

8. CENTRALIZERS: The casing shall be so centered that the completed string will be plumb and true. The casing shall be fitted with proper centering brackets, installed at locations as directed by the Owner, but spaced not over 50 feet apart.

9. PLACING OF GRAVEL: After the well casing has been installed and centered, water shall be introduced into the circulating drilling fluid to properly thin the fluids without endangering the wall structure. The annular space shall then be carefully and completely packed with "gravel."

The gravels may be pumped to the bottom through one or two columns of 2" or 21/2" pipe or tubing. Whether pumped or dumped, a constant check must be made to determine the actual position of the gravel.

A careful record and computation must be made and furnished as the amount of gravel introduced into the annulus of the well to compare it with the total amount required to properly gravel the well.

Fluid circulation shall be continuous with the placing of the gravel by pumping from the inside of the well casing. The method of placement of the gravel shall be approved by the Owner.

10. <u>PROTECTIVE GROUTING</u>: The Contractor shall be required for public health safety to place a cement grout of an approved mixture in the annular space between the well casing and the drilled hole. The Contractor shall place all cement grout in accordance to the State Laws governing such practices as adapted July 15, 19__, and he shall include in his bid price the cost of such work. If local conditions are not sufficiently known prior to drilling operations to specify necessary depth and grouting, the required work may be determined in the field by the Owner and such work shall be considered an extra, payment for which shall be at price agreed to by the Contractor and the Owner in writing before such operation is begun.

11. SWABBING AND JET DEVELOPMENT OF THE WELL: Upon completion of the setting of the well casing and the graveling, the well shall be swabbed at the Contractor's option. Development by jetting the screen and gravel pack as later described is mandatory.

The swab used should have a diameter not less than 1/8 inch smaller than the diameter of pipe to be swabbed. This work should be done by operating a suction swabber immediately after completion of initial graveling. Swabbing should continue opposite the perforated section of the well casing until as much as possible of the drilling fluids, silt and fine sand has been removed and as long as the gravel continues to move in the peripheral area around the well casing.

All material drawn into the well shall be removed often enough to prevent cementing of this material into lower formations. A careful record must be kept and furnished to the Owner as to the quantity and character of material removed from the well as a result of swabbing operations.

If, in the Contractor's opinion, due to some cause over which he has no control, the structural stability of the well is endangered, he may discontinue swabbing, of that particular section of the well casing where danger is imminent.

The Contractor shall develop the well by using a hydraulic jet nozzle to break up bridges in the gravel pack at the slots in the screen. During such an operation, circulation will be maintained by pumping at least 30 percent more water from the well than that being jetted into the well. The pressure of the jet shall not be less than 150 nor more than 200 psi or as directed by the Owner. The tip of the jet nozzle shall be centralized so as not to be more than 2 inches from the well screen. No less than 100 gpm nor more than 200 gpm shall be jetted to the screens through no more than 2 jet nozzles rotating at about 1 rpm.

During development there shall be available up to a 2-inch stream of water which shall be applied to the top of the gravel pack in whatever quantities required. The water is to furnish additional weight to facilitate movement of the gravel. Application of the water may be discontinued, if in the opinion of the Owner, the hole is endangered. Application of the water shall be through a tremie pipe to within a few feet of the gravel pack.

Payment for development of the open-hole portion of the well in the Jordan Sandstone (if drilled) shall be at the same rate as Item 8, Bid Items. Any additional supplier and materials shall be at a negotiated price between the Contractor the Owner.

12. <u>SWAGING</u>: A rigid 15-foot long tapered swage having a diameter 1/2" less than the internal diameter of the casing must pass freely from the top to the bottom of the casing. The taper shall not be more than 2 feet from either end of the swage.

13. DEVELOPING AND TESTING: After completely swabbing, jet developing and sand pumping the well, the Contractor shall furnish and install a test pump with a capacity of 1500 gpm at a head of 300 feet to provide for testing and developing the well. The pump should be driven by a diesel or automotive-type engine or an

electric motor in good condition and of sufficient horsepower to adequately deliver power required for the development of the well. The discharge pipe should be not less than 10 inch I.D. and should be equipped with a sharp-edged orifice plate not less than 8 inches I.D. A gate valve should be installed in the discharge piping close to the pump.

Pumping shall be started at a low rate of flow. The rate of production shall be gradually increased, depending upon the turbidity of the water, and surged until the desired production from the well is obtained.

The well should be developed until it produces no more sand that the amount specified by the Owner. The sand production shall be measured by a device such as the centrifugal sand sampler described in the A.W.W.A. Journal of February, 1954. Turbidity shall be less than 10 on the silica scale described in Standard Methods of Water Analysis. No deviation from this stipulation will be allowed except upon approval of the Owner. In no case, will sand be allowed to exceed 5 parts per million of water.

If in the opinion of the Owner after pumping for 48 hours additional development is necessary, the Contractor shall perform this work at his expense for a reasonable period of time.

The Owner must be notified 72 hours in advance when the Contractor intends to start development. Because of certain conditions, over which the Owner has no control, all development work should be carried out in daylight hours only as far as practical.

Upon completion of the test and development described above, to the satisfaction of the Owner, the Contractor shall remove the test pump and clean the well to bottom.

After the well has been initially pumped in this manner, no further work shall be done on the well for a period of not less than 12 hours or as determined by the Owner.

The test pumping shall be performed for a minimum of 48 hours and shall be directed by the Owner. Pumping may be continued for a period of up to 10 days if determined to be necessary by the Owner. The Contractor shall be paid for the non-operating time the test pump is installed in the well if authorized in writing by the Owner according to Bid Item 12A.

After installation of the permanent well pump, the well pump and all appurtenances will be chlorinated in accordance with the Minnesota State Board of Health specifications.

14. PUMP INSTALLATION AND WELL COMPLETION: The Contractor shall furnish and install a vertical line shaft turbine pump and related appurtenances as shown in Plate 2 and described below. The final design of the pumping equipment will be made after completing and analyzing the aquifer test. The following preliminary design is a guide and represents the probable maximum pumping equipment;

    1. The pump shall be a oil-lubricated vertical line shaft turbine capable of pumping 1500 gpm with a total head (pumping lift plus line pressure) of 600 feet; equivalent to Tait 12BCH;

    2. The bottom of the pump bowls shall be at a depth of about 260 feet below ground surface;

    3. The multiple-stage pump bowl assembly shall not be less than 11 7/8-inch O.D. and not more than 13 7/8-inch O.D.;

    4. The tail pipe shall be 10 feet of 10-inch I.D. pipe (wall thickness not less than 0.25 inches);

    5. The line shaft bearings and bowl bearings shall be bronze;

6. The line shaft shall be not less than 1 15/16 inch in diameter and shall be of No. C1045 chrome plated, turned and polished steel to about 130 feet below the motor, the remainder of the shaft shall be No. 416 stainless steel or better;all couplings shall be of the appropriate material; and shall be enclosed in a 3-inch I.D. schedule 80 oil tube;

7. The column shall be 10-inch I.D. with flange fittings (300 psi);

8. The discharge head shall be close-grain cast iron (minimum 30,000 pounds) and have a 10 inch I.D. discharge flange (300 psi);Tait 1012HA or equivalent;

9. The electric motor shall be 300 hp, 3 phase, 60 cycle, 460 volt, 1770 R.P.M. non-reversible, Nema enclosure, shielded, hollow shaft, equivalent to General Electric or better;

10. The pump base shall be 3 feet square by six feet high of which four feet shall be below grade and shall consist of structurally reinforced concrete as specified and shown in Plate 3 and 4 .

15. <u>APPURTENANCES</u>: The Contractor shall furnish the following appurtenances on the discharge line:

    1. A pressure gage having a dial not less than 3 inch in diameter shall be connected near the discharge head by 1/4-inch cock with level handle and appropriate fittings. The gage shall be graduated in pounds per square inch to at least 300 psi.

    2. A 2-inch or larger automatic air release valve is required to vent air from the column and discharge head upon starting the pump and also to serve to admit air to the column to dissipate the vacuum when the pump is stopped. This valve shall be located at the highest point in the discharge line between the pump and the discharge check valve. The air release valve shall be Consolidated Valve Co. No. 1541H or its equivalent.

    3. A 6-inch I.D. relief valve shall be mounted vertically on a 10 x 10 x 6 reducing tee. The relief valve discharge shall be to an 8 inch I.D. line to outside the pumphouse. Pressure relief valve shall have a 6-inch inlet and outlet flange plates, with the outlet flange on the side of the vertically-mounted valve. The inlet flange shall be rated not less than 250 psi and the side outlet flange can be rated to 125 psi. The relief pressure valve shall be manually adjustable with a side wheel. The pressure relief valve shall be Kunkle No. 218 or its equivalent. The pressure relief valve shall be set at 175 psi.

    4. Check valve shall be 10-inch I.D. and rated not less than 175 psi (non-shock). The check valve shall be Crane Ferrosteel Clearway Swing Check No. 375 or its equivalent.

    5. Gate valve shall be rated not less than 175 psi (non-shock) and shall be a 10-inch I.D. Crane Ferrosteel Gate Valve No. 467 or its equivalent.

    6. The 10-inch I.D. tee shall be a 10 x 10 x 6 inch reducing tee rated for not less than 175 psi.

    7 . All flange fittings shall be of standard design and shall be joined using appropriate gaskets and bolt and nut fasteners as recommended by the manufacturers.

8. The relief valve discharge line shall consist of a 6 inch I.D. 125 psi long-radius 90° elbow with a flange at one end and the other end plain, a plain-end long-radius 90° elbow, and two pieces of standard 6-inch I.D. pipe in the appropriate lengths the total not to exceed 15 feet, the components shall be welded together as one unit.

Payment for the above discharge line appurtenances shall be according to Bid Item 14. The Contractor is not responsible for furnishing nor constructing starter or electric controls.

BID ITEMS

1. MOBILIZATION: Furnish all equipment, personnel, etc.

                                     Lump Sum_____

2. ROTARY DRILLING: No less than 24-inch diameter hole in accordance with attached
                                     Specifications to an estimated depth of 360 feet.

                                     Price/Foot _____

3. CABLE TOOL DRILLING: A. No less than 12-3/4 inch
(At Owner's   diameter hole to accommodate
Option)   the casing in Item 5. Total
                                     estimated quantity of drilling
                                     57 feet.

                                     Price/Foot_____

                                     B. No less than 10-inch diameter
                                     hole. Total estimated quantity
                                     of drilling 63 feet.

                                     Price/Foot_____

4. PIPE: Furnish and install 16-inch O.D. standard pipe casing in accordance with
                                     attached Specifications. Total estimated quantity 300 feet.

                                     Price/Foot_____

5. PIPE: Furnish and install 12-inch I.D. standard pipe casing and a swaged lead
                                     seal all in accordance with attached Specifications. Total estimated
                                     quantity 60 feet. (At Owner's option.)

                                     Price/Foot_____

6. SCREEN: Furnish and install 16-inch O.D. pipe-size stainless steel #304 (U O P
                                     Johnson or equivalent) well screens in slot sizes determined by the
                                     Owner in accordance with attached Specifications. Total estimated
                                     quantity 60 feet.

                                     Price/Foot_____

7. GRAVEL PACK: Install gravel pack at sizes and in a quantity as determined by the
                                     Owner and in accordance with attached specifications.

                                     Price/Yard_____

8. DEVELOPMENT: Develop the test well by jetting and swabbing in accordance with
                                     attached Specifications.

                                     Price/Hour_____

9. STAND-BY:

For the Contractor's time after being notified by written communication to cease operations during normal working hours. Under such conditions, stand-by time will be in effect, all in accordance with attached Specifications.

Price/Hour _____

10. GROUTING:

Install concrete grout in annulus between the wall of the hole and the casing in accordance with attached Specifications and in compliance with Minnesota State Law of July 15, 1974, concerning such practices.

Price/Cubic Yard_____

11. TEST PUMP:

Furnish, install, and remove a line-shaft turbine pump in accordance to the attached Specifications. This item includes prime mover, discharge piping, gate valve, and orifice in accordance to attached Specifications.

Lump Sum_____

12. TEST PUMP RENTAL:

Test pump rental of the equipment in Item 11, in accordance with attached Specifications. A. Non-operating time

Price/Day_____

B. Operating time with Contractor's personnel

Price/Day_____

C. Operating time without Contractor's personnel

Price/Day_____

13. PUMP BASE:

Construct steel-reinforced concrete pump base in accordance to attached Specifications.

Lump Sum_____

14. PUMP:

Furnishing and installing vertical line-shaft turbine pump, motor in accordance with attached Specifications.

Lump Sum_____

15. APPURTENANCES:

Furnishing and installing discharge line appurtenances in accordance to the attached Specifications.

Lump Sum_____

Appendix D-2

EXAMPLE

SPECIFICATIONS FOR CONSTRUCTING

A GROUND-WATER SUPPLY SYSTEM

TECHICAL SPECIFICATIONS

1. DRAWINGS

The drawings accompanying and referred to in these Specifications are as follows:

Drawing 1   Location Plan
Drawing 2   Exploration and Production Borehole Construction
             Diagram
Drawing 3   Details of Borehole Headworks
Drawing 4   Details of jetting tool

2. LOCAL CONDITIONS

The areas are not normally subject to extreme flooding except in the main stream channels, though flash floods can occur and, after rain, access to and between bore sites could be difficult. Under dry conditions, existing access tracks are adequate for most vehicles and tracks will be cleared for access to new borehole locations.

Accommodation and messing facilities are not available at the Facility and the Contractor shall supply his own camp and messing facilities. Temporary camps or caravans may be established in the area of operation on written approval from the Client.

The Contractor shall be totally responsible for his own supply of food. _____ is served by a large supermarket from which perishables can be purchased.
Water is available in town, the Contractor will supply his own method of carting water.

3. CLIMATE

The area is normally dry with the temperature ranging from -5° to 40°C.

4. GEOLOGICAL INFORMATION

From the numerous bores now constructed within the area the following geological log illustrates the expected formations to be encountered:

0- 10 meters -	Kalahari Formation-sands, silcretes and calcretes
10 - 100 meters -	Basalt, weathered to fresh
100 - 200 meters -	Sandstone with minor siltstones, may be fractured
200 - 250 meters -	Mudstone, silt stones, may be hydrating, minor sandstones.

All levels are relative to ground level.

The depths indicated above are approximate only and the Engineer accepts no responsibility as to the accuracy of the values indicated.

5. MATERIALS

5.1   Materials Supplied by the Contractor

The Contractor shall supply and deliver on site, in new and undamaged condition, all materials necessary to complete the Contract without interruption. The materials used shall be paid for at the rates tendered by the Contractor in the Schedule of Quantities and Rates.

5.2   203mm I.D nominal (3mm wall thickness) Steel Surface Casing

The casing shall be plain ended or slip socket, BS43 (SABS 719) or equipment.

5.3   168mm I.D nominal (3mm and 4mm wall thickness) Steel Plain Casing

This casing shall be plain ended or slip socket, BS43 (SABS 719)
or equivalent.

5.4   168mm I.D nominal (3mm and 4mm wall thickness) Slotted
Steel Casing

This casing shall be plain ended or slip socket, BS43 (SABS 719)
or equivalent. It shall be pre-slotted prior to arrival on site.
Slots must be 2mm width and the casing characterized by a minimum
open area of 3% per meter length.

5.5   Other Materials

All other items required in the construction of the bore shall be
constructed in a manner approved by the Engineer or the Technical Supervisor and S.A.B.S. code
of practices.

Because of the nature of the exploration program extra materials may have to be mobilized to site
during the Contract.

6. BOREHOLE CONSTRUCTION

6.1 Depth

The bottom depth of the bore and all other significant depths involved in the design of the bore shall be determined by the Engineer or the Technical Supervisor on site from the cutting samples collected by the Contractor from his strata log and by electrical and gamma ray logging carried out by the Technical Supervisor. It is intended that drilling shall be completed when a 6m sump has been drilled or at some lesser depth as specified by the Engineer or Technical Supervisor.

6.2 Techniques

The Contractor may use any drilling techniques he feels applicable to achieve the DEPTH AND DIAMETER REQUIRED, providing that the techniques used are those specified in his Tender or are approved by the Engineer or Technical Supervisor and that they do not permit formation collapse, hole erosion or involve the use of lost circulation agents, sawdust or any form of plugging that may ultimately affect the production capacity of the water bearing strata intersected.

The Contractor will set the surface casing lm below the depth of intersection of basalt or as directed by the Engineer or the Technical Supervisor on site.

The Technical Supervisor must be present during the placing of the casings in boreholes.

The Contractor should note that 'sloughing' and/or fractured formations may be intersected and the use of special techniques may be required.

The Contractor will be expected to provide coring capabilities at a diameter of 75mm or greater. Core length requirements are a minimum 300mm and core sampling will be as directed. It is anticipated that a minimum of 5 samples will be required over the sections of sandstones intersected per borehole. Cores should be collected using rotary airfoam or air-flush techniques and coring will be paid on an hourly worktime basis as indicated in Item 12a of the Schedules of Rates.

6 . 3 <u>Drilling Media</u>

The Contractor may use any form of drilling media which he feels is applicable to achieve the DEPTH AND DIAMETER REQUIRED, providing that the media used are those specified in his Tender, or are otherwise approved by the Engineer or Technical Supervisor and that they do not permit formation collapse, cause hole erosion or involve the use of native clay, oil, salt or any lost circulation agent, sawdust, cement, or any form of plugging that could affect the production capacity of the water bearing strata intersected.

The Contractor should note that 'sloughing', and/or fractured formations may be intersected and the use of special media may be required .

Low solids, degradable drilling mud may be used where applicable. Where permission is given for the use of Bentonitic muds, they shall be fully hydrated prior to use.
The conditions of the bentonitic drilling mud shall be maintained as follows:

Mud weights:	At all times other than in flowing artesian conditions less than 1,08 kg/1itre.
Viscosity:	Greater than 35 seconds Marsh Funnel where applicable with a maximum of 42 seconds.
Filter cake:	Less than 2m m where practicable with a maximum of 3mm.
Fluid loss:	Less than 5cc where practicable with a maximum of 8cc.
Sand content:	Less than 2 % where practicable with a maximum of 5%.

The conditions of degradable mud shall be maintained as follows:

Mud weight:	At all times other than in flowing artesian conditions, less than 1.08kg/1itre.
Viscosity:	Where practicable less than 42 seconds Marsh Funnel reading.
Sand Content:	Less than 2 ~ with a maximum of 5 %.
pH:	According to manufacturer's specification.

Measurements of the mud conditions shall be collected every one hour or as directed by the Engineer or Technical Supervisor and shall be recorded on the report sheet. Steps should be taken to treat immediately any excessive variation of the preferred value listed.

Two steel tanks with a minimum capacity of 9 cubic meters (approximately 2 000 gallons) shall be used in which to mix and hold all drilling fluid. These tanks should be at least twice as long as they are wide and be fitted with V serrated baffle plates. The flow channel from the bore head to the mud tanks shall be of sufficient length and capacity to allow flocculation and settling of clays and tailings fro m the drilling fluid.

The mud tanks and flow channel shall at all times be kept clean of flocculated and settled material

7. SAMPLING

Representative samples of the strata intersected shall be collected by the Contractor every 2m by whatever method is standard for the drilling technique in use and approved by the Engineer or Technical Supervisor. The Contractor will take every possible precaution to guard against sample contamination due to poor circulation, hole erosion or caving. The sample should be bagged, labeled with the bore number and depth increment and stored in a position where they will not be contaminated by site conditions or drilling operations. The Contractor shall supply stable sample bags and labels as required. During mud drilling, lag time shall be accurately calculated and timed so that the sample collected relates to the depth at which it was cut.

Water samples shall be collected from each bore on the completion of development or at other intervals as directed by the Engineer or Technical Supervisor in containers which will be supplied by the Engineer.

Each water sample bottle should be marked with the well number, depth, date in non-soluble ink.

8. DRILLING AND CONSTRUCTION OF BORES

Bores shall be drilled through the surface formation to an anticipated maximum depth of 10 meters at a diameter of 254mm. The pre-collar will be completed to a depth of 1 or 2 meters into Bedrock.

On completion of the surface drilling, surface casing shall be set and grouted in position using quick setting cement slurry (1,5kg/1itre). Subsequent to the cement setting, drilling shall proceed at a diameter which will permit the insertion of exploration and production casing types to a depth as directed by the Engineer or Technical Supervisor.

On completion of drilling, geophysical logging will be carried out and well development commenced thereafter. Subsequently, the casing string shall be lowered into position and left in tension.

Subsequent to the casing string being positioned in the hole, aquifer development shall proceed as directed by the Engineer or Technical Supervisor.

9. DIMENSIONS OF BOREHOLES

- Surface casing 203m m (8 inch) diameter.
- Production casing to total depth below the surface 168mm (6 5/8 inch) diameter.

10. YIELD MONITORING

When air drilling or air developing, a 90° V notch flow measurement device shall be permanently set up in an approved manner, level and vertical, in the drain line so that continuous monitoring of air lift yields can be obtained. Average yields shall be read and noted every 2m, of penetration and recorded in the drillers' log.

Care shall be taken to ensure that no-floating debris impedes the flow of water over the V. The weir shall at all times be kept clear of a build up of silt.

11. SETTING CASING IN POSITION

All casing strings shall be lowered into position under tension. Any casing string which requires its upper end to be terminated below ground level shall be set in position by being attached to drilling rods by means of a 'j' latch, left hand back off thread or other approved connections and lowered into position .

If a casing string is required to have the upper end terminated below ground level and left in tension, it shall be set in position and suspended by use of a set of tapered support rings.
Under no circumstances will it be permitted to drop casing into position.

12. PLUMBNESS AND ALIGNMENT

All boreholes shall be drilled and cased straight and vertical and all casings and liners shall be set round, plumb and true to line. The Engineer or Technical Supervisor shall have the right to reject any or all drilling or casing which fails to meet this specification and all work and casing rejected will be replaced at the Contractor's expense.

Any delays encountered in running casing, considered to be due to poor hole alignment shall be at the Contractor's expense.

13 . REQUIREMENTS TO TEST

To demonstrate the compliance of the work with this requirement, the Contractor shall furnish all labor, tools and equipment and shall make the tests described herein in the manner prescribed by, and to the satisfaction of the Engineer or Technical Supervisor. Tests for plumbness and alignment must be made after the complete construction of the well and before its acceptance.

14. DESCRIPTION OF TEST

Alignment shall be tested by lowering into the well to a depth as directed by the Engineer or Technical Supervisor, a section of (dummy) pipe 3 meters in length. The outer diameter of the dummy shall not be more than 10 mm smaller than the diameter of that part of the casing being tested.

The dummy shall be suspended on a wire cable and lowered at a maximum rate of 10 meters per minute. The Contractor shall be responsible for bringing his own pre-constructed du m my onto site.
The Engineer or Technical Supervisor, at his discretion, may carry out additional plumbness and alignment tests using specialized equipment to ensure compliance with this clause.

15. REQUIREMENTS FOR PLUMBNESS AND ALIGNMENT

Should the dummy fail to move freely throughout the length of the casing or hole to the required depth or should the well vary from the vertical in excess of 30% of the smallest inside diameter of that part of the well being tested per 30 meter depth, or beyond limitations of this or any other test performed by the Engineer or Technical Supervisor, the plumbness and alignment of the well shall be corrected by the Contractor at his own expense. Should the Contractor fail to correct such faulty alignment of plumbness, the Engineer or Technical Supervisor may refuse to accept the well and no payment shall be made for same.

16. PROTECTION

During the Contract period when work is not in progress, the bores shall be kept capped in such a manner as to prevent the entrance of foreign material. The Contractor shall remove any foreign matter at his own expense. On completion of each bore, the Contractor shall complete the borehole as shown on Drawing 3 .

17. EXPERTISE

The Contractor under this contract is considered to be an expert water well driller and is expected to organize and carry out the work specified hereunder in an expert manner. Drilling problems encountered will be overcome entirely within the framework of the specification and schedule of quantities and rates and no claim for extra payment will be entertained for problems foreshadowed in the specification or due to limitations placed by this specification.

18. ABANDONMENT

The Engineer or Technical Supervisor shall have the right, at any time during the progress of the work, to order the abandonment of a borehole. The Contractor thereupon shall withdraw the casing and screens, if applicable, and salvage or attempt to salvage all such materials as the Engineer or Technical Supervisor may direct and/or up until he revokes such direction, and shall fill or leave the bore to the satisfaction of the Engineer or Technical Supervisor.

Payment shall be made for such abandoned bores at the rate of drilling and other rates as are appropriate or as detailed in this specification.

19. LOST BOREHOLE

Should accident to the plant, behavior of the ground, jamming of the tools, or casing, or any other cause prevent the satisfactory completion of the works, the borehole shall be deemed to be lost and no payment shall be made for that bore nor for any materials not recovered therefrom, nor for any time lost.

Any material provided by the Client which is not recovered from the Lost Bore in good condition, may be at the Contractor's expense, and may be deducted from the Contractor's payment.

In the event of a lost bore, the Contractor shall construct a bore adjacent to the lost bore or at a site indicated by the Engineer or Technical Supervisor. The option of declaring any bore lost shall rest with the Contractor subject to direction from the Engineer,

If the Engineer or Technical Supervisor directs that a replacement bore be re-drilled at a site more than 60m distant, the Contractor shall be paid in full for the move and 10 % of the value of the drilling of the lost bore.

If the Engineer or Technical Supervisor directs that the lost bore be re-drilled close to the lost bore and the Contractor is concerned regarding the possible loss of air or other media the Contractor shall back fill the lost bore with approved material. The top 10 m of the hole shall be cement grouted to provide a complete seal All work and material shall be at the Contractor's expense .

20. DEVELOPMENT:

The water bore on the completion of drilling shall be developed to a maximum yield of water, free of suspended materials. Development will be carried out using water jetting, air surging and back washing, air lift pumping, isolated air pumping and surging, valve surging, simultaneous airlift/high pressure jetting, polyphosphate treatment and such other standard techniques as may be directed by the Engineer or Technical Supervisor.

Any non-standard techniques such as acidizing will be paid for at the development rate, though time spent in making up specialized equipment which holds up progress of rig operation will be paid for under standby rates in the schedule of quantities and rates.

18. ABANDONMENT

The Engineer or Technical Supervisor shall have the right, at any time during the progress of the work, to order the abandonment of a borehole. The Contractor thereupon shall withdraw the casing and screens, if applicable, and salvage or attempt to salvage all such materials as the Engineer or Technical Supervisor may direct and/or up until he revokes such direction, and shall fill or leave the bore to the satisfaction of the Engineer or Technical Supervisor.

Payment shall be made for such abandoned bores at the rate of drilling and other rates as are appropriate or as detailed in this specification.

19. LOST BOREHOLE

Should accident to the plant, behavior of the ground, jamming of the tools, or casing, or any other cause prevent the satisfactory completion of the works, the borehole shall be deemed to be lost and no payment shall be made for that bore nor for any materials not recovered there from, nor for any time lost.

Any material provided by the Client which is not recovered from the Lost Bore in good condition, may be at the Contractor's expense, and may be deducted from the Contractor's payment.

In the event of a lost bore, the Contractor shall construct a bore adjacent to the lost bore or at a site indicated by the Engineer or Technical Supervisor. The option of declaring any bore lost shall rest with the Contractor subject to direction from the Engineer.

If the Engineer or Technical Supervisor directs that a replacement borehole be re-drilled at a site more than 60m distant, the Contractor shall be paid in full for the move and 10% of the value of the drilling of the lost bore.

If the Engineer or Technical Supervisor directs that the lost bore be re-drilled close to the lost bore and the Contractor is concerned regarding the possible loss of air or other media the Contractor shall back fill the lost bore with approved material. The top 10m of the hole shall be cement grouted to provide a complete seal All work and material shall be at the Contractor's expense .

20. DEVELOPMENT:

The water bore on the completion of drilling shall be developed to a maximum yield of water, free of suspended materials. Development will be carried out using water jetting, air surging and back washing, air lift pumping, isolated air pumping and surging, valve surging, simultaneous airlift/high pressure jetting, polyphosphate treatment and such other standard techniques as may be directed by the Engineer or Technical Supervisor.

Any non-standard techniques such as acidizing will be paid for at the development rate, though time spent in making up specialized equipment which holds up progress of rig operation will be paid for under standby rates in the schedule of quantities and rates.

DEVELOPMENT (CONTINUED)

Any borehole which is found by the Engineer or Technical Supervisor to be characterized by an excess of 10% side-wall degradation of permeability following completion of 8 hours airlift development, will be re-developed for a period of 4 hours by the Contractor. Should further hydraulic testing show that the bore is still characterized by an unacceptable level of formation degradation then the Contractor will undertake further remedial action at his own expense. Should the performance of the borehole prove unsatisfactory after this period of remedial action the bore will be declared lost by the Engineer on site.

Payment for the borehole will be negotiable and dependent on the final degree of degradation and the bore's utilization potential

The Contractor should note that it is anticipated that simultaneous airlift/high pressure jetting technique will need to be employed in the development of the bores. The Contractor will be required to bring a jetting tool as outlined in Drawing 4 onto site.

21. PUMP TESTING:

All pumping tests will be carried out by the Engineer or Technical Supervisor.

22. TESTS FOR ACCEPTABILITY

The Contractor shall submit each exploration and production borehole for the Engineer or Technical Supervisor acceptance test.

Boreholes shall only be accepted after passing the following tests and being finally completed.

(i) A multiple stage discharge test with each stage lasting 60 minutes.

(ii) Passing all plumbness and alignment tests and any photographic / video tests deemed necessary.

The Contractor shall submit each borehole for the Engineer's or Technical Supervisor's acceptance test and shall give him at least 12 hours notice of such test .

In order not to unduly delay payment for completed boreholes which have not undergone all tests the Engineer or Technical Supervisor may issue a provisional acceptance. If however, the bore does not pass any test remaining after the provisional acceptance a Final Acceptance will not be issued and the bore will be deemed not to be acceptable and therefore not due for payment.

Any moneys which may have been paid on provisional acceptance shall be deducted from the Contractor's succeeding invoices and final payment.

23 . REPORTS

The Contractor shall provide the following reports to the Engineer or Technical Supervisor:

NAME	DESCRIPTION	SUPPLIED
Strata Log	An accurate record of strata passed through and the depths at which the strata intersected, also progressive measured air lifted yields when drilling with air or air developing.	DAILY
Penetration Log	An accurate record of the penetration rates achieved in minutes per 2 m through the various strata, broken down to not more than 2 meter increments together with weight applied at the bit, type and grade of bit.	DAILY
Drilling Media Log	An accurate record of the components and quantities used in or injected into the drilling media, including for drilling muds, viscosity, weight sand content, water loss, filter cake, pH and temperature. The Contractor will carry out all of above tests each hour and as directed at random by the Engineer of Technical Supervisor.	DAILY
Construction Log	An accurate record of all casing, slotted casing and screen lengths positions run into the borehole	ON COMPLETION OF CONSTRUCTION
Time Log	An accurate record of time spent on all phases of drilling	DAILY and a sum-summary of completion of construction to Engineer of Technical Supervisor for signature.

24 . PAYMENTS

24.1 Drilling

The rates of drilling are based on depth measurement, diameter and penetration rate and are to cover all the costs involved in drilling, including mud mixing, carting of water, injection of mixes, bit sharpening, conditioning of the drilling fluid and cleaning the hole of all bridging, obstructions and backfill ready for geophysical logging, tripping in and out of the hole and all other such works as are associated with the works and are not covered under other allowable payments.

24.2 Supply and Install Casing

This rate is to cover the supply and installation of bore casing into exploration or production bores. It does not cover the running or pulling of casing in bores declared lost or in which the casing cannot be set in position due to misalignment or other operational problems.

No claim for extra payment will be entertained by reason of remoteness, wharfage, insurance etc., or by reason of omission in calculating the Tender rate.

24.3 Standby Time

This time rate is to cover only those items when the rig and crew are waiting on geophysical logging or decisions by the Engineer or Technical Supervisor.

24.4 Work Time

The work time rate is to cover time spent using the rig to carry out any directive by the Engineer or Technical Supervisor for non-standard work not included in the specification. It does not include mud or cement mixing time.

24.5 Bore Development

The bore development time rate is to cover all the time spent on bore development, except where included under other time rates. Contractors will note that time rates do not allow for the manufacture of standard development tools on site. All bore development is subject to the approval of the Engineer or Technical Supervisor.

24.6 Vacation of Project

This item is to cover the removal of all plant, equipment and personnel permanently from the project, and the restoration of the drilling sites to a level and tidy state.

24.7   Supply and Delivery of Materials on Site

The above rate is to cover purchase cost, transport to and delivery and safe storage on site of all materials required for drilling, construction, development and use in the bores and shall be measured and invoiced for each bore. Bulk payment for materials will not be made and all materials remain the property of the Contractor until they are accepted as part of a completed bore by the Engineer Technical Supervisor.

24.8   Moving Borehole Sites

This item of payment is to cover the movement of the rig and ancillary equipment from one bore site to the next. The Contractor shall note that sites are expected to be located within 3km of each other but may be up to 10 kms apart.

25. TIME FOR COMPLETION

The drilling is to be completed by the Contractor as detailed in the contract program, unless otherwise agreed to in writing by the Engineer. The Contractor must mount two 12 hour shifts for six days per week until the works are complete.

26. SHUT DOWN

Notwithstanding any other time rate clause in the specification and the General Conditions of Contract, the Engineer reserves the right to shut down the Contractor's operations without notice, if, in the Engineer's or Technical Supervisor's opinion, the Contractor is failing to carry out the work in accordance with this specification. In this event, the Contractor shall not be liable for payment, but a site conference shall be called immediately between the Contractor's Senior Representative and the Engineer to discuss and resolve this problem.

27. SUPERVISION OF WORKS

The Contractor shall have a skilled senior tool pusher on site at all times to give technical instructions to the drilling crew, manage and organize the Contract and to laissez with the Engineer or Technical Supervisor.

28. SAFETY STANDARDS

All safety standards normal to the Engineer or which are required by site regulations shall be adhered to.

SCHEDULE N0.1 OF QUANTITIES AND RATES
FOR
EXPLORATION DRILLING

ITEM	DESCRIPTION	AMOUNT ESTIMATE	UNIT OF MEASURE	RATE
1.	Establishment	1	Item	
2.	Move and Set-up Between Bores	5	No	
3.	Drill 254mm Air Hammer or Air Rotary	60	m	
4.	Supply and Install 203mm ID. Steel Precollar	62	m	
5.	Cement Grout Steel Precollar to Include Waiting Time for Setting of Accelerated Cement	6	No	
6.	Drill 200mm Air Hammer	1500	m	
7.	Supply 168mm ID. Plain Steel Casing to Site (3mm wall thickness)	700	m	
8.	Supply 168mm ID. Slotted Steel Casing to Site (3mm wall thickness)	800	m	
9.	Install 168mm ID. Plain Steel Casing	700	m	
10.	Install 168mm ID. Slotted Steel Casing	800	m	
11.	Supply and Install Plain 168mm ID. Steel Casing	700	m	
12.	Supply and Install Slotted 168mm ID. Steel Casing	800	m	
13.	Bore Development	72	Hr	
14.	Completion of Bore Headworks	6	No	
15.	Standby Time Awaiting Logging etc.	24	Hr	
16.	Work Time as Directed	50	Hr	
16a	Work Time Involved with Coring	100	Hr	
17.	Vacation of Site	1	Item	
18.	Supply of Materials 14.1 Polyphosphate	Rate	Kg	

TOTAL FOR EXPLORATION BORES
C/F to Form of Tender Page

SCHEDULE NO. 2 OF QUANTITIES AND RATES
F O R
EXPLORATION & PRODUCTION DRILLING

ITEM	DESCRIPTION	ESTIMATE	MEASURE RATE
1.	Establishment	1	Item
2.	Move and Set-up Between Bores	23	No
3.	Drill 254mm Air Hammer or Air Rotary	240	m
4.	Supply and Install 203mm ID. Steel Precollar	252	m
5.	Cement Grout Steel Precollar to Include Waiting Time for Setting of Accelerated Cement	24	No
6.	Drill 200mm Air Hammer	6000	m
7.	Supply 168mm ID. Plain Steel Casing to Site (4mm wall thickness)	2000	m
8.	Supply 168mm ID. Slotted Steel Casing to Site (4mm wall thickness)	4000	m
9.	Install 168mm ID. Plain Steel Casing	2000	m
10.	Install 168mm ID. Slotted Steel Casing	4000	m
11.	Supply and Install Plain 168mm ID. Steel Casing	2000	m
12.	Supply and Install Slotted 168mm ID. Slotted Steel Casing	4000	m
13	Borehole Development	288	Hr
14	Completion of Bore Headworks	24	No
15	Standby Time Awaiting Logging Engineers Instructions.	100	Hr
16	Work Time as Directed	100	Hr
16c	Work Time Involved with Coring	360	Hr
17	Vacation of Site	1	Item
18	Supply of Materials 14.1 Polyphosphate	Rate Only	Kg

TOTAL FOR EXPLORATION & PRODUCTION BORES

C/F to Form of Tender Page...

Appendix D-3
EXAMPLE
SPECIFICATION FOR CONSTRUCTING
A GROUND WATER MONITORING WELL

PART I – INTRODUCTION

Drilling and well construction services are generally subcontracted by the Consultant of Record, on behalf of the Owner, in accordance with prevailing Master Agreements. Therefore, the only contractual issue this specification addresses is the method of subcontractor selection and reimbursement for drilling and well installation services.

Since the consultant of record is ultimately responsible for adherence to this well specification and the quality of the work products and deliverables, the selection of a qualified drilling firm is the consultant's responsibility. The Owner does reserve the right to request documentation of the selected or proposed drilling firm's qualifications and experience and to reject the subcontracting of drilling firms which the Owner deems to be unqualified. Such rejections will constitute the basis for renegotiation of that portion of the contract pertaining to drilling and well installation.

The Owner expects the consultant to procure cost effective drilling and well installation services through the competitive bidding process. The owner reserves the right of prior approval to review and make recommendations to the consultant prior to the consultant's awarding the drilling subcontract.

Drilling and well construction services performed at the Owner's facilities should be performed on principally a footage and materials basis in accordance with a previously submitted and accepted schedule of fees. An example of a fee schedule that would be provided in a bid package is shown on Table A–l. All drilling activities should be reimbursed on a per foot basis. The basic drilling footage rate must include, but is not limited to, any sample and testing protocols which are specified as part of the scope of work, estimated drilling depths, estimated number of borings, decontamination of drilling tools between individual borings, and the cost of any basic safety monitoring equipment or personal protection clothing that is required by the drilling firm or its insurance carrier.

Separate footage rates should be provided for different drilling methods or equipment requirements. Footage surcharge fees should be provided for work performed in adverse weather conditions, in Level A, B, or C personnel protective attire, or for deep borings. Mobilization/demobilization, on–site moves and set–ups, and well construction (including development) should be reimbursed on a per event basis. All well construction materials and expendable items (i.e., drill bits, auger baskets, core boxes, etc.) should be reimbursed on a unit cost basis. Authorized standby is the only item which will be reimbursed on an hourly basis.

To facilitate subcontractor conformance with this reimbursement method, the owner recommends that the consultant provide the bidding firms with a formatted fee schedule, such as that shown on Table A–1, complete with the estimated units. The estimated drilling depths and number of borings should also be provided.

PART II – GENERAL CONDITIONS

SECTION 2.1 – DEFINITIONS

2.1–01 **Agreement**: The contract between the Owner and Contractor including supplements and change orders issued by the Owner's Representative.

2.1–02 **Annular Space:** The space between two concentric tubes or casings, or between the casing and the well hole.

2.1–03 **Aquifer:** A geologic formation, group of formations, or part of a formation that is saturated, and contributes a significant quantity of water to wells or springs.

2.1–04 **Artesian:** A condition in an aquifer where the groundwater is confined under pressure.

2.1–05 **ASTM:** The American Society for Testing and Materials provides guidelines and standards for material testing. The address is 1916 Race Street, Philadelphia, PA 19103.

2.1–06 **Bailer:** A tabular hollow receptacle with a check valve used to facilitate withdrawal of fluid from a well or borehole.

2.1–07 **Bentonite:** A highly plastic absorptive, colloidal natural clay composed largely of sodium montmorillonite and which is sold commercially in dry powder or pelletized form.

2.1–08 **Bid:** The offer or proposal of the bidder submitted on the prescribed form setting forth the prices for the work to be performed.

2.1–09 **Bidder:** Any person, firm, or corporation invited to submit a bid for the work.

2.1–10 **Blow Out:** The inflow of groundwater and soil into the well hole or casing caused by a differential pressure head greater outside the well hole or casing than inside, generally due to a lower water level inside the well hole than that of the surrounding potentiometric level.

2.1–11 **Casing:** Tubular steel, finished in sections with either threaded connections or bevelled edges to be field welded, which is installed to counteract caving of the drilled hole.

2.1–12 **Casing, Flush Joint:** Casing with squared threaded ends such that a fixed inside and outside diameter is maintained across joints.

2.1–13 **Casing, Protective:** Anodized aluminum pipe with aluminum locking lid with provisions for a heavy duty padlock installed at 3 feet above ground surface to protect the PVC well from damage.

2.1–14   **Casing, Surface:** A single section of clean black steel pipe used to stabilize the well hole near the surface during the initial drilling of the hole.

2.1–15   **Cement:** Portland Cement Type 1 meeting ASTM C 150 furnished in 94 pound bags.

2.1–16   **Cement Float Shoe:** A plug or packer constructed of inert materials within the lowermost section of permanent casing fitted with a passageway through which grout is injected under pressure to fill the annular space. After the grout has hardened, the cement float shoe is drilled out.

2.1–17   **Centering Disk:** A flat, perforated disk constructed of PVC which slides over the riser and/or well screen and fits inside the temporary casing or hollow–stem auger to center the riser within the casing.

2.1–18   **Centralizer:** See Centering Disk.

2.1–19   **Change Order:** A written order to the Contractor signed by the Owner's Representative authorizing an addition, deletion, or revision in the work, or an adjustment in the Contract price or the contract time issued after execution of the agreement.

2.1–20   **Conductivity:** See Specific Conductance.

2.1–21   **Cone of Depression:** The zone influenced by withdrawal of water from an aquifer by some artificial or natural means such as a pumped well, leak, or spring.

2.1–22   **Confined Aquifer:** Groundwater under pressure significantly greater than atmospheric pressure; the upper limit of the aquifer being the bottom of a zone of distinctly lower hydraulic conductivity than that of the material in which the confined water occurs.

2.1–23   **Contractor:** The person, firm, or corporation with whom the Owner has executed the agreement.

2.1–24   **Cuttings:** The fragments, particles, or slurry of soil or rock created during the drilling of the well hole.

2.1–25   **D.C.D.M.A.:** The Diamond Core Drill Manufacturer's Association.

2.1–26   **Drawdown:** The difference in elevation between the static water level and the surface of the cone of depression at the time of development.

2.1–27   **Drawings:** Refer to the attached for drawing of single– and multi–cased wells.

2.1–28 **Drilling Fluid:** A water based fluid used in the drilling operation to wash cuttings from the hole, to clean and cool the bit, to reduce friction between the drill stem and sides of the hole, and to seal the sides of the hole to prevent loss of drilling fluids. NOTE: COMMERCIAL DRILLING FLUIDS WITH ADDITIVES ARE NOT TO BE USED.

2.1–29 **Drive Shoe:** A forged steel collar with a cutting edge fastened onto the bottom of the casing to shear off irregularities in the hole as the casing advances, and to protect the lower edge of the casing as it is driven.

2.1–30 **d–15:** The theoretical diameter of the soil particle in millimeters at which 15 percent of the particles are finer and 85 percent are coarser.

2.1–31 **d–85:** The theoretical diameter of the soil particle in millimeters at which 85 percent of the particles are finer and 15 percent are coarser.

2.1–32 **Engineer:** An individual with a degree in civil engineering and having experience in the installation of monitoring wells, who is employed by the consultant.

2.1–33 **Filter:** A clean sand of selected grain size and gradation which is installed in the annular space between the well pipe and the wall of the casing or well hole above the gravel pack and below the bentonite seal.

2.1–34 **Geologist:** An individual with formal training in the science of geology.

2.1–35 **Gravel Pack:** A gravel or coarse sand installed between the well screen and the well hole extending 5 feet above the top of the well screen.

2.1–36 **Ground water:** Naturally occurring water encountered below the ground surface.

2.1–37 **Grout:** A mixture of cement, bentonite, lime, and water which is used to form a seal between the borehole and well casing.

2.1–38 **Hazardous Waste:** A hazardous waste as defined by the Resource Conservation Recovery Act (RCRA) in 40 CFR 261.3.

2.1–39 **Homogeneous:** The property of a material to be essentially uniform in its characteristics of composition, texture, appearance, etc.

2.1–40 **Hydraulic Gradient:** The change in static head per unit of distance in a given direction. If not specified, the direction of flow generally is understood to be that of the maximum rate of decrease in head.

2.1–41   **Jetting:** Water is forced down through the drill rods or well by means of the pressure pump and out through holes in the bit or well screen. This water, being under pressure, creates a quick condition and allows the well or drill rods to sink into the soil or cuttings.

2.1–42   **Leachate:** Contaminated water resulting from the passage of rain, surface water, or ground water through waste.

2.1–43   **Lower Zone:** A readily defined soil strata consisting of a predominate soil type different from the zone(s) above it.

2.1–44   **Measuring Tape:** An electronic water level indicator which utilizes the water as a conductor to indicate submergence of a point containing an energized probe and a neutral wire separated by a short distance.

2.1–45   **Mud Pan:** A metal tub into which the drilling fluid and cuttings are discharged and which serves as a reservoir and settling tank during recirculation of the drilling fluids.

2.1–46   **Oil Trap:** A filter and separator used to remove oil from the compressed air flowing out of the storage tank.

2.1–47   **Owner:** The legal owner of the facility for which the work is being performed.

2.1–48   **Owner's Representative:** The authorized representative of the Owner who is assigned to the project and who has the authority to bind the Owner to an agreement.

2.1–49   **Packer:** A device temporarily placed in a well which plugs or seals a portion of the well at a specific level.

2.1–50   **Perched Ground water:** Ground water in a saturated zone of relatively limited horizontal extent which is separated from the main body of ground water by an unsaturated zone or thick zone of low permeability.

2.1–51   **Permeability:** A measure of the relative ease with which a porous medium can transmit a liquid under a potential gradient. It is a property of the medium that is dependent upon the shape and size of the pores. The rate at which water flows through a soil deposit in response to a differential in hydraulic pressure.

2.1–52   **pH:** The intensity of acidic or alkaline condition of a solution; the symbol for the logarithm of the reciprocal of hydrogen ion concentration in gram–atoms per liter.

2.1–53   **Potentiometric Level:** The level in or above a confined or unconfined aquifer at which the pressure is atmospheric. This level is determined at a location by the static water level in a monitoring well screened in the aquifer.

2.1–54   **Reaming:** The process of enlarging the well hole to remove geologic material from the sides of the well hole.

2.1–55   **Revert(R):** An organic polymer drilling fluid additive of high viscosity manufactured by the Johnson Well Screen Company. NOT TO BE USED UNDER THESE SPECIFICATIONS.

2.1–56   **Riser:** The pipe extending from the well screen to above the ground surface.

2.1–57   **Seal Tamper:** A heavy cylindrical metal section of tubing which is secured to a cable that slips over the riser and fits inside the casing or well hole which is used to tamp the bentonite pellets, gravel pack, or filter.

2.1–58   **Specifications:** The instructions to bidders, the general conditions, the special conditions, and the technical provisions.

2.1–59   **Specific Conductance:** The potential for electrical conductivity of a water sample at 25°C as expressed in micro–ohms per centimeter.

2.1–60   **Standby:** Authorized periods of shut–down whereby drilling and well installation stop by orders of the Owners Representative.

2.1–61   **Static Water Level:** The vertical elevation of the top of a column of water in a monitoring well which is no longer influenced by effects of installation, pumping, or other temporary conditions.

2.1–62   **Stick–up:** The vertical length of the portion of the protective casing which protrudes above the ground surface.

2.1–63   **Subcontractor:** An individual, firm, or corporation employed by the Contractor or any other subcontractor for the performance of a part of the work at the site, other than employees of the Contractor.

2.1–64   **Transmissivity:** The rate at which water of prevailing kinematic viscosity is transmitted through a unit width of an aquifer under a unit hydraulic gradient.

2.1–65   **Tremie Pipe:** A pressurized pipe or tube used to transport the flow of grout from the surface into the annular space beginning at the bottom of the annular space and proceeding upwards. (NOTE: HORIZONTAL OR SIDE DISCHARGE IS REQUIRED.)

2.1–66   **Uniformly Graded:** A particle size distribution of a soil which consists of the majority of particles being of the same appropriate diameter.

2.1–67   **Upper Zone:** A soil strata consisting of a dominant soil type different from the zone immediately below it.

2.1–68   **Utilities:** Service lines or equipment located above, upon, or below the ground surface used for conveyance of electricity, natural gas, petroleum, communications, storm water, waste water, potable water, etc.

2.1–69   **Washout Nozzle:** A device utilized at the end of a string of casing equipped with a check valve through which clear water or grout can be injected to wash out drilling fluids and cuttings from the annular space.

2.1–70   **Water Cement Ratio:** The proportion of the weight of mixing water in pounds to weight of cement in pounds.

2.1–71   **Water Table:** The surface in an unconfined aquifer at which the pressure is atmospheric. This level is determined at a location by the static water level in a monitoring well screened in the aquifer.

2.1–72   **Well Hole:** The open subsurface hole created by conventional drilling methods.

2.1–73   **Well Protector:** See Casing, Protective.

2.1–74   **Well Screen:** Commercially manufactured pipe or cylindrical tubing with slits of a uniform width, orientation, and spacing.

2.1–75   **Zone of Saturation:** The zone below the water table or below the top of a confined aquifer in which all interstices are filled with groundwater.